CELL CYCLE AND GROWTH CONTROL

CELL CYCLE AND GROWTH CONTROL

BIOMOLECULAR REGULATION AND CANCER

SECOND EDITION

Edited by

GARY S. STEIN, PH.D.
Department of Cell Biology and Cancer Center
University of Massachusetts Medical School
Worcester, Massachusetts

ARTHUR B. PARDEE, PH.D.
Department of Biological Chemistry and Molecular Pharmacology
Dana-Farber Cancer Institute
Boston, Massachusetts

WILEY-LISS

A JOHN WILEY & SONS, INC., PUBLICATION

Published by John Wiley & Sons, Inc., Hoboken, New Jersey.
Published simultaneously in Canada.

Library of Congress Cataloging-in-Publication Data:

Cell cycle and growth control: biomolecular regulation and cancer / edited by Gary S. Stein, Arthur B. Pardee.—2nd ed.
p. ; cm.
Rev. ed. of: The molecular basis of cell cycle and growth control. c1999.
Includes bibliographical references and index.
ISBN 0-471-25071-6 (alk. paper : cloth)
1. Cell cycle. 2. Cellular control mechanisms. 3. Cellular signal transduction.
4. Cell differentiation—Molecular aspects.
[DNLM: 1. Cell Cycle—physiology. 2. Cell Death—physiology. 3. Mutagenesis—physiology.
QH 604 C3925 2004] I. Stein, Gary S. II. Pardee, Arthur B. (Arthur Beck), 1921—
III. Molecular basis of cell cycle and growth control.
QH604 .M6 2004
571.8′4—dc22 2003024668

Printed in the United States of America.

10 9 8 7 6 5 4 3 2 1

This volume is dedicated to Dr. Arthur B. Pardee in recognition of his seminal contributions to understanding gene regulation and growth control. His pioneering studies in mammalian cells have provided the underlying principles and experimental approaches that are the foundation for our current understanding of growth control and cell cycle progression.

Dr. Pardee is responsible for establishing a restriction point during the prereplicative phase of the mammalian cell cycle and demonstrating its role as a determinant for regulatory mechanisms requisite for the onset of DNA replication. Over the past several years, the Pardee Laboratory has defined interrelationships between the DNA replication cycle and the mitotic cycle, elucidating important differences between normal and tumor cells. His development of differential display technology has led to the identification of genes aberrantly expressed in cancer, as well as broader applications to genes supporting critical regulatory events. He has then translated these fundamental discoveries, exploiting the vulnerability of transformed and tumor cells to biochemical perturbants and the preferential utilization of signaling pathways in tumors to develop novel approaches to cancer chemotherapy.

The profound biological and clinical importance of Dr. Pardee's characterization of regulatory mechanisms that control cell proliferation is reflective of the highest standards of scientific pursuit. In addition to his consistently outstanding research contributions, he has been an inspirational mentor and valued colleague to all of us in the growth control field.

–The Contributors
February 17, 2003

CONTENTS

PREFACE

Cell cycle and growth control are profoundly relevant to biological regulation of development and tissue renewal. Equally significant is the recognition that aberrations in mechanisms governing proliferation are linked to the onset and progression of tumorogenesis. From an historical perspective, the foundation for our current understanding of cell cycle and growth control has been systematically constructed during the past fifty years through the combined application of cellular, biochemical, molecular and *in vivo* genetic approaches. The discovery that DNA replication and mitotic division are confined to discrete periods, each preceded and followed by complex and interdependent regulatory events that establish competency for proliferation and cell cycle progression, provided a conceptual underpinning for mechanisms mediating growth control.

Initially, somatic cell fusion and nuclear transplantation studies, together with the selective use of growth factors and inhibitors of macromolecular biosynthesis established fundamental parameters of cell cycle regulation. These key elements of cell cycle control include requirements for transcription to initiate DNA replication and mitotic division as well as the restriction point late in G1 when the threshold for growth factor-independent progression to S-phase is traversed. A persuasive platform for assembling the regulatory cascades that control the cell cycle then evolved by exploiting the power of yeast genetics and subsequent validation in mammalian cells and *in vivo* animal models. Valuable insight was attained into checkpoints and surveillance mechanisms that monitor fidelity of growth control and responsiveness of cells to intra- and extracellular physiological cues. With enhanced capabilities to investigate gene expression through genomic and proteomic approaches, we are becoming increasingly aware of compromises in gene expression that account for breaches in fidelity of cell cycle control in transformed and tumor cells. Significance of the delicate balance between cell survival and default to apoptosis is emerging as a fundamental component of biological control and as a viable therapeutic target.

This book was developed with the objective of presenting concepts, experimental strategies and key findings that enhance understanding of cell cycle and growth control as obligatory physiological processes and from the perspective of compromises that occur in cancer. The first two chapters present an overview of the elegantly organized and stringently orchestrated molecular events that determine cell fate within a context of options for proliferation, differentiation and apoptosis. The perspectives of regulation and structure are explored as a basis for addressing the combinatorial assembly and activity of regulatory complexes that are responsive to integrated cascades of signals that connect molecules with phenotypes. Here, intranuclear trafficking is presented as a mechanism to direct regulatory proteins to the right place at the right time for focal assembly of macromolecular complexes that support replication, transcription and repair in nuclear microenvironments. The dynamics of regulatory machinery organization is emphasized in relation to temporal-spatial parameters of cell cycle control.

The chapters that follow expand on the organization of cellular events that incorporate a broad spectrum of catalytic and regulatory proteins, not as a comprehensive catalogue, but as a basis for assembling a blueprint for structure-function interrelationships.

Regulatory cascades are dissected to explain the requirements for passage through the G1, S, G2 and mitotic periods of the cell cycle. There is emphasis on positive and negative control that is required for mitotic events that include the regulated as well as the regulatory activities of centrosomes and the mitotic apparatus. Here, implications for chromosome segregation and factors contributing to chromosome instability and aneuploidy are discussed. S-phase regulatory events are examined with emphasis on the coupling of DNA synthesis with histone gene expression and chromatin remodeling. Subtleties of signaling that discriminate between decisions to progress through the cell cycle or default to apoptosis are reviewed. Consideration is not confined to transcription but extends to regulatory events that impact on translational control during the cell cycle. Here, the common denominator is a necessity to balance and selectively amplify or dampen the multidirectional flow of signals that impact on phenotype-specific control of proliferation. This concept is expanded upon in the chapters that are dedicated to regulation of angiogenesis and metamorphosis. Mechanisms that are central to transformation and tumorigenesis are directly examined in four chapters that address DNA repair, oncogenes, and tumor suppressor genes. Emphasis is on cellular compensation and the decision for survival or default to apoptosis. Genomic instability is considered as a function of telomere structure and function and as a consequence of SV40 immortalization. The implications for apoptotic signaling are evaluated on the basis of responsiveness in normal and cancer cells. A central theme is the boundaries between physiological control and a refractory response to checkpoint signals that sustains incurred genomic damage.

The concluding chapters provide an overview of new dimensions to cancer therapy that are based on regulatory parameters of cell cycle and growth control. Options for therapeutic strategies that selectively target components of signaling pathways that mediate steps in establishing competency for proliferation are presented. The complexities of regulatory cascades controlling cell proliferation, differentiation and apoptosis are expanding appreciation for subtleties of growth control and determinacy of cell fate. Each regulatory parameter is a functional component of biological control that enables cells to respond to a broad spectrum of physiological cues. All perturbations in regulatory mechanisms that occur in tumor cells reflect modifications that are consequential for cell fate and cell survival, proliferation, differentiation, senescence, migration and programmed cell death. And, it is becoming increasingly evident that collectively, the insight we are obtaining into regulatory mechanisms operative in normal and tumor cells will facilitate diagnosis of cancer and the ability to treat the disease by selectively targeting molecular signals that exchange regulatory information between the genome and the extracellular environment.

CONTRIBUTORS

ERIC H. BAEHRECKE, Center for Biosystems Research, University of Maryland Biotechnology Institute, College Park, Maryland

UMA BAI, Vattikuti Urology Institute, Henry Ford Health Sciences Center, Detroit, Michigan

STACEY J. BAKER, Fels Institute for Cancer Research and Molecular Biology, Temple University School of Medicine, Philadelphia, Pennsylvania

EVELYN R. BARRACK, Vattikuti Urology Institute, Henry Ford Health Sciences Center, Detroit, Michigan

SILVIA BENVENUTI, Ludwig Institute for Cancer Research, Royal Free and University College School of Medicine, London, United Kingdom

MINA J. BISSELL, Life Sciences Division, Lawrence Berkeley National Laboratory, Berkeley, California

COREY D. BRAASTAD, Department of Cell Biology and Cancer Center, University of Massachusetts Medical School, Worcester, Massachusetts

CARMEN CARNEIRO, Department of Molecular Biology, Memorial Sloan-Kettering Cancer Center, New York, New York

EUGENIA CIFUENTES, Vattikuti Urology Institute, Henry Ford Health Sciences Center, Detroit, Michigan

JAMES DEGREGORI, Program in Molecular Biology, University of Colorado Health Sciences Center, Denver, Colorado

MICHAEL DEININGER, Center for Hematologic Malignancies, Oregon Health and Science University, Portland, Oregon

DAVID T. DENHARDT, Department of Cell Biology, Rutgers University, Nelson Laboratories, Piscataway, New Jersey

WAFIK S. EL-DEIRY, Howard Hughes Medical Institute, Departments of Medicine, Genetics, Pharmacology, and Cancer Center, University of Pennsylvania, Philadelphia, Pennsylvania

JUDAH FOLKMAN, Department of Surgery, Children's Hospital, Harvard Medical School, Boston, Massachusetts

HEIDE L. FORD, Departments of Obstetrics and Gynecology, Biochemistry, and Molecular Genetics, University of Colorado Health Sciences Center, Denver, Colorado

Colleen Fordyce, UCSF Comprehensive Cancer Center, University of California at San Francisco, San Francisco, California

Antonio Giordano, Sbarro Institute for Cancer Research and Molecular Medicine, Center for Biotechnology, Temple University, Philadelphia, Pennsylvania

Cynthia J. Guidi, Department of Cell Biology, University of Massachusetts Medical School, Worcester, Massachusetts

Edward H. Hinchcliffe, Department of Biological Sciences and Walther Institute for Cancer Research, University of Notre Dame, Notre Dame, Indiana

Anthony N. Imbalzano, Department of Cell Biology, University of Massachusetts Medical School, Worcester, Massachusetts

Parmjit S. Jat, Ludwig Institute for Cancer Research, Royal Free and University College School of Medicine, London, United Kingdom

Roger D. Johnson, Department of Cancer Biology and Department of Cell Biology, University of Massachusetts Medical School, Worcester, Massachusetts

Andrew Koff, Department of Molecular Biology, Memorial Sloan-Kettering Cancer Center, New York, New York

Hans Laufer, Department of Molecular and Cell Biology, University of Connecticut, Storrs, Connecticut; and Marine Biological Laboratory, Woods Hole, MA

Jane B. Lian, Department of Cell Biology and Cancer Center, University of Massachusetts Medical School, Worcester, Massachusetts

Rowena L. Lock, Ludwig Institute for Cancer Research, Royal Free and University College School of Medicine, London, United Kingdom

Valeria Masciullo, Departments of Pathology, Anatomy, and Cell Biology, Thomas Jefferson University, Philadelphia, Pennsylvania

Mani Menon, Vattikuti Urology Institute, Henry Ford Health Sciences Center, Detroit, Michigan

Martin Montecino, Departamento de Biologia Molecular, Facultad de Ciencias Biologicas, Universidad de Concepcion, Barrio Universitario s/n, Concepcion, Chile

Arthur B. Pardee, Department of Biological Chemistry and Molecular Pharmacology, Dana-Farber Cancer Institute, Boston, Massachusetts

Joseph F. Porter, Department of Cell Biology, Rutgers University, Nelson Laboratories, Piscataway, New Jersey

E. Premkumar Reddy, Fels Institute for Cancer Research and Molecular Biology, Temple University School of Medicine, Philadelphia, Pennsylvania

G. Prem-Veer Reddy, Vattikuti Urology Institute, Henry Ford Health Sciences Center, Detroit, Michigan

Robert E. Rhoads, Department of Biochemistry and Molecular Biology, Louisiana State University Health Sciences Center, Shreveport, Louisiana

Conly L. Rieder, Laboratory of Cell Regulation, Division of Molecular Medicine, Wadsworth Center, Albany, New York; and Department of Biomedical Sciences, State University of New York, Albany, New York

Aylin Rizki, Life Sciences Division, Lawrence Berkeley National Laboratory, Berkeley, California

Robert A. Sclafani, Program in Molecular Biology, University of Colorado Health Sciences Center, Denver, Colorado

Greenfield Sluder, Department of Cell Biology, University of Massachusetts Medical Center, Worcester, Massachusetts

Gary S. Stein, Department of Cell Biology and Cancer Center, University of Massachusetts Medical School, Worcester, Massachusetts

Janet L. Stein, Department of Cell Biology and Cancer Center, University of Massachusetts Medical School, Worcester, Massachusetts

Thea D. Tlsty, Department of Pathology, University of California at San Francisco, San Francisco, California

André J. van Wijnen, Department of Cell Biology and Cancer Center, University of Massachusetts Medical School, Worcester, Massachusetts

Shulin Wang, Howard Hughes Medical Institute, Departments of Medicine, Genetics, Pharmacology, and Cancer Center, University of Pennsylvania, Philadelphia, Pennsylvania

S. Kaleem Zaidi, Department of Cell Biology and Cancer Center, University of Massachusetts Medical School, Worcester, Massachusetts

Faith A. Zamamiri-Davis, Department of Biochemistry, St. Jude Children's Research Hospital, Memphis, Tennessee

Gerard P. Zambetti, Department of Biochemistry, St. Jude Children's Research Hospital, Memphis, Tennessee

PART I

CHAPTER 1

CELL FATES

ARTHUR B. PARDEE
Dana-Farber Cancer Institute, Boston, MA 02115

Cells develop phenotypes that are determined by organized and regulated molecular processes. Then diverse fates include proliferation, differentiation, and apoptosis. They proceed along several pathways of molecular signaling that are initiated by external factors, which activate cascades of kinases that bring these signals to the nucleus where they initiate transcriptions. These processes require an organized series of cellular and events in which numerous catalytic and regulatory proteins are involved, which in this book are discussed in detail.

PURPOSE AND ORGANIZATION OF THIS BOOK

A living cell can proceed along alternative pathways to a variety of destinations. These include proliferation to form two daughter cells, irreversible or reversible growth arrest, differentiation to a new type of cell as in development or metamorphosis, and death by necrosis or by programmed cell death (apoptosis). At any time the net number of cells is the result of a balance between proliferation and death. These cell fates may be changed in diseases such as the increased growth and decreased apoptotic death of cancer cells. And also they can be modified by drugs and other extracellular agents.

The purpose of this book is to summarize what has been learned about structural, biochemical, and molecular biological events that are the basis for these cell biological processes and their regulations. The emphasis is on vertebrate cells. Thousands of molecules and reactions have already been reported and organized into functional patterns that connect molecules with phenotypes (Fig. 1.1). In this and the next chapter are provided general concepts and underlying principles of regulation and structure. In the other chapters of this book are presented the mass of information, with references and illustrative examples.

Cell Cycle and Growth Control: Biomolecular Regulation and Cancer, Second Edition, Edited by Gary S. Stein and Arthur B. Pardee
ISBN 0-471-25071-6

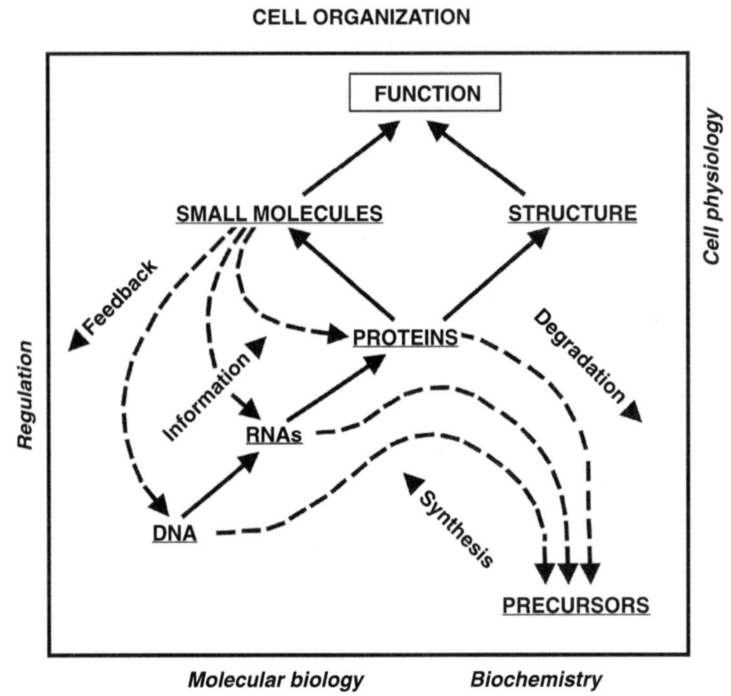

Figure 1.1. Cell molecular and information transfer. The central path of information flow from DNA to cell functions is regulated by feedbacks, indicated on the left. Syntheses from precursors are counterbalanced by degradations, as indicated on the right.

CELL CYCLE BIOLOGY

As an example of a cell fate pathway we outline the general organization of the cell cycle and its biology and biochemistry. By this orderly process one cell grows into two. It is fundamental for the organism's growth and for replacement of cells lost during normal wear and tear (Murray, 1993). Cells from a mature eukaryotic organism can require an interval of a day or more between successive divisions in tissue culture. During this time duplications of all of the myriad molecules that comprise each cell are required, at different times throughout the cycle. The most evident is duplication of deoxyribonucleic acid (DNA), the heredity-carrying material in chromosomes. DNA does not duplicate continously, but only during several hours in midcycle, a period named the S phase for (DNA) synthesis. The cycle is organized, for simplicity, into a sequence of only four major biological and biochemical events which are grouped as gap 1 (G_1 phase) during which a cell prepares for DNA synthesis, DNA synthesis (S phase), preparation for mitosis (G_2 phase), and mitosis (M phase), after which the cell divides and the cycles of the two new cells can commence. For a historical summary, see Baserga (1985).

Quiescence

Most cells in vivo are performing their specialized functions in support of the whole organism. They are quiescent (in G_0 phase), not usually progressing through the cycle, and divide very infrequently. Some cells can remain quiescent for a limited time, an example being fibroblasts whose proliferation resumes after wounding upon stimulation by platelet growth factors. Others such as nerve and muscle cells have become permanently quiescent. Quiescent cells have left the cycle during G_1, and so they contain the unduplicated quantity of DNA, as do G_1 cells. But they differ from G_1 cells in many other properties; in particular, they lack the regulatory molecules required for growth. KI-67 protein is a marker for distinguishing proliferating G_1 from G_0 cells.

GI Phase

Quiescent cells are activated to proliferate by providing suitable conditions. Nutrients including sugars, salts, vitamins, and essential amino acids are needed for their growth (Baserga, 1985). Normal (nontumor) cells also require epidermal growth factor (EGF), insulin-like growth factor (IGF-1), and transferrin. In an organism growth factors and nutrients must be supplied from blood. For cells to grow in tissue culture, a nutrient medium is required that supplies growth factors usually from added serum. Cells again become quiescent if growth factors are removed. These proteins are required to overcome inhibitions created by contacts between receptors on the cell surface with proteins present in the medium such as growth-negative factor TGF-β, in the extracellular matrix, and on other cells with which a cell is in contact at high density.

Cells increase in size in G_1 phase, but they do not exhibit dramatic changes in morphology. But many molecules are synthesized, and molecular processes take place successively during this interval (see below). The time that cells in culture spend traversing this phase is highly variable, for example, from 6 to 24 hours, unlike the rather uniform times they spend in each of the other phases. G_1 culminates in initiation of DNA synthesis. Growth factors initiate a multiple-step cascade of signals that ultimately activate genes to produce messenger ribonucleic acids (mRNA) and proteins.

S Phase

The requirements of growth factors for passage through G_1 phase are lost at the restriction point (R), located shortly before cells start to synthesize DNA (Pardee, 1989). Progression through later phases of the cell cycle depends on internally generated signals. During the 6 to 8 hours of S phase the nuclear DNA comprising possibly 50,000 genes that are located on 23 pairs of chromosomes is replicated. Each gene is duplicated at a definite time. For example, the dihydrofolate reductase gene that is required for synthesis of DNA is replicated in very early S phase.

G_2 Phase, M Phase, and Cell Division (Cytokinesis)

Cells pass through G_2 phase for a few hours after DNA synthesis is completed and before mitosis commences, an interval presumed to be needed to produce the machinery required for mitosis. The complex processes of mitosis then requires less than an hour, during which the nuclear membrane breaks down, duplicated chromosomes condense, are paired, and microtubule proteins segregate them equally between the two daughter cells. These daughter cells then divide, separate, and each can reinitiate its cycle.

BIOCHEMISTRY AND MOLECULAR BIOLOGY OF CYCLE PHASES

Growth Stimulation

The pathways to cell fates are activated by various extracellular and internal molecular signals. Each pathway has is its distinctive molecular basis, a complex set of interactions between very numerous molecules that carry these signals from cell surface into nucleus (Murray and Hunt, 1993; Andreef, 2003). Alternative pathways and their enzymes often perform the same function, a redundancy that provides fail-safe mechanisms. Details of these complex pathways are presented in other chapters this volume.

We illustrate this complexity with as an example an overview of only a part of one pathway. Activation of proliferation by EGF commences when this growth factor binds to its receptor, on the part located on the cell surface. This external stimulus causes the receptor proteins to form a dimer. The entire receptor extends into the cell, and dimerization activates its protein tyrosine kinase portion inside the cell. This in turn initiates signal transduction, a phosphorylation cascade that begins on the membrane's internal surface and ends in the nucleus. Located on the inner surface of the membrane are enzymes and their regulatory noncovalent binding effectors, such as the GTP-binding Ras protein. From there, a cascade of downstream enzymes including kinases B and C carry the signal on to the nucleus, where transcription factors are phosphorylated and form large complexes with accessory proteins that bind to specific promoter and enhancer sequences in DNA of target genes (Naar, 2001).

Steroid hormones also activate transcriptions and initiate growth. These molecules move directly into the nucleus where they activate genes, unlike growth factors that initiate cytoplasmic signaling pathways from the membrane. For example, estrogen activates hormone responsive breast cells by ligating to specific receptor proteins in the nucleus that bind to DNA sequences in promoters and activate growth-stimulating target genes.

Signal Transduction

Numerous genes that are activated during the cycle were discovered by researches with yeast mutants having modified cycle-controls (Hartwell and Kastan, 1994). Activation of G_1 phase in mammalian cells results in expression of at least 100 genes. Biochemistry and molecular biology has identified new key enzymes and regulatory proteins, especially cycle-dependent kinases (cdks) that phosphorylate proteins required for cell cycle progression (Nurse, 2000). A series of regulatory proteins regulate transition through the cycle (Roberts, 1999) by binding to and activating these kinases (Murray, 1993). As a cell proceeds through its cycle, four major cyclins (D, E, A, and B) are produced sequentially, and they activate several cyclin dependent kinases. These complexes catalyze successive stages of cell cycle progression. Cyclin D increases in early to mid G_1 phase and regulates cyclin dependent kinases cdk4 and cdk6 (Sherr, 1996). Cyclin D/cdks trigger synthesis of cyclin E in late G_1 phase, which in turn activates cdk2/cyclin A and DNA synthesis. Cyclins rise and fall during the cycle because of periodic changes in both their synthesis and destruction (Minshull, 1989).

Families of other proteins bind to and block activities of cyclin/cdk complexes. Some named inhibitors of kinases (INK) counterbalance the cyclin's activation of cdks, thereby affecting cycling, development, and tumorigenesis (Sherr, 1996). p27 blocks progression; its level is high in quiescent cells and decreases during late G_1 to release cdk/cyclin activities. Inhibition of cyclins by the cdk inhibitor p21 has been demonstrated to be induced under many conditions that arrest growth.

In addition to the synthesis of cyclins, phosphorylations of these complexes are regulatory. Another kinase, CAK, activates the cyclin-dependent kinases by phosphoryation, and also inhibitory phosphates are removed by phosphatases. Furthermore a major regulatory role during the cell cycle is played by relocalization of cyclin/cdks to the nuclear compartment within a cell. Importantly, proteolytic destruction of these regulatory proteins is vital after a cell passes each phase in the cycle (Koepp, 1999). Proteins targeted for removal, including cyclins, are first specifically labeled with the small ubiquitin protein, and then the proteosome, a biochemical machine composed of many enzymatic subunits, chews them up (Benaroudj, 2001).

Downstream Events

Activated cdks phosphorylate proteins that are essential for progression through the cell cycle. When they phosphorylate the retinoblastoma tumor-suppressor protein pRb, which is absent in retinoblastomas, it releases the E2F1 protein to which it was bound. E2F1 then activates transcriptions of many genes that are necessary for initiating S phase, including those coding for enzymes of DNA synthesis. An example is DNA polymerase-α whose transcription is thereby up regulated at G_1/S phase. These enzymes increase at the beginning of S phase, and also they move from the cytoplasm into the nucleus where they duplicate DNA.

The DNA replication process is initiated at numerous origins of replication, which are sequences in DNA, and is catalyzed by a complex of proteins that includes DNA polymerases. It is closely controlled. In the early S phase cyclins D and E must be degraded by proteasomes. Progression through S phase depends on cyclin A-cdk2 kinase.

After completion of S phase, events in G_2 phase are preparatory for entry into mitosis (M). The maturation-promoting factor (MPF) obtained from mitotic cells was early shown to activate mitosis when introduced into another cell. The cyclin-dependent kinase cdk1 is by itself inactive but has been demonstrated to be essential. It must be activated by binding cyclin B, newly produced in late S and G_2, which forms MPF. It phosphorylates the nuclear membrane protein laminin, which causes breakdown of the nuclear membrane. At the beginning of M phase, after the nuclear membrane is degraded, cyclins A and E2F are removed by proteosome-catalyzed degradation, a process necessary to prevent apoptosis—see below (Lees, 1999). These events are basic to the complex molecular mechanism enabling progression into M phase. To again briefly illustrate the complexity of regulatory mechanisms, this G_2/M checkpoint mechanism is a complex molecular network of phosphorylations and dephosphorylations, catalyzed by several enzymes and proteins. MPF activity is regulated by a variety of proteins that include not only cyclin B but phosphatases, kinases, and also its subcellular localization; cyclin B/cdk1 is rapidly relocated from the cytoplasm to the nucleus at the G_2/M transition.

Thereafter the processes of chromosome condensation, pairing, and segregation in mitosis proceed. The destruction of cyclin B, involving a specialized multiple-subunit anaphase promoting complex, is essential for completion of the cycle. These many phosphorylations are important for the massive morphological changes that are necessary for a cell to divide. Cell separation (cytokinesis) soon follows, but it is not necessary for progression through the next cycle because this is accomplished normally by binucleate cells that can be produced after daughter cell separation is blocked by cytochalasin B.

The cell must prepare for DNA synthesis in its next cycle. Normally only one DNA replication can occur per cycle; DNA synthesis cannot be reinitiated until after mitosis is complete. The retinoblastoma protein pRb is a critical determinant in preventing DNA reduplication. Perhaps related is the breakdown during mitosis of the membrane around the nucleus, which permits interactions between molecules from the nucleus and cytoplasm. Degradation of cyclin B by proteasomes is also necessary to start S phase in the following cycle. This "licensing" of DNA synthesis can be disrupted: cells that have lost the cdk inhibitor p21 undergo multiple rounds of DNA synthesis without mitosis, and this process is also activated by the anticancer agent staurosporin, which eliminates the dependence of DNA synthesis on the prior M phase (Nurse, 2000).

GROWTH DISREGULATION

Proliferative Regulation

The cycle of a normal cell is very closely regulated. Proliferation is determined in G_1 phase by the presence of suitable growth conditions. These controls ensure that a phase of the cell cycle does not begin until the preceding phase has been completed with high fidelity. If a regulation control fails, programmed cell death (apoptosis) or genomic instability can result. In mammalian nontumor cells a surveillance system in G_1 phase is engaged to throw the switch between cell growth and quiescence (Pardee, 1974). A similar regulation point in yeast named START was discovered by Hartwell. These cells cannot pass beyond a specific point in late G_1 phase, named the restriction point (R), if the stimulation by growth factors or nutrients is inadequate, and they remain in or revert to quiescence. The final steps that are needed to pass R require synthesis of an unstable protein, later proposed to be cyclin E. Under inadequate conditions this protein's synthesis does not keep up with its loss, and so it cannot be accumulated to be in excess of the cdk inhibitor p21 and so is insufficient to move the cell into S phase. This G_1 regulatory mechanism is defective in cancer cells, which therefore readily pass through R, and so they proliferate excessively (Pardee, 1974).

DNA Damage-Induced Checkpoints

Uncorrected failures of DNA repair are important in the progression from normal to cancerous mammalian cells. DNA damage results in blocked proliferation. The name checkpoint was proposed for this set of cell cycle controls that are activated after DNA is damaged (Hartwell, 1994). A checkpoint delays entry into the next phase of the cell cycle. A major checkpoint acts upon the G_1 to S transition, and prevents damaged G_1 cells from beginning DNA synthesis until DNA has been repaired, and another is especially evident at the G_2/M interface (Fingert, 1988). Several proteins have been implicated in this checkpoint mechanism, in particular, p53, a tumor suppressor called "the guardian of the genome." It is inactivated by mutation in more than 50% of cancers (Levine, 1997). After DNA is damaged, p53 increases owing to its greater stability; it induces protein p21, which blocks proliferation by inhibiting cyclin/cdk. The ataxia telangiectasia protein (ATM) phosphorylates and increases p53. The gene coding for ATM is mutated in individuals that are very sensitive to X rays and that have a high incidence of tumors.

Mammalian cells in S phase exhibit a dose-dependent reduction in DNA synthesis within several minutes of exposure to DNA damaging agents such as X rays. Less is known about the mechanism of this S phase checkpoint than about those in G_1 or G_2. As little as one double-strand break in DNA activates a G_2/M phase checkpoint control and stops cells at the G_2/M boundary. This is important because it provides time for DNA repair before a cell goes through mitosis. If this interval is shortened by a drug treatment, the cells progress into mitosis without repairing all the damage, which results in death (Fingert, 1988).

Mitosis segregates the duplicated chromosomes between the daughter cells. Accurate segregation depends on proper chromosome alignments on, and attachment to, the mitotic spindle, which is composed of microtubule proteins. A mitotic checkpoint ensures that segregation process occurs correctly by delaying completion of mitosis until all chromosomes are properly attached to the mitotic spindle. This mechanism blocks progression through mitosis if chromosomes are misaligned.

Programmed Cell Death (Apoptosis)

Apoptosis is a terminal cell fate, a highly regulated "suicide" process that eliminates physiologically unneeded or dangerous cells. It may prevent mutations that cause cancer (Sellers, 1999). After a cell is severely damaged the time of checkpoint arrest may be too brief to permit complete repair, and such cells are eliminated by apoptosis. As an example, the cyclin A-kinase complex necessary for S phase progression is inhibited in cells treated with X rays, which can result in apoptosis because of inability of this complex to remove the apoptotic factor E2F (Lees, 1999). Checkpoint genes including p53 are involved in activating apoptosis, and other proteins including NF-κB can prevent apoptosis. Apoptosis is performed by proteases named caspases and by nucleases, activated by a family that includes positively acting Bax and negatively acting Bcl-2 proteins. Various cells have different responses to damage or drugs partly because they express various members of the Bcl-2 family and the modulating proteins.

Tumor Progression

A cancerous cell's regulatory balance is perturbed by additional mutations, which arise though defects of checkpoint regulation and DNA repair. These lead to further errors in repair, replication, and chromosome segregation. The mutations cause further losses of proliferation control, and they block apoptosis, differentiation, and related growth arrest, and limited life span (immortalization). Metastasis follows, the ability of cancer cells to move about in the body and proliferate in unusual environments, and so on (Onn, 2002). Various molecular mechanisms that control cancer cell growth and apoptosis are now being discovered. These differences between cancer and normal cells can provide novel targets for therapy.

MAJOR REGULATORY MECHANISMS

Throughout this book there are detailed discussions and explications of the major molecular pathways that determine cell fates. This chapter concludes with a brief listing of general regulatory mechanisms that apply to cell cycle control, apoptosis, and the other cell fates described in this book. For example, differentiation pathways are activated and regulated by extracellular factors including hormones, retinoic acids, and drugs. These alter expressions of genes that determine the properties of their target cells.

1. Transcription is activated when complexes of proteins bind to specific DNA sequences in a gene's promoter and enhancer regions. An example is binding to DNA of p53, which turns on transcription of many genes, among them ones involved in growth arrest (p21) and then apoptosis. In some cases this functioning depends on covalent bond formation such as protein phosphorylation, or on noncovalent attachment of a small molecule as by retinoic acid's attachment to its receptor proteins.
2. Chromatin structure also regulates transcription. Methylation of the cytosines in CPG islands of DNA favors local histone deacetylations, which is reversed by acetylation. This changes chromatin structure and inactivates transcriptions, Processes of this general kind may be responsible for long term silencing of long DNA regions, as of the entire one of the two X-chromosomes in each female cell.
3. Pre-RNA processing, splicing and export from the nucleus determine the quantity of mRNA available in the cytoplasm to be translated by ribosomes. Of great current interest are the mechanisms by which different splicing of a pre-mRNA produce several mRNAs, and the regulation of these events.
4. Degradation by nucleases limits mRNAs life times, and together with synthesis, rates determine their steady state concentrations.
5. Translation control is an important element in establishing the amount of protein produced from an mRNA. Inequality between a mRNA and its protein has often been observed.
6. Degradation of a protein counteracts its synthesis, and this too can alter the ratio of a protein to its mRNA. The ultimate example of protein degradation is by proteosome action. This process is initiated by a series of three enzymes that specifically identify the proteins to be removed by covalently tagging them with ubiquitins. As an important example, cyclins are degraded by proteosomes after they have served their transient functions in a phase of the cycle.
7. Covalent modifications are among the best known mechanisms of regulation of activities of a protein such as catalysis, ligand binding, and stability. Cleavage by a protease can either activate or inactivate a protein. Protein phosphorylations are frequently identified modifiers of activities. In signal transduction are sequential activations by cascades of kinases, such as MAP kinase kinase kinase, MAP kinase kinase, and MAP kinase. These activities may be positively or negatively modified.
8. The major components of metabolic machinery are catalytic proteins (enzymes) and noncatalytic binding proteins (which might be named "enphores"). A functional molecule's activity is altered by its spe-

cific noncovalent bindings. Regulatory "allosteric" sites of a protein bind small molecules that modify activities of the primary sites on the same or on an associated protein. For example, activation by GTP binding to accessory Ras proteins is involved in signal transduction. Many biosynthetic pathways that produce essential metabolites are closely regulated by feedback mechanisms; an initial enzyme in the pathway is inhibited by its noncovalent binding of the end product metabolite.

9. An example of noncovalent regulation by a large molecule is binding of a growth factor to its receptor on the cell surface, which activates the latter's internal kinase.
10. Control can depend on intracell localization. As one example, NF-κB activates transcriptions when it is moved from cytoplasm to nucleus.

SUMMARY

The several fates of a cell are produced by organized and regulated processes. Recurring principles of regulation are evident. Their molecular mechanisms are similar in diverse organisms. External factors initiate pathways of signaling to create these cell fates. Very many proteins are involved, both catalytic and regulatory. They function in large complexes. Cascades of kinases bring the message to the nucleus, where it initiates transcriptions. The mRNAs produced are translated by cytoplasmic ribosomes to make the cell's machinery. This leads to an organized series of cellular and molecular processes, of which DNA duplication near the middle of the cycle and mitosis at the end stand out.

The ying-yang principle of regulation by opposing dynamic actions is observed throughout biology. Both positively and negatively acting molecules are involved at every level. This is illustrated by proliferation versus apoptosis with cells, by activating cyclin proteins versus inhibitory regulators of cdc kinases in proliferation, by apoptosis action of Bax versus inhibitory Bcl-2, by histone acetylation versus deacetylation, by macromolecules synthesis versus degradation, by enzyme phosphorylation catalyzed by kinases balanced by phosphatases.

The cell cycle must be closely regulated if life is to remain in balance. Problems arise, especially serious being errors in DNA replication and mitosis that can cause mutational insertion of incorrect bases and chromosome rearrangements, respectively. Important safeguards are DNA repair mechanisms, redundant pathways to produce an end result, checkpoints that provide time for repair, and elimination of defective proteins by proteasomes. As the final safeguard, there is apoptosis, causing death of defective and dangerous cells.

REFERENCES

Andreeff M, Goodrich DW, Pardee AB (2003): Cell proliferation and differentiation. In: DW Kufe et al. (eds): Cancer Medicine 6th ed. Hamilton, Ontaraio: Decker, pp 27–40.

Baserga R (1985): The Biology of Cell Reproduction. Cambridge, MA: Harvard University Press.

Benaroudj N, Tarcsa E, Cascio P, Goldberg AL (2001): The unfolding of substrates and ubiquitin-independent protein degradation by proteasomes. Biochimie 8:311–8.

Fingert HJ, Chang JD, Pardee AB (1986): Cytotoxic, cell cycle, and chromosomal effects of methylxanthines in human tumor cells treated with alkylating agents. Cancer Res 46:2463–7.

Hartwell LH, Kastan MB (1994): Cell cycle control and cancer. Science 266:1821–8.

Koepp DM, Harper JW, Elledge SJ (1999): How the cyclin became a cyclin: Regulated proteolysis in the cell cycle. Cell 97:431–4.

Lees JA, Weinberg RA (1999): Tossing monkey wrenches into the clock: new ways of treating cancer. Proc Natl Acad Sci USA 96:4221–3.

Levine AJ (1997): p53, the cellular gatekeeper for growth and division. Cell 88:323–31.

Minshull J, Pines J, Golsteyn R, Standart N, Mackie S, Colman A, Blow J, Ruderman JV, Wu M, Hunt T (1989): The role of cyclin synthesis, modification and destruction in the control of cell division. J Cell Sci Suppl 12:77–97.

Murray AW, Hunt T (1993): The Cell Cycle, An Introduction. New York: Freeman.

Naar AM, Lemon BD, Tjian R (2001): Transcriptional coactivator complexes. An Rev Biochem 70:475–501.

Nurse P (2000): A long twentieth century of the cell cycle and beyond. Cell 100:71–8.

Onn A, Fidler IJ (2002): Metastatic potential of human neoplasms. In vivo 16:423–9.

Pardee AB (1974): A restriction point for control of normal animal cell proliferation. Proc Natl Acad Sci USA 71:1286–90.

Pardee AB (1989): G_1 events and regulation of cell proliferation. Science 246:603–8.

Roberts JM (1999): Evolving ideas about cyclins. Cell 97:129–32.

Sellers WR, Fisher DE (1999): Apoptosis and cancer drug targeting. J Clin Invest 104:1655–61.

Sherr CJ (1996): Cancer cell cycles. Science 274:1672–7.

CHAPTER 2

ARCHITECTURAL ORGANIZATION OF THE REGULATORY MACHINERY FOR TRANSCRIPTION, REPLICATION, AND REPAIR: DYNAMIC TEMPORAL-SPATIAL PARAMETERS OF CELL CYCLE CONTROL

COREY D. BRAASTAD[1], SAYYED K. ZAIDI[1], MARTIN MONTECINO[2], JANE B. LIAN[1], ANDRÉ J. VAN WIJNEN[1], JANET L. STEIN[1], and GARY S. STEIN[1]

[1]Department of Cell Biology and Cancer Center, University of Massachusetts Medical School, Worcester, MA 01655

[2]Departamento de Biologia Molecular, Facultad de Ciencias Biologicas, Universidad de Concepcion, Barrio Universitario s/n, Concepcion, Chile

INTRODUCTION

The regulatory mechanisms that mediate competency for proliferation, cell cycle progression, and exit from the cell cycle must be understood within the context of dynamic modifications in composition, organization, assembly, and activity of the machinery for replication and transcription. Equally important is a stringent requirement for sequence fidelity of genomic DNA as it necessitates editing during the replication and excision/repair of base damage that is incurred in proliferating and postproliferative cells.

We are continually acquiring insight into the complex and interdependent biochemical parameters of cell cycle and growth control. The

Cell Cycle and Growth Control: Biomolecular Regulation and Cancer, Second Edition, Edited by Gary S. Stein and Arthur B. Pardee
ISBN 0-471-25071-6

signaling pathways that govern the activation as well as suppression of genes controlling biological activity necessary for proliferation are also being functionally mapped. However, it is becoming clear that the regulatory parameters of replication and transcription that are operative during proliferation are functionally linked to cellular architecture. Thus the big challenge is to reconcile the extent to which cellular morphology contributes to the biochemistry of growth control.

Initially very subtle, though subsequently striking, modifications occur in cellular morphology, as well as in the localization of regulatory complexes, with transformation and tumor progression. Changes in the biochemistry of cell cycle control and in the temporal-spatial organization of nucleic acids and regulatory proteins are well documented. They reveal that the mechanisms regulating growth and the compromises associated with aberrant replication, repair, and transcription during tumorigenesis reflect breaches in the obligatory interrelationships between the cell's structure and biological control.

In this chapter we focus on the accruing insights into nuclear architecture and cytoarchitecture and their contributions to the subcellular localization and activity of the regulatory machinery for replication, transcription, and repair. We consider the dynamics of the regulatory complex assembly within the three dimensional context of cellular architecture but with an emphasis on the aberrations in transformed and tumor cells. Then we assess the organization of regulatory complexes that are required for genome replication and chromatin remodeling during S phase as well as for DNA repair. We look at the linkages between subcellular placement of genes, regulatory proteins, and structural components of the cell that are compatible with proliferation. Our discussion addresses the mechanisms that mediate the distribution of regulatory complexes to progeny cells for support of postmitotic gene expression. We explore the idea that a sequential and functionally interrelated series of regulatory cycles, requiring the dynamic assembly of architecturally associated regulatory complexes, support the physiological control of cell proliferation and are functionally linked to perturbations in growth regulatory mechanisms in transformed and tumor cells.

INTRANUCLEAR ORGANIZATION OF NUCLEIC ACIDS AND REGULATORY PROTEINS IN FIDELITY OF REPLICATION, REPAIR, AND TRANSCRIPTION

Nuclear Structure–Gene Expression: Architectural Contributions of Nuclear Organization to Biological Control

Physiologically responsive gene expression in an in vivo setting necessitates understanding the temporal and spatial organization, assembly, and activities of the regulatory machinery for transcription. Over the past several decades there has been spectacular progress in identifying and characterizing biochemical components of transcriptional control, and this has yielded insight into the signaling pathways that mediate gene activation and suppression. Concomitantly there have been advances in establishing the regulatory parameters of replication and repair, by which our knowledge of structural and functional components of nuclear morphology has improved. Now the challenge and opportunity is to experimentally establish the obligatory links between nuclear organization and the gene regulatory mechanisms.

The catalog of promoter elements and cognate regulatory proteins that govern gene expression offers essential but insufficient insight into mechanisms that are operative in intact cells. Gene promoters serve as a regulatory infrastructure and thus function as blueprints for responsiveness to the flow of regulatory signals. But the specific genetic information cannot be accessed without an understanding of the transcriptional control of genes in relation to the subnuclear organization of nucleic acids and regulatory proteins. Explanations are lacking for (1) the convergence of multiple regulatory signals and promoter sequences; (2) the integration of regulatory information at independent promoter domains; (3) selective utilization of redundant regulatory pathways; (4) thresholds for initiation or downregulation of transcription with limited intranuclear representation of promoter elements and regulatory factors; (5) mechanisms that render the promoters of cell growth and phenotypic genes competent for protein-DNA and protein-protein interactions in a physiologically responsive manner; (6) the composition, organization, and assembly of sites within the nucleus that support transcription; and (7) the intranuclear trafficking of regulatory proteins to transcriptionally active foci.

Similarly the present repertoire of factors that mediate DNA synthesis and repair does not yet adequately explain replication and maintenance of genome integrity. The fundamental components of key mechanisms remain unresolved. These essential processes require orchestration in a focal assembly of the machinery for replication and repair, and temporal and spatial coordination of regulatory protein recruitment for combinatorial control.

The accumulated evidence is that the architectural organization of nucleic acids and regulatory proteins within the nucleus supports the functional interrelationships between the nuclear structure and gene expression (Fig. 2.1). The components of this nuclear architecture appear to be functionally linked to the organization and sorting of regulatory information in a manner that permits selective access and utilization (Berezney et al., 1996; Gasser, 2002; Lamond and Earnshaw, 1998; Ma et al., 1999; McNeil et al., 1998, 1999; Misteli, 2000; Stein et al., 2000a; Zeng et al., 1997, 1998). At the primary level of nuclear organization—the representation and ordering of genes and promoter elements—the blueprints of alternatives is provided for physiological control. The molecular organization of regulatory elements, the overlap of regulatory sequences within promoter domains, and the multipartite composition of regulatory complexes are among the options for responsiveness. From the context dependency of modularly organized promoter sequences and juxtaposition of regulatory domains, cues emerge for the protein-DNA

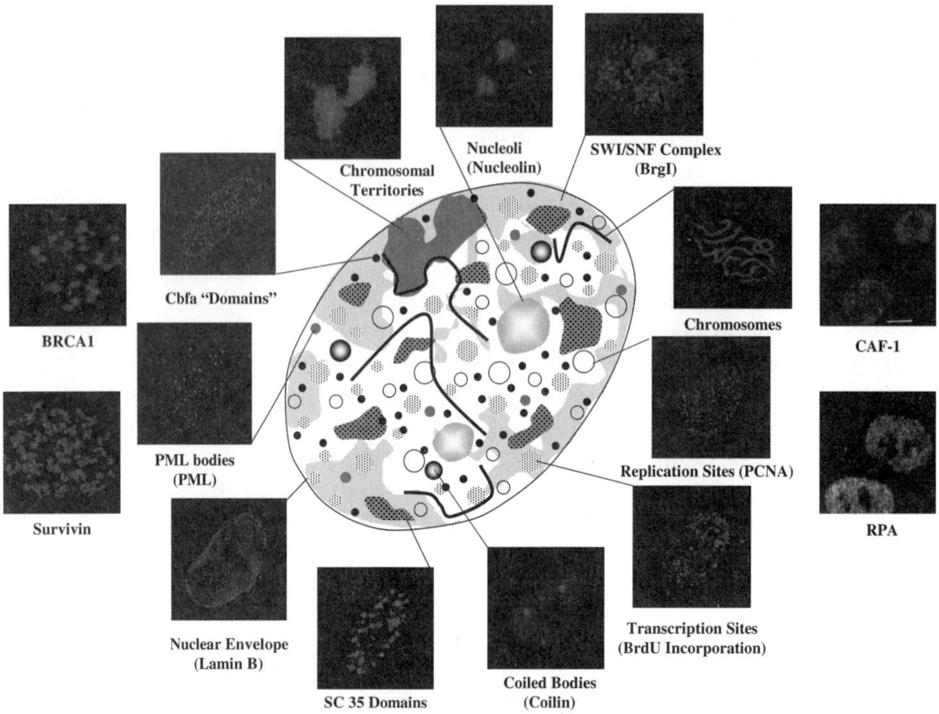

Figure 2.1. Subnuclear compartmentalization of nucleic acids and regulatory proteins into specialized domains. Nuclear functions are organized into distinct, nonoverlapping subnuclear domains. Nuclear matrix, the underlying network of anastomizing network of filaments and fibers provides structural basis for the functional compartmentalization of nuclear functions (*Center*). Immunofluorescence microscopy of the nucleus in situ has revealed the distinct subnuclear distribution of vital nuclear processes, including (but not limited to) DNA replication sites (Ma et al., 1998) and proteins involved in replication such as CAF-1 (Krude, 1995) and RPA (Fortunato and Spector, 1998); DNA damage as shown by BRCA1 (Scully et al., 1997); chromatin remodeling such as mediated by the SWI/SNF complex (Reyes et al., 1997), and Cbfa factors (Zaidi et al., 2001; Zeng et al., 1997); structural parameters of the nucleus such as the nuclear envelope, chromosomes, and chromosomal territories (Ma et al., 1999); Cbfa domains for transcriptional control of tissue-specific genes; and RNA synthesis and processing involving, for example, transcription sites (Wei et al., 1999); SC35 domains (reviewed in Shopland and Lawrence, 2000), coiled bodies (Platani et al., 2000), and nucleoli (Dundr et al., 2000) as well as proteins involved in cell survival such as survivin (Fortugno et al., 2002). Subnuclear PML bodies of unknown function (McNeil et al., 2000) have been examined in numerous cell types. All these domains are associated with the nuclear matrix. (See color insert.)

and protein-protein interactions that dictate the combinatorial assembly and organization of multicomponent regulatory complexes. The chromatin structure and the nucleosome organization reduce the distances between regulatory sequences, facilitate crosstalk between promoter elements, and render elements competent for interactions with positive and negative regulatory factors (Jaskelioff and Peterson, 2003; Peterson, 2002; Peterson and Workman, 2000). At the higher order nuclear architecture, the components include nuclear pores (Mattaj and Englmeier, 1998), the nuclear matrix, and the intranuclear domains that contribute to the bidirectional exchange of regulatory information between the nucleus and cytoplasm (Hieronymus and Silver, 2003; Kau and Silver, 2003) as well as to the subnuclear distribution and activities of genes and regulatory factors (reviewed in Berezney et al., 1996; Misteli, 2000; Penman, 1995). Compartmentalization of regulatory complexes is illustrated by focal organization of PML bodies (Dyck et al., 1994), RUNX bodies (Harrington et al., 2002; Javed et al., 1999; McNeil et al., 1998; Zeng et al., 1997), the nucleolus, chromosomes (Ma et al., 1999), as well as by the punctate intranuclear distribution of sites for replication (Cook, 1999; Leonhardt et al., 1998; Mahadevan et al., 1991), DNA repair (Difilippantonio et al., 2002), transcription (Ciejek et al., 1983; Cook, 1999; Guo et al., 1995; Htun et al., 1996; Kimura et al., 1999; Merriman et al., 1995; Stenoien et al., 1998; van Steensel et al., 1995; Verschure et al., 1999; Wei et al., 1998), steroid and polypeptide modulation of gene expression (reviewed in DeFranco, 2002), and the processing of gene transcripts (Misteli and Spector, 1999; Smith et al., 1999). There is an indication that nuclear structure and function are causally interrelated from evidence that nucleic acids and regulatory proteins in the subnuclear domains are associated with components of nuclear architecture. Thus, rather than a dichotomy in the nuclear architecture with regard to the control of gene expression, there is a mechanism that directs genes and regulatory factors to sites within the nucleus where the regulatory parameters of gene expression establish microenvironments with boundaries between the regulatory complexes.

Compartmentalization of Regulatory Machinery within the Nucleus: Focal Thresholds for Formation of Regulatory Complexes

The compartmentalization of the regulatory machinery for replication and transcription is documented by longstanding biochemical and in situ evidence. Key components of the replication and basal transcription machinery as well as several tissue-specific transcription factor complexes are functionally compartmentalized as specialized, punctate subnuclear domains (DeFranco, 2002; Gasser, 2002; Misteli, 2000; Stein et al., 2000a). It has been demonstrationed that some regulatory domains exhibit similar compartmentalized subnuclear distributions in living cells, and the experimental findings confirm the physiological importance of these nuclear microenvironments (Stein et al., 2000b).

Nuclear Microenvironments. Compartmentalization is particularly evident in the tissue-specific RUNX proteins. This may even be, in part, a characteristic biological constraint in the control of the phenotype-specific transcription in nuclei of intact bone and hematopoietic cells.

Intuitively, and on the basis of experiments, the low representation of promoter regulatory elements and cognate transcription factors suggests that a subnuclear organization of nucleic acids and regulatory proteins supports the threshold concentrations for the activation and repression of gene expression. Over the past several years there has been growing recognition that the organization of nucleic acids and regulatory proteins is functionally linked to the assembly, organization, and activity of gene regulatory machinery. Cellular, molecular, biochemical, and genetic evidence indicates an obligatory relationship between sites within the nucleus where regulatory complexes reside and fidelity of transcriptional control. The biological relevance for the intranuclear distribution of RUNX-containing regulatory complexes is directly reflected by the importance of focal localization of RUNX proteins within the nucleus for tissue-specific transcription (Zeng et al., 1997, 1998) and by aberrant nuclear structure-gene expression interrelationships that are associated with perturbations in skeletal development (Choi et al., 2001) and leukemia (Barseguian et al., 2002; McNeil et al., 1999).

The punctate subnuclear localization and nuclear matrix association of ALL foci (Yano et al., 1997), the glucocorticoid receptor (Tang et al., 1998b), the estrogen receptor (Stenoien et al., 2000), the androgen receptor (van Steensel et al., 1995), and the thyroid hormone receptor are further consistent with compartmentalization and focal concentrations of regulatory machinery for hormone-responsive integration of regulatory signals. Experiments executed in living cells with green fluorescent protein tagged glucocorticoid receptors directly demonstrate agonist-dependent relationships between architectural organization and transcriptional activation (Becker et al., 2002). A striking and clinically relevant example of perturbations in regulatory activity that results from modifications in the intranuclear distribution of receptors is illustrated by PML bodies (Zelent et al., 2001). A limited number of PML bodies contain proteins that mediate physiological control in hematopoietic cells. While, in contrast, chromosomal translocations that involve the RAR locus are characteristic of promyelocytic leukemia, resulting in altered composition, number, and intranuclear localization of PML bodies that appear to be associated with alterations in expression of RAR target genes. Chromosomal rearrangements at the ALL (Pekarsky et al., 2001; Yano et al., 1997) and AML (Rowley, 1999) loci similarly result in altered composition and subnuclear placement of regulatory complexes containing the encoded proteins that are associated with tumor-related changes in gene regulatory mechanisms.

The Nucleolus. For many years the nucleolus has provided a paradigm for compartmentalization of the regulatory machinery for ribosomal gene expression. Biochemical fractionation and characterization of ribosomal genes, transcripts, and nucleolar proteins has been highly informative. The ribosomal genes, which are encoded in several chromosomes at multiple loci, are organized at two sites in normal diploid cells. Dynamic and specific changes in the composition, organization, and activities of nucleolar-associated proteins and nucleic acid-protein interactions occur during the cell cycle. Modifications in the number and structural as well as functional properties occur in transformed and tumor cells. Mechanisms that mediate the structural and functional properties of the nucleolus are revealing nucleolar involvement in regulatory activities that extend beyond ribosomal gene expression.

Replication/Repair Domains. Several lines of evidence implicate compartmentalization of the regulatory machinery for replication in biological activity. Biochemical fractionation has yielded multipartite "replitase" complexes that contain combinatorial components of the enzymology for DNA synthesis and mediators of signaling that interfaces replication with parameters of phenotypic and growth-related regulatory pathways (Reddy and Pardee, 1980; Studzinski et al., 1991). The well-documented punctate organization of replication sites within the nucleus is consistent with focal thresholds (Wei et al., 1998). Recent reports that BRCA foci are linked to DNA repair provide yet another example of architecturally organized regulatory proteins within the nucleus that are compartmentalized (Scully and Livingston, 2000). The striking modification in the representation of BRCA foci following radiation-induced base damage and alterations in the number as well as intranuclear distribution of BRCA foci in tumor cells offers potentially relevant insight into regulatory mechanisms, tumor diagnosis, and therapy (Scully and Livingston, 2000).

Chromosome Territories. Mitotic chromosomes have long been the consummate example of compartmentalized regulatory machinery. Chromosomes are collectively the genetically defined, ordered, and conformationally organized repository of templates for the structural and functional properties of cells, tissues, and organisms. This genetically encoded encyclopedia of information is subdivided and compartmentalized as a series of chromosomes. Each chromosome group, in response to nucleotide sequences, protein-DNA, and protein-protein interactions, as well as RNA, is configured in a manner that is compatible with activation or suppression of genes during interphase as well as with the requirements for chromosome condensation and segregation during mitosis and meosis. Selectivity of the required control for chromosomal compartmentalization is subtly reflected by regions of chromosomes. These regions are organized in a manner that is compatible with accessibility to the regulatory factors that repress expression or render genes competent for transcription. Every chromosome utilizes centromeric sequences for association with the mitotic apparatus that dynamically mediates chromosomal localization and distribution during cell division. A highly specific mechanism for compartmentalization that is invoked for selective inactivation of a single copy of the X chromosome requires both proteins and XIST RNA (Clemson et al., 1996). Thus it is becoming evident that chromosomal compartmentalization, in a manner that is compatible with regulation of gene expression and positioning within the cell, requires mechanisms that are operative for all chromosomes as well as restricted to specific chromosomes.

Recent results provide insight into the positioning of chromosomes during the cell cycle. From the noninvasive labeling of chromosome subsets in living cells there is good evidence that global chromosome positions are heritable through the cell cycle in mammalian cells. By the combined use of approaches that include tracking of labeled chromosomes during segregation and experimental perturbations of chromosomal order, it appears that chromosome-specific timing of chromatid segregation is a determinant for bridging the signal for chromosomal positioning between cell generations (Gerlich et al., 2003). These observations are consistent with an emerging consensus for a mechanism that controls chromosomal compartmentalization in a manner where gene-rich chromosomes are preferentially localized in the nuclear interior while gene-deficient chromosomes predominantly localize in the proximity of the nuclear envelope (Boyle et al., 2001; Croft et al., 1999; Sun et al., 2000; Tanabe et al., 2002). Further support for maintenance of chromosome positions during interphase is provided by Walter and coworkers (Walter et al., 2003). These investigators demonstrate that chromosomal territories are established early in G_1 and are maintained until the completion of G_2. However, Walter et al. (2003) report major changes in chromosome territories during mitosis that modify chromosomal localization from one cell cycle to the next. Thus there is consensus that compartmentalization of chromosomes in interphase nuclei contributes to selective expression of genes. But heritable positioning in somatic cells remains open ended. These are important parameters of subcellular compartmentalization from a fundamental regulatory perspective and relevant to understanding nuclear structure-gene expression interrelationships that relate to tumorigenesis. The proximity of chromosomes may facilitate homologue pairing and, at least in part, account for contributions of nuclear microenvironments to chromosomal translocations (Parada and Misteli, 2002; Sachs et al., 1997). The notion that compartmentalization of chromosomes within the nucleus is conducive to tumor-associated translocations is supported by data from Parada et al. (2002) for the physical proximity of chromosomes undergoing translocations in a mouse lymphoma model.

Architectural Compartmentalization of Gene Expression. Examples that illustrate the involvement of nuclear compartmentalization in biological control are numerous. We have focused on several to emphasize the diverse regulatory activities that occur predominantly in nuclear microenvironments and the extent to which functional interrelationships with components of nuclear architecture are apparent. However, analogous interrelationships between nuclear structure and compartmentalization of gene expression are reflected by intranuclear sites that support processing of gene transcripts (Smith et al., 1999), the subnuclear distribution of the Cajal bodies (Gall, 2000) for S phase specific expression of histone genes, localization of apoptosis-related factors that include survivin (Altieri, 2003), and localization of components for the p53 and RB tumor suppressor mechanisms (Mancini et al., 1994) within the nucleus. The observation that several regulatory domains exhibit the same composition and subnuclear distribution in living cells and in fixed preparations confirms the physiological relevance of these nuclear microenvironments.

The evidence is compelling for compartmentalization of the regulatory domains that are requisite for gene expression, replication, and repair. The results of biochemical, cellular, molecular, and in vivo genetic studies point to a pivotal role of architecturally associated nuclear microenvironments in biological control and perturbations of nuclear microenvironments in tumor cells. However, mechanisms for the organization and assembly of sites within the nucleus that support regulatory activities are minimally understood. To what extent are subnuclear compartments physically associated with nuclear architecture, or is the nuclear scaffold a composite organization of regulatory microenvironments? How is the representation of regulatory proteins within subnuclear compartments modified in response to biological cues? How are regulatory proteins directed to intranuclear foci to support the organization, assembly, and physiologically responsive remodeling of sites within the nucleus that support transcription, replication, and repair? More understanding of regulatory compartmentalization within the nuclear architecture is needed to provide novel options for tumor diagnosis and selective targeting of therapy.

Intranuclear Trafficking of Regulatory Proteins to Subnuclear Sites for Dynamic Assembly and Activities of Cellular Regulatory Machinery

There is a need to understand the targeting and/or recruitment mechanism of regulatory and co-regulatory factors at the subnuclear sites where the machinery for gene activation and suppression is assembled.

Transcription Factor Targeting: Being in the Right Place at the Right Time. The architectural association of osteoblast, myeloid, and lymphoid RUNX transcription factors that mediate tissue-specific transcription (Bae et al., 1993; Banerjee et al., 1996, 1997; Ducy et al., 1997; Merriman et al., 1995; Nuchprayoon et al., 1994) has permitted direct examination of mechanisms for targeting regulatory proteins to transcriptionally active subnuclear domains. Both biochemical and immunofluorescence analyses have shown that RUNX transcription factors exhibit a punctate nuclear distribution that is associated with the nuclear matrix in situ (Zaidi et al., 2001; Zeng et al., 1997, 1998). Taken together, these observations are consistent with the concept that the nuclear matrix is functionally involved in gene localization and in the concentration and subnuclear localization of regulatory factors (Stein et al., 2000a).

The initial indication that nuclear matrix association of RUNX factors is required for maximal activity was provided by the observation that transcriptionally active RUNX proteins associate with the nuclear matrix but inactive C terminally truncated RUNX proteins do not (Zaidi et al., 2001; Zeng et al., 1997, 1998). This localization of RUNX was established by biochemical fractionation and in situ immunofluorescence as well as by green fluorescent protein tagged RUNX proteins (Harrington et al., 2002) in living cells. Colocalization of RUNX1, 2, and 3 at nuclear matrix-associated sites indicates a common intranuclear targeting mechanism may be operative for the family of RUNX transcription (Harrington et al., 2002; Javed et al., 2000; Zeng et al., 1997, 1998). Variations in the partitioning of transcriptionally active and inactive RUNX between subnuclear fractions permitted development of a strategy to identify a region of the RUNX transcription factors that directs the regulatory proteins to nuclear matrix-associated foci. A series of deletions and internal mutations were constructed and assayed for competency to associate with the nuclear matrix by western blot analysis of biochemically prepared nuclear fractions and by in situ immunostaining following transfection into intact cells. Association of osteogenic and hematopoietic RUNX proteins with the nuclear matrix is independent of DNA binding and requires a nuclear matrix targeting signal, a 31 amino acid segment near the C terminus that is distinct from nuclear localization signals (Zeng et al., 1997, 1998). The nuclear matrix targeting signal functions autonomously and is necessary as well as sufficient to direct the transcriptionally active RUNX transcription factors to nuclear matrix-associated sites where gene expression occurs (Zeng et al., 1997, 1998).

Specificity of the RUNX intranuclear targeting signal is directly provided by sequence (Zeng et al., 1997, 1998) and structural (Tang et al., 1999) similarity of the 31 amino acid C terminal regulatory domains in the hematopoietic and osteogenic RUNX transcription factors as well as by the absence of a comparable sequence in other regulatory proteins that are accessible in databases. Additional evidence for specificity of intranuclear targeting signals are unique sequences that support subnuclear trafficking to nuclear matrix-associated sites in the glucocorticoid receptor (DeFranco and Guerrero, 2000), the estrogen receptor (Stenoien et al., 2000), DNA polymerase (Leonhardt et al., 1998), the AML-ETO translocation fusion protein that mediates aberrant gene expression in acute myelogenous leukemia cells with an 8;21 translocation (Barseguian et al., 2002) and in nucleolar proteins.

These findings indicate mechanisms involved in the selective trafficking of proteins to specialized domains within the nucleus where they become components of functional regulatory complexes. At least two trafficking signals appear to be required for subnuclear targeting of RUNX transcription factors: the first supports nuclear import (nuclear localization signal) and the second mediates association with the nuclear matrix (nuclear matrix targeting signal). The multiplicity of determinants for nuclear localization and alternative splicing of RUNX messenger RNA may provide the requisite complexity to support targeting to specific sites within the nucleus in response to diverse biological conditions. Furthermore, because gene expression by RUNX involves contributions by factors and coregulatory proteins that include CBFβ (Banerjee et al., 1996; Gutierrez et al., 2002), ETS-1 (Mao et al., 1999) and C/EBP (Gutierrez et al., 2002; Zhang et al., 1996), Groucho/TLE (Javed et al., 2000; Levanon et al., 1998), HES and SMAD (Zaidi et al., 2001; Zhang et al., 2000c), RUNX may facilitate recruitment of these factors to the nuclear matrix.

Linkage of Aberrant Intranuclear Trafficking with Developmental Arrest and Leukemia. There are biological consequences of perturbations in the subnuclear organization of regulatory complexes. The essential role of RUNX2 in osteogenesis has provided a model to investigate the importance of fidelity of subnuclear localization for tissue differentiation. When the intranuclear targeting signal is deleted by homologous recombination, mice homozygous for the deletion (RUNX2ΔC) do not form bone due to perturbed maturation or arrest of osteoblasts (Choi et al., 2001). Heterozygotes do not develop clavicles but are otherwise normal. These phenotypes are indistinguishable from those of the homozygous and heterozygous null mutants (Komori et al., 1997; Otto et al., 1997), indicating that the intranuclear targeting signal is a critical determinant for function. The expressed truncated RUNX2ΔC protein enters the nucleus and retains normal DNA binding activity, but shows complete loss of intranuclear targeting (Choi et al., 2001). These results establish that the multifunctional N-terminal region of the RUNX2 protein is not sufficient for biological activity. Thus subnuclear localization of RUNX factors in specific foci together with associated regulatory functions is essential for control of RUNX-dependent genes involved in tissue differentiation during embryonic development (Choi et al., 2001). The importance of subnuclear localization of RUNX transcription factors for biological control is further indicated by compromised subnuclear organization and activity of RUNX1 hematopoietic regulatory proteins in acute myelogenous leukemia (McNeil et al., 1999).

Architectural versus Activity-Driven Assembly of Regulatory Foci. It would be presumptuous to propose a single model to account for the specific pathways that direct regulatory factors to sites within the nucleus that support transcription, replication or repair. However, findings suggest that parameters of nuclear architecture functionally interface with components of gene expression and DNA synthesis. The involvement of nuclear matrix-associated regulatory factors with recruitment of regulatory components to modulate replication, transcription, and repair remains to be defined. Working models that serve as frameworks for experimentally addressing components of transcriptional control, replication, or repair within the context of nuclear architecture can be compatible with mechanisms that involve architecturally or activity driven assembly of transcriptionally active intranuclear foci competent to support regulatory activity. The diversity of targeting signals must be established to evaluate the extent to which regulatory discrimination is mediated by encoded intranuclear trafficking signals. It will additionally be important to biochemically and mechanistically define the checkpoints which are operative during subnuclear distribution of regulatory factors, and editing steps which are invoked to ensure structural and functional fidelity of nuclear domains where replication and expression of genes occur. There is emerging recognition that placement of regulatory components of gene expression must be temporally and spatially coordinated to optimally mediate biological control. It is realistic to anticipate that further understanding of mechanisms that dynamically position genes and regulatory factors for establishment and maintenance of cell phenotypes will clarify nuclear structure-function interrelationships that are operative during proliferation and differentiation and are physiologically responsive to modulation of regulatory activity.

Nuclear Architecture and Temporal-Spatial Integration of Physiological Regulatory Signals

Nuclear import, retention, and export support the exchange of regulatory macromolecules between the nucleus and cytoplasm. The entry and exit of nucleic acids and regulatory proteins from the nucleus are becoming increasingly important parameters of biological control within the contexts of modulating the flow of physiological signals and the intranuclear levels of components for assembly, organization, and activity of the machinery for replication, repair, and transcription. There are numerous examples of functional linkage between import of regulatory proteins and modified regulatory mechanisms that are associated with the onset and progression of tumorigenesis. These include but are not restricted to the IGF signaling pathway in hematopoiesis and leukemogenesis (Sun et al., 2003; Tu et al., 2002; Wu et al., 2003), translocation fusion proteins in myeloid leukemias (Wu et al., 2003), RB in osteosarcomas (Thomas et al., 2001), and APC/β catenin signaling in colon cancer (Neufeld et al., 2000b).

The subnuclear compartmentalization of transcription machinery necessitates a mechanistic explanation for directing signaling factors to sites within the nucleus where gene expression occurs under conditions that support integration of regulatory cues. The necessity for architectural compartmentalization of signaling mechanisms is illustrated by gene expression during skeletal development and bone remodeling. The broad spectrum of regulatory signals that control gene expression converge on promoter elements to activate or suppress transcription in a physiologically responsive manner (Fig. 2.2). The interactions of YAP and SMAD coregulatory proteins with C-terminal segments of the RUNX2 transcription factor permits assessment of requirements for recruitment of Src and BMP/TGFβ-mediated signals to skeletal target genes. Recent findings indicate that nuclear import of YAP and SMAD coregulatory factors is agonist dependent. However, there is a stringent

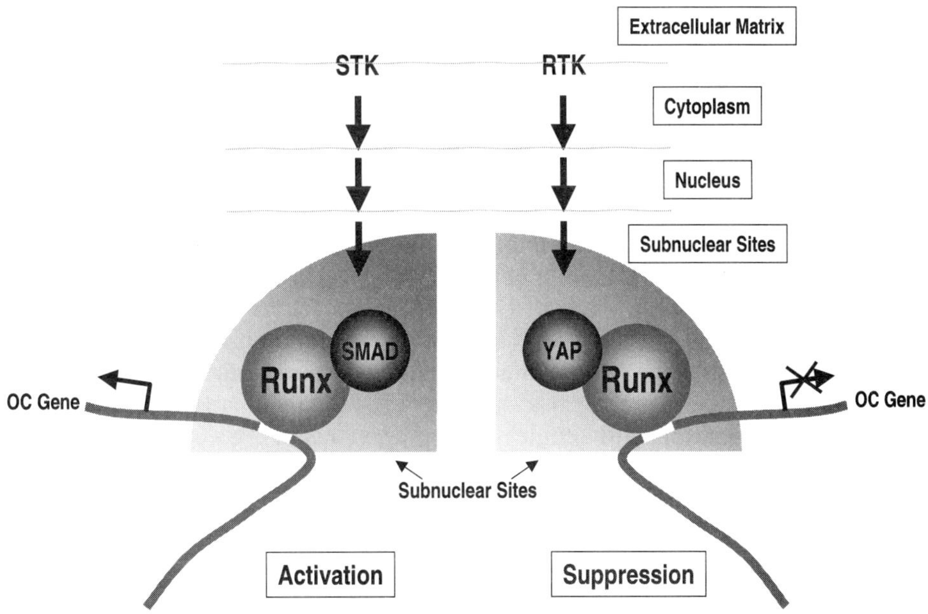

Figure 2.2. Structural and functional integration of regulatory signals at subnuclear sites. Extracellular signaling cascades are triggered by a variety growth factors through the activation of plasma membrane associated receptors. Depicted here are two examples of such receptors: serine threonine kinases (STK) and receptor tyrosine kinases (RTK). Activation of these kinases leads to the phosphorylation of downstream proteins (exemplified here by SMADS, downstream effectors of TGFb/BMP pathway and YAP, a downstream target of Src/Yes tyrosine kinase family) in the cytoplasm. These signaling proteins are then translocated into the nucleus where they interact with several transcription factors such as Runx proteins. In case of SMADS and YAP, Runx transcription factors interact with these proteins in the nucleoplasm and target them to the nuclear matrix-associated sites where these signaling proteins activate (in case of SMADS) or suppress (in case of YAP) Runx target genes. Thus transcription factors functionally and structurally integrate signaling cascades at subnuclear sites where activation or suppression of the target genes takes place.

requirement for fidelity of RUNX subnuclear targeting to recruit these signaling proteins to transcriptionally active or suppressed subnuclear foci. These results demonstrate that the interactions and spatial-temporal organization of RUNX and SMAD as well as YAP coregulatory proteins are essential for assembly of machinery that controls expression of skeletal genes (Zaidi et al., 2001, 2002, 2004). Competency for intranuclear trafficking of RUNX proteins has similarly been functionally linked with the subnuclear localization and activity of TLE/Groucho coregulatory proteins (Javed et al., 2000). These findings are consistent with nuclear matrix-associated proteins serving as a scaffold for interactions with coregulatory proteins that contribute to biological control.

Scaffolding of Regulatory Components for Combinatorial Control of Gene Expression and Replication

Multiple lines of evidence suggest that components of nuclear architecture contribute both structurally and enzymatically to control gene expression and replication during proliferation and differentiation. Sequences have been identified that direct regulatory proteins to nuclear matrix-associated sites that support replication (Leonhardt et al., 1998) and transcription (Zeng et al., 1997, 1998). Insight is thereby provided into mechanisms linked to the assembly and activities of specialized subnuclear domains where replication and transcription occur. In a restricted sense, the foundation has been provided for experimentally addressing intranuclear trafficking of gene regulatory factors and control of association with architectural components of the nucleus to establish and sustain domains that are competent for DNA and RNA synthesis. The unique sequences (Zeng et al., 1997, 1998) and crystal structure for the 31 amino acid nuclear matrix targeting signal of RUNX transcription factors (Tang et al., 1998a) supports specificity for localization at intranuclear sites where the regulatory machinery for gene expression is assembled, rendered operative, and/or suppressed. In a broader context, there is growing appreciation for involvement of nuclear architecture in a dynamic and bidirectional exchange of gene transcripts and regulatory factors between the nucleus and cytoplasm, as well as between regions and structures within the nucleus (Lamond and Earnshaw, 1998; Stein et al., 2000a; Gasser, 2002; Misteli, 2000).

Functional interrelationships between nuclear structure and gene expression are strikingly reflected by dual recognition of regulatory proteins. Included are the RUNX transcription factors for interactions with both promoter elements and coregulatory proteins that modulate the structural and functional properties of targeted genes at microenvironments within the nucleus. Sequence-specific interactions with promoter elements result in placement of RUNX proteins at strategic sites where they provide scaffolds for protein-protein interactions that mediate the organization of machinery for a broad spectrum of regulatory requirements. Among these interactions are histone modifications and chromatin remodeling, which establish competency for transcription factor binding, genomic conformations that interface activities at proximal and upstream promoter domains, and the integration of regulatory cues from signaling pathways that activate of suppress gene expression in a physiologically responsive manner. As a consequence the RUNX proteins are post-translationally modified (e.g., phosphorylated) to further influence the extent to which they engage in biological control (Fig. 2.3).

The complexity of the ALL-1 regulatory protein that assembles as a supercomplex of transcriptional regulatory factors illustrates the potential impact of leukemia-related chromosomal translocations on gene expression (Nakamura et al., 2002). Recent documentation that ALL-1 is a stable complex that includes basal transcription factors, chromatin remodeling factors, and histone modifying factors indicates the scope

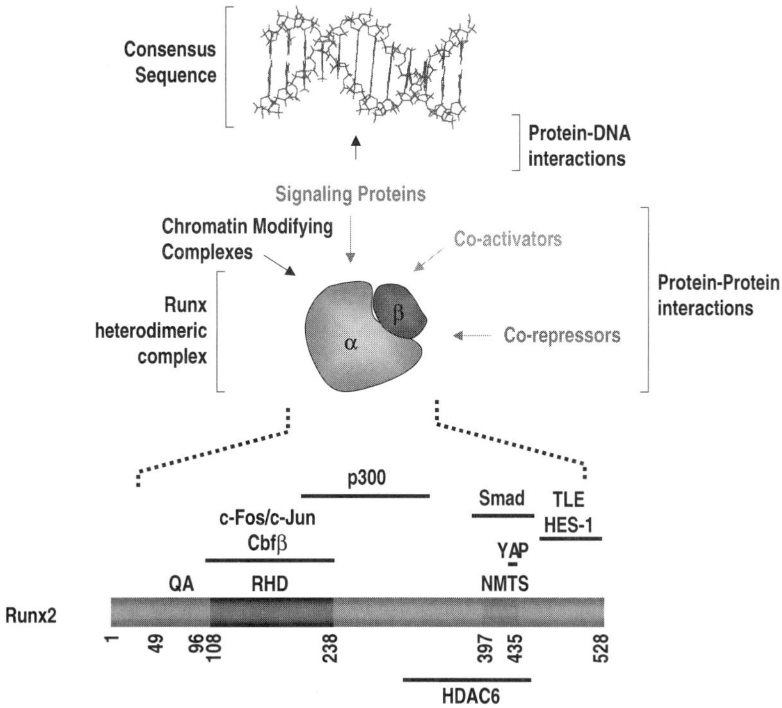

Figure 2.3. Scaffolding nuclear proteins: A mechanism of specificity in gene regulation. Nuclear transcription factors often function as scaffolding proteins that integrate multiple physiological cues on gene promoter elements and subnuclear sites for transcriptional regulation. One such nuclear transcription factor is Runx/Cbfa/AML, a heterodimeric protein complex that is targeted to the nuclear matrix associated sites, interacts with a variety of proteins in the nucleus, and binds DNA in a sequence specific manner. Several transcription factors, co-regulators, and signaling proteins interact with Runx factors at various regions of the proteins as depicted in the bottom panel. Runx factors thus serve as scaffolding proteins; they integrate functions of several co-regulators as well as signaling proteins downstream of key extracellular signaling pathways at the subnuclear regulatory sites and gene promoters. Such a scaffolding function renders tissue specificity in control of gene transcription. (See color insert.)

of combinatorial control that is vulnerable as a consequence of gene rearrangements.

Transcription factors that function as scaffolds for interaction with coregulatory proteins provide an architectural basis for accommodating the combinatorial requirements of biological control. Combinatorial control supports the replication, transcription, and repair by two mechanisms. Context-dependent combinations and permutations of regulatory proteins are assembled into multipartite complexes that increase specificity. Scaffold-associated protein-DNA and protein-protein interactions permit integration of regulatory activities. Nuclear microenvironments are thereby organized, with gene promoters as focal points, where threshold concentrations of regulatory macromolecules are attained. The complexity that is achieved by these architecturally organized oligomeri factors can maximize options for responsiveness to diverse regulatory requirements for transient and long-term biological control.

TEMPORAL-SPATIAL PARAMETERS OF CELL CYCLE CONTROL

Nucleolar Cycle: Programmed Remodeling of Regulatory Machinery of Ribosomal Biogenesis

Cell cycle-dependent transitions in structural and functional properties of the nucleolus are well documented and reflect proliferation dependent requirements for protein synthesis. Nucleoli are the most obvious example of nonmembrane-bound structure within the membrane-bound nucleus. Visible since the early days of microscopy, nucleoli have been heavily studied since. Early drug inhibitory experiments and more recent knockdown studies have revealed that the structure of the nucleolus depends on transcription by RNA polymerase I (RNAP I), and to a lesser extent, RNAP II (Hadjiolov, 1985; Oakes et al., 1998). Consistent with RNAP I transcribing pre-ribosomal RNAs, ribosomal biogenesis has been found to be the predominant function of the nucleolus (Olson et al., 2002; Schwarzacher and Mosgoeller, 2000).

Much is known about the ultrastructure and protein composition of the nucleolus. Although rDNA genes are required for nucleolar structure, they are clearly not sufficient for full formation of functional and dynamic nucleoli (Scheer and Hock, 1999). Two ultrastructures found in all nucleoli are the granular component (GC) and dense fibrillar component (DFC), while a third structure, the fibrillar center (FC) is missing from yeast. In higher eukaryotes, however, the FC forms a nearly spherical center around which the DFC is wrapped. The DFC is comprised of nascent and intermediate pre-rRNP particles and active rRNA gene transcription by RNAP I occurs near its boundary with the FC. The GC plays a role in the processing of ribosome subunits as the assemblage continues and finishes (Olson et al., 2002).

A striking feature of the structure of the nucleolus is its ability to cycle and recycle. As cells begin to enter early mitosis, ribosomal gene transcription begins to wane and the nucleolus begins to structurally disassemble. The disassembly is clearly an organized process as nucleolar organizer regions (NORs) are formed around chromosomal loci representing tandemly repeated rRNA genes (Huang et al., 1997). Interestingly, each NOR is competent to construct a nucleolus upon re-entry of interphase (Hernandez-Verdun et al., 2002). Additionally nucleolar position within daughter cells is maintained from mother cells during division, likely via association of NORs with rDNA loci whose position is generally conserved within an interphase nuclei (Gerlich et al., 2003).

These NORs, by definition of their association with chromosomal loci, are conserved from mother to daughter cells in a common structural mechanism known as the perichromosomal space, or layer (Huang et al., 1997).

As nucleolar structure changes during the cell cycle, changes also occur during tumorigenesis. Scoring for the number and size of silver-stained NORs (AgNORs), which contain two proteins involved in rRNA transcription and processing, is part of a repertoire of tumor pathology done to determine the malignance of some cancers (Roussel and Hernandez-Verdun, 1994; Trere, 2000). The strong growth phenotype, and therefore high ribosomal biogenesis levels, of cancers seems to correlate with AgNOR counts.

Tumor-suppressor regulation is linked to the nucleolus. Cells are likely to have evolved this link as an efficient way to functionally relate ribosome biogenesis with cell cycle progression via cell cycle checkpoint controls (Tsai and McKay, 2002). The tumor-suppressor protein p53, a protein mutated in over half of all tumors, is under direct inhibitory control by Mdm2 which is sequestered and attenuated in the nucleolus by $p19^{ARF}$ binding (Lohrum et al., 2000a; Lohrum et al., 2000b; Tao and Levine, 1999). $p19^{ARF}$ contains a complex nucleolar localization signal (NoLS), which results in its continual retention within the nucleolus (Rizos et al., 2000). Many tumors have been found to overcome the growth blocks imposed by the p53/ARF proteins by mutating and/or inactivating the transcription of these genes (Tsai and McKay, 2002).

Nucleolar ARF has also been shown to sequester and inactivate levels of both the nucleoplasmic transcriptional factors E2F-1, -2, and -3 and its cytoplasmic functional partner DP1 (Datta et al., 2002). ARF seems only to sequester free forms of the proteins, and not the heterodimeric form, thereby regulating well-characterized E2F-mediated progression of the cell cycle into early S phase. The nucleolus may contribute to sensing and mediating the balance of protein turnover versus ribosomal de novo synthesis of proteins that is vital to cellular survival (Dantuma and Masucci, 2002; Hadjiolov, 1985; Merker and Grune, 2000; Szweda et al., 2002).

Cell Structure and Gene Expression at the G_1-S Phase Transition

Fidelity of chromatin organization is critical for proper control of gene expression during the cell cycle, as well as for the orchestrated separation of a full complement of intact chromosomes at mitosis. The mitotic condensation of DNA near the end of each cell division represents one of the most compelling illustrations of a cell cycle-dependent structural modification in chromatin. Apart from the structural cycle of chromatin condensation/decondensation that occurs once during each cell division, cell cycle-dependent remodeling of chromatin occurs at selected gene loci to accommodate the stage-specific expression of genes required for the cell cycle. At each of these loci, there are reversible modifications in nucleosomal organization that, by increasing accessibility of gene regulatory elements, support formation of a promoter architecture capable of integrating the activities of different transcription factors. The local chromatin structure of promoters must remain flexible to signals that terminate, attenuate, or sustain transcriptional initiation. Recent evidence suggests that cells may prolong spatial integrity of promoter architecture by stabilizing factors and co-regulators at specific subnuclear foci that dynamically assimilate and discharge their resident proteins. The entry and exit of transcription factors at gene regulatory foci represent cyclical events in the milisecond time scale while the subnuclear foci remain stable for hours and exhibit cell cycle-dependent modifications relative to mitosis.

Stringent Requirement for Coupling Histone Gene Expression with DNA Replication. The core histones H2A, H2B, H3, and H4 are the key proteins that support the structural integrity of the genome and regulate accessibility of promoters to cognate transcription factors. Two each of the four core histone subtypes form the histone octamer, which packages approximately 0.2 kb of DNA into nucleosomes. Nucleosomal DNA permits folding of DNA into higher order chromatin structures to accommodate the inclusion of the linear genome within the limited dimensions of the nucleus. During each S phase, newly synthesized DNA must be immediately packaged into nucleosomes. This structural requirement is reflected by the stringent functional coupling between histone gene expression and DNA replication. The coupling defines an S phase-related cyclical event, which involves the de novo synthesis of histone mRNA and protein as well as the stoichiometric ordering of histone octamers and the precise incorporation of these octamers at nascent DNA near progressing DNA replication forks.

Temporal-Spatial Identity of Programmed Gene Expression at the R Point versus G_1-S Phase Transition. The onset of de novo synthesis of histone mRNAs is temporally restricted to the G_1/S phase transition by both transcriptional and post-transcriptional mechanisms. Recent data indicate that the transcriptional activation of histone genes at the G_1/S phase transition (S point) is temporally, functionally, and spatially distinct from transcriptional mechanisms at the restriction point (R point). The spatial distinction in R-point versus S-point control is the localization of clustered histone gene loci at Cajal bodies, which is in part modulated during the cell cycle. The functional differences between R-point related genes, which prepare the cells for onset of DNA synthesis, and S-point related genes, which are required for subsequent events, are consistent with the temporal distinction between the two cell cycle transitions.

The temporal-functional aspects of gene expression in late G_1 and early S phase have been elucidated in considerable detail. Passage beyond the G_1/S boundary depends initially on the activation of the cyclin/cyclin-dependent kinase (CDK) cascade by growth factors and the induction of the cyclin E/CDK2 kinase complex at the R point (Dou et al., 1993; Harper and Adams, 2001; Morgan, 1997; Murray and Hunt, 1993; Pardee, 1974; Paulovich et al., 1997). At the R point, when cell cycle progression becomes growth factor independent, cells prepare for the onset of DNA replication by modulating the expression of genes that are directly or indirectly required for DNA synthesis. Many of the genes that are activated at the R-point are controlled by the E2F class of factors, which regulate expression of genes encoding enzymes involved in nucleotide metabolism (i.e., thymidine kinase and dihydrofolate reductase) (Dou et al., 1993; Nevins, 2001; Trimarchi and Lees, 2002). Subsequently, at the onset of S phase, de novo synthesis of histone proteins is required to package nascent DNA into chromatin immediately upon initiation of DNA synthesis (Osley, 1991; Stein et al., 1996). The exquisitely stringent coupling between histone biosynthesis and DNA replication is illustrated by the coordinate transcriptional activation of the 14 distinct human genes encoding histone H4, the most highly conserved nucleosomal protein (Green et al., 1984; Lichtler et al., 1982; Osley, 1991; Pauli et al., 1987). However, the cell cycle regulatory mechanisms that control transcription of the genes for histone H4 and other histones (i.e., H1, H2A, H2B, and H3) function independently of E2F at the onset of S phase (Osley, 1991; Ramsey-Ewing et al., 1994; van Wijnen et al., 1996). Thus, gene regulatory mechanisms controlling histone genes and E2F-dependent genes are temporally and functionally distinct (Fig. 2.4A).

Histone Gene Expression as a Paradigm for Transcriptional Control at the Initiation of S Phase. The human histone H4 gene promoter has been used extensively as a paradigm to define the key gene regulatory factors that control transcription and ultimately the stoichiometric biosynthesis of the four classes of histone protein (H4, H3, H2B, and H2A) that together form nucleosomes (e.g., Last et al., 1998, 1999a,b; Ramsey-Ewing et al., 1994; Shakoori et al., 1995; van den Ent et al., 1993, 1994; van der Meijden et al., 1998; van Wijnen et al., 1997). Thus far, at least three functionally distinct histone gene transcription factors have been identified that (1) activate histone gene transcription in proliferating cells, (2) enhance mRNA synthesis at the G_1/S phase transition, or (3) suppress transcription (Stein et al., 1992, 1996). For example, the histone H4 gene contains two sites of in vivo genomic protein/DNA interactions, designated sites I and II (Pauli et al., 1987). Site I represents an enhancer element of basal histone gene transcription and interacts with a series of transcription factors (i.e., SP-1/HiNF-C, YY1/HiNF-I, HMG-I/HiNF-A, and ATF factors) that together stimulate histone H4 gene transcription (Birnbaum et al., 1995a,b; Guo et al., 1997; Last et al., 1999a). Site II mediates cell cycle control at the G_1/S phase transition and has been shown to interact with at least three distinct DNA binding proteins (i.e., IRF2/HiNF-M, CDP-cut/HiNF-D, and HiNF-P) (van Wijnen et al., 1991, 1992, 1994, 1996; Vaughan et al., 1995, 1998). Point mutational analyses have revealed that each of these proteins contributes to control of histone H4 gene transcription during the cell cycle

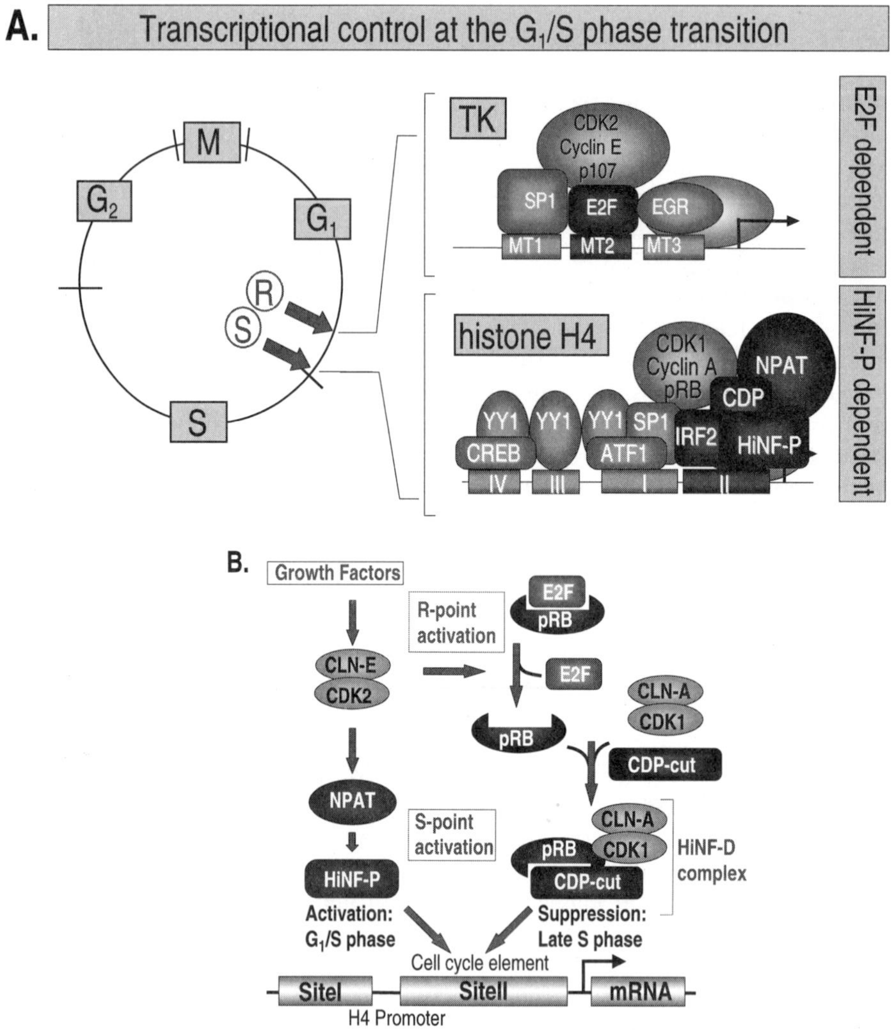

Figure 2.4. Transcriptional mechanisms for cell cycle control of S phase related gene expression. (**A**) Schematic illustration of the promoters of the E2F-dependent thymidine kinase (TK) gene and the E2F-independent histone H4 gene. Indicated are key regulatory elements of the TK gene (MT1, MT2, and MT3) and H4 gene (site I and site II) as well as the corresponding cognate factors that regulate transcription in a cell cycle dependent or constitutive manner. Upregulation of the nucleotide metabolism related TK gene occurs at the restriction (R) point, and this event precedes the activation of histone gene expression at the G_1/S phase transition (S point). (**B**) Model for the interactions of HiNF-P and HiNF-D (CDP/pRB/cyclin A/CDK1 complex) with the site II cell cycle element of the H4 gene that integrates temporally distinct cell cycle regulatory signals. The growth factor-dependent activation of cyclin E/CDK2 kinase complexes releases E2F from pRB at the restriction (R) point. Concomitant activation of NPAT by cyclin E/CDK2 supports the subsequent HiNF-P dependent induction of the histone H4 gene at the G_1/S phase transition. The formation of the gene suppressive HiNF-D complex, which contains pRB, the homeodomain protein CDP/cut, cyclin A, and CDK1, occurs after the cyclin E/CDK2-dependent hyperphosphorylation of pRB protein when cells progress through later stages of S phase.

(Aziz et al., 1998a,b). Furthermore co-regulatory factors that interact with histone gene transcription factors may contribute to transcriptional control of histone gene expression (Staal et al., 2000).

One of the most critical factors that regulates histone H4 gene transcription is Histone Nuclear Factor P (HiNF-P), which binds to the principal cell cycle regulatory domain, site II (Dailey et al., 1986, 1987, 1988; Holthuis et al., 1990; Langdahl et al., 1997; Pauli et al., 1987; Ramsey-Ewing et al., 1994; Stein et al., 1996; van Wijnen et al., 1991, 1996; Vaughan et al., 1995), through a specific recognition motif that is phylogenetically conserved among multiple histone H4 genes in metazoan species (Aziz et al., 1998a; Ramsey-Ewing et al., 1994; van Wijnen et al., 1992; Vaughan et al., 1998). The HiNF-P-dependent activation of H4 genes is functionally linked to NPAT (nuclear protein mapped to the ATM locus), which is a direct downstream target of the cyclin E/CDK2 signaling pathway (Imai et al., 1997; Zhao et al., 1998a). NPAT is essential for normal mammalian development and enhances histone gene transcription (Di Fruscio et al., 1997; Ma et al., 2000; Zhao et al., 2000), but NPAT does not bind directly to DNA. Recent findings have demonstrated that HiNF-P is the mediator that transduces the NPAT/cyclin E/CDK2 signal at the histone H4 gene promoter. Thus HiNF-P is the final link in the intricate signaling cascade that is initiated with the growth factor-dependent induction of cyclin E/CDK2 kinase activity at the R point and culminates in the NPAT-mediated activation of histone H4 genes through HiNF-P at the G_1/S phase transition (Fig. 2.4B).

Biological characterization of all three principal factors HiNF-P, -M, and -D, which interact with the site II cell cycle element in histone H4 genes, has provided insight into the physiological role of these factors in E2F-independent mechanisms mediating cell growth control. Deregulated expression of HiNF-M/IRF-2 causes cell cycle defects, resulting in polyploidy and apoptosis (Xie et al., 2002). Genetic inactivation of CDP-cut, the DNA binding subunit of the HiNF-D complex, causes several developmental abnormalities (e.g., in cells of the skin and the immune system) that are attributable to in vivo defects in cell growth and differentiation (Ellis et al., 2001; Luong et al., 2002; Sinclair et al., 2001). Antisense inhibition of HiNF-P activity impedes S phase progression in actively proliferating cells. Hence all three site II binding proteins have different functional roles in cell growth control.

Eukaryotic cells have developed complex mechanisms to mediate E2F-dependent regulation of genes involved in nucleotide metabolism (e.g., TK and DHFR) at the growth factor-related restriction point in anticipation of DNA replication. The functions of individual E2F transcription factors are partially redundant, and these factors promote either proliferation or exit from the cell cycle depending on the biological context (Trimarchi and Lees, 2002). The E2F-independent activation of DNA replication dependent histone H4 genes at the G_1/S phase transition appears to involve the intricately regulated functions of the principal site II binding activities.

Replication Cycle: Architectural Control of DNA Synthesis

Maintenance of the eukaryotic genome requires precisely coordinated replication of the entire genome each time a cell divides. Complete and accurate DNA replication during S phase synthesis of the cell cycle is integral to sustaining the genetic integrity of all organisms. A wealth of data originating from biochemical fractionation of replicating cells indicates that the orchestrated activity of multi-component protein complexes (named "replitase"; Noguchi et al., 1983) or replisomes (reviewed in Bell and Dutta, 2002) ensures the fidelity of DNA replication. The complexity of the eukaryotic replication process, namely simultaneous firing of multiple replication origins, and enzymology of the process necessitates the in situ organization and association of the proteins involved in genome duplication.

In Situ Assessment of Replication Domains. Biochemical characterization of the protein complexes involved in DNA replication during 1970s and 1980s provided fundamental insight into the enzymology of the whole process (Reddy and Pardee, 1980; Takahashi, 1987). Early studies designed to analyze DNA replication in cultured eukaryotic cells revealed that adjacent replication origins are synchronously activated or "fired" during the S phase of the cell cycle (reviewed in Kaftory and Fry, 1978). However, the resolution of these experiments provided only a minimal size estimate of the number of origins in a cluster. The direct visualization of intact replication domains by BrdU labeling (Manders et al., 1992) provided the means to examine the structural organization of the replication clusters. The results obtained from these microscopic observations revealed that DNA replication initiated at a discrete and limited number of locations within the nucleus. These findings lead to a structural model for DNA replication whereby the formation of each replication domain resulted from the aggregation of at least 10 replication origins around a central ring containing all the proteins required for faithful duplication of DNA (Nakayasu and Berezney, 1989). Such an architectural organization of DNA replication accommodates the requirement of coordinate "firing" of multiple replication origins within a specified window of time (the S phase) and argues for the temporal-spatial distribution of protein complexes that are involved in the process of replication.

Replication Sites: The Focal Thresholds of Proteins to Support DNA Replication. A growing body of literature indicates that the protein complexes assigned to the task of DNA replication are organized as punctate sites within the nucleus (reviewed in Berezney and Wei, 1998; Getzenberg et al., 1991; Leonhardt et al., 2000). These sites colocalize with nascent DNA and undergo a cyclic assembly and disassembly; that is, proteins that form these replication sites exhibit a disperse pattern of distribution in all but the S phase of the cell cycle. During S phase these proteins organize as large distinct subnuclear domains and contain newly synthesized DNA (reviewed in Newport and Yan, 1996; Pasero and Schwob, 2000). As cells progress through the S phase and enter the G_2 phase (Gap-2), these subnuclear domains that support DNA replication disperse as smaller numerous punctate foci.

In situ immunofluorescence microscopy has revealed that several proteins, including proliferating cell nuclear antigen (PCNA), DNA ligase I, chromatin assembly factor-1 (CAF-1), DNA polymerase σ, and DNA methyltransferase, are present in these foci (Hozak et al., 1994; Krude, 1995; Leonhardt et al., 1992; Montecucco et al., 1995). These punctate sites provide architecturally organized nuclear microenvironments for the optimal concentrations of replication proteins. High-resolution electron microscopic observations of the movement of DNA during replication and the localization of DNA polymerase σ and PCNA (Hozak et al., 1994) provide compelling evidence that the nuclear substructure serves as an organizer of DNA replication sites. Indeed, anchorage of DNA polymerase σ and PCNA during DNA replication led Peter Cook and colleagues to propose a model of "replication factories" as architecturally organized subnuclear sites through which DNA moves during the duplication process. The "replication factories" model, while providing the basis for a relationship between nuclear structure and function, fails to accommodate the kinetics of DNA replication as well as structural obstacles offered by DNA packaged into a nucleosomal array. Nonetheless, a common theme emerges from all of the experimental observations described above, namely the nuclear substructure functions as an architectural platform for replication proteins to organize into focal thresholds. These focal thresholds can be directly observed in situ by immunofluorescence microscopy and serve to facilitate protein-protein and protein-DNA interactions.

Cyclical Parameters of Replication Sites: S Phase Specific Nuclear Microenvironment. The eukaryotic cell cycle has evolved as an intricate interplay between growth factor-dependent signal transduction pathways and nuclear proteins that execute the replication program in a timely and precise manner. One can therefore describe the various phases of the cell cycle as an example of "division of labor" at the molecular level 2.4. S phase is characterized by, and is dedicated to, duplication of the genome. Several proteins engaged in DNA replication are involved in other cellular processes; for example, PCNA is also involved in DNA repair, replication protein A (RPA) is required for recombination events (Celis and Madsen, 1986; He et al., 1995; Longhese et al., 1994; Shivji et al., 1992; Toschi and Bravo, 1988). While DNA repair or recombination, broadly speaking, are linked to the synthesis of DNA, both are fundamentally different from DNA replication, as the cell can require either of these events under stress, regardless of the phase of the cell cycle. These bi-functional proteins are present therefore throughout the cell cycle to ensure the faithful execution of cell survival mechanisms. This apparent paradox raises an important question: How do the proteins involved in DNA repair and recombination, which can occur at any

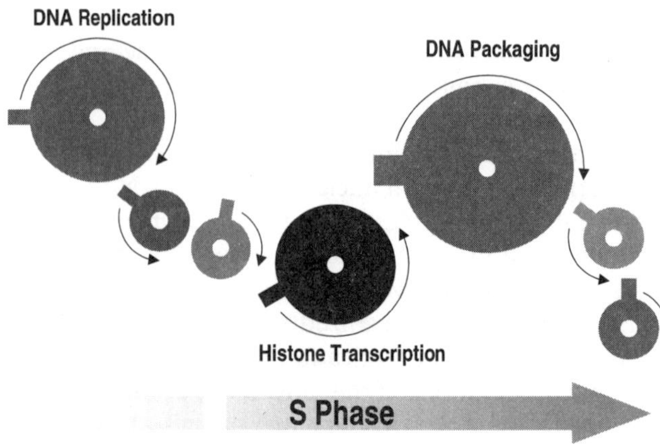

Figure 2.5. Integration of independent cycles in the S phase: A requirement for progression of the S phase. Individual but interdependent structural cycles within the cell cycle can be explained with a simplified chain of events in the S phase. The major nuclear event in the S phase is the duplication of the genome. However, this genome must be packaged into chromatin by histone proteins. The genome packaging into the chromatin requires synthesis of histone genes. Thus the S phase can then be divided into three individual cycles: DNA replication, histone gene transcription, and DNA packaging. According to the "interlinked structure-cycle model," signals that dictate initiation of DNA replication also induce histone gene expression, thus integrating two independent processes. Such integrated cycles are operative for the assembly and activities of regulatory compartments that mediate competency for proliferation and cell cycle progression (e.g., cyclin degradation cycle).

place within the genome and at any time during the cell cycle, assemble to carry out a time-dependent process such as DNA replication that requires a specific number of replication origins? Do replication sites containing these proteins follow a cyclical pattern of assembly and reassembly as DNA replication itself is a cyclical process?

Immunofluorescence microscopy provides a powerful tool to address some of these compelling questions by direct visualization of protein dynamics. Experiments with synchronized cells suggest an ordered transition of replication foci throughout the cell cycle and during S phase (Manders et al., 1992; Nakayasu and Berezney, 1989; O'Keefe et al., 1992; van Dierendonck et al., 1989). These observations made in fixed cells are further supported in live cells by the use of enhanced green fluorescent protein (EGFP) fused replication proteins such as PCNA and DNA methyltransferase (Leonhardt et al., 2000; Liu et al., 1998). These studies provide direct evidence of the cyclical nature of assembly and reassembly of replication foci. Replication proteins are assembled in early S phase as large subnuclear domains. These domains, while remaining "fixed" throughout S phase, continuously undergo waves of assembly and disassembly throughout S phase, indicating that the proteins involved in different steps of DNA replication associate with these sites in a sequential manner. As cells complete DNA duplication and exit S phase, the few large replication sites disperse and result in numerous smaller foci. Thus the replication machinery provides an excellent example of architecturally organized cycles of reorganization to facilitate protein-protein interactions and carry out DNA replication in a timely manner.

The observations described above, and cyclical assembly of replication sites, raises another interesting question: What triggers such an ordered protein assembly to ensure DNA replication during S phase? A little is known about the precise mechanisms that regulate the synchronous process of replication site assembly. It is safe to speculate that like many other cell cycle related events, this assembly is triggered by growth factor signaling. Another possible explanation is provided by the microscopic observations of RPA (Adachi and Laemmli, 1992). RPA is associated with replication origins throughout the cell cycle and is capable of interacting with several other proteins involved in DNA replication. Therefore RPA can potentially provide an interface between replication origins and proteins required to initiate the process. RPA's activity may be regulated by physiological cues that ensure passage of cells through pre–S phase check points. However, our lack of understanding of the mechanisms involved in assembly of replication sites does not undermine the significance of their synchronous and cyclical organization.

Replication of the Viral Genome in Mammalian Cells: An Exception to the Rule. Viruses utilize the cellular machinery to replicate and propagate. The initial steps of viral infection involve replication of viral genomes within host cells. Cells have developed several mechanisms to cope with the requirement of the viral genome to utilize cellular machinery. One of the best studied mechanisms is the formation of approximately 10 nuclear dots or domains (hence named as ND10) in response to interferon signaling, which in turn is activated by viral infection (reviewed in Maul, 1998). The ND10 domains contain several proteins including Sp100 and PML and serve as cellular defense mechanism. Compatible with such mechanisms is the deposition of herpes virus, adenovirus and papovirus genomes at the periphery of ND10. However, these DNA viruses begin their transcription at these sites and eventually utilize them for the replication of their genomes. Thus ND10 domains function as replication sites in an asynchronous and cell cycle independent manner and provide an exception to the rule that all replication sites are assembled sequentially and cyclically to accommodate DNA replication within the cell.

The Chromosome Cycle: Temporal-Spatial Packaging of the Genome to Accommodate Gene Expression and Chromosome Segregation

There is a cyclical series of stringently controlled biochemical events to establish and sustain physiological responsiveness during the cell cycle and heritable chromosome and chromatin architecture in progeny cells following cell division.

The Nucleosome: Primary Level of Genome Organization. Heritable DNA within each cell has been found to have exquisite structure as well as periodicity on several levels (Fig. 2.6). The first level of DNA structure, referred to as chromatin, mediates the complexities of packing a multi-billion subunit polymer into heritable chromosomes. Yet chromatin must maintain flexibility of access to the myriad factors that must cooperate and coordinate to produce the building blocks of a living cell. This has been achieved through the course of evolution by the packaging material, histones, a highly integrated component of all gene regulatory mechanisms. A histone core particle is an octamer of small proteins that acts as an axis about which DNA wraps twice to form a nucleosome, the basic chromatin subunit.

The 30-nm Fiber: Transcriptionally Active or Suppressed Chromatin Template. The second level of chromatin structure is a thermodynamically favorable (stable) nucleosomal array at physiological ion conditions, which forms a 10-nanometer fiber with some regularity along its length. Early studies on the 30-nm chromatin fiber suggested the presence of non-core linker histone proteins, H1 and H1 variant H5, which may repress chromatin structure via additional compaction (Sun et al., 1989). The exact histone or cofactor composition status of the 30-nm fiber is no longer considered definitely established. However, it is accepted that active transcription is generally from a 30-nm structure (Horn and Peterson, 2002).

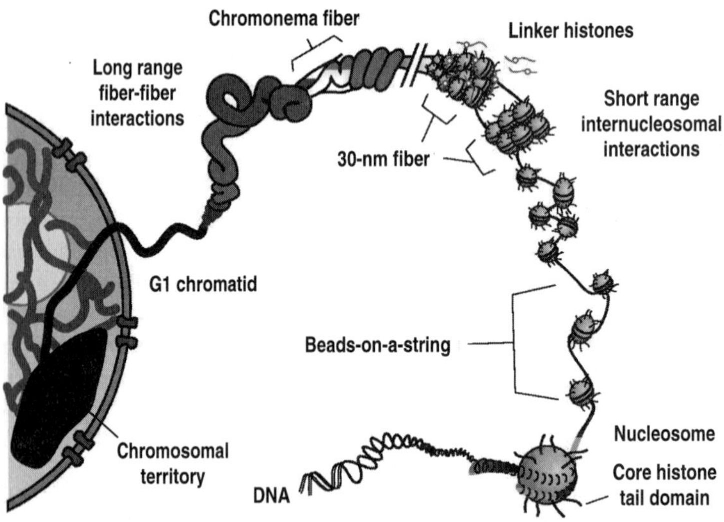

Figure 2.6. Levels of chromatin organization. Depicted are renditions of the levels of chromatin organization, including nucleosomal (primary), 30-nm fiber (secondary), chromonema (tertiary), and chromosomal territories (quaternary). (With permission, from the *Annual Review of Biophysics and Biomolecular Structure, Volume 31*(2002) by Annual Reviews, www.annualreviews.org)

Just as the majority of the genome is devoid of actively transcribed genes, the majority of the chromatin in the in vivo genome is without precise nucleosomal array structure. The most thermodynamically favorable primary structure is in contrast to many examples of exquisite order and periodicity. Histone cores are post-translationally modified by methylases, acetyltransferases, and kinases. These modifications are transient and cyclic, resulting in a tightly regulated steady state alterations that affect chromatin structure, and therefore transcription factor accessibility and binding. Consequently transcriptional activity of the locus is influenced. Nucleosomal structure and order at numerous loci have been shown to be precisely positioned, in a way that this phasing determines the transcriptional activity of the loci (Lohr et al., 1977). Phasing of some enhancer arrays is sufficiently precise that rotational settings of nucleotide sequences wrapped around histone cores differing by a single nucleotide can determine efficacious factor binding (McPherson et al., 1993, 1996; Shim et al., 1998). Construction of characteristic chromatin arrays is dynamic and often periodic. The majority of regulatory loci within proliferating cells undergo chromatin changes as those cells cycle through mitosis, exemplifying ordered periodic structure. Differentiated cells also exemplify dynamic and periodic chromatin structure within the context of tissue- and/or temporal-specific regulation of gene expression.

Chromonema: Higher Order Chromatin Structure. A third level of chromatin structure, referred to as chromonema (Spector and Triemer, 1981), pertains to nonlinear structure with looping and condensation beyond 30-nm fibers and longer-range condensation and congression of chromatin in preparation for cell division. It is likely that chromonema fibers greater than 30-nm, and perhaps even greater than 100-nm, are common substrates for transcription and/or phased arrays (Horn and Peterson, 2002). Mediators of general chromonema structure include the activities of multi-partite protein complexes, condensin, and cohesin. The condensin and cohesin core complexes consist of structural maintenance of chromosomes (SMC) protein subfamilies –2/–4, and –1/–3 respectively (Losada and Hirano, 2001). Condensin complexes help to mediate and scaffold higher order chromatin structure as an intramolecular DNA crosslinker. Cohesin complexes maintain physical linkages between sister chromatids through G_2. The chromonema cycle can be defined by the periodic structural alterations by condensin and cohesin activities.

Additional layers of structure are associated with condensed chromosomes during the processes of nuclear breakdown during early mitosis (Hernandez-Verdun and Gautier, 1994; Moyne and Garrido, 1976). These accumulations form the perichromosomal space, which is a cyclic mechanism for the conservation of material transfer between mother and daughter cells. The ordered breakdowns of most subnuclear domains continue downstream to this perichromosomal-associated mechanism of transfer.

Periodicity and order of another kind exists on large portions of chromatin with regard to replication by DNA-dependent DNA polymerases. The duration of the S phase of the cell cycle is typically about one third of the total, corresponding to an average of about 8 hours in actively proliferating cells. The physical duplication of the approximately three billion nucleotides in a human genome occurring over these 8 hours results in the differential timing of replication of different portions, or replication zones, of the genome. Classic staining patterns of metaphase chromosomes as light and dark alternating bands has been found to generally be representative of early versus late replication zones (Zink et al., 1999). The temporally ordered pattern of replication of these zones is programmed, highly regulated, and maintained through generations of cell divisions (Visser et al., 1998).

An exception to the normal condensation/decondensation chromosome cycle in a proliferating or differentiated cell is imprinting and X-chromosome inactivation (Kelsey and Reik, 1998; Reik and Walter, 2001). It has become clear that while the normal chromosome cycle is not followed, imprinting is yet another example of regulated chromatin structure that is not simply thermodynamically favorable. Both imprinting and X-chromosome inactivation are ultrastructurally similar to condensed mitotic chromatin, yet the mechanisms are completely different, as exemplified by the different modifications predominant in imprinted versus mitotic chromosomes being CpG dinucleotide methylation (Bird et al., 1982; Bird, 1984; Bird and Wolffe, 1999) and histone H3 lysine 9 methylation (Boggs et al., 2002; Peters et al., 2002) versus histone H3 serine 10 phosphorylation (de la Barre et al., 2000). Imprinting occurs on one parental allele, resulting in transcription from only the nonimprinted allele (McGrath and Solter, 1984; Surani et al., 1984). X-chromosome inactivation is the process by which abundance of transcription from individual loci on the X-chromosome can be normalized in female (XX) versus male (XY) cells. Cells containing two X-chromosomes decondense only the one X-chromosome, specifically maternal or paternal depending on the organism, while the other remaining compact X-chromosome localizes to, and tethers, the nuclear periphery. Both of these processes are both highly structural and regulated, are not cyclic, and are maintained.

Chromosomal Territories: Consistently Positioned Genomic Niches. A fourth level (dimension) of chromatin structure is defined by chromosomal territories in decondensed interphase nuclei and ordered alignment of condensed mitotic chromosomes at the mitotic plate (Schardin et al., 1985). As imaging techniques continue to improve, we are able to analyze, in real time, the constituents of chromosomal neighborhoods from one interphase to the next through cell division. Imaging of chromosome painting, double chromosome labeling through fluorescent protein-tagged histone incorporation, and photo-bleaching have been combined for startling pictures of conservation of gene position (Bickmore and Chubb, 2003; Chubb et al., 2002; Gerlich et al., 2003; Haberma et al., 2001; Walter et al., 2003).

Suggestive of remarkable order and maintenance of subnuclear structure, it appears as though complex, but consistent, chromosomal neighborhoods are transmitted through the condensation and metaphase alignment processes to daughter cells. Even at this early stage of study, there are clear data to demonstrate a remarkable level of consistency of association of hemispheres of chromosomal territories between mother and daughter interphase nuclei. The level of complexity associated with the study of chromosomal neighborhoods is still pushing the limits of current assays and analysis; however, the inherent structure is beginning to be revealed. Nucleolar association and maintenance at rDNA gene clusters on multiple chromosomal loci exemplifies additional structural clues to the persistence of chromosomal neighborhoods.

Protein Metabolism and Distribution Cycle: Conservation versus Turnover

Precise progression of the cell cycle requires tightly controlled transcriptional and post-translational mechanisms to ensure availability of regulatory proteins at various cell cycle stages. Transcriptional regulatory mechanisms such as histone acetylation and methylation and DNA methylation, control synthesis of proteins prior to their temporal roles in the cell cycle. Post-translational modifications include, but are not restricted to, phosphorylation and ubiquitination, which render proteins active or inactive, or serve to "tag" proteins for proteasome-mediated degradation. Additional architecturally linked mechanisms redistribute proteins in various subcellular and subnuclear compartments, thereby altering their activity.

Selective and Periodic Protein Turnover. Selective and periodic degradation of cyclins, inhibitors of cyclin-dependent kinases (CDKI) and anaphase inhibitors is responsible for several major cell cycle transitions. The different cyclins, specific for the G_1, S, or M phases of the cell cycle, accumulate and activate Cdks at the appropriate times during the cell cycle and then are degraded, causing kinase inactivation.

Mitotic Cyclins. Though all cyclins are degraded by ubiquitin-mediated processes, the mechanisms that result in their ligation to ubiquitin moiety, and the mode by which these mechanisms are connected to the cell cycle regulatory phosphorylation network, are different for mitotic and G_1 cyclins. In general, mitotic cyclins are ubiquitinated by ubiquitin ligases, whose activity is regulated in a cell cycle-dependent manner. The proteolysis of the G_1 cyclins is controlled by phosphorylation of cyclins. Cyclin B-Cdk1 forms the major mitotic kinase M phase promoting factor (MPF), which is responsible for entry of cells into mitosis. Later, the MPF activates a self-regulatory loop that degrades its cyclin subunit (reviewed in Hershko, 1997). MPF inactivation, caused by the degradation of cyclin B, is required for exit from mitosis (Surana et al., 1993). The activity of cyclosome, the complex that carries out degradation of cyclin B, and some S phase specific cyclins such as cyclin A, is in part regulated by reversible phosphorylation by MPF (Hershko et al., 1994; Lahav-Baratz et al., 1995; Sudakin et al., 1995). The activation of cyclosome by MPF-mediated phosphorylation results in degradation of cyclin B, thus rendering Cdk1 inactive. It is not known how the activity of cyclosome is turned off by G_1 cyclins, but some phosphorylation event by G_1 cyclin/Cdk complexes likely inhibits cyclosome directly or indirectly (reviewed in Hershko and Ciechanover, 1998).

G_1 Cyclins. Mechanisms that regulate turnover of mammalian G_1 phase specific cyclins are emerging. In general, G_1 cyclins are targeted for degradation by phosphorylation. For example, mammalian G_1 specific cyclins D1 and E are phosphorylated on specific single threonine residues. Mutation of these residues slows down the rapid degradation of these cyclins (Diehl et al., 1997; Won and Reed, 1996). The proteasome system that recognizes and degrades phosphorylated G_1 cyclins remains to be identified.

CDK Inhibitors. The activities of some CDKs are controlled tightly by fluctuations in the levels of their negative regulatory proteins, CDKIs. Thus a cyclin/CDK complex cannot act until the inhibitor removed by selective proteolysis. Many CDKIs have been identified in mammalian cells, and these can be divided into two families based on sequence similarities: The KIP/CIP family contains p21, p27, and p57, and the INK family includes p15, p16, p18, and p19. All mammalian CDKIs inhibit G_1 cyclin/Cdk complexes with different specificities and thus mediate cell cycle arrest in response to a variety of growth inhibitory conditions. For example, p21 is induced by DNA damage (reviewed in Elledge, 1996), p27 levels are increased greatly in cells arrested by deprivation of growth factors or contact inhibition (reviewed in Sherr and Roberts, 1999), and p18 levels are elevated in terminal differentiation resulting in permanent cell cycle arrest (Franklin and Xiong, 1996). Mammalian CDKIs are highly unstable proteins and their levels are modulated by alterations in rates of their degradation. The best studied example is p27, whose levels are elevated in quiescent cells in part from decreased degradation (Hengst and Reed, 1996; Pagano et al., 1995). Growth stimulation results in rapid degradation of p27 by proteasome system (Pagano et al., 1995). An additional mechanism by which p27 is "tagged" for degradation is its phosphorylation by the cyclin E/Cdk2 complex (Sheaff et al., 1997). The proteasome system that targets p27 for degradation remains to be identified.

Regulation of Proteasome Activity: Dynamic Redistribution during Cell Cycle. Proteasome machinery utilizes an ATP-dependent proteolysis mechanism and is comprised of a central catalytic machine called the 26S proteasome, and ubiquitin ligases, proteins that ligate ubiquitin moieties to targets (reviewed in Tanaka and Tsurumi, 1997). In situ immunofluorescence microscopy reveals that the proteasome subunits undergo dynamic redistribution during cell cycle, suggesting that subcellular localization plays a role in regulation of proteasome activity.

The yeast 26S proteasome subunit localizes to the nuclear envelope and rough endoplasmic reticulum in the interphase nucleus (Enenkel et al., 1999). As cells enter mitosis, the 26S proteasome is redistributed and localized to the chromosomal periphery, where it degrades proteins to facilitate mitotic progression. In higher eukaryotes, live cell microscopy using EGFP-labeled proteasome subunit LMP2 shows that the proteasome is distributed both in the cytoplasm and nucleus (Reits et al., 1997). FRAP studies reveal that LMP2 is highly mobile in both subcellular compartments. Furthermore proteasomes slowly and unidirectionally move into the nucleus from the cytoplasm. During cell division the proteasome is diffused rapidly throughout the cell due the absence of a selective barrier. Following cell division, the newly formed nuclear envelope offers a barrier for proteasome diffusion and slows its mobility. The dynamic redistribution of the proteasome has been implicated in selective degradation of misfolded or unnecessary proteins (Enenkel et al., 1999; Reits et al., 1997).

Thus the proteasome serves as a major defining component of several mechanisms in place to ensure periodic and cyclical availability of cell cycle regulatory proteins required for faithful cell cycle progression.

Histones: Stable, Segregated, Modified Mediators of Chromatin Remodeling. Chromatin structure and nucleosome organization provide architectural linkages between gene organization and components of transcriptional control. Changes in chromatin organization have been documented under many biological conditions in which modifications in gene expression are necessary for the execution of physiological control. Thus the chromatin template undergoes dynamic changes during the different stages of the cell cycle. These changes include the reorganization that occurs during DNA replication and cell cycle progression, as well as spatially and temporally coordinated gene expression. Over the past several years there have been major advances in the ability to assess the molecular mechanisms that mediate chromatin remodeling. This is, to a significant extent, attributable to an increased understanding of the enzymatic control of nucleosome structure and organization.

Post-Translational Modifications: Physiologically Responsive Switches. Post-translational modifications of histone proteins have been associated with the physiological control of chromatin structure for the past three decades. Covalent modifications occur at the N-terminal "tails" of the core histones as a result of various enzymatic pathways. Several reports indicate that these modifications (acetylation, methylation, phosphorylation, and ubiquitination) modulate the role of core histone tails in chromatin compaction (Fischle et al., 2003).

Acetylation of the amino-termini of nucleosomal histones has been directly correlated with transcriptional activity (Narlikar et al., 2002). This modification is catalyzed within the cells by proteins containing histone acetyl transferase activity (HAT), among which we find p300/CBP (Ogryzko et al., 1996), TAFII250 (Mizzen et al., 1996), P/CAF (Yang et al., 1996), and others. Histone acetylation promotes chromatin decondensation, thereby facilitating the accessibility of transcriptional activators to gene promoter regulatory elements.

Histone acetylation is physiologically reversed by the activity of a family of enzymes designated as histone deacetyl transferases (HDACs), which remove the acetyl groups from the histone N-termini, thereby inducing chromatin compaction and transcriptional repression. HDACs form large multiprotein complexes with co-repressor molecules, which together are target to promoter regulatory regions by sequence-specific factors functioning as transcriptional repressors (Narlikar et al., 2002).

In contrast to acetylation and phosphorylation, methylation of histones H3 and H4 N-terminal "tails" appears to be biochemically stable and irreversible, as no enzyme with demethylase activity has been reported. Based on this, it has been proposed by several authors that once methylated, a histone will remain associated with a promoter until physiological histone turnover or DNA replication changes the methylated histone with an unmodified one.

Lysine 9 of histone H3 (K9-H3) is targeted by both acetylation and methylation (Fischle et al., 2003). Histone H3 lys9 acetylation is associated with transcriptionally active sequences, while methylation of histone lys9 generally accompanies transcriptional silencing. Thus HDAC activity serves as an intermediary between the actions of HATs and histone methylases (HMTases) by creating the substrate site for methylation upon removal of the acetate group from K9-H3. The competition between acetylation and methylation of K9 has the potential then to generate a switch that determines the on–off states to which the histone are associated (Fischle et al., 2003). It has been recently described that nucleosomal histones surrounding a MEF2 target site in the myogenin gene promoter, which is transcriptionally active only in differentiating muscle, are differentially modified by acetylation and methylation during myogenesis (Zhang et al., 2002). High levels of histone methylation are observed at this site in undifferentiated myoblasts. Upon differentiation, the level of histone methylation is decrased at this MEF2 element, with a concomitant increase in histone acetylation. As histone methylation appears to be irreversible, the reduction in histone methylation observed during myogenesis may be due to the exchange of methylated histones with unmodified or acetylated histones through a process dependent on DNA synthesis (Bannister et al., 2002).

Recent results also indicate that maintaining an acetylated state may be required to prevent transcriptional silencing through the cell cycle, especially in genes that are necessary during cell cycle progression. It has been reported that during differentiation of HL-60 promyelocytic leukemia cells, postproliferative downregulation of histone H4/n gene transcription is not associated with a decrease in acetylated histones H3 and H4 (Hovhannisyan et al., 2003). In addition micrococcal nuclease,

DNase I, and restriction enzymes show similar cleavage sites and levels of sensitivity at the H4/n locus in both proliferating and differentiated HL-60 cells. Thus the chromatin structure of the H4/n gene locus remains in an open state even after transcription ceases. The cells, by keeping the histones H3 and H4 associated with the H4/*n* gene in an acetylated state, are preventing these histones from being methylated and thus maintain the gene poised for expression (Hovhannisyan et al., 2003).

Interestingly it has also been found that the expression of other cell cycle-related genes is regulated by differential histone methylation. It was recently reported that the activity of the cyclin E gene promoter is repressed by K9-H3 methylation in the G_1 phase of the cell cycle (Nielsen et al., 2001). As this promoter is activated at the G_1/S transition, K9-H3 methylation needs to be reversed to allow cyclin E gene expression. This could be achieved by the replacement of the modified histones with unmodified variants such the histone H3.3 (Bannister et al., 2002; Fischle et al., 2003). These variants, unlike the cell cycle-dependent histones, are continuously expressed throughout the cell cycle.

Chromatin Remodeling Complexes: ATP-Dependent Regulators of Chromatin Structure. A family of SWI/SNF-related protein complexes has been described in eukaryotic cells (Neely and Workman, 2002) that promotes transcription by altering chromatin structure in an ATP-dependent manner. The alterations render DNA sequences containing regulatory elements accessible for binding cognate transcription factors. All of the members in this family of chromatin remodeling complexes include a catalytic subunit that contains an ATPase activity (Neely and Workman, 2002) that is critical for modifying nucleosomal organization. In humans, as well as in all mammals studied, the hSWI/SNF complex may include one of the other two different catalytic subunits, BRG1 or hBRM. These two ATPases are clearly present in separate complexes although they are found associated with a similar group of subunits (Neely and Workman, 2002). The BRG1-containing complex was found to be present throughout the entire cell cycle and appears to be regulated by phosphorylation of two subunits, hSWI3 and BRG1 (Muchardt et al., 1996; Sif et al., 1998). It has been proposed that inactivation of hSWI/SNF by phosphorylation during mitosis facilitates formation of a repressed chromatin structure at this stage of the cell cycle. The hBRM-containing complex is also phosphorylated, but appears to be targeted for degradation during mitosis, indicating that this complex is regulated differently by phosphorylation events (Muchardt et al., 1996; Sif et al., 1998). Interestingly it has been recently reported that in yeast, the ySWI/SNF complex plays a more global role in the transcriptional activation of genes expressed in late mitosis (Horn and Peterson, 2002). This finding has led to the suggestion that ATP-dependent remodeling may lead to a localized disruption of mitotic condensation, thus promoting the expression of genes required for the progress of this stage.

There are numerous reports indicating that hSWI/SNF complexes may also function as regulators of cell cycle progression. Thus it has been shown that both BRG1 and BRM are able to interact with the tumor suppressor retinoblastoma protein (Rb), forming a hSWI/SNF-Rb complex that represses the expression of genes such as those encoding for cyclins and cyclin-dependent kinases during cell cycle, and in particular during S phase (Dunaief et al., 1994; Strober et al., 1996; Trouche et al., 1997; Zhang et al., 2000b). Thus hSWI/SNF interacts with an Rb-HDAC complex, and together they inhibit the expression of the cyclin E gene, blocking the exit of the cells from G_1 phase (Zhang et al., 2000b). Moreover BRG1 is required for the Rb-dependent G_1 phase arrest (Dunaief et al., 1994; Strobeck et al., 2000; Strober et al., 1996; Trouche et al., 1997; Zhang et al., 2000b). BRCA-1, another tumor suppressor, was also recently found associated to the hSWI/SNF complex, interacting directly with the BRG1 protein. Taken together these results indicate that the hSWI/SNF complex interacts with tumor suppressors and together regulates cell cycle progression.

Accordingly it is predictable that altered interactions between components of hSWI/SNF and tumor suppressor proteins may lead to tumorigenesis. Already it has been reported that mutations in the hSWI/SNF subunit hSNF5/INI 1 are associated with malignant rhabdoid tumors and with rhabdo myosarcomas, both very aggressive pediatric cancers (Versteege et al., 1998). In addition the gene encoding for BRG1 is mutated in several cancer cell lines, further indicating its role as a tumor suppressor (Wong et al., 2000).

Processing Cycle: Dynamic Redistribution of Nuclear Proteins Supporting Gene Expression

The rapid and dynamic turnover of cell cycle regulatory proteins, discussed above, is only one of the several mechanisms that are in place to ensure progression of the cell cycle. Most cell cycle regulatory proteins are ubiquitous in their expression. Eukaryotic cells, however, also express phenotypic and tissue-specific transcription factors. Regulatory and regulated mechanisms control developmental and temporal expression and the activity of lineage-specific proteins, which are often nuclear and represented in limited amounts. Several lines of evidence suggest that transcription factors are present in multi-protein complexes and are organized at transcriptionally active subnuclear sites (Berezney et al., 1996; Gasser, 2002; Lamond and Earnshaw, 1998; Ma et al., 1999; McNeil et al., 1998, 1999; Misteli, 2000; Stein et al., 2000a; Zaidi et al., 2001; Zeng et al., 1997, 1998). An accumulating body of knowledge suggests that some transcription factors serve as "scaffolding proteins" that are associated with the nuclear matrix; that is, they interact with several co-regulatory proteins temporally or simultaneously to form large protein complexes, whose activities are defined by the composition of the complex (Stein et al., 2000b). For example, transcription factors can interact with co-activators such as acetyl transferase p300 on some promoters resulting in gene activation, while simultaneously present with co-repressors such as histone deacetylases on different promoters to repress gene expression. A biologically relevant question arises from such complexity: What is the fate of these multi-protein, nuclear-matrix-associated complexes required for phenotype maintenance during mitosis when essentially the entire nuclear structure is remodeled? Do cells re-synthesize these tissue-specific proteins, whose expression is tightly regulated both at transcriptional and post-translational levels, in the G_1 phase of the cell cycle? Is it possible that these tissue-restricted proteins "bypass" the usual fate of other ubiquitously expressed regulatory factors?

Answers to some of these fundamental questions come from accumulated knowledge of the dynamics of tissue-specific proteins during the cell cycle. For example, it has been recently shown that the general transcription factors are excluded from the chromosomes during mitosis and re-enter the nucleus upon completion of anaphase concomitant with re-assembly of the nuclear envelope (Prasanth et al., 2003). Recent reports have shown that hematopoiesis related transcription factors ALL-1 and Runx1, as well as the bone tissue-specific Runx2 transcription factors remain throughout mitosis (Ennas et al., 1997; Zaidi et al., 2003). These transcription factors retain their punctate subnuclear organization during mitosis and a subset of these foci associate with chromosomes during mitosis. In addition, it has been shown that Runx proteins partition equally to, and resume their subnuclear organization in, the daughter cells, thus rendering them equivalently competent for phenotypic gene expression (Zaidi et al., 2003b). It is noteworthy that this behavior is specific for transcription factors, as the RNA processing proteins that enter into the nucleus after transcription factors (Prasanth et al., 2003) are not equally partitioned to daughter nuclei and their subnuclear organization is not resumed (Zaidi et al., 2003b). These findings are indicative of mechanisms that are in place to sustain and equally partition tissue-specific transcription factors to the daughter nuclei to maintain phenotypic properties of cells.

Checkpoint Cycles: Physiological Safeguards against Tumorigenesis

The cellular equivalents of an emergency ripcord or panic button are devices called cell cycle checkpoints. These devices are centered on cellular DNA and the protection, conservation, and maintenance of the fidelity of the genome for progeny cells. Checkpoint pathways are inhibitory in nature. As such, lack of expression of a checkpoint allele results in loss of dependence from the checkpoint. The structural basis of the majority of cancers is due to loss of checkpoint alleles, resulting in an inability to protect, conserve, and maintain genome fidelity.

There are three major cell cycle checkpoints, each of which acts as a sensor for particular structural characteristics. The mitotic spindle checkpoint senses misalignment of chromosomes at the mitotic plate, the S phase replication checkpoint senses DNA replication failure, and the multifaceted DNA damage checkpoint senses various forms of genotoxic insult.

The Centrosome Cycle: The Architecture-Dependent Cycle of the Spindle Checkpoint. The spindle checkpoint is highly structural and cyclic (Gardner and Burke, 2000; Musacchio and Hardwick, 2002). The actual checkpoint device is a complex of proteins, called the mitotic checkpoint complex (MCC; Sudakin et al., 2001) that occupies the mitotic-specific chromosomal structure of a kinetochore. The MCC inhibits progression of mitosis until certain crucial genomic integrity criteria are verified. The mitotic inhibitory nature of the MCC is conserved in function and structure between budding and fission yeast and higher eukaryotes, and it is mainly composed of Mad (mitotic arrest defective) and Bub (budding uninhibited in benzimidazole) proteins. The MCC has at least three important duties: it acts as a sensor, a signal transducer, and an inhibitor. First, the MCC is a sensor of chromosomes that are not aligned to the mitotic plate. Second, the MCC relays signals from the sensors by way of transducers. Third, the MCC must effect inhibition of cell cycle progression based on sensor signal transduction. The MCC has been shown to sense even a single unattached kinetochore, and effect that signal to cause inhibition of mitotic progression (Rieder et al., 1995).

The MCC is able to control mitotic progression and exit by directly inhibiting the activity of a structure called the anaphase promoting complex/cyclosome (APC/C: Harper et al., 2002; Morgan, 1999; Zachariae and Nasmyth, 1999). The APC/C is an E3 ubiquitin ligase whose activity is required to target 26S proteasome-mediated degradation in two mitotic pathways (Gmachl et al., 2000). One of the APC/C targets is securin (Pds1) whose targeted degradation lifts the inhibition of a protease, separase (Esp1), whose activity is essential to sister chromatid separation and therefore progression from metaphase to anaphase. Other APC/C targets are B-type cyclin-dependent kinases (i.e., cdc2/cdk1), whose targeted degradation is required to exit mitosis.

APC/C undergoes a mitotic-specific modification that renders it able to be inhibited by the MCC at unattached kinetochores (Sudakin et al., 2001). Subsequent to the mitotic APC/C modification, MCC specifically inhibits the cdc20 component of the APC/C (Hwang et al., 1998). This inhibition of the cdc20-APC/C by the MCC seems to be a structural phenomenon that may include the exchange of cdc20 between the two complexes. The MCC protein Mad2 has been shown to structurally behave as a molecular safety belt (Sironi et al., 2002), wrapping around Mad1 (MCC-associated) and cdc20 proteins with similar efficiencies.

Much remains to be resolved by way of molecular interactions. However, it is clear that the spindle checkpoint is highly ordered and structural in nature. The cyclic and dynamic nature of the kinetochore-MCC-APC/C-spindle is a wonderful example of highly regulated cyclic architecture and structure within the cell cycle.

Common Architectural Features of Replication and DNA Damage Checkpoints. The progression of cells through S phase is monitored by the replication checkpoint. It can be described as a subset of the DNA damage checkpoint, which is active throughout the cell cycle. An architectural feature of these checkpoints is the striking focal nucleation of various regulatory, scaffold, repair, and maintenance proteins at specific genomic loci in response to checkpoint activation (Qin and Li, 2003). Structural preservation of replication intermediates demonstrates highly ordered, reversible and cyclic architecture (Kelly and Brown, 2000). The composition of the focal complexes triggered by the DNA damage checkpoint is completely dependent on the phase of the cell cycle and ploidy state of the cell at that point (Carr, 2002; Hendrickson, 1997). These fascinating observations will be considered with regard to periodic structural phenomena and fidelity-maintenance biochemical activities.

Structural Cycles of the Replication Checkpoint. The replication checkpoint cycle is intimately tied to the structural cycle of the replication complex. The replication checkpoint is readily separable from the DNA damage checkpoint by treatment of cells with hydroxyurea (HU), which inhibits ribonucleotide reductase thereby causing replication fork stalling due to a lack of nucleotide precursors (Yarbro, 1992). Activation of the replication checkpoint with HU results in three main downstream architectural phenomena (Kelly and Brown, 2000). First, there is active delay in mitotic chromosome segregation; the consequence is cell mortality if replication is incomplete (Enoch and Nurse, 1990). Second, structural replication intermediates are stabilized by inhibition of replication fork elongation, re-initiation, and recombination, all of which might cause unresolved intermediates and render the arrest irreversible (Stewart et al., 1997). Third, there is transcriptional induction of ribonucleotide reductase and other genes important for recovery (Desany et al., 1998; Elledge et al., 1993; Zhao et al., 1998b).

Several proteins have been found to be important for the integrity and downstream effectiveness of the replication checkpoint. These proteins are conserved between budding and fission yeasts and higher eukaryotes and have been called the "checkpoint rads" (radiation-hypersensitive) in budding yeast, consisting of Rad1, Rad3, Rad9, Rad17, Rad26, and Hus1 (Al-Khodairy et al., 1994; Al-Khodairy and Carr, 1992; Enoch et al., 1992). Rad1/9/17 and Hus1 have mammalian homologues with the same name, which are likely to form an architectural complex resembling the PCNA sliding clamp protein (Thelen et al., 1999). The sliding clamp is likely an integral part of the replication complex architecture itself. A role for the replication architecture itself in the replication checkpoint is supported by several lines of evidence. For example, budding yeast polymerase α (pol 1) mutants are unable to activate the replication checkpoint (D'Urso et al., 1995), as are polymerase ε (pol 2) relatively to a lesser degree (Navas et al., 1995).

Rad3 is homologous to each member of the phosphotidlyinositol 3-kinase (PI3-kinase) family in higher eukaryotes, which includes ATM (mutated in ataxia telangectasia), ATR (AT- and Rad3-related) and DNA-PKcs (DNA-dependent protein kinase catalytic subunit/*scid*). These PI3-kinases are involved in the DNA damage response as well as the replication checkpoint (Abraham, 2001). In this way "checkpoint rad" signaling leads directly downstream to pathways that are shared between replication and DNA damage checkpoints (Carr, 2002). This fact may not be surprising, since replication fork stalling is not only a consequence of low nucleotide pools but also is a consequence of encountering unreplicable DNA damage.

Fidelity of Nuclear Architecture at DNA Damage Checkpoints. DNA damage checkpoints are active in all phases of the cell cycle. Damage detected during the G_1 and S phases of the cell cycle delays entry into S or slows progression through S, respectively, likely to provide the opportunity for repair mechanisms to act and allow smooth replication of a lesion-free genome. Likewise damage detected during G_2 and possibly M phases of the cell cycle delays entry into mitosis and exit from mitosis, respectively, allowing for repair of chromosomal substrates to permit mitotic segregation. As with the spindle checkpoint, there are three components to the DNA checkpoint: (1) a damage sensor, (2) transduction of sensor signaling, and (3) downstream effectors of cell cycle inhibition and damage repair mechanisms.

Sensing and Signaling DNA Damage. DNA damage comes in many forms, all of which must be detected to signal arrest and repair. DNA damage can occur intrinsically through reactive oxygen species that are generated as metabolic by-products, through spontaneous disintegration of chemical bonds within DNA or exogenous induction of modifications through chemical insult, UV or ionizing radiation (IR) exposure (Hoeijmakers, 2001). Different types of genotoxic events lead to different forms of DNA damage, including thymidine crosslinks, bulky group adducts, single-strand breaks, double-strand breaks, and other phosphodiester and nitrogenous base modifications (Cadet et al., 1997).

Biochemical sensing of these different types of damage fall into two main categories: (1) sensing of DNA double-strand breaks (DSBs) and (2) sensing of all other aberrant DNA structures, lesions, and stalled replication forks (Fig. 2.7). ATM is likely to detect DSBs through its intrinsic DSB-end binding activity, while ATR and its associated coiled-coil cofactor, ATRIP/Rad26, seems to broadly detect most other types of damage (Abraham, 2001).

ATM and ATR/ATRIP transduce damage sensor signals via phosphorylation of many proteins important for the downstream control of cell cycle inhibition and DNA repair mechanisms. The effected signaling pathways are completely dependent on cell cycle position. Targets of ATM-dependent phosphorylation include important cell cycle signal transduction proteins and also proteins directly implicated in DNA repair, including p53 (tumor-suppressor: Lu and Lane, 1993), MDM2 (p53 repressor: Maya et al., 2001), CHK2 (checkpoint kinase with FHA:

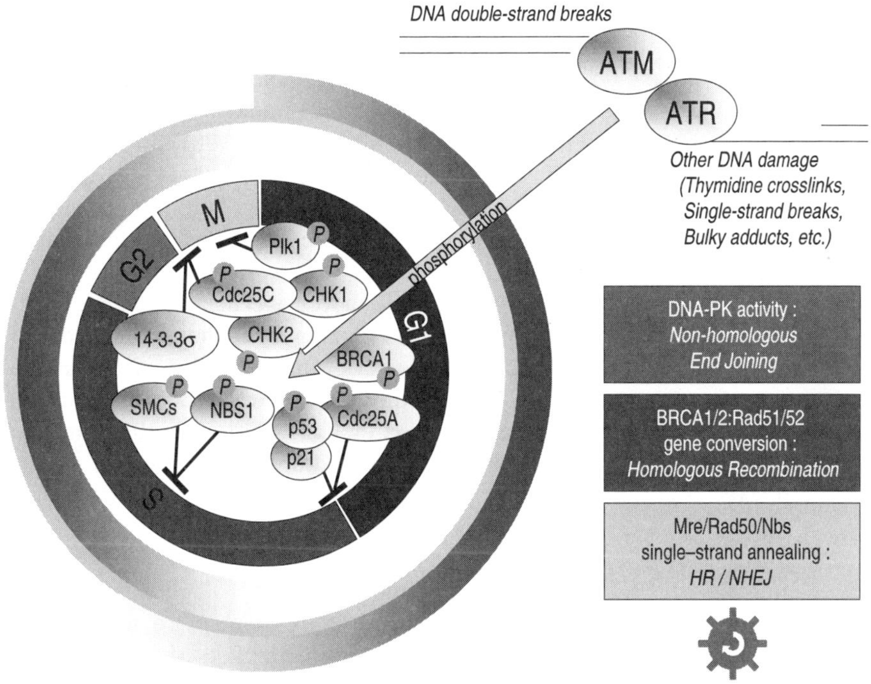

Figure 2.7. Cell cycle and DNA damage checkpoints, cycle arrest, and DNA repair. Diagrammed is the canonical cell cycle, represented by G_1 (gap 1), S (DNA synthesis), G_2 (gap2), and M (mitotic) phases. DNA damage is sensed by mechanisms including the ATM/ATR kinases. Cell cycle checkpoint activation initiates signaling cascades that include the proteins shown within the cell cycle diagram, effecting phase-specific cell cycle arrest as indicated. DNA repair complexes (DNA-PK, BRCA and MRN) are graphically represented around the cell cycle diagram in terms of their phase-specific activities.

Matsuoka et al., 1998), BRCA1 (BRCT domain: Koonin et al., 1996), 53BP1 (BRCT domain: Iwabuchi et al., 1994), NBS (forkhead-associated domain: Carney et al., 1998; Varon et al., 1998), SMC1/3 (Cohesin complex: Losada and Hirano, 2001), and perhaps Plks (Polo-like kinases: Smits et al., 2000). Similarly targets of ATR-dependent damage signal transduction overlap some ATM targets and minimally include CHK1 (checkpoint kinase: Liu et al., 2000), CHK2, and BRCA1. ATM/ATR-dependent signaling results in arrest in each phase of the cell cycle by activating specific cell cycle inhibitory molecules.

Phase-Specific Arrest. Checkpoint activation in G_1 or S phase results in G_1/S boundary or S progression arrest via at least two potential effector inhibitors: the CDK inhibitor p21WAF1/CIP1 (el-Deiry et al., 1993; Harper et al., 1993) and the phosphatase CDC25A (Falck et al., 2001). Both p21 and CDC25A inhibit CDK2 activity, which is essential for multiple S phase requirements (Harper and Adams, 2001). Checkpoint arrest at the G2/M boundary and during progression of M is effected via 14-3-3σ-mediated sequestration and nuclear exclusion of CHK1-phosphorylated CDC25C phosphatase (Peng et al., 1997). Active nuclear CDC25C is required to dephosphorylate Tyr14 and Tyr15 of CDC2, an event essential for G2/M transition (Jin et al., 1998). An M phase specific arrest may occur in both lower and higher eukaryotes, consisting of putative ATM-directed inactivation of Plk1 (Polo-like kinase), a kinase that activates the APC and promotes centromere maturation (Nigg, 1998).

Checkpoint activation and cell cycle arrest is crucial to cellular survival in the case of repairable DNA damage. Further evaluation of damage is required and repair versus programmed cell death decision-making is downstream of successfully executed checkpoint control. Arrest in any phase of the cell cycle is thought to provide the temporal opportunity for these downstream evaluation, direction, and action pathways.

DNA Repair Cycle

Cellular mechanisms to sense and signal DNA damage are consequentially followed by repair of damage. In higher eukaryotes three main complexes of proteins carry out repair of the most crucial DNA damage. Interestingly the activities of these complexes are regulated by cell cycle position and, perhaps more important, the ploidy state of the genome.

Among the many forms of damage that can cause genomic instability, DNA double-strand breaks (DSBs) appear to be the most insidious (reviewed in Jackson, 2002; Schar, 2001). DSBs can occur spontaneously during DNA replication, are formed transiently during meiosis and lymphoid V(D)J recombination, and can arise in patients exposed to chemo- or radiotherapeutic agents (Hoeijmakers, 2001). Improper repair of DSBs results in gross chromosomal rearrangements involving translocations, inversions, and fusions, which invariably lead to oncogenic transformation or cell death (Norbury and Hickson, 2001).

Ploidy-Specific Repair. Repairs of DSBs by the three cell cycle regulated complexes utilize different strategies depending upon the ploidy state of the genome (Fig. 2.7). During G_1, the genome is represented as 2N (diploid), and homologous chromosomes are not necessarily adjacent or available for recombination events between homologous alleles. Consequently, during G_1 and early S phase, higher eukaryotes predominantly depend on a DSB repair pathway called nonhomologous DNA end joining (NHEJ) mediated by the DNA-PK complex (Barnes, 2001; Hoeijmakers, 2001). The remainder of the cell cycle is characterized by a 4N (tetrapolid) genome, where sister chromatids are a template from which DSBs can be efficiently repaired without error. Consequently, during late S, G_2, and early M phases, higher eukaryotes predominantly depend on a DSB repair pathway called homologous recombination (HR) mediated by the BRCA/Rad51 complex (reviewed in Thompson and Schild, 2002; van den Bosch et al., 2002). A third complex forms the minor DSB repair pathway and is composed of Mre11/Rad50/Nbs (MRN) proteins (Carney et al., 1998). The MRN complex is active throughout the cell cycle and represents properties of both NHEJ and HR by joining DSBs with short stretches of base pairing, or microhomologies, thereby not depending on the ploidy state of the genome (Norbury and Hickson, 2001).

Repair via Nonhomologous End Joining in Diploid Cells. NHEJ uses limited sequence homology to rejoin ends in a manner that is often error prone. In mammalian cells, NHEJ is the preferred mechanism of DSB repair (Barnes, 2001; Karran, 2000). The two major complexes that appear to be critical to the normal repair of DSBs in mammalian cells are the MRN complex and the DNA-PK complex.

In mammals, the gene products that comprise the mammalian DNA-PK complex minimally include Ku70 (XRCC6), Ku86 (XRCC5), and the DNA-dependent protein kinase catalytic subunit (DNA-PK_{cs}; XRCC7) (Smith and Jackson, 1999). This DNA-PK complex, along with DNA ligase IV (Adachi et al., 2001) and the DNA ligase IV associated factor, XRCC4 (Sibanda et al., 2001; Wang et al., 2001b), are required for the rejoining of DSBs (Critchlow et al., 1997; Grawunder et al., 1997, 1998; Wang et al., 2001a, b). Moreover the DNA-PK complex is required for most, if not all, NHEJ DSB repair (Karran, 2000; Norbury and Hickson, 2001). The DNA-PK complex is also critical to the formation of a normal immune system through the regulated process of V(D)J recombination, where intermediates are generated that are biochemically equivalent to DSBs (Hendrickson et al., 1988, 1991; Smith and Jackson, 1999).

The ~465 kDa DNA-PK_{cs}:XRCC7 (DNA-dependent protein kinase catalytic subunit) protein is the product of the severe combined immune deficiency (*scid*) gene, which is a member of the phosphotidlyinositol 3-kinase (PI3-kinase) family (Hartley et al., 1995; Poltoratsky et al., 1995). Mutation of the genes in this protein kinase subfamily, such as ATM (ataxia telangectasia-, or AT-, mutated; see Khanna et al., 2001; Lavin and Shiloh, 1997) and ATR (AT-related; see Nghiem et al., 2001, 2002; Tibbetts et al., 1999), of the PI-3 lipid kinase superfamily often results in chromosomal instability syndromes in mammals (Durocher and Jackson, 2001; Shiloh, 2001). The importance of DNA-PK_{cs}'s kinase activity is not yet clear insofar as it pertains to signal transduction leading to NHEJ. However, signaling the completion of repair to many disparate pathways might be one role. The potential role of the ~465 kDa DNA-PK_{cs} protein as a scaffolding structure has been made clear by cryo-EM imaging, revealing the potential for a sheltered, rigid microenvironment in which DNA ends might be internalized along with accessory factors that include ligase IV or XRCC-4 (Chiu et al., 1998).

Ku was originally discovered as an autoantigen recognized by the antisera of patients with autoimmune diseases (Mimori and Hardin, 1986). Ku is a heterodimeric DNA end-binding complex composed of 70 and

86 kDa subunits (Ku70 and Ku86, respectively; reviewed in Smith and Jackson, 1999; Tuteja and Tuteja, 2000) that binds in a sequence nonspecific fashion to virtually all double-stranded DNA ends, including telomeres (Falzon et al., 1993; Haber, 1999; Mimori and Hardin, 1986). Ku86/70 has DNA-dependent ATPase (Cao et al., 1994) and helicase (Tuteja et al., 1994) activities in addition to end-binding activity (Tuteja and Tuteja, 2000). The binding of Ku to broken DNA ends is required to prevent unnecessary DNA degradation (Liang and Jasin, 1996) and juxtapose the DNA ends (Bliss and Lane, 1997; Pang et al., 1997; Walker et al., 2001). The binding of Ku to free DNA ends recruits and activates DNA-PK_{cs} (Gottlieb and Jackson, 1993; Suwa et al., 1994), DNA ligase IV (McElhinny et al., 2000; Teo and Jackson, 2000), and XRCC4 (Gao et al., 1998; Li et al., 1995).

Activity of the DNA-PK complex and the IR sensitivity of *scid* cells fluctuates during the cell cycle (Hendrickson, 1997). Wild-type cells demonstrate DNA-PK activity in the G_1 and early S phases of the cell cycle, a profile that parallels the IR hypersensitivity in *scid* cells (Lee et al., 1997). The complex does not appear to be controlled by protein quantity because the levels of the DNA-PK components remain constant throughout the cell cycle (Lee et al., 1997). The regulatory mechanism is undefined and may include relief of repression or stimulation of activation via posttranslational modifications and/or protein interactions (Hendrickson, 1997).

Repair via Homologous Recombination in Tetraploid Cells. Homologous recombination ensures relatively error-free repair by using an undamaged sister chromatid, homologous chromosome, or duplicate gene elsewhere in the genome as a template (Kuzminov, 2001; Paques and Haber, 1999). In higher eukaryotes, sister chromatids are available only in the late S and G_2 phases of the cell cycle. Higher eukaryotes utilize this pathway predominately for meiotic recombination and for a minority of exogenously induced DSBs. HR is classically represented by two competing mechanisms of action; one is gene conversion (Norbury and Hickson, 2001) and the other is single-strand annealing (SSA. Lin et al., 1984; Norbury and Hickson, 2001).

The participants in gene conversion include RAD51/RAD54 (mitotic), Dmc1/Tid1 (meiotic), RAD52, BRCA1, and BRCA2 (Haber, 2000; Hoeijmakers, 2001; Norbury and Hickson, 2001; Paques and Haber, 1999). RAD51 (Lambert and Lopez, 2001; Shibata et al., 2001; Yu et al., 2001) and DMC1 (Dresser et al., 1997; Harada et al., 2001; Masson and West, 2001) are homologues of the bacterial RecA strand exchange and single-stranded binding protein, with RAD51 being utilized primarily during mitosis and DMC1 during meiosis. Similarly RAD54 (Ristic et al., 2001; Solinger and Heyer, 2001; Swagemakers et al., 1998) and Tid1 (Shinohara et al., 1997; Shinohara et al., 2000) are homologues that act as binding partners for RAD51 and DMC1 during mitosis and meiosis, respectively (Arbel et al., 1999; Shinohara et al., 1997). RAD52 is also a RAD51 interacting protein that appears to target RAD51 to DNA ends

(Parsons et al., 2000; Shinohara et al., 1998; Stasiak et al., 2000). Whether RAD52 also interacts with DMC1 or there is a meiosis-specific equivalent of RAD52 is not currently known.

Initial resection of the free double-stranded DNA ends enables RAD52 and then RAD51 to interact with the resected region and free end. These interactions strongly potentiate the progression of gene conversion. The RAD51/RAD52 bound single-strand is recombinogenic and can invade a homologous sequence. In this way gene conversion via strand invasion typically occurs between two alleles of a gene.

Strand invasion is thought to establish a structure very similar to a replication fork or Holliday junction (HJ) (Haber, 2000; Paques and Haber, 1999). The structural similarity may be extensive in that there is even evidence of utilization of leading and lagging strand synthesis at gene conversion loci (Holmes and Haber, 1999). The precise mechanism of action of gene conversion is still a matter of some debate. Gene conversion results in the repair of DSBs with no loss of genetic information. An exception is allelic differences that may have existed between homologous chromosomes.

Importantly, the gene products of the BRCA1 and BRCA2 breast cancer susceptibility genes are capable of forming complexes with a large number of proteins through their BRCT and RING domains. These structural and architectural complexes have been associated with a multitude of biochemical roles, including transcription-coupled repair and nucleation of repair complexes at sites of DNA damage (Haber, 2000; Kerr and Ashworth, 2001; Paques and Haber, 1999; Venkitaraman, 1999).

Repair via Single-Strand Annealing throughout the Cell Cycle. Single-strand annealing (SSA) is an alternative form of HR. SSA and gene conversion are competing mechanisms. The predominant pathway appears to vary with the organism. The proteins controlling the process of SSA are MRE11, whose important biochemical activities include a $3' \rightarrow 5'$ double-stranded exonuclease activity (Stewart et al., 1999; Yamaguchi-Iwai et al., 1999), RAD50, a potential ATP-dependent DNA-binding protein (Luo et al., 1999; Stewart et al., 1999), and NBS/Xrs1, a putative enabler of signal transduction activities (Carney et al., 1998; Dong et al., 1999). These proteins form the MRN complex, and each is required for SSA to occur (Karran, 2000; Norbury and Hickson, 2001). Additional helicase, exonuclease, and kinase activities are associated with the MRN complex, although precisely which activity goes with which protein is not yet well characterized. The mechanism of MRN complex action within SSA is also not well defined, although it may act at the sites of the lesions because it localizes in nuclear foci following genotoxic insult (Maser et al., 1997; Nelms et al., 1998).

Single-strand annealing (SSA) may occur if initial resection of the free double-stranded DNA ends exposes complementary homologous sequences. SSA is relatively inefficient between short homologies (~30 base pairs) or relatively efficient between long homologies (~200–400 base pairs) on either side of the original DSB (Sugawara et al., 2000).

Annealing of homologous sequences results in the exclusion of the intervening single-stranded tails. These excluded single-stranded tails are trimmed. DNA synthesis is then employed to fill-in the intervening gaps, and subsequent ligation of the newly synthesized sequence to the resected template completes the reaction (Haber, 2000). SSA is the only type of HR that results in the loss of genetic information, with the loss occurring between the two stretches of homology.

Architectural Organization of Repair. At least two of the three major DSB repair complexes are structurally and architecturally ordered into nuclear foci. BRCA and MRN complexes rapidly form foci at sites of DNA damage (Maser et al., 1997; Scully et al., 1997), likely through ATM/ATR signaling and additionally via interaction with phosphorylated H2AX core histone variant (Kobayashi et al., 2002; Paull et al., 2000). H2AX integration into nucleosomes adjacent to sites of DSBs occurs very quickly (1–10 min) prior to repair foci formation (Rogakou et al., 1998). Additionally there are reports of a large architecturally organized complex designated the BASC (BRCA1-associated genome surveillance complex), consisting of a very large number of functionally related complexes (Wang et al., 2000). The BASC has been reported to contain the BRCA/Rad51 and MRN complexes as well as ATM and a large number of other repair-associated activities.

Tumor-Suppressor Gene Cycle

Tumor-suppressor proteins have disparate roles in the regulation of cellular growth factor responsiveness, DNA damage/repair, and cell cycle checkpoints. Mutation or silencing of tumor-suppressor genes is common in cancers, and a germline mutation in one allele of a tumor-suppressor is the basis of many hereditary cancers.

So-called uncontrolled cell division in tumors refers in part to tumor-suppressor protein "control" of cell division. From the perspective of cancer biology, tumor-suppressor genes are functionally informative. However, from a cell biology perspective, a cell must regulate the activity of these tumor-suppressor (cycle suppressor) proteins during the course of "controlled" cell division. The regulation of tumor-suppressor proteins has been found to be highly structural, compartmental, and architectural. These features of regulation are periodic during the cell cycle. The cyclic and structural aspects of tumor suppressor protein regulation considered here include nuclear-cytoplasmic shuttling, subnuclear sequestration, and focal concentrations of activities in response to signaling mechanisms. Each of these strategies are misregulated in some tumors.

Regulation of Tumor-Suppressor Proteins through Scheduled Nucleocytoplasmic Shuttling. Of the field of tumor-suppressor proteins, many of the key proteins screened and studied to date include both nuclear transport and export signals (Fabbro and Henderson, 2003). These subcellular localization signals help to mediate the regulated process of nuclear membrane translocation via signal recognition, nuclear pore docking, nuclear pore transport, and protein release. Regulation of some tumor-suppressor proteins via orderly shuttling is an ideal mechanism due to nuclear localization of crucial targets.

Among the tumor-suppressor proteins regulated by nucleocytoplasmic shuttling are p53 ("guardian of the genome": Liang and Clarke, 1999; Zhang and Xiong, 2001), p73 (p53-related: Inoue et al., 2002), BRCA1 (BRCT-domain containing: Rodriguez and Henderson, 2000; Wilson et al., 1999), APC (adenomatous polyposis coli: Galea et al., 2001; Neufeld et al., 2000a; Rosin-Arbesfeld et al., 2000; Zhang et al., 2000a), PML (promyelocytic leukemia gene: Daniel et al., 1993; Henderson and Eleftheriou, 2000), p130 (Rb-related: Chestukhin et al., 2002), VHL (von Hippel-Lindau gene: Lee et al., 1996, 1999), Smad4 (growth receptor-nucleus signaling: Pierreux et al., 2000), Beclin (Bcl2-interacting coiled-coil protein: Liang et al., 2001), and INI1/hSNF5 (SWI/SNF component: Craig et al., 2002).

The detailed regulation of nucleocytoplasmic shuttling of each of these tumor-suppressors is different from one another, indicative of independently evolved (versus common) mechanisms. The import pathways minimally include usage of either Importin-α/β, BARD1 and B56α, while export pathways nearly exclusively utilize CRM1-dependent nuclear export (Fabbro and Henderson, 2003).

Nucleolar Sequestration of Tumor-Suppressors and their Regulatory Factors. A common regulatory mechanism for the activity of certain tumor-suppressor proteins is subnuclear targeting, or sequestration. The nucleolus is a common architectural subnuclear target of sequestration for the regulation and homeostatic maintenance of a wide array of proteins (Olson et al., 2002).

Structural subnuclear targeting has been found to directly regulate tumor-suppressor proteins but also to act indirectly within a relevant regulatory pathway. The tumor-suppressor itself may be sequestered in the nucleolus, as might be the case with BRCA1 in some tumor cells (Tulchin et al., 1998), and the candidate tumor-suppressor ING1 upon cellular conditions of UV-induced DNA damage (Scott et al., 2001). Alternatively, regulatory proteins can be sequestered to the nucleolus, as is the case with nucleolar-localized protein $p14^{ARF}$ (Lindstrom et al., 2000; Rizos et al., 2000) and its sequestration of MDM2 to the nucleolus (Lohrum et al., 2000b; Tao and Levine, 1999) to relieve inhibition on p53 (Lohrum et al., 2000a; Tsai and McKay, 2002).

$p14^{ARF}$ has also been found to exert a sequestration influence on some members of the E2F family of transcription factors (Datta et al., 2002). E2F transcription factor activities are required for multiple cell cycle progression pathways in the G_1/S transition, including DNA replication (Farnham et al., 1993). Classic E2F regulation inhibits activity while complexed with the Rb tumor-suppressor in a manner dependent on Rb phosphorylation status (Chellappan et al., 1991; Lee et al., 2002;

Weintraub et al., 1995). Rb is phosphorylated by multiple cyclin-dependent kinases such that Rb phosphorylation, and thus E2F activity, is periodic within the cell cycle (Nevins, 1992; Stevaux and Dyson, 2002). The involvement of nucleolar sequestration within this cyclic pathway must be further defined.

Intranuclear Compartmentalization of Tumor-Suppressors. Some tumor-suppressor proteins undergo structural and architectural associations in a cell cycle-specific manner and/or following signaling of specific stimuli. This is particularly relevant to the activity of many tumor-suppressor proteins that have roles in the cellular DNA damage response. Rb localizes with replication origins (foci) during S phase following DNA damage in a protein phosphatase 2A (PP2A)-dependent manner, presumably to suppress inappropriate replication initiation at those origins (Burger, 2002). Similarly p53 is recruited by the BLM helicase (defective in Bloom's Syndrome; Elledge, 1996) to replication origins upon hydroxyurea (HU) treatment, which results in stalling of replication forks and triggering of the replication checkpoint (Sengupta et al., 2003).

The best example of focal concentrations of a tumor-suppressor is BRCA1, which exhibits germ-line mutations in over 50% of patients with inherited breast cancers and 90% with breast and ovarian cancer susceptibility (Couch and Weber, 1996). Clearly, BRCA1 serves an important tumor-suppressor role (Chen et al., 1999; Scully and Livingston, 2000). BRCA1 interacts with a large number of proteins, in part through two BRCT (*BR*CA1 *c*arboxyl-*t*erminus) domains that interact with BRCT domains on other proteins. BRCA1 also contains an amino-terminal RING domain that binds BARD1 (*B*RCA1 *a*ssociated *R*ING *d*omain protein), which masks the BRCA1 nuclear export signal, thereby acting as a nuclear chaperone (Fabbro et al., 2002). BRCA1-BARD1 form discrete nuclear foci during S phase in addition to DNA damage inducible foci important for replication and DNA repair (Jin et al., 1997; Scully et al., 1997). By these protein interaction motifs, BRCA1 seems to play a central role in many of the structural and architectural responses to DNA damage as well as normal progression through S phase.

Clearly, there is much to be gained by further examining tumor-suppressor protein function, and structural and architectural cycles. Continued discovery of patterns and consistencies including nucleocytoplasmic shuttling, subnuclear targeting, sequestration, and nuclear focal concentrations are important to continued understanding of regulation and activity of tumor-suppressors during the cell cycle.

Proliferation/Differentiation Cell Cycle Control

Multicellular organisms are characterized by the presence of highly specialized cells that are programmed to differentiate into specific lineages, consequently giving rise to various organs with different functions. These specialized cells require exit from active proliferation to execute their differentiation programs. Once cells exit from the cell cycle (proliferative stage), they enter quiescent or G_0 stage in which a completely new set of genes is induced while genes related to cell cycle progression are turned off (reviewed in Malumbres and Barbacid, 2001). Exit from the cell cycle and entry into G_0 also result in major architectural changes in cellular and nuclear parameters concomitant with the initiation of differentiation program.

Stem Cell Maintenance: Asymmetric Fate and Architecture. Asymmetric division in cells is a highly structural and architectural feat of cell kinetics (Hawkins and Garriga, 1998). The importance of asymmetric cell division has long been recognized in relation to maintenance of stem cells, including the well-characterized gametic and therapeutically relevant hematopoietic stem cells (Wolpert, 1988). Maintenance of stem cells during the production of a more differentiated cell obligates asymmetric division ipso facto. The canonical decision of proliferation versus differentiation is one that is split in the case of stem cell maintenance (Sherley, 2002). This type of asymmetric division is not only important in early development, it is relevant throughout development and in the continued homeostasis of an adult organism (Brummendorf et al., 1999; Knoblich, 2001). Tissue remodeling and wound healing are dependent on recruitment, proliferation, and differentiation of stem cells. The fates of the progeny cells can be significantly different, just as the biology and biochemistry associated with proliferation versus differentiation are quite different. In particular, these differences include regulation of gene expression, translation, and protein turnover ratios in staggering complexities that result in a differentiated cell phenotypically distinct from the parental cell (Brummendorf et al., 1998, 1999; Mantel et al., 2001; Seery and Watt, 2000; Shen et al., 2002).

A small group of proteins has been shown to be potentially important for asymmetric cell kinetics. These proteins include the tumor suppressors p53, p63, and Pten, and the p53-regulated genes inosine-5′-monophosphate dehydrogenase (IMPDH) and $p21^{cip1/waf1}$ (Sherley, 2002). Gene products and pathways contributing to asymmetric cell division are not likely to be limited to these few proteins, and much progress is being made in characterizing important mechanisms of stem cell propagation and maintenance. Future findings are essential not only to in vivo biology but also in vitro cultures of embryonic and nonembryonic stem cells of all types.

Hematopoiesis: Distinct Nuclear Architecture upon Differentiation. Granulocytes, monocytes platelets, red blood cells, and a variety of lymphocyte types are produced primarily in marrow tissue in the adult mammal. This production proceeds from primitive multipotent stem or progenitor cells through progressively lineage-restricted progenitors to morphologically recognizable dividing and finally nondividing end cells (reviewed in Quesenberry et al., 1999). Cells of hematopoietic lineage therefore serve as a paradigm for studying the link between nuclear architecture and differentiation programs. The hematopoietic stem cell exhibits a round nuclear shape, while cells of different lineages originating from the stem cells possess diverse nuclear shapes; for example, platelets and lymphocytes retain round nuclear shapes but exhibit great variation in nuclear size, monocytes are characterized by kidney bean-shaped nucleus, eosinophils have horse shoe-shaped nuclei, and multi-lobed nuclei are present in basophiles and neutrophiles. These microscopic observations are often accompanied by variations in gene expression profiles. Thus the nuclear architecture is equally involved in regulation of differentiation programs in lineage-committed cells.

A different scenario for the involvement of nuclear structure and gene regulation is presented during the differentiation of monocytes into osteoclasts, the bone-resorbing cells, as well as during the differentiation of pre-myoblasts into mature myotubes. In these cases mononuclear cells fuse together to form multi-nucleated, lineage-committed cells. Similar to hematopoietic differentiation, osteoclastogenesis and myogenesis involve changes in gene expression profiles that accompany alterations in nuclear architecture (reviewed in Duong and Rodan, 2001; Wigmore and Evans, 2002).

Leukemiogenesis: Alterations in Subnuclear Organization of Transcription Factors Accompany Disease. Differentiation of pluripotent cells into different lineages is often marked by the induction of "master regulator" transcription factors. Gene ablation studies have provided valuable insight into biological activities of such transcription factors. Runx1 (also known as AML1 or Cbfa2) has been shown to be required for definitive hematopoiesis (Wang et al., 1996). At the cellular level, Runx1 is present in nuclei as distinct subnuclear foci that are transcriptionally active (Zeng et al., 1998). Importantly, a 31 amino acid segment, named the nuclear matrix-targeting signal (NMTS), is responsible for targeting of Runx1 to subnuclear sites (Zeng et al., 1997, 1998). Runx1 is a frequent target of chromosomal translocations that are involved in the development of leukemiogenesis (reviewed in Speck et al., 1999). Interestingly, the most frequent translocation in human acute myelogenous leukemia t(8;21) results in a fusion protein that lacks the Runx1 NMTS and is targeted to different subnuclear compartment than wild type Runx1 (Barseguian et al., 2002; McNeil et al., 1999). Thus alterations in subnuclear organization and distribution are accompanied by the development of pathological condition, suggesting a link between nuclear architecture and pathogenesis (reviewed in Stein et al., 2000b).

Dynamic Changes in the Nuclear Envelope during the Cell Cycle

The functional complexity of the eukaryotic nucleus is often supported by, and explained in terms of, architecturally distinguishable features that include the nuclear matrix, chromatin, and the nucleolus. The nuclear envelope, one of such architectural parameters of the eukaryotic nucleus, is characterized by the presence of a "nuclear rim" formed by nuclear lamins. A primary function of the nuclear envelope is to separate nuclear transcription of genes from cytoplasmic translation of messenger RNA. Another important function of the nuclear envelope is to regulate the nucleocytoplasmic transport of macromolecules through nuclear pores, which in turn is pivotal to temporal-spatial regulation of gene expression. Nuclear transcription is regulated in a temporal-spatial manner throughout the cell cycle. Transcription is transiently silenced as cells enter mitosis and is restored in progeny cells. The equivalent partitioning of chromosomes into the progeny cells necessitates gross structural changes in nuclear morphology and requires the disintegration of the nuclear envelope. However, the mitotic or M phase (mitotic) of the cell cycle is less than an hour, and the nuclear envelope must be reassembled around newly segregated genomes to ensure the integrity of progeny cells.

Nuclear lamins, the primary subunit of the nuclear envelope, belong to intermediate filament family of proteins. Lamins polymerize to form a two-dimensional lattice and connect to endoplasmic reticulum in the cytoplasm, chromatin, and inner membrane integral proteins inside the nucleus (reviewed in Gant and Wilson, 1997). In situ immunofluorescence of nuclear lamins in fixed mitotic cells, together with the application of biochemical assays, has revealed a sequence of events leading to the disassembly of the nuclear envelope at the onset of mitosis and re-assembly as cells enter telophase. At the onset of mitosis, cyclin-dependent kinase-mediated phosphorylation of nuclear lamins results in reversible depolymerization of nuclear lamins (Gerace and Blobel, 1980), thus disassembling the nuclear envelope. In situ immunofluorescence microscopy shows that nuclear lamins are enclosed in tubules and vesicles during prophase and metaphase, which are released into the "cytoplasm" of the mitotic cell. During anaphase-telophase transition the membrane vesicles fuse to form the nuclear envelope and lamins repolymerize to provide the required structural integrity (reviewed in Gant and Wilson, 1997).

The microscopic observations of nuclear lamins in fixed cells have been supported by recent studies in live cells (reviewed in Lippincott-Schwartz, 2002). Studies using several nuclear envelope proteins (e.g., Lamin B receptor: Ellenberg et al., 1997; and nuclear pore complex (NPC) proteins, e.g., POM121, Nup153 and Lamin B1: Daigle et al., 2001) fused with EGFP have revealed distinct behavior for various nuclear envelope components in live cells. For example, Lamin B receptor (LBR) is synthesized in the endoplasmic reticulum during interphase and then targeted to the inner nuclear membrane independent of cell division. Fluorescence recovery after photo-bleaching (FRAP) of the EGFP-LBR chimeric protein shows that the LBR is highly mobile in the ER fraction, and its targeting to the inner nuclear membrane results in immobilization of the receptor. High-resolution confocal microscopy of mitotic cells reveals that the LBR becomes highly mobile during mitosis, and is redistributed to the ER where it colocalizes with the ER markers

(Ellenberg et al., 1997). In contrast, the NPC proteins are not mobile during the interphase. As cells enter mitosis, a large array of NPC proteins slowly and synchronously moves suggesting that NPC proteins are interconnected. During mitosis all the NPC proteins are completely mobile and dispersed and rapidly redistribute to form an immobile pool around chromatin during anaphase-telophase transition. The recruitment of Lamin B1 to the NPC follows that of nucleoporins such as POM121 and Nup153 (Daigle et al., 2001). Thus components of the nuclear envelope and nuclear pore complex follow a cyclical and sequential pattern during each cell cycle and assembly, and re-assembly of the nuclear envelope must be completed within the mitotic time frame to ensure the integrity of the eukaryotic nucleus.

Several lines of evidence suggest that the nuclear lamins are required for DNA replication to proceed through S phase. In addition to organizing into nuclear envelope, the nuclear lamins are also present in the interior of the nucleus as intranuclear foci that, in case of lamin B, colocalize with replication sites as well as with replication proteins such as PCNA and replication fork complex (RFC) (Moir et al., 1994; Spann et al., 1997). Furthermore immunodepletion of nuclear lamins results in cellular extracts that are incompetent for DNA replication (Newport et al., 1990). These findings suggest an "architecturally linked" crosstalk between two distinct cycles within the cell cycle, namely the replication cycle and the nuclear envelope cycle. It is appropriate to suggest that the integration of these two pathways, and perhaps several others, is required for precise and faithful progression of the cell cycle.

Apoptosis: A Graceful Exit from Cycles

All cells die. Yet not all cells die equally. Cell death is an important evolutionary consideration. A unicellular organism has a much different "evolutionary view" of cell death than a multicellular organism. As such, one would expect that a unicellular organism, which is not reliant, nor relies on other similar organisms, would struggle to survive absolutely despite any circumstance. In a completely contrasting evolutionary strategy, a multicellular organism is a homeostatic environment in which cells die and divide to maintain the organism as a whole. It is now clear that cells from multicellular organisms do not simply look out for themselves, and rather regulate themselves to preserve the function of the whole.

Cell death can be accidental, as in the case of a wound or injury of some kind. However, cell death has been found to be important to normal organismal homeostasis, development, and elimination of cancerous cells. Accidental injuries damage membranes and cellular architecture, spilling carefully packaged noxious contents into the extracellular matrix, producing an inflammatory response. Homeostatic cell death, on the other hand, avoids damaging neighboring cells and prevents an inflammatory response via a mechanism known as programmed cell death (PCD), or apoptosis (Kerr et al., 1972). Apoptosis is a highly regulated mechanism by which cells can systematically shut down growth signaling, cell cycling, DNA repair, transcription, translation, energy production, and metabolism while remaining membrane encapsulated (Earnshaw et al., 1999).

Nuclear Architectural Modifications through Apoptosis. Apoptosis is a re-structural event, exquisite examples of which are the nuclear architectural changes (Martelli et al., 2001). There are several classic nuclear phenotypic hallmarks of apoptosis, each of which has structural and architectural ramifications and is consistent between most systems examined to date. Chromatin condenses and collapses during early stages of apoptosis, and this is associated with endonuclease activity (CAD: Counis and Torriglia, 2000) and possibly histone H3, H2B and/or H2AX phosphorylation (Ajiro, 2000; Rogakou et al., 2000; Waring et al., 1997). Chromatin condensation can be seen in characteristic crescent shapes on one side of the nuclear membrane. Nuclear shrinkage and/or blebbing of ribonucleotide-filled membrane bound buds seems to be a structural characteristic of the protease (caspases: Earnshaw et al., 1999) mediated cleavage of the nucleoporin (Nup: Buendia et al., 1999), lamin B receptor (LBR: Duband-Goulet et al., 1998) and lamin-associated polypeptide $\alpha 2$ (LAP: Gotzmann et al., 2000), and actin cytoskeleton (Coleman and Olson, 2002). The buds have structure whose composition is not yet clear, as particular buds solely contain DNA or RNA (Halicka et al., 2000). Nuclear pores are proteolytically released from chromatin via S/MAR component cleavage (Martelli et al., 2001), and redistribute away from the crescent of condensed chromatin (Falcieri et al., 1994).

These nuclear re-structuring events during apoptosis serve to package the cell into membrane-bound and noninflammatory orbs that can cleared by the immune system via phagocytosis. Much of the restructuring relies upon caspase proteolysis of existing structural and architectural associated proteins. Far from completely degrading these structural proteins, caspase specificity of cleavage often results in a protein with altered function. The characteristic structural phenotypes, which are seen as apoptosis progresses, is in part due to re-structuring of cellular architecture due to those altered functions.

CONCLUDING REMARKS: CHALLENGES AND OPPORTUNITIES FOR INSIGHTS INTO BIOLOGICAL REGULATION

Competency for proliferation and cell cycle progression require the stringent execution of regulatory cascades that are governed by the temporal-spatial integration of physiological cues. There is growing evidence that subnuclear localization of nucleic acids and regulatory proteins is necessary for gene expression, replication, and repair. Consequently there is a necessity to regulate the organization and compartmentalization as well as the mitotic distribution of the machinery for transcription and DNA synthesis that is requisite for fidelity of activity.

A similar architectural perspective of the regulatory machinery for cell survival and apoptosis should be established.

Compelling support is emerging for mechanisms that mediate the assembly of regulatory complexes at sites within the nucleus where threshold levels are attained for physiological responsiveness. Examples of architecturally organized regulatory complexes that contribute to cell cycle and growth control are numerous. It is well recognized that chromosomes, nucleoli, sites of replication, repair, and transcription reflect context-dependent specialization of niches within the nucleus that can profoundly influence biological control.

Traditional molecular, biochemical, and genetic approaches, together with high-throughput analyses, have yielded insight into pathways that regulate proliferation and aberrations that are incurred with compromised control during the onset and progression of tumorigenesis. But, as the databases of regulatory macromolecules expand, the challenge is to understand combinatorial mechanisms as they are operative in intact cells where a broad spectrum of regulatory factors are assembled to regulate the cell cycle. The capacity of regulatory proteins to function as scaffolds and substrates further illustrates options inherent in the regulatory circuitry of signaling networks and pathways that contribute to options for responsiveness to both intrinsic and extrinsic cues. A perspective that can be explored experimentally is that sequentially and functionally interrelated regulatory cycles, each requiring the dynamic assembly of architecturally associated macromolecular complexes, support physiological control of proliferation and contribute to perturbations in growth regulatory mechanisms in transformed cells and cancer.

A prominent role for the execution of cell cycle and growth regulatory mechanisms within the three-dimensional context of nuclear architecture is becoming increasingly evident. Further characterizing regulatory components of proliferation that are embedded in nuclear structure–gene expression interrelationships is formidable but compelling. Equivalently relevant is the necessity to further elucidate the interfacing of cell cycle and growth control with regulation of apoptosis and cell survival. Here the common denominator is emerging linkages between nuclear architecture and biological activity. The outcome will be expanded insight into fundamental parameters of proliferation and novel options for therapy that are predicated on functional interrelationships between nuclear structure and biological control.

ACKNOWLEDGMENTS

The authors thank Elizabeth Bronstein for editorial assistance with the preparation of this manuscript. Results presented in this chapter were in part supported by the National Institutes of Health grants R01-GM32010, PO1-AR48818, PO1-CA82834.

REFERENCES

Abraham RT (2001): Cell cycle checkpoint signaling through the ATM and ATR kinases. Genes Dev 15:2177–96.

Adachi N, Ishino T, Ishii Y, Takeda S, Koyama H (2001): DNA ligase IV-deficient cells are more resistant to ionizing radiation in the absence of Ku70: Implications for DNA double-strand break repair. Proc Nat Acad Sci USA 98: 12109–13.

Adachi Y, Laemmli UK (1992): Identification of nuclear pre-replication centers poised for DNA synthesis in *Xenopus* egg extracts: Immunolocalization study of replication protein A. J Cell Biol 119:1–15.

Ajiro K (2000): Histone H2B phosphorylation in mammalian apoptotic cells: An association with DNA fragmentation. J Biol Chem 275:439–43.

Al-Khodairy F, Carr AM (1992): DNA repair mutants defining G_2 checkpoint pathways in *Schizosaccharomyces pombe*. EMBO J 11:1343–50.

Al-Khodairy F, Fotou E, Sheldrick KS, Griffiths DJ, Lehmann AR, Carr AM (1994): Identification and characterization of new elements involved in checkpoint and feedback controls in fission yeast. Mol Biol Cell 5:147–60.

Altieri DC (2003): Survivin and apoptosis control. Adv Cancer Res 88:31–52.

Arbel A, Zenvirth D, Simchen G (1999): Sister chromatid-based DNA repair is mediated by RAD54, not by DMC1 or TID1. EMBO J 18:2648–58.

Aziz F, van Wijnen AJ, Stein JL, Stein GS (1998a): HiNF-D (CDP-*cut*/CDC2/cyclin A/pRB-complex) influences the timing of IRF-2 dependent cell cycle activation of human histone H4 gene transcription at the G_1/S phase transition. J Cell Physiol 177:453–64.

Aziz F, van Wijnen AJ, Vaughan PS, Wu S, Shakoori AR, Lian JB, Soprano KJ, Stein JL, Stein GS (1998b): The integrated activities of IRF-2 (HiNF-M) CDP/cut (HiNF-D) and H4TF-2 (HiNF-P) regulate transcription of a cell cycle controlled human histone H4 gene: mechanistic differences between distinct H4 genes. Mol Biol Rep 25:1–12.

Bae SC, Yamaguchi-Iwai Y, Ogawa E, Maruyama M, Inuzuka M, Kagoshima H, Shigesada K, Satake M, Ito Y (1993): Isolation of PEBP2αB cDNA representing the mouse homolog of human acute myeloid leukemia gene, AML1. Oncogene 8:809–14.

Banerjee C, Hiebert SW, Stein JL, Lian JB, Stein GS (1996): An AML-1 consensus sequence binds an osteoblast-specific complex and transcriptionally activates the osteocalcin gene. Proc Natl Acad Sci USA 93:4968–73.

Banerjee C, McCabe LR, Choi J-Y, Hiebert SW, Stein JL, Stein GS, Lian JB (1997): Runt homology domain proteins in osteoblast differentiation: AML-3/CBFA1 is a major component of a bone specific complex. J Cell Biochem 66:1–8.

Bannister AJ, Schneider R, Kouzarides T (2002): Histone methylation: Dynamic or static? Cell 109:801–6.

Barnes DE (2001): Non-homologous end joining as a mechanism of DNA repair. Curr Biol 11:455–7.

Barseguian K, Lutterbach B, Hiebert SW, Nickerson J, Lian JB, Stein JL, van Wijnen AJ, Stein GS (2002): Multiple subnuclear targeting signals of the leukemia-related AML1/ETO and ETO repressor proteins. Proc Natl Acad Sci USA 99:15434–9.

Becker M, Baumann C, John S, Walker DA, Vigneron M, McNally JG, Hager GL (2002): Dynamic behavior of transcription factors on a natural promoter in living cells. EMBO Rep 3:1188–94.

Bell SP, Dutta A (2002): DNA replication in eukaryotic cells. An Rev Biochem 71:333–74.

Berezney R, Mortillaro M, Ma H, Meng C, Samarabandu J, Wei X, Somanathan S, Liou WS, Pan SJ, Cheng PC (1996): Connecting nuclear architecture and genomic function. J Cell Biochem 62:223–6.

Berezney R, Wei X (1998): The new paradigm: Integrating genomic function and nuclear architecture. J Cell Biochem Suppl 30–31:238–42.

Bickmore WA, Chubb JR (2003): Chromosome position: Now, where was I? Curr Biol 13:357–9.

Bird AP (1984): DNA methylation—How important in gene control? Nature 307:503–4.

Bird AP, Macleod D, Taggart MH (1982): Studies of the relationship between DNA methylation and transcription of the ribosomal RNA genes. Adv Exp Med Biol 158:375–80.

Bird AP, Wolffe AP (1999): Methylation-induced repression—Belts, braces, and chromatin. Cell 99:451–4.

Birnbaum MJ, van Wijnen AJ, Odgren PR, Last TJ, Suske G, Stein GS, Stein JL (1995a): Sp1 trans-activation of cell cycle regulated promoters is selectively repressed by Sp3. Biochem 34:16503–8.

Birnbaum MJ, Wright KL, van Wijnen AJ, Ramsey-Ewing AL, Bourke MT, Last TJ, Aziz F, Frenkel B, Rao BR, Aronin N, Stein GS, Stein JL (1995b): Functional role for Sp1 in the transcriptional amplification of a cell cycle regulated histone H4 gene. Biochem 34:7648–58.

Bliss TM, Lane DP (1997): Ku selectively transfers between DNA molecules with homologous ends. J Biol Chem 272:5765–73.

Boggs BA, Cheung P, Heard E, Spector DL, Chinault AC, Allis CD (2002): Differentially methylated forms of histone H3 show unique association patterns with inactive human X chromosomes. Nat Genet 30:73–6.

Boyle S, Gilchrist S, Bridger JM, Mahy NL, Ellis JA, Bickmore WA (2001): The spatial organization of human chromosomes within the nuclei of normal and emerin-mutant cells. Hum Mol Genet 10:211–19.

Brummendorf TH, Dragowska W, Lansdorp PM (1999): Asymmetric cell divisions in hematopoietic stem cells. An NY Acad Sci 872:265–72.

Brummendorf TH, Dragowska W, Zijlmans J, Thornbury G, Lansdorp PM (1998): Asymmetric cell divisions sustain long-term hematopoiesis from single-sorted human fetal liver cells. J Expr Med 188:1117–24.

Buendia B, Santa-Maria A, Courvalin JC (1999): Caspase-dependent proteolysis of integral and peripheral proteins of nuclear membranes and nuclear pore complex proteins during apoptosis. J Cell Sci 112:1743–53.

Burger MM (2002): UICC study group on basic and clinical cancer research: Mechanisms of metastasis. J Cell Biochem 87:253–7.

Cadet J, Berger M, Douki T, Ravanat JL (1997): Oxidative damage to DNA: formation, measurement, and biological significance. Rev Physiol Biochem Pharmacol 131:1–87.

Cao QP, Pitt S, Leszyk J, Baril EF (1994): DNA-dependent ATPase from HeLa cells is related to human Ku autoantigen. Biochem 33:8548–57.

Carney JP, Maser RS, Olivares H, Davis EM, Le Beau M, Yates JR3, Hays L, Morgan WF, Petrini JH (1998): The hMre11/hRad50 protein complex and Nijmegen breakage syndrome: Linkage of double-strand break repair to the cellular DNA damage response. Cell 93:477–86.

Carr AM (2002): DNA structure dependent checkpoints as regulators of DNA repair. DNA Rep 1:983–94.

Celis JE, Madsen P (1986): Increased nuclear cyclin/PCNA antigen staining of non S-phase transformed human amnion cells engaged in nucleotide excision DNA repair. FEBS Lett 209:277–83.

Chellappan SP, Hiebert S, Mudryj M, Horowitz JM, Nevins JR (1991): The E2F transcription factor is a cellular target for the RB protein. Cell 65:1053–61.

Chen Y, Lee WH, Chew HK (1999): Emerging roles of BRCA1 in transcriptional regulation and DNA repair. J Cellular Physiol 181:385–92.

Chestukhin A, Litovchick L, Rudich K, DeCaprio JA (2002): Nucleocytoplasmic shuttling of p130/RBL2: novel regulatory mechanism. Mol Cell Biol 22: 453–68.

Chiu CY, Cary RB, Chen DJ, Peterson SR, Stewart PL (1998): Cryo-EM imaging of the catalytic subunit of the DNA-dependent protein kinase. J Mol Biol 284: 1075–81.

Choi J-Y, Pratap J, Javed A, Zaidi SK, Xing L, Balint E, Dalamangas S, Boyce B, van Wijnen AJ, Lian JB, Stein JL, Jones SN, Stein GS (2001): Subnuclear targeting of Runx/Cbfa/AML factors is essential for tissue-specific differentiation during embryonic development. Proc Natl Acad Sci USA 98:8650–5.

Chubb JR, Boyle S, Perry P, Bickmore WA (2002): Chromatin motion is constrained by association with nuclear compartments in human cells. Curr Biol 12:439–45.

Ciejek EM, Tsai MJ, O'Malley BW (1983): Actively transcribed genes are associated with the nuclear matrix. Nature 306:607–9.

Clemson CM, McNeil JA, Willard HF, Lawrence JB (1996): XIST RNA paints the inactive X chromosome at interphase: Evidence for a novel RNA involved in nuclear/chromosome structure. J Cell Biol 132:259–75.

Coleman ML, Olson MF (2002): Rho GTPase signalling pathways in the morphological changes associated with apoptosis. Cell Death Diff 9:493–504.

Cook PR (1999): The organization of replication and transcription. Science 284: 1790–5.

Couch FJ, Weber BL (1996): Mutations and polymorphisms in the familial early-onset breast cancer (BRCA1) gene. Breast Cancer Information Core. Hum Mut 8:8–18.

Counis MF, Torriglia A (2000): DNases and apoptosis. Biochem Cell Biol 78: 405–14.

Craig E, Zhang ZK, Davies KP, Kalpana GV (2002): A masked NES in INI1/hSNF5 mediates hCRM1-dependent nuclear export: Implications for tumorigenesis. EMBO J 21:31–42.

Critchlow SE, Bowater RP, Jackson SP (1997): Mammalian DNA double-strand break repair protein XRCC4 interacts with DNA ligase IV. Curr Biol 7: 588–98.

Croft JA, Bridger JM, Boyle S, Perry P, Teague P, Bickmore WA (1999): Differences in the localization and morphology of chromosomes in the human nucleus. J Cell Biol 145:1119–31.

D'Urso G, Grallert B, Nurse P (1995): DNA polymerase alpha, a component of the replication initiation complex, is essential for the checkpoint coupling S phase to mitosis in fission yeast. J Cell Sci 108:3109–18.

Daigle N, Beaudouin J, Hartnell L, Imreh G, Hallberg E, Lippincott-Schwartz J, Ellenberg J (2001): Nuclear pore complexes form immobile networks and have a very low turnover in live mammalian cells. J Cell Biol 154:71–84.

Dailey L, Hanly SM, Roeder RG, Heintz N (1986): Distinct transcription factors bind specifically to two regions of the human histone H4 promoter. Proc Natl Acad Sci USA 83:7241–5.

Dailey L, Roberts SB, Heintz N (1987): RNA polymerase II transcription factors H4TF-1 and H4TF-2 require metal to bind specific DNA sequences. Mol Cell Biol 7:4582–4.

Dailey L, Roberts SB, Heintz N (1988): Purification of the human histone H4 gene-specific transcription factors H4TF-1 and H4TF-2. Genes Dev 2:1700–12.

Daniel MT, Koken M, Romagne O, Barbey S, Bazarbachi A, Stadler M, Guillemin MC, Degos L, Chomienne C, de The H (1993): PML protein expression in hematopoietic and acute promyelocytic leukemia cells. Blood 82:1858–67.

Dantuma NP, Masucci MG (2002): Stabilization signals: A novel regulatory mechanism in the ubiquitin/proteasome system. FEBS Lett 529:22–6.

Datta A, Nag A, Raychaudhuri P (2002): Differential regulation of E2F1, DP1, and the E2F1/DP1 complex by ARF. Mol Cell Biol 22:8398–408.

de la Barre AE, Gerson V, Gout S, Creaven M, Allis CD, Dimitrov S (2000): Core histone N-termini play an essential role in mitotic chromosome condensation. EMBO J 19:379–91.

DeFranco DB (2002): Navigating steroid hormone receptors through the nuclear compartment. Mol Endocrinol 16:1449–55.

DeFranco DB, Guerrero J (2000): Nuclear matrix targeting of steroid receptors: Specific signal sequences and acceptor proteins. Crit Rev Eukaryot Gene Expr 10:39–44.

Desany BA, Alcasabas AA, Bachant JB, Elledge SJ (1998): Recovery from DNA replicational stress is the essential function of the S-phase checkpoint pathway. Genes Dev 12:2956–70.

Di Fruscio M, Weiher H, Vanderhyden BC, Imai T, Shiomi T, Hori TA, Jaenisch R, Gray DA (1997): Proviral inactivation of the Npat gene of Mpv 20 mice results in early embryonic arrest. Mol Cell Biol 17:4080–6.

Diehl JA, Zindy F, Sherr CJ (1997): Inhibition of cyclin D1 phosphorylation on threonine-286 prevents its rapid degradation via the ubiquitin-proteasome pathway. Genes Dev 11:957–72.

Difilippantonio MJ, Petersen S, Chen HT, Johnson R, Jasin M, Kanaar R, Ried T, Nussenzweig A (2002): Evidence for replicative repair of DNA double-strand breaks leading to oncogenic translocation and gene amplification. J Expr Med 196:469–80.

Dong Z, Zhong Q, Chen PL (1999): The Nijmegen breakage syndrome protein is essential for Mre11 phosphorylation upon DNA damage. J Biol Chem 274: 19513–6.

Dou QP, Levin AH, Zhao S, Pardee AB (1993): Cyclin E and cyclin A as candidates for the restriction point protein. Cancer Res 53:1493–7.

Dresser ME, Ewing DJ, Conrad MN, Dominguez AM, Barstead R, Jiang H, Kodadek T (1997): DMC1 functions in a *Saccharomyces cerevisiae* meiotic pathway that is largely independent of the RAD51 pathway. Genetics 147: 533–44.

Duband-Goulet I, Courvalin JC, Buendia B (1998): LBR, a chromatin and lamin binding protein from the inner nuclear membrane, is proteolyzed at late stages of apoptosis. J Cell Sci 111:1441–51.

Ducy P, Zhang R, Geoffroy V, Ridall AL, Karsenty G (1997): Osf2/Cbfa1: A transcriptional activator of osteoblast differentiation. Cell 89:747–54.

Dunaief JL, Strober BE, Guha S, Khavari PA, Alin K, Luban J, Begemann M, Crabtree GR, Goff SP (1994): The retinoblastoma protein and BRG1 form a complex and cooperate to induce cell cycle arrest. Cell 79:119–30.

Dundr M, Misteli T, Olson MO (2000): The dynamics of postmitotic reassembly of the nucleolus. J Cell Biol 150:433–46.

Duong LT, Rodan GA (2001): Regulation of osteoclast formation and function. Rev Endocr Metab Disord 2:95–104.

Durocher D, Jackson SP (2001): DNA-PK, ATM and ATR as sensors of DNA damage: Variations on a theme? Curr Opin in Cell Biol 13:225–31.

Dyck JA, Maul GG, Miller WH, Chen JD, Kakizuka A, Evans RM (1994): A novel macromolecular structure is a target of the promyelocyte-retinoic acid receptor oncoprotein. Cell 76:333–43.

Earnshaw WC, Martins LM, Kaufmann SH (1999): Mammalian caspases: structure, activation, substrates, and functions during apoptosis. An Rev Biochem 68:383–424.

el-Deiry WS, Tokino T, Velculescu VE, Levy DB, Parsons R, Trent JM, Lin D, Mercer WE, Kinzler KW, Vogelstein B (1993): WAF1, a potential mediator of p53 tumor suppression. Cell 75:817–25.

Elledge SJ (1996): Cell cycle checkpoints: Preventing an identity crisis. Science 274:1664–72.

Elledge SJ, Zhou Z, Allen JB, Navas TA (1993): DNA damage and cell cycle regulation of ribonucleotide reductase. BioEssays 15:333–9.

Ellenberg J, Siggia ED, Moreira JE, Smith CL, Presley JF, Worman HJ, Lippincott-Schwartz J (1997): Nuclear membrane dynamics and reassembly in living cells: Targeting of an inner nuclear membrane protein in interphase and mitosis. J Cell Biol 138:1193–1206.

Ellis T, Gambardella L, Horcher M, Tschanz S, Capol J, Bertram P, Jochum W, Barrandon Y, Busslinger M (2001): The transcriptional repressor CDP (Cutl1) is essential for epithelial cell differentiation of the lung and the hair follicle. Genes Dev 15:2307–19.

Enenkel C, Lehmann A, Kloetzel PM (1999): GFP-labelling of 26S proteasomes in living yeast: insight into proteasomal functions at the nuclear envelope/rough ER. Mol Biol Rep 26:131–5.

Ennas MG, Sorio C, Greim R, Nieddu M, Scarpa A, Orlandini S, Croce CM, Fey GH, Marschalek R (1997): The human ALL-1/MLL/HRX antigen is predominantly localized in the nucleus of resting and proliferating peripheral blood mononuclear cells. Cancer Res 57:2035–41.

Enoch T, Carr AM, Nurse P (1992): Fission yeast genes involved in coupling mitosis to completion of DNA replication. Genes Dev 6:2035–46.

Enoch T, Nurse P (1990): Mutation of fission yeast cell cycle control genes abolishes dependence of mitosis on DNA replication. Cell 60:665–73.

Fabbro M, Henderson BR (2003): Regulation of tumor suppressors by nuclear-cytoplasmic shuttling. Expr Cell Res 282:59–69.

Fabbro M, Rodriguez JA, Baer R, Henderson BR (2002): BARD1 induces BRCA1 intranuclear foci formation by increasing RING-dependent BRCA1 nuclear import and inhibiting BRCA1 nuclear export. J Biol Chem 277:21315–24.

Falcieri E, Gobbi P, Cataldi A, Zamai L, Faenza I, Vitale M (1994): Nuclear pores in the apoptotic cell. Histochem J 26:754–63.

Falck J, Mailand N, Syljuasen RG, Bartek J, Lukas J (2001): The ATM-Chk2-Cdc25A checkpoint pathway guards against radioresistant DNA synthesis. Comment. Nature 410:842–7.

Falzon M, Fewell JW, Kuff EL (1993): EBP-80, a transcription factor closely resembling the human autoantigen Ku, recognizes single- to double-strand transitions in DNA. J Biol Chem 268:10546–52.

Farnham PJ, Slansky JE, Kollmar R (1993): The role of E2F in the mammalian cell cycle. Biochim Biophys Acta 1155:125–31.

Fischle W, Wang Y, Allis CD (2003): Histone and chromatin cross-talk. Curr Opin Cell Biol 15:172–83.

Fortugno P, Wall NR, Giodini A, O'Connor DS, Plescia J, Padgett KM, Tognin S, Marchisio PC, Altieri DC (2002): Survivin exists in immunochemically distinct subcellular pools and is involved in spindle microtubule function. J Cell Sci 115:575–85.

Fortunato EA, Spector DH (1998): p53 and RPA are sequestered in viral replication centers in the nuclei of cells infected with human cytomegalovirus. J Virol 72:2033–9.

Franklin DS, Xiong Y (1996): Induction of p18INK4c and its predominant association with CDK4 and CDK6 during myogenic differentiation. Mol Biol Cell 7:1587–99.

Galea MA, Eleftheriou A, Henderson BR (2001): ARM domain-dependent nuclear import of adenomatous polyposis coli protein is stimulated by the B56 alpha subunit of protein phosphatase 2A. J Biol Chem 276:45833–9.

Gall JG (2000): Cajal bodies: The first 100 years. An Rev Cell Dev Biol 16:273–300.

Gant TM, Wilson KL (1997): Nuclear assembly. An Rev Cell Dev Biol 13:669–95.

Gao Y, Sun Y, Frank KM, Dikkes P, Fujiwara Y, Seidl KJ, Sekiguchi JM, Rathbun GA, Swat W, Wang J, Bronson RT, Malynn BA, Bryans M, Zhu C, Chaudhuri J, Davidson L, Ferrini R, Stamato T, Orkin SH, Greenberg ME, Alt FW (1998): A critical role for DNA end-joining proteins in both lymphogenesis and neurogenesis. Cell 95:891–902.

Gardner RD, Burke DJ (2000): The spindle checkpoint: Two transitions, two pathways. Trends Cell Biol 10:154–8.

Gasser SM (2002): Visualizing chromatin dynamics in interphase nuclei. Science 296:1412–16.

Gerace L, Blobel G (1980): The nuclear envelope lamina is reversibly depolymerized during mitosis. Cell 19:277–87.

Gerlich D, Beaudouin J, Kalbfuss B, Daigle N, Eils R, Ellenberg J (2003): Global chromosome positions are transmitted through mitosis in mammalian cells. Cell 112:751–64.

Getzenberg RH, Pienta KJ, Ward WS, Coffey DS (1991): Nuclear structure and the three-dimensional organization of DNA. J Cell Biochem 47:289–99.

Gmachl M, Gieffers C, Podtelejnikov AV, Mann M, Peters JM (2000): The RING-H2 finger protein APC11 and the E2 enzyme UBC4 are sufficient to ubiquitinate substrates of the anaphase-promoting complex. Proc Nat Acad Sci USA 97:8973–8.

Gottlieb TM, Jackson SP (1993): The DNA-dependent protein kinase: requirement for DNA ends and association with Ku antigen. Cell 72:131–42.

Gotzmann J, Vlcek S, Foisner R (2000): Caspase-mediated cleavage of the chromosome-binding domain of lamina-associated polypeptide 2 alpha. J Cell Sci 113(Pt 21):3769–80.

Grawunder U, Wilm M, Wu X, Kulesza P, Wilson TE, Mann M, Lieber MR (1997): Activity of DNA ligase IV stimulated by complex formation with XRCC4 protein in mammalian cells. Comment. Nature 388:492–5.

Grawunder U, Zimmer D, Fugmann S, Schwarz K, Lieber MR (1998): DNA ligase IV is essential for V(D)J recombination and DNA double-strand break repair in human precursor lymphocytes. Mol Cell 2:477–84.

Green L, Van Antwerpen R, Stein J, Stein G, Tripputi P, Emanuel B, Selden J, Croce C (1984): A major human histone gene cluster on the long arm of chromosome 1. Science 226:838–40.

Guo B, Odgren PR, van Wijnen AJ, Last TJ, Nickerson J, Penman S, Lian JB, Stein JL, Stein GS (1995): The nuclear matrix protein NMP-1 is the transcription factor YY1. Proc Natl Acad Sci USA 92:10526–30.

Guo B, Stein JL, van Wijnen AJ, Stein GS (1997): ATF1 and CREB transactivate a cell cycle regulated histone H4 gene at a distal nuclear matrix associated promoter element. Biochem 36:14447–55.

Gutierrez S, Javed A, Tennant D, van Rees M, Montecino M, Stein GS, Stein JL, Lian JB (2002): CCAAT/enhancer-binding proteins (C/EBP) β and δ activate osteocalcin gene transcription and synergize with Runx2 at the C/EBP element to regulate bone-specific expression. J Biol Chem 277:1316–23.

Haber JE (1999): Sir-Ku-itous routes to make ends meet. Cell 97:829–32.

Haber JE (2000): Partners and pathways; repairing a double-strand break. Trends Genet 16:259–64.

Haberma FA, Cremer M, Walter J, Kreth G, von Hase J, Bauer K, Wienberg J, Cremer C, Cremer T, Solovei I (2001): Arrangements of macro- and microchromosomes in chicken cells. Chromosome Res 9:569–84.

Hadjiolov AA (1985): The nucleolus and ribosome biogenesis. (in press).

Halicka HD, Bedner E, Darzynkiewicz Z (2000): Segregation of RNA and separate packaging of DNA and RNA in apoptotic bodies during apoptosis. Expr Cell Res 260:248–56.

Harada H, Nagai H, Tsuneizumi M, Mikami I, Sugano S, Emi M (2001): Identification of DMC1, a novel gene in the TOC region on 17q25.1 that shows loss of expression in multiple human cancers. J Hum Genet 46:90–5.

Harper JW, Adami GR, Wei N, Keyomarsi K, Elledge SJ (1993): The p21 Cdk-interacting protein Cip1 is a potent inhibitor of G1 cyclin-dependent kinases. Cell 75:805–16.

Harper JW, Adams PD (2001): Cyclin-dependent kinases. Chem Rev 101: 2511–26.

Harper JW, Burton JL, Solomon MJ (2002): The anaphase-promoting complex: It's not just for mitosis any more. Genes Dev 16:2179–206.

Harrington KS, Javed A, Drissi H, McNeil S, Lian JB, Stein JL, van Wijnen AJ, Wang Y-L, Stein GS (2002): Transcription factors RUNX1/AML1 and RUNX2/Cbfa1 dynamically associate with stationary subnuclear domains. J Cell Sci 115:4167–76.

Hartley KO, Gell D, Smith GC, Zhang H, Divecha N, Connelly MA, Admon A, Lees-Miller SP, Anderson CW, Jackson SP (1995): DNA-dependent protein kinase catalytic subunit: a relative of phosphatidylinositol 3-kinase and the ataxia telangiectasia gene product. Cell 82:849–56.

Hawkins N, Garriga G (1998): Asymmetric cell division: from A to Z. Genes Dev 12:3625–38.

He Z, Henricksen LA, Wold MS, Ingles CJ (1995): RPA involvement in the damage-recognition and incision steps of nucleotide excision repair. Nature 374:566–9.

Henderson BR, Eleftheriou A (2000): A comparison of the activity, sequence specificity, and CRM1-dependence of different nuclear export signals. Expr Cell Res 256:213–24.

Hendrickson EA (1997): Cell-cycle regulation of mammalian DNA double-strand-break repair. Am J Hum Genet 61:795–800.

Hendrickson EA, Qin XQ, Bump EA, Schatz DG, Oettinger M, Weaver DT (1991): A link between double-strand break-related repair and V(D)J recombination: the scid mutation. Proc Natl Acad Sci USA 88:4061–5.

Hendrickson EA, Schatz DG, Weaver DT (1988): The scid gene encodes a trans-acting factor that mediates the rejoining event of Ig gene rearrangement. Genes Dev 2:817–29.

Hengst L, Reed SI (1996): Translational control of p27Kip1 accumulation during the cell cycle. Science 271:1861–4.

Hernandez-Verdun D, Gautier T (1994): The chromosome periphery during mitosis. BioEssays 16:179–85.

Hernandez-Verdun D, Roussel P, Gebrane-Younes J (2002): Emerging concepts of nucleolar assembly. J Cell Sci 115:2265–70.

Hershko A (1997): Roles of ubiquitin-mediated proteolysis in cell cycle control. Curr Opin Cell Biol 9:788–99.

Hershko A, Ciechanover A (1998): The ubiquitin system. An Rev Biochem 67: 425–79.

Hershko A, Ganoth D, Sudakin V, Dahan A, Cohen LH, Luca FC, Ruderman JV, Eytan E (1994): Components of a system that ligates cyclin to ubiquitin and their regulation by the protein kinase cdc2. J Biol Chem 269:4940–6.

Hieronymus H, Silver PA (2003): Genome-wide analysis of RNA-protein interactions illustrates specificity of the mRNA export machinery. Nat Genet 33: 155–61.

Hoeijmakers JH (2001): Genome maintenance mechanisms for preventing cancer. Nature 411:366–74.

Holmes AM, Haber JE (1999): Double-strand break repair in yeast requires both leading and lagging strand DNA polymerases. Cell 96:415–24.

Holthuis J, Owen TA, van Wijnen AJ, Wright KL, Ramsey-Ewing A, Kennedy MB, Carter R, Cosenza SC, Soprano KJ, Lian JB, et al. (1990): Tumor cells exhibit deregulation of the cell cycle histone gene promoter factor HiNF-D. Science 247:1454–7.

Horn PJ, Peterson CL (2002): Molecular biology. Chromatin higher order folding—Wrapping up transcription. Science 297:1824–7.

Hovhannisyan H, Cho B, Mitra P, Montecino M, Stein GS, van Wijnen AJ, Stein JL (2003): Maintenance of open chromatin and selective genomic occupancy at the cell-cycle-regulated histone H4 promoter during differentiation of HL-60 promyelocytic leukemia cells. Mol Cell Biol 23:1460–9.

Hozak P, Jackson DA, Cook PR (1994): Replication factories and nuclear bodies: the ultrastructural characterization of replication sites during the cell cycle. J Cell Sci 107 (Pt 8):2191–202.

Htun H, Barsony J, Renyi I, Gould DL, Hager GL (1996): Visualization of glucocorticoid receptor translocation and intranuclear organization in living cells with a green fluorescent protein chimera. Proc Natl Acad Sci USA 93: 4845–50.

Huang S, Deerinck TJ, Ellisman MH, Spector DL (1997): The dynamic organization of the perinucleolar compartment in the cell nucleus. J Cell Biol 137: 965–74.

Hwang LH, Lau LF, Smith DL, Mistrot CA, Hardwick KG, Hwang ES, Amon A, Murray AW (1998): Budding yeast Cdc20: A target of the spindle checkpoint. Comment. (erratum appears in Science 1998 May 29;280(5368):1331). Science 279:1041–4.

Imai T, Sugawara T, Nishiyama A, Shimada R, Ohki R, Seki N, Sagara M, Ito H, Yamauchi M, Hori T (1997): The structure and organization of the human NPAT gene. Genomics 42:388–92.

Inoue T, Stuart J, Leno R, Maki CG (2002): Nuclear import and export signals in control of the p53-related protein p73. J Biol Chem 277:15053–60.

Iwabuchi K, Bartel PL, Li B, Marraccino R, Fields S (1994): Two cellular proteins that bind to wild-type but not mutant p53. Proc Natl Acad Sci USA 91: 6098–102.

Jackson SP (2002): Sensing and repairing DNA double-strand breaks. Carcinogenesis 23:687–96.

Jaskelioff M, Peterson CL (2003): Chromatin and transcription: Histones continue to make their marks. Nat Cell Biol 5:395–9.

Javed A, Guo B, Hiebert S, Choi J-Y, Green J, Zhao S-C, Osborne MA, Stifani S, Stein JL, Lian JB, van Wijnen AJ, Stein GS (2000): Groucho/TLE/R-Esp proteins associate with the nuclear matrix and repress RUNX (CBFα/AML/PEBP2α) dependent activation of tissue-specific gene transcription. J Cell Sci 113:2221–31.

Javed A, Gutierrez S, Montecino M, van Wijnen AJ, Stein JL, Stein GS, Lian JB (1999): Multiple Cbfa/AML sites in the rat osteocalcin promoter are required for basal and vitamin D responsive transcription and contribute to chromatin organization. Mol Cell Biol 19:7491–500.

Jin P, Hardy S, Morgan DO (1998): Nuclear localization of cyclin B1 controls mitotic entry after DNA damage. J Cell Biol 141:875–85.

Jin Y, Xu XL, Yang MC, Wei F, Ayi TC, Bowcock AM, Baer R (1997): Cell cycle-dependent colocalization of BARD1 and BRCA1 proteins in discrete nuclear domains. Proc Natl Acad Sci USA 94:12075–80.

Kaftory A, Fry M (1978): Highly efficient copying of single-stranded DNA by eukaryotic cell chromatin. Nucl Acids Res 5:2679–93.

Karran P (2000): DNA double strand break repair in mammalian cells. Curr Opin Genet Dev 10:144–50.

Kau TR, Silver PA (2003): Nuclear transport as a target for cell growth. Drug Discov Today 8:78–85.

Kelly TJ, Brown GW (2000): Regulation of chromosome replication. An Rev Biochem 69:829–80.

Kelsey G, Reik W (1998): Analysis and identification of imprinted genes. Methods 14:211–34.

Kerr JF, Wyllie AH, Currie AR (1972): Apoptosis: A basic biological phenomenon with wide-ranging implications in tissue kinetics. Brit J Cancer 26: 239–57.

Kerr P, Ashworth A (2001): New complexities for BRCA1 and BRCA2. Curr Biol 11:668–76.

Khanna KK, Lavin MF, Jackson SP, Mulhern TD (2001): ATM, a central controller of cellular responses to DNA damage. Cell Death Diff 8:1052–65.

Kimura H, Tao Y, Roeder RG, Cook PR (1999): Quantitation of RNA polymerase II and its transcription factors in an HeLa cell: Little soluble holoenzyme but significant amounts of polymerases attached to the nuclear substructure. Mol Cell Biol 19:5383–92.

Knoblich JA (2001): Asymmetric cell division during animal development. Nat Rev Mol Cell Biol 2:11–20.

Kobayashi J, Tauchi H, Sakamoto S, Nakamura A, Morishima K, Matsuura S, Kobayashi T, Tamai K, Tanimoto K, Komatsu K (2002): NBS1 localizes to gamma-H2AX foci through interaction with the FHA/BRCT domain. Curr Biol 12:1846–51.

Komori T, Yagi H, Nomura S, Yamaguchi A, Sasaki K, Deguchi K, Shimizu Y, Bronson RT, Gao Y-H, Inada M, Sato M, Okamoto R, Kitamura Y, Yoshiki S, Kishimoto T (1997): Targeted disruption of *Cbfa1* results in a complete lack of bone formation owing to maturational arrest of osteoblasts. Cell 89:755–64.

Koonin EV, Altschul SF, Bork P (1996): BRCA1 protein products . . . Functional motifs. Nat Genet 13:266–8.

Krude T (1995): Chromatin assembly factor 1 (CAF-1) colocalizes with replication foci in HeLa cell nuclei. Expr Cell Res 220:304–11.

Kuzminov A (2001): DNA replication meets genetic exchange: chromosomal damage and its repair by homologous recombination. Proc Natl Acad Sci USA 98:8461–8.

Lahav-Baratz S, Sudakin V, Ruderman JV, Hershko A (1995): Reversible phosphorylation controls the activity of cyclosome-associated cyclin-ubiquitin ligase. Proc Natl Acad Sci USA 92:9303–7.

Lambert S, Lopez BS (2001): Role of RAD51 in sister-chromatid exchanges in mammalian cells. Oncogene 20:6627–31.

Lamond AI, Earnshaw WC (1998): Structure and function in the nucleus. Science 280:547–53.

Langdahl BL, Knudsen JY, Jensen HK, Gregersen N, Eriksen EF (1997): A sequence variation: 713–8delC in the transforming growth factor-beta 1 gene has higher prevalence in osteoporotic women than in normal women and is associated with very low bone mass in osteoporotic women and increased bone turnover in both osteoporotic and normal women. Bone 20:289–94.

Last TJ, Birnbaum M, van Wijnen AJ, Stein GS, Stein JL (1998): Repressor elements in the coding region of the human histone H4 gene interact with the transcription factor CDP/*cut*. Gene 221:267–77.

Last TJ, van Wijnen AJ, Birnbaum MJ, Stein GS, Stein JL (1999a): Multiple interactions of the transcription factor YY1 with human histone H4 gene regulatory elements. J Cell Biochem 72:507–16.

Last TJ, van Wijnen AJ, de Ridder MC, Stein GS, Stein JL (1999b): The homeodomain transcription factor CDP/*cut* interacts with the cell cycle regulatory element of histone H4 genes packaged into nucleosomes. Mol Biol Rep 26:185–94.

Lavin MF, Shiloh Y (1997): The genetic defect in ataxia-telangiectasia. An Rev Immunol 15:177–202.

Lee C, Chang JH, Lee HS, Cho Y (2002): Structural basis for the recognition of the E2F transactivation domain by the retinoblastoma tumor suppressor. Genes Dev 16:3199–212.

Lee S, Chen DY, Humphrey JS, Gnarra JR, Linehan WM, Klausner RD (1996): Nuclear/cytoplasmic localization of the von Hippel-Lindau tumor suppressor gene product is determined by cell density. Proc Natl Acad Sci USA 93: 1770–5.

Lee S, Neumann M, Stearman R, Stauber R, Pause A, Pavlakis GN, Klausner RD (1999): Transcription-dependent nuclear-cytoplasmic trafficking is required for the function of the von Hippel-Lindau tumor suppressor protein. Mol Cell Biol 19:1486–97.

Lee SE, Mitchell RA, Cheng A, Hendrickson EA (1997): Evidence for DNA-PK-dependent and -independent DNA double-strand break repair pathways in mammalian cells as a function of the cell cycle. Mol Cell Biol 17:1425–33.

Leonhardt H, Page AW, Weier HU, Bestor TH (1992): A targeting sequence directs DNA methyltransferase to sites of DNA replication in mammalian nuclei. Cell 71:865–73.

Leonhardt H, Rahn HP, Cardoso MC (1998): Intranuclear targeting of DNA replication factors. J Cell Biochem Suppl 30–31:243–9.

Leonhardt H, Rahn HP, Weinzierl P, Sporbert A, Cremer T, Zink D, Cardoso MC (2000): Dynamics of DNA replication factories in living cells. J Cell Biol 149:271–80.

Levanon D, Goldstein RE, Bernstein Y, Tang H, Goldenberg D, Stifani S, Paroush Z, Groner Y (1998): Transcriptional repression by AML1 and LEF-1 is mediated by the TLE/Groucho corepressors. Proc Natl Acad Sci USA 95: 11590–5.

Li Z, Otevrel T, Gao Y, Cheng HL, Seed B, Stamato TD, Taccioli GE, Alt FW (1995): The XRCC4 gene encodes a novel protein involved in DNA double-strand break repair and V(D)J recombination. Cell 83:1079–89.

Liang F, Jasin M (1996): Ku80-deficient cells exhibit excess degradation of extrachromosomal DNA. J Biol Chem 271:14405–11.

Liang SH, Clarke MF (1999): A bipartite nuclear localization signal is required for p53 nuclear import regulated by a carboxyl-terminal domain. J Biol Chem 274:32699–703.

Liang XH, Yu J, Brown K, Levine B (2001): Beclin 1 contains a leucine-rich nuclear export signal that is required for its autophagy and tumor suppressor function. Cancer Res 61:3443–9.

Lichtler AC, Sierra F, Clark S, Wells JR, Stein JL, Stein GS (1982): Multiple H4 histone mRNAs of HeLa cells are encoded in different genes. Nature 298: 195–8.

Lin FL, Sperle K, Sternberg N (1984): Model for homologous recombination during transfer of DNA into mouse L cells: role for DNA ends in the recombination process. Mol Cell Biol 4:1020–34.

Lindstrom MS, Klangby U, Inoue R, Pisa P, Wiman KG, Asker CE (2000): Immunolocalization of human p14(ARF) to the granular component of the interphase nucleolus. Expr Cell Res 256:400–10.

Lippincott-Schwartz J (2002): Cell biology: Ripping up the nuclear envelope. Nature 416:31–2.

Liu Q, Guntuku S, Cui XS, Matsuoka S, Cortez D, Tamai K, Luo G, Carattini-Rivera S, DeMayo F, Bradley A, Donehower LA, Elledge SJ (2000): Chk1 is an essential kinase that is regulated by Atr and required for the G(2)/M DNA damage checkpoint. Genes Dev 14:1448–59.

Liu Y, Oakeley EJ, Sun L, Jost JP (1998): Multiple domains are involved in the targeting of the mouse DNA methyltransferase to the DNA replication foci. Nucl Acids Res 26:1038–45.

Lohr D, Tatchell K, van Holde KE (1977): On the occurrence of nucleosome phasing in chromatin. Cell 12:829–36.

Lohrum MA, Ashcroft M, Kubbutat MH, Vousden KH (2000a): Contribution of two independent MDM2-binding domains in p14(ARF) to p53 stabilization. Curr Biol 10:539–42.

Lohrum MA, Ashcroft M, Kubbutat MH, Vousden KH (2000b): Identification of a cryptic nucleolar-localization signal in MDM2. Nat Cell Biol 2:179–81.

Longhese MP, Plevani P, Lucchini G (1994): Replication factor A is required in vivo for DNA replication, repair, and recombination. Mol Cell Biol 14: 7884–90.

Losada A, Hirano T (2001): Shaping the metaphase chromosome: Coordination of cohesion and condensation. BioEssays 23:924–35.

Lu X, Lane DP (1993): Differential induction of transcriptionally active p53 following UV or ionizing radiation: defects in chromosome instability syndromes? Cell 75:765–78.

Luo G, Yao MS, Bender CF, Mills M, Bladl AR, Bradley A, Petrini JH (1999): Disruption of mRad50 causes embryonic stem cell lethality, abnormal embryonic development, and sensitivity to ionizing radiation. Proc Natl Acad Sci USA 96:7376–81.

Luong MX, van der Meijden CM, Xing D, Hesselton R, Monuki ES, Jones SN, Lian JB, Stein JL, Stein GS, Neufeld EJ, van Wijnen AJ (2002): Genetic ablation of the CDP/Cux protein C terminus results in hair cycle defects and reduced male fertility. Mol Cell Biol 22:1424–37.

Ma H, Samarabandu J, Devdhar RS, Acharya R, Cheng PC, Meng C, Berezney R (1998): Spatial and temporal dynamics of DNA replication sites in mammalian cells. J Cell Biol 143:1415–25.

Ma H, Siegel AJ, Berezney R (1999): Association of chromosome territories with the nuclear matrix. Disruption of human chromosome territories correlates with the release of a subset of nuclear matrix proteins. J Cell Biol 146:531–42.

Ma T, Van Tine BA, Wei Y, Garrett MD, Nelson D, Adams PD, Wang J, Qin J, Chow LT, Harper JW (2000): Cell cycle-regulated phosphorylation of p220(NPAT) by cyclin E/Cdk2 in Cajal bodies promotes histone gene transcription. Genes Dev 14:2298–313.

Mahadevan LC, Willis AC, Barratt MJ (1991): Rapid histone H3 phosphorylation in response to growth factors, phorbol esters, okadaic acid, and protein synthesis inhibitors. Cell 65:775–83.

Malumbres M, Barbacid M (2001): To cycle or not to cycle: a critical decision in cancer. Nat Rev Cancer 1:222–31.

Mancini MA, Shan B, Nickerson JA, Penman S, Lee WH (1994): The retinoblastoma gene product is a cell cycle-dependent, nuclear matrix-associated protein. Proc Natl Acad Sci USA 91:418–22.

Manders EM, Stap J, Brakenhoff GJ, van Driel R, Aten JA (1992): Dynamics of three-dimensional replication patterns during the S-phase, analysed by double labelling of DNA and confocal microscopy. J Cell Sci 103 (Pt 3):857–62.

Mantel C, Hendrie P, Broxmeyer HE (2001): Steel factor regulates cell cycle asymmetry. Stem Cells 19:483–91.

Mao S, Frank RC, Zhang J, Miyazaki Y, Nimer SD (1999): Functional and physical interactions between AML1 proteins and an ETS protein, MEF: Implications for the pathogenesis of t(8;21)- positive leukemias. Mol Cell Biol 19: 3635–44.

Martelli AM, Zweyer M, Ochs RL, Tazzari PL, Tabellini G, Narducci P, Bortul R (2001): Nuclear apoptotic changes: An overview. J Cellular Biochem 82: 634–46.

Maser RS, Monsen KJ, Nelms BE, Petrini JH (1997): hMre11 and hRad50 nuclear foci are induced during the normal cellular response to DNA double-strand breaks. Mol Cell Biol 17:6087–96.

Masson JY, West SC (2001): The Rad51 and Dmc1 recombinases: A non-identical twin relationship. Trends Biochem Sci 26:131–6.

Matsuoka S, Huang M, Elledge SJ (1998): Linkage of ATM to cell cycle regulation by the Chk2 protein kinase. Science 282:1893–7.

Mattaj IW, Englmeier L (1998): Nucleocytoplasmic transport: The soluble phase. An Rev Biochem 67:265–306.

Maul GG (1998): Nuclear domain 10, the site of DNA virus transcription and replication. BioEssays 20:660–7.

Maya R, Balass M, Kim ST, Shkedy D, Leal JF, Shifman O, Moas M, Buschmann T, Ronai Z, Shiloh Y, Kastan MB, Katzir E, Oren M (2001): ATM-dependent phosphorylation of Mdm2 on serine 395: Role in p53 activation by DNA damage. Genes Dev 15:1067–77.

McElhinny SA, Snowden CM, McCarville J, Ramsden DA (2000): Ku recruits the XRCC4-ligase IV complex to DNA ends. Mol Cell Biol 20:2996–3003.

McGrath J, Solter D (1984): Completion of mouse embryogenesis requires both the maternal and paternal genomes. Cell 37:179–83.

McNeil S, Guo B, Stein JL, Lian JB, Bushmeyer S, Seto E, Atchison ML, Penman S, van Wijnen AJ, Stein GS (1998): Targeting of the YY1 transcription factor to the nucleolus and the nuclear matrix in situ: the C-terminus is a principal determinant for nuclear trafficking. J Cell Biochem 68:500–10.

McNeil S, Javed A, Harrington KS, Lian JB, Stein JL, van Wijnen AJ, Stein GS (2000): Leukemia-associated AML1/ETO (8;21) chromosomal translocation protein increases the cellular representation of PML bodies. J Cell Biochem 79:103–12.

McNeil S, Zeng C, Harrington KS, Hiebert S, Lian JB, Stein JL, van Wijnen AJ, Stein GS (1999): The t(8;21) chromosomal translocation in acute myelogenous leukemia modifies intranuclear targeting of the AML1/CBFalpha2 transcription factor. Proc Natl Acad Sci USA 96:14882–7.

McPherson CE, Horowitz R, Woodcock CL, Jiang C, Zaret KS (1996): Nucleosome positioning properties of the albumin transcriptional enhancer. Nucl Acids Res 24:397–404.

McPherson CE, Shim EY, Friedman DS, Zaret KS (1993): An active tissue-specific enhancer and bound transcription factors existing in a precisely positioned nucleosomal array. Cell 75:387–98.

Merker K, Grune T (2000): Proteolysis of oxidised proteins and cellular senescence. Expr Gerontol 35:779–86.

Merriman HL, van Wijnen AJ, Hiebert S, Bidwell JP, Fey E, Lian J, Stein J, Stein GS (1995): The tissue-specific nuclear matrix protein, NMP-2, is a member of the AML/CBF/PEBP2/runt domain transcription factor family: Interactions with the osteocalcin gene promoter. Biochem 34:13125–32.

Mimori T, Hardin JA (1986): Mechanism of interaction between Ku protein and DNA. J Biol Chem 261:10375–9.

Misteli T (2000): Cell biology of transcription and pre-mRNA splicing: Nuclear architecture meets nuclear function. J Cell Sci 113:1841–9.

Misteli T, Spector DL (1999): RNA polymerase II targets pre-mRNA splicing factors to transcription sites in vivo. Mol Cell 3:697–705.

Mizzen CA, Yang XJ, Kokubo T, Brownell JE, Bannister AJ, Owen-Hughes T, Workman J, Wang L, Berger SL, Kouzarides T, Nakatani Y, Allis CD (1996): The TAF(II)250 subunit of TFIID has histone acetyltransferase activity. Cell 87:1261–70.

Moir RD, Montag-Lowy M, Goldman RD (1994): Dynamic properties of nuclear lamins: Lamin B is associated with sites of DNA replication. J Cell Biol 125:1201–12.

Montecucco A, Savini E, Weighardt F, Rossi R, Ciarrocchi G, Villa A, Biamonti G (1995): The N-terminal domain of human DNA ligase I contains the nuclear localization signal and directs the enzyme to sites of DNA replication. EMBO J 14:5379–86.

Morgan DO (1997): Cyclin-dependent kinases: engines, clocks, and microprocessors. An Rev Cell Dev Biol 13:261–91.

Morgan DO (1999): Regulation of the APC and the exit from mitosis. Nature Cell Biol 1:47–53.

Moyne G, Garrido J (1976): Ultrastructural evidence of mitotic perichromosomal ribonucleoproteins in hamster cells. Expr Cell Res 98:237–47.

Muchardt C, Reyes J-C, Bourachot B, Legouy E, Yaniv M (1996): The hbrm and BRG-1 proteins, components of the human SNF/SWI complex, are phosphorylated and excluded from the condensed chromosomes during mitosis. EMBO J 15:3394–402.

Murray AW, Hunt T (1993): The Cell Cycle, An Introduction. New York: Oxford University Press.

Musacchio A, Hardwick KG (2002): The spindle checkpoint: Structural insights into dynamic signalling. Nat Rev Mol Cell Biol 3:731–41.

Nakamura T, Mori T, Tada S, Krajewski W, Rozovskaia T, Wassell R, Dubois G, Mazo A, Croce CM, Canaani E (2002): ALL-1 is a histone methyltransferase that assembles a supercomplex of proteins involved in transcriptional regulation. Mol Cell 10:1119–28.

Nakayasu H, Berezney R (1989): Mapping replicational sites in the eucaryotic cell nucleus. J Cell Biol 108:1–11.

Narlikar GJ, Fan HY, Kingston RE (2002): Cooperation between complexes that regulate chromatin structure and transcription. Cell 108:475–87.

Navas TA, Zhou Z, Elledge SJ (1995): DNA polymerase epsilon links the DNA replication machinery to the S phase checkpoint. Cell 80:29–39.

Neely KE, Workman JL (2002): The complexity of chromatin remodeling and its links to cancer. Biochim Biophys Acta 1603:19–29.

Nelms BE, Maser RS, MacKay JF, Lagally MG, Petrini JH (1998): In situ visualization of DNA double-strand break repair in human fibroblasts. Science 280:590–2.

Neufeld KL, Nix DA, Bogerd H, Kang Y, Beckerle MC, Cullen BR, White RL (2000a): Adenomatous polyposis coli protein contains two nuclear export signals and shuttles between the nucleus and cytoplasm. Proc Natl Acad Sci USA 97:12085–90.

Neufeld KL, Zhang F, Cullen BR, White RL (2000b): APC-mediated downregulation of beta-catenin activity involves nuclear sequestration and nuclear export. EMBO Rep 1:519–23.

Nevins JR (1992): E2F: A link between the Rb tumor suppressor protein and viral oncoproteins. Science 258:424–9.

Nevins JR (2001): The Rb/E2F pathway and cancer. Hum Mol Genet 10:699–703.

Newport J, Yan H (1996): Organization of DNA into foci during replication. Curr Opin Cell Biol 8:365–8.

Newport JW, Wilson KL, Dunphy WG (1990): A lamin-independent pathway for nuclear envelope assembly. J Cell Biol 111:2247–59.

Nghiem P, Park PK, Kim Ys YS, Desai BN, Schreiber SL (2002): ATR is not required for p53 activation but synergizes with p53 in the replication checkpoint. J Biol Chem 277:4428–34.

Nghiem P, Park PK, Kim Y, Vaziri C, Schreiber SL (2001): ATR inhibition selectively sensitizes G1 checkpoint-deficient cells to lethal premature chromatin condensation. Proc Natl Acad Sci USA 98:9092–7.

Nielsen SJ, Schneider R, Bauer UM, Bannister AJ, Morrison A, OíCarroll D, Firestein R, Cleary M, Jenuwein T, Herrera RE, Kouzarides T (2001): Rb targets histone H3 methylation and HP1 to promoters. Nature 412:561–5.

Nigg EA (1998): Polo-like kinases: positive regulators of cell division from start to finish. Curr Opin Cell Biol 10:776–83.

Noguchi H, Reddy PV, Pardee AB (1983): Rapid incorporation of label from ribonucleoside disphosphates into DNA by a cell-free high molecular weight fraction from animal cell nuclei. Cell 32:443–51.

Norbury CJ, Hickson ID (2001): Cellular responses to DNA damage. An Rev Pharmacol Toxicol 41:367–401.

Nuchprayoon I, Meyers S, Scott LM, Suzow J, Hiebert SW, Friedman AD (1994): PEBP2/CBF, the murine homolog of the human myeloid AML1 and PEBP2β/CBFβ proto-oncoproteins, regulates the murine myeloperoxidase and neutrophil elastase genes in immature myeloid cells. Mol Cell Biol 14: 5558–68.

OíKeefe RT, Henderson SC, Spector DL (1992): Dynamic organization of DNA replication in mammalian cell nuclei: spatially and temporally defined replication of chromosome-specific alpha-satellite DNA sequences. J Cell Biol 116:1095–110.

Oakes M, Aris JP, Brockenbrough JS, Wai H, Vu L, Nomura M (1998): Mutational analysis of the structure and localization of the nucleolus in the yeast *Saccharomyces cerevisiae*. J Cell Biol 143:23–34.

Ogryzko VV, Schiltz RL, Russanova V, Howard BH, Nakatani Y (1996): The transcriptional coactivators p300 and CBP are histone acetyltransferases. Cell 87:953–9.

Olson MO, Hingorani K, Szebeni A (2002): Conventional and nonconventional roles of the nucleolus. Internatl Rev Cytol 219:199–266.

Osley MA (1991): The regulation of histone synthesis in the cell cycle. An Rev Biochem 60:827–61.

Otto F, Thornell AP, Crompton T, Denzel A, Gilmour KC, Rosewell IR, Stamp GWH, Beddington RSP, Mundlos S, Olsen BR, Selby PB, Owen MJ (1997): *Cbfa1*, a candidate gene for cleidocranial dysplasia syndrome, is essential for osteoblast differentiation and bone development. Cell 89:765–71.

Pagano M, Tam SW, Theodoras AM, Beer-Romero P, Del Sal G, Chau V, Yew PR, Draetta GF, Rolfe M (1995): Role of the ubiquitin-proteasome pathway in regulating abundance of the cyclin-dependent kinase inhibitor p27. Science 269:682–5.

Pang D, Yoo S, Dynan WS, Jung M, Dritschilo A (1997): Ku proteins join DNA fragments as shown by atomic force microscopy. Cancer Res 57:1412–15.

Paques F, Haber JE (1999): Multiple pathways of recombination induced by double-strand breaks in *Saccharomyces cerevisiae*. Microbiol Mol Biol Rev 63:349–404.

Parada L, Misteli T (2002): Chromosome positioning in the interphase nucleus. Trends Cell Biol 12:425–32.

Parada LA, McQueen PG, Munson PJ, Misteli T (2002): Conservation of relative chromosome positioning in normal and cancer cells. Curr Biol 12:1692–7.

Pardee AB (1974): A restriction point for control of normal animal cell proliferation. Proc Natl Acad Sci USA 71:1286–90.

Parsons CA, Baumann P, Van Dyck E, West SC (2000): Precise binding of single-stranded DNA termini by human RAD52 protein. EMBO J 19:4175–81.

Pasero P, Schwob E (2000): Think global, act local–how to regulate S phase from individual replication origins. Curr Opin Genet Dev 10:178–86.

Pauli U, Chrysogelos S, Stein G, Stein J, Nick H (1987): Protein-DNA interactions in vivo upstream of a cell cycle- regulated human H4 histone gene. Science 236:1308–11.

Paull TT, Rogakou EP, Yamazaki V, Kirchgessner CU, Gellert M, Bonner WM (2000): A critical role for histone H2AX in recruitment of repair factors to nuclear foci after DNA damage. Curr Biol 10:886–95.

Paulovich AG, Toczyski DP, Hartwell LH (1997): When checkpoints fail. Cell 88:315–21.

Pekarsky Y, Hallas C, Croce CM (2001): Molecular basis of mature T-cell leukemia. JAMA 286:2308–14.

Peng CY, Graves PR, Thoma RS, Wu Z, Shaw AS, Piwnica-Worms H (1997): Mitotic and G2 checkpoint control: regulation of 14–3-3 protein binding by phosphorylation of Cdc25C on serine-216. Comment. Science 277:1501–5.

Penman S (1995): Rethinking cell structure. Proc Natl Acad Sci USA 92:5251–7.

Peters AH, Mermoud JE, O'Carroll D, Pagani M, Schweizer D, Brockdorff N, Jenuwein T (2002): Histone H3 lysine 9 methylation is an epigenetic imprint of facultative heterochromatin. Nat Genet 30:77–80.

Peterson CL (2002): Chromatin remodeling enzymes: Taming the machines. Third in review series on chromatin dynamics. EMBO Rep 3:319–22.

Peterson CL, Workman JL (2000): Promoter targeting and chromatin remodeling by the SWI/SNF complex. Curr Opin Genet Dev 10:187–92.

Pierreux CE, Nicolas FJ, Hill CS (2000): Transforming growth factor beta-independent shuttling of Smad4 between the cytoplasm and nucleus. Mol Cell Biol 20:9041–54.

Platani M, Goldberg I, Swedlow JR, Lamond AI (2000): In vivo analysis of Cajal body movement, separation, and joining in live human cells. J Cell Biol 151: 1561–74.

Poltoratsky VP, Shi X, York JD, Lieber MR, Carter TH (1995): Human DNA-activated protein kinase (DNA-PK) is homologous to phosphatidylinositol kinases. J Immunol 155:4529–33.

Prasanth KV, Sacco-Bubulya PA, Prasanth SG, Spector DL (2003): Sequential entry of components of gene expression machinery into daughter nuclei. Mol Biol Cell 14:1043–57.

Qin J, Li L (2003): Molecular anatomy of the DNA damage and replication checkpoints. Radiation Research 159:139–48.

Quesenberry PJ, Stewart FM, Zhong S, Habibian H, McAuliffe C, Reilly J, Carlson J, Dooner M, Nilsson S, Peters S, Stein G, Stein J, Emmons R, Benoit B, Bertoncello I, Becker P (1999): Lymphohematopoietic stem cell engraftment. An NY Acad Sci 872:40–5.

Ramsey-Ewing A, van Wijnen AJ, Stein GS, Stein JL (1994): Delineation of a human histone H4 cell cycle element in vivo: the master switch for H4 gene transcription. Proc Natl Acad Sci USA 91:4475–9.

Reddy PV, Pardee AB (1980): Multienzyme complex for metabolic channeling in mammalian DNA replication. Proc Natl Acad Sci USA 77:3312–16.

Reik W, Walter J (2001): Genomic imprinting: Parental influence on the genome. Nat Rev Genet 2:21–32.

Reits EA, Benham AM, Plougastel B, Neefjes J, Trowsdale J (1997): Dynamics of proteasome distribution in living cells. EMBO J 16:6087–94.

Reyes J-C, Muchardt C, Yaniv M (1997): Components of the human SWI/SNF complex are enriched in active chromatin and are associated with the nuclear matrix. J Cell Biol 137:263–74.

Rieder CL, Cole RW, Khodjakov A, Sluder G (1995): The checkpoint delaying anaphase in response to chromosome monoorientation is mediated by an inhibitory signal produced by unattached kinetochores. J Cell Biol 130:941–8.

Ristic D, Wyman C, Paulusma C, Kanaar R (2001): The architecture of the human Rad54-DNA complex provides evidence for protein translocation along DNA. Proc Natl Acad Sci USA 98:8454–60.

Rizos H, Darmanian AP, Mann GJ, Kefford RF (2000): Two arginine rich domains in the p14ARF tumour suppressor mediate nucleolar localization. Oncogene 19:2978–85.

Rodriguez JA, Henderson BR (2000): Identification of a functional nuclear export sequence in BRCA1. J Biol Chem 275:38589–96.

Rogakou EP, Nieves-Neira W, Boon C, Pommier Y, Bonner WM (2000): Initiation of DNA fragmentation during apoptosis induces phosphorylation of H2AX histone at serine 139. J Biol Chem 275:9390–5.

Rogakou EP, Pilch DR, Orr AH, Ivanova VS, Bonner WM (1998): DNA double-stranded breaks induce histone H2AX phosphorylation on serine 139. J Biol Chem 273:5858–68.

Rosin-Arbesfeld R, Townsley F, Bienz M (2000): The APC tumour suppressor has a nuclear export function. Nature 406:1009–12.

Roussel P, Hernandez-Verdun D (1994): Identification of Ag-NOR proteins, markers of proliferation related to ribosomal gene activity. Exprl Cell Res 214:465–72.

Rowley JD (1999): The role of chromosome translocations in leukemogenesis. Semin Hematol 36:59–72.

Sachs RK, Chen AM, Brenner DJ (1997): Review: Proximity effects in the production of chromosome aberrations by ionizing radiation. Int J Radiat Biol 71:1–19.

Schar P (2001): Spontaneous DNA damage, genome instability, and cancer—When DNA replication escapes control. Comment. Cell 104:329–32.

Schardin M, Cremer T, Hager HD, Lang M (1985): Specific staining of human chromosomes in Chinese hamster x man hybrid cell lines demonstrates interphase chromosome territories. Hum Genet 71:281–7.

Scheer U, Hock R (1999): Structure and function of the nucleolus. Curr Opin Cell Biol 11:385–90.

Schwarzacher HG, Mosgoeller W (2000): Ribosome biogenesis in man: Current views on nucleolar structures and function. Cytogenet Cell Genet 91:243–52.

Scott M, Boisvert FM, Vieyra D, Johnston RN, Bazett-Jones DP, Riabowol K (2001): UV induces nucleolar translocation of ING1 through two distinct nucleolar targeting sequences. Nucl Acids Res 29:2052–8.

Scully R, Chen J, Ochs RL, Keegan K, Hoekstra M, Feunteun J, Livingston DM (1997): Dynamic changes of BRCA1 subnuclear location and phosphorylation state are initiated by DNA damage. Cell 90:425–35.

Scully R, Livingston DM (2000): In search of the tumour-suppressor functions of BRCA1 and BRCA2. Nature 408:429–32.

Seery JP, Watt FM (2000): Asymmetric stem-cell divisions define the architecture of human oesophageal epithelium. Curr Biol 10:1447–50.

Sengupta S, Linke SP, Pedeux R, Yang Q, Farnsworth J, Garfield SH, Valerie K, Shay JW, Ellis NA, Wasylyk B, Harris CC (2003): BLM helicase-dependent transport of p53 to sites of stalled DNA replication forks modulates homologous recombination. EMBO J 22:1210–22.

Shakoori AR, van Wijnen AJ, Cooper C, Aziz F, Birnbaum M, Reddy GP, Grana X, De Luca A, Giordano A, Lian JB, Stein JL, Quesenberry P, Stein GS (1995): Cytokine induction of proliferation and expression of CDC2 and cyclin A in FDC-P1 myeloid hematopoietic progenitor cells: Regulation of ubiquitous and cell cycle-dependent histone gene transcription factors. J Cell Biochem 59:291–302.

Sheaff RJ, Groudine M, Gordon M, Roberts JM, Clurman BE (1997): Cyclin E-CDK2 is a regulator of p27Kip1. Genes Dev 11:1464–78.

Shen Q, Zhong W, Jan YN, Temple S (2002): Asymmetric Numb distribution is critical for asymmetric cell division of mouse cerebral cortical stem cells and neuroblasts. Development 129:4843–53.

Sherley JL (2002): Asymmetric cell kinetics genes: The key to expansion of adult stem cells in culture. Stem Cells 20:561–72.

Sherr CJ, Roberts JM (1999): CDK inhibitors: Positive and negative regulators of G_1-phase progression. Genes Dev 13:1501–12.

Shibata T, Nishinaka T, Mikawa T, Aihara H, Kurumizaka H, Yokoyama S, Ito Y (2001): Homologous genetic recombination as an intrinsic dynamic property of a DNA structure induced by RecA/Rad51-family proteins: A possible advantage of DNA over RNA as genomic material. Proc Natl Acad Sci USA 98:8425–32.

Shiloh Y (2001): ATM and ATR: Networking cellular responses to DNA damage. Curr Opin Genet Dev 11:71–7.

Shim EY, Woodcock C, Zaret KS (1998): Nucleosome positioning by the winged helix transcription factor HNF3. Genes Dev 12:5–10.

Shinohara A, Shinohara M, Ohta T, Matsuda S, Ogawa T (1998): Rad52 forms ring structures and co-operates with RPA in single-strand DNA annealing. Genes Cells 3:145–56.

Shinohara M, Gasior SL, Bishop DK, Shinohara A (2000): Tid1/Rdh54 promotes colocalization of rad51 and dmc1 during meiotic recombination. Proc Natl Acad Sci USA 97:10814–9.

Shinohara M, Shita-Yamaguchi E, Buerstedde JM, Shinagawa H, Ogawa H, Shinohara A (1997): Characterization of the roles of the *Saccharomyces cerevisiae* RAD54 gene and a homologue of RAD54, RDH54/TID1, in mitosis and meiosis. Genetics 147:1545–56.

Shivji KK, Kenny MK, Wood RD (1992): Proliferating cell nuclear antigen is required for DNA excision repair. Cell 69:367–74.

Shopland LS, Lawrence JB (2000): Seeking common ground in nuclear complexity. J Cell Biol 150:F1–F4.

Sibanda BL, Critchlow SE, Begun J, Pei XY, Jackson SP, Blundell TL, Pellegrini L (2001): Crystal structure of an Xrcc4-DNA ligase IV complex. Nat Struct Biol 8:1015–9.

Sif S, Stukenberg PT, Kirschner MW, Kingston RE (1998): Mitotic inactivation of a human SWI/SNF chromatin remodeling complex. Genes Dev 12:2842–51.

Sinclair AM, Lee JA, Goldstein A, Xing D, Liu S, Ju R, Tucker PW, Neufeld EJ, Scheuermann RH (2001): Lymphoid apoptosis and myeloid hyperplasia in CCAAT displacement protein mutant mice. Blood 98:3658–67.

Sironi L, Mapelli M, Knapp S, De Antoni A, Jeang KT, Musacchio A (2002): Crystal structure of the tetrameric Mad1-Mad2 core complex: implications of a 'safety belt' binding mechanism for the spindle checkpoint. EMBO J 21:2496–506.

Smith GC, Jackson SP (1999): The DNA-dependent protein kinase. Genes Dev 13:916–34.

Smith KP, Moen PT, Wydner KL, Coleman JR, Lawrence JB (1999): Processing of endogenous pre-mRNAs in association with SC-35 domains is gene specific. J Cell Biol 144:617–29.

Smits VA, Klompmaker R, Arnaud L, Rijksen G, Nigg EA, Medema RH (2000): Polo-like kinase-1 is a target of the DNA damage checkpoint. Nat Cell Biol 2:672–6.

Solinger JA, Heyer WD (2001): Rad54 protein stimulates the postsynaptic phase of Rad51 protein-mediated DNA strand exchange. Proc Natl Acad Sci USA 98:8447–53.

Spann TP, Moir RD, Goldman AE, Stick R, Goldman RD (1997): Disruption of nuclear lamin organization alters the distribution of replication factors and inhibits DNA synthesis. J Cell Biol 136:1201–12.

Speck NA, Stacy T, Wang Q, North T, Gu TL, Miller J, Binder M, Marin-Padilla M (1999): Core-binding factor: A central player in hematopoiesis and leukemia. Cancer Res 59:1789s–93s.

Spector DL, Triemer RE (1981): Chromosome structure and mitosis in the dinoflagellates: An ultrastructural approach to an evolutionary problem. Biosystems 14:289–98.

Staal A, Enserink JM, Stein JL, Stein GS, van Wijnen AJ (2000): Molecular characterization of Celtix-1, a bromodomain protein interacting with the transcription factor interferon regulatory factor 2. J Cell Physiol 185:269–79.

Stasiak AZ, Larquet E, Stasiak A, Muller S, Engel A, Van Dyck E, West SC, Egelman EH (2000): The human Rad52 protein exists as a heptameric ring. Curr Biol 10:337–40.

Stein GS, Montecino M, van Wijnen AJ, Stein JL, Lian JB (2000a): Nuclear structure-gene expression interrelationships: Implications for aberrant gene expression in cancer. Cancer Res 60:2067–76.

Stein GS, Stein JL, van Wijnen AJ, Lian JB (1996): Transcriptional control of cell cycle progression: the histone gene is a paradigm for the G_1/S phase and proliferation/differentiation transitions. Cell Biol Int 20:41–9.

Stein GS, van Wijnen AJ, Stein JL, Lian JB, Montecino M, Choi J-Y, Zaidi K, Javed A (2000b): Intranuclear trafficking of transcription factors: implications for biological control. J Cell Sci 113:2527–33.

Stenoien D, Sharp ZD, Smith CL, Mancini MA (1998): Functional subnuclear partitioning of transcription factors. J Cell Biochem 70:213–21.

Stenoien DL, Mancini MG, Patel K, Allegretto EA, Smith CL, Mancini MA (2000): Subnuclear trafficking of estrogen receptor-alpha and steroid receptor coactivator-1. Mol Endocrinol 14:518–34.

Stevaux O, Dyson NJ (2002): A revised picture of the E2F transcriptional network and RB function. Curr Opin Cell Biol 14:684–91.

Stewart E, Chapman CR, Al-Khodairy F, Carr AM, Enoch T (1997): $rqh1^+$, a fission yeast gene related to the Bloom's and Werner's syndrome genes, is required for reversible S phase arrest. EMBO J 16:2682–92.

Stewart GS, Maser RS, Stankovic T, Bressan DA, Kaplan MI, Jaspers NG, Raams A, Byrd PJ, Petrini JH, Taylor AM (1999): The DNA double-strand break repair gene hMRE11 is mutated in individuals with an ataxia-telangiectasia-like disorder. Cell 99:577–87.

Strobeck MW, Knudsen KE, Fribourg AF, DeCristofaro MF, Weissman BE, Imbalzano AN, Knudsen ES (2000): BRG-1 is required for RB-mediated cell cycle arrest. Proc Natl Acad Sci USA 97:7748–53.

Strober BE, Dunaief JL, Guha, Goff SP (1996): Functional interactions between the hBRM/hBRG1 transcriptional activators and the pRB family of proteins. Mol Cell Biol 16:1576–83.

Studzinski GP, Shankavaram UT, Moore DC, Reddy PV (1991): Association of c-Myc protein with enzymes of DNA replication in high molecular weight fractions from mammalian cells. J Cell Physiol 147:412–19.

Sudakin V, Chan GK, Yen TJ (2001): Checkpoint inhibition of the APC/C in HeLa cells is mediated by a complex of BUBR1, BUB3, CDC20, and MAD2. Comment. J Cell Biol 154:925–36.

Sudakin V, Ganoth D, Dahan A, Heller H, Hershko J, Luca FC, Ruderman JV, Hershko A (1995): The cyclosome, a large complex containing cyclin-selective ubiquitin ligase activity, targets cyclins for destruction at the end of mitosis. Mol Biol Cell 6:185–97.

Sugawara N, Ira G, Haber JE (2000): DNA length dependence of the single-strand annealing pathway and the role of *Saccharomyces cerevisiae* RAD59 in double-strand break repair. Mol Cell Biol 20:5300–9.

Sun H, Tu X, Prisco M, Wu A, Casiburi I, Baserga R (2003): Insulin-like growth factor I receptor signaling and nuclear translocation of insulin receptor substrates 1 and 2. Mol Endocrinol 17:472–86.

Sun HB, Shen J, Yokota H (2000): Size-dependent positioning of human chromosomes in interphase nuclei. Biophys J 79:184–90.

Sun JM, Wiaderkiewicz R, Ruiz-Carrillo A (1989): Histone H5 in the control of DNA synthesis and cell proliferation. Science 245:68–71.

Surana U, Amon A, Dowzer C, McGrew J, Byers B, Nasmyth K (1993): Destruction of the CDC28/CLB mitotic kinase is not required for the metaphase to anaphase transition in budding yeast. EMBO J 12:1969–78.

Surani MA, Barton SC, Norris ML (1984): Development of reconstituted mouse eggs suggests imprinting of the genome during gametogenesis. Nature 308: 548–50.

Suwa A, Hirakata M, Takeda Y, Jesch SA, Mimori T, Hardin JA (1994): DNA-dependent protein kinase (Ku protein-p350 complex) assembles on double-stranded DNA. Proc Natl Acad Sci USA 91:6904–8.

Swagemakers SM, Essers J, de Wit J, Hoeijmakers JH, Kanaar R (1998): The human RAD54 recombinational DNA repair protein is a double-stranded DNA-dependent ATPase. J Biol Chem 273:28292–7.

Szweda PA, Friguet B, Szweda LI (2002): Proteolysis, free radicals, and aging. Free Rad Biol Med 33:29–36.

Takahashi M (1987): A model for the spatio-temporal organization of DNA replication in mammalian cells. J Theor Biol 129:91–115.

Tanabe H, Muller S, Neusser M, von Hase J, Calcagno E, Cremer M, Solovei I, Cremer C, Cremer T (2002): Evolutionary conservation of chromosome territory arrangements in cell nuclei from higher primates. Proc Natl Acad Sci USA 99:4424–9.

Tanaka K, Tsurumi C (1997): The 26S proteasome: subunits and functions. Mol Biol Rep 24:3–11.

Tang L, Guo B, Javed A, Choi J-Y, Hiebert S, Lian JB, van Wijnen AJ, Stein JL, Stein GS, Zhou GW (1999): Crystal structure of the nuclear matrix targeting signal of the transcription factor AML-1/PEBP2αB/CBFα2. J Biol Chem 274: 33580–6.

Tang L, Guo B, van Wijnen AJ, Lian JB, Stein JL, Stein GS, Zhou GW (1998a): Preliminary crystallographic study of the glutathione S-transferase fused with the nuclear matrix targeting signal of the transcription factor AML-1/CBFα2. J Struct Biol 123:83–5.

Tang Y, Getzenberg RH, Vietmeier BN, Stallcup MR, Eggert M, Renkawitz R, DeFranco DB (1998b): The DNA-binding and τ2 transactivation domains of the rat glucocorticoid receptor constitute a nuclear matrix targeting signal. Mol Endocrinol 12:1420–31.

Tao W, Levine AJ (1999): P19(ARF) stabilizes p53 by blocking nucleo-cytoplasmic shuttling of Mdm2. Proc Natl Acad Sci USA 96:6937–41.

Teo SH, Jackson SP (2000): Lif1p targets the DNA ligase Lig4p to sites of DNA double-strand breaks. Curr Biol 10:165–8.

Thelen MP, Venclovas C, Fidelis K (1999): A sliding clamp model for the Rad1 family of cell cycle checkpoint proteins. Cell 96:769–70.

Thomas DM, Carty SA, Piscopo DM, Lee JS, Wang WF, Forrester WC, Hinds PW (2001): The retinoblastoma protein acts as a transcriptional coactivator required for osteogenic differentiation. Mol Cell 8:303–16.

Thompson LH, Schild D (2002): Recombinational DNA repair and human disease. Mut Res 509:49–78.

Tibbetts RS, Brumbaugh KM, Williams JM, Sarkaria JN, Cliby WA, Shieh SY, Taya Y, Prives C, Abraham RT (1999): A role for ATR in the DNA damage-induced phosphorylation of p53. Genes Dev 13:152–7.

Toschi L, Bravo R (1988): Changes in cyclin/proliferating cell nuclear antigen distribution during DNA repair synthesis. J Cell Biol 107:1623–28.

Trere D (2000): AgNOR staining and quantification. Micron 31:127–31.

Trimarchi JM, Lees JA (2002): Sibling rivalry in the E2F family. Nat Rev Mol Cell Biol 3:11–20.

Trouche D, Le Chalony C, Muchardt C, Yaniv M, Kouzarides T (1997): RB and hbrm cooperate to repress the activation functions of E2F1. Proc Natl Acad Sci USA 94:11268–73.

Tsai RY, McKay RD (2002): A nucleolar mechanism controlling cell proliferation in stem cells and cancer cells. Genes Dev 16:2991–3003.

Tu X, Batta P, Innocent N, Prisco M, Casaburi I, Belletti B, Baserga R (2002): Nuclear translocation of insulin receptor substrate-1 by oncogenes and Igf-I: Effect on ribosomal RNA synthesis. J Biol Chem 277:44357–65.

Tulchin N, Ornstein L, Bleiweiss IJ, Mendlowitz M, Weinstein E, Dikman SH (1998): Nucleolar localization of BRCA1 protein in human breast cancer. Intl J Oncol 13:513–18.

Tuteja N, Tuteja R, Ochem A, Taneja P, Huang NW, Simoncsits A, Susic S, Rahman K, Marusic L, Chen J (1994): Human DNA helicase II: A novel DNA unwinding enzyme identified as the Ku autoantigen. EMBO J 13:4991–5001.

Tuteja R, Tuteja N (2000): Ku autoantigen: a multifunctional DNA-binding protein. Critical Reviews in Biochemistry & Molecular Biology 35:1–33.

van den Bosch M, Lohman PH, Pastink A (2002): DNA double-strand break repair by homologous recombination. Biol Chem 383:873–92.

van den Ent FM, van Wijnen AJ, Last TJ, Bortell R, Stein JL, Lian JB, Stein GS (1993): Concerted control of multiple histone promoter factors during cell density inhibition of proliferation in osteosarcoma cells: Reciprocal regulation of cell cycle-controlled and bone-related genes. Cancer Res 53:2399–409.

van den Ent FM, van Wijnen AJ, Lian JB, Stein JL, Stein GS (1994): Cell cycle controlled histone H1, H3, and H4 genes share unusual arrangements of recognition motifs for HiNF-D supporting a coordinate promoter binding mechanism. J Cell Physiol 159:515–30.

van der Meijden CMJ, Vaughan PS, Staal A, Albig W, Doenecke D, Stein JL, Stein GS, van Wijnen AJ (1998): Selective expression of specific histone H4 genes reflects distinctions in transcription factor interactions with divergent H4 promoter elements. Biochim Biophys Acta 1442:82–100.

van Dierendonck JH, Keyzer R, van de Velde CJ, Cornelisse CJ (1989): Subdivision of S-phase by analysis of nuclear 5-bromodeoxyuridine staining patterns. Cytometry 10:143–50.

van Steensel B, Jenster G, Damm K, Brinkmann AO, van Driel R (1995): Domains of the human androgen receptor and glucocorticoid receptor involved in binding to the nuclear matrix. J Cell Biochem 57:465–78.

van Wijnen AJ, Aziz F, Grana X, De Luca A, Desai RK, Jaarsveld K, Last TJ, Soprano K, Giordano A, Lian JB, et al. (1994): Transcription of histone H4, H3, and H1 cell cycle genes: promoter factor HiNF-D contains CDC2, cyclin A, and an RB-related protein. Proc Natl Acad Sci USA 91:12882–6.

van Wijnen AJ, Cooper C, Odgren P, Aziz F, De Luca A, Shakoori RA, Giordano A, Quesenberry PJ, Lian JB, Stein GS, Stein JL (1997): Cell cycle-dependent modifications in activities of pRb-related tumor suppressors and proliferation-specific CDP/cut homeodomain factors in murine hematopoietic progenitor cells. J Cell Biochem 66:512–23.

van Wijnen AJ, Ramsey-Ewing AL, Bortell R, Owen TA, Lian JB, Stein JL, Stein GS (1991): Transcriptional element H4-site II of cell cycle regulated human H4 histone genes is a multipartite protein/DNA interaction site for factors HiNF-D, HiNF-M, and HiNF-P: Involvement of phosphorylation. J Cell Biochem 46:174–89.

van Wijnen AJ, van den Ent FM, Lian JB, Stein JL, Stein GS (1992): Overlapping and CpG methylation-sensitive protein-DNA interactions at the histone H4 transcriptional cell cycle domain: Distinctions between two human H4 gene promoters. Mol Cell Biol 12:3273–87.

van Wijnen AJ, van Gurp MF, de Ridder MC, Tufarelli C, Last TJ, Birnbaum M, Vaughan PS, Giordano A, Krek W, Neufeld EJ, Stein JL, Stein GS (1996): CDP/*cut* is the DNA-binding subunit of histone gene transcription factor HiNF-D: A mechanism for gene regulation at the G_1/S phase cell cycle transition point independent of transcription factor E2F. Proc Natl Acad Sci USA 93:11516–21.

Varon R, Vissinga C, Platzer M, Cerosaletti KM, Chrzanowska KH, Saar K, Beckmann G, Seemanova E, Cooper PR, Nowak NJ, Stumm M, Weemaes CM, Gatti RA, Wilson RK, Digweed M, Rosenthal A, Sperling K, Concannon P, Reis A (1998): Nibrin, a novel DNA double-strand break repair protein, is mutated in Nijmegen breakage syndrome. Cell 93:467–76.

Vaughan PS, Aziz F, van Wijnen AJ, Wu S, Harada H, Taniguchi T, Soprano KJ, Stein JL, Stein GS (1995): Activation of a cell-cycle-regulated histone gene by the oncogenic transcription factor IRF-2. Nature 377:362–5.

Vaughan PS, van der Meijden CMJ, Aziz F, Harada H, Taniguchi T, van Wijnen AJ, Stein JL, Stein GS (1998): Cell cycle regulation of histone H4 gene transcription requires the oncogenic factor IRF-2. J Biol Chem 273:194–9.

Venkitaraman AR (1999): Breast cancer genes and DNA repair. Comment. Science 286:1100–2.

Verschure PJ, van Der Kraan I, Manders EM, van Driel R (1999): Spatial relationship between transcription sites and chromosome territories. J Cell Biol 147:13–24.

Versteege I, Sevenet N, Lange J, Rousseau-Merck MF, Ambros P, Handgretinger R, Aurias A, Delattre O (1998): Truncating mutations of hSNF5/INI1 in aggressive paediatric cancer. Nature 394:203–6.

Visser AE, Eils R, Jauch A, Little G, Bakker PJ, Cremer T, Aten JA (1998): Spatial distributions of early and late replicating chromatin in interphase chromosome territories. Expr Cell Res 243:398–407.

Walker JR, Corpina RA, Goldberg J (2001): Structure of the Ku heterodimer bound to DNA and its implications for double-strand break repair. Nature 412:607–14.

Walter J, Schermelleh L, Cremer M, Tashiro S, Cremer T (2003): Chromosome order in HeLa cells changes during mitosis and early G_1, but is stably maintained during subsequent interphase stages. J Cell Biol 160:685–97.

Wang H, Zeng ZC, Bui TA, DiBiase SJ, Qin W, Xia F, Powell SN, Iliakis G (2001a): Nonhomologous end-joining of ionizing radiation-induced DNA double-stranded breaks in human tumor cells deficient in BRCA1 or BRCA2. Cancer Res 61:270–7.

Wang H, Zeng ZC, Perrault AR, Cheng X, Qin W, Iliakis G (2001b): Genetic evidence for the involvement of DNA ligase IV in the DNA-PK-dependent pathway of non-homologous end joining in mammalian cells. Nucl Acids Res 29:1653–60.

Wang Q, Stacy T, Binder M, Marin-Padilla M, Sharpe AH, Speck NA (1996): Disruption of the Cbfa2 gene causes necrosis and hemorrhaging in the central nervous system and blocks definitive hematopoiesis. Proc Natl Acad Sci USA 93:3444–9.

Wang Y, Cortez D, Yazdi P, Neff N, Elledge SJ, Qin J (2000): BASC, a super complex of BRCA1-associated proteins involved in the recognition and repair of aberrant DNA structures. Genes Dev 14:927–39.

Waring P, Khan T, Sjaarda A (1997): Apoptosis induced by gliotoxin is preceded by phosphorylation of histone H3 and enhanced sensitivity of chromatin to nuclease digestion. J Biol Chem 272:17929–36.

Wei X, Samarabandu J, Devdhar RS, Siegel AJ, Acharya R, Berezney R (1998): Segregation of transcription and replication sites into higher order domains. Science 281:1502–5.

Wei X, Somanathan S, Samarabandu J, Berezney R (1999): Three-dimensional visualization of transcription sites and their association with splicing factor-rich nuclear speckles. J Cell Biol 146:543–58.

Weintraub SJ, Chow KN, Luo RX, Zhang SH, He S, Dean DC (1995): Mechanism of active transcriptional repression by the retinoblastoma protein. Nature 375:812–15.

Wigmore PM, Evans DJ (2002): Molecular and cellular mechanisms involved in the generation of fiber diversity during myogenesis. Int Rev Cytol 216: 175–232.

Wilson CA, Ramos L, Villasenor MR, Anders KH, Press MF, Clarke K, Karlan B, Chen JJ, Scully R, Livingston D, Zuch RH, Kanter MH, Cohen S, Calzone FJ, Slamon DJ (1999): Localization of human BRCA1 and its loss in high-grade, non-inherited breast carcinomas. Nat Genet 21:236–40.

Wolpert L (1988): Stem cells: A problem in asymmetry. J Cell Sci Suppl 10:1–9.

Won KA, Reed SI (1996): Activation of cyclin E/CDK2 is coupled to site-specific autophosphorylation and ubiquitin-dependent degradation of cyclin E. EMBO J 15:4182–93.

Wong AK, Shanahan F, Chen Y, Lian L, Ha P, Hendricks K, Ghaffari S, Iliev D, Penn B, Woodland AM, Smith R, Salada G, Carillo A, Laity K, Gupte J, Swedlund B, Tavtigian SV, Teng DH, Lees E (2000): BRG1, a component of the SWI-SNF complex, is mutated in multiple human tumor cell lines. Cancer Res 60:6171–7.

Wu A, Sciacca L, Baserga R (2003): Nuclear translocation of insulin receptor substrate-1 by the insulin receptor in mouse embryo fibroblasts. J Cell Physiol 195:453–60.

Xie RL, van Wijnen AJ, van der Meijden CM, Stein JL, Stein GS (2002): Forced expression of the interferon regulatory factor 2 oncoprotein causes polyploidy and cell death in FDC-P1 myeloid hematopoietic progenitor cells. Cancer Res 62:2510–15.

Yamaguchi-Iwai Y, Sonoda E, Sasaki MS, Morrison C, Haraguchi T, Hiraoka Y, Yamashita YM, Yagi T, Takata M, Price C, Kakazu N, Takeda S (1999): Mre11 is essential for the maintenance of chromosomal DNA in vertebrate cells. EMBO J 18:6619–29.

Yang XJ, Ogryzko VV, Nishikawa J, Howard BH, Nakatani Y (1996): A p300/CBP-associated factor that competes with the adenoviral oncoprotein E1A. Nature 382:319–24.

Yano T, Nakamura T, Blechman J, Sorio C, Dang CV, Geiger B, Canaani E (1997): Nuclear punctate distribution of ALL-1 is conferred by distinct elements at the N terminus of the protein. Proc Natl Acad Sci USA 94:7286–91.

Yarbro JW (1992): Mechanism of action of hydroxyurea. Sem Oncol 19:1–10.

Yu X, Jacobs SA, West SC, Ogawa T, Egelman EH (2001): Domain structure and dynamics in the helical filaments formed by RecA and Rad51 on DNA. Proc Natl Acad Sci USA 98:8419–24.

Zachariae W, Nasmyth K (1999): Whose end is destruction: Cell division and the anaphase-promoting complex. Genes Dev 13:2039–58.

Zaidi SK, Javed A, Choi J-Y, van Wijnen AJ, Stein JL, Lian JB, Stein GS (2001): A specific targeting signal directs Runx2/Cbfa1 to subnuclear domains and contributes to transactivation of the osteocalcin gene. J Cell Sci 114:3093–102.

Zaidi SK, Sullivan AJ, Medina R, Ito Y, van Wijnen AJ, Stein JL, Lian JB, Stein GS (2004): Tyrosine phosphorylation regulates Runx2 dependent subnuclear targeting of YAP and its ability to suppress transcription. EMBO J., 23(4):790–9.

Zaidi SK, Sullivan AJ, van Wijnen AJ, Stein JL, Stein GS, Lian JB (2002): Integration of Runx and Smad regulatory signals at transcriptionally active subnuclear sites. Proc Natl Acad Sci USA 99:8048–53.

Zaidi SK, Young DW, Pockwinse SH, Javed A, Lian JB, Stein JL, van Wijnen AJ, Stein GS (2003): Mitotic partitioning of the tissue specific Runx/Cbfa/AML transcription factor to progeny cells. Proc Natl Acad Sci USA 100:14852–14857.

Zelent A, Guidez F, Melnick A, Waxman S, Licht JD (2001): Translocations of the RARalpha gene in acute promyelocytic leukemia. Oncogene 20:7186–203.

Zeng C, McNeil S, Pockwinse S, Nickerson JA, Shopland L, Lawrence JB, Penman S, Hiebert SW, Lian JB, van Wijnen AJ, Stein JL, Stein GS (1998): Intranuclear targeting of AML/CBFα regulatory factors to nuclear matrix-associated transcriptional domains. Proc Natl Acad Sci USA 95:1585–9.

Zeng C, van Wijnen AJ, Stein JL, Meyers S, Sun W, Shopland L, Lawrence JB, Penman S, Lian JB, Stein GS, Hiebert SW (1997): Identification of a nuclear matrix targeting signal in the leukemia and bone-related AML/CBFα transcription factors. Proc Natl Acad Sci USA 94:6746–51.

Zhang CL, McKinsey TA, Olson EN (2002): Association of class II histone deacetylases with heterochromatin protein 1: Potential role for histone methylation in control of muscle differentiation. Mol Cell Biol 22:7302–12.

Zhang DE, Hetherington CJ, Meyers S, Rhoades KL, Larson CJ, Chen HM, Hiebert SW, Tenen DG (1996): CCAAT enhancer-binding protein (C/EBP) and AML1 (CBFα2) synergistically activate the macrophage colony-stimulating factor receptor promoter. Mol Cell Biol 16:1231–40.

Zhang F, White RL, Neufeld KL (2000a): Phosphorylation near nuclear localization signal regulates nuclear import of adenomatous polyposis coli protein. Proc Natl Acad Sci USA 97:12577–82.

Zhang HS, Gavin M, Dahiya A, Postigo AA, Ma D, Luo RX, Harbour JW, Dean DC (2000b): Exit from G_1 and S phase of the cell cycle is regulated by repressor complexes containing HDAC-Rb-hSWI/SNF and Rb-hSWI/SNF. Cell 101: 79–89.

Zhang Y, Xiong Y (2001): A p53 amino-terminal nuclear export signal inhibited by DNA damage-induced phosphorylation.[comment]. Science 292:1910–15.

Zhang YW, Yasui N, Ito K, Huang G, Fujii M, Hanai J, Nogami H, Ochi T, Miyazono K, Ito Y (2000c): A RUNX2/PEBP2αA/CBFA1 mutation displaying impaired transactivation and Smad interaction in cleidocranial dysplasia. Proc Natl Acad Sci USA 97:10549–54.

Zhao J, Dynlacht B, Imai T, Hori T, Harlow E (1998a): Expression of NPAT, a novel substrate of cyclin E-CDK2, promotes S- phase entry. Genes Dev 12: 456–61.

Zhao J, Kennedy BK, Lawrence BD, Barbie DA, Matera AG, Fletcher JA, Harlow E (2000): NPAT links cyclin E-Cdk2 to the regulation of replication-dependent histone gene transcription. Genes Dev 14:2283–97.

Zhao X, Muller EG, Rothstein R (1998b): A suppressor of two essential checkpoint genes identifies a novel protein that negatively affects dNTP pools. Mol Cell 2:329–40.

Zink D, Bornfleth H, Visser A, Cremer C, Cremer T (1999): Organization of early and late replicating DNA in human chromosome territories. Expr Cell Res 247:176–88.

PART II

CHAPTER 3

CELL CYCLE REGULATORY CASCADES

HEIDE L. FORD, ROBERT A. SCLAFANI, and JAMES DEGREGORI
Departments of Obstetrics and Gynecology, Biochemistry, and Molecular Genetics, University of Colorado Health Sciences Center, Denver, CO, 80262

INTRODUCTION

Definition of Cell Cycle Phases and the Concept of Coordination of Growth and Division

Cells are the basic units of life. Therefore the regulation of cell number is of major importance to both unicellular and multicellular organisms. Eukaryotic cells have evolved mechanisms of coordinating the replication and segregation of their genetic material with cellular growth by distributing these events to specific phases of a temporal cycle known as the cell division cycle or simply, "the cell cycle" (Fig. 3.1). The main goal of the cell cycle is to produce two cells that have equal amounts of genetic material (chromosomes) and a proper cell size.

Replication of the chromosomes occurs in the S (DNA synthesis) phase, while segregation of the newly replicated sister chromatids occurs in M (mitosis) phase (see Chapters 5 and 6 for detailed descriptions of both S and M phases). The G_1 (gap 1) and G_2 (gap 2) phases are the gaps between the S and M phases, with G_1 preceding S phase and G_2 preceding M phase. As we will see in this chapter, regulation of chromosome replication and segregation occurs in these two gap phases. All phases except M phase can also be grouped together and called interphase, which is the intervening phase between mitoses.

Most eukaryotic cells coordinate cell growth and division in G_1 of the cell cycle with some exceptions such as *Schizosaccharomyces pombe* fission yeast (see below) and epidermal cells. Before they can continue into the cell cycle, cells wait in G_1 until they reach a critical mass or size.

Cell Cycle and Growth Control: Biomolecular Regulation and Cancer, Second Edition, Edited by Gary S. Stein and Arthur B. Pardee
ISBN 0-471-25071-6

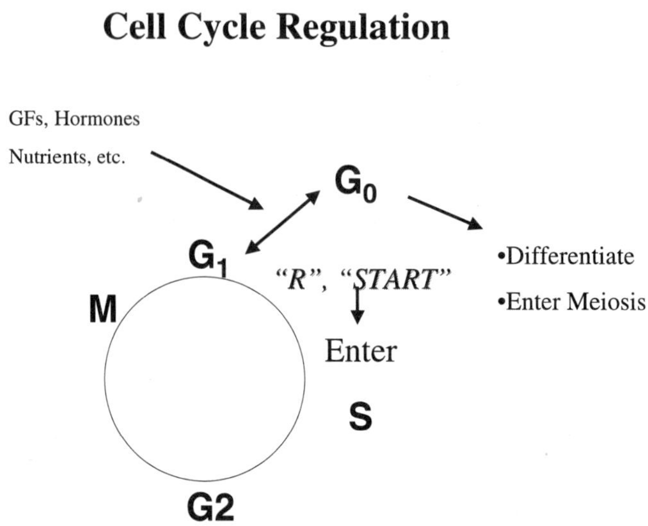

Figure 3.1. Cell cycle phases. The cell cycle is shown as a circle. Chromosomal DNA replication occurs in S phase and segregation of newly, replicated daughter chromosomes occurs in M phase. G_1 and G_2 mark the gap periods that precede S and M phases, respectively. R indicates the restriction point, and START would be a similar point in yeast cells. Cells can enter S phase or exit the cell cycle to enter G_0, from which they can differentiate or enter meiosis.

In this way cells are prevented from dividing if they are too small. This prevents the production of very small inviable cells and maintains proper cell size. An exception to this regulation occurs in the early cleavage divisions of an embryo, in which the cells get smaller after each division.

Definition of the "Restriction Point" and Analogy to "START" in Yeast

How is this regulation accomplished? At a specific point in G_1 phase, communication between the outside and the inside of the cell occurs. If nutrients and growth factors are present and the cell has attained the critical mass necessary, then the cell is allowed to pass this regulatory point known experimentally as the "restriction" point or "R." Normally cells grow and divide asynchronously with a population having different amounts of cells in all four phases, typically 40% G_1, 40% S, and 20% G_2/M for mammalian cells in culture. "R" was defined experimentally by first synchronizing cells in G_1 by removal of serum, which contains nutrients and growth factors, and then adding back the serum to produce a synchronous cycling population. Serum was removed from this synchronous population and progression into S phase was monitored. If serum was removed early in G_1 phase when the cells were too small, they could not proceed into S phase and eventually left the cell cycle and went to a resting phase known as G_0 (Fig. 3.1). However, if the serum was removed later in G_1 phase after they had attained the critical mass, they could enter S phase and even complete the cell cycle. The point at which the cells attain this critical size is known as R. This point is also a point of commitment in that cells must complete the cell cycle after passing it. It is additionally a point of regulation as cells can go to different fates from this point. For example, some cells exit the cell cycle at R, enter G_0, and then differentiate into nondividing neurons. A point similar to R has been defined experimentally and the regulatory proteins important for establishing R in a number of different eukaryotic organisms will be discussed below.

Model Organisms as a Way to Study the Cell Cycle

A number of model organisms have been used to study the cell cycle. Both fission and budding yeasts (*Schizosaccharomyces pombe* and *Saccharomyces cerevisiae*, respectively) have been used because of their sophisticated molecular genetics. This allowed investigators to isolate mutants, called cdc (cell division cycle) mutants, that were defective in transiting from one cell cycle phase to another, and to then use these mutants to identify the gene products. Both systems have different advantages. With *S. cerevisiae*, it was easy to identify cells in different cell cycle phases cytologically. In addition, cell division is unequal, producing a small daughter cell and large mother cell (Fig. 3.2). Cells in G_1 phase are unbudded, cells in early S phase have a small bud, while cells in G_2/M phase have a large bud (Fig. 3.2). Because cdc gene products were expected to be essential for cell division, conditional temperature-sensitive (ts) cdc mutants were isolated which arrested in a specific cell cycle phase under restrictive conditions, such that at 37°C the gene product does not function. Normally, populations of yeast, like mammalian cells, are asynchronous with cells at all stages of the cell cycle. Cdc mutants have the majority of cells at a specific cell cycle stage and thus have a uniform cytology. For example, the *CDC28* gene of *S. cerevisiae* encodes the major Cdk (cyclin-dependent kinase), which is part of a family of protein kinase enzymes that regulate the cell cycle by phosphorylating critical target genes (see below). Therefore *cdc28*-ts mutants arrest in G_1 phase as unbudded cells.

Like mammalian cells, *S. cerevisiae* also coordinate size and division in G_1 phase at a point called "START" (analogous to R) in that cells that have passed START can complete the cell cycle in the absence of nutrients. Daughter cells are too small to enter the cell cycle and must grow in G_1 phase to a critical size before they can START the cycle. Cells of *S. cerevisiae* also must be at START to enter a different developmental fate such as the meiotic (germ-line) cell cycle. This makes logical sense as it would be disastrous for cells to enter meiosis in the middle of S phase with partially replicated DNA. In this case chromosome instability would result in cell death.

In *S. pombe*, mutants were isolated on the basis of their reduced size. This yeast coordinates growth and division in G_2 phase, where they are

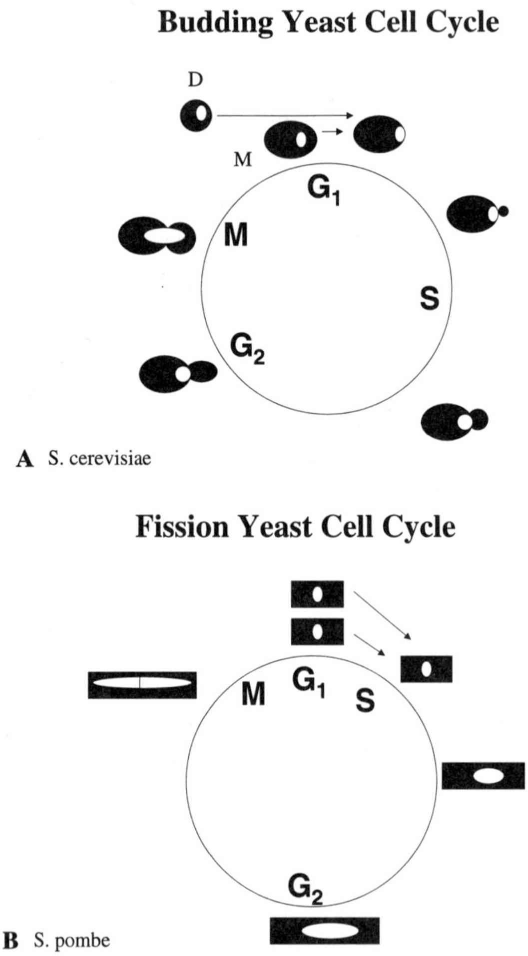

Figure 3.2. The budding and fission yeast cell cycles. The cytology of (**A**) *S. cerevisiae* budding yeast and (**B**) *S. pombe* fission yeast cells proceeding through the cell cycle is depicted. The nucleus is shown as a white image on a black background. Smaller daughter (D) and larger mother (M) budding yeast cells are shown in G_1 phase. The lengths of cell cycle phases are drawn approximately to scale.

held and prevented from entering M phase until they reach a critical size. Mutants in *cdc2*, the major Cdk, or in genes that regulate the Cdk, such as *Wee1*, can enter M phase early, thereby producing smaller cells, referred to as "Wee" because they were isolated in Scotland. The comparison between these two yeasts is particularly informative and reveals much about how the cell cycle works. Although we see that the two yeast cells regulate size and division in different cell cycle phases, the same Cdk enzyme is used. Later it was found that the Cdk enzyme is also used in the other gap regulatory phases, that is, in G_2 in *S. cerevisiae* and in G_1 in *S. pombe*. Thus Cdk is used for regulation in both gap phases of the cell cycle. However, a different form of the cyclin (regulatory subunit)/Cdk complex is used in the G_1 and G_2 phases.

No discussion of model organisms for studying the cell cycle is complete without a discussion of MPF (maturation promoting factor) in frogs (*Xenopus laevis*). MPF was originally identified as a cytoplasmic factor that when injected would cause germ-line oocytes to mature into eggs in the absence of hormones. Oocytes are normally arrested in G_2 of meiosis I and become mature eggs by completion of meiosis I and II, which are essentially the result of two sequential G_2 to M transitions. The discovery that MPF is a CDK enzyme (consisting of a cyclin/Cdk complex) provided the first biochemical evidence that active CDK can drive the cell cycle. Again, as we saw with the two yeasts, the same CDK is responsible even though the physiology is different, that is, G_2 to M transition in a different cell cycle (meiosis instead of mitosis). This points out the universality of CDK enzymes as regulators of cell cycle progression in all eukaryotic cells.

CDKs as Regulators of the "Cycling" of Cell Cycle

These CDK enzymes are responsible for the "cycling" of the cell cycle. Simply, the CDK is composed of a protein kinase subunit called Cdk that becomes active when bound to a regulatory subunit called cyclin. Thus the CDK enzyme is a heterodimeric complex of a Cdk subunit and a cyclin subunit, which is referred to as "cyclin/Cdk" or "CDK" (see below). It is the level of cyclin protein that fluctuates or cycles during the cell cycle, and thereby regulates, in part, the activity of the Cdk. It should be noted that while yeast have one major Cdk, metazoans have numerous Cdks, as will be further discussed below.

What process ensures that cells always go from S phase to the next M phase to the next S phase, and so on? The whole cell cycle can be divided into two phases (Fig. 3.3): one with low CDK activity (G_1) and one with high CDK activity (S, G_2, M). The process of DNA replication is regulated by forming a pre-replication complex (pre-RC; see Chapter 5 for S phase discussion) in G_1 phase while CDK activity is low, and then activating the pre-RC in S phase when CDK activity is high. Importantly, high CDK activity is needed for pre-RC activation but also inhibits pre-RC formation. Thus the pre-RC can only be assembled when CDK activity is low and can only be activated when CDK activity is high. In this way high CDK activity and pre-RC formation can never co-exist. This ensures that re-replication of the DNA cannot take place ("the once and only once rule"; see Chapter 5). It also ensures that S phase is dependent on a prior M phase; that is, cells must go through M phase to reduce CDK activity to allow for pre-RC formation in G_1 phase, and then high CDK activity produced by production of cyclins E and A ensures S phase completion (see below and Chapter 5 on S phase regulation). The CDK enzyme is then destroyed upon exit from M phase and the cell begins a new cell cycle (see below). In this way cells go from S to M to S to M, and so on.

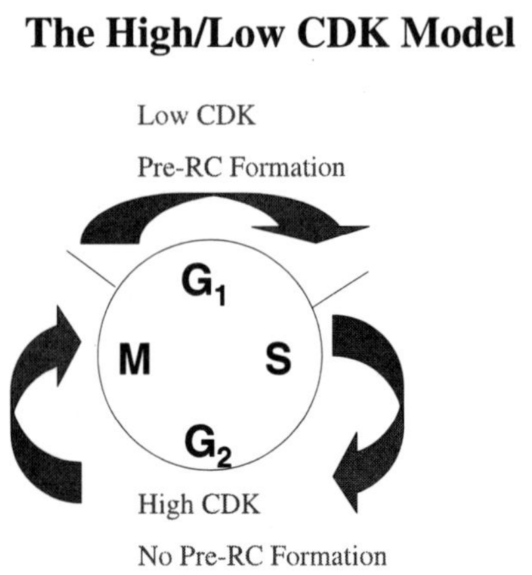

Figure 3.3. High/low CDK model for the cell cycle. A model in which there are two states: G_1 phase in which the pre-RC (pre-replication complex) assembly occurs in low CDK (Cyclin-dependent kinase) activity and the combined $S/G_2/M$ phases in which high CDK activity activates the pre-RC to produce DNA replication but blocks pre-RC formation.

The Checkpoint Concept as a Surveillance Mechanism

The concept of a checkpoint control mechanism was suggested from mammalian studies and fully gleaned from the phenotype of yeast cdc mutants. As described above, yeast cells with temperature-sensitive mutations in important genes arrest in the cell cycle with a uniform cytology. For instance, mutations in enzymes needed for DNA replication, such as DNA polymerase arrest in S phase, and do not enter M phase. How do the cells know that DNA was not made? What prevents a mutant that cannot make DNA from entering M phase? The hypothesis is that the cells have a surveillance or checkpoint mechanism that prevents future cell cycle events from happening if the prior event is blocked. Furthermore cells that have damage in their DNA also activate a checkpoint and do not enter M phase. Support for this hypothesis was provided by showing that yeast mutants defective in the checkpoint have exactly this phenotype; that is, they do not know that DNA replication is blocked or that the DNA is damaged, so they enter M phase and even divide, which can result in cell death. In the case of DNA damage, the cell cycle arrest is transient, allowing time for the DNA to be repaired before proceeding into M phase. This ensures efficient DNA repair before mitosis. This is consistent with earlier observations in mammalian cells that showed that when the DNA is damaged, the cells arrest transiently in G2 phase, and that argued that this arrest may provide time for repair processes that are critical for survival after DNA damage. If caffeine is added, the G2 arrest does not occur and the cells die at a higher rate. Therefore cells have two checkpoints for monitoring the DNA: a DNA replication and a DNA damage checkpoint.

Cells also have evolved checkpoint mechanisms for preventing exit from mitosis when the spindle apparatus is defective (see below). Again, studies of the yeast system were informative. If yeast cells are treated with microtubule poisons such as Nocadazole or Benomyl, the cells sense the defect, and block in mitosis. As seen for the DNA checkpoint above, mutations in this "spindle checkpoint" cause the cells to divide and die.

Clearly, cells have evolved elaborate checkpoint mechanisms for maintenance of the genome. If these mechanisms are subverted by mutation, then chromosomal instability will occur and result either in cell death or in the proliferation of cells with a multitude of defects, some of which may produce transformed or cancerous cells. In the next section, we will explore the evidence that defects in checkpoints can result in cancer.

Checkpoint Control and Cancer

The process of DNA checkpoint control (Fig. 3.4) has been described as analogous to a signal transduction pathway. In this analogy, if DNA replication stops for any reason and/or the DNA is damaged, a signal is detected by "sensor" proteins and then sent by "transducer" proteins to "effector" proteins, which block the cell cycle and elicit DNA repair. DNA checkpoint control can occur in G_1, S phase, or at the G_2/M transition.

Again, as seen above, yeast mutants defective in the response helped to define the basic pathway. These studies helped to define the sensors, transducers, and effectors. Mutants in *rad9* cannot sense the damage, while *chk2* mutants cannot transduce the signal generated by the Rad9 protein. Many of the transducers are protein kinase enzymes that will phosphorylate effectors to regulate the response.

Mammalian cells have additional regulators for this checkpoint pathway, many of which are found mutated in cancer cells. The study of several familial syndromes, in which a greatly increased level of cancer occurs because of mutations in a single gene, has been informative in this regard. The p53 gene or the Chk2 gene is mutated in Li-Fraumeni syndrome, which presents with increased incidence of sarcomas and leukemias. Familial ATM (ataxia-telangiectasia mutated) mutations give rise to lymphomas.

Notably the p53 protein, which is mutated in more than 50% of all tumors, is known as the "guardian" of the genome (see below). The idea is that loss of p53 by mutation results in the accumulation of many additional mutations resulting in tumor progression. Loss of p53 can occur either in the germ-line and be inherited as in Li-Fraumeni syndrome or in adult somatic cells in the acquired cases. When the DNA is damaged, p53 levels increase and then p53 acts a transcription factor to increase expression of a number of important genes. One of these genes is p21, a CDK inhibitor (see Chapter 7), which then arrests the cell cycle and allows for DNA repair to occur. In this pathway, DNA damage is sensed by the ATM protein kinase, which then phosphorylates p53 and thereby

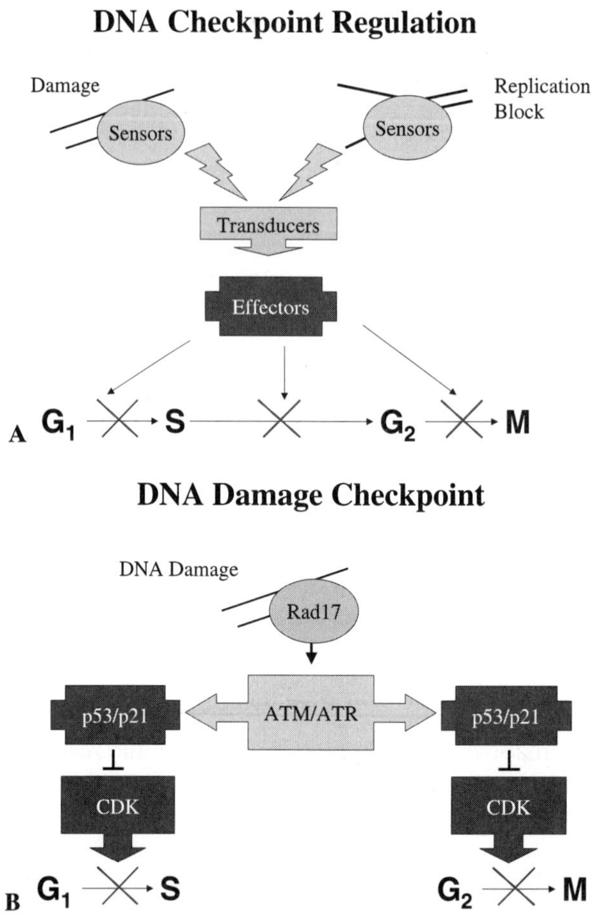

Figure 3.4. (**A**) DNA checkpoint regulation. The DNA checkpoint is depicted as a signal transduction cascade in which DNA damage or DNA replication blocks are detected by sensor proteins and the signal is transmitted by transducer proteins to effector molecules, which arrest the cell cycle or cause repair of the DNA damage lesions. (**B**) DNA damage checkpoint. The DNA damage checkpoint is depicted similarly to that in (**A**), but specific proteins known to be important for response to DNA damage are shown. In this case ATM/ATR protein kinases are activated by DNA damage signal sent by Rad17 protein. These kinases phosphorylate p53, which acts as transcription factor to produce p21. The resultant p21 protein inhibits CDK activity, and thereby blocks the G_1 to S or G_2 to M phase transition.

results in transcription of p21 (Fig. 3.4B). Thus p53 "guards" the genome by stopping cell cycle progression by inhibiting the CDK enzyme either at G_1/S or at G_2/M when the DNA is damaged. Other examples of checkpoint regulation will be described below.

Finally, although it is important to arrest the cell cycle during the checkpoint response, it is also important to stabilize DNA replication forks and to repair the DNA. Thus many of the effectors of the checkpoint pathway are believed to be molecules important for these two processes as well.

Summary

In this section the basic premise for cell cycle regulation was covered, highlighting work from model organisms that demonstrated how cell division and growth are coupled, and also underscoring the importance of the cyclin/Cdk complexes in cell cycle progression. The checkpoint concept as a surveillance mechanism was introduced, and dysregulation of these checkpoints as a contributing factor to tumorigenesis was addressed. Subsequent sections of this chapter will be devoted to the regulatory cascades that govern cycling of normal cells (with some reference to tumorigenesis). We will cover all stages of the cell cycle, with the majority of our emphasis on the transitions from G_1 to S phase, and from G_2 to mitosis.

G_1/S TRANSITION

Introduction to the Retinoblastoma Protein Rb as a Tumor Suppressor and Key Regulator of Proliferation

In order to maintain control of cell numbers, whether during tissue development or tissue homeostasis, the decision of a cell to enter the cell cycle must be tightly controlled by extracellular cues. These cues include diffusible growth factors, contact with extracellular matrix, and interactions with other cells (discussed in more detail in Chapter 4). For simplicity, we will refer to all of these signals generically as "growth factors." As you learned in the previous section, these growth factors control the cell cycle by ultimately impinging on the activities of key components of the cell cycle regulatory machinery such as the cyclin-dependent kinases (Cdks). A key component of the regulation of cell cycle entry in mammalian cells is the retinoblastoma protein, pRb, which functions as a barrier to inappropriate cell cycle progression. The Rb gene was originally cloned as the gene mutated in patients with hereditary retinoblastoma (retinal tumors), and is now known to be mutated in about 30% of human cancers. Readers are referred to Chapter 18 for a more in-depth discussion of roles for Rb in tumor suppression. In this section, we will discuss how cyclin-dependent kinases, in response to growth factor-mediated signaling, phosphorylate Rb, relieving Rb's restraint of cell cycle progression. We will discuss how Rb controls the transcription of a variety of genes required for the progression of cells out of G_1 and through S phase via its association with the E2F family of transcription factors. Our discussion will range from the genetic analysis of Rb and E2F function using model organisms to a more biochemical understanding of the mechanism underlying Rb/E2F control of transcription.

History: Isolation as the Retinoblastoma Susceptibility Protein and Early Studies

Prior to the isolation of the retinoblastoma (Rb) gene as the gene whose deletion or loss-of-function mutation led to the development of retinoblastoma tumors in children, many cancer biologists believed that gain-of-function mutations in oncogenes primarily contributed to tumorigenesis. A loss-of-function, or recessive, mutation results in the loss or reduction of biological activity of the mutated gene, and mutation of both alleles of the gene is generally required for the phenotype. In contrast, a gain-of-function, or dominant, mutation usually confers either altered or increased activity on the encoded protein, such that even when encoded together with the unaltered (wild-type) allele, the mutant protein confers the phenotype. Rb was the founding member of a now large and still expanding class of genes, termed tumor suppressors, whose loss-of-function mutation contributes to tumorigenesis. A key to how Rb functions as a tumor suppressor was uncovered when pRb was shown to associate with viral oncogenic proteins, such as the adenoviral E1A protein, and this association was shown to prevent pRb from limiting cellular proliferation. pRb was subsequently shown to associate with the cellular transcription factor E2F, and E1A binding to pRb was shown to sequester pRb from E2F, resulting in increased E2F-dependent transcription. As will be described below, E2F activity plays critical roles in G_1 to S phase, as well as S and M phase progression, by regulating the transcription of a large number of cell cycle control factors. Although the E1A studies told us how adenovirus, with the goal of stimulating cell cycle progression into S phase in order to achieve the replication of its own genome, could relieve Rb-mediated inhibition of E2F-dependent transcription of cell cycle progression genes, it was still unclear how Rb was regulated during normal cell cycle entry.

Rb Family Members

Like most genes in vertebrates, Rb is a member of a gene family encoding structurally and functionally similar proteins, which in addition to pRb include the p107 and p130 proteins. Like pRb, p107 and p130 associate with viral oncoproteins like E1A, are regulated during the cell cycle by cyclin/Cdk-dependent phosphorylation, and associate with and inhibit E2F transcription factors. However, the p107 and p130 genes appear to be less frequently mutated in human cancers relative to Rb. There are other differences. The p130 protein is expressed in quiescent (G_0 phase, or "out of the cell cycle") cells, and following growth factor stimulation and cell cycle progression, p130 protein disappears as the result of regulated protein degradation. In contrast, Rb and p107 levels increase in late G_1 phase (discussed further below). Also, as discussed below, the three Rb family members differentially associate with different subsets of the E2F family. While Rb family members clearly play overlapping roles in regulating E2F and the cell cycle, there is clearly specificity in their actions as well.

All human tumor mutations in Rb described to date prevent Rb from inhibiting E2F-dependent transcription and thus cell cycle progression, highlighting the essential function of E2F inhibition in tumor suppression by Rb. In addition all three Rb family members have been shown to promote cellular differentiation. Thus the increased expression of Rb family members can promote cellular differentiation and mutation of Rb, p107, or p130 can prevent differentiation both in cell culture and in mice, with different effects caused by mutation of different Rb family members. In fact certain Rb mutant genes, both engineered and cloned from retinoblastoma patients, disrupt the ability of Rb to inhibit E2F-dependent transcription but retain the ability to promote differentiation. Such partially penetrant Rb mutants were isolated from patients with low risk retinoblastoma, suggesting that the mutations resulted in partial disruption of tumor suppression by Rb. Thus both the promotion of differentiation and the inhibition of E2F contribute to tumor suppression by Rb, and these properties can be separated genetically, indicating that different parts of the Rb protein are responsible for the different functions. Although the regulation of differentiation is clearly an important aspect of Rb function, this chapter will focus on cell cycle regulation by Rb, which appears to be largely mediated via association with E2F.

Regulation of Rb by Cyclin/Cdks

A clue to how Rb is regulated during the cell cycle came from the observations that pRb is heavily phosphorylated starting in late G_1 of the cell cycle until mitosis. E2F was found to associate only with hypophosphorylated pRb (less phosphorylated than hyperphosphorylated pRb), and numerous studies have confirmed that hypophosphorylated Rb is the active form of Rb that negatively regulates E2F and cell cycle entry (Fig. 3.5). The phosphorylation of Rb (as well as p107 and p130) during G_1 progression is largely carried out by CDKs. Specifically, in mid-G_1, Rb is first phosphorylated by cyclin D-dependent kinases, which are composed of one of three different D type cyclin proteins (the regulatory subunit) with either Cdk4 or Cdk6 (the catalytic kinase subunit). As stated above, the kinase subunits of cyclin/Cdks are absolutely dependent on association with a cyclin for activity.

D type cyclin/Cdks are highly responsive to growth factor stimulation at several levels, including the synthesis of their subunits, the association with inhibitory proteins (cyclin kinase inhibitors or CKIs), the assembly of the subunits, and the stability of cyclin D, all of which will be discussed in more detail in Chapter 4. CKIs of the Ink4 family ($p16^{Ink4a}$, $p15^{Ink4b}$, $p18^{Ink4c}$, and $p19^{Ink4d}$) specifically associate with Cdk4 and Cdk6, blocking the kinase active site and preventing association with cyclins. In contrast, CKIs of the CIP/KIP family ($p21^{CIP1}$, $p27^{Kip1}$, and $p57^{Kip2}$) associate with and inhibit all cyclin/Cdk complexes. Growth factors can decrease CKI expression (e.g., p15 and p27), which together with increased cyclin/Cdk expression results in active complex assemblies. In addition cyclin D/Cdk4,6 complex association with p21 and p27 following growth factor activation is required for cyclin E/Cdk2 activation, by sequestering these

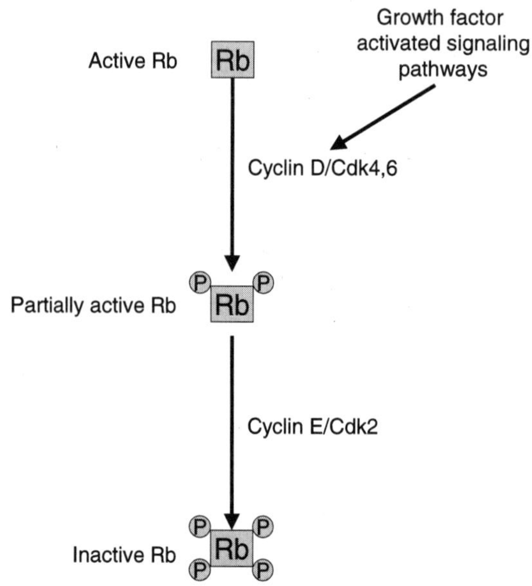

Figure 3.5. Cyclin-dependent kinases sequentially inactivate Rb. Growth factor-activated signaling pathways lead to the activation of Cyclin D/Cdk complexes, which phosphorylate Rb (or other Rb family members) on specific serine and threonine residues, resulting in the partial inactivation of Rb. Cyclin E/Cdk2 activated in late G_1 further phosphorylates Rb, resulting in hyperphosphorylated Rb that can no longer inhibit E2F dependent transcription and cell cycle progression.

CKIs away from cyclin E/Cdk2. Interestingly, at low stochiometries, p21 and p27 are actually required for the assembly of cyclin D/Cdk complexes. Most important for our discussion, it appears that cyclin D/Cdk dependent phosphorylation of Rb is not sufficient to fully relieve Rb-mediated repression of E2F, and in late G_1 increased cyclin E/Cdk2-mediated phosphorylation of Rb results in complete inactivation of Rb. Cyclin D and cyclin E dependent Cdks phosphorylate distinct sites (serine and threonine amino acids) on Rb. Thus the sequential and combined phosphorylation of Rb by cyclin D and cyclin E dependent kinases contributes to full inactivation of Rb. The dephosphorylation of Rb is also important to reactivate Rb, either following mitosis or in response to growth factor withdrawal, and appears to be mediated by the combined action of phosphatases together with the inactivation of cyclin-dependent kinases.

Rb and the Restriction Point (R)

In the first part of this chapter, you learned about the functional definition of the restriction (R) point, the point where a cell no longer needs growth factor stimulation in order to continue G_1 progression into S phase. The control of Rb, E2F, and/or cyclin E activities may represent, biochemically, the R point. Certainly the inactivation of Rb, increased E2F-dependent transcription, and cyclin E/Cdk2 activation coincide temporally with the R point in late G_1. In addition overexpression of E2F or ectopic cyclin E/Cdk2 activation are each sufficient to drive a quiescent cell into S phase in the absence of growth factors. Furthermore the inactivation of all three Rb family members results in inappropriate G_1 to S phase progression in the absence of growth factors. Cyclin E overexpression or Rb inactivation may also uncouple the R point from the requirement that a cell grow to a particular size prior to R, allowing cells to enter S phase at a smaller size. Since Rb inactivation, E2F upregulation, and cyclin E/Cdk2 activation are all mutually dependent, it is difficult to ascribe the R point to only one of these events.

Rb Control of E2F Transcriptional Activity

The G_1 CDK-Rb-E2F Pathway and Proliferation Control. Growth factor-dependent activation of cyclin/Cdk-mediated Rb phosphorylation and E2F activation, referred to as the CDK-Rb-E2F pathway, is a prerequisite for cell cycle entry and progression. In fact, as discussed further in Chapter 18, deregulation of this key pathway, by mutation of CDK inhibitors, increased expression of cyclin/Cdk subunits, or mutation of Rb, occurs in virtually all human cancers and thus appears to be a prerequisite for tumorigenesis. While different cell types respond to diverse extracellular signals controlling cell cycle entry, all of the growth factor-activated signaling pathways that lead to cell cycle entry appear to ultimately result in Cdk activation, Rb phosphorylation, and increased E2F dependent transcription (Fig. 3.6). For example, while a T lymphocyte proliferates in response to antigen and an epidermal cell divides in response to epidermal growth factor (EGF), both antigen and EGF stimulation activate the CDK-Rb-E2F pathway.

Active Repression of Transcriptional Targets by Rb/E2F. Overexpression of E2F proteins increases the transcription of target genes, which together with other experimental data, indicates that E2Fs can function as bona fide transcriptional activators. Rb association with E2F masks the transcriptional activation domain of E2F. In the discussion below, "Rb" refers generally to all three Rb family members. However, Rb does much more than simply prevent E2F from activating transcription. In fact it is now clear that the Rb/E2F complex functions to actively repress transcription (Fig. 3.7). Thus the elimination of E2F DNA binding to promoters usually results in increased transcription from these promoters, suggesting that a major function of E2F is to recruit Rb to promoters for transcriptional repression. Rb functions as a transcriptional repressor by recruiting various cofactors, many of which are involved in remodeling chromatin. DNA in the nucleus is organized into higher ordered structures together with proteins (primarily histones), and this DNA-associated protein structure is referred to as chromatin. Modifications of histones and other chromatin proteins can influence chromatin structure,

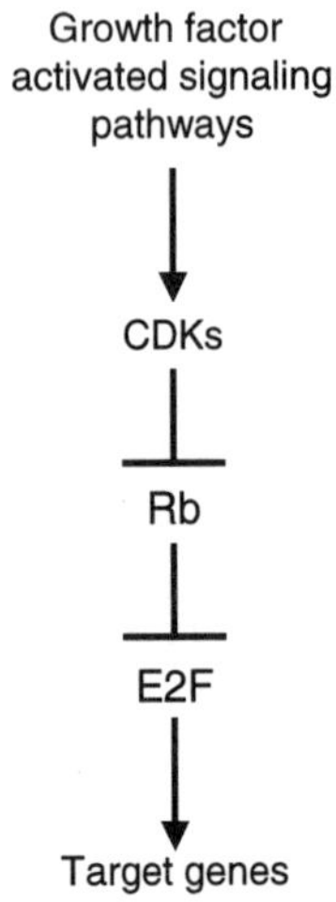

Figure 3.6. The CDK-Rb-E2F pathway. Growth factor stimulation of a number of cellular receptors activates signaling pathways that activate CDKs. Activated CDKs phosphorylate Rb, relieving Rb/E2F repression of target gene transcription and releasing transcriptionally active E2F. The CDK-Rb-E2F pathway plays a central role in G_1 to S phase progression. Note that arrows denote activation, while the blunt arrows denote inhibition. The Rb shown represents all three Rb family members. E2F refers to E2F/DP heterodimers.

interconverting open or closed chromatin states. The open state is more accessible to transcription factors, and thus open chromatin is generally associated with active transcription. Rb recruits factors that induce a closed chromatin state that does not support transcription. For example, Rb/E2F recruits histone deacetylases (HDACs) to E2F target promoters, which function to remove acetyl groups from histone proteins at the promoter. Acetylated histones are associated with open chromatin, and thus deacetylation of chromatin contributes to transcriptional repression. Chapter 8 provides more details on how chromatin modifications modulate gene expression. The analysis of endogenous E2F-regulated promoters in cells reveals decreased histone acetylation and increased Rb/E2F promoter association in quiescent, unstimulated cells, and increased acetylation with increased free, transcriptionally activating E2F (not Rb) associated with promoters in late G_1 following growth factor stimulation.

In summary, Rb/E2F complexes contribute to the maintenance of cell quiescence by actively repressing, through an alteration of promoter chromatin structure, the expression of genes that promote cell cycle progression. The phosphorylation of Rb relieves this repression and also allows E2F-dependent activation of these genes, reversing the repressive chromatin state. The repressive chromatin state is reversed both by the elimination of HDAC recruitment as well as by the E2F-dependent recruitment of histone acetyl transferases (HATs), which acetylate histones (Fig. 3.7). Thus Rb/E2F and E2F differentially regulate target

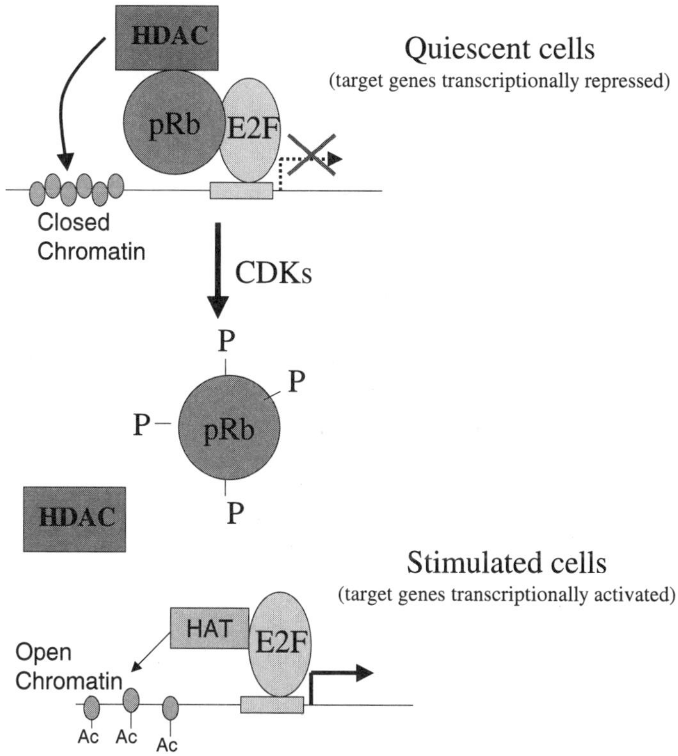

Figure 3.7. E2F-mediated gene repression and activation. In quiescent cells, pRb recruitment of HDAC and other corepressors to the promoter actively represses gene expression by promoting a closed chromatin conformation, in part by deacetylation of histones. Following growth factor stimulation and CDK activation, phosphorylated Rb is no longer able to bind E2F or recruit corepressors, and E2F is now free to promote transcription, in part by recruitment of histone acetyl transferases (HATs). The pRb shown represents all three Rb family members.

genes required for cell cycle progression, with cyclin/Cdk-mediated phosphorylation of Rb regulating the switch from repression to activation and stimulating the E2F-dependent transcription of genes that promote entrance into S phase.

E2F Transcriptional Targets and Their Role in Cell Cycle Progression

A Central Role for E2F in Control of G_1 to S Phase Transitions. Cell cycle progression is regulated at multiple levels. Post-transcriptional regulation, the regulation of protein levels and activities independent of transcriptional control, clearly plays a major role in cell cycle transitions. For example, cyclin-dependent kinases are highly regulated during the cell cycle by the control of cyclin protein stability. Also modifications of pro-

E2F Targets

Nucleotide synthesis
- thymidine kinase
- thymidylate synthase
- ribonucleotide reductatase
- dihydrofolate reductase

DNA replication
- PCNA
- DNA polymerase α
- Cdc6
- Mcm2, 3, 4, 5, 6, 7
- Dbf4

Cell cycle regulators
- Cyclin E
- Cdk2
- E2F1,2, 3
- Cyclin A
- Cyclin B
- Cdc2
- Cyclin A

Inhibitors
- p18Ink4c
- p19Ink4d
- Rb
- p107
- p21

Figure 3.8. E2F target genes. E2Fs regulate the expression of a large number of genes that play critical roles in cell cycle progression. Representative genes are shown. Some target genes are required for the synthesis of the nucleotide pool necessary for DNA replication. Other target genes are directly involved in DNA replication. Finally cell cycle regulators coordinate and control cell cycle progression, from G_1 phase through mitosis.

teins, such as by phosphorylation, control the activities of key components of the cell cycle machinery. An additional level of control is provided by the regulated transcription of genes that are required for cell cycle progression, and E2Fs play a major role in this regulation. The expression of E2F regulated genes is generally increased during late G_1 and/or in S phase of the cell cycle. These genes play various critical roles involved in cell cycle progression, and include genes involved in cell cycle regulation and DNA replication (see Fig. 3.8 for representative target genes). In terms of DNA replication, E2F targets include the enzymes required for deoxynucleotide synthesis, components of the complex that recognize origins of replication, components of the DNA polymerase holoenzyme, and the cyclins and Cdks that regulate origin firing (for a detailed discussion of the mechanics of S phase, see Chapter 5). Importantly, while E2F is required to regulate the transcriptional increase of both cyclin/Cdk subunits and replication components, post-translational control of these activities, including Cdk-dependent phosphorylation of replication components, is required for their proper regulation during the cell cycle. It is also critical to stress that the cyclin E/Cdk2-dependent regulation of targets other than Rb is important for S phase, contributing to the control of origin firing, histone synthesis, and centrosome duplication. Although less studied, E2F also functions to regulate G_2 to M phase progression by regulating the transcription of genes such as cyclin B and Cdc2. The multi-layered regulation of cell cycle progression, including E2F-dependent transcriptional regulation, functions to ensure ordered progression through the cell cycle that is dependent on proper environmental cues.

Thus E2F-dependent transcriptional control coupled with multilayered post-transcriptional control coordinates the generation of waves of different Cyclin/Cdk activities as a cell progresses from quiescence through the cell cycle (Fig. 3.9). Growth factor dependent activation of cyclin D/Cdk4,6 complexes initiates Rb phosphorylation and E2F acti-

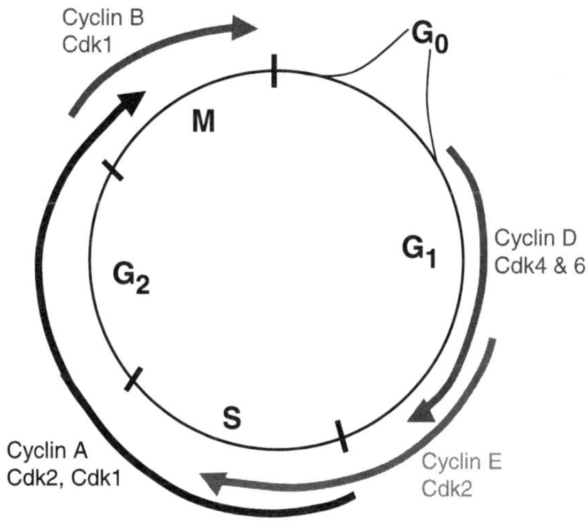

Figure 3.9. Waves of specific CDK activities during the cell cycle. The activities of Cdks associated with specific Cyclins are depicted by arrows. For each Cyclin, multiple family members can contribute to the activity (e.g., Cyclin D has three family members, Cyclins D1, D2, and D3). Growth factor-dependent signaling pathways control the accumulation of Cyclin D/Cdk proteins. The activities of Cyclin E, A, and B associated kinases are in part determined by the regulated accumulation of the subunits via E2F dependent control of transcription. The abrupt loss of Cyclin E, A, and B kinase activities at specific points in the cell cycle is largely the result of the regulated degradation of the Cyclin subunits.

vation, contributing to increased cyclin E/Cdk2 activity, which further increases E2F activity and coordinates DNA replication. Cyclin A is activated in a second wave of E2F-dependent transcription in S phase, and cyclin A associated Cdk2 and Cdc2 kinases are required for appropriate S and G_2/M phase progression. Finally, E2F-dependent upregulation of cyclin B and Cdc2 together with a host of post-transcriptional controls contribute to the activation of cyclin B/Cdc2 at the G2/M boundary, which is required for mitosis. The proper coordination of each cyclin/Cdk wave is essential for ensuring appropriate cell cycle entry, accurate replication of the genome once and only once per cell cycle, and equal segregation of sister chromosomes into the two daughter cells.

E2F Family and Specific Functions. E2F transcription factor activity is composed of various heterodimers, each formed from one E2F subunit and one DP subunit. There are six known genes encoding E2F family members and two known DP family members. Dimers of two E2Fs or two DP proteins are not known to naturally occur, such that E2F/DP heterodimers appear to represent all E2F transcriptional activity. E2F proteins all share similar domains required for DNA binding and heterodimerization, and all but E2F6 possess a C-terminal domain involved both in Rb family member binding as well as transcriptional activation. E2F1, E2F2, and E2F3 appear to only associate with Rb, while E2F4 can

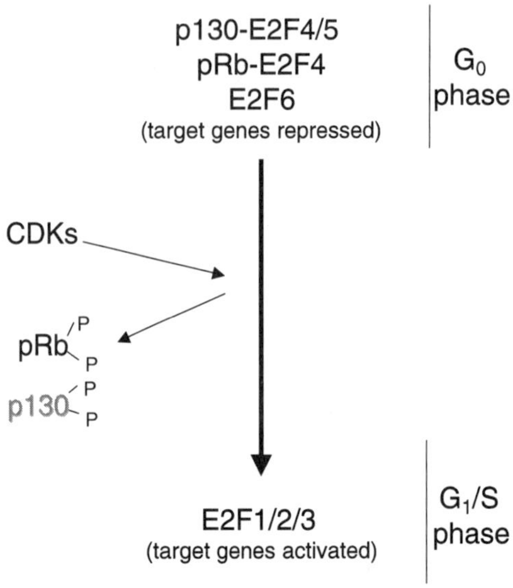

Figure 3.10. Differential roles for E2F family members in progression from G_0 (quiescence) to S phase. In quiescent cells, E2F4 or E2F5 in complex with p130 and E2F4 in complex with Rb function to repress E2F target gene expression. E2F6 also functions as part of a transcriptional repressor independent of Rb. During G_1 progression, CDK phosphorylation and inactivation of Rb and p130 together with p130 degradation result in relief of repression, the accumulation of E2F1, 2, and 3, and transcriptional activation of E2F target genes.

associate with Rb, p107, and p130 and E2F5 appears to predominantly associate with p130. Thus the E2F and Rb families can be distinguished by their associations with one another. In addition, while E2F4, 5, and 6 are expressed throughout the cell cycle, E2F1, 2, and 3 are transcriptionally upregulated in late G_1 phase, coincident with increased E2F-dependent transcription. In part, E2F1, 2, and 3 upregulation results from growth factor-dependent activation of the Myc transcription factor, which directly activates the transcription of these E2Fs. Current evidence favors a model whereby E2F4 and E2F5 function primarily as Rb family member associated transcriptional repressors in quiescent cells, and E2F1, 2, and 3 function primarily as transcriptional activators in late G_1 and S phase (Fig. 3.10). E2F6 appears to function as a transcriptional repressor independent of Rb.

Specific roles for E2F family members have been revealed by genetic studies in both mice and flies. The *Drosophila Melanogaster* (fruit fly) genome encodes for only two E2Fs, and mutation of either of these E2Fs reveals their opposing functions in the regulation of target genes and cell cycle progression. The null mutation of dE2F1 (*Drosophila* E2F1) reduces E2F dependent transcription (i.e., the levels of known E2F targets were greatly downregulated) and decreases proliferation. In contrast, null mutation of dE2F2 increased E2F-dependent transcription. Thus dE2F1 appears to be analogous to mammalian E2F1, 2, and 3, and dE2F2 appears to be analogous to mammalian E2F4 and 5. Mice can be engineered with mutations (or "knockouts") in chosen genes using gene-targeting technology. Mice deficient for each E2F have been created, revealing distinct and often complex roles for E2Fs in controlling proliferation and development. Mouse cells lacking all E2F1, 2, and 3 exhibit decreased expression of E2F target genes and virtually no cell cycle progression, confirming positive roles for these E2Fs as transcriptional activators and cell cycle promoters. Mouse cells lacking E2F4 and 5 show inappropriate cell cycle entry in some contexts, consistent with roles for these E2Fs in Rb-dependent transcriptional repression and the maintenance of quiescence. Importantly, the clean separation of E2Fs into cell cycle promoting and inhibiting groups as presented above is clearly an oversimplification. For example, in some cell types, E2F2 can clearly function to limit cell cycle progression. Still, in general, the simple dichotomy within the E2F family appears to hold true. Finally, based on the overexpression of E2Fs and on gene knockout mice, roles for E2F1 and in some cases E2F3 in promoting apoptosis have been demonstrated, perhaps serving as a barrier to inappropriate E2F activation and proliferation.

E2F-Mediated Positive and Negative Feedback Loops. The CDK-Rb-E2F pathway, like many cell regulatory pathways, is regulated by positive and negative feedback loops that are activated by the CDK-Rb-E2F pathway. Positive feedback loops amplify signaling, enforcing the cell fate outcome. Negative feedback loops inhibit signaling, often functioning to limit the duration of the signal. E2F1, 2, and 3 are themselves E2F regulated, and thus the transcription of these E2Fs substantially increase in late G_1, resulting in more E2F-dependent transcription (Fig. 3.11). In

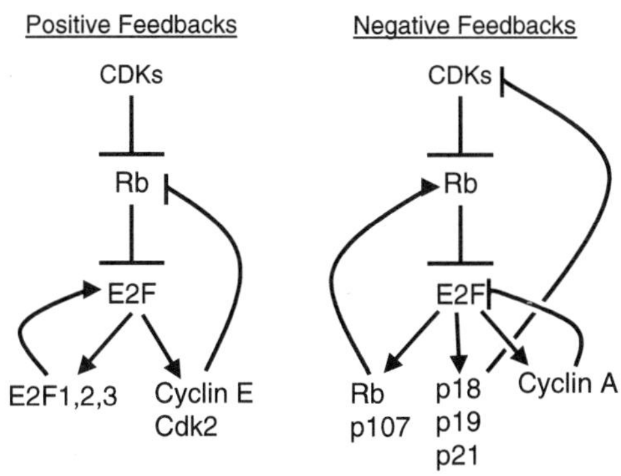

Figure 3.11. Feedback loops in the CDK-Rb-E2F pathway. E2F-dependent activation of E2Fs 1–3 and cyclin E/Cdk2 expression functions as a positive feedback loop by increasing E2F transcriptional activity and CDK-mediated inactivation of Rb, respectively. E2F activation of Rb family members and CDK inhibitors function as negative feedback loops by potentiating Rb-mediated inhibition of E2F activity. E2F activation of cyclin A and cyclin A/Cdk2-mediated phosphorylation of E2F/DP results in decreased E2F DNA binding.

addition cyclin E and Cdk2 are also E2F regulated with increased expression in late G_1. Increased expression of these kinase subunits contributes to increased cyclin E/Cdk2 activity, increased phosphorylation of Rb, and thus increased E2F-dependent transcription (which increases cyclin E and Cdk2 transcription, etc.). Hence the increased E2F-dependent transcription of E2Fs 1–3, cyclin E, and Cdk2 amplifies E2F activation, ensuring G_1 to S phase progression.

Cells also have an interest in preventing spurious CDK-Rb-E2F activation, which could amplify into inappropriate cell cycle progression. Also it is important for cells to inactivate E2F following G_1 to S progression in order to allow for proper progression into the G_2 and M phases. Not surprisingly then, E2F also activates negative feedback loops. The Rb and p107 genes are under E2F control and are upregulated in late G_1, functioning to limit E2F activation in the absence of sufficient CDK activation. A cell receiving insufficient signaling and thus insufficient CDK activation would presumably not be able to inactivate the increased levels of Rb and p107, preventing inappropriate cell cycle progression. Several CDK inhibitors, including $p21^{CIP1}$, $p18^{INK4C}$, and $p19^{INK4D}$, are also E2F regulated, functioning either to limit CDK activation or to downregulate cyclin/Cdks after their cell cycle job has been completed. An additional negative feedback loop involves E2F-dependent upregulation of cyclin A expression. Cyclin A/Cdk2-mediated phosphorylation of the E2F1,2,3/DP heterodimers decreases DNA binding, down-regulating E2F-dependent transcription as a cell progresses through S phase. In sum, positive and negative feedback loops promote and limit Cdk and E2F activation as a means to carefully regulate E2F activation and cell cycle entry.

Summary

The deregulation or mutation of a gene in cancer often highlights its critical role in normal proliferation control. The CDK-Rb-E2F pathway is almost invariably deregulated in human tumors, and indeed this pathway plays a critical role in regulating entry into and progression through the cell cycle. CDK activity functions to convert Rb/E2F repressor complexes into E2F transcriptional activators, resulting in the upregulation of a variety of genes required for cell cycle progression. Given the seminal role of E2F activation in the maintenance of quiescence as well as cell cycle entry, it is not surprising that the CDK-Rb-E2F pathway is both highly regulated and highly complex, with multiple family members of each pathway component differentially contributing to cell cycle control.

THE G_2/M TRANSITION

As outlined above, cell replication is controlled by regulating the timing of two major events within the cell cycle: DNA replication and mitosis.

While the section above addressed the regulatory cascades in G_1 and S phase, this section focuses on events important in the G_2/M transition, and in mitosis itself. In mitosis the nucleus is divided to produce two daughter nuclei, each genetically equivalent and containing the diploid number of chromosomes. There are four main stages of mitosis that will be briefly outlined (prophase, metaphase, anaphase, and telophase), and are discussed in more detail in Chapter 6. In prophase the chromosomes condense and the nuclear envelope begins to break down. At this point DNA replication has occurred, and thus each chromosome has duplicated (yielding a 4N DNA content). Each copy of the duplicated chromosome is called a sister chromatid, and they are joined at the centromere. Next, during metaphase, the fully condensed chromosomes align in the center of the cell. In anaphase, the sister chromatids that were still held together during metaphase separate and move to opposite poles of the mitotic apparatus, segregating one sister chromatid to each daughter cell. Finally, in telophase, the nuclear envelope that had broken down early in mitosis, reforms around the segregated chromosomes and the chromosomes decondense. Following telophase, the cytoplasm is divided through a process called cytokinesis, resulting in two daughter cells.

Similar to other processes already described in this chapter, events that lead up to and through mitosis require the action of heterodimeric protein kinases containing both a catalytic (cyclin-dependent kinase) and regulatory (cyclin) subunit. These kinases phosphorylate target proteins, either activating or repressing their activities, thereby coordinating the progression through mitosis.

Introduction to Mitotic Cyclins and Cdks

During late S and G_2, cells prepare for mitosis in part by increasing the levels of two regulatory subunits of the cyclin-dependent kinases (Cdks), cyclins A and B. Cyclin B is the main mitotic cyclin, and several forms of this cyclin have been identified. For simplicity, we will generally refer to the family of B type cyclins as cyclin B. Cyclin A, which is mostly involved in S phase events, is also necessary for cells to enter mitosis.

The cyclin B/Cdk1 complex was first identified as the maturation-promoting factor (MPF), an activity discovered in frog (*Xenopus laevis*) eggs that was capable of inducing meiosis in immature G_2 oocytes (see model organisms above). Although first discovered in *Xenopus*, this activity was found in mitotic cells from all species examined. For example, when cytoplasm from mitotically arrested mammalian somatic cells was injected into interphase cells, the interphase cells entered mitosis. Such experiments, as well as cell-fusion experiments, demonstrated that MPF was a diffusible factor that promoted the entry of cells into mitosis.

Some years later the MPF factors were determined to be cyclin B and Cdk1. Through experiments first carried out in sea urchin embryos, cyclin B levels were found to oscillate throughout the cell cycle, peaking in early mitosis, falling during anaphase, and then accumulating slowly during interphase until reaching a peak early in the next mitosis. Subsequent experiments in the frog system would conclusively demonstrate that the cyclin B component of MPF was the crucial protein required to regulate MPF activity.

Identification of the catalytic subunit of MPF came from studies performed in fission yeast (see model organisms above). An absence of activity of one of these genes, Cdc2, prevented entry into mitosis, whereas an excess of its activity caused early entrance into mitosis. This gene was subsequently isolated from numerous organisms, including human (the mammalian counterpart is referred to as cyclin-dependent kinase 1, or Cdk1, and will be used throughout the rest of this chapter), demonstrating the high conservation of cell cycle events throughout evolution. When MPF purified from *Xenopus* was tested for protein kinase activity, Cdk1 was found to be the catalytic component of this activity, and to work in concert with the regulatory component, cyclin B. As is true with other cyclins, Cdk1 must be bound to cyclin B to be catalytically active.

Cyclin B/Cdk1 Regulation

To ensure tight regulation of mitotic entrance and exit, cyclin B/Cdk1 activity must be exquisitely controlled. This is done at multiple levels and will be described below.

Cyclin B Synthesis, mRNA, and Protein Stability. As described previously, the levels of cyclin B protein oscillate throughout the cell cycle. This occurs, in part, because of both transcriptional control and regulation of mRNA stability. Cyclin B begins to be synthesized at the end of S-phase, and its mRNA is believed to be more stable in G_2 as compared to G_1. In addition cyclin B protein levels are controlled throughout the cell cycle via proteolytic degradation (see exit from mitosis). Together, these mechanisms ensure that cyclin B levels are tightly controlled throughout the cell cycle.

Phosphorylation of Cdk1. Once cyclin B has accumulated and can associate with Cdk1, the complex is further regulated by a number of phosphorylation and dephosphorylation events on the kinase. These events, however, can only occur if Cdk1 is bound to cyclin B. In order to be active, mammalian Cdk1 must be phosphorylated on threonine 161 (Thr161) and dephosphorylated on tyrosine 15 (Tyr15) and threonine 14 (Thr14).

The activating phosphorylation on Thr161 occurs in the T-loop, a flexible region of the kinase that, in its inactive state, blocks access of protein substrates to the active ATP-bound site. By analogy to cyclin A/Cdk2 (for which the three dimensional structure is known), this T-loop is believed to change in position once cyclin B has bound Cdk1, allowing for minimal activity. The activity is increased upon phosphorylation of the activating threonine in the T-loop (which increases in parallel with Cyclin binding), and presumably this causes additional conformational changes that greatly increase the affinity of the kinase for its substrates.

This activating phosphorylation is carried out by the Cdk-activating kinase (CAK). CAK is itself a cyclin/Cdk complex, composed of cyclin H and Cdk7, as well as a third component that allows for stabilization of the complex.

However, the cyclinB/Cdk1 complex, even if phosphorylated on Thr161, can be held in an inactive state by inhibitory phosphorylations on Thr14 and Tyr15, as is observed during the G_2 period of the cell cycle. These sites are within the region of the kinase that binds to ATP, and thus their phosphorylations inhibit catalytic activity. Phosphorylation of Thr14 inhibits Cdk1 activity by interfering with ATP binding, whereas phosphorylation on Tyr15 interferes with transfer of the phosphate to a bound substrate. These phosphorylation events, in particular, that on Tyr15, play very important roles in controlling the initiation of mitosis, and as will be outlined later in this chapter, are critical in engaging the G_2 checkpoint in response to DNA damage.

The Wee kinase described above (see model organisms) is one of a family of kinases (Wee1/Mik1) responsible for the inactivating phosphorylation on Tyr15 of Cdk1. Thus mutants in these kinases allow for premature entrance into mitosis, whereas overexpression of Wee1 (or related kinases) increases the length of G_2. The Myt1 kinase, which is related to Wee1, phosphorylates both Thr14 and Tyr15, with preference for Thr14. Together then, these kinases hold cyclinB/Cdk1 in an inactive state.

In late G_2 these phosphorylation events are counteracted by the dual specificity phosphatases from the Cdc25 family. Members of the Cdc25 family can dephosphorylate both Thr14 and Tyr15 of Cdk1, fully activating the cyclinB/Cdk1 complex and triggering the initiation of mitosis (Fig. 3.12).

Subcellular Localization as a Control Mechanism of CyclinB/Cdk1 Activity. While cyclin B/Cdk1 activity is regulated by multiple phosphorylation events on the Cdk subunit, it is additionally regulated by its location within the cell. Cyclin B1 is localized to the cytoplasm during S phase and G_2, and moves to the nucleus at the onset of mitosis. This subcellular localization of cyclin B1 is affected by a cytoplasmic retention signal (CRS) in the molecule as well as by continuous nuclear export during interphase. Phosphorylation of cyclin B1 at the G_2/M transition not only masks the CRS, thus allowing nuclear entry, but also inhibits interaction of cyclin B1 with the CRM nuclear export factor, thus inhibiting its export from the nucleus. So the overall activity of cyclin B1/Cdk1 is regulated by phosphorylation events on both the catalytic and the regulatory subunits.

Cyclin-Dependent Kinase Inhibitors (CKIs). These inhibitors are most known for their activity against the G_1 cyclins, and will be further discussed in Chapter 7. However, one particular CKI, p21, has been implicated in the G_2/M transition and is able to inhibit Cdk1 kinase activity. p21 has also been implicated in the G_1 and G_2 checkpoint responses to

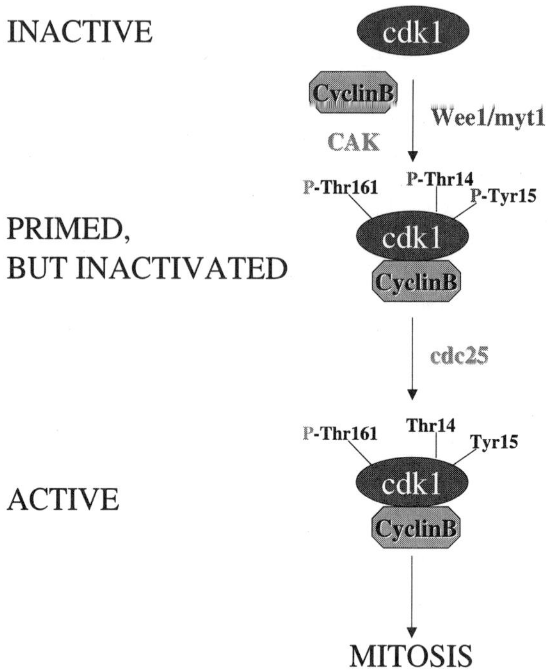

Figure 3.12. Phosphorylation events regulating Cdk1 activity. Cyclin B is the regulatory subunit of Cdk1. When Cdk1 is not bound to cyclin B, it remains inactive. The cyclin-activating kinase (CAK) phosphorylates cdk1 at Thr161, an event that is promoted by cyclin B binding to Cdk1, and stimulates its activity. However, if Cdk1 is also phosphorylated by Wee1/Myt1 (on Thr14 and Tyr15), the kinase remains inactive. In late G_2, the Cdc25 dual specificity phosphatases remove the inhibitory phosphorylations on Thr14 and Tyr15, allowing for Cdk1 activity and promoting entrance into mitosis. Kinases/phosphatases and their respective phosphorylation events that are activating for Cdk1 are listed in ■, whereas those that are inhibitory for Cdk1 are listed in ■.

DNA damage (see above, checkpoint control and cancer). The role of p21 in the DNA damage-induced G_2 checkpoint will be more thoroughly discussed below.

Targets of CyclinB/Cdk1

At mitosis, the architecture of the cell changes dramatically. Among other changes, the nuclear envelope disassembles, chromosomes condense, and actin microfilaments and microtubules are reorganized. This is believed to occur, in part, because of phosphorylation of a number of target proteins by cyclinB/Cdk1. Many of these targets remain unidentified, however, numerous targets have been discovered that are important for the mitotic process. For example, cyclin B/Cdk1 is known to phosphorylate lamin subunits, resulting in nuclear breakdown. CyclinB/Cdk1 is also important in phosphorylating Eg5, a motor protein required for establishing the bipolar spindle during mitosis. CyclinB/Cdk1 may additionally be involved in downregulating transcription during mitosis by inhibiting TFIIIB, a component of the polymerase III associated transcription complex.

Exit from Mitosis

For cells to exit mitosis, the processes of sister chromatid separation, spindle disassembly, and cytokinesis are required. These processes are initiated and coordinated by ubiquitin-dependent proteolysis of a number of critical regulatory proteins. For example, while cyclinB/Cdk1 activity is critical for entry into mitosis, its inactivation in anaphase and telophase is just as critical for mitotic exit. This inactivation occurs through proteolysis of cyclinB and is carried out though the ubiquitin pathway.

In the ubiquitin proteolysis pathway, two successive steps are required. First ubiquitin molecules are covalently attached to the substrate, and then the poly-ubiquitinated substrate is degraded by the 26S proteasome, leaving ubiquitin to be recycled and reused to tag additional proteins for degradation. The addition of ubiquitin occurs through a three-step mechanism, involving three enzymes, El–E3. The El enzyme, or ubiquitin-activating enzyme, forms a thioester bond with the ubiquitin molecule, thus activating it. Ubiquitin is then transferred to the E2 enzyme, or ubiquitin-conjugating enzyme, which works together with an E3 enzyme, or ubiquitin ligase enzyme, to covalently attach ubiquitin to lysine residues on the specific substrate to be degraded. The E3 proteins therefore provide the specificity of the system by recognizing the substrate to be tagged (Fig. 3.13).

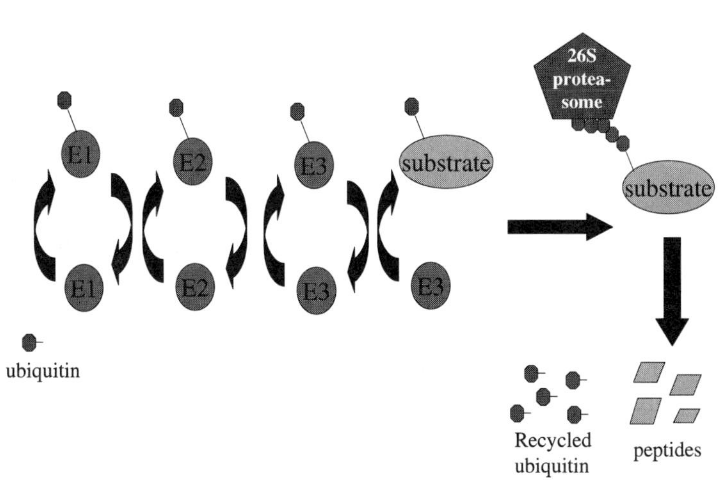

Figure 3.13. The ubiquitin-proteasome pathway. Ubiquitin is transferred to the substrate molecule via three enzymes (E1–E3). Once multiple ubiquitin molecules have been transferred, the substrate is recognized by the 26S proteasome, which degrades the substrate into peptides and releases the ubiquitin to be recycled.

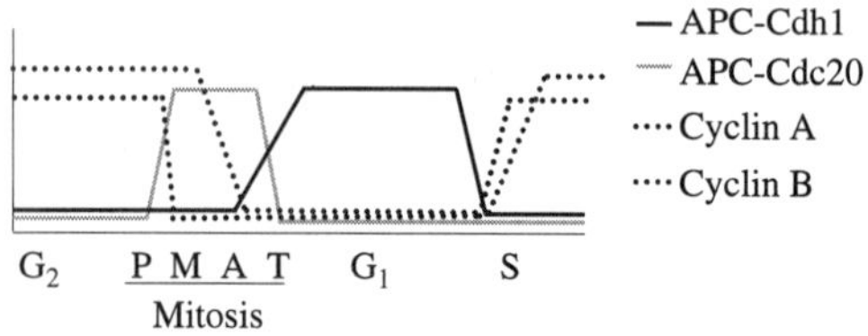

Figure 3.14. Role of the anaphase promoting complex (APC) in regulating mitotic cyclin levels. In prometaphase, APC (an E3 ubiquitin ligase) is activated by Cdc20 in a mechanism dependent on cyclin B/Cdk1 activity. This in turn initiates the degradation of cyclin A. Proteolysis of cyclin B also occurs in response to APC-Cdc20 but does not occur until metaphase. Degradation of the mitotic cyclins (which had been inhibiting Cdh1 activity) then allows for the activation of APC-Cdh1, which continues to keep mitotic cyclins low by targeting them for degradation, and also degrades Cdc20. APC-Cdh1 remains active until the end of G_1, when S phase Cdk activity causes its inactivation and allows mitotic cyclin protein levels to rise again. Solid lines represent protein activity, and dotted lines represent protein levels. P, prophase; M, metaphase; A, anaphase; T, telophase.

The E3 that is critical for ubiquitin-dependent proteolysis in mitosis is the anaphase promoting complex (APC). APC consists of at least 11 subunits, and only becomes fully active after binding to Cdc20, Cdh1, or related activators. APC is first activated at the onset of prometaphase by Cdc20. This activation is dependent on Cdk1 activity, and initiates the degradation of cyclin A. Proteolysis of CyclinB and other substrates is also carried out by APC-Cdc20, and is outlined in more detail below. In most species, Cdc20 is itself degraded during anaphase via APC-Cdh1, whose activity is activated at this stage by the degradation of cyclins A and B. Cdh1 thus keeps the APC active, and mitotic cyclin levels down, until the end of the next G_1 phase, when APC-Cdh1 is inactivated and mitotic cyclin levels are thus able to increase once again (Fig. 3.14). Cyclin degradation can be prevented through the mutation of sequences in their N-terminus called destruction boxes. Such stabilized cyclins prevent mitotic exit, demonstrating the importance of APC-mediated degradation in promoting cell cycle progression.

As mentioned above, APC activity is necessary for cyclin degradation. However, additional substrates of APC exist that are also critical for progression through mitosis. One such substrate is securin, whose destruction is essential to initiate anaphase and to regulate sister chromatid separation. While the mechanism-regulating sister chromatid separation is not identical in all eukaryotes, it is highly conserved and therefore the common components are outlined.

Central to APC's role in initiating anaphase is the destruction of securin, a molecule that prevents the separation of sister chromatids. In a metaphase chromosome, sister chromatids are attached to microtubules via a complex of proteins at the centromere, called the kinetochore. The kinetechore microtubules are attached at opposite ends to the spindle poles but are unable to pull the sister chromatids apart due to the fact that the sister chromatids are attached at their centromeres and at multiple positions along the chromosome arm by the cohesin protein complexes.

Securin inhibits anaphase by binding to the separase protein, and preventing it from cleaving cohesins. At the metaphase/anaphase transition, securin is degraded through an APC-dependent mechanism, thus releasing separase and allowing cleavage of the cohesin complex. This in turn allows the separation of sister chromatids and the onset of anaphase.

In summary, the APC is a critical component in mitotic exit. It not only triggers sister chromatid separation via the destruction of inhibitors of anaphase (securins) as described above but thereafter promotes additional mitotic events, such as spindle disassembly and mitotic exit via degradation of the mitotic cyclins.

Polo-like Kinases

While the above-mentioned material focuses primarily on cyclinB/Cdk1 and its role in mitosis, it should be noted that other kinases are also critical in numerous aspects of the G_2/M transition. One such family of kinases is the polo-like kinases. The founding member of this family is the *Drosophila* polo, but homologues have been identified in yeast, *Xenopus,* and mammalian cells. All polo-like kinases contain a region in the C-terminal noncatalytic domain called the polo box. Mutation of this box disrupts protein localization and as well as mitotic function.

There are three polo-like kinases in mammalian cells: PLK1, SNK, and Fnk/Prk. The functional mammalian homologue of *Drosophila* polo may be PLK1. It regulates a variety of mitotic events including the onset of mitosis, via activation of Cdc25c, and the DNA damage checkpoint, via its inactivation, and thus inhibition of Cdc25c activation. It is also known to activate the anaphase-promoting complex (APC), thus participating in mitotic exit, and to be involved in centrosome duplication and maturation. Taken together, the polo-like kinases clearly play numerous functions in both entrance into and exit from mitosis. More detailed reviews of polo-like kinases are listed at the end of this chapter.

DNA Damage-Induced G_2 Checkpoint

To ensure that the integrity of the genome is maintained, cell cycle progression must be prevented in the event of DNA damage. This occurs through the establishment of checkpoints, as introduced earlier (checkpoint concept as a surveillance mechanism). In response to DNA damage, cells can arrest in G_1, S, or G_2, depending on the phase in which the damage is sensed. In some instances, when DNA damage is very severe, cells will apoptose rather than arrest.

Studies on G_2 checkpoint regulation have identified a hierarchical signal transduction pathway consisting of sensors, signal transducers, and effectors that ultimately regulate Cdk1, thereby controlling mitotic entry. While it is generally thought of as a linear pathway, it should be noted that the pathway is more like a network of pathways that act together to carry out the checkpoint response to DNA damage.

Targets of the G_2 DNA Damage Checkpoint. As stated above, the main target of the G_2 arrest is Cdk1, which when inactivated, prevents cells from entering mitosis. Dephosphorylation of Tyr15 of Cdk1 is necessary for its activation. When the G_2 checkpoint is engaged, this dephosphorylation is prevented by the inactivation of the Cdc25c phosphatase via phosphorylation at serine 216 by upstream kinases Chk1 and Chk2. Phosphorylation of Cdc25c creates a binding site for the 14-3-3 proteins, which then sequester Cdc25c in the cytoplasm and prevent it from dephosphorylating and activating Cdk1. It should be noted, however, that this is not the sole event regulating Cdk1 activity in response to DNA damage, as expression of a nonphosphorylatable Cdc25c, with an alanine in place of the serine at 216, leads to only a modest effect on the G_2 DNA damage checkpoint.

DNA damage also regulates cyclin B levels, which decrease transiently after irradiation. In addition cyclin B localization is affected by DNA damage through a sequestration mechanism that is similar to what is observed with Cdc25c. The 14-3-3σ protein is increased after irradiation in a p53-dependent manner (see Chapter 19 for a discussion on the p53 tumor suppressor). p53 is itself a target of a stabilizing phosphorylation in response to DNA damage by Chk kinases and by the proximal kinases described below (ATM and ATR). When p53 is stabilized, it induces 14-3-3σ which then sequesters cyclin B in the cytoplasm, further inhibiting Cdk1 activity.

p53 has also been shown to upregulate expression of the cyclin-dependent kinase inhibitor, p21, in response to DNA damage (see Chapter 7 for a thorough discussion on cell cycle inhibitors) as well as GADD45. Initially p21 was believed to be primarily involved in the G_1 arrest, however, it is now known that p21 also plays a role in sustaining the G_2 arrest, possibly by inhibiting CAK-mediated Cdk1 activation. GADD45 binds to and dissociates the cyclin B/Cdk1 complex, further inhibiting its activity.

Sensors and Proximal Signal Transducers of the G_2 Checkpoint. Little is known about the sensors of DNA damage, and a more detailed discussion of these can be found in references regarding the G_2 checkpoint listed below. Signal transducers upstream of the Chk family of kinases include the phospho-inositide kinase (PIK)-related protein kinase ATM, originally cloned as a gene mutated in ataxia telangectasia, and a related kinase ATR. These two kinases play a central role in the DNA damage response, with ATM primarily involved in the response to irradiation and ATR primarily involved in the response to other genotoxic stress. Both kinases phosphorylate a number of target proteins important in arresting the cell cycle, including the Chk kinases and p53. In addition they have been shown to phosphorylate such targets as the tumor-suppressor protein BRCA1, which is involved in double-strand break repair (for a discussion on tumor-suppressor genes, see Chapter 17).

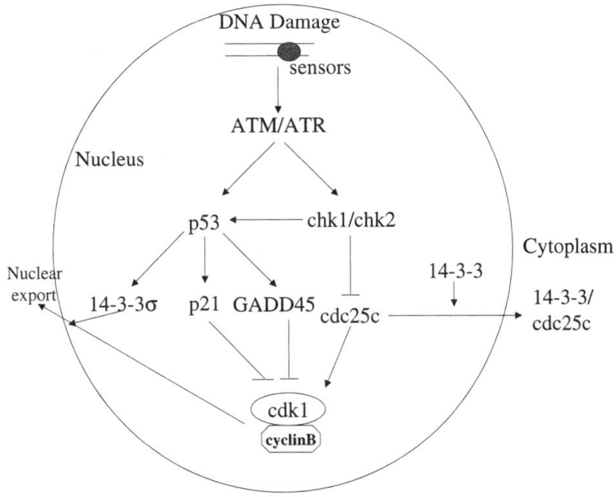

Figure 3.15. DNA damage-induced G_2 checkpoint. DNA damage is first recognized by sensor molecules on the DNA, which activate signal transducers such as ATM/ATR. These kinases then phosphorylate targets proteins, initiating two cascades that result in the inhibition of Cdk1 activity. One of these cascades is rapid and signals through the kinases Chk1 and Chk2, kinases that both phosphorylate the Cdc25c phosphatase, causing its interaction with 14-3-3 proteins and sequestration in the cytoplasm. This inhibits the activating dephosphorylation events on Cdk1. The other pathway involves phosphorylation and stabilization of p53 (which occurs through ATM/ATR and Chk1/Chk2), which causes the transcriptional activation of a number of target genes whose protein products inhibit Cdk1 activity.

Summary of G_2 Checkpoint. Many, but not all, of the mechanisms engaged to inhibit Cdk1 activity in response to DNA damage are listed above. ATM/ATR thus initiate two cascades that act in parallel to inactivate cyclinB/Cdk1 activity. The first, and presumably more rapid, involves the inactivation of Cdc25c by the Chk kinases. The second involves a mechanism to stabilize p53, which results in the transcriptional activation of numerous genes whose protein products act in multiple ways to inhibit Cdk1 activity (Fig. 3.15). By preventing Cdk1 activity in response to DNA damage, entrance into mitosis can be delayed until the damagc is repaired. This in turn protects the genomic integrity of the cell, ensuring that damaged DNA is not passed on to daughter cells. If this checkpoint is not maintained, the mutation rate of the cell will increase, which could ultimately result in tumorigenesis.

Spindle Assembly Checkpoint

The spindle assembly checkpoint prevents cells from entering anaphase until the chromosomes all have their kinetochores attached to spindle microtubules, ensuring that none will be left behind during mitosis. If the chromosomes are not properly aligned and cell division ensues, daughter cells may not receive one copy of each chromosome, which can result in aneuploidy. Such an event could be lethal during development, and could also lead to cancer.

The kinetochore is central to the spindle checkpoint. When the kinetochore is not bound to spindle microtubules or not under appropriate tension, it generates a checkpoint signal. This signal is a diffusible "wait anaphase" signal, which inhibits the APC from degrading proteins that are required for the onset of anaphase, such as the securins. Once all the kinetochores bind microtubules, this signal is inhibited and APC is activated via Cdc20, allowing for sister chromatid separation. It is important to note that all the kinetochores need to be bound, and that one unbound kinetochore will inhibit entrance into anaphase. Exactly how the "wait anaphase" signal is generated and how the checkpoint monitors microtubule attachment or tension at the kinetochore to release anaphase inhibition is still much debated.

What is known is that numerous proteins are bound to unattached kinetochores, including several members of the Mad (mitotic arrest deficient) and Bub (budding uninhibited by benzimidazole) families. Mad1 to 3 and Bub 1 and 3 were discovered using genetic screens in budding yeast to identify mutants that do not arrest in mitosis following drug-induced inhibition of spindle microtubule assembly, and several of their homologues have been identified in higher organisms.

It is believed that kinetochores are the sites at which two main proteins, Mad2 and BubRl (a mammalian protein kinase with homology to both Mad3 and Bub 1), gain their ability to interact with and inhibit Cdc20. These proteins can bind to unattached kinetochores or kinetochores under low tension, which promotes their inhibitory interaction with Cdc20. When kinetochore tension increases, or microtubule attachment occurs, the presence of these proteins on the kinetochores is lost, APC is activated via Cdc20 binding, and anaphase ensues. Many additional proteins are involved in this checkpoint, including kinases such as Mps1 and MAPK, and kinetochore motors such as CENP-E (see Chapter 6 for more thorough discussion). Continued investigations are rapidly shedding light on the complex interactions that regulate the spindle assembly checkpoint (Fig. 3.16).

CONCLUSION

The G_2/M transition involves a complex set of regulatory cascades that center around the activity of cyclinB/Cdk1. A central theme that arises out of this chapter is how similar the regulation is between each cell cycle phase, and how logical the molecular controls are. In each instance, kinases (the cyclin-dependent kinases) are controlled by regulatory molecules (the cyclins and the cyclin-dependent kinase inhibitors). Once activated, these kinases phosphorylate a number of target proteins that allow progression into the subsequent phase of the cell cycle. Together, these regulatory cascades ensure proper progression throughout the cell cycle. In addition each phase "monitors" its proper progression through

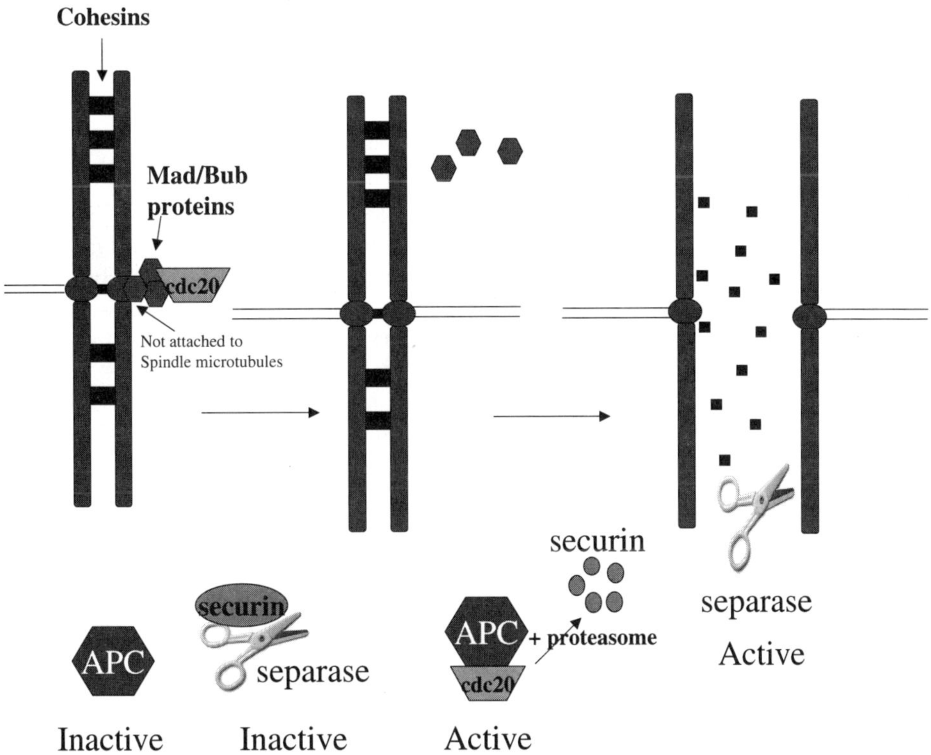

Figure 3.16. Spindle assembly checkpoint. When sister chromatids are not bound to spindle microtubules, the kinetochore proteins at the centromere bind to numerous members of the Mad and Bub family. This binding allows for interaction of at least two of the members (Mad2 and BubR1) with Cdc20, which is dependent on the Mads/Bubs cycling on and off of the kinetochore in mechanism that is not depicted here. Interaction of Mad2 or BuBR1 with Cdc20 inhibits its ability to associate with and activate the APC. When the spindle microtubules bind at the kinetochore, the presence of the Mad/Bub complex on the kinetochores is lost, APC is activated via Cdc20 binding, and securin is degraded. This releases separase, which cleaves the cohesins that were keeping the sister chromatids together and allows for their segregation.

the use of checkpoints, which maintain the integrity of the genome. It is no wonder, then, that many of the molecules involved in these regulatory cascades are altered in cancers, which exhibit uncontrolled proliferation and genomic instability.

REFERENCES

Introduction

Forsburg SL, Nurse P (1991): Cell cycle regulation in the yeasts *Saccharomyces cerevisiae* and *Schizosaccharomyces pombe*. In: GE Palade, BM Alberts, JA

Spudich (eds): Annual Review of Cell Biology, vol. 7. Palo Alto, CA: Annual Reviews, pp 227–56.

Lau CC, Pardee AB (1982): Mechanism by which caffeine promotes the lethality of nitrogen mustard. Proc Natl Acad Sci USA 79:2942–6.

Maller JL (1991): Mitotic control. Curr Opin Cell Biol 3:269–75.

Murray A, Hunt T (1993): The Cell Cycle: An Introduction. New York: Freeman.

Nasmyth K (1996): Viewpoint: Putting the cell cycle in order. Science 274:1643–5.

Nyberg KA, Michelson RJ, Putnam CW, Weinert TA (2002): Toward maintaining the genome: DNA damage and replication checkpoints. Ann Rev Genet 36:617–56.

Planas-Silva MD, Weinberg RA (1997): The restriction point and control of cell proliferation. Curr Opin Cell Biol 9:768–72.

Rossow PW, Riddle VGH, Pardee AB (1979): Synthesis of labile, serum-dependent protein in early G_1 controls animal cell growth. Proc Natl Acad Sci USA 76:4446–50.

Weinert TA, Hartwell LH (1988): The *RAD9* gene controls the cell cycle response to DNA damage in *Saccharomyces cerevisiae*. Science 241:317–22.

Zetterberg A, Larsson O, Wiman KG (1995): What is the restriction point? Curr Opin Cell Biol 7:835–42.

General G_1/S Transition and S Phase

Bartek J, Lukas J (2001): Pathways governing G_1/S transition and their response to DNA damage. FEBS Lett 490:117–22.

DeSalle LM, Pagano M (2001): Regulation of the G_1 to S transition by the ubiquitin pathway. FEBS Lett 490:179–89.

Ekholm SV, Reed SI (2000): Regulation of G(1) cyclin-dependent kinases in the mammalian cell cycle. Curr Opin Cell Biol 12:676–84.

Ford HL, Pardee AB (1998): The S-phase: Beginning, middle, and end; a perspective. J Cell Biochem Suppl 30–31:1–7.

Reed SI (1997): Control of the G_1/S transition. Cancer Surveys 29:7–23.

Rb and E2F

DeGregori J (2002): The genetics of the E2F family of transcription factors: shared functions and unique roles. Biochim Biophys Acta 1602:131–50.

Dyson N (1998): The regulation of E2F by pRB-family proteins. Genes Dev 12:2245–62.

Ferreira R, Naguibneva I, Pritchard LL, Ait-Si-Ali S, Harel-Bellan A (2001): The Rb/chromatin connection and epigenetic control: Opinion. Oncogene 20: 3128–33.

Trimarchi JM, Lees JA (2002): Sibling rivalry in the E2F family. Nat Rev Mol Cell Biol 3:11–20.

History of Mitotic Cyclins/CDKS

Doree M, Hunt T (2002): From Cdc2 to Cdk1: When did the cell cycle kinase join its cyclin partner? J Cell Sci 115 (Pt 12):2461–4.

Dunphy WG, Brizuela L, Beach D, Newport J (1988): The *Xenopus* Cdc2 protein is a component of MPF, a cytoplasmic regulator of mitosis. Cell 54:423–31.

Evans T, Rosenthal ET, Youngblom J, Distel D, Hunt T (1983): Cyclin: A protein specified by maternal mRNA in sea urchin eggs that is destroyed at each cleavage division. Cell 33:389–96.

Gautier J, Norbury C, Lohka M, Nurse P, Maller JL (1988): Purified maturation-promoting factor contains the product of *Xenopus* homolog of the fission yeast cell cycle control gene $cdc2^+$. Cell 54:433–9.

Gautier J, Minshull J, Lohka M, Glotzer M, Hunt T, Maller JL (1990): Cyclin is a component of maturation-promoting factor from *Xenopus*. Cell 60:487–94.

Gerhart J, Wu M, Kirschner MJ (1984): Cell cycle dynamics of an M-phase specific cytoplasmic factor in *Xenopus laevis* oocytes and eggs. J Cell Biol 98:1247–55.

Masui Y, Markert CL (1971): Cytoplasmic control of nuclear behavior during meiotic maturation of frog oocytes. J Exp Zool 177:129–45.

Nurse P, Bissett Y (1981): Gene required in G_1 for commitment to cell cycle and in G_2 for control of mitosis in fission yeast. Nature 292:558–60.

Lee MG, Nurse P (1987): Complementation used to clone a human homologue of the fission yeast cell cycle control gene cdc2. Nature 327:31–5.

CyclinB/CDK1 Regulation

Jackman MR, Pines JN (1997): Cyclins and the G_2/M Transition. Cancer Surveys 29:47–73.

Smits VAJ, Medema RH (2001): Checking out the G_2/M Transition. Biochim Biophys Acta 1519:1–12.

Proteolytic Events in the Cell Cycle and Exit from Mitosis

Ciechanover A, Orian A, Schwartz AL (2000): Ubiquitin-mediated proteolysis: Biological regulation via destruction. Bioessays 22:442–51.

Hershko A (1997): Roles of ubiquitin-mediated proteolysis in cell cycle control. Curr Opin Cell Biol 9:788–99.

King RW, Deshaies RJ, Peters JM, Kirschner MW (1996): How proteolysis drives the cell cycle. Science 274:1652–59.

Peters JM (2002): The anaphase-promoting complex: Proteolysis in mitosis and beyond. Mol Cell 9:931–43.

Polo-like Kinases

Dai W, Wang Q, Traganos F (2002): Polo-like kinases and centrosome regulation. Oncogene 21:6195–200.

Donaldson MM, Tavares AA, Hagan IM, Nigg EA, Glover DM (2001): The mitotic roles of Polo-like kinase. J Cell Sci 114 (Pt 13):2357–8.

Clover DM, Hagan IM, Tavares AA (1998): Polo-like kinases: A team that plays throughout mitosis. Genes Dev 12(24):3777–87.

Nigg EA (1998): Polo-like kinases: Positive regulators of cell division from start to finish. Curr Opin Cell Biol 10:776–83.

DNA Damage-Induced G_2 Checkpoint and Spindle Assembly Checkpoint

Abraham RT (2002): Cell cycle checkpoint signaling through the ATM and ATR kinases. Genes Dev 15:2177–96.

Clarke DJ, Gimenez-Abian JF (2000): Checkpoints controlling mitosis. Bioessays 22:351–63.

Melo J, Toczyski D (2002): A unified view of the DNA-damage checkpoint. Curr Opin Cell Biol 14:237–45.

O'Connell MJ, Walworth NC, Walworth AM (2000): The G_2-phase DNA-damage checkpoint. Trends Cell Biol 10:296–303.

Rudner AD, Murray AW (1996): The spindle assembly checkpoint. Curr Opin Cell Biol 8:773–80.

Shah JV, Cleveland DW (2000): Waiting for anaphase: Mad2 and the spindle assembly checkpoint. Cell 103:997–1000.

Zhou BB, Elledge SJ (2000): The DNA damage response: putting checkpoints in perspective. Nature 408:433–9.

CHAPTER 4

MEMBRANE RECEPTORS AND SIGNAL TRANSDUCTION PATHWAYS IN G_1: REGULATION OF LIVER REGENERATION AND T CELL PROLIFERATION

JOSEPH F. PORTER and DAVID T. DENHARDT
Cell Biology, Rutgers University, Nelson Laboratories, 604 Allison Road, Piscataway, NJ 08854

INTRODUCTION

Cycle: "an interval of time in which a certain succession of events or phenomena is completed, and then returns again and again, uniformly and continually in the same order" (*Webster's International Dictionary of the English Language*, 1903 edition). As applied to the conventional view of the cell cycle, composed of the G_1, S, G_2, and M phases, this definition is as good today as it was 100 years ago, at least for exponentially growing cells in culture with a continuously replenished medium. But is the cell cycle truly a cycle under normal physiological conditions?

At mitosis, the cell divides to become two cells, neither of which is precisely identical to the cell that began the cycle. There are several reasons for this. One is that as the cell proceeds through its replicative process, it is impacted upon by environmental signals, arising for example from changes in culture conditions or the hormonal milieu, that modulate signal transduction pathways and modify gene expression. Another is that during gene duplication and the reassortment of chromosomes into the daughter cells, there are random modifications in the structure and function of the genomes, for example, as the result of recombination or epigenetic changes affecting gene expression. Changes in DNA methylation, and possibly DNA damage, may also distinguish the two daughter cells. Finally some cytoplasmic constituents may not be equally distributed, and telomeres may become shorter. Despite these caveats,

Cell Cycle and Growth Control: Biomolecular Regulation and Cancer, Second Edition, Edited by Gary S. Stein and Arthur B. Pardee
ISBN 0-471-25071-6

however, we can still consider the cells under approximate steady state conditions to cycle through the G_1, S, G_2, and M phases. But what about non–steady state conditions? We will address this question after first considering some of the events that regulate the passage of the cell into and through G_1.

SIGNAL TRANSDUCTION PATHWAYS

The cell cycle will not repeat if the requisite exogenous and endogenous signals are not compatible with continued cell proliferation, which is a tightly regulated process requiring a cascade of intracellular events to occur in order for the process to proceed. The tightest regulation of the cell cycle occurs at a late G_1 checkpoint, referred to as the restriction point (review: Denhardt, 1999). This checkpoint is regulated at the intersection of several signal transduction pathways. The two most important are the Ras-Raf-MEK-ERK pathway and (for cells engaging an extracellular matrix) the integrin-FAK/Src pathway. These two pathways synergize to produce a sustained level of ERK activity, which in turn promotes up-regulation of cyclin D1 and down-regulation of the p16/p21 cyclin-dependent kinase (CDK) inhibitors (Zhu et al., 1996; Aktas et al., 1997; Assoian and Schwartz, 2001; Hulleman and Boonstra, 2001; Takuwa and Takuwa, 2001). The differential regulation of the various cyclins, cyclin-dependent kinases, and CDK inhibitors orchestrates progression through the cell cycle. Here we examine the Ras and FAK pathways on an individual level in terms of function and how they interact with each other to guide cells through the G_1/S cell cycle checkpoint.

The Ras-Raf-MEK-ERK Pathway

The Ras-Raf-MEK-ERK pathway is the most well-known signal transduction pathway. Ras functions as a GTP "switch" and is the main activator of the Ras-Raf-MEK-ERK pathway as well as other signal transduction pathways in the cell (Denhardt, 1996; Campbell et al., 1998; Gille and Downward, 1999; Liebmann, 2001). Ras cycles between an inactive GDP-bound state and an active GTP-bound state. The cycling between these two states is regulated by guanine nucleotide exchange factors (GEFs) and GTPase-activating proteins (GAPs). Ras is activated by (among other stimuli) receptor tyrosine kinases that are activated by growth factors. These RTKs (epidermal growth factor receptor being one of the best studied examples) will, when engaged by growth factors, phosphorylate themselves on tyrosines in their cytoplasmic domains. The phosphotyrosine moieties will then bind Src homology 2 (SH2)-containing adapter proteins such as Shc, which is in turn phosphorylated. Phosphotyrosine residues bind different SH-2 domains as a function of the local amino acid sequence within which the phosphotyrosine is embedded. As pictured in Figure 4.1, growth factor receptor 2 (Grb 2) can bind to certain phosphotyrosines either in the RTK or in Shc. Asso-

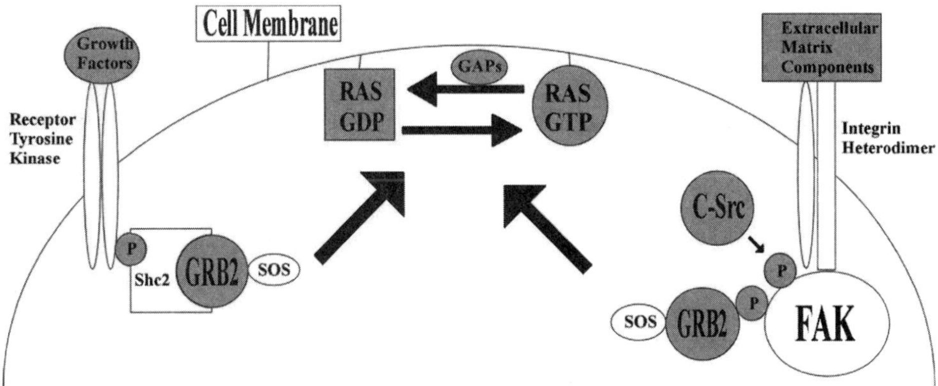

Figure 4.1. Interactions of receptor tyrosine kinases (RTK) and integrin-FAK complexes leading to Ras activation. Growth factor binding to the extracellular domain of an RTK causes autophosphorylation of the intracellular domain. Tyrosine phosphorylation of the intracellular domain creates recognition sites for Shc2, which will in turn bind Grb2/SOS. SOS activates Ras, forcing it to release the bound GDP and allowing GTP, present in excess, to bind. This exchange converts Ras from an inactive form to an active form. Additionally integrins, when engaged with elements of the extracellular matrix, will bind FAK, which when bound will phosphorylate itself. This autophosphorylation will allow cSrc to bind and phosphorylate FAK, providing a docking site for Grb2.

ciated with Grb2, the GEF son of sevenless (SOS) is thus relocated from the cytosol to the inner plasma membrane where it can engage membrane-bound inactive Ras, causing it to release bound GDP, which is then replaced by the more abundant GTP. The Shc, Grb2, and SOS proteins thus transmit the mitogenic signal from the cell surface receptor to Ras. Activated Ras is inactivated by a GAP (e.g., p120 GAP or NF1-GAP), which enhances the intrinsic GTPase ability of Ras, forming Ras-GDP. The relative activity of the GEFs and GAPs thus determines the overall activity of the Ras-Raf-MEK-ERK pathway.

The Ras protein contains several domains that are essential for its activity. There are several amino acids identified as being essential for Ras GTPase activity. They are Gly^{12}, Gly^{13}, Ala^{59}, and Gln^{61} (Scheffzek et al., 1998; Macaluso et al., 2002). Ras mutations in human cancers are most frequently found in these amino acids. Gln^{61} acts to stabilize the negative charge on the bound GTP molecule and carries one water molecule used to hydrolyze the GTP to GDP. Upon binding of GAP, the negative charge of the GTP is stabilized by GAP allowing for Gln^{61}, carrying one water molecule, to become repositioned to the active site of Ras where it can then catalyze hydrolysis of the GTP. Mutations in Gly^{12}, Gly^{13}, or Ala^{59} interfere with the conformational change induced by the binding of the GAP, and hence interfere with the repositioning of the Gln^{61} residue to the active site of Ras. Thus the amino acid residues at positions 12, 13, 59, and 61 are essential for Ras GTPase activity, and when mutated, they desensitize Ras to the action of GAP, resulting in a permanently active (oncogenic) GTP-bound Ras.

One Ras domain, residues 32–40, interacts with downstream Ras effector proteins (Campbell et al., 1998; Webb et al., 1998; Macaluso et al., 2002). Several studies show the importance of this domain for the activity of the Ras-Raf-MEK-ERK pathway. This domain undergoes a conformational change when Ras binds GTP allowing the domain to interact with downstream Ras effectors such as Raf. Specifically, the amino acids at position 37 and 40 have been shown, when mutated, to completely abolish the ability of Ras to interact with and activate Raf and hence activate MEK and ERK as well (Webb et al., 1998).

The second Ras domain essential for its activity is located at the carboxyl terminus of the protein (Macaluso et al., 2002). There are several post-translational modifications in this region that render this portion of the Ras molecule hydrophobic. A cysteine residue located at position 186 is first farnesylated, followed by a cleavage of the amino acid residue downstream of it and methylation of the resulting free carboxyl group. Cys^{181} and Cys^{184} are then palmitoylated. These post-translational modifications cause Ras to associate with the cytoplasmic side of the plasma membrane, allowing it to interact more efficiently with membrane-associated proteins.

Raf is the second signal transduction molecule in the Ras-Raf-MEK-ERK pathway. Raf comprises a family of serine/threonine kinases that has three members, A-Raf, B-Raf, and C-Raf or Raf-1. Raf-1 is the best studied of the three forms and will be discussed here. Raf-1, when not interacting with Ras-GTP, is kept in an inactive state by the adapter protein 14-3-3 (Kolch, 2000; Dhillon and Kolch, 2002). As illustrated in Figure 4.2, the adapter protein 14-3-3 interacts with two inhibitory phosphorylated sites on the Raf-1 molecule, ser^{259} and ser^{621}. The interaction with 14-3-3 appears to allow the cysteine-rich domain (CRD) to interact with the kinase domain of Raf-1, inhibiting its kinase activity. Raf-1 is indirectly phosphorylated and activated by active GTP-bound Ras. Raf-1 will bind to Ras and phosphatidylserine through its Ras binding domain (RBD), localizing Raf-1 to the cytoplasmic side of the plasma membrane. The binding of activated Ras to Raf-1 dissociates the 14-3-3 adapter protein from the phosphoserine residue at position 259 in Raf-1. The dissociation of the 14-3-3 adapter protein from the phosphoserine residue allows dephosphorylation of the phosphoser^{259} residue to occur through the action of protein phosphatase 2A (PP2A). Additionally the CRD domain of Raf-1 is presumed to dissociate from the Raf-1 kinase domain upon binding of Ras. At this point Raf-1 is in a state primed for activation by other kinases, which phosphorylate it at several sites within its kinase domain.

Stimulation of Ras results in the phosphorylation of Raf-1 at ser^{338} and tyr^{341} within the kinase domain as well as thr^{491}, ser^{494}, and ser^{499} within the activation loop (Kolch, 2000; Dhillon and Kolch, 2002). The first two phosphorylation sites within the kinase domain synergistically act to activate Raf-1 kinase activity. When these sites are both mutated,

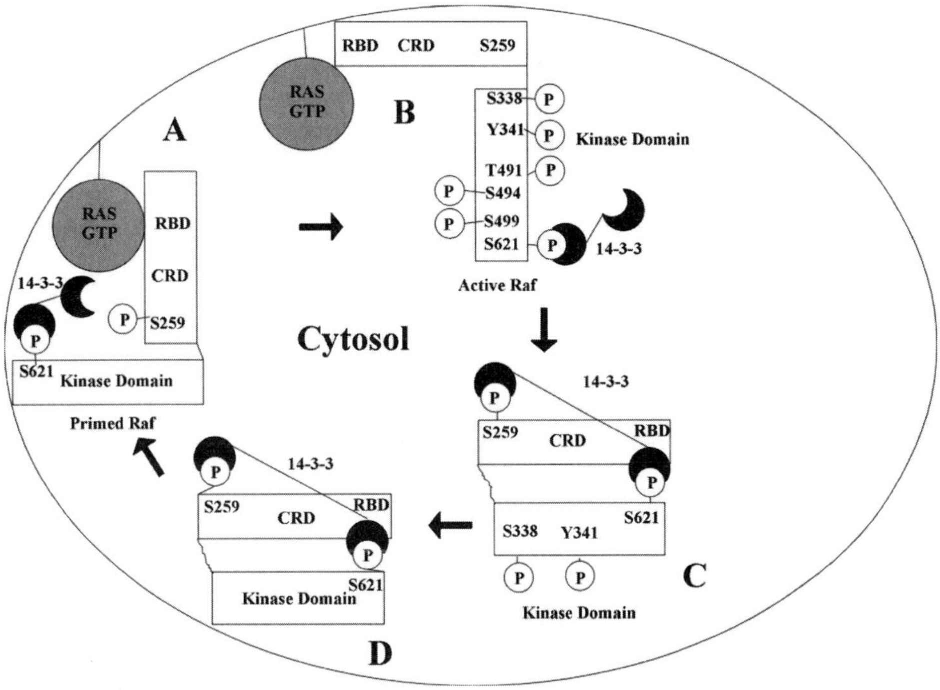

Figure 4.2. The Raf-1 cycle. The adapter protein 14-3-3 is represented by the two solid black semicircles connected by a line. Raf-1 is represented by two rectangles connected by a line. One of these rectangles is the kinase domain, and the other is the Ras binding domain (RBD) and the cysteine rich domain (CRD). **(A)** Priming of Raf-1. Active Ras binds to the RBD, forcing 14-3-3 to dissociate from ser^{259} and exposing the kinase domain to other kinases. **(B)** Activation of Raf. Raf-1 becomes phosphorylated at several sites within the kinase domain. **(C)** Initial deactivation of Raf-1. Raf-1 is partially dephosphorylated, allowing 14-3-3 to bind serine 259, causing a conformational change in Raf-1 such that the phosphorylation sites within the kinase domain are rendered unavailable. **(D)** Complete inactivation of Raf-1. Raf-1, as it occurs in the cytosol, is almost completely dephosphorylated and kept in a conformation by 14–3-3 such that the phosphorylation sites within the kinase domain are sequestered.

Raf-1 kinase activation by mitogens is almost completely abolished. Additionally, replacing the tyr^{341} residue with a phosphomimetic aspartic acid residue results in a highly active Raf-1, whereas the same replacement of ser^{338} results in only modest activation of Raf-1. The phosphorylation of ser^{338} is correlated with Raf-1 kinase activation as well as activation of downstream MEK and ERK kinases. However, the phosphorylation of ser^{338} does not correlate to the magnitude of activation of Raf-1 or its downstream effectors. These observations of ser^{338} phosphorylation imply that phosphorylation of this site is required for activation of the pathway but is not the only requirement. Phosphorylation of tyr^{341} has been observed to relieve repression of the kinase domain by the regulatory domain. Kinases that have been described either directly or indirectly to phosphorylate these sites include the Rac/Cdc42-activated kinase PAK for ser^{338} and members of the Src family of tyrosine kinases for tyr^{341} (Li et al., 2001).

Three other phosphorylation sites exist within the activation loop of Raf-1 (Kolch, 2000; Dhillon and Kolch, 2002). Two of these sites, thr^{491} and ser^{494}, are phosphorylated in a mitogen-dependent manner and are a factor in Raf-1 activation. Replacement of these two sites with aspartic acid enhances Raf-1 activation. The third phosphorylation site, ser^{499}, is involved in protein kinase C activation of Raf-1; mutation of the ser^{499} site eliminates Raf-1 autophosphorylation but does not interfere with MEK activation. The existence of these multiple phosphorylation sites within Raf-1 and the indication that the phosphorylation sites are involved in activating Raf-1 to differing levels of activity, implies that different mitogens could activate Raf-1 to different extents. This differential phosphorylation would taylor a Raf-1 activation level specific for the mitogen activating the pathway and the biological response specific for that mitogen. Multiple pathways converging on Raf-1 could synergize to produce maximal activation. Raf-1 can also be regulated negatively through the process of phosphorylation. Ser^{43} is unphosphorylated in active Raf-1; when phosphorylated, possibly by an MAPK, Raf-1 losses its affinity for binding Ras-GTP and is released from Ras. The release of Raf-1 allows Ras GAP to gain access to Ras-GTP, thereby downregulating Ras signaling.

In addition to regulation by phosphorylation, Raf-1 is also regulated through interaction with a number of other proteins. One of these proteins is known as suppressor of Ras-8 (SUR-8), which has been shown to form a complex with Ras-GTP and the kinase domain of Raf-1 (Kolch, 2000; Dhillon and Kolch, 2002). SUR-8 has been shown to enhance Raf-1 activation and is hypothesized to act as a physical link between Ras-GTP and the kinase domain of Raf-1, allowing Ras-GTP to directly signal to Raf-1. A second protein is known as kinase suppressor of Ras (KSR), which is believed to act as a scaffolding protein for the Ras-Raf-MEK-ERK pathway (Roy et al., 2002). KSR binds simultaneously to both Raf-1 and MEK1/2, facilitating activation of MEK1/2 by Raf-1. KSR has been shown to bind ERK1/2 also, setting the stage for efficient activation of ERK1/2 by MEK1/2. KSR itself is also regulated by phosphorylation of a serine residue located at position 392. Phosphorylation of this site allows the binding of the adapter protein 14-3-3, which acts to confine KSR in the cytoplasm.

Additional proteins that may be associated, directly or indirectly, with Raf-1 include heat shock protein 90 (Hsp90) and raf kinase inhibitor protein (RKIP). Hsp90 has been shown to prevent Raf-1 degradation. When Hsp90 activity is inhibited Raf-1 is ubiquitinated and degraded. It is hypothesized that Hsp90 serves as a chaperone for Raf-1 allowing it to maintain its structure and biological activity (Schulte et al., 1997). RKIP, on the other hand, serves as a negative regulator of Raf-1 activity. RKIP abolishes the interaction between Raf-1 and MEK1/2, thereby preventing downstream signaling from Raf-1. It is further hypothesized that KSR may be able to counteract this inhibitory effect of RKIP, permitting Raf-1 to phosphorylate MEK1/2 (Yeung et al., 2000). MEK1/2 is a dual specificity kinase capable of phosphorylating specific threonine and tyrosine residues of ERK1/2 in a defined domain.

The MEKs are regulated by their C-terminal regions, which appear to determine their cellular distribution and their ability to interact with Raf-1 and activate ERK1/2 (Cha et al., 2001). This C-terminal region contains a proline-rich region and multiple phosphorylation sites that presumably act in the regulation of MEK1/2. C-terminal deletion mutants were constructed to investigate how MEK-1 was regulated by its C-terminal region. These mutant proteins were found to be anomalously associated with membrane-bound compartments instead of being homogeneously distributed throughout the cytosol. These same C-terminal mutant proteins also could not be phosphorylated by constitutively active Raf-1 and lacked the ability to phosphorylate ERK1/2. The MEK kinases also associate with other regulatory proteins. An adapter protein called MP1 interacts with MEK-1 and ERK-1, bringing them into close proximity and enabling MEK-1 to phosphorylate and activate ERK-1 (Kolch, 2000; Dhillon and Kolch, 2002). MP1 appears to interact preferentially with MEK-1 and ERK-1, facilitating the activation of ERK-1 over ERK-2. The physiological consequences of this preferential activation of ERK-1 remain unknown.

ERK 1 and 2 are a pair of serine/threonine kinases that phosphorylate and regulate numerous proteins, including various transcription factors. ERK1/2, as mentioned earlier in this section, is brought into close proximity to MEK1/2 with the help of scaffolding proteins. The current paradigm hypothesizes that ERK1/2 are "guided" to their intracellular targets by way of docking domains (Barsyte-Lovejoy et al., 2002). These docking domains are located on the intracellular targets with which the ERKs interact. The ERKs themselves contain reciprocal docking domains, which interact with the docking domains on the target proteins. One type of docking domain has been characterized on some transcription factor targets of the ERKs. This docking domain consists of a number of submotifs, including a stretch of basic amino acids, an LXL submotif, and a stretch of hydrophobic amino acids. These submotifs can be subtly modified to ensure that they are activated by only one type of MAPK (e.g., p38 or ERK1/2).

Besides being directed by docking domains to recognize specific targets, ERKs are regulated in other ways. ERKs can be dephosphorylated by mitogen-activated protein kinase phosphatase-3 (MKP-3), which acts as a dual specificity phosphatase (Nichols et al., 2000). Binding of MKP-3 to ERKs is required for its activation and subsequent dephosphorylation of the ERKs. A series of studies done with p38/ERK chimera molecules indicates that MKP-3 binds to the C-terminal domain of the ERKs. Additionally the binding site for MKP-3 overlaps with the domain that grants substrate specificity to the ERKs. Consistent with this observation is the fact that some known ERK1/2 substrates, such as Elk-1 and $p90^{rsk}$, inhibit ERK-dependent activation of MKP-3.

The Ras-Raf-MEK-ERK pathway has been shown to integrate growth factor stimulation with the regulation of the G_1/S restriction point and progression through the cell cycle (Aktas et al., 1997). One line of evidence points to the regulation of CDK4, CDK6, cyclin D1 and $p27^{KIP1}$ by Ras. Activation of CDK4 and CDK6 by the binding of cyclin D1 and down-regulation of the CDK inhibitor $p27^{KIP1}$ promotes cell cycle progression. A dominant negative mutant of Ras was used to prove that induction of cyclin D1 gene expression and down-regulation of $p27^{KIP1}$ gene expression were the result of Ras signaling. Additionally this same study found that overexpression of cyclin D1 eliminated the need for active Ras for progression through the cell cycle. An increase in the activities of both CDK4 and CDK6 is achieved through the ability of Ras both to increase cyclin D1 and to decrease $p27^{KIP1}$ activities, thus driving the cells out of G_1 and into S phase if conditions are suitable.

Integrin Signaling

In order for a cell to progress through the G_1/S checkpoint, a sustained level of ERK activity must be maintained. Growth factors alone can only cause a transient increase in ERK activation, particularly in adherent cells. In order for a sustained level of ERK activity to be maintained, it has been shown, as illustrated in Figure 4.1, that integrin engagement with the extracellular matrix is essential (Schwartz and Assoian, 2001). Integrins are a large group of heterodimers where each integrin is made of one α and one β chain. There are 16 different α and 8 different β chains that can interact to form at least 22 known integrins (Hulleman and Boonstra, 2001). Integrins, in general, have a long extracellular domain and a short cytoplasmic domain with no intrinsic kinase activity. The cytoplasmic domain of integrins is what typically interacts with other intracellular proteins, which themselves are kinases or adapter proteins.

Several of the 22 known integrins have been shown to enhance cell proliferation when they are engaged with components of the extracellular matrix and the cells are exposed to growth factors. Integrin α5β1, for example, has been shown in fibroblasts to increase cyclin D1 expression by causing a sustained ERK activity in cells treated with growth factors (Roovers et al., 1999). Integrin αvβ3 has been shown in at least two separate cases to promote passage through the G_1/S restriction point. In one case fibroblasts were shown to proliferate in response to platelet-derived growth factor β (PDGFβ) when the αvβ3 integrin was engaged with vitronectin (Schneller et al., 1997). The second case involves vascular smooth muscle cells that are exposed to epidermal growth factor (EGF) with the αvβ3 integrin bound to tenascin-C (Jones et al., 1997). These cells were shown to proliferate when exposed to EGF with the αvβ3 integrin engaged with tenascin-C.

In order for integrins to cause a sustained ERK activation in cells treated with growth factors, the integrin signal cascade must somehow synergize with the Ras-Raf-MEK-ERK signaling pathway. Focal adhesion kinase (FAK), which interacts with the cytoplasmic domain of the engaged integrin, is a key player. When associated with the activated integrin, FAK autophosphorylates itself at tyr^{397} (Zhao et al., 1998). This phosphorylation allows FAK to bind to Src through Src's SH2 domain. Src in turn phosphorylates FAK at tyr^{925}. This second phosphorylation event allows FAK to bind the Grb2/SOS complex and thereby contribute to further activation of Ras signaling pathways, leading to increased ERK activity as described above. Integrin-mediated adhesion has been shown to promote the nuclear translocation of ERK 1/2 and the phosphorylation of Elk-1 (Aplin et al., 2001). This increased ERK activity leads, as already noted, to increased expression of cyclin D1 and decreased expression of the CDK inhibitor p21. Both an increased expression of cyclin D1 and a decreased expression of p21 are seen in fibroblasts transfected with constitutively active FAK. Adherence of cells to an extracellular matrix and engagement by growth factors are both essential for maximal activation of cyclin E-cdk2 and phosphorylation of Rb (Zhu et al., 1996).

ENTRANCE INTO THE CELL CYCLE: G_1 REGULATION

The most widely studied model of the cell cycle is based on the stimulation of serum-deprived quiescent (nonreplicating, contact-inhibited, out-of-cycle, G_0) fibroblasts in cell culture to re-enter the cell cycle in response to serum stimulation. Because of the substantial impact of serum factors on gene expression, this is not a good model for continuously cycling cells (Hofbauer and Denhardt, 1991; Cooper, 2003). Two aspects of the response to serum-stimulation have been defined: first the induction of competence, by PDGF, for example, which enables the cell to then respond to progression signals, such as provided by platelet-poor plasma, to progress into S phase (Olashaw and Pledger, 2002). Transformed cells, in contrast to primary cells and to untransformed but immortal cells, often do not require these signals and proliferate with less dependence on exogenous growth factors. Cells may not continue through the replicative cycle for many reasons, for example the inability to pass a checkpoint as the result of DNA damage or a missing signal.

As detailed throughout this book, and in an earlier extensive review (Denhardt, 1999), many of the physiological signals that control proliferation act in the G_1 phase of the cycle, stimulating the cells to pass through what is known as the "restriction point," proceeding to replicate their DNA and undergo mitosis. Cells that have ceased replicating because of the absence of required stimuli are defined as being in the G_0 phase, an out-of-cycle phase that may in some cases last a very long time. An example of cells in the G_0 phase are confluent, nontransformed fibroblasts in cell culture that are unable to proceed into S phase as the result of contact inhibition. These contact-inhibited cells are in a quiescent, nonproliferating phase but will re-enter the cell cycle when replated at a lower density. Many of the intracellular events occurring as the cells re-enter the cycle, phosphorylation of Rb, loss of E2F inhibition, and activation of cyclin-dependent kinases are described in detail elsewhere in this volume.

This widely accepted understanding of how the cell cycle is regulated has been challenged by Cooper (1999). Cooper, quite rightly, takes issue with some of the experimental results on which the conventional model is based, particularly with regard to methods used to synchronize the cell cycle (Cooper, 2003). It is likely that some of the biochemical processes exhibiting an apparent cell cycle dependence in synchronized cells are the consequence of the synchronization process itself—possibly even Rb phosphorylation (Cooper and Shayman, 2001). The alternative view that Cooper vigorously postulates is that there are no G_1-specific controls regulating the division cycle; instead, he proposes that a triggering substance begins to accumulate at the start of one S phase, and that when it reaches a critical level, the next S phase is initiated. The identity of this cell cycle regulator and the factors that control its continuous accumulation in parallel with the increase in cell mass remain to be defined. Although this "continuum model" is consistent with many aspects of the kinetics of the cell cycle, as documented in Cooper's publications, it will remain only a hypothesis until the underlying biochemistry is clarified. Cooper argues that out-of-cycle (G_0) cells do not exist—that they cannot be distinguished experimentally from slow-growing cells with a G_1 phase amount of DNA that are slowly accumulating the putative trigger substance. Although Cooper has successfully challenged some of the underlying support for the current paradigm of cell cycle control, in our view the model elaborated throughout this volume provides a satisfactory framework on which to build our understanding of cell cycle control.

In the following paragraphs we describe two physiologically relevant examples of cell cycle regulation controlled in part by membrane receptors. In both cases cells remain quiescent for long periods (in G_0 if you will), but when appropriately stimulated, they initiate proliferation via mechanisms, using cyclins and cyclin-dependent kinases, that appear very similar to those defined in serum-stimulated quiescent fibroblasts. Once proliferation is underway, with many of the regulatory proteins present and appropriately phosphorylated, the cells continue to proliferate under their own momentum until conditions no longer support cell replication.

Liver Homeostasis

In the healthy adult mammal the size of the liver is strictly regulated. Excise part of the liver and the remainder regenerates the missing portion; transplant in excess tissue, and an equivalent portion of the liver regresses (Zimmermann, 2002). How is this accomplished? Generally, liver regrowth entails substantially increased proliferation of several cell types, including hepatocytes, cholangiocytes, Kupffer cells, and sinusoidal endothelial cells, followed by the regeneration of relatively normal liver structure. The mechanisms regulating these processes are not fully understood, but they do appear to involve several cytokines, including hepatocyte growth factor/scatter factor (HGF/SF), IL6, TNFα as well as several other less well-characterized factors (e.g., augmenter of liver regeneration, hepassocin, hepatopoietin, and hepatic stimulator substance) signaling through cell surface receptors to control liver cell growth and proliferation. In cases of severe liver damage, proliferation of liver stem cells (oval cells) is also elicited (Vessey and Hall, 2001). Other cell types in the body have not been reported to initiate proliferation in response to partial hepatectomy, so there must be something special about the set of factors engaging the unique constellation of receptors on the surface of liver cells that controls their proliferation with great selectivity.

Hepatocytes in the normal liver are largely in the G_0 phase, though there is a slow rate of turnover (~0.1%) such that the entire liver is renewed about once a year, presumably representing replacement of senescent/necrotic/apoptotic cells (Vessey and Hall, 2001). HGF, which engages the cMet receptor, is a significant stimulator of hepatocyte proliferation. However, given that cMet is also found on most epithelial cells and that HGF is produced by cells in other organs (kidney, spleen, lung), it can only be part of the process. Thus HGF, which is also required in early development, should not be considered an uniquely liver-specific factor (Zimmermann, 2002). Liver regeneration after partial hepatectomy begins with the activation (or priming) of many of the hepatocytes to enter early G_1, initially in response to IL-6, TNFα and likely other unidentified factors. Production of both IL-6 and TNFα by the liver is rapidly increased; the importance of both cytokines is reflected in the fact that IL-6-deficient and TNF-receptor-1-deficient mice are impaired in their ability to support liver regeneration. Early intracellular events include activation of the NFκB, STAT3, AP-1, c-Fos, c-Jun, c-Myc, and CEBP transcription factors.

Progression through G_1 of primed, committed hepatocytes requires HGF and TGFα signaling, after which the replicative process proceeds autonomously under the control of the cyclins and cyclin-dependent kinases (Rozga, 2002). TGFα (whose synthesis is stimulated by TNFα and HGF) binds to the EGF receptor, and EGFR kinase activation appears essential for the mitogenic activity of HGF (Scheving et al., 2002). Survivin (TIAP), a member of the inhibitor of apoptosis protein (IAP) family, is strongly induced during liver regeneration; it enhances Rb phosphorylation and cell cycle progression (Deguchi et al., 2002). Expression is highest in the G_2/M phase, where it is thought to act to maintain cell viability during mitosis.

Proliferation ceases when the original liver mass is regained, presumably the consequence of the re-establishment of the necessary cytokine/hormone balance. Two factors implicated in the termination of liver regeneration are TGFβ and activin, which suppress cell growth at a distinct set point (Zimmermann, 2002). In mice lacking the ability to produce Skp2, an F-box protein of the SCF ubiquitin ligase complex that targets $p27^{Kip1}$, a cyclin-dependent kinase inhibitor, for degradation, restoration of liver mass after partial hepatectomy is accomplished by an increase in hepatocyte polyploidy and cell mass (Minamishima et al.,

2002). These results suggest that cell proliferation and cell growth are independently regulated, and that the loss of proliferative ability can be compensated by an increase in cell size. Lack of Skp2 apparently suppressed cell division in this model, presumably because of an increase in $p27^{Kip1}$ levels.

Fas/FasL-mediated apoptosis is a major regulator of liver cell homeostasis, as evidenced, for example, by the fact that Fas-deficient mice exhibit excessive liver growth (Desbarats and Newell, 2000). Fas (CD95), and also TNFR1, are cell surface receptors that typically trigger apoptosis when engaged by their cognate ligand, FasL and TNF-α respectively. The FADD (Fas-associated death domain) protein binds the activated receptor, mediating apoptosis via caspase 8. Interestingly, in the livers of mice subjected to partial hepatectomy, Fas stimulates cell growth and liver regeneration, apparently as follows: Cytokines generated in response to liver damage modify Fas signaling by augmenting the action of FLIP (FLICE-inhibitory protein) and reducing the extent of Fas-induced apoptosis. FLIP inhibits Fas-induced apoptosis by binding the FADD and preventing its association with caspase 8/FLICE (Seino et al., 2001).

Immune Cells

The survival and proliferation of T and B cells recognizing specific antigens are subject to complex regulatory controls integrating signals delivered to cell surface receptors by various cytokines and antigen-presenting cells. T cells develop in the thymus and are responsible for cell-mediated immunity and aspects of humoral immunity, functioning to kill pathogens and abnormal cells. B cells develop in the bone marrow and produce antibodies to foreign antigens. Mature T and B cells, capable of responding to specific antigens, are the result of intricate differentiation and selection processes occurring primarily in the thymus, marrow, and spleen. Because more is known about T cells, the rest of this discussion will focus on them, specifically the cells (naïve, memory) that are stimulated to proliferate upon encountering their target antigen.

Mature T cells circulating between the blood, lymph and lymphoid organs are quiescent. Similar to quiescent fibroblasts that require competence and progression signals as described above, T cells require two types of signals to become fully active and to proliferate (Appleman et al., 2000). The first (competence) signal is the result of the engagement of the T cell receptor/CD3 complex by its cognate antigen presented by an antigen-presenting cell (macrophage, dendritic cell) together with co-stimulation of CD28 by other surface molecules on the antigen-presenting cell. This activates several signal transduction pathways that enhance cdk/cyclin activities, that propel the cell further into G_1, and that stimulate IL-2 production and expression of the IL-2 receptor. IL-2 transcription is dependent in part on JNK activation via a Rac-dependent pathway stimulated by PKC-θ and calcineurin in the activated T cell (Werlen et al., 1998). Autocrine signaling resulting from the interaction of IL-2 with its receptor then provides the second required (progression) signal that results in robust T cell proliferation. However, optimal TCR and CD28 engagement can elicit IL-2-independent cell cycle progression also (Colombetti et al., 2002). If the T cell does not progress through the cell cycle and fails to proliferate in response to its cognate antigen, then it becomes anergic, unresponsive to subsequent stimulation.

Activation of T cells has been studied extensively using nonspecific activators such as concanavalin A or antibody crosslinking. These act nonspecifically (i.e., independently of antigen) on the T cell receptor and up-regulate genes such as the α subunit of the IL-2 receptor and Jak3, which together allow the T cell to become responsive to IL-2 (Ellery and Nicholls, 2002). Osteopontin (early T cell activation gene 1) is also expressed at high levels by activated T cells; its functions include supporting cell survival, costimulating (with anti-CD3) T cell proliferation, and regulating autoimmunity by modulating Th1/Th2 ratios (Ashkar et al., 2000; O'Regan et al., 2000; Denhardt et al., 2001; Chabas et al., 2001).

An army of signal transduction intermediates mediate signaling downstream of the TCR/CD3 complex (Cantrell, 2002). Among the first events after engagement of the receptor are activation of the Src family kinases p59fyn and p56lck, leading to phosphorylation of immunoreceptor tyrosine activation motifs (ITAMs) in the TCR complex that provide protein tyrosine phosphate docking sites for the SH-2-containing protein ZAP-70/Syk, which in turn phosphorylates tyrosines in the adaptors SLP76 and LAT (linker for activation of T cells). LAT is a 37 kDa integral membrane protein that, when phosphorylated by ZAP-70/Syk, recruits PLC-γ1, Grb2 and Gads to the plasma membrane (Zhang et al., 1998; 2000). Gads nucleates multi-protein complexes that are required for tyrosine kinase-dependent signaling in immune cells: it may also represent a point of modulation for these pathways through the activation of caspase-dependent signaling events. The importance of PKC signaling in lymphocyte activation is evidenced by the ability of phorbol esters to mimic many aspects of antigen receptor triggering (Cantrell, 2002). One of the negative regulators of T cell activation and mitogenic signaling is cAMP, whose levels are increased, for example, by prostaglandin E or HIV infection. PKA, the cAMP-dependent protein kinase, colocalizes with the TCR/CD3 complex and inhibits Lck-mediated tyrosine phosphorylation by activating Csk, the c-src family kinase that negatively regulates Lck (Vang et al., 2001).

Phospholipase-Cγ1, activated by tyrosine phosphoryation, cleaves the membrane phosphoinositide PtdIns(3,4,5)P_3 to generate inositol trisphosphate and diacylglycerol (DAG), which mobilize calcium from intracellular stores and activate protein kinase C family members respectively. DAG also binds and activates serine protein kinase D and RasGRP (guanine nucleotide releasing protein) (Ebinu et al., 2000). Ras is activated not only by RasGRP but also by PKC-dependent inhibition of RasGAP proteins and activation of SOS via the adaptor Grb2 engaged by tyrosine-phosphorylated proteins. Targets of the Ras/Raf/MEK/Erk pathway include the transcription factors AP-1 and NFAT. Substantial

NFAT activation and translocation to the nucleus depend strongly on dephosphorylation by calcineurin, which is stimulated by increases in intracellular calcium levels. A critical but poorly understood aspect of regulation of these protein phosphorylation cascades is the role of protein phosphatases in reversing the action of the protein kinases. Protein scaffolds (e.g., SLP-76, LAT) and adaptors (e.g., Grb2, Gads) play key roles in mediating signal transduction pathways, making them more efficient and directing the signals toward specific downstream targets. However, somewhat discouragingly, Burack et al. (2002) conclude in a recent review that "so many molecules have been shown to interact with the various scaffolds and so many molecules have been shown to interact with multiple scaffolds that building some sort of model of a highly specific structure seems difficult (p. 314)."

Boussiotis and colleagues have recently shed considerable light on the role CD28 plays as a co-stimulator of cell cycle progression and T cell expansion using primary peripheral blood human T lymphocytes stimulated by crosslinking with rabbit anti-mouse Ig after the cells were exposed to anti-CD3 and/or anti-CD28 (Appleman et al., 2000, 2002). They found that TCR/CD3 activated ERK1/2 via an MEK pathway likely controlled by Ras, whereas CD28 activated PI3K. Products of PI3K include PtdIns(3,4,5)P_3 and PtIns(3,4)P_2, which bind pleckstrin homology domains and induce relocalization of proteins, in this case TEC family protein kinases, to defined areas of the plasma membrane. Also downstream of TCR/CD3 are PDK1 and PKB/c-Akt, which phosphorylate a number of proteins including the ribosomal S6 kinase and proteins controlling both cell cycle progression and cell survival (Cantrell, 2002). The GTPases Rac and Rho are also regulated by PI3K signals. Cell proliferation was stimulated only when both the RAS and PI3K pathways were activated.

Passage of the cells from their quiescent state into the early stages of G_1 required first the degradation of the cdk inhibitor $p27^{kip1}$ by an ubiquitin-dependent, proteasome-mediated pathway, likely initiated by Erk1/2 phosphorylation of one or more of the proteins regulating ubiquitination. Degradation of $p27^{kip1}$, which much evidence suggests plays a key role in maintaining T cell quiescence, releases cyclinD2/cdk4-cdk6 and cyclin E/cdk2 from inhibition, allowing them to phosphorylate target proteins such as Rb. As discussed elsewhere in this volume, Rb phosphorylation is one of the key events leading into S phase. Expression of IL-2 and its receptor are up-regulated in late G_1, providing the necessary stimulus to propel the cells through the G_1/S transition. Proliferation will continue, driven by IL-2, until (in real life) the antigenic stimulus is eliminated or the cells become replicatively senescent and die by apoptosis.

Activation induced cell death is a major regulator of T cell numbers at the end of an immune response when encounters with antigen become less frequent. IL-2, which stimulated the proliferation of antigen-stimulated cells, now functions to sensitize the cells to apoptosis, possibly by increasing Fas/FasL expression (Thome and Tschopp, 2001; Budd, 2002). As discussed above with regard to liver homeostasis, TNF receptor family members such as Fas (CD95/APO-1) signal via FADD and caspase-8 to drive cells into apoptosis. Inhibiting this apoptotic pathway is FLIP, which via its death effector domains interacts with a number of molecules involved in the apoptotic response. Several studies have implicated FLIP in a proliferative response, possibly because of its ability to activate NF-κB and AP-1.

One of the important lessons learned recently regarding cell signaling mechanisms is that the same receptor (TCR in this case) can differentially stimulate intracellular signal transduction pathways as a function of the affinity/avidity of the specific ligand (Werlen et al., 2000; Mariathasan et al., 2001; Werlen et al., 2003). During the maturation of thymocytes, ligands that bind strongly to the TCR cause rapid and abundant phosphorylation of downstream mediators, whereas weakly binding ligands deliver a weaker but more prolonged stimulus. Positive selection (= cell survival) correlates with a sustained low-level activation of ERK induced by low-affinity ligands, whereas negative selection (= cell death) is associated with a strong transient activation of ERK induced by high affinity ligands. Kinetic parameters that determine the specificity of the TCR-generated signal include the off-rate of the bound ligand, the rate of co-receptor recruitment, the efficiency of TCR-signalosome formation, and the kinetics of ERK, JNK, and p38 activation. Regulatory subtleties such as these will undoubtedly be operative in many other situations also, allowing essentially the same set of signal transduction pathways to orchestrate different outcomes, even in the same cell type. Clearly much remains to be learned about the complexities of signal transduction pathways and how they control cell behavior.

ACKNOWLEDGMENT

Research in the authors' laboratory has been supported by grants from the National Institutes of Health and the Charles and Johanna Busch Biomedical Research Fund. We thank Heide Ford, Guy Werlen, and Arthur Zimmermann for comments on parts of the manuscript.

REFERENCES

Aktas H, Cai H, Cooper G (1997): Ras links growth factor signaling to the cell cycle machinery via regulation of cyclin D1 and the cdk inhibitor p27KIP1. Mol Cell Biol 17:3850–7.

Aplin A, Stewart S, Assoian R, Juliano R (2001): Integrin-mediated adhesion regulates ERK nuclear translocation and phosphorylation of Elk-1. J Cell Biol 153:273–81.

Appleman LJ, Berezovskaya A, Grass I, Boussiotis VA (2000): CD28 costimulation mediates T cell expansion via IL-2-independent and IL-2-dependent regulation of cell cycle progression. J Immunol 164:144–51.

Appleman LJ, van Puijenbroek AA, Shu KM, Nadler LM, Boussiotis VA (2002): CD28 costimulation mediates down-regulation of p27kip1 and cell cycle progression by activation of the PI3K/PKB signaling pathway in primary human T cells. J Immunol 168:2729–36.

Ashkar S, Weber GF, Panoutsakopoulou V, Sanchirico ME, Jansson M, Zawaideh S, Rittling SR, Denhardt DT, Glimcher MJ, Cantor H (2000): Eta-1 (osteopontin): An early component of type-1 (cell-mediated) immunity. Science 287:860–4.

Assoian R, Schwartz M (2001): Coordinate signaling by integrins and receptor tyrosine kinases in the regulation of G_1 phase cell cycle progression. Curr Opin Genetics Dev 11:48–53.

Barsyte-Lovejoy D, Galanis A, Sharrocks AD (2002): Specificity determinants in MAPK signaling to transcription factors. J Biol Chem 277:9896–903.

Budd RC (2002): Death receptors couple to both cell proliferation and apoptosis. J Clin Invest 109:437–41.

Burack WR, Cheng AM, Shaw AS (2002): Scaffolds, adaptors and linkers of TCR signaling: Theory and practice. Curr Opin Immunol 14:312–6.

Campbell S, Khosravi-Far R, Rossman K, Clark G, Der C (1998): Increasing complexity of Ras signaling. Oncogene 17:1395–413.

Cantrell DA (2002): T-cell antigen receptor signal transduction. Immunology 105:369–74.

Cha H, Lee EK, Shapiro P (2001): Identification of a C-terminal region that regulates mitogen-activated protein kinase kinase-1 cytoplasmic localization and ERK activation. J Biol Chem 276:48494–501.

Chabas D, Baranzini SE, Mitchell D, Bernard CC, Rittling SR, Denhardt DT, Sobel RA, Lock C, Karpuj M, Pedotti R, Heller R, Oksenberg JR, Steinman L (2001): The influence of the proinflammatory cytokine, osteopontin, on autoimmune demyelinating disease. Science 294:1731–5.

Colombetti S, Benigni F, Basso V, Mondino A (2002): Clonal anergy is maintained independently of T cell proliferation. J Immunol 169:6178–86.

Cooper S (1999): The continuum model and G_1-control of the mammalian cell cycle. In: L Meijer, S Guidet, M Philippe (eds): Progress in Cell Cycle Research, Vol. 4. New York: Plenum Press, pp 27–39.

Cooper S (2003): Reappraisal of serum starvation, the restriction point, G_0 and G_1 phase arrest points. FASEB J 17:333–40.

Cooper S, Shayman JA (2001): Revisiting retinoblastoma protein phosphorylation during the mammalian cell cycle. Cell Mol Life Sci 58:580–95.

Deguchi M, Shiraki K, Inoue H, Okano H, Ito T, Yamanaka T, Sugimoto K, Sakai T, Ohmori S, Murata K, Furusaka A, Hisatomi H, Nakano T (2002): Expression of survivin during liver regeneration. Biochem Biophys Res Commun 297:59–64.

Denhardt DT (1996): Signal-transducing protein phosphorylation cascades mediated by Ras/Rho proteins in the mammalian cell: The potential for multiplex signalling. Biochem J 318:729–47.

Denhardt DT (1999): Signal transduction pathways and regulation of the mammalian cell cycle: Cell type-dependent integration of external signals. In: GS Stein, R Baserga, A Giordano, DT Denhardt (eds): The Molecular Basis of Cell Cycle and Growth Control. New York: Wiley-Liss, pp 225–304.

Denhardt DT, Noda M, O'Regan AW, Pavlin D, Berman JS (2001): Osteopontin as a means to cope with environmental insults: regulation of inflammation, tissue remodeling, and cell survival. J Clin Invest 107:1055–61.

Desbarats J, Newell MK (2000): Fas engagement accelerates liver regeneration after partial hepatectomy. Nat Med 6:920–3.

Dhillon A, Kolch W (2002): Untying the regulation of Raf-1 kinase. Arch Biochem Biophys 404:3–9.

Ebinu JO, Stang SL, Teixeira C, Bottorff DA, Hooton J, Blumberg PM, Barry M, Bleakley RC, Ostergaard HL, Stone JC (2000): RasGRP links T-cell receptor signaling to Ras. Blood 95:3199–203.

Ellery JM, Nicholls PJ (2002): Alternate signalling pathways from the interleukin-2 receptor. Cytokine Growth Factor Rev 13:27–40.

Gille H, Downward J (1999): Multiple Ras effector pathways contribute to G_1 cell cycle progression. J Biol Chem 274:2033–40.

Hofbauer R, Denhardt DT (1991): Cell cycle-regulated and proliferation stimulus-responsive genes. Crit Rev Eukaryot Gene Expr 1:247–300.

Hulleman E, Boonstra J (2001): Regulation of G_1 phase progression by growth factors and the extracellular matrix. Cell Mol Life Sci 58:80–93.

Jones P, Crack J, Rabinovitch M (1997): Regulation of tenascin-C, a vascular smooth muscle cell survival factor that interacts with the alphavbeta3 integrin to promote epidermal growth factor receptor phosphorylation and growth. J Cell Biol 139:279–93.

Kolch W (2000): Meaningful relationships: The regulation of the Ras/Raf/MEK/ERK pathway by protein interactions. Biochem J 351:289–305.

Li W, Chong H, Guan KL (2001): Function of the Rho family GTPases in Ras-stimulated Raf activation. J Biol Chem 276:34728–37.

Liebmann C (2001): Regulation of MAP kinase activity by peptide receptor signaling pathway: Paradigms of multiplicity. Cellular Signal 13:777–85.

Macaluso M, Russo G, Cinti C, Bazan V, Gebbia N, Russo A (2002): Ras family genes: An interesting link between cell cycle and cancer. J Cell Physiol 192: 125–30.

Mariathasan S, Zakarian A, Bouchard D, Michie AM, Zuniga-Pflucker JC, Ohashi PS (2001): Duration and strength of extracellular signal-regulated kinase signals are altered during positive versus negative thymocyte selection. J Immunol 167:4966–73.

Minamishima YA, Nakayama K, Nakayama K (2002): Recovery of liver mass without proliferation of hepatocytes after partial hepatectomy in Skp2-deficient mice. Cancer Res 62:995–9.

Nichols A, Camps M, Gillieron C, Chabert C, Brunet A, Wilsbacher J, Cobb M, Pouyssegur J, Shaw JP, Arkinstall S (2000): Substrate recognition domains within extracellular signal-regulated kinase mediate binding and catalytic activation of mitogen-activated protein kinase phosphatase-3. J Biol Chem 275:24613–21.

Olashaw N, Pledger WJ (2002): Paradigms of growth control: Relation to Cdk activation. Sci STKE May 28;2002(134):RE7.

O'Regan AW, Hayden JM, Berman JS (2000): Osteopontin augments CD3-mediated interferon-gamma and CD40 ligand expression by T cells, which results in IL-12 production from peripheral blood mononuclear cells. J Leukoc Biol 68:495–502.

Roovers K, Davey G, Zhu X, Bottazzi M, Assoian R (1999): Alpha5Beta1 integrin controls cyclin D1 expression by sustaining mitogen-activated protein kinase activity in growth factor-treated cells. Mol Biol Cell 10:3197–204.

Roy F, Laberge G, Douziech M, Ferland-McCollough D, Therrien M (2002): KSR is a scaffold required for activation of the ERK/MAPK module. Genes Dev 16:27–38.

Rozga J (2002): Hepatocyte proliferation in health and in liver failure. Med Sci Monit 8:RA32–8.

Scheffzek K, Ahmadian M, Wittinghofer A (1998): GTPase-activating proteins: Helping hands to compliment an active site. Trends Biochem Sci 7:257–62.

Scheving LA, Stevenson MC, Taylormoore JM, Traxler P, Russell WE (2002): Integral role of the EGF receptor in HGF-mediated hepatocyte proliferation. Biochem Biophys Res Commun 290:197–203.

Schneller M, Vuori K, Ruoslahti E (1997): AlphavBeta3 integrin associates with activated insulin and PDGFBeta receptors and potentiates the biological activity of PDGF. EMBO J 16:5600–7.

Schulte TW, An WG, Neckers LM (1997): Geldanamycin-induced destabilization of Raf-1 involves the proteasome. Biochem Biophys Res Commun 239:655–9.

Schwartz M, Assoian R (2001): Integrins and cell proliferation: regulation of cyclin dependent kinases via cytoplasmic signaling pathways. J Cell Sci 114: 2553–60.

Seino K, Setoguchi Y, Ogino T, Kayagaki N, Akiba H, Nakano H, Taniguchi H, Takada Y, Yuzawa K, Todoroki T, Fukuchi Y, Yagita H, Okumura K, Fukao K (2001): Protection against Fas-mediated and tumor necrosis factor receptor 1-mediated liver injury by blockade of FADD without loss of nuclear factor-kappaB activation. Ann Surg 234:681–8.

Takuwa N, Takuwa Y (2001): Regulation of cell cycle molecules by the Ras effector system. Mol Cell Endo 177:25–33.

Thome M, Tschopp J (2001): Regulation of lymphocyte proliferation and death by FLIP. Nat Rev Immunol 1:50–8.

Vang T, Torgersen KM, Sundvold V, Saxena M, Levy FO, Skalhegg BS, Hansson V, Mustelin T, Tasken K (2001): Activation of the COOH-terminal Src kinase (Csk) by cAMP-dependent protein kinase inhibits signaling through the T cell receptor. J Exp Med 193:497–507.

Vessey CJ, de la Hall PM (2001): Hepatic stem cells: a review. Pathology 33: 130–41.

Webb C, Van Aelst L Wigler M, Woude G (1998): Signaling pathways in Ras-mediated tumorigenicity and metastasis. Proc Natl Acad Sci USA 95:8773–8.

Werlen G, Hausmann B, Naeher D, Palmer E (2003): Signaling life and death in the thymus: Timing is everything. Science 299:1859–63.

Werlen G, Hausmann B, Palmer E (2000): A motif in the alphabeta T-cell receptor controls positive selection by modulating ERK activity. Nature 406:422–6.

Werlen G, Jacinto E, Xia Y, Karin M (1998): Calcineurin preferentially synergizes with PKC-theta to activate JNK and IL-2 promoter in T lymphocytes. EMBO J 17:3101–11.

Zhang W, Sloan-Lancaster J, Kitchen J, Trible RP, Samelson LE (1998): LAT: The ZAP-70 tyrosine kinase substrate that links T cell receptor to cellular activation. Cell 92:83–92.

Yeung K, Janosch P, McFerran B, Rose DW, Mischak H, Sedivy JM, Kolch W (2000): Mechanism of suppression of the Raf/MEK/extracellular signal-regulated kinase pathway by the raf kinase inhibitor protein. Mol Cell Biol 20:3079–85.

Zhang W, Trible RP, Zhu M, Liu SK, McGlade CJ, Samelson LE (2000): Association of Grb2, Gads, and phospholipase C-gamma 1 with phosphorylated LAT tyrosine residues. Effect of LAT tyrosine mutations on T cell antigen receptor-mediated signaling. J Biol Chem 275:23355–61.

Zhao J, Reiske H, Guan J (1998): Regulation of the cell cycle by focal adhesion kinase. J Cell Biol 143:1997–2008.

Zhu X, Ohtsubo M, Bohmer R, Roberts J, Assoian R (1996): Adhesion-dependent cell cycle progression linked to the expression of cyclin D1, activation of cyclin E-cdk2, and phosphorylation of the retinoblastoma protein. J Cell Biol 133:391–403.

Zimmermann A (2002): Liver regeneration: the emergence of new pathways. Med Sci Monit 8:RA53–63.

CHAPTER 5

ONSET OF DNA SYNTHESIS AND S PHASE

G. PREM-VEER REDDY, EUGENIA CIFUENTES, UMA BAI, MANI MENON, and EVELYN R. BARRACK
Vattikuti Urology Institute, Henry Ford Health Sciences Center, Detroit, MI 48202

INTRODUCTION

The onset of DNA replication marks cell entry into S phase. Cellular processes leading to the initiation, and subsequent termination, of DNA synthesis represent regulatory events necessary for cell entry into, and progression through, S phase. The DNA must be duplicated fully, accurately, and only once per cell cycle. A defect in any of these processes is debilitating, leading to cell death or promiscuous proliferation.

Initial glimpses into the complexity of cell cycle-based differences in the ability of cells to initiate DNA synthesis has come from mammalian cell fusion experiments of Rao and Johnson (1970). When synchronized HeLa cells in S phase were fused with cells in G_1 phase, the G_1 cells abruptly resumed DNA synthesis and entered into S phase. However, when S phase cells were fused with G_2 phase cells, the G_2 cells did not synthesize DNA until after mitosis. On the other hand, neither G_1 nor G_2 phase cells prevented S phase cells from completing DNA replication and subsequent passage through G_2 and M phases. These classic observations (Rao and Johnson, 1970) revealed that (1) The onset of DNA replication in G_1 cells requires specific factors whose expression and/or activation is restricted to cells that have entered into S phase; otherwise, the DNA itself is competent to replicate at any point in G_1. (2) G_1 and G_2 cells do not contain any inhibitors capable of preventing S phase cells from completing DNA replication and passing through G_2 and M phases; thus the commitment of cells to enter into S phase is the rate-limiting step in their ability to transit through a full cycle of cell division. (3) DNA replicated once during S phase becomes inaccessible to factors in S phase cells for its re-replication at any time prior to nuclear division; this sug-

Cell Cycle and Growth Control: Biomolecular Regulation and Cancer, Second Edition, Edited by Gary S. Stein and Arthur B. Pardee
ISBN 0-471-25071-6

gests that the activity of nuclear factors necessary for the initiation of DNA replication is being turned over with one round of replication during S phase and the restoration of these factors again in the nucleus would require breakdown of the nuclear envelope during mitosis. Two of the most fundamental questions about cell cycle control then became: What are the factors in S phase cells that are capable of initiating DNA replication in G_1 phase cells? What are the nuclear factors that render DNA incapable of re-replicating following one round of replication during S phase?

Answers to these questions have emerged from subsequent studies in yeast, *Saccharomyces cerevisiae* (*S. cerevisiae*) and *Schizosaccharomyces pombe* (*S. pombe*), frog (*Xenopus laevis*) egg extracts. In the last two decades we have witnessed the identification and characterization of cyclins and cyclin-dependent kinases (Cdks) associated with the entry into and progression through S phase, proteins involved in the initiation of DNA replication, proteins that prevent DNA from replicating more than once per cell cycle, origins of DNA replication, and nuclear architecture facilitating spatial, structural, and functional organization of chromatin and enzymes of DNA synthesis. Although most of these discoveries have come from studies with unicellular organisms and frog eggs, important details of the regulation of DNA replication and S phase seem to be universal to proliferating cells in higher organisms. A bird's-eye view of each of these discoveries, as they pertain to the progression of mammalian cells from G_1 into S phase, is presented in this chapter.

SIGNALING PATHWAYS IN CELL CYCLE PROGRESSION FROM G_1 INTO S PHASE

Growth factor-induced extracellular mitogenic stimuli are required for the transition of cells from G_1 into S phase (Pardee, 1989). As depicted in Figure 5.1, growth factor binding to receptors on the cell membrane result in activation of extracellular signal-related kinases 1/2 (ERK 1/2) through a cascade of kinase and phosphatase reactions involving small guanine nucleotide-binding proteins, such as Ras. The mitogen activated Ras-Raf-MAPK pathway phosphorylates and activates transcription factors required for the expression of cell cycle regulatory genes, such as cyclins that regulate cyclin-dependent kinases required for the progression of cells from G_1 into S phase. Ras activation occurs at multiple points during the progression of cells from G_0/G_1 into S phase by a variety of growth factors including insulin-like growth factor-I (IGF-I) (Dobrowolski et al., 1994; Lu et al., 1989).

In addition to their role in expression of cell cycle regulatory proteins, growth factors, such as IGF-I, acting in late G_1, stimulate membrane-bound phospholipase C (PLC), which converts phosphatidylinositol 4,5-bisphosphate (PIP_2) to diacylglycerol (DG) and inositol 1,4,5-trisphosphate (IP_3), and activates phosphatidylinositol-3-OH kinase

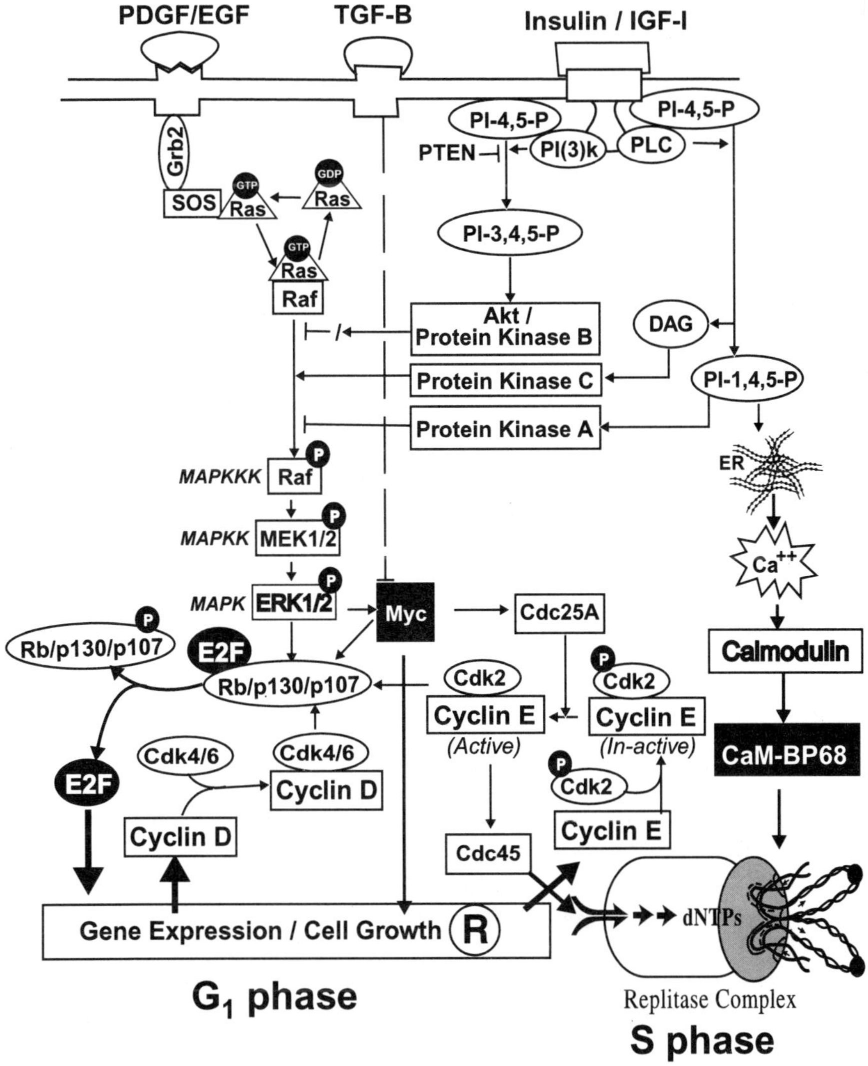

Figure 5.1. A simplified view of signaling pathways emanating from growth factor/receptor interaction that govern the transition of mammalian cells from G_1 into S phase and the assembly of replication machinery (replitase complex) for DNA synthesis.

(PI(3)K) and diacylglycerol (DAG)-dependent forms of protein kinase C (PKC). Activation of PI(3)K and PKC in late G_1 is essential for the entry of cells into S phase (Jones and Kazlauskas, 2001). IP_3 formed by PLC activation also releases Ca^{2+} from intracellular stores, leading to the activation of several proteins and enzymes, including a calcium-receptor protein called calmodulin (CaM). CaM is implicated to play a pivotal role in progression of cells from G_1 into S phase (Means, 1994; Reddy et al., 1994).

Thus a sequence of events resulting from mitogenic activation of the Ras-Raf-MAPK pathway and membrane-bound PLC converges in late G_1 to trigger the entry of cells into S phase. The role of mitogen-stimulated periodic changes in expression and/or activities of cyclins and cyclin-dependent kinases, and PLC-mediated activation of Ca^{++}/CaM in progression of cells from G_1 into S phase, are discussed below.

CELL CYCLE REGULATORS AFFECTING THE PROGRESSION OF CELLS FROM G_1 INTO S PHASE

Cyclins and Cyclin-Dependent Kinases

Cyclins and cyclin-dependent kinases (Cdks) have emerged as key regulators of cell cycle progression. They are for the most part evolutionarily conserved from yeast to humans. However, the increase in genome size and its complex nuclear organization, as well as the need to respond to extracellular mitogenic or inhibitory stimuli at the tissue level, may have contributed to the evolutionary development of additional cell cycle regulators to sustain a timely and controlled proliferation of cells in higher organisms. For example, cdc2/cdc28 kinase, which is common for induction of both S phase and mitosis in yeast, is involved in regulation of only mitosis in mammalian cells and is designated as Cdk1. There are at least three other Cdks involved in the progression of mammalian cells through G_1 and entry into S phase; Cdk4 and Cdk6 for progression through early to late G_1, and Cdk2 for entry into and progression through S phase (see Reddy, 1999, and references therein for details). The regulatory subunits of these kinases are cyclins, which affect the transition of cells from G_1 into S phase, as summarized below (Fig. 5.2) (see Ford et al., Chapter 3, for details).

The D-type cyclins in association with Cdk4 or Cdk6 are involved in progression of mammalian cells not only through G_1, but also in triggering entry into S phase (Baldin et al., 1993; Quelle et al., 1993; Sherr, 1993). Overexpression of cyclin D1, in either cycling or quiescent cells stimulated to enter into S, results in a significant reduction in the G_1 period (Quelle et al., 1993; Resnitzky et al., 1994). Nonetheless, the overall rate of proliferation is unchanged. Similarly, in cyclin D1 transgenic mice, the proliferative rate is unchanged, but in combination with other oncogenes can induce tumorigenesis (Bodrug et al., 1994; Lovec et al., 1994). By contrast, cells lacking cyclin D1 undergo cell division and complete normal prenatal development (Sicinski et al., 1995), suggesting that either cyclin D1 is not directly responsible for the transition of mammalian cells into S phase, or other cyclins expressed later in the cell cycle are able to fulfill the function of cyclin D1 to trigger cell entry into S phase.

Cyclins E and A, in association with Cdk2, are more directly involved in the entry and progression of cells through S phase. Both of these

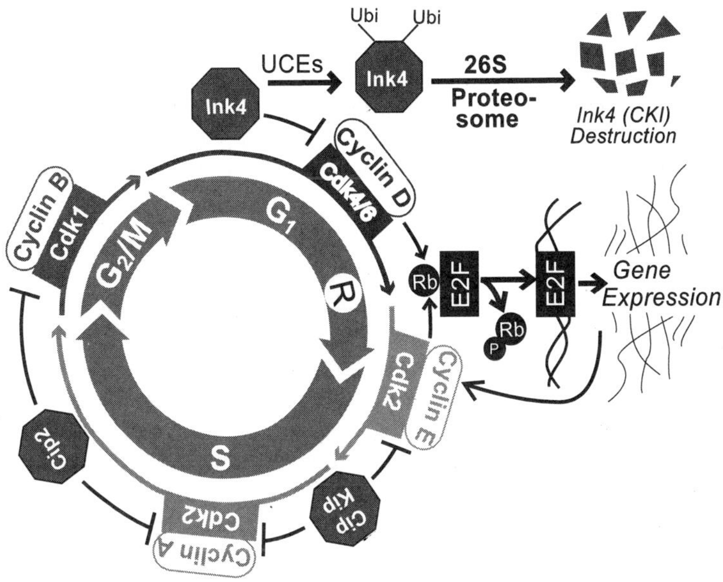

Figure 5.2. Expression and sequential activation of CDKs and CKIs required for the progression of cells from G_1 into S phase. Ubiquitin- and 26S proteosome-dependent degradation shown for Ink4 is also responsible for timely destruction of other CKIs and cyclins during cell cycle. Cyclin D/Cdk4/6 and cyclin E/Cdk2 phosphorylation of Rb allows E2F to be transcriptionally active and express the genes necessary for progression of cells from G_1 into S phase. UCEs, ubiquitin conjugating enzymes; Ubi, ubiquitin.

proteins increase to maximal levels when cells pass through S phase; cyclin E increases at the beginning of S phase (Dulic et al., 1992; Koff et al., 1992), and cyclin A increases during S and G_2 phases (Pines and Hunter, 1990; Tsai et al., 1991). Induction of cyclin E expression in quiescent mammalian cells allows progression into S phase with a modest reduction in the time required for transit through G_1 (Ohtsubo and Roberts, 1993; Resnitzky et al., 1994). In fact, there is an absolute requirement for cyclin E for commitment to S phase, since other cyclins do not have the same effect.

Only cyclin A, which has some structural and functional resemblance to B-type cyclins, can regulate more than one step in the mammalian cell cycle. It has been implicated in the control of S phase (Girard et al., 1991; Pagano et al., 1992; Strausfeld et al., 1996; Zindy et al., 1992) and mitosis (Lehner and O'Farrell, 1989; Minshull et al., 1989; Strausfeld et al., 1996), and also in preventing nuclear DNA from replicating more than once per cell cycle (Sauer et al., 1995). Variation in the level of cyclin A/Cdk2 activity during the cell cycle seems to determine its functional specificity to induce either S phase or mitosis, a low level promotes passage through S phase, and a high level induces mitosis (Strausfeld et al., 1996). Cyclin A/Cdk2 activity seems to be critical for the entry of cells into S phase, because inhibition of cyclin A function prevents the cells from entering into S phase (Girard et al., 1991; Pagano et al., 1992; Zindy et al., 1992). The labile nature of both cyclin E and cyclin A, and the increase in the level of their activity at the beginning of S phase (Dou et al., 1993), are consistant with their role in allowing cells to pass through the restriction (R) point in G_1 to enter into S phase (Pardee, 1974, 1989).

Cyclin/Cdk Inhibitors

Periodic changes in the cyclins (Fig. 5.2) do not fully account for the stringent temporal order of cell cycle progression. Family of low molecular weight proteins that inhibit the activity of Cdks exert a second tier of regulation. These Cdk inhibitors (CKIs) play a pivotal role in regulating cell cycle progression from G_1 into S phase and also in preventing DNA from re-replicating prior to nuclear division. There are four classes of CKIs in mammalian cells: p27Kip1, Ink4, Pic1, and Cip2/Cdi1. Each of these CKIs can bind and inactivate multiple cyclin/Cdk complexes.

The initial understanding of CKI involvement in controlling the entry of mammalian cells into S phase came from the work of Koff et al. (1993). They found that the extracts of the cells arrested in G_1 by transforming growth factor-β (TGF-β) contained normal amounts of cyclin E and Cdk2 but failed to exhibit cyclin E/Cdk2 activity. These studies revealed the presence of an inhibiting factor in extracts of TGF-β-treated cell extracts that is capable of blocking cyclin E/Cdk2 activity in the extracts of untreated cells. This inhibitory factor is identified to be a heat-stable protein with apparent molecular weight of 27 kDa that exists in an inactive form in untreated cells and is referred to as p27Kip1 (Polyak et al., 1994). p27Kip1 levels are found to decline as quiescent macrophage cells (Kato et al., 1994) and T cells (Firpo et al., 1994) are induced to enter into S phase following growth factor/cytokine stimulation.

Ink4 inactivates both Cdk4 and Cdk6 associated with cyclin D (Guan et al., 1994; Hannon and Beach, 1994; Xiong et al., 1993) and arrests cells in G_1. Cyclin D/Cdk4- or cyclin D/Cdk6-dependent phosphorylation of retinoblastoma protein (Rb) is essential for the progression of cells through G_1 (Quelle et al., 1993; Resnitzky et al., 1994). Ink4 fails to induce G_1 arrest in cells that lack functional Rb (Koh et al., 1995; Lukas et al., 1995), suggesting Ink4 involvement in the control of cyclin D/Cdk4- or Cdk6-dependent phosphorylation of RB. Ink4 regulation of these processes seems to play an important role in normal cell proliferation, as a number of primary tumors and tumor cell lines contain mutations in the Ink4 gene (Hunter and Pines, 1994). Similarly Cip1, also known as Waf1 (El-Deiry et al., 1993), Sdi1 (Noda et al., 1994), Cap20 (Gu et al., 1993), or Pic1 (Hunter, 1993), inhibits the kinase activity of cyclin A/Cdk2, cyclin E/Cdk2, and cyclin D1/Cdk4 strongly and that of cyclin B/Cdc2 weakly. It is found associated with cyclins A, E, and D1 under in vitro conditions (Xiong et al., 1993). Cip1/Pic1 induction is linked more directly to a block in the entry of X-ray treated cells into S phase. DNA damage caused by irradiation activates wild-type (but not mutant) p53, which in turn binds to the Cip1/Pic1 promoter, leading to transcription. An increase in abundance of Cip1/Pic1 following p53 activation blocks entry of irradiated cells into S phase by inactivating cyclin E/Cdk2 and/or cyclin D/Cdk4 (Deng et al., 1995). Cip2 binds tightly to Cdk2 but not to Cdk4. Overexpression of wild-type, but not mutant, Cip2/Cdi1 in HeLa cells leads to arrest in G_1; in normal cycling cells its mRNA and protein reach a maximum in late G_1 (Gyuris et al., 1993), suggesting a negative regulatory role for this protein in the control of cell entry into S phase.

Most of the CKIs identified to date exhibit a broad specificity in binding to Cdks. This has made it difficult to assign the negative regulatory function exclusively to any one particular CKI in the control of S phase onset in mammalian cells. Furthermore, even when a potential role in blocking S phase is assigned to a particular CKI, it is hard to establish whether its regulation during the cell cycle contributes to the G_1/S checkpoint control or to the growth factor-induced signal transduction pathways governing $G_1 \rightarrow S$ transition at the restriction (R) point. However, the observations that Ink4 and Cip1/Pic1 in transformed cells with normal checkpoint controls do not respond to growth inhibitory signals, such as TGF-β and cell-cell contact, and that they fail to inhibit G_1 cyclin/Cdk activity, indicate possible involvement of these two CKIs in signal transduction pathways controlling the transition of cells from G_1 into S phase (Nasmyth and Hunt, 1993).

Proteolysis in Progression of Cells from G1 into S Phase

Degradation of CKIs at the end of G_1 by ubiquitin-dependent mechanism is an essential step in the onset of DNA replication. CKIs are marked for degradation by an ubiquitin-conjugating enzyme system consisting of E1 (ubiquitin-activating enzyme), E2 (ubiquitin-conjugating enzyme) and E3 (ubiquitin-ligating enzyme) (Ciechanover, 1994). While E1 enzyme initiates the first step in the reaction, a variety of E2 enzymes in conjunction with E3 enzymes seem to determine the specificity for the proteins targeted for ubiquitination. Once the target proteins are ubiquitinated, they are readily degraded by 26S proteosomes in an ATP-dependent reaction (Hilt and Wolf, 1996). This ubiquitin-dependent proteolysis (Fig. 5.2) is also responsible for the destruction of cyclins, contributing to periodic changes in their levels during the cell cycle (Deshaies et al., 1995; Glotzer et al., 1991; Luca et al., 1991; Seufert et al., 1995; Yaglom et al., 1995). *Cis*-acting signals in cyclins, such as "destruction box," consisting of short stretches of highly conserved amino acids, or "PEST" sequences, consisting of regions rich in proline, aspartic acid, glutamic acid, serine, and threonine, targets them for ubiquitin-dependent proteolysis. It is not known whether such *cis*-acting signals are also present in CKIs to make them susceptible to ubiquitination. Specific phosphorylated states of these proteins seem to determine their susceptibility to ubiquitination (Deshaies et al., 1995; Yaglom et al., 1995). Phosphorylation of CKIs by G_1 cyclin/Cdk complexes during late G_1 triggers ubiquitin-dependent degradation, thereby allowing S phase cyclin/Cdks to activate the progression of cells into S phase.

CDKS IN REGULATION OF DNA SYNTHESIS

Cyclin/Cdk-Mediated Expression of the Enzymes and Proteins Associated with DNA Synthesis

While the identification of specific substrates of cyclin/Cdks that are directly involved in the events leading to the onset of DNA synthesis has been a subject of intense investigation in recent years (see below), active cyclin/Cdk complexes are known to regulate the expression of a number of enzymes and proteins associated with DNA replication. This is mediated by changing the phosphorylated state of Rb and Rb-like proteins (p130, p107) that control the activity of a family of heterodimeric transcriptional regulators called E2Fs (La Thangue, 1994). E2Fs induce expression of cell cycle and DNA synthesis regulatory genes by binding to their promoter sequences. These genes include those encoding thymidine kinase, dihydrofolate reductase, thymidylate synthase, DNA polymerase-α, Cdc2, cyclin E, cyclin A, and *C-myc* (Dalton, 1992; Dou et al., 1992; Geng et al., 1996; Karlseder et al., 1996; Lam and Watson, 1993; Means et al., 1992; Mudryj et al., 1990; Ogris et al., 1993; Pearson et al., 1991; Reed et al., 1992; Sherr, 1996). Hypo-phosphorylated Rb represses the expression of these genes by binding and inactivating E2F/DP-1 hetero-dimeric transcription factors. Phosphorylation of Rb causes Rb to be released from E2F/DP-1 complexes, allowing E2F/DP-1 complexes to be transcriptionally active. Cyclin D/Cdk4 (Baldin et al., 1993; Lukas et al., 1994; Quelle et al., 1993) in G_1 and cyclin E/Cdk2 (Beijersbergen et al., 1995; Hinds et al., 1992) during the transition of cells from G_1 into S phosphorylate Rb, leading to E2F/DP-1-dependent induction of the enzymes/proteins required for DNA replication.

Cyclin A/Cdk2 is also capable of phosphorylating Rb to allow E2F transactivation of genes and premature entry of cells into S phase (Hinds et al., 1992; Resnitzky et al., 1995; Resnitzky and Reed, 1995). Cyclin A/Cdk2 is also implicated in suppressing gene expression at the end of S phase by binding to E2F/DP-1 DNA complexes and phosphorylating DP-1; this causes release of E2F/DP-1 from DNA, thereby inhibiting the expression of E2F-target genes (Krek et al., 1994). These opposing dual roles of cyclin A/Cdk2 in gene expression seem to be facilitated by the periodic changes in cyclin A levels; low levels, as at the beginning of S phase, promote gene expression by phosphorylating Rb, and high levels, toward the end of S phase, suppress E2F-target gene expression by phosphorylating its DNA-binding subunit, DP-1. Thus gene expression facilitated by periodic changes in the level of cyclins E and A during late G_1 and early S phase represents an important step in the ability of cells to enter into S phase and initiate DNA replication.

Association of Cyclin/Cdks with Enzymes of DNA Replication

Direct interaction between cyclin/Cdk complexes and the enzymes of DNA replication may determine the ability of cells to enter into, and/or progress through, S phase. This is indicated from the observation that cyclin A/Cdk2 is specifically co-localized with discrete sites of DNA replication in the nuclei of S phase cells (Cardoso et al., 1993). This raises the possibility that S phase cyclin/Cdks may play a role in the assembly or activation of complex of enzymes of DNA replication machinery. Consistent with such a possibility is the observation that cyclin A/Cdk2, but not cyclin B/Cdk1, in HeLa cells is associated with a high molecular weight nuclear fraction consisting of DNA polymerase-α and proliferating cell nuclear antigen (PCNA) (Jaumot et al., 1994). Furthermore $p21^{Cip1}$, an inhibitor of most cyclin/Cdks induced following DNA damage, directly inhibits DNA replication by binding to and inhibiting PCNA associated with DNA polymerase-δ (Li et al., 1994; Waga et al., 1994; Waga and Stillman, 1994).

ORIGINS OF DNA REPLICATION

A full understanding of the role that cell cycle regulators play in promoting the onset of DNA synthesis and entry of cells into S phase requires identification and characterization of DNA and chromatin where replication initiates. A limited duration of S phase in which a large amount of DNA has to duplicate in eukaryotes necessitates simultaneous replication of their DNA at multiple sites on each chromosome. Otherwise, in human cells, for example, a single replication fork extending at a rate of 5 kb per minute on each chromosome would require more than 18 days to fully duplicate around 3×10^6 kb DNA in 23 chromosomes during a single S phase. Pulse labeling and autoradiography experiments have indicated that the long DNA fibers, ranging in length from 500–1800 μm (Cairns, 1966; Huberman and Riggs, 1966) to more than 2 cm (Sasaki and Norman, 1966), in mammalian chromosomes replicate in separate tandemly joined units of about 30 μm (Cairns, 1966; Huberman and Riggs, 1968). A unit of DNA replicated from single initiation site is termed replicon (Jacob et al., 1963). In a functional analogy to the operon model, the replicon model postulates that the initiation of DNA replication is determined by the binding of *trans*-acting proteins ("initiators") to the *cis*-acting DNA sequences ("replicators") in a DNA template. Binding of the initiators to the replicators facilitates localized unwinding of the DNA, thereby allowing the replication machinery to initiate DNA replication. The regions or the segments of DNA in a chromosome at which DNA replication is initiated are referred to as "origins of replication" ("ori"). Although specific DNA sequences and structures interacting with replication proteins have been identified to serve as the origins of replication in prokaryotes (Bramhill and Kornberg, 1988), animal viruses (Challberg and Kelly, 1989; DePamphilis, 1987; Stillman,

1989), and budding yeast *S. cerevisiae* (Deshpande and Newlon, 1992; Marahrens and Stillman, 1992; Rivier and Rine, 1992; Walker et al., 1991), similar origins of replication in multicellular eukaryotes (metazoans) could not be established. In metazoans, instead of specific short sequences, relatively large stretches of DNA seem to facilitate initiation of DNA synthesis.

In budding yeast, *S. cerevisiae*, *a*utonomously *r*eplicating *s*equence (ARS) elements with origin function were identified by a selective screening process in which segments of yeast DNA sequences in plasmids were tested for their ability to promote extrachromosomal replication of plasmids (Stinchcomb et al., 1979). ARS elements containing two functional domains, A and B, are about 150 bp long. Domain A contains a short (11 bp) conserved core consensus sequence required for origin function, and the domain B contains several stimulatory elements (Brewer and Fangman, 1987; Brewer and Fangman, 1988; Broach et al., 1983; Huberman et al., 1988; Linskens and Huberman, 1988). Not all ARS elements exhibit origin function within a given cell. For example, ARS in tandemly repeated ribosomal DNA are used in approximately 20% of cell cycles on the average (Fangman and Brewer, 1991). However, ARS 501, which is replicated late in S phase, is activated in almost every cycle (Ferguson et al., 1991). Furthermore, not all replication origins within a given cell are activated at the same time during S phase. There are some origins that are activated in early S phase (e.g., ARS1), and there are others that are activated in late S phase (e.g., ARS 501). These differences in the timing of their activation in S phase seem to be determined by the context of their location in the chromosome. For instance, moving ARS 501 from its normal telomeric location to a circular, but not to a linear, plasmid causes it to activate in early S phase, whereas placing a copy of ARS1 near a telomere converts it to activate in late S phase (Ferguson and Fangman, 1992). Replication origins, comparable to those found in *S. cerevisiae*, are also present in *S. pombe*. But the origin sequences in *S. pombe* are much longer, being on the order of 500 to 1000 bp, and appear to be more diffuse and functionally less efficient than those in *S. cerevisiae* (Caddle and Calos, 1994; Clyne and Kelly, 1995; Dubey et al., 1996; Wohlgemuth et al., 1994).

Although it is recognized that in mammalian cells, as in yeast, the initiation of DNA replication occurs mostly at intergenic regions in chromosomes, the sequence and the structural identity of replicators in mammalian cells remains elusive. In comparison to discrete origins of 100 to 200 bp in *S. cerevisiae* chromosome, initiation of DNA replication in mammalian cells is known to occur in large zones, ranging in size from 0.5 to 55 kb. A number of approaches involving the analysis of nascent DNA and replication intermediates led to the identification of some genomic regions in mammalian cells that contained preferential sites for initiation of DNA replication. These include dihydrofolate reductase (DHFR) (Burhans et al., 1990; Dijkwel and Hamlin, 1992; Dijkwel and Hamlin, 1995; Leu and Hamlin, 1989; Vaughn et al., 1990), carbamoyl phosphate synthetase-aspartate transcarbamylase-dihydrooratase

(CAD) (Kelly et al., 1995) and rhodopsin (Gale et al., 1992) gene loci in hamster cells, histone gene repeats (Shinomiya and Ina, 1993), DNA polymerase α gene (Shinomiya and Ina, 1994) and chorion gene (Orr-Weaver, 1991) in *Drosophila* cultured cells, DNA puff II/9A gene in the fungus fly *Sciara coprophila* (Bielinsky et al., 2001; Liang et al., 1993) adenosine deaminase (ADA) region of mouse genome (Carroll et al., 1993; Virta-Pearlman et al., 1993), and c-myc (Vassilev and Johnson, 1990), β-globin (Kitsberg et al., 1993), rRNA (Little et al., 1993; Yoon et al., 1995), and lamin B2 (Abdurashidova et al., 2000) genes in human cells.

One of the most extensively studied among these has been DHFR locus. Chinese hamster ovary cells subjected to selective pressure to develop resistance against methotrexate yielded a cell line called CHOC 400 that contained over 1000 copies of the gene encoding DHFR, target enzyme of methotrexate (Looney and Hamlin, 1987; Milbrandt et al., 1981). Several mapping methods applied to the amplified DHFR domain in CHOC 400 cells have indicated that the replication is initiated at a preferred site down stream of DHFR gene. However, the length of the DNA containing potential origin of replication in this region varied depending on the method employed for its mapping. Initial attempts to map origins of replication in DHFR domain by analyzing early radio-labeled restriction fragments in synchronized CHOC 400 cells entering into S phase have indicated that the replication begins within 28 kb region downstream of DHFR gene (Heintz and Hamlin, 1982). Using a sensitive method of quantitative analysis of the early labeled restriction fragments, the replication in DHFR domain was found to initiate at two preferred sites referred to as "ori-β" and "ori-α" (Leu and Hamlin, 1989). These sites are located 22 kb apart in the intergenic region between DHFR and 2BE2121 genes. A similar restriction fragment analysis of the DNA labeled with radioactive thymidine during the first 2 minutes of cell entry into S phase has narrowed the region in which replication is initiated to a 4.3 kb Xba1 restriction fragment surrounding ori-β (Burhans et al., 1986a, 1986b). Treatment of the cells with AraC, an inhibitor of DNA replication, limited replication to this region, further indicating the presence of an initiation site within 4.3 kb region of DHFR domain (Burhans et al., 1986b). In subsequent studies using an analogous lagging strand assay, in which radiolabeled short nascent DNA representing Okazaki fragments synthesized in permeabilized CHO cells were examined for hybridization to "plus" and "minus" template strands of DNA in the ori-β locus, a 0.45 kb region located 17 kb downstream from DHFR gene was shown to contain initiation site (Burhans et al., 1990). However, in contrast to the observations with radiolabeled nascent DNA analysis, the two-dimensional gel electrophoresis analysis of replication intermediates consistently revealed a delocalized image of initiation sites in a 55 kb region encompassing ori-β and ori-α (Dijkwel and Hamlin, 1992; Vaughn et al., 1990).

These differences in the length of the region containing potential initiation sites identified in DHFR domain by two-dimensional gel analysis method and radiolabeled nascent DNA analysis method are suggested to have resulted possibly from the differences in the stability of replication intermediates analyzed in these two different methods (Dijkwel and Hamlin, 1996). Furthermore the relatively higher sensitivity of 2-D gel analysis methods may have contributed to the detection of trace amounts of individual initiation sites that are not easily detected by the other relatively less-sensitive methods (Burhans and Huberman, 1994). Although trace amounts of replication bubbles, representing individual initiation sites, are seen throughout the 55 kb initiation zone of DHFR domain, a quantitative analysis of replication intermediates containing bubbles indicated their abundance essentially in the 12 kb region surrounding ori-β (Dijkwel and Hamlin, 1992).

Genetically Determined Origins of Replication

Although specific DNA sequences, analogous to those characterized as replicators in simple prokaryotes and yeast *S. cerevisiae*, have not been found in metazoans, origins of replication in higher eukaryotes, as in lower organisms, seem to be conserved and genetically determined. This is implicit in the observation that the replication is initiated at the same specific site in a genomic region when the locus containing that region is present in either two copies per cell or over 1000 copies per cell as observed in the case of hamster DHFR (Dijkwel and Hamlin, 1992, 1995; Handeli et al., 1989; Vassilev et al., 1990) and mouse ADA (Virta-Pearlman et al., 1993) gene loci. This is further reinforced by the observation that the ori regions of hamster DHFR domain (Handeli et al., 1989) and *Drosophila* chorian gene (Orr-Weaver, 1991) retain ability to initiate DNA replication when they are translocated to other chromosomal sites. Similarly activity of ori region in Syrian hamster CAD gene (Kelly et al., 1995) or Chinese hamster DHFR domain (Gilbert et al., 1993) is maintained when they are transfected into chinese hamster cells or incubated with replication-competent protein extract of *Xenopus* oocytes, respectively. Most important, the deletion of an 8 kb region containing initiation sites from human β-globin gene cluster abolishes its bidirectional replication (Kitsberg et al., 1993).

Chromosomal Context of Origin Function

Initiation of replication in the human β-globin gene locus occurs in 8 kb region that is located 50 kb downstream of the locus control region (LCR). The sequences covering LCR are also required for initiation of replication as the deletion of LCR abolishes initiation in the entire locus (Aladjem et al., 1995). These observations indicate that the initiation of replication at specific sites in metazoan chromosomes is determined not only by the conserved sequences at which replication is initiated but also by their interaction with other sequence elements located at a distance in the chromosome. In order for an origin of replication to be active, it seems necessary that its location in the chromosome should permit its interaction with other regulatory sequences or the factors associated with such sequences. Alternatively, it is possible that transcriptional activity in the genes located at a distance from initiation sites may change chromosomal architecture in such a way that the initiation sites become accessible to the replication machinery. These possibilities are also reflected in the observation that the deletion of the promoter in DHFR gene locus abrogates initiation of replication in ori-β region located several kilobases downstream of DHFR gene (Dijkwel and Hamlin, 1995). However, an assertion that transcriptional activity per se is responsible for the initiation of replication in the genomic region is confounded by the observation that the transcriptional activation of genes during embryo development, represses initiation of replication within the transcribed regions. For instance, in early stages of *Xenopus* embryo development, replication in both the intragenic and intergenic regions of ribosomal RNA gene (rDNA) locus is initiated at every 9 to 12 kb interval. However, in late-blastula stage, when rDNA gene becomes transcriptionally active, initiation of replication within the transcribed region gets repressed while that in nontranscribed intergenic regions continues to persist in ensuing divisions of embryo development (Hyrien et al., 1995). These observations suggest that it is the chromatin remodeling that occurs to facilitate gene transcription during embryo development, rather than the transcriptional activity itself may specify the sites at which replication is initiated.

Relationship between Transcription and Replication in an Ori Region

DNA replication in eukaryotes is localized both in time and space, with specific regions of chromosomal DNA replicating at specific intervals and at limited number of sites within the nucleus during S phase. As mentioned above, depending on the location of yeast ARS elements on the chromosome, they replicate either early or late in S phase (Ferguson et al., 1991). In metazoans transcriptionally active regions were found to replicate early in S phase, whereas transcriptionally inactive regions replicate late in S phase (Holmquist et al., 1982; Yunis et al., 1977). Furthermore the same gene in different cell types may replicate either early or late in S phase, depending on whether or not it is being expressed in a given cell type. For example, the genomic region containing cystic fibrosis (CF) gene replicates early in S phase in the cells expressing the CF gene, whereas it replicates late in S phase in the cells that do not express the gene (Selig et al., 1992). Cell fusion studies have also revealed a tight coordination between transcriptional activity and the timing of replication in the β-globin gene locus (Dhar et al., 1989). When β-globin gene expression in mouse hepatoma cells is repressed following their fusion with mouse erythroleukemia (MEL) cells, replication of β-globin gene locus is shifted to a later time in S phase. Similarly, when β-globin gene is activated in human fibroblasts by their fusion with MEL cells, replication of the entire locus is shifted to an earlier time in S phase. Although it is evident from these observations that the transcription and replication occur coordinately in specific genomic regions, the causal relationship between these processes remains to be determined.

Conceivable Models for the Functional Origins of Replication in Metazoans

These observations, taken together, raise the possibility that while replication in metazoans is initiated at multiple sites in a broad region, most of the initiations become futile, and only those at selected sites within the ori region will be effective in allowing bi-directional replication of a replicon. Several models have been proposed to explain how a functional origin of replication is manifested in intergenomic regions of multicellular eukaryotes. In one model, it is suggested that the DNA replication initiates at a number of sites in a broad region and extend unidirectionally toward a specific site from where the replication becomes bi-directional (Linskens and Huberman, 1990). In a second model, DNA primers are suggested to first synthesize at multiple sites in a large uncoiled region of duplex DNA before bi-directional replication is initiated at a fixed site (Benbow et al., 1992). In a third model, it is proposed that the replication is initiated at multiple sites on naked DNA, but their elongation is suppressed by the organization of DNA into chromatin; only selected initiation sites in chromatin that are associated with nuclear structure are capable of further unwinding and promoting DNA replication (Burhans and Huberman, 1994; DePamphilis, 1993a, b, c). These models cannot be validated without a full understanding of the structural and functional organization of the DNA and its interaction with proteins and enzymes associated with DNA replication in the context of chromatin and nuclear architecture.

INITIATORS OF DNA SYNTHESIS AT THE ORIGINS OF REPLICATION

The idea put forth some 40 years ago by Jacob et al. (1963), that trans-acting factors, "initiators," necessary for DNA synthesis assemble at the origins of DNA replication, is now confirmed. This assembly of initiators is an orderly process involving stepwise recruitment of proteins into a complex that is capable of unwinding DNA at the origins, guiding the formation of replication machinery capable of initiating DNA synthesis, and bringing replication competence to the cells. Such a complex of initiators at the origins is referred to as pre-*r*eplication *c*omplex (pre-RC). Identifying individual components necessary for the assembly of pre-RC and their regulation during the progression of cells from G_1 into S phase has been a subject of intense investigation in the last few years. Current understanding of individual initiators is discussed below, in the order in which they are recruited to ori regions to form pre-RC.

Origin Recognition Complex

Since replication in *S. cerevisiae* initiates specifically at ARS elements (Brewer and Fangman, 1987, 1988; Stinchcomb et al., 1979), biochemical and genetic studies were designed to identify proteins that bind specifically to these elements. Using a DNase protection footprinting assay, Bell and Stillman (1992) first reported a multiprotein complex in yeast nuclear extract that binds in a sequence-specific manner to A and B domains of ARS1 element. Purified multiprotein complex, referred to as *o*rigin *r*ecognition *c*omplex (ORC) consists of six protein subunits (Orc1 to Orc6) ranging in molecular weight from 120 to 50 KDa. In vivo studies also revealed binding of proteins to A and B domains of the ARS1 element in a pattern similar to that seen in vitro with ORC (Diffley and Cocker, 1992). All six ORC subunits are essential for initiation of DNA replication and cell viability (Foss et al., 1993; Liang et al., 1995; Loo et al., 1995; Micklem et al., 1993).

Homologues of ORC subunits have been identified in several eukaryotes including *S. pombe*, *Drodophila melanogaster*, *Xenopus leavis*, and humans (Carpenter et al., 1996; Dhar and Dutta, 2000; Gavin et al., 1995; Gossen et al., 1995; Kelly and Brown, 2000; Leatherwood et al., 1996; Muzi-Falconi and Kelly, 1995). Structural and functional interaction between individual subunits of ORC, and their binding to chromatin is essential for DNA replication and entry of cells into S phase in higher eukaryotes including human cells, just as it is in *S. cerevisiae* (Dhar et al., 2001; Landis et al., 1997). In analogy to the ARS1 element in *S. cerevisiae*, Bielinsky et al. (2001) reported a distinct 80 bp sequence to which ORC binds in a metazoan replication origin. However, despite significant structural and functional homology of ORC proteins between *S. cerevisiae* and other eukaryotic cells, an ORC binding consensus sequence similar to that in the ARS1 element of *S. ceravisiae*, remains to be established in higher eukaryotes. In most eukaryotes, including *S. pombe*, ORC binding seems to be facilitated mostly by AT-rich elements in the ori region (Austin et al., 1999; Chuang and Kelly, 1999).

In *S. cerevisae* there is a tight stoichiometric association of all six ORC subunits in pre-RC, and their binding to DNA does not fluctuate during the cell cycle; they bind to DNA not just during S phase but throughout the cell cycle (Aparicio et al., 1997; Diffley and Cocker, 1992; Diffley et al., 1994). However, in higher eukaryotes individual ORC proteins exhibit cell cycle-dependent differences in their ability to bind to chromatin. In hamster cells, Orc1 dissociates easily from chromatin during mitosis and early G_1 and binds stably to the functional pre-RC at the origins during mid G_1 (Natale et al., 2000). In human cells there is a similar dissociation of Orc1/Orc6 and Orc1/Orc2 complexes from chromatin during S phase, and reassociation is required for entry of cells into G_1 phase after mitosis (Dhar and Dutta, 2000; Kreitz et al., 2001). Studies with recombinant human ORC subunits revealed an orderly assembly of individual protein subunits to form ORC; first a core complex of Orc2, Orc3, and Orc4 is formed to which Orc1, Orc5, and Orc6 bind, Orc6 showing the weakest affinity to the complex (Vashee et al., 2001). Thus an orderly assembly, and varied binding abilities, of individual subunits of ORC to chromatin may play an important role in regulating the entry into, and/or exit from, S phase in higher eukaryotes.

Although ORC binding marks the site at which DNA replication can initiate, ORC has no direct role in either the activation or initiation of origins. Recruitment of additional initiation factors necessary for the activation of origins is indicated from the observation that the DNase protected footprint of ORC bound to DNA extends as the cells exit mitosis and progress through G_1 phase (Cocker et al., 1996; Diffley et al., 1994). ORC at the origins serves to recruit "origin loading factors," Cdc6/Cdc18 and Cdt1, essential for loading a complex of six membered *m*ini-*c*hromosome *m*aintenance (MCM) proteins capable of activating origins to initiate DNA synthesis (Cocker et al., 1996; Kearsey et al., 2000; Maiorano et al., 2000a, b; Nishitani et al., 2000).

Cdc6/Cdc18

Cdc6 in *S. cerevisiae*, and its homologue Cdc18 in *S. pombe*, are required for the formation of functional pre-RC and the initiation of DNA replication (Bueno and Russell, 1992; Cocker et al., 1996; Dahmann et al., 1995; Kelly et al., 1993; Liang et al., 1995; Muzi Falconi et al., 1996; Piatti et al., 1995). Genetic studies revealed functional interaction of Cdc6 with ORC subunits Orc2, Orc5, and Orc6 (Li and Herskowitz, 1993; Loo et al., 1995). Biochemical studies have shown physical interaction of Cdc6/Cdc18 with Orc2 in both *S. cerevisiae* and *S. pombe* (Leatherwood et al., 1996; Liang et al., 1995). There is considerable sequence homology between Cdc6/Cdc18 and one of the ORC subunits Orc1, and it has an essential ATP-binding motif (Bell et al., 1995; Gavin et al., 1995; Muzi-Falconi and Kelly, 1995). These observations suggest the possibility that Orc1 homology regions in Cdc6 may play a role in its interaction with Orc2 and with other subunits of ORC in pre-RC. Cdc6/Cdc18 homologues have been identified in *Xenopus* and mammals (Carpenter et al., 1996; Coleman et al., 1996; Williams et al., 1997). As indicated from studies in yeast, in *Xenopus leavis* also Cdc6 binding to chromatin requires Orc2. Furthermore Cdc6 in *Xenopus* egg extract is shown to be required for initiation, but not for elongation, of replication forks.

Cdt1

Initially Cdt1 was identified as a target of the Cdc10/Sct1 transcription factor required for the progression of *S. Pombe* from G_1 into S phase (Hofmann and Beach, 1994). Deletion of Cdt1, just as that of Cdc18, prevents cells from initiating DNA synthesis, and its excessive expression potentiates re-replication by inducing origins to fire more persistently (Yanow et al., 2001). Cdt1, like Cdc6/Cdc18, is evolutionarily conserved in vertebrates with homologues identified in *Xenopus leavis*, *Drosophila*, humans (Maiorano et al., 2000a, b; Nishitani et al., 2001; Whittaker et al.,

2000; Wohlschlegel et al., 2000), and *S. cerevisiae* (Tanaka and Diffley, 2002). *S. cerevisae* depleted of Cdt1 fail to initiate DNA synthesis by failing to load Mcm2 onto chromatin (Devault et al., 2002).

The levels of both Cdc6/Cdc18 and Cdt1 oscillate with the cell cycle, reaching a peak in late G_1 phase, and are transcriptionally regulated (Blow and Tada, 2000; Drury et al., 1997; Lopez-Girona et al., 1998; Maiorano et al., 2000a; Muzi Falconi et al., 1996; Nishitani et al., 2000; Nishitani and Nurse, 1995; Piatti et al., 1995; Whittaker et al., 2000). In human cells, as in *S. Pombe*, Cdt1 peaks at G_1/S transition and disappears after the onset of DNA synthesis (Nishitani et al., 2001; Wohlschlegel et al., 2000). In *S. cervisiae* it is present in the nucleus only during G_1 phase and gets excluded from the nucleus for the rest of the cell cycle (Tanaka and Diffley, 2002). In fission yeast and metazoans, Cdc6/Cdc18 and Cdt1 act synergistically to recruit the MCM complex and exhibit ORC-dependent binding to chromatin during G_1 phase, and physically interact with each other (Maiorano et al., 2000a, b; Nishitani et al., 2000). While they play a critical role in recruitment of MCM complex, retention of MCM proteins in pre-RC is unaffected by the removal of Cdc6/Cdc18 and Cdt1 from the pre-RC, suggesting their role mainly in the assembly, but not in the maintenance, of functional pre-RC (Hua and Newport, 1998; Maiorano et al., 2000b; Nishitani et al., 2000; Rowles et al., 1999).

Minichromosome Maintenance Proteins

Genes encoding minichromosome maintenance proteins in *S. cerevisiae* were first identified in a genetic screening of mutants defective in maintaining plasmids (minichromosomes) containing an ARS element (Maine et al., 1984). A high frequency of minichromosome loss in these mutants is due to the defect in initiation of DNA replication (Maiti and Sinha, 1992; Yan et al., 1993). A family of six MCM gene products, Mcm2 to Mcm7, is required for viability as well as for entry of cells into S phase (Chong et al., 1996; Gibson et al., 1990; Hennessy et al., 1991; Maiorano et al., 1996). Cells defective in these genes are arrested at nonpermissive conditions with partially replicated DNA. These proteins seem to play an important role in both the initiation and elongation of DNA replication forks (Tye, 1999).

Homologues of these Mcm proteins have been identified in *S. pombe* (Coxon et al., 1992; Forsburg and Nurse, 1994; Miyake et al., 1993; Takahashi et al., 1994) and in higher eukaryotes, including mouse (Thommes et al., 1992), human (Hu et al., 1993; Todorov et al., 1994), *Drosophila* (Treisman et al., 1995), and *Xenopus* (Kubota et al., 1995; Madine et al., 1995a). Mouse Mcm3 protein was initially identified as P1 protein that copurified with DNA polymerase-α (Thommes et al., 1992). Microinjection of antibodies against P1 protein into mouse cells (Thommes et al., 1992) and those against BM28 (human Mcm2) protein into human cells (Todorov et al., 1994) prevented transition from G_1 into S phase. Cells in imaginal discs and the central nervous system of mutant *Drosophila* embryos defective in MCM2 delayed or failed to replicate

DNA (Treisman et al., 1995). *Xenopus* egg extract immunodepleted of Mcm3 is also incapable of supporting chromatin replication (Chong et al., 1995; Kubota et al., 1995; Madine et al., 1995a).

MCM proteins are recruited to the origins of DNA replication during G_1 phase (Labib et al., 2000; Tanaka et al., 1997). These proteins exhibit nuclear localization throughout the cell cycle (Kimura et al., 1994; Thommes et al., 1992; Todorov et al., 1994). Furthermore Mcm3 and other proteins of MCM complex are able to enter freely into nuclei without requiring the breakdown of nuclear membrane (Madine et al., 1995a). Once inside the nuclei, however, their binding to chromatin shows an absolute requirement for the breakdown of nuclear membrane (Madine et al., 1995b), suggesting the requirement of cytoplasmic factors, such as Cdc6/Cdc18 and Cdt1 (see below), to gain access to the nuclei in order for Mcm proteins to bind to chromatin. After the initiation of DNA replication in *Xenopus leavis*, Cdc6 dislodges from chromatin and remains associated with the nuclear envelope. However, Mcm3 is released, not at the time of initiation per se, but as the replication forks begin to extend bidirectionally and is dispersed in the nuclear compartment in a soluble form (Chong et al., 1995; Kubota et al., 1995; Madine et al., 1995a).

Some Mcm proteins, Mcm2, Mcm4, Mcm6, and Mcm7, contain zinc finger motifs, which are suggested to play a role in their binding to DNA and interaction with one another (Kearsey and Labib, 1998; Tye, 1999; You et al., 2002). All six Mcm proteins have a 240 amino acid conserved region with a DNA-dependent ATPase motif that includes Walker motifs A and B characteristic of ATPase and helicase. Stoichiometric amounts of each of these proteins interact with one another to form a heterohexameric complex of about 600 kDa with a globular structure (Adachi et al., 1997; Brown and Kelly, 1998; Kubota et al., 1997; Thommes et al., 1997). Although a hexameric complex of all six Mcm proteins is required for the activation of origins (Maiorano et al., 2000b; Prokhorova and Blow, 2000), these proteins, either individually or in hexameric complex form, displayed neither ATPase nor helicase activity. However, in trimeric complexes of Mcm4/6/7, Mcm4 and Mcm7 are shown to contain helicase activity and Mcm6 is reported to play an essential role in ATP binding. Interestingly, addition of Mcm2 to this trimeric complex abrogated helicase activity (Ishimi, 1997; You et al., 1999). It is proposed that a coordinated action of two trimeric subcomplexes, a catalytic Mcm-4-6-7 and a regulatory Mcm-2-3-5, may constitute helicase activity in the MCM complex (Schwacha and Bell, 2001). Of 15 different pairwise combinations of six Mcm recombinant proteins (Mcm2–Mcm7), only 3 pairs of Mcm proteins, Mcm3/7, Mcm4/7, and Mcm2/6, are shown to exhibit ATPase activity (Davey et al., 2003). The physiological significance of activities in dimeric or trimeric complexes of Mcm proteins isolated from HeLa cells in the initiation of DNA synthesis remains to be determined.

Even though the recruitment of Mcm proteins to the origins is dependent on ORC, Cdc6/Cdc18, and Cdt1 at the origins, none of these proteins seem to interact directly with Mcm proteins. Furthermore, once the

Mcm proteins are recruited to the origins, removal of ORC, Cdc6/Cdc18 and Cdt1 from pre-RC has little effect on the retention or initiation function of Mcm proteins at the origins (Donovan et al., 1997; Hua and Newport, 1998; Maiorano et al., 2000a; Rowles et al., 1999). Mcm proteins seem to assemble initially at or near the origins where ORC is bound during G_1 phase but as the cells enter into S phase they are distributed over a large region surrounding ORC (Alexandrow et al., 2002; Edwards et al., 2002; Schaarschmidt et al., 2002). Based on structural similarities, Cdc6 is suggested to facilitate loading of Mcm proteins onto chromatin in a manner similar to a superfamily of loading factors such as replication factor C (RF-C) that loads sliding-clamp protein *p*roliferating-*c*ell *n*uclear *a*ntigen (PCNA), the processesivity factor associated with DNA polymerase δ, onto DNA (Perkins and Diffley, 1998). This may imply that MCM complex may play a processesive role in elongation of replication forks.

In vitro and in vivo studies revealed a direct interaction between Mcm2 and histone acetyltransferase HBO1 (Burke et al., 2001). The role of HBO1 in DNA replication remains to be determined. However, acetylation of Mcm3 by an MCM3 acetylating protein (MCM3AP) in human cells has been shown to inhibit initiation of DNA replication (Takei et al., 2002). Furthermore it is reported that mouse P1 (Mcm3) (Kimura et al., 1994) and human BM28 (Mcm2) (Todorov et al., 1995) proteins undergo periodic changes in their phosphorylated states and in their intranuclear distribution during the cell cycle. It is observed that Mcm proteins in G_1 phase are hyperphosphorylated, and following the onset of S phase they gradually become underphosphorylated. These differences in phosphorylation states of Mcm proteins may determine their affinity to Cdc6 and loading onto chromatin (Hendrickson et al., 1996; Lei et al., 1996).

Mcm10

In addition to the heterohexameric complex of Mcm2–Mcm7, another Mcm protein, Mcm10, that bears no sequence homology to Mcm2–7, is shown to be present at the origins of DNA replication, and it interacts with all six subunits of the hexomeric complex (Homesley et al., 2000; Kawasaki et al., 2000; Merchant et al., 1997). In contrast to the Mcm complex components, Mcm10 binds to chromatin constitutively during all phases of the cell cycle, and its binding to chromatin is unaffected by the removal of ORC from origins. Furthermore its binding to pre-RC is Mcm2–7 dependent (Wohlschlegel et al., 2002). Mcm10 presence on chromatin seems to be critical for the stability of pre-RC, since its removal, unlike that of ORC, Cdc6/Cdc18, and Cdt1, disassociates Mcm proteins from the origins (Homesley et al., 2000). Mutations in MCM10 (mcm10-1) result in pausing of elongation of replication forks at the origins that failed to initiate. This defect is corrected in double mutants that carried a second mutation in MCM7 (mcm7-1), suggesting a direct interaction between Mcm10 and Mcm7 that is essential for initiation and elongation of replication forks in *S. cerevisiae* (Homesley et al., 2000). Mcm10 is also reported to interact with Mcm2, Mcm3, Mcm4, and Mcm6 (Merchant et al., 1997).

Cdc45

S. cerevisiae with a CDC45 gene mutation fail to initiate DNA replication. Interestingly, this defect in CDC45 mutants, like that of MCM10 mutants, is suppressed by a second mutation in MCM5 or MCM7 gene alleles (Hennessy et al., 1991), suggesting a physical and functional interaction between Cdc45 and Mcm proteins. Accordingly Cdc45 co-immunoprecipitates with Mcm2, Mcm5, and Mcm7 and its binding to pre-RC is Cdc6, Mcm2, and S phase CDK dependent (Hopwood and Dalton, 1996; Zou et al., 1997; Zou and Stillman, 1998). Cdc45, like Mcm proteins, relocates from the origins of replication to the interorigin regions during S phase, possibly to associate with elongation machinery (Aparicio et al., 1997). Cdc45 homologues have been identified in *Xenopus leavis* and in human cells (Mimura and Takisawa, 1998). Cdc45 depletion abrogates DNA replication activity in *Xenopus* extracts. Temporal association of Cdc45 with chromatin coincides with that of DNA polymerase α, and they both physically interact with each other in *Xenopus* extracts. These observations suggest a critical role of Cdc45 in recruitment of DNA polymerase-α-primase to the origins of DNA replication and in intiation of DNA synthesis.

Other Proteins Associated with Pre-RC

In addition to the proteins described above, mutations in *Drosophila* E2F1, DP, and RB genes are shown to affect ORC and initiation of replication at the chorion gene origin of replication. E2F1, DP, and Rb proteins are also shown to complex with ORC at chorion origin of replication in vivo (Bosco et al., 2001). The functional significance of interaction of transcription factors with ORC at the origins in regulation of DNA replication remains to be determined. However, this report raises the possibility that transcription factors through their interaction with the components of pre-RC may coordinate gene transcription during the replicative process in S phase. Unlike in yeast, in mammalian cells replication and transcription occurs simultaneously at thousands of origins and genes, respectively, during S phase. Since origins of DNA replication and gene promoter sequences are interspersed, their coordinated activation or repression may require crosstalk between these two important cellular processes. It is conceivable that the interaction of transcription factors with pre-RC components may allow a temporal coordination of these two processes during S phase. We have recently observed that in androgen-sensitive prostate epithelial (LNCaP) cells, androgen receptor (AR) colocalizes with replication foci containing BrdU incorporated DNA, and co-immunoprecipitates with Cdc6 in chromatin preparations (Cifuentes, Bai, and Reddy, manuscript in preparation). AR, in addition to its role in transcriptional regulation of androgen-responsive genes, plays a critical role in transition of LNCaP cells from G_1 into S phase (Cifuentes et al., 2003). Thus it is expected from these early findings that depending on the cell type, additional factors involved in transcriptional regulation and hormonal control of cell proliferation could also be associated with pre-RC.

Assembly and Activation of Pre-RC

As described above, ORC, Cdc6/Cdc18, Cdt1, and MCM complex assemble at the origins of DNA replication to form pre-RC and activation of pre-RC depends on its recruitment of Cdc45. An orderly assembly of each of these components into pre-RC at the origins is temporally separated from the initiation of DNA synthesis (Fig. 5.3). Pre-RC assembly starts immediately after anaphase and continues throughout G_1 phase, and is activated as the cells enter into S phase following the recruitment of Cdc45. Once the cells enter into S phase, they cannot form any new pre-RC until after mitosis. This temporal separation of assembly and activation of pre-RC is critical for ensuring that any segment of cellular DNA is not replicated more than once per cell cycle. This overall process of an orderly assembly of pre-RC in G_1 phase and the restrictions on its reassembly during S and G_2/M phase are under stringent control of cyclin-dependent kinases (CDKs), a 25 kDa protein called geminin, and Dbf4-dependent Cdc7 kinase (DDK).

As cells exit mitosis, the level of mitotic CDKs, such as cyclin B/Cdk1 or Cdc2, decreases and ORC at the origins becomes available for Cdc6/Cdc18 binding. In mammalian cells, mitotic CDKs, if present, will prevent the binding of Cdc6/Cdc18 to ORC (Diffley, 1996; Fujita et al., 1998, 1999; Tanaka et al., 1997), and target it for rapid proteosome-dependent degradation (Jallepalli et al., 1997; Mendez and Stillman, 2000). In *S. cerevisiae*, Cdc6 may also play a role in mitotic CDK inactivation during the exit from mitosis (Calzada et al., 2001). Following mitosis, Cdc6/Cdc18 levels increase and translocate into nuclei for binding to ORC.

Cyclin/Cdks (CDKs), in addition to their role in regulation of gene expression as described above (Fig. 5.2), seem to control regulatory events leading to a stepwise recruitment of pre-RC components to the site of DNA replication. At permissive levels, cyclin E/Cdk2, in cooperation with Cdc6, stimulates the recruitment of Mcm2 and functional assembly of pre-RC. Cyclin A/Cdk2 could not be substituted for this function in late-G_1. However, once pre-RC is assembled cyclin A, but not cyclin E, activates DNA synthesis in late-G_1 (Coverley et al., 2002). Activation of CDKs at the G_1/S boundary requires the destruction of CDK-inhibitors (CKIs). CKIs, such as Sic1, are marked for ubiquitin-dependent degradation by G_1 CDKs (Montagnoli et al., 1999; Verma et al., 1997). In *Xenopus*, ubiquitin-dependent destruction of CKIs is spatially constrained to chromatin bound pre-RC, and occurs independent of its phosphorylation by cyclin E/Cdk2 (Furstenthal et al., 2001).

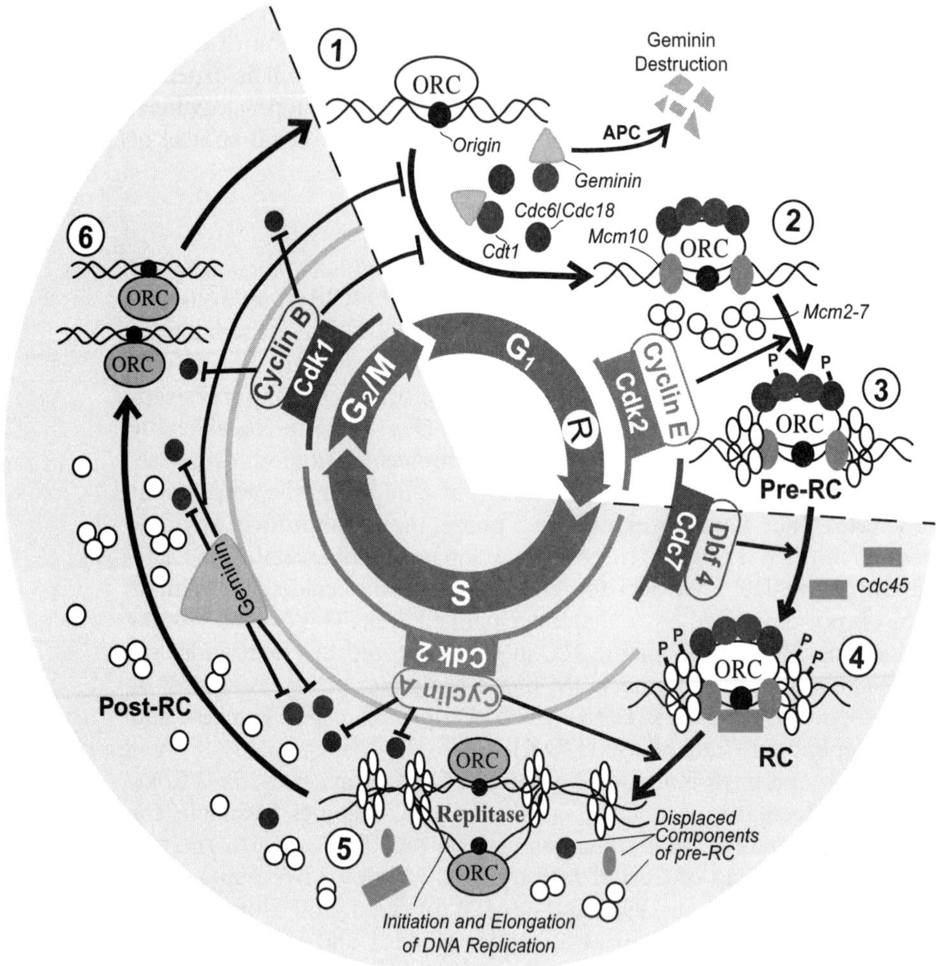

Figure 5.3. Cell cycle regulatory events controlling an orderly assembly and activation of pre-RC required for the transition of cells from G_1 into S phase. (1) ORC localized to the sites at which DNA replication initiates (origins). (2) MCM loading factors, Cdc6/Cdc18 and Cdt1, are recruited to the origins. This requires anaphase-promoting complex (APC)- and 26S proteosome-dependent destruction of geminin and mitotic CDK (cyclin B/Cdk1), respectively. (3) MCM proteins are recruited to the origins completing the assembly of pre-RC. This requires cyclin E/Cdk2-dependent phosphorylation of Cdc6/Cdc18 and Cdt1. (4) Cdc45 joins pre-RC to unwind DNA at the origins. This requires DDK (Cdc7/Dbf4) phosphorylation of MCM and ORC protein subunits. (5) DNA strand separation allows recruitment of DNA polymerase-α/primase and other enzymes of DNA replication machinery required for initiation and elongation of DNA replication. Cyclin A/Cdk2, geminin, and cyclin B/Cdk1 play a critical role in preventing displaced Cdc6/Cdc18 and Cdt1 from their reassociation with ORC at the origins until the completion of mitosis. (6) Anaphase separation of daughter DNA strands. Destruction of mitotic cyclins and geminin is required for reassembly of pre-RC. Pre-RC assembly (*clear area*) is temporally separated from initiation and elongation of DNA replication forks (post-RC) (*shaded area*).

Unlike in yeast, binding of Cdc6/Cdc18 alone to ORC is not sufficient for effective recruitment of MCM complex to pre-RC. Cdt1 synergizes Cdc6/Cdc18 to recruit MCM complex. Just as mitotic CDK destruction is essential for Cdc6/Cdc18 recruitment, anaphase-promoting complex (APC)-dependent destruction of 25 kDa protein called geminin is critical for Cdt1 binding to DNA (Yanagi et al., 2002). Geminin is absent in G_1 phase cells but accumulates during S and G_2/M, and disappears at the time of metaphase-anaphase transition (McGarry and Kirschner, 1998). Geminin also prevents the recruitment of Mcm proteins to the replication origins by binding to Cdt1 (Tada et al., 2001; Wohlschlegel et al., 2000). Thus geminin as a negative regulator of Cdt1 plays a critical role in loading of Mcm proteins to pre-RC and also ensures that the Cdt1 released from pre-RC after initiation of DNA synthesis is prevented from being active again until after mitosis.

Following the assembly of ORC, Cdc6/Cdc18, Cdt1, and MCM complex containing pre-RC, is joined by Cdc45 required for the activation of pre-RC and the loading of DNA polymerase-α and primase at the origins (Walter and Newport, 2000). Joining of Cdc45 with Mcm proteins in pre-RC requires a protein kinase consisting of a catalytic subunit encoded by CDC7 and a regulatory subunit encoded by DBF4 in budding yeast. Previously Cdc7 kinase activity was shown to be required for the initiation of replication (Hereford and Hartwell, 1974; Jackson et al., 1993; Kitada et al., 1992). Cdc7 kinase activity fluctuates with the cell cycle and its activation at the G_1/S boundary is dependent on its interaction with a regulatory protein Dbf4 (Jackson et al., 1993; Yoon and Campbell, 1991). Cells lacking either Cdc7 or Dbf4 fail to initiate DNA replication, even though they contain normal S phase-promoting CDK activity and are able to transit through the START point (R point) in late G_1 phase. It is shown that Cdc7 binds to ORC, and Dbf4, like Orc6, interacts with the origins of replication (Dowell et al., 1994). From these observations it is possible that Dbf4 may target Cdc7 kinase to ORC, allowing its interaction with, and phosphorylation of, Mcm proteins in pre-RC. Phosphorylation of Mcm proteins by Dbf4-Cdc7 (Dbf4-dependent kinase, DDK) is essential for the activation of pre-RC, without which origins cannot be activated. Mcm2 is a target of DDK during the initiation of DNA synthesis (Lei et al., 1997). Phosphorylation of MCM complex may unveil its helicase activity required for strand separation and ORC displacement at the origins to allow Cdc45 binding to the origins. Homologues of Cdc7-related kinases are present in *Xenopus* and humans (Sato et al., 1997).

It is conceived from these observations that the phosphorylation of pre-RC components, Cdc6/Cdc18, Mcm proteins and/or ORC, by S phase-promoting CDKs and DDK may lead to the recruitment of the enzymes of DNA replication, including DNA polymerase-α-primase, to the origins of replication. Initiation of replication would then displace Cdc6/Cdc18, and subsequently Mcm proteins as the replication forks extend, from ORC. The phosphorylated state of the displaced Cdc6/Cdc18 may also target for ubiquitin-dependent proteolysis as described above (Fig. 5.2). Such degradation of displaced Cdc6/Cdc18 would ensure that no new pre-RCs are assembled to re-initiate DNA replication from the same origins during the remainder of S phase and prior to the passage of cells through mitosis. This is consistent with the observation that mutations in genes encoding the components of ubiquitin-dependent protein degradation complex would lead to re-initiation of replication at the origins within a single cell cycle. Furthermore overexpression of Cdc18 in fission yeast leads to a continuous accumulation of DNA due to re-initiation of replication at each origin within the same S phase (Muzi Falconi et al., 1996; Nishitani and Nurse, 1995). Thus timely degradation of phosphorylated Cdc6/Cdc18 displaced from the origins of replication by ubiquitin-dependent proteolysis is essential for limiting the initiation of replication at an origin of replication to once per cell cycle.

Inactivation of mitotic CDK by various methods, including overexpression of Rum1 in fission yeast (Moreno and Nurse, 1994), or Sic1 in budding yeast (Dahmann et al., 1995), also leads to re-replication of DNA without intervening mitosis. However, this re-replication due to CDK inactivation seems to result in the accumulation of DNA in full genome increments, rather than in a continuous increase, which, as described above, occurs if re-initiation takes place at each origin within a single S phase. These observations indicate that in the absence of mitotic CDK, cells lose the controls that limit their re-entry into S phase before mitosis but retain the ability to prevent re-initiation of replication at each origin in a single S phase. Thus active mitotic CDK plays an essential role in preventing the onset of S phase. In vivo experiments in budding yeast revealed a direct correlation between the inhibition of mitotic CDK and the assembly of pre-RC on chromatin (Dahmann et al., 1995). This observation may imply that high mitotic CDK activity in cells may prevent nuclear accumulation of Cdc6/Cdc18 required for the assembly of pre-RC. Therefore mitotic CDK must be inactivated, which normally occurs at the end of mitosis, in order for pre-RC to re-assemble during G_1 phase. Once again, an increase in S phase cyclin CDK, which occurs in direct relation to the increase in cell size, would trigger the entry of cells into S phase by activating pre-RC to initiate DNA replication.

ENZYMES OF DNA SYNTHESIS IN THE REPLICATION COMPLEXES

Pre-RC with helicase activity, described above, allows the assembly of DNA polymerase-α-primase at the origins to initiate DNA replication. Elongation of initiated DNA replication strands is then facilitated by the concerted action of a number of enzymes and proteins at fixed sites within the nuclei. These sites, referred to as replication machineries or replication factories, have been the subject of both immunocytochemical and biochemical studies. Autoradiographic (Pardoll et al., 1980) and fluorescent immunocytochemical (Adachi and Laemmli, 1992; Cardoso et al., 1993; Kill et al., 1991; Nakamura et al., 1984) studies have been useful in establishing that DNA replication occurs at fixed sites within the nuclei and also in the identification of some of the components of the replication machinery. Biochemical studies allowed the identification and characterization of the enzymes in mega-complexes, representing replication machineries/factories, isolated from the nuclei of proliferating cells. Isolated mega-complexes capable of replicating DNA in vitro have been suggested to increase the functional efficiency of the enzymes required for DNA synthesis (for a review, see Reddy and Fager, 1993).

In vitro measurements indicate the formidable task for DNA polymerases at replication forks to sustain the rapid rate of DNA synthesis under conditions in which deoxynucleoside triphosphate (dNTP) pools in cells are well below the Km required for their activity. Furthermore the rate of DNA replication fork movement (the rate of DNA synthesis) in eukaryotes (about 80 nucleotides/s/replication fork, or about 4×10^6 nucleotides/s/cell) is so rapid that the entire pool of dNTPs in a cell will be depleted within one minute of the initiation of DNA replication (Reddy, 1989). Thus the kinetics of enzyme reaction and the supply of dNTPs to meet the demands of their utilization during DNA replication warrant coordinated activation of, and interaction between, the enzymes of DNA replication and DNA precursor synthesis. Furthermore, considering that the sole purpose of deoxynucleotides generated by ribonucleotide reductase is to serve as substrates for DNA replication, it is likely that this and other enzymes of dNTP de novo synthesis are localized in close proximity to DNA replication in S phase cells. There is a growing body of evidence for such functional and physical interactions in prokaryotes as well as in eukaryotes (Chiu et al., 1982; Noguchi et al., 1983; Reddy and Mathews, 1978; Reddy and Pardee, 1980; Wheeler et al., 1996).

Physical Interaction between the Enymes of DNA Synthesis

A number of enzymes required for DNA synthesis in synchronized mammalian cells are shown to relocate from cytosol to the nucleus as cells transit from G_1 into S phase and assemble into a mega-complex called replitase (Fig. 5.4) (Reddy and Pardee, 1980). Enzymes of deoxynucleotide metabolism including ribonucleotide reductase, thymidylate synthase, thymidylate kinase, and nucleoside diphosphate kinase, in nuclear extracts of regenerating rat liver or Novikoff tumor cells are reported to co-sediment with DNA polymerase-α on sucrose density gradients (Baril et al., 1974). Such sedimentation of enzymes associated with dNTP synthesis and DNA replication is observed in variety of mammalian cells including Chinese hamster embryo fibroblast (CHEF/18) cells (Noguchi et al., 1983; Reddy and Pardee, 1980, 1982), mouse FM3A cells (Ayusawa et al., 1983), BHK fibroblast cells (Harvey and Pearson, 1988), and human lymphoblasts (Wickremasinghe and Hoffbrand, 1983; Wickremasinghe et al., 1982, 1983). Multi-enzyme complexes containing

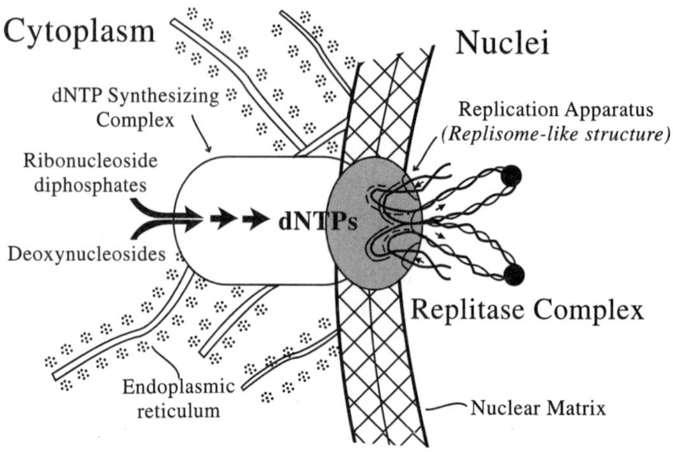

Figure 5.4. Hypothetical model of replitase complex in which dNTP synthesizing complex is juxtaposed with replication apparatus attached to the nuclear matrix. According to this model, DNA precursors (dNTPs) are compartmentalized in the microvicinity of DNA replication to facilitate a rapid rate of DNA synthesis. Endoplasmic reticulum (ER) may play a role translocation of enzymes of DNA replication from cytoplasm to the sites of DNA replication on nuclear matrix.

the enzymes of both dNTP synthesis and DNA replication were also observed in mammalian cells infected with herpes simplex virus-1 (Harvey and Pearson, 1988; Jong et al., 1984; Sclafani and Fangman, 1984) and adenovirus (Arens et al., 1977; Yamashita et al., 1977). In yeast, the replication of 2 micron extra-chromosomal plasmid DNA is also reported to be facilitated by multienzyme complexes of about 2 million daltons (Jazwinski and Edelman, 1984). Similar complexes characterized for the presence of enzymes of DNA replication, but not DNA precursor synthesis, were isolated from breast cancer epithelial cells (Coll et al., 1996) and Hela cells (Frouin et al., 2002).

Enzyme activities that are found associated with the replitase complex include: DNA polymerase-α-primase, 3′ to 5′ exonuclease, DNA topoisomerase II, thymidylate synthase, dihydrofolate reductase, ribonucleotide reductase, nucleoside diphosphate kinase, dCMP and dTMP kinases, and thymidine kinase (Hammond et al., 1989; Noguchi et al., 1983; Reddy, 1982; Reddy and Pardee, 1980). In addition insulin/IGF-I-regulated CaM-BP68 (Subramanyam et al., 1990), and S phase-specific cyclin A, Cdk2, and PCNA are found associated with the complex (Jaumot et al., 1994). Isolated replication complexes from breast cancer epithelial cells and HeLa cells are shown to contain cell cycle regulatory proteins and PCNA (Coll et al., 1996; Frouin et al., 2002). Also, androgen receptor (AR), which is required for the transition of androgen-sensitive prostate cancer epithelial (LNCaP) cells from G_1 into S phase (Cifuentes et al., 2003), is found in a complex that contains DNA polymerase activity (Reddy, manuscript in preparation). Biological implications and functional role of these cell cycle regulatory proteins and hormone receptor with isolated replication complexes from cancer cells remains to be determined.

A direct role of Ca^{++}/CaM in DNA synthesis is evident from the observation that CaM-specific monoclonal antibodies inhibit DNA replication in permeabilized S phase cells (Reddy et al., 1992a). In Chinese hamster embryo fibroblasts (Subramanyam et al., 1990) and hematopoietic progenitor cells (Reddy and Quesenberry, 1996; Reddy et al., 1992b, 1994), expression and nuclear localization of a specific calmodulin-binding protein of 68 kDa, called CaM-BP68, is associated with specific growth factor/cytokine-dependent progression from G_1 into S phase. Furthermore purified CaM-BP68 stimulates DNA replication in permeabilized density-arrested hematopoietic progenitor cells (Reddy et al., 1994). CaM-BP68 is associated with replitase complex in Chinese hamster embryo fibroblasts (Subramanyam et al., 1990) and the DNA polymerase-α-primase complex in HeLa cells (Cao et al., 1995).

Functional Interaction between the Enzymes of DNA Synthesis

Isolated replitase complex is shown to support DNA synthesis in vitro, and to facilitate deoxynucleoside triphosphate (dNTP) compartmentation in the micro-vicinity of DNA replication (Noguchi et al., 1983). In vitro (Noguchi et al., 1983; Wickremasinghe et al., 1982, 1983), in situ (Ayusawa et al., 1983; Reddy et al., 1982, 1986), and in vivo (Reddy, 1989) studies have led to the understanding that channeling and functional compartmentation of deoxynucleotides in the nucleus of mammalian cells are facilitated by interaction between enzymes of DNA precursor synthesis and replication in such complexes. Interactions between DNA precursor synthesizing and DNA-replicating enzymes in replitase complex are allosteric in nature—in the sense that the functional state of one enzyme affects the activity of a second enzyme within the complex, resulting in their coordinated activation or cross-inhibition. As a consequence of such interactions, the in vivo catalytic activity of enzymes such as thymidylate synthase (TS) and DNA polymerase is confined to S phase, even though the enzyme levels, as measured in vitro, remain relatively constant throughout the cell cycle (Reddy, 1982). Furthermore Reddy and Pardee (1983) showed that a variety of antimetabolites cross inhibit TS in vivo but not in vitro. For instance, hydroxyurea (HU), which inhibits ribonucleotide reductase, shows an in vivo block of TS; TS, when isolated in soluble form, is not affected by HU. Similarly, in vivo, aphidicolin, an inhibitor of DNA polymerase-α, also blocks TS; in vitro, there is no inhibition. Cross-inhibition is a function of allosteric interaction between the enzymes of DNA synthesis in the replitase complex, rather than a consequence of deoxynucleotide pool disruption (Chiba et al., 1984; Nicander and Reichard, 1983); this has been established by systematic analysis of in vivo deoxynucleotide pool composition, under conditions where cross-inhibition occurs (Plucinski et al., 1990).

NUCLEAR CONTEXT IN THE CONTROL OF DNA REPLICATION

It appears that chromatin and nuclear architecture, rather than specific DNA sequences, specify the sites at which DNA is replicated in metazoan nuclei. Autoradiographic analysis of pulse-labeled DNA reveals that chromosomal DNA replicates in clusters of synchronously initiated replicons and that different clusters initiate at different times during S phase (Dubey and Raman, 1987; Hand, 1978). Replication in a cluster of replicons is initiated at discrete sites within the nucleus (Nakamura et al., 1986; Nakayasu and Berezney, 1989). Each S phase nucleus contains about 100 to 300 such sites. The number of replication forks at each site ranges from 20 to 40 in cultured mammalian cells (Hassan and Cook, 1993; Nakamura et al., 1986).

Replication Foci Attached to the Nuclear Structure as Detected in S Phase Cells

The nuclear matrix structure that remains after extraction of cells with DNase I and 0.2 M ammonium sulfate retains the ability to incorporate biotinylated-dUMP into DNA, and the nascent DNA synthesized on templates attached to the nuclear matrix exhibit a pattern of replication foci similar to that seen in intact calls labeled with BrdU (Nakayasu and Berezney, 1989). These observations point to a physical association between template DNA, the enzymes of DNA replication, and the nuclear matrix. Earlier studies also showed that nascent DNA is firmly attached to the nuclear matrix (Berezney and Coffey, 1975; Dijkwel et al., 1979; Jackson and Cook, 1986a; McCready et al., 1980; Mirkovitch et al., 1984; Pardoll et al., 1980; Vogelstein et al., 1980). Furthermore gel electrophoretic analysis reveals that nascent DNA and replication forks partition with the nuclear matrix (Vaughn et al., 1990). Attachment of replication origins to the nuclear matrix is indicated from the observations that the radiolabel incorporated into DNA at the onset of S phase stays in close proximity to the matrix during G_2 and the next S phase, whereas the radiolabel incorporated at a later time in S phase is chased into surrounding DNA loops away from the matrix (Aelen et al., 1983; Carri et al., 1986; Dijkwel et al., 1986).

Segments of DNA physically associated with the nuclear matrix are referred to as matrix-attached regions (MARs). Although no consensus sequence has been discerned for MARs, MAR activity may reside in certain sequence motifs (Nakagomi et al., 1994). MARs are often located in the vicinity of replication origins (Amati and Gasser, 1988, 1990; Dijkwel and Hamlin, 1988). A potential role of MARs is to retain *cis*-regulatory sequences in a nuclear subcompartment that is accessible to *trans*-acting factors required for replication and transcription (Mirkovitch et al., 1984). For a MAR to be functionally associated with the matrix, a specific protein or a complex of proteins that recognizes

MAR sequence motifs must be present in the nuclear matrix structure. A number of proteins that interact with MARs have been identified (Dickinson et al., 1992; Nakagomi et al., 1994; Romig et al., 1992). However, a potential role of these proteins in anchoring MARs to the matrix or in initiation of DNA replication is not known.

Localization of Enzymes and Proteins at Sites of DNA Replication

Incorporation of biotinylated-dUTP into DNA in isolated nuclear matrix structures (Nakayasu and Berezney, 1989) indicates that in addition to template DNA, matrix preparations contain enzymes of DNA replication at the sites where DNA is replicated. Several enzymes, including DNA polymerase-α and primase, are found associated with the nuclear matrix preparation in a cell cycle and DNA replication dependent manner (Collins and Chu, 1987; Jackson and Cook, 1986b; Mikhailov and Tsanev, 1983; Nakayasu and Berezney, 1989; Nishizawa et al., 1984; Smith and Berezney, 1980; Tubo and Berezney, 1987a; Tubo and Berezney, 1987b; Wood and Collins, 1986). Replication enzymes and proteins that are associated with a discrete granular structure and exhibit a punctate, rather than diffuse, distribution in the nucleus include DNA polymerase-α (Bensch et al., 1982; Nakamura et al., 1984, 1986; Yamamoto et al., 1984), DNA ligase I (Lasko et al., 1990), PCNA (Bravo and Macdonald-Bravo, 1987; Kill et al., 1991; Kitada et al., 1992; Madsen and Celis, 1985), single-stranded DNA-binding protein RP-A (Adachi and Laemmli, 1992; Wilcock and Lane, 1991), and two essential S phase protein kinases, cyclin A and Cdk-2 (Cardoso et al., 1993).

Based on electron microscopic analysis, both DNA polymerase-α and PCNA are associated with dense structures in which DNA is replicated (Hozak et al., 1994). These structures, referred to as replication factories, are attached to the nucleoskeleton. These factories appear at the end of G_1 phase and increase in size and decrease in number as S phase progresses (Hozak et al., 1994). Immunofluorescence studies reveal a similar nucleoskeletal localization pattern of the specific calmodulin-binding protein CaM-BP68 (Reddy, 1999), which is tightly associated with the DNA polymerase α-primase complex (Cao et al., 1995) and is involved in DNA replication (Reddy et al., 1994).

DNA methyltransferase, which methylates deoxycytidine residues, also associates with replication foci only during S phase (Leonhardt et al., 1992). This activity was reported earlier to be associated with the replitase complex isolated from S phase nuclei (Noguchi et al., 1983). A specific role of DNA methyltransferase in replication is not clear. In somatic cells B-type lamins also localize to replication foci (Moir et al., 1994). Immunodepletion of lamin B3 from *Xenopus* egg extract allows normal assembly of the nuclear envelope but prevents DNA replication (Jenkins et al., 1993; Meier et al., 1991; Newport et al., 1990); this suggests a direct role of B-type lamins in DNA replication (Hutchison et al., 1994).

Replicative Process at Fixed Sites within the Nuclei

A hypothetical model in which DNA is replicated by its spooling through a complex of immobilized enzymes is presented in Figure 5.5. According to this model several adjacent replicons that replicate synchronously are arranged in loops by the binding of a replication origin in each replicon to the replication apparatus, a replisome-like structure. The replication apparatus is a component of the replitase complex containing the enzymes of DNA replication and DNA precursor synthesis (Fig. 5.4). A group of replication apparatuses in a cluster of replicons may represent a replication factory (Hozak et al., 1993). Once replication is initiated, it extends bidirectionally 5′ to 3′ at replication forks that remain associated with the replication apparatus. As DNA from both sides of a replication origin spools through the complex of enzymes (as indicated by arrows in Fig. 5.5), two adjacent replication forks continue to extend until replication of a replicon comes close to completion. Two daughter strands generated by this process loop out from around the site where the replication origin was initially bound to the replication apparatus. This model accommodates most of the biochemical (existence of replication enzymes in mega complexes), biophysical (physical and functional organization of replicating DNA that is over 100 cm in length inside a nuclear compartment that is no more than 10 μm in diameter), and structural (association of enzymes of DNA replication with replication foci attached to the nucleoskeleton) aspects of DNA replication in mammalian cells.

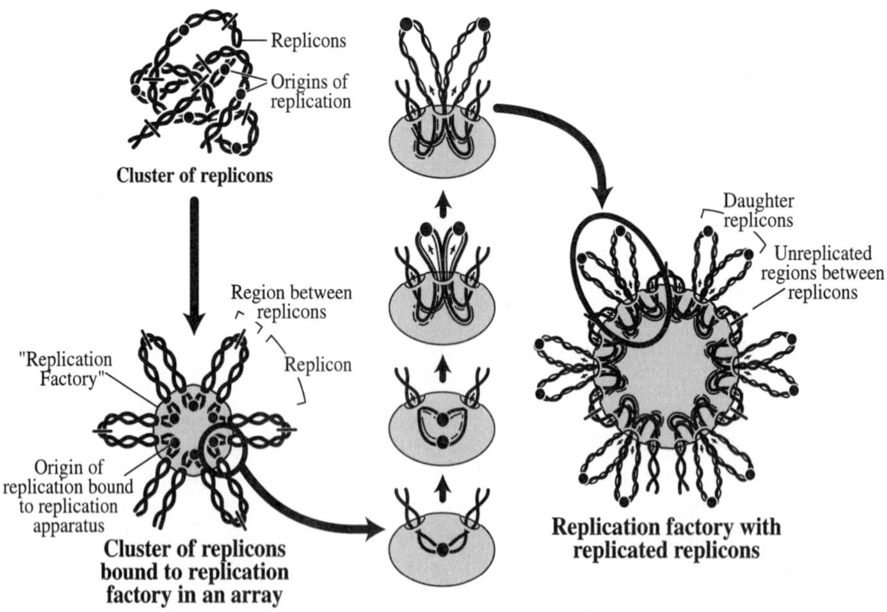

Figure 5.5. A hypothetical model depicting nuclear organization of a cluster of replicons during their replication by immobilized complexes of enzymes in a replication machinery or factory (reproduced from Reddy, 1999).

However, the model falls short of explaining how the stretches of DNA between two replicons in a replication factory and that between replication factories are fully replicated. A possible solution to this puzzle may lie in the observation that several adjacent replication factories merge as the S phase progresses (Hozak et al., 1993) and telomerase acting in concert with the replication machinery may also complete the replicative process (Harrington, 2003; Ray et al., 2002).

Role of Nuclear Membrane in Limiting Replication to Once per Cell Cycle

Cell fusion studies of Rao and Johnson (1970) indicate that the G_2 nucleus must go through mitosis before it is able to replicate again. Breakdown of nuclear membrane seems to be the primary event during mitosis that confers re-replicating ability to DNA in G_2 nuclei. This is based on the observation that G_2 nuclei incubated with fresh *Xenopus* egg extract are able to replicate again, without undergoing mitosis, only if their nuclear membrane is transiently permeabilized with non-ionic detergents or lysolecithin (Blow and Laskey, 1988; Coverley et al., 1993; Leno et al., 1992). Collectively these observations indicate that in order for the DNA in G2 nuclei to replicate again, it must be exposed to a cytoplasmic replication-initiating factor that is incapable of entering into nucleus unless its membrane breaks down at mitosis. This replication factor is referred to as "licensing factor" constituting the components of pre-RC described above (Blow and Laskey, 1988). According to the model proposed by Blow and Laskey (1988), binding of this factor to chromatin in reassembled mature nuclei allows initiation of replication at its binding sites. Once replication is initiated during S phase, the licensing factor bound to chromatin at initiation sites is inactivated or destroyed, making it incapable of binding to chromatin again. This ensures that DNA does not replicate more than once during each cell cycle. For DNA to replicate again, cytoplasmic licensing factor, represented by Cdt1, must gain access to the DNA, which requires breakdown of the nuclear membrane at mitosis. Thus the nuclear membrane seems to play an essential role in both, providing a structure on which DNA is replicated inside the nucleus and ensuring that DNA does not replicate more than once per cell cycle.

SUMMARY

Cellular preparation for entry into S phase begins following the completion of anaphase and continues through the entire G_1 phase. This preparation in mammalian cells is sensitive to the extracellular stimuli generated by growth promoting, as well as growth inhibitory, factors such as peptide growth factors/cytokines and hormones. Signals emanating from growth factor-receptor interactions control the expression and activation of proteins and enzymes required for the assembly of replication machinery at specific sites on chromatin (origins of DNA replication) and within the nuclei (replication foci). This involves first an orderly assembly of ORC, Cdc6/Cdc18, Cdt1, Mcm2-7, and Mcm10 to form pre-RC and then the activation of pre-RC by the joining of Cdc45 and DNA polymerase-α-primase at the origins of DNA replication. This orderly assembly of pre-RC during G_1 and its activation in S phase is governed by a coordinated action of CDKs and DDK, and 26S proteosome and anaphase-promoting complex (APC)-dependent destruction of CKIs, cyclins, and geminin. Activated pre-RC at the origins of DNA replication become the site at which replication machinery, containing the enzymes of both DNA precursor (dNTP) synthesis and DNA replication, present in the replitase complex, assembles to ensure rapid and faithful elongation of replication forks. Nuclear architecture, serving as the launching pad for DNA replication, seems to play a dynamic role in the overall assembly and functional stability of replication machinery and also in ensuring that none of the genome in a cell is duplicated more than once per cell cycle. This chapter provides a bird's-eye view of regulatory events controlling the onset of DNA synthesis, and entry of cells into S phase, as they are currently understood. Our understanding, however, is likely to change with the discovery of new factor or post-translational modification events that are associated with replication machinery. Further characterization of physical and functional interactions of E2F, Rb, DNA methyltransferase, and B-type lamins that are known to be associated with pre-RC or replication foci may also provide insights into the regulatory events coordinating DNA replication with transcription, and help in defining the role of nuclear architecture in the assembly of replication machinery at the sites of DNA replication during S phase.

ACKNOWLEDGMENTS

We gratefully acknowledge the support of NIH grant DK57864.

REFERENCES

Abdurashidova G, Deganuto M, Klima R, Riva S, Biamonti G, Giacca M, Falaschi A (2000): Start sites of bidirectional DNA synthesis at the human lamin B2 origin. Science 287:2023–6.

Adachi Y, Laemmli UK (1992): Identification of nuclear pre-replication centers poised for DNA synthesis in *Xenopus* egg extracts: immunolocalization study of replication protein A. J Cell Biol 119:1–15.

Adachi Y, Usukura J, Yanagida M (1997): A globular complex formation by Nda1 and the other five members of the MCM protein family in fission yeast. Genes Cells 2:467–79.

Aelen JM, Opstelten RJ, Wanka F (1983): Organization of DNA replication in Physarum polycephalum. Attachment of origins of replicons and replication forks to the nuclear matrix. Nucl Acids Res 11:1181–95.

Aladjem MI, Groudine M, Brody LL, Dieken ES, Fournier RE, Wahl GM, Epner EM (1995): Participation of the human beta-globin locus control region in initiation of DNA replication. Science 270:815–19.

Alexandrow MG, Ritzi M, Pemov A, Hamlin JL (2002): A potential role for mini-chromosome maintenance (MCM) proteins in initiation at the dihydrofolate reductase replication origin. J Biol Chem 277:2702–8.

Amati B, Gasser SM (1990): *Drosophila* scaffold-attached regions bind nuclear scaffolds and can function as ARS elements in both budding and fission yeasts. Mol Cell Biol 10:5442–54.

Amati BB, Gasser SM (1988): Chromosomal ARS and CEN elements bind specifically to the yeast nuclear scaffold. Cell 54:967–78.

Aparicio OM, Weinstein DM, Bell SP (1997): Components and dynamics of DNA replication complexes in *S. cerevisiae*: Redistribution of MCM proteins and Cdc45p during S phase. Cell 91:59–69.

Arens M, Yamashita T, Padmanabhan R, Tsuruo T, Green M (1977): Adenovirus deoxyribonucleic acid replication: Characterization of the enzyme activities of a soluble replication system. J Biol Chem 252:7947–54.

Austin RJ, Orr-Weaver TL, Bell SP (1999): *Drosophila* ORC specifically binds to ACE3, an origin of DNA replication control element. Genes Dev 13:2639–49.

Ayusawa D, Shimizu K, Koyama H, Takeishi K, Seno T (1983): Unusual aspects of human thymidylate synthase in mouse cells introduced by DNA-mediated gene transfer. J Biol Chem 258:48–53.

Baldin V, Lukas J, Marcote MJ, Pagano M, Draetta G (1993): Cyclin D1 is a nuclear protein required for cell cycle progression in G_1. Genes Dev 7:812–21.

Baril EF, Baril B, Elford H, Luftig R (1974): DNA polymerases and a possible multienzyme complex for DNA biosynthesis in eukaryotes. In: MK Kolber (ed): Mechanisms and DNA regulation of Replication. New York: Plenum Press, pp 275–91.

Beijersbergen RL, Carlee L, Kerkhoven RM, Bernards R (1995): Regulation of the retinoblastoma protein-related p107 by G_1 cyclin complexes. Genes Dev 9:1340–53.

Bell SP, Mitchell J, Leber J, Kobayashi R, Stillman B (1995): The multidomain structure of Orc1p reveals similarity to regulators of DNA replication and transcriptional silencing. Cell 83:563–8.

Bell SP, Stillman B (1992): ATP-dependent recognition of eukaryotic origins of DNA replication by a multiprotein complex. Nature 357:128–34.

Benbow RM, Zhao J, Larson DD (1992): On the nature of origins of DNA replication in eukaryotes. Bioessays 14:661–70.

Bensch KG, Tanaka S, Hu SZ, Wang TS, Korn D (1982): Intracellular localization of human DNA polymerase alpha with monoclonal antibodies. J Biol Chem 257:8391–6.

Berezney R, Coffey DS (1975): Nuclear protein matrix: Association with newly synthesized DNA. Science 189:291–3.

Bielinsky AK, Blitzblau H, Beall EL, Ezrokhi M, Smith HS, Botchan MR, Gerbi SA (2001): Origin recognition complex binding to a metazoan replication origin. Curr Biol 11:1427–31.

Blow JJ, Laskey RA (1988): A role for the nuclear envelope in controlling DNA replication within the cell cycle. Nature 332:546–8.

Blow JJ, Tada S (2000): Cell cycle. A new check on issuing the licence. Nature 404:560–1.

Bodrug SE, Warner BJ, Bath ML, Lindeman GJ, Harris AW, Adams JM (1994): Cyclin D1 transgene impedes lymphocyte maturation and collaborates in lymphomagenesis with the *myc* gene. EMBO J 13:2124–30.

Bosco G, Du W, Orr-Weaver TL (2001): DNA replication control through interaction of E2F-RB and the origin recognition complex. Nat Cell Biol 3:289–95.

Bramhill D, Kornberg A (1988): A model for initiation at origins of DNA replication. Cell 54:915–18.

Bravo R, Macdonald-Bravo H (1987): Existence of two populations of cyclin/proliferating cell nuclear antigen during the cell cycle: Association with DNA replication sites. J Cell Biol 105:1549–54.

Brewer BJ, Fangman WL (1987): The localization of replication origins on ARS plasmids in *S. cerevisiae*. Cell 51:463–71.

Brewer BJ, Fangman WL (1988): A replication fork barrier at the 3′ end of yeast ribosomal RNA genes. Cell 55:637–43.

Broach JR, Li YY, Feldman J, Jayaram M, Abraham J, Nasmyth KA, Hicks JB (1983): Localization and sequence analysis of yeast origins of DNA replication. Cold Spring Harb Symp Quant Biol 47 (Pt 2):1165–73.

Brown GW, Kelly TJ (1998): Purification of Hsk1, a minichromosome maintenance protein kinase from fission yeast. J Biol Chem 273:22083–90.

Bueno A, Russell P (1992): Dual functions of CDC6: A yeast protein required for DNA replication also inhibits nuclear division. EMBO J 11:2167–76.

Burhans WC, Huberman JA (1994): DNA replication origins in animal cells: A question of context? Science 263:639–40.

Burhans WC, Selegue JE, Heintz NH (1986a): Replication intermediates formed during initiation of DNA synthesis in methotrexate-resistant CHOC 400 cells are enriched for sequences derived from a specific, amplified restriction fragment. Biochemistry 25:441–9.

Burhans WC, Selegue JE, Heintz NH (1986b): Isolation of the origin of replication associated with the amplified Chinese hamster dihydrofolate reductase domain. Proc Natl Acad Sci USA 83:7790–4.

Burhans WC, Vassilev LT, Caddle MS, Heintz NH, DePamphilis ML (1990): Identification of an origin of bidirectional DNA replication in mammalian chromosomes. Cell 62:955–65.

Burke TW, Cook JG, Asano M, Nevins JR (2001): Replication factors MCM2 and ORC1 interact with the histone acetyltransferase HBO1. J Biol Chem 276:15397–408.

Caddle MS, Calos MP (1994): Specific initiation at an origin of replication from Schizosaccharomyces pombe. Mol Cell Biol 14:1796–805.

Cairns J (1966): Autoradiography of HeLa cell DNA. J Mol Biol 15:372–3.

Calzada A, Sacristan M, Sanchez E, Bueno A (2001): Cdc6 cooperates with Sic1 and Hct1 to inactivate mitotic cyclin-dependent kinases. Nature 412:355–8.

Cao QP, McGrath CA, Baril EF, Quesenberry PJ, Reddy GP (1995): The 68 kDa calmodulin-binding protein is tightly associated with the multiprotein DNA polymerase alpha-primase complex in HeLa cells. Biochemistry 34:3878–83.

Cardoso MC, Leonhardt H, Nadal-Ginard B (1993): Reversal of terminal differentiation and control of DNA replication: Cyclin A and Cdk2 specifically localize at subnuclear sites of DNA replication. Cell 74:979–92.

Carpenter PB, Mueller PR, Dunphy WG (1996): Role for a *Xenopus* Orc2-related protein in controlling DNA replication. Nature 379:357–60.

Carri MT, Micheli G, Graziano E, Pace T, Buongiorno-Nardelli M (1986): The relationship between chromosomal origins of replication and the nuclear matrix during the cell cycle. Exp Cell Res 164:426–36.

Carroll SM, DeRose ML, Kolman JL, Nonet GH, Kelly RE, Wahl GM (1993): Localization of a bidirectional DNA replication origin in the native locus and in episomally amplified murine adenosine deaminase loci. Mol Cell Biol 13: 2971–81.

Challberg MD, Kelly TJ (1989): Animal virus DNA replication. An Rev Biochem 58:671–717.

Chiba P, Bacon PE, Cory JG (1984): Studies directed toward testing the "channeling" hypothesis—ribonucleotides—DNA in leukemia L1210 cells. Biochem Biophys Res Commun 123:656–62.

Chiu CS, Cook KS, Greenberg GR (1982): Characteristics of a bacteriophage T4-induced complex synthesizing deoxyribonucleotides. J Biol Chem 257: 15087–97.

Chong JP, Mahbubani HM, Khoo CY, Blow JJ (1995): Purification of an MCM-containing complex as a component of the DNA replication licensing system. Nature 375:418–21.

Chong JP, Thommes P, Blow JJ (1996): The role of MCM/P1 proteins in the licensing of DNA replication. Trends Biochem Sci 21:102–6.

Chuang RY, Kelly TJ (1999): The fission yeast homologue of Orc4p binds to replication origin DNA via multiple AT-hooks. Proc Natl Acad Sci USA 96:2656–61.

Ciechanover A (1994): The ubiquitin-proteasome proteolytic pathway. Cell 79:13–21.

Cifuentes E, Croxen R, Menon M, Barrack ER, Reddy GPV (2003): Synchronized prostate cancer cells for studying androgen regulated events in cell cycle progression from G_1 into S phase. J Cell Physiol 195:337–45.

Clyne RK, Kelly TJ (1995): Genetic analysis of an ARS element from the fission yeast *Schizosaccharomyces pombe*. EMBO J 14:6348–57.

Cocker JH, Piatti S, Santocanale C, Nasmyth K, Diffley JF (1996): An essential role for the Cdc6 protein in forming the pre-replicative complexes of budding yeast. Nature 379:180–2.

Coleman TR, Carpenter PB, Dunphy WG (1996): The *Xenopus* Cdc6 protein is essential for the initiation of a single round of DNA replication in cell-free extracts. Cell 87:53–63.

Coll JM, Sekowski JW, Hickey RJ, Schnaper L, Yue W, Brodie AM, Uitto L, Syvaoja JE, Malkas LH (1996): The human breast cell DNA synthesome: Its purification from tumor tissue and cell culture. Oncol Res 8:435–47.

Collins JM, Chu AK (1987): Binding of the DNA polymerase alpha-DNA primase complex to the nuclear matrix in HeLa cells. Biochemistry 26:5600–7.

Coverley D, Downes CS, Romanowski P, Laskey RA (1993): Reversible effects of nuclear membrane permeabilization on DNA replication: Evidence for a positive licensing factor. J Cell Biol 122:985–92.

Coverley D, Laman H, Laskey RA (2002): Distinct roles for cyclins E and A during DNA replication complex assembly and activation. Nat Cell Biol 4:523–8.

Coxon A, Maundrell K, Kearsey SE (1992): Fission yeast cdc21$^+$ belongs to a family of proteins involved in an early step of chromosome replication. Nucl Acids Res 20:5571–7.

Dahmann C, Diffley JF, Nasmyth KA (1995): S-phase-promoting cyclin-dependent kinases prevent re-replication by inhibiting the transition of replication origins to a pre-replicative state. Curr Biol 5:1257–69.

Dalton S (1992): Cell cycle regulation of the human cdc2 gene. EMBO J 11: 1797–804.

Davey MJ, Indiani C, O'Donnell M (2003): Reconstitution of the Mcm2–7p heterohexamer, subunit arrangement, and ATP site architecture. J Biol Chem 278:4491–9.

Deng C, Zhang P, Harper JW, Elledge SJ, Leder P (1995): Mice lacking p21CIP1/WAF1 undergo normal development, but are defective in G_1 checkpoint control. Cell 82:675–84.

DePamphilis M (1987): Replication of simian virus 40 and polyoma virus chromosomes. In: Molecular Aspects of Papovaviruses. Boston: Martinus Nijhoff, pp 1–40.

DePamphilis ML (1993a): Eukaryotic DNA replication: Anatomy of an origin. An Rev Biochem 62:29–63.

DePamphilis ML (1993b): Origins of DNA replication that function in eukaryotic cells. Curr Opin Cell Biol 5:434–41.

DePamphilis ML (1993c): Origins of DNA replication in metazoan chromosomes. J Biol Chem 268:1–4.

Deshaies RJ, Chau V, Kirschner M (1995): Ubiquitination of the G_1 cyclin Cln2p by a Cdc34p-dependent pathway. EMBO J 14:303–12.

Deshpande AM, Newlon CS (1992): The ARS consensus sequence is required for chromosomal origin function in *Saccharomyces cerevisiae*. Mol Cell Biol 12:4305–13.

Devault A, Vallen EA, Yuan T, Green S, Bensimon A, Schwob E (2002): Identification of Tah11/Sid2 as the ortholog of the replication licensing factor Cdt1 in *Saccharomyces cerevisiae*. Curr Biol 12:689–94.

Dhar SK, Delmolino L, Dutta A (2001): Architecture of the human origin recognition complex. J Biol Chem 276:29067–71.

Dhar SK, Dutta A (2000): Identification and characterization of the human ORC6 homolog. J Biol Chem 275:34983–8.

Dhar V, Skoultchi AI, Schildkraut CL (1989): Activation and repression of a beta-globin gene in cell hybrids is accompanied by a shift in its temporal replication. Mol Cell Biol 9:3524–32.

Dickinson LA, Joh T, Kohwi Y, Kohwi-Shigematsu T (1992): A tissue-specific MAR/SAR DNA-binding protein with unusual binding site recognition. Cell 70:631–45.

Diffley JF (1996): Once and only once upon a time: Specifying and regulating origins of DNA replication in eukaryotic cells. Genes Dev 10:2819–30.

Diffley JF, Cocker JH (1992): Protein-DNA interactions at a yeast replication origin. Nature 357:169–72.

Diffley JF, Cocker JH, Dowell SJ, Rowley A (1994): Two steps in the assembly of complexes at yeast replication origins in vivo. Cell 78:303–16.

Dijkwel PA, Hamlin JL (1988): Matrix attachment regions are positioned near replication initiation sites, genes, and an interamplicon junction in the amplified dihydrofolate reductase domain of Chinese hamster ovary cells. Mol Cell Biol 8:5398–409.

Dijkwel PA, Hamlin JL (1992): Initiation of DNA replication in the dihydrofolate reductase locus is confined to the early S period in CHO cells synchronized with the plant amino acid mimosine. Mol Cell Biol 12:3715–22.

Dijkwel PA, Hamlin JL (1995): The Chinese hamster dihydrofolate reductase origin consists of multiple potential nascent-strand start sites. Mol Cell Biol 15:3023–31.

Dijkwel PA, Hamlin JL (1996): Sequence and context effects on origin function in mammalian cells. J Cell Biochem 62:210–22.

Dijkwel PA, Mullenders LH, Wanka F (1979): Analysis of the attachment of replicating DNA to a nuclear matrix in mammalian interphase nuclei. Nucl Acids Res 6:219–30.

Dijkwel PA, Wenink PW, Poddighe J (1986): Permanent attachment of replication origins to the nuclear matrix in BHK-cells. Nucl Acids Res 14:3241–9.

Dobrowolski S, Harter M, Stacey DW (1994): Cellular *ras* activity is required for passage through multiple points of the G_0/G_1 phase in BALB/c 3T3 cells. Mol Cell Biol 14:5441–9.

Donovan S, Harwood J, Drury LS, Diffley JF (1997): Cdc6p-dependent loading of Mcm proteins onto pre-replicative chromatin in budding yeast. Proc Natl Acad Sci USA 94:5611–16.

Dou QP, Levin AH, Zhao S, Pardee AB (1993): Cyclin E and cyclin A as candidates for the restriction point protein. Cancer Res 53:1493–7.

Dou QP, Markell PJ, Pardee AB (1992): Thymidine kinase transcription is regulated at G_1/S phase by a complex that contains retinoblastoma-like protein and a cdc2 kinase. Proc Natl Acad Sci USA 89:3256–60.

Dowell SJ, Romanowski P, Diffley JF (1994): Interaction of Dbf4, the Cdc7 protein kinase regulatory subunit, with yeast replication origins in vivo. Science 265:1243–6.

Drury LS, Perkins G, Diffley JF (1997): The Cdc4/34/53 pathway targets Cdc6p for proteolysis in budding yeast. EMBO J 16:5966–76.

Dubey DD, Kim SM, Todorov IT, Huberman JA (1996): Large, complex modular structure of a fission yeast DNA replication origin. Curr Biol 6:467–73.

Dubey DD, Raman R (1987): Factors influencing replicon organization in tissues having different S-phase durations in the mole rat, *Bandicota bengalensis*. Chromosoma 95:285–9.

Dulic V, Lees E, Reed SI (1992): Association of human cyclin E with a periodic G_1-S phase protein kinase. Science 257:1958–61.

Edwards MC, Tutter AV, Cvetic C, Gilbert CH, Prokhorova TA, Walter JC (2002): MCM2–7 complexes bind chromatin in a distributed pattern surrounding the origin recognition complex in *Xenopus* egg extracts. J Biol Chem 277: 33049–57.

El-Deiry WS, Tokino T, Velculescu VE, Levy DB, Parsons R, Trent JM, Lin D, Mercer WE, Kinzler KW, Vogelstein B (1993): WAF1, a potential mediator of p53 tumor suppression. Cell 75:817–25.

Fangman WL, Brewer BJ (1991): Activation of replication origins within yeast chromosomes. An Rev Cell Biol 7:375–402.

Ferguson BM, Brewer BJ, Reynolds AE, Fangman WL (1991): A yeast origin of replication is activated late in S phase. Cell 65:507–15.

Ferguson BM, Fangman WL (1992): A position effect on the time of replication origin activation in yeast. Cell 68:333–9.

Firpo EJ, Koff A, Solomon MJ, Roberts JM (1994): Inactivation of a Cdk2 inhibitor during interleukin 2-induced proliferation of human T lymphocytes. Mol Cell Biol 14:4889–901.

Forsburg SL, Nurse P (1994): The fission yeast cdc19$^+$ gene encodes a member of the MCM family of replication proteins. J Cell Sci 107:2779–88.

Foss M, McNally FJ, Laurenson P, Rine J (1993): Origin recognition complex (ORC) in transcriptional silencing and DNA replication in *S. cerevisiae*. Science 262:1838–44.

Frouin I, Montecucco A, Biamonti G, Hubscher U, Spadari S, Maga G (2002): Cell cycle-dependent dynamic association of cyclin/Cdk complexes with human DNA replication proteins. EMBO J 21:2485–95.

Fujita M, Yamada C, Goto H, Yokoyama N, Kuzushima K, Inagaki M, Tsurumi T (1999): Cell cycle regulation of human CDC6 protein. Intracellular localization, interaction with the human mcm complex, and CDC2 kinase-mediated hyperphosphorylation. J Biol Chem 274:25927–32.

Fujita M, Yamada C, Tsurumi T, Hanaoka F, Matsuzawa K, Inagaki M (1998): Cell cycle- and chromatin binding state-dependent phosphorylation of human MCM heterohexameric complexes. A role for cdc2 kinase. J Biol Chem 273: 17095–101.

Furstenthal L, Swanson C, Kaiser BK, Eldridge AG, Jackson PK (2001): Triggering ubiquitination of a CDK inhibitor at origins of DNA replication. Nat Cell Biol 3:715–22.

Gale JM, Tobey RA, D'Anna JA (1992): Localization and DNA sequence of a replication origin in the rhodopsin gene locus of Chinese hamster cells. J Mol Biol 224:343–58.

Gavin KA, Hidaka M, Stillman B (1995): Conserved initiator proteins in eukaryotes. Science 270:1667–71.

Geng Y, Eaton EN, Picon M, Roberts JM, Lundberg AS, Gifford A, Sardet C, Weinberg RA (1996): Regulation of cyclin E transcription by E2Fs and retinoblastoma protein. Oncogene 12:1173–80.

Gibson SI, Surosky RT, Tye BK (1990): The phenotype of the minichromosome maintenance mutant mcm3 is characteristic of mutants defective in DNA replication. Mol Cell Biol 10:5707–20.

Gilbert DM, Miyazawa H, Nallaseth FS, Ortega JM, Blow JJ, DePamphilis ML (1993): Site-specific initiation of DNA replication in metazoan chromosomes and the role of nuclear organization. Cold Spring Harb Symp Quant Biol 58: 475–85.

Girard F, Strausfeld U, Fernandez A, Lamb NJ (1991): Cyclin A is required for the onset of DNA replication in mammalian fibroblasts. Cell 67:1169–79.

Glotzer M, Murray AW, Kirschner MW (1991): Cyclin is degraded by the ubiquitin pathway. Nature 349:132–8.

Gossen M, Pak DT, Hansen SK, Acharya JK, Botchan MR (1995): A Drosophila homolog of the yeast origin recognition complex. Science 270:1674–7.

Gu Y, Turck CW, Morgan DO (1993): Inhibition of CDK2 activity in vivo by an associated 20K regulatory subunit. Nature 366:707–10.

Guan KL, Jenkins CW, Li Y, Nichols MA, Wu X, O'Keefe CL, Matera AG, Xiong Y (1994): Growth suppression by p18, a p16INK4/MTS1- and p14INK4B/

MTS2-related CDK6 inhibitor, correlates with wild-type pRb function. Genes Dev 8:2939–52.

Gyuris J, Golemis E, Chertkov H, Brent R (1993): Cdi1, a human G_1 and S phase protein phosphatase that associates with Cdk2. Cell 75:791–803.

Hammond RA, Miller MR, Gray MS, Reddy GPV (1989): Association of 3′—5′ exodeoxyribonuclease activity with DNA replitase complex from S-phase Chinese hamster embryo fibroblast cells. Exp Cell Res 183:284–93.

Hand R (1978): Eucaryotic DNA: Organization of the genome for replication. Cell 15:317–25.

Handeli S, Klar A, Meuth M, Cedar H (1989): Mapping replication units in animal cells. Cell 57:909–20.

Hannon GJ, Beach D (1994): p15INK4B is a potential effector of TGF-beta-induced cell cycle arrest. Nature 371:257–61.

Harrington L (2003): Biochemical aspects of telomerase function. Cancer Lett 194:139–54.

Harvey G, Pearson CK (1988): Search for multienzyme complexes of DNA precursor pathways in uninfected mammalian cells and in cells infected with herpes simplex virus type I. J Cell Physiol 134:25–36.

Hassan AB, Cook PR (1993): Visualization of replication sites in unfixed human cells. J Cell Sci 105:541–50.

Heintz NH, Hamlin JL (1982): An amplified chromosomal sequence that includes the gene for dihydrofolate reductase initiates replication within specific restriction fragments. Proc Natl Acad Sci USA 79:4083–7.

Hendrickson M, Madine M, Dalton S, Gautier J (1996): Phosphorylation of MCM4 by cdc2 protein kinase inhibits the activity of the minichromosome maintenance complex. Proc Natl Acad Sci USA 93:12223–8.

Hennessy KM, Lee A, Chen E, Botstein D (1991): A group of interacting yeast DNA replication genes. Genes Dev 5:958–69.

Hereford LM, Hartwell LH (1974): Sequential gene function in the initiation of *Saccharomyces cerevisiae* DNA synthesis. J Mol Biol 84:445–61.

Hilt W, Wolf DH (1996): Proteasomes: destruction as a programme. Trends Biochem Sci 21:96–102.

Hinds PW, Mittnacht S, Dulic V, Arnold A, Reed SI, Weinberg RA (1992): Regulation of retinoblastoma protein functions by ectopic expression of human cyclins. Cell 70:993–1006.

Hofmann JF, Beach D (1994): cdt1 is an essential target of the Cdc10/Sct1 transcription factor: Requirement for DNA replication and inhibition of mitosis. EMBO J 13:425–34.

Holmquist G, Gray M, Porter T, Jordan J (1982): Characterization of Giemsa dark- and light-band DNA. Cell 31:121–9.

Homesley L, Lei M, Kawasaki Y, Sawyer S, Christensen T, Tye BK (2000): Mcm10 and the MCM2-7 complex interact to initiate DNA synthesis and to release replication factors from origins. Genes Dev 14:913–26.

Hopwood B, Dalton S (1996): Cdc45p assembles into a complex with Cdc46p/Mcm5p, is required for minichromosome maintenance, and is essential for chromosomal DNA replication. Proc Natl Acad Sci USA 93:12309–14.

Hozak P, Hassan AB, Jackson DA, Cook PR (1993): Visualization of replication factories attached to nucleoskeleton. Cell 73:361–73.

Hozak P, Jackson DA, Cook PR (1994): Replication factories and nuclear bodies: the ultrastructural characterization of replication sites during the cell cycle. J Cell Sci 107:2191–202.

Hu B, Burkhart R, Schulte D, Musahl C, Knippers R (1993): The P1 family: a new class of nuclear mammalian proteins related to the yeast Mcm replication proteins. Nucl Acids Res 21:5289–93.

Hua XH, Newport J (1998): Identification of a preinitiation step in DNA replication that is independent of origin recognition complex and cdc6, but dependent on cdk2. J Cell Biol 140:271–81.

Huberman JA, Riggs AD (1966): Autoradiography of chromosomal DNA fibers from Chinese hamster cells. Proc Natl Acad Sci USA 55:599–606.

Huberman JA, Riggs AD (1968): On the mechanism of DNA replication in mammalian chromosomes. J Mol Biol 32:327–41.

Huberman JA, Zhu JG, Davis LR, Newlon CS (1988): Close association of a DNA replication origin and an ARS element on chromosome III of the yeast, *Saccharomyces cerevisiae*. Nucl Acids Res 16:6373–84.

Hunter T (1993): Braking the cycle. Cell 75:839–41.

Hunter T, Pines J (1994): Cyclins and cancer. II: Cyclin D and CDK inhibitors come of age. Cell 79:573–82.

Hutchison CJ, Bridger JM, Cox LS, Kill IR (1994): Weaving a pattern from disparate threads: Lamin function in nuclear assembly and DNA replication. J Cell Sci 107:3259–69.

Hyrien O, Maric C, Mechali M (1995): Transition in specification of embryonic metazoan DNA replication origins. Science 270:994–7.

Ishimi Y (1997): A DNA helicase activity is associated with an MCM4, –6, and –7 protein complex. J Biol Chem 272:24508–13.

Jackson AL, Pahl PM, Harrison K, Rosamond J, Sclafani RA (1993): Cell cycle regulation of the yeast Cdc7 protein kinase by association with the Dbf4 protein. Mol Cell Biol 13:2899–908.

Jackson DA, Cook PR (1986a): Replication occurs at a nucleoskeleton. EMBO J 5:1403–10.

Jackson DA, Cook PR (1986b): A cell-cycle-dependent DNA polymerase activity that replicates intact DNA in chromatin. J Mol Biol 192:65–76.

Jacob F, Brenner S, Cuzin F (1963): On the regulation of DNA replication in bacteria. Cold Spring Harb Symp Quant Biol 28:329–48.

Jallepalli PV, Brown GW, Muzi-Falconi M, Tien D, Kelly TJ (1997): Regulation of the replication initiator protein p65cdc18 by CDK phosphorylation. Genes Dev 11:2767–79.

Jaumot M, Grana X, Giordano A, Reddy GPV, Agell N, Bachs O (1994): Cyclin/cdk2 complexes in the nucleus of HeLa cells. Biochem Biophys Res Commun 203:1527–34.

Jazwinski SM, Edelman GM (1984): Evidence for participation of a multiprotein complex in yeast DNA replication in vitro. J Biol Chem 259:6852–7.

Jenkins H, Holman T, Lyon C, Lane B, Stick R, Hutchison C (1993): Nuclei that lack a lamina accumulate karyophilic proteins and assemble a nuclear matrix. J Cell Sci 106:275–85.

Jones SM, Kazlauskas A (2001): Growth-factor-dependent mitogenesis requires two distinct phases of signalling. Nat Cell Biol 3:165–72.

Jong AY, Kuo CL, Campbell JL (1984): The CDC8 gene of yeast encodes thymidylate kinase. J Biol Chem 259:11052–9.

Karlseder J, Rotheneder H, Wintersberger E (1996): Interaction of Sp1 with the growth- and cell cycle-regulated transcription factor E2F. Mol Cell Biol 16:1659–67.

Kato JY, Matsuoka M, Polyak K, Massague J, Sherr CJ (1994): Cyclic AMP-induced G_1 phase arrest mediated by an inhibitor (p27Kip1) of cyclin-dependent kinase 4 activation. Cell 79:487–96.

Kawasaki Y, Hiraga S, Sugino A (2000): Interactions between Mcm10p and other replication factors are required for proper initiation and elongation of chromosomal DNA replication in *Saccharomyces cerevisiae*. Genes Cells 5: 975–89.

Kearsey SE, Labib K (1998): MCM proteins: Evolution, properties, and role in DNA replication. Biochim Biophys Acta 1398:113–36.

Kearsey SE, Montgomery S, Labib K, Lindner K (2000): Chromatin binding of the fission yeast replication factor mcm4 occurs during anaphase and requires ORC and cdc18. EMBO J 19:1681–90.

Kelly RE, DeRose ML, Draper BW, Wahl GM (1995): Identification of an origin of bidirectional DNA replication in the ubiquitously expressed mammalian CAD gene. Mol Cell Biol 15:4136–48.

Kelly TJ, Brown GW (2000): Regulation of chromosome replication. An Rev Biochem 69:829–80.

Kelly TJ, Martin GS, Forsburg SL, Stephen RJ, Russo A, Nurse P (1993): The fission yeast cdc18+ gene product couples S phase to START and mitosis. Cell 74:371–82.

Kill IR, Bridger JM, Campbell KH, Maldonado-Codina G, Hutchison CJ (1991): The timing of the formation and usage of replicase clusters in S-phase nuclei of human diploid fibroblasts. J Cell Sci 100:869–76.

Kimura H, Nozaki N, Sugimoto K (1994): DNA polymerase alpha associated protein P1, a murine homolog of yeast MCM3, changes its intranuclear distribution during the DNA synthetic period. EMBO J 13:4311–20.

Kitada K, Johnston LH, Sugino T, Sugino A (1992): Temperature-sensitive cdc7 mutations of *Saccharomyces cerevisiae* are suppressed by the DBF4 gene, which is required for the G_1/S cell cycle transition. Genetics 131:21–9.

Kitsberg D, Selig S, Keshet I, Cedar H (1993): Replication structure of the human beta-globin gene domain. Nature 366:588–90.

Koff A, Giordano A, Desai D, Yamashita K, Harper JW, Elledge S, Nishimoto T, Morgan DO, Franza BR, Roberts JM (1992): Formation and activation of a cyclin E-cdk2 complex during the G_1 phase of the human cell cycle. Science 257:1689–94.

Koff A, Ohtsuki M, Polyak K, Roberts JM, Massague J (1993): Negative regulation of G_1 in mammalian cells: inhibition of cyclin E-dependent kinase by TGF-beta. Science 260:536–9.

Koh J, Enders GH, Dynlacht BD, Harlow E (1995): Tumour-derived p16 alleles encoding proteins defective in cell-cycle inhibition. Nature 375:506–10.

Kreitz S, Ritzi M, Baack M, Knippers R (2001): The human origin recognition complex protein 1 dissociates from chromatin during S phase in HeLa cells. J Biol Chem 276:6337–42.

Krek W, Ewen ME, Shirodkar S, Arany Z, Kaelin WG, Jr., Livingston DM (1994): Negative regulation of the growth-promoting transcription factor E2F-1 by a stably bound cyclin A-dependent protein kinase. Cell 78:161–72.

Kubota Y, Mimura S, Nishimoto S, Masuda T, Nojima H, Takisawa H (1997): Licensing of DNA replication by a multi-protein complex of MCM/P1 proteins in *Xenopus* eggs. EMBO J 16:3320–31.

Kubota Y, Mimura S, Nishimoto S, Takisawa H, Nojima H (1995): Identification of the yeast MCM3-related protein as a component of *Xenopus* DNA replication licensing factor. Cell 81:601–9.

La Thangue NB (1994): DRTF1/E2F: an expanding family of heterodimeric transcription factors implicated in cell-cycle control. Trends Biochem Sci 19:108–14.

Labib K, Tercero JA, Diffley JF (2000): Uninterrupted MCM2-7 function required for DNA replication fork progression. Science 288:1643–7.

Lam EW, Watson RJ (1993): An E2F-binding site mediates cell-cycle regulated repression of mouse B-*myb* transcription. EMBO J 12:2705–13.

Landis G, Kelley R, Spradling AC, Tower J (1997): The k43 gene, required for chorion gene amplification and diploid cell chromosome replication, encodes the *Drosophila* homolog of yeast origin recognition complex subunit 2. Proc Natl Acad Sci USA 94:3888–92.

Lasko DD, Tomkinson AE, Lindahl T (1990): Mammalian DNA ligases: Biosynthesis and intracellular localization of DNA ligase I. J Biol Chem 265: 12618–22.

Leatherwood J, Lopez-Girona A, Russell P (1996): Interaction of Cdc2 and Cdc18 with a fission yeast ORC2-like protein. Nature 379:360–3.

Lehner CF, O'Farrell PH (1989): Expression and function of *Drosophila* cyclin A during embryonic cell cycle progression. Cell 56:957–68.

Lei M, Kawasaki Y, Tye BK (1996): Physical interactions among Mcm proteins and effects of Mcm dosage on DNA replication in *Saccharomyces cerevisiae*. Mol Cell Biol 16:5081–90.

Lei M, Kawasaki Y, Young MR, Kihara M, Sugino A, Tye BK (1997): Mcm2 is a target of regulation by Cdc7-Dbf4 during the initiation of DNA synthesis. Genes Dev 11:3365–74.

Leno GH, Downes CS, Laskey RA (1992): The nuclear membrane prevents replication of human G2 nuclei but not G1 nuclei in *Xenopus* egg extract. Cell 69: 151–8.

Leonhardt H, Page AW, Weier HU, Bestor TH (1992): A targeting sequence directs DNA methyltransferase to sites of DNA replication in mammalian nuclei. Cell 71:865–73.

Leu TH, Hamlin JL (1989): High-resolution mapping of replication fork movement through the amplified dihydrofolate reductase domain in CHO cells by in-gel renaturation analysis. Mol Cell Biol 9:523–31.

Li JJ, Herskowitz I (1993): Isolation of ORC6, a component of the yeast origin recognition complex by a one-hybrid system. Science 262:1870–4.

Li R, Waga S, Hannon GJ, Beach D, Stillman B (1994): Differential effects by the p21 CDK inhibitor on PCNA-dependent DNA replication and repair. Nature 371:534–7.

Liang C, Spitzer JD, Smith HS, Gerbi SA (1993): Replication initiates at a confined region during DNA amplification in Sciara DNA puff II/9A. Genes Dev 7:1072–84.

Liang C, Weinreich M, Stillman B (1995): ORC and Cdc6p interact and determine the frequency of initiation of DNA replication in the genome. Cell 81:667–76.

Linskens MH, Huberman JA (1988): Organization of replication of ribosomal DNA in Saccharomyces cerevisiae. Mol Cell Biol 8:4927–35.

Linskens MH, Huberman JA (1990): The two faces of higher eukaryotic DNA replication origins. Cell 62:845–7.

Little RD, Platt TH, Schildkraut CL (1993): Initiation and termination of DNA replication in human rRNA genes. Mol Cell Biol 13:6600–13.

Loo S, Fox CA, Rine J, Kobayashi R, Stillman B, Bell S (1995): The origin recognition complex in silencing, cell cycle progression, and DNA replication. Mol Biol Cell 6:741–56.

Looney JE, Hamlin JL (1987): Isolation of the amplified dihydrofolate reductase domain from methotrexate-resistant Chinese hamster ovary cells. Mol Cell Biol 7:569–77.

Lopez-Girona A, Mondesert O, Leatherwood J, Russell P (1998): Negative regulation of Cdc18 DNA replication protein by Cdc2. Mol Biol Cell 9:63–73.

Lovec H, Grzeschiczek A, Kowalski MB, Moroy T (1994): Cyclin D1/bcl-1 cooperates with *myc* genes in the generation of B-cell lymphoma in transgenic mice. Embo J 13:3487–95.

Lu KH, Levine RA, Campisi J (1989): c-ras-Ha gene expression is regulated by insulin or insulinlike growth factor and by epidermal growth factor in murine fibroblasts. Mol Cell Biol 9:3411–7.

Luca FC, Shibuya EK, Dohrmann CE, Ruderman JV (1991): Both cyclin A delta 60 and B delta 97 are stable and arrest cells in M- phase, but only cyclin B delta 97 turns on cyclin destruction. EMBO J 10:4311–20.

Lukas J, Muller H, Bartkova J, Spitkovsky D, Kjerulff AA, Jansen-Durr P, Strauss M, Bartek J (1994): DNA tumor virus oncoproteins and retinoblastoma gene mutations share the ability to relieve the cell's requirement for cyclin D1 function in G_1. J Cell Biol 125:625–38.

Lukas J, Parry D, Aagaard L, Mann DJ, Bartkova J, Strauss M, Peters G, Bartek J (1995): Retinoblastoma-protein-dependent cell-cycle inhibition by the tumour suppressor p16. Nature 375:503–6.

Madine MA, Khoo CY, Mills AD, Laskey RA (1995a): MCM3 complex required for cell cycle regulation of DNA replication in vertebrate cells. Nature 375: 421–4.

Madine MA, Khoo CY, Mills AD, Musahl C, Laskey RA (1995b): The nuclear envelope prevents reinitiation of replication by regulating the binding of MCM3 to chromatin in *Xenopus* egg extracts. Curr Biol 5:1270–9.

Madsen P, Celis JE (1985): S-phase patterns of cyclin (PCNA) antigen staining resemble topographical patterns of DNA synthesis: A role for cyclin in DNA replication? FEBS Lett 193:5–11.

Maine GT, Sinha P, Tye BK (1984): Mutants of *S. cerevisiae* defective in the maintenance of minichromosomes. Genetics 106:365–85.

Maiorano D, Lemaitre JM, Mechali M (2000b): Stepwise regulated chromatin assembly of MCM2-7 proteins. J Biol Chem 275:8426–31.

Maiorano D, Moreau J, Mechali M (2000a): XCDT1 is required for the assembly of pre-replicative complexes in *Xenopus laevis*. Nature 404:622–5.

Maiorano D, Van Assendelft GB, Kearsey SE (1996): Fission yeast cdc21, a member of the MCM protein family, is required for onset of S phase and is located in the nucleus throughout the cell cycle. EMBO J 15:861–72.

Maiti AK, Sinha P (1992): The mcm2 mutation of yeast affects replication, rather than segregation or amplification of the two micron plasmid. J Mol Biol 224: 545–58.

Marahrens Y, Stillman B (1992): A yeast chromosomal origin of DNA replication defined by multiple functional elements. Science 255:817–23.

McCready SJ, Godwin J, Mason DW, Brazell IA, Cook PR (1980): DNA is replicated at the nuclear cage. J Cell Sci 46:365–86.

McGarry TJ, Kirschner MW (1998): Geminin, an inhibitor of DNA replication, is degraded during mitosis. Cell 93:1043–53.

Means AL, Slansky JE, McMahon SL, Knuth MW, Farnham PJ (1992): The HIP1 binding site is required for growth regulation of the dihydrofolate reductase gene promoter. Mol Cell Biol 12:1054–63.

Means AR (1994): Calcium, calmodulin and cell cycle regulation. FEBS Lett 347:1–4.

Meier J, Campbell KH, Ford CC, Stick R, Hutchison CJ (1991): The role of lamin LIII in nuclear assembly and DNA replication, in cell-free extracts of *Xenopus* eggs. J Cell Sci 98:271–9.

Mendez J, Stillman B (2000): Chromatin association of human origin recognition complex, cdc6, and minichromosome maintenance proteins during the cell cycle: Assembly of prereplication complexes in late mitosis. Mol Cell Biol 20: 8602–12.

Merchant AM, Kawasaki Y, Chen Y, Lei M, Tye BK (1997): A lesion in the DNA replication initiation factor Mcm10 induces pausing of elongation forks through chromosomal replication origins in *Saccharomyces cerevisiae*. Mol Cell Biol 17:3261–71.

Micklem G, Rowley A, Harwood J, Nasmyth K, Diffley JF (1993): Yeast origin recognition complex is involved in DNA replication and transcriptional silencing. Nature 366:87–9.

Mikhailov VS, Tsanev R (1983): Intranuclear localization of DNA polymerases alpha and beta in regenerating rat liver. Int J Biochem 15:855–9.

Milbrandt JD, Heintz NH, White WC, Rothman SM, Hamlin JL (1981): Methotrexate-resistant Chinese hamster ovary cells have amplified a 135-kilobase-pair region that includes the dihydrofolate reductase gene. Proc Natl Acad Sci USA 78:6043–7.

Mimura S, Takisawa H (1998): Xenopus Cdc45-dependent loading of DNA polymerase alpha onto chromatin under the control of S-phase Cdk. EMBO J 17:5699–707.

Minshull J, Blow JJ, Hunt T (1989): Translation of cyclin mRNA is necessary for extracts of activated *Xenopus* eggs to enter mitosis. Cell 56:947–56.

Mirkovitch J, Mirault ME, Laemmli UK (1984): Organization of the higher-order chromatin loop: specific DNA attachment sites on nuclear scaffold. Cell 39: 223–32.

Miyake S, Okishio N, Samejima I, Hiraoka Y, Toda T, Saitoh I, Yanagida M (1993): Fission yeast genes nda1$^+$ and nda4$^+$, mutations of which lead to S-phase block, chromatin alteration and Ca2$^+$ suppression, are members of the CDC46/MCM2 family. Mol Biol Cell 4:1003–15.

Moir RD, Montag-Lowy M, Goldman RD (1994): Dynamic properties of nuclear lamins: Lamin B is associated with sites of DNA replication. J Cell Biol 125: 1201–12.

Montagnoli A, Fiore F, Eytan E, Carrano AC, Draetta GF, Hershko A, Pagano M (1999): Ubiquitination of p27 is regulated by Cdk-dependent phosphorylation and trimeric complex formation. Genes Dev 13:1181–9.

Moreno S, Nurse P (1994): Regulation of progression through the G1 phase of the cell cycle by the rum1$^+$ gene. Nature 367:236–42.

Mudryj M, Hiebert SW, Nevins JR (1990): A role for the adenovirus inducible E2F transcription factor in a proliferation dependent signal transduction pathway. EMBO J 9:2179–84.

Muzi Falconi M, Brown GW, Kelly TJ (1996): cdc18$^+$ regulates initiation of DNA replication in *Schizosaccharomyces pombe*. Proc Natl Acad Sci USA 93:1566–70.

Muzi-Falconi M, Kelly TJ (1995): Orp1, a member of the Cdc18/Cdc6 family of S-phase regulators, is homologous to a component of the origin recognition complex. Proc Natl Acad Sci USA 92:12475–9.

Nakagomi K, Kohwi Y, Dickinson LA, Kohwi-Shigematsu T (1994): A novel DNA-binding motif in the nuclear matrix attachment DNA-binding protein SATB1. Mol Cell Biol 14:1852–60.

Nakamura H, Morita T, Masaki S, Yoshida S (1984): Intracellular localization and metabolism of DNA polymerase alpha in human cells visualized with monoclonal antibody. Expr Cell Res 151:123–33.

Nakamura H, Morita T, Sato C (1986): Structural organizations of replicon domains during DNA synthetic phase in the mammalian nucleus. Expr Cell Res 165:291–7.

Nakayasu H, Berezney R (1989): Mapping replicational sites in the eucaryotic cell nucleus. J Cell Biol 108:1–11.

Nasmyth K, Hunt T (1993): Cell cycle. Dams and sluices. Nature 366:634–5.

Natale DA, Li CJ, Sun WH, DePamphilis ML (2000): Selective instability of Orc1 protein accounts for the absence of functional origin recognition complexes during the M-G(1) transition in mammals. EMBO J 19:2728–38.

Newport JW, Wilson KL, Dunphy WG (1990): A lamin-independent pathway for nuclear envelope assembly. J Cell Biol 111:2247–59.

Nicander B, Reichard P (1983): Dynamics of pyrimidine deoxynucleoside triphosphate pools in relationship to DNA synthesis in 3T6 mouse fibroblasts. Proc Natl Acad Sci USA 80:1347–51.

Nishitani H, Lygerou Z, Nishimoto T, Nurse P (2000): The Cdt1 protein is required to license DNA for replication in fission yeast. Nature 404:625–8.

Nishitani H, Nurse P (1995): p65cdc18 plays a major role controlling the initiation of DNA replication in fission yeast. Cell 83:397–405.

Nishitani H, Taraviras S, Lygerou Z, Nishimoto T (2001): The human licensing factor for DNA replication Cdt1 accumulates in G_1 and is destabilized after initiation of S-phase. J Biol Chem 276:44905–11.

Nishizawa M, Tanabe K, Takahashi T (1984): DNA polymerases and DNA topoisomerases solubilized from nuclear matrices of regenerating rat livers. Biochem Biophys Res Commun 124:917–24.

Noda A, Ning Y, Venable SF, Pereira-Smith OM, Smith JR (1994): Cloning of senescent cell-derived inhibitors of DNA synthesis using an expression screen. Expr Cell Res 211:90–8.

Noguchi H, Reddy GPV, Pardee AB (1983): Rapid incorporation of label from ribonucleoside disphosphates into DNA by a cell-free high molecular weight fraction from animal cell nuclei. Cell 32:443–51.

Ogris E, Rotheneder H, Mudrak I, Pichler A, Wintersberger E (1993): A binding site for transcription factor E2F is a target for *trans* activation of murine thymidine kinase by polyomavirus large T antigen and plays an important role in growth regulation of the gene. J Virol 67:1765–71.

Ohtsubo M, Roberts JM (1993): Cyclin-dependent regulation of G_1 in mammalian fibroblasts. Science 259:1908–12.

Orr-Weaver TL (1991): *Drosophila* chorion genes: Cracking the eggshell's secrets. Bioessays 13:97–105.

Pagano M, Pepperkok R, Verde F, Ansorge W, Draetta G (1992): Cyclin A is required at two points in the human cell cycle. Embo J 11:961–71.

Pardee AB (1974): A restriction point for control of normal animal cell proliferation. Proc Natl Acad Sci USA 71:1286–90.

Pardee AB (1989): G_1 events and regulation of cell proliferation. Science 246: 603–8.

Pardoll DM, Vogelstein B, Coffey DS (1980): A fixed site of DNA replication in eucaryotic cells. Cell 19:527–36.

Pearson BE, Nasheuer HP, Wang TS (1991): Human DNA polymerase alpha gene: Sequences controlling expression in cycling and serum-stimulated cells. Mol Cell Biol 11:2081–95.

Perkins G, Diffley JF (1998): Nucleotide-dependent prereplicative complex assembly by Cdc6p, a homolog of eukaryotic and prokaryotic clamp-loaders. Mol Cell 2:23–32.

Piatti S, Lengauer C, Nasmyth K (1995): Cdc6 is an unstable protein whose de novo synthesis in G_1 is important for the onset of S phase and for preventing a "reductional" anaphase in the budding yeast Saccharomyces cerevisiae. EMBO J 14:3788–99.

Pines J, Hunter T (1990): Human cyclin A is adenovirus E1A-associated protein p60 and behaves differently from cyclin B. Nature 346:760–3.

Plucinski TM, Fager RS, Reddy GPV (1990): Allosteric interaction of components of the replitase complex is responsible for enzyme cross-inhibition. Mol Pharmacol 38:114–20.

Polyak K, Lee MH, Erdjument-Bromage H, Koff A, Roberts JM, Tempst P, Massague J (1994): Cloning of p27Kip1, a cyclin-dependent kinase inhibitor and a potential mediator of extracellular antimitogenic signals. Cell 78:59–66.

Prokhorova TA, Blow JJ (2000): Sequential MCM/P1 subcomplex assembly is required to form a heterohexamer with replication licensing activity. J Biol Chem 275:2491–8.

Quelle DE, Ashmun RA, Shurtleff SA, Kato JY, Bar-Sagi D, Roussel MF, Sherr CJ (1993): Overexpression of mouse D-type cyclins accelerates G_1 phase in rodent fibroblasts. Genes Dev 7:1559–71.

Rao PN, Johnson RT (1970): Mammalian cell fusion: studies on the regulation of DNA synthesis and mitosis. Nature 225:159–64.

Ray S, Karamysheva Z, Wang L, Shippen DE, Price CM (2002): Interactions between telomerase and primase physically link the telomere and chromosome replication machinery. Mol Cell Biol 22:5859–68.

Reddy GPV, Klinge EM, Pardee AB (1986): Ribonucleotides are channeled into a mixed DNA-RNA polymer by permeabilized hamster cells. Biochem Biophys Res Commun 135:340–6.

Reddy GPV (1982): Catalytic function of thymidylate synthase is confined to S phase due to its association with replitase. Biochem Biophys Res Commun 109:908–15.

Reddy GPV (1989): Compartmentation of deoxypyrimidine nucleotides for nuclear DNA replication in S phase mammalian cells. J Mol Recognit 2:75–83.

Reddy GPV (1999): DNA replication and S phase. In GS Stein, R Baserga, A Giordano, DT Denhardt (eds): The Molecular Basis of Cell Cycle and Growth Control. New York: Wiley-Liss, pp 80–154.

Reddy GPV, Fager RS (1993): Replitase: A complex integrating dNTP synthesis and DNA replication. Crit Rev Eukaryot Gene Expr 3:255–77.

Reddy GPV, Mathews CK (1978): Functional compartmentation of DNA precursors in T4 phage-infected bacteria. J Biol Chem 253:3461–7.

Reddy GPV, Pardee AB (1980): Multienzyme complex for metabolic channeling in mammalian DNA replication. Proc Natl Acad Sci USA 77:3312–16.

Reddy GPV, Pardee AB (1982): Coupled ribonucleoside diphosphate reduction, channeling, and incorporation into DNA of mammalian cells. J Biol Chem 257:12527–31.

Reddy GPV, Pardee AB (1983): Inhibitor evidence for allosteric interaction in the replitase multienzyme complex. Nature 304:86–8.

Reddy GPV, Quesenberry PJ (1996): Stem cell factor enhances interleukin-3 dependent induction of 68-kD calmodulin-binding protein and thymidine kinase activity in NFS-60 cells. Blood 87:3195–202.

Reddy GPV, Reed WC, Deacon DH, Quesenberry PJ (1992b): Growth factor-dependent proliferative stimulation of hematopoietic cells is associated with the modulation of cytoplasmic and nuclear 68-Kd calmodulin-binding protein. Blood 79:1946–55.

Reddy GPV, Reed WC, Deacon DH, Quesenberry PJ (1994): Growth factor modulated calmodulin-binding protein stimulates nuclear DNA synthesis in hemopoietic progenitor cells. Biochemistry 33:6605–10.

Reddy GPV, Reed WC, Sheehan E, Sacks DB (1992a): Calmodulin-specific monoclonal antibodies inhibit DNA replication in mammalian cells. Biochemistry 31:10426–30.

Reed SI, Dulic V, Lew DJ, Richardson HE, Wittenberg C (1992): G_1 control in yeast and animal cells. Ciba Found Symp 170:7–15; Discussion 15–9.

Resnitzky D, Gossen M, Bujard H, Reed SI (1994): Acceleration of the G_1/S phase transition by expression of cyclins D1 and E with an inducible system. Mol Cell Biol 14:1669–79.

Resnitzky D, Hengst L, Reed SI (1995): Cyclin A-associated kinase activity is rate limiting for entrance into S phase and is negatively regulated in G_1 by p27Kip1. Mol Cell Biol 15:4347–52.

Resnitzky D, Reed SI (1995): Different roles for cyclins D_1 and E in regulation of the G_1-to-S transition. Mol Cell Biol 15:3463–9.

Rivier DH, Rine J (1992): An origin of DNA replication and a transcription silencer require a common element. Science 256:659–63.

Romig H, Fackelmayer FO, Renz A, Ramsperger U, Richter A (1992): Characterization of SAF-A, a novel nuclear DNA binding protein from HeLa cells with high affinity for nuclear matrix/scaffold attachment DNA elements. EMBO J 11:3431–40.

Rowles A, Tada S, Blow JJ (1999): Changes in association of the *Xenopus* origin recognition complex with chromatin on licensing of replication origins. J Cell Sci 112 (Pt 12):2011–8.

Sasaki MS, Norman A (1966): DNA fibres from human lymphocyte nuclei. Expr Cell Res 44:642–5.

Sato N, Arai K, Masai H (1997): Human and Xenopus cDNAs encoding budding yeast Cdc7-related kinases: in vitro phosphorylation of MCM subunits by a putative human homologue of Cdc7. EMBO J 16:4340–51.

Sauer K, Knoblich JA, Richardson H, Lehner CF (1995): Distinct modes of cyclin E/cdc2c kinase regulation and S-phase control in mitotic and endoreduplication cycles of *Drosophila* embryogenesis. Genes Dev 9:1327–39.

Schaarschmidt D, Ladenburger EM, Keller C, Knippers R (2002): Human Mcm proteins at a replication origin during the G_1 to S phase transition. Nucl Acids Res 30:4176–85.

Schwacha A, Bell SP (2001): Interactions between two catalytically distinct MCM subgroups are essential for coordinated ATP hydrolysis and DNA replication. Mol Cell 8:1093–104.

Sclafani RA, Fangman WL (1984): Yeast gene CDC8 encodes thymidylate kinase and is complemented by herpes thymidine kinase gene TK. Proc Natl Acad Sci USA 81:5821–5.

Selig S, Okumura K, Ward DC, Cedar H (1992): Delineation of DNA replication time zones by fluorescence in situ hybridization. EMBO J 11:1217–25.

Seufert W, Futcher B, Jentsch S (1995): Role of a ubiquitin-conjugating enzyme in degradation of S- and M-phase cyclins. Nature 373:78–81.

Sherr CJ (1993): Mammalian G_1 cyclins. Cell 73:1059–65.

Sherr CJ (1996): Cancer cell cycles. Science 274:1672–7.

Shinomiya T, Ina S (1993): DNA replication of histone gene repeats in *Drosophila melanogaster* tissue culture cells: Multiple initiation sites and replication pause sites. Mol Cell Biol 13:4098–106.

Shinomiya T, Ina S (1994): Mapping an initiation region of DNA replication at a single-copy chromosomal locus in *Drosophila melanogaster* cells by two-dimensional gel methods and PCR-mediated nascent-strand analysis: multiple replication origins in a broad zone. Mol Cell Biol 14:7394–403.

Sicinski P, Donaher JL, Parker SB, Li T, Fazeli A, Gardner H, Haslam SZ, Bronson RT, Elledge SJ, Weinberg RA (1995): Cyclin D1 provides a link between development and oncogenesis in the retina and breast. Cell 82: 621–30.

Smith HC, Berezney R (1980): DNA polymerase alpha is tightly bound to the nuclear matrix of actively replicating liver. Biochem Biophys Res Commun 97:1541–7.

Stillman B (1989): Initiation of eukaryotic DNA replication in vitro. An Rev Cell Biol 5:197–245.

Stinchcomb DT, Struhl K, Davis RW (1979): Isolation and characterisation of a yeast chromosomal replicator. Nature 282:39–43.

Strausfeld UP, Howell M, Descombes P, Chevalier S, Rempel RE, Adamczewski J, Maller JL, Hunt T, Blow JJ (1996): Both cyclin A and cyclin E have S-phase promoting (SPF) activity in Xenopus egg extracts. J Cell Sci 109(Pt 6): 1555–63.

Subramanyam C, Honn SC, Reed WC, Reddy GPV (1990): Nuclear localization of 68 kDa calmodulin-binding protein is associated with the onset of DNA replication. J Cell Physiol 144:423–8.

Tada S, Li A, Maiorano D, Mechali M, Blow JJ (2001): Repression of origin assembly in metaphase depends on inhibition of RLF-B/Cdt1 by geminin. Nat Cell Biol 3:107–13.

Takahashi K, Yamada H, Yanagida M (1994): Fission yeast minichromosome loss mutants mis-cause lethal aneuploidy and replication abnormality. Mol Biol Cell 5:1145–58.

Takei Y, Assenberg M, Tsujimoto G, Laskey R (2002): The MCM3 acetylase MCM3AP inhibits initiation, but not elongation, of DNA replication via interaction with MCM3. J Biol Chem 277:43121–5.

Tanaka S, Diffley JF (2002): Interdependent nuclear accumulation of budding yeast Cdt1 and Mcm2-7 during G_1 phase. Nat Cell Biol 4:198–207.

Tanaka T, Knapp D, Nasmyth K (1997): Loading of an Mcm protein onto DNA replication origins is regulated by Cdc6p and CDKs. Cell 90:649–60.

Thommes P, Fett R, Schray B, Burkhart R, Barnes M, Kennedy C, Brown NC, Knippers R (1992): Properties of the nuclear P1 protein, a mammalian homologue of the yeast Mcm3 replication protein. Nucl Acids Res 20:1069–74.

Thommes P, Kubota Y, Takisawa H, Blow JJ (1997): The RLF-M component of the replication licensing system forms complexes containing all six MCM/P1 polypeptides. EMBO J 16:3312–9.

Todorov IT, Attaran A, Kearsey SE (1995): BM28, a human member of the MCM2-3-5 family, is displaced from chromatin during DNA replication. J Cell Biol 129:1433–45.

Todorov IT, Pepperkok R, Philipova RN, Kearsey SE, Ansorge W, Werner D (1994): A human nuclear protein with sequence homology to a family of early S phase proteins is required for entry into S phase and for cell division. J Cell Sci 107(Pt 1):253–65.

Treisman JE, Follette PJ, O'Farrell PH, Rubin GM (1995): Cell proliferation and DNA replication defects in a *Drosophila* MCM2 mutant. Genes Dev 9: 1709–15.

Tsai LH, Harlow E, Meyerson M (1991): Isolation of the human cdk2 gene that encodes the cyclin A- and adenovirus E1A-associated p33 kinase. Nature 353:174–7.

Tubo RA, Berezney R (1987a): Pre-replicative association of multiple replicative enzyme activities with the nuclear matrix during rat liver regeneration. J Biol Chem 262:1148–54.

Tubo RA, Berezney R (1987b). Nuclear matrix-bound DNA primase: Elucidation of an RNA priming system in nuclear matrix isolated from regenerating rat liver. J Biol Chem 262:6637–42.

Tye BK (1999): MCM proteins in DNA replication. An Rev Biochem 68:649–86.

Vashee S, Simancek P, Challberg MD, Kelly TJ (2001): Assembly of the human origin recognition complex. J Biol Chem 276:26666–73.

Vassilev L, Johnson EM (1990): An initiation zone of chromosomal DNA replication located upstream of the c-*myc* gene in proliferating HeLa cells. Mol Cell Biol 10:4899–904.

Vassilev LT, Burhans WC, DePamphilis ML (1990): Mapping an origin of DNA replication at a single-copy locus in exponentially proliferating mammalian cells. Mol Cell Biol 10:4685–9.

Vaughn JP, Dijkwel PA, Hamlin JL (1990): Replication initiates in a broad zone in the amplified CHO dihydrofolate reductase domain. Cell 61:1075–87.

Verma R, Annan RS, Huddleston MJ, Carr SA, Reynard G, Deshaies RJ (1997): Phosphorylation of Sic1p by G_1 Cdk required for its degradation and entry into S phase. Science 278:455–60.

Virta-Pearlman VJ, Gunaratne PH, Chinault AC (1993): Analysis of a replication initiation sequence from the adenosine deaminase region of the mouse genome. Mol Cell Biol 13:5931–42.

Vogelstein B, Pardoll DM, Coffey DS (1980): Supercoiled loops and eucaryotic DNA replicaton. Cell 22:79–85.

Waga S, Hannon GJ, Beach D, Stillman B (1994): The p21 inhibitor of cyclin-dependent kinases controls DNA replication by interaction with PCNA. Nature 369:574–8.

Waga S, Stillman B (1994): Anatomy of a DNA replication fork revealed by reconstitution of SV40 DNA replication in vitro. Nature 369:207–12.

Walker SS, Malik AK, Eisenberg S (1991): Analysis of the interactions of functional domains of a nuclear origin of replication from *Saccharomyces cerevisiae*. Nucl Acids Res 19:6255–62.

Walter J, Newport J (2000): Initiation of eukaryotic DNA replication: Origin unwinding and sequential chromatin association of Cdc45, RPA, and DNA polymerase alpha. Mol Cell 5:617–27.

Wheeler LJ, Ray NB, Ungermann C, Hendricks SP, Bernard MA, Hanson ES, Mathews CK (1996): T4 phage gene 32 protein as a candidate organizing factor for the deoxyribonucleoside triphosphate synthetase complex. J Biol Chem 271:11156–62.

Whittaker AJ, Royzman I, Orr-Weaver TL (2000): *Drosophila* double parked: A conserved, essential replication protein that colocalizes with the origin recognition complex and links DNA replication with mitosis and the down-regulation of S phase transcripts. Genes Dev 14:1765–76.

Wickremasinghe RG, Hoffbrand AV (1983): Inhibition by aphidicolin and dideoxythymidine triphosphate of a multienzyme complex of DNA synthesis from human cells. FEBS Lett 159:175–9.

Wickremasinghe RG, Yaxley JC, Hoffbrand AV (1982): Solubilization and partial characterization of a multienzyme complex of DNA synthesis from human lymphoblastoid cells. Eur J Biochem 126:589–96.

Wickremasinghe RG, Yaxley JC, Hoffbrand AV (1983): Gel filtration of a complex of DNA polymerase and DNA precursor-synthesizing enzymes from a human lymphoblastoid cell line. Biochim Biophys Acta 740:243–8.

Wilcock D, Lane DP (1991): Localization of p53, retinoblastoma and host replication proteins at sites of viral replication in herpes-infected cells. Nature 349:429–31.

Williams RS, Shohet RV, Stillman B (1997): A human protein related to yeast Cdc6p. Proc Natl Acad Sci USA 94:142–7.

Wohlgemuth JG, Bulboaca GH, Moghadam M, Caddle MS, Calos MP (1994): Physical mapping of origins of replication in the fission yeast *Schizosaccharomyces pombe*. Mol Biol Cell 5:839–49.

Wohlschlegel JA, Dhar SK, Prokhorova TA, Dutta A, Walter JC (2002): *Xenopus* Mcm10 binds to origins of DNA replication after Mcm2-7 and stimulates origin binding of Cdc45. Mol Cell 9:233–40.

Wohlschlegel JA, Dwyer BT, Dhar SK, Cvetic C, Walter JC, Dutta A (2000): Inhibition of eukaryotic DNA replication by geminin binding to Cdt1. Science 290:2309–12.

Wood SH, Collins JM (1986): Preferential binding of DNA primase to the nuclear matrix in HeLa cells. J Biol Chem 261:7119–22.

Xiong Y, Zhang H, Beach D (1993): Subunit rearrangement of the cyclin-dependent kinases is associated with cellular transformation. Genes Dev 7: 1572–83.

Yaglom J, Linskens MH, Sadis S, Rubin DM, Futcher B, Finley D (1995): p34Cdc28-mediated control of Cln3 cyclin degradation. Mol Cell Biol 15: 731–41.

Yamamoto S, Takahashi T, Matsukage A (1984): Tight association of DNA polymerase alpha with granular structures in the nuclear matrix of chick embryo cell: immunocytochemical detection with monoclonal antibody against DNA polymerase alpha. Cell Struct Funct 9:83–90.

Yamashita T, Arens M, Green M (1977): Adenovirus deoxyribonucleic acid replication. Isolation of a soluble replication system and analysis of the in vitro DNA product. J Biol Chem 252:7940–6.

Yan H, Merchant AM, Tye BK (1993): Cell cycle-regulated nuclear localization of MCM2 and MCM3, which are required for the initiation of DNA synthesis at chromosomal replication origins in yeast. Genes Dev 7:2149–60.

Yanagi K, Mizuno T, You Z, Hanaoka F (2002): Mouse geminin inhibits not only Cdt1-MCM6 interactions but also a novel intrinsic Cdt1 DNA binding activity. J Biol Chem 277:40871–80.

Yanow SK, Lygerou Z, Nurse P (2001): Expression of Cdc18/Cdc6 and Cdt1 during G_2 phase induces initiation of DNA replication. EMBO J 20:4648–56.

Yoon HJ, Campbell JL (1991): The CDC7 protein of *Saccharomyces cerevisiae* is a phosphoprotein that contains protein kinase activity. Proc Natl Acad Sci USA 88:3574–8.

Yoon Y, Sanchez JA, Brun C, Huberman JA (1995): Mapping of replication initiation sites in human ribosomal DNA by nascent-strand abundance analysis. Mol Cell Biol 15:2482–9.

You Z, Ishimi Y, Masai H, Hanaoka F (2002): Roles of Mcm7 and Mcm4 subunits in the DNA helicase activity of the mouse Mcm4/6/7 complex. J Biol Chem 277:42471–9.

You Z, Komamura Y, Ishimi Y (1999): Biochemical analysis of the intrinsic Mcm4-Mcm6-mcm7 DNA helicase activity. Mol Cell Biol 19:8003–15.

Yunis JJ, Kuo MT, Saunders GF (1977): Localization of sequences specifying messenger RNA to light-staining G-bands of human chromosomes. Chromosoma 61:335–44.

Zindy F, Lamas E, Chenivesse X, Sobczak J, Wang J, Fesquet D, Henglein B, Brechot C (1992): Cyclin A is required in S phase in normal epithelial cells. Biochem Biophys Res Commun 182:1144–54.

Zou L, Mitchell J, Stillman B (1997): CDC45, a novel yeast gene that functions with the origin recognition complex and Mcm proteins in initiation of DNA replication. Mol Cell Biol 17:553–63.

Zou L, Stillman B (1998): Formation of a preinitiation complex by S-phase cyclin CDK-dependent loading of Cdc45p onto chromatin. Science 280:593–6.

CHAPTER 6

THE PROGRESSION AND REGULATION OF MITOTIC EVENTS

GREENFIELD SLUDER[1], EDWARD H. HINCHCLIFFE[2], and CONLY L. RIEDER[3,4]
[1]Department of Cell Biology, University of Massachusetts Medical Center, Worcester, MA 01605
[2]Department of Biological Sciences and Walther Institute for Cancer Research, University of Notre Dame, Notre Dame, IN 46556
[3]Laboratory of Cell Regulation, Division of Molecular Medicine, Wadsworth Center, Albany, NY 12201-0509
[4]Department of Biomedical Sciences, State University of New York, Albany, NY 12222

INTRODUCTION

The purpose of the cell cycle is the formation of two genetically identical daughter cells. Mitosis is the division process that makes two cells out of one; it is the culmination of the cell cycle and the reason for all the events of growth and duplication. In this chapter we review the events of mitosis in animal cells and discuss some of the regulatory mechanisms that ensure the equal segregation of the genome.

PHASES OF MITOSIS

Historically mitosis has been separated into five phases: prophase, prometaphase, metaphase, anaphase, and telophase (Schrader, 1953). Over the years these stages, which are based on the structure and position of the chromosomes, have served as convenient labels to indicate how far the cell has progressed through mitosis, and define in a shorthand fashion what events are occurring at a particular time. However, it is important to keep in mind that the definitions of these stages are based

Cell Cycle and Growth Control: Biomolecular Regulation and Cancer, Second Edition, Edited by Gary S. Stein and Arthur B. Pardee
ISBN 0-471-25071-6

on morphology as seen by the light microscope, which does not reveal the underlying biochemical events at work. Today we know that many of these morphological events begin well before they become visible in the light microscope and thus, fall into more than one of the traditional stages. In assigning stages to mitosis one is slicing the flow of events into separate pieces based on the particular criteria used, be they morphological or biochemical. As a result the terminology used to traditionally describe the stages of mitosis can lead to ambiguity, especially when comparing mitosis in different experimental systems. Thus, in the light of new advances in our understanding of mitotic events, some of the classical terms are losing their precise meaning when used as undefined labels. Nevertheless, the classical terminology is still useful when used properly and consequently is still in widespread use today. For a thoughtful examination of how the traditional stages of mitosis fit with recent advances in our understanding of the molecular transitions that control the progress of mitosis, the reader is referred to Pines and Rieder (2001).

Prophase

Traditionally mitosis is defined to start in prophase (Figs. 6.1A, 6.2A), which begins when the light microscopist can first detect the presence of condensing chromosomes in the nucleus, and ends with nuclear envelope breakdown (NEB). Although the initiation of prophase is generally defined as the first visible signs of chromosome condensation, it is not clear that this phase has a sharply defined beginning. There is evidence that the chromosome condensation cycle is a continuum (Mazia, 1961; Pederson, 1972; Pederson and Robbins, 1972), beginning in S phase and reaching its greatest extent in early anaphase (Bajer, 1959, 1965). This means that the traditional term prophase does not have a precise meaning as a cell cycle transition but rather is a handy label to indicate that chromosome condensation is well underway and the cell will soon undergo NEB if not perturbed.

Chromosome condensation is an important event in the cell's preparations for mitosis because it compacts and decatenates the long and intertwined strands of DNA into discrete bodies that can align on the spindle and allows sister chromosomes to separate later in anaphase. Chromosome condensation has been estimated to involve an approximately 10,000- to 20,000-fold linear compaction of the DNA (Li et al., 1998). A thorough discussion of the mechanism of chromosome condensation is beyond the scope of this chapter, and for a review of the current understanding of this process, the reader is referred to Swedlow and Hirano (2003).

As prophase progresses, the chromosomes become progressively more condensed, the nucleoli dissipate, and the extensive interphase cytoplasmic microtubule array becomes reorganized into two focal arrays, known as asters, centered on the centrosomes (Fig. 6.2A). Ultimately these astral arrays, which are nucleated by the replicated centrosomes, separate to form the spindle poles and supply the microtubules

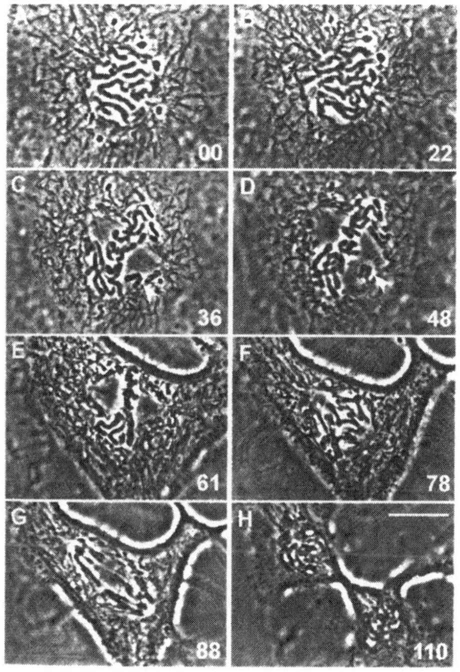

Figure 6.1. Sequential phase-contrast images of a living PtK_1 cell in the process of mitosis. (**A**) By *late prophase* the chromosomes are condensed within the nucleus and the nucleolar organizers have dissipated. (**B**) *Prometaphase* is initiated when the nuclear envelope breaks down to allow the chromosomes to interact with the centrosomes to form the spindle. (**C**) By *mid-prometaphase* all of the chromosomes have acquired a bipolar alignment, but one (white arrowhead) is still monooriented. (**D**) By *late prometaphase* this last monooriented chromosome (white arrowhead) has become bioriented and is congressing to the spindle equator. (**E**) At *metaphase* all of the chromosomes are positioned on the spindle equator, at approximately equal distances between the two poles. (**F**) As *anaphase* begins, the chromatids separate and move toward their respective poles. (**G**) During *late anaphase* the two spindle poles move farther apart in a process known as *anaphase B*, which additionally separates the two genomes. (**H**) During *telophase* the cytokinesis pinches the cell into two daughter cells, and a nuclear envelope re-forms around the two groups of chromosomes. Time in minutes is at the lower right corner of each picture. Bar in H = 15 µm.

used to construct the mitotic apparatus. However, the extent to which the aster are formed and separated by the time of NEB varies greatly between cells, even for neighboring cells in the same culture (reviewed in Rieder, 1990). In some cases the duplicated centrosomes remain close together with little evidence of astral microtubule assembly. In others, both asters are well developed and have separated to opposite sides of the nucleus before the end of prophase (Fig. 6.2A).

The force-producing mechanism for spindle pole separation has been the subject of much debate. Some favor the proposal that forces exerted

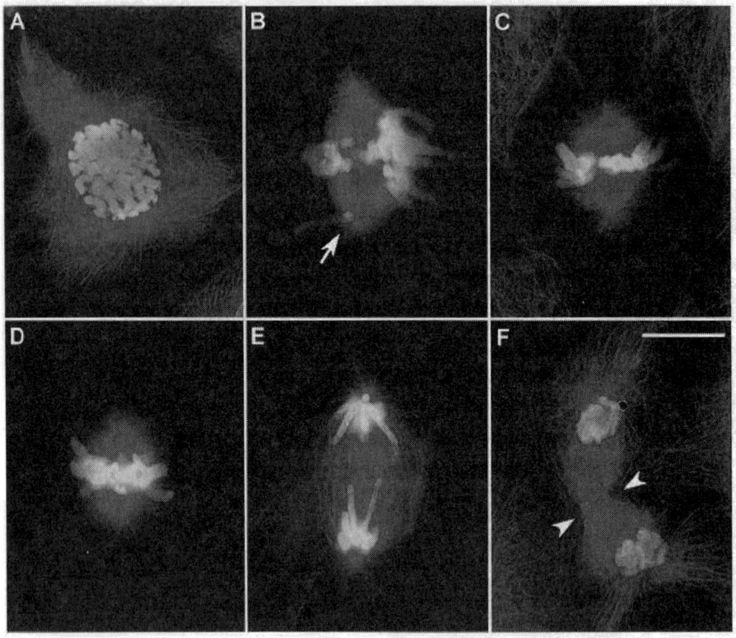

Figure 6.2. Gallery of fluorescent micrographs depicting glutaraldehyde-fixed and lysed PtK_1 cells in various stages of mitosis. The microtubules (red) were stained using indirect immunofluorescent methods, while the chromosomes (green) were stained with the DNA probe Hoechst 33342. (**A**) A *late prophase* cell in which the two centrosomes and their associated radial arrays of astral microtubules have separated to opposite sides of the nucleus. Note that there are still many cytoplasmic microtubules in this cell that are not associated with the asters. (**B**) A *mid-prometaphase* cell that contains one monooriented chromosome (white arrow), and several congressing chromosomes. By this time, all of the cytoplasmic microtubules have disassembled, and most of the astral microtubules have been incorporated into the spindle. (**C**) A *metaphase* cell in which all of the chromosomes are aligned on the spindle equator. (**D**) A cell just entering *anaphase* in which the chromatids are disjoining. Note the compact nature of the spindle (cf. C, D). (**E**) A *late anaphase* cell in which the two groups of chromosomes are already at the spindle poles, which themselves are moving farther apart (anaphase B). (**F**) A *telophase* cell in which the two groups of well-separated chromosomes are reforming nuclei, and in which cytokinesis (between the white arrowheads) is almost complete. The prominent bundle of microtubules between the two nuclei participates in cytokinesis and is known as the mid-body. Bar in F = 15 μm. (Figure courtesy of Dr. Alexey Khodjakov) (See color insert.)

between the two overlapping and antiparallel astral microtubule arrays push the poles apart, which is clearly the mechanism for spindle pole separation in yeast and diatoms (reviewed in Hogan and Cande, 1990). However, in vertebrate somatic cells, the asters continue to separate with normal kinetics even when their arrays of microtubules no longer overlap (Waters et al., 1993). In such cells the force-generating mechanism for centrosome separation is intrinsic to each aster, and the centrosomes pull themselves apart, perhaps as their associated arrays of astral microtubules interact with minus end directed motor molecules (e.g., cytoplasmic dynein) that are anchored in the cytoplasm or at the cell cortex (reviewed in Ault and Rieder, 1994). Recent work indicates that cells use multiple mechanisms to separate their asters. There is a balance of inward and outward forces generated by microtubule dynamics, cortical dynein, and a number of microtubule based motors located on the interpolar microtubules (Whitehead et al., 1996; Sharp et al., 2000; Brust-Mascher and Scholey, 2002).

Cell cycle progress through prophase into mitosis is driven by cyclin-dependent kinases (Cdk), primarily Cdk1 complexed with cyclin A and cyclin B (see Chapter 3 by H. Ford et al. in this volume for a thorough discussion of Cdk cascades in cell cycle regulation). Starting in S phase and continuing into prophase, the cyclin B proteins are synthesized and accumulate in the cytoplasm where they associate with Cdk1 whose activity drives the cell into mitosis. However, once this CdK1/cyclin B2 complex forms, its activity is inhibited by the phosphorylations on Thr-14 and Tyr-15 of Cdk1 by the Wee1 and Myt1 kinases located in the nucleus and cytoplasm (reviewed in Jackman and Pines, 1997; Smits and Medema, 2001). Then in late G_2/prophase these inhibitory phosphorylations are removed by members of the Cdc25 phosphastase family. A rapid rise in Cdk1-cyclin B activity is promoted by a positive feedback loop in which activating phosphorylation of Cdc25 phosphatases by Cdk1-cyclin B in turn increases the rate at which more Cdk1-cyclin B is activated.

Entry into mitosis is not necessarily a unitary event; both the nuclear and the cytoplasmic compartments must be coordinately brought into mitosis. Work with binucleate sea urchin zygotes has revealed that control of nuclear envelope breakdown is under local nuclear control and that cytoplasmic and nuclear entry into mitosis can be uncoupled (Sluder et al., 1995; Hinchcliffe et al., 1999). Thoughout G_2 and early prophase Cdk1-cyclin B enters the nucleus where Wee1 can put inhibitory phosphorylations on the Cdk1 and the kinase complex is actively exported from the nucleus (reviewed in Jackman and Pines, 1997; Smits and Medema, 2001). Then in late prophase the Cdk1/cyclin B2 complex translocates into the nucleus (Pines and Hunter, 1991; Gallant and Nigg, 1992). This nuclear import is thought to occur when the cytoplasmic retention signal (CRS) associated with cyclin B2 subunit becomes masked due to phosphorylation of Cyclin B (e.g., Li et al., 1997) and the interaction of the kinase complex with the nuclear export factor is inhibited.

In recent years, as an understanding of the cyclin-dependent kinases that control the cell cycle has increased, the beginning of mitosis has been increasingly defined as NEB. Using NEB as the marker for the start of mitosis has the advantage in that it is a discrete irreversible event that can be used for timing studies, and it integrates morphological changes in the cell with distinct biochemical events. NEB is controlled, in part, by the activity of Cdk1-cyclin B2, which allows the nuclear envelope to vesiculate by hyper-phosphorylating the lamin proteins that form a meshwork on the inner face of the nuclear envelope (reviewed in Gerace and Foisner, 1994; Fields and Thompson, 1995). Recent morphological analysis of NEB indicates that the growth of astral microtubules into the nuclear envelope may speed the process of NEB by providing mechanical disruption and facilitating the movement of pieces of the nuclear envelope toward the poles of the forming spindle (Terasaki et al., 2001; Beaudouin et al., 2002; Lenart et al., 2003). However, NEB is not strictly dependent on mechanical disruption of the nuclear envelope by microtubules since it occurs at the normal time when microtubule assembly is completely inhibited (Sluder, 1979; Rieder and Palazzo, 1992).

In addition to the nuclear events that lead to chromosome condensation and ultimately NEB, some cytoplasmic components, such as the duplicated centrosomes, also undergo extensive biochemical modifications during prophase. Immunological evidence reveals that some Cdk1-cyclin B1 accumulates at the centrosome during G_2 and may be activated there (Bailly et al., 1992; Pockwinse et al., 1997; Jackman et al., 2003). This centrosome-associated Cdk1/cyclin B1 is then activated near the G_2/M boundary by Cdc25B, whose concentration increases during prophase (Gabrielli et al., 1996). Possibly the activation of centrosome-associated Cdk1/cyclin B1 leads to the accumulation of the microtubule nucleating complexes at the centrosome and hyper-phosphorylation of some centrosomal proteins that may promote microtubule nucleation (see Vandre et al., 1984; and Fig. 6.2A). Throughout this period there is a conversion of the relatively stable interphase microtubule array into two independent radial arrays of dynamically unstable microtubules through the action of a number of accessory proteins that influence microtubule tip stability (see Scholey et al., 2003).

Prometaphase

The breakdown of the nuclear envelope (NEB) occurs over a 1 to 2 minute interval and it signals the start of prometaphase, the stage when the spindle forms (Figs. 6.1B–D, 6.2B). Three essential events must be accomplished during this phase if the division is to be normal: the cell must establish a bipolar spindle axis; the daughter chromatids of each replicated chromosome must become connected to opposing spindle poles (i.e., bioriented); and the chromosomes must become aligned at or near the spindle equator.

For animal cells spindle bipolarity is determined by the two radial arrays of centrosomal microtubules (asters) as they separate. As noted above, this may occur prior to NEB, or it may occur after NEB. Regardless, nearly all of the microtubules used to construct the spindle are derived from the centrosomes when they are present (Sluder and Rieder, 1985; reviewed in Brinkley, 1985). This is clearly demonstrated by the fact that cells with only one centrosome assemble a monopolar spindle, at least initially (Mazia et al., 1960; Bajer, 1982; Sluder and Begg, 1985), and that those with more than two centrosomes typically form multipolar spindles (Heneen, 1975; Sluder et al., 1997).

It is important to note, however, that spindle assembly does not occur only by this centrosome-based mechanism. There is a chromosome-based spindle assembly pathway that is revealed when centrosomes are naturally not present (e.g., in some female meiotic systems) or when centrosomes are experimentally removed from dividing cells (reviewed in Compton, 2000; Scholey et al., 2003). Earlier findings, some dating back almost 40 years, indicated that male and female meiotic cells of insects can form bipolar acentrosomal spindles (Dietz, 1964; Steffen et al., 1986). More recent work with Xenopus egg extracts has revealed that bipolar spindles will assemble from initially randomly oriented microtubules assembled in the vicinity of chromatin, be it chromosomes or beads coated with DNA fragments (Heald et al., 1996; 1997; reviewed in Karsenti and Vernos, 2001). Spindle assembly starts with the spontaneous assembly of randomly organized microtubules in the immediate vicinity of the chromatin. This is promoted by the guanine nucleotide-exchange factor RCC1 on the chromosomes that produces a spatial gradient of Ran-GTP centered on the chromatin (reviewed by Walczak, 2001). These microtubules are then bundled into antiparallel arrays by bipolar kinesins, and the minus ends are moved distal to the chromosomes by chromokinesins (a class of kinesins bound to the chromosomes). Finally, minus end directed motor molecules, such as cytoplasmic dynein, move to and crosslink the minus ends of the microtubules to form a somewhat focused spindle pole (Walczak et al., 1998; Karsenti and Vernos, 2001). Also the microtubule bundling protein NuMA, which accumulates at the polar ends of the spindle, may contribute to the focused anchorage of spindle microtubules (see Keating et al., 1997) and keep the centrosomes, when present, attached to the ends of the spindle (Heald et al., 1997).

Significantly, mammalian somatic cells that normally have centrosomes also have this chromosome-based spindle assembly pathway. When the centrosomes of African green monkey cells are laser ablated during prophase or microsurgically removed before mitosis, the cells assemble a functional bipolar spindle at mitosis (Khodjakov et al., 2000; Hinchcliffe et al., 2001). Importantly, this latter study also found that the time from nuclear envelope breakdown to nuclear envelope reformation in acentrosomal cells was almost three times as long at that in normal cells. This indicates that even though centrosomes may not be totally necessary for spindle assembly, their presence accelerates spindle assembly and alignment of chromosomes. Thus centrosomes, when present, promote the timeliness and fidelity of the mitotic process. In summary, it appears that spindle pole formation in higher animal cells is the result of the cooperative action of two mechanisms: the microtubule motor protein bundling/rearrangement of cytoplasmic microtubules and centrosomes.

Kinetochores are paired complex macromolecular assemblies that form on opposite sides of the primary constriction, or centromeric region, of each chromosome (reviewed in Rieder, 1982; Yen and Schaar, 1996). A chromosome becomes attached to the forming spindle when its "sister" kinetochores become associated with astral microtubules growing from the spindle poles (Rieder and Alexander, 1990). This attachment process is remarkably dynamic and depends on the properties of the microtubule ends as well as the ability of kinetochores to interact with these astral microtubules. At NEB astral microtubules grow into the volume previously occupied by the nucleus that now contains the condensed chromosomes. Although there is net growth of these microtubules, the growing tip of each is dynamically unstable (reviewed in Cassimeris et al., 1987); for each microtubule there is a variable period of growth followed by rapid shortening, either back to the centrosome or more likely to some intermediate length. Thereafter the tip grows again. The result is that the volume occupied by the chromosomes is constantly being probed by the tips of growing astral microtubules.

Chromosome attachment to the spindle is accomplished by the ability of kinetochores to capture the ends or walls of astral microtubules, thereby forming the kinetochore fibers of the spindle (reviewed in Rieder and Alexander, 1990; Mitchison, 1990; Skibbens et al., 1993). These fibers, in turn, serve as the scaffold on which the poleward (P) forces are produced to move the chromosomes. Normally, due to the stochastic nature of kinetochore fiber formation, the attachment of sister kinetochores to the spindle is asynchronous. As a rule, the first kinetochore to attach is the one located closest to and facing a spindle pole at NEB (reviewed in Rieder, 1990). This attachment "monoorients" the chromosome and allows the kinetochore to move toward that pole (see Rieder and Alexander, 1990; Khodjakov et al., 1996; Figs. 6.1C, 6.2B). Once near the pole, monooriented chromosomes begin to undergo continuous oscillatory movements toward and away from the pole, which reflects the directionally unstable nature of the attached kinetochore (reviewed in Khodjakov and Rieder, 1996). When moving toward the pole (P motion), the kinetochore is translocated by forces produced primarily at, or acting on, the kinetochore in concert with the coordinated disassembly of the ends of microtubules at the kinetochore. During away-from-the pole (AP) motion the kinetochore appears to be in a "neutral" non-force-producing state that allows it to be pushed AP, while its associated microtubules elongate, by the action of a polar ejection force that acts along the length of the chromosome. Some think that this polar ejection force, or "polar wind," is produced by the growth of astral microtubules that contact the chromosomes and push them away from the pole (Rieder et al., 1986; Khodjakov and Rieder, 1996; Ault et al., 1991; Cassimeris et al., 1994; reviewed in Rieder and Salmon, 1994). However, recent work indicates that the polar ejection force is also due to chromokinesin microtubule plus end motors that are associated with the chromosome arms, thereby sliding the chromosome toward the microtubule plus ends located in the spindle midzone (Wang and Aldler, 1995; Tokai et al., 1996; Antonio et al., 2000; Funabiki and Murray, 2000).

Regardless of the mechanism, as the chromosome moves AP the microtubules associated with the following kinetochore lengthen by the addition of tubulin subunits at that kinetochore.

The bipolar attachment of a monooriented chromosome occurs as the previously unattached kinetochore captures and stabilizes microtubules from the more distant aster (McEwen et al., 1997). Again, this is a stochastic process that relies on the chance encounter of a microtubule wall or tip with the kinetochore, and it may be facilitated by the constant positional changes of the monooriented chromosome (Rieder, 1990). When the unattached kinetochore eventually captures one or more microtubules, the now "bioriented" chromosome rapidly initiates movement to the spindle equator. During this "congression" process both sister kinetochores remain directionally unstable and continue show transient periods of P and AP motions. However, net changes in chromosome position occur because, once bioriented, the motilities of sister kinetochores become coordinated to allow for changes in chromosome position, and this coordination is thought to be mediated by a tension sensing mechanism that acts across the centromere (see Skibbens et al., 1995).

Over a variable period of time all of the chromosomes become attached to the spindle in a bipolar fashion and move to the midpoint or equator of the spindle. The aggregate of chromosomes positioned near or on the spindle equator forms the "metaphase plate." The establishment of this equilibrium position for any given chromosome is thought to be due to a balance between poleward pulling forces on the sister kinetochores, which is not necessarily "on" or equal at any given time, and the action of polar ejection forces, whose strength in each half spindle drop off from the pole to the equator as the density of the growing astral microtubules falls off (reviewed in Rieder and Salmon, 1994; Khodjakov and Rieder, 1996; McEwen et al., 1997). Thus, as a chromosome moves away from the metaphase plate toward a spindle pole, it encounters a progressively stronger force pushing it away from that pole. The poleward-moving leading kinetochore, now under greater tension, has a higher probability of becoming directionally unstable and changing from P movement to AP movement. As a consequence the chromosome moves back toward the metaphase plate. Although photographs of living or fixed cells might suggest that chromosomes at the metaphase plate cease moving, time lapse cinematography reveals that individually, they constantly oscillate back and forth across the metaphase plate, rarely making large excursions. In addition the size of the chromosomes determines whether or not the chromosomes are evenly distributed through the metaphase plate. During spindle formation there is a tendency for the larger chromosomes to be excluded from the spindle, and to be positioned at the periphery—with the kinetochores just within the spindle and the chromosome arms projecting into the cytoplasm. When viewed with the microscope along the axis of the spindle, cells with predominantly large chromosomes (e.g., newt lung cells and rat kangaroo cells) have a metaphase plate that looks like a ring of chromosomes. On the other hand, cells with very small chromosomes (e.g., HeLa, LLC-PK, and CHO cells) have a metaphase plate that is solidly packed with chromosomes.

Metaphase

When all the chromosomes are bioriented and positioned near the spindle equator the cell is considered to be in the metaphase stage of mitosis (Figs. 6.1E, 6.2C). This stage has been traditionally defined solely by morphological criteria, namely the alignment of all chromosomes on the metaphase plate. Also during metaphase the distance between the spindle poles decreases as the spindle becomes progressively more compacted (Fig. 6.2C–D). By morphological criteria metaphase represents the culmination of spindle assembly events occurring during prometaphase. As a consequence the classical cytological terms "metaphase arrest" and "metaphase block" have lost any real meaning when applied to cells treated with agents that prevent spindle microtubule assembly; such cells are arrested in prometaphase of mitosis and are not necessarily poised to initiate anaphase onset (reviewed in Rieder and Palazzo, 1992).

Anaphase

The anaphase stage of mitosis starts when the sister chromatids, of each replicated chromosome, disjoin to form two independent chromosomes, each of which immediately begins moving toward its attached spindle pole at 1 to 2 μm per minute. The initial disjunction of sister chromatids, as opposed to actual chromosome movement, does not depend on pulling forces generated by the spindle; when microtubule assembly is completely prevented, chromatid disjunction still occurs as the cell undergoes a delayed metaphase-anaphase transition (Eigsti and Dustin, 1955; Sluder, 1979; reviewed in Bajer and Mole-Bajer, 1972; Rieder and Palazzo, 1992). Also the P motion of the newly disjoined anaphase chromosomes does not appear to arise from the sudden activation of P force producers that begin to act on the kinetochore only during anaphase. The directionally unstable sister kinetochores on a metaphase chromosome undergo constant tension-related switches between P and AP activity state, and disjunction of the chromatids suddenly relieves the tension on both sister kinetochores which then allows them to switch into a P state of motion. When one kinetochore on a metaphase chromosome is destroyed by laser microsurgery the chromosome moves toward the other spindle pole with the same kinetics of an anaphase chromosome (reviewed in Rieder et al., 1995). Anaphase ends when chromosome P motion is completed. At some point in mid to late anaphase the process of cytokinesis (Figs. 6.1H, 6.2F), which pinches the cytoplasm in two between the separating groups of chromosomes, is also initiated but not always apparent (reviewed in Rappaport, 1969; White and Borisy, 1983; Salmon, 1989; Oegema and Mitchison, 1997).

The term "metaphase-anaphase transition" is widely used today to represent entry into the anaphase portion of the cell cycle. However, this term implies far more than chromatid disjunction and subsequent P chromosome motion; it refers to a fundamentally important transition in the cell cycle that commits the cell to finish mitosis and enter the next cell cycle. At one time chromatid disjunction and exit from mitosis were thought to be triggered by the same mechanistic pathway (the sudden inactivation of Cdk1/cyclin B2). Now, however, we know that chromatid disjunction can occur even when the cell is prevented from exiting mitosis by the expression of a nondegradable form of cyclin B (which
et al., 1993; Wheatley et al., 1997;
es that chromatid disjunction and
eparate but normally coordinated
95; Straight et al., 1996). That is,
ase transition can occur with the
teins even though the cell does not

ion, the destruction of cyclin B and
involves proteolysis of a specific
inction of the anaphase-promoting
ivation of the APC/C occurs when
quires or captures spindle micro-
e APC/C, ceases to be inhibited by
scussed later in this chapter). This
called the "cyclosome," is an E3
quitination of specific proteins. In
by proteosomes (reviewed in King
alysis of lysed HeLa cells suggest
ly with the spindle (Tugendreich et

iromatid disjunction (reviewed in
somes are held together, primarily
iesin complex of proteins. Disjunc-
e, becomes activated and degrades
plex. Activation of separase occurs
ein that inhibits separase, is proteolytically degraded upon activation of the APC/C. However, other levels of control over chromatid disjunction exit, at least under experimental conditions. First, polo-like kinase (PLK) phosphorylation of cohesion complex subunits is required for their dissociation from the chromosomes and thus, chromatid disjunction. Second, phosphorylation of separase at Ser1126 by high levels of Cdk1-cyclin B activity inactivates this protease in a securin-independent fashion. Whether or not this inhibitory site is phosphorylated by Cdk1-cyclin B under physiological conditions remains to be determined. In any case, the dependency of the metaphase-anaphase transition on proteolytic events ensures that this critical cell cycle transition is irreversible for both chromatid disjunction and exit from mitosis. Although the assembly of the actin-based cyto-

kinetic apparatus is initiated at or shortly after the metaphase-anaphase transition, the actual furrowing process is not apparent until later (Fig. 6.1F–H).

During anaphase each chromosome moves to its respective pole (anaphase A: Fig. 6.1F) and the poles themselves move further apart (anaphase B: Figs. 6.1G, 6.2E). These two motions act additively to increase the distance between the two separating groups of chromosomes. Although anaphase A and B movements usually start simultaneously upon chromatid disjunction (as in vertebrates), in some organisms they begin at different times (reviewed in Mazia, 1961), suggesting that in some cases they can be independently regulated. Anaphase A involves two coordinated events: the movement of chromosomes through the cytoplasm by P forces that act at the kinetochore, and the shortening of the microtubules attached to the kinetochore via tubulin subunit loss at the kinetochore. The mechanism(s) by which the motive force is generated for chromosome P motion remains a subject of lively debate and may differ, to various extents, between organisms (reviewed in McIntosh and Pfarr, 1991; Sawin and Endow, 1993; Rieder and Salmon, 1994). Mechanisms that are capable of providing the force for P chromosome movement include minus end directed motor molecules associated with the kinetochore that act on the microtubules associated with, and disassembling at, the kinetochore (Rieder and Alexander, 1990; McIntosh and Pfarr, 1991; Thrower et al., 1996; Brown et al., 1996); the disassembly of microtubule subunits at the kinetochore occurs while the kinetochore "hangs on" to the shortening end (Koshland et al., 1988; Steffen and Linck, 1992; reviewed in Inoue and Salmon, 1995). In addition the slow depolymerization of kinetochore microtubule minus ends at the spindle pole, occurring throughout mitosis, contributes a force-producing component for chromosome P motion, particularly in late anaphase (Mitchison and Salmon, 1992; Sawin and Mitchison, 1994; Waters et al., 1996a). Since very little force is required to move even large chromosomes through the cytoplasm at the slow (~1–2 μm/min) speeds normally seen in anaphase (reviewed in Nicklas, 1988), any of these mechanisms could, in principle, provide the requisite forces for anaphase A.

The motive force for separating the spindle poles during anaphase B could come from two processes acting singly or in combination. In yeast and diatoms microtubule plus end directed motors (e.g., members of the kinesin superfamily), which are anchored to a matrix in the spindle midzone and crosslink adjacent antiparallel microtubules, push the poles apart by working against the overlapping pole-to-pole microtubules (Hogan and Cande, 1990; Hogan et al., 1993; reviewed in Ault and Rieder, 1994). However, recent work suggests that the separation of the centrosomes during anaphase B in vertebrate somatic cells occurs also from a pulling mechanism intrinsic to each pole (Waters et al., 1993; Wheatley et al., 1997). Indeed, in these cells the overlapping antiparallel microtubules that connect the two centrosomes during metaphase detache from the centrosomes during anaphase, so the centrosomes are no longer connected as in yeast and diatoms (Mastronarde et al., 1993).

The pulling forces that act on these centrosomes, to effect anaphase B, are presumably produced by the interaction of astral microtubules with minus end directed microtubule motors anchored to the cell cortex or cytoplasmic structures such as the endoplasmic reticulum (see Vaisberg et al., 1993; Shaw et al., 1997).

Telophase

Telophase, the last stage of mitosis, consists of a series of events that complete cell division and take the cell out of mitosis. The events of telophase require the inactivation of Cdk1/cyclin B, and are clearly separable from those controlling chromatid disjunction, as discussed earlier. Shortly after the completion of anaphase B a nuclear envelope reforms around both masses of separated daughter chromosomes. In those cell in which all of the chromosomes come into close contact during late anaphase (e.g., vertebrates), a single nuclear envelope simply forms around the single mass of chromosomes. However, in other systems (e.g., sea urchin zygotes), in which the chromosomes are still separated and not touching at the end of anaphase, a nuclear envelope forms around each individual chromosome. These micronuclei (called karyomeres) then aggregate and fuse into a single nucleus. During this final "telophase" stage of mitosis (Figs. 6.1H, 6.2F) the cell also begins to cleave between the separated nuclei in a process known as cytokinesis. The cleavage apparatus is composed of a circumferential band of actin and myosin that coordinately contracts and disassembles so that the constricting furrow is not sterically constrained from completing cell division by a mass of actomyosin (see Fishkind and Wang, 1993; Oegema and Mitchison, 1997). Cytokinesis is not a simple unitary event; it appears to consist of a number of overlapping processes that involve a wide variety of mechanisms. Although the details of these processes is beyond the scope of this chapter, the general outlines are as follows: First, the equatorial cell cortex is signaled or stimulated to begin assembling the actin-based cleavage apparatus (reviewed in Rappaport, 1986, 1990). Even though the entire cortex is competent to assemble a cleavage apparatus, only the equatorial cortex is stimulated to do so. Exactly how this happens is a mystery but appears to require an activity provided by the microtubules of the two asters and/or the central spindle. Second, the ring of actin filaments contacts to bring the cortex in close proximity to the tightly bundled remnant of the central spindle where it must be held to keep the furrow from regressing. Finally, the separation of the daughter cells must be completed. This involves both the severing or disassembly of the bundle of microtubules in the furrow neck (discussed further below) and the joining of membranes to close the furrow and make two intact cells. This last process appears to involve mechanisms that participate in vesicle fusion with the plasma membrane and wound repair (O'Halloran, 2000; Sisson et al., 2000; Skop et al., 2001).

Before the final completion of cleavage the two daughter cells remain connected by a midbody that is composed of tightly bundled antiparallel microtubules embedded in a densely staining matrix material. Traditionally the completion of cleavage, seen as the rupturing of the midbody, was thought to be mediated by the cells crawling apart leading to the rupture of the midbody with one daughter cell inheriting the midbody apparatus, a process termed "traction mediated cytofission." Although this can occur, particularly for cells growing sparsely on artificial two-dimensional substrates, this may not be the normal process for cells in a tissue. Recent work has raised the possibility that abscission of the midbody may be a distinct and highly regulated event. Piel et al. (2001) observed that one or both mother (older) centrioles move into close proximity to the midbody and remain there for a variable period of time. Narrowing of the midbody and abscission are temporally correlated with the rapid movement of the centriole(s) away from the midbody and back to the central region of the cell. These observations make the intriguing suggestion that perhaps the centrioles participate in a signaling pathway that triggers a specific midbody abscission mechanism. This could be necessary because cells in tissues may not have the ability to separate widely and the midbody may have significant structural integrity due to the tightly packed microtubules and residual actin filaments. However, it should be noted that the successful completion of cleave does not require the presence of centrioles; for mammalian somatic cells lacking centrioles approximately 60% complete cleavage (Piel et al., 2001). This last observation does not necessarily disprove the notion that centriole-based signaling is important for the completion of cleavage; rather, it may indicate that centrioles are important for the timeliness and fidelity of the cleavage process.

After cleavage is complete, the daughter cells flatten out again and the centrosome inherited by each cell reassumes its role as the nucleation center for the cytoplasmic microtubule complex.

ERRORS AND QUALITY CONTROL MECHANISMS

The purpose of mitosis is the generation of genetically identical daughter cells. However, there are a number of things that can go wrong shortly before and during the mitotic process. Some errors, such as the initial monopolar attachment of chromosomes, are a normal part of mitosis and can be corrected. Others, such as DNA damage, perturbations of the microtubule cytoskeleton before mitosis, and later cleavage failure, are not normal, but the cell has ways of coping. Finally, some errors, such as the presence of supernumerary centrosomes, cannot be resolved. Although these various errors may not have lethal consequences for a cell, at least in the short term, they can have disastrous long-term consequences for the organism, such as the genesis of transformed cells with aggressive growth characteristics (discussed in Khodjakov and Rieder, 2001). Most organisms have evolved quality control mechanisms, called "checkpoints," to deal with a number of these errors. Checkpoints are signal transduction pathways that block progression of the cell cycle until the specific event being monitored is completed (concepts reviewed in Hartwell and Wienert, 1989; Murray, 1992). It is important to note that checkpoint pathways are separate and distinct from the chain of molecular events that drive the progression of the cell cycle. If a checkpoint pathway is disabled, say by mutation, the cell cycle will proceed in a normal fashion, but chance errors in the event being monitored are apt to not be corrected in time. This "relief of dependence" is a key measure of whether a cell cycle arrest is truly due to a checkpoint or not. Importantly, mutations in checkpoint pathways are found in many human cancer cells (Orr-Weaver and Weinberg, 1999) and are believed to contribute to tumor genesis (Cahill et al., 1999). Below we discuss in general terms a number of commonly observed problems and how the cell deals with some but not all of them.

Control of the G_2/M Transition

In the early 1950s Bullough and coworkers noted that mitosis was delayed in mouse epidermal preparations by starvation, insulin, respiratory poisons or shock and, more importantly, that the block occurred "at only one point in the process of mitosis, that of the change from the resting cell to the prophase" (Green and Bullough, 1950). It was also evident that "none of the (insults) has any action on the passage of a mitosis once it has begun" (Bullough and Johnson, 1951). Shortly thereafter the term "antephase" was coined to delineate that period, in late G2, just prior to the onset of chromosome condensation (reviewed in Pines and Ricder, 2001). It is now well established that a wide variety of insults arrest cells in antephase, but that if the cell has already passed through a point of no return (Mazia, 1961) prior to the treatment, entry into mitosis cannot be delayed. Clearly, passage through this point represents the functional termination of interphase and the beginning of mitosis.

When antephase ends appears to differ depending on the organism. In vertebrates that contain many small chromosomes (mice, humans, chickens, monkeys, etc.) the duration of visible prophase is relatively short (<15 min), and the commitment point appears to fall just before chromosome condensation is evident at the light microscope level. By contrast, in cells that contain large chromosomes (newts, rat kangaroos, Indian muntjac, etc.), chromosome condensation is apparent up to one hour before nuclear envelope breakdown (NEB). In these cells the commitment to mitosis does not occur until late prophase, approximately when the nucleoli begin to dissipate. When these cells are stressed during early to mid prophase, the process of chromosome condensation is arrested and/or reversed. As the chromosomes decondense, those nuclear protcins that were phosphorylated in preparation for mitosis (e.g., the MPM2 epitopes) are dephosphorylated (Rieder and Cole, 1998). This reversion itself is reversible since, after a period of adaptation or when the stress is relieved, the cells ultimately re-enter prophase and proceed through mitosis (Rieder, 1981; Rieder and Cole, 2000)

Importantly, when a wide variety of cells, both plant and animal, containing large chromosomes are stressed during *late prophase*, about 10 to 15 minutes prior to NEB, they proceed into prometaphase and complete mitosis, usually without further delay (Carlson and Hollaender, 1948; Carlson, 1950, 1969; Gaulden and Perry, 1958; Ducoff and Ehret, 1959; Klasterska et al., 1977; Onuki, 1972; Rieder, 1981; Jiang and Liang, 1989). The commitment to mitosis temporally coincides with the sudden accumulation and activation of Cdk1-cyclin B in the nucleus (Toyoshima et al., 1998; Hagting et al., 1999; Clute and Pines, 1999; Jin et al., 1998), and it is likely that this event defines the point of no return.

The control of the antephase to mitosis transition is an important transition in the cell cycle, because any agent or stress that interferes with the integrity of the genome (DNA) enhances the potential for loss or gain of genetic information in the daughter cells resulting from mitosis. Thus cells have evolved at least three of checkpoint control pathways that regulate the antephase/M transition. The first G_2/M checkpoint to be discovered, and therefore best studied, monitors DNA integrity and is triggered by DNA damage. In humans this control uses the ATM/ATR serine/threonine kinases to ultimately prohibit the activation of Cdk1-cyclin B via two separate and independent signal transduction pathways. In both pathways the damage to DNA activates ATM/ATR kinase pathways: the current evidence suggests that ATM responds primarily to double-strand DNA breaks, while ATR also responds to UV damage and replication arrest (reviewed in Durocher and Jackson, 2001). In one pathway, activated ATM/ATR activates the Chk1/Chk2 kinase, which then blocks the function of the Cdc25C phosphatase that removes the inhibitory phosphorylations on Cdk1-cyclin B thereby activating it. In the other route, the activation of ATM/ATR leads to the phosphorylation of the p53 transcription factor, which induces the synthesis of p21, a Cdk2 inhibitor.

In addition vertebrate cells in G_2 possess a feedback pathway that monitors the integrity of their cytoplasmic microtubule complex. When cells in antephase are suddenly exposed to drugs that disrupt microtubules, they transiently arrest in the cell cycle for several hours before finally adapting and entering mitosis, albeit a dysfunctional one (Rieder and Cole, 2000). Importantly, in humans this delay of mitosis is a general feature of normal but not tumor cells (Jha et al., 1994), implying that the latter have lost this checkpoint pathway. Indeed, cells lacking a functional copy of the Chfr (checkpoint with FHA and ring finger) gene do not exhibit a G_2 delay when their microtubules are disassembled, while those possessing a functional copy do (Scolnick and Halazonetis, 2000). The report that taxol, which stabilizes microtubules, does not delay cells in antephase (Rieder and Cole, 2000) suggests that the event monitored has to do with the dynamic properties of microtubules and not their ability to support bi-directional transport. Recently Cfhr has been characterized as a unique ubiquitin ligase (Chaturvedi et al., 2002) that exerts its effect on the cell cycle by targeting the polo-like kinase (Plk1) for proteolysis (Kang et al., 2002). Perhaps this checkpoint pathway works by promoting the destruction of Plk1, whose activity is required for activation of Cdc25C and thus activation of Cdk1-cyclin B. However, because triggering the Chfr pathway induces chromosome decondensation in mid-prophase cells, it is likely that the targets of the checkpoint also include Cdk1-cyclin and the aurora B kinases, whose activity drive the early stages of chromosome condensation (Furuno et al., 1999; Giet and Glover, 2001; Van Hooser et al., 1998).

Finally, progression through antephase is also guarded by an interwoven and complex series of signal transduction pathways that are mediated by the mitogen-activated protein kinases (MAPKs), in particular the p38 (Hog1 in yeast) stress kinase pathway (see Bulavin et al., 2002). This kinase is activated by a wide variety of insults including, for example, heat or osmotic shocks (Dmitrieva et al., 2001), UV (Bulavin et al., 2001) or g- irradiations (Wang et al., 2000), and drugs that induce genotoxic stress like the histone deacetylase (Qiu et al., 2000) or topoisomerase II inhibitors (Pandey et al., 1996; Kharbanda et al., 1995; Goldstone et al., 2001). Evidence is accumulating that this pathway "senses" changes in the topology or structure of chromatin (Pandey et al., 1996; Yoshida et al., 2000) via the DNA-PK kinase (Kharbanda et al., 1997). Once activated, it initiates a signal cascade, not inhibited by caffeine (i.e., it does not involve the ATM/ATR kinase (Goldstone et al., 2001; Jha et al., 2002), that ultimately prevents activation of Cdk1-cyclin B (Bulavin et al., 2002). When the p38 pathway is triggered by topoisomerase II inhibitors, mid-prophase cells become "locked" in prophase within minutes (Mikhailov et al., 2004). The speed of the arrest and the condensed stage of the prophase chromosomes make it highly unlikely that the arrest is mediated by transcription factors like p53, which take several hours to exert their effect. The fact that this arrest can be overridden by the pyridinyl imidazole SB203580, which specifically inhibits the p38 kinase, again provides the relief of dependence fiduciary indicating the existence of a bond fide checkpoint control. Although the mechanism(s) by which this pathway inhibits the antephase/M transition remain vague, it may work by inhibiting Cdk2-cyclin A (Goldstone et al., 2001), whose activity is required for Cdk1-cyclin B activation.

Although there are reports that the DNA-damage checkpoint continues to operate during prometaphase in vertebrates (Smits et al., 2000), the delay in the metaphase-anaphase transition seen in response to DNA damage is likely due to problems in kinetochore attachment and thus the Mad-2 mediated spindle assembly checkpoint (Mikhailov et al., 2002; see below). Similarly there is no evidence that Chfr plays a role in maintaining a mitotic arrest in response to lack of microtubule function. Mitotic arrest occurs whether the cell contains or lacks a functional copy of Chfr. Finally, there is a recent report that the p38 pathway becomes active, and contributes to a mitotic arrest in 3T3 cells, when microtubule assembly is disrupted by nocodazole (Takenaka et al., 1998). However, we find that inhibiting p38 (with SB203580), in a wide variety of cells, does not relieve a nocodazole or colcemid induced mitotic block (A. Mikhailov and C. L. Rieder, personal observation).

Kinetochore Attachment to the Spindle

To equally segregate chromosomes, the cell must ensure the attachment of all sister chromatids to opposite spindle poles before the cell initiates the metaphase-anaphase transition. The problem the cell faces is that incomplete chromosome attachment (i.e., monoorientation) to the spindle is a normal part of the mitotic process (concepts discussed in Nicklas, 1989; Rieder et al., 1994; Nicklas and Ward, 1994). As discussed earlier, a chromosome first attaches to the spindle when one of its kinetochores captures microtubules from one aster; this leads to monoorientation of the chromosome to that spindle pole. Since astral microtubule density drops with distance from the more distant pole and relatively few dynamically unstable microtubules grow sufficiently long to span the full interpolar distance, the distal kinetochore on the monooriented chromosome remains unattached for a highly variable period of time—in normal PtK cells this is typically 7 minutes to one hour and as much as 3 hours (Rieder et al., 1994). Given this enormous variability in the amount of time required for the completing the bipolar attachment of all chromosomes, the cell would risk unequal chromosome distribution if the time of the metaphase-anaphase transition were determined by an invariant timing mechanism. One daughter would inherit an extra copy of a chromosome and the other daughter would lack that chromosome.

This essential coordination between chromosome attachment to the spindle and anaphase onset is not left to chance; in almost all higher eukaryotic cells the metaphase-anaphase transition is subject to a "checkpoint" control that delays anaphase onset in response to perturbations in spindle microtubule assembly and chromosome attachment to the spindle (Sluder, 1979; Sluder and Begg, 1983; Hoyt et al., 1991; Li and Murray, 1991; reviewed in Wells, 1996). Broadly speaking, the checkpoint for the metaphase-anaphase transition (commonly called the "spindle assembly" checkpoint) consists of a detector that monitors bipolar chromosome attachment to the spindle and a pathway that targets the machinery that executes the sequence of molecular events of the metaphase-anaphase transition (see Hardwick and Murray, 1995).

The first direct demonstration for the existence of a checkpoint for the metaphase-anaphase transition that monitors chromosome attachment to the spindle in mammalian somatic cells came from the detailed analysis of the temporal relationship between chromosome attachment and the duration of mitosis, defined as the time from nuclear envelope breakdown to anaphase onset (Rieder et al., 1994). The duration of mitosis was variable, ranging from 23 to 198 minutes. Importantly, this variability was found to be entirely due to the range of times required for the last monooriented chromosome to establish connections to both poles of the spindle. Once the last chromosome attached to the spindle and started to congress to the metaphase plate, anaphase onset occurred on average 23 minutes later, regardless of how long the cell had a monooriented chromosome. In these cells even a single monooriented chromosome delayed the onset of anaphase for up to three hours.

We now know that this is due to an inhibitory signal from the unattached kinetochore as first suggested by McIntosh (1991); the cell is not counting attached kinetochores or monitoring chromosome monoorientation per se. This information came from a study from Rieder et al. (1995), which used a laser microbeam to selectively destroy the unattached kinetochore on the last monooriented chromosome in PtK1 cells. Normally these cells enter anaphase on average 23 minutes after the last kinetochore attaches. However, when the unattached kinetochore on the last monooriented chromosome is destroyed, cells initiate anaphase on average 17 minutes later, even though the irradiated chromosome remains monooriented at one of the spindle poles. Recent work has provided direct evidence that inhibitor production ceases once the last kinetochore attaches to the spindle, and its concentration ultimately falls below the threshold that permits the metaphase-anaphase transition to occur. By following the fluorescence of GFP cyclin B1 through mitosis, Clute and Pines (1999) provided evidence that cyclin B degradation begins as soon as the last kinetochore attaches but chromosome disjuction occurs several minutes later when cyclin B levels are low. Presumably the APC/C complexes become increasingly activated as soon as the last kinetochore attaches and several minutes are needed for the degradation of the securins and Scc1 subunits of the cohesin complexes that hold the chromatids together.

Initially it was thought that the inhibitory activity produced by the unattached kinetochore(s) is freely diffusible throughout the cell, because the entire cell must be arrested in mitosis. However, an investigation of mitosis in PtK1 cells containing two spindles in a common cytoplasm provided the surprising result that the inhibitory influence is functionally restricted to the spindle (Rieder et al., 1997; also see Sluder et al., 1994). The rationale for this approach was that the variability in the time for the completion of kinetochore attachment for any one spindle should produce instances where one spindle had completed chromosome congression while the other still had one or more unattached kinetochores. These workers found that multiple unattached kinetochores in one spindle did not inhibit anaphase onset in the neighboring spindle that was "mature," that is, in which all chromosomes had established bipolar attachments. As in normal cells with only one spindle, anaphase started in the "mature" spindle on average 24 minutes after the last monooriented chromosome established bipolar connections. Thus the inhibitory influence produced by an unattached kinetochore acts locally only at the level of the individual spindle, and therefore the target of the inhibitor is in the spindle not the cytoplasm. For example, at least 7 unattached kinetochores on a monopolar spindle approximately 20 microns away from a bipolar spindle did not prevent anaphase onset in the latter. In comparison, a single unattached kinetochore near one of the poles on a large multipolar spindle PtK1 spindle inhibits anaphase onset in all parts of the spindle, even those located greater than 20 microns from the monooriented chromosome (Sluder et al., 1997). Thus the inhibitory influence can propagate greater distances within a spindle than between adjacent-independent spindles. Together these observations suggest that the inhibitor of the metaphase-anaphase transition produced by unattached kinetochores becomes structurally associated with the spindle containing such kinetochores.

This study also revealed that the molecular changes of the metaphase-anaphase transition propagate throughout the cell and are dominant over the inhibitory activity of unattached kinetochores (also see Sluder et al., 1994). The key finding was that the lagging or "immature" spindles initiated anaphase on average 9 minutes after the leaders even though the former contained one or more unattached kinetochores. This observation suggests that the molecular events of the metaphase-anaphase transition are initiated within the spindle not the cytoplasm. In this regard there are intriguing observations for Drosophila embryos that the degradation of cyclin B starts in the asters and propagates to the central spindle (Wakefield et al., 2000; Raff et al., 2002). The generality of this phenomenon is not yet certain because this spatial specificity of cyclin B degradation was not observed in mammalian cells (Clute and Pines, 1999).

Although the checkpoint for the metaphase-anaphase transition in some cell types, such as HeLa, leads to a permanent arrest in mitosis until the cells undergo apoptosis, it is important to note that many other kinds of cells will, after a long delay (typically 4–6 hours), show synchronous chromosome disjunction, and progress into the next G_1 regardless of whether unattached kinetochores are present (reviewed in Rieder and Palazzo, 1992). Little information is available concerning how cells "adapt" to the checkpoint, or why the checkpoint in some cells is leakier than others. Perhaps the inhibitory influence produced by unattached kinetochores acts to greatly slow, but not stop, the progression of molecular reactions that trigger the execution of the metaphase-anaphase transition. Alternatively, cells may have evolved specific compensatory mechanisms that in time allow the completion of mitosis, albeit defective, by down-regulating the checkpoint.

The aspect of kinetochore attachment that is monitored by the checkpoint includes the extent to which the kinetochore is saturated with attached microtubules (Rieder et al., 1994; McEwen et al., 1997; Waters et al., 1996b) and tension across the centromere due to poleward directed forces exerted by motor molecules in the sister kinetochores (McIntosh, 1991; Li and Nicklas, 1995, 1997). Although the relative roles of microtubule occupancy at the kinetochore and tension at the kinetochore have been vigorously debated, experimentally distinguishing between these two possibilities has been difficult because the phenomena are interrelated; the accumulation of microtubules at the kinetochore promotes poleward forces, and the resulting tension at the kinetochore stabilizes microtubule attachment, leading to the stable capture of more microtubules by the kinetochore (Nicklas and Staehly, 1967; Nicklas, 1967; Nicklas and Koch, 1969; Nicklas and Ward, 1994). Current thinking arguably favors the notion that microtubule occupancy is the parameter being monitored and that tension profoundly influences the stability of microtubule attachment to the kinetochore (Nicklas et al., 2001; reviewed in Zhou et al., 2002).

The details of the signal transduction pathway of the checkpoint that originates in the unattached kinetochore and blocks the metaphase-anaphase transition is still being actively investigated. Since a detailed review of the current understanding of this pathway is beyond the scope of this chapter, the reader is referred to reviews by Zhou et al. (2002), Yu (2002), and Irniger (2002) for a more comprehensive treatment of this subject. Genetic analysis of yeast mutants has identified a number of genes that code for proteins involved with the checkpoint: *mad1*, *mad2*, *mad3*, *bub1*, and *bub3*. Another kinase important for the checkpoint, Mps1, appears to act immediately upstream of Mad1 (Weiss and Winey, 1996; Stucke et al., 2002). Recent work with the vertebrate homologs of these proteins has led to the formulation of the following model. Unattached kinetochores serve as sites for the transient binding and activation of Mad2 and its association with other checkpoint proteins, which then move as a complex poleward along microtubules (Yao et al., 2000; Howell et al., 2000; 2001). Somewhere within the spindle or even at the kinetochore the inhibitory complex binds to and inactivates Cdc20, which is an activator of the APC/C. There is also evidence that the diffusible inhibitory signal originating at the unattached kinetochore may consist of more than one complex, such as BubR1-Bub3-Mad2-Cdc20 and BubR1-Bub3-Cdc20. In any case, keeping Cdc20 from activating the APC/C prevents the destruction cyclin B and securin, which are respectively required for exit from mitosis and release of sister chromatids from each other (Zacharie and Nasmyth, 1999; Naysmith et al., 2000). Once the last kinetochore attaches to the spindle, the assembly of these inhibitory complexes ceases and the activation of the APC/C leads to the metaphase-anaphase transition.

Extra Centrosomes and Cleavage Failure Defects

The presence of extra centrosomes (often called "centrosome amplification") represents a serious problem for the organism because cells do not have a checkpoint that aborts mitosis in response to extra spindle poles or when the spindle lacks a bipolar symmetry (Sluder et al., 1997). In PtK1 cells that formed tripolar or tetrapolar spindles the interval between the bipolar attachment of the last monooriented chromosome and anaphase onset was normal, regardless of the number of spindle poles. As long as a cell assembles spindle microtubules when it comes into mitosis, the completion of kinetochore attachment is the event that limits when it will initiate the metaphase-anaphase transition. The fact that spindle multipolarity can lead to aneuploidy raises the question why there is no checkpoint to monitor this important parameter. A possible answer is that checkpoint mechanisms offer a direct selective advantage only for defects that the cell can ultimately resolve. Cells are not able to correct for the presence of too many spindle poles, and thus a checkpoint for the metaphase-anaphase transition that monitors bipolar spindle symmetry would serve no functional purpose.

Cells with more than the normal two centrosomes are prone to assemble multipolar spindles that unequally distribute chromosomes to the several daughter cells. Multipolar mitoses are dangerous, because chromosome loss or gain produces genetic imbalances that can lead to the evolution of cells with unregulated growth characteristics and diminished apoptotic response to cellular damage (see Orr-Weaver and Weinberg, 1998; Rieder et al., 2001; Brinkley, 2001). Indeed, the demonstration that the cells of many aggressive human tumors show centrosome amplification (Pihan et al., 1998; 2001; Lingle et al., 1998) implicates the presence of extra centrosomes in the genesis of the transformed phenotype during cancer progression (Sato et al., 1999; Lingle et al., 2002; D'Assoro et al., 2002; reviewed in Kramer et al., 2002). Whether or not centrosome amplification is a leading event that causes cancer is currently a matter of debate, but at a minimum it should contribute to the genesis of the transformed phenotype by producing genomic instability (see Brinkley, 2001).

Unfortunately, centrosome amplification is not necessarily self-limiting through the loss of chromosomes and consequent loss of daughter cell viability. We have found that *populations* of cells with multiple centrosomes propagate efficiently (through spindle pole bundling in some cells that gives bipolar mitoses) but have a significant mitotic error rate that can generate random genotypes, thereby driving the evolution of the transformed state (Sluder and Nordberg unpublished). In effect, extra centrosomes are a "mistake machine" that compromises the essential fidelity of the mitotic process.

The mechanism by which supernumerary centrosomes arise in cells, particularly those that are p53 deficient, is not fully known. There are two schools of thought that are not mutually exclusive. One camp holds that centrosome amplification is simply the consequence of cleavage failure that produces a polyploid cell that has two centrosomes (Meraldi et al., 2002). If such a cell were to commit to another cell cycle, both centrosomes would duplicate at S phase, and at mitosis the cell would be predisposed to assemble a multipolar spindle that would distribute chromosomes at random to produce multiple daughter cells that are genetically unbalanced, as discussed above. Importantly, the presence of extra chromosomes should increase the chances that some daughter cells will have enough genetic information to remain viable and proliferate. Subsequent multipolar divisions, bundling of spindle poles in some divisions, and selection for the fastest growing progeny will lead to a population with a significant percentage of cells showing extra centrosomes (Borel et al., 2002). Additional support for the cleavage failure school comes from reports that the targeted inactivation of p53 in human cells produces supernumerary centrosomes in conjunction with multinucleation (Duensing et al., 2000; Bunz et al., 2002).

However, cleavage failure is normally a rare event, even for cells growing in two dimensions on artificial substrates. Time lapse analysis of dividing cells reveals that cleavage failure ranges from 0% with NIH3T3 cells to approximately 2% for BSC1 cells (Piel et al., 2001; Sluder and Nordberg unpublished). Whether or not normal cells proliferating in a tissue show a significant rate of cleavage failure has not been systematically explored. Nevertheless, this may well be a real problem in the organism because mammals appear to have evolved a checkpoint mechanism to deal with cleavage failure. A number of workers have found that mammalian cells have a mechanism that arrests the cell cycle in G_1, in a p53 dependent fashion, when they become polyploid (Andreassen et al., 2001). To be of tangible benefit to the organism, this arrest must be enduring and/or predispose the polyploid cells to apoptosis. Presently the cellular parameter(s) being monitored and the targets of this putative checkpoint are not known.

Cleavage failure may not be the only source of centrosome amplification. Others posit that reduplication of the centrosome within a cell cycle occurs in p53 deficient cells and is an important source of centrosome amplification (Fukasawa et al., 1996; Tarapore et al., 2001). This position draws its published support from the finding that >90% of early passage p53–/– mouse embryo fibroblasts (MEFs) have 2N/4N DNA content yet >30% have extra centrosomes. Additionally many of the 2N cells contained more than 2 centrosomes (Tarapore and Fukasawa, 2002). There appears to be a link between loss of p53 and centrosome amplification (Levine et al., 1991; Weber et al., 1998; Carroll et al., 1999; Tarapore et al., 2001). In addition, restoring p53 to p53–/– cells leads to the restoration of a normal centrosome complement (Tarapore et al., 2001). Taken together, these data support the notion that centrosome amplification may have more than one source, particularly in tumor cells lacking a functional p53 gene.

CONCLUSION

Mitosis in the higher animal cell consists of a highly conserved sequence of events. Although these have been divided into five phases based on morphological criteria, we now know that many of the events begin and start to end before this becomes apparent at the morphological level. As a consequence these five phases are losing their precise definitions and must be used carefully and knowledgeably. Nevertheless, the terminology has become embedded in our thinking and can be used as a convenient shorthand way to indicate how far a cell has progressed in the mitotic process and what it is doing at a particular time.

The overarching purpose of mitosis is to take one cell and produce two genetically identical daughter cells. The penalties for mistakes include genomic instability, genetic imbalances, and the acquisition of unregulated growth characteristics. Although this may not present a short-term problem for the cell or its progeny, genetic imbalances can have disastrous consequences for the organism. As a consequence higher animals have evolved quality control mechanisms that can detect common naturally occurring mistakes and stop the progression of mitosis until they are remedied. The fact that the cells of some aggressive tumors are defective for one or more of these checkpoint mechanisms provides functional proof of their importance to the organism.

ACKNOWLEDGMENTS

The authors would like to thank Dr. Alexey Khodjakov for help in preparing the figures. Work cited from our laboratories was supported by: NIH GM 30758 to G. Sluder, NIH GM 40198 to C. L. Rieder, and NIH NCRR-01219, awarded by the DHHS/PHS, which supports the Wadsworth Center Biological Microscopy and Image Reconstruction Facility as a National Biotechnological Resource. E. H. Hinchcliffe is supported by an American Cancer Society Research Scholar Award.

REFERENCES

Andreassen PR, Lacroix FB, Lohez OD, Margolis RL (2001): Neither p21WAF1 nor 14-3-3sigma prevents G_2 progression to mitotic catastrophe in human colon carcinoma cells after DNA damage, but p21WAF1 induces stable G_1 arrest in resulting tetraploid cells. Cancer Res 61:7660–8.

Antonio C, Ferby I, Wilhelm H, Jones M, Karsenti E, Nebreda AR, Vernos I (2000): Xkid, a chromokinesin required for chromosome alignment on the metaphase plate. Comment. Cell 102:425–35.

Ault JG, DeMarco AJ, Salmon ED, Rieder CL (1991): Studies on the ejection properties of asters: astral microtubule turnover influences the oscillatory behavior and positioning of monooriented chromosomes. J Cell Sci 99:701–10.

Ault JG, Rieder CL (1994): Centrosome and kinetochore movement during mitosis. Curr Opin Cell Biol 6:41–9.

Bailly E, Pines J, Hunter T, Bornens M (1992): Cytoplasmic accumulation of cyclin B1 in human cells: Association with a detergent-resistant compartment and with the centrosome. J Cell Sci 101:529–45.

Bajer A (1959): Change of length and volume of mitotic chromosomes in living cells. Hereditas 45:579–96.

Bajer A (1965): Subchromatid structure of chromosomes in the living state. Chromosoma 17:291–302.

Bajer A, Mole-Bajer J (1972): Spindle Dynamics and Chromosome Movements. New York: Academic Press.

Bajer AS (1982): Functional autonomy of monopolar spindle and evidence for oscillatory movement in mitosis. J Cell Biol 93:33–48.

Beaudouin J, Gerlich D, Daigle N, Eils R, Ellenberg J (2002): Nuclear envelope breakdown proceeds by microtubule-induced tearing of the lamina. Cell 108: 83–96.

Borel F, Lohez OD, Lacroix FB, Margolis RL (2002): Multiple centrosomes arise from tetraploidy checkpoint failure and mitotic centrosome clusters in p53 and RB pocket protein-compromised cells. Proc Natl Acad Sci USA 99: 9819–24.

Brinkley BR (1985): Microtubule organizing centers. An Rev Cell Biol 1:145–72.

Brinkley BR (2001): Managing the centrosome numbers game: From chaos to stability in cancer cell division. Trends Cell Biol 11:18–21.

Brown KD, Wood KW, Cleveland DW (1996): The kinesin-like protein CENP-E is kinetochore-associated throughout poleward chromosome segregation during anaphase-A. J Cell Sci 109:961–9.

Brust-Mascher I, Scholey JM (2002): Microtubule flux and sliding in mitotic spindles of *Drosophila* embryos. Mol Biol Cell 13:3967–75.

Bulavin DV, Amundson SA, Fornace AJ (2002): p38 and Chk1 kinases: Different conductors for the G(2)/M checkpoint symphony. Curr Opin Genet Dev 12:92–7.

Bulavin DV, Higashimoto Y, Popoff IJ, Gaarde WA, Basrur V, Potapova O, Appella E, Fornace AJ, Jr (2001): Initiation of a G_2/M checkpoint after ultraviolet radiation requires p38 kinase. Nature 411:102–7.

Bullough W, Johnson M (1951): The energy relations of mitotic activity in adult mouse epidermis. Pro Royal Soc (Ser B) B138:562–75.

Bunz F, Fauth C, Speicher MR, Dutriaux A, Sedivy JM, Kinzler KW, Vogelstein B, Lengauer C (2002): Targeted inactivation of p53 in human cells does not result in aneuploidy. Cancer Res 62:1129–33.

Cahill DP, Kinzler KW, Vogelstein B, Lengauer C (1999): Genetic instability and Darwinian selection in tumours. Trends Cell Biol 9:M57–60.

Carlson JG (1950): Effects of radiation on mitosis. J Cell Comp Physiol 35: 89–101.

Carlson JG (1969): X-ray-induced prophase delay and reversion of selected cells in certain avian and mammalian tissues in culture. Radiat Res 37:15–30.

Carlson JG, Hollaender A (1948): Mitotic effects of ultraviolet radiation of the 2250 Å region, with special reference to the spindle and cleavage. J Cell Comp Physiol 31:149–73.

Carroll PE, Okuda M, Horn HF, Biddinger P, Stambrook PJ, Gleich LL, Li YQ, Tarapore P, Fukasawa K (1999): Centrosome hyperamplification in human cancer: Chromosome instability induced by p53 mutation and/or Mdm2 overexpression. Oncogene 18:1935–44.

Cassimeris L, Rieder CL, Salmon ED (1994): Microtubule assembly and kinetichore directional instability in vertebrate monopolar spindles: implications for the mechanism of chromosome congression. J Cell Sci 107:285–97.

Cassimeris LU, Walker RA, Pryer NK, Salmon ED (1987): Dynamic instability of microtubules. Bioessays 7:149–54.

Chaturvedi P, Sudakin V, Bobiak ML, Fisher PW, Mattern MR, Jablonski SA, Hurle MR, Zhu Y, Yen TJ, Zhou BB (2002): Chfr regulates a mitotic stress pathway through its RING-finger domain with ubiquitin ligase activity. Cancer Res 62:1797–801.

Clute P, Pines J (1999): Temporal and spatial control of cyclin B1 destruction in metaphase. Nat Cell Biol 1:82–7.

Compton DA (2000): Spindle assembly in animal cells. An Rev Biochem 69: 95–114.

D'Assoro AB, Lingle WL, Salisbury JL (2002): Centrosome amplification and the development of cancer. Oncogene 21:6146–53.

Dietz R (1964): The dispensability of the centrioles in the spermatocyte division of *Pales ferruginea* (Nematocera). Proc Oxford Chromosome Conf 1:161–6.

Dmitrieva NI, Bulavin DV, Fornace AJ, Jr, Burg MB (2002): Rapid activation of G2/M checkpoint after hypertonic stress in renal inner medullary epithelial (IME) cells is protective and requires p38 kinase. Proc Natl Acad Sci USA 99:184–9.

Ducoff H, Ehret C (1959): Mitogenesis. Chicago: University of Chicago Press.

Duensing S, Lee LY, Duensing A, Basile J, Piboonniyom S, Gonzalez S, Crum CP, Munger K (2000): The human papillomavirus type 16 E6 and E7 oncoproteins cooperate to induce mitotic defects and genomic instability by uncoupling centrosome duplication from the cell division cycle. Proc Natl Acad Sci USA 97:10002–7.

Durocher D, Jackson SP (2001): DNA-PK, ATM and ATR as sensors of DNA damage: Variations on a theme? Curr Opin Cell Biol 13:225–31.

Eigsti O, Dustin P (1955): Colchicine in Agriculture, Medicine, Biology, and Chemistry. Ames, IA: Iowa State College Press.

Fields AP, Thompson LJ (1995): The regulation of mitotic nuclear envelope breakdown: A role for multiple lamin kinases. Prog Cell Cycle Res 1:271–86.

Fishkind DJ, Wang YL (1993): Orientation and three-dimensional organization of actin filaments in dividing cultured cells. J Cell Biol 123:837–48.

Fukasawa K, Choi T, Kuriyama R, Rulong S, Vande Woude GF (1996): Abnormal centrosome amplification in the absence of p53. Science 271: 1744–7.

Funabiki H, Murray AW (2000): The *Xenopus* chromokinesin Xkid is essential for metaphase chromosome alignment and must be degraded to allow anaphase chromosome movement. Comment. Cell 102:411–24.

Furuno N, den Elzen N, Pines J (1999): Human cyclin A is required for mitosis until mid prophase. J Cell Biol 147:295–306.

Gabrielli BG, De Souza CP, Tonks ID, Clark JM, Hayward NK, Ellem KA (1996): Cytoplasmic accumulation of cdc25B phosphatase in mitosis triggers centrosomal microtubule nucleation in HeLa cells. J Cell Sci 109:1081–93.

Gallant P, Nigg EA (1992): Cyclin B2 undergos cell cycle-depentant nuclear translocation and, when expressed as a non-destructable mutant, causes mitotic arrest in HeLa cells. J Cell Biol 117:213–24.

Gaulden M, Perry R (1958): Influence of the nucleolus on mitosis as revealed by ultraviolet microbeam irradiation. Proc Natl Acad Sci USA 44:553–9.

Gerace A, Foisner F (1994): Integral membrane proteins and dynamic organization of the nuclear envelope. Trends Cell Biol 4:127–31.

Giet R, Glover DM (2001): *Drosophila* aurora B kinase is required for histone H3 phosphorylation and condensin recruitment during chromosome condensation and to organize the central spindle during cytokinesis. J Cell Biol 152: 669–82.

Goldstone S, Pavey S, Forrest A, Sinnamon J, Gabrielli B (2001): Cdc25-dependent activation of cyclin A/cdk2 is blocked in G2 phase arrested cells independently of ATM/ATR. Oncogene 20:921–32.

Green H, Bullough W (1950): Mitotic activity in the shock state. British J Expr Pathol 31:175–82.

Hagting A, Jackman M, Simpson K, Pines J (1999): Translocation of cyclin B1 to the nucleus at prophase requires a phosphorylation-dependent nuclear import signal. Curr Biol 9:680–9.

Hardwick KG, Murray AW (1995): Mad1p, a phosphoprotein component of the spindle assembly checkpoint in budding yeast. J Cell Biol 131:709–20.

Hartwell LH, Weinert TA (1989): Checkpoints: Controls that ensure the order of cell cycle events. Science 246:629–34.

Heald R, Tournebize R, Blank T, Sandaltzopoulos R, Becker P, Hyman A, Karsenti E (1996): Self-organization of microtubules into bipolar spindles around artificial chromosomes in *Xenopus* egg extracts. Nature 382:420–5.

Heald R, Tournebize R, Habermann A, Karsenti E, Hyman A (1997): Spindle assembly in *Xenopus* egg extracts: Respective roles of centrosomes and microtubule self-organization. J Cell Biol 138:615–28.

Heneen WK (1975): Kinetochores and microtubules in multipolar mitosis and chromosome orientation. Expr Cell Res 91:57–62.

Hinchcliffe EH, Cassels GO, Rieder CL, Sluder G (1998): The coordination of centrosome reproduction with nuclear events during the cell cycle in the sea urchin zygote. J Cell Biol 140:1417–26.

Hinchcliffe EH, Li C, Thompson EA, Maller JL, Sluder G (1999): Requirement of Cdk2-cyclin E activity for repeated centrosome reproduction in *Xenopus* egg extracts. Science 283:851–4.

Hinchcliffe EH, Miller FJ, Cham M, Khodjakov A, Sluder G (2001): Requirement of a centrosomal activity for cell cycle progression through G_1 into S phase. Comment. Science 291:1547–50.

Hogan CJ, Cande WZ (1990): Antiparallel microtubule interactions: Spindle formation and anaphase B. Cell Motil Cytoskeleton 16:99–103.

Hogan CJ, Wein H, Wordeman L, Scholey JM, Sawin KE, Cande WZ (1993): Inhibition of anaphase spindle elongation in vitro by a peptide antibody that recognizes kinesin motor domain. Proc Natl Acad Sci USA 90:6611–15.

Holloway S, Glotzer M, King RW, Murray AW (1993): Anaphase is initiated by proteolysis rather than by the inactivation of maturation-promoting factor. Cell 73:1393–402.

Holloway SL (1995): Sister chromatid separation in vivo and in vitro. Curr Opin Genet Dev 5:243–8.

Howell BJ, Hoffman DB, Fang G, Murray AW, Salmon ED (2000): Visualization of Mad2 dynamics at kinetochores, along spindle fibers, and at spindle poles in living cells. J Cell Biol 150:1233–50.

Howell BJ, McEwen BF, Canman JC, Hoffman DB, Farrar EM, Rieder CL, Salmon ED (2001): Cytoplasmic dynein/dynactin drives kinetochore protein transport to the spindle poles and has a role in mitotic spindle checkpoint inactivation. J Cell Biol 155:1159–72.

Hoyt MA, Totis L, Roberts BT (1991): *S cerevisiae* genes required for cell cycle arrest in response to loss of microtubule function. Cell 66:507–17.

Inoue S, Salmon ED (1995): Force generation by microtubule assembly/disassembly in mitosis and related movements. Mol Biol Cell 6:1619–40.

Irniger S (2002): Cyclin destruction in mitosis: A crucial task of Cdc20. FEBS Lett 532:7–11.

Jackman M, Lindon C, Nigg EA, Pines J (2003): Active cyclin B1-Cdk1 first appears on centrosomes in prophase. Nat Cell Biol 5:143–8.

Jackman M, Pines J (1997): Cyclins and the G_2/M transition. Cancer Surv 29: 47–73.

Jha MN, Bamburg JR, Bedford JS (1994): Cell cycle arrest by Colcemid differs in human normal and tumor cells. Cancer Res 54:5011–15.

Jha MN, Bamburg JR, Bernstein BW, Bedford JS (2002): Caffeine eliminates gamma-ray-induced G_2-phase delay in human tumor cells but not in normal cells. Radiat Res 157:26–31.

Jiang RL, Liang HN (1989): Mitotic reversion in prophase of PTK1 cells induced by argon laser microirradiation. Cell Biophys 14:271–82.

Jin P, Hardy S, Morgan DO (1998): Nuclear localization of cyclin B1 controls mitotic entry after DNA damage. J Cell Biol 141:875–85.

Kang D, Chen J, Wong J, Fang G (2002): The checkpoint protein Chfr is a ligase that ubiquitinates Plk1 and inhibits Cdc2 at the G_2 to M transition. J Cell Biol 156:249–59.

Karsenti E, Vernos I (2001): The mitotic spindle: A self-made machine. Science 294:543–7.

Keating TJ, Peloquin JG, Rodionov VI, Momcilovic D, Borisy GG (1997): Microtubule release from the centrosome. Proc Natl Acad Sci USA 94: 5078–83.

Kharbanda S, Pandey P, Jin S, Inoue S, Bharti A, Yuan ZM, Weichselbaum R, Weaver D, Kufe D (1997): Functional interaction between DNA-PK and c-Abl in response to DNA damage. Nature 386:732–5.

Kharbanda S, Ren R, Pandey P, Shafman TD, Feller SM, Weichselbaum RR, Kufe DW (1995): Activation of the c-Abl tyrosine kinase in the stress response to DNA-damaging agents. Nature 376:785–8.

Khodjakov A, Cole RW, Bajer AS, Rieder CL (1996): The force for poleward chromosome motion in Haemanthus cells acts along the length of the chromosome during metaphase but only at the kinetochore during anaphase. J Cell Biol 132:1093–104.

Khodjakov A, Cole RW, Oakley BR, Rieder CL (2000): Centrosome-independent mitotic spindle formation in vertebrates. Curr Biol 10:59–67.

Khodjakov A, Rieder CL (1996): Kinetochores moving away from their associated pole do not exert a significant pushing force on the chromosome. J Cell Biol 135:315–27.

Khodjakov A, Rieder CL (2001): Centrosomes enhance the fidelity of cytokinesis in vertebrates and are required for cell cycle progression. J Cell Biol 153: 237–42.

King RW, Deshaies RJ, Peters J-M, Kirschner MW (1996): How proteolysis drives the cell cycle. Science 274:1652–9.

Klasterska I, Natarajan AT, Ramel C (1977): A highly radiation-sensitive stage in the late prophase of mouse spermatocyte meiosis. Hereditas 87:99–106.

Koshland DE, Mitchison TJ, Kirschner MW (1988): Polewards chromosome movement driven by microtubule depolymerization in vitro. Nature 331: 499–504.

Kramer A, Neben K, Ho AD (2002): Centrosome replication, genomic instability and cancer. Leukemia 16:767–75.

Lenart P, Rabut G, Daigle N, Hand AR, Terasaki M, Ellenberg J (2003): Nuclear envelope breakdown in starfish oocytes proceeds by partial NPC disassembly followed by a rapidly spreading fenestration of nuclear membranes. J Cell Biol 160:1055–68.

Levine DS, Sanchez CA, Rabinovitch PS, Reid BJ (1991): Formation of the tetraploid intermediate is associated with the development of cells with more than four centrioles in the elastase-simian virus 40 tumor antigen transgenic mouse model of pancreatic cancer [published erratum appears in Proc Natl Acad Sci USA 1991 Sep 15;88(18):8282]. Proc Natl Acad Sci USA 88:6427–31.

Li G, Sudlow G, Belmont AS (1998): Interphase cell cycle dynamics of a late-replicating, heterochromatic homogeneously staining region: Precise choreography of condensation/decondensation and nuclear positioning. J Cell Biol 140:975–89.

Li J, Meyer AN, Donoghue DJ (1997): Nuclear localization of cyclin B1 mediates its biological activity and is regulated by phosphorylation. Proc Natl Acad Sci USA 94:502–7.

Li R, Murray AW (1991): Feedback control of mitosis in budding yeast. Cell 66:519–31.

Li X, Nicklas RB (1995): Mitotic forces control a cell-cycle checkpoint. Nature 373:630–2.

Li X, Nicklas RB (1997): Tension-sensitive kinetochore phosphorylation and the chromosome distribution checkpoint in praying mantid spermatocytes. J Cell Sci 110:537–45.

Lingle WA, Lutz WH, Ingle JN, Maihle NJ, Salisbury JL (1998): Centrosome hypertrophy in human breast tumors: Implications for genomic stability and cell polarity. Proc Natl Acad Sci USA 95:2950–5.

Lingle WL, Barrett SL, Negron VC, D'Assoro AB, Boeneman K, Liu W, Whitehead CM, Reynolds C, Salisbury JL (2002): Centrosome amplification drives chromosomal instability in breast tumor development. Proc Natl Acad Sci USA 99:1978–83.

Mastronarde DN, McDonald KL, Ding R, McIntosh JR (1993): Interpolar spindle microtubules in PTK cells. J Cell Biol 123:1475–89.

Mazia D (1961): Mitosis and physiology of cell division. In: J Brachet, A Mirsky (eds): The Cell, Biochemistry, Physiology, Morphology. New York: Academic Press, pp 77–412.

Mazia D, Harris P, Bibring T (1960): The multiplicity of the mitotic centers and the time-course of their duplication and separation. Biophys Biochem Cytol 7:1–20.

McEwen BF, Heagle AB, Cassels GO, Buttle KF, Rieder CL (1997): Kinetochore fiber maturation in PtK1 cells and its implications for the mechanisms of chromosome congression and anaphase onset. J Cell Biol 137:1567–80.

McIntosh JR (1991): Structural and mechanical control of mitotic progression. Cold Spring Harbor Symp Quant Biol 56:613–19.

McIntosh JR, Pfarr CM (1991): Mitotic motors. J Cell Biol 115:577–85.

Meraldi P, Honda R, Nigg EA (2002): Aurora-A overexpression reveals tetraploidization as a major route to centrosome amplification in p53–/– cells. EMBO J 21:483–92.

Mikhailov A, Cole RW, Shinohara M, Rieder CL (2004): Modifying chromatin structure during terminal G2 delays entry into mitosis by triggering the p38 MAPK checkpoint pathway. J Cell Biol in press.

Mikhailov A, Cole RW, Rieder CL (2002): DNA damage during mitosis in human cells delays the metaphase/anaphase transition via the spindle-assembly checkpoint. Curr Biol 12:1797–806.

Mitchison TJ (1990): Mitosis The kinetochore in captivity. Nature 348:14–15.

Mitchison TJ, Salmon ED (1992): Poleward kinetochore fiber movement occurs during both metaphase and anaphase-A in newt lung cell mitosis. J Cell Biol 119:569–82.

Murray AW (1992): Creative blocks: cell-cycle checkpoints and feedback controls. Nature 359:599–604.

Nasmyth K (2002): Segregating sister genomes: The molecular biology of chromosome separation. Science 297:559–65.

Nasmyth K, Peters JM, Uhlmann F (2000): Splitting the chromosome: Cutting the ties that bind sister chromatids. Comment. Science 288:1379–85.

Nicklas RB (1967): Chromosome micromanipulation II Induced reorientation and the experimental control of segregation in meiosis. Chromosoma 21: 17–50.

Nicklas RB (1988): The forces that move chromosomes in mitosis. An Rev Biophys Chem 17:431–49.

Nicklas RB (1989): The motor for poleward chromosome movement in anaphase is in or near the kinetochore. J Cell Biol 109:2245–55.

Nicklas RB, Koch CA (1969): Chromosome micromanipulation III Spindle fiber tension and the reorientation of mal-oriented chromosomes. J Cell Biol 43: 40–50.

Nicklas RB, Staehly CA (1967): Chromosome micromanipulation I: The mechanics of chromosome attachment to the spindle. Chromosoma 21:1–16.

Nicklas RB, Ward SC (1994): Elements of error correction in mitosis: Microtubule capture, release, and tension. J Cell Biol 126:1241–53.

Nicklas RB, Waters JC, Salmon ED, Ward SC (2001): Checkpoint signals in grasshopper meiosis are sensitive to microtubule attachment, but tension is still essential. J Cell Sci 114:4173–83.

Oegema K, Mitchison TJ (1997): Rappaport rules: Cleavage furrow induction in animal cells. Proc Natl Acad Sci USA 94:4817–20.

O'Halloran TJ (2000): Membrane traffic and cytokinesis. Traffic 1:921–6.

Onuki Y, Rounds DE, Olson RS, Berns MW (1972): Laser microbeam irradiation of the juxtanucleolar region of prophase nucleolar chromosomes. Expr Cell Res 71:132–44.

Orr-Weaver TL, Weinberg RA (1998): A checkpoint on the road to cancer. Nature 392:223–4.

Pandey P, Raingeaud J, Kaneki M, Weichselbaum R, Davis RJ, Kufe D, Kharbanda S (1996): Activation of p38 mitogen-activated protein kinase by c-Abl-dependent and -independent mechanisms. J Biol Chem 271:23775–9.

Pederson T (1972): Chromatin structure and the cell cycle. Proc Natl Acad Sci USA 69:2224–8.

Pederson T, Robbins E (1972): Chromatin structure and the cell division cycle Actinomycin binding in synchronized HeLa cells. J Cell Biol 55:322–7.

Piel M, Nordberg J, Euteneuer U, Bornens M (2001): Centrosome-dependent exit of cytokinesis in animal cells. Science 291:1550–3.

Pihan GA, Purohit A, Wallace J, Knecht H, Woda B, Quesenberry P, Doxsey SJ (1998): Centrosome defects and genetic instability in malignant tumors. Cancer Res 58:3974–85.

Pihan GA, Purohit A, Wallace J, Malhotra R, Liotta L, Doxsey SJ (2001): Centrosome defects can account for cellular and genetic changes that characterize prostate cancer progression. Cancer Res 61:2212–19.

Pines J, Hunter T (1991): Human cyclins A and B1 are differentially located in the cell and undergo cell cycle-dependant nuclear transport. J Cell Biol 115: 1–17.

Pines J, Rieder CL (2001): Re-staging mitosis: A contemporary view of mitotic progression. Nat Cell Biol 3:E3–6.

Pockwinse SM, Krockmalnic G, Doxsey SJ, Nickerson J, Lian JB, van Wijnen AJ, Stein JL, Stein GS, Penman S (1997): Cell cycle independent interaction of CDC2 with the centrosome, which is associated with the nuclear matrix-intermediate filament scaffold. Proc Natl Acad Sci USA 94:3022–7.

Qiu L, Burgess A, Fairlie DP, Leonard H, Parsons PG, Gabrielli BG (2000): Histone deacetylase inhibitors trigger a G_2 checkpoint in normal cells that is defective in tumor cells. Mol Biol Cell 11:2069–83.

Raff JW, Jeffers K, Huang JY (2002): The roles of Fzy/Cdc20 and Fzr/Cdh1 in regulating the destruction of cyclin B in space and time. J Cell Biol 157: 1139–49.

Rappaport R (1969): Aster-equatorial surface relations and furrow establishment. J Expr Zool 171:59–68.

Rappaport R (1986): Establishment of the mechanism of cytokinesis in animal cells. Int Rev Cytol 105:245–81.

Rappaport R (1990): Role of the mitotic apparatus in furrow initiation. An NY Acad Sci 582:15–21.

Rieder CL (1981): Effect of hypothermia (20–25 degrees C) on mitosis in PtK1 cells. Cell Biol Int Rep 5:563–73.

Rieder CL (1982): The formation, structure, and composition of the mammalian kinetochore and kinetochore fiber. Int Rev Cytol 79:1–58.

Rieder CL (1990): Formation of the astral mitotic spindle: Ultrastructural basis for the centrosome-kinetochore interaction. Electron Microsc Rev 3:269–300.

Rieder CL, Alexander SP (1990): Kinetochores are transported poleward along a single astral microtubule during chromosome attachment to the spindle in newt lung cells. J Cell Biol 110:81–95.

Rieder CL, Cole R (2000): Microtubule disassembly delays the G_2-M transition in vertebrates. Curr Biol 10:1067–70.

Rieder CL, Cole RW (1998): Entry into mitosis in vertebrate somatic cells is guarded by a chromosome damage checkpoint that reverses the cell cycle when triggered during early but not late prophase: The vertebrate cell kinetochore and its roles during mitosis. J Cell Biol 142:1013–22.

Rieder CL, Cole RW, Khodjakov A, Sluder G (1995): The checkpoint delaying anaphase in response to chromosome monoorientation is mediated by an inhibitory signal produced by unattached kinetochores. J Cell Biol 130:941–8.

Rieder CL, Davison EA, Jensen LC, Cassimeris L, Salmon ED (1986): Oscillatory movements of monooriented chromosomes and their position relative to the spindle pole result from the ejection properties of the aster and half-spindle. J Cell Biol 103:581–91.

Rieder CL, Faruki S, Khodjakov A (2001): The centrosome in vertebrates: More than a microtubule-organizing center. Trends Cell Biol 11:413–19.

Rieder CL, Khodjakov A, Paliulis LV, Fortier TM, Cole RW, Sluder G (1997): Mitosis in vertebrate somatic cells with two spindles: Implications for the metaphase/anaphase transition checkpoint and cleavage (see Comments). Proc Natl Acad Sci USA 94:5107–12.

Rieder CL, Palazzo RE (1992): Colcemid and the mitotic cycle. J Cell Sci 102: 387–92.

Rieder CL, Salmon ED (1994): Motile kinetichores and polar ejection forces dictate chromosome position on the vertebrate mitotic spindle. J Cell Biol 124:223–33.

Rieder CL, Schultz A, Cole R, Sluder G (1994): Anaphase onset in vertebrate somatic cells is controlled by a checkpoint that monitors sister kinetochore attachment to the spindle. J Cell Biol 127:1301–10.

Salmon ED (1989): Cytokinesis in animal cells. Curr Opin Cell Biol 1:541–7.

Sato N, Mizumoto K, Nakamura M, Nakamura K, Kusumoto M, Niiyama H, Ogawa T, Tanaka M (1999): Centrosome abnormalities in pancreatic ductal carcinoma. Clin Cancer Res 5:963–70.

Sawin KE, Endow SA (1993): Meiosis, mitosis and microtubule motors. Bioessays 15:399–407.

Sawin KE, Mitchison TJ (1994): Microtubule flux in mitosis is independent of chromosomes, centrosomes, and antiparallel microtubules. Mol Biol Cell 5: 217–26.

Scholey JM, Brust-Mascher I, Mogilner A (2003): Cell division. Comment. Nature 422:746–52.

Schrader F (1953): Mitosis: The Movements of Chromosomes in Cell Division. 2nd ed. New York: Columbia University Press.

Scolnick DM, Halazonetis TD (2000): Chfr defines a mitotic stress checkpoint that delays entry into metaphase. Comment. Nature 406:430–5.

Sharp DJ, Rogers GC, Scholey JM (2000): Microtubule motors in mitosis. Nature 407:41–7.

Shaw SL, Yeh E, Maddox P, Salmon ED, Bloom K (1997): Astral microtubule dynamics in yeast: a microtubule-based searching mechanism for spindle orientation and nuclear migration into the bud. J Cell Biol 139:985–94.

Sisson JC, Field C, Ventura R, Royou A, Sullivan W (2000): Lava lamp, a novel peripheral golgi protein, is required for *Drosophila melanogaster* cellularization. Comment. J Cell Biol 151:905–18.

Skibbens RV, Rieder CL, Salmon ED (1995): Kinetochore motility after severing between sister centromeres using laser microsurgery: Evidence that kinetochore directional instability and position is regulated by tension. J Cell Sci 108:2537–48.

Skibbens RV, Skeen VP, Salmon ED (1993): Directional instability of kinetochore motility during chromosome congression and segregation in mitotic newt lung cells: A push-pull mechanism. J Cell Biol 122:859–75.

Skop AR, Bergmann D, Mohler WA, White JG (2001): Completion of cytokinesis in *C. elegans* requires a brefeldin A-sensitive membrane accumulation at the cleavage furrow apex. Curr Biol 11:735–46.

Sluder G (1979): Role of spindle microtubules in the control of cell cycle timing. J Cell Biol 80:674–91.

Sluder G, Begg DA (1983): Control mechanisms of the cell cycle: Role of the spatial arrangement of spindle components in the timing of mitotic events. J Cell Biol 97:877–86.

Sluder G, Begg DA (1985): Experimental analysis of the reproduction of spindle poles. J Cell Sci 76:35–51.

Sluder G, Miller FJ, Thompson EA, Wolf DE (1994): Feedback control of the metaphase-anaphase transition in sea urchin zygotes: Role of maloriented chromosomes. J Cell Biol 126:189–98.

Sluder G, Rieder CL (1985): Experimental separation of pronuclei in fertilized sea urchin eggs: Chromosomes do not organize a spindle in the absence of centrosomes. J Cell Biol 100:897–903.

Sluder G, Thompson EA, Miller FJ, Hayes J, Rieder CL (1997): The checkpoint control for anaphase onset does not monitor excess numbers of spindle poles or bipolar spindle symmetry. J Cell Sci 110:421–9.

Sluder G, Thompson EA, Rieder CL, Miller FJ (1995): Nuclear envelope breakdown is under nuclear not cytoplasmic control in sea urchin zygotes. J Cell Biol 129:1447–58.

Smits VA, Klompmaker R, Arnaud L, Rijksen G, Nigg EA, Medema RH (2000): Polo-like kinase-1 is a target of the DNA damage checkpoint. Nat Cell Biol 2:672–6.

Smits VA, Medema RH (2001): Checking out the G(2)/M transition. Biochim Biophys Acta 1519:1–12.

Steffen W, Fuge H, Dietz R, Bastmeyer M, Muller G (1986): Aster-free spindle poles in insect spermatocytes: evidence for chromosome-induced spindle formation? J Cell Biol 102:1679–87.

Steffen W, Linck RW (1992): Evidence for a non-tubulin spindle matrix and for spindle components immunologically related to tektin filaments. J Cell Sci 101:809–22.

Straight AF, Belmont AS, Robinett CC, Murray AW (1996): GFP tagging of budding yeast chromosomes reveals that protein-protein interactions can mediate sister chromatid cohesion. Curr Biol 6:1599–608.

Stucke VM, Sillje HH, Arnaud L, Nigg EA (2002): Human Mps1 kinase is required for the spindle assembly checkpoint but not for centrosome duplication. EMBO J 21:1723–32.

Swedlow JR, Hirano T (2003): The making of the mitotic chromosome: Modern insights into classical questions. Mol Cell 11:557–69.

Takenaka K, Moriguchi T, Nishida E (1998): Activation of the protein kinase p38 in the spindle assembly checkpoint and mitotic arrest. Science 280:599–602.

Tarapore P, Fukasawa K (2002): Loss of p53 and centrosome hyperamplification. Oncogene 21:6234–40.

Tarapore P, Horn HF, Tokuyama Y, Fukasawa K (2001): Direct regulation of the centrosome duplication cycle by the p53-p21Waf1/Cip1 pathway. Oncogene 20:3173–84.

Terasaki M, Campagnola P, Rolls MM, Stein PA, Ellenberg J, Hinkle B, Slepchenko B (2001): A new model for nuclear envelope breakdown. Mol Biol Cell 12:503–10.

Thrower DA, Jordan MA, Wilson L (1996): Modulation of CENP-E organization at kinetochores by spindle microtubule attachment. Cell Motil Cytoskeleton 35:121–33.

Tokai N, Fujimoto-Nishiyama A, Toyoshima Y, Yonemura S, Tsukita S, Inoue S, Yamamoto T (1996): KID, a novel kinesin-like DNA binding protein, is localized to chromosomes and the mitotic spindle. EMBO J 15:457–67.

Toyoshima F, Moriguchi T, Wada A, Fukuda M, Nishida E (1998): Nuclear export of cyclin B1 and its possible role in the DNA damage-induced G_2 checkpoint. EMBO J 17:2728–35.

Tugendreich S, Tomkiel J, Earnshaw W, Hieter P (1995): CDC27Hs colocalizes with CDC16Hs to the centrosome and mitotic spindle and is essential for the metaphase to anaphase transition. Cell 81:261–8.

Vaisberg E, Koonce M, McIntosh J (1993): Cytoplasmic dynein plays a role in mammalian mitotic spindle formation. J Cell Biol 123:849–58.

Van Hooser A, Goodrich DW, Allis CD, Brinkley BR, Mancini MA (1998): Histone H3 phosphorylation is required for the initiation, but not maintenance, of mammalian chromosome condensation. J Cell Sci 111:3497–506.

Vandre DD, Davis FM, Rao PN, Borisy GG (1984): Phosphoproteins are components of mitotic microtubule organizing centers. Proc Natl Acad Sci USA 81:4439–43.

Wakefield JG, Huang JY, Raff JW (2000): Centrosomes have a role in regulating the destruction of cyclin B in early *Drosophila* embryos. Curr Biol 10: 1367–70.

Walczak CE (2001): Ran hits the ground running. Comment. Nat Cell Biol 3: E69–70.

Walczak CE, Vernos I, Mitchison TJ, Karsenti E, Heald R (1998): A model for the proposed roles of different microtubule-based motor proteins in establishing spindle bipolarity. Curr Biol 8:903–13.

Wang S-Z, Adler R (1995): Chromokinesin: A DNA-binding, kinesin-like nuclear protein. J Cell Biol 128:761–8.

Wang X, McGowan CH, Zhao M, He L, Downey JS, Fearns C, Wang Y, Huang S, Han J (2000): Involvement of the MKK6-p38gamma cascade in gamma-radiation-induced cell cycle arrest. Mol Cell Biol 20:4543–52.

Waters JC, Cole RW, Rieder CL (1993): The force-producing mechanism for centrosome separation during spindle formation in vertebrates is intrinsic to each aster. J Cell Biol 122:361–72.

Waters JC, Mitchison TJ, Rieder CL, Salmon ED (1996a): The kinetochore microtubule minus-end disassembly associated with poleward flux produces a force that can do work. Mol Biol Cell 7:1547–58.

Waters JC, Skibbens RV, Salmon ED (1996b): Oscillating mitotic newt lung cell kinetochores are, on average, under tension and rarely push. J Cell Sci 109: 2823–31.

Weber RG, Bridger JM, Benner A, Weisenberger D, Ehemann V, Reifenberger G, Lichter P (1998): Centrosome amplification as a possible mechanism for numerical chromosome aberrations in cerebral primitive neuroectodermal tumors with TP53 mutations. Cytogenet Cell Genet 83:266–9.

Weiss E, Winey M (1996): The Saccharomyces cerevisiae spindle pole body duplication gene MPS1 is part of a mitotic checkpoint. J Cell Biol 132:111–23.

Wells W (1996): The spindle assembly checkpoint: Aiming for a perfect mitosis, every time. Trends Cell Biol 6:228–34.

Wheatley SP, Hinchcliffe EH, Glotzer M, Hyman AA, Sluder G, Wang Y (1997): CDK1 inactivation regulates anaphase spindle dynamics and cytokinesis in vivo. J Cell Biol 138:385–93.

White JG, Borisy GG (1983): On the mechanisms of cytokinesis in animal cells. J Theor Biol 101:289–316.

Whitehead CM, Winkfein RJ, Rattner JB (1996): The relationship of HsEg5 and the actin cytoskeleton to centrosome separation. Cell Motil Cytoskeleton 35: 298–308.

Yao X, Abrieu A, Zheng Y, Sullivan KF, Cleveland DW (2000): CENP-E forms a link between attachment of spindle microtubules to kinetochores and the mitotic checkpoint. Nat Cell Biol 2:484–91.

Yen TJ, Schaar BT (1996): Kinetochore function: molecular motors, switches and gates. Curr Opin Cell Biol 8:381–8.

Yoshida K, Weichselbaum R, Kharbanda S, Kufe D (2000): Role for Lyn tyrosine kinase as a regulator of stress-activated protein kinase activity in response to DNA damage. Mol Cell Biol 20:5370–80.

Yu H (2002): Regulation of APC-Cdc20 by the spindle checkpoint. Curr Opin Cell Biol 14:706–14.

Zachariae W, Nasmyth K (1999): Whose end is destruction: Cell division and the anaphase-promoting complex. Genes Dev 13:2039–58.

Zhou J, Yao J, Joshi HC (2002): Attachment and tension in the spindle assembly checkpoint. J Cell Sci 115:3547–55.

CHAPTER 7

CELL CYCLE INHIBITORY PROTEINS

CARMEN CARNEIRO and ANDREW KOFF
Department of Molecular Biology, Memorial Sloan-Kettering Cancer Center, New York, NY 10021

INTRODUCTION

From the beginning and through its life the cells of an organism are facing the decision to proliferate or not to proliferate. They need to proliferate in order to build up or repair tissues and organs, and they often withdraw from the cell cycle to differentiate. Most of the cells in an adult are quiescent, but unless they are "terminally" differentiated, they can re-enter the cell cycle. Proliferative fate is governed by mitogenic and anti-mitogenic signals that come to cells in different flavors: extracellular factors and interactions with other cells, all contribute to the cellular milieu. Any failure to choose the right proliferative fate can have severe consequences.

Studies on cell cycle control focus on the progression of cells through G_1 into S phase. Cells that fail to progress withdraw from the cell cycle into a nonproliferative quiescent state. We now recognize that the decision of a cell to withdraw from the cell cycle, to not proliferate, has its own fundamental importance. Defects in this decision affect development and cause disease. New clues of how this program is actively engaged versus the consequence of a cell simply not proliferating are emerging. Nowhere is this more important than in developmental biology, where cell fate is intertwined with appropriate proliferation decisions and a number of signal pathways converge on the cell cycle machinery.

In this chapter we describe what is known about a particular group of proteins with an important role controlling the decision of cells to exit the cell cycle, the CDK-inhibitory proteins.

Cell Cycle and Growth Control: Biomolecular Regulation and Cancer, Second Edition, Edited by Gary S. Stein and Arthur B. Pardee
ISBN 0-471-25071-6

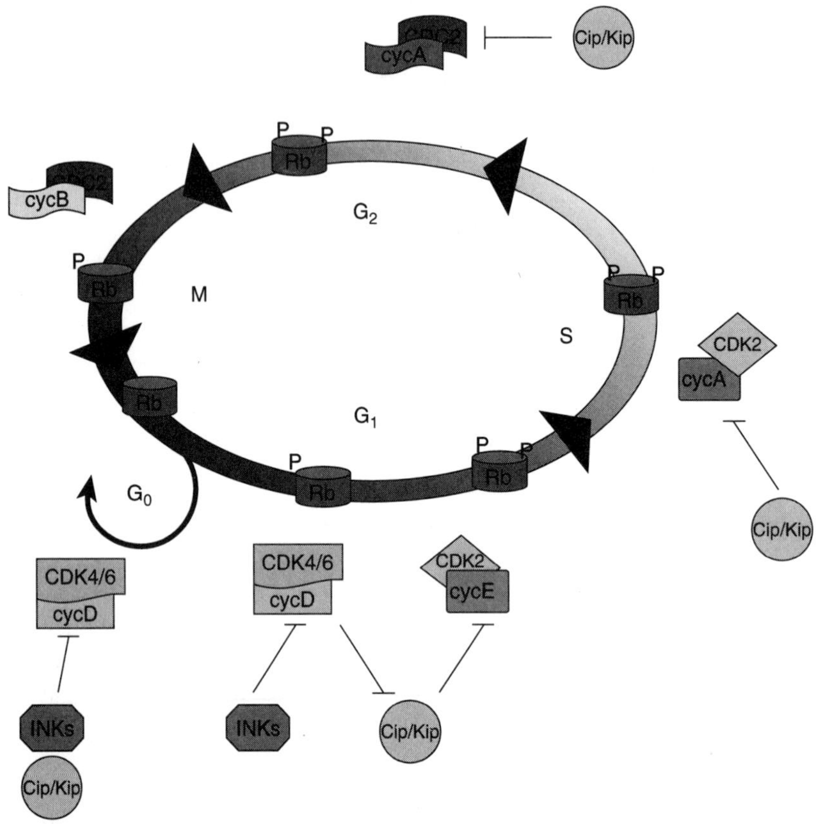

Figure 7.1. Control of cell cycle progression by the cyclin-CDK complex. The progression through each phase of the cell cycle is controlled by a specific cyclin-CDK complex. The binding to CDK inhibitors blocks the activity of these complexes.

HOW THE CELL CYCLE KEEPS ON GOING

The activation of the cyclin-dependent kinases controls the transition from one phase of the cell cycle to the next. Each cdk is a holoenzyme complex formed by a regulatory subunit, the cyclin, and a catalytic subunit, the cyclin-dependent kinase (CDK). Each particular phase and transition within the cell cycle can be identified by its "own" signature of cyclin and their associated kinase activities (Fig. 7.1).

Cyclin-CDK Complexes That Govern the G_1 to S Transition

Upon entering G_1 phase, the first complex that becomes active is a D-type cyclin (D1, D2, and D3) and one of its two catalytic partners CDK4 or CDK6. D-type cyclins are unstable, and their induction and expression are dependent on persistent mitogenic stimulation (Matsushime et

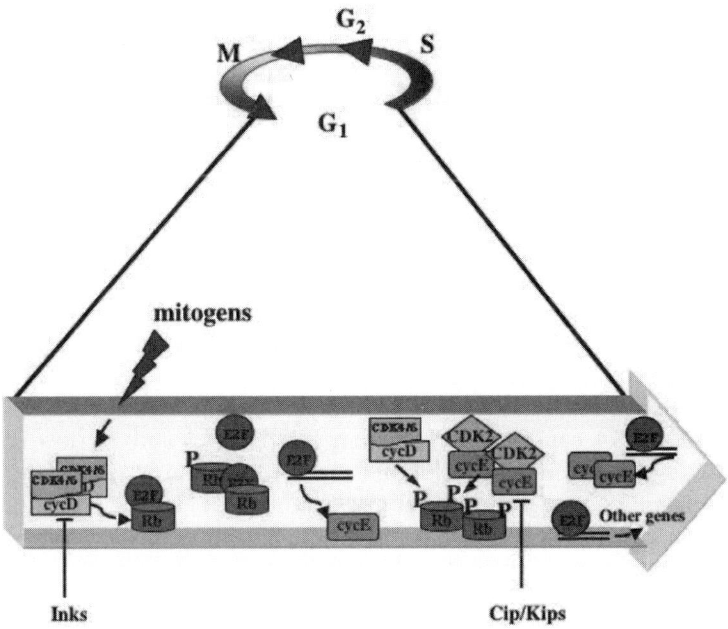

Figure 7.2. Progression through G_1. Mitogenic signals induce cyclin D-CDK4/6 complex formation, which initiates pRb phosphorylation in early to mid G_1. Complete pRb phosphorylation requires cyclin E-CDK2 kinase activity in late G_1. Inactivation of pRb releases the E2F transcription factors, which begin to transcribe the several genes necessary for S phase. Among them, newly synthesized cyclin E generates more cyclin E-CDK2 activity, reinforcing the commitment into S phase.

al., 1994; Sherr and Roberts, 1999) (Fig. 7.2). They are expressed in a cell-type specific manner (Sherr, 1993) and their patterns of expression often overlap. However, it is not always clear whether they have a redundant function regulating progression through G_1. CDK4 and CDK6 are stable and their levels are constant through the cell cycle. They are co-expressed in a number of cell types, but in some cases such as pancreatic β-islet cells (Rane et al., 1999) and mouse embryo fibroblasts (Tsutsui et al., 1999), it is clear that the function of CDK4 can not be compensated by CDK6.

As cells progress from mid to late G_1 a second cyclin-CDK complex appears, cyclin E-CDK2 (Fig. 7.2). Cyclin E and CDK2 are expressed in all cell types, and neither their accumulation nor assembly is dependent on persistent mitogenic stimulation.

Cyclin-D and cyclin-E associated kinase activities phosphorylate and inactivate the retinoblastoma gene product, pRb in a sequential manner. Cyclin D-CDK4/6 complexes initially phosphorylate Rb in mid G_1. The later phosphorylation by the cyclin E-associated kinase further disrupts the pocket domain of pRb dissociating the pRb-E2F complex and releasing the E2F transcription factors (Harbour et al., 1999) (Fig. 7.2). E2F activity is essential for the transcription of several genes necessary for S phase (reviewed in Dyson, 1998).

These complexes are not redundant. Cyclin D activity is required for S phase entry in Rb positive cells and dispensable in Rb negative cells (Koh et al., 1995; Lukas et al., 1995; Medema et al., 1995), whereas cyclin E activity is necessary in both Rb positive and negative cells (Ohtsubo et al., 1995). The co-expression of both cyclins decreases the duration of G_1 phase even further than when either is expressed alone (Resnitzky et al., 1994).

Several findings suggest that the main objective of G_1 necessary to progress into the S phase is to accumulate cyclin E associated activity. First, in mice loss of cyclin D can be compensated by cyclin E (Geng et al., 1999). Second, introduction of catalytically inactive forms of CDK2 but not CDK4 causes G_1 arrest (van den Heuvel and Harlow, 1993). Finally, cyclin E gene transcription is one of the main targets of the new released E2F activity (Botz et al., 1996; Ohtani et al., 1995). This generates a feedback loop: more cyclin E message means more protein that in turn generates more cyclin E-cdk2 activity and more pRb phosphorylation, reinforcing cell progression into the cell cycle. The positive feedback loop may account, at least in part, for extracellular growth factors' independence after commitment to S phase, as pRb phosphorylation is maintained now by a mitogen-independent complex.

Then, what is the role of the cyclin D-associated kinase? Unlike cyclin E, cyclin D expression and associated kinase activity is highly dependent of mitogens. Cyclin D expression, its assembly with CDKs and its turnover are mitogen regulated, mostly via events that induce Ras activation (reviewed in Sherr and Roberts, 1999). Thus D-type cyclins serve as a link between the signals coming from outside the cell to activate an internal machinery that is independent from the exterior, the CDK2 activity. Cyclin D-associated kinase activity makes cyclin E regulation sensitive to mitogens by affecting E2F activity via Rb. As we will discuss later, it can also affect CDK2 activity by titrating CDK inhibitors.

Putting Brakes to the Cell Cycle: The Cell Cycle Inhibitors

Cells need to have machinery that can stop their proliferation in response to anti-mitogenic signals or keep them in a quiescence state in absence of a mitogenic stimulation. As mentioned before, the activation of cyclin-dependent kinases is the main event necessary for driving cells into the S phase. They are also the main targets for anti-proliferative signals. The control of cyclin-dependent kinase activity is exerted at multiple levels. First, signals can control the accumulation of the cyclin; second, they can control the assembly of the cyclin-CDK complex; third, they can control the phosphorylation and dephosphorylation on specific residues; and fourth, they can control the availability of CDK inhibitory proteins. CDK inhibitors associate with either CDKs or cyclin-CDK complexes and block their activation or ability to phosphorylate substrates.

The remaining parts of this chapter will be focus on the cell cycle inhibitory proteins, from their structure and regulation to the biological consequences of their absence, their possible redundancy, and their cooperation to mediate effects.

Two Families with Two Different Mechanisms of Action

Cell cycle inhibitors have been classified into two different families: Ink4 and Cip/Kip, based on their structural similarities.

The Ink4 Family. To date, this family group contains four proteins: p16Ink4a (Serrano et al., 1993), p15Ink4b (Hannon and Beach, 1994), p18Ink4c (Guan et al., 1994; Hirai et al., 1995) and p19Ink4d (Chan et al., 1995; Hirai et al., 1995). In humans, p16Ink4a and p15Ink4b are located on the short arm of chromosome 9 (Hannon and Beach, 1994; Kamb et al., 1994), p18Ink4c maps to chromosome 1 (Guan et al., 1994), and p19Ink4d to chromosome 19 (Chan et al., 1995). The members of this family share a common structural motif, the ankyrin repeat. There are four repeats in p16Ink4a and p15Ink4b and five in p18Ink4c and p19Ink4d (rev. in Ortega et al., 2002).

The Ink4a proteins were named for their ability to bind and inhibit CDK4 (inhibitor of CDK4). They also can bind CDK6 (Chan et al., 1995; Hannon and Beach, 1994; Hirai et al., 1995; Serrano et al., 1993). They compete with the D-type cyclins for binding to the CDK subunit. The structural basis of this interaction is well established. Although the Ink and the cyclin binding sites on the CDK do not overlap, Ink binding causes an allosteric change that alters the cyclin binding site (Pavletich, 1999). Another consequence of Ink association is the distortion of the ATP binding site, resulting in reduced affinity for ATP (Russo et al., 1998).

The ability of Ink4 proteins to arrest cells in G_1 is largely dependent on the presence of a functional pRb. Ectopically expressed p16Ink4a is unable to arrest either Rb null cells (Lukas et al., 1995; Medema et al., 1995) or cells lacking the two other Rb-related pocket proteins p107 and p130 (Bruce et al., 2000). This is may be the result of how much CDK activity has to be inhibited and how much inhibitor is present. Cyclin E-cdk2 activity is positively controlled by E2F and cells that lack pRb or its related pocket-proteins have higher E2F activity and thus higher amounts of cyclin E, which like in the knockout mice can overcome the requirement for cyclin D.

The Cip/Kip Family. The Cip/Kip family (Cdk interacting protein/kinase inhibitory protein) is currently formed by three proteins: p21Cip1, p27Kip1, and p57Kip2. All share a homologous inhibitory domain through which they bind to and inhibit the cyclin-CDK complex. The members of the Cip/Kip family show a broad spectrum of activity, and in vitro they can inhibit both CDK4/CDK6 and CDK2-containing complexes. In vivo they preferentially bind to and inhibit the CDK2 complex.

The explanation of this is still unclear. CDK2 activity is inhibited by equimolar concentrations of p21 (Hengst and Reed, 1998), but cyclin D-CDK4-/Kip complexes show no significant inhibition at a 1:1:1 ratio (Blain et al., 1997; Zhang et al., 1994). Not only is cyclin D-CDK4 activity not affected by the presence of the inhibitor, they *need* the Kips for their function both in vivo (Cheng et al., 1999; Soos et al., 1996) and in vitro (Blain et al., 1997). However, too much Kip eventually inhibits the activity (Blain et al., 1997). The binding of the inhibitor stabilizes the complex, increases the stability of the D-type cyclins, and directs the complex to the nucleus (Cheng et al., 1999; LaBaer et al., 1997). An understanding of the molecular mechanisms accounting for this assembly/inhibition function remains of utmost importance (as discussed later).

How do the Cip/Kip proteins inhibit CDK-associated activity? The development of crystallographic models showed us the interaction between these proteins at the molecular level. p27 is able to "mimic" ATP and interacts with the CDK to block the ATP binding site. In addition a conformational change on the ATP and substrate binding cleft is observed (Russo et al., 1996).

Thus the two families of inhibitors use two different mechanisms. The Ink proteins inhibit CDK activity by preventing the cyclin-CDK interaction and the Cip/Kip family binds to cyclin-CDK complex and disrupts the ATP binding and substrate access.

Deep Look Inside: What Do We Know Today about CKIs

In addition to their mechanism of action, other aspects of CKIs have to be considered. An overwhelming amount of data has provided us with information about their regulation, their expression pattern, their possible roles, and the consequences of their absence.

While large amounts of information regarding the biochemical interactions between cell cycle regulatory proteins comes from studies using cell culture systems, some questions like the functional importance of these proteins or the functional differences between them cannot be answered in a cell culture system. In the past years the development of genetically altered mice has been used to address some of these issues. At the present time knockout strains for all the cell cycle inhibitors as well as combinations of some of them have been engineered (Table 7.1). Besides providing us with some ideas about the critical role of these regulators in differentiation and tumorigenesis, the mouse system allows us to ask if some of the data obtained in vitro was of biologic relevance. They also serve as a source from which cell can be obtained to carry out new studies on proliferation and differentiation.

Members of the Ink Family. The founder of this family, p16Ink4a, was identified in a two-hybrid interaction screening using CDK4 as bait (Serrano et al., 1993). p16 has attracted more attention than the other members of the family for several reasons. First, its frequent loss of func-

TABLE 7.1. Mouse Strains Locking One or Combinations of Two CDK Inhibitors and Their Phenotype Consequences

Knockout Gene	Phenotypic Consequence	Reference
$p15^{-/-}$	No developmental defects or tumor predisposition	Latres et al. (2000), Roussel (1999)
$p16^{-/-}$	Tumor free or slight tumor predisposition (depending on the strain)	Krimpenfort et al. (2001), Sharpless et al. (2001)
$p18^{-/-}$	Organomegalia and gigantism; pituitary hiperplasia	Franklin et al. (1998)
$p19^{-/-}$	Testicular atrophy	Zindy et al. (2001)
$p21^{-/-}$	No developmental defects or tumor predisposition	Brugarolas et al. (1995), Deng et al. (1995)
$p27^{-/-}$	Organomegalia and gigantism; pituitary hiperplasia or adenoma; female infertility	Fero et al. (1996), Kiyokawa et al. (1996), Nakayama et al. (1996)
$p57^{-/-}$	Embryonic and neonatal lethality; altered proliferation and apoptosis in some tissues	Yan et al. (1997), Zhang et al. (1997)
$p15^{-/-}; p18^{-/-}$	Infertility	Latres et al. (2000)
$p18^{-/-}; p19^{-/-}$	Male infertility	Zindy et al. (2001)
$p18^{-/-}; p27^{-/-}$	Increased organomegalia; earlier onset of pituitary adenomas	Franklin et al. (1998)
$p19^{-/-}$ $p27^{-/-}$	Inapropriate proliferation on postmitotic neurons Postnatal lethality at day 18	Zindy et al. (1999)
$p27^{-/-}; p57^{-/-}$	Skeletal muscle differentiation failure; altered lung development	Zhang et al. (1999)
$p21^{-/-}$ $p57^{-/-}$	Increased neonatal mortality; increase in lens defects; placental alterations	Zhang et al. (1998)

tion in different human cancers (reviewed in Ruas and Peters, 1998). In fact, among all the CDK inhibitors, p16Ink4a is the only one that can be considered as a tumor suppressor by the criteria of LOH. This tumor-suppressor function is supported by the studies developed using p16-deficient mice (see below). Second, the Ink4a locus encodes not only p16Ink4a but also another tumor-suppressor gene, p14ARF (p19ARF in mice). The way the two proteins are encoded is intriguing. The initiation codons in two alternative promoters located in exons 1α and 1β splice to the same sequence within exon 2 but are read in different reading frames to give two final products, p16Ink4a and p19ARF, with no sequence relationship to one another (Quelle et al., 1995).

In mouse, p16Ink4a is only detected after birth, and both the mRNA and protein levels increase with the age (Zindy et al., 1997a). This also occurs in culture, where there is a progressive increase in p16Ink4a protein levels as cells are continually passaged (Alcorta et al., 1996; Zindy et al., 1997a). In culture, this increase is related to a phenomenon known as replicative cellular senescence (reviewed in Serrano, 1997), and what that means in tissues is not clear. At present there is some controversy about whether this represents just an artifact caused by the culture conditions or a real response that can be found inside a tissue (Sherr and DePinho, 2000; Tang et al., 2001). Oncogenic stress also induces p16Ink4a (reviewed in Serrano, 1997), and at least in mouse cells, p16Ink4a has been implicated as a mediator of JunB growth inhibitory activity (Passegue and Wagner, 2000).

Inactivation of the entire Ink4a locus, both p16Ink4a and p19Arf, by knocking out exons 2 and 3 was first reported by Serrano and coworkers (Serrano et al., 1996). The Ink4a$^{\Delta 2,3}$ mice were born at the expected Mendelian ratio, and they grew without any gross developmental defects. However, they were predisposed to develop tumors, both spontaneously or induced (Ortega et al., 2002; Serrano et al., 1996). This was not an unexpected finding, as clinical data had suggested a role for p16Ink4a as a tumor suppressor. Therefore the observed phenotype was thought to reflect a direct consequence of p16 loss. Such interpretation had to be revised after the generation of the p19ARF knockout mice. Surprisingly, these animals had a very similar phenotype to the p16Ink4$^{\Delta 2,3}$ and showed also a tumor predisposition (Kamijo et al., 1997). Moreover these tumors expressed both p16Ink4a protein and mRNA, and this suggests that lost of p19ARF and not p16Ink4a might be the cause of the p16Ink4$^{\Delta 2,3}$ mouse phenotype.

The generation of "pure" p16Ink4a knockout mice was recently reported by two groups using two different strategies (Krimpenfort et al., 2001; Sharpless et al., 2001; reviewed in Sherr, 2001). In one of the strains, replacement of the wild-type gene by a mutated form resulted in animals that expressed an unstable p16Ink4a protein without ability to inhibit cyclin D-CDK complexes (Krimpenfort et al., 2001). The second p16Ink4a knockout strain was generated by removing the exon 1α (Sharpless et al., 2001). The null animals generated through the deletion showed some predisposition to spontaneously develop tumors, but they could not recapitulate the tumor incidence observed in the p19ARF strain, confirming the strong tumor-suppressor role of p19ARF in mice. As will be discussed later, a different picture emerged when the role of these two proteins in tumor development was explored in humans.

The second member of the family, p15Ink4b, was first identified in human keratinocytes treated with transforming growth factor-β (Hannon and Beach, 1994). p15Ink4b gene is located adjacent to p16Ink4a. Like p16, p15Ink4b expression is only detected after birth (Zindy et al., 1997a) and is not normally expressed during the cell cycle. p15Ink4b expression is induced in culture by TGF-β (Hannon and Beach, 1994) in a pathway involving the transcription factors Smad2, Smad3, Smad4, and Sp1 (Feng et al., 2000). Induction is repressed by c-myc, which physically binds to the Smad-Sp1 complexes on the p15 promoter inhibiting their transcriptional activity (Feng et al., 2002).

Mice lacking p15Ink4b protein were born at the expected Mendelian ratio and did not exhibit any gross abnormality during development

(Latres et al., 2000). The incidence of both spontaneously and induced tumors was low, suggesting that p15Ink4b has limited tumor-suppressing activities.

p18Ink4c (Guan et al., 1994; Hirai et al., 1995) and p19Ink4d (Chan et al., 1995; Hirai et al., 1995) are less studied members of the family. p18Ink4c and p19Ink4d are expressed mostly during embryonic development but can also be detected in the adult life in tissues like brain and testis (Roussel, 1999; Zindy et al., 1997b). p18Ink4c and p19Ink4d mRNA and protein levels increase at the G_1/S transition (Hirai et al., 1995), and after G_2, p19Ink4d is rapidly degraded by the ubiquitin-proteasome dependent machinery (Thullberg et al., 2000). These observations suggest a possible role for p18Ink4c and p19Ink4d in controlling cell cycle arrest coupled to, at least in some cases, differentiation programs (Zindy et al., 1999).

The p18Ink4c null mice are, among all the Ink knockouts, the only ones that display some developmental alteration (Franklin et al., 1998; Latres et al., 2000). The Mendelian ratio of these animals at birth is normal, but they are larger than their wild-type littermates and display widespread organomegaly. They show a high incidence of spontaneous pituitary intermediate lobe hyperplasia, which progresses slowly to pituitary adenoma. This phenotype is remarkably similar to that reported in p27Kip1 null mice (see below). p18Ink4c null animals also display a low incidence of other neoplasias such as testicular tumors and pheochromocytomas (Franklin et al., 1998; Latres et al., 2000; Ortega et al., 2002).

The only phenotype observed on the p19Ink4d null mice is testicular atrophy, but without affecting their ability to breed (Zindy et al., 2000). There is no effect on spontaneous or carcinogen-induced tumorigenesis.

Members of the Cip/Kip Family. p21Cip was the first CDK inhibitor identified. It was discovered almost simultaneously by different groups as a mediator of p53-induced arrest (el-Deiry et al., 1993), as a CDK2-associated protein (Gu et al., 1993; Harper et al., 1993; Xiong et al., 1993), and as a gene whose expression is induced in senescence cells (Noda et al., 1994). Such a variety of actions was also reflected in its multiple names: Sdi1 (for senescent cell-derived inhibitor), Waf1 (for wild-type p53-activated fragment), and Cip1 (for CDK-interacting protein).

p21 expression is mainly controlled at a transcriptional level by both p53-dependent and p53-independent mechanisms (reviewed in Gartel and Tyner, 1999). In some cases, post-transcriptional regulation, such as mRNA stabilization by UVC (Gorospe et al., 1998), or post-translational stabilization through the interaction with the transcription factor C/EBPα [Timchenko, 1997; Timchenko, 1996] has been observed.

Its role in cell cycle control, DNA damage response, senescence, differentiation, and DNA replication is mediated by its interaction with a large number of proteins. p21Cip has two cyclin-CDK binding domains. One is homologous to the other family members, and there is another cyclin binding site at the C-terminus, in a region that overlaps with its

PCNA-binding domain (Chen et al., 1996). In proliferating cells, the cyclin-CDK-p21 complex also contains PCNA, perhaps linking the control of the cyclin-CDK activity to DNA synthesis. In vitro p21 binding prevents PCNA-dependent DNA replication but not PCNA-dependent excision repair (reviewed in Dotto, 2000). Whether the amount of p21 ever reaches the level necessary to do this, in vivo, is controversial. p21Cip association with PCNA can be inhibited by the binding of the transcription factor c-myc to C-terminal region (Kitaura et al., 2000).

p21Cip is most clearly involved in p53-dependent G_1 arrest after DNA damage. The amount of p21 increases after the exposure to DNA damaging agents (Dulic et al., 1994; el-Deiry et al., 1994). p21 null mouse embryo fibroblasts arrest in G_1 after γ irradiation. However, these cells show an intermediate phenotype between *p53*$^{-/-}$ and wild type (Brugarolas et al., 1995; Deng et al., 1995), suggesting that p21 is not the only protein involved in this p53-dependent response. c-myc binds to the p21 promoter after DNA damage, blocking p21 induction by p53, and it may play a big role in the decision between an apoptotic or cell cycle arrest response to induction of p53 (Seoane et al., 2002).

p21Cip has been linked to senescence, as its levels increase in primary fibroblast that express oncogenic Ras (Serrano et al., 1997) and as part of cellular response to stress (reviewed in Dotto, 2000). However, p21 null cells were shown to undergo senescence and mount responses to stress, suggesting that this relationship needs further clarification (Pantoja and Serrano, 1999).

The already long list of protein interactions and biological functions where p21 is involved is still growing. It is worth mentioning that the role of p21 in some of these is likely to be dependent on the system used to identify them. Thus, although p21 usually acts as a negative regulator of the cell cycle, in some instances it has been observed to be induced after mitogenic stimulation (Michieli et al, 1994; Nourse et al, 1994; Halaban et al, 1998). It is possible that these noninhibitory functions of p21 may be reflecting its "assembly factor" role previously discussed. p21 is generally induced during terminal differentiation both in vivo and in vitro (reviewed in Dotto, 2000), but it participates in a non-growth-arrest function in terminally differentiated keratinocytes (Di Cunto et al., 1998). A third example where we can find a dual function for p21 is its role in apoptosis. An increase of p21 is generally linked to an induction of the apoptotic process (reviewed in Dotto, 2000), but in colorectal carcinona and melanoma cells p21 expression seems to protect cells from p53-induced apoptosis (Gorospe et al., 1997; Polyak et al., 1996; Seoane et al., 2002).

Mice lacking p21Cip did not show any developmental defect or tumor predisposition (Brugarolas et al., 1995; Deng et al., 1995). The consequences of p21 loss are only related with its function within the p53 pathway.

p27Kip1 (for kinase inhibitor protein 1) is the second member of the family. It was initially identified as a protein associated with the cyclin

E-CDK2 complexes in cells treated with transforming growth factor-β (TGF-β), Lovastatin, and in contact-inhibited cells (Hengst et al., 1994; Koff et al., 1993; Polyak et al., 1994).

The highest amount of p27 is found in quiescent cells, decreasing as the cell enters in G_1, reaching its lowest amount in S phase, and being maintained at this low level through the rest of the cell cycle. These changes are the result of complex regulation that can be exerted at different levels: transcription (Dijkers et al., 2000; Gardner et al., 2001; Hirano et al., 2001; Inoue et al., 1999; Servant et al., 2000; Yang et al., 2001), protein synthesis (Hengst and Reed, 1996; Millard et al., 1997; Vidal et al., 2002), and sequestration (Soos et al., 1996) and degradation (Harper, 2001; Malek et al., 2001; Nguyen et al., 1999; Pagano et al., 1995). From all these mechanisms, CDK2-independent proteolysis and synthesis rate are the major contributors to p27 threshold levels between G_0 noncycling and G_1 cycling cells. After the cell has entered S phase, p27 levels reach a nadir because of CDK2-dependent proteolysis.

As we mentioned, p27 protein levels increase when the cell enters in a quiescent state (reviewed in Philipp-Staheli et al., 2001). A large number of antimitogenic stimuli, including contact inhibition (Polyak et al., 1994), TGF-β (Polyak et al., 1994), cAMP (Kato et al., 1994), rapamycin (Nourse et al., 1994), and IL6 (Kortylewski et al., 1999), arrest cells and induce p27 accumulation. In some of these, the induction of p27 contributes to growth arrest since cells lacking p27 are unable to arrest as efficiently. The addition of growth factors as estrogens, IL-2, PDGF, or serum correlates with a decrease on p27 expression (reviewed in Philipp-Staheli et al., 2001). In some cases cells lacking p27 require less mitogenic signal to remain in cell cycle.

The status of p27 can affect apoptosis, although whether it protects or promotes depends on the cell type and cellular context. For example, p27 overexpression induces apoptosis in some cancer cell lines (Katayose et al., 1997), but in others p27 can prevent the apoptotic effects of drugs or DNA-damaging agents (Katayose et al., 1997). Recently it has been found that absence of p27 desensitizes $Rb^{-/-}$ pituitary tumor cells to response to apoptotic signals, suggesting a new mechanism by which p27 can contribute to tumor formation (Carneiro et al., 2003).

The p27 knockout mouse was generated independently by three laboratories. Two of the strains show a completely lack of the protein (Fero et al., 1996; Nakayama et al., 1996), whereas the third one expresses a truncated form of the protein that lacks the cyclin-CDK inhibitory domain (Kiyokawa et al., 1996). The most apparent phenotype observed in the three lines was an increase in body size that can be detected as early as three weeks after birth. This size increase is a consequence of multiorgan hypercellularity and is dosage dependent: the p27 heterozygous mouse carrying the truncated allele expresses equal amount of functional and nonfunctional protein, thus reducing the amount of p27 by half, and it has an intermediate phenotype. In addition to that, the organs that in the wild-type animals express the highest amount of p27 are the ones that show the biggest increase in size in the knockout. Another relevant feature in these animals is the development of pituitary intermediate lobe hyperplasia, and in some cases adenoma, with almost 100% penetrance (Fero et al., 1996; Kiyokawa et al., 1996; Nakayama et al., 1996). p27 knockouts also show female infertility (Fero et al., 1996; Kiyokawa et al., 1996; Nakayama et al., 1996) and deafness (Chen and Segil, 1999; Lowenheim et al., 1999). All the phenotypes of the p27 null animals reflect a defect in antimitogenic responses, and confirm the important contribution of p27 to the cell decision between proliferation and cell cycle withdraw. This was demonstrated on a variety of cellular systems: tissue culture cells, including oligodendrocytes (Casaccia-Bonnefil et al., 1997) and osteoblasts (Drissi et al., 1999), and in the animal, including luteal cells (Tong et al., 1998), hair cells of the organ of Corti (Chen and Segil, 1999; Lowenheim et al., 1999), and hematopoietic progenitor cells in the bone marrow (Cheng et al., 2000).

The third member of the family, p57Kip2, was cloned simultaneously by two different groups (Lee et al., 1995; Matsuoka et al., 1995). The gene that encodes p57Kip2 is genomically imprinted, and the paternal allele is transcriptionally repressed and methylated in mouse (Hatada and Mukai, 1995). In human the paternal allele is expressed at low levels in most of the tissues except in the developing brain and some embryonal tissues, where its levels are comparable to the maternal allele (Matsuoka et al., 1995). The gene is located in a chromosomal region implicated in sporadic cancers, the Beckwith-Wiedemann syndrome, and Wilm's tumors, pointing to a possible role of p57Kip2 as a tumor-supressor gene. A functional interaction between p57Kip2 and another imprinted gene, IGF-II, in the development of the Beckwith-Wiedemann syndrome has been suggested (Grandjean et al., 2000). p57Kip2 is the most structurally diverse member of the family. It shares more similarity to p27 than to p21, both at the C-terminus and N-terminus, were the CDK inhibitory domain is located. The internal domain of p57 is unique, consisting of a proline-rich region and an acidic repeat region in mouse and a proline-alanine repeat region in human. p57Kip2 is implicated in differentiation of myogenic cells where it regulates MyoD expression (Reynaud et al., 1999, 2000). Recently it has been reported that p57Kip2 expression can be transcriptionaly induced by the β isoform of p73, but not by p53 (Blint et al., 2002).

The p57Kip2 null mouse phenotype suggests an important role during development. Within the knockouts for the Kip/Cip inhibitors, the p57Kip2 null mice are the only ones that exhibit a severe phenotype, with a high percentage of animals dying at day 1 after birth (Yan et al., 1997; Zhang et al., 1997). This perinatal death is the consequence of several developmental abnormalities that include cleft palate and abdominal defects. These defects occur as a result of increased apoptosis, endochondral bone ossification defects with incomplete differentiation, and inappropriate S phase entry in lens fiber cells, also with an increase in apoptosis.

Redundancy or Compensatory Roles of CDK Inhibitors

With the exception of p57Kip2, absence of a single CDK inhibitor does not correlate with the development of a severe phenotype, suggesting the existence of compensatory mechanisms, or alternatively, redundancy between inhibitors. Compensation or redundancy between proteins can be exerted in different ways, and the functional implications of each of them are different (Vidal and Koff, 2000). The development of double knockouts had provided us with a useful tool to study redundancy or compensation between CDK inhibitors (Table 7.1).

Combined loss of p21Cip1 and p57Kip2 revealed a phenotypic redundancy of these two inhibitors in some tissues. Thus *p21*$^{-/-}$*p57*$^{-/-}$ mice showed a profound defect in skeletal muscle formation as a consequence of a failure in myotubes formation and an increase on proliferation and apoptotic rates of myoblasts (Zhang et al., 1999). Neither of these phenotypes was previously observed on the single mutants. The generation of double knockouts for p18Ink4c and p19Ink4d had also revealed a phenotypic redundancy between those inhibitors. Mice lacking both proteins are sterile due to a delayed exit of spermatogonia from the mitotic cell cycle, suggesting a collaboration between both proteins in regulating spermatogenesis (Zindy et al., 2001).

Simultaneous loss of two inhibitors with the same phenotype like p27Kip1 and p18Ink4c resulted in acceleration of the pituitary tumor development (Franklin et al., 1998). *p27*$^{-/-}$*p18*$^{-/-}$ mice also develop hyperplasia or adenoma in some organs, mostly endocrine, with a higher frequency than in the single null strains (Franklin et al., 2000). In addition to that, some organs were even more enlarged (Franklin et al., 1998). This suggests that both proteins are collaborating on the same pathway or are controlling different pathways that cooperate to control cell proliferation. A functional collaboration in controlling body size was not found when p18Ink4c mice were crossed into a p21Cip1 null background, although they did cooperate to increase the incidence of pituitary adenomas when compared with the single nulls (Franklin et al., 2000). In addition to the effect on the pituitary, these animals also develop a unique tumor profile, different from that detected in the *p27*$^{-/-}$*p18*$^{-/-}$ mice. This suggests an influence of the cell type on the functional collaboration between distinct CDK inhibitors. Finally, the cross between p19Ink4d and p27Kip1 null animals shows again a cooperation between Cip and Ink proteins. The *p19*$^{-/-}$*p27*$^{-/-}$ mice die very soon after birth with bradykinesia, proprioceptive abnormalities, and seizures as a result of inappropriate proliferation of postmitotic neurons in all parts of the brain (Zindy et al., 1999). This suggests that postmitotic neurons are maintained in a quiescent state as a result of a cooperation between these two inhibitors. The previously found lens defect on the p57Kip2 null mice was slightly more severe when crossed into a p27 null background (Zhang et al., 1998).

Cell Cycle Inhibitors and Cancer

One of the characteristics that all tumor cells display is a decreased responsiveness to antimitogenic signals that control their growth. The data obtained from the analyses of the different phenotype show that deletion of the CDK inhibitors, either alone or in combination, does not cause a loss of proliferation control and cancer. In a few cases mutations or deletions in the p15Ink4b, p18Ink4c, and p19Ink4d genes can be found in human tumors (reviewed in Ortega et al., 2002). At the present just two of the CDK inhibitors, p27Cip1 and p16Ink4a, are considered tumor-suppressor genes.

p27Kip1 is not what we would call a classical tumor suppressor. As we mentioned, p27 null mice are not predisposed to a general increase in tumor development, although they do develop pituitary adenomas (Fero et al., 1996; Kiyokawa et al., 1996; Nakayama et al., 1996) and benign prostate hyperplasia (Cordon-Cardo et al., 1998). Interestingly, both *p27*$^{-/-}$ and *p27*$^{+/-}$ mice are predisposed to tumor formation after being expose to ionizing radiation or chemical carcinogens (Fero et al., 1998). The genetical and biochemical analysis of the tumors arising in the carcinogen-treated *p27*$^{+/-}$ animals revealed that the wild-type allele is not mutated and the protein is not expressed. In the classical tumor-suppressor genes such as pRb, p19ARF, and p53, the tumors that arise in the heterozygous animals show frequently the loss of the remaining wild-type allele (Harvey et al., 1993; Jacks et al., 1992; Kamijo et al., 1999; Williams et al., 1994), consistent with the Knudson's "two-hit" model (Knudson, 1971).

Reducing p27 levels in the absence of two other cell cycle-related genes, p18Ink4c and Rb, increases tumor aggressiveness. We already mentioned that combined loss of p27Kip1 and p18Ink4c causes an early appearance of pituitary tumors (Franklin et al., 1998). pRb heterozygous mice display the same tumor spectrum as p27 null animals, with adenocarcinoma of the pituitary intermediate lobe. These tumors showed loss of the remaining wild-type allele (Harrison et al., 1995; Hu et al., 1994). The development of pituitary tumors in the *Rb*$^{+/-}$ mice occurs after a long latency period and reflects the time necessary to overcome the apoptosis induced by antiproliferative signals that control abnormal growth (Nikitin and Lee, 1996). *Rb*$^{+/-}$*p27*$^{-/-}$ mice develop more aggressive pituitary tumors with an earlier onset (Park et al., 1999). This is consistent with a model where loss of response to antimitogenic signals (*p27*$^{-/-}$) provides an additional advantage over the already altered proliferation (*Rb*$^{-/-}$) shortening the latency period. Loss of p27 provides this advantage by desensitizing these cells to the apoptotic signals (Carneiro et al., 2003). Exacerbation of the tumor development after loss of p27 can also be found in other mouse models such as Pten (Di Cristofano et al., 2001), GHRH (Teixeira et al., 2000), Inhibin (Cipriano et al., 2001), and APC

(Philipp-Staheli et al., 2002), although in these systems its contribution is more associated with an increase in proliferation.

With regard to its role in human tumors, an increasing number of studies point to p27 as protein with prognostic significance. Mutation or homozygous deletion of the p27 gene in human tumors is rare, although some exceptions can be found (Komuro et al., 1999). Reduction of p27 is correlated with increased aggressiveness and decreased patient survival in a wide variety of tumor types (reviewed in Philipp-Staheli et al., 2001; Slingerland and Pagano, 2000). In some cases this appears to reflect an increase in proteasome-mediated p27 degradation (Loda et al., 1997; Piva et al., 1999), a regulatory process evident in cycling cells. However, the loss of p27 may not be simply consequential since the prognostic significance of low p27 is not equivalent to increased proliferation. Recently it had been described that p27 cellular localization plays an important role in certain types of tumors. Thus, in some carcinomas of the breast, thyroid or colon, p27 levels are normal but the protein is localized in the cytoplasm (Liang et al., 2002; Shin et al., 2002; Viglietto et al., 2002). p27 cytoplasm localization also correlates with poor prognosis (Liang et al., 2002). The underlying mechanism appears to involve Akt-dependent phosphorylation of p27 within the nuclear localization signal, preventing its entry inside the nucleus and its binding to CDK2.

Although with some differences between the two knockout strains, "pure" p16Ink4a mice did not showed a high tumor predisposition (Krimpenfort et al., 2001; Sharpless et al., 2001). One of the strains developed a broader spectrum of tumors when treated with carcinogens. The remaining wild-type allele was silenced in some of the aggressive tumors (Sharpless et al., 2001). The incidence of spontaneous tumors on the strain that carried the mutation within the exon 1α was very low. However, simultaneously deletion of one p19ARF allele in these animals provoked the development of several types of tumors, including melanomas, sarcomas, and lymphomas. The frequency of these tumors was increased when these animals were treated with carcinogenes (Krimpenfort et al., 2001), suggesting that lost of p16 cooperated with p19Arf heterozygosity in tumor formation.

In contrast to its weak role as tumor-suppressor gene in mice, several findings show that p16Ink4a is a strong tumor suppressor in humans. Ink4a/Arf and p15Ink4b loci are encoded in chromosomal region 9p21. After p53, alterations involving this chromosomal region are probably the most frequently found in human cancers (Kamb et al., 1994; Nobori et al., 1994; Ruas and Peters, 1998). Homozygous deletions as well as loss of expression due to promoter hypermethylation are the most common ways of p16Ink4a function inactivation, although point mutations are also frequently found in pancreatic cancers and melanomas (Ruas and Peters, 1998). In addition p16Ink4a-specific germ-line mutations had been identified in several studies carried out in kindred with familial melanoma and pancreatic carcinoma (reviewed in Rocco and Sidransky, 2001; Ruas and Peters, 1998). In some tumors, specific alteration of exon 1α selectively targets p16Ink4a, but a large number of tumor deletions or mutations within the p16/p19-shared sequence of exon 2 also occur. However none of these mutations affect the ability of p19ARF to cause G_1 arrest, a function that resides within the N-terminal domain encoded by exon 1β (Quelle et al., 1997). In human tumors, alterations that exclusively affect p19ARF are rare. Thus, in marked contrast to what happens in mouse, p16Ink4a has a more predominant role over p19ARF in human cancer.

To Cycle or Not to Cycle, How Cells Decide

We have described the essential pieces of the cell machinery needed to respond to the mitogenic and antimitogenic signals. But, of course, this is a dynamic process with proteins that are continuously synthesized and degraded, that are being held together, or that are changing partners. So there has to be a coordination, a sequence of events so that the signal can be interpreted and executed in the correct way. The decision to proliferate has to be made before crossing the point of no return, the restriction point. Once the cell has passed that point, the commitment to enter into S phase and to cycle is irreversible.

Once they appear in early to mid G_1, the first mission of the cyclin D-CDK4/6 complexes is to start phosphorylating pRB. But at the same time they carry out a second important function, they sequester p21Cip and p27Kip molecules. Remember that cyclin E levels are being increased by the transcriptional activity of the E2F factors being released from the pRb repression. Thus unbound p21 and p27 can still inhibit the activity of the new synthesized cyclin E-CDK2 complexes. However, when the inhibitory activity is sequestered by the cyclin D-CDK4/6 complexes, the cyclin E-CDK2 complexes can facilitate its own activation by inducing p27 degradation. To do that, cyclin E-CDK2 phosphorylates p27Kip on a particular threonine residue (Thr-187), which targets p27 to ubiquitination-mediated proteolysis (Harper, 2001; Sheaff et al., 1997; Vlach et al., 1997).

Until now, we have described how the Cip/Kip inhibitors regulate the response to mitogenic signals. Where do the Ink proteins fit in all this process? As we said before, Ink proteins are inhibitors of the CDK4 activity, so an increase in Ink levels will affect pRb phosphorylation. The treatment of the mink lung cell line Mv1Lu with TGF-β has shed light on an interesting mechanism where, as a response to an antimitogenic signal, the two classes of inhibitors, Ink and Cip/Kip, cooperate to affect CDK activity and induce cell cycle arrest. In these cells, TGF-β treatment induces p15Ink4b accumulation, which binds to the cyclin D-CDK4 complexes. This provokes a redistribution of p27Cip from the cyclin D-CDK4 to the cyclin E-CDK2 complexes and does not require new p27 synthesis (Reynisdottir and Massague, 1997; Reynisdottir et al., 1995). A similar mechanism operates to mobilize p21Cip after p16 induction (McConnell et al., 1999).

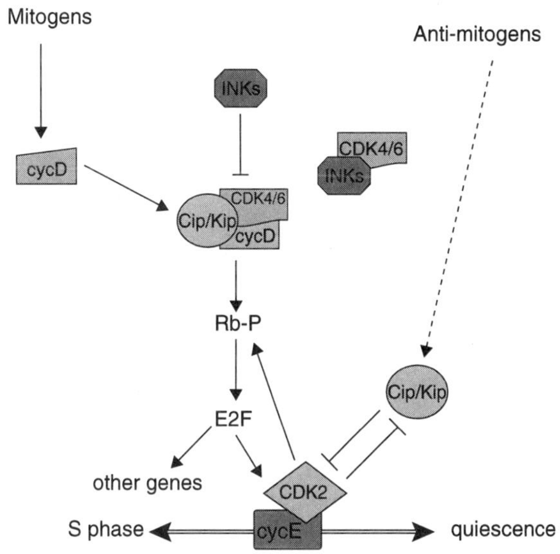

Figure 7.3. Balance model. Mitogenic signals activate cyclin D complexes that induce pRb phosphorylation and inactivation. Release of Rb-bounded E2F allows transcription of genes necessaries for S phase. Antimitogenic signals inhibit cyclin E-associated kinase activity through p27Kip1. Binding to p27 facilitates cyclin D-CDK4/6 assembly, and this negatively regulates p27Kip inhibitory activity. Once all inhibitory activity has been sequestered, cyclin E-CDK2 complexes can facilitate its own activation by inducing p27 degradation. The balance between signals that induce and those that inhibit cyclin E-associated activity determines whether there will occur progression to S phase or growth arrest.

The final decision between progression to S phase or growth arrest is determined by the balance between the signals that activate and those that inhibit cyclin E activity (Vidal and Koff, 2000) (Fig. 7.3).

Beyond the Cell Cycle: New Roles for the CKIs

There is no doubt that CDK inhibitors play an important role in cell cycle arrest, but is that their only function? In the last few years it has become clear that they also participate in other processes once the cell has achieved arrest. Among them, a great attention is been paid to the role of CDK inhibitors in differentiation and to the fact that they may contribute to the differentiation process, perhaps using mechanisms different than those used to induce growth arrest.

Increased expression of cell cycle inhibitors is observed during differentiation in several cell types such as keratinocytes (Hauser et al., 1997), oligodendrocyte progenitor cells (Casaccia-Bonnefil et al., 1997; Ghiani et al., 1999; Tang et al., 1998), and retinal progenitor cells (Dyer and Cepko, 2000).

As we mentioned earlier, lost of p57kip2 has severe consequences in development, namely bone ossification defects, which suggests the role of p57 in chondrocytes differentiation (Yan et al., 1997; Zhang et al., 1997). Combined lost of p21 and p57 alters skeletal muscle differentiation (Zhang et al., 1999) and implicates p21 in the process. The role of Ink inhibitors in differentiation is less clear, although some examples can be found, like p19Ink4d cooperation with p27Kip1, that maintain differentiated neurons in a quiescent state (Zindy et al., 1999).

In addition to the increase in cells that is induced at terminal differentiation, a function for p27 in this process is suggested by the observation that the inability to increase p27 prevents differentiation (reviewed in Philipp-Staheli et al., 2001). It is important to note that although p27 plays an important role in differentiation, it is not a decisive factor. Indeed, in absence of p27, the differentiation process is only delayed because cells fail to withdraw from the cell cycle in a timely fashion. However, differentiation is ultimately achieved, indicating that the cell has other ways to exit from the cell cycle (Fero et al., 1996; Kiyokawa et al., 1996; Nakayama et al., 1996). We can speculate that the delay reflects the period of time that the cell needs to activate an alternative pathway to replace p27 function or to wait for other processes to act.

Besides mice, studies carried out in *Xenopus* had provided us with additional evidence implicating CDK inhibitors in differentiation. In *Xenopus* only one CDK inhibitor, that shares structural and functional characteristics with p21Cip1, p27Kip1, and p57Kip2 (Shou and Dunphy, 1996; Su et al., 1995), p27Xic1. In this system it has been demonstrated that an increase in p27Xic expression promotes the differentiation of Muller cells of the retina (Ohnuma et al., 1999). Recently p27Xic1 has been also implicated in the induction of both muscle and neuron differentiation. These activities are separable from its role in cell cycle regulation, as a truncated form of the protein that retained the CDK inhibitory domain but lacks the N-terminus was unable to promote differentiation (Vernon et al., 2003; Vernon and Philpott, 2003).

Cellular systems derived from knockout mice have also confirmed a role of cell cycle inhibitors in differentiation. Thus oligodendrocyte derived from both p21 and p27 null mice failed to differentiate when placed in a differentiation media, but only absence of p27 prevented growth arrest (Zezula et al., 2001). p21 null oligodendrocytes successfully exit the cell cycle, indicating a role of p21 in differentiation that is independent of its role in regulating cell cycle exit. Interestingly the differentiation defect observed in the p21 null oligodendrocytes is completely complemented when conditions inhibit cyclinD-CDK4/6 kinase activity (Zezula et al., 2001). It is possible to speculate that in the absence of p21, the cyclin D-CDK4/6 complex can interfere in a differentiation program, perhaps by affecting the differentiation-promoting function of pRb (Kaelin, 1997).

The search of molecular mechanisms accounting for the assembly/inhibitory function of CDK inhibitors may provide us with a better understanding of their function in proliferation control and their role beyond the cell cycle.

REFERENCES

Alcorta DA, Xiong Y, Phelps D, Hannon G, Beach D, Barrett JC (1996): Involvement of the cyclin-dependent kinase inhibitor p16 (INK4a) in replicative senescence of normal human fibroblasts. Proc Natl Acad Sci USA 93: 13742–7.

Blain SW, Montalvo E, Massague J (1997): Differential interaction of the cyclin-dependent kinase (Cdk) inhibitor p27Kip1 with cyclin A-Cdk2 and cyclin D2-Cdk4. J Biol Chem 272:25863–72.

Blint E, Phillips AC, Kozlov S, Stewart CL, Vousden KH (2002): Induction of p57(KIP2) expression by p73beta. Proc Natl Acad Sci USA 99:3529–34.

Botz J, Zerfass-Thome K, Spitkovsky D, Delius H, Vogt B, Eilers M, Hatzigeorgiou A, Jansen-Durr P (1996): Cell cycle regulation of the murine cyclin E gene depends on an E2F binding site in the promoter. Mol Cell Biol 16:3401–9.

Bruce JL, Hurford RK, Jr., Classon M, Koh J, Dyson N (2000): Requirements for cell cycle arrest by p16INK4a. Mol Cell 6:737–42.

Brugarolas J, Chandrasekaran C, Gordon JI, Beach D, Jacks T, Hannon GJ (1995): Radiation-induced cell cycle arrest compromised by p21 deficiency. Nature 377:552–7.

Carneiro C, Jiao MS, Hu M, Shaffer D, Park M, Pandolfi PP, Cordon-Cardo C, Koff A (2003): p27 deficiency desensitizes cells to signals that trigger apoptosis during pituitary tumor development. Oncogene 22:361–9.

Casaccia-Bonnefil P, Tikoo R, Kiyokawa H, Friedrich V, Jr., Chao MV, Koff A (1997): Oligodendrocyte precursor differentiation is perturbed in the absence of the cyclin-dependent kinase inhibitor p27Kip1. Genes Dev 11:2335–46.

Chan FK, Zhang J, Cheng L, Shapiro DN, Winoto A (1995): Identification of human and mouse p19, a novel CDK4 and CDK6 inhibitor with homology to p16ink4. Mol Cell Biol 15:2682–8.

Chen IT, Akamatsu M, Smith ML, Lung FD, Duba D, Roller PP, Fornace AJ, Jr., O'Connor PM (1996): Characterization of p21Cip1/Waf1 peptide domains required for cyclin E/Cdk2 and PCNA interaction. Oncogene 12:595–607.

Chen P, Segil N (1999): p27(Kip1) links cell proliferation to morphogenesis in the developing organ of Corti. Development 126:1581–90.

Cheng M, Olivier P, Diehl JA, Fero M, Roussel MF, Roberts JM, Sherr CJ (1999): The p21(Cip1) and p27(Kip1) CDK "inhibitors" are essential activators of cyclin D-dependent kinases in murine fibroblasts. EMBO J 18:1571–83.

Cheng T, Rodrigues N, Dombkowski D, Stier S, Scadden DT (2000): Stem cell repopulation efficiency but not pool size is governed by p27(kip1). Nat Med 6:1235–40.

Cipriano SC, Chen L, Burns KH, Koff A, Matzuk MM (2001): Inhibin and p27 interact to regulate gonadal tumorigenesis. Mol Endocrinol 15:985–96.

Cordon-Cardo C, Koff A, Drobnjak M, Capodieci P, Osman I, Millard SS, Gaudin PB, Fazzari M, Zhang ZF, Massague J, Scher HI (1998): Distinct altered patterns of p27KIP1 gene expression in benign prostatic hyperplasia and prostatic carcinoma. J Natl Cancer Inst 90:1284–91.

Deng C, Zhang P, Harper JW, Elledge SJ, Leder P (1995): Mice lacking p21CIP1/WAF1 undergo normal development, but are defective in G_1 checkpoint control. Cell 82:675–84.

Di Cristofano A, De Acetis M, Koff A, Cordon-Cardo C, Pandolfi PP (2001): Pten and p27KIP1 cooperate in prostate cancer tumor suppression in the mouse. Nat Genet 27:222–4.

Di Cunto F, Topley G, Calautti E, Hsiao J, Ong L, Seth PK, Dotto GP (1998): Inhibitory function of p21Cip1/WAF1 in differentiation of primary mouse keratinocytes independent of cell cycle control. Science 280:1069–72.

Dijkers PF, Medema RH, Pals C, Banerji L, Thomas NS, Lam EW, Burgering BM, Raaijmakers JA, Lammers JW, Koenderman L, Coffer PJ (2000): Forkhead transcription factor FKHR-L1 modulates cytokine-dependent transcriptional regulation of p27(KIP1). Mol Cell Biol 20:9138–48.

Dotto GP (2000): p21(WAF1/Cip1): more than a break to the cell cycle? Biochim Biophys Acta 1471:M43–56.

Drissi H, Hushka D, Aslam F, Nguyen Q, Buffone E, Koff A, van Wijnen A, Lian JB, Stein JL, Stein GS (1999): The cell cycle regulator p27kip1 contributes to growth and differentiation of osteoblasts. Cancer Res 59:3705–11.

Dulic V, Kaufmann WK, Wilson SJ, Tlsty TD, Lees E, Harper JW, Elledge SJ, Reed SI (1994): p53-dependent inhibition of cyclin-dependent kinase activities in human fibroblasts during radiation-induced G_1 arrest. Cell 76:1013–23.

Dyer MA, Cepko CL (2000): p57(Kip2) regulates progenitor cell proliferation and amacrine interneuron development in the mouse retina. Development 127:3593–605.

Dyson N (1998): The regulation of E2F by pRB-family proteins. Genes Dev 12:2245–62.

el-Deiry WS, Harper JW, O'Connor PM, Velculescu VE, Canman CE, Jackman J, Pietenpol JA, Burrell M, Hill DE, Wang Y, et al. (1994): WAF1/CIP1 is induced in p53-mediated G_1 arrest and apoptosis. Cancer Res 54:1169–74.

el-Deiry WS, Tokino T, Velculescu VE, Levy DB, Parsons R, Trent JM, Lin D, Mercer WE, Kinzler KW, Vogelstein B (1993): WAF1, a potential mediator of p53 tumor suppression. Cell 75:817–25.

Feng XH, Liang YY, Liang M, Zhai W, Lin X (2002): Direct interaction of c-Myc with Smad2 and Smad3 to inhibit TGF-beta-mediated induction of the CDK inhibitor p15(Ink4B). Mol Cell 9:133–43.

Feng XH, Lin X, Derynck R (2000): Smad2, Smad3 and Smad4 cooperate with Sp1 to induce p15(Ink4B) transcription in response to TGF-beta. EMBO J 19:5178–93.

Fero ML, Randel E, Gurley KE, Roberts JM, Kemp CJ (1998): The murine gene p27Kip1 is haplo-insufficient for tumour suppression. Nature 396:177–80.

Fero ML, Rivkin M, Tasch M, Porter P, Carow CE, Firpo E, Polyak K, Tsai LH, Broudy V, Perlmutter RM, Kaushansky K, Roberts JM (1996): A syndrome of multiorgan hyperplasia with features of gigantism, tumorigenesis, and female sterility in p27(Kip1)-deficient mice. Cell 85:733–44.

Franklin DS, Godfrey VL, Lee H, Kovalev GI, Schoonhoven R, Chen-Kiang S, Su L, Xiong Y (1998): CDK inhibitors p18(INK4c) and p27(Kip1) mediate two separate pathways to collaboratively suppress pituitary tumorigenesis. Genes Dev 12:2899–911.

Franklin DS, Godfrey VL, O'Brien DA, Deng C, Xiong Y (2000): Functional collaboration between different cyclin-dependent kinase inhibitors suppresses tumor growth with distinct tissue specificity. Mol Cell Biol 20:6147–58.

Gardner LB, Li Q, Park MS, Flanagan WM, Semenza GL, Dang CV (2001): Hypoxia inhibits G_1/S transition through regulation of p27 expression. J Biol Chem 276:7919–26.

Gartel AL, Tyner AL (1999): Transcriptional regulation of the p21(WAF1/CIP1) gene. Expr Cell Res 246:280–9.

Geng Y, Whoriskey W, Park MY, Bronson RT, Medema RH, Li T, Weinberg RA, Sicinski P (1999): Rescue of cyclin D1 deficiency by knockin cyclin E. Cell 97:767–77.

Ghiani CA, Eisen AM, Yuan X, DePinho RA, McBain CJ, Gallo V (1999): Neurotransmitter receptor activation triggers p27(Kip1) and p21(CIP1) accumulation and G_1 cell cycle arrest in oligodendrocyte progenitors. Development 126:1077–90.

Gorospe M, Cirielli C, Wang X, Seth P, Capogrossi MC, Holbrook NJ (1997): p21(Waf1/Cip1) protects against p53-mediated apoptosis of human melanoma cells. Oncogene 14:929–35.

Gorospe M, Wang X, Holbrook NJ (1998): p53-dependent elevation of p21Waf1 expression by UV light is mediated through mRNA stabilization and involves a vanadate-sensitive regulatory system. Mol Cell Biol 18:1400–7.

Grandjean V, Smith J, Schofield PN, Ferguson-Smith AC (2000): Increased IGF-II protein affects p57kip2 expression in vivo and in vitro: Implications for Beckwith-Wiedemann syndrome. Proc Natl Acad Sci USA 97:5279–84.

Gu Y, Turck CW, Morgan DO (1993): Inhibition of CDK2 activity in vivo by an associated 20K regulatory subunit. Nature 366:707–10.

Guan KL, Jenkins CW, Li Y, Nichols MA, Wu X, O'Keefe CL, Matera AG, Xiong Y (1994): Growth suppression by p18, a p16INK4/MTS1- and p14INK4B/MTS2-related CDK6 inhibitor, correlates with wild-type pRb function. Genes Dev 8:2939–52.

Hannon GJ, Beach D (1994): p15INK4B is a potential effector of TGF-beta-induced cell cycle arrest. Nature 371:257–61.

Harbour JW, Luo RX, Dei Santi A, Postigo AA, Dean DC (1999): Cdk phosphorylation triggers sequential intramolecular interactions that progressively block Rb functions as cells move through G_1. Cell 98:859–69.

Harper JW (2001): Protein destruction: adapting roles for Cks proteins. Curr Biol 11:R431–5.

Harper JW, Adami GR, Wei N, Keyomarsi K, Elledge SJ (1993): The p21 Cdk-interacting protein Cip1 is a potent inhibitor of G_1 cyclin-dependent kinases. Cell 75:805–16.

Harrison DJ, Hooper ML, Armstrong JF, Clarke AR (1995): Effects of heterozygosity for the Rb-1t19neo allele in the mouse. Oncogene 10:1615–20.

Harvey M, McArthur MJ, Montgomery CA, Jr., Butel JS, Bradley A, Donehower LA (1993): Spontaneous and carcinogen-induced tumorigenesis in p53-deficient mice. Nat Genet 5:225–9.

Hatada I, Mukai T (1995): Genomic imprinting of p57KIP2, a cyclin-dependent kinase inhibitor, in mouse. Nat Genet 11:204–6.

Hauser PJ, Agrawal D, Flanagan M, Pledger WJ (1997): The role of p27kip1 in the in vitro differentiation of murine keratinocytes. *Cell* Growth Differ 8: 203–11.

Hengst L, Dulic V, Slingerland JM, Lees E, Reed SI (1994): A cell cycle-regulated inhibitor of cyclin-dependent kinases. Proc Natl Acad Sci USA 91:5291–5.

Hengst L, Reed SI (1996): Translational control of p27Kip1 accumulation during the cell cycle. Science 271:1861–4.

Hengst L, Reed SI (1998): Inhibitors of the Cip/Kip family. Curr Top Microbiol Immunol 227:25–41.

Hirai H, Roussel MF, Kato JY, Ashmun RA, Sherr CJ (1995): Novel INK4 proteins, p19 and p18, are specific inhibitors of the cyclin D-dependent kinases CDK4 and CDK6. Mol Cell Biol 15:2672–81.

Hirano M, Hirano K, Nishimura J, Kanaide H (2001): Transcriptional up-regulation of p27(Kip1) during contact-induced growth arrest in vascular endothelial cells. Expr Cell Res 271:356–67.

Hu N, Gutsmann A, Herbert DC, Bradley A, Lee WH, Lee EY (1994): Heterozygous Rb-1 delta 20/+mice are predisposed to tumors of the pituitary gland with a nearly complete penetrance. Oncogene 9:1021–7.

Inoue T, Kamiyama J, Sakai T (1999): Sp1 and NF-Y synergistically mediate the effect of vitamin D(3) in the p27(Kip1) gene promoter that lacks vitamin D response elements. J Biol Chem 274:32309–17.

Jacks T, Fazeli A, Schmitt EM, Bronson RT, Goodell MA, Weinberg RA (1992): Effects of an Rb mutation in the mouse. Nature 359:295–300.

Kaelin WG, Jr. (1997): Recent insights into the functions of the retinoblastoma susceptibility gene product. Cancer Invest 15:243–54.

Kamb A, Shattuck-Eidens D, Eeles R, Liu Q, Gruis NA, Ding W, Hussey C, Tran T, Miki Y, Weaver-Feldhaus J, et al. (1994): Analysis of the p16 gene (CDKN2) as a candidate for the chromosome 9p melanoma susceptibility locus. Nat Genet 8:23–6.

Kamijo T, van de Kamp E, Chong MJ, Zindy F, Diehl JA, Sherr CJ, McKinnon PJ (1999): Loss of the ARF tumor suppressor reverses premature replicative arrest but not radiation hypersensitivity arising from disabled atm function. Cancer Res 59:2464–9.

Kamijo T, Zindy F, Roussel MF, Quelle DE, Downing JR, Ashmun RA, Grosveld G, Sherr CJ (1997): Tumor suppression at the mouse INK4a locus mediated by the alternative reading frame product p19ARF. Cell 91:649–59.

Katayose Y, Kim M, Rakkar AN, Li Z, Cowan KH, Seth P (1997): Promoting apoptosis: a novel activity associated with the cyclin- dependent kinase inhibitor p27. Cancer Res 57:5441–5.

Kato JY, Matsuoka M, Polyak K, Massague J, Sherr CJ (1994): Cyclic AMP-induced G1 phase arrest mediated by an inhibitor (p27Kip1) of cyclin-dependent kinase 4 activation. Cell 79:487–96.

Kitaura H, Shinshi M, Uchikoshi Y, Ono T, Iguchi-Ariga SM, Ariga H (2000): Reciprocal regulation via protein-protein interaction between c-Myc and p21(cip1/waf1/sdi1) in DNA replication and transcription. J Biol Chem 275: 10477–83.

Kiyokawa H, Kineman RD, Manova-Todorova KO, Soares VC, Hoffman ES, Ono M, Khanam D, Hayday AC, Frohman LA, Koff A (1996): Enhanced growth of mice lacking the cyclin-dependent kinase inhibitor function of p27(Kip1). Cell 85:721–32.

Knudson AG, Jr. (1971): Mutation and cancer: Statistical study of retinoblastoma. Proc Natl Acad Sci USA 68:820–3.

Koff A, Ohtsuki M, Polyak K, Roberts JM, Massague J (1993): Negative regulation of G1 in mammalian cells: inhibition of cyclin E-dependent kinase by TGF-beta. Science 260:536–9.

Koh J, Enders GH, Dynlacht BD, Harlow E (1995): Tumour-derived p16 alleles encoding proteins defective in cell-cycle inhibition. Nature 375:506–10.

Komuro H, Valentine MB, Rubnitz JE, Saito M, Raimondi SC, Carroll AJ, Yi T, Sherr CJ, Look AT (1999): p27KIP1 deletions in childhood acute lymphoblastic leukemia. Neoplasia 1:253–61.

Kortylewski M, Heinrich PC, Mackiewicz A, Schniertshauer U, Klingmuller U, Nakajima K, Hirano T, Horn F, Behrmann I (1999): Interleukin-6 and oncostatin M-induced growth inhibition of human A375 melanoma cells is STAT-dependent and involves upregulation of the cyclin-dependent kinase inhibitor p27/Kip1. Oncogene 18:3742–53.

Krimpenfort P, Quon KC, Mooi WJ, Loonstra A, Berns A (2001): Loss of p16Ink4a confers susceptibility to metastatic melanoma in mice. Nature 413: 83–6.

LaBaer J, Garrett MD, Stevenson LF, Slingerland JM, Sandhu C, Chou HS, Fattaey A, Harlow E (1997): New functional activities for the p21 family of CDK inhibitors. Genes Dev 11:847–62.

Latres E, Malumbres M, Sotillo R, Martin J, Ortega S, Martin-Caballero J, Flores JM, Cordon-Cardo C, Barbacid M (2000): Limited overlapping roles of P15(INK4b) and P18(INK4c) cell cycle inhibitors in proliferation and tumorigenesis. EMBO J 19:3496–506.

Lee MH, Reynisdottir I, Massague J (1995): Cloning of p57KIP2, a cyclin-dependent kinase inhibitor with unique domain structure and tissue distribution. Genes Dev 9:639–49.

Liang J, Zubovitz J, Petrocelli T, Kotchetkov R, Connor MK, Han K, Lee JH, Ciarallo S, Catzavelos C, Beniston R, Franssen E, Slingerland JM (2002): PKB/Akt phosphorylates p27, impairs nuclear import of p27 and opposes p27-mediated G_1 arrest. Nat Med 8:1153–60.

Loda M, Cukor B, Tam SW, Lavin P, Fiorentino M, Draetta GF, Jessup JM, Pagano M (1997): Increased proteasome-dependent degradation of the cyclin-dependent kinase inhibitor p27 in aggressive colorectal carcinomas (see Comments). Nat Med 3:231–4.

Lowenheim H, Furness DN, Kil J, Zinn C, Gultig K, Fero ML, Frost D, Gummer AW, Roberts JM, Rubel EW, Hackney CM, Zenner HP (1999): Gene disruption of p27(Kip1) allows cell proliferation in the postnatal and adult organ of corti. Proc Natl Acad Sci USA 96:4084–8.

Lukas J, Parry D, Aagaard L, Mann DJ, Bartkova J, Strauss M, Peters G, Bartek J (1995): Retinoblastoma-protein-dependent cell-cycle inhibition by the tumour suppressor p16. Nature 375:503–6.

Malek NP, Sundberg H, McGrew S, Nakayama K, Kyriakidis TR, Roberts JM (2001): A mouse knock-in model exposes sequential proteolytic pathways that regulate p27Kip1 in G_1 and S phase. Nature 413:323–7.

Matsuoka S, Edwards MC, Bai C, Parker S, Zhang P, Baldini A, Harper JW, Elledge SJ (1995): p57KIP2, a structurally distinct member of the p21CIP1

Cdk inhibitor family, is a candidate tumor suppressor gene. Genes Dev 9: 650–62.

Matsushime H, Quelle DE, Shurtleff SA, Shibuya M, Sherr CJ, Kato JY (1994): D-type cyclin-dependent kinase activity in mammalian cells. Mol Cell Biol 14:2066–76.

McConnell BB, Gregory FJ, Stott FJ, Hara E, Peters G (1999): Induced expression of p16(INK4a) inhibits both CDK4- and CDK2- associated kinase activity by reassortment of cyclin-CDK-inhibitor complexes. Mol Cell Biol 19:1981–9.

Medema RH, Herrera RE, Lam F, Weinberg RA (1995): Growth suppression by p16ink4 requires functional retinoblastoma protein. Proc Natl Acad Sci USA 92:6289–93.

Millard SS, Yan JS, Nguyen H, Pagano M, Kiyokawa H, Koff A (1997): Enhanced ribosomal association of p27(Kip1) mRNA is a mechanism contributing to accumulation during growth arrest. J Biol Chem 272:7093–8.

Nakayama K, Ishida N, Shirane M, Inomata A, Inoue T, Shishido N, Horii I, Loh DY (1996): Mice lacking p27(Kip1) display increased body size, multiple organ hyperplasia, retinal dysplasia, and pituitary tumors. Cell 85:707–20.

Nguyen H, Gitig DM, Koff A (1999): Cell-free degradation of p27(kip1), a G_1 cyclin-dependent kinase inhibitor, is dependent on CDK2 activity and the proteasome. Mol Cell Biol 19:1190–201.

Nikitin A, Lee WH (1996): Early loss of the retinoblastoma gene is associated with impaired growth inhibitory innervation during melanotroph carcinogenesis in $Rb^{+/-}$ mice. Genes Dev 10:1870–9.

Nobori T, Miura K, Wu DJ, Lois A, Takabayashi K, Carson DA (1994): Deletions of the cyclin-dependent kinase-4 inhibitor gene in multiple human cancers. Nature 368:753–6.

Noda A, Ning Y, Venable SF, Pereira-Smith OM, Smith JR (1994): Cloning of senescent cell-derived inhibitors of DNA synthesis using an expression screen. Expr Cell Res 211:90–8.

Nourse J, Firpo E, Flanagan WM, Coats S, Polyak K, Lee MH, Massague J, Crabtree GR, Roberts JM (1994): Interleukin-2-mediated elimination of the p27Kip1 cyclin-dependent kinase inhibitor prevented by rapamycin. Nature 372:570–3.

Ohnuma S, Philpott A, Wang K, Holt CE, Harris WA (1999): p27Xic1, a Cdk inhibitor, promotes the determination of glial cells in *Xenopus* retina. Cell 99:499–510.

Ohtani K, DeGregori J, Nevins JR (1995): Regulation of the cyclin E gene by transcription factor E2F1. Proc Natl Acad Sci USA 92:12146–50.

Ohtsubo M, Theodoras AM, Schumacher J, Roberts JM, Pagano M (1995): Human cyclin E, a nuclear protein essential for the G1-to-S phase transition. Mol Cell Biol 15:2612–24.

Ortega S, Malumbres M, Barbacid M (2002): Cyclin D-dependent kinases, INK4 inhibitors and cancer. Biochim Biophys Acta 1602:73–87.

Pagano M, Tam SW, Theodoras AM, Beer-Romero P, Del Sal G, Chau V, Yew PR, Draetta GF, Rolfe M (1995): Role of the ubiquitin-proteasome pathway in regulating abundance of the cyclin-dependent kinase inhibitor p27 (see Comments). Science 269:682–5.

Pantoja C, Serrano M (1999): Murine fibroblasts lacking p21 undergo senescence and are resistant to transformation by oncogenic Ras. Oncogene 18:4974–82.

Park MS, Rosai J, Nguyen HT, Capodieci P, Cordon-Cardo C, Koff A (1999): p27 and Rb are on overlapping pathways suppressing tumorigenesis in mice. Proc Natl Acad Sci USA 96:6382–7.

Passegue E, Wagner EF (2000): JunB suppresses cell proliferation by transcriptional activation of p16(INK4a) expression. EMBO J 19:2969–79.

Pavletich NP (1999): Mechanisms of cyclin-dependent kinase regulation: structures of Cdks, their cyclin activators, and Cip and INK4 inhibitors. J Mol Biol 287:821–8.

Philipp-Staheli J, Kim KH, Payne SR, Gurley KE, Liggitt D, Longton G, Kemp CJ (2002): Pathway-specific tumor suppression. Reduction of p27 accelerates gastrointestinal tumorigenesis in Apc mutant mice, but not in Smad3 mutant mice. Cancer Cell 1:355–68.

Philipp-Staheli J, Payne SR, Kemp CJ (2001): p27(Kip1): regulation and function of a haploinsufficient tumor suppressor and its misregulation in cancer. Expr Cell Res 264:148–68.

Piva R, Cancelli I, Cavalla P, Bortolotto S, Dominguez J, Draetta GF, Schiffer D (1999): Proteasome-dependent degradation of p27/kip1 in gliomas. J Neuropathol Expr Neurol 58:691–6.

Polyak K, Hamilton SR, Vogelstein B, Kinzler KW (1996): Early alteration of cell-cycle-regulated gene expression in colorectal neoplasia. Am J Pathol 149: 381–7.

Polyak K, Kato JY, Solomon MJ, Sherr CJ, Massague J, Roberts JM, Koff A (1994): p27Kip1, a cyclin-Cdk inhibitor, links transforming growth factor-beta and contact inhibition to cell cycle arrest. Genes Dev 8:9–22.

Quelle DE, Cheng M, Ashmun RA, Sherr CJ (1997): Cancer-associated mutations at the INK4a locus cancel cell cycle arrest by p16INK4a but not by the alternative reading frame protein p19ARF. Proc Natl Acad Sci USA 94:669–73.

Quelle DE, Zindy F, Ashmun RA, Sherr CJ (1995): Alternative reading frames of the INK4a tumor suppressor gene encode two unrelated proteins capable of inducing cell cycle arrest. Cell 83:993–1000.

Rane SG, Dubus P, Mettus RV, Galbreath EJ, Boden G, Reddy EP, Barbacid M (1999): Loss of Cdk4 expression causes insulin-deficient diabetes and Cdk4 activation results in beta-islet cell hyperplasia. Nat Genet 22:44–52.

Resnitzky D, Gossen M, Bujard H, Reed SI (1994): Acceleration of the G_1/S phase transition by expression of cyclins D1 and E with an inducible system. Mol Cell Biol 14:1669–79.

Reynaud EG, Leibovitch MP, Tintignac LA, Pelpel K, Guillier M, Leibovitch SA (2000): Stabilization of MyoD by direct binding to p57(Kip2). J Biol Chem 275:18767–76.

Reynaud EG, Pelpel K, Guillier M, Leibovitch MP, Leibovitch SA (1999): p57(Kip2) stabilizes the MyoD protein by inhibiting cyclin E-Cdk2 kinase activity in growing myoblasts. Mol Cell Biol 19:7621–9.

Reynisdottir I, Massague J (1997): The subcellular locations of p15(Ink4b) and p27(Kip1) coordinate their inhibitory interactions with cdk4 and cdk2. Genes Dev 11:492–503.

Reynisdottir I, Polyak K, Iavarone A, Massague J (1995): Kip/Cip and Ink4 Cdk inhibitors cooperate to induce cell cycle arrest in response to TGF-beta. Genes Dev 9:1831–45.

Rocco JW, Sidransky D (2001): p16(MTS-1/CDKN2/INK4a) in cancer progression. Expr Cell Res 264:42–55.

Roussel MF (1999): The INK4 family of cell cycle inhibitors in cancer. Oncogene 18:5311–17.

Ruas M, Peters G (1998): The p16INK4a/CDKN2A tumor suppressor and its relatives. Biochim Biophys Acta 1378:F115–77.

Russo AA, Jeffrey PD, Patten AK, Massague J, Pavletich NP (1996): Crystal structure of the p27Kip1 cyclin-dependent-kinase inhibitor bound to the cyclin A-Cdk2 complex. Nature 382:325–31.

Russo AA, Tong L, Lee JO, Jeffrey PD, Pavletich NP (1998): Structural basis for inhibition of the cyclin-dependent kinase Cdk6 by the tumour suppressor p16INK4a. Nature 395:237–43.

Seoane J, Le HV, Massague J (2002): Myc suppression of the p21(Cip1) Cdk inhibitor influences the outcome of the p53 response to DNA damage. Nature 419:729–34.

Serrano M (1997): The tumor suppressor protein p16INK4a. Expr Cell Res 237: 7–13.

Serrano M, Hannon GJ, Beach D (1993): A new regulatory motif in cell-cycle control causing specific inhibition of cyclin D/CDK4. Nature 366:704–7.

Serrano M, Lee H, Chin L, Cordon-Cardo C, Beach D, DePinho RA (1996): Role of the INK4a locus in tumor suppression and cell mortality. Cell 85:27–37.

Serrano M, Lin AW, McCurrach ME, Beach D, Lowe SW (1997): Oncogenic *ras* provokes premature cell senescence associated with accumulation of p53 and p16INK4a. Cell 88:593–602.

Servant MJ, Coulombe P, Turgeon B, Meloche S (2000): Differential regulation of p27(Kip1) expression by mitogenic and hypertrophic factors: Involvement of transcriptional and posttranscriptional mechanisms. J Cell Biol 148:543–56.

Sharpless NE, Bardeesy N, Lee KH, Carrasco D, Castrillon DH, Aguirre AJ, Wu EA, Horner JW, DePinho RA (2001): Loss of p16Ink4a with retention of p19Arf predisposes mice to tumorigenesis. Nature 413:86–91.

Sheaff RJ, Groudine M, Gordon M, Roberts JM, Clurman BE (1997): Cyclin E-CDK2 is a regulator of p27Kip1. Genes Dev 11:1464–78.

Sherr CJ (1993): Mammalian G_1 cyclins. Cell 73:1059–65.

Sherr CJ (2001): The INK4a/ARF network in tumour suppression. Nat Rev Mol Cell Biol 2:731–7.

Sherr CJ, DePinho RA (2000): Cellular senescence: Mitotic clock or culture shock? Cell 102:407–10.

Sherr CJ, Roberts JM (1999): CDK inhibitors: Positive and negative regulators of G_1-phase progression. Genes Dev 13:1501–12.

Shin I, Yakes FM, Rojo F, Shin NY, Bakin AV, Baselga J, Arteaga CL (2002): PKB/Akt mediates cell-cycle progression by phosphorylation of p27(Kip1) at threonine 157 and modulation of its cellular localization. Nat Med 8:1145–52.

Shou W, Dunphy WG (1996): Cell cycle control by *Xenopus* p28Kix1, a developmentally regulated inhibitor of cyclin-dependent kinases. Mol Biol Cell 7:457–69.

Slingerland J, Pagano M (2000): Regulation of the cdk inhibitor p27 and its deregulation in cancer. J Cell Physiol 183:10–17.

Soos TJ, Kiyokawa H, Yan JS, Rubin MS, Giordano A, DeBlasio A, Bottega S, Wong B, Mendelsohn J, Koff A (1996): Formation of p27-CDK complexes during the human mitotic cell cycle. Cell Growth Differ 7:135–46.

Su JY, Rempel RE, Erikson E, Maller JL (1995): Cloning and characterization of the *Xenopus* cyclin-dependent kinase inhibitor p27XIC1. Proc Natl Acad Sci USA 92:10187–91.

Tang DG, Tokumoto YM, Apperly JA, Lloyd AC, Raff MC (2001): Lack of replicative senescence in cultured rat oligodendrocyte precursor cells. Science 291:868–71.

Tang XM, Strocchi P, Cambi F (1998): Changes in the activity of cdk2 and cdk5 accompany differentiation of rat primary oligodendrocytes. J Cell Biochem 68:128–37.

Teixeira LT, Kiyokawa H, Peng XD, Christov KT, Frohman LA, Kineman RD (2000): p27Kip1-deficient mice exhibit accelerated growth hormone-releasing hormone (GHRH)-induced somatotrope proliferation and adenoma formation. Oncogene 19:1875–84.

Thullberg M, Bartek J, Lukas J (2000): Ubiquitin/proteasome-mediated degradation of p19INK4d determines its periodic expression during the cell cycle. Oncogene 19:2870–6.

Tong W, Kiyokawa H, Soos TJ, Park MS, Soares VC, Manova K, Pollard JW, Koff A (1998): The absence of p27Kip1, an inhibitor of G_1 cyclin-dependent kinases, uncouples differentiation and growth arrest during the granulosa->luteal transition. Cell Growth Differ 9:787–94.

Tsutsui T, Hesabi B, Moons DS, Pandolfi PP, Hansel KS, Koff A, Kiyokawa H (1999): Targeted disruption of CDK4 delays cell cycle entry with enhanced p27(Kip1) activity. Mol Cell Biol 19:7011–19.

van den Heuvel S, Harlow E (1993): Distinct roles for cyclin-dependent kinases in cell cycle control. Science 262:2050–4.

Vernon AE, Devine C, Philpott A (2003): The cdk inhibitor p27(Xic1) is required for differentiation of primary neurones in *Xenopus*. Development 130:85–92.

Vernon AE, Philpott A (2003): A single cdk inhibitor, p27(Xic1), functions beyond cell cycle regulation to promote muscle differentiation in *Xenopus*. Development 130:71–83.

Vidal A, Koff A (2000): Cell-cycle inhibitors: three families united by a common cause. Gene 247:1–15.

Vidal A, Millard SS, Miller JP, Koff A (2002): Rho activity can alter the translation of p27 mRNA and is important for RasV12 induced transformation in a manner dependent on p27 status. J Biol Chem 277:16433–40.

Viglietto G, Motti ML, Bruni P, Melillo RM, D'Alessio A, Califano D, Vinci F, Chiappetta G, Tsichlis P, Bellacosa A, Fusco A, Santoro M (2002): Cytoplasmic relocalization and inhibition of the cyclin-dependent kinase inhibitor p27(Kip1) by PKB/Akt-mediated phosphorylation in breast cancer. Nat Med 8:1136–44.

Vlach J, Hennecke S, Amati B (1997): Phosphorylation-dependent degradation of the cyclin-dependent kinase inhibitor p27. EMBO J 16:5334–44.

Williams BO, Remington L, Albert DM, Mukai S, Bronson RT, Jacks T (1994): Cooperative tumorigenic effects of germline mutations in Rb and p53. Nat Genet 7:480–4.

Xiong Y, Hannon GJ, Zhang H, Casso D, Kobayashi R, Beach D (1993): p21 is a universal inhibitor of cyclin kinases. Nature 366:701–4.

Yan Y, Frisen J, Lee MH, Massague J, Barbacid M (1997): Ablation of the CDK inhibitor p57Kip2 results in increased apoptosis and delayed differentiation during mouse development. Genes Dev 11:973–83.

Yang W, Shen J, Wu M, Arsura M, FitzGerald M, Suldan Z, Kim DW, Hofmann CS, Pianetti S, Romieu-Mourez R, Freedman LP, Sonenshein GE (2001): Repression of transcription of the p27(Kip1) cyclin-dependent kinase inhibitor gene by c-Myc. Oncogene 20:1688–702.

Zezula J, Casaccia-Bonnefil P, Ezhevsky SA, Osterhout DJ, Levine JM, Dowdy SF, Chao MV, Koff A (2001): p21cip1 is required for the differentiation of oligodendrocytes independently of cell cycle withdrawal. EMBO Rep 2:27–34.

Zhang H, Hannon GJ, Beach D (1994): p21-containing cyclin kinases exist in both active and inactive states. Genes Dev 8:1750–8.

Zhang P, Liegeois NJ, Wong C, Finegold M, Hou H, Thompson JC, Silverman A, Harper JW, DePinho RA, Elledge SJ (1997): Altered cell differentiation and proliferation in mice lacking p57KIP2 indicates a role in Beckwith-Wiedemann syndrome. Nature 387:151–8.

Zhang P, Wong C, DePinho RA, Harper JW, Elledge SJ (1998): Cooperation between the Cdk inhibitors p27(KIP1) and p57(KIP2) in the control of tissue growth and development. Genes Dev 12:3162–7.

Zhang P, Wong C, Liu D, Finegold M, Harper JW, Elledge SJ (1999): p21(CIP1) and p57(KIP2) control muscle differentiation at the myogenin step. Genes Dev 13:213–24.

Zindy F, Cunningham JJ, Sherr CJ, Jogal S, Smeyne RJ, Roussel MF (1999): Postnatal neuronal proliferation in mice lacking Ink4d and Kip1 inhibitors of cyclin-dependent kinases. Proc Natl Acad Sci USA 96:13462–7.

Zindy F, den Besten W, Chen B, Rehg JE, Latres E, Barbacid M, Pollard JW, Sherr CJ, Cohen PE, Roussel MF (2001): Control of spermatogenesis in mice by the cyclin D-dependent kinase inhibitors p18(Ink4c) and p19(Ink4d). Mol Cell Biol 21:3244–55.

Zindy F, Quelle DE, Roussel MF, Sherr CJ (1997a): Expression of the p16INK4a tumor suppressor versus other INK4 family members during mouse development and aging. Oncogene 15:203–11.

Zindy F, Soares H, Herzog KH, Morgan J, Sherr CJ, Roussel MF (1997b): Expression of INK4 inhibitors of cyclin D-dependent kinases during mouse brain development. Cell Growth Differ 8:1139–50.

Zindy F, van Deursen J, Grosveld G, Sherr CJ, Roussel MF (2000): INK4d-deficient mice are fertile despite testicular atrophy. Mol Cell Biol 20:372–8.

CHAPTER 8

CHROMATIN REMODELING AND CANCER

CYNTHIA J. GUIDI and ANTHONY N. IMBALZANO
Department of Cell Biology, University of Massachusetts Medical School, Worcester, MA 01655

OVERVIEW

The human genome encodes for over 30,000 genes, with only a fraction of these genes expressed in a given cell. It is critical to the viability of a cell that the proper genes be activated or repressed at the appropriate time. An important level of regulation is provided by chromatin structure. When DNA is packaged into chromatin structure, the transcriptional machinery is unable to access regulatory sequences, and thus gene activation generally is repressed. These repressive effects of chromatin can be overcome by the action of proteins known as chromatin-remodeling enzymes. These enzymes can be divided generally into two groups: those that chemically modify chromatin and those that utilize the energy derived from ATP hydrolysis to alter chromatin structure. Constituents of each of these groups play significant roles in gene regulation. As such, the chromatin-remodeling enzymes themselves must be properly regulated. Misregulation of many of the chromatin remodeling enzymes has been associated with defects in cellular proliferation and tumorigenesis.

CHROMATIN STRUCTURE

The core particle of chromatin structure is the nucleosome. Combined data obtained from micrococcal nuclease digestion, as well as X-ray and electron crystallography at 7 Å resolution, indicate that the nucleosome consists of approximately 146 base pairs of DNA wrapped in 1.8 helical turns around the histone octamer (van Holde, Shaw et al., 1975; Finch, Lutter et al., 1977; Noll and Kornberg, 1977). The octamer itself has a

Cell Cycle and Growth Control: Biomolecular Regulation and Cancer, Second Edition, Edited by Gary S. Stein and Arthur B. Pardee
ISBN 0-471-25071-6

$(H3)_2(H4)_2$ tetramer at its center with an H2A-H2B dimer at each end of the DNA path. Each histone has a polypeptide chain fold known as the "histone fold" (Arents and Moudrianakis, 1995). The histone fold is formed by a long, central α-helix that is flanked on either side by shorter helices and loops that interact with DNA. At the amino terminal end of each histone are 15 to 30 residues that comprise the "histone tail." The histone tails appear unstructured at this resolution.

The 2.8 Å resolution crystal structure shows that the phosphodiester backbones of the DNA strands on the inner surface of the superhelix contact the octamer every ten base pairs, where the minor groove of the double helix faces inward (Luger, Mäder et al., 1997). The amino-terminal tails of both H2B and H3 pass through the gap in the DNA superhelix formed by aligned minor grooves to the outside of the core particle. The H2A and H4 tails pass across the superhelix on the flat faces of the particle to the outside as well. The position of the tails suggests that they are exposed. The 16 to 25 amino terminal residues of H4 tail extend into the adjacent nucleosome to interact with the negatively charged face of the H2A-H2B dimer. This interaction may mediate higher order folding.

The 1.9 Å resolution X-ray crystal structure of a nucleosome core particle containing 147 base pairs of DNA shows that water molecules and ions play in important role in nucleosome structure (Davey, Sargent et al., 2002). The water molecules serve as hydrogen bond bridges between the histone proteins and DNA. It has been suggested that these bonds diminish the requirement for sequence specificity in nucleosome positioning. Monovalent anions are located in proximity to the DNA phosphodiester backbone and may partially neutralize the electrostatic interaction between histones and DNA. Divalent cations, bound at specific sites in the nucleosome, contribute to histone-histone and histone-DNA interactions between adjacent nucleosomes. As with the histone tail of histone H4, these divalent cations may participate in higher order folding.

The DNA between adjacent nucleosomes is called linker DNA. The histone H1 binds to the linker DNA near one end of the core DNA inside the chromatin fiber (Zhou, Gerchman et al., 1998). Chromatin fibers are composed of arrays of nucleosomes, linker histones, and transacting factors. In vitro, nucleosomal arrays adopt an extended 10 nm diameter or 30 nm diameter fiber, depending on ionic strength of the medium. Tail-less chromatin fibers can neither fold into 30 nm fibers nor form fiber-fiber associations, suggesting that the tails play an important role in higher order chromatin structure (Carruthers and Hansen, 2000). The 30 nm fiber is the basic component of both interphase chromatin and mitotic chromosomes; however, the mechanism by which these fibers are packed into the highly condensed, organized structure of the mitotic chromosome is not well understood. Recent data indicate that a macromolecular complex called condensin is required for proper chromosome condensation, but how this complex functions is unclear (Hirano and Mitchison, 1994; Cubizollez, Legagneux et al., 1998; Sutani, Yuasa et al., 1999). Furthermore the core histone tails, but not histone H1, are required for mitotic chromosome condensation (de la Barre, Gerson et al., 2000).

Chromatin structure generally inhibits the function of transcriptional machinery. The packaging of promoters in nucleosomes prevents the initiation of transcription by bacterial and eukaryotic RNA polymerases in vitro (Knezetic and Luse, 1986; Lorch, LaPointe et al., 1987). In vivo, when H4 synthesis is inhibited, several TATA-containing promoters are activated in the absence of their normal activation mechanisms (Han and Grunstein, 1988). Furthermore, in DNA microarray analysis, nucleosome loss results in activation of 15% of yeast genes, not including the nearly 40% of the yeast genome that is constitutively active (Grunstein, 1990; Wyrick, 1999).

CHEMICAL MODIFICATION OF CHROMATIN STRUCTURE

The core histone tails, and in some cases the histone H1 tail, are susceptible to a wide range of post-translational modifications, including acetylation, methylation, phosphorylation, ubiquitination, glycosylation, and ADP-ribosylation. The effects of these modifications on gene expression are varied. In the following sections, histone acetylation/deacetylation, methylation, phosphorylation, and ubiquitination, as well as their links to cellular proliferation and tumorigenesis, are discussed.

HISTONE ACETYLATION

Hyperacetylation of histone tails has been correlated with increased gene activity (Gross and Garrard, 1988; Hebbes, Thorne et al., 1988; Hebbes, Thorne et al., 1992; Grunstein, 1997; Struhl, 1998). The regions of the histone tails that are acetylated are conserved, often invariant, lysine residues. Mutation of acetylatable lysines in histone H4 of *Saccharomyces cerevisiae* shows that these residues are required for activation of regulated genes. It is believed that the changes in the charge of histone tails resulting from acetylation weakens histone : DNA contacts, alters histone : histone interactions between neighboring nucleosomes, or disrupts histone : regulatory protein interactions, or a combination of all three (Hecht, Laroche et al., 1995; Luger, Mäder et al., 1997; Luger and Richmond, 1998; Tse, Sera et al., 1998; Wolffe and Hayes, 1999).

Histone acetyl transferases, or HATs, are responsible for the acetylation of histones. They can be divided into two categories: A-type and B-type. B-type HATs are cytoplasmic and likely catalyze acetylation events linked to transport of newly synthesized histones from cytoplasm to nucleus for deposition onto newly replicated DNA (Ruiz-Carrillo, Wangh et al., 1975; Allis, Chicoine et al., 1985). A-type HATs include nuclear HATs that likely catalyze transcription-related acetylation events (Brownell, Zhou et al., 1996). The A-type HAT proteins can be divided, based on sequence, into distinct families that show high

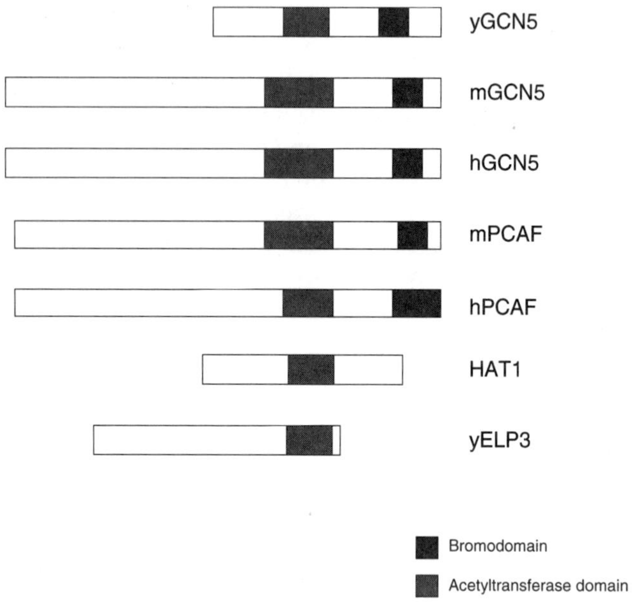

Figure 8.1. Schematic representation and comparison of the members of the GNAT family of acetyl transferase enzymes.

sequence similarity within families but poor to no sequence similarity between families. These families include the GNAT superfamily, the MYST family, the p300/CBP family, the basal transcription factors, and the nuclear receptor cofactors (Roth, Denu et al., 2001).

The GNAT superfamily encompasses the GCN5-related N-acetyltransferases (Neuwald and Landsman, 1997) (Fig. 8.1). They contain limited sequence homology within four, 15 to 35 residue motifs (named A–D). This family includes the prototype GCN5/PCAF, as well as Hat1, Elp3, and Hpa2. The first description linking histone acetyltransferase activity to gene activation came in 1996 with the finding that the *Tetrahymena* histone acetyltransferase A had homology with the yeast GCN5, a known transcriptional activator (Brownell and Allis, 1995; Brownell, Zhou et al., 1996).

The MYST family is named for the founding members: MOZ, Ybf2/Sas3, Sas2, and Tip60 (Fig. 8.2). It also includes Esa1, MOF, and Hbo1. Many members of the MYST family contain chromodomains (chromatin organization modifier), protein-protein interaction domains often found in heterochromatin-associated proteins (Jones, Cowell et al., 2000). It is possible that these domains serve to target members of the MYST family to chromatin targets. The MYST family has been linked to cancer via the founding member, MOZ (monocytic leukemia zinc finger protein). As its name implies, *MOZ* is an oncogene, whose translocations are involved in certain cases of monocytic leukemia (Borrow, Shearman et al., 1996; Carapeti, Aguiar et al., 1998; Carapeti, Aguiar et al., 1999). *MOZ* is the human homologue of yeast Ybf2/Sas3, the catalytic subunit

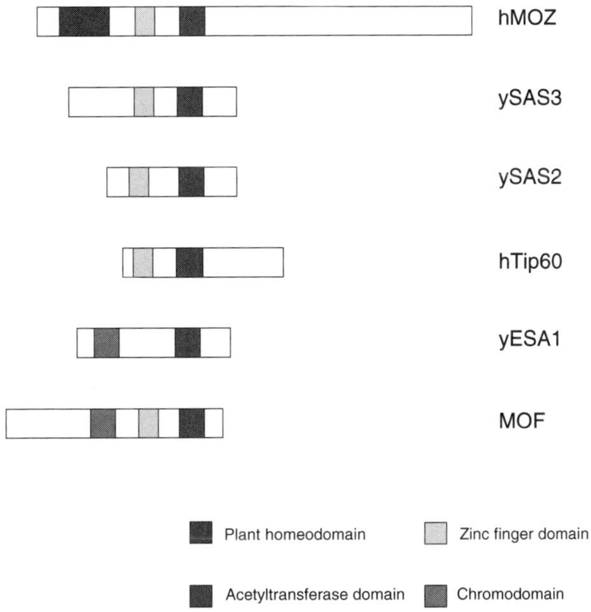

Figure 8.2. Schematic representation and comparison of the members of the MYST family of acetyl transferase enzymes.

of NuA3, a yeast HAT complex that specifically acetylates histone H3 (Reifsnyder, Lowell et al., 1996; Grant, Duggan et al., 1997; John, Howe et al., 2000). Although MOZ has not been demonstrated to possess HAT activity, the sequence's similarity to Sas3 suggests that it is likely a HAT.

P300 and CBP were isolated independently as factors that interact with adenovirus E1A protein (p300) or with the phosphorylated form of the transcription factor CREB (CBP) (Chrivia, Kwok et al., 1993; Eckner, Ewen et al., 1994). Both share sequence similarities, and their function is interchangeable in vitro (Arany, Sellers et al., 1994; Arany, Newsome et al., 1995; Lundblad, Kwok et al., 1995). Each contains three putative zinc finger regions, a bromodomain (a domain that interacts with acetyl-lysine residues), a HAT domain, and at least two independent regions that interact with multiple transcription factors. They are transcriptional co-activators; they do not bind DNA directly. They interact with many factors including, but not limited to, c-jun, c-myb, c-fos, TFIID, MyoD, nuclear hormone receptors, and E2F-1 (Ferreri, Gill et al., 1994; Bannister, Oehler et al., 1995; Janknecht, Cahill et al., 1995; Dai, Akimaru et al., 1996; Janknecht and Hunter, 1996; Kamei, Xu et al., 1996; Oelgeschlager, Janknecht et al., 1996; Yuan, Condorelli et al., 1996; Sartorelli, Huang et al., 1997; Martinez-Balbas, Bauer et al., 2000). Their HAT activities are required for their functions in transcriptional activation (Bannister and Kouzarides, 1996; Ogryzko, Schiltz et al., 1996; Martinez-Balbas, Bauer et al., 2000).

The first line of evidence linking misregulation of HAT activity to cancer came from the finding that the adenoviral E1A oncoprotein targets p300/CBP (Arany, Sellers et al., 1994; Eckner, Ewen et al., 1994). Overexpression of E1A prevents binding of p300/CBP to PCAF and induces entry of cells into S phase (Yang, Ogryzko et al., 1996). The transforming activity of E1A depends on its ability to interact with and sequester p300/CBP, as excess p300/CBP inhibits E1A-mediated cell immortilization.

The gene encoding CBP has been shown to be involved in chromosomal translocations in certain leukemias. In acute myeloid leukemia, the t(8;16)(p11;p13) translocation results in the fusion of *CBP* to the human oncogene *MOZ* (Borrow, Shearman et al., 1996). This fusion creates a protein with two HAT domains. Recruitment of this protein by CBP or MOZ cofactors may bring inappropriate HAT activity to target promoters. In addition two inversions within chromosome 8 that are associated with leukemia fuse *MOZ* to transcriptional intermediary factor 2 (*TIF2*), a p300/CBP interacting protein with intrinsic HAT activity (Carapeti, Aguiar et al., 1998; Carapeti, Aguiar et al., 1999). The resulting fusions retain the HAT domains of both proteins.

The t(11;16)(q23;p13) chromosomal translocation, found in many leukemias, fuses *CBP* to *MLL/ALL-1* (Sobulo, Borrow et al., 1997). *MLL/ALL-1* is the human homologue of *Drosophila* trithorax, a protein that functions during development in maintenance of open chromatin configuration for proper expression of homeotic genes. Additionally a *MLL-p300* translocation has been described in a patient with AML (Ida, Kitabayashi et al., 1997).

There is evidence suggesting that *CBP* is a bona fide tumor suppressor. *CBP* heterozygosity is associated with Rubenstein-Taybi Syndrome (RTS), a human disorder characterized by cranial and digital malformation, mental retardation, hematopoietic abnormalities, and higher risk for developing certain types of cancer (Miller and Rubinstein, 1995; Petrij, Giles et al., 1995). *CBP* has been targeted in mouse knockout experiments (Oike, Hata et al., 1999; Kung, Rebel et al., 2000). *CBP* heterozygous mice display various developmental defects and develop a high incidence of hematological malignancies, including histiocytic sarcomas and myelogenous and lymphocytic leukemias. Tumorigenesis is correlated with loss of heterozygosity in transformed cells.

The histone acetyltransferase p300 also may be a tumor suppressor. A number of human tumors, including glioblastomas, colorectal cancers, and breast cancer, show loss of heterozygosity of *p300* (Muraoka, Konishi et al., 1996; Gayther, Batley et al., 2000). In a study examining a variety of primary tumors or tumor cell lines for mutations in *p300*, ten of 193 were shown to have loss of function mutations (Gayther, Batley et al., 2000). *p300* has been targeted in mouse knockout experiments (Yao, Oh et al., 1998). However, there have been no reported cases of malignancy in *p300* heterozygous mice.

Overexpression of some histone acetyltransferases has been correlated with cancer. The nuclear hormone cofactor ACTR is overexpressed in several breast and ovarian cancers (Anzick, Kononen et al., 1997). Although it is unclear if this is a cause or effect of these cancers, it is possible that overexpression of ACTR leads to increased activation of target genes, which in turn may lead to increased cellular proliferation. Additionally the RNA polymerase III transcription factor TFIIIC2, which is an acetyltransferase, is overexpressed in ovarian tumors, contributing to the abnormal abundance of pol III transcripts in these tumors (Winter, Sourvinos et al., 2000).

HISTONE DEACETYLATION

While histone acetylation is associated with gene activation, histone deacetylation is associated with gene repression. In fact many gene products that were known to act as corepressors were later found to have deacetylase activity. The link between histone deacetylation and gene repression first was demonstrated by the isolation of the human histone deacetylase HDAC1, which has sequence highly similar to the yeast Rpd3, a known negative regulatory protein (Taunton, Hassig et al., 1996). Histone deacetylases (HDACs) are categorized, based on homology, into two classes. The first class includes the yeast HDACs Rpd3, Hos1, and Hos2 as well as the mammalian histone deacetylases HDAC1–3, and 9. The second class consists of yeast Hda1 and mammalian HDAC4–8, and 10. Most HDACs are associated in multisubunit complexes; substrate specificity is regulated by components of these complexes.

The mammalian HDAC1 and HDAC2 have been shown to play important roles in cellular growth arrest (Davie and Chadee, 1998; Luo, Postigo et al., 1998; Koipally, Renold et al., 1999). The multiprotein complex SIN3-HDAC consists of both HDAC1 and HDAC2, along with the scaffolding protein SIN3 and at least eight other proteins (Alland, Muhle et al., 1997; Heinzel, Lavinsky et al., 1997; Nagy, Kao et al., 1997). This co-repressor complex has been shown to associate with the basic helix-loop-helix-zipper protein Mad and is required for Mad-induced transcriptional repression. The repression mediated by this complex prevents the activation of target genes such as E2F and cdc25, leading to growth arrest in a wide range of cells. HDAC activity appears to be required for the ability of Mad to induce growth arrest, as inhibitors of deacetylase activity partially overcome this effect.

The SIN3-HDAC complex also plays an important role in retinoblastoma tumor suppressor protein (Rb)-mediated repression (reviewed in Harbour and Dean, 2000). Rb controls cellular proliferation by repressing transcription of genes required for progression through G_1 and S of the cell cycle. Rb is recruited to target genes via its interaction with the E2F family of transcription factors. Rb represses E2F-mediated transactivation by two mechanisms; it blocks the E2F transactivation domain and it actively represses E2F promoters. The deacetylase activity of the SIN3-HDAC complex helps to repress E2F-regulated genes.

Certain forms of leukemia are associated with misregulation of SIN3-HDAC activity. RAR is a transcriptional regulator that responds to retinoids and is important for the differentiation of cells into many lineages, especially myeloid lineages (Chambon, 1996). RARs recruit the SIN3-HDAC complex, via N-CoR (nuclear receptor corepressor) or SMRT (silencing mediator for retinoid and thyroid receptors), to promoters containing RARE (retinoic acid response element) sequences. In the presence of retinoic acid (RA), the SIN3-HDAC complex is released from RAR allowing the TIF2-CBP HAT complex to bind to a domain on RAR that is masked in the absence of ligand (Alland, Muhle et al., 1997; Heinzel, Lavinsky et al., 1997; Nagy, Kao et al., 1997; Nagy, 1999). In this manner, retinoic acid is able to induce genes containing RARE sequences.

Chromosomal translocations resulting in the fusion of the *RAR* gene to the gene encoding PML have been associated with some cases of human acute promyelocytic leukemia (APL) (Grignani, De Matteis et al., 1998; Lin, Magy et al., 1998). The normal function of PML is unclear; however, it is known to homodimerize and to interact with HDACs (Melnick and Licht, 1999). PML-RAR fusion proteins retain the regions of RAR required for DNA and ligand binding, as well as the regions of PML required for HDAC interaction and homodimerization. Leukemogenesis is believed to result from the dimerization of the fusion proteins and subsequent stronger association with HDACs. HDAC association is maintained at physiological levels of RA but released at high levels of ligand. Patients with PML-RAR translocations often go into remission after treatment with pharmacological doses of retinoic acid. In other forms of APL, *RAR* is fused to PLZF (promyelocytic leukemia zinc finger) (Grignani, De Matteis et al., 1998; Lin, Magy et al., 1998). The normal function of PLZF is not known, though it is able to homodimerize and interacts with SIN3-HDAC. The PLZF-RAR fusion protein retains these known abilities of PLZF. The SIN3 protein is not released from PLZF-RAR even at high concentrations of RA, and patients with this translocation are resistant to treatment with pharmacological does of retinoic acid. Interestingly, inhibitors of histone deacetylase activity have been shown to dramatically potentiate retinoid-induced gene activation of RA-sensitive and restore retinoid response of RA-resistant APL cell lines (Grignani, De Matteis et al., 1998; Lin, Magy et al., 1998). This finding suggests that the RAR fusion proteins mediate leukemogenesis through aberrant chromatin acetylation.

HISTONE METHYLATION

Lysine histone methyltransferases contain a conserved methyltransferase domain termed a SET [Su(var)3–9, Enhancer-of-zeste, Trithorax] domain (reviewd in Kouzarides, 2002; Schneider, Bannister et al., 2002). To date, not all SET-domain containing proteins have been shown to have methyltransferase activity, though lack of detectable activity may be due to inappropriate assay conditions. The effect of histone methylation on gene activation is varied. The lysine histone methyltransferases are divided into four families: SUV39, SET1, SET2, and RIZ (Kouzarides, 2002; Schneider, Bannister et al., 2002).

The SUV39 subfamily includes Suv39h1, Suv39h2, EuHMTase1, G9a, ESET, and CLLL8. Su(var)3–9 originally was identified in a genetic screen as a suppressor of position effect variegation in *Drosophila melanogaster*. The SET domain of Su(var)3–9 is the founding member of the SUV39 subfamily of SET domains. The mouse homologues are Suv39h1 and Suv39h2 (Rea, 2000). Though mice deficient for either gene are phenotypically normal, double-knockout mice of *Suv39h1/h2* display dramatic genomic instability (Peters, O'Carroll et al., 2001). They are predisposed to cancer and approximately one-third of the mice develops late-onset B-cell lymphoma. A common feature of these tumors is non-segregated chromosomes that are linked via acrocentric regions. These knockout mice have a greatly reduced level of H3 K9 methylation, suggesting that the methyltransferase activity of Suv39h1/h2 is important for suppressing tumorigenesis. The human SUV39H1/2 methyltransferase has been linked to oncogenesis via its interaction with Rb (Nielsen, Schneider et al., 2001). This interaction is required for correct regulation of the gene encoding cyclin E, which is important in cell cycle regulation (Owa, 2001). Many human cancers have mutations in Rb and some of these Rb mutants fail to bind SUV39H1 (Nielsen, Schneider et al., 2001). It is possible that the interaction between Rb and SUV39H1 plays a significant role in tumor suppression.

The SET1 subfamily includes hSET7 and ySET1, both of which have been shown to possess a H3 K4-specific methyltransferase activity (Roguev, Schaft et al., 2001; Wang, Cao et al., 2001; Yang, Xia et al., 2002). Other members of this subfamily have not been shown to have methyltransferase activity. These include the polycomb (PcG) proteins EZH1 and EZH2. Polycomb genes are a group of genes required to repress homeotic (hox) gene activity. MLL1–3 and ALR, trithorax (trxG) genes that are required to maintain hox gene activity, also belong to the SET1 subfamily. There are many links between members of the SET1 subfamily and cancer. *MLL1* is translocated in many leukemias (Ziemin-van der Poel, McCabe et al., 1991; Zeleznik-Le, Harden et al., 1994; Ayton and Cleary, 2001). In fact over 30 different chromosomal fusions of this region have been observed, and all of these fusions lack the SET domain. Additionally deletions in exon 8 of *MLL1* have been observed in acute lymphoblastic leukemias (Lochner, Siegler et al., 1996). A partial duplication of *MLL1* has been documented in acute myeloblastic leukemia and gastric carcinoma cell lines (Schichman, Caligiuri et al., 1994). It is unclear if these mutations in *MLL1* result in tumorigenesis due to loss of function of the normal *MLL1* product, a gain of function of the fusion proteins, or a combination of both. Another MLL gene product, MLL2 is amplified in some solid tumor cell lines (Huntsman, Chin et al., 1999). Chromosomal aberrations of the third MLL gene, *MLL3*, are associated with hematological neoplasia and holoprosencephaly, a congenital malformation of the brain and face (Tan and Chow, 2001). The polycomb gene *EZH2* is upregulated in tumor cell lines (Visser, Gunster et al., 2001). It is localized to a region crucial for malignant myeloid disorders (Cardoso, Mignon et al., 2000), and its SET domain interacts with XNP, which is mutated in different inherent disorders, including ATR-X syndrome (Cardoso, Timsit et al., 1998).

The SET2 subfamily includes NSD1–3, HIF1, AND ASH1. The founding member of this subfamily, the *S. cerevisiae* SET2 protein, has intrinsic histone methyltransferase activity specific for H3 K36 (Strahl, Grant et al., 2002). Members of the mammalian nuclear receptor-binding SET-domain containing (NSD) family contain a SET domain that is highly related to that of ySET2; however, NSD proteins have yet to be shown to possess methyltransferase activity. NSD1 can enhance androgen receptor (AR)-mediated transactivation in prostate cancer, though it is unclear if this is a cause or result of oncogenesis (Wang, Yeh et al., 2001). In the t[5,11](q35;p15.5) translocation in acute myeloid leukemia, *NSD1* is fused to the *NUP98* gene, which encodes a nucleoporin that plays a role in nuclear trafficking (Jaju, Fidler et al., 2001). In addition truncations in the SET domain of *NSD1* have been identified in individuals with Sotos syndrome, a familial disorder linked with a predisposition to cancers such as Wilm's tumor, hepatocarcinomas, mixed paratoid tumors, and osteochondromas (Kurotaki, Imaizumi et al., 2002). *NSD2* maps to a region deleted in the Wolf-Hirschhorn syndrome (WHS) critical region (Stec, Wright et al., 1998). Deletions in this region cause WHS, which is characterized by mental retardation and developmental defects. *NSD2* often is found fused to the *IgH* gene in multiple myeloma (Stec, Wright et al., 1998; Malgeri, Baldini et al., 2000). The third member of the NSD family, *NSD3*, is amplified in several breast cancer cell lines and in primary breast carcinomas (Stec, van Ommen et al., 2001). In addition this gene also is found fused to *NUP98* in acute myeloid leukemia (Rosati, La Starza et al., 2002).

The RIZ subfamily includes RIZ, BLIMP-1, MEL1, PFM1, and MDS1-EVI1. The SET domain of the RIZ protein was the founding member of this subfamily. None of the proteins in this subfamily have been shown to possess methyltransferase activity. The *RIZ* gene encodes for two proteins, RIZ1 and RIZ2, via the use of two alternative promoters (Abbondanza, Medici et al., 2000). RIZ2 is identical to RIZ1 except that it lacks the first 200 amino acids, including the SET domain. RIZ1 expression is reduced or lost in many types of cancer including breast cancer, lung cancer, osteosarcomas, hepatoma, neuroblastoma, and colorectal cancer (Huang, 1999; Abbondanza, Medici et al., 2000). Frameshift mutations in *RIZ* have been found in 37% of primary tumors of the colon, stomach, endometrium, and pancreas (Buyse, Shao et al., 1995; Piao, Fang et al., 2000). Furthermore mice deficient for RIZ1 are prone to develop diffuse B-cell lymphomas and a broad spectrum of unusual tumors (Steele-Perkins, Fang et al., 2001). It is interesting to note that *RIZ1*-deficient mice present with similar tumors as mice deficient for the Suv39h1/h2 methyltransferases. The *MDS1-EVI1* gene encodes for two products: the SET-domain-containing MDS1-EVI1 and the EVI1 protein that lacks the SET domain (Fears, Mathieu et al., 1996). Certain chromosomal rearrangements cause disruption of the MDS-EVI1 protein and activation of the EVI1 protein leading to myeloid leukemia. Furthermore EVI1 is overexpressed in solid tumors and leukemia (Fears, Mathieu et al., 1996). Another RIZ subfamily member, *BLIMP-1*, is deleted in B-cell non-Hodgkin lymphoma (Keller and Maniatis, 1991; Mock, Liu et al., 1996). *MEL1* is transcriptionally activated by translocation in acute myeloid leukemia (Mochizuki, Shimizu et al., 2000). Lastly, *PFM1* maps to a tumor suppressor locus on chromosome 12 (Yang and Huang, 1999).

Clearly, a large number of the SET-domain-containing proteins play an important role in cell cycle regulation. In fact misregulation of a number of these proteins has been linked to a variety of cancers. In some cases tumorigenesis has been linked to the diminished methyltransferase activity of the disrupted gene products. However, not all of the SET-domain proteins have been shown to possess methyltransferase activity. As such, it is not clear if a methyltransferase activity of all of the described factors is required for their normal activity.

HISTONE PHOSPHORYLATION

The core histones and histone H1 undergo phosphorylation on specific serine and threonine residues. Phosphorylation of H3 and H1 are cell cycle regulated, with the highest level of phosphorylation occurring during M phase (Gurley, D'Anna et al., 1978; Paulson and Taylor, 1982; Goto, Tomono et al., 1999; Wei, Yu et al., 1999). Phosphorylation of H1 has been associated with transcriptional activation of the MMTV promoter (Lee and Archer, 1998). Phosphorylation of H3 also has been shown to play a role in the transcriptional induction of immediate early genes in mammalian cells (Mahadevan, Willis et al., 1991; Chadee, Hendzel et al., 1999). H3 residues within the promoter of c-fos and c-myc are rapidly phosphorylated in serum-starved cells when the Ras-mitogen activated protein kinase (MAPK) pathway is stimulated by growth factors. Furthermore the mitotic phosphorylation of H3 also is associated with chromosomal condensation. The condensation of chromosomes during mitosis is essential for the proper transmission of parental genetic information to daughter cells. The aurora kinase family is involved in histone H3 phosphorylation (Hsu, Sun et al., 2000). Members of the aurora kinase family are overexpressed in a variety of cancers including colorectal cancers and invasive ductal carcinomas of the breast (Bischoff, Anderson et al., 1998; Tatsuka, Katayama et al., 1998; Zhou, Kuang et al., 1998; Tanaka, Kimura et al., 1999). The mechanism by which overexpression of aurora kinase family members leads to tumorigenesis is unclear; however, this finding points to the importance of proper regulation of histone phosphorylation in maintaining normal cellular proliferation.

HISTONE UBIQUITINATION

Histones H2A, H2B, H3, and the linker histone H1 can be reversibly ubiquitinated. The carboxyl-terminus of the ubiquitin molecule is covalently attached via an isopeptide bond to the ε-amino group of lysine. Approximately 5% to 15% of total H2A and about 1.5% of total H2B present in mammalian cells are monoubiquitinated (Levinger and Varshavsky, 1980; West and Bonner, 1980; Kleinschmidt and Martinson, 1981; Levinger, Barsoum et al., 1981). Ligation of ubiquitin moieties to short-lived proteins tags them for degradation by the 26S proteasome; however, mono-ubiquitinated histones do not appear to be tagged for degradation in vivo (Seale, 1981; Wu, Kohn et al., 1981). The biological significance of histone ubiquitination is unclear. Studies suggesting that ubiquitinated histone H2A is associated with transcriptional activation are contrasted by those that suggest ubiquitination of histone H2A result in gene repression (for a review, see Jason, Moore et al., 2002). To date, only one tentative link between misregulation of histone ubiquitination and cancer has been published. The levels of ubiquitinated H2A were found to be highly upregulated in SV-40 transformed human fibroblasts and keratinocytes, suggesting that this modification may play an important role in cell cycle control (Vassilev, Rasmussen et al., 1995).

ATP-DEPENDENT CHROMATIN REMODELING

ATP-dependent chromatin remodeling complexes use the energy of ATP hydrolysis to alter chromatin structure. Every ATP-dependent chromatin-remodeling complex contains an ATPase subunit that is highly conserved across species. Each of the ATPase subunits belongs to the SWI2/SNF2 superfamily of proteins. Based on the homology of the ATPase subunit, these complexes can be classified into three subfamilies: the SWI2/SNF2 subfamily, the ISWI subfamily, and the CHD subfamily (Fig. 8.3). Members of each of these subfamilies, and, where applicable, their links to human cancers are discussed.

SWI2/SNF2 SUBFAMILY

The SWI2/SNF2 subfamily includes *S. cerevisiae* SWI/SNF, RSC (remodels the structure of chromatin), and INO80.com; *Drosophila* Brahma; and mammalian SWI/SNF. The activity of the ATPase subunit of each of these complexes is stimulated by both DNA and nucleosomes (Côté, Quinn et al., 1994; Imbalzano, Kwon et al., 1994; Kwon, Imbalzano et al., 1994; Cairns, Lorch et al., 1996; Du, Nasir et al., 1998; Phelan, Sif et al., 1999). The ATPase subunits also share a C-terminal bromodomain and two other conserved regions of unknown function (Workman and Kingston, 1998).

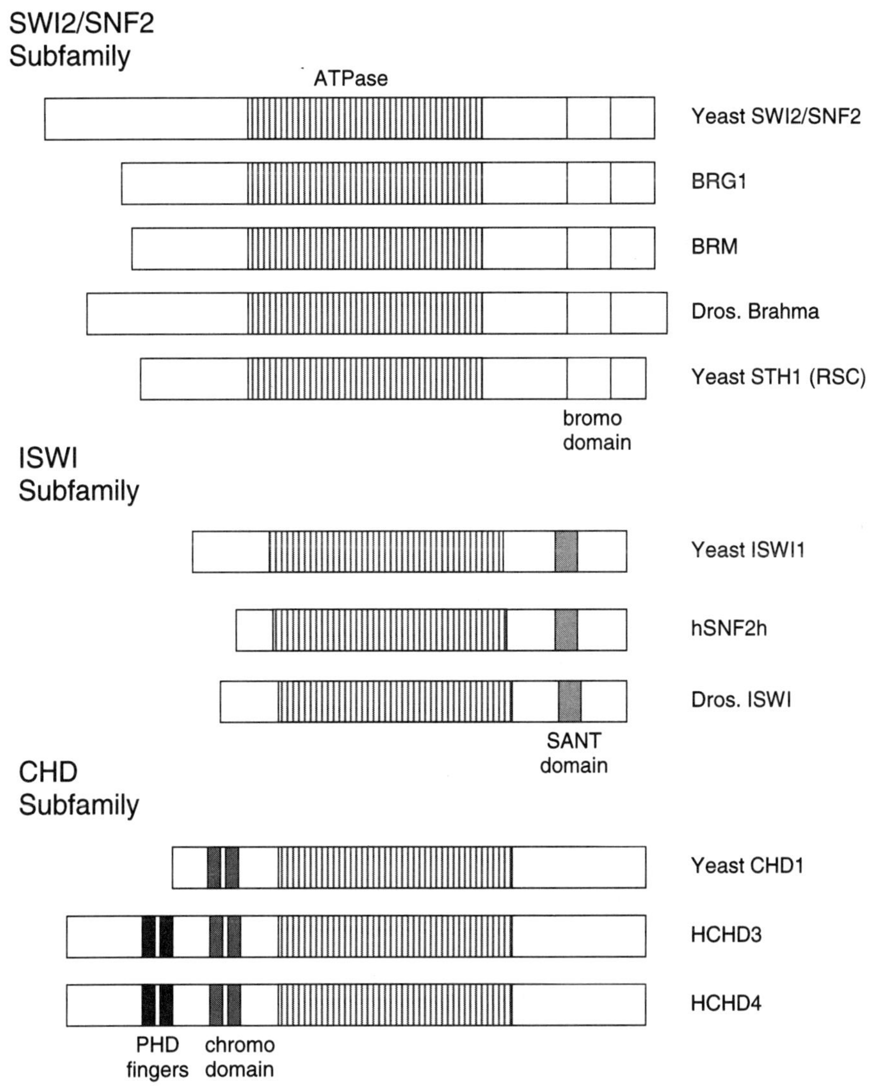

Figure 8.3. Schematic representation and comparison of the different SNF2 family ATPases that are the catalytic subunits of ATP dependent chromatin remodeling enzymes.

Yeast SWI/SNF Complex

The SWI/SNF complex first was identified in yeast (Cairns, Kim et al., 1994; Peterson, Dingwall et al., 1994). It is comprised of 11 subunits, with the core ATPase subunit encoded by the SWI2/SNF2 gene. None of the members of the yeast SWI/SNF complex are required for viability; however, several of its components originally were isolated as being required for mating type switching (SWI) and sucrose fermentation (SNF) (Neigeborn and Carlson, 1984; Stern, Jensen et al., 1984; Breeden and Nasmyth, 1987). These phenotypes are due to the fact that SWI/SNF is required for induction of the mating type switch gene, HO and the SUC2 invertase that is required for sucrose fermentation. The first hint that SWI/SNF plays a role in chromatin remodeling came from the discovery that several mutations that suppressed swi/snf phenotypes corresponded to genes encoding histones and nonhistone components of chromatin structure (Kruger and Herskowitz, 1991; Hirschhorn, Brown et al., 1992; Kruger, Peterson et al., 1995). SWI/SNF later was shown to alter the DNase I digestion pattern of in vitro assembled mononucleosomes, giving credence to the idea that it could directly alter chromatin structure. In addition the activity of SWI/SNF can facilitate the binding of a number of transcription factors and restriction nucleases to nucleosomal DNA templates (Côté, Quinn et al., 1994; Burns and Peterson, 1997; Logie and Peterson, 1997; Utley, Côté et al., 1997). Data obtained from DNA microarray expression analysis indicate that approximately 5% of yeast genes that are constitutively expressed are dependent on the ATPase activity of SWI/SNF. Interestingly, SWI/SNF appears to be involved in the repression of just as many genes as it activates (Holstege, Jennings et al., 1998; Sudarsanam, Iyer et al., 2000).

Mammalian SWI/SNF Complexes

Mammalian SWI/SNF complexes contain one of two SWI2/SNF2 ATPase homologues, BRM (SNF2α) or BRG1 (SNF2β) (Wang, Côte et al., 1996). The mammalian SWI/SNF complex is composed of 8 to 12 subunits, with its composition differing slightly between cell types. Like its yeast counterpart, mammalian SWI/SNF complexes are able to disrupt the DNase I digestion pattern of in vitro assembled mononucleosomes and increase the accessibility of some transcription factors to nucleosomal templates (Imbalzano, Kwon et al., 1994). Components of mammalian SWI/SNF complexes have been implicated in a variety of cellular processes, including gene activation and repression, development and differentiation, cell cycle regulation, and recombination and repair (Muchardt and Yaniv, 1993; Chiba, Muramatsu et al., 1994; Dunaief, Strober et al., 1994; Trouche, Le Chalony et al., 1997; Fryer and Archer, 1998; Murphy, Hardy et al., 1999; Shanahan, Seghezzi et al., 1999; Agalioti, Lomvardas et al., 2000; Bochar, Wang et al., 2000; de la Serna, Carlson et al., 2000; Strobeck, Knudsen et al., 2000; Zhang, Gavin et al., 2000; de la Serna, Carlson et al., 2001). Furthermore members of the SWI/SNF complex are targets of viral regulatory proteins (Kalpana, Marmon et al., 1994; Miller, Cairns et al., 1996; Lee, Sohn et al., 1999; Wu, Krumm et al., 2000; Lee, Lim et al., 2002). As SWI/SNF plays such a diverse role in cellular regulation, one might expect misregulation of SWI/SNF activity to result in events such as tumorigenesis.

SWI/SNF constituents associate with a number of known tumor suppressors. Both BRG1 and BRM have been shown to interact with the Rb tumor suppressor and facilitate the repression of certain gene expression events required for entry into S phase. In fact BRG1 or BRM is required for Rb-dependent G_1 arrest (Dunaief, Strober et al., 1994; Strobeck, Knudsen et al., 2000; Zhang, Gavin et al., 2000). BRG1 also has been shown to interact with the breast cancer susceptibility gene product, BRCA1 (Bochar, Wang et al., 2000). The ATPase activity of BRG1 is required for the ability of BRCA1 to stimulate p53-mediated transcription. Furthermore BRG1 interacts directly with p53 in co-immunoprecipitation experiments (Lee, Kim et al., 2002). This interaction appears to facilitate activation of some p53-responsive genes. It is unclear if any of these interactions are important in suppressing human cancers. To date, no mutations have been identified in any cancers that specifically disrupt any of these interactions, though few cancer cell lines or primary tumors have been screened for such mutations.

Misexpression of *BRG1* and *BRM* has been found in a number of human tumor cell lines and primary tumors. Expression of *BRG1* and *BRM* is down-regulated or absent in tumor cell lines derived from various tissues, including prostate, lung, and breast (Wong, Shanahan et al., 2000). In another study both alleles of *BRG1* were found to be mutated in 2 out of 22 breast carcinoma cell lines examined (DeCristofaro, Betz et al., 2001). On the contrary, BRG1 was found to be overexpressed in approximately 60% of gastric carcinomas examined (Sentani, Oue et al., 2001).

Results from mouse knockout experiments reveal variable roles for *Brg1* and Brm in tumorigenesis. Mice lacking Brm are viable but show mild proliferative effects, suggesting a role for Brm in the control of cellular proliferation (Reyes, Barra et al., 1998). Mice lacking Brg1 are early embryonic lethal (Bultman, Gebuhr et al., 2000). Furthermore a small percentage of mice heterozygous for *Brg1* develop apocrine tumors. However, loss of heterozygosity in the tumors has yet to been demonstrated.

SNF5/INI1 is a core subunit of all mammalian SWI/SNF complexes purified to date (Wang, Côte et al., 1996). It originally was identified based on its homology to the yeast Snf5 protein and by a yeast two-hybrid screen as a protein that interacts with HIV-1 integrase (integrase interactor **1**) (Kalpana, Marmon et al., 1994; Muchardt, Sardet et al., 1995). Bi-allelic deletions or truncating mutations of *INI1* have been shown to be associated with most cases of malignant rhabdoid tumor, a rare but aggressive pediatric cancer of the soft tissues (Versteege, Sevenet et al., 1998; Biegel, Zhou et al., 1999; DeCristofaro, BLBetz et al., 1999; Rousseau-Merck, Versteege et al., 1999; Biegel, Tan et al., 2002; Uno, Takita et al., 2002). Mutations in *INI1* also have been found in other neuronal tumors such as choroid plexus carcinomas, medullablastomas, and central primitive neuroectodermal tumors (Sévenet, Lellouch-Tubiana et al., 1999; Biegel, Fogelgren et al., 2000). Furthermore deletions of *INI1* have been reported in chronic phase and blast crisis of chronic myeloid leukemia (Grand, Kulkarni et al., 1999). Recent studies indicate that germ-line mutations in *INI1* predispose afflicted individuals to some of these cancers (Sévenet, Lellouch-Tubiana et al., 1999; Taylor, Gokgoz et al., 2000). Mice lacking Ini1, like those lacking Brg1, are early embryonic lethal (Klochendler-Yeivin, Fiette et al., 2000; Roberts, Galusha et al., 2000; Guidi, Sands et al., 2001). Approximately 30% of mice heterozygous for *Ini1* develop undifferentiated or poorly differentiated sarcomas, with variable rhabdoid features, of soft tissues. In these cases, tumor occurrence has been correlated with loss of heterozygosity at the *Ini1* locus.

It is unclear if BRG1 and INI1 function independently as tumor suppressors or function cooperatively via an activity of the SWI/SNF complex. Heterozygous disruption of *Brg1* and *Ini1* in mouse models results in divergent phenotypes. However, while disruption of Ini1 may affect both Brg1- and Brm-containing complexes, it is possible that Brm is able to partially compensate for Brg1 deficiency. Clearly, Brm is unable to compensate for the absence of Brg1 in early development. This may be due to the fact that during early mouse embryonic development, Brg1 and Brm show differences in their levels of expression as well as localization at the blastocyst stage (LeGouy, Thompson et al., 1998). On the contrary, the level of Brm message is comparable to that of Brg1 in adult tissues and many cell lines. In human tumor cell lines lacking BRG1, BRM is able to compensate for BRG1 function in cell cycle arrest mediated by Rb (Strobeck, Reisman et al., 2002). Thus it is possible that the presence of Brm in Brg1-heterozygous mice is sufficient to maintain the putative tumor suppressor function of SWI/SNF.

While it is possible that the ability of BRG1 and INI1 to function as tumor suppressors depends on their role in the SWI/SNF complex, recent data suggest that INI1 has functions distinct from those of BRG1 and BRM. As mentioned above, cell cycle arrest mediated by Rb depends on the presence of functional BRG1 or BRM. On the contrary, INI1 is not required for the ability of Rb to induce arrest (Betz, Strobeck et al., 2002; Versteege, Medjkane et al., 2002). When a constitutively active Rb is introduced into human tumor cell lines lacking INI1, the cells arrest in G_1. Therefore it is possible that the tumor-suppressor function of Ini1 is distinct from its function as a member of the SWI/SNF complex.

RSC Complex

The yeast RSC complex contains the ATPase Sth1, a protein that shares high homology with Swi2/Snf2 (Cairns, Lorch et al., 1996). This complex consists of 15 subunits, some of which share homology with other members of the yeast SWI/SNF complex. Rsc8/Swh3, Rsc6, and Sfh1 are homologues to SWI/SNF subunits Swi3, Swp73, and Snf5, respectively. Unlike the yeast SWI/SNF constituents, members of the RSC complex are required for mitotic growth (Cao, Cairns et al., 1997). The RSC complex catalyzes the transfer of histone octamers from one strand of DNA to another (Lorch, Zhang et al., 1999). It is also able to increase the accessibility of restriction nucleases to nucleosomal templates (Lorch, Cairns et al., 1998). The remodeled state persists after removal of RSC and ATP, and can be reversed upon re-addition of RSC and ATP.

It is unclear if mammalian cells contain a complex homologous to yeast RSC. The BAF180 subunit of the SWI/SNF-B complex shares homology with three yeast RSC complex subunits, Rsc1, Rsc2, and Rsc4 (Xue, Canman et al., 2000). This has led some to propose that SWI/SNF-B is the mammalian homologue of yeast RSC (Neely and Workman, 2002). Furthermore the mammalian SWI/SNF-B complex localizes to the kinetochores of mitotic chromosomes, suggesting that this complex may play a similar to RSC in cell cycle progression.

Ino80.com

The Ino80 protein was identified in yeast based on its homology to Swi2/Snf2 (Shen, Mizuguchi et al., 2000). Ino80 also has homologues in Drosophila (dINO80) and humans (hINO80). The yeast Ino80 associates with approximately 12 proteins in a complex called Ino80.com. This complex possesses a 3′ to 5′ DNA helicase activity, though it has yet to be determined if Ino.com is able to alter chromatin structure. Ino80-null mutants are viable but are sensitive to hydroxyurea, methyl methane-sulfonate, ultraviolet light, and ionizing radiation, suggesting a role for Ino80.com in DNA damage response.

ISWI SUBFAMILY

Members of the ISWI subfamily contain a subunit that shares homology with the *Drosophila* ISWI (imitation switch) protein. These subunits are homologous to Swi2/Snf2 only in their ATPase domain. The ATPase activity is stimulated by nucleosomal DNA.

In *Drosophila*, three ISWI-containing complexes have been identified: NURF (nucleosome remodeling factor), CHRAC (chromatin accessibility complex), and ACF (ATP-utilizing chromatin assembly and remodeling factor) (Tsukiyama, Daniel et al., 1995; Tsukiyama and Wu, 1995; Ito, Bulger et al., 1997; Varga-Weisz, Wilm et al., 1997). Aside from ISWI, the constituents of these complexes vary. All share the ability to regularly space nucleosome arrays in an ATP-dependent fashion; however, only CHRAC has been shown to increase the accessibility of restriction enzymes to chromatin templates. *Drosophila* ISWI is essential for cell viability. Interestingly, null and dominant-negative mutations in ISWI resulted in alteration of the structure of the male X-chromosome, suggesting that this factor plays a role in higher order chromatin structure (Deuring, Fanti et al., 2000).

Two homologues of *Drosophila* ISWI, Isw1p and Isw2p, have been identified in yeast cells (Tsukiyama, Palmer et al., 1999; Gelbart, Rechsteiner et al., 2001). These two subunits are present in distinct complexes. Like *Drosophila* ISWI, Isw1p and Isw2p possess an ATPase activity that is stimulated by nucleosomal DNA. Isw1p- and Isw2p-containing complexes have an ATP-dependent nucleosome remodeling and spacing acitivity.

In humans, the *Drosophila* ISWI-homologue, hSnf2H, has been purified in four, apparently distinct, complexes: RSF (remodeling and spacing factor), WCRF, ACF, and hCHRAC (LeRoy, Orphanides et al., 1998; Bochar, Savard et al., 2000; LeRoy, Loyola et al., 2000; Poot, Dellaire et al., 2000). Like their homologues these complexes have an ATPase activity that is stimulated by nucleosomal DNA. Furthermore they remodel and space nucleosomes in an ATP-dependent manner. The RSF complex also has been shown to stimulate transcriptional initiation from a promoter within a nucleosome template. The WCRF and ACF complexes contain WSTF (Williams syndrome transcription factor) protein, which has been found to be mutated in the developmental disorder Williams syndrome.

CHD SUBFAMILY

CHD (chromo-helicase-DNA-binding) proteins have a SWI2/SNF2-like helicase/ATPase domain, a DNA-binding domain, and a chromodomain. This subfamily includes *S. cerevisiae* Chd1, human NURD complexes, Xenopus Mi-2 complex, and *Drosophila* Mi-2 complex.

The yeast Chd1 has not been found to assemble into a complex, but rather appears to dimerize (Tran, Steger et al., 2000). Chd1 has an ATPase activity that is stimulated by DNA and nucleosomes. Chd1 is able to alter, to some extent, the DNase I digestion pattern of in vitro assembled mononucleosomes. Yeast strains bearing chd1-null deletions are viable; however, chd1-null mutants are synthetically lethal with swi2-null mutants, suggesting that Chd1 and SWI/SNF may share redundant functions.

In human cells, a complex possessing both ATP-dependent chromatin remodeling activity and histone deacetylase activity was purified simultaneously by three groups. These complexes were named NURD nucleosome remodeling and histone deacetylation), NuRD, and NRD (nucleosome remodeling and deacetylating) (Tong, Hassig et al., 1998; Xue, Wong et al., 1998; Zhang, LeRoy et al., 1998). It is unclear whether these are identical complexes or separate, highly related complexes. They contain one or both of two human CHD proteins, CHD3/Mi-2α and/or CHD4/Mi-2β. CHD3/Mi-2α and CHD4/Mi-2β are highly related proteins that are autoantigens in dermatomyositis, a human disease that predisposes 15% to 30% of those afflicted to cancer (Ge, Nilasena et al., 1995; Seelig, Moosbrugger et al., 1995). Recombinant Mi-2 protein was found to have ATPase activity similar to that of intact NuRD complex (Wang and Zhang, 2001). The histone deacetylase activity of these complexes is provided by HDAC1 and HDAC2. These complexes also contain either MTA1 or MTA2 (metastasis-associated protein), whose expression correlates with the metastatic potential of several human cancer cell lines and tissues (Toh, Pencil et al., 1994). The NuRD complex has been shown to contain two alternatively spliced forms of MBD3 (methyl-CpG binding domain). Furthermore this complex interacts with

MBD2, a protein that is believed to link NuRD to methylated DNA. MBD2 also has been identified as NY-CO-41, a human cancer antigen that is recognized by autoantibodies from some colon cancer patients (Wade, Gegonne et al., 1999).

SUMMARY

The chromatin remodeling complexes include a large, and continually growing, number of factors. While it is not clear how many of these complexes function in vivo, it has become apparent that they are important for a variety of cellular processes, and in many instances, cell viability. As described in the sections above, a multitude of chromatin remodeling enzymes are disrupted in a wide range of cancers. It also is likely that some of the more recently discovered enzymes will also be found to play a role in oncogenesis. In summary, the findings reviewed here further signify the necessity to maintain proper regulation of chromatin structure to maintain a healthy cellular environment.

REFERENCES

Abbondanza C, Medici N, et al. (2000): The retinoblastoma-interacting zinc-finger protein RIZ is a downstream effector of estrogen action. Proc Natl Acad Sci USA 97(7):3130–5.

Agalioti T, Lomvardas S, et al. (2000): Ordered recruitment of chromatin modifying and general transcription factors to the IFN-β promoter. Cell 103: 667–78.

Alland L, Muhle R, et al. (1997): Role for N-CoR and histone deacetylase in Sin3-mediated transcriptional repression. Nature 387:49–55.

Allis CD, Chicoine LG, et al. (1985): Deposition-related histone acetylation in micronuclei of conjugating Tetrahymena. Proc Natl Acad Sci USA 82(23): 8048–52.

Anzick SL, Kononen J, et al. (1997): AIB1, a steroid receptor coactivator amplified in breast and ovarian cancer. Science 277:965–8.

Arany Z, Newsome D, et al. (1995): A family of transcriptional adaptor proteins targeted by the E1A oncoprotein. Nature 374(6517):81–4.

Arany Z, Sellers WR, et al. (1994): E1A-associated p300 and CREB-associated CBP belong to a conserved family of coactivators. Cell 77(6):799–800.

Arents G, Moudrianakis EN (1995): The histone fold: a ubiquitous architectural motif utilized in DNA compaction and protein dimeriztion. Proc Natl Acad Sci USA 93:11174–9.

Ayton PM, Cleary ML (2001): Molecular mechanisms of leukemogenesis mediated by MLL fusion proteins. Oncogene 20(40):5695–707.

Bannister AJ, Kouzarides T (1996): The CBP co-activator is a histone acetyltransferase. Nature 384:641–3.

Bannister AJ, Oehler T, et al. (1995): Stimulation of c-Jun activity by CBP: c-Jun residues Ser63/73 are required for CBP induced stimulation in vivo and CBP binding in vitro. Oncogene 11(12):2509–14.

Betz BL, Strobeck MW, et al. (2002): Re-expression of hSNF5/INI1/BAF47 in pediatric tumor cells leads to G_1 arrest associated with induction of p16ink4a and activation of RB. Oncogene 21(34):5193–203.

Biegel JA, Fogelgren B, et al. (2000): Mutations of the INI1 rhabdoid tumor suppressor gene in medulloblastomas and primitive neuroectodermal tumors of the central nervous system. Clin Cancer Res 6(7):2759–63.

Biegel JA, Tan L, et al. (2002): Alterations of the hSNF5/INI1 gene in central nervous system atypical teratoid/rhabdoid tumors and renal and extrarenal rhabdoid tumors. Clin Cancer Res 8(11):3461–7.

Biegel JA, Zhou JY, et al. (1999): Germ-line and acquired mutations of INI1 in atypical teratoid and rhabdoid tumors. Cancer Res 59(1):74–9.

Bischoff JR, Anderson L, et al. (1998): A homologue of *Drosophila* aurora kinase is oncogenic and amplified in human colorectal cancers. EMBO J 17:3052–65.

Bochar DA, Savard J, et al. (2000): A family of chromatin remodeling factors related to Williams syndrome transcription factor. Proc Natl Acad Sci USA 97(3):1038–43.

Bochar DA, Wang L, et al. (2000): BRCA1 is associated with a human SWI/SNF-related complex: linking chromatin remodeling to breast cancer [In process citation]. Cell 102(2):257–65.

Borrow J, Shearman AM, et al. (1996): The t(7;11)(p15;p15) translocation in acute myeloid leukaemia fuses the genes for nucleoporin NUP98 and class I homeoprotein HOXA9. Nat Genet 12(2):159–67.

Breeden L, Nasmyth K (1987): Cell cycle control of the yeast HO gene: *cis*- and *trans*-acting regulators. Cell 48:389–97.

Brownell JE, Allis CD (1995): An activity gel assay detects a single, catalytically active histone acetyltransferase subunit in *Tetrahymena* macronuclei. Proc Natl Acad Sci USA 92:6364–8.

Brownell JE, Zhou J, et al. (1996): *Tetrahymena* histone acetyltransferase A: A homolog to yeast Gcn5p linking histone acetylation to gene activation. Cell 84:843–51.

Bultman S, Gebuhr T, et al. (2000): A *Brg1* null mutation in the mouse reveals functional differences among mammalian SWI/SNF complexes. Mol Cell 6: 1287–95.

Burns LG, Peterson CL (1997): The yeast SWI/SNF complex facilitates binding of a transcriptional activator to nucleosomal site in vivo. Mol Cell Biol 17: 4811–9.

Buyse IM, Shao G, et al. (1995): The retinoblastoma protein binds to RIZ, a zinc-finger protein that shares an epitope with the adenovirus E1A protein. Proc Natl Acad Sci USA 92(10):4467–71.

Cairns BR, Kim Y-J, et al. (1994): A multisubunit complex containing the *SWI1/ADR6*, *SWI2/SNF2*, *SWI3*, *SNF5*, and *SNF6* gene products isolated from yeast. Proc Natl Acad Sci USA 91:1950–4.

Cairns BR, Lorch Y, et al. (1996): RSC, an essential, abundant, chromatin-remodeling complex. Cell 87:1249–60.

Cao Y, Cairns BR, et al. (1997): Sfh1p, a component of a novel chromatin-remodeling complex, is required for cell cycle progression. Mol Cell Biol 17(6):3323–34.

Carapeti M, Aguiar RC, et al. (1998): A novel fusion between MOZ and the nuclear receptor coactivator TIF2 in acute myeloid leukemia. Blood 91(9):3127–33.

Carapeti M, Aguiar RC, et al. (1999): Consistent fusion of MOZ and TIF2 in AML with inv(8)(p11q13). Cancer Genet Cytogenet 113(1):70–2.

Cardoso C, Mignon C, et al. (2000): The human EZH2 gene: genomic organisation and revised mapping in 7q35 within the critical region for malignant myeloid disorders. Eur J Hum Genet 8(3):174–80.

Cardoso C, Timsit S, et al. (1998): Specific interaction between the XNP/ATR-X gene product and the SET domain of the human EZH2 protein. Hum Mol Genet 7(4):679–84.

Carruthers LM, Hansen JC (2000): The core histone N-termini function independently of linker histones during chromatin condensation. J Biol Chem 275:37285–90.

Chadee DN, Hendzel MJ, et al. (1999): Increased Ser-10 phosphorylation of histone H3 in mitogen-stimulated and oncogene-transformed mouse fibroblasts. J Biol Chem 274(35):24914–20.

Chambon P (1996): A decade of molecular biology of retinoic acid receptors. FASEB J 10:940–54.

Chiba H, Muramatsu M, et al. (1994): Two human homologues of *Saccharomyces cerevisiae* SWI2/SNF2 and *Drosophila Brahma* are transcriptional coactivators cooperating with the estrogen receptor and the retinoic acid receptor. Nucl Acids Res 22(10):1815–20.

Chrivia JC, Kwok RP, et al. (1993): Phosphorylated CREB binds specifically to the nuclear protein CBP. Nature 365(6449):855–9.

Côté J, Quinn J, et al. (1994): Stimulation of GAL4 derivative binding to nucleosomal DNA by the yeast SWI/SNF complex. Science 265:53–60.

Cubizollez F, Legagneux V, et al. (1998): pEg7, a new *Xenopus* protein required for mitotic chromosome condensation in egg extracts. J Cell Biol 143:1437–46.

Dai P, Akimaru H, et al. (1996): CBP as a transcriptional coactivator of c-Myb. Genes Dev 10(5):528–40.

Davey CA, Sargent DF, et al. (2002): Solvent mediated interactions in the structure of the nucleosome core particle at 1.9 a resolution. J Mol Biol 319(5):1097–113.

Davie JR, Chadee DN (1998): Regulation and regulatory parameters of histone modifications. J Cellular Biochem 30:203–13.

de la Barre AE, Gerson V, et al. (2000): Core histone amino-termini play an essential role in mitotic chromosome condensation. EMBO J 19:379–91.

de la Serna IL, Carlson KA, et al. (2000): Mammalian SWI-SNF complexes contribute to activation of the hsp70 gene. Mol Cell Biol 20(8):2839–51.

de la Serna IL, Carlson KA, et al. (2001): Mammalian SWI/SNF complexes promote MyoD-mediated muscle differentiation. Nat Gene 27:1–4.

DeCristofaro MF, Betz BL, et al. (2001): Characterization of SWI/SNF protein expression in human breast cancer cell lines and other malignancies. J Cellular Physiol 186:136–45.

DeCristofaro MF, Betz BL, et al. (1999): Alteration of hSNF5/INI1/BAF47 detected in rhabdoid cell lines and primary rhabdomyosarcomas but not Wilms' tumors. Oncogene 18:7559–65.

Deuring R, Fanti L, et al. (2000): The ISWI chromatin-remodeling protein is required for gene expression and the maintenance of higher order chromatin structure in vivo. Mol Cell 5(2):355–65.

Du J, Nasir I, et al. (1998): Sth1p, a saccharomyces cerevisiae Snf2p/Swi2p homolog, is an essential ATPase in RSC and differs from Snf/Swi in its interactions with histones and chromatin-associated proteins [In process citation]. Genetics 150(3):987–1005.

Dunaief JL, Strober BE, et al. (1994): The retinoblastoma protein and BRG1 form a complex and cooperate to induce cell cycle arrest. Cell 79:119–30.

Eckner R, Ewen ME, et al. (1994): Molecular cloning and functional analysis of the adenovirus E1A-associated 300-kD protein (p300) reveals a protein with properties of a transcriptional adaptor. Genes Dev 8(8):869–84.

Fears S, Mathieu C, et al. (1996): Intergenic splicing of MDS1 and EVI1 occurs in normal tissues as well as in myeloid leukemia and produces a new member of the PR domain family. Proc Natl Acad Sci USA 93(4):1642–7.

Ferreri K, Gill G, et al. (1994): The cAMP-regulated transcription factor CREB interacts with a component of the TFIID complex. Proc Natl Acad Sci USA 91(4):1210–3.

Finch JT, Lutter LC, et al. (1977): Structure of nucleosome core particles of chromatin. Nature 269:29–36.

Fryer CJ, Archer TK (1998): Chromatin remodelling by the glucocorticoid receptor requires the BRG1 complex. Nature 393:88–91.

Gayther SA, Batley SJ, et al. (2000): Mutations truncating the EP300 acetylase in human cancers. Nat Genet 24(3):300–3.

Ge Q, Nilasena DS, et al. (1995): Molecular analysis of a major antigenic region of the 240-kD protein of Mi-2 autoantigen. J Clin Invest 96:1730–7.

Gelbart ME, Rechsteiner T, et al. (2001): Interactions of Isw2 chromatin remodeling complex with nucleosomal arrays: Analyses using recombinant yeast histones and immobilized templates. Mol Cell Biol 21(6):2098–106.

Goto H, Tomono Y, et al. (1999): Identificaion of a novel phosphorylation site on histone H3 coupled with mitotic chromsome condensation. J Biol Chem 274:25543–9.

Grand F, Kulkarni S, et al. (1999): Frequent deletion of hSNF5/INI1, a component of the SWI/SNF complex, in chronic myeloid leukemia. Cancer Res 59(16):3870–4.

Grant PA, Duggan L, et al. (1997): Yeast Gcn5 functions in two multisubunit complexes to acetylate nucleosomal histones: Characterization of an Ada complex and the SAGA (Spt/Ada) complex. Genes Dev 11:1640–50.

Grignani F, De Matteis S, et al. (1998): Fusion proteins of the retinoic acid receptor-alpha recruit histone deacetylase in promyelocytic leukaemia. Nature 391:815–18.

Gross DS, Garrard WT (1988): Nuclease hypersensitive sites in chromatin. An Rev Biochem 57:159–97.

Grunstein M (1990): Histone function in transcription. An Rev Cell Biol 6:643–78.

Grunstein M (1997): Histone acetylation in chromatin structure and transcription. Nature 389:349–52.

Guidi CJ, Sands AT, et al. (2001): Disruption of Ini1 leads to peri-implantation lethality and tumorigenesis in mice. Mol Cell Biol 21(10):3598–603.

Gurley LR, D'Anna JA, et al. (1978): Histone phosphorylation and chromatin structure during mitosis in Chinese hamster cells. Eur J Biochem 84(1):1–15.

Han M, Grunstein M (1988): Nucleosome loss activates yeast downstream promoters in vivo. Cell 55:1137–45.

Harbour JW, Dean DC (2000): The Rb/E2F pathway: Expanding roles and emerging paradigms. Genes Dev 14:2393–409.

Hebbes TR, Thorne AW, et al. (1992): Histone acetylation and globin gene switching. Nucl Acids Res 20(5):1017–22.

Hebbes TR, Thorne AW, et al. (1988): A direct link between core histone acetylation and transcriptionally active chromatin. EMBO J 7:1395–1402.

Hecht A, Laroche T, et al. (1995): Histone H3 and H4 N-termini interact with SIR3 and SIR4 proteins: A molecular model for the formation of heterochromatin in yeast. Cell 80:583–92.

Heinzel T, Lavinsky RM, et al. (1997): A complex containing N-CoR, mSin3 and histone deacetylase mediates transcription al repression. Nature 387:43–8.

Hirano T, Mitchison TJ (1994): A heterodimeric coil-coil protein required for mitotic chromosome condensation *in vitro*. Cell 79:449–58.

Hirschhorn JN, Brown SA, et al. (1992): Evidence that SNF2/SWI2 and SNF5 activate transcription in yeast by altering chromatin structure. Genes Dev 6:2288–98.

Holstege FC, Jennings EG, et al. (1998): Dissecting the regulatory circuitry of a eukaryotic genome. Cell 95(5):717–28.

Hsu JY, Sun ZW, et al. (2000): Mitotic phosphorylation of histone H3 is governed by Ipl1/aurora kinase and Glc7/PP1 phosphatase in budding yeast and nematodes. Cell 102:279–91.

Huang S (1999): The retinoblastoma protein-interacting zinc finger gene RIZ in 1p36-linked cancers. Front Biosci 4:D528–32.

Huntsman DG, Chin SF, et al. (1999): MLL2, the second human homolog of the Drosophila trithorax gene, maps to 19q13.1 and is amplified in solid tumor cell lines. Oncogene 18(56):7975–84.

Ida K, Kitabayashi I, et al. (1997): Adenoviral E1A-associated protein p300 is involved in acute myeloid leukemia with t(11;22)(q23;q13). Blood 90(12): 4699–704.

Imbalzano AN, Kwon H, et al. (1994): Facilitated binding of TATA-binding protein to nucleosomal DNA. Nature 370:481–5.

Ito T, Bulger M, et al. (1997): ACF, an ISWI-containing and ATP-utilizing chromatin assembly and remodeling factor. Cell 90(1):145–55.

Jaju RJ, Fidler C, et al. (2001): A novel gene, NSD1, is fused to NUP98 in the t(5;11)(q35;p15.5) in de novo childhood acute myeloid leukemia. Blood 98(4):1264–7.

Janknecht R, Cahill MA, et al. (1995): Signal integration at the c-fos promoter. Carcinogenesis 16(3):443–50.

Janknecht R, Hunter T (1996): Transcription: A growing coactivator network. Nature 383(6595):22–3.

Jason LJM, Moore SC, et al. (2002): Histone uiquitination: A tagging tail unfolds? Bioessays 24:166–74.

John S, Howe L, et al. (2000): The something about silencing protein, Sas3, is the catalytic subunit of NuA3, a yTAF(II)30-containing HAT complex that interacts with the Spt16 subunit of the yeast CP (Cdc68/Pob3)-FACT complex. Genes Dev 14(10):1196–208.

Jones DO, Cowell IG, et al. (2000): Mammalian chromodomain proteins: Their role in genome organisation and expression. Bioessays 22(2):124–37.

Kalpana GV, Marmon S, et al. (1994): Binding and stimulation of HIV-1 integrase by a human homolog of yeast transcription factor SNF5. Science 266: 2002–6.

Kamei Y, Xu L, et al. (1996): A CBP integrator complex mediates transcriptional activation and AP-1 inhibition by nuclear receptors. Cell 85(3):403–14.

Keller AD, Maniatis T (1991): Identification and characterization of a novel repressor of beta-interferon gene expression. Genes Dev 5(5):868–79.

Kleinschmidt AM, Martinson HG (1981): Structure of nucleosome core particles containing uH2A (A24). Nucl Acids Res 9:2423–31.

Klochendler-Yeivin A, Fiette L, et al. (2000): The murine SNF5/INI1 chromatin remodeling factor is essential for embryonic development and tumor suppression. EMBO Rep 1(6):500–6.

Knezetic JA, Luse DS (1986): The presence of nucleosomes on a DNA template prevents initiation by RNA polymerase II in vitro. Cell 45:95–104.

Koipally J, Renold A, et al. (1999): Repression by Ikaros and Aiolos is mediated through histone deacetylase complexes. EMBO J 18:3090–100.

Kouzarides T (2002): Histone methylation in transcriptional control. Curr Opin Genet Dev 12:198–209.

Kruger W, Herskowitz I (1991): A negative regulator of HO transcription, SIN1 (SPT2), is a nonspecific DNA binding protein related to HMG1. Mol Cell Biol 11:4135–46.

Kruger W, Peterson CL, et al. (1995): Amino acid substitutions in the structured domains of histones H3 and H4 partially relieve the requirement of the yeast SWI/SNF complex for transcription. Genes Dev 9:2770–9.

Kung AL, Rebel VI, et al. (2000): Gene dose-dependent control of hematopoiesis and hematologic tumor suppression by CBP. Genes Dev 14(3):272–7.

Kurotaki N, Imaizumi K, et al. (2002): Haploinsufficiency of NSD1 causes Sotos syndrome. Nat Genet 30(4):365–6.

Kwon H, Imbalzano AN, et al. (1994): Nucleosome disruption and enhancement of activator binding by a human SWI/SNF complex. Nature 370:477–81.

Lee D, Kim JW, et al. (2002): SWI/SNF complex interacts with tumor suppressor p53 and is necessary for the activation of p53-mediated transcription. J Biol Chem 277(25):22330–7.

Lee D, Lim C, et al. (2002): The viral oncogene HPV E7 deregulates transcriptional silencing by BRG-1 via molecular interactions. J Biol Chem 277(50): 48842–8.

Lee D, Sohn H, et al. (1999): Interaction of E1 and hSNF5 proteins stimulates replication of human papillomavirus DNA. Nature 399(6735):487–91.

Lee HL, Archer TK (1998): Prolonged glucocorticoid exposure dephosphor lates histone H1 and inactivates the MMTV promoter. EMBO J 17(5):1454–66.

LeGouy E, Thompson EM, et al. (1998): Differential preimplantation regulation of two mouse homologues of the yeast SWI2 protein. Dev Dyn 212(1):38–48.

LeRoy G, Loyola A, et al. (2000): Purification and characterization of a human factor that assembles and remodels chromatin. J Biol Chem 275(20):14787–90.

LeRoy G, Orphanides G, et al. (1998): Requirement of RSF and FACT for transcription of chromatin templates in vitro [see Comments]. Science 282(5395):1900–4.

Levinger L, Barsoum J, et al. (1981): Two-dimenstiolnal hybridization mapping of nucleosomes: Comparison of DNA and protein patterns. J Mol Biol 146:287–304.

Levinger L, Varshavshy A (1980): High-resolution fractionation of nucleosomes: Minor particles, "whiskers" and separation of nucleosomes containing and lacking A24 semihistone. Proc Natl Acad Sci USA 77:3244–8.

Lin RJ, Magy L, et al. (1998): Role of the histone deacetylase complex in acute promyelocytic leukaemia. Nature 391:811–14.

Lochner K, Siegler G, et al. (1996): A specific deletion in the breakpoint cluster region of the ALL-1 gene is associated with acute lymphoblastic T-cell leukemias. Cancer Res 56(9):2171–7.

Logie C, Peterson CL (1997): Catalytic activity of the yeast SWI/SNF complex on reconstituted nucleosome arrays. EMBO J 16:6772–82.

Lorch Y, Cairns BR, et al. (1998): Activated RSC-nucleosome complex and persistently altered form of the nucleosome. Cell 94(1):29–34.

Lorch Y, LaPointe JW, et al. (1987): Nucleosomes inhibit the initiation of transcription but allow chain elongation with the displacement of histones. Cell 49:203–10.

Lorch Y, Zhang M, et al. (1999): Histone octamer transfer by a chromatin-remodeling complex. Cell 96(3):389–92.

Luger K, Mäder AW, et al. (1997): Crystal structure of the nucleosome core particle at 2.8 A resolution. Nature 389:251–60.

Luger K, Richmond TJ (1998): The histone tails of the nucleosome. Curr Opin in Genet Dev 8:140–6.

Lundblad JR, Kwok RP, et al. (1995): Adenoviral E1A-associated protein p300 as a functional homologue of the transcriptional co-activator CBP. Nature 374(6517):85–8.

Luo RX, Postigo AA, et al. (1998): Rb interacts with histone deacetylase to repress transcription. Cell 92:463–73.

Mahadevan LC, Willis AC, et al. (1991): Rapid histone H3 phosphorylation in response to growth factors, phorbol esters, okadaic acid, and protein synthesis inhibitors. Cell 65(5):775–83.

Malgeri U, Baldini L, et al. (2000): Detection of t(4;14)(p16.3;q32) chromosomal translocation in multiple myeloma by reverse transcription-polymerase chain reaction analysis of IGH-MMSET fusion transcripts. Cancer Res 60(15):4058–61.

Martinez-Balbas MA, Bauer UM, et al. (2000): Regulation of E2F1 activity by acetylation. EMBO J 19(4):662–71.

Melnick A, Licht JD (1999): Deconstructing a disease: RAR alpha, its fusion partners, and their roles in the pathogenesis of acute promyelocytic leukemia. Blood 93:3167–215.

Miller ME, Cairns BR, et al. (1996): Adenovirus E1A specifically blocks SWI/SNF-dependent transcriptional activation. Mol Cell Biol 16(10):5737–43.

Miller RW, Rubinstein JH (1995): Tumors in Rubinstein-Taybi syndrome. Am J Med Genet 56(1):112–15.

Mochizuki N, Shimizu S, et al. (2000): A novel gene, MEL1, mapped to 1p36.3 is highly homologous to the MDS1/EVI1 gene and is transcriptionally activated in t(1;3)(p36;q21)-positive leukemia cells. Blood 96(9):3209–14.

Mock BA, Liu L, et al. (1996): The B-lymphocyte maturation promoting transcription factor BLIMP1/PRDI-BF1 maps to D6S447 on human chromosome 6q21–q22.1 and the syntenic region of mouse chromosome 10. Genomics 37(1):24–8.

Muchardt C, Sardet C, et al. (1995): A human protein with homology to *Saccharomyces cerevisiae* SNF5 interacts with the potential helicase hbrm. Nucl Acids Res 23(7):1127–32.

Muchardt C, Yaniv M (1993): A human homologue of *Saccharomyces cerevisiae SNF2/SWI2* and *Drosophila brm* genes potentiates transcriptional activation by the glucocorticoid receptor. EMBO J 12:4279–90.

Muraoka M, Konishi M, et al. (1996): p300 gene alterations in colorectal and gastric carcinomas. Oncogene 12(7):1565–9.

Murphy DJ, Hardy S, et al. (1999): Human SWI-SNF component BRG1 represses transcription of the c-*fos* gene. Mol Cell Biol 19(4):2724–33.

Nagy L, Kao H-Y, et al. (1997): Nuclear receptor repression mediated by a complex containing SMRT, mSin3A, and histone deacetylase. Cell 89:373–80.

Nagy LEA (1999): Mechanism of corepressor binding and release from nuclear hormone receptors. Genes Dev 13:3209–16.

Neely KE, Workman JL (2002): The complexity of chromatin remodeling and its links to cancer. Biochim Biophys Acta 1603(1):19–29.

Neigeborn L, Carlson M (1984): Genes affecting the regulation of *SUC2* gene expression by glucose repression in *Saccharomyces cerevisiae*. Genetics 108: 845–58.

Neuwald AF, Landsman D (1997): GCN5-related histone N-acetyltransferases belong to a diverse superfamily that includes the yeast SPT10 protein. Trends Biochem Sci 22(5):154–5.

Nielsen SJ, Schneider R, et al. (2001): Rb targets histone H3 methylation and HP1 to promoters. Nature 412(6846):561–5.

Noll M, Kornberg RD (1977): Action of micrococcal nuclease on chromatin and the location of H1. J Mol Biol 109:393–404.

Oelgeschlager M, Janknecht R, et al. (1996): Interaction of the co-activator CBP with Myb proteins: Effects on Myb-specific transactivation and on the cooperativity with NF-M. EMBO J 15(11):2771–80.

Ogryzko VV, Schiltz RL, et al. (1996): The transcriptional coactivators p300 and CBP are histone acetyltransferases. Cell 87:953–9.

Oike Y, Hata A, et al. (1999): Truncated CBP protein leads to classical Rubinstein-Taybi syndrome phenotypes in mice: implications for a dominant-negative mechanism. Hum Mol Genet 8(3):387–96.

Owa TEA (2001): Cell cycle regulation in the G_1 phase: A promising target for the development of new chemotherapeutic anticancer agents. Curr Med Chem 8:1487–503.

Paulson JR, Taylor SS (1982): Phosphorylation of histones 1 and 3 and nonhistone high mobility group 14 by an endogenous kinase in HeLa metaphase chromosomes. J Biol Chem 257:6064–72.

Peters AH, O'Carroll D, et al. (2001): Loss of the Suv39h histone methyltransferases impairs mammalian heterochromatin and genome stability. Cell 107(3):323–37.

Peterson CL, Dingwall A, et al. (1994): Five *SWI/SNF* gene products are components of a large multiprotein complex required for transcriptional enhancement. Proc Natl Acad Sci USA 91:2905–8.

Petrij F, Giles RH, et al. (1995): Rubinstein-Taybi syndrome caused by mutations in the transcriptional co-activator CBP. Nature 376(6538):348–51.

Phelan ML, Sif S, et al. (1999): Reconstitution of a core chromatin remodeling complex from SWI/SNF subunits. Mol Cell 3(2):247–53.

Piao Z, Fang W, et al. (2000): Frequent frameshift mutations of RIZ in sporadic gastrointestinal and endometrial carcinomas with microsatellite instability. Cancer Res 60(17):4701–4.

Poot RA, Dellaire G, et al. (2000): HuCHRAC, a human ISWI chromatin remodelling complex contains hACF1 and two novel histone-fold proteins. EMBO J 19(13):3377–87.

Rea SEA (2000): Regulation of chromatin structure by site-specific histone H3 methyltransferases. Nature 406:593–9.

Reifsnyder C., Lowell J, et al. (1996): Yeast SAS silencing genes and human genes associated with AML and HIV-1 Tat interactions are homologous with acetyltransferases. Nat Genet 14(1):42–9.

Reyes JC, Barra J, et al. (1998): Altered control of cellular proliferation in the absence of mammalian brahma (SNF2alpha). EMBO J 17(23):6979–91.

Roberts CWM, Galusha SA, et al. (2000): Haploinsufficiency of Snf5 (integrase interactor 1) predisposes to malignant rhabdoid tumors in mice. PNAS 97(25):13796–800.

Roguev A, Schaft D, et al. (2001): The *Saccharomyces cerevisiae* Set1 complex includes an Ash2 homologue and methylates histone 3 lysine 4. EMBO J 20(24):7137–48.

Rosati R, La Starza R, et al. (2002): NUP98 is fused to the NSD3 gene in acute myeloid leukemia associated with t(8;11)(p11.2;p15). Blood 99(10):3857–60.

Roth SY, Denu JM, et al. (2001): Histone acetyltransferases. An Rev Biochem 70:81–120.

Rousseau-Merck MF, Versteege I, et al. (1999): hSNF5/INI1 inactivation is mainly associated with homozygous deletions and mitotic recombinations in rhabdoid tumors. Cancer Res 59(13):3152–6.

Ruiz-Carrillo A, Wangh LJ, et al. (1975): Processing of newly synthesized histone molecules. Science 190(4210):117–28.

Sartorelli V, Huang J, et al. (1997): Molecular mechanisms of myogenic coactivation by p300: direct interaction with the activation domain of MyoD and with the MADS box of MEF2C. Mol Cell Biol 17(2):1010–26.

Schichman SA, Caligiuri MA, et al. (1994): ALL-1 partial duplication in acute leukemia. Proc Natl Acad Sci USA 91(13):6236–9.

Schneider R, Bannister AJ, et al. (2002): Unsafe SETs: histone lysine methyltransferases and cancer. Trends Biochem Sci 27(8):396–402.

Seale RL (1981): Rapid turnover of the histone-ubiquitin conjugate, protein A24. Nucl Acids Res 9:3151–8.

Seelig HP, Moosbrugger I, et al. (1995): The major dermatomyositis-specific Mi-2 autoantigen is a presumed helicase involved in transcriptional activation. Arthritis Rheum 38:1389–99.

Sentani K, Oue N, et al. (2001): Increased expression but not genetic alteration of BRG1, a component of the SWI/SNF complex, is associated with the advanced stage of human gastric carcinomas. Pathobiology 69:315–20.

Sévenet N, Lellouch-Tubiana A, et al. (1999): Spectrum of hSNF5/INI1 somatic mutations in human cancer and genotype-phenotype correlations. Hum Mol Genet 8(13):2359–68.

Shanahan F, Seghezzi W, et al. (1999): Cyclin E associates with BAF155 and BRG1, components of the mammalian SWI-SNF complex, and alters the ability of BRG1 to induce growth arrest. Mol Cell Biol 19(2):1460–9.

Shen X, Mizuguchi G, et al. (2000): A chromatin remodelling complex involved in transcription and DNA processing. Nature 406(6795):541–4.

Sobulo OM, Borrow J, et al. (1997): MLL is fused to CBP, a histone acetyltransferase, in therapy-related acute myeloid leukemia with a t(11;16)(q23;p13.3). Proc Natl Acad Sci USA 94(16):8732–7.

Stec I, van Ommen GJ, et al. (2001): WHSC1L1, on human chromosome 8p11.2, closely resembles WHSC1 and maps to a duplicated region shared with 4p16.3. Genomics 76(1–3):5–8.

Stec I, Wright TJ, et al. (1998): WHSC1, a 90kb SET domain-containing gene, expressed in early development and homologous to a *Drosophila* dysmorphy gene maps in the Wolf-Hirschhorn syndrome critical region and is fused to IgH in t(4;14) multiple myeloma. Hum Mol Genet 7(7):1071–82.

Steele-Perkins G, Fang W, et al. (2001): Tumor formation and inactivation of RIZ1, an Rb-binding member of a nuclear protein-methyltransferase superfamily. Genes Dev 15(17):2250–62.

Stern MJ, Jensen R, et al. (1984): Five *SWI* genes are required for expression of the *HO* gene in yeast. J Mol Biol 178:853–68.

Strahl BD, Grant PA, et al. (2002): Set2 is a nucleosomal histone H3-selective methyltransferase that mediates transcriptional repression. Mol Cell Biol 22(5):1298–306.

Strobeck MW, Knudsen KE, et al. (2000): BRG-1 is required for Rb mediated cell cycle arrest. Proc Natl Acad Sci USA 97:7748–53.

Strobeck MW, Reisman DN, et al. (2002): Compensation of BRG-1 function by Brm: insight into the role of the core SWI-SNF subunits in retinoblastoma tumor suppressor signaling. J Biol Chem 277(7):4782–9.

Struhl K (1998): Histone acetylation and transcriptional regulatory mechanisms. Genes Dev 12:599–606.

Sudarsanam P, Iyer VR, et al. (2000): Whole-genome expression analysis of snf/swi mutants of *Saccharomyces cerevisiae*. Proc Natl Acad Sci USA 97(7): 3364–9.

Sutani T, Yuasa T, et al. (1999): Fission yeast condensin complex: essential roles of non-SMC subunits for condensation and Cdc2 phosphorylation of Cut3/SMC4. Genes Dev 13:2271–83.

Tan YC, Chow VT (2001): Novel human HALR (MLL3) gene encodes a protein homologous to ALR and to ALL-1 involved in leukemia, and maps to chromosome 7q36 associated with leukemia and developmental defects. Cancer Detect Prev 25(5):454–69.

Tanaka T, Kimura M, et al. (1999): Centrosomal kinase AIK1 is overexpressed in invasive ductal carcinoma of the breast. Cancer Res 59:2041–4.

Tatsuka M, Katayama H, et al. (1998): Multinuclearity and increased ploidy caused by overexpression of the aurora- and Ipl1-like midbody-associated protein mitotic kinase in human cancer cells. Cancer Res 58:4811–16.

Taunton J, Hassig CA, et al. (1996): A mammalian histone deacetylase related to the yeast transcriptional regulator Rpd3p. Science 272:408–11.

Taylor MD, Gokgoz N, et al. (2000): Familial posterior fossa brain tumors of infancy secondary to germline mutation of the hSNF5 gene. Am J Hum Genet 66:1403–6.

Toh Y, Pencil SD, et al. (1994): A novel candidate metastasis-associated gene, mta1, differentially expressed in highly metastatic mammary adenocarcinoma cell lines: cDNA cloning, expression and protein analyses. J Biol Chem 269: 22958–63.

Tong JK, Hassig CA, et al. (1998): Chromatin deacetylation by an ATP-dependent nucleosome remodelling complex. Nature 395(6705):917–21.

Tran HG, Steger DJ, et al. (2000): The chromo domain protein chd1p from budding yeast is an ATP-dependent chromatin-modifying factor. EMBO J 19(10):2323–31.

Trouche D, Le Chalony C, et al. (1997): Rb and hbrm cooperate to repress the activation functions of E2F1. Proc Natl Acad Sci USA 94:11268–73.

Tse C, Sera T, et al. (1998): Disruption of higher-order folding by core histone acetylation dramatically enhances transcription of nucleosomal arrays by RNA polymerase III [In process citation]. Mol Cell Biol 18(8):4629–38.

Tsukiyama T, Daniel C, et al. (1995): ISWI, a member of the SWI2/SNF2 ATPase family, encodes the 140kDa subunit of the nucleosome remodeling factor. Cell 83(6):1021–6.

Tsukiyama T, Palmer J, et al. (1999): Characterization of the imitation switch subfamily of ATP-dependent chromatin-remodeling factors in Saccharomyces cerevisiae. Genes Dev 13(6):686–97.

Tsukiyama T, Wu C (1995): Purification and properties of an ATP-dependent nucleosome remodeling factor. Cell 83(6):1011–20.

Uno K, Takita J, et al. (2002): Aberrations of the hSNF5/INI1 gene are restricted to malignant rhabdoid tumors or atypical teratoid/rhabdoid tumors in pediatric solid tumors. Genes Chromosomes Cancer 34(1):33–41.

Utley PT, Côté J, et al. (1997): SWI/SNF stimulates the formation of disparate activator-nulceosome complexes but is partially redundant with cooperative binding. J Biol Chem 272:12642–9.

van Holde KE, Shaw BR, et al. (1975): Organization and Expression of the Eukaryotic Genome. Amsterdam: North Holland, American Elsevier.

Varga-Weisz PD, Wilm M, et al. (1997): Chromatin-remodelling factor CHRAC contains the ATPases ISWI and topoisomerase II. Nature 388:598–602.

Vassilev AP, Rasmussen HH, et al. (1995): The levels of ubiquitinated histone H2A are highly upregulated in transformed human cells: Partial colocalization of uH2A cluster and PCNA/cyclin foci in a fraction of cells in S-phase. J Cell Sci 108:1205–15.

Versteege I, Medjkane S, et al. (2002): A key role of the hSNF5/INI1 tumour suppressor in the control of the G_1-S transition of the cell cycle. Oncogene 21(42):6403–12.

Versteege I, Sevenet N, et al. (1998): Truncating mutations of hSNF5/INI1 in aggressive paediatric cancer. Nature 394(6689):203–6.

Visser HP, Gunster MJ, et al. (2001): The Polycomb group protein EZH2 is upregulated in proliferating, cultured human mantle cell lymphoma. Br J Haematol 112(4):950–8.

Wade PA, Gegonne A, et al. (1999): Mi-2 complex couples DNA methylation to chromatin remodelling and histone deacetylation [see Comments]. Nat Genet 23(1):62–6.

Wang H, Cao R, et al. (2001): Purification and functional characterization of a histone H3-lysine 4-specific methyltransferase. Mol Cell 8(6):1207–17.

Wang HB, Zhang Y (2001): Mi2, an auto-antigen for dermatomyositis, is an ATP-dependent nucleosome remodeling factor. Nucl Acids Res 29(12): 2517–21.

Wang W, Côte J, et al. (1996): Purification and biochemical heterogeneity of the mammalian SWI-SNF complex. EMBO J 15(19):5370–82.

Wang X, Yeh S, et al. (2001): Identification and characterization of a novel androgen receptor coregulator ARA267-alpha in prostate cancer cells. J Biol Chem 276(44):40417–23.

Wei Y, Yu L, et al. (1999): Phosphorylation of histone H3 is required for proper chromosome condensation and segregation. Cell 97(1):99–109.

West MHP, Bonner WM (1980): Histone 2B can be modified by the attachment of ubiquitin. Nucl Acids Res 8:4671–80.

Winter AG, Sourvinos G, et al. (2000): RNA polymerase III transcription factor TFIIIC2 is overexpressed in ovarian tumors. Proc Natl Acad Sci USA 97(23):12619–24.

Wolffe AP, Hayes JJ (1999): Chromatin disruption and modification. Nucl Acids Res 27(3):711–20.

Wong AK, Shanahan F, et al. (2000): BRG1, a component of the SWI-SNF complex, is mutated in multiple human tumor cell lines. Cancer Res 60(21): 6171–7.

Workman JL, Kingston RE (1998): Alteration of nucleosome structure as a mechanism of transcriptional regulation. An Rev Biochem 67:545–79.

Wu DY, Krumm A, et al. (2000): Promoter-specific targeting of human SWI-SNF complex by Epstein-Barr virus nuclear protein 2. J Virol 74(19):8893–903.

Wu RS, Kohn KW, et al. (1981): Metabolism of ubiquitinated histones. J Biol Chem 256:5916–20.

Wyrick JJEA (1999): Chromosomal landscape of nucleosome-dependent gene expression and silencing in yeast. Nature 402:418–21.

Xue Y, Canman JC, et al. (2000): The human SWI/SNF-B chromatin-remodeling complex is related to yeast Rsc and localizes at kinetochores of mitotic chromosomes. Proc Natl Acad Sci USA 97:13015–20.

Xue Y, Wong J, et al. (1998): NURD, a novel complex with both ATP-dependent chromatin-remodeling and histone deacetylase activities [In process citation]. Mol Cell 2(6):851–61.

Yang L, Xia L, et al. (2002): Molecular cloning of ESET, a novel histone H3-specific methyltransferase that interacts with ERG transcription factor. Oncogene 21(1):148–52.

Yang X-J, Ogryzko VV, et al. (1996): A p300/CBP-associated factor that competes with the adenoviral oncoprotein E1A. Nature 382:319–24.

Yang XH, Huang S (1999): PFM1 (PRDM4), a new member of the PR-domain family, maps to a tumor suppressor locus on human chromosome 12q23–q24.1. Genomics 61(3):319–25.

Yao TP, Oh SP, et al. (1998): Gene dosage-dependent embryonic development and proliferation defects in mice lacking the transcriptional integrator p300. Cell 93(3):361–72.

Yuan W, Condorelli G, et al. (1996): Human p300 protein is a coactivator for the transcription factor MyoD. J Biol Chem 271(15):9009–13.

Zeleznik-Le NJ, Harden AM, et al. (1994): 11q23 translocations split the "AT-hook" cruciform DNA-binding region and the transcriptional repression domain from the activation domain of the mixed-lineage leukemia (MLL) gene. Proc Natl Acad Sci USA 91(22):10610–14.

Zhang HS, Gavin M, et al. (2000): Exit from G_1 and S phase of the cell cycle is regulated by repressor complexes containing HDAC-Rb-hSWI/SNF and Rb-hSWI/SNF. Cell 101(1):79–89.

Zhang Y, LeRoy G, et al. (1998): The dermatomyositis-specific autoantigen Mi2 is a component of a complex containing histone deacetylase and nucleosome remodeling activities. Cell 95(2):279–89.

Zhou H, Kuang J, et al. (1998): Tumour amplified kinase STK15/BTAK induces centrosome amplification, aneuploidy and transformation. Nat Genet 20: 189–93.

Zhou YB, Gerchman SE, et al. (1998): Position and orientation of the globular domain of linker histone H5 on the nucleosome. Nature 395(6700):402–5.

Ziemin-van der Poel S, McCabe NR, et al. (1991): Identification of a gene, MLL, that spans the breakpoint in 11q23 translocations associated with human leukemias. Proc Natl Acad Sci USA 88(23):10735–9.

CHAPTER 9

EXTRACELLULAR MATRIX: TISSUE-SPECIFIC REGULATOR OF CELL PROLIFERATION

AYLIN RIZKI and MINA J. BISSELL
Life Sciences Division, Lawrence Berkeley National Laboratory, Berkeley, CA 94720

INTRODUCTION

There are two broad categories of extracellular matrix (ECM) in tissues: interstitial/stromal matrix and basement membrane (BM). The interstitial matrix is the loose material around the epithelial cells that is separated from the cells by a basement membrane in many solid tissues. BM is most commonly found lining epithelial cell layers in tissues such as skin and breast (Fuchs et al., 1997; Ronnov-Jessen et al., 1996). Both the composition and the ultrastructure of BM exhibit tissue specificity, as well as temporal regulation during development (Jones and Jones, 2000; Miosge, 2001; Streuli, 1999; Tsai, 1998). Studies of the ultrastructural composition of basement membranes in vivo suggest that the relative arrangement of various ECM components are not only tissue specific but can also be different in certain parts of the same tissue (Lin and Bissell, 1993; Miosge, 2001). For example, ultrastructurally identical basement membranes, such as those found in the proximal and distal tubules of the kidney, have been shown to have a different molecular arrangement when examined by electron microscopy that allows observation of component orientation in tissue samples (Miosge, 2001).

One manifestation of tissue specificity is observed in the form of gene expression patterns, including expression of genes involved in cell cycle regulation. Establishment of tissue-specific gene expression patterns is not simply a result of which ECM molecules surround the cells in the adult tissue. Developmental processes (both during embryogenesis and postbirth, as is the case for the mammary gland) that produce a differentiated tissue involve sequential and interrelated gene regulatory

Cell Cycle and Growth Control: Biomolecular Regulation and Cancer, Second Edition, Edited by Gary S. Stein and Arthur B. Pardee
ISBN 0-471-25071-6

events that are temporally regulated and that result in an integrated pattern of gene expression in the differentiated tissue (Bissell et al., 1999; Boudreau and Jones, 1999; Lohi et al., 1997; Wiseman and Werb, 2002). Therefore, to fully solve the puzzle of how ECM affects tissue-specific gene expression or cell cycle progression, the information context needs to include the history of the cell and its surrounding ECM (both of which change during development), as well as of the molecular characteristics of the cell-ECM interactions (Bissell and Ram, 1989; Werb and Chin, 1998; Wiseman and Werb, 2002).

Developmental events create a particular imprint in each tissue of both specific ECM molecules and ECM receptors. Thus how a cell behaves in response to its surrounding ECM is dependent not only on the level and composition of the ECM, but also on the cell-surface receptors that recognize and respond to it. The best-studied ECM receptors are the integrin family (Hynes, 1987, 1992; Miranti and Brugge, 2002). However, increasingly other receptors such as syndecans and dystroglycan have been shown to play a role in ECM-mediated signaling (Carey, 1997; Couchman and Woods, 1999; Rapraeger, 2000; Zimmermann and David, 1999). Multiple receptors can recognize a single type of ECM molecule, and a single receptor type may respond to multiple ECM components (Ashkenas et al., 1996; Boudreau and Jones., 1999; Watt, 2002). In addition how a cell responds to ECM is dependent on its growth factor and cytokine context. This is due to the extensive and reciprocal cross-talk between ECM, growth factor, and cytokine receptors (Damsky and Werb, 1992; Danen and Yamada, 2001; Schwartz et al., 1995). Besides cell cycle progression and differentiation, cell-ECM interactions regulate other cellular events such as apoptosis (Boudreau et al., 1995; Howlett et al., 1995). Not surprisingly, disruption of cell-ECM interactions, either by misregulated receptor function or by aberrant ECM composition and arrangement, results in tumorigenesis (Bissell and Radisky, 2001; Radisky et al., 2002; Simpson et al., 1994; Sternlicht et al., 2000; Sternlicht et al., 1999; Talhouk et al., 1992).

TISSUE SPECIFICITY OF ECM AND ITS RECEPTORS

Tissue-specific effects of ECM on cell proliferation are dependent on the molecular composition of the matrix surrounding the cells, as well as the ECM receptor makeup of the particular cell type within a tissue. Here we briefly discuss the function of some of the main ECM component families and their receptors with emphasis on tissue-distribution and tissue-specific diseases associated with these molecules. The ECM components we focus on are collagens, laminins, nidogens, glycosaminoglycans, and proteoglycans; ECM receptors include integrins, dystroglycan, and syndecans. Examples of genetic diseases associated with aberrant ECM or ECM receptor components are listed in Table 9.1.

TABLE 9.1. ECM Component Mutations in Human Disease

Gene[a]	Disease[b]	Reference[c]
Laminins		
LAMA2	Merosin-deficient muscular dystrophy	Kuang et al. (1998)
LAMA3	Herlitz type junctional epidermolysis bullosa (skin)	Vidal et al. (1995)
LAMB3	Herlitz type junctional epidermolysis bullosa (skin)	Kon et al. (1998)
LAMC3	Herlitz type junctional epidermolysis bullosa (skin)	Kon et al. (1998)
Collagens (fibrillar)		
COL1A1	Osteogenesis imperfecta (bone, muscle)	Ward et al. (2001)
COL1A2	Osteogenesis imperfecta (bone, muscle)	Trummer et al. (2001)
COL1A2	Ehlers-Danlos syndrome (connective tissue)	Byers et al. (1997)
COL1A2	Marfan syndrome (bone, ocular, cardiovascular)	Dalgleish et al. (1986)
COL2A1	Spondyloepiphyseal dysplasia (bone, retina)	Tiller et al. (1995)
COL2A1	Kniest dysplasia (bone, cartilage)	Spranger et al. (1994)
COL2A1	Achondrogenesis-hypochondrogenesis (bone, cartilage)	Godfrey and Hollister (1988)
COL2A1	Osteoarthiritis with mild condrodysplasia (bone, joint)	Ritvaniemi et al. (1995)
COL2A1	Stickler syndrome (joint, hearing, eye, cleft)	Brown et al. (1995)
COL2A1	Wagner syndrome (eye)	Richards et al. (2000)
COL3A1	Ehlers-Danlos syndrome (connective tissue)	Smith et al. (1997)
COL3A1 (1999)	Familial aortic aneurysms (endothelial)	van Keulen et al.
COL6A1	Bethlem myopathy (muscle)	Jobsis et al. (1996)
COL6A2	Bethlem myopathy (muscle)	Jobsis et al. (1996)
COL6A3	Bethlem myopathy (muscle)	Jobsis et al. (1996)
COL10A1	Metaphyseal chondrodysplasia (bone)	McIntosh et al. (1995)
COL10A1	Spondylomethaphyseal dysplasia (bone)	Ikegawa et al. (1998)
COL11A1	Stickler syndrome (joint, hearing, eye, cleft)	Snead et al. (1996)
COL11A1	Marshall syndrome (hearing, eye, facial skeletal defect)	Meisler et al. (1998)
Collagens (BM)		
COL4A3	Alport syndrome (kidney)	Kashtan (1995)
COL4A3	Benign hematuria (kidney)	Badenas et al. (2002)
COL4A4	Alport syndrome (kidney)	Kashtan (1995)
COL4A4	Benign hematuria (kidney)	Lemmink et al. (1996)
COL4A5	Alport syndrome (kidney)	Kashtan (1995)
COL4A6	Alport syndrome (kidney)	Kashtan (1995)
COL8A2	Fuchs endothelial corneal dystrophy	Biswas et al. (2001)

[a] A gene is listed more than once if it has multiple associated diseases. These genes were selected because mutations in them are associated with genetic diseases.
[b] The affected tissues are listed in parentheses.
[c] Additional information and references are available at the NIH web site OMIM (Online Mendelian Inheritance In Man), at *http://www.ncbi.nlm.nih.gov:80/entrez/query.fcgi?db=OMIM.*

Collagens

Collagens are the most abundant components of the extracellular matrix as well as of interstitial/stromal matrices. Indeed, collagens are the most abundant proteins in mammals, being a major component of the skin and bone. Most collagens are heterotrimeric molecules of alpha chains (more than 25 of which have been described so far) folded into a coiled-coil triple helix. The triple helix can be homotrimeric, as in the case of type II, type III, type IV, and type VII. Collagens can also be heterotrimeric containing either two or three different alpha chains. For example, collagen type XI is heterotrimeric, containing an alpha 1 (COL11A1), an alpha 2 (COL11A2), and an alpha 3 (COL11A3) chain; collagen type I contains two alpha 1 chains (COL1A1) and one alpha 2 chain (COL1A2). More than 20 different types of collagens have been described based on the structural domains of their alpha chain components. However, within one collagen type, there may be significant variation among molecules because of the large number of alpha chain isoforms that exist. For example, the basement membrane collagen, collagen type IV, can be composed of helix combinations that are compiled from six different gene products (COL4A1 through COL4A6), and their many alternatively spliced variants, in a somewhat tissue-specific manner (Myllyharju and Kivirikko, 2001).

Collagens can be organized into fibrillar structures in connective tissues and in interstitial matrices of soft tissues, or they organize into sheet-like structures in basement membranes. Some nonfibrillar collagens are found associated with fibrils, and sometimes also with the basement membrane and are thought to stabilize interactions between the basement membrane and the interstitial stromal matrix. Fibril-forming collagens include collagen types I, II, III, V, VI, IX, X, and XI (Kadler, 1994, 1995). Fibril-associated collagens are collagen types VII, XII, XV, XVI, XIX, and XXI. Basement membrane collagens are collagen types IV and VIII (Fukai et al., 1994). Functions of collagens XIII, XIV, XXII, XXIII, and XXVI have not been well characterized. Collagen type XVII is an unusual collagen since it is a transmembrane protein (Uitto and Pulkkinen, 1996).

A comprehensive overview of the tissue-specific distribution and functions of this vast number of collagens is not within the scope of this review. However, examples below indicate that certain tissues feature some collagens more prominently, and that diseases associated with each collagen type hints at the tissue specificity of expression (Olsen, 1995; Tryggvason, 1995). Examples of tissue-specific functions and diseases of fibrillar collagen are as follows: Collagen type I is a fibrillar collagen, most commonly found in connective tissues such as bone, cartilage, and skin (Cremer et al., 1998). Mutations in collagen type I have been associated with bone diseases such as osteogenesis imperfecta, Ehlers-Danlos syndrome, and idiopathic osteoporosis, as well as causing a particular type of skin tumor called dermafibrosarcoma protuberans (Kuivaniemi et al., 1997). Collagen type II is a fibrillar collagen found in cartilage and in the vitreous humor of the eye (Cremer et al., 1998; Kuivaniemi et al., 1997). Accordingly mutations in collagen type II are associated with connective tissue disorders such as achondrogenesis, chondrodysplasia, early onset familial osteoarthritis, SED congenita, Langer-Saldino achondrogenesis, Kniest dysplasia, Stickler syndrome type I, and spondyloepimetaphyseal dysplasia Strudwick type. Collagen type III is also a fibrillar collagen that is associated with flexible connective tissues such as that of the skin, lung, and vasculature, often found associated with collagen type I. Accordingly mutations in this collagen are found associated with blood vessel abnormalities, such as aortic and arterial aneurysms, as well as with connective tissue disorders such as Ehlers-Danlos syndrome (Kuivaniemi et al., 1997). Type V collagen is found in tissues containing type I collagen and appears to regulate the assembly of heterotypic fibers composed of both type I and type V collagen. Collagen VI is a major structural component of muscle microfibrils, and mutations in the genes that code for the collagen VI subunits result in the autosomal dominant disorder, Bethlem myopathy. Type X collagen is short fibrillar collagen that is expressed by chondrocytes during ossification. Mutations in collagen type X result in bone diseases such as Schmid-type metaphyseal chondrodysplasia (SMCD) and Japanese-type spondylometaphyseal dysplasia (SMD) (Kuivaniemi et al., 1997). Collagen type XI is a fibrillar collagen which is a minor constituent of cartilage and mutations in this collagen are associated with type II Stickler syndrome and with Marshall syndrome, both connective tissue disorders (Cremer et al., 1998; Kuivaniemi et al., 1997).

Examples of tissue-specific functions and diseases of basement membrane collagens are as follows: Collagen type IV is found in many basement membranes. There are 6 alpha chains that can give rise to a possible 56 combinations of collagen type IV. Furthermore there are multiple isoforms of many of the subunits, providing a large number of possible collagen type IV combinations for establishment of tissue specificity (Kuhn, 1995; Petitclerc et al., 2000). Mutations in the alpha 1 chain of collagen type IV are associated with type II autosomal Alport syndrome (hereditary glomerulonephropathy) and with familial benign hematuria (thin basement membrane disease). Both diseases affect the kidneys, suggesting that the alpha 3, alpha 4, alpha 5 chains (COL4A3, COL4A4, COL4A5) are prominently featured in the kidney basement membranes (Kashtan, 2000). Type VIII collagen is the main component of the corneal epithelium. This collagen has only two alpha subunits but it can exist either as a homo- or a heterotrimer, as a combination of these two subunits. Mutations in collagen type VIII cause corneal endothelial dystrophy, consistent with the specificity/prominence of its function in the cornea (Meek and Fullwood, 2001).

Laminins

Laminins are a family of heterotrimeric extracellular basement membrane glycoproteins. They are composed of three chains, alpha, beta, and gamma, which form a cruciform structure with three short arms (each from a different chain) and a long arm (composed of all three chains) (Cheng et al., 1997; Yurchenco et al., 1992). Each laminin subunit has multiple functional domains and is encoded by a distinct gene. Expression of laminin encoding genes is regulated at multiple levels, giving rise to different isoforms. Laminin alpha, beta, and gamma chain isoforms can combine to give rise to different laminins. A total of 11 laminins have been described so far (and they were named in the order of their discovery: laminin-1, laminin-2, etc.); however, a very large number of permutations of the 5 alpha, 4 beta, and 3 gamma subunits and their isoforms is possible (Ekblom et al., 1998).

The tissue-specific distribution, and possible distinct functions of the different alpha, beta, and gamma chains and their isoforms remain largely unknown. Some laminin components have wide tissue distributions. For example, laminin beta 1 is expressed in most tissues that produce a basement membrane, and is one of the three chains constituting laminin 1 (alpha 1/LAMA1, beta 1/LAMB1, gamma 1/LAMC1). This was the first laminin isolated from Engelbreth-Holm-Swarm (EHS) tumor (a basement membrane gel isolated from EHS tumors is widely used in reconstitution experiments in culture and is referred to as a laminin-rich basement membrane, lrBM, in the rest of this chapter) (Friedman et al., 1989; Grant et al., 1985; Kleinman et al., 1982; Li et al., 1987). However, some chains are more prominent in certain tissues, reflecting tissue specificity of BM composition. For example, laminin alpha 2 (LAMA2) is prominently expressed in striated muscle (laminins that contain LAMA2 are called merosins), and the significance of LAMA2 function in muscle is exemplified by the causal relationship between mutations in this gene and congenital merosin-deficient muscular dystrophy (Kuang et al., 1998; Vachon et al., 1996; Wewer and Engvall, 1996). Laminin 5 (alpha 3/LAMA3, beta 3/LAMB3, gamma 2/LAMC2) has a significant role in skin keratinocyte function, as shown by the observation that mutations in any one of its three subunits causes Herlitz type junctional epidermolysis bullosa (Kon et al., 1998; Vidal et al., 1995). Another example of laminin chains that show much restricted tissue distribution is beta 2/LAMB2. It is enriched in the basement membrane of muscles at the neuromuscular junctions, kidney glomerulus and vascular smooth muscle (Hunter et al., 1989; Virtanen et al., 1995).

Nidogens/Entactins

Nidogens are sulfated glycoproteins that are found in basement membranes. Two mammalian nidogens have so far been identified, nidogen 1 and nidogen 2. Their function is to bridge the laminin network with the collagen IV network and provide stability of the basement membranes. Nidogen is essential for both embryonic development and maintenance of proper differentiation of adult tissues. Since there are only two members of the family, nidogen is a less likely candidate for establishment and maintenance of tissue specificity (Dziadek, 1995; Mayer et al., 1998; Timpl et al., 1984). However, distribution of the two members of this family does appear to show some tissue specificity: nidogen 2 is more prominent in endothelial cells but nidogen 1 is more widely distributed (Schymeinsky et al., 2002).

Proteoglycans and Glycosaminoglycans

Proteoglycans have important structural, as well as signaling functions (Perrimon and Bernfield, 2001). They play a part in providing shape and biomechanical strength to organs and tissues mainly because of the nature of their glycosaminoglycan (GAG) chains. Structural diversity within GAG chains ensures that each protein-GAG interaction is as specific as necessary. A single proteoglycan, even if it carries a single GAG chain, can bind multiple proteins, suggesting the possibility of functional diversity (Delehedde et al., 2001). The core proteins of cell surface proteoglycans may be transmembrane, such as syndecan (an ECM receptor), or GPI-anchored, such as glypican, or matrix proteins, such as perlecan. Perlecan is a heparin sulfate proteoglycan and is a major component of basement membranes (Oldberg et al., 1990; Yanagishita, 1993).

Many growth factors and morphogens (fibroblast growth factors, hepatocyte growth factor/scatter factor, members of the midkine family, and wnts), and matrix proteins (collagen, fibronectin, and laminin) interact with proteoglycans when they signal. The GAG-protein interactions serve to regulate the signal output of growth factor receptor tyrosine kinases and hence cell fate (Park et al., 2000). In addition GAGs coordinate stromal and epithelial development, and they are active participants in mediating cell-cell and cell-matrix interactions (Kresse and Schonherr, 2001; Lander et al., 1996).

An example of a functionally diverse GAG chain (that actually is not attached to protein) is hyaluronic acid/hyaluronan (HA). HA serves a variety of functions, including space filling, lubrication of joints, provision of a matrix through which cells can migrate, and intracellular signaling (Toole, 2001). HA is actively produced during wound healing and tissue repair to provide a framework for growth of blood vessels and fibroblasts (Toole et al., 2002). In addition the interaction of HA with the leukocyte receptor CD44 is important in tissue-specific homing by leukocytes (Turley et al., 2002), and overexpression of HA receptors has been correlated with tumorigenicity and tumor metastasis (Isacke and Yarwood, 2002; Toole, 2002).

Integrins

Integrins are the most extensively studied receptors that transmit ECM signals. The name integrin stems from the observation that integrins are the integrators of extracellular (ECM) and intracellular (cytoskeletal) signals (Boudreau and Jones, 1999; Giancotti and Ruoslahti, 1999; Hynes, 1987). Integrins are heterodimeric receptors of alpha and beta subunits that interact noncovalently to form transmembrane receptors. Currently 17 α and 8 β subunits have been identified. So far, at least 20 heterodimeric receptors have been identified, many of which are tissue specific (Schwartz and Ingber, 1994; Schwartz et al., 1995). Substrates of integrins are various ECM components including laminins, collagens, fibronectin, and vitronectin. A specific integrin heterodimer can have multiple substrates. For example, $\alpha v\beta 3$ receptor shows strong binding affinity for fibronectin, collagen, tenscin-C, thrombospondin, and fibrinogen, as well as vitronectin. A certain type of ECM molecule can interact with more than one kind of integrin heterodimer in a cell context dependent manner (Giancotti and Ruoslahti, 1999). For example, laminin binds $\alpha 3\beta 1$, $\alpha 6\beta 1$, and $\alpha 6\beta 4$ integrins with high affinity. The number of diseases that have been shown to have mutations in integrin subunits is limited so far. This could be partly because such mutations may be lethal. The most prominent defect identified so far is epidermolysis bullosa, a skin disease that is strongly associated with mutations in integrin beta 4 (ITGB4 gene) (Pulkkinen et al., 1998).

In addition to transmitting biochemical signals from ECM proteins, integrins are also involved in sensing and transducing mechanochemical signals (Alenghat and Ingber, 2002; Chen et al., 1997; Ingber, 2002). Integrin signaling has effects on cell adhesion, the cytoskeleton, proliferation, growth, differentiation, and apoptosis pathways. How these pathways are connected is discussed in some detail below, with emphasis on integrin-mediated signaling effects on cell cycle progression.

Non-integrin ECM Receptors

A number of transmembrane cell surface receptor families function in transmitting ECM mediated signals. Among them are syndecans and glypicans. In addition dystroglycan has recently been identified as an ECM signal transmitting receptor. Syndecans are expressed on cells that are dependent on adhesion for proliferation and their distribution is highly cell type and tissue specific. Syndecans are involved in formation

---

Figure 9.1. Integrin signaling exhibits dynamic reciprocity. **(A)** ECM-to-nucleus signaling. Integrins are activated and cluster either by crosslinking or occupancy of ECM molecules. Interaction of the cytoplasmic tail of integrins with talin, paxillin, and vinculin or plectin has two main consequences: the actin/keratin cytoskeleton is reorganized, and activation of focal adhesion kinase (FAK) or Shc initiates downstream signaling pathways, such as the Ras-Raf-MEK-ERK (MAPK cascade), the PI-3 kinase, and the JNK pathway. The clustered integrin/ECM and many downstream targets that accumulate at the adhesion site form a focal adhesion complex or a hemidesmosome. Downstream effects of the activation of the MAPK cascade, the PI-3 kinase, and the JNK pathway include transcriptional modulation of a number of genes including those involved in cell cycle regulation. **(B)** Nucleus-to-ECM signaling. The affinity of membrane-bound integrins for ECM can be altered by change in nuclear functions. Transcriptional changes result in altered expression of cell surface receptors such as integrins, as well as in the composition of ECM via regulation of ECM components and ECM-degrading enzymes (blue). These growth status regulated changes in the nucleus are initiated by the MAPK cascade, or G-protein, PtdIns(4,5)P2 and PKC. MAPKs and PKC signals can also affect the affinity of integrins for their ligands.

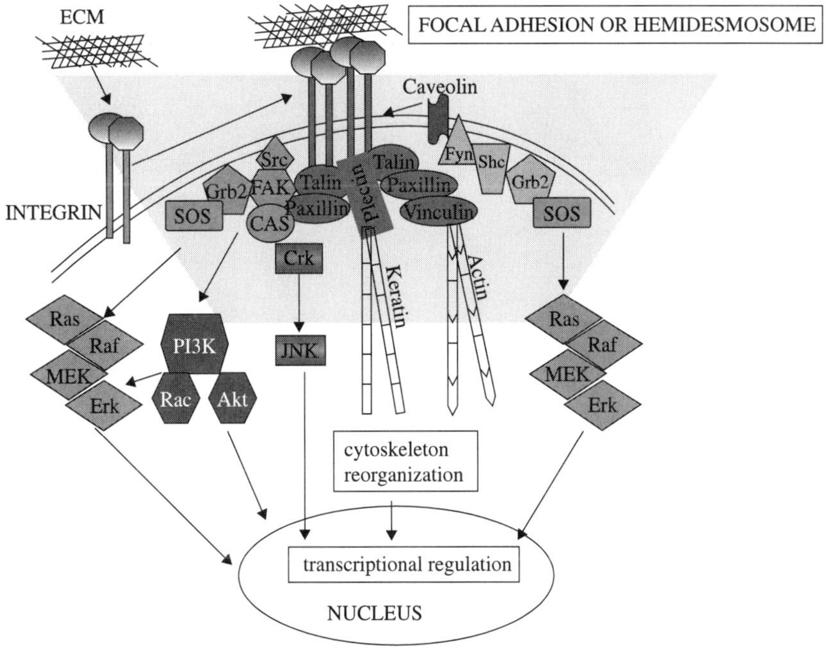
ECM
FOCAL ADHESION OR HEMIDESMOSOME
Caveolin
INTEGRIN
Src
Grb2
FAK
Talin
Plectin
Talin
Fyn
Shc
Grb2
SOS
CAS
Paxillin
Paxillin
Vinculin
SOS
Crk
Keratin
Actin
Ras
Raf
PI3K
JNK
Ras
Raf
MEK
Erk
Rac
Akt
MEK
Erk
cytoskeleton reorganization
transcriptional regulation
NUCLEUS

A

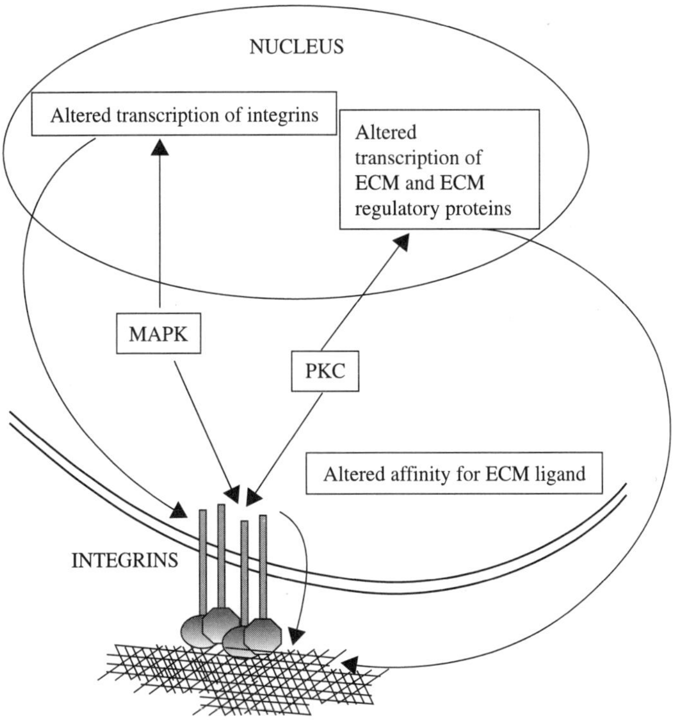
NUCLEUS
Altered transcription of integrins
Altered transcription of ECM and ECM regulatory proteins
MAPK
PKC
Altered affinity for ECM ligand
INTEGRINS
ECM

B

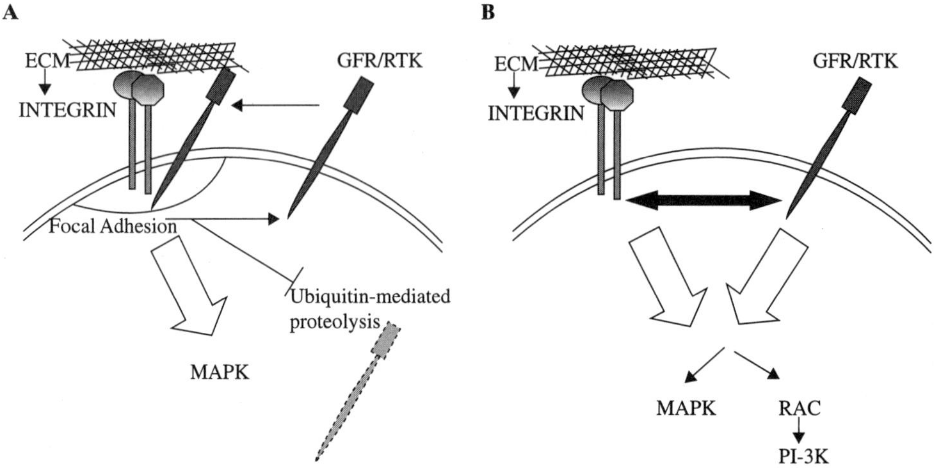

Figure 9.2. Cooperation of integrin and RTK/GFR signaling. **(A)** ECM activated integrin signaling results in focal adhesion complex formation followed by recruitment of growth factor receptors (GFRs). Such recruitment activates GFRs and their downstream targets such as the MAPK pathway. Alternatively, stability of GFRs can be increased by blocking their ubiquitin-mediated degradation upon attachment to ECM. **(B)** Sustained activation of the MAPK pathway, as well as activation of RAC-mediated signaling can be dependent on both integrin and GFR signaling.

of focal adhesion complexes, as are integrins (see below), and syndecans can also modulate integrin function (Carey, 1997; Couchman and Woods, 1999; Rapraeger, 2000; Zimmermann and David, 1999).

Glypicans are a distinct family of transmembrane heparan sulfate proteoglycans that contain a core protein anchored to the cell surface via a glycosyl phosphatidylinositol linkage (GPI-linked). Some members of the glypican family of integral membrane proteins are putative cell surface coreceptors for growth factors, ECM proteins, proteases, and antiproteases (Lander et al., 1996), and have been implicated in regulation of cell cycle progression (Delehedde et al., 2001).

Dystroglycan is both an ECM receptor and an organizer of the ECM (Ekblom et al., 1998; Henry et al., 2001). Originally isolated in skeletal muscle, it has now been shown to function in many tissues (Durbeej et al., 1998b). Dystroglycan substrates identified up to now include laminins, and proteoglycans. So far, only one dystroglycan gene has been identified. It has wide tissue distribution and multiple functions (Durbeej and Ekblom, 1997; Durbeej et al., 1998a,b; Muschler et al., 2002).

TISSUE-SPECIFIC EFFECTS OF ECM ON CELL PROLIFERATION IN MOUSE MODELS

Most knockout mouse models of ECM components and ECM receptors are not conditional; therefore the final phenotype observed reflects the changes throughout development. For this reason, interpretations of such models for tissue specificity of ECM effects on cell cycle progression need to be made with caution. Furthermore ability of another ECM component to substitute for the mutated gene can obscure interpretations. However, using conditional loss-of-function mutations in mice allows observation of tissue-specific developmental defects for certain ECM or receptor components. Tissue-specific conditional knockouts in the adult mouse are the best sources of information for determining cell type specificity of ECM or its receptors in terms of their roles in cell cycle control. Only a few examples of conditional knockouts exist for ECM and its receptors, namely for β1 integrin and dystroglycan (Brakebusch et al., 2000; Hirsch et al., 1996; Raghavan et al., 2000). An overview of ECM and ECM receptor components that have been knocked out in mice is shown in Table 9.2.

Examination of homozygous knockouts of ECM components and their receptors shows that loss of a single subunit is usually not embryonic lethal, suggesting that other BM components can substitute for function in most tissues. Nidogen-1 and nidogen-2 knockout mice demonstrate an extreme case of this. Targeted homozygous disruption of neither nidogen-1 nor nidogen-2 results in the disruption of BM. Since nidogen is necessary for the integrity of basement membranes, this suggests that nidogen-1 and nidogen-2 can substitute for each other's function with no or very little tissue specific requirements (Murshed et al., 2000; Schymeinsky et al., 2002). In other ECM knockouts, some tissues and cell types show gross developmental abnormalities, some of which are due to changes in the regulation of cell proliferation by ECM. For example, disruption of the alpha 3 chain of laminin 5 result in abnormalities in the survival of epithelial cells (especially in the skin) as well as extreme blistering of the skin, similar to a skin disease known as junctional epidermolysis bullosa-gravis in humans. LAMA3 was shown to be necessary for the proper formation and stabilization of hemidesmosomes in the epidermis, and the skin abnormalities could be attributed to the hemidesmosome defects (Ryan et al., 1999). A few knockouts do show embryonic lethality, suggesting that the mutated genes are essential in early development and not replaceable by similar components. For example, mice homozygous null for the gamma 1 chain of laminin (LAMC1), beta 1 integrin (ITGB1), dystroglycan (DAG1) show embryonic lethality (Smyth et al., 1999; Stephens et al., 1995; Williamson et al., 1997). Conditional knockouts of integrin beta 1 and dystroglycan have been produced for multiple tissues (Brakebusch et al., 2000; Cohn et al., 2002; Hirsch et al., 1996; Moore et al., 2002; Raghavan et al., 2000). Conditional knockouts are more informative about cell proliferation effects especially when the knockout event occurs either after completion of the development of the tissue or in later stages of development. For example, disrupting beta 1 integrin in skin, using a keratin 14 promoter that is turned on very late in development, results in failure of basement membrane assembly and maintenance, and impairment of epidermal proliferation (Raghavan et al., 2000).

TABLE 9.2. ECM and ECM Receptor Mutations in Mice

Gene	Homozygous Mutant Phenotype	Reference
Laminins		
LAMA2	Murine muscular dystrophy	Xu et al. (1994)
LAMA3	Neonatal lethality	Ryan et al. (1999)
LAMA4	Nervous system developmental defects	Patton et al. (2001)
LAMB2	Aberrant neuromuscular development	Noakes et al. (1995)
LAMB3	Junctional epidermolysis bullosa (skin blistering)	Kuster et al. (1997)
LAMC1	Embyonic lethality	Smyth et al. (1999)
Collagens		
COLA1	Dermal fibrosis, impaired uterine involution	Liu et al. (1995)
COLA2	Glomerulopathy (kidney)	Phillips et al. (2002)
COL3A1	Cardiovascular defects	Liu et al. (1997)
COL4A3	Alport syndorome (kidney)	Cosgrove et al. (1996)
COL5A2	Spinal deformities, skin and eye abnormalities	Andrikopoulos et al. (1995)
COL10A1	Defects in bone development and hematopoiesis	Gress and Jacenko (2000)
COL11A1	Skeletal development defects	Li et al. (1995)
COL11A2	Hearing loss	McGuirt et al. (1999)
COL15A1	Skeletal myopathy, cardiovascular defects	Eklund et al. (2001)
Nidogens		
NID1	No detectable defects	Murshed et al. (2000)
NID2	No detectable defects	Schymeinsky et al. (2002)
Proteoglycans		
HSPG2	Cartilage defects, cardiac and brain BM defects	Costell et al. (1999)
AGC	Bone and cartilage phenotypes	Wai et al. (1998)
FN1	Mesoderm, neural tube, vascular development defects	George et al. (1993)
Integrins		
ITGA1	Derived MEFs have adhesion defects in vitro	Gardner et al. (1996)
ITGA2	Kidney and lung development defects	Kreidberg et al. (1996)
ITGA4	Placental and cardiac development defects	Yang et al. (1995)
ITGA5	Embryonic mesodermal defects	Yang et al. (1993)
ITGA6	Epidermolysis bullosa (skin blisters), neonatal death	Georges-Labouesse et al. (1996)
ITGA7	Homozygous null shows muscle and tendon organization defects, including a novel form of muscular dystropy	Mayer et al. (1997), Miosge et al. (1999)
ITGA8	Homozygous null has defects in kidney morphogenesis	Muller et al. (1997)
ITGA9	Homozygous mutants have fatal bilateral chylothorax (lymphatic system and thoracic duct defects)	Huang et al. (2000)
ITGAE	Cutaneous inflammatory disorder	Schon et al. (2000)
ITGAM	Neutrophil adhesion, migration, apoptosis defects	Coxon et al. (1996), Ding et al. (1999)
ITGAV	Lethality preceded by aberrant vasculogenesis, angiogenesis, and organogenesis	Bader et al. (1998)
ITGB1	Peri-implantation lethality and inner cell mass failure Their ES cells have migration and adhesion defects	Fassler et al. (1995), Stephens et al. (1995)

TABLE 9.2. *Continued*

Gene	Homozygous Mutant Phenotype	Reference
ITGB1	Conditional null in skin has epidermal proliferation, basement membrane formation, and hair follicle invagination defects	Brakebusch et al. (2000), Raghavan et al. (2000)
ITGB1	Chimeric mice, with null hematopoietic cells display impaired migration of hematopoietic stem cells	Hirsch et al. (1996)
ITGB3	Homozygous mutants of the cytoplasmic domain have impaired platelet function	Law et al. (1999)
ITGB3	Osteoclast abnormalities	McHugh et al. (2000)
ITGB3	Heterozygous mutants have placental defects and reduced survival	Hodivala-Dilke et al. (1999)
ITGB4	Homozygous nulls have defects in hemidesmosome formation, cell adhesion, and cell survival, followed by lethality soon after birth	Dowling et al. (1996), Frei et al. (1999)
ITGB4	Homozygous deletion of the cytoplasmic domain results in cell cycle and adhesion defects, and lethality postbirth	Murgia et al. (1998)
ITGB6	Juvenile baldness and asthma-like syndromes due to infiltration of skin and lung epithelium by macrophages and lymphocytes	Huang et al. (1996)
ITGB7	Defects in gut-associated lymphoid tissue formation possibly due to defective lymphocyte attachment	Wagner et al. (1996)
ITGA8	Homozygous nulls die soon after birth due to kidney defects, with ear neuroepithelial deformities	Littlewood Evans and Muller (2000)
Syndecans		
SDC1	Decreased Wnt-1 induced mammary tumorigenesis	Alexander et al. (2000)
SDC3	Feeding behavior defects due to altered hypothalamic functions (predominantly neural)	Reizes et al. (2001)
SDC4	Homozygous mutants display delayed healing of skin wounds and defective angiogenesis	Echtermeyer et al. (2001)
Glypicans		
GPC3	Defects in limb patterning, skeletal development, and kidney branching morphogenesis, renal cystic dysplasia, ventral wall defects	Cano-Gauci et al. (1999), Grisaru et al. (2001), Paine-Saunders et al. (2000)
Dystroglycan		
DAG1	Early embryonic lethality associated with defects in the extra-embryonic basement membrane	Williamson et al. (1997)
DAG1	Conditional mutants with homozygous deletion in the brain display congenital muscular dystrophy syndromes	Moore et al. (2002)
DAG1	Conditional mutants with homozygous deletion in skeletal muscle display muscle regeneration defects	Cohn et al. (2002)

Note: The *Homozygous Mutant Phenotype* column shows the phenotype of mice carrying a targeted homozygous null mutation of the indicated gene, unless otherwise indicated to be a conditional mutation in a certain tissue or a heterozygous mutation.

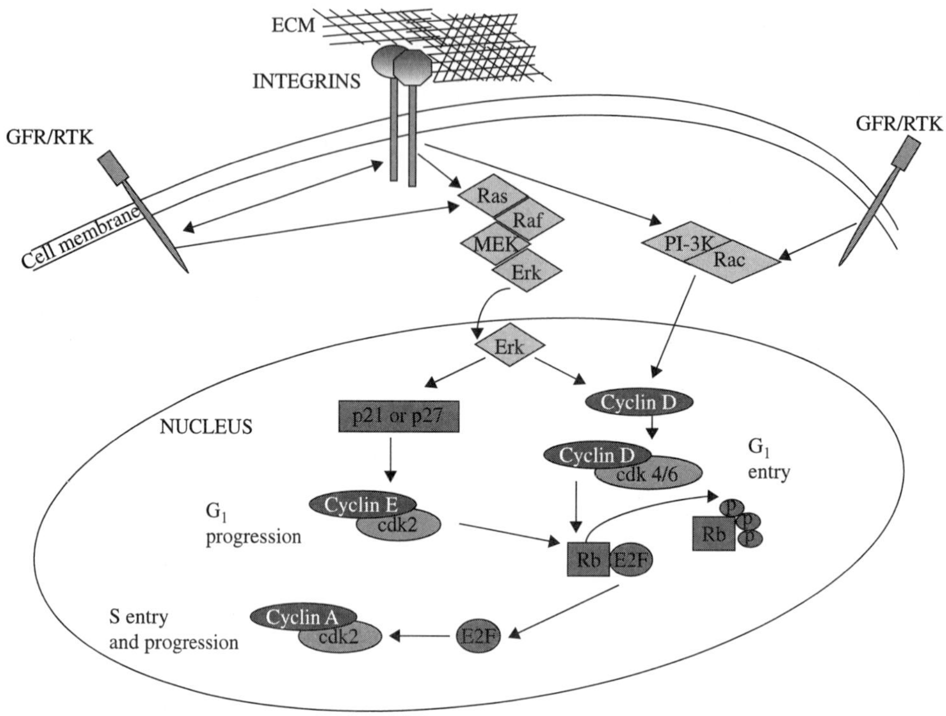

Figure 9.3. ECM-initiated signals regulate G_1 and S phase entry and progression. Sustained activation of Erk via integrin-mediated and/or RTK initiated activation of the MAPK pathway results in translocation of Erk into the nucleus. Erk regulates cyclin D1 expression positively and p21 and p27 negatively. Activation of cyclin D1-cdk4/6 complexes allow G_1 entry. Progression through G_1 is controlled by cyclin E-cdk2. Cyclin E-cdk2 activation requires downregulation of p21 or p27, which is accomplished by Erk-mediated signaling as well. Activation of cyclin E-cdk2 and of cyclin D-cdk4/6 free up the E2F transcription factor by phosphorylating Rb. E2F is needed for transcription of cyclin A. Thus S phase entry and progression that is dependent on cyclin A-cdk2 activation is also regulated by ECM, since downstream targets of cyclin D1 activation and p21 or p27 inactivation include cyclin A. In addition to the MAPK pathway, PI-3K or Rac can also upregulate cyclin D1 transcription through activation of JNK via Akt (a downstream target of PI-3K) or via JNK (a downstream target of Rac).

EFFECTS OF CELL-ECM INTERACTIONS ON CELL CYCLE PROGRESSION IN CULTURED CELLS

Overview

For most cells derived from multicellular organisms, proliferation ex vivo requires simulating at least part of the cellular microenvironment within the tissue. Many nonmalignant cells derived from solid, multicellular organs require adhesion to a substratum in order to proliferate. ECM components, such as fibronectin, laminin, and collagen, have been used as thin monolayer coats onto which cells can attach and grow (Ashkenas et al., 1996). Although plastic that is treated to simulate a charge distribution that allows cellular attachment has been widely used, presence of ECM provides additional biochemical signals that can promote or inhibit cell proliferation (Cambier et al., 2000; Giancotti and Ruoslahti, 1999; Schwartz, 2001). The ECM effect on cell cycle is dependent on cell type, cell-surface receptor composition, as well as the nature of the ECM molecules.

A classic example of positive regulation of growth by ECM, which was later followed by many other examples, is the demonstration that mouse bone marrow cells, when cultured on a complex ECM derived from marrow, exhibit a dramatically increased ability to proliferate, compared to the controls (Campbell et al., 1985). Negative regulation of growth by ECM in nontumorigenic adherent cells is perhaps best demonstrated by use of three-dimensional (3D) cultures of laminin rich reconstituted basement membrane (lrBM). For example, mammary epithelial cells plated in 3D lrBM proliferate for only a limited number of divisions, arrest growth, and form differentiated colonies resembling tissue structures in vivo, in both shape and function (Fig. 9.4) (Petersen et al., 1992), and specific integrins such as $\alpha v\beta 8$ exert negative growth control in epithelial cells (Cambier et al., 2000). Loss of dependence on adherence or acquisition of anchorage-independent growth (as measured by growth in soft agar) usually accompanies transformation to malignancy. Interestingly, tumorigenic counterparts of cell types that differentiate within 3D ECM cultures exhibit loss of growth regulation as well as their ability to differentiate in response to ECM-mediated signals in the same assay.

Control of Cell Proliferation by Integrin Signaling

ECM-mediated signals that are transmitted via integrins are generally involved in controlling events in G_1 phase progression. Of the G_1 phase cyclins and CDKs (cyclin D, cyclin E, CDK2 and CDK4 and their inhibitors), integrin signaling has the most significant effect on the induction of cyclin D1 and repression of the CDK inhibitors p21 and p27 (Assoian 1997; Roovers and Assoian 2000) (Fig. 9.3). Signaling through integrins is bi-directional (Bissell et al., 1982; Chen et al., 1994). ECM signals are transmitted through integrins to their downstream effectors that regulate expression and activity of the cell cycle regulators in the nucleus. For the rest of this review we call these events ECM-to-nucleus signaling. Growth signals from the nucleus reciprocally regulate the extracellular levels and binding potential of ECM molecules and integrins by regulating their structure and function. We refer to such events as nucleus-to-ECM signaling. While we have focused on integrins as the main transducers of such bi-directional flow of information, or dynamic reciprocity, between the nucleus and cell surface receptors, this phenomenon is also observed for growth factor, cytokine, and other receptors.

ECM-to-Nucleus Signaling. In response to ECM signaling, integrins and many associated proteins cluster and associate with the cellular cytoskeleton to promote filament assembly or disassembly, followed by an intricate series of signaling events that result in changes in a number

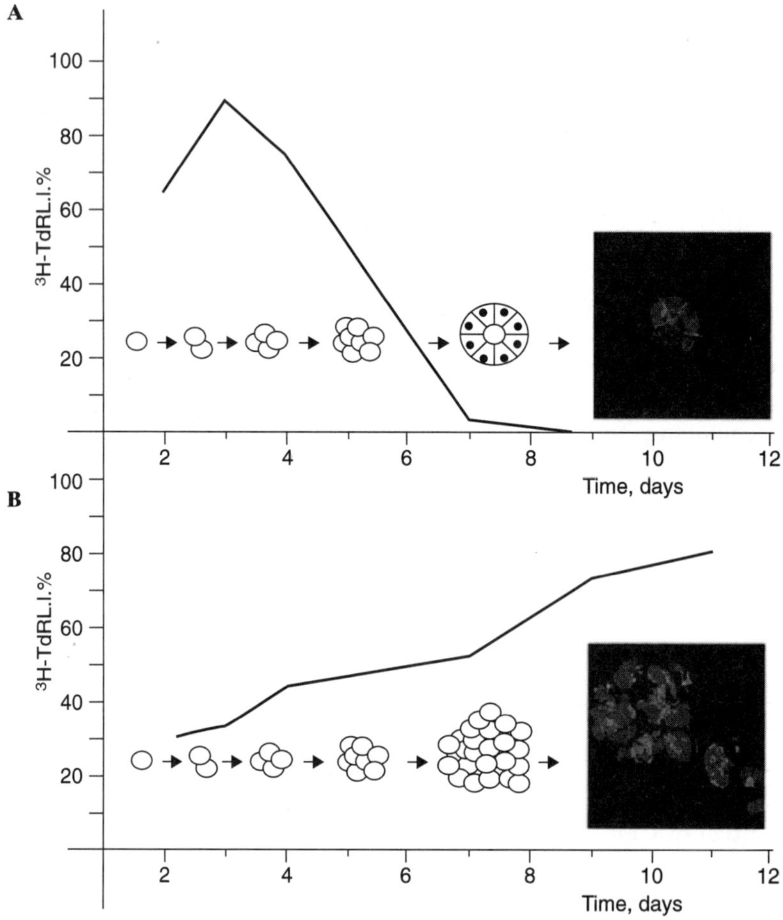

Figure 9.4. Nontumorigenic and tumorigenic mammary epithelial cells differ in their ability to proliferate and differentiate in lrBM. Tritiated tymidine incorporation was measured as a function of time in primary "normal" or nontumorigenic mammary epithelial cell lines **(A)**, and in tumorigenic lines **(B)**. Initially both normal and tumorigenic cells are able to divide. However, normal cells growth arrest and differentiate **(A)**, but tumorigenic cells continue proliferating **(B)** (Reprinted with modifications by permission of Cancer Biology from Weaver et al. 1995). (See color insert.)

of nuclear responses including transcription of cell cycle regulators (Fig. 9.1A). The overview below describes the main components of such signaling events, each of which is a potential point of regulation for cell cycle progression.

Reorganization of actin or keratin filaments into large stress fibers produces signals that feed back into the integrin clusters and causes enhanced integrin clustering and increased binding to the ECM. This region of clustering and signaling is called either a focal adhesion if actin reorganization is involved or a hemidesmosome if the keratin cytoskeleton is involved as in the case of $\alpha6\beta4$ signaling (Jones et al., 1998; Nievers et al., 1999). Although for the rest of this section we focus on focal adhesion-mediated signaling, similar principles also apply to hemidesmosome-mediated signaling. Integrin clustering can further be affected by lateral association of integrins with other membrane proteins, such as caveolin-1 and urokinase plasminogen activator (uPAR), an ECM degrading enzyme (Wary et al., 1996; Wei et al., 1999). Integrin clustering activates some protein tyrosine kinases, including focal adhesion kinase (FAK), Src-family kinases, Abl, as well as a serine-threonine kinase called integrin-linked kinase (ILK) (Giancotti and Ruoslahti, 1999; Wary et al., 1996, 1999). First, interaction of the cytoplasmic tail of integrins with cytoskeletal proteins such as talin and paxillin results in recruitment of FAK to the nascent focal adhesions (Chen et al., 1995; Lewis and Schwartz, 1995; Miyamoto et al., 1995; Parsons et al., 1994; Schaller et al., 1995). This is followed by autophosphorylation of FAK, creating a binding site for Src homology 2 (SH2) domain proteins (Schaller et al., 1994; Schlaepfer et al., 1994). The SH2 domain kinase then phosphorylates a number of focal adhesion proteins, including cytoskeletal proteins such as paxillin and tensin, docking proteins such as $p130^{CAS}$ that recruits adapter proteins such as Crk and Nck (Vuori et al., 1996). FAK can activate phosphoinositide 3-OH kinase (PI 3-kinase) either through activation of Src or directly (Chen et al., 1996). The interaction of Src and FAK can be reciprocal; that is, Src can phosphorylate FAK at the same tyrosine that is autophosphorylated by FAK. This creates a binding site for the adapter complex containing Grb2 and Ras guanosine 5′-triphosphate exchange factor mSOS (Schlaepfer et al., 1994). These signals are transmitted to the MAPK pathway through activation cascade of Ras, Raf, MEK, and ERK sequentially. The MAPKs are also downstream targets of growth factor receptor tyrosine kinases, providing a link between integrin and growth factor receptor signaling. The downstream effects of MAPK signaling include controlling cyclin D1 expression (needed for cell cycle entry), as well as controlling integrin and ECM molecule expression in a feedback loop (nucleus-to-ECM signaling).

Another feedback loop is observed between mitotic signals and FAK. FAK is further phosphorylated on serine when cells enter mitosis. This results in dissociation of FAK from Src and $p130^{CAS}$ (Yamakita et al., 1999). Loosening of focal contacts may allow the cells to decrease adhesion to the substratum and help them divide and spread out. A parallel pathway of integrin-mediated activation starts with activation of a Src family kinase, Fyn, by some integrins. In this pathway a membrane-bound receptor (e.g., caveolin-1) acts as an adapter linking the integrin alpha subunit to Fyn. When ECM binding activates integrins, Fyn becomes activated and its Src homology 3 (SH3) domain interacts with Shc, which in turn phosphorylates it on tyrosine. This targets Shc for the adapter GrbB2-mSOS complex. The SOS complex then transduces signals to the MAPK pathways through Ras, Raf, MEK, and ERK (Schlaepfer et al., 1994), which in turn induce changes in nuclear events that result in enhanced cell proliferation.

Both activation of FAK and Shc (and perhaps other yet unidentified molecules) can contribute to the Ras-ERK MAPK signaling cascade (Howe et al., 2002; Hughes et al., 1997). In some cell types Shc is responsible for the high-level activation of ERK, following cell adhesion to a substratum. FAK most likely has a more significant role in sustaining the activation signal (Pozzi et al., 1998; Wary et al., 1999). Both activation of FAK and Shc can be regulated positively and negatively by tyrosine phosphatases. Some examples include receptor-type protein phosphatase alpha and cytosolic phosphatase SHP-2, which remove the negative regulatory phosphate in Src kinases and therefore amplify both FAK and Shc signaling (Oh et al., 1999; Su et al., 1999). PTP-PEST and PTP-1B are examples of cytoplasmic tyrosine phosphatases that dephosphorylate $p130^{CAS}$; and this inhibits some of the downstream signals of FAK (Garton et al., 1997; Garton and Tonks, 1999; Liu et al., 1998). The phosphatases themselves can be regulated by signals that are integrin-ECM activated. For example, PTP-PEST is anchored to the endoplasmic reticulum and needs to be recruited to the focal adhesions. This occurs when integrin-mediated adhesion activates a protease called calpain, which in turn cleaves the PTP-PEST extension and anchors it to the endoplasmic reticulum, allowing the phosphatase to localize to focal adhesions (Rock et al., 1997). Since phosphatases can have multiple specificities, they can affect cell proliferation by regulating two related pathways. For example, PTEN (a tumor-suppressor gene-encoded protein) dephophorylates PI 3-kinase generated inositol lipids (Li et al., 1997; Maehama and Dixon, 1998; Myers et al., 1998; Stambolic et al., 1998). PTEN can also dephosphorylate FAK and Shc and suppress integrin signaling. Inhibition of PI 3-kinase results in downmodulation of Ras Raf Erk (MAPK) signaling (Gu et al., 1998; Tamura et al., 1998) (Fig. 9.1A). This, in turn, can downmodulate active integrin expression. Therefore a phosphatase like PTEN can inhibit focal adhesion formation via multiple mechanisms. Inhibition of adhesion, as well as down-regulation of the PI 3-kinase survival pathway, can cooperate to reduce the cell's ability to proliferate.

Nucleus-to-ECM Signaling. This can manifest itself in multiple forms (Fig. 9.1B). In many instances, proliferation-related signals that alter nuclear functions result in altered functional expression of cell surface receptors or of the ECM components produced by the cell, thereby changing both the ECM composition surrounding the cell and its ability to respond to the changed microenvironment.

For example, the MAPK-dependent differentiation and growth arrest of PC12 cells is accompanied by up-regulation of $\alpha 1\beta 2$ integrin expression, which helps maintain an elongated morphology via its collagen/laminin interactions (Rossino et al., 1990). A similar nucleus-to-ECM signaling event is observed in erythroleukemia cells that are dependent on MAPK for differentiation; MAPK-mediated growth arrest and differentiation in these cells is accompanied by up-regulation of receptor $\alpha IIb\beta 3$ (Woods et al., 2001). Furthermore expression of some ECM components is dependent on the MAPK-initiated differentiation. Therefore growth arrest related changes in the nucleus result both in increased ligand concentration for integrin binding and in expression of the integrin receptors (Jones et al., 1999).

An interesting mechanism by which the proliferation status of cells results in nuclear signals that manifest themselves in the form of altered cell-ECM interactions comes from the studies of the Ets transcription factor PEA3 (polyome enhancer activator 3). PEA3 is a direct downstream target of ERK and is essential for the transcriptional activation of many integrins, including αIIb and αv integrins. PEA3 also plays a significant role in the expression of proteases that degrade ECM, such as uPA (urokinase plasminogen activator), collagenases, and stromelysin-1 (Boudreau and Jones, 1999; Crawford and Matrisian, 1996; Wasylyk et al., 1991).

It is intriguing to note that MAPK activation, when it reflects a change in proliferation, can also lead to suppression of high affinity ligand binding of many integrins including β1, β3, and α6, by a mechanism that does not involve transcriptional or translational regulation of integrins or their ligands directly. In a screen for suppressors of integrin activation, H-Ras and its effector kinase Raf-1 were identified as negative regulators of integrin activation (Hughes et al., 1997). H-Ras inhibited the activation of integrins with three distinct alpha and beta subunit cytoplasmic domains. Suppression was independent of both transcription and protein synthesis. Furthermore suppression correlated with activation of the ERK MAP kinase pathway. Therefore the effect of ECM/adhesion on cell proliferation via integrin activation is tightly interconnected with proliferation signals that alter nuclear events. Such nuclear signals regulate integrin composition, affinity of integrins for their ligands, and the composition of the ligands as demonstrated by the transcriptional effects on ECM molecules themselves, as well as on the expression of ECM degrading enzymes.

Cooperation of Integrins and Growth Factor Receptors. Multiple mechanisms exist for crosstalk between integrin-mediated and growth factor/receptor tyrosine kinase (RTK) signaling that result in a cumulative effect on cell proliferation (Fig. 9.3). Following integrin activation by ECM, some RTKs are recruited to the focal adhesion complexes. In various cell types, epidermal growth factor receptor (EGFR), platelet-derived growth factor receptor (PDGFR), and fibroblast growth factor receptor (FGFR) have been found recruited to focal adhesion complexes (Miyamoto et al., 1996; Plopper and Ingber, 1993; Plopper et al., 1995). This recruitment results in phosphorylation and activation of growth factor receptors and their downstream targets (Miyamoto et al., 1996). Cell adhesion-mediated signals can also affect stability of growth factor receptors as exemplified by an increase in the number of PDGF receptors by blocking ubiquitin-mediated degradation of the receptor following adherence to a fibronectin substratum (Baron and Schwartz, 2000). The downstream MAPK cascade of Ras, Raf, MEK, and ERK can also be differentially affected by adhesion-mediated signaling. There are reported instances where activation of Raf and its downstream targets were promoted by adhesion, but Ras activation was not (Lin et al., 1997;

Roovers and Assoian, 2000). Furthermore integrin-mediated adhesion can be necessary for activation of MEK by Raf and maintenance of active ERK requires adhesion (Howe et al., 2002; Renshaw et al., 1999; Roovers and Assoian, 2000; Roovers et al., 1999). Pathways that are sensitive to both integrin and RTK-initiated signals include the Rac pathway. Growth factor induction of Rac is integrin dependent as demonstrated by observations that RTKs can activate Rac in suspended cells, but the activated Rac is not targeted to the plasma membrane and does not interact and activate its downstream targets such as PI 3-kinase and Akt (del Pozo et al., 2000; Khwaja et al., 1997).

It should be noted that the crosstalk between integrins and RTKs is not only dependent on the presence or absence of integrin-mediated adhesion but also on the nature of the substrata. Experiments with mammary epithelial cells grown either as monolayers on 2D plastic substratum or in 3D laminin-rich reconstituted basement membrane (lrBM) suggest that the composition of the substrata has profound consequences for integrin and growth factor receptor crosstalk. Tumorigenic mammary epithelial cells are unable to growth arrest in 3D lrBM but their normal counterparts stop growth and differentiate into in vivo like structures called acini (Petersen et al., 1992; Weaver et al., 1995) (Fig. 9.4). However, blocking cell-ECM interactions by inhibiting β1 integrin or inactivating EGF receptor signaling in tumorigenic cells in 3D lrBM results in reacquisition of an ability to growth arrest and differentiate (Wang et al., 1998; Weaver et al., 1997) (Fig. 9.5A and B). In 3D lrBM, inhibiting one receptor (β1 integrin or EGFR) results in down-regulation of the expression of the other (EGFR or β1 integrin), but this is not the case when cells are grown in monolayer 2D cultures (Fig. 9.5C). Conversely, both EGFR (ErbB1) and ErbB2 can induce increased proliferation in MCF-10A cells in monolayer, but in 3D lrBM, once the acini are formed, only ErbB2 results in uncontrolled proliferation that fills the acinar lumen with cells and ErbB1 has no effect (Muthuswamy et al., 2001). Taken together with the effect of the MAPK pathway on nucleus-to-ECM signaling that can determine integrin composition, integrin affinity, and ECM molecule composition (see above), these observations suggest that the role of context should be taken into account when interpreting signaling events.

Cell Cycle Regulatory Targets of Integrin-Mediated Signaling. Integrin-mediated signaling can induce either cell cycle progression or exit from the cell cycle followed by differentiation. Positive regulation of cell proliferation by integrins has been observed in cell types that are dependent on adhesion for cell cycle entry. The most common and most highly studied role of integrin signaling on promoting cell cycle entry is in promoting progression to G_1 and the G_1/S transition. The G_1 phase is also controlled by growth factor signals; therefore the crosstalk between integrins and growth factors can manifest itself in the form of co-regulation of cell cycle progression. Although overcoming G_2/M arrest can contribute to cell cycle progression, the role of ECM signaling on the G_2/M transition has not been given much attention. One reason for

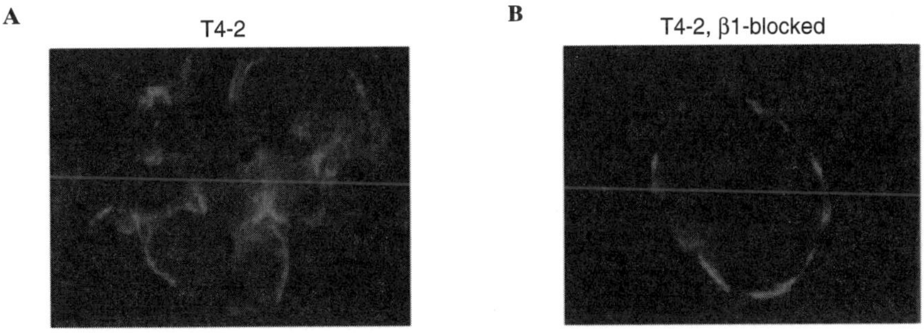

C

	3D lr BM				2D plastic		
	S1	fr4-2	T4-2 "treated"		T4-2	T4-2 "treated"	
Inhibitor Added:	–	–	β1 Integrin Inhibitor	EGFR Inhibitor	–	β1 Integrin Inhibitor	EGFR Inhibitor
β1 Integrin total levels	+	+++	+	+	+++	+++	+++
EGFR total levles	+	++++	+	+	++++	++++	++++
EGFR* activated	+	++++	+	+	++++	++++	+

Figure 9.5. Reverted tumorigenic mammary epithelial cells exhibit crosstalk between β1 integrin and EGFR in 3D lrBM but not in 2D monolayer cultures. Treatment of tumorigenic human mammary epithelial cells by inhibitory antibodies against β1 integrin or EGFR results in reversion to differentiated acinar phenotype as shown by re-establishment of basal α6β4 integrin expression. (Reprinted with permission of *Journal of Cell Biology* from Weaver et al., 1997) **(A)** α6 integrin is mislocalized in tumorigenic T4-2 cells (orange). (See color insert.) **(B)** α6 integrin is relocalized to the basal side after tumorigenic T4-2 cells are reverted to a differentiated, nondividing structure in lrBM by treating with inhibitory β1 integrin antibodies (orange). **(C)** Inhibiting β1 integrin in T4-2 cells results in downregulation of the level and activity of EGFR to levels similar to what is observed for non-tumorigenic S1 cells when T4 is grown in 3D lrBM but not in monolayer 2D cultures. Likewise inhibition of EGFR results in β1 integrin down-regulation in 3D but not in 2D. (Reprinted with permission of *Cancer Research* from Bissell et al., 1999).

this is that several studies have shown that only G_1 phase is subject to control by the ECM (and growth factors) for the cell types and the substrata studied (Fang et al., 1996; Sherr, 1994; Zhu et al., 1996). Therefore more thorough studies using a larger number of cell types and substrata are needed to determine if ECM signaling under normal growth conditions may control G_2/M transition. In addition, whether ECM plays a role in the G_2/M progression when cells are arrested in response to external or internal stimuli, such as DNA damaging agents, has not yet been determined.

The mammalian cell cycle is controlled by cyclins, cyclin-dependent kinases (cdks) that are activated by cyclins, cdk inhibitors, and cyclin activators. There are many cdks that function at different stages of the cell cycle. Progression through G_1 phase is modulated by the binding of cyclin D family (D1, D2, D3) to cdk4 or its homolog cdk6 (cdk4/6). Cyclin E binding to cdk2 is also involved in G_1 phase progression. Binding of cyclin A to cdk2 is required for progression through S phase. Activated Cdk4/6 and cdk2 are required for phosphorylation of the retinoblastoma protein (Rb), which activates transcription of genes regulated by E2F family of transcription factors. One of the functions of E2Fs is to regulate expression of cyclin A and therefore S phase entry (Weinberg, 1995). This finalizes the known points of cell cycle regulation initiated by adhesion or growth factors. Progression through G_2/M phase of the cell cycle is activated by binding of cyclin B to cdk1 (= cdc2) (Assoian, 1997; Heichman and Roberts, 1994; Sherr, 1994); however, G_2/M transition does not appear to be one of the main points of regulation initiated by ECM or growth factors. Progression through G_1 or G_2/M can be inhibited by either the INK4 family (p15, p16, p18, and p19) or the p21 family (p21cip1, p27kip1, and p53kip2) of cdk inhibitors. The INK4 inhibitors bind and inhibit cdk4/6, resulting in G_1 arrest. The p21 family of cdk inhibitors can bind to, and inhibit, either cdk4/6 or cdk2, preventing progression to G_1 or S phases of the cell cycle.

The effect of ECM-mediated cell cycle control is mainly manifested by regulation of G_1 and S phase progression for both positive and negative regulatory functions (Schwartz, 2001) (Fig. 9.3). In cases where integrin-mediated adhesion exhibits a positive effect on cell proliferation, activation of the Ras-Raf-MEK-ERK cascade by either ECM-mediated signaling or by growth factors results in increased cyclin D1 expression or down-regulation of the cdk inhibitors p21 and p27 (Roovers and Assoian, 2000). At least in some cases, ERK activity by itself is not sufficient for cyclin D1 expression, cells need to be attached to a substratum (Weber et al., 1997). For example, in suspended CCL39 cells, which require adhesion for proliferation, sustained ERK activity is not sufficient for cyclin D1 activation (Le Gall et al., 1998). It has been reported that translocation of ERK from the cytoplasm to the nucleus is also adhesion dependent (Aplin et al., 2001; Assoian and Schwartz, 2001). It is indicated that other downstream targets that are regulated by both integrin and growth factor receptor/RTK signaling, such as PI 3-kinase, play a role in the transcriptional activation and stability of cyclin D1 (Danilkovitch et al., 2000; Gille and Downward, 1999; Takuwa et al.,

1999). Cyclin D1 regulation by integrin signaling can also be manifested at the level of translation, at least in some cell types, in a manner that is dependent on activation of Rac, a signaling molecule that is translocated to the plasma membrane via events mediated by integrins (Huang et al., 1998; Schwartz and Assoian, 2001). Therefore, in addition to the MAPK signaling cascade, PI 3-kinase and Rac-mediated events can control the transcription and translation of cyclin D1.

Integrin signaling can also mediate cell cycle progression by regulating the cell cycle inhibitors p21 and p27. These cdk inhibitors are normally upregulated in early G_1 phase, but their levels drop as cells progress through G_1 to allow exit from G_1 and entry into the S phase. Induction of p21 in adherence-dependent cells is strongly regulated by ERK activity in early G_1 and the down-regulation of p21 in late G_1 is impaired when integrin signaling is inhibited. However, involvement of ERK with p21 appears to be indirect and through the small GTP binding protein Rho. Regulation of Rho activity by cytoskeletal changes, such as induction of actin stress fibers, has been well studied; however, the impact of integrin signaling on Rho remains somewhat controversial. One reason for the controversy is the difficulty of separating the effects of ECM on cell shape (known to induce Rac) from its effects on integrin signaling. A recent report indicates that activation of the downstream effector of integrin signaling, FAK, can negatively regulate Rho activity (Ren et al., 1999, 2000; Schwartz and Assoian, 2001). Effects of Rho-mediated regulation of cdk inhibitors is likely to turn out to be a cell type, cell surface receptor, and ECM-specific effect, as are many of the signaling cascades described above.

Dependence of cell cycle regulation by ECM on cell type, receptor composition, and substratum identity is best demonstrated by the observations that integrins can also regulate cell cycle progression negatively. For example, expression of αvβ8 results in increased expression of the cdk inhibitor p21 in a lung carcinoma cell line that has lost αvβ8 with subsequent growth arrest when grown on a monolayer of vitronectin, an αvβ8 ligand (Cambier et al., 2000). Many lung tumors have lost their αvβ8 and introduction of this integrin to tumorigenic lung epithelial cells results in loss of tumorigenicity in nude mice (Cambier et al., 2000).

CONCLUSION

A large number of studies on the effect of ECM on proliferation of cultured cells, data on tissue-specific diseases that are associated with specific ECM signaling components, and knockout mice studies provide clues that control of cell proliferation within a tissue is dependent on the nature of the ECM and its receptors in the tissues. A more comprehensive understanding of contribution of ECM to the tissue-specific regulation of cell cycle progression will require more systematic approaches, including conditional knockouts in different tissues as well as development of culture systems that better mimic physiological tissue environment.

ACKNOWLEDGMENTS

We thank the numerous investigators, including the past and present members of the Bissell laboratory, who contributed to the field of extracellular matrix research and apologize to those colleagues whose work we were unable to cite due to space limitations. The authors' work was supported by funds from the U.S. Department of Energy, Office of Biological and Environmental Research, the National Cancer Institute, and by an Innovator Award from the U.S. Department of Defense Breast Cancer Research Program (to M.J.B.), and by a Postdoctoral Fellowship from the California Breast Cancer Research Program (to A.R.).

REFERENCES

Alenghat FJ, Ingber DE (2002): Mechanotransduction: All signals point to cytoskeleton, matrix, and integrins. Sci STKE 2002:PE6.

Alexander CM, Reichsman F, Hinkes MT, et al (2000): Syndecan-1 is required for Wnt-1-induced mammary tumorigenesis in mice. Nat Genet 25:329–32.

Andrikopoulos K, Liu X, Keene DR, Jaenisch R, Ramirez F (1995): Targeted mutation in the col5a2 gene reveals a regulatory role for type V collagen during matrix assembly. Nat Genet 9:31–6.

Aplin AE, Stewart SA, Assoian RK, Juliano RL (2001): Integrin-mediated adhesion regulates ERK nuclear translocation and phosphorylation of Elk-1. J Cell Biol 153:273–82.

Ashkenas J, Muschler J, Bissell MJ (1996): The extracellular matrix in epithelial biology: Shared molecules and common themes in distant phyla. Dev Biol 180: 433–44.

Assoian RK (1997): Control of the G_1 phase cyclin-dependent kinases by mitogenic growth factors and the extracellular matrix. Cytokine Growth Factor Rev 8:165–70.

Assoian RK, Schwartz MA (2001): Coordinate signaling by integrins and receptor tyrosine kinases in the regulation of G_1 phase cell-cycle progression. Curr Opin Genet Dev 11:48–53.

Badenas C, Praga M, Tazon B, et al (2002): Mutations in the COL4A4 and COL4A3 genes cause familial benign hematuria. J Am Soc Nephrol 13: 1248–54.

Bader BL, Rayburn H, Crowley D, Hynes RO (1998): Extensive vasculogenesis, angiogenesis, and organogenesis precede lethality in mice lacking all alpha v integrins. Cell 95:507–19.

Baron V, Schwartz M (2000): Cell adhesion regulates ubiquitin-mediated degradation of the platelet-derived growth factor receptor beta. J Biol Chem 275: 39318–23.

Bissell MJ, Hall HG, Parry G (1982): How does the extracellular matrix direct gene expression? J Theor Biol 99:31–68.

Bissell MJ, Radisky D (2001): Putting tumours in context. Nat Rev Cancer 1: 46–54.

Bissell MJ, Ram TG (1989): Regulation of functional cytodifferentiation and histogenesis in mammary epithelial cells: Role of the extracellular matrix. Environ Health Perspect 80:61–70.

Bissell MJ, Weaver VM, Lelievre SA, Wang F, Petersen OW, Schmeichel KL (1999): Tissue structure, nuclear organization, and gene expression in normal and malignant breast. Cancer Res 59:1757s–63s; Discussion 1763s–64s.

Biswas S, Munier FL, Yardley J, et al (2001): Missense mutations in COL8A2, the gene encoding the alpha2 chain of type VIII collagen, cause two forms of corneal endothelial dystrophy. Hum Mol Genet 10:2415–23.

Boudreau N, Sympson CJ, Werb Z, Bissell MJ (1995): Suppression of ICE and apoptosis in mammary epithelial cells by extracellular matrix. Science 267: 891–3.

Boudreau NJ, Jones PL (1999): Extracellular matrix and integrin signalling: the shape of things to come. Biochem J 339 (Pt 3):481–8.

Brakebusch C, Grose R, Quondamatteo F, et al (2000): Skin and hair follicle integrity is crucially dependent on beta 1 integrin expression on keratinocytes. EMBO J 19:3990–4003.

Brown DM, Vandenburgh K, Kimura AE, Weingeist TA, Sheffield VC, Stone EM (1995): Novel frameshift mutations in the procollagen 2 gene (COL2A1) associated with Stickler syndrome (hereditary arthro-ophthalmopathy). Hum Mol Genet 4:141–2.

Byers PH, Duvic M, Atkinson M, et al (1997): Ehlers-Danlos syndrome type VIIA and VIIB result from splice-junction mutations or genomic deletions that involve exon 6 in the COL1A1 and COL1A2 genes of type I collagen. Am J Med Genet 72:94–105.

Cambier S, Mu DZ, O'Connell D, et al (2000): A role for the integrin alphav-beta8 in the negative regulation of epithelial cell growth. Cancer Res 60: 7084–93.

Campbell A, Wicha MS, Long M (1985): Extracellular matrix promotes the growth and differentiation of murine hematopoietic cells in vitro. J Clin Invest 75:2085–90.

Cano-Gauci DF, Song HH, Yang H, et al (1999): Glypican-3-deficient mice exhibit developmental overgrowth and some of the abnormalities typical of Simpson-Golabi-Behmel syndrome. J Cell Biol 146:255–64.

Carey DJ (1997): Syndecans: Multifunctional cell-surface co-receptors. Biochem J 327 (Pt 1):1–16.

Chen CS, Mrksich M, Huang S, Whitesides GM, Ingber DE (1997): Geometric control of cell life and death. Science 276:1425–8.

Chen HC, Appeddu PA, Isoda H, Guan JL (1996): Phosphorylation of tyrosine 397 in focal adhesion kinase is required for binding phosphatidylinositol 3-kinase. J Biol Chem 271:26329–34.

Chen HC, Appeddu PA, Parsons JT, Hildebrand JD, Schaller MD, Guan JL (1995): Interaction of focal adhesion kinase with cytoskeletal protein talin. J Biol Chem 270:16995–9.

Chen YP, O'Toole TE, Shipley T, et al (1994): "Inside-out" signal transduction inhibited by isolated integrin cytoplasmic domains. J Biol Chem 269:18307–10.

Cheng YS, Champliaud MF, Burgeson RE, Marinkovich MP, Yurchenco PD (1997): Self-assembly of laminin isoforms. J Biol Chem 272:31525–32.

Cohn RD, Henry MD, Michele DE, et al (2002): Disruption of DAG1 in differentiated skeletal muscle reveals a role for dystroglycan in muscle regeneration. Cell 110:639–48.

Cosgrove D, Meehan DT, Grunkemeyer JA, et al (1996): Collagen COL4A3 knockout: A mouse model for autosomal Alport syndrome. Genes Dev 10: 2981–92.

Costell M, Gustafsson E, Aszodi A, et al (1999): Perlecan maintains the integrity of cartilage and some basement membranes. J Cell Biol 147:1109–22.

Couchman JR, Woods A (1999): Syndecan-4 and integrins: Combinatorial signaling in cell adhesion. J Cell Sci 112 (Pt 20):3415–20.

Coxon A, Rieu P, Barkalow FJ, et al (1996): A novel role for the beta 2 integrin CD11b/CD18 in neutrophil apoptosis: A homeostatic mechanism in inflammation. Immunity 5:653–66.

Crawford HC, Matrisian LM (1996): Mechanisms controlling the transcription of matrix metalloproteinase genes in normal and neoplastic cells. Enzyme Protein 49:20–37.

Cremer MA, Rosloniec EF, Kang AH (1998): The cartilage collagens: a review of their structure, organization, and role in the pathogenesis of experimental arthritis in animals and in human rheumatic disease. J Mol Med 76:275–88.

Dalgleish R, Williams G, Hawkins JR (1986): Length polymorphism in the pro alpha 2(I) collagen gene: An alternative explanation in a case of Marfan syndrome. Hum Genet 73:91–2.

Damsky CH, Werb Z (1992): Signal transduction by integrin receptors for extracellular matrix: cooperative processing of extracellular information. Curr Opin Cell Biol 4:772–81.

Danen EH, Yamada KM (2001): Fibronectin, integrins, and growth control. J Cell Physiol 189:1–13.

Danilkovitch A, Donley S, Skeel A, Leonard EJ (2000): Two independent signaling pathways mediate the antiapoptotic action of macrophage-stimulating protein on epithelial cells. Mol Cell Biol 20:2218–27.

del Pozo MA, Price LS, Alderson NB, Ren XD, Schwartz MA (2000): Adhesion to the extracellular matrix regulates the coupling of the small GTPase Rac to its effector PAK. EMBO J 19:2008–14.

Delehedde M, Lyon M, Sergeant N, Rahmoune H, Fernig DG (2001): Proteoglycans: Pericellular and cell surface multireceptors that integrate external stimuli in the mammary gland. J Mammary Gland Biol Neoplasia 6:253–73.

Ding ZM, Babensee JE, Simon SI, et al (1999): Relative contribution of LFA-1 and Mac-1 to neutrophil adhesion and migration. J Immunol 163:5029–38.

Dowling J, Yu QC, Fuchs E (1996): Beta4 integrin is required for hemidesmosome formation, cell adhesion and cell survival. J Cell Biol 134:559–72.

Durbeej M, Ekblom P (1997): Dystroglycan and laminins: Glycoconjugates involved in branching epithelial morphogenesis. Expr Lung Res 23:109–18.

Durbeej M, Henry MD, Campbell KP (1998a): Dystroglycan in development and disease. Curr Opin Cell Biol 10:594–601.

Durbeej M, Henry MD, Ferletta M, Campbell KP, Ekblom P (1998b): Distribution of dystroglycan in normal adult mouse tissues. J Histochem Cytochem 46:449–57.

Dziadek M (1995): Role of laminin-nidogen complexes in basement membrane formation during embryonic development. Experientia 51:901–13.

Echtermeyer F, Streit M, Wilcox-Adelman S, et al (2001): Delayed wound repair and impaired angiogenesis in mice lacking syndecan-4. J Clin Invest 107: R9–R14.

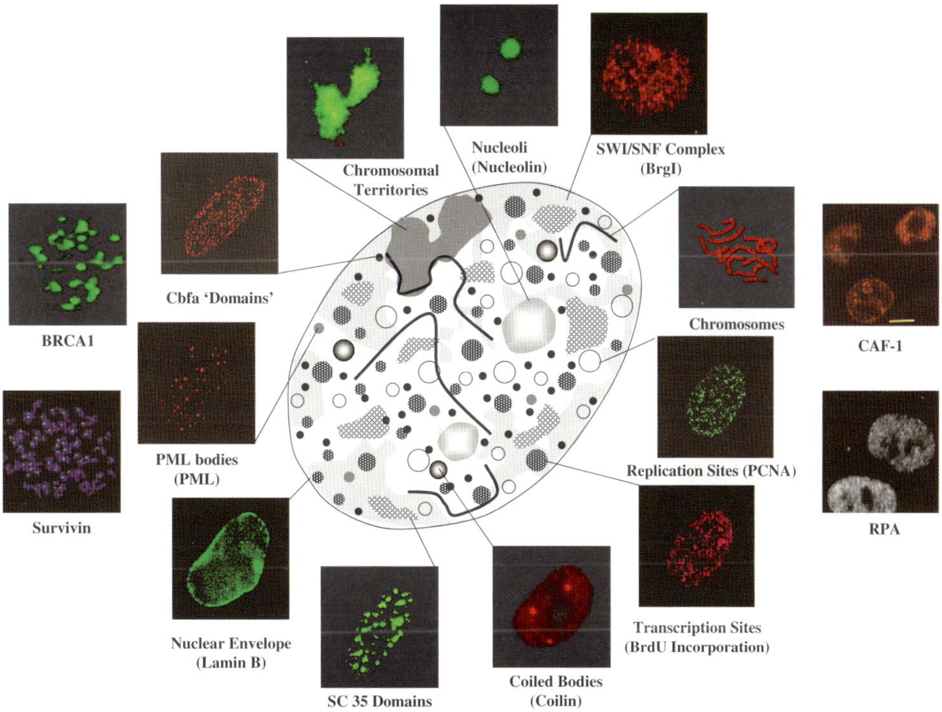

Figure 2.1. Subnuclear compartmentalization of nucleic acids and regulatory proteins into specialized domains. See text for full caption.

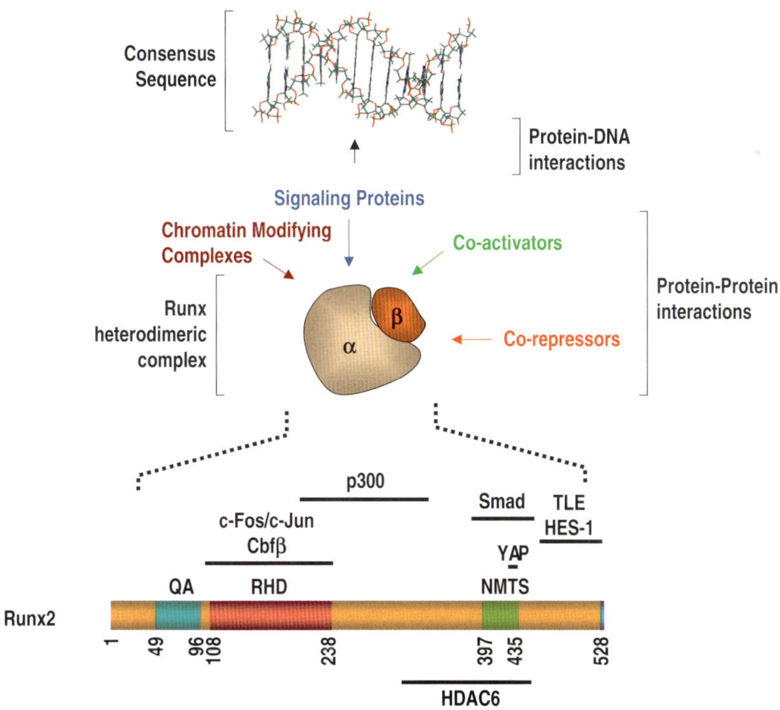

Figure 2.3. Scaffolding nuclear proteins: A mechanism of specificity in gene regulation. See text for full caption.

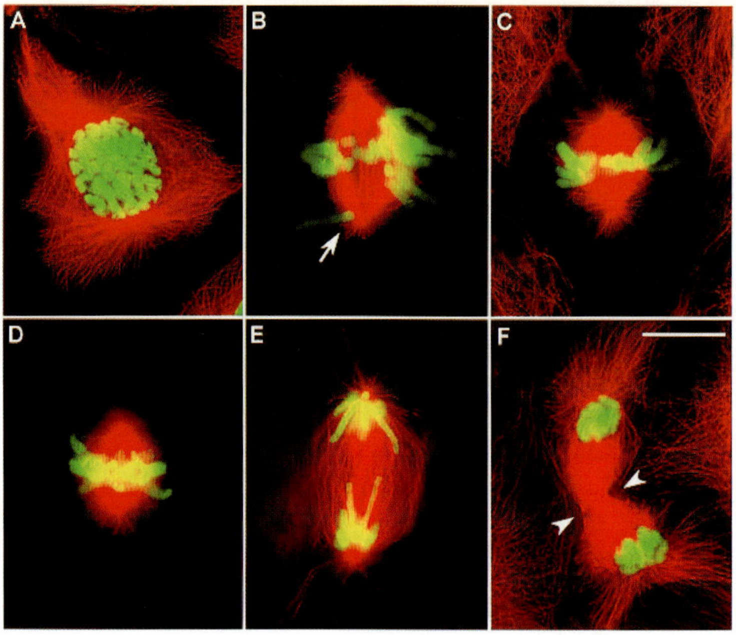

Figure 6.2. Gallery of fluorescent micrographs depicting glutaraldehyde-fixed and lysed PtK_1 cells in various stages of mitosis. See text for full caption.

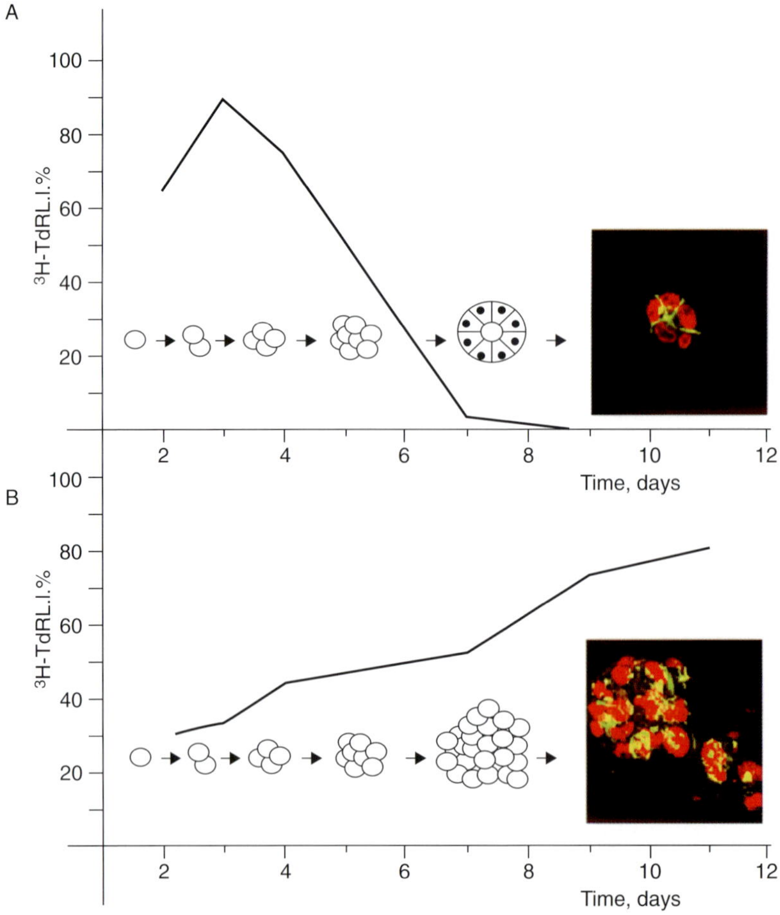

Figure 9.4. Nontumorigenic and tumorigenic mammary epithelial cells differ in their ability to proliferate and differentiate in lrBM. See text for full caption.

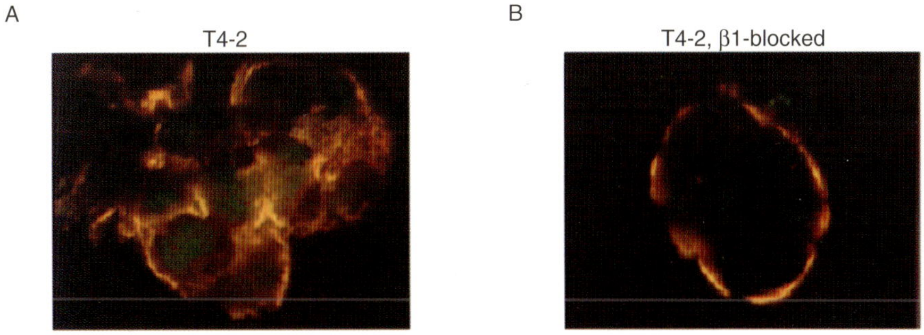

Figure 9.5. Reverted tumorigenic mammary epithelial cells exhibit crosstalk between β1 integrin and EGFR in 3D lrBM but not in 2D monolayer cultures. See text for full caption.

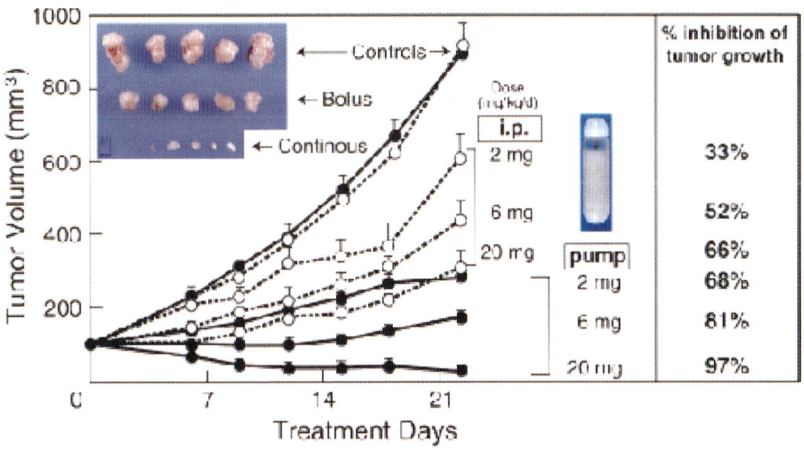

Figure 10.1. Continuous versus bolus administration of human endostatin to SCID mice bearing human pancreatic cancer that is $p53^{-/-}$. See text for full caption.

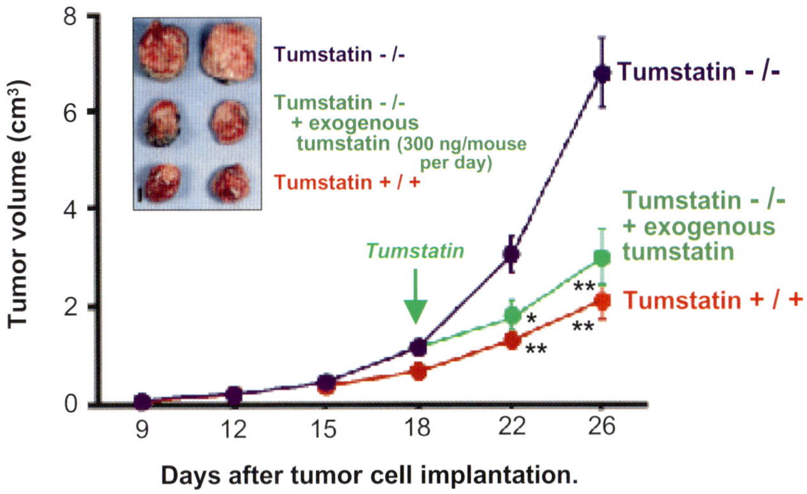

Figure 10.4. In mice depleted of the endogenous angiogenesis inhibitor tumstatin, tumors grow 300% to 400% more rapidly than in wild-type mice. See text for full caption.

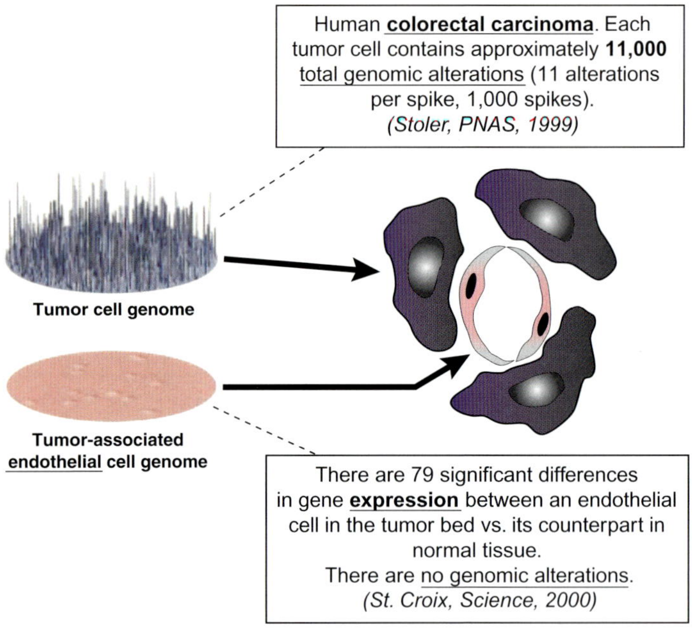

Figure 10.8. Tumor cells are genetically unstable and contain thousands of genomic alterations. See text for full caption.

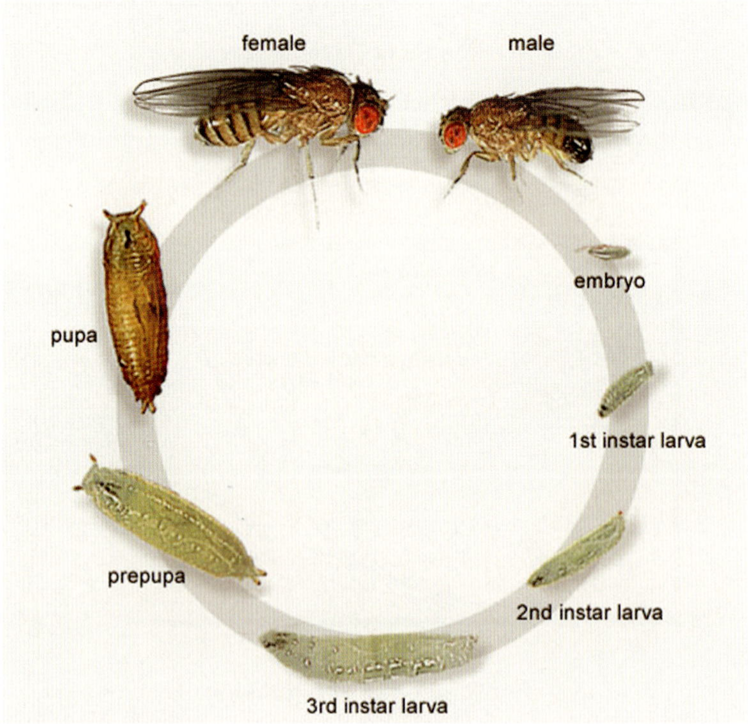

Figure 11.1. Developmental stages of the fruit fly *Drosophila melanogaster*. See text for full caption.

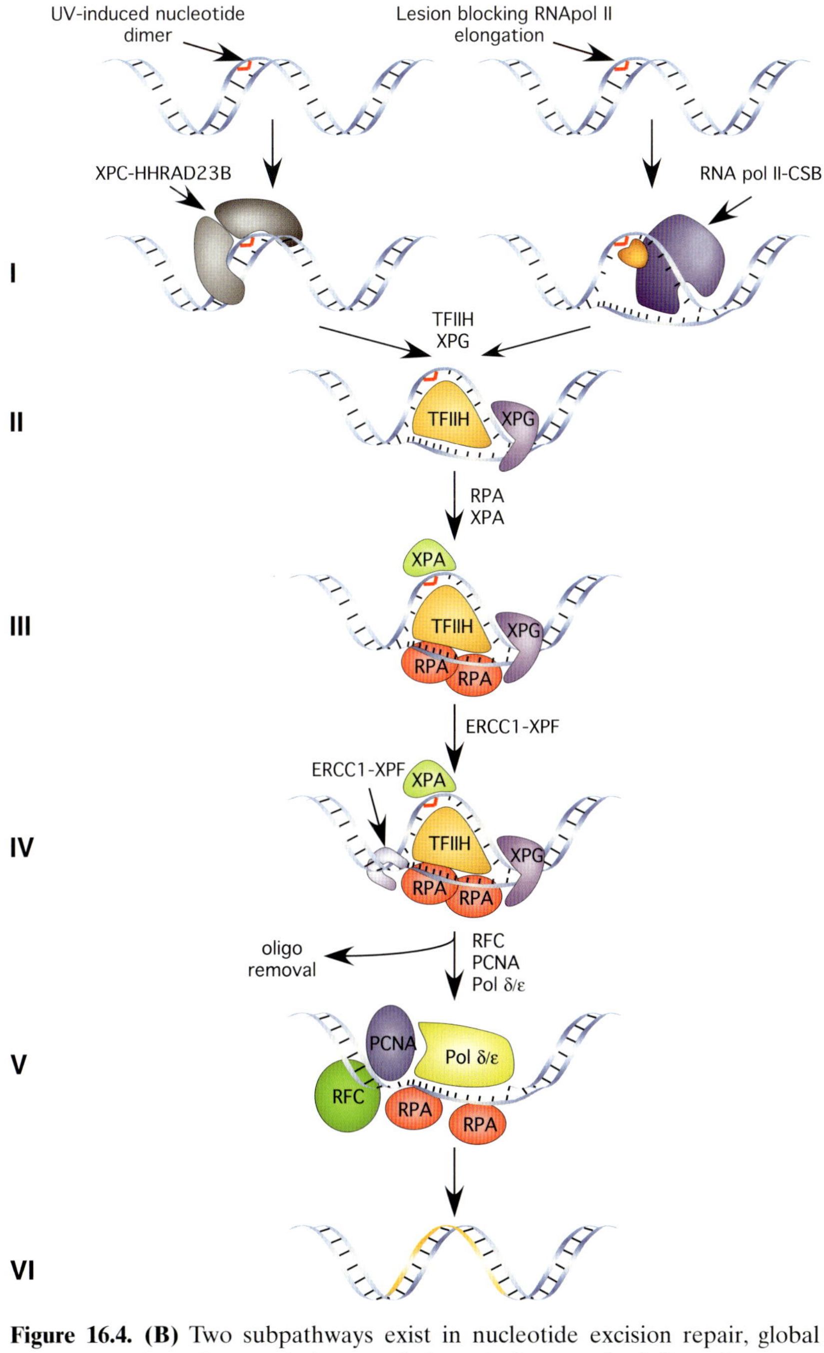

Figure 16.4. (B) Two subpathways exist in nucleotide excision repair, global genome repair, and transcription coupled repair. See text for full caption.

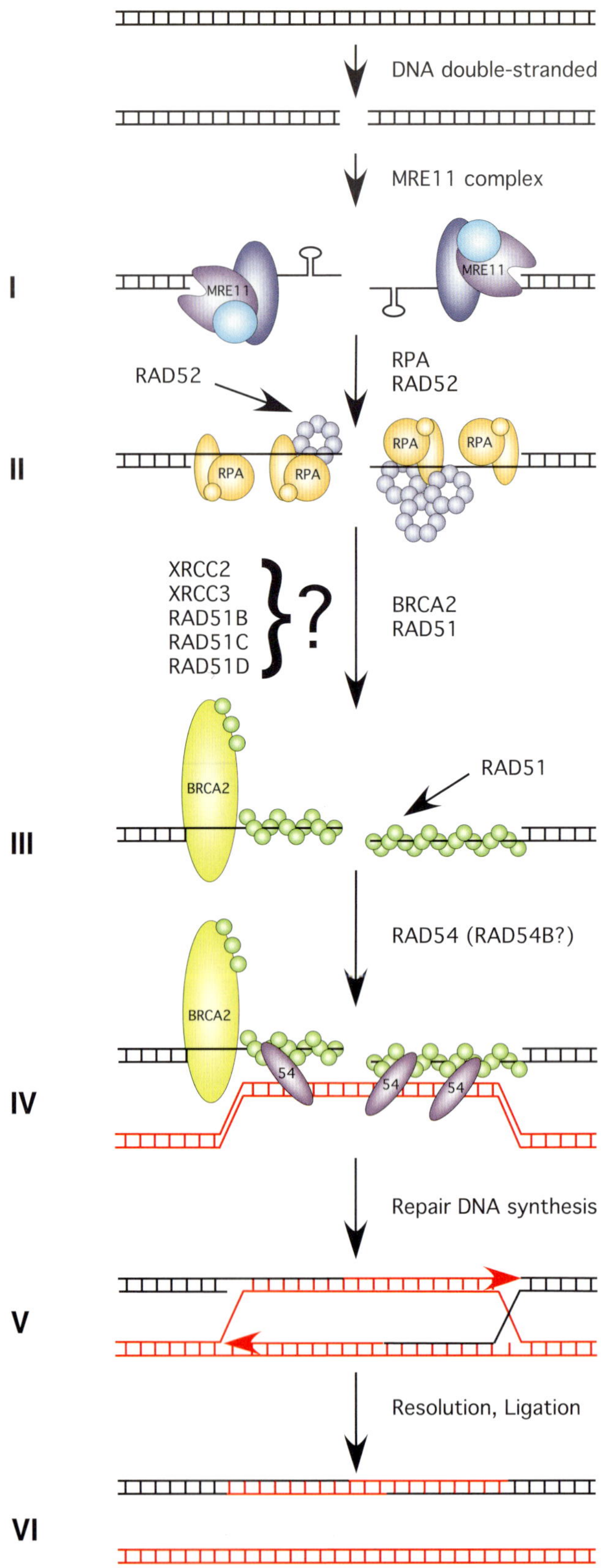

Figure 16.8. Schematic representation of the homologous recombination mechanism. See text for full caption.

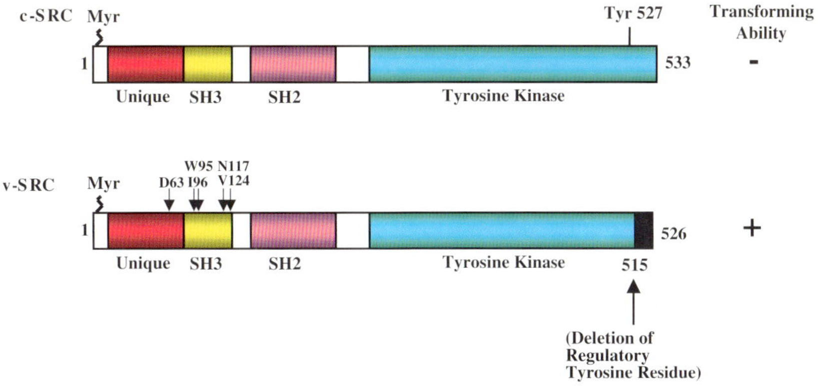

Figure 17.2. Activation of the Src oncoprotein. See text for full caption.

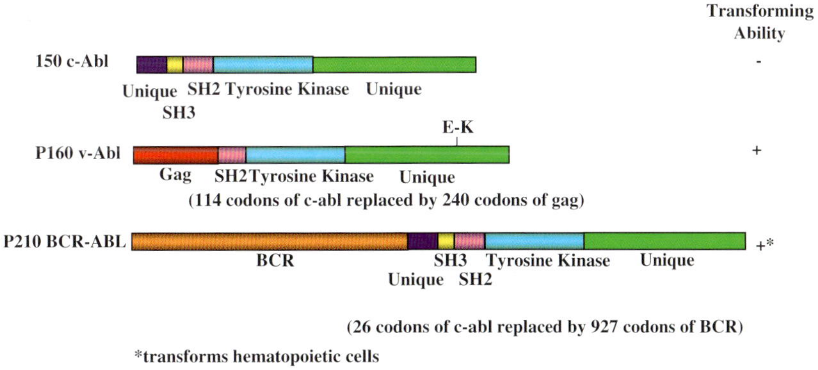

Figure 17.3. Activation of the Abl oncoprotein. See text for full caption.

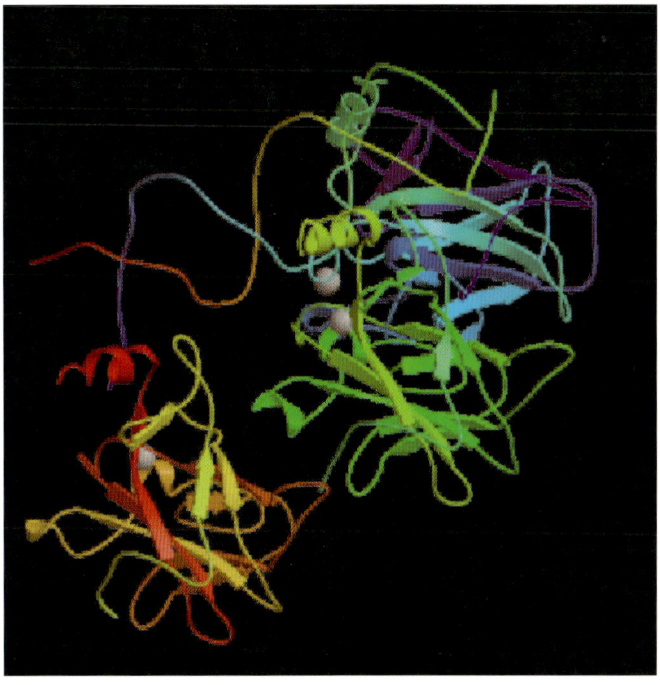

Figure 19.6. Structure of wild-type p53 bound to DNA. Protein Data Bank ID: 1TUP (see Web Resources).

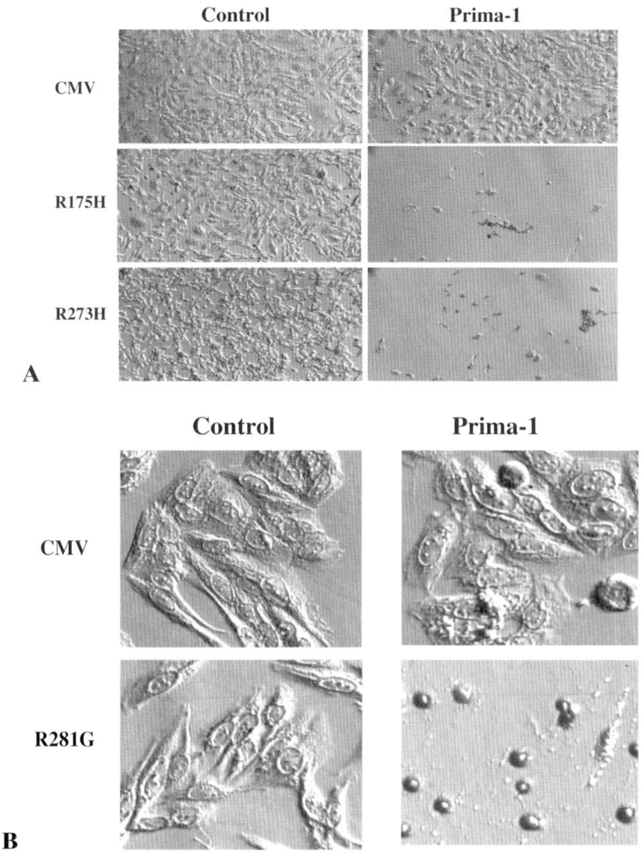

Figure 19.7. Prima-1 reactivates mutant p53. (**A**) Restoration of wild-type p53 activity to mutant p53 by Prima-1 in mouse 10(3) cells. Murine (10)3 fibroblasts lacking endogenous p53 were engineered to express only the selectable marker (CMV) or either the human mutant p53-R175H or R273H. Cells were grown under normal culture conditions (control) or treated with Prima-1 (10 μM) for 48 hours and stained for morphological analysis. Note that cells lacking p53 maintained viability after Prima-1 treatment (upper right panel), whereas cells expressing mutant p53 underwent apoptosis (middle and lower right panels) (unpublished data). (**B**) Restoration of wild-type p53 activity to mutant p53 by Prima-1 in Saos-2 cells. Human osteosarcoma Saos-2 cells lacking endogenous p53 were engineered to express only the selectable marker (CMV) or human mutant p53-R281G. Cells were grown under normal culture conditions (Control) or treated with Prima-1 (75 μM) for 48 hours and stained for morphological analysis. Note that cells lacking p53 maintained viability after Prima-1 treatment (upper right panel) whereas cells expressing mutant p53 underwent apoptosis (lower right panel) (unpublished data).

Ekblom M, Falk M, Salmivirta K, Durbeej M, Ekblom P (1998): Laminin isoforms and epithelial development. An NY Acad Sci 857:194–211.

Eklund L, Piuhola J, Komulainen J, et al (2001): Lack of type XV collagen causes a skeletal myopathy and cardiovascular defects in mice. Proc Natl Acad Sci USA 98:1194–9.

Fang F, Orend G, Watanabe N, Hunter T, Ruoslahti E (1996): Dependence of cyclin E-CDK2 kinase activity on cell anchorage. Science 271:499–502.

Fassler R, Pfaff M, Murphy J, et al (1995): Lack of beta 1 integrin gene in embryonic stem cells affects morphology, adhesion, and migration but not integration into the inner cell mass of blastocysts. J Cell Biol 128:979–88.

Frei R, Dowling J, Carenini S, Fuchs E, Martini R (1999): Myelin formation by Schwann cells in the absence of beta4 integrin. Glia 27:269–74.

Friedman SL, Roll FJ, Boyles J, Arenson DM, Bissell DM (1989): Maintenance of differentiated phenotype of cultured rat hepatic lipocytes by basement membrane matrix. J Biol Chem 264:10756–62.

Fuchs E, Dowling J, Segre J, Lo SH, Yu QC (1997): Integrators of epidermal growth and differentiation: Distinct functions for beta 1 and beta 4 integrins. Curr Opin Genet Dev 7:672–82.

Fukai N, Apte SS, Olsen BR (1994): Nonfibrillar collagens. Methods Enzymol 245:3–28.

Gardner H, Kreidberg J, Koteliansky V, Jaenisch R (1996): Deletion of integrin alpha 1 by homologous recombination permits normal murine development but gives rise to a specific deficit in cell adhesion. Dev Biol 175:301–13.

Garton AJ, Burnham MR, Bouton AH, Tonks NK (1997): Association of PTP-PEST with the SH3 domain of p130cas; a novel mechanism of protein tyrosine phosphatase substrate recognition. Oncogene 15:877–85.

Garton AJ, Tonks NK (1999): Regulation of fibroblast motility by the protein tyrosine phosphatase PTP-PEST. J Biol Chem 274:3811–8.

George EL, Georges-Labouesse EN, Patel-King RS, Rayburn H, Hynes RO (1993): Defects in mesoderm, neural tube and vascular development in mouse embryos lacking fibronectin. Development 119:1079–91.

Georges-Labouesse E, Messaddeq N, Yehia G, Cadalbert L, Dierich A, Le Meur M (1996): Absence of integrin alpha 6 leads to epidermolysis bullosa and neonatal death in mice. Nat Genet 13:370–3.

Giancotti FG, Ruoslahti E (1999): Integrin signaling. *Science* 285:1028–32.

Gille H, Downward J (1999): Multiple ras effector pathways contribute to G(1) cell cycle progression. J Biol Chem 274:22033–40.

Godfrey M, Hollister DW (1988): Type II achondrogenesis-hypochondrogenesis: Identification of abnormal type II collagen. Am J Hum Genet 43:904–13.

Grant DS, Kleinman HK, Leblond CP, Inoue S, Chung AE, Martin GR (1985): The basement-membrane-like matrix of the mouse EHS tumor: II. Immunohistochemical quantitation of six of its components. Am J Anat 174:387–98.

Gress CJ, Jacenko O (2000): Growth plate compressions and altered hematopoiesis in collagen X null mice. J Cell Biol 149:983–93.

Grisaru S, Cano-Gauci D, Tee J, Filmus J, Rosenblum ND (2001): Glypican-3 modulates BMP- and FGF-mediated effects during renal branching morphogenesis. Dev Biol 231:31–46.

Gu J, Tamura M, Yamada KM (1998): Tumor suppressor PTEN inhibits integrin- and growth factor-mediated mitogen-activated protein (MAP) kinase signaling pathways. J Cell Biol 143:1375–83.

Heichman KA, Roberts JM (1994): Rules to replicate by. Cell 79:557–62.

Henry MD, Satz JS, Brakebusch C, et al. (2001): Distinct roles for dystroglycan, beta1 integrin and perlecan in cell surface laminin organization. J Cell Sci 114:1137–44.

Hirsch E, Iglesias A, Potocnik AJ, Hartmann U, Fassler R (1996): Impaired migration but not differentiation of haematopoietic stem cells in the absence of beta1 integrins. Nature 380:171–5.

Hodivala-Dilke KM, McHugh KP, Tsakiris DA, et al. (1999): Beta3-integrin-deficient mice are a model for Glanzmann thrombasthenia showing placental defects and reduced survival. J Clin Invest 103:229–38.

Howe AK, Aplin AE, Juliano RL (2002): Anchorage-dependent ERK signaling—Mechanisms and consequences. Curr Opin Genet Dev 12:30–5.

Howlett AR, Bailey N, Damsky C, Petersen OW, Bissell MJ (1995): Cellular growth and survival are mediated by beta 1 integrins in normal human breast epithelium but not in breast carcinoma. J Cell Sci 108 (Pt 5):1945–57.

Huang S, Chen CS, Ingber DE (1998): Control of cyclin D1, p27(Kip1), and cell cycle progression in human capillary endothelial cells by cell shape and cytoskeletal tension. Mol Biol Cell 9:3179–93.

Huang XZ, Wu JF, Cass D, et al. (1996): Inactivation of the integrin beta 6 subunit gene reveals a role of epithelial integrins in regulating inflammation in the lung and skin. J Cell Biol 133:921–8.

Huang XZ, Wu JF, Ferrando R, et al. (2000): Fatal bilateral chylothorax in mice lacking the integrin alpha9beta1. Mol Cell Biol 20:5208–15.

Hughes PE, Renshaw MW, Pfaff M, et al. (1997): Suppression of integrin activation: A novel function of a Ras/Raf-initiated MAP kinase pathway. Cell 88:521–30.

Hunter DD, Shah V, Merlie JP, Sanes JR (1989): A laminin-like adhesive protein concentrated in the synaptic cleft of the neuromuscular junction. Nature 338:229–34.

Hynes RO (1987): Integrins: A family of cell surface receptors. Cell 48:549–54.

Hynes RO (1992): Integrins: Versatility, modulation, and signaling in cell adhesion. Cell 69:11–25.

Ikegawa S, Nishimura G, Nagai T, Hasegawa T, Ohashi H, Nakamura Y (1998): Mutation of the type X collagen gene (COL10A1) causes spondylometaphyseal dysplasia. Am J Hum Genet 63:1659–62.

Ingber D (2002): Mechanical signaling. An NY Acad Sci 961:162–3.

Isacke CM, Yarwood H (2002): The hyaluronan receptor, CD44. Int J Biochem Cell Biol 34:718–21.

Jobsis GJ, Keizers H, Vreijling JP, et al. (1996): Type VI collagen mutations in Bethlem myopathy, an autosomal dominant myopathy with contractures. Nat Genet 14:113–15.

Jones FS, Jones PL (2000): The tenascin family of ECM glycoproteins: structure, function, and regulation during embryonic development and tissue remodeling. Dev Dyn 218:235–59.

Jones JC, Hopkinson SB, Goldfinger LE (1998): Structure and assembly of hemidesmosomes. Bioessays 20:488–94.

Jones PL, Jones FS, Zhou B, Rabinovitch M (1999): Induction of vascular smooth muscle cell tenascin-C gene expression by denatured type I collagen is dependent upon a beta3 integrin-mediated mitogen-activated protein kinase pathway and a 122-base pair promoter element. J Cell Sci 112 (Pt 4): 435–45.

Kadler K (1994): Extracellular matrix. 1: Fibril-forming collagens. Protein Profile 1:519–638.

Kadler K (1995): Extracellular matrix 1: Fibril-forming collagens. Protein Profile 2:491–619.

Kashtan CE (1995): Clinical and molecular diagnosis of Alport syndrome. Proc Assoc Am Physicians 107:306–13.

Kashtan CE (2000): Alport syndrome: Abnormalities of type IV collagen genes and proteins. Ren Fail 22:737–49.

Khwaja A, Rodriguez-Viciana P, Wennstrom S, Warne PH, Downward J (1997): Matrix adhesion and Ras transformation both activate a phosphoinositide 3-OH kinase and protein kinase B/Akt cellular survival pathway. EMBO J 16:2783–93.

Kleinman HK, McGarvey ML, Liotta LA, Robey PG, Tryggvason K, Martin GR (1982): Isolation and characterization of type IV procollagen, laminin, and heparan sulfate proteoglycan from the EHS sarcoma. Biochemistry 21: 6188–93.

Kon A, Pulkkinen L, Hara M, et al. (1998): Laminin 5 genes and Herlitz junctional epidermolysis bullosa: novel mutations and polymorphisms in the LAMB3 and LAMC2 genes: Mutations in brief no. 190. Online. Hum Mutat 12:288.

Kreidberg JA, Donovan MJ, Goldstein SL, et al. (1996): Alpha 3 beta 1 integrin has a crucial role in kidney and lung organogenesis. Development 122: 3537–47.

Kresse H, Schonherr E (2001): Proteoglycans of the extracellular matrix and growth control. J Cell Physiol 189:266–74.

Kuang W, Xu H, Vachon PH, et al. (1998): Merosin-deficient congenital muscular dystrophy: Partial genetic correction in two mouse models. J Clin Invest 102:844–52.

Kuhn K (1995): Basement membrane (type IV) collagen. Matrix Biol 14:439–45.

Kuivaniemi H, Tromp G, Prockop DJ (1997): Mutations in fibrillar collagens (types I, II, III, and XI), fibril-associated collagen (type IX), and network-forming collagen (type X) cause a spectrum of diseases of bone, cartilage, and blood vessels. Hum Mutat 9:300–15.

Kuster JE, Guarnieri MH, Ault JG, Flaherty L, Swiatek PJ (1997): IAP insertion in the murine LamB3 gene results in junctional epidermolysis bullosa. Mamm Genome 8:673–81.

Lander AD, Stipp CS, Ivins JK (1996): The glypican family of heparan sulfate proteoglycans: Major cell-surface proteoglycans of the developing nervous system. Perspect Dev Neurobiol 3:347–58.

Law DA, DeGuzman FR, Heiser P, Ministri-Madrid K, Killeen N, Phillips DR (1999): Integrin cytoplasmic tyrosine motif is required for outside-in alphaIIbbeta3 signalling and platelet function. Nature 401:808–11.

Le Gall M, Grall D, Chambard JC, Pouyssegur J, Van Obberghen-Schilling E (1998): An anchorage-dependent signal distinct from p42/44 MAP kinase activation is required for cell cycle progression. Oncogene 17:1271–7.

Lemmink HH, Nillesen WN, Mochizuki T, et al. (1996): Benign familial hematuria due to mutation of the type IV collagen alpha4 gene. J Clin Invest 98: 1114–18.

Lewis JM, Schwartz MA (1995): Mapping in vivo associations of cytoplasmic proteins with integrin beta 1 cytoplasmic domain mutants. Mol Biol Cell 6: 151–60.

Li J, Yen C, Liaw D, et al. (1997): PTEN, a putative protein tyrosine phosphatase gene mutated in human brain, breast, and prostate cancer. Science 275:1943–7.

Li ML, Aggeler J, Farson DA, Hatier C, Hassell J, Bissell MJ (1987): Influence of a reconstituted basement membrane and its components on casein gene expression and secretion in mouse mammary epithelial cells. Proc Natl Acad Sci USA 84:136–40.

Li Y, Lacerda DA, Warman ML, et al. (1995): A fibrillar collagen gene, Col11a1, is essential for skeletal morphogenesis. Cell 80:423–30.

Lin CQ, Bissell MJ (1993): Multi-faceted regulation of cell differentiation by extracellular matrix. Faseb J 7:737–43.

Lin TH, Chen Q, Howe A, Juliano RL (1997): Cell anchorage permits efficient signal transduction between *ras* and its downstream kinases. J Biol Chem 272: 8849–52.

Littlewood Evans A, Muller U (2000): Stereocilia defects in the sensory hair cells of the inner ear in mice deficient in integrin alpha8beta1. Nat Genet 24:424–8.

Liu F, Sells MA, Chernoff J (1998): Protein tyrosine phosphatase 1B negatively regulates integrin signaling. Curr Biol 8:173–6.

Liu X, Wu H, Byrne M, Jeffrey J, Krane S, Jaenisch R (1995): A targeted mutation at the known collagenase cleavage site in mouse type I collagen impairs tissue remodeling. J Cell Biol 130:227–37.

Liu X, Wu H, Byrne M, Krane S, Jaenisch R (1997): Type III collagen is crucial for collagen I fibrillogenesis and for normal cardiovascular development. Proc Natl Acad Sci USA 94:1852–6.

Lohi J, Leivo I, Franssila K, Virtanen I (1997): Changes in the distribution of integrins and their basement membrane ligands during development of human thyroid follicular epithelium. Histochem J 29:337–45.

Maehama T, Dixon JE (1998): The tumor suppressor, PTEN/MMAC1, dephosphorylates the lipid second messenger, phosphatidylinositol 3,4,5-trisphosphate. J Biol Chem 273:13375–8.

Mayer U, Kohfeldt E, Timpl R (1998): Structural and genetic analysis of laminin-nidogen interaction. An NY Acad Sci 857:130–42.

Mayer U, Saher G, Fassler R, et al. (1997): Absence of integrin alpha 7 causes a novel form of muscular dystrophy. Nat Genet 17:318–23.

McGuirt WT, Prasad SD, Griffith AJ, et al. (1999): Mutations in COL11A2 cause non-syndromic hearing loss (DFNA13). Nat Genet 23:413–19.

McHugh KP, Hodivala-Dilke K, Zheng MH, et al. (2000): Mice lacking beta3 integrins are osteosclerotic because of dysfunctional osteoclasts. J Clin Invest 105:433–40.

McIntosh I, Abbott MH, Francomano CA (1995): Concentration of mutations causing Schmid metaphyseal chondrodysplasia in the C-terminal noncollagenous domain of type X collagen. Hum Mutat 5:121–5.

Meek KM, Fullwood NJ (2001): Corneal and scleral collagens—A microscopist's perspective. Micron 32:261–72.

Meisler MH, Griffith AJ, Warman M, Tiller G, Sprunger LK (1998): Gene symbol: COL11A1. Disease: Marshall syndrome. Hum Genet 102:498.

Miosge N (2001): The ultrastructural composition of basement membranes in vivo. Histol Histopathol 16:1239–48.

Miosge N, Klenczar C, Herken R, Willem M, Mayer U (1999): Organization of the myotendinous junction is dependent on the presence of alpha7beta1 integrin. Lab Invest 79:1591–9.

Miranti CK, Brugge JS (2002): Sensing the environment: A historical perspective on integrin signal transduction. Nat Cell Biol 4:E83–90.

Miyamoto S, Teramoto H, Coso OA, et al. (1995): Integrin function: Molecular hierarchies of cytoskeletal and signaling molecules. J Cell Biol 131:791–805.

Miyamoto S, Teramoto H, Gutkind JS, Yamada KM (1996): Integrins can collaborate with growth factors for phosphorylation of receptor tyrosine kinases and MAP kinase activation: Roles of integrin aggregation and occupancy of receptors. J Cell Biol 135:1633–42.

Moore SA, Saito F, Chen J, et al. (2002): Deletion of brain dystroglycan recapitulates aspects of congenital muscular dystrophy. Nature 418:422–5.

Muller U, Wang D, Denda S, Meneses JJ, Pedersen RA, Reichardt LF (1997): Integrin alpha8beta1 is critically important for epithelial-mesenchymal interactions during kidney morphogenesis. Cell 88:603–13.

Murgia C, Blaikie P, Kim N, Dans M, Petrie HT, Giancotti FG (1998): Cell cycle and adhesion defects in mice carrying a targeted deletion of the integrin beta4 cytoplasmic domain. EMBO J 17:3940–51.

Murshed M, Smyth N, Miosge N, et al. (2000): The absence of nidogen 1 does not affect murine basement membrane formation. Mol Cell Biol 20:7007–12.

Muschler J, Levy D, Boudreau R, Henry M, Campbell K, Bissell MJ (2002): A role for dystroglycan in epithelial polarization: loss of function in breast tumor cells. Cancer Res 62:7102–9.

Muthuswamy SK, Li D, Lelievre S, Bissell MJ, Brugge JS (2001): ErbB2, but not ErbB1, reinitiates proliferation and induces luminal repopulation in epithelial acini. Nat Cell Biol 3:785–92.

Myers MP, Pass I, Batty IH, et al. (1998): The lipid phosphatase activity of PTEN is critical for its tumor supressor function. Proc Natl Acad Sci USA 95: 13513–18.

Myllyharju J, Kivirikko KI (2001): Collagens and collagen-related diseases. An Med 33:7–21.

Nievers MG, Schaapveld RQ, Sonnenberg A (1999): Biology and function of hemidesmosomes. Matrix Biol 18:5–17.

Noakes PG, Gautam M, Mudd J, Sanes JR, Merlie JP (1995): Aberrant differentiation of neuromuscular junctions in mice lacking s-laminin/laminin beta 2. Nature 374:258–62.

Oh ES, Gu H, Saxton TM, et al. (1999): Regulation of early events in integrin signaling by protein tyrosine phosphatase SHP-2. Mol Cell Biol 19:3205–15.

Oldberg A, Antonsson P, Hedbom E, Heinegard D (1990): Structure and function of extracellular matrix proteoglycans. Biochem Soc Trans 18:789–92.

Olsen BR (1995): New insights into the function of collagens from genetic analysis. Curr Opin Cell Biol 7:720–7.

Paine-Saunders S, Viviano BL, Zupicich J, Skarnes WC, Saunders S (2000): Glypican-3 controls cellular responses to Bmp4 in limb patterning and skeletal development. Dev Biol 225:179–87.

Park PW, Reizes O, Bernfield M (2000): Cell surface heparan sulfate proteoglycans: Selective regulators of ligand-receptor encounters. J Biol Chem 275: 29923–6.

Parsons JT, Schaller MD, Hildebrand J, Leu TH, Richardson A, Otey C (1994): Focal adhesion kinase: Structure and signalling. J Cell Sci Suppl 18:109–13.

Patton BL, Cunningham JM, Thyboll J, et al. (2001): Properly formed but improperly localized synaptic specializations in the absence of laminin alpha4. Nat Neurosci 4:597–604.

Perrimon N, Bernfield M (2001): Cellular functions of proteoglycans—An overview. Semin Cell Dev Biol 12:65–7.

Petersen OW, Ronnov-Jessen L, Howlett AR, Bissell MJ (1992): Interaction with basement membrane serves to rapidly distinguish growth and differentiation pattern of normal and malignant human breast epithelial cells. Proc Natl Acad Sci USA 89:9064–8.

Petitclerc E, Boutaud A, Prestayko A, et al. (2000): New functions for non-collagenous domains of human collagen type IV: Novel integrin ligands inhibiting angiogenesis and tumor growth in vivo. J Biol Chem 275:8051–61.

Phillips CL, Pfeiffer BJ, Luger AM, Franklin CL (2002): Novel collagen glomerulopathy in a homotrimeric type I collagen mouse (oim). Kidney Int 62:383–91.

Plopper G, Ingber DE (1993): Rapid induction and isolation of focal adhesion complexes. Biochem Biophys Res Commun 193:571–8.

Plopper GE, McNamee HP, Dike LE, Bojanowski K, Ingber DE (1995): Convergence of integrin and growth factor receptor signaling pathways within the focal adhesion complex. Mol Biol Cell 6:1349–65.

Pozzi A, Wary KK, Giancotti FG, Gardner HA (1998): Integrin alpha1beta1 mediates a unique collagen-dependent proliferation pathway in vivo. J Cell Biol 142:587–94.

Pulkkinen L, Kim DU, Uitto J (1998): Epidermolysis bullosa with pyloric atresia: novel mutations in the beta4 integrin gene (ITGB4). Am J Pathol 152:157–66.

Radisky D, Muschler J, Bissell MJ (2002): Order and disorder: The role of extracellular matrix in epithelial cancer. Cancer Invest 20:139–53.

Raghavan S, Bauer C, Mundschau G, Li Q, Fuchs E (2000): Conditional ablation of beta1 integrin in skin: Severe defects in epidermal proliferation, basement membrane formation, and hair follicle invagination. J Cell Biol 150:1149–60.

Rapraeger AC (2000): Syndecan-regulated receptor signaling. J Cell Biol 149: 995–8.

Reizes O, Lincecum J, Wang Z, et al. (2001): Transgenic expression of syndecan-1 uncovers a physiological control of feeding behavior by syndecan-3. Cell 106:105–16.

Ren XD, Kiosses WB, Schwartz MA (1999): Regulation of the small GTP-binding protein Rho by cell adhesion and the cytoskeleton. EMBO J 18:578–85.

Ren XD, Kiosses WB, Sieg DJ, Otey CA, Schlaepfer DD, Schwartz MA (2000): Focal adhesion kinase suppresses Rho activity to promote focal adhesion turnover. J Cell Sci 113 (Pt 20):3673–8.

Renshaw MW, Price LS, Schwartz MA (1999): Focal adhesion kinase mediates the integrin signaling requirement for growth factor activation of MAP kinase. J Cell Biol 147:611–8.

Richards AJ, Martin S, Yates JR, et al. (2000): COL2A1 exon 2 mutations: Relevance to the Stickler and Wagner syndromes. Br J Ophthalmol 84:364–71.

Ritvaniemi P, Korkko J, Bonaventure J, et al. (1995): Identification of COL2A1 gene mutations in patients with chondrodysplasias and familial osteoarthritis. Arthritis Rheum 38:999–1004.

Rock MT, Brooks WH, Roszman TL (1997): Calcium-dependent signaling pathways in T cells: Potential role of calpain, protein tyrosine phosphatase 1b, and p130Cas in integrin-mediated signaling events. J Biol Chem 272: 33377–83.

Ronnov-Jessen L, Petersen OW, Bissell MJ (1996): Cellular changes involved in conversion of normal to malignant breast: Importance of the stromal reaction. Physiol Rev 76:69–125.

Roovers K, Assoian RK (2000): Integrating the MAP kinase signal into the G_1 phase cell cycle machinery. Bioessays 22:818–26.

Roovers K, Davey G, Zhu X, Bottazzi ME, Assoian RK (1999): Alpha5beta1 integrin controls cyclin D1 expression by sustaining mitogen-activated protein kinase activity in growth factor-treated cells. Mol Biol Cell 10:3197–204.

Rossino P, Gavazzi I, Timpl R, et al. (1990): Nerve growth factor induces increased expression of a laminin-binding integrin in rat pheochromocytoma PC12 cells. Expr Cell Res 189:100–8.

Ryan MC, Lee K, Miyashita Y, Carter WG (1999): Targeted disruption of the LAMA3 gene in mice reveals abnormalities in survival and late stage differentiation of epithelial cells. J Cell Biol 145:1309–23.

Schaller MD, Hildebrand JD, Shannon JD, Fox JW, Vines RR, Parsons JT (1994): Autophosphorylation of the focal adhesion kinase, pp125FAK, directs SH2-dependent binding of pp60src. Mol Cell Biol 14:1680–8.

Schaller MD, Otey CA, Hildebrand JD, Parsons JT (1995): Focal adhesion kinase and paxillin bind to peptides mimicking beta integrin cytoplasmic domains. J Cell Biol 130:1181–7.

Schlaepfer DD, Hanks SK, Hunter T, van der Geer P (1994): Integrin-mediated signal transduction linked to Ras pathway by GRB2 binding to focal adhesion kinase. Nature 372:786–91.

Schon MP, Schon M, Warren HB, Donohue JP, Parker CM (2000): Cutaneous inflammatory disorder in integrin alphaE (CD103)-deficient mice. J Immunol 165:6583–9.

Schwartz MA (1997): Integrins, oncogenes, and anchorage independence. J Cell Biol 139:575–8.

Schwartz MA, Assoian RK (2001): Integrins and cell proliferation: Regulation of cyclin-dependent kinases via cytoplasmic signaling pathways. J Cell Sci 114: 2553–60.

Schwartz MA, Assoian RK (2001): Integrins and cell proliferation: Regulation of cyclin-dependent kinases via cytoplasmic signaling pathways. J Cell Sci 114: 2553–60.

Schwartz MA, Ingber DE (1994): Integrating with integrins. Mol Biol Cell 5:389–93.

Schwartz MA, Schaller MD, Ginsberg MH (1995): Integrins: Emerging paradigms of signal transduction. An Rev Cell Dev Biol 11:549–99.

Schymeinsky J, Nedbal S, Miosge N, et al. (2002): Gene structure and functional analysis of the mouse nidogen-2 gene: Nidogen-2 is not essential for basement membrane formation in mice. Mol Cell Biol 22:6820–30.

Sherr CJ (1994): G_1 phase progression: Cycling on cue. Cell 79:551–5.

Simpson DG, Terracio L, Terracio M, Price RL, Turner DC, Borg TK (1994): Modulation of cardiac myocyte phenotype in vitro by the composition and orientation of the extracellular matrix. J Cell Physiol 161:89–105.

Smith LT, Schwarze U, Goldstein J, Byers PH (1997): Mutations in the COL3A1 gene result in the Ehlers-Danlos syndrome type IV and alterations in the size and distribution of the major collagen fibrils of the dermis. J Invest Dermatol 108:241–7.

Smyth N, Vatansever HS, Murray P, et al. (1999): Absence of basement membranes after targeting the LAMC1 gene results in embryonic lethality due to failure of endoderm differentiation. J Cell Biol 144:151–60.

Snead MP, Yates JR, Williams R, Payne SJ, Pope FM, Scott JD (1996): Stickler syndrome type 2 and linkage to the COL11A1 gene. An NY Acad Sci 785: 331–2.

Spranger J, Menger H, Mundlos S, Winterpacht A, Zabel B (1994): Kniest dysplasia is caused by dominant collagen II (COL2A1) mutations: Parental somatic mosaicism manifesting as Stickler phenotype and mild spondyloepiphyseal dysplasia. Pediatr Radiol 24:431–5.

Stambolic V, Suzuki A, de la Pompa JL, et al. (1998): Negative regulation of PKB/Akt-dependent cell survival by the tumor suppressor PTEN. Cell 95: 29–39.

Stephens LE, Sutherland AE, Klimanskaya IV, et al. (1995): Deletion of beta 1 integrins in mice results in inner cell mass failure and peri-implantation lethality. Genes Dev 9:1883–95.

Sternlicht MD, Bissell MJ, Werb Z (2000): The matrix metalloproteinase stromelysin-1 acts as a natural mammary tumor promoter. Oncogene 19: 1102–13.

Sternlicht MD, Lochter A, Sympson CJ, et al. (1999): The stromal proteinase MMP3/stromelysin-1 promotes mammary carcinogenesis. Cell 98:137–46.

Streuli C (1999): Extracellular matrix remodelling and cellular differentiation. Curr Opin Cell Biol 11:634–40.

Su J, Muranjan M, Sap J (1999): Receptor protein tyrosine phosphatase alpha activates Src-family kinases and controls integrin-mediated responses in fibroblasts. Curr Biol 9:505–11.

Takuwa N, Fukui Y, Takuwa Y (1999): Cyclin D1 expression mediated by phosphatidylinositol 3-kinase through mTOR-p70(S6K)-independent signaling in growth factor-stimulated NIH 3T3 fibroblasts. Mol Cell Biol 19:1346–58.

Talhouk RS, Bissell MJ, Werb Z (1992): Coordinated expression of extracellular matrix-degrading proteinases and their inhibitors regulates mammary epithelial function during involution. J Cell Biol 118:1271–82.

Tamura M, Gu J, Matsumoto K, Aota S, Parsons R, Yamada KM (1998): Inhibition of cell migration, spreading, and focal adhesions by tumor suppressor PTEN. Science 280:1614–17.

Tiller GE, Polumbo PA, Weis MA, et al. (1995): Dominant mutations in the type II collagen gene, COL2A1, produce spondyloepimetaphyseal dysplasia, Strudwick type. Nat Genet 11:87–9.

Timpl R, Fujiwara S, Dziadek M, Aumailley M, Weber S, Engel J (1984): Laminin, proteoglycan, nidogen and collagen IV: Structural models and molecular interactions. Ciba Found Symp 108:25–43.

Toole BP (2001): Hyaluronan in morphogenesis. Semin Cell Dev Biol 12:79–87.

Toole BP (2002): Hyaluronan promotes the malignant phenotype. Glycobiology 12:37R–42R.

Toole BP, Wight TN, Tammi MI (2002): Hyaluronan-cell interactions in cancer and vascular disease. J Biol Chem 277:4593–6.

Trummer T, Brenner R, Just W, Vogel W, Kennerknecht I (2001): Recurrent mutations in the COL1A2 gene in patients with osteogenesis imperfecta. Clin Genet 59:338–43.

Tryggvason K (1995): Molecular properties and diseases of collagens. Kidney Int Suppl 49:S24–8.

Tsai LH (1998): Stuck on the ECM. Trends Cell Biol 8:292–5.

Turley EA, Noble PW, Bourguignon LY (2002): Signaling properties of hyaluronan receptors. J Biol Chem 277:4589–92.

Uitto J, Pulkkinen L (1996): Molecular complexity of the cutaneous basement membrane zone. Mol Biol Rep 23:35–46.

Vachon PH, Loechel F, Xu H, Wewer UM, Engvall E (1996): Merosin and laminin in myogenesis; specific requirement for merosin in myotube stability and survival. J Cell Biol 134:1483–97.

van Keulen CJ, van de Akker E, Pals G, Rauwerda JA (1999): The role of type III collagen in the development of familial abdominal aortic aneurysms. Eur J Vasc Endovasc Surg 18:65–70.

Vidal F, Baudoin C, Miquel C, et al. (1995): Cloning of the laminin alpha 3 chain gene (LAMA3) and identification of a homozygous deletion in a patient with Herlitz junctional epidermolysis bullosa. Genomics 30:273–80.

Virtanen I, Laitinen L, Korhonen M (1995): Differential expression of laminin polypeptides in developing and adult human kidney. J Histochem Cytochem 43:621–8.

Vuori K, Hirai H, Aizawa S, Ruoslahti E (1996): Introduction of p130cas signaling complex formation upon integrin-mediated cell adhesion: A role for Src family kinases. Mol Cell Biol 16:2606–13.

Wagner N, Lohler J, Kunkel EJ, et al. (1996): Critical role for beta7 integrins in formation of the gut-associated lymphoid tissue. Nature 382:366–70.

Wai AW, Ng LJ, Watanabe H, Yamada Y, Tam PP, Cheah KS (1998): Disrupted expression of matrix genes in the growth plate of the mouse cartilage matrix deficiency (cmd) mutant. Dev Genet 22:349–58.

Wang F, Weaver VM, Petersen OW, et al. (1998): Reciprocal interactions between beta1-integrin and epidermal growth factor receptor in three-dimensional basement membrane breast cultures: A different perspective in epithelial biology. Proc Natl Acad Sci USA 95:14821–6.

Ward LM, Lalic L, Roughley PJ, Glorieux FH (2001): Thirty-three novel COL1A1 and COL1A2 mutations in patients with osteogenesis imperfecta types I-IV. Hum Mutat 17:434.

Wary KK, Dans M, Mariotti A, Giancotti FG (1999): Biochemical analysis of integrin-mediated Shc signaling. Methods Mol Biol 129:35–49.

Wary KK, Mainiero F, Isakoff SJ, Marcantonio EE, Giancotti FG (1996): The adaptor protein Shc couples a class of integrins to the control of cell cycle progression. Cell 87:733–43.

Wasylyk C, Gutman A, Nicholson R, Wasylyk B (1991): The c-Ets oncoprotein activates the stromelysin promoter through the same elements as several non-nuclear oncoproteins. EMBO J 10:1127–34.

Watt FM (2002): Role of integrins in regulating epidermal adhesion, growth and differentiation. EMBO J 21:3919–26.

Weaver VM, Howlett AR, Langton-Webster B, Petersen OW, Bissell MJ (1995): The development of a functionally relevant cell culture model of progressive human breast cancer. Semin Cancer Biol 6:175–84.

Weaver VM, Petersen OW, Wang F, et al. (1997): Reversion of the malignant phenotype of human breast cells in three-dimensional culture and in vivo by integrin blocking antibodies. J Cell Biol 137:231–45.

Weber JD, Raben DM, Phillips PJ, Baldassare JJ (1997): Sustained activation of extracellular-signal-regulated kinase 1 (ERK1) is required for the continued expression of cyclin D1 in G_1 phase. Biochem J 326 (Pt 1):61–8.

Wei Y, Yang X, Liu Q, Wilkins JA, Chapman HA (1999): A role for caveolin and the urokinase receptor in integrin-mediated adhesion and signaling. J Cell Biol 144:1285–94.

Weinberg RA (1995): The retinoblastoma protein and cell cycle control. Cell 81: 323–30.

Werb Z, Chin JR (1998): Extracellular matrix remodeling during morphogenesis. An NY Acad Sci 857:110–18.

Wewer UM, Engvall E (1996): Merosin/laminin-2 and muscular dystrophy. Neuromuscul Disord 6:409–18.

Williamson RA, Henry MD, Daniels KJ, et al. (1997): Dystroglycan is essential for early embryonic development: Disruption of Reichert's membrane in Dag1-null mice. Hum Mol Genet 6:831–41.

Wiseman BS, Werb Z (2002): Stromal effects on mammary gland development and breast cancer. Science 296:1046–9.

Woods D, Cherwinski H, Venetsanakos E, et al. (2001): Induction of beta3-integrin gene expression by sustained activation of the Ras-regulated Raf-MEK-extracellular signal-regulated kinase signaling pathway. Mol Cell Biol 21:3192–205.

Xu H, Wu XR, Wewer UM, Engvall E (1994): Murine muscular dystrophy caused by a mutation in the laminin alpha 2 (Lama2) gene. Nat Genet 8:297–302.

Yamakita Y, Totsukawa G, Yamashiro S, et al. (1999): Dissociation of FAK/p130(CAS)/c-Src complex during mitosis: Role of mitosis-specific serine phosphorylation of FAK. J Cell Biol 144:315–24.

Yanagishita M (1993): Function of proteoglycans in the extracellular matrix. Acta Pathol Jpn 43:283–93.

Yang JT, Rayburn H, Hynes RO (1993): Embryonic mesodermal defects in alpha 5 integrin-deficient mice. Development 119:1093–105.

Yang JT, Rayburn H, Hynes RO (1995): Cell adhesion events mediated by alpha 4 integrins are essential in placental and cardiac development. Development 121:549–60.

Yurchenco PD, Cheng YS, Colognato H (1992): Laminin forms an independent network in basement membranes. J Cell Biol 117:1119–33.

Zhu X, Ohtsubo M, Bohmer RM, Roberts JM, Assoian RK (1996): Adhesion-dependent cell cycle progression linked to the expression of cyclin D1, activation of cyclin E-cdk2, and phosphorylation of the retinoblastoma protein. J Cell Biol 133:391–403.

Zimmermann P, David G (1999): The syndecans, tuners of transmembrane signaling. Faseb J 13 Suppl:S91–S100.

CHAPTER 10

ANGIOGENESIS AND BLOOD SUPPLY

JUDAH FOLKMAN
Dept of Surgery, Children's Hospital, Harvard Medical School, Vascular Biology Program, Boston, MA 02115

TUMOR ANGIOGENESIS: EXPERIMENTAL BASIS

Hypothesis: Tumor Growth Is Angiogenesis Dependent

In 1971 I proposed a hypothesis that tumor growth is angiogenesis dependent (Folkman, 1971). In its simplest terms the hypothesis stated that in the absence of neovascularization, a tumor would remain dormant at a size of less than a few millimeters diameter. This paper introduced the term "anti-angiogenesis" to describe a novel potential therapy that would prevent recruitment of new blood vessels by a tumor. These ideas were based on experiments that I performed with Frederick Becker in the early 1960s at the Naval Medical Research Institute in Bethesda, Maryland. Tumors implanted into isolated canine thyroid glands perfused with hemoglobin solutions, stopped growing at 1 to 2 mm diameter, but remained viable and resumed growth to reach a size of cubic centimeters when transplanted to syngeneic mice. Tumors in the mice were highly vascularized but remained avascular in the isolated thyroid glands (Folkman et al., 1963, 1966, 1972; Gimbrone et al., 1972; Folkman, 1976). This hypothesis was not widely accepted for more than a decade. The idea was resisted mainly because tumor hyperemia was conveniently explained as vasodilation of preexisting host vessels resulting from tumor metabolites or from necrotic tumor products (Coman and Sheldon, 1946). Three publications between 1939 and 1947 had previously alluded to the possibility that tumor vessels were new (Ide et al., 1939; Algire et al., 1945; Algire and Legalais, 1947), but these suggestions were not taken seriously by the scientific community, nor followed up, because the entrenched notion of simple vasodilation led to the belief that the inflammatory products that supposedly caused it were a side effect of tumor growth and not a requirement (Day, 1964; Folkman, 1985). Another

Cell Cycle and Growth Control: Biomolecular Regulation and Cancer, Second Edition, Edited by Gary S. Stein and Arthur B. Pardee
ISBN 0-471-25071-6

obstacle to the acceptance of a hypothesis of angiogenesis dependence of tumors was the requirement for a diffusible substance or substances to be released by the tumor that could stimulate blood vessel growth toward the tumor. No such substance existed. In fact, at the time the hypothesis was first proposed (Folkman, 1971), there were no bioassays available to purify a putative angiogenic factor from a tumor. Vascular endothelial cells had never been cultured in vitro, and it was widely assumed that this goal would remain elusive.

Development of Bioassays to Study Angiogenesis

Throughout the 1970s bioassays were developed that made it possible to study the process of angiogenesis, to purify angiogenic factors, to identify the first angiogenesis inhibitors, and to uncover mechanisms of tumor dormancy when angiogenesis was blocked (reviewed in Folkman and Kalluri, 2003). These included (1) the shell-less chick chorioallantoic membrane (Auerbach et al., 1974), (2) induction of corneal neovascularization in the rabbit (Gimbrone et al., 1974) and mouse cornea (Muthukkaruppan and Auerbach, 1979), (3) the development of polymer pellets that could be implanted into the cornea and that could provide sustained release of angiogenic proteins (Langer and Folkman, 1976), (4) in vitro growth and cloning of endothelial cells from the umbilical vein (Gimbrone et al., 1973) and microvascular endothelium from capillary beds (Folkman et al., 1979), and (5) the induction of angiogenesis in vitro (Folkman and Haudenschild, 1980). These methods have stood the test of time and are still employed today in laboratories engaged in angiogenesis research. More recently developed methods include (1) the in vitro culture of rings of aorta from mice from which capillary sprouts grow (Bonanno and Nicosia, 1992) and from the chick embryo (Muthukkaruppan et al., 2000), (2) subcutaneous implantation in mice of matrigel containing an angiogenic protein such as vascular endothelial growth factor (VEGF) (Akhtar et al., 2002), and (3) transparent chambers in the skin or skull used in mice for video analysis of tumor angiogenesis (Boucher et al., 1996).

Identification of Positive Regulators of Angiogenesis

This methodology led to the identification of molecules that stimulate angiogenesis. These are mainly proteins which are expressed for brief periods (days or weeks) during the physiological functions of development, reproduction and repair (Table 10.1). Tumors express these proteins during induction of angiogenesis, but tumor angiogenesis usually persists, in contrast to physiological angiogenesis. Tumors most commonly overexpress VEGF. About 60% of human breast cancers express mainly VEGF when they are first diagnosed (Relf et al., 1997). However, tumors may express multiple pro-angiogenic proteins. Some human breast cancers express up to six angiogenic proteins (e.g., VEGF, PLGF, PD-ECGF, TGF-beta, bFGF, and pleiotrophin). Tumors can also induce

TABLE 10.1. Angiogenic Proteins Commonly Expressed by Human Tumors

	Most commonly produced by human tumors	
VEGF	45,000	Vascular endothelial growth factor
bFGF	18,000	Basic fibroblast growth factor
aFGF	16,400	Acidic fibroblast growth factor
PDGF	40,000	Platelet-derived growth factor
PD-ECGF	45,000	Platelet-derived endothelial growth factor
IL-8	40,000	Interleukin-8
HGF	92,000	Hepatocyte growth factor
EGF	6,000	Epidermal growth factor
Angiogenin	14,100	
Others:		
TNF-alpha	17,000	Tumor necrosic factor alpha
TGF-beta	25,000	Transforming growth factor beta
TGF-alpha	5,500	Transforming growth factor alpha
Proliferin	35,000	
PLGF	25,000	Placental growth factor

Source: Adapted from Folkman, and Kalluri 2003, (with permission of the publisher).

stromal cells to express angiogenic proteins such as VEGF (Fukumura et al., 1998). Tumor expression of angiogenic promoters induces an increase of circulating endothelial precursor endothelial cells from the bone marrow (Monestiroli et al., 2001). However, it is unclear how a small matrigel pellet containing only 50 nanograms of VEGF and implanted subcutaneously can stimulate bone marrow to release endothelial cells, which then home to the matrigel and form new blood vessels.

Pharmacologic Evidence That Tumor Growth Is Angiogenesis Dependent

Because no angiogenesis inhibitors existed in the late 1970s, our laboratory set out to discover such molecules to provide pharmacologic evidence that tumor growth is angiogenesis dependent. Over the next 23 years, 11 angiogenesis inhibitors were discovered (Table 10.2). Five of these are currently in phase II clinical trials (Table 10.2).

The first angiogenesis inhibitors immediately revealed their broad anticancer spectrum in experimental animals in contrast to conventional cytotoxic chemotherapeutic agents. For example, TNP-470 (Takeda Neoplastic Product-470), a synthetic analogue of fumagillin (Ingber et al., 1990; Folkman, 1998), inhibits methionine aminopeptidase-2 potently and selectively in endothelial cells (Griffith et al., 1997; Sin et al., 1997). In the endothelial cell cycle pathway, TNP-470 blocks cdk2, inhibits cdc, and inhibits Rb-phosphorylation. It inhibits endothelial cell proliferation

TABLE 10.2. Angiogenesis Inhibitors Identified in the Folkman Laboratory from 1980 to 2003

Angiogenesis Inhibitors (1980–2003)		
1980	Interferon α/β, new activity	(Brouty-Boye, D. and Zetter, B.R. *Science 208:516–518, 1980*)
1982	Platelet factor 4 Protamine	(Taylor, S. and Folkman, J. *Nature 297:307–312, 1982*)
1985	Angiostatic steroids	(Crum, R. et al. (Folkman) *Science 230:1375–1378, 1985*)
1990	TNP-470 a fumagillin analogue	(Ingber, D. et al. (Folkman) *Nature 348:555–557, 1990*)
1994	Angiostatin	(O'Reilly, M. et al. (Folkman) *Cell 79:315–328, 1994*)
1994	Thalidomide	(D'Amato, R.J. et al. (Folkman) *Proc Natl cad Sci USA 91:4082–4085, 1994*)
1994	2-methoxyestradiol	(D'Amato, R.J. et al. (Folkman) *Proc Natl Acad Sci USA 91:3964–3968, 1994*)
1997	Endostatin	(O'Reilly, M. et al. (Folkman) *Cell 88:277–285, 1997*)
1999	Cleaved antithrombin III	(O'Reilly, M. et al. (Folkman) Science *285:1926–1928, 1999*)
2002	3-amino thalidomide	(Lentzsch, S. et al. (D'Amato) *Cancer Res 62:2300–2305, 2002)*
2003	DBP-*maf*	(Kisker, O. et al. (Folkman) *Neoplasia 5:32–40, 2003)*

in vitro at concentrations three logs below concentrations that inhibit tumor cell proliferation (see Ingber et al., 1990; Sin et al., 1997). Many investigators reported inhibition of 50 different tumors from 40% to 100% (complete regression) in tumors and metastases implanted in mice, rats, hamsters, and rabbits and human tumors implanted in immunodeficient mice. TNP-470 has the widest antitumor spectrum in preclinical studies of a known anticancer agent (Table 10.3).

Endostatin, a 20 kD internal fragment of collagen XVIII, was the first endogenous angiogenesis inhibitor to be isolated as a cryptic fragment of a basement membrane (O'Reilly et al., 1997; Boehm et al., 1997; for review see, Folkman and Kalluri 2003, pp 173). The alpha5beta1 integrin has been identified as a functional receptor for endostatin (Sudhakar et al., 2003; Wickstrom et al., 2002). Endostatin interferes with phosphorylation of serine 1177 in endothelial nitric oxide synthase (eNOS) (Urbich et al., 2002). Endostatin also inhibits the VEGF receptor (Kim et al., 2002), binds and inhibits metalloproteinase-2 (MMP-2) (Lee et al., 2002), blocks endothelial cell motility by binding to the $alpha_5$ and $alpha_v$ integrins (Rehn et al., 2001), and inhibits cyclin D1 in endothelial cells (Hanai et al., 2002). Endostatin also has a wide spectrum of anti-tumor activity in tumor-bearing mice (Table 10.4).

TABLE 10.3. Spectrum of Tumor Types Inhibited by TNP-470 in Animals

Experimental tumors inhibited by TNP-470			
Human tumors	% inhibition	Mouse tumors	% inhibition
Gastric carcinoma	43	TBJ neuroblastoma	60
Embryonal rhadomyosarcoma	47	Retinoblastoma	60
Ovarian carcinoma	60	Lewis lung carcinoma	64
Choriocarcinoma	60	C-1300 neuroblastoma	67
Colon carcinoma	61	Mammary carcinoma	70
Meningioma (benign)	63	B16 melanoma	71
Medulloblastoma	66	Colon 38 carcinoma	75
Prostate carcinoma	67	Renal cortical adenocarcinoma	77
MDA-MC-231 breast carcinoma	72	MCA-105 fibrosarcoma	83
T98G glioblastoma	72	Hemangioendothelioma	90
Meningioma (malignant)	77	M5076 reticulum cell sarcoma	91
MCF-7 breast carcinoma	80		
U87 glioma	95		
Breast carcinoma	96		
Neurofibrosarcoma	97		
Neurofibroma	100		
Acoustic neuroma	100		

Metastatic tumors	Location	% inhibition
H59 carcinoma	Lung	50
MCA-105 fibrosarcoma (mouse)	Lung	67
Lewis lung carcinoma-L1 (mouse)	Lung	69
Fibrosarcoma A5653HM (rat)	Lymph nodes	69
GCH-1 choriocarcinoma (human)	Lung	73
TBJ neuroblastoma (mouse)	Lymph nodes	81
C-1300 neuroblastoma (mouse)	Lymph nodes	82
B16 B16 melanoma (mouse)	Lung	87
AH-130 hepatoma (rat)	Liver	89
Bomirski Ab melanoma (hamster)	Lung	89
B16 F10 melanoma (mouse)	Lung	90
M27 lung carcinoma (mouse)	Lung	90
Hepatocellular carcinoma (human)	Liver	90
VX-2 carcinoma (rabbit)	Liver	92
Renal adenocarcinoma (mouse)	Lung & liver	92
Colon adenocarcinoma (human)	Liver	93
LM8 osteosarcoma (mouse)	Lung	97
M5076 reticulum sarcoma (mouse)	Liver	98
Breast carcinoma (human)	Lymph nodes	100
NUC-1 choriocarcinoma (human)	Lung	100
M5076 reticulum sarcoma (mouse)	Lung	100
Az-H5c gastric carcinoma (human)	Liver	100
MT-5 gastric carcinoma	Liver	100

Source: as reported by different laboratories all using a similar dose of 30 mg/kg subcutaneously every other day; modified from Folkman (1988).

Note: For references to each tumor type see (Folkman 1998 with permission of the publisher).

TABLE 10.4. Wide Spectrum of Tumors Inhibited in Mice by Recombinant Murine or Human Endostatin

Endostatin protein: Experimental anti-cancer	
1. **Fibrosarcoma, Lewis lung, B-16 melanoma**	*O'Reilly Cell 1997: Nature 1997*
2. **Spontaneous pancreatic islet carcinoma**	*Bergers Science 1999*
3. **Spontaneous mammary carcinoma**	*Yokoyama, Cancer Res. 2000*
4. **Spontaneous murine breast cancer**	*Perletti, Cancer Res. 2000*
5. **Ovarian cancer**	*Yokoyama, Cancer Res, 2000*
6. **SQ20B radio-resistant ca.** (+/– short radioRx)	*Hanna, Cancer Jour. 2000*
7. **Human pancreatic cancer**	*Kisker, Cancer Res 2001*
8. **Lewis lung carcinoma**	*Huang Cancer Res, 2001*
9. **Human neuroblastoma**	*Kuroiwa, Int. J. Mol. Med 2001*
10. **Human lung cancer**	*Boehle, Int J. Cancer 2001*
11. **Lymphoma** (+/– angiostatin)	*Scappaticci, Angiogenesis 2001*
12. **Murine neuroblastoma** (+/– GFP)	*Davidoff Cancer Gene Ther 2001*
13. **Murine gliomas** (released from cells in beads)	*Read, Nature Biotechnology 2001*
14. **Gliosarcoma**	*Joki, Nature Biotechnology 2001*
15. **Glioblastoma** (released from cells in beads)	*Sorensen, Neuro-oncol 2000*
16. **Thyroid**	*Ye, Endocrinology 2002*
17. **Leukemia**	*Iversen, Leukemia 2002*
18. **Liver metastases** (+/– chemotherapy)	*te Velde, Brit J. Surg 2002*
19. **Rat malignant glioma**	*Sorensen, Neoplasia, 2002*

Studies with endostatin revealed another general principle about anti-angiogenic therapy: continuous administration of endostatin inhibited tumor growth approximately 10 times more effectively than once a day bolus dosing (Fig. 10.1) (Kisker et al., 2001). In fact continuous administration of endostatin over 24 hours caused tumor regression (>97% inhibition) in a $p53^{-/-}$ human pancreatic cancer in SCID mice, whereas the same dose administered as a once/day bolus only inhibited tumor growth by approximately 68% (Fig. 10.1). A compelling pharmacologic proof that tumor growth is angiogenesis dependent was demonstrated by Kim et al., who showed that a specific anti-VEGF antibody administered to mice bearing a tumor that secreted only VEGF caused significant inhibition of tumor growth (Kim et al., 1993).

Genetic Evidence That Tumor Growth Is Angiogenesis Depeendent

Douglas Hanahan hybridized the large T antigen of the SV40 oncogene to the rat insulin promoter, which was then expressed in transgenic mice (RIP-Tag mice) (Hanahan, 1985). The oncogene was expressed in every beta cell of the pancreatic islets and only in beta cells. Approximately 50% of the islets began to grow (hyperplasia), doubled their size, and then stopped expanding at approximately 5 weeks. At approximately 6 to 8 weeks, 10% of the islets became angiogenic. From these angiogenic

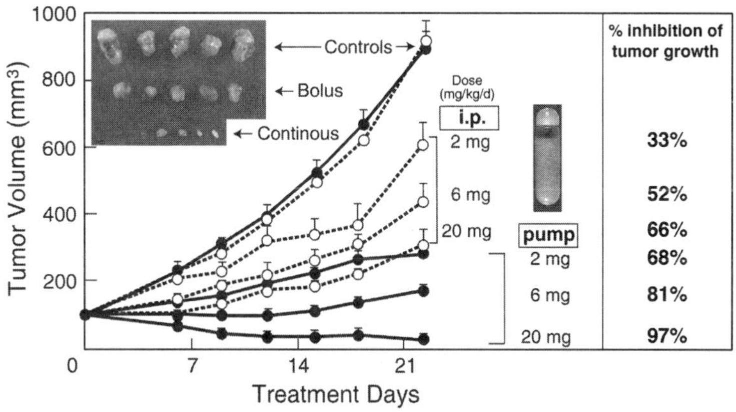

Figure 10.1. Continuous versus bolus administration of human endostatin to SCID mice bearing human pancreatic cancer that is $p53^{-/-}$. An intraperitoneal microosmotic pump releasing endostatin at 1 microliter/hour provided continuous dosing and yielded a constant blood level of 200 ng/ml to 250 ng/ml (normal = <20 ng/ml). In contrast, the same 24 hour dose administered as a bolus intraperitoneal injection once a day resulted in a rapid high peak blood level of endostatin level followed by a rapid fall below therapeutic concentrations within a few hours (pharmacokinetic data not shown here, but reported in Kisker et al., 2001). Continuous dosing regresses tumors (>97% inhibition), whereas bolus dosing does not. The implication of this study is that to therapeutically inhibit tumor growth, or to cause tumor regression, endothelial cells in the tumor vascular bed must be continuously exposed to the therapeutic levels of inhibitor, to counteract (or to titrate against) stimulators of endothelial growth and migration from tumor cells and from local stromal cells, which constantly bathe the endothelium. (From Folkman and Kalluri, 2003, with permission of the publisher.) (See color insert.)

islets, 3% to 4% grew rapidly into large tumors and killed the mice by about 14 weeks. This system was employed to develop a model of the "angiogenic switch," in which the onset of tumor angiogenesis is the result of a shift in the net balance between expressed positive and negative regulators of angiogenesis (Hanahan and Folkman, 1996). In subsequent experiments it was demonstrated that VEGF was the predominant pro-angiogenic protein and that it was highly expressed in the pre-angiogenic islets and also in the angiogenic tumors (Inoue et al., 2002). The VEGF receptor was highly expressed on microvascular endothelium in the pre-angiogenic islets as well as in the angiogenic tumors. When VEGF was deleted from the pre-angiogenic islets by Cre-lox technology, there was a 90% reduction in the number of angiogenic islets by 10 weeks and a 95% reduction in total tumor burden by 16 weeks compared to the tumor burden of the wild-type mice at 14 weeks who were dying of a large tumor burden (Fig. 10.2). This is one line of genetic evidence that pro-angiogenic proteins play a role in the angiogenic switch.

In a different experiment Arbiser (Arbiser et al., 1997) transfected endothelial cells with the SV40-T oncogene. The cells were immortalized

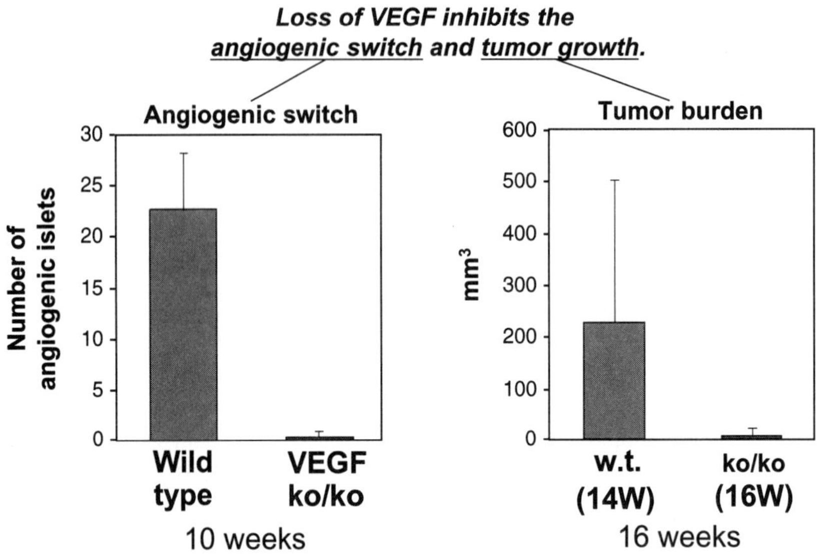

Figure 10.2. In transgenic mice that develop carcinomas of the beta cells in the pancreatic islets, deletion of the gene of VEGF only in these cells (by the cre-lox system) significantly decreases the number of angiogenic islets and final tumor burden. This is one example of genetic evidence that tumor growth is angiogenesis dependent. (Inoue et al., 2002)

but were poorly tumorigenic when injected subcutaneously into mice. Small tumors (<0.5 mm) formed but remained non-angiogenic and failed to expand further. When these cells were then transfected by a second oncogene, *ras*, VEGF expression and metalloproteinases 2 and 9 were all up-regulated, while tissue inhibitors of metalloproteinases were down-regulated. Large highly angiogenic angiosarcomas grew rapidly.

One of the most compelling genetic proofs that tumors are angiogenesis dependent, was reported by Lyden (Lyden et al., 1999). Id1 and Id3 are transcription factors active during development. If both genes are deleted, the embryo dies. However, if one aliele of Id1 and both alleles of Id2 are deleted a relatively normal mouse is born. However, when tumor cells are injected subcutaneously, small avascular tumors appear but fail to expand for most tumor types (Fig. 10.3). However, if wild-type bone marrow cells are injected intravenously into these mice, endothelial precursor cells carrying the normal Id1 and Id3 genes home to the tiny avascular tumor, which then becomes highly angiogenic and grows rapidly (Lyden et al., 2001). These experiments provide evidence that endogenous angiogenesis inhibitors defend against the angiogenesis switch in tumors.

At least 12 well-characterized negative regulators of angiogenesis are either expressed by host cells—such as interferon-alpha (Folkman et al., 1997), pigment epithelial derived factor (Dawson et al., 1999), 2-methoxyestradiol (D'Amato et al., 1994; Fotsis et al., 1994), platelet

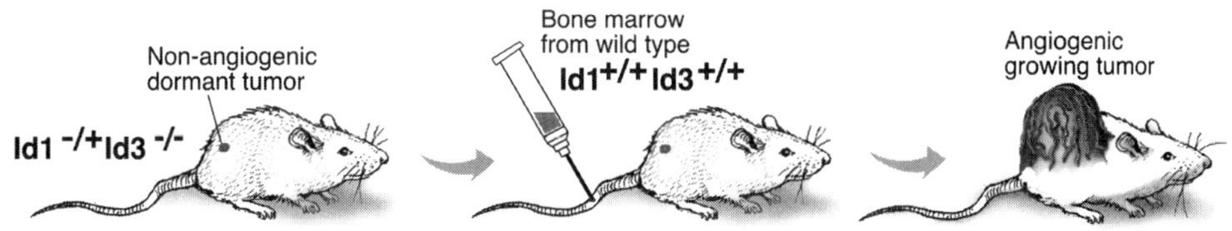

Figure 10.3. Subcutaneous implantation of tumor cells into transgenic mice in which one allele of Id1 and two alleles of Id3 have been deleted develop small tumors that are generally unable to induce angiogenesis (Lyden et al., 1999). Intravenous injection of wild-type bone marrow provides a source of endothelial cells that traffic to the tumor and permit it to become angiogenic and grow rapidly. This is another genetic proof that tumor growth is angiogenesis dependent. (Lyden et al., 2001)

factor 4 (Taylor and Folkman, 1982; Maione et al., 1990), tetrahydrocortisol (Crum et al., 1985; Williams 1999), and thrombospondin-1 (Volpert et al., 1995; Rodriguez-Manzaneque et al., 2001)—or are generated by enzymes secreted by host cells or by tumor cells—angiostatin (O'Reilly et al., 1994), arresten (Colorado et al., 2000), cleaved anti-thrombin III (O'Reilly et al., 1999), endostatin (O'Reilly et al., 1997), tumstatin (Maeshima et al., 2001), and vitamin D binding protein-*maf* (Kisker et al., 2003). Three of these endogenous angiogenesis inhibitors illustrate general mechanisms by which these inhibitors may guard against pathological angiogenesis and especially tumor angiogenesis.

Thrombospondin. Thrombospondin-1 inhibits angiogenesis and tumor growth (Lawler et al., 2002; Tanaka et al., 2002; Reiher et al., 2002). Transfection of thrombospondin-1, or thrombospondin-1 and -2, potently inhibited growth of a human tumor in mice by inhibition of angiogenesis. Tumor proliferation rates were not different from controls (Streit et al., 1999). This response is similar to the response of other tumors where angiogenesis is blocked. Tumor cell proliferation does not decrease significantly, but tumor cell apoptosis increases significantly (Holmgren et al., 1995; Achilles et al., 2001). Thrombospondin-1 expression in host cells and in pre-angiogenic tumor cells is increased by wild-type p53 (Dameron et al., 1994). In fact, p53 also inhibits angiogenesis by three other mechanisms: degradation of hypoxia-inducible factor-1 (HIF-1) (Ravi et al., 2000), repression of VEGF expression (Zhang et al., 2000), and down-regulation of expression of an FGF binding protein (Sherif et al., 2001). Thrombospondin-1 expression is repressed by the cooperation of *ras* and *myc*, which leads to increased tumor angiogenesis (Watnick et al., 2003). Thrombospondin-1 is increased in plasma, and angiogenesis and tumor growth are potently inhibited in mice (95%), when a continuous low dose (metronomic therapy) (Browder et al., 2000; Klement 2000; Hanahan et al., 2000) of a chemotherapeutic agent (cyclophosphamide) is administered to tumor-bearing mice (Bocci et al., 2003). In contrast, metronomic therapy with cyclophosphamide had no effect on angiogenesis or tumor growth in thrombospondin-1 null mice bearing the same tumor. This provocative finding suggests that it may be fruitful to determine whether any other chemotherapeutic agents, when administered on a metronomic low dose schedule, will increase expression of thrombospondin-1 or other endogenous angiogenesis inhibitors. The same idea may be extended to noncytotoxic anticancer drugs. For example, Herceptin increases thrombospondin-1 expression by fivefold.

Endostatin. The anti-angiogenic and anti-tumor activity of endostatin that is observed against tumors in animals and humans generally requires serum levels of approximately 250 nanograms/ml or above (normal = 20 ± 11 ng/ml), when the recombinant protein is administered as a subcutaneous injection once or twice daily. However, lower circulating levels may act as a physiological defense against pathologic angiogenesis. For example, the incidence of solid tumors is lower in people with Down syndrome than in any other humans as determined by a very recent study of 17,897 age-matched patients with Down syndrome (Yang et al., 2002). Down patients have virtually no breast cancer, prostate cancer, colon cancer, or other solid tumors. They have a mild form of erythroleukemia and the same incidence of testicular cancer as normal controls. Their overall incidence of cancer is less than 10% of the normal human population, and it would be even less if testicular cancer were not included. This clinical observation has been known for at least 50 years, but it was simply explained that these patients did not live long enough. However, now most Down patients are living until middle age or later, and there has still been no satisfactory explanation for why they appear to be the most protected against cancer of all humans. Recently it has been found that Down patients have an extra copy of collagen XVIII on chromosome 21, and as a result, a significantly higher level of circulating endostatin (39 ± 20 ng/ml) than normal controls (20 ± 11 ng/ml) (Zorick et al., 2001). While these results are only correlative data, they are considerably strengthened by three additional new findings. Testicular cancer employs a novel angiogenic protein Bv8 to induce angiogenesis, and the receptor Bv8R is restricted to testicular endothelium (LeCouter et al., 2003). It is not clear that endostatin, for which a functional receptor is $alpha_5beta_1$, can block Bv8. Furthermore atheromas are virtually nonexistent in patients with Down syndrome (Chadefaux et al., 1988) and diabetic retinopathy is also extremely rare in Down patients despite a similar incidence of diabetes as in patients without Down syndrome (Fulcher et al., 1990). Endostatin has been shown to inhibit atheroma formation by 85% in ApoE null mice, by its ability to inhibit neovascularization in arterial plaques (Moulton et al., 1999). Therefore it is possible that the elevated blood level of endostatin is acting to prevent angiogenesis in these three pathological states.

Additional evidence that the endogenous anti-angiogenic activity of endostatin is defending against tumor growth is the finding that a polymorphism of endostatin, in which an asparagine is substituted for aspartic acid at amino acid 104 (from the N-terminus), increases the risk of prostate cancer 2.5-fold (Iughetti et al., 2001). Other forms of pathologic angiogenesis are also prevented by physiologic levels of endostatin. For example, children born with Knobloch syndrome are blind because of failure of development of retinal vessels (vitreo-retinal degeneration and retinal detachment; Sertie et al., 2002). Recent reports show that these patients have a mutation in the alpha1collagen XVIII short chain that leaves the eye without endostatin. Normally the lens is vascularized by hyaloid vessels during fetal life and becomes avascular by 30 days after birth. However, in the absence of endostatin the hyaloid vessels fail to regress in the vitreous. It is thought that this elevates oxygen in the eye, which inhibits VEGF expression and therefore prevents development of retinal vasculature. This mechanism is supported by collagen XVIII null mice, which also have delayed regression of blood vessels in the vitreous and retinal detachment (Fukai et al., 2002).

Tumstatin. After endostatin was discovered to be a proteolytic cryptic fragment in collagen XVIII, Kalluri and colleagues discovered that tumstatin (and three other proteins) could be isolated from collagen IV. Tumstatin is a peptide of 232 amino acids in the NC1 domain of the alpha3 chain of collagen IV. Like endostatin, tumstatin is highly conserved. Collagen and IV and XVIII are the only collagens present in all vertebrates including *C. elegans*. When the alpha 3 chain was deleted, the normal blood level of tumstatin of 336 ± 28 ng/ml fell to 0. Tumors grew 300% to 400% faster in the knockout mice than in wild-type mice (Hamano et al., 2003). This is the first demonstration that tumors in wild-type mice do not grow at ceiling rates, but harbor a capacity for more rapid growth once the endogenous angiogenesis inhibitor brakes are removed (Fig. 10.4). When tumstatin was replaced at physiological levels, namely at 300 ng/day, which because of a half-life of 14 hours achieves a blood level of approximately 336 ng/ml, tumors returned to the same slower growth rate of the wild-type tumors. This experiment fulfills the classic paradigm of a tumor-suppressor protein, like p53, except that tumstatin is purely anti-angiogenic and has no other known function at this writing. Furthermore a functional receptor for tumstatin has been identified as the alpha$_v$beta$_3$ integrin (Sudhaker et al., 2003). Tumstatin or a 25 amino acid peptide significantly inhibited DNA synthesis in endothelial cells from wild-type mice but not in endothelial cells from alpha$_v$beta$_3$ null mice (Hamano et al., 2003). Tumstatin also suppressed angiogenesis in beta$_3^{+/+}$ mice but not in beta$_3^{-/-}$ mice. In contrast, endostatin blocked DNA synthesis in endothelial cells, which either expressed the alpha$_v$beta$_3$ integrin or did not express this integrin. Endostatin also blocked angiogenesis in both beta$_3^{+/+}$ mice and in beta$_3^{-/-}$ mice. Endostatin does not require alpha$_v$beta$_3$ for its anti-endothelial or functions. Metalloproteinase 9 effectively cleaves tumstatin from type IV collagen. In metalloproteinase-null mice, tumstatin blood levels decreased by 40%, and tumors grew 70% faster in the metalloproteinase-null mice than in the wild-type mice. When both tumor sets had reached 2 cm^3 (i.e., 20 days for

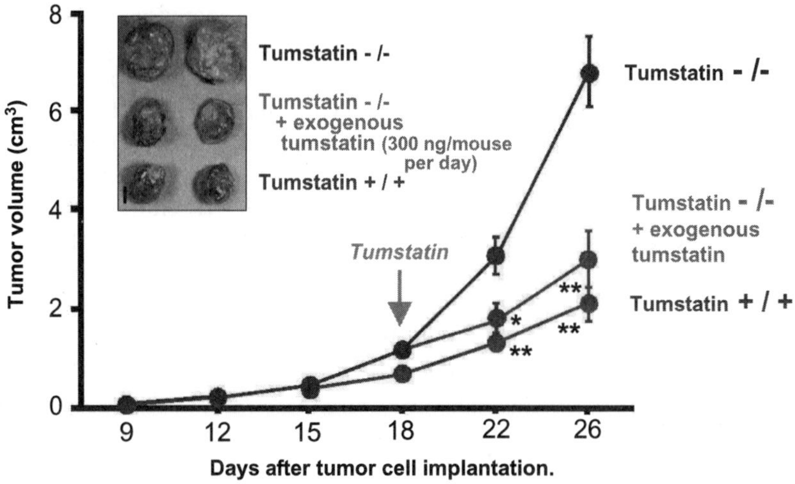

Figure 10.4. In mice depleted of the endogenous angiogenesis inhibitor tumstatin, tumors grow 300% to 400% more rapidly than in wild-type mice. This experiment shows that tumstatin is an endogenous inhibitor of angiogenesis and that tumors in wild-type mice do not grow at ceiling rates. It provides additional genetic proof that tumor growth is angiogenesis dependent. (Hamano et al., 2003.) (See color insert.)

the metalloproteinase-null mice and 28 days for the wild-type-mice), metalloproteinase-null mice were treated with replacement doses of tumstatin to obtain physiologic levels of tumstatin. At that point the tumors in the metalloproteinase-null mice slowed to the same growth rate as the tumors in the wild-type mice (Hamano et al., 2003). Therefore tumstatin experiments show that it is possible to delete an endogenous angiogenesis inhibitor, tumstatin (a ligand), delete its receptor, or delete the enzyme that releases the ligand from its matrix. All predictably show the same effect: increased angiogenesis followed by increased tumor growth. If tumstatin is restored to physiological levels, tumor growth is slowed to rates of tumors in wild-type mice. Wound healing and pregnancy are unaffected in mice treated with either endostatin or tumstatin.

In summary, when the results from experiments with thrombospondin, endostatin, and tumstatin are taken together, they provide genetic evidence that a normal physiological function of an endogenous angiogenesis inhibitor may be to defend against pathological angiogenesis.

Role of Oncogenes in the Switch to the Angiogenic Phenotype

Oncogene products appear to override endogenous inhibitors of angiogenesis during the switch to the angiogenic phenotype. Oncogenes were initially discovered because of their ability to induce proliferation of pre-

TABLE 10.5. Pro-angiogenic Oncogenes

Oncogene	Implicated pro-angiogenic activity
K-ras, H-ras	VEGF upregulation, TSP-1 downregulation
v-src	VEGF upregulation, TSP-1 downregulation
c-myb	TSP-2 downregulation
N-myc	angiogenic properties in neuroblastoma
c-myc	angiogenic properties in epidermis
HER-2	VEGF upregulation
EGFR	VEGF, bFGF, IL-8 upregulation
PyMT	TSP-1 downregulation
c-fos	VEGF expression
trkB	VEGF downregulation
HPV-16	secretion of VEGF and IFN-a
v-p3k	VEGF production and angiogenesis
ODC	novel angiogenic factor
PTTG1	VEGF and bFGF upregulation
E2a-Pbx1	induction of mouse angiogenin-3
bcl-2	VEGF upregulation

Source: Assembled from Rak et al. (2000), Fernandez et al. (2000), and Folkman and Kalluri (2003).

neoplastic or neoplastic cells in vitro and to protect these cells from apoptosis. Recent experiments from many laboratories reveal that oncogenes are also pro-angiogenic (Rak et al., 1995, 2000) (Table 10.5) (For references for each oncogene, see the review in Kerbel and Folkman, 2002.) For example, *ras* oncogene up-regulates VEGF and down-regulates thrombospondin-1. *Src* oncogene has similar pro-angiogenic activity. EGFR up-regulates VEGF, bFGF, and IL-8. Bcl-2 up-regulates VEGF. When Bcl-2 was transfected into prostate cancer cells that were already angiogenic, there was a marked rise in angiogenic activity as determined by increased microvessel density in tumor-bearing mice (Fernandez et al., 2001).

From these studies an important lesson for clinicians is that when an oncogene is used as a target to develop an anticancer drug, the drug may inhibit tumor cell production of one or more pro-angiogenic factors, such as VEGF or bFGF, and may also restore the physiological level of an endogenous angiogenesis inhibitor, such as thrombospondn-1. For example, an anticancer drug that targets tyrosine kinase in the signaling pathway of the EGF receptor in a cancer cell may be a potent angiogenesis inhibitor. Such a drug could have a broader spectrum of anticancer activity than would be predicted only from its label as an "EGF receptor tyrosine kinase inhibitor" (Fig. 10.5).

Direct and Indirect Angiogenesis Inhibitors

It is also possible that certain limitations accrue to an anticancer drug that targets an oncogene or its product, or the receptor for that product

Drugs developed by screening for molecules which block the **EGF receptor tyrosine kinase**.

Anti-angiogenic mechanism:
Inhibits tumor production of:
VEGF
bFGF
TGF-alpha
(possibly IL-8)

ZD 1839	Iressa	AstraZeneca (approved - Japan)
ZD 6474		AstraZeneca
OSI 774	Tarceva	Owned by Genentech
CI 1033		Pfizer
PKI 1666		Novartis
IMC 225	Erbitux	Imclone

Figure 10.5. Anticancer drugs that target oncogenes or their products or their receptors, may inhibit angiogenic proteins. (Folkman and Kalluri, 2003)

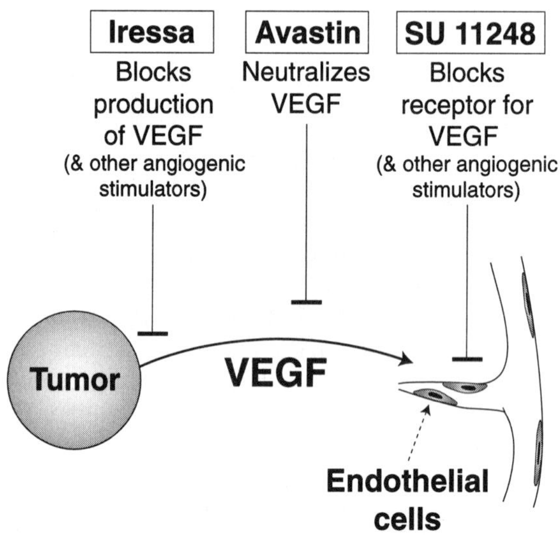

Figure 10.6. "Indirect" angiogenesis inhibitors target tumor cell production of an angiogenic protein, the protein itself, or its receptor on endothelial cells. (From Folkman and Kalluri, 2003)

on vascular endothelium. This type of anticancer drug may be classified as an "indirect" angiogenesis inhibitor (Fig. 10.6). (Folkman and Kalluri, 2003). For example, an EGF receptor tyrosine kinase inhibitor could suppress production of VEGF and bFGF by a tumor. But over time, if mutations occurred so that the tumor began to produce redundant pro-angiogenic proteins such as PLGF (placenta like growth factor) or PD-ECGF (platelet derived endothelial growth factor), the tumor could appear to have become drug resistant. Also treatment of all patients

Types of Angiogenesis Inhibitors

	Indirect	Direct
Inhibitor	*Iressa*	*Endostatin*
Mechanism	Inhibits synthesis by tumor cells of angiogenic proteins: **bFGF, VEGF,** and **TGF-α.**	Inhibits endothelial cells from responding to multiple angiogenic proteins, e.g., **bFGF, VEGF, IL-8, PDGF.**

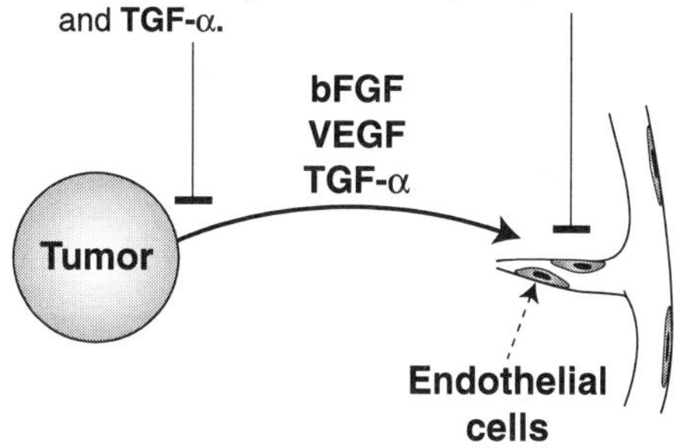

Figure 10.7. "Direct" angiogenesis inhibitors prevent endothelial cells from responding to a spectrum of angiogenic proteins. (From Folkman and Kalluri, 2003)

without stratifying them for VEGF-producing tumors or bFGF-producing tumors could lead to failure of a good drug because of a poorly designed clinical trial. In contrast a "direct" angiogenesis inhibitor targets vascular endothelium directly, and prevents it from responding to a wide spectrum of pro-angiogenic stimuli. Endostatin and TNP-470 are examples of direct angiogenesis inhibitors (Fig. 10.7). Direct angiogenesis inhibitors are less susceptible to loss of efficacy because of the emergence of redundant pro-angiogenic proteins in a tumor.

Genetic Stability of Vascular Endothelial Cells

The fundamental basis for the clinical use of angiogenesis inhibitors against cancer is that these therapeutic agents are directed against a genetically stable target—the vascular endothelial cell (Fig. 10.8) (Folkman et al., 2000, 2001). For the past 50 years the genetically unstable tumor cell has been the major target of anticancer cytotoxic chemotherapy. However, the high rate of mutations in most tumor cells increases the risk of emergence of resistant tumor cells. For example, a single colorectal carcinoma cell may contain approximately 11,000 genomic alterations. (Stoler et al., 1999). In contrast, there are no genomic alterations in endothelial cells. When endothelial cells are recruited to a tumor, they may undergo temporary changes in gene expression, but these appear to reverse when the angiogenic stimulus from a tumor or its stroma is interrupted (St Croix et al., 2000). A provocative recent example of one of these induced changes in gene

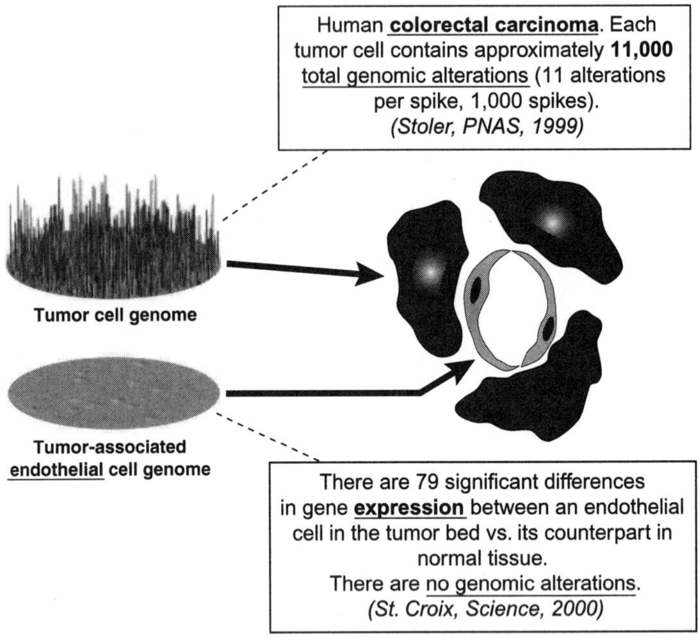

Figure 10.8. Tumor cells are genetically unstable and contain thousands of genomic alterations. In contrast, endothelial cells are genetically stable, and when recruited to a tumor vascular bed, they may undergo temporary changes in gene expression but no genomic alterations. (Folkman et al., 2000, with permission of the publisher.) (See color insert.)

expression is the demonstration that conditioned medium from human glioblastoma can induce vascular endothelial cells to express telomerase (Falchetti et al., 2003).

This finding sheds light on a central question in tumor angiogenesis research. If endothelial cells in the vascular bed of a tumor are stimulated to undergo a high proliferation rate, why don't we see more evidence of endothelial cell senescence? It is now clear that the microvascular endothelial cell recruited to a tumor has become an important second target in cancer therapy. As angiogenesis inhibitors are becoming more available, it is recognized that treating both the cancer cell and the endothelial cell in a tumor, or even treating the endothelial cell alone, may be more effective than treating the tumor cell alone.

Antiangiogenic Chemotherapy

If, as Kisker and coworkers (2001) showed, antiangiogenic therapy is best administered on a schedule that permits continuous exposure of the activated microvascular endothelium in the tumor bed to angiogenesis inhibitor(s), would administration of conventional chemotherapy on a similar schedule improve its efficacy? Could a drug-resistant tumor be inhibited by the same chemotherapeutic agent if it were administered on a dose schedule that would optimize its antiangiogenic activity? Timothy

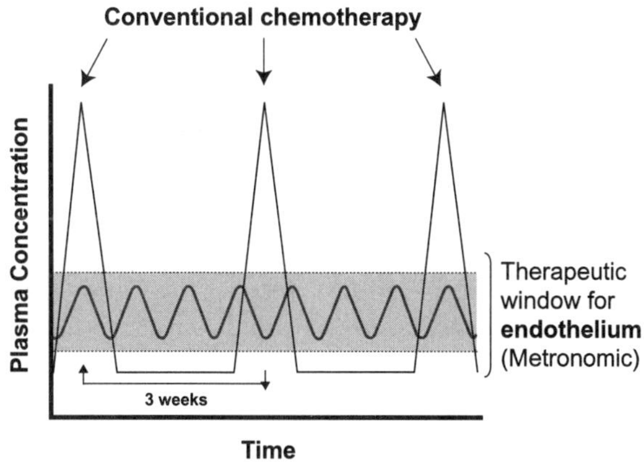

Figure 10.9. Conventional cytotoxic chemotherapy is administered at maximum tolerated dose with treatment-free intervals of a week or weeks, to rescue bone marrow and to permit return of gastrointestinal function. This off-therapy interval may allow regrowth of microvascular endothelial cells in the tumor bed. In contrast, when chemotherapy is administered at lower doses and more frequently, its anti-angiogenic activity is optimized. This is called metronomic therapy (or antiangiogenic chemotherapy, or low-dose chemotherapy). In experimental animals, metronomic therapy provides effective tumor inhibition even when the tumor cells themselves are resistant to the chemotherapeutic agent. (See Browder T., et al., 2000, and Klement G., et al., 2000.)

Browder in the Folkman laboratory answered this question by mouse experiments in which drug-resistant tumors that had failed to respond to "conventional scheduling" of cyclophosphamide (i.e., maximum tolerated doses administered for 3 days in 21 day cycles) regressed when mice were treated on an antiangiogenic (metronomic) schedule of cyclophosphamide (Fig. 10.9) administered every 6 days at a total lower dose (Browder et al., 2000; Folkman, 2003). Endothelial cell apoptosis in the tumor bed *preceded* apoptosis of tumor cells by 4 to 5 days. This report was confirmed and extended by Robert Kerbel's laboratory (Klement et al., 2000), using etoposide instead of cyclophosphamide. In a subsequent editorial by Douglas Hanahan (Hanahan et al., 2000), this approach was termed "metronomic chemotherapy." These results may explain why some patients who receive long-term maintenance or palliative chemotherapy experience stable disease beyond the time when the tumor would have been expected to develop drug resistance. Patients who are on continuous infusion of 5-fluorouracil, weekly paclitaxel, or daily etoposide, have shown improved outcomes despite the fact that the tumor had already become drug resistant in some of these patients (Kakolyris et al., 1998; Folkman and Kalluri, 2003). Furthermore angiogenesis inhibitors may be combined with low dose (metronomic) chemotherapy more effectively, and with less toxicity than if conventional maximum tolerated dose chemotherapy is added to antiangiogenic therapy.

Distinction between Antivascular Therapy and Anti-angiogenic Therapy

Antivascular therapy of tumors differs from anti-angiogenic therapy in at least two aspects: First, antivascular therapy causes death and/or sloughing of endothelial cells in the tumor vasculature within hours or a few days after the agent is administered intravenously. Examples are antibody-directed toxins (Burrows and Thorpe, 1993) and drugs such as combretastatin (Hori and Saito, 2003). Second, the tumor necrosis, which is induced by antivascular therapy, is rapid, patients often experience pain in their tumors, and tumor regression may occur early. In contrast, with anti-angiogenic therapy, where new microvessels are prevented from growing or are caused to involute, stable disease is usually observed first and tumor regression months later.

In animal studies and in clinical trials antivascular therapy rapidly causes a large crater of central necrosis in the tumor. However, this may be followed by a rim of rapid tumor growth. It is likely that the peripheral regrowth of tumor after combretastatin therapy is due to a surge of hypoxia-inducible factor-1 alpha (HIF-1 alpha). Therefore, if an angiogenesis inhibitor is employed in combination with antivascular therapy, or is administered after antivascular therapy, it may be prudent to choose an angiogenesis inhibitor that suppresses expression of HIF-1 alpha. Endostatin down-regulates expression of HIF-1 alpha in human endothelial cells (Abdollahi et al., unpublished data 2003). 2-methoxyestradiol inhibits expression of HIF-1 alpha in tumor cells (Mabjeesh et al., 2003).

Role of Angiogenesis in Tumor Progression

Tumors can become more locally aggressive or more metastatic by several angiogenic mechanisms: (1) induction of stromal angiogenesis (Fukumura et al., 1998), (2) mobilization of endothelial precursor cells from the bone marrow (Asahara et al., 1997) and (Rafii et al., 2002), (3) expression of redundant angiogenic promoters (e.g., breast cancers, which can express up to six angiogenic proteins) (Relf et al., 1997), (4) induction of telomerase expression in endothelial cells by neighboring tumor cells (Falchetti et al., 2003), and (5) possible loss of endogenous angiogenesis inhibitors by proteinuria induced by circulating VEGF from tumor burden (Kalluri and Folkman, unpublished data 2003). Another interesting possible mechanism of tumor progression is phagocytosis of apoptotic bodies of neighboring tumor cells, which horizontally transfers genetic information and may increase aneuploidy (Bergsmedh et al., 2001). It is unclear whether increased angiogenic activity can be horizontally transferred by this route. However, it has been shown that tumors with a mutated or missing p53 can vertically transmit genetic information acquired by phagocytosis of apoptotic bodies (Bergsmedh et al., 2001).

TABLE 10.6. Some Examples of Angiogenesis Inhibitors in Clinical Trials in the United States and Europe

Drugs which are *exclusively* anti-angiogenic.	Drugs which *include* anti-angiogenic activity.
In clinical trial:	***FDA approved:***
Angiostatin	Celecoxib (Celebrex)
Avastin (Anti-VEGF antibody) (Bevacizumab)	Herceptin
Endostatin	Iressa (Inhibits VEGF production by tumor)
Tetrahydrocortisol	Rosiglitazone
TNP-470	Taxol (Ultra low continuous dose)
Thrombospondin peptide	Velcade (proteasome inhibitor PS-341)
VEGF-Trap	Zoledronic acid (Bisphosphonate)
Vitaxin	***In clinical trial:***
	C225 (Erbitux)
Not yet in clinical trial:	Combretastatin
Arresten	Interferon-alpha
Canstatin	NM-3
Cleaved Antithrombin III	OSI774 (Tarceva)
DBP-maf	PTK787
PEDF	SU 5416
Tumstatin	SU 6668
	SU 11248
	Thalidomide (& 3-amino thalidomide)
	2-Methoxyestradiol

Note: The inhibitors can be categorized into those that exclusively inhibit angiogenesis and have no other activity, and those that have other activities besides inhibiting angiogenesis.

Angiogenesis Inhibitors: Translation to the Clinic

Approximately 30 angiogenesis inhibitors are currently in clinical trial in the United States and Europe. Some of these are listed in Table 10.6. They can be divided into two categories: (1) drugs that are exclusively anti-angiogenic (i.e., Avastin an anti-VEGF antibody, VEGF-Trap, endostatin, and a peptide of thrombospondin) and (2) drugs that inhibit angiogenesis but have other activities. Several of these drugs received FDA approval for another activity before they were found to have anti-angiogenic activity (i.e., celecoxib, rosiglitazone, and zolendronic acid). The anti-angiogenic activity for some inhibitors is known to basic scientists but not widely known to clinicians. For example, Velcade is a proteasome inhibitor that received FDA approval in 2003 for use in patients with multiple myeloma. However, pre-clinical studies reveal that it is a potent angiogenesis inhibitor (Sunwoo et al., 2001; Nawrocki et al., 2002; LeBlanc et al., 2002). Clinicians who know this drug only as a proteosome inhibitor for multiple myeloma may miss the opportunity to design clinical trials that take advantage of the drug's anti-angiogenic activity against a wider spectrum of tumor types.

In 1988 low-dose interferon alpha was the first angiogenesis inhibitor to be used in humans at 2 to 3 million units/meter^2/day) (White et al., 1989; Folkman, 1989). Its antiangiogenic activity was based on its anti-motility activity against endothelial cells in vitro (Brouty-Boye and Zetter, 1980) and on subsequent studies in tumor-bearing mice (Sidky et al., 1987; Dvorak et al., 1989). It was subsequently used successfully to treat life-threatening hemangiomas and hemangioendotheliomas that had failed to respond to steroid therapy (Ezekowitz et al., 1992; Ohlms et al., 1994; Folkman et al., 1997; Ezekowitz et al., 1991; Deb et al., 2002). Hemangiomas overexpressed bFGF (Takahashi et al., 1994) and interferon alpha was found to down-regulate expression of bFGF in human tumor cells (Singh et al., 1995) and to have a biphasic effect as an angiogenesis inhibitor. Low doses were more effective than high doses in pre-clinical studies (Slaton et al., 1999). More recently low-dose interferon alpha has brought about complete and durable regression of recurrent high-grade giant cell tumors of the mandible, maxilla, and pelvis (Kaban et al., 1999, 2002; Folkman, 2002), and also angioblastomas (Marler et al., 2002), all of which had failed conventional therapies. Interferon alpha may also inhibit VEGF transcription (Von Marschall et al., 2003). From 10 years of experience with low-dose interferon alpha therapy in patients, certain principles have emerged that may apply generally to anti-angiogenic therapy. While only 8 months of low-dose interferon alpha caused durable complete regression of high-grade giant cell tumors, other tumors (i.e., angioblastoma of the palate metastatic to the tonsils; Marler et al., 2002) required more than 3.5 years of therapy for complete durable regression. Low doses administered daily were more effective than every other day dosing. This is consistent with experimental evidence that continuous exposure of tumor vasculature to an angiogenesis inhibitor is more effective than bolus dosing. Interferon alpha at low doses was well tolerated. Drug resistance was not observed. Patients were selected for those whose tumors produced bFGF as their major or sole angiogenic protein. In contrast to breast cancers, high-grade giant cell tumors and angioblastomas do not appear to form mutations that secrete a spectrum of other angiogenic proteins. Therefore it is unlikely that interferon alpha would be effective in breast cancer. The important lesson from this experience is that when an indirect angiogenesis inhibitor is used, patients may need to be stratified for those whose tumors produce the angiogenic protein that the inhibitor blocks. When a direct angiogenesis inhibitor is used (i.e., TNP-470), a wider spectrum of tumors may be treatable.

Avastin, a humanized monoclonal anti-VEGF antibody, has recently completed a successful phase III clinical trial for colorectal cancer (Hurwitz et al., 2003). It was a randomized, placebo controlled, blinded trial of more than 900 patients with metastatic colorectal cancer. Overall survival of patients treated with Avastin plus conventional chemotherapy was significantly increased compared to patients treated with chemotherapy alone. In addition, time to progression, response rate, tumor regression, and duration of response were all significantly improved. These are the first positive phase III results with an anti-angiogenic therapy for cancer.

CURRENT AND FUTURE DIRECTIONS OF CLINICAL APPLICATION OF ANGIOGENESIS INHIBITORS

Currently it is estimated that more than 100,000 cancer patients are receiving angiogenesis inhibitors (Table 10.2) in clinical trials in the United States and in other countries. It would be impossible to summarize these in this space. However, it is possible to foresee that combinations of angiogenesis inhibitors once they become FDA approved to be more commonly used with each other or in combination with conventional therapies.

Other angiogenesis inhibitors are currently in phase II clinical trials including thalidomide, 2-methoxyestradiol, and endostatin, as well as those listed in Table 10.6. As angiogenesis inhibitors become more widely used in anticancer therapy, it will be important to determine: (1) Can the harsh side-effects of conventional chemotherapy be reduced? (2) Can the risk of drug resistance be reduced? (3) Can cancer eventually be converted to a chronic manageable disease, like heart disease or diabetes? (Ezzell, 1998).

NORMAL VESSELS

Today's modern understanding of the vascular system at the biological and molecular levels emerged largely from early studies of tumor angiogenesis. Many of the positive and negative regulators of physiological blood vessel growth and quiescence were first discovered as mediators of tumor angiogenesis. For this reason, I conclude this chapter with a brief section on normal endothelium and quiescent vasculature. Below I highlight a few major differences between resting vascular endothelium and activated endothelium recruited to a tumor bed. More extensive current reviews of normal vascular endothelium and vasculature may be found in Voest and D'Amore (2001) and Tomanek (2002).

One major difference between normal microvasculature and tumor microvasculature is the configuration of normal cells around each microvessel. Normal tissue cells generally live in close apposition to a capillary blood vessel (Fig. 10.10). For example, every liver cell lives adjacent to a capillary blood vessel. Certain cells such as beta cells in the islets of the pancreas, or adipocytes, have a capillary blood vessel on each side. In contrast, tumor cells surround a capillary blood vessel in multiple layers to form a microcylinder with a radius that is limited by the oxygen diffusion. Capillary blood vessels in normal tissues are densely packed (Fig. 10.11). For example, in the human heart there are 2500 millimeters of capillary blood vessels packed into each cubic millimeter of myocardial tissue (Hoppeler and Kayar, 1988).

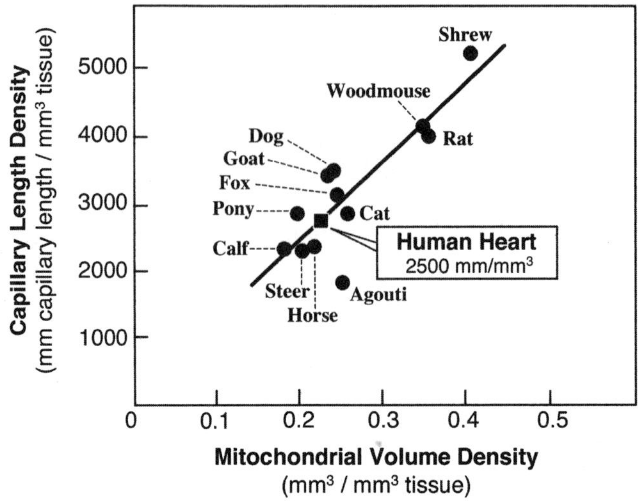

Figure 10.10. Capillary blood vessel length in cubic millimeters of tissue for different species. (From Hoppeler and Krayar, 1988)

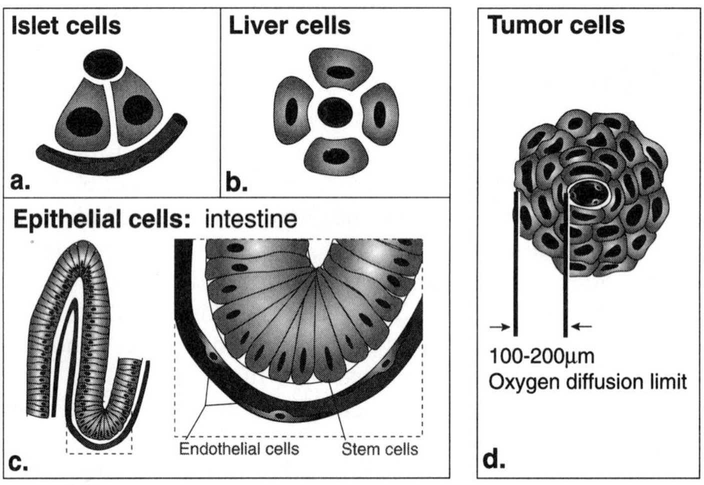

Figure 10.11. Diagram to illustrate that normal cells generally live in close apposition to a capillary blood vessel. For example, every liver cell lives adjacent to a capillary blood vessel. Certain cells like islet cells, or adipocytes, have a capillary blood vessel on each side. In contrast, tumor cells surround a capillary blood vessel in multiple layers to form a microcylinder with a radius that is limited by the oxygen diffusion distance (~100–200 microns). At this writing it is not clear what mediates the attraction of tumor cells to capillary blood vessels.

Another difference between normal endothelium and endothelium in a tumor bed is in rate of proliferation. Proliferation of endothelial cells in a Lewis lung carcinoma in mice revealed a labeling index of 40% (Holmgren et al., 1995). In a variety of experimental tumors, endothelial turnover times range from 1 day to approximately 10 days (Darland and D'Amore, 2001). In contrast, endothelial turnover times in normal tissues range from 80 days in the lung to 8000 days in the brain (Hobson and Denekamp, 1980). In an adult human, endothelial cells cover a surface area of approximately $1000\,m^2$.

Endothelial cells are among the most highly sensitive of all cells to anchorage dependence (Folkman et al., 1978), now called anoikas. After endothelial cells are removed from their in vitro substratum for as little as 10 hours they begin to undergo apoptosis. The in vivo counterpart of this phenomenon is the requirement for endothelial cells to bind to fibronectin and other extracellular matrix molecules through integrins expressed on the endothelial cell surface (Hynes, 2002). For example, if the adhesion of the integrin $alpha_v beta_3$ to extracellular matrix is interrupted, endothelial cells undergo apoptosis and newly formed tumor vessels will involute (Brooks et al., 1994).

Endothelial cells are also among the most tightly regulated by contact inhibition. Confluent layers of endothelial cells in vitro shut off DNA synthesis and cannot be stimulated to undergo another round of DNA synthesis by addition of serum, as is possible with fibroblasts and other cell types (Gimbrone et al., 1974). Tie-1 and Tie-2 receptors are expressed by endothelial cells and hematopoietic cells in all organisms so far examined including zebrafish (for review, see Suri and Yancopoulos, 2002).

Expression of angiopoietin-1 by tissue cells surrounding capillary blood vessels, and subsequent binding of angiopoietin-1 to the Tie-2 receptor of endothelial cells, stabilizes the vessels, keeps them in a quiescent state, and induces their endothelium to recruit pericytes, which appose themselves to the capillary vessel and further inhibit endothelial DNA synthesis (Fig. 10.12). In contrast, when angiopoietin-1 expression is decreased or when angiopoietin-2 expressed by endothelial cells (e.g., under activation by tumor cells) blocks the Tie-2 receptor from binding angiopoietin-1, capillary vessels become unstable, pericytes can migrate away, and endothelial cells can proliferate or undergo apoptosis, depending on the balance of positive and negative regulators of endothelial growth in the neighborhood. Thus, in the switch to the angiogenic phenotype by tumors, or during the angiogenic switch in wounds, the angiopoietins play an important permissive regulatory role that orchestrates the actions of endothelial mitogens, motility factors, and apoptotic factors in the vascular environment.

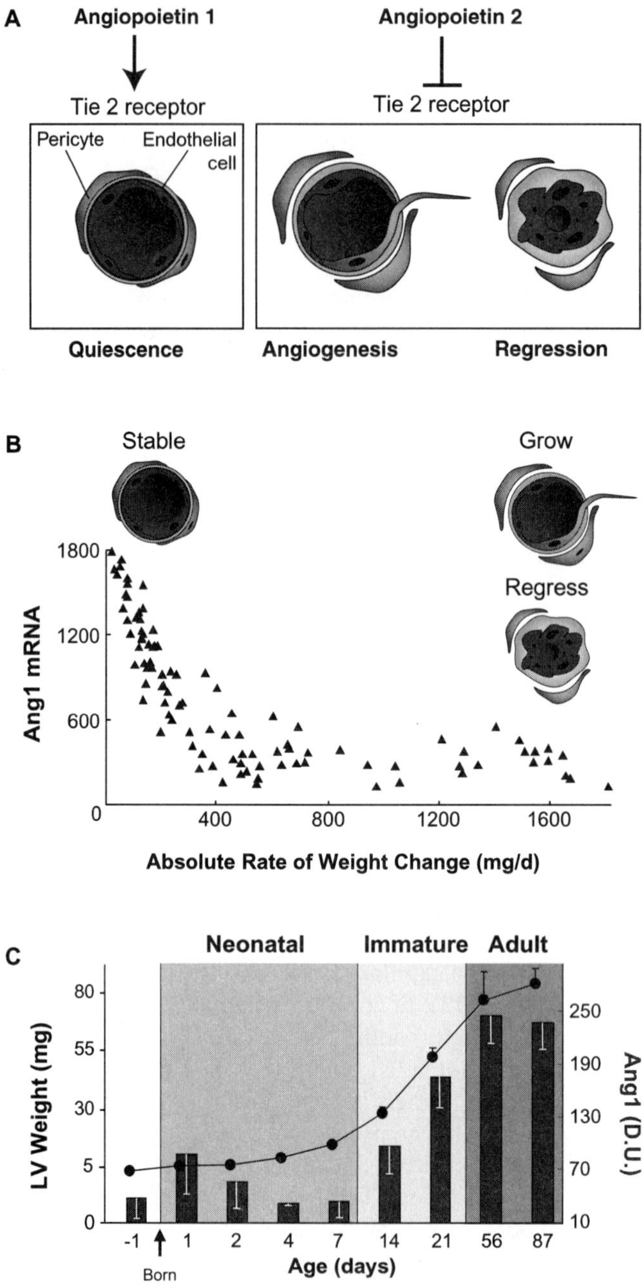
A
Angiopoietin 1
Tie 2 receptor
Pericyte
Endothelial cell
Quiescence
Angiopoietin 2
Tie 2 receptor
Angiogenesis
Regression
B
Stable
Grow
Regress
Ang1 mRNA
1800
1200
600
0
400
800
1200
1600
Absolute Rate of Weight Change (mg/d)
C
Neonatal
Immature
Adult
LV Weight (mg)
80
55
30
5
0
Ang1 (D.U.)
250
190
130
70
10
-1
1
2
4
7
14
21
56
87
Born
Age (days)

Figure 10.12. (**A**) Diagram to illustrate that tissue cells express angiopoietin that binds to the Tie-2 receptor on endothelium and induces endothelium to become quiescent. Binding of angiopoietin-1 to the Tie-2 receptor also induces endothelium to recruit pericytes, which further suppress endothelial proliferation. When endothelial cells are stimulated (i.e., by tumor cells or other cell types), angiopoietin-2 is upregulated, blocks the Tie-2 receptor from binding by angiopoietin-1, and permits endothelial cells to either undergo proliferation or apoptosis, depending on the balance of positive and negative regulators of growth in the neighborhood. (**B**) In leptin-deficient mice, as weight gain increases because of increased proliferation of adipocytes, endothelial cells in the neighboring microvessels also proliferate inversely to angiopoietin-1 expression. When angiogpoitein-1 expression is decreased, new vessels can grow or regress. When angiopoietin-1 expression is high, vessels remain stable. Each adipocyte is surrounded by at least two microvessels, which control growth of adipocyte total mass (Rupnick et al., 2002). (**C**) Neonatal heart versus adult heart. As angiopoietin-1 expression increases, myocardial tissue mass slows its growth. (Rupnick et al., unpublished, 2003)

This work was supported by the National Institutes of Health grants R01 CA37395 and P01 CA45548 and P01 CA64481 and a grant from the Breast Cancer Research Foundation, New York.

REFERENCES

Achilles EG, Fernandez A, Allred EN, Kisker O, Udagawa T, Beecken WD, Flynn E, Folkman J (2001): Heterogeneity of angiogenic activity in a human liposarcoma: A proposed mechanism for "no take" of human tumors in mice. J Natl Cancer Inst 93:1075–81.

Akhtar N, Dickerson EB, Auerbach R (2002): The sponge/Matrigel angiogenesis assay. Angiogenesis 5:75–80.

Algire GH, Chalkely HW, Legallais FY, Park H (1945): Vascular reactions of normal and malignant tumors in vivo. I. Vascular reactions of mice to wounds and to normal and neoplastic transplants. J Natl Cancer Inst 6:73–85.

Algire GH, Legallais FY (1947): Growth rate of transplanted tumors in relation to latent period and host vascular reaction. Cancer Res 7:724.

Algire GH, Chalkely HW, Legallais FY, Park H (1945): Vascular reactions of normal and malignant tumors in vivo: I. Vascular reactions of mice to wounds and to normal and neoplastic transplants. J Natl Cancer Inst 6:73–85.

Algire GH, Legallais FY, Park H (1947): Vascular reactions of normal and malignant tissues in vivo. II. The vascular reaction of normal and neoplastic tissues of mice to a bacterial polysaccharide from serratia marcesscens (bacillus prodigiosus) culture filtrates. J Natl Cancer Inst 8:53–62.

Arbiser JL, Moses MA, Fernandez CA, Ghiso N, Cao Y, Klauber N, Frank D, Brownlee M, Flynn E, Parangi S, Byers HR, Folkman J (1997): Oncogenic H-ras stimulates tumor angiogenesis by two distinct pathways. Pro Natl Acad Sci USA 94:861–6.

Asahara T, Murohara T, Sullivan A, Silver M, van der Zee R, Li T, Witzenbichler B, Schatteman G, Isner JM (1997): Isolation of putative progenitor endothelial cells for angiogenesis. Science 275:964–7.

Auerbach R, Kubai L, Knighton D, Folkman J (1974): A simple procedure for the long-term cultivation of chicken embryos. Dev Biol 41:391–4.

Bergers G, Javaherian K, Lo KM, Folkman J, Hanahan D (1999): Effects of angiogenesis inhibitors on multistage carcinogenesis in mice. Science 284:808–12.

Bergsmedh A, Szeles A, Henriksson M, Bratt A, Folkman MJ, Spetz AL, Holmgren L (2001): Horizontal transfer of oncogenes by uptake of apoptotic bodies. Proc Natl Acad Sci USA 98:6407–11.

Bocci G, Francia G, Man S, Lawler J, Kerbel RS (2003): Thrombospondin 1, a mediator of the antiangiogenic effects of low-dose metronomic chemotherapy. Proc Natl Acad Sci USA 100:12917–22.

Boehle AS, Kurdow R, Schulze M, Kliche U, Sipos B, Soondrum K, Ebrahimnejad A, Dohrmann P, Kalthoff H, Henne-Bruns D, Neumaier M (2001): Human endostatin inhibits growth of human non–small-cell lung cancer in a murine xenotransplant model. Int J Cancer 94:420–8.

Boehm T, Folkman J, Browder T, O'Reilly MS (1997): Antiangiogenic therapy of experimental cancer does not induce acquired drug resistance. Nature 390:404–7.

Bonanno E, Nicosia RF (1992): Ring formation in cultures of rat aortic smooth muscle cells. In vitro Cell Dev Biol 28A:703–4.

Boucher Y, Leunig M, Jain RK (1996): Tumor angiogenesis and interstitial hypertension. Cancer Res 56:4264–6.

Brooks PC, Montgomery AM, Rosenfeld M, Reisfeld RA, Hu T, Klier G, Cheresh DA (1994): Integrin alpha v beta 3 antagonists promote tumor regression by inducing apoptosis of angiogenic blood vessels. Cell 79:1157–64.

Brouty-Boye D, Zetter BR (1980): Inhibition of cell motility by interferon. Science 208:516–8.

Browder T, Butterfield CE, Kraling BM, Shi B, Marshall B, O'Reilly MS, Folkman J (2000): Antiangiogenic scheduling of chemotherapy improves efficacy against experimental drug-resistant cancer. Cancer Res 60:1878–86.

Burrows FJ, Thorpe PE (1993): Eradication of large solid tumors in mice with an immunotoxin directed against tumor vasculature. Proc Natl Acad Sci USA 90:8996–9000.

Chadefaux B, Ceballos I, Hamet M, Coude M, Poissonnier M, Kamoun P, Allard D (1988): Is absence of atheroma in Down syndrome due to decreased homocysteine levels? Lancet 2:741.

Colorado PC, Torre A, Kamphaus G, Maeshima Y, Hopfer H, Takahashi, K, Volk R, Zamborsky ED, Herman S, Sarkar PK, Ericksen MB, Dhanabal M, Simons M, Post M, Kufe DW, Weichselbaum RR, Sukhatme VP, Kalluri R (2000): Anti-angiogenic cues from vascular basement membrane collagen. Cancer Res 60:2520–6.

Coman DR, Sheldon WF (1946): The significance of hyperemia around tumor implants. Am J Pathol 22:821–31.

Crum R, Szabo S, Folkman J (1985): A new class of steroids inhibits angiogenesis in the presence of heparin or a heparin fragment. Science 230:1375–8.

D'Amato RJ, Lin CM, Flynn E, Folkman J, Hamel E (1994): 2-Methoxyestradiol, an endogenous mammalian metabolite, inhibits tubulin polymerization by interacting at the colchicine site. Proc Natl Acad Sci USA 91:3964–8.

Dameron KM, Volpert OV, Tainsky MA, Bouck N (1994): Control of angiogenesis in fibroblasts by p53 regulation of thrombospondin-1. Science 265: 1582–4.

Darland DC, D'Amore PA (2001): Endothelial Cells and Pericytes in Tumor Vasculature. In: EE Voest, PA D'Amore (eds): Tumor Angiogenesis and Microcirculation. New York: Dekker, pp 1–8.

Davidoff AM, Leary MA, Ng CY, Spurbeck WW, Frare P, Vanhove M, Nienhuis AW, Vanin EF (2001): Autocrine expression of both endostatin and green fluorescent protein provides a synergistic antitumor effect in a murine neuroblastoma model. Cancer Gene Ther 8:537–45.

Dawson DW, Volpert OV, Gillis P, Crawford SE, Xu H, Benedict W, Bouck NP (1999): Pigment epithelium-derived factor: A potent inhibitor of angiogenesis. Science 285:245–8.

Day ED (1964): Vascular relationships of tumor and host. Prog Exp Tumor Res 4:57–97.

Deb G, Donfrancesco A, Ilari I, De Sio L, Milano GM, Ghitti C, Fontana G, Sandri A, Helson L (2002): Hemangioendothelioma: successful therapy with interferon-alpha: A study in Association with the Italian Pediatric Haematology/Oncology Society (AIEOP). Med Pediatr Oncol 38:118–9.

Dvorak HF, Gresser I (1989): Microvascular injury in pathogenesis of interferon-induced necrosis of subcutaneous tumors in mice. J Natl Cancer Inst 81:497–502.

Ezekowitz RA, Mulliken JB, Folkman J (1992): Interferon alfa-2a therapy for life-threatening hemangiomas of infancy. N Engl J Med 326:1456–63.

Ezekowitz A, Mulliken J, Folkman J (1991): Interferon alpha therapy of haemangiomas in newborns and infants. Br J Haematol Suppl 79:67–8.

Ezzell C (1998): Starving tumors of their lifeblood. Sci Am 279:33–4.

Falchetti ML, Pierconti F, Casalbore P, Maggiano N, Levi A, Larocca LM, Pallini R (2003): Glioblastoma induces vascular endothelial cells to express telomerase in vitro. Cancer Res 63:3750–4.

Fernandez A, Udagawa T, Schwesinger C, Beecken W, Achilles-Gerte E, McDonnell T, D'Amato R (2001): Angiogenic potential of prostate carcinoma cells overexpressing Bcl-2. J Natl Cancer Inst 93:208–13.

Folkman J (1971): Tumor angiogenesis: Therapeutic implications. N Engl J Med 285:1182–6.

Folkman J, Long DM Jr, Becker FF (1963): Growth and metastasis of tumor in organ culture. Cancer 16:452–67.

Folkman J, Cole P, Zimmerman S (1966): Tumor behavior in isolated perfused organs: in vitro growth and metastases of biopsy material in rabbit thyroid and canine intestinal segment. Ann Surg 164:491–502.

Folkman J (1976): The vascularization of tumors. Sci Am 234:58–64, 70–53.

Folkman J (1985): Tumor angiogenesis. Adv Cancer Res 43:175–203.

Folkman J, Kalluri R (2003): Tumor Angiogenesis. In: DW Kufe, RE Pollock, RR Weichselbaum, RC Bast Jr, JF Holland, E Frei, III (eds): Cancer Medicine, 6th ed. Hamilton, Ontario: BC Decker, pp 161–94.

Folkman J, Haudenschild CC, Zetter BR (1979): Long-term culture of capillary endothelial cells. Proc Natl Acad Sci USA 76:5217–21.

Folkman J, Haudenschild C (1980): Angiogenesis in vitro. Nature 288:551–6.

Folkman J (1998): Tumor angiogenesis. In: SA Wells Jr, PA Sharp (eds): Accomplishments in Cancer Research, Philadelphia: JB Lippincott Williams and Wikins, pp 32–44.

Folkman J (1989): Successful treatment of an angiogenic disease. N Engl J Med 320:1211–2.

Folkman J, Mulliken JB, Ezekowitz RA (1997): Antiangiogenic therapy of haemangiomas with interferon. In: R Stuart-Harris, R Penny (eds): The Clinical Applications of the Interferons. London: Chapman & Hall Medical, pp 255–65.

Folkman J (2002): Discussion. Antiangiogenic therapy with interferon alpha for giant cell lesions of the jaws. J Oral Maxillofac Surg 60:111–3.

Folkman J, Hahnfeldt P, Hlatky L (2000): Cancer: Looking outside the genome. Nat Rev Mol Cell Biol 1:76–9.

Folkman J (2003): Fundamental concepts of the angiogenic process. Curr Mol Med 3:643–51.

Folkman J, Moscona A (1978): Role of cell shape in growth control. Nature 273:345–9.

Fotsis T, Zhang Y, Pepper MS, Adlercreutz H, Montesano R, Nawroth PP, Schweigerer L (1994): The endogenous oestrogen metabolite 2-methoxy-oestradiol inhibits angiogenesis and suppresses tumour growth. Nature 368:237–9.

Fukai N, Eklund L, Marneros AG, Oh SP, Keene DR, Tamarkin L, Niemela M, Ilves M, Li E, Pihlajaniemi T, Olsen BR (2002): Lack of collagen XVIII/endostatin results in eye abnormalities. EMBO J 21:1535–44.

Fukumura D, Xavier R, Sugiura T, Chen Y, Park EC, Lu N, Selig M, Nielsen G, Taksir T, Jain RK, Seed B (1998): Tumor induction of VEGF promoter activity in stromal cells. Cell 94:715–25.

Fulcher T, Griffin M, Crowley S, Firth R, Acheson R, O'Meara N (1998): Diabetic retinopathy in Down's syndrome. Br J Ophthalmol 82:407–9.

Gimbrone MA Jr, Leapman SB, Cotran RS, Folkman J (1972): Tumor dormancy in vivo by prevention of neovascularization. J Exp Med 136:261–76.

Gimbrone MA Jr, Cotran RS, Leapman SB, Folkman J (1974): Tumor growth and neovascularization: An experimental model using the rabbit cornea. J Natl Cancer Inst 52:413–27.

Gimbrone MA Jr, Cotran RS, Folkman J (1973): Endothelial regeneration: Studies with human endothelial cells in culture. Ser Haematol 6:453–5.

Gimbrone MA Jr, Cotran RS, Folkman J (1974): Human vascular endothelial cells in culture. Growth and DNA synthesis. J Cell Biol 60:673–84.

Griffith EC, Su Z, Turk BE, Chen S, Chang YH, Wu Z, Biemann K, Liu JO (1997): Methionine aminopeptidase (type 2) is the common target for angiogenesis inhibitors AGM-1470 and ovalicin. Chem Biol 4:461–71.

Hamano Y, Zeisberg M, Sugimoto H, Lively JC, Maeshima Y, Yang C, Hynes RO, Werb Z, Sudhakar A, Kalluri R (2003): Physiological levels of tumstatin, a fragment of collagen IV alpha3 chain, are generated by MMP-9 proteolysis and suppress angiogenesis via alphaVbeta3 integrin Cancer Cell 3:589–601.

Hanahan D (1985): Heritable formation of pancreatic beta-cell tumours in transgenic mice expressing recombinant insulin/simian virus 40 oncogenes. Nature 315:115–22.

Hanahan D, Folkman J (1996): Patterns and emerging mechanisms of the angiogenic switch during tumorigenesis. Cell 86:353–64.

Hanahan D, Bergers G, Bergsland E (2000): Less is more, regularly: Metronomic dosing of cytotoxic drugs can target tumor angiogenesis in mice. J Clin Invest 105:1045–7.

Hanai J, Dhanabal M, Karumanchi SA, Albanese C, Waterman M, Chan B, Ramchandran R, Pestell R, Sukhatme VP (2002): Endostatin causes G_1 arrest of endothelial cells through inhibition of cyclin D1. J Biol Chem 277: 16464–9.

Hanna NN, Seetharam S, Mauceri HJ, Beckett MA, Jaskowiak NT, Sallourm RM, Hari D, Dhanabal M, Ramchandran R, Kalluri R, Sukhatme VP, Kufe DW, Weichselbaum RR (2000): Antitumor interaction of short-course endostatin and ionizing radiation. Cancer J 6:287–93.

Hobson B, Denekamp J (1984): Endothelial proliferation in tumours and normal tissues: Continuous labelling studies. Br J Cancer 49:405–13.

Holmgren L, O'Reilly MS, Folkman J (1995): Dormancy of micrometastases: Balanced proliferation and apoptosis in the presence of angiogenesis suppression. Nat Med 1:149–53.

Hoppeler H, Kayar SR (1988): Capillarity and oxidative capacity of muscles. News Physiol Sci 3:113–60.

Hori K, Saito S (2003): Microvascular mechanisms by which the combretastatin A-4 derivative AC7700 (AVE8062) induces tumour blood flow stasis. Br J Cancer 89:1334–44.

Huang X, Wong MK, Zhao Q, Zhu Z, Wang KZ, Huang N, Ye C, Gorelik E, Li M (2001): Soluble recombinant endostatin purified from *Escherichia coli*: Antiangiogenic activity and antitumor effect. Cancer Res 61:478–81.

Hurwitz H, Fehrenbacher L, Cartwright T, Hainsworth J, Heim W, Berlin J, Griffing S, Novotny W, Holmgren E, Kabbinavar F (2003): Bevacizumab (a monoclonal antibody to vascular endothelial growth factor) prolongs survival in first-line colorectal cancer (CRC): Results of a phase III trial of bevacizumab in combination with bolus IFL (irinotecan, 5-fluorouracil, leucovorin) as first-line therapy in subjects with metastatic CRC. Proc Am Soc Clin Oncol, LB abstract 3646.

Hynes RO (2002): Integrins: bi-directional, allosteric, signaling machines. Cell 110:673–7.

Ide AG, Baker NH, Warren SL (1939): Vascularization of the Brown-Pearce rabbit epithelioma transplant as seen in the transparent ear chamber. Am J Roentgenol 42:891–9.

Ingber D, Fujita T, Kishimoto S, Sudo K, Kanamaru T, Brem H, Folkman J (1990): Synthetic analogues of fumagillin that inhibit angiogenesis and suppress tumour growth. Nature 348:555–7.

Inoue M, Hager JH, Ferrara N, Gerber HP, Hanahan D (2002): VEGF-A has a critical, nonredundant role in angiogenic switching and pancreatic beta cell carcinogenesis. Cancer Cell 1:193–202.

Iughetti P, Suzuki O, Godoi PH, Alves VA, Sertie AL, Zorick T, Soares F, Camargo A, Moreira ES, di Loreto C, Moreira-Filho CA, Simpson A, Oliva G, Passos-Bueno MR (2001): A polymorphism in endostatin, an angiogenesis inhibitor, predisposes for the development of prostatic adenocarcinoma. Cancer Res 61:7375–8.

Iversen PO, Sorensen DR, Benestad HB (2002): Inhibitors of angiogenesis selectively reduce the malignant cell load in rodent models of human myeloid leukemias. Leukemia 16:376–81.

Izumi Y, Xu L, di Tomaso E, Fukumura D, Jain RK (2002): Tumour biology: Herceptin acts as an anti-angiogenic cocktail. Nature 416:279–80.

Joki T, Machluf M, Atala A, Zhu J, Seyfried NT, Dunn IF, Abe T, Carroll RS, Black PM (2001): Continuous release of endostatin from microencapsulated engineered cells for tumor therapy. Nat Biotechnol 19:35–9.

Kaban LB, Mulliken JB, Ezekowitz RA, Ebb D, Smith PS, Folkman J (1999): Antiangiogenic therapy of a recurrent giant cell tumor of the mandible with interferon alfa-2a. Pediatrics 103:1145–9.

Kaban LB, Troulis MJ, Ebb D, August M, Hornicek FJ, Dodson TB (2002): Antiangiogenic therapy with interferon alpha for giant cell lesions of the jaws. J Oral Maxillofac Surg 60:1103–11.

Kakolyris S, Samonis G, Koukourakis M, Vlachonicolis I, Chalkiadakis G, Kalbakis K, Souglakos I, Agelaki S, Toloudis P, Georgoulias V (1998): Treatment of non–small-cell lung cancer with prolonged oral etoposide. Am J Clin Oncol 21:505–8.

Kerbel R, Folkman J (2002): Clinical translation of angiogenesis inhibitors. Nat Rev Cancer 2:727–39.

Kim YM, Hwang S, Pyun BJ, Kim TY, Lee ST, Gho YS, Kwon YG (2002): Endostatin blocks vascular endothelial growth factor-mediated signaling via direct interaction with KDR/Flk-1. J Biol Chem 277:27872–9.

Kim KJ, Li B, Winer J, Armanini M, Gillett N, Phillips HS, Ferrara N (1993): Inhibition of vascular endothelial growth factor-induced angiogenesis suppresses tumour growth in vivo. Nature 362:841–4.

Kisker O, Becker CM, Prox D, Fannon M, D'Amato R, Flynn E, Fogler WE, Sim BK, Allred EN, Pirie-Shepherd SR, Folkman J (2001): Continuous administration of endostatin by intraperitoneally implanted osmotic pump improves the efficacy and potency of therapy in a mouse xenograft tumor model. Cancer Res 61:7669–74.

Kisker O, Onizuka S, Becker CM, Fannon M, Flynn E, D'Amato R, Zetter B, Folkman J, Ray R, Swamy N, Pirie-Shepherd S (2003): Vitamin D binding protein-macrophage activating factor (DBP-maf) inhibits angiogenesis and tumor growth in mice. Neoplasia 5:32–40.

Kisker O, Onizuka S, Banyard J, Komiyama T, Becker Cm, Achilles EG, Barnes CM, O'Reilly MS, Folkman J, Pirie-Shepherd SR (2001): Generation of multiple angiogenesis inhibitors by human pancreatic cancer. Cancer Res 61:7298–304.

Klement G, Baruchel S, Rak J, Man S, Clark K, Hicklin DJ, Bohlen P, Kerbel RS (2000): Continuous low-dose therapy with vinblastine and VEGF receptor-2 antibody induces sustained tumor regression without overt toxicity. J Clin Invest 105:R15–24.

Kuroiwa M, Ikeda H, Hongo T, Tsuchida Y, Hirato J, Kaneko Y, Suzuki N, Obana K, Makino SI (2001): Effects of recombinant human endostatin on a human neuroblastoma xenograft. Int J Mol Med 8:391–6.

Langer R, Folkman J (1976): Polymers for the sustained release of proteins and other macromolecules. Nature 263:797–800.

Lawler J (2002): Thrombospondin-1 as an endogenous inhibitor of angiogenesis and tumor growth. J Cell Mol Med 6:1–12.

LeBlanc R, Catley LP, Hideshima T, Lentzsch S, Mitsiades CS, Mitsiades N, Neuberg D, Goloubeva O, Pien CS, Adams J, Gupta D, Richardson PG, Munshi NC, Anderson KC (2002): Proteasome inhibitor PS-341 inhibits human myeloma cell growth in vivo and prolongs survival in a murine model. Cancer Res 62:4996–5000.

LeCouter J, Lin R, Tejada M, Frantz G, Peale F, Hillan KJ, Ferrara N (2003): The endocrine-gland-derived VEGF homologue Bv8 promotes angiogenesis in the testis: Localization of Bv8 receptors to endothelial cells. Proc Natl Acad Sci USA 100:2685–90.

Lee SJ, Jang JW, Kim YM, Lee HI, Jeon JY, Kwon YG, Lee ST (2002): Endostatin binds to the catalytic domain of matrix metalloproteinase-2. FEBS Lett 519:147–52.

Lyden D, Young AZ, Zagzag D, Yan W, Gerald W, O'Reilly R, Bader BL, Hynes RO, Zhuang Y, Manova K, Benezra R (1999): ld1 and ld3 are required for neurogenesis, angiogenesis and vascularization of tumour xenografts. Nature 401:670–7.

Lyden D, Hattori K, Dias S, Costa C, Blaikie P, Butros L, Chadburn A, Heissig B, Marks W, Witte L, Wu Y, Hicklin D, Zhu Z, Hackett NR, Crystal RG, Moore MA, Hajjar KA, Manova K, Benezra R, Rafii S (2001): Impaired recruitment of bone-marrow-derived endothelial and hematopoietic precursor cells blocks tumor angiogenesis and growth. Nat Med 7:1194–201.

Mabjeesh NJ, Escuin D, LaVallee TM, Pribluda VS, Swartz GM, Johnson MS, Willard MT, Zhong H, Simons JW, Giannakakou P (2003): 2ME2 inhibits tumor growth and angiogenesis by disrupting microtubules and dysregulating HIF. Cancer Cell 3:363–75.

Maeshima Y, Manfredi M, Reimer C, Holthaus KA, Hopfer H, Chandamuri BR, Kharbanda S, Kalluri R (2001): Identification of the anti-angiogenic site within vascular basement membrane-derived tumstatin. J Biol Chem 276:15240–8.

Maione TE, Gray GS, Petro J, Hunt AJ, Donner AL, Bauer SI, Carson HF, Sharpe RJ (1990): Inhibition of angiogenesis by recombinant human platelet factor-4 and related peptides. Science 247:77–9.

Marler JJ, Rubin JB, Trede NS, Connors S, Grier H, Upton J, Mullken JB, Folkman J (2002): Successful antiangiogenic therapy of giant cell angioblastoma with interferon alfa 2b: Report of 2 cases. Pediatrics 109:E37 (on-line only).

Monestiroli S, Mancuso P, Burlini A, Pruneri G, Dell'Agnola C, Gobbi A, Martinelli G, Bertolini F (2001): Kinetics and viability of circulating endothelial cells as surrogate angiogenesis marker in an animal model of human lymphoma. Cancer Res 61:4341–4.

Moulton KS, Heller E, Konerding MA, Flynn E, Palinski W, Folkman J (1999): Angiogenesis inhibitors endostatin or TNP-470 reduce intimal neovascularization and plaque growth in apolipoprotein E-deficient mice. Circulation 99:1726–32.

Muthukkaruppan V, Shinners BL, Lewis RL, Park S, Baechler BJ, Auerbach R (2000): The chick embryo aortic arch assay: a new, rapid, quantifiable in vitro method for testing the efficacy of angiogenic and anti-angiogenic factors in a three-dimensional, serum-free organ culture system. Proc Am Assoc Cancer Res 41:65, abstract 415.

Muthukkaruppan V, Auerbach R (1979): Angiogenesis in the mouse cornea. Science 205:1416–8.

Nawrocki ST, Bruns CJ, Harbison MT, Bold RJ, Gotsch BS, Abbruzzese JL, Elliott P, Adams J, McConkey DJ (2002): Effects of the proteasome inhibitor PS-341 on apoptosis and angiogenesis in orthotopic human pancreatic tumor xenografts. Mol Cancer Ther 1:1243–53.

Ohlms LA, Jones DT, McGill TJ, Healy GB (1994): Interferon alfa-2a therapy for airway hemangiomas. Ann Otol Rhinol Laryngol 103:1–8.

O'Reilly MS, Boehm T, Shing Y, Fukai N, Vasios G, Lane WS, Flynn E, Birkhead JR, Olsen BR, Folkman J (1997): Endostatin: An endogenous inhibitor of angiogenesis and tumor growth. Cell 88:277–85.

O'Reilly MS, Holmgren L, Shing Y, Chen C, Rosenthal RA, Moses M, Lane WS, Cao Y, Sage EH, Folkman J (1994): Angiostatin: A novel angiogenesis inhibitor that mediates the suppression of metastases by a Lewis lung carcinoma. Cell 79:315–28.

O'Reilly MS, Pirie-Shepherd S, Lane WS, Folkman J (1999): Antiangiogenic activity of the cleaved conformation of the serpin antithrombin. Science 285:1926–8.

Perletti G, Concari P, Giardini R, Marras E, Piccinini F, Folkman J, Chem L (2000): Antitumor activity of endostatin against carcinogen-induced rat primary mammary tumors. Cancer Res 60:1793–6.

Rafii S, Lyden D, Benezra R, Hattori K, Heissig B (2002): Vascular and haematopoietic stem cells: Novel targets for anti-angiogenesis therapy? Nat Rev Cancer 2:826–35.

Rak J, Filmus J, Finkenzeller G, Grugel S, Marme D, Kerbel RS (1995): Oncogenes as inducers of tumor angiogenesis. Cancer Metastasis Rev 14:263–77.

Rak J, Yu JL, Klement G, Kerbel RS (2000): Oncogenes and angiogenesis: Signaling three-dimensional tumor growth. J Investig Dermatol Symp Proc 5:24–33.

Rak J, Mitsuhashi Y, Sheehan C, Tamir A, Viloria-Petit A, Filmus J, Mansour SJ, Ahn NG, Kerbel RS (2000): Oncogenes and tumor angiogenesis: Differential modes of vascular endothelial growth factor up-regulation in *ras*-transformed epithelial cells and fibroblasts. Cancer Res 60:490–8.

Ravi R, Mookerjee B, Bhujwalla ZM, Sutter CH, Artemov D, Zeng Q, Dillehay LE, Madan A, Semenza GL, Bedi A (2000): Regulation of tumor angiogenesis by p53-induced degradation of hypoxia-inducible factor 1alpha. Genes Dev 14:34–44.

Read TA, Sorensen DR, Mahesparan R, Enger PO, Timpl R, Olsen BR, Hjelstuen MH, Haraldseth O, Bjerkvig R (2001): Local Endostatin treatment of gliomas administered by microencapsulated producer cells. Nat Biotechnol 19:29–34.

Rehn M, Veikkola T, Kukk-Valdre E, Nakamura H, Illmonen M, Lombardo C, Pihlajaniemi T, Alitalo K, Vuori K (2001): Interaction of endostatin with integrins implicated in angiogenesis. Proc Natl Acad Sci USA 98:1024–9.

Reiher FK, Volpert OV, Jimenez B, Crawford SE, Dinney CP, Henkin J, Haviv F, Bouck NP, Campbell SC (2002): Inhibition of tumor growth by systemic treatment with thrombospondin-1 peptide mimetics. Int J Cancer 98:682–9.

Relf M, LeJeune S, Scott PA, Fox S, Smith K, Leek R, Moghaddam A, Whitehouse R, Bicknell R, Harris AL (1997): Expression of the angiogenic factors vascular endothelial cell growth factor, acidic and basic fibroblast growth factor, tumor growth factor beta-1, platelet-derived endothelial cell growth factor, placenta growth factor, and pleiotrophin in human primary breast cancer and its relation to angiogenesis. Cancer Res 57:963–9.

Rodriguez-Manzaneque JC, Lane TF, Ortega MA, Hynes RO, Lawler J, Iruela-Arispe ML (2001): Thrombospondin-1 suppresses spontaneous tumor growth and inhibits activation of matrix metalloproteinase-9 and mobilization of vascular endothelial growth factor. Proc Natl Acad Sci USA 98:12485–90.

Rupnick MA, Panigrahy D, Zhang CY, Dallabrida SM, Lowell BB, Langer R, Folkman MJ (2002): Adipose tissue mass can be regulated through the vasculature. Proc Natl Acad Sci USA 99:10730–5.

Scappaticci FA, Contreras A, Smith R, Bonhoure L, Lum B, Cao Y, Engleman EG, Nolan GP (2001): Statin-AE: A novel angiostatin-endostatin fusion protein with enhanced antiangiogenic and antitumor activity. Angiogenesis 4:263–8.

Sertie AL, Sossi V, Camargo AA, Zatz M, Brahe C, Passos-Bueno MR (2000): Collagen XVIII, containing an endogenous inhibitor of angiogenesis and tumor growth, plays a critical role in the maintenance of retinal structure and in neural tube closure (Knobloch syndrome). Hum Mol Genet 9:2051–8.

Sherif ZA, Nakai S, Pirollo KF, Rait A, Chang EH (2001): Downmodulation of bFGF-binding protein expression following restoration of p53 function. Cancer Gene Ther 8:771–82.

Sidky YA, Borden EC (1987): Inhibition of angiogenesis by interferons: Effects on tumor-and lymphocyte-induced vascular responses. Cancer Res 47: 5155–61.

Sin N, Meng L, Wang MQ, Wen JJ, Bornmann WG, Crews CM (1997): The anti-angiogenic agent fumagillin covalently binds and inhibits the methionine aminopeptidase, MetAP-2. Proc Natl Acad Sci USA 94:6099–103.

Singh RK, Gutman M, Bucana CD, Sanchez R, Llansa N, Fidler IJ (1995): Interferons alpha and beta down-regulate the expression of basic fibroblast growth factor in human carcinomas. Proc Natl Acad Sci USA 92:4562–6.

Slaton JW, Perrotte P, Inoue K, Dinney CP, Fidler IJ (1999): Interferon-alpha-mediated down-regulation of angiogenesis-related genes and therapy of bladder cancer are dependent on optimization of biological dose and schedule. Clin Cancer Res 5:2726–34.

Sorensen DR, Read TA, Porwol T, Olsen BR, Timpl R, Sasaki T, Iversen PO, Benestad HB, Sim BK, Bjerkvig R (2002): Endostatin reduces vascularization, blood flow, and growth in a rat gliosarcoma. Neuro-oncol 4:1–8.

Sorensen DR, Leirdal M, Iversen PO, Sioud M (2002): Combination of endostatin and a protein kinase Calpha DNA enzyme improves the survival of rats with malignant glioma. Neoplasia 4:474–9.

St Croix B, Rago C, Velculescu V, Traverso G, Romans KE, Montgomery E, Lal A, Riggins GJ, Lengauer C, Vogelstein B, Kinzler KW (2000): Genes expressed in human tumor endothelium. Science 289:1197–202.

Stoler DL, Chen N, Basik M, Kahlenberg MS, Rodriguez-Bigas MA, Petrelli NJ, Anderson GR (1999): The onset and extent of genomic instability in sporadic colorectal tumor progression. Proc Natl Acad Sci USA 96:1512–6.

Streit M, Riccardi L, Velasco P, Brown LF, Hawighorst T, Bornstein P, Detmar M (1999):Thrombospondin-2: A potent endogenous inhibitor of tumor growth and angiogenesis. Proc Natl Acad Sci USA 96:14888–93.

Sudhakar A, Sugimoto H, Yang C, Lively J, Zeisberg M, Kalluri R (2003): Human tumstatin and human endostatin exhibit distinct antiangiogenic activities mediated by alpha v beta 3 and alpha 5 beta 1 integrins. Proc Natl Acad Sci USA 100:4766–71.

Sunwoo JB, Chen Z, Dong G, Yeh N, Crowl Bancroft C, Sausville E, Adams J, Elliott P, Van Waes C (2001): Novel proteasome inhibitor PS-341 inhibits activation of nuclear factor-kappa B, cell survival, tumor growth, and angiogenesis in squamous cell carcinoma. Clin Cancer Res 7:1419–28.

Suri C, Yancopoulos GD (2002): The ties that bind: Emerging concepts about the structure and function of angiopoietins and their receptors in angiogenesis. In: RJ, Tomanek (ed): Assembly of the Vasculature and Its Regulation. Boston: Birkhauser, pp 55–66.

Takahashi K, Mulliken JB, Kozakewich HP, Rogers RA, Folkman J, Ezekowitz RA (1994): Cellular markers that distinguish the phases of hemangioma during infancy and childhood. J Clin Invest 93:2357–64.

Tanaka K, Sonoo H, Kurebayashi J, Nomura T, Ohkubo S, Yamamoto Y, Yamamoto S (2002): Inhibition of infiltration and angiogenesis by thrombospondin-1 in papillary thyroid carcinoma. Clin Cancer Res 8:1125–31.

Taylor S, Folkman J (1982): Protamine is an inhibitor of angiogenesis. Nature 297:307–12.

te Velde EA, Vogten JM, Gebbink MF, van Gorp JM, Voest EE, Borel Rinkes IH (2002): Enhanced antitumour efficacy by combining conventional chemotherapy with angiostatin or endostatin in a liver metastasis model. Br J Surg 89:1302–9.

Tomanek RJ (2002): Assembly of the Vasculature and Its Regulation. Boston: Birkhauser.

Urbich C, Reissner A, Chavakis E, Dernbach E, Haendeler J, Fleming I, Zeiher AM, Kaszkin M, Dimmeler S (2002): Dephosphorylation of endothelial nitric oxide synthase contributes to the anti-angiogenic effects of endostatin. FASEB J 16:706–8.

Voest EE, D'Amore PA (2001): Tumor Angiogenesis and Microcirculation. New York: Dekker.

Volpert OV, Stellmach V, Bouck N (1995): The modulation of thrombospondin and other naturally occurring inhibitors of angiogenesis during tumor progression. Breast Cancer Res Treat 36:119–26.

Von Marschall Z, Scholz A, Cramer T, Schafer G, Schirner M, Oberg K, Wiedenmann B, Hocker M, Rosewicz S (2003): Effects of interferon alpha on vascular endothelial growth factor gene transcription and tumor angiogenesis. J Natl Cancer Inst 95:437–48.

Watnick RS, Cheng Y-N, Rangarajan A, Ince TA, Weinberg RA (2003): Ras modulates Myc activity to repress thrombospondin-1 expression and increase tumor angiogenesis. Cancer Cell 3:210–31.

White CW, Sondheimer HM, Crouch EC, Wilson H, Fan LL (1989): Treatment of pulmonary hemangiomatosis with recombinant interferon alfa-2a. N Engl J Med 320:1197–200.

Wickstrom SA, Alitalo K, Keski-Oja J (2002): Endostatin associates with integrin alpha5beta1 and caveolin-1, and activates Src via a tyrosyl phosphatase-dependent pathway in human endothelial cells. Cancer Res 62: 5580–9.

Williams JI (1999): A new angiostatic steroid. In: BA, Teicher (ed): Antiangiogenic Agents in Cancer Therapy. NJ: Totowa, Humana Press, pp 153–74.

Yang Q, Rasmussen SA, Friedman JM (2002): Mortality associated with Down's syndrome in the USA from 1983 to 1997: a population-based study. Lancet 359:1019–25.

Ye C, Feng C, Wang S, Liu X, Lin Y, Li M (2002): Antiangiogenic and antitumor effects of endostatin on follicular thyroid carcinoma. Endocrinology 143: 3522–8.

Yokoyama Y, Green JE, Sukhatme VP, Ramakrishnan S (2000): Effect of endostatin on spontaneous tumorigenesis of mammary adenocarcinoma in a transgenic mouse model. Cancer Res 60:4362–5.

Zhang L, Yu D, Hu M, Xiong S, Lang A, Ellis LM, Pollock RE (2000): Wild-type p53 suppresses angiogenesis in human leiomyosarcoma and synovial sarcoma by transcriptional suppression of vascular endothelial growth factor expression. Cancer Res 60:3655–61.

Zorick TS, Mustacchi Z, Bando SY, Zatz M, Moreira-Filho CA, Olsen B, Passos-Bueno MR (2001): High serum endostatin levels in Down syndrome: implications for improved treatment and prevention of solid tumours. Eur J Hum Genet 9:811–4.

CHAPTER 11

REGULATION OF CELL GROWTH, DIFFERENTIATION, AND DEATH DURING METAMORPHOSIS

HANS LAUFER[1] and ERIC H. BAEHRECKE[2]
[1]Department of Molecular and Cell Biology, University of Connecticut, Storrs, CT 06269-3125 and Marine Biological Laboratory, Woods Hole, MA 02543
[2]Center for Biosystems Research, University of Maryland Biotechnology Institute, College Park, MD 20742

INTRODUCTION

Many, perhaps most, life forms undergo metamorphic changes in form, shape, or substance. Typically one thinks of tadpoles turning into frogs or insect larvae turning into pupae and then into butterflies, moths, or flies. However the phenomenon is much broader. For instance, V. B. Wigglesworth (1972) pointed out that metamorphosis is part of a more comprehensive phenomenon termed polymorphism. This includes casts among social insects that not only produce larvae, pupae, and adults but different kinds of adults, namely kings, queens, soldiers, and workers.

Holometabolous insects have a complete metamorphosis (several larval stages, whose number is genetically determined, one pupal stage, and an adult reproductive stage; Fig. 11.1). However, hemimetabolous insects have only immature nymphoid stages that resemble the final adult stage, except that they are neither winged nor sexually mature. Examples of this kind of development are roaches, grasshoppers, and locusts, which are only reproductive in their final winged stage. The more primitive ametabolous insects such as silverfish or firebrats have a series of nymphal stages followed by a wingless adult that resembles the earlier stages. Thus they do not have a metamorphosis.

The varieties in life cycles are extensive. There are species of invertebrates such as some starfish, which metamorphose through a swimming

Cell Cycle and Growth Control: Biomolecular Regulation and Cancer, Second Edition, Edited by Gary S. Stein and Arthur B. Pardee
ISBN 0-471-25071-6

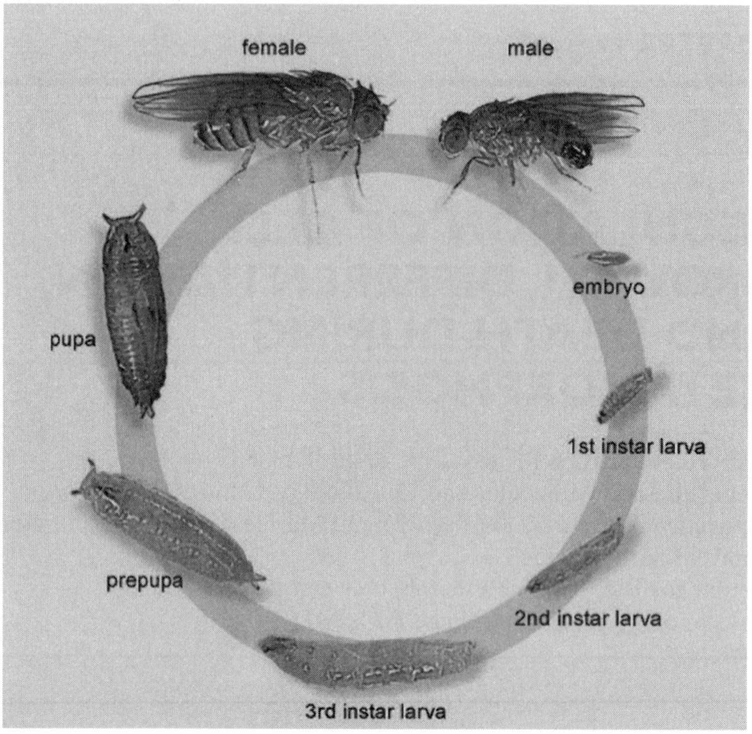

Figure 11.1. Developmental stages of the fruit fly *Drosophila melanogaster*. A female (top left) and a male adult fly (top right). In the circle below starting at right, an embryo, 3 instars, a prepupa, and a pupal stage. (Source: Weigmann et al., 2003.) (See color insert.)

larval stage (*Asterias forbesi*) while other species occupying similar waters may have "direct development" which does not present a larval stage (e.g., a local starfish species to Woods Hole, *Henricia sanguinolenta*). This is a direct development in which eggs hatch into immature starfish.

Many crustaceans hatch from their egg at the nauplius stage, which utilize appendages formed from head segments for mobility that are homologous to body segment appendages (e.g., *Artemia salina*, the brine shrimp, and true shrimp; Fig. 11.2), while others such as the common Louisiana crayfish (*Procambarus clarkii*) hatch from eggs as juveniles. Some zoologists maintain that these latter forms undergo the nauplius stage, and beyond, during embryonic development in the egg before hatching. These varying stages of the life cycle are apparently evolutionary adaptations, and through evolution, they seem to result in new life forms and the occupation of new ecological niches. For example, the four-legged amphibious frog can move to explore a terrestrial environment, leaving the aquatic environment behind. Other frogs, like *Xenopus laevis*, the African "clawed toad," will metamorphose into frog that remains in the aquatic environment for its entire life. Most salamanders

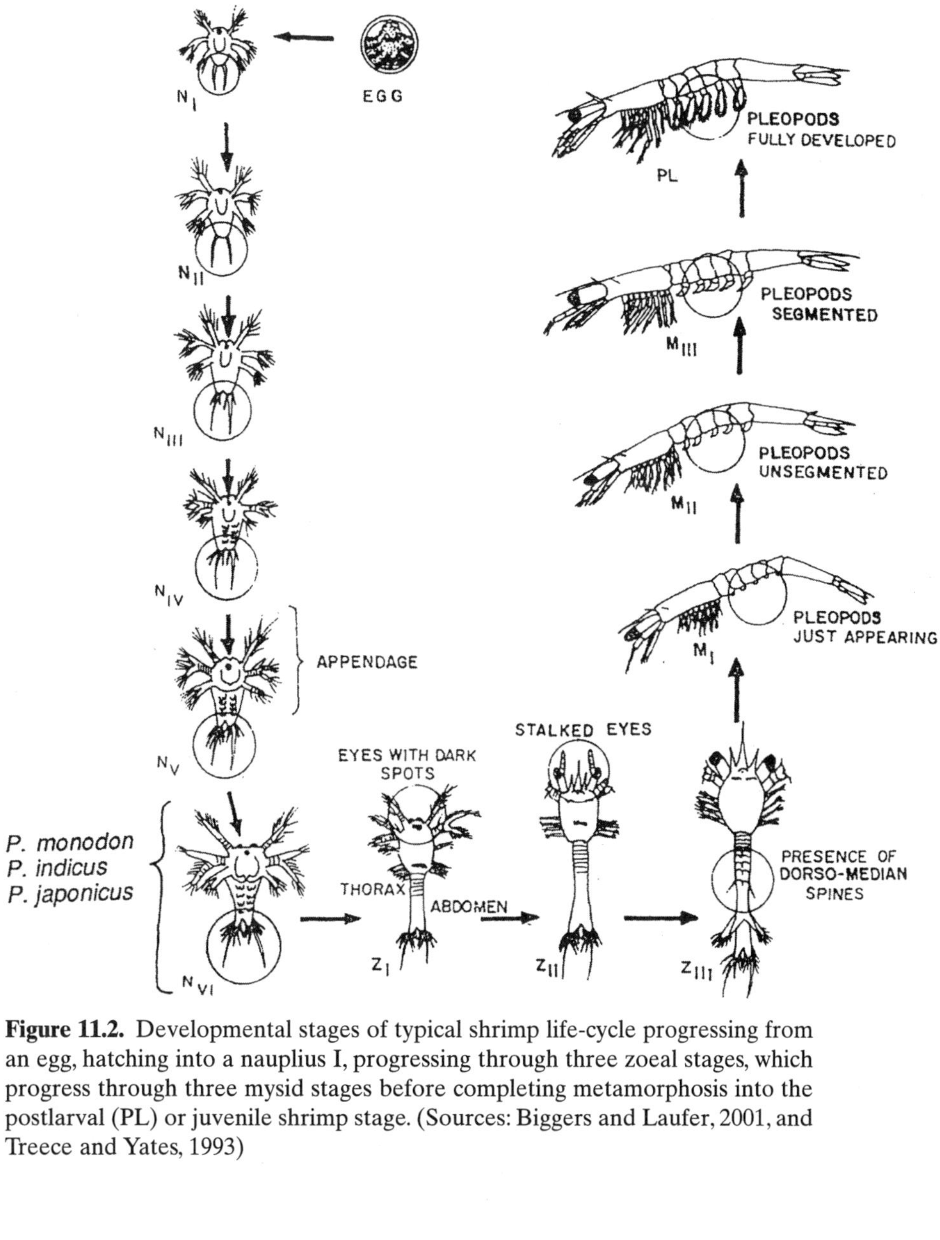

Figure 11.2. Developmental stages of typical shrimp life-cycle progressing from an egg, hatching into a nauplius I, progressing through three zoeal stages, which progress through three mysid stages before completing metamorphosis into the postlarval (PL) or juvenile shrimp stage. (Sources: Biggers and Laufer, 2001, and Treece and Yates, 1993)

have larval forms with gills that are resorbed at metamorphosis. Most adult salamanders become terrestrial, but others like *Necturus* the "mud puppy," and the Mexican axolotl, *Ambystoma mexicanum*, retain their larval gills into the adult stage, a phase called neoteny, and they remain aquatic (Prahlad and Delanney, 1965). The latter species fails to secrete thyrotropin from the pituitary gland to activate the thyroid gland to produce thyroxin, which is needed for metamorphosis of frogs and salamanders. In the case of *Triturus viridescens*, some populations (e.g., Cape Cod population) may retain their neotenous gills, while most populations elsewhere metamorphose into adult morphotypes. These variations seem to be remnants of evolutionary adaptations in transition.

Such residues of polymorphic evolutionary forms express themselves even in humans, occasionally seen in such oddities as webbed hands, webbed toes, absence of wisdom teeth, vestigial small toes, remnants of a coxygial tail, and so on.

The ability to occupy and exploit new niches, such as flying in insects, permits feeding on nectar and adaptations to pollinating flowers. It permits in colder climates migrations for overwintering in warmer climates as in the case of the monarch butterfly. However, other insects have adapted other strategies for overwintering in harsh climates by digging into the ground or production of eggs that overwinter in cold soil or in which the pupal stage overwinters in a state of arrested development, termed diapause. One of the most studied and extreme case of adaptation to adversity is seen in the brine shrimp, *Artemia salina*, cysts that can be dehydrated for hundreds of years and then revive within a few hours to hatch into an active nauplius larva, which are produced by immersion of the cysts into saline solution.

Parasitic organisms undergo complex changes in form during their life, and this is associated with changes in their environment and diet. For example, insects in many orders including Lepidoptera (moths and butterflies), Coleoptera (beetles), and Hymenoptera (bees and wasps) undergo multiple metamorphoses (also known as hypermetamorphosis) (Clausen, 1940). These insects change their larval body pattern in association with changing from an herbivorous plant-eating stage to a parasitic stage, and this occurs prior to their metamorphosis into an adult life-stage. Complex life-histories and multiple changes in body form are not restricted to parasitic insects, as parasites such as malaria have different morphologies to dwell in different hosts for parts of their lives. In the case of malaria, part of its existence is in a mosquito in which it grows, propagates, and undergoes extensive changes in form and function. When the *sporozoite* stage cells are expelled into an animal host during a feeding incident, the parasite is injected into the blood stream, from which it invades a red blood cell and becomes a *trophozoite*, an amoeba-like organism. After growth, the *trophozoite* becomes a *schizont* capable of multiple divisions of 15 to 24 *merozoites*, which attack more red blood cells. Some cells undergo further replication as *schizonts* while others mature into gametocytes, a sexual stage, which if they end up in a blood meal through ingestion by the right kind of mosquito of the *Anopheles* group, they produce male and female reproduction gametes capable of forming a zygote. The fertile egg develops into a worm-like structure called an *ookinete*. In the mosquito gut the parasite forms an *oocyst* and during the next 6 to 7 days undergoes sporulation into hundreds of *spororzoites* some of which find their way to the salivary gland from which they can infect another host (Hegner, 1933). *Sporozoites* metamorphose into the miracidium stage and proliferate in red blood cells, resulting in a disease state for the host. It is estimated, by methods of microarray analysis, that the malaria parasite genome is composed of about 7,000 genes that change dramatically during development (BenMamoun et al., 2001).

SIGNALS THAT CONTROL METAMORPHOSIS

Many signals have been implicated as regulators of animal development. In the context of metamorphosis, signals trigger large changes in body pattern and serve as a sort of timing mechanism. A pre-pattern is often established early in development and then elaborated during metamorphosis, such as during development of the insect wing. Growth factors, nitric oxide, and hormones including thyroid, steroid, and isoprenoid hormones (Fig. 11.3) are among the signals that regulate metamorphosis.

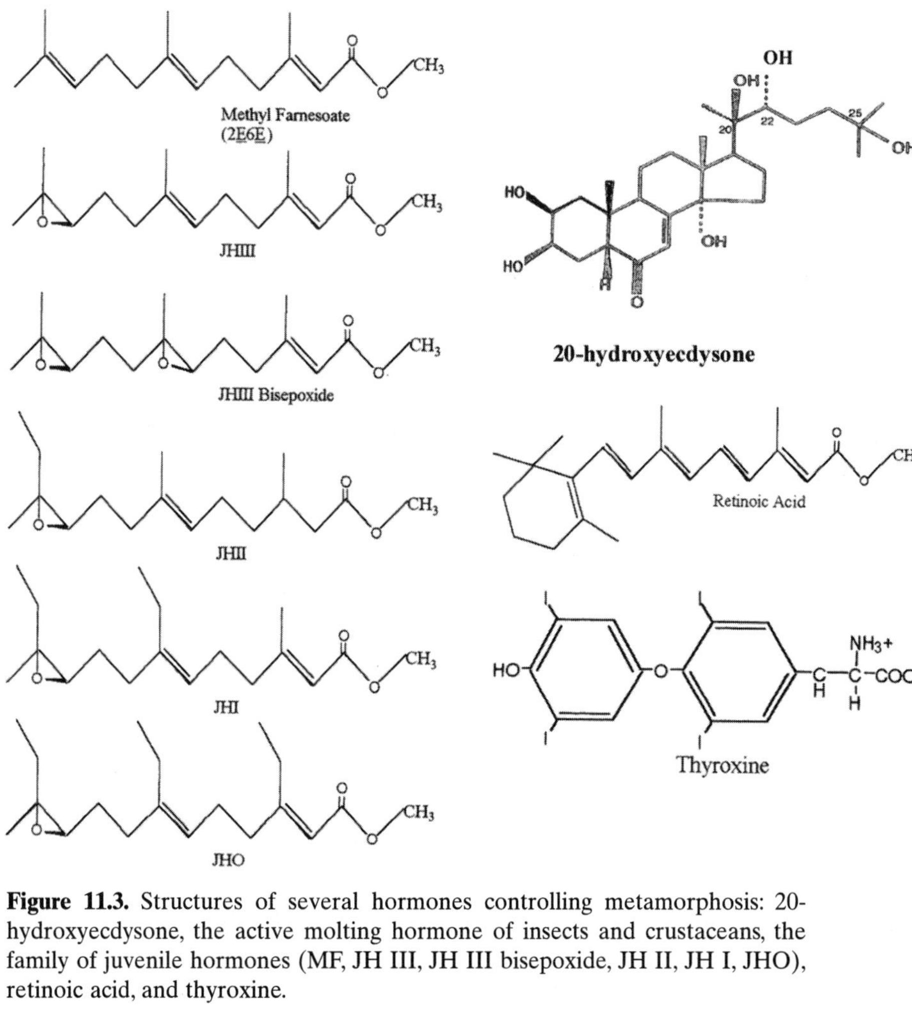

Figure 11.3. Structures of several hormones controlling metamorphosis: 20-hydroxyecdysone, the active molting hormone of insects and crustaceans, the family of juvenile hormones (MF, JH III, JH III bisepoxide, JH II, JH I, JHO), retinoic acid, and thyroxine.

GENETIC AND MOLECULAR CONTROL OF A METAMORPHOSING SYSTEM

The Insect Wing

An intensively studied and well-understood example of metamorphic development is the formation of wings in *Drosophila melanogaster*. They arise from a cluster of imaginal disc cells set aside in the early forming larva with great precision as to final position and location in the second thoracic segment. The formation and patterning of wings involves more than 200 genes and gene products in complex interactions (reviewed by Carroll et al., 2001; Davidson, 2001; Gilbert 2003). A single homeotic mutation called *ultra bithorax* (UBX), discovered by Calvin Bridges in 1915, provides significant insight into the genetic control of body pattern. This mutation transforms the halteres, a set of balancing organs homologous to wings, on the third thoracic segment, into a second pair of wings. A second mutant *antennapedia* transforms the *Drosophila* antenna into legs. Such mutations resulting in a homologous structure in an unusual position are called homeotic mutants. In time most of these mutations reveal themselves to be homeotic mutants affecting homologous structures of different locations along various body segments such as limbs, antennae, wings, and so on. Studies of the mutations show that they are arranged in a linear fashion on chromosomes forming complexes. The proteins produced by homeotic genes share a similar block of amino acids, a "homeobox," that binds to specific regulatory sites on DNA and in turn regulates transcription of target genes. Furthermore, while homeotic mutants act in the formation of the adult fly, similar homeotic or "hox" genes, as these are called, not only act during the larval segment formation setting up the body-segmented organization but also function in regulating body pattern and limb formation in vertebrate animals.

Body pattern is established in the *Drosophila* embryo, which is segregated into segments during early development. Many genes function in the establishment of the dorsal/ventral axis (i.e., *dorsal*, *cactus*, *snail*, *sog*, and *zen*), while the actions of bicoid and nanos proteins, which are maternally derived, establish anterior and posterior gradients. The accumulation of the gap proteins such as Krüppel in the mid-embryo nuclei is followed by the accumulation of the pair-rule protein Hairy, and the segment polarity protein Engrailed, which are expressed in blocks of cells and determine the 14 larval and adult body segments along the anterior-posterior body axis.

The wing disc is also divided into anterior and posterior compartments by protein determinants. Engrailed protein is a transcription factor that is expressed in the posterior compartment of the wing imaginal disc in the first-instar larva. A short-range paracrine factor named Hedgehog is activated by the action of the *engrailed* gene. The *hedgehog* gene product activates the *decapentaplegic* (Dpp) gene in cells that are in the midline of the imaginal disc. The Dpp protein is a long-range paracrine factor that diffuses into both the anterior and posterior compartments of the wing disc, forming a gradient with the highest concentration in the dorsal ventral axes central of the disc. Dpp is also a transcription factor that activates the factors *spalt* and *omb* at the midline of the disc, and *omb* is also activated nearer the periphery where Dpp is in lower concentration (reviewed by Carroll et al., 2001). By the second larval instar the dorsal ventral axis of the wing is established. The *apterous* gene is turned on in cells destined to become the dorsal surface of the wing (Blair, 1993; Diaz-Benjumea and Cohen, 1993) while the *vestigial* gene is expressed in the prospective ventral portion of the wing blade. The *wingless* gene is turned on at the dorso-ventral boundary of the wing disc. The wingless protein is a growth factor promoting the growth and extension of the margin of the wing disc. The wingless protein also activates the *distal-less* gene at the outer margin (Neumann and Cohen, 1996). This complex regulatory hierarchy is used to establish a pre-pattern that is maintained during larval growth and until the wing is further differentiated during metamorphosis to the adult.

Regulation of Metamorphosis by Nitric Oxide

The intercellular messenger nitric oxide (NO) has been shown to be a naturally produced regulator of metamorphosis in frogs, fish, and also several invertebrate species, and is known to prevent cells from undergoing programmed cell death. One species for which this has been demonstrated is the marine gastropod *Ilyanassa obsoleta* (Froggett and Leise, 1996; Thavaradhara and Leise, 2001). Similar to most marine invertebrate larvae, swimming larvae of *Ilyanassa* will settle and metamorphose on suitable substrates in response to specific chemical cues present in the environment. These chemical cues are picked up by apical sensory ganglia (ASG), which is a sensory organ containing 25 neurons. During larval metamorphosis of *Ilyanassa* into the juvenile form, the apical sensory ganglia are lost by programmed cell death, since they are no longer needed (Gifondorwa and Leise, 2001). This programmed cell death of the apical sensory ganglia has been demonstrated by Leise et al. (2001) to be regulated by serotonin and nitric oxide. Endogenous NO production was found to be necessary for the maintenance of the larval state, whereas exogenous application of serotonin was found to induce metamorphosis. This inductive effect of serotonin can be inhibited by exogenous NO. Endogenous nitric oxide synthetase (NOS), along with NADPH diaphorase, an indicator of NOS activity were demonstrated to be present in the ASG (Thavaradhara and Leise, 2001), and inhibitors of endogenous NOS were found to stimulate larval metamorphosis.

In the tobacco hornworm moth *Manduca sexta*, NO coordinates ecdysone-dependent neural cell proliferation of the optic anlage and development of the eye during metamorphosis (Champlin and Truman, 1998, 2000). When ecdysone (Fig. 11.3) levels drop below a critical threshold level, the neural cell precursors undergo proliferative arrest in the

G_2 phase of the cell cycle (Champlin and Truman, 1998, 2000) and increases in ecdysone over the threshold level results in rapid resumption of development of the eye disc and brain. NO conversely inhibits proliferation of neural cells of the optic anlage of *Manduca* by causing an arrest in the cell cycle in the G_2 phase. NOS activity is stimulated by subthreshold levels of ecdysone, causing cell cycle arrest, and is inhibited by suprathreshold levels of ecdysone, resulting in resumption of the cell cycle.

HORMONAL CONTROL OF METAMORPHOSIS

Regulation of Frog Metamorphosis by Thyroid Hormone and Thyroxine

One remarkable example of metamorphosis in amphibians is the transformation of tadpoles into adult frogs. This metamorphosis, which is marked by development of the adult form involving tissue remodeling and loss of the tail through programmed cell death, is under regulation by thyroid hormone (TH) (Fig. 11.3). The thyroid hormone, thyroxin (T4), produced by the thyroid gland, is rendered more active by a type III deiodinase to tri-iodothyroxine (T3) and is regulated by corticotrophin releasing hormone (CRH), a neuropeptide from the pituitary gland. Thyroid-stimulating hormone (TSH) from the pituitary gland controls amphibian metamorphosis by T3, which involves many physiological, biochemical, and morphological changes including the growth of limbs, tail resorption, and air breathing through lungs, as well as a change in diet from plant eating to mostly insects (reviewed in Gilbert et al., 1996).

In eliciting the metamorphic response, T3 causes genetic reprogramming, resulting in the formation of adult tissues and resorption and apoptosis of larval tissues. In accomplishing this, T3 binds with nuclear receptors TH receptor alpha (TR-A), which has been implicated to function in the formation of adult tissues, and TR-B has been implicated in causing larval tissue resorption and apoptosis. Using GC-1, a selective agonist of TR-B, Furlow et al. (2000) found that this agonist activated *Xenopus* TR-B 20-fold more than *Xenopus* TR-A, compared with the natural ligand T3, and caused tail and gill resorption. Further analysis showed that GC-1 activated a subset of TH-response genes in the tail, including several protease genes.

In bringing about this change in gene expression, Sachs and Shi (2000) have found that T3-mediated gene regulation also involves chromatin remodeling by histone acetylation. In their studies, TR in the absence of TH ligand repressed transcription by causing histone deacetylation. Unliganded TR was found to directly bind to TH response elements. In the presence of TH during metamorphosis, TH binds with TR and promotes histone acetylation, causing chromatin remodeling and allowing TH response gene activation. In the intestine and tail, Sachs and Shi

(2000) found that TH response gene activation by TH was associated with histone H4 acetylation.

Amphibian metamorphosis also involves selective neuronal cell death. One protein that appears to play an essential role in the regulation of this process is the anti-apoptotic protein xR11, which is a homologue of the Bcl-XL protein that plays a role in postnatal mammalian neuronal cell survival (Coen et al., 2001). Overexpression of xR11 in vivo in Xenopus tadpoles was found to significantly prolong Rohon-Beard neuron survival in the tail during metamorphosis, whereas in controls these neurons disappeared. The overexpressed xR11 protein also counteracted in a dose-dependent manner the pro-apoptotic effects of the co-expressed Bax protein.

Metamorphosis in Insects and Crustacea

V. B. Wigglesworth (1972) pointed out that the fertilized insect egg contained all of the genetic information to produce the several larval, pupal, and adult stages of a metamorphosing insect. Furthermore he indicated that the transformations that insects go through are under hormonal control. All of the changes that are manifested are affected at the level of target cells. That is, some cells are discarded such as larval appendages at metamorphosis, while others such as epidermal cells of most insects are transformed into pupal and adult cells. The hormones affecting metamorphosis in insects are the molting hormones, which are a group of steroids including ecdysone and 20-hydroxyecdysone (Fig. 11.3).

Ecdysone is secreted by the prothoracic gland of most insects. In *Drosophila melanogaster*, the fruit fly, the ecdysone secreting tissue is part of the ring gland. Ecdysone is converted to 20-hydroxyecdysone (20HE) by internal tissues and is the more active form of the hormone. There are also several other active forms of the molecule (Hoffmann and Hetru, 1983), as there are other sources of ecdysone (Fallon et al., 1974).

To exert their effects, steroid hormones such as 20HE and thyroid hormones, enter target cells and bind to nuclear receptor proteins. These receptors have been conserved in all higher animals, and often act as homo- or hetero-dimers (described more below). Once ligand is bound, these hormone receptor complexes bind to DNA and activate target gene transcription.

Insects and crustacea grow by shedding their exoskeletons, through a process called molting (Fig. 11.1, 11.2). Molting is controlled by the molting hormone(s), ecdysones, which are a family of steroids produced by the prothoracic gland of insects and the y-organ of crustacea. The various members of this steroid family are metabolic products of tissues. By convention, we refer here to the class of active compounds as ecdysone(s). The mechanism of ecdysone action was proposed by Clever and Karlson (1960) to be in the nucleus and at the chromosome (gene) level, and is the model for all steroid hormones including those of vertebrates. Ecdysone activated chromosomal puffs–puff I 18c in *Chironomus tentaus* was increased in activity within an hour of treatment with ecdysone (Clever and Karlson, 1960). This region of the chromosome was activated to synthesize RNA in response to hormone treatment. Laufer and colleagues showed that JH-active compounds activated larval specific chromosomal puff loci in *Chironomus thumni* salivary glands in vivo (Laufer and Holt, 1970; Laufer and Calvet, 1972) and in vitro (Vafopoulou et al., 1989) in short-term incubations, suggesting a nuclear action for JH as well. Ecdysone is regulated in turn by neuropeptides from the brain, the prothoracotropic hormone (PTTH) of insects, and molt-inhibiting hormone (MIH) from the sinus gland-x-organ complex of the crustacean eyestalk complex. The molting process leads to the shedding of the exoskeleton and results in increased size and growth of the organism.

The quality of the molt, whether the organism remains larval or progresses to an alternate form, depends on the presence or absence of juvenile hormone (JH). These are a family of isoprenoids (Fig. 11.3) secreted by the corpus allatum (corpora allata) (CA) of insects and the mandibular organs of crustaceans (Laufer et al., 1987). The secretion of JH in the larva maintains the larval state. Bouniol (1938) produced additional larval stages in the silk moth *Bombyx mori* which normally has five larval instars, by implanting CA, while removal of the CA in the early fourth larval stage resulted in the production of a smaller than normal pupa and adult with the next two molts. Williams (1961) showed that adding JH to a pupal *Cecropia* silkmoth, *Hyalophora cecropia*, produced a second pupal stage, which the silkworm would never do under normal conditions, showing both the flexibility and programmability of the organism's cellular responses.

JH Active Compounds in Early Larval Development

The action of JH on late larval insect development has been studied and described for many years (reviews by Gilbert et al., 1996; Truman and Riddiford, 1999). Addition of JH to a late insect larva at the critical prepupal stage at the point of decreased JH hormone concentration, when there is a small ecdysone peak, determines the subsequent events for metamorphosis. This addition of JH active material to a crustacean reverses the process and also inhibits progression of a larval to juvenile form (hence the name "status quo" hormone).

However, there have been instances reported where JH active compounds actually promote metamorphosis. For instance, JH active compounds stimulate the metamorphosis of the cipris larvae of a crustacean barnacle (Ramanofsky et al., 1974; Gomez et al., 1973; Tighe-Ford 1977). Laufer and Landau (1991) reported finding the JH active compound, methyl farnesoate, synthesized by barnacles. In addition JH active compounds promote the metamorphic settlement and formation of a worm from the metatrochophore larva of the annelid *Capitella capitata* (in 30–60 min)(Biggers and Laufer, 2001). The mechanism for this transformation involves a protein kinase C pathway (PKC) and potassium and calcium ion channels. Suggesting that JH active compounds may have additional modes of action than affecting only nuclear sites (see review by Davey, 2000).

Davey and colleagues have provided evidence that in several insects, including *Locusta migratoria*. JH acts at the level of the cell membrane to open up channels between follicle cells for yolk protein uptake form the blood by the egg cell. JH binds to cell surface membrane to activate a protein kinase C signaling cascade involving phosphorylation of a 100 kDa protein, which is a subunit of a Na^{+}/K^{+} ATPase (Davari, 1999; Davey 1996). This would be similar to other bioactive compounds such as retinoids (Fig. 11.3), which, in the case of vision, retinoids have a signal transduction mechanism, generating an action potential, while in the induction of limbs may function through classical nuclear receptors (Laufer, 1988). Nemec et al. (1993) showed that retinoids may act as juvenoids, suggesting a possibly broad recognition by JH receptors. Jones and Sharp (1997) and Jones et al. (2001) provide evidence that JH in insect cells utilizes nuclear ultraspiracle (USP) receptors, either as homo or heterodimers. Crustacea differ from classical insect development (which have larvae, pupae, and adult stages), as many shrimp have multiple larval stages starting with six naupilar stages and then progress through three zoeal stages and three mysid stages before reaching the postlarval metamorphosis (Fig. 11.2). Some of the early stage transformations occur in the presence of methyl farnesoate, and may even depend on it (Laufer and Biggers, 2001). In a similar fashion Truman and Riddiford (1999) proposed that early prolarvae of insects may be developing in the presence of JH, and may be dependent on JH for their progressive developmental alterations.

Endocrine Control

JH synthesis and secretion is controlled by neurosecretory cells from the brain, in the form of activators by allatotropins, and inhibitors, allatostatins in insects. Corpora allata of *Manduca sexta*, the tobacco hornworm adults, are stimulated to secrete JH isoforms and are in turn activated to synthesize and secrete JH by allatotropins (Kataoka et al., 1989). A series of allatostatins have been discovered that control the level of JH produced in larval stages. These are pentoadecapeptides in the case of *Manduca* (Tobe and Stay, 1965); others are 13 amino acid long neuropeptides from the central nervous system of *Diploptera*, a roach. Allatostatins seem to function by turning off the CAs at premetamorphosis at the "critical" stages at the time of pupal commitment. Thus they are important actors in metamorphosis.

In crustacea the glands homologous to the CA are the mandibular organs (MOs), which synthesize methyl farnesoate (MF) (Fig. 11.3), an unepoxidated form of JH III (Laufer et al., 1987). MF seems to be a JH of crustaceans, since it has been found in every crustacean studied to date (more than 30 species) (Laufer and Biggers, 2001). Also MF has the same dual functions in crustaceans that JH has in insects; that is, it maintains larval and juvenile characteristics in young crustaceans and it acts as a gonadotropin, stimulating gonad maturation in mature crustaceans (Laufer and Biggers, 2001).

The hormone regulating MF synthesis by the MO is MOIH (mandibular organ inhibiting hormone), which is produced by the eyestalk sinus gland x-organ (Liu et al., 1997). MOIH is a neuropeptide 72 amino acids in length and is a member of the crustacean hyperglycemic hormone (CHH) family, which also includes MIH, the crustacean molt-inhibiting hormone. Many of the CHHs have multiple functions. That is, they are very similar in structure and have overlapping functional activities, and most regulate glucose mobilization in the blood (hyperglycemic hormone activity) (Liu and Laufer, 1996).

Riddiford showed that there was a small peak of ecdysone in the late fifth larval stage of *Manduca sexta*, the tobacco hornworm, which if it occurred in the absence of JH would determine the progress to the pupal stage. At this time genes of the broad complex are activated and result in the production of pupal characteristics (Zhou and Riddiford, 2002). Addition of JH to pupae will re-activate the broad complex genes, enabling a second pupa to be formed at the next molt. In the absence of JH, the adult morphology would be generated. JH effects several isoforms of the broad complex, which lead to the differential activation of specific cuticle genes such as Edg78E Edg84D, and PcP (Henikoff et al., 1986; Fechtel et al., 1988; Apple and Fristrom 1991), while genes affecting adult cuticle such as Acp65A, Edg78E, Edg84D, and APF are repressed (Henikoff and Eghtedarzadeh, 1987; Zhou and Riddiford, 2002).

While *Drosophila melanogaster* is an ideal organism for genetic and molecular biology studies of metamorphosis, it is a dipteran that differs from most other holometabolous insects, in that its prospective adult cells are restricted to imaginal discs. This means that most of the larval structures, including the larval skin or hypodermis is discarded during adult development. However, in most other insects the larval hypodermis is transformed into pupal and adult hypodermis, and transplanted skin pieces can be altered in either a forward or backward direction when implanted into an appropriate host (Piepho, 1939a, b).

Drosophila Metamorphosis as a System to Study the Mechanisms of Development

The genetic mechanisms that regulate metamorphosis have been most thoroughly studied in the fruit fly *Drosophila melanogaster*. This is because *Drosophila* possesses many attributes for genetic and genomic studies, including a small fully sequenced genome and a rapid life cycle, so it has been used as a model for genetic studies since the early 1900s (Rubin and Lewis, 2000). In addition 62% of the genes that have been implicated in human disease possess related genes in *Drosophila* (Fortini et al., 2000; Rubin, 2000). Therefore *Drosophila* metamorphosis provides an ideal system to identify genes that function in normal growth and development, and to study the mechanisms that regulate abnormal development when genes are perturbed.

Drosophila development consists of embryonic, three larval instar, prepupal, pupal, and adult stages (Fig. 11.1). Increases in the steroid hormone 20-hydroxyecdysone (ecdysone) punctuate each of these developmental stages (Riddiford, 1993). During the onset of metamorphosis, increases in ecdysone titer trigger cellular changes that are required to transform a larva into an adult. At the end of the third larval instar, an increase in ecdysone induces the formation of a prepupa (Handler, 1982; Pak and Gilbert, 1987; Richards, 1981). The ecdysone titer decreases in the mid-prepupal stage (Handler, 1982; Pak and Gilbert, 1987), and increases again 12 hours following puparium formation (Handler, 1982; Sliter and Gilbert, 1992). This rise in ecdysone triggers future adult head eversion, which marks the beginning of pupation (Handler, 1982; Sliter and Gilbert, 1992). The ecdysone titer then decreases at the onset of pupation, and another large rise in ecdysone titer occurs during pupal development peaking approximately 30 hours after puparium formation (Pak and Gilbert, 1987).

Drosophila metamorphosis involves the destruction of most of the larval tissues, and differentiation and morphogenesis of the tissues that form the adult fly. Many tissues initiate metamorphic changes in synchrony with the rise in ecdysone titer at the larval-to-prepupal transition. Imaginal discs undergo morphogenesis to form future adult appendages, and adult flight muscles form in the anterior region of the prepupa (Bodenstein, 1965; Robertson, 1936). While these adult structures are forming, several tissues are destroyed including the larval midgut (Bodenstein, 1965; Lee et al., 2002; Robertson, 1936). Cell- and tissue-specific changes are also induced when the ecdysone titer rises at the prepupal-to-pupal transition. Procuticle is deposited on adult appendages (Doctor et al., 1985), the larval muscles are destroyed in the abdomen, the larval foregut epithelium is replaced by the adult foregut, and the larval salivary glands die when adult salivary glands initiate morphogenesis (Bodenstein, 1965; Robertson, 1936). While many changes associated with the transformation of a larva into an adult fly occur during the prepupal stage, additional details of adult formation are elaborated during the three days between pupation and adult eclosion. These include the maturation of the adult neurons (Truman et al., 1993) and patterning of the adult eye (Wolff and Ready, 1993).

The stage- and cell-specific changes that are triggered by ecdysone indicate that fluctuations in this hormone alone are not sufficient to determine the nature of the cellular response. Unfortunately, detailed studies of the mechanisms underlying ecdysone-regulated responses have been restricted to a limited number of cells and tissues. Studies of ecdysone-triggered imaginal disc evagination have served as a useful model for morphogenesis (Fristrom and Fristrom, 1993; von Kalm et al., 1995). The nervous system, larval midgut, and larval salivary glands have been useful for studies of steroid regulation of cell death in *Drosophila* (Jiang et al., 1997; Lee and Baehrecke, 2001; Lee et al., 2002a,b; Robinow et al., 1993). While all of these studies indicate that patterning of the response to ecdysone is a critical component of proper growth and development, limited connections have been made between the initial systemic steroid signal that triggers the cell response and the factors that determine if the cell is competent to respond to the hormone.

Steroids Signal by Triggering a Genetic Regulatory Hierarchy

The mechanisms of steroid signaling have been most thoroughly studied in *Drosophila* larval salivary glands because of the giant polytene chromosomes that form ecdysone-induced puffs reflecting a transcriptional regulatory hierarchy. Changes in chromosome puffing (decondensation of chromatin) accompany the changes in ecdysone titer during development. A series of elegant studies led to a model for genetic regulation of chromosome puffing (Ashburner et al., 1974; Becker, 1959; Clever, 1964). According to this model, the ecdysone receptor complex directly induces a small set of early puff genes, and the protein products of these genes then repress their own activity and induce a large set of secondary late response genes.

The isolation and characterization of the ecdysone receptor and ecdysone-regulated puff genes have supported the model that was proposed based on chromosome puffing (Andres and Thummel, 1992; Ashburner, 1990). The *EcR* (Koelle et al., 1991) and *usp* (Henrich et al., 1990; Oro et al., 1990; Shea et al., 1990) genes both encode members of the nuclear hormone receptor family of proteins, and heterodimerize to form the ecdysone receptor (Thomas et al., 1993; Yao et al., 1992). This receptor complex binds to DNA, and activates transcription of early puff genes. Early puffs and the genes encoded by these genetic loci are not properly induced in *EcR* and *usp* mutants (Bender et al., 1997; Hall and Thummel, 1998). The characterization of the *BR-C*, *E74*, and *E75* early puff genes provided further support of the model for ecdysone signaling based on chromosome puffs (Burtis et al., 1990; DiBello et al., 1991; Segraves and Hogness, 1990). These early puff genes are complicated and encode multiple isoforms of transcription factor proteins by alternative promoter usage and splicing. *BR-C* encodes zinc finger proteins, *E74* encodes members of the ETS family of DNA binding proteins, and *E75* encodes nuclear hormone receptor family member zinc finger proteins. E74 and E75 proteins bind to both early and late puff chromosome loci (Hill et al., 1993; Urness and Thummel, 1990). Late puff genes have not been extensively characterized, but the isolation of the *L71* late genes (Restifo and Guild, 1986; Wright et al., 1996) have been useful for testing the tenets of the steroid signaling model that was based on chromosome puffing. *BR-C* and *E74* mutations impact transcription of late target genes (Fletcher and Thummel, 1995). Furthermore BR-C and E74 proteins bind to glue and *L71* gene regulatory elements, providing a direct link between these DNA binding proteins and the regulation of target gene transcription (Crossgrove et al., 1996; Urness and Thummel, 1995; von Kalm et al., 1994).

The steroid regulatory hierarchy is activated by different pulses of ecdysone during development (Fig. 11.3). The increase in ecdysone titer at the end of the third larval instar regulates the transcription of glue and late genes in the salivary gland (Crowley and Meyerowitz, 1984; Hansson and Lambertsson, 1989; Restifo and Guild, 1986; Wright et al., 1996). The ecdysone titer then drops to a low level in midprepupae, enabling the induction of the nuclear hormone receptor *ßFTZ-F1* (Woodard et al., 1994). *ßFTZ-F1* serves as competence factor that enables the reinduction of the *BR-C* and *E74* early genes, and the stage-specific induction of *E93* by the pulse of ecdysone 12 hours after puparium formation in salivary glands (Broadus et al., 1999; Woodard et al., 1994). Like *BR-C* and *E74A*, *E93* also regulates the transcription of target late genes (Lee et al., 2000), and these downstream genes appear to function more directly in programmed cell death. While late puffs are observed at this stage of development (Ashburner, 1967), none of these late genes have been identified based on puffing. However, targets of the early genes that are induced at this stage have been identified (Jiang et al., 2000; Lee et al., 2000, 2002b, 2003).

Genetic Regulation of Programmed Cell Death during *Drosophila* Metamorphosis

Programmed cell death plays an important role during animal development by functioning in the formation of structures, deletion of structures, control of cell numbers, and elimination of abnormal cells (Baehrecke, 2002). Several cell types undergo programmed cell death during *Drosophila* metamorphosis including those in the central nervous system (Truman et al., 1993, 1994), eye (Baker Brachmann and Cagan, 2003; Wolff and Ready, 1991, 1993), midgut (Jiang et al., 1997; Lee et al., 2002a), and salivary glands (Jiang et al., 1997; Lee and Baehrecke, 2001). Of these cell types, some of the most detailed understanding of the mechanisms that regulate cell death have emerged from studies of salivary glands.

The ecdysone titer begins to rise 10 hours after puparium formation, and reaches its peak level 12 hours after puparium formation when it triggers the synchronous death of larval salivary gland cells (Bodenstein, 1965; Robertson, 1936). This cell death is preceded by markers of apoptosis including DNA fragmentation and nuclear acridine orange staining (Jiang et al., 1997). While these data suggest that salivary glands die by apoptosis, their morphology indicates that they die by autophagic programmed cell death. Following the rise in ecdysone titer 12 hours after puparium formation, salivary gland cells become round in shape, large cytoplasmic vacuoles appear to fragment, and a second class of vacuoles appears that is associated with the plasma membrane (Lee and Baehrecke, 2001). By 14 hours after puparium formation, salivary glands possess autophagic vacuoles that contain cytoplasmic structures including mitochondria (Lee and Baehrecke, 2001). The cellular changes that occur following the rise in ecdysone titer are accompanied by changes in the tubulin and actin cytoskeleton, and accumulation of acid phosphatase activity (Jochova et al., 1997). Salivary glands are destroyed 16 hours following puparium formation. These changes in cell morphology indicate that the production and movement of vacuoles, reorganization of the cytoplasm and cell shape, and cytoplasmic degradation by proteases may all be involved in the autophagic destruction of salivary glands.

Several ecdysone-regulated genes function in programmed cell death. The *EcR*, *usp*, *ßFTZ-F1*, *BR-C*, *E74*, and *E93* genes have been implicated in the regulation of programmed cell death in a variety of tissues including the midgut, nervous system, and salivary glands (Broadus et al., 1999; Hall and Thummel, 1998; Jiang et al., 2000; Lee et al., 2000; Restifo and White, 1992; Robinow et al., 1993; Truman et al., 1994). *EcR*, *usp*, *ßFTZ-F1*, *BR-C*, and *E74* are pleiotropic, however, and function in cell responses other than death including the proper formation of adult cells (Bender et al., 1997; Broadus et al., 1999; Fletcher et al., 1995). Significantly *E93* appears to function more specifically in the destruction of larval tissues (Lee et al., 2000). *E93* encodes a novel nuclear protein that is expressed in larval midgut and salivary gland cells immediately prior to ecdysone induction of their death. Furthermore *E93* mutants have defects in larval salivary gland cell destruction, and expression of E93 is sufficient to induce programmed cell death. *E93* mutants possess decreased transcription of the *BR-C* and *E74* genes, and each of these early genes impact the transcription of programmed cell death genes prior to the stage that larval salivary glands initiate destruction (Jiang et al., 2000; Lee et al., 2000, 2002b).

Several lines of evidence indicate that the evolutionarily conserved apoptosis machinery functions in salivary gland cell death during *Drosophila* development. Markers of apoptosis including DNA fragmentation foreshadow programmed cell death of salivary glands (Jiang et al., 1997; Lee and Baehrecke, 2001; Lee et al., 2002b), and their destruction is prevented by expression of the caspase inhibitor p35 (Jiang et al., 1997; Lee and Baehrecke, 2001; Lee et al., 2002b). Significantly the pro-apoptotic genes *rpr* and *hid* (*W*) increase in transcription following the rise in ecdysone that triggers salivary autophagic cell death, and this change is concurrent with a decrease in the inhibitor of apoptosis *diap2* (Jiang et al., 1997). At the same time, the fly *Apaf-1* homologue and caspase co-factor *ark* (*dark*, *dapaf1*, *hac1*), the caspases *dronc* (*Nc*), *drice* (*Ice*) and *dream* (*strica*), the Bcl-2 family member *buffy* (*dborg-2*), the CD36 realtive *crq*, and the DNase *rep4* all exhibit dramatic increases in transcription just before salivary gland autophagic cell death (Dorstyn et al., 1999; Lee et al., 2002b, 2003). Animals with mutations in the transcription regulators *BR-C*, *E74A*, and *E93* prevent proper transcription of apoptosis genes. Both *BR-C* and *E93* mutant salivary glands have dramatically reduced levels of *rpr*, *hid*, *dronc*, *dream*, and *crq* (Lee et al., 2002b, 2003). While *buffy* and *drice* RNA levels are also decreased in *E93* mutant salivary glands, *drice* is not altered and *buffy* is ectopically expressed in *BR-C* mutants (Lee et al., 2003). *E74A* mutants exhibit reduced levels of *hid*, *crq*, *buffy*, *drice*, and *dream* (Lee et al., 2002b, 2003). Studies of apoptosis gene promoters indicate that *rpr* is a direct target of the ecdysone receptor (Jiang et al., 2000) and that *dronc* is a direct target of BR-C (Cakouros et al., 2002). These data suggest that the ecdysone-regulated early genes *BR-C*, *E74A*, and *E93* have overlapping and distinct target apoptosis genes that they regulate. Significantly, it appears that proper transciption of apoptosis genes is essential for autophagic cell death, since animals with mutations in *BR-C*, *E74A*, and *E93* prevent destruction of salivary glands.

While caspases and their cofactors play an important role in cell death of salivary glands, accumulating evidence points to the role of numerous other factors in the destruction of this tissue. Eleven of the 19 genes that are required for yeast autophagy encode related genes in flies and humans (hereafter refered to as *apg*-related genes) (Baehrecke, 2002; Reggiori and Klionsky, 2002). Nine of these 11 *apg*-related genes are transcribed in dying *Drosophila* glands, and the 7 genes that are most similar to *apg2* (*CG1241*), *apg3* (*CG6877*), *apg4* (*CG6194*), *apg5* (*CG1643*), *apg7* (*CG5489*), *apg9* (*CG3615*), and *aut10/cvt18* (*CG7986*) exhibit increased transcription following the rise in steroid hormone that triggers autophagic programmed cell death of this tissue (Gorski et al., 2003; Lee et al., 2003). These yeast genes, and presumably these similar fly genes, are involved in two autophagy ubiquitin conjugation systems (Ohsumi, 2001). Mutations in the ecdysone-regulated transcription factor genes that prevent proper salivary gland cell death also inhibit proper transcription of the *apg*-related genes (Lee et al., 2003). Animals with mutations in *BR-C* exhibit ectopic transcription of *CG6194* (*apg4*-like) and *CG1643* (*apg5*-like), and decreased transcription of *CG5489* (*apg7*-like) in salivary glands at the stage that they would normally die. While animals with mutations in *E74A* only exhibit decreased transcription of *CG6194*, salivary glands dissected from *E93* mutants have decreased levels of *CG1241* (*apg2*-like), *CG6194*, *CG1643*, and *CG5489*. Yeast with mutations in *apg4*, *apg5*, and *apg7* are defective in autophagic vacuole formation (Tsukada and Ohsumi, 1993). Therefore it is striking that *E93* mutants exhibit decreased *CG6194*, *CG1643*, and *CG5489* RNA levels, since *E93* mutant salivary gland and midgut cells are defective in the formation of autophagic vacuoles (Lee and Baehrecke, 2001; Lee et al., 2002a,b).

Genome-wide studies of dying salivary gland cells have identified numerous interesting genes that are induced just prior to autophagic programmed cell death. For example, several signaling molecules and transcription regulators increase in transcription in dying salivary glands. The *Drosophila* gene *CG8304* encodes a serine/threonine kinase that is most similar to the death-associated protein kinase (DAPk) in humans, and transcription of this gene increases 36-fold in dying salivary glands (Lee et al., 2003). DAPk mediates membrane blebbing and formation of vesicles in dying human cell lines (Inbal et al., 2002), and suggests that *CG8304* may have a related function in *Drosophila*. Steroids usually function by activating transcription, and therefore it is not surprising that transcription regulators are induced in dying salivary glands including the co-repressors (*Smarter* and *CG4756*), the NFkB regulator *cactus*, the NFkB family member *dif*, several other genes encoding DNA binding proteins (*bun*, *sox14*, *CG8319*), and components of transcription complexes (*trap95* and *hsf*) (Gorski et al., 2003; Lee et al., 2003).

Salivary gland cells exhibit dynamic changes in cell shape and vacuole localization just before autophagic cell destruction, suggesting that cytoplasmic reorganization and proteolysis are important for their death. Consistent with this notion, genes that encode motor proteins, including *ctp*, *ck*, and *dlc90F*, and members of the Rho, Rac, and Rab families of small guanosine triphosphates (GTPases), exhibited increases in transcription just before salivary gland autophagic cell death (Gorski et al., 2003; Lee et al., 2003). While cell remodeling is known to play an important role in phagocyte engulfment of apoptotic cells (Hengartner, 2001), the role of these factors in autophagic cell death is less clear. In addition to caspases, several cysteine, serine, and metallo-proteases increase in transcription in dying salivary glands, and these changes in RNA level are complemented by decreased transcription of cysteine, serine, and metallo-protease inhibitors (Gorski et al., 2003; Lee et al., 2003). Furthermore up-regulation of the matrix metalloprotease *mmp1* is abolished in mutants that prevent salivary gland cell death (Lee et al., 2003). These proteases likely complement caspase function during autophagic cell death. As mutations in *mmp2* prevent proper destruction of *Drosophila* midguts (Page-McCaw et al., 2003), this tissue dies by autophagic cell death (Lee et al., 2002a,b), and inhibition of caspases does not completely prevent autophagic changes in dying cells (Lee and Baehrecke, 2001).

Genetic Regulation of Cell Proliferation and Growth during *Drosophila* Metamorphosis

The proper size and shape of tissues occurs through the regulation of growth during development. *Drosophila* has remarkably uniform tissue sizes, and tissue growth is directly related to the number of cells produced by regulation of the cell cycle minus the number of cells that are destroyed by programmed cell death. Like humans, *Drosophila* can acquire mutations that result in neoplastic growth typical of cancer (Gateff, 1978). Therefore studies of *Drosophila* metamorphosis have been useful for the identification of genes that are required for normal growth, and for understanding the mechanisms that occur in neoplasia.

Studies of developing adult tissues (imaginal discs) during metamorphosis have been useful for the identification of genes that are required for normal growth and development. The first mutations that result in neoplasia were isolated by Calvin Bridges in the 1930s, and were named lethal (2) giant larvae (*l(2)gl*). As the name implies, animals that are homozygous for the *l(2)gl* mutation form large larvae. These giant larvae contain imaginal discs and brains with cells that overproliferate, and this converts the tissue into an invasive malignant neuroblastoma following transplantation into a wild-type animal (Gateff and Schneiderman, 1974). *l(2)gl* encodes a cadherin which is similar to a family of calcium-dependent cell adhesion molecules (Klämbt et al., 1989).

Recent genetic studies of growth in *Drosophila* imaginal discs have taken advantage of increases in cell proliferation and growth in clones of homozygous mutant cells in the adult eye. This approach has the distinct advantage that it enables the identification of genes that function in growth that may not be easily identified in a homozygous lethal animal. For example, *archipelago* (*ago*) mutant cells have a proliferative advantage over their wild-type neighbors (Moberg et al., 2001). These *ago* mutant cells have persistently elevated cyclin E levels, and proliferate when wild-type neighbors become quiescent; cyclin E degradation plays a critical role in the regulation of exit from the cell cycle. AGO encodes a protein with an F-box and seven WD repeats, and the most similar human gene was mutated in several human cancer cell lines (Moberg et al., 2001). Similarly mutations in *salvador* (*sav*) also result in increased growth compared to wild-type neighboring cells, and these mutant cells also possess elevated levels of cyclin E (Tapon et al., 2002). Unlike *ago* mutant cells, however, *sav* mutant cells have impaired programmed cell death. SAV encodes a protein with WW domains, and interacts with WARTS (also known as LATS), which encodes a serine-threonine kinase. Since mutations in *warts/lats* also cause cell overgrowth that resembles neoplasia (Justice et al., 1995; Xu et al., 1995), it is thought that these genes act in the same pathway. Significantly the human orthologue of SAV, named hWW45, is mutated in several human cancer cell lines (Tapon et al., 2002).

Drosophila metamorphosis also provides an excellent system to study orthologues of human genes that have been associated with human disorders. The human disease tuberous sclerosis is caused by mutations in *TSC1* and *TSC2*, and result in tumorous growths called hamartomatous (Young and Povey, 1998). In *Drosophila*, mutations in either *TSC1* or *TSC2* cause increase cell growth and size (Potter et al., 2001; Tapon et al., 2001). Both *TSC1* and *TSC2* mutant cells spend reduced time in the G_1 phase of the cell cycle, and have elevated levels of cyclins E and A. Finally genetic experiments indicate that *TSC1* and *TSC2* function together in the insulin-signaling pathway. These studies provide an elegant example of how studies of metamorphosis can provide important insights into the regulation of growth, and defects associated with human diseases.

ACKNOWLEDGMENTS

The research from the laboratory of H. Laufer on crustacean development was supported by grants from the Sea Grant College Program, NOAA, and the Department of Environmental Protection, State of Connecticut; research on chironomid development was supported by grants form the National Science Foundation. The research of Dr. Baehrecke was supported by NIH grant GM59136. We acknowledge the assistance and discussion of Dr. William Biggers of Wilkes College for some aspects of this chapter.

REFERENCES

Andres AJ, Thummel CS (1992): Hormones, puffs, and flies: The molecular control of metamorphosis by ecdysone. Trends Genet 8:132–8.

Apple RT, Fristrom JW (1991): 20-Hydroxyecdysone is required for, and negatively regulates, transcription of *Drosophila* pupal cuticle protein genes. Devel Biol 146:569–82.

Ashburner M (1967): Patterns of puffing activity in the salivary gland chromosomes of *Drosophila.* I. Autosomal puffing patterns in a laboratory stock of *Drosophila melanogaster*. Chromosoma 21:398–428.

Ashburner M (1990): Puffs, genes, and hormones revisited. Cell 61:1–3.

Ashburner M, Chihara C, Meltzer P, Richards G (1974): Temporal control of puffing activity in polytene chromosomes. Cold Spring Harbor Symp Quant Biol 38:655–62.

Baehrecke EH (2002): How death shapes life during development. Nature Rev Mol Cell Biol 3:779–87.

Baker Brachmann C, Cagan RL (2003): Patterning the fly eye: the role of apoptosis. Trends Genet 19:91–6.

Becker HJ (1959): Die Puffs der Speicheldrüsenchromosomen von *Drosophila melanogaster*. I. Beobachtungen zur verhalten des Puffmusters in Normalstamm und bei zwei Mutanten, *giant* und *lethal-giant-larvae*. Chromosoma 10: 654–8.

Bender M, Imam FB, Talbot WS, Ganetzky B, Hogness DS (1997): *Drosophila* ecdysone receptor mutations reveal functional differences among receptor isoforms. Cell 91:777–88.

BenMamoun C, Gluzman IY, Hott C, MacMillan SK, Amarakone AS, Anderson DL, Carlton MR, Dame JB, Chakrabarti D, Martin RK, Brownstein BH, Goldberg DE (2001): Co-ordinated programme of gene expression during asexual intraerythrocytic development of the human malaria parasite *Plasmodium falciparum* revealed by microarray analysis. Mol Microbiol 39:26–36.

Blair SS (1993): Mechanisms of compartment formation: Evidence that non-proliferating cells do not play a role in defining the D/V lineage restriction in the developing wing of *Drosophila*. Development 119:339–51.

Bodenstein D (1965): The postembryonic development of *Drosophila.* In: Demerec M (ed): Biology of *Drosophila*, New York: Hafner, pp 275–367.

Bounhiol JJ (1938): Recherches expérimentales sur le déterminisme de la métamorphose chez les L'epidoptères. Bull Biol Suppl 24:1–199.

Broadus J, McCabe JR, Endrizzi B, Thummel CS, Woodard CT (1999): The *Drosophila* ßFTZ-F1 orphan nuclear receptor provides competence for stage-specific responses to the steroid hormone ecdysone. Mol Cell 3:143–9.

Burtis KC, Thummel CS, Jones CW, Karim FD, Hogness DS (1990): The *Drosophila* 74EF early puff contains *E74*, a complex ecdysone-inducible gene that encodes two *ets*-related proteins. Cell 61:85–99.

Cakouros D, Daish T, Martin D, Baehrecke EH, Kumar S (2002): Ecdysone-induced expression of the caspase DRONC during hormone-dependent programmed cell death in *Drosophila* is regulated by Broad-Complex. J Cell Biol 157:985–95.

Carrol SB, Grenier JK, Weatherbee SD (2001): From DNA to Diversity. Malden MA: Blackwell Science.

Champlin DT, Truman JW (2000): Ecdysteroid coordinates optic lobe neurogenesis via a nitric oxide signaling pathway. Development 127:3543–51.

Champlin DT, Truman JW (1998): Ecdysteroids control cell proliferation during optic lobe neurogenesis of the moth, *Manduca Sexta*. Development 125: 269–77.

Clausen CP (1940): Entomophagous Insects. New York: McGraw-Hill.

Clever U (1964): Actinomycin and puromycin: Effects on sequential gene activation by ecdysone. Science 146:794–5.

Clever U, Karlson P (1960): Induktion von Puff-Veränderugen in den Speicheldrüsen-chromosomen von *Chironomus tentaus* durch Ecdyson. Expr Cell Res 20:623–6.

Coen L, Aurent N, du-Pasquier D, Le-Mevel S, Brown S, Tata J, Mazabraud A, Demeneix BA (2001): *Xenopus* Bcl-XL selectively protects Rohon-Beard neurons from metamorphic degeneration. Proc Natl Acad Sci USA 98: 7869–74.

Costlow JD (1968): Metamorphosis in crustacea. In: W Etkin, LI Gilbert (eds): Metamorphosis. New York: Appleton, pp 3–41.

Crossgrove K, Bayer CA, Fristrom JW, Guild GM (1996): The *Drosophila broad-complex* early gene directly regulates late gene transcription during the ecdysone-induced puffing cascade. Dev Biol 180:745–58.

Crowley TE, Meyerowitz EM (1984): Steroid regulation of RNAs transcribed from the *Drosophila* 68C polytene chromosome puff. Dev Biol 102:110–21.

Davari ED (1999): The effect of JH on isozymes of Na^+/K^+—ATPase in ovarian follicles of *Locusta migratoria*. M Sc Thesis. York University, Toronto.

Davey KG (1996): Hormonal control of the follicular epithelium during vitellogenin uptake. Invert Reprod Dev 30:249–54.

Davey KG (2000): The modes of action of juvenile hormones: Some questions we ought to ask. Insect Biochem Mol Biol 30:663–9.

Davidson EH (2001): Genomic Regulatory Systems. San Diego, CA: Academic Press.

Diaz-Benjumea EJ, Cohen SM (1993): Interaction between dorsal and ventral cells in the imaginal disc directs wing development in *Drosophila*. Cell 75: 741–52.

DiBello PR, Withers DA, Bayer CA, Fristrom JW, Guild GM (1991): The *Drosophila Broad-Complex* encodes a family of related proteins containing zinc fingers. Genetics 129:385–97.

Doctor J, Fristrom D, Fristrom JW (1985): The pupal cuticle of *Drosophila*: Biphasic synthesis of pupal cuticle proteins in vivo and in vitro in response to 20-hydroxyecdysone. J Cell Biol 101:189–200.

Dorstyn L, Colussi PA, Quinn LM, Richardson H, Kumar S (1999): DRONC, an ecdysone-inducible *Drosophila* caspase. Proc Natl Acad Sci USA 96:4307–12.

Fallon AM, Hagedorn HH, Wyatt GR, Laufer H (1974): Activation of vitellogenin synthesis in the mosquito *Aedes aegypti* by ecdysone. J Insect Physiol 20:1815–23.

Fechtal K, Natzle JE, Brown EE, Fristrom JW (1988): Prepupal differentiation of *Drosophila* imaginal discs: Identification of four genes whose transcripts accumulate in response to a pulse of 20-hydroxyecdysone. Genetics 120: 465–74.

Fletcher JC, Burtis KC, Hogness DS, Thummel CS (1995): The *Drosophila E*74 gene is required for metamorphosis and plays a role in the polytene chromosome puffing response to ecdysone. Development 121:1455–65.

Fletcher JC, Thummel CS (1995): The ecdysone-inducible *Broad-Complex* and *E*74 early genes interact to regulate target gene transcription and *Drosophila* metamorphosis. Genetics 141:1025–35.

Fortini ME, Skupski MP, Boguski MS, Hariharan IK (2000): A survey of human disease gene counterparts in the *Drosophila* genome. J Cell Biol 150:F23–30.

Fristrom D, Fristrom JW (1993): The metamorphic development of the adult epidermis. In: M Bate, A Martinez Arias (eds): The Development of *Drosophila melanogaster*. Cold Spring Harbor: Cold Spring Harbor Laboratory Press, pp 843–97.

Froggett SJ, Leise EM (1996): Nitric oxide inhibits metamorphosis induction in a larval gastropod mollusk. Am Zool 36:13A.

Furlow JD, Lim W, Ermis DJ, Chiellini G, Scanlan TS (2000): Molecular mechanisms underlying thyroid hormone induced gene expression cascades during amphibian metamorphosis. Am Zool 40:1022–3.

Gateff E (1978): Malignant neoplasms of genetic origin in *Drosophila melanogaster*. Science 200:1448–59.

Gateff E, Schneiderman HA (1974): Developmental capacities of benign and malignant neoplasms of *Drosophila*. Wilhelm Roux Arch Entw Mech 176: 23–65.

Gifondorwa DJ, Leise EM (2001): Does the apical ganglion undergo a form of programmed cell death during metamorphosis in *Ilyanassa obsolete.* Am Zool 41:1453–4.

Gilbert LI, Tata JR, Atkinson BG, eds (1996): Metamorphosis. San Diego, CA: Academic Press.

Gilbert SF (2003): Development Biology, 7th ed. Sunderland, MA: Sinauer Associates.

Gomez ED, Faulkner D, Newman W, Ireland C (1973): Juvenile hormone mimics: Effects on cirriped crustacean metamorphosis. Science 179:813–14.

Gorski SM, Chittaranjan S, Pleasance ED, Freeman JD, Anderson CL, Varhol RJ, Coughlin SM, Zuyderduyn SD, Jones SJ, Marra MA (2003): A SAGE approach to discovery of genes involved in autophagic cell death. Curr Biol 13:358–63.

Hall BL, Thummel CS (1998): The RXR homolog Ultraspiracle is an essential component of the *Drosophila* ecdysone receptor. Development 125:4709–17.

Handler AM (1982): Ecdysteroid titres during pupal and adult development in *Drosophila melanogaster.* Dev Biol 93:73–82.

Hansson L, Lambertsson A (1989): Steroid regulation of glue protein genes in *Drosophila melanogaster*. Hereditas 110:61–7.

Hegner RW (1933): Invertebrate Zoology. New York: Macmillan.

Hengartner MO (2001): Apoptosis: Coralling the corpses. Cell 104:325–8.

Henikoff S, Eghtedarzadeh MK (1982): Conserved arrangement of nested genes at the *Drosophila Gant* locus. Genetics 117:711–25.

Henikoff S, Keene MA, Fechtel K, Fristrom JW (1986): Gene within a gene: Nested *Drosophila* genes encode unrelated proteins on opposite DNA strands. Cell 44:33–4.

Henrich VC, Sliter TJ, Lubahn DB, MacIntyre A, Gilbert LI (1990): A steroid/thyroid hormone receptor superfamily member in *Drosophila melanogaster* that shares extensive sequence similarity with a mammalian homologue. Nucl Acids Res 18:4143–8.

Hill RJ, Segraves WA, Choi D, Underwood PA, Macavoy E (1993): The reaction with polytene chromosomes of antibodies raised against *Drosophila* E75 protein. Insect Biochem Mol Biol 23:99–104.

Hinsch G (1972): Some factors controlling reproduction in the spider crab, *Libinia emarginata.* Biol Bull 143:358–66.

Hoffmann JA, Hetru C (1983): Ecdysone In: R Downer, H Laufer (eds): Endocrinology of Insects. New York: Liss, pp 65–88.

Inbal B, Bialik S, Sabanay I, Shani G, Kimchi A (2002): DAP kinase and DRP-1 mediate membrane blebbing and the formation of autophagic vesicles during programmed cell death. J Cell Biol 157:455–68.

Jiang C, Baehrecke EH, Thummel CS (1997): Steroid regulated programmed cell death during *Drosophila* metamorphosis. Development 124:4673–83.

Jiang C, Lamblin AFJ, Steller H, Thummel CS (2000): A steroid-triggered transcriptional hierarchy controls salivary gland cell death during *Drosophila* metamorphosis. Mol Cell 5:445–55.

Jochova J, Zakeri Z, Lockshin RA (1997): Rearrangement of the tubulin and actin cytoskeleton during programmed cell death in *Drosophila* salivary glands. Cell Death Differ 4:140–9.

Jones G, Sharp PA (1997): Ultraspiracle: An invertebrate nuclear receptor for juvenile hormones. Proc Natl Acad Sci USA 94:13499–503.

Jones G, Wozniak M, Chu Y-X, Dhar S, Jones D (2001): Juvenile hormone III dependent conformational changes of the nuclear receptor ultraspiracle. Insect Biochem Mol Biol 32:33–49.

Justice RW, Zilian O, Woods DF, Noll M, Bryant PJ (1995): The *Drosophila* tumor suppressor gene warts encodes a homolog of human myotonic dystrophy kinase and is required for the control of cell shape and proliferation. Genes Dev 9:534–46.

Kim J, Sebring A, Esch JJ, Kraus ME, Vornerk K, Magee J, Carroll SB (1996): Integration of positional information and identity by *Drosophila vestigial gene*. Nature 382:133–8.

Klämbt C, Muller S, Lutzelschwab R, Rossa R, Totzke F, Schmidt O (1989): The *Drosophila melanogaster l(2)gl* gene encodes a protein homologous to the cadherin cell-adhesion molecule family. Dev Biol 133:425–36.

Koelle MR, Talbot WS, Segraves WA, Bender MT, Cherbas P, Hogness DS (1991): The *Drosophila* EcR gene encodes an ecdysone receptor, a new member of the steroid receptor superfamily. Cell 67:59–77.

Laufer H, Calvet JP (1972): Hormonal effects on chromosomal puffs and insect development. Gen Comp Endocrinol Suppl 3:137–48.

Laufer H, Borst D, Baker FC, Carrasco C, Sinkus M, Reuter CC, Tsai LW, Schooley DA (1987): The identification of a crustacean juvenile hormone. Science 235:202–5.

Laufer H (1988): Comparisons of terpinoids in vertebrates and invertebrates. In: F Sehnal, A Zabza, DL Denlinger (eds): Endocrinological Frontiers in Physiological Insect Ecology. Wrodaw, Poland: Wrodaw Technical University Press, pp 907–18.

Laufer H, Biggers WJ (2001): Unifying concepts learned from methyl farnesoate for invertebrate reproduction and post-embryonic development. Am Zool 41:442–57.

Lee C-Y, Baehrecke EH (2001): Steroid regulation of autophagic programmed cell death during development. Development 128:1443–55.

Lee C-Y, Clough EA, Yellon P, Teslovich TM, Stephan DA, Baehrecke EH (2003): Genome-wide analyses of steroid- and radiation-triggered programmed cell death in *Drosophila*. Curr Biol 13:350–7.

Lee C-Y, Cooksey BAK, Baehrecke EH (2002a): Steroid regulation of midgut cell death during *Drosophila* development. Dev Biol 250:101–11.

Lee C-Y, Simon CR, Woodard CT, Baehrecke EH (2002b): Genetic mechanism for the stage- and tissue-specific regulation of steroid-triggered programmed cell death in *Drosophila*. Dev Biol 252:138–48.

Lee C-Y, Wendel DP, Reid P, Lam G, Thummel CS, Baehrecke EH (2000): *E*93 directs steroid-triggered programmed cell death in *Drosophila*. Mol Cell 6: 433–43.

Laufer H, Holt TKH (1970): Juvenile hormone effects on chromosomal puffing and development in *Chironomus thummi*. J Expr Zool 173:341–52.

Laufer H, Landau M (1991): Endocrine control of reproduction in shrimp and other crustacea. In: P DeLoach, WJ Dougherty, MA Davidson (eds): Frontiers in Shrimp Research. Amsterdam: Elsevier, pp 65–81.

Leise EM, Thavaradhara K, Durham NR, Turner BE (2001): Serotonin and nitric oxide regulate metamorphosis in the marine snail *Ilyanassa obsolete*. Am Zool 41:258–67.

Liu L, Laufer H (1996): Isolation and characterization of sinus gland neuropeptides with both mandibular organ inhibiting and hyperglycemic effects from the spider crab *Libinia emarginata*. Arch Insect Biochem Physiol 32: 375–85.

Liu L, Laufer H, Gogarten PJ, Wang M (1997): cDNA cloning of a mandibular organ inhibiting hormone from the spider crab *Libinia ermarginata*. Invert Neuroscience 3:199–204.

Moberg KH, Bell DW, Wahrer DC, Haber DA, Hariharan IK (2001): Archipelago regulates cyclin E levels in *Drosophila* and is mutated in human cancer cell lines. Nature 413:311–16.

Nemec V, Kodovik D, Matolin S, Laufer H (1993) Juvenile hormone-like effects of retinoic acid in insect metamorphosis, embryogenesis and reproduction. J Insect Physiol 39:1083–93.

Neumann CJ, Cohen SM (1996): Distinct mitogenic and cell fate specification functions of wingless in different regions of the wing. Development 122: 1781–9.

Ohsumi Y (2001): Molecular dissection of autophagy: two ubiquitin-like systems. Nat Rev Mol Cell Biol 2:211–16.

Oro AE, McKeown M, Evans RM (1990): Relationship between the product of the *Drosophila ultraspiracle* locus and vertebrate retinoid X receptor. Nature 347:298–301.

Page-McCaw A, Serano J, Sante JM, Rubin GM (2003): *Drosophila* matrix metalloproteinases are required for tissue remodeling, but not embryonic development. Dev Cell 4:95–106.

Pak MD, Gilbert LI (1987): A developmental analysis of ecdysteroids during the metamorphosis of *Drosophila melanogaster*. J Liq Chrom 10:2591–611.

Pecasse F, Beck Y, Ruiz C, Richards G (2000): Kruppel-homolog, a stage-specific modulator of the prepupal ecdysone response, is essential for *Drosophila* metamorphosis. Dev Biol 221:53–67.

Piepho H (1939a): Raupenhäutungen bereits verpuppter Hautstücke bei der Wachsmotte *Galleria mellonella* L. Naturwissenschaften. 27:301–2.

Piepho H (1939b): Über den Determinations-zustand der Vorpuppen Hypodermis bei der Wacksmotte *Galleria mellonella L.* Biol Zbl 59:314–26.

Potter CJ, Huang H, Xu T (2001): *Drosophila* Tsc1 functions with Tsc2 to antagonize insulin signaling in regulating cell growth, cell proliferation, and organ size. Cell 105:357–68.

Prahlad KV, Delanney LF (1965): A study of induced metamorphosis in the axolotl. J Expr Zool 160:137–46.

Ramanofsky M, Faulkner DJ, Ireland C (1974): Effects of juvenile hormone on cirriped metamorphosis. Biochem Res Comm 60:172–8.

Reggiori F, Klionsky DJ (2002): Autophagy in the eukaryotic cell. Eukaryot. Cell 1:11–21.

Restifo LL, Guild GM (1986): An ecdysone-responsive puff site in *Drosophila* contains a cluster of seven differentially regulated genes. J Mol Biol 188:517–28.

Restifo LL, White K (1992): Mutations in a steroid hormone-regulated gene disrupt the metamorphosis of internal tissues in *Drosophila*: Salivary glands, muscle, and gut. Wilhelm Roux's Arch Dev Biol 201:221–34.

Richards G (1981): The radioimmune assay of ecdysteroid titres in *Drosophila melanogaster.* Mol Cell Endocrinol 21:181–97.

Riddiford LM (1993): Hormones and *Drosophila* development. In: M Bate, A Martinez Arias (eds): The Development of *Drosophila melanogaster*. Cold Spring Harbor: Cold Spring Harbor Laboratory Press, pp 899–940.

Robertson CW (1936): The metamorphosis of *Drosophila melanogaster*, including an accurately timed account of the principal morphological changes. J Morphol 59:351–99.

Robinow S, Talbot WS, Hogness DS, Truman JW (1993): Programmed cell death in the *Drosophila* CNS is ecdysone-regulated and coupled with a specific ecdysone receptor isoform. Development 119:1251–9.

Rubin GM, Lewis EB (2000): A brief history of *Drosophila's* contributions to genome research. Science 287:2216–18.

Rubin GM (2000): Comparative genomics of the eukaryotes. Science 287: 2204–15.

Sachs LM, Shi YB (2000): Targeted chromatin binding and histone acetylation in vivo by thyroid hormone receptor during amphibian development. Proc Natl Acad Sci USA 97:13138–43.

Segraves WA, Hogness DS (1990): The *E*75 ecdysone-inducible gene responsible for the 75B early puff in *Drosophila* encodes two new members of the steroid receptor superfamily. Genes Dev 4:204–19.

Shea MJ, King DL, Conboy MJ, Mariani BD, Kafatos FC (1990): Proteins that bind to *Drosophila* chorion *cis*-regulatory elements: A new C_2H_2 zinc finger protein and a C_2C_2 steroid receptor-like component. Genes Dev 4:1128–40.

Sliter TJ, Gilbert LI (1992): Developmental arrest and ecdysteroid deficiency resulting from mutations at the *dre*4 locus of *Drosophila*. Genetics 130: 555–68.

Stowers RS, Garza D, Rascle A, Hogness DS (2000): The L63 gene is necessary for the ecdysone-induced 63E late puff and encodes CDK proteins required for *Drosophila* development. Dev Biol 221:23–40.

Tapon N, Harvey KF, Bell DW, Wahrer DC, Schiripo TA, Haber DA, Hariharan IK (2002): *Salvador* promotes both cell cycle exit and apoptosis in *Drosophila* and is mutated in human cancer cell lines. Cell 110:467–78.

Tapon N, Ito N, Dickson BJ, Treisman JE, Hariharan IK (2001): The *Drosophila* tuberous sclerosis complex gene homologs restrict cell growth and cell proliferation. Cell 105:345–55.

Thavaradhara K, Leise EM (2001): Localization of nitric oxide synthase-like immunoreactivity in the developing nervous system of the snail *Ilyanassa obsolete*. J Neurocytol 30:449–56.

Thomas HE, Stunnenberg HG, Stewart AF (1993): Heterodimerization of the *Drosophila* ecdysone receptor with retinoid X receptor and *ultraspiracle*. Nature 362:471–5.

Tighe-Ford JD (1977): Effects of juvenile hormone analogues on larval metamorphosis in the barnacle *Eliminus modestus* Darwin. (Crustacea: Cirripedia). J Expr Mar Biol Ecol 25:163–76.

Treece GD, Yates ME (1993): Laboratory Mmanual for the Culture of Panaeid Shrimp Larvae. Marine Advisory Service, Sea Grant College Program. Texas A&M University TAMU-SG-88–202 (R).

Truman JW, Riddiford LM (1999): The origins of insect metamorphosis. Nature 401:447–52.

Truman JW, Talbot WS, Fahrbach SE, Hogness DS (1994): Ecdysone receptor expression in the CNS correlates with stage-specific responses to ecdysteroids during *Drosophila* and *Manduca* development. Development 120:219–34.

Truman JW, Taylor BJ, Awad TA (1993): Formation of the adult nervous system. In: M Bate, A Martinez Arias (eds): The Development of *Drosophila melanogaster*. Cold Spring Harbor: Cold Spring Harbor Laboratory Press, pp 1245–75.

Tsukada M, Ohsumi Y (1993): Isolation and characterization of autophagy-defective mutants of *Saccharomyces cerevisiae*. FEBS Lett. 1–2:169–74.

Tobe SS, Stay B (1985): Structure and regulation of the corpus allatum. Adv Insect Physiol 18:305–432.

Urness LD, Thummel CS (1995): Molecular analysis of a steroid-induced regulatory hierarchy: The *Drosophila* E74A protein directly regulates *L*71–6 transcription. EMBO J 14:6239–46.

Urness LD, Thummel CS (1990): Molecular interactions within the ecdysone regulatory hierarchy: DNA binding properties of the *Drosophila* ecdysone-inducible *E74A* protein. Cell 63:47–61.

Vafopoulou X, Steel CGH, Laufer H (1989): Selective stimulation of puffing activity *in vitro* with a juvenile hormone analogue by polytene chromosomes in the salivary gland of *Chironomus thummi* (*Diptera*). Can J Zool 67:2528–32.

von Kalm L, Crossgrove K, Von Seggern D, Guild GM, Beckendorf SK (1994): The *Broad-Complex* directly controls a tissue-specific response to the steroid hormone ecdysone at the onset of *Drosophila* metamorphosis. EMBO J 13: 3505–16.

von Kalm L, Fristrom D, Fristrom J (1995): The making of a fly leg: A model for epithelial morphogenesis. Bioessays 17:693–702.

Weigmann K, Klapper R, Strasser T, Rickert C, Technau GM, Jäckie H, Janning W, Klämbt C (2003): Flymove- a new way to look at development of *Drosophila* trends in Genetics 19:310–11.

Wigglesworth VB (1972): The Principles of Insect Physiology, 7th ed. London: Chapman and Hall.

Williams CM (1961): The juvenile hormone II: Its role in the endocrine control of molting, pupation, and adult development in the *Cecropia* silkworm. Biol Bull 121:572–85.

Wolff T, Ready DF (1991): Cell death in normal and rough eye mutants of *Drosophila*. Development 113:825–39.

Wolff T, Ready DF (1993): Pattern formation in the *Drosophila* retina. In: M Bate, A Martinez Arias (eds): The Development of *Drosophila melanogaster*. Cold Spring Harbor: Cold Spring Harbor Laboratory Press, pp 1277–325.

Woodard CT, Baehrecke EH, Thummel CS (1994): A molecular mechanism for the stage-specificity of the Drosophila prepupal genetic response to ecdysone. Cell 79:607–15.

Wright LG, Chen T, Thummel CS, Guild GM (1996): Molecular characterization of the 71E late puffin *Drosophila melanogaster* reveals a family of novel genes. J Mol Biol 255:387–400.

Xu T, Wang W, Zhang S, Stewart RA, Yu W (1995): Identifying tumor suppressors in genetic mosaics: the Drosophila lats gene encodes a putative protein kinase. Development 121:1053–63.

Yao T-P, Segraves WA, Oro AE, McKeown M, Evans RM (1992): *Drosophila ultraspiracle* modulates ecdysone receptor function via heterodimer formation. Cell 71:63–72.

Young J, Povey S (1998): The genetic basis of tuberous sclerosis. Mol Med Today 4:313–19.

Zhou X, Riddiford LM (2002): Broad specifies pupal development and mediates the status quo action of juvenile hormone on the pupal-adult transformation in *Drosophila and Manudca.* Development 129:2259–69.

CHAPTER 12

TRANSLATIONAL CONTROL AND THE CELL CYCLE

ROBERT E. RHOADS
Department of Biochemistry and Molecular Biology, Louisiana State University Health Sciences Center, Shreveport, LA 71130-3932

INTRODUCTION

It is not surprising that two processes so central to cellular function as protein synthesis and cell division are interrelated at many levels. As is reviewed in this chapter, there are specific translational signals that promote the progression from one stage of the cycle to the next, in the absence of which cells remain arrested. Conversely, the characteristics of protein synthesis, such as the overall rate of protein synthesis, whether the process is limited by initiation or elongation, and whether initiation favors cap-dependent or cap-independent pathways, are strongly influenced by the stage of the cell cycle. The ability of the protein synthesis machinery to translate different types of mRNAs varies throughout the cell cycle, independently of transcriptional control and the spectrum of mRNAs present in the cell.

Cell growth and cell division are distinct but coupled processes (Conlon and Raff, 1999; Schmelzle and Hall, 2000), and protein synthesis plays separate and independent roles in both. Cell growth requires synthesis of new proteins, which in turn is controlled by a variety of signaling pathways responsive to hormonal stimuli, extracellular growth factors, and availability of nutrients. Progression through the cell cycle halts at specific checkpoints in G_1 or G_0 if the cell has not grown to a minimum size. However, protein synthesis is not merely a passive process necessary to achieve a cell of sufficient volume to divide. Rather, specific signals are transmitted from the protein synthesis machinery to the cell cycle machinery.

Mitogens are not the same as growth factors, the former stimulating cell cycle progression and the latter stimulating cell growth. Yet the terms are often used synonymously because growth factors can promote

Cell Cycle and Growth Control: Biomolecular Regulation and Cancer, Second Edition, Edited by Gary S. Stein and Arthur B. Pardee
ISBN 0-471-25071-6

mitogenesis if cell growth is limiting for cell cycle progression. The link between these two processes is protein synthesis. Agents possessing both growth factor and mitogenic activity regulate protein synthesis in multiple ways. Some lead to cell growth through a stimulation of global protein synthesis, thereby bringing the cell closer to the threshold size required for entry into S phase. Some lead to mitogenesis through synthesis of specific proteins needed for the $G_1 \rightarrow S$ transition or biosynthetic reactions occurring in S phase and beyond. Interestingly it is possible to separate the two end points by interruption of specific signaling pathways leading to the protein synthesis machinery.

Because of this intimate relationship between cell growth and cell division, this chapter deals with the relationship of protein synthesis to both processes but with emphasis on cell division. In the area of cell growth, the discussion is limited to anabolic process, even though there is a great deal known about the starvation response and nutrient-responsive transcriptional factors (Hinnebusch, 2000). In the area of cell division, two aspects of translational control have been intensively investigated, the progression from G_1 into S, and the progression from G_2 into M. Relatively little is known about the requirement for, or characteristics of, protein synthesis at other stages of the cell cycle (Burke and Church, 1991). Interestingly the mechanisms for both $G_1 \rightarrow S$ and $G_2 \rightarrow M$ progression involve mRNA-specific translational control, rather than global translational control, and both involve the same protein synthesis factor. The translational regulation of a number of proteins is discussed, but not their transcriptional regulation unless it is directly affected by translational events. Finally, as cancer has been called a disease of the cell cycle, this chapter also reviews current information on the deregulation of translational processes affecting the cell cycle in malignantly transformed cells and naturally occurring human cancers.

MECHANISM OF EUKARYOTIC PROTEIN SYNTHESIS

The three stages of protein synthesis are catalyzed by three groups of proteins: initiation, elongation, and release factors (Hershey and Merrick, 2000). The most highly regulated phase of protein synthesis is initiation (Gingras et al., 1999; Rhoads, 1999). A different class of initiation factors catalyzes each of the steps of initiation. A ternary complex of eIF2·GTP·Met-$tRNA_i$ binds to the 40S ribosomal subunit to form the 43S initiation complex (Fig. 12.1). The next step, recruitment of mRNA to the 43S initiation complex to form the 48S initiation complex, is rate limiting for initiation under normal conditions. *Cis*-acting elements in mRNA that stimulate this process include the 5′-terminal m^7G-containing cap, the 3′-terminal poly(A) tract, and, in a small subset of viral and cellular mRNAs, an internal ribosome entry sequence (IRES; Jackson, 2000). *Trans*-acting factors include eIF3, poly(A)-binding protein (PABP), and the eIF4 proteins. eIF3 is 520kDa multimer that is required for both Met-$tRNA_i$ and mRNA binding. PABP is a 70kDa

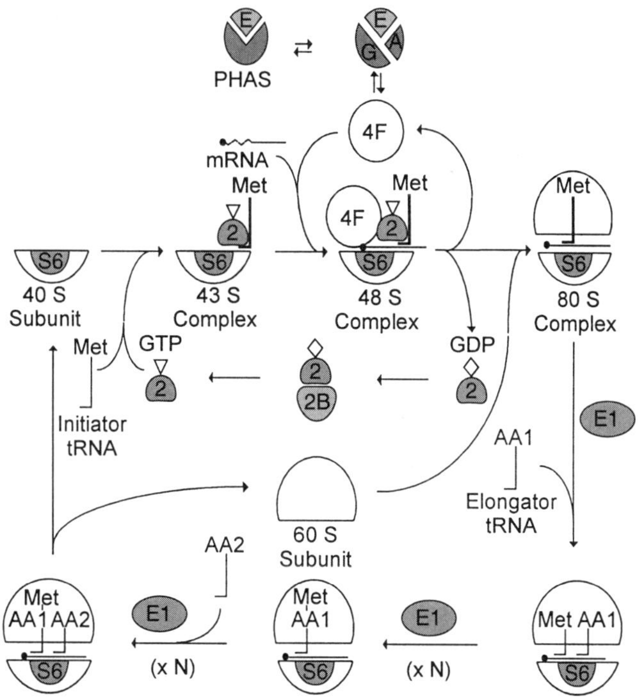

Figure 12.1. Eukaryotic protein synthesis and sites of action for initiation and elongation factors. Factors are abbreviated as 2, eIF2; 2B, eIF2B; A, eIF4A; E, eIF4E; 4F, eIF4F; G, eIF4G, E1, eEF1; E2, eEF2; and S6, ribosomal protein S6. The triangle represents GTP and the diamond, GDP.

protein that specifically binds poly(A) and homo-oligomerizes. The eIF4 factors consist of eIF4A, a 46 kDa RNA helicase; eIF4B, a 70 kDa RNA-binding and annealing protein that stimulates eIF4A; eIF4E, a 25 kDa cap-binding protein; and eIF4G, a 185 kDa protein that specifically binds to and co-localizes all of the other proteins involved in mRNA recruitment on the 40 S ribosomal subunit.

The 48 S complex consists of the eIF4 factors plus PABP, eIF3 and the eIF2·GTP·Met-tRNA$_i$ ternary complex all bound to the 40 S ribosomal subunit (Fig. 12.2). It scans until the first AUG in good sequence context is encountered. Scanning requires ATP hydrolysis by eIF4A and the presence of eIF1 and eIF1A. Then the GTPase-activating protein eIF5, together with eIF5B, stimulates GTP hydrolysis by eIF2. The initiation factors are replaced by the 60 S subunit to form the 80 S complex, and the first peptide bond is formed. The released eIF2·GDP is recycled to eIF2·GTP by the guanine nucleotide exchange factor eIF2B.

The first elongator aminoacyl-tRNA is brought to the A-site by eEF1 (Fig. 12.1). This is followed by a cycle of GTP hydrolysis and exchange, analogous to that of eIF2. Translocation is catalyzed by eEF2, again with a GTP hydrolysis cycle. This frees up the acceptor site on the ribosome so that the next incoming aminoacyl-tRNA can bind. When the ribosome

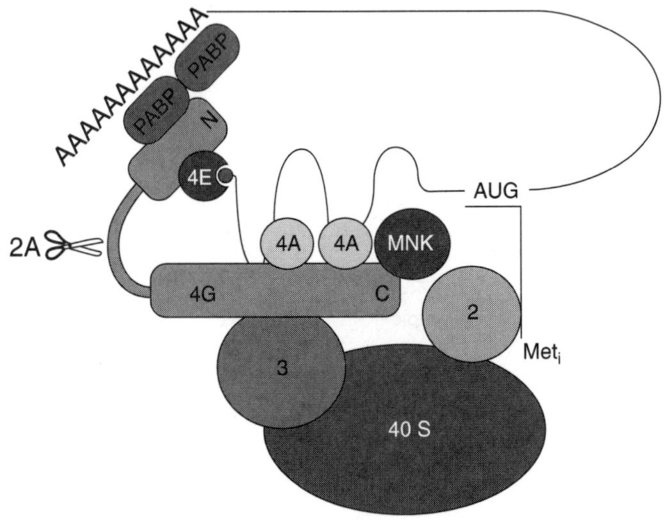

Figure 12.2. Model for the 48S initiation complex. "2A" indicates the site of cleavage in eIF4G by picornaviral protease 2A. The initiation factors are defined as in Figure 12.1, except that "3" indicates eIF3, "4A" indicates eIF4A, "4E" indicated eIF4E, and "4G" indicated eIF4G. "N" and "C" designate the NH_2- and COOH-termini of eIF4G, respectively. The wavy line conveys secondary structure in mRNA.

reaches a termination codon, the release factor eRF1 catalyzes termination, recognizing all three termination codons and structurally mimicking tRNA (Welch et al., 2000). Then eRF3 releases eRF1 from the ribosome in a reaction that utilizes its GTPase function.

REGULATION OF PROTEIN SYNTHESIS FACTOR ACTIVITY

The activities and availabilities of several initiation and elongation factors are regulated by three basic mechanisms: intracellular level, phosphorylation, and sequestration in inactive complexes with modulatory proteins (Rhoads, 1999; Gingras et al., 1999). This section will concentrate only on those factors and mechanisms that are implicated in cell growth and cell cycle progression. In this section is discussed the modifications that affect factor activities, enzymes responsible for those modifications, and signaling pathways leading to changes in protein synthesis in response to growth factors, hormones, cytokines, and nutrients.

eEF1

The elongation factor eEF1 is phosphorylated and activated by multipotential S6 kinase, which phosphorylates the α, β, and γ subunits (Chang and Traugh, 1997). This causes a 2–2.6-fold increase in eEF1 activity in vitro. Serum starvation of 3T3-L1 cells decreases phosphorylation but insulin restores it. Similarly insulin doubles the rate of elongation in serum-starved cells. As noted below, eEF1 is also a substrate for phosphorylation by mitosis promoting factor.

eEF2

By contrast to eEF1, the activity of eEF2 is *inhibited* by phosphorylation, causing a decrease in the rate of polypeptide elongation (Nairn and Palfrey, 1987; Ryazanov et al., 1988; Sitikov et al., 1988). The 100kDa elongation factor is phosphorylated by eEF2 kinase on Thr-56 and Thr-58, the phosphorylation of the latter site requiring prior phosphorylation of the former (Redpath et al., 1993; Nairn and Palfrey, 1996). This enzyme, apparently a dedicated kinase, is activated by Ca^{++}/calmodulin and contains a calmodulin-binding domain. eEF2 kinase is also activated by cAMP, independent of Ca^{++} (Diggle et al., 1998). There is evidence that insulin regulates eEF2 through a rapamycin-sensitive pathway; in CHO-IR cells, insulin causes decreased phosphorylation of eEF2, and this is associated with stimulation of the rate of polypeptide chain elongation (Redpath et al., 1996). Rapamycin blocks the activation of elongation, the decrease in eEF2 phosphorylation, and the decrease in eEF2 kinase activity.

eIF2B

eIF2 is a heterotrimeric initiation factor. The α subunit is phosphorylated and inhibited by a variety of kinases (Hinnebusch, 2000; Kaufman, 2000; Chen, 2000; Ron and Harding, 2000), but this generally occurs in response to cellular stresses and is not discussed further here. The other major mechanism by which eIF2 is regulated is through the guanine-nucleotide exchange factor eIF2B. Phosphorylation of the ε subunit of mammalian eIF2B by glycogen synthase kinase-3 (GSK-3) decreases its nucleotide exchange activity (Welsh et al., 1998; Jefferson et al., 1999; Fig. 12.3). GSK-3 inactivates eIF4Bε by phosphorylation on Ser-535, but it requires a priming phosphorylation at Ser-539 by the kinase DYRK (Woods et al., 2001). *Drosophila* eIF2Bε is also phosphorylated and inactivated by GSK-3 and casein kinase II (Williams et al., 2001).

GSK-3 regulates numerous cellular processes besides phosphorylation of glycogen synthase, including insulin-stimulated protein synthesis (Cross et al., 1995). Phosphorylation of GSK-3 at Ser-9 *inactivates* the enzyme and is correlated with the activation of protein synthesis. This is catalyzed by PKB (Moule et al., 1997; Hajduch et al., 1998; van Weeren et al., 1998; Fig. 12.3). However, insulin may activate eIF2B through additional routes, since constitutively active protein kinase Cζ (ΔPKCζ) stimulates general protein synthesis in 32D cells in an insulin-dependent manner without activating S6K1 (labeled $p70^{S6K}$ in Fig. 12.3), suggesting that PKB is not involved (Mendez et al., 1997). Ectopic expression of ΔPKCζ restores insulin-stimulated general protein synthesis to 32D cells containing a form of IRS-1 lacking all 18 Tyr residues; that is, IRS-1

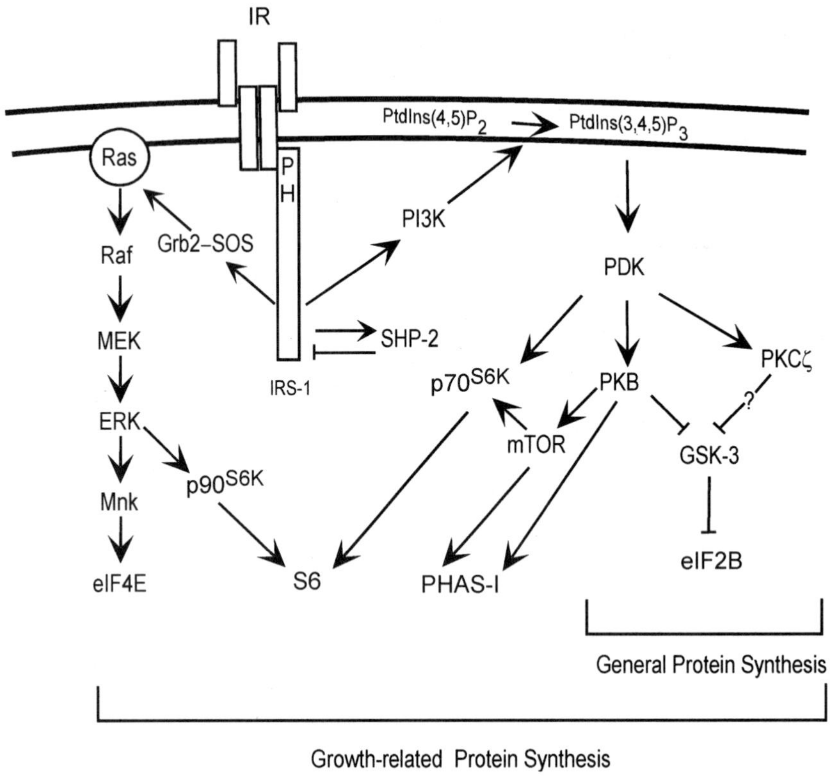

Figure 12.3. Model for the signaling pathways leading to general and growth-regulated protein synthesis. Pathways differ for different growth factors and cells. The figure is based largely on insulin stimulation of 32D cells stably expressing insulin receptor (IR) and IRS-1. "PtdIns" indicates phosphatidylinositol, and "PH," pleckstrin homology. The 70kDa S6 kinase is labeled "p70^{S6K} in the figure to distinguish it from p90^{S6K}, but in the text it is referred to as "S6K1."

cannot be phosphorylated by insulin receptor. This is accompanied by a commensurate decrease in GSK-3 activity and increase in eIF2B activity. Also, in 32D cells completely lacking IRS-1 but containing ΔPKCζ, the activities of GSK-3 and eIF2B are fully responsive to insulin (Mendez et al., 2001). In L6 myotubes, on the other hand, insulin-induced activation of GSK-3 is inhibited by a dominant-negative form of PKB but not PKCλ (another atypical PKC; Takata et al., 1999). Similarly, in CHO cells, insulin-induced phosphorylation of PHAS-I (see below) is inhibited by a dominant-negative form of PKB but not PKCλ.

There are at least 10 isoforms of PKC (α–ζ) that differ in responsiveness to phospholipids and Ca^{++} (Dekker and Parker, 1994). Classical PKCs (α, β and γ) are activated and eventually down-regulated by phorbol esters, but atypical isoforms (λ and ζ) are not. Insulin activates both classical and atypical isoforms (Bandyopadhyay et al., 1997; Mendez et al., 1997). PKCζ is activated downstream of phosphatidylinositol-3-kinase (PI3K) (Herrera-Velit et al., 1997) via direct phosphorylation at Thr-410 by phosphatidylinositol-dependent kinase-1 (PDK1) (Le Good et al., 1998; Dong et al., 1999; Fig. 12.3).

PKCζ activation is necessary and sufficient to activate maturation in oocytes and to produce deregulation of growth control in mouse fibroblasts (Berra et al., 1993). Furthermore use of a dominant kinase-defective mutant of PKCζ shows that it is required for mitogenic activation in oocytes and fibroblasts. Ectopic expression of ΔPKCζ in mouse 32D cells eliminates the requirement of IRS-1 for general protein synthesis but not for insulin-stimulated mitogenesis (Mendez et al., 1997). PDK1 appears to represent a bifurcation in the insulin signaling pathway, one branch leading through PKCζ to general protein synthesis and one, through PKB and mTOR, to growth-regulated protein synthesis and cell cycle progression (Fig. 12.3).

eIF4E

eIF4E levels, availability, and activity are controlled by at least three mechanisms that are relevant to cell cycle and cell growth control: phosphorylation, sequestration by PHAS, and transcription.

eIF4E Phosphorylation. Initial interest in eIF4E phosphorylation came from the observation that the rate of phosphate incorporation into eIF4E correlates with cap-dependent protein synthesis in a large number of systems (Rhoads, 1993). Extracellular effectors such as growth factors, cytokines, and hormones increase both protein synthesis and the rate of eIF4E phosphorylation, whereas viruses, heat shock, and other stresses decrease both. Phosphorylation was originally thought to increase its affinity for caps (Minich et al., 1994). RNA-Sepharose separates the phosphorylated from nonphosphorylated forms of eIF4E, and the former preparation has higher affinity for m^7GTP. However, this study was conducted before the discovery of PHAS-I, and it is possible that the higher affinity was due to the presence of PHAS-I. Subsequent work has shown that eIF4E phosphorylation *decreases* its affinity for the cap (Scheper et al., 2002).

eIF4E is phosphorylated on Ser-209 (Joshi et al., 1995; Flynn and Proud, 1995) by Mnk1 and 2 (Fukunaga and Hunter, 1997; Waskiewicz et al., 1997, 1999). These MAPK-interacting Ser/Thr kinases were identified as binding partners and substrates of ERK1 and 2. Mnk1 is phosphorylated and activated in vitro by ERK1 and p38 MAP kinases (Fig. 12.3) upon stimulation of mammalian cells with polypeptide growth factors, phorbol esters, fetal calf serum, anisomycin, UV irradiation, TNFα, IL-1β, or osmotic shock. Expression of dominant-negative Mnk1 reduces mitogen-induced eIF4E phosphorylation, while expression of activated Mnk1 increases basal eIF4E phosphorylation (Waskiewicz et al., 1999). Interestingly Mnk1 also binds to the COOH-terminus of eIF4G and co-purifies with eIF4G and eIF4E (Waskiewicz et al., 1999; Pyronnet et al., 1999; Fig. 12.2). Ectopic expression of constitutively active Mnk1 also induces extensive phosphorylation of eIF4E in cells overexpressing PHAS-I, suggesting that this pathway is independent of other mitogenic signals that release eIF4E from PHAS-I (Waskiewicz et al., 1999).

Determining the effect of eIF4E phosphorylation on translational initiation and overall protein synthesis has been more elusive, with studies in various systems showing an increase, a decrease, or no change. Several reports have failed to find a difference between phosphorylated and nonphosphorylated eIF4E in a variety of assays related to protein synthesis (McKendrick et al., 2001; Scheper and Proud, 2002). Another study found that expression of active forms of Mnk1 and Mnk2 in 293 cells diminishes cap-dependent translation relative to cap-independent translation in a transient reporter assay and reduces in vitro protein synthesis (Knauf et al., 2001). By contrast, a transgenic study in *Drosophila* showed that phosphorylation of eIF4E is critical for growth (Lachance et al., 2002). *Drosophila* expressing an unphosphorylatable variant of eIF4E in an *eIF4E* mutant background have reduced viability. Those that live develop more slowly and are smaller than controls.

PHAS-I. A second regulatory mechanism for eIF4E involves sequestration into an inactive complex (Fig. 12.1). A cDNA for one of the most prominent insulin-stimulated phosphoproteins in rat adipocytes was cloned (Hu et al., 1994) and named PHAS-I (phosphorylated heat- and acid-stable substrate regulated by insulin). The human cDNA for this protein was cloned and named 4E-BP1 (Pause et al., 1994). Both names are used in current literature; for simplicity the original name is used here. Multiple proteins similar in structure and function to PHAS have been described, called PHAS-I, PHAS-II, or PHAS–III (4E-BP1, 4E-BP2, or 4E-BP3).

PHAS phosphorylation regulates the availability of eIF4E for translation. In extracts of 3T3-L1 adipocytes, eIF4E binds PHAS-I unless the cells are previously treated with insulin (Lin et al., 1994). Nonphosphorylated PHAS-I binds eIF4E and inhibits protein synthesis. Similarly interaction of PHAS-I and -II with eIF4E inhibits translation, but treatment of cells with insulin causes PHAS-I to become hyperphosphorylated and dissociate from eIF4E, allowing eIF4E to bind eIF4G and thereby relieving translational inhibition (Pause et al., 1994; Fig. 12.1). PHAS phosphorylation levels are increased by many types of extracellular stimuli, including mitogens, growth factors, hormones, cytokines, and G-protein-coupled receptor agonists (reviewed in Gingras et al., 2001). Conversely, PHAS phosphorylation is decreased by environmental or nutritional stresses.

PHAS-I is phosphorylated at multiple sites. Rat PHAS-I is phosphorylated at Thr-36, Thr-45, Ser-64, Thr-69, and Ser-82 (Fadden et al., 1997). All sites are increased in rat adipocytes by insulin and decreased by rapamycin. Phosphorylation of Thr-36 alone is insufficient to prevent eIF4E binding. In human PHAS-I, phosphorylation of Ser-65 and Thr-70 is strongly enhanced by addition of serum. Changing Thr-37 and/or

Thr-46 to Ala prevents phosphorylation of Ser-65 and Thr-70, suggesting that phosphorylation of these residues serves as a priming event (Gingras et al., 2001). Phosphorylation of Ser-65 and Thr-70 in PHAS-I is carried out by mTOR, a kinase inhibited by rapamycin (see below). Treatment of cells with rapamycin prevents mitogen-stimulated PHAS-I phosphorylation, and immunoprecipitates of mTOR phosphorylate PHAS-I on the "priming" sites (von Manteuffel et al., 1996; Mendez et al., 1996; Brunn et al., 1997). Also treatment of cells with the phosphatase inhibitor calcyculin prevents PHAS-I phosphorylation, suggesting that a PP2A-type phosphatase directed toward Ser-65 and Thr-70 is activated following rapamycin treatment (Gingras et al., 2001). This is particularly relevant to TOR action in budding yeast (see below).

Transcription of the eIF4E Gene. Finally, the total intracellular protein levels of eIF4E are controlled at the transcriptional level. eIF4E gene transcription is increased by a variety of mitogenic stimuli (Rosenwald et al., 1993; Jones et al., 1996). The eIF4E gene promoter contains an E-box recognized by the transcription factor c-Myc.

Ribosomal Protein S6

S6 is a protein present in the small ribosomal subunit (Fig. 12.1) that is phosphorylated coincident with stimulation of protein synthesis under a variety of conditions (Fig. 12.3). The major activity responsible for phosphorylation of S6 is $p70^{S6K}$ (S6K1), and this S6 kinase has been most studied in the context of mitogenic responses. S6K1 is activated by hierarchical phosphorylation of seven Ser/Thr sites (Pullen and Thomas, 1997, Alessi et al., 1998). Phosphorylation of COOH-terminal sites occurs first, making Thr-389 available for phosphorylation by a rapamycin-sensitive pathway (Burnett et al., 1998; Weng et al., 1998). Rapamycin principally affects only one early site in the phosphorylation hierarchy (Pullen and Thomas, 1997; Weng et al., 1998). Upon addition of rapamycin or wortmannin to insulin-treated cells, the decrease in activity of S6K1 is closely correlated with the disappearance of Thr-412 phosphorylation. Immunoprecipitates of mTOR phosphorylate S6K1 in vitro (Fig. 12.3), and treatment of HEK293 cells with serum stimulates mTOR kinase activity with the same kinetics as the in vivo phosphorylation of S6K1 (Burnett et al., 1998; Isotani et al., 1999), but it remains unclear whether this phosphorylation is physiologically relevant in vivo. S6K1 activation also requires direct phosphorylation by PDK1 at Thr-229 in the catalytic domain (Pullen et al., 1998; Alessi et al., 1998). This phosphorylation is strongly dependent on S6K1 being already phosphorylated on Thr-412.

Other kinases have been identified that phosphorylate S6. Multipotential S6 kinase is activated and becomes membrane-associated in response to insulin but by an unknown pathway (Chang and Traugh, 1997). This is accompanied by phosphorylation of S6 in ribosomes and a twofold increase in the rate of elongation in vitro. $p90^{S6K}$ is activated in response to growth factors by forming complexes with ERK1 and 2 and undergoing phosphorylation (Smith et al., 1999; Fig. 12.3). Additional S6 kinases continue to be discovered (Shima et al., 1998). An enzyme named S6K2 was discovered because targeted gene disruption of S6K1 led to a reduced animal size, but S6 phosphorylation continued, suggesting a second enzyme (Park et al., 2002). Activation of S6K2 requires signaling from both PI3K and mTOR. RLPK is yet another recently discovered S6 kinase (New et al., 1999).

mTOR

Target of rapamycin (mTOR, FRAP, RAFT-1), a 290-kDa protein kinase that phosphorylates Ser/Thr-Pro motifs, is upstream of both PHAS and S6K1 (Gingras et al., 2001). The mammalian protein shares 45% identity with yeast TOR proteins (see below). mTOR belongs to a larger protein family called PIKKs (phosphatidylinositotal kinase-related kinases), all of which may be involved in cell-cycle control. Despite its similarity to lipid kinases, mTOR functions as a protein kinase.

As the name implies, the kinase activity of mTOR is inhibited by the immunosuppressant rapamycin (Abraham and Wiederrecht, 1996). Rapamycin acts by binding to the protein FKBP12, the complex binding in turn to mTOR in a configuration by which rapamycin simultaneously occupies two hydrophobic pockets, one in FKBP12 and the other in mTOR. This explains why rapamycin blocks phosphorylation of PHAS-I and inhibits cap-dependent initiation of translation in 32D (Mendez et al., 1996) and NIH 3T3 (Beretta et al., 1996) cells. Variants of mTOR that do not respond to rapamycin protect both S6K1 and PHAS-I against rapamycin-induced dephosphorylation (Hara et al., 1997).

mTOR is activated downstream of PI3K in response to growth factors, cytokines, and hormones, but the precise mechanism is not clear (Fig. 12.3). Evidence has been presented for activation by PKB (Scott et al., 1998). The COOH terminus of mTOR contains a consensus site for phosphorylation by PKB. In 3T3-L1 adipocytes, insulin increases the ^{32}P content of mTOR and the phosphorylation of PHAS-I in immune complexes of mTOR. Activation of PKB mimics the effect of insulin. However, other studies have failed to detect significant alteration in mTOR autokinase activity or mTOR-associated PHAS-I kinase activity in response to insulin (Hara et al., 1997, 1998). Also only Thr-37 and Thr-46 of PHAS-I are phosphorylated by mTOR in vitro (Gingras et al., 1999). As noted above, phosphorylation of Thr-37 and Thr-46 is required for subsequent phosphorylation of several COOH-terminal serum-sensitive sites. These results suggest that mTOR-mediated priming phosphorylation of PHAS-I is a prerequisite for growth factor-induced PKB-mediated full phosphorylation. New light has been shed on this area by the discovery of a downstream effector of mTOR, the protein raptor, based on the observation that nutrients activate downstream effectors of mTOR without changing the in vitro kinase activity of mTOR (Kim et al., 2002; Hara et al., 2002). mTOR forms a complex with raptor to stimulate S6K1 phosphorylation in response to nutrients. Raptor also binds both S6K1 and PHAS-I and is necessary for the phosphorylation of the latter by mTOR.

The phosphorylation S6K1 and PHAS-I is also sensitive to the presence of amino acids, leading to the model of mTOR as a nutrition-sensitive "gatekeeper" (Gingras et al., 2001). Upon withdrawal of amino acids from the medium, CHO-IR cells become unresponsive to all agonists, accompanied by rapid deactivation of S6K1 and dephosphorylation of PHAS-I (Hara et al., 1998). Re-addition of amino acids quickly restores the phosphorylation and responsiveness to insulin. A S6K1 variant that is unresponsive to rapamycin is also unresponsive to amino acid withdrawal. These effects are not mediated through PI3K or MAPK pathways. Also, in pancreatic β-cells, amino acids are required for glucose or exogenous insulin to stimulate phosphorylation of PHAS-I (Xu et al., 1998). Rapamycin inhibits the ability of branched-chain amino acids to stimulate the phosphorylation of PHAS-I and S6K1, suggesting the involvement of mTOR. Furthermore, in human T-cells, downstream signaling events of mTOR are dependent on the amino acid concentration (Iiboshi et al., 1999). In cells expressing a temperature-sensitive variant of histidyl-tRNA synthetase, the suppression of S6K1 activity by amino acid depletion is blocked at the restrictive temperature but reversed by excess histidine, suggesting that the accumulation of uncharged tRNA deactivates S6K1. These studies suggest that mTOR plays a role as nutritional checkpoint, permitting intracellular signals to be transmitted to the translational apparatus only when sufficient amino acids are present.

mRNA-SPECIFIC TRANSLATIONAL CONTROL AND THE CELL CYCLE

Signaling pathways do not stimulate translation of all mRNAs equally. In response to growth and mitogenic stimuli, global translation rates do not change nearly as dramatically as the translation rates of a small population of mRNAs. For instance, a DNA-array study of ribosome loading of 1200 mRNAs in resting and mitogenically activated fibroblasts showed that fewer than 1% of mRNAs are translationally regulated in response to serum (Zong et al., 1999). This is explained in part by the fact that messenger RNAs differ widely in translational efficiency. Although many mechanisms have been described for mRNA-specific translational control, some of them relate directly to cell cycle progression and are discussed below. Three of these factors contributing to low efficiency of translation, a highly structured 5′-untranslated region (5′-UTR), upstream AUGs, and poor sequence context for the initiating AUG, are found in the 5′-UTRs of mRNAs for scarce proteins (Kochetov et al., 1998). These mRNAs, which have been termed "growth-regulated" (Baserga, 1990), are poorly translated in quiescent cells but are preferentially recruited to ribosomes after a mitogenic signal (Kaspar et al., 1990; Manzella et al., 1991; Nielsen et al., 1995). They encode a disproportionate share of proteins involved in cell growth and cell cycle progression, such as proto-oncoproteins, growth factors, growth factor receptors, enzymes involved in DNA synthesis, and cyclins (De Benedetti and Harris, 1999). In this section are described *cis*-acting factors that contribute to differences in mRNA translational efficiency and the corresponding *trans*-acting factors that elicit changes in translational efficiency in response to mitogens and growth factors.

RNA Secondary Structure

The presence of extensive base-pairing in the 5′-UTR of an mRNA reduces its translational efficiency. For instance, two naturally occurring *c-myc* transcripts differ in the length and secondary structure of their 5′-UTRs, and the shorter transcript is translated in vitro 10-fold more efficiently (Darveau et al., 1985), adding support to the hypothesis that translation of full-length c-*myc* mRNA is normally repressed, whereas in Burkitt lymphomas that have deletions in the 5′-UTR, translation of the c-*myc* mRNA is more efficient (Saito et al., 1983). Similarly, when oligonucleotide linkers that form hairpin loops are inserted into the region of the herpes thymidine kinase gene corresponding to the 5′-UTR of the mRNA, translational efficiency of the mRNAs is decreased both in vivo and in vitro as the number of linkers is increased (Pelletier and Sonenberg, 1985). The translation of these mRNAs, however, is preferentially stimulated when components of the mRNA unwinding machinery (i.e., initiation factors of the eIF4 group) are overexpressed or activated (Rhoads, 1999; Gingras et al., 1999). In NIH 3T3 cells, expression of a reporter from transiently transfected vectors is inhibited when an increasing number of self-complementary *Bam*HI linkers is introduced into the 5′-UTR, but ectopic overexpression of eIF4E relieves this inhibition (Koromilas et al., 1992). Examples of increased translation of natural mRNAs with high 5′-UTR secondary structure by ectopic expression of eIF4E are discussed below in this chapter. eIF4A is another member of the unwinding machinery (see Figs. 12.1, 12.2). In vitro translation of mRNAs containing stable secondary structures in the 5′-UTR is more susceptible to inhibition by a dominant-negative variant of eIF4A (Svitkin et al., 2001). Another member of the eIF4 group of factors, eIF4B, stimulates the unwinding activity of eIF4A. In vitro translation of β-galacosidase reporter mRNAs with varying degrees of RNA secondary structure in the 5′-UTR in extracts of *Saccharomyces* containing a disruption of the gene encoding eIF4B shows preferential inhibition for mRNAs with more stable secondary structure (Altmann et al., 1993).

uORFs

The presence of open-reading frames (uORFs) upstream of the major open-reading frame in an mRNA can have a profound effect on translational efficiency (Morris, 1997). Three types of uORFs can exist: in-frame overlapping, out-of-frame overlapping, and freestanding. If the AUG initiating the uORF is in good sequence context (Kozak, 1997), a freestanding uORF will severely inhibit initiation of the major downstream ORF. The most extensively studied uORF of this type is in S-adenosylmethionine decarboxylase (AdoMetDC) mRNA. This mRNA contains a 21-nucleotide (nt) uORF 14 nt downstream of the cap site. Translation of AdoMetDC mRNA is enhanced in polyamine-depleted Swiss 3T3 fibroblasts; the mRNA is normally on high polysomes, but adding putrescine to the medium causes a shift to low polysomes (White, 1990). Pertinent to the derangement of protein synthesis in neoplastically transformed cells (see below), regulation of translation by uORFs can change as the cellular milieu changes. AdoMetDC mRNA is on small polysomes in cells of T lymphocyte origin but large polysomal in non-lymphoid cells, such as fibroblasts and the adrenal carcinoma line Y1 (Hill and Morris, 1992). The regulation depends on the amino acid sequence encoded by the uORF, not the nucleotide sequence of the ORF (Hill and Morris, 1993). The uORF of AdoMetDC mRNA encodes the peptide MAGDIS, and regulation is abolished by changing any of the last three amino acid residues. This suggests that the hexapeptide interacts with the translating ribosome to suppress translation of AdoMetDC mRNA specifically.

5′-TOP Sequences

Another subset of mRNAs containing a *cis*-acting modulator of translational efficiency is the 5′-terminal oligopyrimidine tract (5′-TOP) mRNAs (Meyuhas and Hornstein, 2000). These contain an uninterrupted stretch of 5 to 14 pyrimidine residues immediately following the cap. The 5′-TOP mRNAs encode many translational components including eEF1α, eEF2, PABP, and all vertebrate ribosomal proteins characterized to date.

5′-TOP mRNAs are recruited to polysomes in a growth-dependent fashion (Kaspar et al., 1992; Jefferies and Thomas, 1994). For instance, the mRNA for ribosomal protein L32 moves from mRNPs to polysomes following serum activation of quiescent Swiss 3T3 cells, but actin mRNA shows no translational control (Kaspar et al., 1990). Labeling of eIF4E with ^{32}P-orthophosphate is increased by the same stimulus and with the same time course. Selective translation of 5′-TOP mRNAs is also observed during development in *Xenopus* (Mariottini and Amaldi, 1990) and *Drosophila* (Patel and Jacobs-Lorena, 1992). 5′-TOP mRNAs fail to be translated upon withdrawal of growth factors, in nutrient deprivation, upon contact inhibition of cultured cells, and at the initiation of a differentiation program. There is no obvious consensus among the 5′-TOP sequences, but mutation of the sequence leads to constitutive translation (Levy et al., 1991; Kaspar et al., 1992).

Translation of 5′-TOP mRNAs is selectively inhibited by rapamycin (Terada et al., 1994; Jefferies et al., 1994; Redpath et al., 1996). Initial evidence suggested that 5′-TOP mRNAs require activation of S6K1 for translation (Jefferies et al., 1997; Kawasome et al., 1998), but more recent studies argue against this. In murine erythroleukemia cells or other hematopoietic cells, S6 phosphorylation is not detectable, apparently due to the action of a phosphatase acting downstream of S6K1 (Barth-Baus et al., 2002). Despite the absence of changes in S6 phosphorylation, translation of 5′-TOP mRNAs is repressed during differentiation. In human embryonic kidney 293 cells, translation of 5′-TOP mRNAs is rapidly repressed by amino acid withdrawal, and this depends on the 5′-TOP motif. However, neither phosphorylation of S6 nor activation of S6K1 is sufficient to relieve this repression (Tang et al., 2001). Also inhibition of S6K1 activity and S6 phosphorylation by overexpression of a dominant-negative kinase fails to suppress the translational activation of 5′-TOP mRNAs in cells re-fed amino acids. Furthermore 5′-TOP mRNAs are translationally regulated by amino acids in embryonic stem cells lacking both alleles of the S6K1 gene. Rapamycin leads to complete repression of S6K1 activity but only partial and delayed repression of 5′-TOP mRNA translation. By contrast, interference with the PI3K pathway blocks translational activation of 5′-TOP mRNAs. These studies indicate that the translational regulation of 5′-TOP mRNAs requires PI3K but not S6K1.

IRESes

Initiation of nearly all eukaryotic mRNAs proceeds by a cap-dependent mechanism whereby the AUG nearest the 5′-end serves as the initiation codon (Kozak, 1992). Yet other modes of initiation codon selection are used in special cases (Jackson, 2000). IRESes direct ribosomes to internal AUGs (Jang et al., 1988; Pelletier et al., 1988). Internal initiation has been demonstrated for picornaviruses and certain other viruses (Jackson, 2000), but some cellular mRNAs are also translated by internal initiation (Table 12.1).

Most insight into the mechanism of internal initiation has been obtained with picornaviral IRESes (Jackson, 2000). By contrast, relatively little is known about cellular (non-viral) IRESes. The 5′-UTRs of BiP (Macejak and Sarnow, 1991), *Ultrabithorax* (Ye et al., 1997), and *Antennapedia* (Oh et al., 1992) mRNAs are, respectively, 220 nt, 968 nt, and 252 nt. In the case of the 318-nt 5′-UTR of FGF2 mRNA, the IRES resides in a 165-nt segment (Vagner et al., 1995). In the 1022-nt 5′-UTR of PDGF2 mRNA, the IRES is present in a 395-nt segment (Bernstein et al., 1997). The role of *trans*-acting factors in the utilization of cellular IRESes is not well established, although some results are beginning to emerge. For example, IRES-driven translation for the anti-apoptotic protein XIAP is modulated by hnRNPs C1 and C2 (Holcik et al., 2003).

Pertinent to the topic of translational control during mitosis (see below), IRES-dependent translation frequently occurs under conditions when cap-dependent translation is depressed. It is well documented that cap-dependent translation is inhibited during picornavirus infection while IRES-driven translation continues (Jackson, 2000). This can be demonstrated in vitro by cleavage of eIF4G by the 2A proteases of picor-

TABLE 12.1. mRNAs Containing Cellular (Non-viral) IRESes and Their Preferential Translation

mRNA	Function	Species and/or Cell	Preferred Conditions for Translation	References
GRP78/BiP	Endoplasmic reticulum-resident chaperone	HeLa (human)	Picornavirus infection, glucose starvation	Sarnow, (1989), Macejak and Sarnow (1991), Johannes and Sarnow (1998)
Antennapedia	Homeobox protein	*Drosophila*	Development	Oh et al. (1992)
FGF-2	Growth factor	COS-7 (monkey)		Vagner et al. (1995)
PDGF-2/c-*sis*	Growth factor	K562 cells (human)	Megakaryocytic differentiation	Bernstein et al. (1997)
Ultrabithorax	Homeobox protein	*Drosophila*	Development	Ye et al. (1997)
VEGF	Growth factor	Rat glial, 293, NIH 3T3 (mouse)	Hypoxia	Stein et al. (1998)
c-Myc	Transcription factor	HeLa, COS-7, HepG2	Met starvation, poliovirus infection	Nanbru et al. (1997), Stoneley et al. (1998), Johannes and Sarnow (1998)
eIF4G	Translation initiation factor	HeLa	Poliovirus infection	Johannes and Sarnow (1998)
XIAP	Anti-apoptotic factor	Human lung, ovary, and kidney tumor cell lines	Cellular stress	Holcik et al. (2003)
Kv1.4	Cardiac voltage-gated potassium channel	NIH 3T3		Negulescu et al. (1998)
FMR1	Fragile X syndrome	SK-N-MC neuroepithelial cells	Translation in dendrites?	Chiang (2001)
Rbm3	Cold stress-induced protein	Rat glioma, human neuroblastoma, NIH 3T3	Hypothermia	Chappell et al. (2001)
ODCase	Polyamine biosynthesis	HeLa	Late G2 phase of the cell cycle	Pyronnet et al. (2000)
cat-1	Cationic amino acid transporter	Rat C6 glioma	Amino acid starvation	Fernandez et al. (2001)

naviruses (see Fig. 12.2). This separates the N-terminal one-third of eIF4G, containing binding sites for eIF4E and PABP, from the C-terminal two-thirds, containing binding sites for eIF3 and eIF4A (Lamphear et al., 1995). Translation of capped mRNAs is drastically inhibited, whereas initiation of IRES-containing and uncapped mRNAs is either unaffected or even stimulated (Liebig et al., 1993; Ohlmann et al., 1995; Borman et al., 1997). Inhibition of cap-dependent translation with rapamycin via the mTOR pathway stimulates the translation of mRNAs containing an IRES (Jackson, 2000). Inhibition of cap-dependent translation by deprivation of nutrients, again via the TOR pathway, also promotes cap-independent translation. Starved *Saccharomyces* have the capacity to support internal initiation of a reporter mRNA (Paz et al., 1999). In C6 rat glioma cells, amino acid starvation increases translation of the IRES-containing cat-1 mRNA (Fernandez et al., 2001). IRES-driven translation also occurs during hypothermia (Chappell et al., 2001). Finally the IRES-mediated translation of XIAP mRNA is increased in response to cellular stress (Holcik et al., 2003).

Cytoplasmic Poly(A) Addition

Cytoplasmic addition of poly(A) plays a key role in $G_2 \rightarrow M$ progression in maturing oocytes (see below). A long 3′-poly(A) tract stimulates initiation of translation synergistically with the 5′-cap (Galili et al., 1988; Munroe and Jacobson, 1990; Gallie, 1991; Gallie and Tanguay, 1994; Preiss and Hentze, 1998). The role of the poly(A) tract in translation is mediated by PABP, as demonstrated by variety of experimental approaches (Hensold et al., 1996; Campbell et al., 1994; Sieliwanowicz, 1987; Munroe and Jacobson, 1990; Grossi de Sa et al., 1988; Tarun and Sachs, 1995). Synergistic enhancement of translation by the cap and the poly(A) tract, mediated by eIF4E and PABP, respectively, requires binding of both factors to the N-terminus of eIF4G in *Saccharomyces*, wheat germ, *Xenopus*, and mammals (Tarun and Sachs, 1996; Le et al., 1997; Imataka et al., 1998; Kessler and Sachs, 1998; see Fig. 12.2).

Global versus mRNA-Specific Protein Synthesis Regulation

As noted in the Introduction, protein synthesis affects cell cycle progression in two distinct ways: the rate of protein synthesis sets the timetable for reaching the critical cell size for the $G_1 \rightarrow S$ transition, and synthesis of specific proteins needed for cell cycle progression is controlled at the translational level (discussed in more detail below). A single growth factor can promote both of these activities but through separate pathways. This can be illustrated by the best characterized growth factor and mitogen, insulin. The relationship between insulin-stimulated protein synthesis and insulin-stimulated cell proliferation is complex. Rapamycin inhibits DNA synthesis and cell cycle progression through G_1 (Terada et al., 1993; Lane et al., 1993; Chung et al., 1992). Activation of the rapamycin-sensitive pathway is required for growth-regulated but not general protein synthesis (Fig. 12.3). Rapamycin specifically prevents increased translation of highly cap-dependent, growth-regulated mRNAs such as c-*myc* by both insulin (Mendez et al., 1996) and other growth factors (Beretta et al., 1996; Jefferies et al., 1994; Nielsen et al., 1995; Terada et al., 1994). However, rapamycin has little or no effect on synthesis of housekeeping proteins such as actin (Mendez et al., 1996). Conversely, expression of ΔPKCζ permits insulin-stimulated *general* protein synthesis but not activation of the rapamycin-sensitive branch of the pathway, as evidenced by the absence of insulin-stimulated S6K1 activation (Mendez et al., 1997; Fig. 12.3). ΔPKCζ permits insulin-stimulated actin but not c-Myc synthesis. Consistent with this model, ΔPKCζ replaces IRS-1 for protein synthesis but not for DNA synthesis.

Insulin stimulates both the rapamycin-sensitive (Mendez et al., 1996) and rapamycin-insensitive (Mendez et al., 1997) branches through PI3K. On the other hand, the IRS-1$^{Y608-658}$ variant, which lacks the Tyr residues necessary to activate PI3K, stimulates protein synthesis well (Mendez et al., 1996) but DNA synthesis only poorly (Myers et al., 1996). Furthermore IR variants that bind and activate PI3K directly mediate protein synthesis but not proliferation in response to insulin (Yenush et al., 1996). A certain level of PI3K may be required for proliferation, but PI3K-independent pathways mediated by IRS-1 must be limiting for this effect.

TRANSLATIONAL SIGNALS AFFECTING PASSAGE FROM G_1 TO S

The role of translation in G_1→S progression is best understood in *Saccharomyces*, although critical links remain to be established. Many additional factors seem to be involved in mammalian cells, but striking similarities to *Saccharomyces* hint of a conserved central mechanism. The following section will review what is known about the role translation in G_1→S progression in both yeast and mammalian cells.

Budding Yeast

G_1 Cyclins. In budding yeast, G_1 cells increase in cell mass until they reach a critical size, at which point (called START) they enter S phase, bud, and duplicate their spindle pole bodies (North, 1991). START is the point at which the mating pheromones cause cell cycle arrest, allowing cells of complementary mating type to undergo conjugation. At START, cells must integrate both external environmental signals (nutrients and pheromones) and internal signals (size) to determine whether to commit to mitosis, differentiation, or stationary phase. Cyclins made and functioning in G_1 were first discovered by their essential role in START. The G_1 cyclins, Cln1, Cln2, and Cln3, are redundant for function and act to promote the G_1→S transition (Richardson et al., 1989). Cln2 peaks during G_1, decreases thereafter, and is rapidly lost following exposure of cells to mating pheromone (Wittenberg et al., 1990). Cln2 abundance can be explained by the G_1-specific accumulation of the *CLN2* transcript. The Cln2 polypeptide interacts with p34^{CDC28} to form an active protein kinase complex required for the G_1→S transition.

Cln3 is an upstream activator of Cln1 and Cln2 (Tyers et al., 1993). Cln3 is a much rarer protein than Cln1 or Cln2 and has a much weaker histone H1 kinase activity. *CLN1* and *CLN2* mRNAs fluctuate periodically in the cell cycle, peaking in the G_1 phase, whereas *CLN3* mRNA is constant through the cell cycle. Artificial expression of *CLN3* accelerates START and induces at least five other cyclin genes. Despite its weak biochemical activity, Cln3 is essential for expression of cyclin genes in a *cln1 cln2* strain. Cln3, but not Cln1 or Cln2, is necessary for activation of the transcription factors SBF and MBF, which are required for transcriptional activation of Cln1 and Cln2 synthesis (Dirick et al., 1995). Cln1 and Cln2 then trigger the proteolysis of the cyclin B-p34^{CDC28} inhibitor p40^{SIC1}, which initiates S phase. Thus the accumulation of Cln1 and Cln2 kinases, which starts the yeast cell cycle, is set in motion by prior activation of SBF- and MBF-mediated transcription by Cln3-p34^{CDC28} kinase.

When nutrients are limiting, haploid yeast cells do not proceed to START in late G_1 but instead exit the mitotic cell cycle in early G_1 and enter a stationary or G_0 phase (Barbet et al., 1996). Cln3 is degraded rapidly by ubiquitin-dependent mechanisms in cells deprived of a nitrogen source (Gallego et al., 1997). Translation of *CLN3* mRNA is repressed eightfold under nitrogen-deprivation conditions. Because of their transcriptional dependence on SBF, the G_1 cyclins Cln1 and Cln2 become undetectable in starved cells. This provides the molecular basis for the G_1 arrest caused by nitrogen deprivation. Because of the post-translational regulation of its levels, Cln3 is a good candidate for coordinating the control of early cell cycle events.

eIF4E. Two classes of START mutants have been identified (reviewed in Brenner et al., 1988). Upon shifting to the nonpermissive temperature, class I mutants (*cdc28*, *cdc36*, *cdc37*, and *cdc39*) resemble mating-pheromone-arrested cells, while class II mutants (*cdc25*, *cdc33*, and *cdc35*) resemble nutritionally arrested cells. The G_1 period of exponentially growing *cdc33-1* cells is approximately twice as long as that of *CDC33 cells*, indicating that the *cdc33-1* mutation affects the G_1 phase. The eIF4E gene of yeast was cloned by Altmann et al. (Altmann et al., 1987). Subsequently *cdc33-1* was shown to encode a temperature-sensitive form of eIF4E (Brenner et al., 1988). The *cdc33-1* strain arrests with unbudded cells at the nonpermissive temperature. The cdc33-1 form of eIF4E has a single Gly→Asp substitution that decreases the affinity of eIF4E for m^{7}GDP and the rate of protein synthesis (Altmann and Trachsel, 1989). *CDC33* may also play a regulatory function during DNA synthesis by controlling the synthesis of RNR2, the small subunit of ribonucleotide reductase (Abid et al., 1999).

TOR. The TOR pathway regulates a variety of processes contributing to cell growth, including initiation of mRNA translation, ribosome synthesis, expression of metabolism-related genes, autophagy, and cytoskeletal reorganization (reviewed in Gingras et al., 2001; Schmelzle and Hall, 2000). *TOR2* encodes an essential 292-kDa PI3K homologue (Kunz et al., 1993). Disruption of the *TOR1* gene is not lethal but leads to a slight decrease in growth rate. Disruption of *TOR2* is lethal and is accompanied by random arrest throughout the cell cycle. Rapamycin treatment or inactivation of both *TOR* genes results in a severe decrease in translation initiation and arrest in the early G_1 phase of the cell cycle (Kunz et al., 1993; Barbet et al., 1996). *TOR2* mutations confer resistance to rapamycin-induced G_1 arrest. Loss of TOR activity causes a starvation response, eliciting activation or repression of the same subset of genes that are modulated following a switch from a rich to a poor carbon or nitrogen source (Barbet et al., 1996).

Expression of Cln3 from an *UBI4* 5′-UTR overcomes the G_1 arrest caused by TOR inhibition by rapamycin (Barbet et al., 1996). *UBI4* mRNA is normally translated during starvation conditions. Expression of Cln3 in this fashion confers starvation sensitivity. These results suggest that the block in translation initiation is a direct result of loss of TOR activity and is the cause of G_1 cell cycle arrest. Supporting this interpretation is the fact that *CLN3* mRNA translation is down-regulated during the G_1 arrest caused by nitrogen deprivation (Gallego et al., 1997). Interestingly growth-dependent regulation of *CLN3* mRNA translation is controlled by an uORF in the 5′-UTR that specifically represses *CLN3* expression during conditions of diminished protein synthesis or slow growth (Polymenis and Schmidt, 1997). Inactivation of the uORF accelerates the completion of START and entry into the cell cycle. This suggests that translational regulation of *CLN3* provides a mechanism for coupling cell growth and division.

Cln3 Bypasses an eIF4E Defect. The nature of this translational regulation was brought into more focus by the observation that *CLN3* expression is sufficient to restore $G_1 \rightarrow S$ progression in *cdc33-1* mutants (Danaie et al., 1999). Constructs carrying the 5′-UTR of *CLN3* mRNA fused to *lazZ* exhibit weak reporter activity, which is significantly decreased in a *cdc33-1* mutant. This implies that *CLN3* mRNA is an inefficiently translated mRNA that is sensitive to perturbations in the translation machinery. A *cdc33-1* strain expressing stable Cln3 or a hybrid *UBI4 5′-CLN3* mRNA display less dependence on eIF4E. Induction of the hybrid *UBI4 5′-CLN3* mRNA in a *cdc33-1* mutant previously arrested in G_1 causes entry into a new cell cycle. This suggests that eIF4E activity in the G_1 phase is critical to produce sufficient Cln3 for G_1 progression.

How Does TOR Affect Translation Initiation? The preceding results indicate that Cln3 is downstream of eIF4E, which is in turn downstream of TOR. Whereas in metazoans there are substrates for the mTOR kinase that relate to protein synthesis initiation (as discussed above), the connection in the case of yeast TOR1 and TOR2 has been elusive. One possibility is that TOR regulates protein synthesis through protein phosphatases. The protein Tap42 associates with and functions positively with both Sit4, a type 2A-related protein phosphatase, and the type 2A phosphatase catalytic subunit PP2A (Di Como and Arndt, 1996). Tap42 is directly phosphorylated by TOR2, and this phosphorylation promotes formation of the Tap42-PP2A complex. Mutations in *TAP42* can confer almost complete resistance to rapamycin. Tap42-Sit4 and Tap42-PP2A complex formation is regulated by nutrient growth signals and the rapamycin-sensitive TOR pathway. The *tap42-11* mutant is defective in translation at the non-permissive temperature. Another study showed that inactivation of either Cdc55 or Tpd3, which regulate type 2A phosphatase activity, results in rapamycin resistance, and this resistance correlates with an increased association of Tap42 with PP2A (Jiang and Broach, 1999). TOR-phosphorylated Tap42 may compete with Cdc55 or Tpd3 for binding to the phosphatase 2A catalytic subunit. Recently a protein, Tip41, was identified that negatively regulates the TOR pathway by binding and inhibiting Tap42 (Jacinto et al., 2001). The binding of Tip41 to Tap42 is stimulated upon rapamycin treatment via Sit4-dependent dephosphorylation of Tip41, suggesting that Tip41 is part of a feedback loop that rapidly amplifies Sit4 phosphatase activity under TOR-inactivating conditions. These results suggest that Tap42, Sit4, and PP2A are components of the TOR signaling pathway. Despite these advances the mechanism by which Tap42 signals to activate protein synthesis is unknown.

Another possible route linking TOR to translational initiation is via Eap1 (Cosentino et al., 2000). Eap1 is a functional homologue of mammalian PHAS. It binds eIF4E, prevents binding of eIF4E to eIF4G, and inhibits cap-dependent translation in vitro. Interestingly targeted disruption of the *EAP1* gene confers partial resistance to growth inhibition by rapamycin, suggesting that Eap1 may pay a role in the TOR signaling pathway for cell growth.

Mammals

The elegant genetic tools that have elucidated such clear links between translational control and the cell cycle in *Saccharomyces* are not available in mammalian cells. Nonetheless, there are intriguing parallels between yeast and mammalian cells involving signaling pathways, protein synthesis factors, and components of the cell cycle machinery.

In Swiss 3T3 cells inhibited to varying degrees with cycloheximide, cell growth is proportional to protein synthesis rate (Rossow, 1979). The elongation of the cell cycle occurs entirely in the portion of G_1 preceding the "restriction point," which is analogous to START in budding yeast. Furthermore, as documented below, there are similar relationships between mTOR, eIF4E, and G_1 cyclin mRNA translation, although the biochemical mechanisms governing them may be different. The information in mammalian cells is much more fragmentary than in *Saccharomyces*. However, it is useful to have the firm, clear connections between successive events established in *Saccharomyces* to serve as a framework for the partial information in mammalian cells.

Mammalian Cell Growth and Division Rates are Altered by eIF4E Levels. Overexpression of eIF4E three- to ninefold over the endogenous level, by way of an episomally replicating vector, leads to accelerated cell growth of HeLa cells and densely packed, multilayered foci (De Benedetti and Rhoads, 1990). Microinjection of eIF4E, but not eEF1α, eEF1H, eIF2, or eIF4A, into NIH 3T3 cells induces the cells to enter the S phase (Smith et al., 1990). HeLa cells transformed to express antisense RNA against eIF4E mRNA from an inducible promoter are morphologically similar to control cells but grow four- to sevenfold more slowly (De Benedetti et al., 1991). Induction of antisense RNA, however, is lethal. Both eIF4E mRNA and protein levels are reduced in proportion to the degree of antisense RNA expression, as are the rates of protein synthesis and polysome size.

Translationally Controlled Proteins Involved in $G_1 \rightarrow S$ Progression. Since cells must reach a minimum size before they are poised at the restriction point, any protein that contributes to cell growth is involved in to $G_1 \rightarrow S$ progression, albeit indirectly. This group therefore includes proteins encoded by 5′-TOP mRNAs, for instance, many translational components (as discussed above). Translation of 5′-TOP mRNAs is preferentially stimulated upon growth factor or mitogen exposure, provided there is nutrient sufficiency, via the mTOR pathway. Other proteins, discussed below, are more directly involved with either the $G_1 \rightarrow S$ transition or S phase itself. Interestingly forced expression of many of these proteins shortens the overall cell cycle, particularly G_1. Another unifying feature is that many of these proteins are preferentially synthesized in eIF4E-overexpressing cells.

Cyclin D1. Cyclin D1 is necessary for the $G_1 \rightarrow S$ transition (Sherr, 1993). Several studies have shown that eIF4E strongly affects intracellular levels of the cyclin D1 protein. Overexpression of cyclin D1 mRNA in NIH 3T3 cells does not increase cyclin D1 protein, but raising intracellular levels of eIF4E causes an increase in cyclin D1 without an accompanying increase in cyclin D1 mRNA (Rosenwald et al., 1993). Interestingly there is no significant effect of eIF4E on the polysomal distribution of cyclin D1 mRNA (Rosenwald et al., 1995). However, the total amount of cyclin D1 mRNA associated with polysomes is increased without an increase in total cyclin D1 mRNA (Rousseau et al., 1996). Whereas in the parental NIH 3T3 cell line, a large proportion of cyclin D1 mRNA is confined to the nucleus, in the eIF4E overexpressing cells, the vast majority of the mRNA is present in the cytoplasm.

eIF4E promotes the nucleo-cytoplasmic transport of cyclin D1 mRNA. The promyelocytic leukemia protein PML is organized into nuclear bodies that mediate suppression of oncogenic transformation and growth (Cohen et al., 2001). eIF4E directly binds PML, which modulates eIF4E activity by reducing its affinity for the cap. eIF4E requires cap binding for transport of cyclin D1 mRNA and subsequent transformation activity. PML causes a reduction in cyclin D1 protein levels and consequent inhibition of transformation. PML co-localizes with eIF4E in nuclear bodies (Topisirovic et al., 2002). Treatment of NIH 3T3 cells with γ-interferon or cadmium leads to dispersal of PML from nuclear bodies but leaves the eIF4E. This results in increased cyclin D1 mRNA transport and increased cyclin D1 synthesis.

The signaling pathways leading to increased cyclin D1 levels are complex. In human colon carcinoma and breast cancer cell lines, the induction of cyclin D1 and D3 expression by serum, or repression by herbimycin A, occurs prior to a change in cyclin D mRNA level (Muise-Helmericks et al., 1998). These effects are mediated by PI3K and mTOR, based on inhibitor studies. Another study showed that IGF-1 and estradiol act synergistically to stimulate expression of cyclin D1, hyperphosphorylation of pRb, DNA synthesis, and cell division in MCF7 cells (Hamelers et al., 2002). IGF-I induces nuclear accumulation of cyclin D1 by a PI3K-dependent mechanism. Inhibition of GSK-3β also causes nuclear accumulation of cyclin D1 and cell cycle progression in estradiol-treated cells. Conversely, overexpression of constitutively active GSK-3β blocks IGF-I-induced accumulation of cyclin D1 as well as S phase transition.

c-Myc. The transcription factor c-Myc is also strongly implicated in cellular growth control, particularly $G_1 \rightarrow S$ progression (reviewed in Schmidt, 1999). c-Myc functions to drive cell-cycle progression, induce apoptosis, block differentiation, and, when deregulated, promote cellular transformation. The 5′-UTR of mouse *c-myc* mRNA has a negative effect on translational efficiency in vitro, the restrictive element being confined to a 240-nt sequence (Parkin et al., 1988). As mentioned previously, it was subsequently shown that *c-myc* mRNA contains an IRES (Table 12.1). Ectopic overexpression of eIF4E in CHO and CREF cells results in a prominent increase in the synthesis of a few proteins, including c-Myc, but no change in *c-myc* mRNA (De Benedetti et al., 1994). Wortmannin and rapamycin abolish PHAS phosphorylation and the stimulation of *c-myc* mRNA translation that occurs upon addition of serum to immortalized B-cells (West et al., 1998), suggesting a mechanism involving an increase in available eIF4E via serum → PDGFR → PI3K → PDK1 → PKB → mTOR → PHAS-I (Fig. 12.3).

Knowledge of the downstream targets of transcriptional activation by c-Myc is critical to an understanding of how it promotes $G_1 \rightarrow S$ progression. Evidence has been presented that the ornithine decarboxylase (ODCase) gene (Bello-Fernandez et al., 1993) and eIF4E gene (Jones et al., 1996) are regulated by c-Myc. However, a more recent study, utilizing a cDNA microarray approach to identify downstream targets of c-Myc, applied more stringent criteria to distinguish genes whose regulation was dependent on c-Myc per se from those regulated in response to activation of general mitogenic or apoptotic programs (Watson et al., 2002). Only 15 target genes were uncovered out of more than 6000, and these did not include eIF4E or ODCase.

ODCase. ODCase is the rate-limiting enzyme in synthesis of polyamines, which are needed for S phase. Its activity is near zero in quiescent, nonproliferating cells but is dramatically induced by a wide variety of growth factors and mitogens. Its relation to cell cycle progression is illustrated by the fact that ectopic expression of ODCase in NIH 3T3 cells induces cell transformation and anchorage-independent growth, whereas expression of antisense RNA is associated with reduced cell proliferation (Auvinen et al., 1992).

ODCase mRNA has an unusually long and highly structured 5′-UTR. Its translational efficiency is increased in mitogen-activated lymphocytes (White et al., 1987). Plasmids containing sequences representing various portions of the 5′-UTR of rat ODCase mRNA express a reporter protein in mouse fibroblasts in an insulin-dependent manner, the expression directly correlated with the extent of 5′-UTR secondary structure and phosphorylation of eIF4E (Manzella et al., 1991). ODCase activity is much greater in NIH 3T3 cells ectopically overexpressing eIF4E (Shantz and Pegg, 1994). The increase in the amount of ODCase and the translation initiation rate, as judged by the size of ODCase mRNA-containing polysomes, is directly proportional to the degree of eIF4E overexpression (Rousseau et al., 1996). In malignantly transformed cells, the mRNA undergoes alternative splicing to relieve the translational inhibition (Pyronnet et al., 1996). Conversely, in CREF-T24 cells expressing the activated Ras oncogene, expression of antisense RNA against eIF4E reduces the malignant phenotype (Rinker-Schaeffer et al., 1993) and translation of ODCase mRNA (Graff et al., 1997).

Signaling to ODCase translation involves mTOR, whether via a G-protein-coupled receptor or amino acid signals. Gastrin increases ODCase mRNA translation in AR4–2J tumor cells via a mechanism involving the 5′-UTR and inhibited by rapamycin (Pyronnet et al., 1998). Leu stimulates rapamycin-sensitive synthesis of ODCase and eEF1A, a 5′-TOP mRNA-encoded protein, in L6 myoblasts (Kimball et al., 1999).

FGF-2. Fibroblast growth factor 2 (FGF-2) belongs to a family of multifunctional molecules that act as potent mitogens for a variety of mesoectodermal lineages. FGF-2 is also a powerful angiogenic factor, involved in both normal and pathologic processes such as wound repair and tumor vascularization (De Benedetti and Harris, 1999). FGF-2 mRNA contains a long and inhibitory 5′-UTR in addition to at least two CUG codons upstream of, and in frame with, the AUG start codon. These CUGs are used for translation initiation to generate several different polypeptides. Cells ectopically overexpressing eIF4E produce and secrete large amounts of FGF-2, particularly the largest, CUG1-initiated form (Kevil et al., 1995). FGF-2 mRNA is found on large polysomes in eIF4E overexpressing cells but small polysomes in control cells. Breast carcinomas expressing elevated eIF4E also exhibit the large FGF-2 isoforms, which could play a role in tumor angiogenesis. Addition of purified eIF4E also increases initiation from the CUG1 and AUG codons in a rabbit reticulocyte lysate translation system.

VEGF. Vascular endothelial growth factor (VEGF) is a specific cytokine for vascular endothelia (De Benedetti and Harris, 1999) and contains a highly structured 5′-UTR as well as an IRES (Table I). CHO cells ectopically overexpressing eIF4E have 130-fold higher levels of secreted VEGF with no difference in the level of VEGF mRNA. VEGF mRNA is associated with the heavy polysomal region in eIF4E-overexpressing cells but the light region in control cells.

Ribonucleotide Reductase. Ribonucleotide reductase synthesizes deoxyribonucleotide diphosphates, a specific and rate-limiting step in S phase. In Chinese hamster ovary cells overexpressing eIF4E, the small subunit of ribonucleotide reductase is elevated 20- to 40-fold (Abid et al., 1999). These cells display an altered cell cycle with shortened S phase, similar to cells selected for RNR2 overexpression with hydroxyurea.

Thymidylate Synthase. Thymidylate synthase is also essential for S phase. An inhibitor of thymidylate synthase, ICI D1694, causes a 10- to 40-fold increases in thymidylate synthase in human mammary epithelial cells with no change in thymidylate synthase mRNA (Keyomarsi et al., 1993). This is blocked by cycloheximide, suggesting that a translational block is relieved when the enzyme is inhibited. The effect is overcome by addition of 5,10-methylenetetrahydrofolate, indicating that the drug has its effect by binding to the folate substrate site of thymidylate synthase. The mRNA for thymidylate synthase contains a 36-nt binding site for thymidylate synthase in the 5′-UTR, indicating that synthesis is regulated by a negative feedback mechanism (Chu et al., 1993).

Blocking the mTOR Pathway Inhibits Cell Growth and Proliferation. As was noted in the previous section, the translational expression of several proteins implicated in $G_1 \rightarrow S$ progression in mammalian cells can be blocked by inhibitors of PI3K or mTOR. These inhibitors also block $G_1 \rightarrow S$ progression itself, as reviewed in the present section. These correlations do not have the same cause-and-effect power as the observations in *Saccharomyces* cited above, but the results are highly suggestive. These two effects of PI3K and mTOR inhibitors can be accommodated in a model whereby increasing the activity and availability of eIF4E yields products (cyclins, transcription factors, etc.) that promote G_1 progression.

PI3K is stimulated by insulin and a variety of other growth factors. Treatment of 3T3-L1 adipocytes with LY294002, a specific PI3K inhibitor, prevents insulin-stimulated PI3K activation, S6K1 activation, and DNA synthesis (Cheatham et al., 1994). Similarly rapamycin inhibits progression through the G_1 phase of the cell cycle. In activated T cells, rapamycin blocks cell-cycle progression at a stage prior to events characteristic of the middle to late G_1 phase of the cycle (Terada et al., 1993) as well as ribosomal protein synthesis (Terada et al., 1995). Because of its inhibitory effects on lymphocyte proliferation, rapamycin is a potent immunosuppressant and effectively prevents allograft rejection (reviewed in Gingras et al., 2001). In mouse T cells, rapamycin inhibits the IL-2-induced expression of cyclin A and the appearance of active cyclin A-p34^{CDC2} complexes (Morice et al., 1993). The rapamycin inhibition of $G_1 \rightarrow S$ is mediated by effects on cyclin D1 mRNA (Hashemolhosseini et al., 1998). In serum-starved NIH 3T3 cells, rapamycin treatment delays the accumulation of cyclin D1 mRNA during progression through G_1 and also affects stability of the transcript. The combined transcriptional and post-transcriptional effects of the drug ultimately result in decreased levels of cyclin D1 protein.

mTOR activation of S6K1 can be uncoupled from $G_1 \rightarrow S$ progression. In Swiss 3T3 cells, the peptide bombesin is a potent mitogen and stimulates S6K1 activation (Withers et al., 1997). Both DNA synthesis and S6K1 activation are inhibited by rapamycin. However, the combination of insulin and bombesin stimulates maximum DNA synthesis despite persistent inhibition of S6K1 by rapamycin throughout G_1. Rapamycin also inhibits synthesis of cyclin D1 in bombesin-stimulated cells, but not in cells stimulated with bombesin plus insulin. This suggests that a rapamycin-insensitive pathway leads to cyclin D1 expression and $G_1 \rightarrow S$ progression.

Commonalities among Species

Exciting new insights are emerging from the study of knockout, dominant-negative, and constitutively active mutations in *Drosophila* of the signaling components involved in growth control (PI3K, S6K, PKB, IRS, TOR, PTEN, PHAS, and Myc) (reviewed in Schmelzle and Hall, 2000 and Gingras et al., 2001). However, since these results deal more with changes in cell, organ, and organism size than with the molecular mechanisms linking translational control and cell cycle progression, they are not be considered further. It is striking, however, that in all three organisms, *Saccharomyces*, *Drosophila*, and mammals, signals from growth factors and nutrients converge and are integrated by TOR. Similarly all regulate expression of G_1 cyclins at the post-transcriptional level, and eIF4E is involved in this regulation, although the molecular mechanisms may differ.

TRANSLATIONAL SIGNALS AFFECTING PASSAGE FROM G_2 TO M

Translational control also occurs at the $G_2 \rightarrow M$ transition. Unlike G_1, where protein synthesis is rapid, there is a general depression of overall protein synthesis in M phase. Despite this depression, the translation of certain mRNAs is enhanced during M phase. This section also discusses a specialized case of $G_2 \rightarrow M$ progression in maturing oocytes.

Protein Synthesis in M Phase

The observation that protein synthesis is reduced in M phase was made in one of the first studies investigating the relationship between protein synthesis and the cell cycle (Fan and Penman, 1970). The rate of protein synthesis in CHO cells arrested in metaphase with colcemid is approximately 30% that of interphase cells. There is a precipitous drop in polysome size, indicating a reduction in the rate of initiation of protein synthesis. However, the elongation phase occurs at a normal rate. Polysome loading can be increased in both interphase and metaphase cells by treatment with cycloheximide, indicating that the decrease in protein synthesis is not due to a decrease in mRNA levels. Another early study showed in synchronized HeLa cells that the fraction of ribosomes in polysomes and the radioactive labeling of polypeptides in polysomes varies widely throughout the cell cycle (Eremenko and Volpe, 1975). There are two peaks of incorporation into protein, the higher one occurring at the G_1/S boundary and the lower one in G_2. Valleys occur in M phase (deepest), the S/G_2 boundary (intermediate), and in G_1 (shallowest). During M and G_2, the content of large polysomes is high, but their labeling is relatively low, suggesting an elongation block. Binding of Met-tRNA$_i$ to the 40 S ribosome (see Fig. 12.1) is the same for HeLa cells in S and M phase both in vivo and in vitro, indicating the block in initiation occurs after 43 S initiation complex formation (Tarnowka and Baglioni, 1979).

The decline in the rate of protein synthesis during mitosis may be explained, at least in part, by eEF2 phosphorylation. As noted above, phosphorylation of eEF2 inhibits its activity. The phosphorylated form of eEF2 increases dramatically during mitosis in human amnion cells (Celis et al., 1990). Such a mechanism would be at variance with the observation that elongation is not inhibited (Fan and Penman, 1970), but the latter study was performed with isolated polysomes in vitro, perhaps after dephosphorylation of eEF2.

Cap-dependent, but not cap-independent, translation is reduced in mitosis. Addition of eIF4F stimulates the translation of endogenous mRNA in extracts from mitotic but not interphase cells (Bonneau and Sonenberg, 1987). Cap-independent poliovirus RNA translation is not affected, but cap-dependent vesicular stomatitis virus mRNA translation is. eIF4E from mitotic cells is metabolically labeled with ^{32}P to a lesser extent than from interphase cells. In another study, arrest of 293 cells in metaphase causes cellular protein synthesis to be inhibited more than 90% but has little effect on cap-independent, late adenovirus mRNA translation (Huang and Schneider, 1991). Furthermore phosphorylation of eIF4E is decreased in late adenovirus-infected cells.

Specific Increase in IRES-Driven Translation during M Phase

As noted above, IRES-dependent translation frequently occurs under conditions when cap-dependent translation is depressed, suggesting that this type of translation may occur during M phase (see also Table 12.1). Despite the fact that ODCase mRNA has a long and structured 5′-UTR, its translation is considered to be cap dependent (Manzella et al., 1991;

Pyronnet et al., 1996). Synthesis of ODCase peaks twice during the cell cycle, at the G_1/S and G_2/M transitions (Fredlund et al., 1995). The latter occurs at a stage when cap-dependent initiation is reduced. This apparent paradox is explained by the finding that ODC mRNA contains an IRES, which functions exclusively at G_2/M, as does *c-myc* mRNA (Pyronnet et al., 2000). The latter may be regulated by the translocation of hnRNP C, which stimulates translation from the *c-myc* IRES, from the nucleus to the cytosol at G_2/M (Kim et al., 2003).

Recruitment of mRNAs to Polysomes in G_2 Phase by Cytoplasmic Polyadenylation

The synthesis and destruction of cyclin B drives mitosis (Pines, 1993). During G_2, human cyclin B mRNA increases to four times that present in G_1, but cyclin B protein increases to 20 times, indicating translational control (Pines and Hunter, 1989). This cyclin activates $p34^{CDC2}$ and is then abruptly degraded at M. Most of our understanding of translational control at the G_2/M interface, however, comes from studies in *Xenopus* oocytes (reviewed in Mendez and Richter, 2001a). Cyclin B1 mRNA translation is regulated during the meiotic divisions of *Xenopus* oocytes, which are similar to the mitotic divisions of somatic cells. Oocytes are arrested at the end of meiotic prophase I, which resembles G_2 of the somatic cell cycle. Because of the wealth of molecular detail known about translational control in the maturing *Xenopus* oocyte, this system is primarily discussed here, even though it is unclear whether the same mechanisms are utilized by somatic cells.

During vertebrate oogenesis, gene expression is governed primarily by translational control (Dworkin and Dworkin-Rastl, 1990; Curtis et al., 1995). The developing oocyte accumulates maternal components, including mRNAs, proteins, and ribosomes, that will be required during the rapid development that follows fertilization (Davidson, 1986). The stored maternal mRNAs encode cell cycle regulatory proteins such as c-Mos, Cdk2, and cyclins A, B1, and B2, which play a role in early embryonic development (Gabrielli et al., 1993; Sheets et al., 1994; Stebbins-Boaz et al., 1996). These translationally dormant mRNAs reside in stable ribonucleoprotein complexes and contain short 3′ poly(A) tracts (Dworkin and Dworkin-Rastl, 1985; Kelso-Winemiller and Winkler, 1991; Richter, 1988). Reentry of quiescent oocytes into the meiotic cell cycle is triggered in vivo by progesterone, causing dissolution of the nucleus and sending the cells through two meiotic cycles to arrest once again in metaphase (Maller, 1985). Some of the stored maternal mRNAs are recruited to the protein synthetic machinery, while many of the housekeeping mRNAs cease to be translated (Richter, 1991; Wormington, 1993).

Modification of the 3′ poly(A) tracts of stored maternal mRNAs plays an important role in their recruitment to the ribosome (Bachvarova, 1992; Richter et al., 1990). Progesterone activates a cytoplasmic poly(A) polymerase that adds 100 to 300 A residues to these mRNAs (Fox et al., 1989). The poly(A) elongation is required for recruitment onto ribosomes; failure to polyadenylate inhibits both the translational activation of these mRNAs and progression of the meiotic cell cycle (Kuge and Inoue, 1992; McGrew et al., 1989). Later during maturation, housekeeping mRNAs such as actin and ribosomal protein messages lose their poly(A) tracts and cease to be translated (Varnum and Wormington, 1990). Thus, for most maternal mRNAs, there is strict correlation between the presence of a long poly(A) tract and efficient translation.

Regulated translation of mRNAs for B cyclins is required for progression to M phase. B cyclins are a part of *Xenopus* mitosis-promoting factor (MPF, also known as maturation-promoting factor) (Gautier et al., 1990). Interestingly so are the β and γ subunits of eEF1 (Belle et al., 1989). eEF1 is a major phosphorylation substrate of MPF (Janssen et al., 1991).

B cyclins are synthesized during G_2 as a result of cytoplasmic polyadenylation. Cytoplasmic polyadenylation requires two elements in the 3′-UTR, the hexanucleotide AAUAAA and a cytoplasmic polyadenylation element (CPE) (Mendez and Richter, 2001a). The CPE is bound by a protein, CPEB, which contains both zinc finger and RRM motifs. Polyadenylation is mediated by the phosphorylation of CPEB by the Aurora kinase Eg2, whose activity oscillates with the cell cycle (Andresson and Ruderman, 1998).

Another protein, maskin, appears to contribute to the translational inhibition in the absence of polyadenylation (Stebbins-Boaz et al., 1999). Maskin interacts simultaneously with both CBEB and eIF4E and hence is tethered to the 3′ UTR of the cyclin B mRNA. The interaction between maskin and eIF4E is mediated by an eIF4E-binding motif that is present in both eIF4G and PHAS-I, indicating that maskin and eIF4G compete for binding to eIF4E. The interaction between eIF4E and maskin appears to be regulated by the polyadenylation state of cyclin B mRNA during the cell cycle.

CPEB, once phosphorylated by Eg2, recruits the polyadenylation machinery (Mendez et al., 2000). This is followed by the binding of PABP to the newly elongated poly(A) tail, the interaction of PABP with eIF4G (Fig. 12.2), and replacement of the maskin:eIF4E inhibitory complex with the eIF4E:eIF4G complex that promotes cap-dependent translation. Phosphorylation of CPEB by Eg2 regulates translation of c-*mos* mRNA as well (Mendez et al., 2000). Cdc2-catalyzed CPEB phosphorylation and its turnover via PEST box-mediated destruction during oocyte maturation is necessary for meiotic progression (Mendez et al., 2002). Interestingly CPEB, maskin, and cyclin B1 mRNA are localized at the mitotic apparatus of animal pole blastomeres in embryos, suggesting local translational control of cell division (Groisman et al., 2000). CPEB interacts with microtubules and is involved in the localization of cyclin B1 mRNA to the mitotic apparatus stabilizing the spindle and centrosome.

DEREGULATION OF PROTEIN SYNTHESIS AND THE CELL CYCLE IN CANCER

Cancer has been called a disease of the cell cycle. A number of observations have been made concerning alterations in the protein synthesis machinery, or its signaling pathways, that result in malignant transformation of cells. Conversely, the malignant state of some transformed cells can be attenuated by interference with translational factors or signaling pathways governing their activity, which holds promise for novel approaches to cancer intervention. This section reviews protein synthesis factors and signaling pathways that meet two criteria: they affect both cell cycle progression and malignant transformation of cells.

One modulator of initiation factor activity that will not be discussed in detail is the double-stranded RNA-activated protein kinase PKR, which phosphorylates the α subunit of eIF2 (Kaufman, 2000). Ectopic expression of a functionally defective form of human PKR in NIH 3T3 cells causes malignant transformation, suggesting that PKR may function as a suppressor of cell proliferation and tumorigenesis (Koromilas et al., 1992). However, it is not known whether PKR action is mediated through a direct change in cell cycle progression.

This section is divided into three parts: work in experimental systems, which yields much mechanistic detail but may not be directly relevant to human cancer; observations in naturally occurring cancers which, while relevant, tend to be more descriptive than mechanistic; and implications for cancer diagnosis and treatment.

Experimental Manipulation of the Expression of Protein Synthesis Factors

eIF4E. Overexpression of eIF4E has a profound effect on cell growth and phenotype. In HeLa cells, inducible overexpression of eIF4E by 3- to 9-fold over the endogenous level causes formation of densely packed, multilayered foci reminiscent of transformed cells (De Benedetti and Rhoads, 1990). In NIH 3T3 and Rat 2 fibroblasts, overexpression of eIF4E causes malignant transformation as determined by three criteria: formation of foci on a monolayer of cells, anchorage-independent growth, and tumor formation in nude mice (Lazaris-Karatzas et al., 1990). Microinjection of eIF4E into NIH 3T3 cells produces a cell morphology characteristic of oncogenically transformed cells that is not observed with eEF1α, eEF1H, eIF2, or eIF4A (Smith et al., 1990). eIF4E overexpression from transfected vectors also prevents apoptosis in growth-factor-restricted fibroblasts (Polunovsky et al., 1996). Ras is activated in NIH 3T3 cells in which eIF4E is overexpressed ectopically (Lazaris-Karatzas et al., 1992). Overexpression of GAP, the negative regulator of Ras, causes reversion of the transformed phenotype as do neutralizing antibodies or a dominant-negative form of Ras. In NIH 3T3 cells that overexpress eIF4E from a tetracycline-inducible promoter, fivefold overexpression of eIF4E causes both PHAS-I and S6K1 to become dephosphorylated, but phosphorylation of PKB is not affected (Khaleghpour et al., 1999). This suggests that eIF4E participates in a negative feedback loop designed to maintain translational homeostasis, perhaps compensating for elevated eIF4E by complexing it with PHAS-I.

Conversely, reduction in eIF4E expression attenuates the transformed cell phenotype. Reducing eIF4E levels in HeLa cells by expression of antisense RNA complementary to eIF4E mRNA causes a decrease in eIF4E levels, protein synthesis, and cell growth rate (De Benedetti et al., 1991). In rat embryo fibroblasts transformed with T24 Ras (CREF T24 cells), expression of eIF4E antisense RNA (CREF T24/AS cells) reduces eIF4E levels to between 30% and 50% of normal and decreases both cell growth and protein synthesis (Rinker-Schaeffer et al., 1993). CREF T24 have a spindle-shaped, retractile appearance whereas CREF T24/AS grow in ordered, parallel patterns and exhibit contact inhibition. The efficiency of growth in soft agar is 11-fold lower for CREF T24/AS, and the latency period for tumor formation in nude mice increases from 8 to 23 days. Cell lines established from CREF T24/AS-derived tumors partially regain the eIF4E levels characteristic of CREF T24. CREF T24/AS cells also have delayed and reduced invasiveness and decreased experimental metastasis (Graff et al., 1995). They express the metastasis-associated 92-kDa collagenase type-IV and exon-6 variant of the CD44 adhesion molecule. Reduced eIF4E levels correlate inversely with increased levels of the putative metastasis-suppressor protein nm23. All of these parameters (growth rate, protein synthesis rate, metastasis, and expression of CD44, type IV collagenase, and nm23) are reversed in cell lines established from CREF T24/AS-derived tumors and lung metastases. Similarly, reducing eIF4E levels with antisense RNA in MDA-435, a cell line derived from breast cancer, suppresses tumorigenic and angiogenic properties in nude mice, concomitant with loss of FGF-2 synthesis (Nathan et al., 1997).

eIF4G. Similar to eIF4E, the transfection of NIH 3T3 cells with a vector expressing human eIF4G causes foci on a monolayer of cells, anchorage-independent growth, and formation of tumors in nude mice (Fukuchi-Shimogori et al., 1997). There is no difference in the amount of eIF4E in the eIF4G-overexpressing cells, indicating that it is not the cause of malignant transformation.

PHAS-I. Overexpression of PHAS-I and PHAS-II in NIH 3T3 cells transformed by eIF4E or v-*src* causes a partial reversion of the transformed phenotype (doubling time, growth in soft agar, tumor growth rate, and latency period; Rousseau et al., 1996). Enforced expression of PHAS-I sensitizes CREF cells transformed with V12 Ras to apoptosis in a manner that is dependent on its ability to sequester eIF4E from a translationally active complex with eIF4G-1 and the co-expression of oncogenic Ras (Polunovsky et al., 2000). Also, peptides containing the eIF4E-binding motif of PHAS-I linked to the penetratin peptide-carrier sequence, which mediates the rapid transport of peptides across cell membranes, can be introduced into MRC5 cells (Herbert et al., 2000). This leads to rapid, dose-dependent cell death with characteristics of apoptosis. Ala substitutions at key positions in the peptides impair their binding to eIF4E and markedly reduce their ability to induce apoptosis.

Signaling Molecules Upstream of Translation. Cells can be malignantly transformed by overexpression of a variety of signaling molecules, or their constitutively active variants, upstream of protein synthesis factors. These include PI3K, PKB, PTEN, $\alpha 6\beta 4$ integrin, and others (Gingras et al., 2001). However, as signaling pathways branch to many endpoints, the transforming ability of these intermediates cannot be unambiguously linked to protein synthesis factors and hence is not considered here.

Naturally Occurring Cancers

Complementary to the observation that cells in culture can be malignantly transformed by overexpression of eIF4E, naturally occurring cancers, and cell lines derived from them, are found to overexpress eIF4E.

Breast Cancer. In a survey of 38 breast carcinomas, eIF4E was elevated 3- to 10-fold in virtually all of the carcinomas but not in fibroadenomas (Kerekatte et al., 1995). In a series of breast cancer cell lines, eIF4E protein was elevated 10-fold, and eIF4E mRNA was elevated 5- to 10-fold, suggesting a transcriptional rather than translational mechanism (Anthony et al., 1996). This idea is strengthened by the observation that the eIF4E gene is amplified in breast cancer tissues (Sorrells et al., 1997) and eIF4E mRNA levels are elevated in a broad spectrum of transformed cell lines (Miyagi et al., 1995). eIF4E is markedly increased in vascularized malignant ductules of invasive breast carcinomas, whereas necrotic and avascular ductal carcinomas in situ display significantly lower levels (Nathan et al., 1997). Interestingly eIF4E expression is significantly increased in invasive ductal carcinomas and in islets of viable cells in the centers of poorly vascularized regions (DeFatta et al., 1999). Expression of eIF4E is increased by hypoxia and, presumably, in hypoxic areas of these lesions. Overexpression of eIF4E may be induced by hypoxia and in turn afford cancer cells a critical advantage to survive hypoxia by elaborating angiogenic factors such as FGF-2 and VEGF (as mentioned above). This would then mark the transition toward the vascular phase of cancer progression.

Bladder Cancer. Expression of eIF4E and FGF-2 isoforms is also elevated during vascularization of bladder cancers (Crew et al., 2000). As noted above, VEGF mRNA is particularly responsive to elevated eIF4E levels. The VEGF protein-to-mRNA ratio is elevated as much as 15-fold in invasive bladder cancers, suggesting translational regulation. Expression of eIF4E is greater in tumors than normal bladder ($p < 0.0001$) and correlates with VEGF protein-to-mRNA ratios. In superficial tumors, high expression of eIF4E is associated with a poor prognosis and reduced survival, suggesting that eIF4E may have a role as a prognostic indicator.

Colon Cancer. An elevation of eIF4E levels in colon adenomas and carcinomas is accompanied by elevation of cyclin D1 levels (Rosenwald et al., 1999). Synthesis of only a small fraction of the total proteins is preferentially upregulated, as judged by 2D electrophoresis. In another study, 87 cases with lesions in the colorectum with a variety of histopathological diagnoses were randomly selected (Berkel et al., 2001). A statistically significant relationship was found between the level of eIF4E expression and histological type of lesion, the highest being in colorectal adenocarcinomas. Carcinomatous lesions were found to have a 43 times higher chance of having a high level of eIF4E expression compared with normal tissue ($p < 0.0001$). In a multivariate analysis, histological type was the only variable that showed a significant relationship with eIF4E expression; no effect was found due to age, gender, race, history of polyps, or family history.

Head-and-Neck Cancer. In a study of 26 squamous cell cancers (HNSCC), eIF4E expression was high in contrast to its low expression in benign lesions (Nathan et al., 1997). This was also true of histologically cancer-negative margins in a majority of patients. Cancer recurred in 5 of the 12 patients with eIF4E-positive margins as opposed to none of the 11 patients with eIF4E-negative margins ($p < 0.02$). An immunohistochemical analysis performed on 115 specimens revealed a significant correlation in eIF4E expression from low-grade dysplasia through tumor (Nathan et al., 1999). For VEGF and FGF-2, a significant increase was only seen between the tumor group and dysplastic group; no significant increase was noted between low-grade and high-grade dysplasia. There was also a significant increase in mean vessel density from low- to high-grade dysplasia, supporting a role for eIF4E in angiogenesis.

Lung Cancer. In a study of 143 examples of atypical adenomatous hyperplasia and adenocarcinomas of the human peripheral lung, eIF4E levels were elevated in both tumoral and stromal tissues and were significantly associated ($p < 0.001$) with histological grade (Seki et al., 2002). eIF4E expression in adenocarcinomas was 3.4- to 7.4-fold higher than in normal lung.

Other Initiation Factors. Considerably less has been published on the elevation of initiation factors other than eIF4E in naturally occurring cancers. Nearly one-third of squamous cell carcinomas of the lung have an amplification of the genome at 3q26.1–q26.3 (Brass et al., 1997), the same region where the eIF4G-1 gene is located (Yan and Rhoads, 1995). Among the genes detected in a tumor cDNA expression library was that of eIF4G-1. eIF4A-1 mRNA is overexpressed in human melanoma cells (Eberle et al., 1997). Various eIF3 subunits are overexpression in a broad spectrum of cancers (Hershey and Miyamoto, 2000).

Implications for Cancer Therapy

Knowledge of the involvement of protein synthesis factors, particularly eIF4E, in development and progression of malignant cells offers several possibilities for cancer treatment, as discussed in detail below.

Cancer Screening. First, eIF4E levels can be used to predict severity and recurrence of cancers. Tumorigenesis is proposed to be a multi-step process preceded by clinically evident precancerous lesions. Earlier detection through molecular, rather than histological, changes can enable earlier treatment. Even after surgery, a better predictor of clinical outcome can direct more targeted therapies. Several studies have shown that eIF4E level provides a superior indicator of cancer severity and progression. Sevenfold or higher eIF4E elevation is a strong, independent prognostic indicator of breast cancer recurrence and death (Li et al., 1997, 1998). In a prospective study of 191 women with stage I to III breast cancer, the subset of patients with 14-fold or more eIF4E overexpression had a 7.2-fold higher risk of cancer recurrence ($p < 0.002$) (Li et al., 2002). In another prospective study, eIF4E levels were measured on all newly diagnosed HNSCC patients who underwent surgical resection for their disease (Nathan et al., 1999). All 65 patients had elevated levels of eIF4E in the tumors, and 20 subsequently had local-regional recurrences. Elevated eIF4E in the margins was an independent prognostic factor ($p = 0.009$) for recurrence. In histologically tumor-free surgical margins, elevated levels of eIF4E predict a significantly increased risk of recurrence and may identify patients who could benefit from additional therapy. In a similar retrospective analysis of 31 patients who underwent surgery for HNSCC, eIF4E in the surgical margins was an independent prognostic factor (Franklin et al., 1999). eIF4E was also seen in premalignant lesions with increasing grades of dysplasia from biopsies of patients with head-and-neck lesions, suggesting a possible role of eIF4E in the angiogenic conversion of premalignant lesions to invasive cancer.

Cancer Chemotherapy. It may be possible to interfere pharmacologically with signaling pathways for protein synthesis factors involved in aggressive cell growth. Rapamycin at low doses reverses the transformation of chicken embryo fibroblasts expressing p3k or Akt (Aoki et al., 2001). Because of these anti-proliferative effects, rapamycin is used clinically to block immune rejection of transplanted organs. CCI-779, an ester derivative of rapamycin formulated for intravenous use, has strong anti-tumor potential and favorable pharmaceutical and toxicological characteristics in preclinical studies (Hidalgo and Rowinsky, 2000; Yu et al., 2001; Dudkin et al., 2001). In nude mouse xenografts, CCI-779 inhibits growth of tumors derived from sensitive but not resistant lines, resulting in a decrease in D-type cyclin and c-Myc levels and an increase in $p27^{KIP-1}$ (Yu et al., 2001). SDZ RAD, another rapamycin analogue, has an antiproliferative effect on EBV-transformed B lymphocytes in culture and in a mouse model, blocking these cells in G_1 and inducing apoptosis (Majewski et al., 2000). These agents appear to be well tolerated at doses that have anti-tumor activity against several types of refractory neoplasms. It is likely that rapamycin is acting in these studies through an inhibition of eIF4E. A study of cell lines derived from childhood tumors found that some, such as Rh30, are highly sensitive to growth inhibition by rapamycin, whereas others, such as Rh1, have high intrinsic resistance (Hosoi et al., 1998). Rapamycin blocks serum induction of c-Myc in Rh30 but not in Rh1 cells. Rh30 cells selected for acquired resistance to rapamycin showed 10-fold less PHAS-I bound to eIF4E (Dilling et al., 2002). Similarly several colon carcinoma cell lines with intrinsic rapamycin resistance had low PHAS-to-eIF4E ratios, suggesting that PHAS-to-eIF4E ratio is a determinant of rapamycin resistance and controls certain aspects of the malignant phenotype.

Cancer Gene Therapy. Finally protein synthesis factors involved in cell cycle progression may be targets of gene therapy. The high levels of eIF4E in breast cancer cell lines were exploited to selectively kill cancer cells (DeFatta et al., 2002a). The authors constructed a vector (BK-UTK) encoding a highly structured 5′-UTR upstream of the coding region for herpes thymidine kinase (HTK). Because eIF4E enables translation of mRNAs with high secondary structure (as discussed above), various breast cancer cell lines efficiently synthesized HTK from the translationally regulated mRNA, whereas normal cells did not. Only cancer cells were killed upon administration of gancyclovir, a predrug that is converted to a toxic nucleotide analogue by HTK. By altering the expression of eIF4E, it was possible to modulate the sensitivity of various cell lines to gancyclovir. In a companion study, the BK-UTK vector and a control vector lacking the 5′-UTR (BK-TK) were used to treat experimental tumors in mice derived from a murine breast cancer line (DeFatta et al., 2002b). Both vectors reduced subcutaneous tumors and lung metastases following administration of gancyclovir, but whereas the BK-TK vector was highly toxic, resulting in severe weight loss, degeneration of various organs, and early death of mice following systemic vector delivery, the BK-UTK vector increased mean survival time without toxicity.

ACKNOWLEDGMENTS

This work was supported by Grant RO1 GM20818 from the National Institutes of Health. The author is grateful for the bibliographic, editorial, and graphic assistance of Lisa Whittington, Tolvert Fowler and Dr. Nadejda Korneeva.

REFERENCES

Abid MR, Li Y, Anthony C, De Benedetti A (1999): Translational regulation of ribonucleotide reductase by eukaryotic initiation factor 4E links protein synthesis to the control of DNA replication. J Biol Chem 274:35991–8.

Abraham RT, Wiederrecht GJ (1996): Immunopharmacology of rapamycin. Ann Rev Immunol 14:483–510.

Alessi DR, Kozlowski MT, Weng QP, Morrice N, Avruch J (1998): 3-Phosphoinositide-dependent protein kinase 1 (PDK1) phosphorylates and activates the p70 S6 kinase in vivo and in vitro. Curr Biol 8:69–81.

Altmann M, Handschin C, Trachsel H (1987): mRNA cap binding protein: Cloning of the gene encoding protein synthesis initiation factor eIF-4E from *Saccharomyces cerevisiae*. Mol Cell Biol 7:998–1003.

Altmann M, Muller PP, Wittner B, Ruchti F, Lanker S, Trachsel H (1993): A *Saccharomyces cerevisiae* homologue of mammalian translation initiation factor 4B contributes to RNA helicase activity. EMBO J 12:3997–4003.

Altmann M, Trachsel H (1989): Altered mRNA cap recognition activity of initiation factor 4E in the yeast cell cycle division mutant cdc33. Nucl Acids Res 17:5923–31.

Andresson T, Ruderman JV (1998): The kinase Eg2 is a component of the *Xenopus* oocyte progesterone-activated signaling pathway. EMBO J 17:5627–37.

Anthony B, Carter P, De Benedetti A (1996): Overexpression of the proto-oncogene/translation factor 4E in breast-carcinoma cell lines. Int J Cancer 65: 858–63.

Aoki M, Blazek E, Vogt PE (2001): A role of the kinase mTOR in cellular transformation induced by the oncoprotein P3k and Akt. Proc Natl Acad Sci USA 98:136–41.

Auvinen M, Paasinen A, Andersson LC, Holtta E (1992): Ornithine decarboxylase activity is critical for cell transformation. Nature 360:355–8.

Bachvarova RF (1992): A maternal tail of poly(A): The long and short of it. Cell 69:895–7.

Bandyopadhyay G, Standaert ML, Zhao L, Yu B, Avignon A, Galloway L, Karnam P, Moscat J, Farese RV (1997): Activation of protein kinase C (α, β, and ζ) by insulin in 3T3/L1 cells. J Biol Chem 272:2551–8.

Barbet NC, Schneider U, Helliwell SB, Stansfield I, Tuite MF, Hall MN (1996): TOR controls translation initiation and early G_1 progression in yeast. Mol Biol Cell 7:25–42.

Barth-Baus D, Stratton CA, Parrott l, Myerson H, Meyuhas L, Templeton DJ, Landreth GE, Hensold JO (2002): S6 phosphorylation-independent pathways regulate translation of 5′-terminal oligopyrimidine tract-containing mRNAs in differentiating hematopoietic cells. Nucl Acids Res 30:1919–28.

Baserga R (1990): The cell cycle: myths and realities. Cancer Res 50:6769–71.

Belle R, Derancourt J, Poulhe R, Capony J-P, Ozon R, Mulner-Lorillon O (1989): A purified complex from *Xenopus* oocytes contains a p47 protein, an in vivo substrate of MPF, and a p30 protein respectively homologous to elongation factors EF-1γ and EF-1β. FEBS Lett 255:101–4.

Bello-Fernandez C, Packham G, Cleveland JL (1993): The ornithine decarboxylase gene is a transcriptional target of c-Myc. Proc Natl Acad Sci USA 90: 7804–8.

Beretta L, Gingras AC, Svitkin YV, Hall MN, Sonenberg N (1996): Rapamycin blocks the phosphorylation of 4E-BP and inhibits cap-dependent initiation of translation. EMBO J 15:658–64.

Berkel HJ, Turbat-Herrera EA, Shi R, De Benedetti A (2001): Expression of the translation initiation factor eIF4E in the polyp-cancer sequence in the colon. Cancer Epidem Biomark Prevention 10:663–6.

Bernstein J, Sella O, Le SY, Elroy-Stein O (1997): PDGF2/c-sis mRNA leader contains a differentiation-linked internal ribosomal entry site (D-IRES). J Biol Chem 272:9356–62.

Berra E, Diaz-Meco MT, Dominguez I, Municio MM, Sanz L, Lozano J, Chapkin RS, Moscat J (1993): Protein kinase C ζ isoform is critical for mitogenic signal transduction. Cell 74:555–63.

Bonneau A-M, Sonenberg N (1987): Involvement of the 24-kDa cap-binding protein in regulation of protein synthesis in mitosis. J Biol Chem 262:11134–9.

Borman AM, Kirchweger R, Ziegler E, Rhoads RE, Skern T, Kean KM (1997): eIF4G and its proteolytic cleavage products: Effect on initiation of protein synthesis from capped, uncapped and IRES-containing mRNAs. RNA 3: 186–96.

Brass N, Keckel D, Sahin U, Pfreundschuh M, Sybrecht GW, Meese E (1997): Translation initiation factor eIF-4γ is encoded by an amplified gene and induces an immune response in squamous cell lung carcinoma. Hum Mol Gen 6:33–9.

Brenner C, Nakayama N, Goebl M, Tanaka K, Toh-e A, Matsumoto K (1988): CDC33 encodes mRNA cap-binding protein eIF-4E of *Saccharomyces cerevisiae*. Mol Cell Biol 8:3556–9.

Brunn GJ, Hudson CC, Sekulic A, Williams JM, Hosoi H, Houghton PJ, Lawrence JC, Abraham RT (1997): Phosphorylation of the translational repressor PHAS-I by the mammalian target of rapamycin. Science 277:99–101.

Burke DJ, Church D (1991): Protein synthesis requirements for nuclear division, cytokinesis, and cell separation in *Saccharomyces cerevisiae*. Mol Cell Biol 11:3691–8.

Burnett PE, Barrow RK, Cohen NA, Snyder SH, Sabatini DM (1998): RAFT1 phosphorylation of the translational regulators p70 S6 kinase and 4E-BP1. Proc Natl Acad Sci USA 95:1432–7.

Campbell LH, Borg KT, Haines RT, Moon RT, Schoenberg DR, Arrigo SJ (1994): Human immunodeficiency virus type 1 is required for binding of poly(A)-binding protein to Rev-dependent RNAs. J Virol 68:5433–8.

Celis JE, Madsen P, Ryazanov AG (1990): Increased phosphorylation of elongation factor 2 during mitosis in transformed human amnion cells correlates with a decreased rate of protein synthesis. Proc Natl Acad Sci USA 87: 4231–5.

Chang YW, Traugh JA (1997): Phosphorylation of elongation factor 1 and ribosomal protein S6 by multipotential S6 kinase and insulin stimulation of translational elongation. J Biol Chem 272:28252–7.

Chappell SA, Owens GC, Mauro VP (2001): A 5′ leader of Rbm3, a cold stress-induced mRNA, mediates internal initiation of translation with increased efficiency under conditions of mild hypothermia. J Biol Chem 276: 36917–22.

Cheatham B, Vlahos CJ, Cheatham L, Wang L, Blenis J, Kahn CR (1994): Phosphatidylinositol 3-kinase activation is required for insulin stimulation of pp70 S6 kinase, DNA synthesis, and glucose transporter translocation. Mol Cell Biol 14:4902–11.

Chen J-J (2000): Heme-regulated eIF2α kinase. In: N Sonenberg, JWB Hershey, MB Mathews (eds): Translational Control of Gene Expression. Cold Spring Harbor: Cold Spring Harbor Laboratory Press, pp 529–46.

Chiang P-W (2001): The 5-untranslated region of the *FMR1* message facilitates translation by internal ribosome entry. J Biol Chem 276:37916–21.

Chu E, Voeller D, Koeller DM, Drake JC, Takimoto CH, Maley GF, Maley F, Allegra CJ (1993): Identification of an RNA binding site for human thymidylate synthase. Proc Natl Acad Sci USA 90:517–21.

Chung J, Kuo CJ, Crabtree GR, Blenis J (1992): Rapamycin-FKBP specifically blocks growth-dependent activation of signaling by the 70 kd S6 protein kinase. Cell 69:1227–36.

Cohen N, Sharma M, Kentsis A, Perez JM, Strudwick S, Borden KLB (2001): PML RING suppresses oncogenic transformation by reducing the affinity of eIF4E for mRNA. EMBO J 20:4547–59.

Conlon I, Raff M (1999): Size control in animal development. Cell 96:235–44.

Cosentino GP, Schmelzle T, Haghighat A, Helliwell SB, Hall MN, Sonenberg N (2000): Eap1p, a novel eukaryotic translation initiation factor 4E-associated protein in *Saccharomyces cerevisiae*. Mol Cell Biol 20:4604–13.

Crew JP, Fuggle S, Bicknell R, Cranston DW, De Benedetti A, Harris AL (2000): Eukaryotic initiation factor-4E in superficial and muscle invasive bladder cancer and its correlation with vascular endothelial growth factor expression and tumour progression. Br J Cancer 82:161–6.

Cross DAE, Allessi DR, Cohen P, Andjelkovich M, Hemmings BA (1995): Inhibition of glycogen synthase kinase-3 by insulin mediated by protein kinase B. Nature 378:785–9.

Curtis D, Lehmann R, Zamore PD (1995): Translational regulation in development. Cell 81:171–8.

Danaie P, Altmann M, Hall MN, Trachsel H, Helliwell SB (1999): *CLN*3 expression is sufficient to restore G_1-to-S-phase progression in *Saccharomyces cerevisiae* mutants defective in translation initiation factor eIF4E. Biochem J 340: 135–41.

Darveau A, Pelletier J, Sonenberg N (1985): Differential efficiencies of in vitro translation of mouse c-*myc* transcripts differing in the 5′ untranslated region. Proc Natl Acad Sci USA 82:2315–19.

Davidson EH (1986): Gene activity in early development. Orlando, Academic Press, Inc.

De Benedetti A, Harris AL (1999): eIF4E expression in tumors: Its possible role in progression of malignancies. Int J Biochem Cell Biol 31:59–72.

De Benedetti A, Joshi B, Graff JR, Zimmer SG (1994): CHO cells transformed by the translation factor eIF4E display increased c-Myc expression, but require overexpression of Max for tumorigenicity. Mol Cell Differ 2:347–71.

De Benedetti A, Joshi-Barve S, Rinker-Schaeffer C, Rhoads RE (1991): Expression of antisense RNA against initiation factor eIF-4E mRNA in HeLa cells results in lengthened cell division times, diminished translation rates, and reduced levels of both eIF-4E and the p220 component of eIF-4F. Mol Cell Biol 11:5435–45.

De Benedetti A, Rhoads RE (1990): Overexpression of eukaryotic protein synthesis initiation factor 4E in HeLa cells results in aberrant growth and morphology. Proc Natl Acad Sci USA 87:8212–16.

DeFatta RJ, Chervenak RP, De Benedetti A (2002b): A cancer gene therapy approach through translational control of a suicide gene. Cancer Gene Ther 9:505–12.

DeFatta RJ, Li Y, De Benedetti A (2002a): Selective killing of cancer cells based on translational control of a suicide gene. Cancer Gene Ther 9:573–8.

DeFatta RJ, Turbat-Herrera E, Li BDL, Anderson W, De Benedetti A (1999): Elevated expression of eIF4E in confined early breast cancer lesions: Possible role of hypoxia. Int J Cancer 80:516–22.

Dekker LV, Parker PJ (1994): Protein kinase C—A question of specificity. Trends Biochem Sci 19:73–7.

Di Como CJ, Arndt KT (1996): Nutrients, via the Tor proteins, stimulate the association of Tap42 with type 2A phosphatases. Genes Dev 10:1904–16.

Diggle TA, Redpath NT, Heesom KJ, Denton RM (1998): Regulation of protein synthesis elongation-factor-2 kinase by cAMP in adipocytes. Biochem J 336: 525–9.

Dilling MB, Germain GS, Dudkin L, Jayaraman AL, Zhang X, Harwood FC, Houghton PJ (2002): 4E-binding proteins, the suppressors of eukaryotic initiation factor 4E, are down-regulated in cells with acquired or intrinsic resistance to rapamycin. J Biol Chem 277:13907–17.

Dirick L, Bohm T, Nasmyth K (1995): Roles and regulation of Cln-Cdc28 kinases at the start of the cell cycle of *Saccharomyces cerevisiae*. EMBO J 14:4803–13.

Dong LQ, Zhang RB, Langlais P, He H, Clark M, Zhu L, Liu F (1999): Primary structure, tissue distribution, and expression of mouse phosphoinositide-dependent protein kinase-1, a protein kinase that phosphorylates and activates protein kinase Cζ. J Biol Chem 274:8117–22.

Dudkin L, Dilling MB, Cheshire PJ, Harwood FC, Hollingshead M, Arbuck SG, Travis R, Sausville EA, Houghton PJ (2001): Biochemical correlates of mTOR inhibition by the rapamycin ester CCI-779 and tumor growth inhibition. Clin Cancer Res 7:1758–64.

Dworkin MB, Dworkin-Rastl E (1985): Changes in RNA titers and polyadenylation during oogenesis and oocyte maturation in *Xenopus laevis*. Dev Biol 112:451–7.

Dworkin MB, Dworkin-Rastl E (1990): Functions of maternal mRNA in early development. Mol Reprod Dev 26:261–97.

Eberle J, Krasagakis K, Orfanos CE (1997): Translation initiation factor eIF-4AI mRNA is consistently overexpressed in human melanoma cells in vitro. Int J Cancer 71:396–401.

Eremenko T, Volpe P (1975): Polysome translational state during the cell cycle. Eur J Biochem 52:203–10.

Fadden P, Haystead TAJ, Lawrence JC, Jr. (1997): Identification of phosphorylation sites in the translational regulator, PHAS-I, that are controlled by insulin and rapamycin in rat adipocytes. J Biol Chem 15:10240–7.

Fan H, Penman S (1970): Regulation of protein synthesis in mammalian cells. J Mol Biol 50:665–70.

Fernandez R, Yaman I, Mishra R, Merrick WC, Snider MD, Lamer WH, Hatzoglou M (2001): Internal ribosome entry site-mediated translation of a mammalian mRNA is regulated by amino acid availability. J Biol Chem 276:12285–91.

Flynn A, Proud CG (1995): Serine 209, Not Serine 53, Is the major site of phosphorylation in initiation factor eIF-4E in serum-treated Chinese hamster ovary cells. J Biol Chem 270:21684–8.

Fox CA, Sheets MD, Wickens MP (1989): Poly(A) addition during maturation of frog oocytes: Distinct nuclear and cytoplasmic activities and regulation by the sequence UUUUUAU. Genes Dev 3:2151–62.

Franklin S, Pho T, Abreo F, Nassar R, De Benedetti A, Stucker F, Nathan CA (1999): Immunohistochemical detection of the proto-oncogene eIF4E in larynx and hypopharynx cancers. Arch Otolaryngol Head Neck Surg 125: 177–82.

Fredlund JO, Johansson MC, Dahlberg E, Oredsson SM (1995): Ornitihine decarboxylase and S-adenosylmethionine decarboxylase expression during the cell cycle of Chinese hamster ovary cells. Expr Cell Res 216:86–92.

Fukuchi-Shimogori T, Itsuko I, Kashiwagi K, Mashiba H, Ekimoto H, Igarashi K (1997): Malignant transformation by overproduction of translation initiation factor eIF4G. Cancer Res 57:5041–4.

Fukunaga R, Hunter T (1997): MNK1, a new MAP kinase-activated protein kinase, isolated by a novel expression screening method for identifying protein kinase substrates. EMBO J 16:1921–33.

Gabrielli BG, Roy LM, Maller JL (1993): Requirement for cdk2 in cytostatic factor-mediated metaphase II arrest. Science 259:1766–9.

Galili G, Kawata EE, Smith LD, Larkins BA (1988): Role of the 3′-poly(A) sequence in translational regulation of mRNAs in *Xenopus laevis* oocytes. J Biol Chem 263:5764–70.

Gallego C, Gari E, Colomina N, Herrero E, Aldea M (1997): The Cln3 cyclin is down-regulated by translational repression and degradation during the G_1 arrest caused by nitrogen deprivation in budding yeast. EMBO J 16:7196–206.

Gallie DR (1991): The cap and poly(A) function synergistically to regulate mRNA translation efficiency. Genes Dev 5:2108–16.

Gallie DR, Tanguay R (1994): Poly(A) binds to initiation factors and increases cap-dependent translation in vitro. J Biol Chem 269:17166–73.

Gautier J, Minshull J, Lohka M, Glotzer M, Hunt T, Maller JL (1990): Cyclin is a component of maturation-promoting factor from *Xenopus*. Cell 60:487–94.

Gingras A-C, Gygi SP, Raught B, Polakiewicz RD, Abraham RT, Hoekstra MF, Aebersold R, Sonenberg N (1999): Regulation of 4E-BP1 phosphorylation: A novel two-step mechanism. Genes Dev 13:1422–37.

Gingras A-C, Raught B, Gygi SP, Niedzwiecka A, Miron M, Burley SK, Polakiewicz RD, Wyslouch-Cieszynska A, Aebersold R, Sonenberg N (2001): Hierarchical phosphorylation of the translation inhibitor 4E-BP1. Genes Dev 15:2852–64.

Gingras A-C, Raught B, Sonenberg N (1999): eIF4 initiation factors: effectors of mRNA recruitment to ribosomes and regulators of translation. An Rev Biochem 68:913–63.

Gingras A-C, Raught B, Sonenberg N (2001): Regulation of translation initiation by FRAP/mTOR. Genes Dev 15:807–26.

Graff JR, Boghaert ER, De Benedetti A, Chan SK, Zimmer SG (1995): Reduction of the levels of initiation factor 4E mediates the malignant phenotype of *ras*-transformed rat fibroblasts. Int J Cancer 60:255–63.

Graff JR, De Benedetti A, Olson JW, Tamez P, Casero RA, Jr., Zimmer S (1997): Translation of ODC mRNA and polymine transport are suppressed in *ras*-transformed CREF cells by depleting translation initiation factor 4E. Biochem Biophys Res Comm 240:15–20.

Groisman I, Huang Y-S, Mendez R, Cao Q, Theurkauf W, Richter JD (2000): CPEB, maskin, and cyclin B1 mRNA at the mitotic apparatus: Implications for local translational control of cell division. Cell 103:435–47.

Grossi de Sa MF, Standart N, Martins de Sa C, Akhayat O, Huesca M, Scherrer K (1988): The poly(A)-binding protein facilitates in vitro translation of poly(A)-rich mRNA. Eur J Biochem 176:521–6.

Hajduch E, Alessi DR, Hemmings BA, Hundal HS (1998): Constitutive activation of protein kinase Bα by membrane targeting promotes glucose and system A amino acid transport, protein synthesis, and inactivation of glycogen synthase kinase 3 in L6 muscle cells. Diabetes 47:1006–13.

Hamelers IHL, van Schaik RFMA, Skpkema J, Sussenbach JS, Steenbergh PH (2002): Insulin-like growth factor I triggers nuclear accumulation of cyclin D1 in MCF-7S breast cancer cells. J Biol Chem 277:47645–52.

Hara K, Maruki Y, Long X, Yoshino K-i, Oshioro N, Hidayat S, Tokunaga C, Avruch J, Yonezawa K (2002): Raptor, a binding partner of target of rapamycin (TOR), mediates TOR action. Cell 110:177–89.

Hara K, Yonezawa K, Kozlowski MT, Sugimoto T, Andrabi K, Weng QP, Kasuga M, Nishimoto I, Avruch J (1997): Regulation of eIF-4E BP1 phosphorylation by mTOR. J Biol Chem 272:26457–63.

Hara K, Yonezawa K, Weng QP, Kozlowski MT, Belham C, Avruch J (1998): Amino acid sufficiency and mTOR regulate p70 S6 kinase and eIF-4E BP1 through a common effector mechanism. J Biol Chem 273:14484–94.

Hashemolhosseini S, Nagamine Y, Morley SJ, Desrivieres S, Mercep L, Ferrari S (1998): Rapamycin inhibition of the G_1 to S transition is mediated by effects on cyclin D1 mRNA and protein stability. J Biol Chem 273:14424–9.

Hensold JO, Barth-Baus D, Stratton CA (1996): Inducers of erythroleukemic differentiation cause messenger RNAs that lack poly(A)-binding protein to accumulate in translationally inactive, salt-stable 80S ribosomal complexes. J Biol Chem 271:23246–54.

Herbert TP, Fåhraeus R, Prescott AR, Lane DP, Proud CG (2000): Rapid induction of apoptosis by peptides that bind initiation factor eIF4E. Curr Biol 10:793–6.

Herrera-Velit P, Knutson KL, Reiner NE (1997): Phosphatidylinositol 3-kinase-dependent activation of protein kinase C-ζ in bacterial lipopolysaccharide-treated human monocytes. J Biol Chem 272:16445–52.

Hershey JWB, Merrick WC (2000): Pathway and mechanism of initiation of protein synthesis. In: N Sonenberg, Hershey JWB, Mathews MB (eds): Translational Control of Gene Expression. Cold Spring Harbor: Cold Spring Harbor Laboratory Press, pp 33–88.

Hershey JWB, Miyamoto S (2000): Translational control and cancer. In: N Sonenberg, JWB Hershey, MB Mathews (eds): Translational Control of Gene Expression. Cold Spring Harbor: Cold Spring Harbor Laboratory Press, pp 637–54.

Hidalgo M, Rowinsky EK (2000): The rapamycin-sensitive signal transduction pathway as a target for cancer therapy. Oncogene 19:6680–6.

Hill JR, Morris DR (1992): Cell-specific translation of S-adenosylmethionine decarboxylase messenger RNA: Regulation by the 5′ transcript leader. J Biol Chem 267:21886–93.

Hill JR, Morris DR (1993): Cell-specific translational regulation of S-adenosylmethionine decarboxylase mRNA. J Biol Chem 268:726–31.

Hinnebusch AG (2000): Mechanism and regulation of initiator methionyl-tRNA binding to ribosomes. In: N Sonenberg, JWB Hershey, MB Mathews (eds): Translational Control of Gene Expression. Cold Spring Harbor: Cold Spring Harbor Laboratory Press, pp 185–244.

Holcik M, Gordon BW, Korneluk RG (2003): The internal ribosome entry site-mediated translation of antiapoptotic protein XIAP is modulated by the heterogeneous nuclear ribonucleoproteins C1 and C2. Mol Cell Biol 23:280–8.

Hosoi H, Dilling MB, Liu LN, Danks MK, Shikata T, Sekulic A, Abraham RT, Lawrence JC, Jr, Houghton PJ (1998): Studies on the mechanism of resistance to rapamycin in human cancer cells. Mol Pharm 54:815–24.

Hu C, Pang S, Kong X, Velleca M, Lawrence JC (1994): Molecular cloning and tissue distribution of PHAS-I, an intracellular target for insulin and growth factors. Proc Natl Acad Sci USA 91:3730–4.

Huang J, Schneider RJ (1991): Adenovirus inhibition of cellular protein synthesis involves inactivation of cap-binding protein. Cell 65:271–80.

Iiboshi Y, Pabst PJ, Kawasome H, Hosoi H, Abraham RT, Houghton PJ, Terada N (1999): Amino acid-dependent control of $p70^{s6k}$. J Biol Chem 274:1092–9.

Imataka H, Gradi A, Sonenberg N (1998): A newly identified N-terminal amino acid sequence of human eIF4G binds poly(A) binding protein and functions in poly(A)-dependent translation. EMBO J 17:7480–9.

Isotani S, Hara K, Tokunaga C, Inoue H, Avruch J, Yonezawa K (1999): Immunopurified mammalian target of rapamycin phosphorylates and activates p70 S6 kinase α in vitro. J Biol Chem 274:34493–8.

Jacinto E, Guo B, Arndt KT, Schmelzle T, Hall MN (2001): TIP41 interacts with TAP42 and negatively regulates the TOR signaling pathway. Cell 8:1017–26.

Jackson RJ (2000): Comparative view of initiation site selection mechanisms. In: N Sonenberg, JWB Hershey, MB Mathews (eds): Translational Control of Gene Expression. Cold Spring Harbor: Cold Spring Harbor Laboratory Press, pp 127–84.

Jang SK, Kräusslich H-G, Nicklin MJH, Duke GM, Palmenberg AC, Wimmer E (1988): A segment of the 5′-nontranslated region of encephalomyocarditis virus RNA directs internal entry of ribosomes during in vitro translation. J Virol 62:2636–43.

Janssen GMC, Morales J, Schipper A, Labbe J-C, Mulner-Lorillon O, Belle R, Möller W (1991): A major substrate of maturation promoting factor identified as elongation factor 1 βγδ in *Xenopus* laevis. J Biol Chem 266:14885–8.

Jefferies BJ, Fumagalli S, Dennis PB, Reinhard C, Pearson RB, Thomas G (1997): Rapamycin suppresses 5′TOP mRNA translation through inhibition of $p70^{s6k}$. EMBO J 16:3693–704.

Jefferies HBJ, Reinhard C, Kozma SC, Thomas G (1994): Rapamycin selectively represses translation of the "polypyrimidine tract" mRNA family. Proc Natl Acad Sci USA 91:4441–5.

Jefferies HBJ, Thomas G (1994): Elongation factor-1α is selectively translated following mitogenic stimulation. J Biol Chem 269:4367–72.

Jefferson LS, Fabian JR, Kimball SR (1999): Glycogen synthase kinase-3 is the predominant insulin-regulated eukaryotic initiation factor 2B kinase in skeletal muscle. Int J Biochem Cell Biol 31:191–200.

Jiang Y, Broach JR (1999): Tor proteins and protein phosphatase 2A reciprocally regulate Tap42 in controlling cell growth in yeast. EMBO J 18:2782–92.

Johannes G, Sarnow P (1998): Cap-independent polysomal association of natural mRNAs encoding c-myc, BiP and eIF4G conferred by internal ribosome entry sites. RNA 4:1500–13.

Jones RM, Branda J, Johnston KA, Polymenis M, Gadd M, Rustgi A, Callanan L, Schmidt EV (1996): An essential E box in the promoter of the gene encoding the mRNA cap-binding protein (eukaryotic initiation factor 4E) is a target for activation by c-*myc*. Mol Cell Biol 16:4754–64.

Joshi B, Cai A-L, Keiper BD, Minich WB, Mendez R, Beach CM, Stepinski J, Stolarski R, Darzynkiewicz E, Rhoads RE (1995): Phosphorylation of eukaryotic protein synthesis initiation factor 4E at Ser-209. J Biol Chem 270:14597–603.

Kaspar RL, Kakegawa T, Cranston H, Morris DR, White MW (1992): A regulatory cis element and a specific binding factor invovled in the mitogenic control of murine ribosomal protein L32 translation. J Biol Chem 267:508–14.

Kaspar RL, Rychlik W, White MW, Rhoads RE, Morris DR (1990): Simultaneous cytoplasmic redistribution of ribosomal protein L32 mRNA and phosphorylation of eukaryotic initiation factor 4E after mitogenic stimulation of Swiss 3T3 cells. J Biol Chem 265:3619–22.

Kaufman RJ (2000): Double-stranded RNA-activated protein kinase PKR. In: N Sonenberg, JWB Hershey, MB Mathews (eds): Translational Control of Gene Expression. Cold Spring Harbor: Cold Spring Harbor Laboratory Press, pp 503–28.

Kawasome H, Papst P, Webb S, Keller GM, Johnson GL, Gelfand EW, Terada N (1998): Targeted disruption of $p70^{s6k}$ defines its role in protein synthesis and rapamycin sensitivity. Proc Natl Acad Sci USA 95:5033–8.

Kelso-Winemiller LC, Winkler MM (1991): "Unmasking" of stored maternal mRNAs and the activation of protein synthesis in sea urchins. Development 111:623–33.

Kerekatte V, Smiley K, Hu B, Smith A, Gelder F, De Benedetti A (1995): The protooncogene/translation factor eIF-4E—A survey of its expression in breast carcinomas. Int J Cancer 65:27–31.

Kessler SH, Sachs AB (1998): RNA recognition motif 2 of yeast PAB1p is required for its functional interaction with eukaryotic translation initiation factor 4G. Mol Cell Biol 18:51–7.

Kevil C, Carter P, Hu B, De Benedetti A (1995): Translational enhancement of FGF-2 by eIF-4 factors, and alternate utilization of CUG and AUG codons for translation initiation. Oncogene 11:2339–48.

Keyomarsi K, Samet J, Molnar G, Pardee AB (1993): The thymidylate synthase inhibitor, ICI D1694, overcomes translational detainment of the enzyme. J Biol Chem 268:15142–9.

Khaleghpour E, Pyronnet S, Gingras A-C, Sonenberg N (1999): Translational homeostasis: Eukaryotic translation initiation factor 4E contol of 4E-binding protein 1 and p70 S6 kinase activities. Mol Cell Biol 19:4302–10.

Kim D-H, Sarbassov DD, Ali SM, King JE, Latek RR, Erdjument-Bromage H, Tempst P, Sabatini DM (2002): mTOR interacts with raptor to form a nutrient-sensitive complex that signals to the cell growth machinery. Cell 110: 163–75.

Kim JH, Paek KY, Choi K, Kim T-D, Hahm B, Kim K-T, Jang SK (2003): Heterogeneous nuclear ribonucleoprotein C modulates translation of c-*myc* mRNA in a cell cycle phase-dependent manner. Mol Cell Biol 23:708–20.

Kimball SR, Shantz LM, Horetsky RL, Jefferson LS (1999): Leucine regulates translation of specific mRNAs in L6 myoblasts through mTOR-mediated changes in availability of eIF4E and phosphorylation of ribosomal protein S6. J Biol Chem 274:11647–52.

Knauf U, Tschopp C, Gram H (2001): Negative regulation of protein translation by mitogen-activated protein kinase-interacting kinases 1 and 2. Mol Cell Biol 21:5500–11.

Kochetov AV, Ischenko IV, Vorobiev DG, Kel AE, Babenko VN, Kisselev LL, Kolchanov NA (1998): Eukaryotic mRNAs encoding abundant and scarce proteins are statistically dissimilar in many structural features. FEBS Lett 440: 351–5.

Koromilas AE, Lazaras-Karatzas A, Sonenberg N (1992): mRNAs containing extensive secondary structure in their 5′ non-coding region translate efficiently in cells overexpressing initiation factor eIF-4E. EMBO J 11:4153–8.

Koromilas AE, Roy S, Barber GN, Katze MG, Sonenberg N (1992): Malignant transformation by a mutant of the IFN-inducible dsRNA-dependent protein kinase. Science 257:1685–9.

Kozak M (1992): Regulation of translation in eukaryotic systems. An Rev Cell Biol 8:197–225.

Kozak M (1997): At least six nucleotides preceding the initiator AUG codon enhance translation in mammalian cells. J Mol Biol 196:947–50.

Kuge H, Inoue A (1992): Maturation of *Xenopus laevis* oocyte by progesterone requires poly(A) tail elongation of mRNA. Expr Cell Res 202:52–8.

Kunz J, Henriquez R, Schneider U, Deuter-Reinhard M, Movva NR, Hall MN (1993): Target of rapamycin in yeast, TOR2, is an essential phosphatidylinositol kinase homolog required for G_1 progression. Cell 73:585–96.

Lachance PED, Miron M, Raught B, Sonenberg N, Lasko P (2002): Phosphorylation of eukaryotic translation initiation factor 4E is critical for growth. Mol Cell Biol 22:1656–63.

Lamphear BJ, Kirchweger R, Skern T, Rhoads RE (1995): Mapping of functional domains in eIF4G with picornaviral proteases. Implications for cap-dependent and cap-independent translational initiation. J Biol Chem 270: 21975–83.

Lane HA, Fernandez A, Lamb NJ, Thomas G (1993): $p70^{S6K}$ function is essential for G_1 progression. Nature 363:170–2.

Lazaris-Karatzas A, Montine KS, Sonenberg N (1990): Malignant transformation by a eukaryotic initiation factor subunit that binds to mRNA 5′ cap. Nature 345:544–7.

Lazaris-Karatzas A, Smith MR, Frederickson RM, Jaramillo ML, Liu Y-L, Kung H-F, Sonenberg N (1992): Ras mediates translation initiation factor 4E-induced malignant transformation. Genes Dev 6:1631–42.

Le Good JA, Ziegler WH, Parekh DB, Allessi DR, Cohen P, Parker PJ (1998): Protein kinase C isotypes controlled by phosphoinositide 3-kinase through the protein kinase PDK1. Science 281:2042–5.

Le H, Tanguay RL, Balasta ML, Wei CC, Browning K, Metz AM, Goss DJ, Gallie DR (1997): Translation initiation factors eIF-iso4G and eIF-4B interact with the poly(A)-binding protein and increase its RNA binding activity. J Biol Chem 272:16247–55.

Levy S, Avni D, Hariharan N, Perry RP, Meyuhas O (1991): Oligopyrimidine tract at the 5′ end of mammalian ribosomal protein mRNAs is required for their translational control. Proc Natl Acad Sci USA 88:3319–23.

Li BD, Gruner JS, Abreo F, Johnson LW, Yu H, Nawas A, McDonald JC, De Benedetti A (2002): Prospective study of eukaryotic initiation factor 4E protein elevation and breast cancer outcome. An Surg 235:732–9.

Li BD, Liu L, Dawson M, De Benedetti A (1997): Overexpression of eukaryotic initiation factor 4E (eIF4E) in breast carcinoma. Cancer 79:2385–90.

Li BDL, McDonald J, Nassar R, De Benedetti A (1998): Clinical outcome in stage 1 to 3 breast carcinomas and eIF4E overexpression. An Surg Soc 227:756–62.

Liebig H-D, Ziegler E, Yan R, Hartmuth K, Klump H, Kowalski H, Blaas D, Sommergruber W, Frasel L, Lamphear B, Rhoads RE, Kuechler E, Skern T (1993): Purification of two picornaviral 2A proteinases: interaction with eIF-4γ and influence on in vitro translation. Biochemistry 32:7581–8.

Lin T, Kong X, Haystead TAJ, Pause A, Belsham G, Sonenberg N, Lawrence JC (1994): PHAS-I as a link between mitogen-activated protein kinase and translation initiation. Science 266:653–6.

Macejak DG, Sarnow P (1991): Internal initiation of translation mediated by the 5′ leader of a cellular mRNA. Nature 353:90–4.

Majewski M, Korecka M, Kossev P, Li S, Goldman J, Moore J, Silberstein LE, Nowell PC, Schuler W, Shaw LM, Wasik MA (2000): The imunosuppressive macrolide RAD inhibits growth of human Epstein-Barr virus-transformed B lymphocytes in vitro and in vivo. Proc Natl Acad Sci USA 97:4285–90.

Maller JL (1985): Regulation of amphibian oocyte maturation. Cell Differ 16: 211–21.

Manzella JM, Rychlik W, Rhoads RE, Hershey JWB, Blackshear PJ (1991): Insulin induction of ornithine decarboxylase. Importance of mRNA secondary structure and phosphorylation of eucaryotic initiation factors eIF-4B and eIF-4E. J Biol Chem 266:2383–9.

Mariottini P, Amaldi F (1990): The 5′ untranslated region of mRNA for ribosomal protein S19 is involved in its translational regulation during *Xenopus* development. Mol Cell Biol 10:816–22.

McGrew LL, Dworkin-Rastl E, Dworkin MB, Richter JD (1989): Poly(A) elongation during *Xenopus* oocyte maturation is required for translational recruitment and is mediated by a short sequence element. Genes Dev 3: 803–15.

McKendrick L, Morley SJ, Pain VM, Jagus R, Joshi B (2001): Phosphorylation of eukaryotic initiation factor 4E (eIF4E) at Ser209 is not required for protein synthesis in vitro and in vivo. Eur J Biochem 268:5375–85.

Mendez R, Barnard D, Richter JD (2002): Differential mRNA translation and meiotic progression require cdc-2-mediated CPEB destruction. EMBO J 21: 1833–44.

Mendez R, Hake LE, Andresson T, Littlepage LE, Ruderman JV, Richter JD (2000): Phosphorylation of CPE binding factor by Eg2 regulates translation of c-mos-mRNA. Nature 404:302–7.

Mendez R, Kollmorgen G, White MF, Rhoads RE (1997): Requirement of protein kinase Cζ for stimulation of protein synthesis by insulin. Mol Cell Biol 17:5184–92.

Mendez R, Murthy EG, Ryan K, Manley JL, Richter JD (2000): Phosphorylation of CPEB by Eg2 mediates the recruitment of CPSF into an active cytoplasmic polyadenylation complex. Mol Cell 6:1253–9.

Mendez R, Myers MG, White MF, Rhoads RE (1996): Stimulation of protein synthesis, eukaryotic translation initiation factor 4E phosphorylation, and PHAS-I phosphorylation by insulin requires insulin receptor substrate-1 and phosphatidylinositol 3-kinase. Mol Cell Biol 16:2857–64.

Mendez R, Richter JD (2001a): Translational control by CPEB: A means to the end. Mol Cell Biol 2:521–9.

Mendez R, Welsh G, Kleijn M, Myers MG, White MF, Proud CG, Rhoads RE (2001b): Regulation of protein synthesis by insulin through IRS-1, in Rhoads (ed) Prog Nucl Acids Res and Mol Biol 26, Springer–Verlag, Berlin, pp 49–93.

Meyuhas O, Hornstein E (2000): Translational control of TOP mRNAs. In: N Sonenberg, JWB Hershey, MB Mathews (eds): Translational Control of Gene Expression. Cold Spring Harbor: Cold Spring Harbor Laboratory Press, pp 671–94.

Minich WB, Balasta ML, Goss DJ, Rhoads RE (1994): Chromatographic resolution of in vivo–phosphorylated and nonphosphorylated translation initiation factor eIF-4E: Demonstration of increased cap affinity of the phosphorylated form. Proc Natl Acad Sci USA 91:7668–72.

Miyagi Y, Sugiyama A, Asai A, Okazaki T, Kuchino Y, Kerr J (1995): Elevated levels of eukaryotic translation initiation factor eIF-4E mRNA in a broad spectrum of transformed cell lines. Cancer Lett 91:247–52.

Morice WG, Wiederrecht G, Brunn GJ, Siekierka JJ, Abraham RT (1993): Rapamycin inhibition of interleukin-2-dependent $p33^{cdk2}$ and $p34^{cdc2}$ kinase activation in T lymphocytes. J Biol Chem 268:22737–45.

Morris DR (1997): cis-Acting mRNA structures in gene-specific translational control. In mRNA Metabolism and Post-transcriptional Gene Regulation. New York: Wiley-Liss, pp 165–80.

Moule SK, Welsh GI, Edgell NJ, Foulstone EJ, Proud CG, Denton RM (1997): Regulation of protein kinase B and glycogen synthase kinase-3 by insulin and β-adrenergic agonists in rat epididymal fat cells. J Biol Chem 272:7713–19.

Muise-Helmericks RC, Grimes HL, Bellacosa A, Malstrom SE, Tsichlis PN, Rosen N (1998): Cyclin D expression is controlled post-transcriptionally via a phosphatidylinositol 3-kinase/Akt-dependent pathway. J Biol Chem 273: 29864–72.

Munroe D, Jacobson A (1990): mRNA poly(A) tail, a 3′ enhancer of translation initiation. Mol Cell Biol 10:3441–55.

Myers MG, Zhang Y, Aldaz GAI, Grammer T, Glasheen EM, Yenush L, Wang LM, Sun XJ, Blenis J, Pierce JH, White MF (1996): YMXM-motifs and signaling by an IRS-1 molecule without tyrosine phosphorylation sites. Mol Cell Biol 16:4147–55.

Nairn A, Palfrey HC (1996): Regulation of protein synthesis by calcium. In: JWB Hershey, MB Mathews, N Sonenberg (eds): Translational Control. Cold Spring Harbor: Cold Spring Harbor Laboratory Press, pp 295–318.

Nairn AC, Palfrey HC (1987): Identification of the mayor Mr 100,000 substrate for calmodulin-dependent protein kinase III in mammalian cells as elongation factor-2. J Biol Chem 262:17299–303.

Nanbru C, Lafon I, Audigier S, Gensac MC, Vagner S, Huez G, Prats AC (1997): Alternative translation of the proto-oncogene c-*myc* by an internal ribosome entry site. J Biol Chem 272:32061–6.

Nathan CO, Carter P, Liu L, Li B, Abreo F, Tudor A, Zimmer S, De Benedetti A (1997): Elevated expression of eIF4E and FGF-2 isoforms during vascularization of breast carcinomas. Oncogene 15:1087–95.

Nathan CO, Franklin S, Abreo F, Nassar R, De Benedetti A, Williams J, Stucker F (1999): Expression of eIF4E during head and neck tumorigenesis: Possible role in angiogenesis. Laryngoscope 109:1253–8.

Nathan CO, Franklin S, Abreo FW, Nassar R, De Benedetti A, Glass J (1999): Analysis of surgical margins with the molecular marker eIF4E: a prognostic factor in patients with head and neck cancer. J Clin Oncol 17:2909–14.

Nathan CO, Liu L, Li BDL, Abreo F, De Benedetti A (1997): Detection of the proto-oncogene eIF4E in surgical margins may predict recurrence in head and neck cancer. Oncogene 15:579–84.

Negulescu D, Leong LE-C, Chandy KG, Semler BL, Gutman GA (1998): Translation initiation of a cardiac voltage-gated potassium channel by internal ribosome entry. J Biol Chem 273:20109–13.

New L, Zhao M, Li Y, Bassett WW, Feng Y, Ludwig S, Padova FD, Gram H, Han J (1999): Cloning and characterization of a novel RSK-related protein kinase. J Biol Chem 274:1026–32.

Nielsen FC, Ostergaard L, Nielsen J, Christiansen J (1995): Growth-dependent translation of IGF-II mRNA by a rapamycin-sensitive pathway. Nature 377: 358–62.

North G (1991): Starting and stopping. Nature 351:604.

Oh S-K, Scott MP, Sarnow P (1992): Homeotic gene *Antennapedia* messenger RNA contains 5′-noncoding sequences that confer translational initiation by internal ribosome binding. Genes Dev 6:1643–53.

Ohlmann T, Rau M, Morley SJ, Pain VM (1995): Proteolytic cleavage of initiation factor eIF-4γ in the reticulocyte lysate inhibits translation of capped mRNAs but enhances that of uncapped mRNAs. Nucl Acids Res 23:334–40.

Park I-H, Bachmann R, Shirazi H, Chen J (2002): Regulation of ribosomal S6 kinase 2 by mammalian target of rapamycin. J Biol Chem 277:31423–9.

Parkin N, Darveau A, Nicholson R, Sonenberg N (1988): *cis*-Acting translational effects of the 5′ noncoding region of *c-myc* mRNA. Mol Cell Biol 8:2875–83.

Patel RC, Jacobs-Lorena M (1992): *cis*-Acting sequences in the 5′-untranslated region of the ribosomal protein A1 mRNA mediate its translational regulation during early embryogenesis of *Drosophila*. J Biol Chem 267:1159–64.

Pause A, Belsham GJ, Gingras A, Donze O, Lin T, Lawrence JC, Sonenberg N (1994): Insulin-dependent stimulation of protein synthesis by phosphorylation of a regulator of 5′-cap function. Nature 371:762–7.

Paz I, Abramovitz L, Choder M (1999): Starved *Saccharomyces cerevisiae* cells have the capacity to support internal initiation of translation. J Biol Chem 274:21741–5.

Pelletier J, Kaplan G, Racaniello VR, Sonenberg N (1988): Cap-independent translation of poliovirus mRNA is conferred by sequence elements within the 5′ noncoding region. Mol Cell Biol 8:1103–12.

Pelletier J, Sonenberg N (1985): Insertion mutagenesis to increase secondary structure within the 5′ noncoding region of a eukaryotic mRNA reduces translational efficiency. Cell 40:515–26.

Pines J (1993): Cyclins and cyclin-dependent kinases—Take your partners. Trends Biochem Sci 18:195–7.

Pines J, Hunter T (1989): Isolation of a human cyclin cDNA: Evidence for cyclin mRNA and protein regulation in the cell cycle and for interaction with $p34^{cdc2}$. Cell 58:833–46.

Polunovsky VA, Gingras A-C, Sonenberg N, Peterson M, Tan A, Rubins JB, Manivel JC, Bitterman PB (2000): Translational control of the antiapoptotic function of Ras. J Biol Chem 275:24776–80.

Polunovsky VA, Rosenwald IB, Tan AT, White J, Chiang L, Sonenberg N, Bitterman PB (1996): Translational control of programmed cell death: Eukaryotic translation initiation factor 4E blocks apoptosis in growth-factor-restricted fibroblasts with physiologically expressed or deregulated Myc. Mol Cell Biol 16:6573–81.

Polymenis M, Schmidt EV (1997): Coupling of cell division to cell growth by translational control of the G_1 cyclin CLN3 in yeast. Genes Dev 11:2522–31.

Preiss T, Hentze MW (1998): Dual function of the messenger RNA cap structure in poly(A)-tail-promoted translation in yeast. Nature 392:516–20.

Pullen N, Dennis PB, Andjelkovic M, Dufner A, Kozma SC, Hemmings BA, Thomas G (1998): Phosphorylation and activation of $p70^{s6k}$ by PDK1. Science 279:707–10.

Pullen N, Thomas G (1997): The modular phosphorylation and activation of $p70^{s6k}$. FEBS Lett 410:78–82.

Pyronnet S, Gingras A-C, Bouisson M, Kowalski-Chauvel A, Seva C, Vaysse N, Sonenberg N, Pradayrol L (1998): Gastrin induces phsophorylation of eIF4E binding protien 1 and translation initiation of ornithine decarboxylase mRNA. Oncogene 16:2219–7.

Pyronnet S, Imataka H, Gingras A-C, Fukunaga R, Hunter T, Sonenberg N (1999): Human eukaryotic translation initiation factor 4G (eIF4G) recruits Mnk1 to phosphorylate eIF4E. EMBO J 18:270–9.

Pyronnet S, Pradayrol L, Sonenberg N (2000): A cell cycle-dependent internal ribosome entry site. Mol Cell 5:607–16.

Pyronnet S, Vagner S, Bouisson M, Prats A-C, Vaysse N, Pradayrol L (1996): Relief of ornithine decarboxylase messenger RNA translational repression induced by alternative splicing of its 5′ untranslated region. Cancer Res 56: 1742–5.

Redpath NT, Foulstone EJ, Proud CG (1996): Regulation of translation elongation factor-2 by insulin via rapamycin-sensitive signalling pathway. EMBO J 15:2291–7.

Redpath NT, Price NT, Severinov KV, Proud CG (1993): Regulation of elongation factor-2 by multisite phosphorylation. Eur J Biochem 213:689–99.

Rhoads RE (1993): Regulation of eukaryotic protein synthesis by initiation factors. J Biol Chem 268:3017–20.

Rhoads RE (1999): Minireview: Signal transduction pathways that regulate eukaryotic protein synthesis. J Biol Chem 274:30337–40.

Richardson HE, Wittenberg C, Cross F, Reed SI (1989): An essential G1 function for cyclin-like proteins in yeast. Cell 59:1127–33.

Richter JD (1988): Information relay from gene to protein: the mRNP connection. Trends Biochem Sci 13:483–6.

Richter JD (1991): Translational control during early development. Bioessays 13: 179–83.

Richter JD, Paris J, McGrew LL (1990): Maternal mRNA expression in early development: Regulation at the 3′ end. Enzyme 44:129–46.

Rinker-Schaeffer CW, Graff JR, De Benedetti A, Zimmer SG, Rhoads RE (1993): Decreasing the level of translation initiation factor 4E with antisense RNA causes reversal of *ras*-mediated transformation and tumorigenesis of cloned rat embryo fibroblasts. Int J Cancer 55:841–7.

Ron D, Harding HP (2000): PERK and translational control by stress in the endoplasmic reticulum. In: N Sonenberg, JWB Hershey, MB Mathews (eds): Translational Control of Gene Expression. Cold Spring Harbor: Cold Spring Harbor Laboratory Press, pp 547–60.

Rosenwald EB, Lazaris-Karatzas A, Sonenberg N, Schmidt EV (1993): Elevated levels of cyclin D1 protein in response to increased expression of eukaryotic initiation factor 4E. Mol Cell Biol 13:7358–63.

Rosenwald IB, Chen JJ, Wang S, Savas L, London IM, Pullman J (1999): Upregulation of protein synthesis initiation factor eIF4E is an early event during colon carcinogenesis. Oncogene 18:2507–17.

Rosenwald IB, Kaspar R, Rousseau D, Gehrke L, Leboulch P, Chen JJ, Schmidt EV, Sonenberg N, London IM (1995): Eukaryotic translation initiation factor 4E regulates expression of cyclin D1 at transcriptional and posttranscriptional levels. J Biol Chem 270:21176–80.

Rosenwald IB, Rhoads DB, Callanan LD, Isselbacher KJ, Schmidt EV (1993): Increased expression of eukaryotic translation initiation factors eIF-4E and eIF-2 alpha in response to growth induction by c-*myc*. Proc Natl Acad Sci USA 90:6175–8.

Rossow PW (1979): Synthesis of labile, serum-dependent protein in early G_1 controls animal cell growth. Proc Natl Acad Sci USA 76:4446–50.

Rousseau D, Gingras A-C, Pause A, Sonenberg N (1996): The eIF4E-binding proteins 1 and 2 are negative regulators of cell growth. Oncogene 13:2415–20.

Rousseau D, Kaspar R, Rosenwald IB, Gehrke L, Sonenberg N (1996): Translation initiation of ornithine decarboxylase and nucleocytoplasmic transport of cyclin D1 mRNA are increased in cells overexpressing eukaryotic initiation factor 4E. Proc Natl Acad Sci USA 93:1065–70.

Ryazanov AG, Shestakova EA, Natapov PG (1988): Phosphorylation of elongation factor 2 by EF-2 kinase affects rate of translation. Nature 334:170–3.

Saito H, Hayday AC, Wiman K, Hayward WS, Tonegawa S (1983): Activation of the *c-myc* gene by translocation: A model for translational control. Proc Natl Acad Sci USA 80:7476–80.

Sarnow P (1989): Translation of glucose-regulated protein 78/immunoglobulin heavy-chain binding protein mRNA is increased in poliovirus-infected cells at a time when cap-dependent translation of cellular mRNAs is inhibited. Proc Natl Acad Sci USA 86:5795–9.

Scheper GC, Proud CG (2002): Does phosphorylation of the cap-binding protein eIF4E play a role in translation initiation? Eur J Biochem 269:5350–9.

Scheper GC, van Kollenburg B, Hu J, Luo I, Goss DJ, Proud CG (2002): Phosphorylation of eukaryotic initiation factor 4E markedly reduces its affinity for capped mRNA. J Biol Chem 277:3303–9.

Schmelzle T, Hall MN (2000): TOR, a central controller of cell growth. Cell 103: 253–62.

Schmidt EV (1999): The role of *c-myc* in cellular growth control. Oncogene 18: 2988–996.

Scott PH, Brunn GJ, Kohn RA, Lawrence JC (1998): Evidence of insulin-stimulated phosphorylation and activation of the mammalian target of rapamycin mediated by a protein kinase B signaling pathway. Proc Natl Acad Sci USA 95:7772–7.

Seki N, Takasu T, Mandai K, Nakata M, Saeki H, Heike Y, Takata I, Segawa Y, Hanafusa T, Eguchi K (2002): Expression of eukaryotic initiation factor 4E in atypical adenomatous hyperplasia and adenocarcinoma of the human peripheral lung. Clin Cancer Res 8:3046–53.

Shantz LM, Pegg AE (1994): Overproduction of ornithine decarboxylase caused by relief of translational repression is associated with neoplastic transformation. Cancer Res 54:2313–16.

Sheets MD, Fox CA, Hunt T, Vande Woude G, Wickens M (1994): The 3′-untranslated regions of c-mos and cyclin mRNAs stimulate translation by regulating cytoplasmic polyadenylation. Genes Dev 8:926–38.

Sherr CJ (1993): Mammalian G_1 cyclins. Cell 73:1059–65.

Shima H, Pende M, Chen Y, Fumagalli S, Thomas G, Kozma SC (1998): Disruption of the $p70^{s6k}$/ $p85^{s6k}$ gene reveals a small mouse phenotype and a new functional S6 kinase. EMBO J 17:6649–59.

Sieliwanowicz B (1987): The influence of poly(A)-binding proteins on translation of poly(A)+ RNA in a cell-free system from embryo axes of dry pea seeds. Biochim Biophys Acta 908:54–9.

Sitikov AS, Simonenko PN, Shestakova EA, Ryazanov AG, Ovchinnikov LP (1988): cAMP-dependent activation of protein synthesis correlates with dephosphorylation of elongation factor 2. FEBS Lett 228:327–31.

Smith JA, Poteet-Smith CE, Malarkey K, Sturgill TW (1999): Identification of an extracellular signal-regulated kinase docking site in ribosomal S6 kinase, a sequence critical for activation by ERK *in vivo*. J Biol Chem 274:2893–8.

Smith MR, Jaramillo M, Liu Y, Dever TE, Merrick WC, Kung H, Sonenberg N (1990): Translation initiation factors induce DNA synthesis and transform NIH 3T3 cells. New Biol 2:648–54.

Sorrells DL, Destin RB, Meschonat C, Rhoads R, De Benedetti A, Gao M, Williams BJ, Li BDL (1997): Detection of eIF4E gene amplification in breast cancer by competitive PCR. An Surg Oncol 5:232–7.

Stebbins-Boaz B, Cao Q, de Moor CH, Mendez R, Richter JD (1999): Maskin is a CPEB-associated factor that transiently interacts with eIF4E. Mol Cell 4: 1017–27.

Stebbins-Boaz B, Hake LE, Richter JD (1996): CPEB controls the cytoplasmic polyadenylation of cyclin, Cdk2 and c-mos mRNAs and is necessary for oocyte maturation in *Xenopus*. EMBO J 15:2582–92.

Stein I, Itin A, Einat P, Skaliter R, Grossman Z, Keshet E (1998): Translation of vascular endothelial growth factor mRNA by internal ribosome entry: Implications for translation under hypoxia. Mol Cell Biol 18:3112–9.

Stoneley M, Paulin FEM, LeQuesne JPC, Chapell SA, Willis AE (1998): C-myc 5′ untranslated region contains an internal ribosome entry segment. Oncogene 16:423–8.

Svitkin YV, Pause A, Haghighat A, Pyronnet S, Witherell G, Belsham GJ, Sonenberg N (2001): The requirement for eukaryotic initiation factor 4A (eIF4A) in translation is in direct proportion to the degree of mRNA 5′ secondary structure. RNA 7:382–94.

Takata M, Ogawaa W, Kitamura T, Hino Y, Kuroda S, Kotani K, Klip A, Gingras A-C, Sonenberg N, Kasuga M (1999): Requirement for Akt (protein kinase B) in insulin-induced activation of glycogen synthase and phosphorylation of 4E-BP1 (PHAS-1). J Biol Chem 274:20611–18.

Tang H, Hornstein E, Stolovich M, Levy G, Livingstone M, Templeton DJ, Avruch J, Meyuhas O (2001): Amino acid-induced translation of TOP mRNAs is fully dependent on phosphatidylinositol 3-kinase-mediated signaling, is partially inhibited by rapamycin, and is independent of S6K1 and rpS6 phosphorylation. Mol Cell Biol 21:8671–83.

Tarnowka MA, Baglioni C (1979): Regulation of protein synthesis in mitotic HeLa cells. J Cell Physiol 99:359–68.

Tarun SZ, Sachs AB (1995): A common function for mRNA 5′ and 3′ ends in translation initiation in yeast. Genes Dev 9:2997–3007.

Tarun SZ, Sachs AB (1996): Association of the yeast poly(A) tail binding protein with translation initiation factor eIF-4G. EMBO J 15:7168–77.

Terada N, Lucas JJ, Szepesi A, Franklin RA, Domenico J, Gelfand EW (1993): Rapamycin blocks cell cycle progression of activated T cells prior to events characteristic of the middle to late G_1 phase of the cycle. J Cell Physiol 154: 7–15.

Terada N, Patel HR, Takase K, Kohno K, Nairn AC (1994): Rapamycin selectively inhibits translation of mRNAs encoding elongation factors and ribosomal proteins. Proc Natl Acad Sci USA 91:11477–81.

Terada N, Takase K, Pabst P, Nairn AC, Gelfand EW (1995): Rapamycin inhibits ribosomal protein synthesis and induces G_1 prolongation in mitogen-activated T lymphocytes. J Immunol 155:3418–26.

Topisirovic I, Capili AD, Borden KLB (2002): Gamma interferon and cadmium treatments modulate eukaryotic initiation factor 4E-dependent mRNA transport of cyclin D1 in a PML-dependent manner. Mol Cell Biol 22:6183–98.

Tyers M, Tokiwa G, Futcher B (1993): Comparison of the Saccharomyces cerevisiae G_1 cyclins: Cln3 may be an upstream activator of Cln1, Cln2 and other cyclins. EMBO J 12:1955–68.

Vagner S, Gensac MC, Maret A, Bayard F, Amalric F, Prats H, Prats AC (1995): Alternative translation of human fibroblast growth factor 2 mRNA occurs by internal entry of ribosomes. Mol Cell Biol 15:35–44.

van Weeren PC, Bruyn MT, de Vries-Smits MM, van Lint J, Burgering BM (1998): Essential role for protein kinase B (PKB) in insulin-induced glycogen synthase kinase 3 inactivation. J Biol Chem 273:13150–6.

Varnum SM, Wormington WM (1990): Deadenylation of maternal mRNAs during Xenopus oocyte maturation does not require specific *cis*-sequences: A default mechanism for translational control. Genes Dev 4:2278–86.

von Manteuffel SR, Gingras A-C, Ming X-F, Sonenberg N, Thomas G (1996): 4E-BP1 phosphorylation is mediated by the FRAP-p70^{s6k} pathway and is independent of mitogen-activated protein kinase. Proc Natl Acad Sci USA 93:4076–80.

Waskiewicz AJ, Flynn A, Proud CG, Cooper JA (1997): Mitogen-activated protein kinases activate the serine/threonine kinases Mnk1 and Mnk2. EMBO J 16:1909–20.

Waskiewicz AJ, Johnson JC, Penn B, Mahalingam M, Kimball SR, Cooper JA (1999): Phosphorylation of the cap-binding protein eukaryotic translation initiation factor 4E by protein kinase Mnk1 in vivo. Mol Cell Biol 19:1871–80.

Watson JD, Oster SK, Shago M, Khosravi F, Penn LZ (2002): Identifying genes regulated in a Myc-dependent manner. J Biol Chem 277:36921–30.

Welch EM, Wang W, Peltz SW (2000): Translation termination: It's not the end of the story. In: N Sonenberg, JWB Hershey, MB Mathews (eds): Translational Control of Gene Expression. Cold Spring Harbor: Cold Spring Harbor Laboratory Press, pp 467–86.

Welsh GI, Miller CM, Loughlin AJ, Price NT, Proud CG (1998): Regulation of eukaryotic initiation factor eIF2B: glycogen synthase kinase-3 phosphorylates a conserved serine which undergoes dephosphorylation in response to insulin. FEBS Lett 421:125–30.

Weng Q-P, Kozlowski MT, Belham C, Zhang A, Comb MJ, Avruch J (1998): Regulation of the p70 S6 kinase by phosphorylation in vivo. J Biol Chem 273:16621–9.

West MJ, Stoneley M, Willis AE (1998): Translational induction of the *c-myc* oncogene via activation of the FRAP/TOR signaling pathway. Oncogene 17: 769–80.

White MW (1990): Specific regulation by endogenous polyamines of translational initiation of *S*-adenosylmethionine decarboxylase mRNA in Swiss 3T3 fibroblasts. Biochem J 268:657–60.

White MW, Kameji T, Pegg AE, Morris DR (1987): Increased efficiency of translation of ornithine decarboxylase mRNA in mitogen-activated lymphocytes. Eur J Biochem 170:87–92.

Williams DD, Pavitt GD, Proud CG (2001): Characterization of the initiation factor eIF2B and its regulation in *Drosophila melanogaster*. J Biol Chem 276: 3733–42.

Withers DJ, Seufferlein T, Mann D, Garcia B, Jones N, Rozengurt E (1997): Rapamycin dissociates $p70^{S6K}$ activiation from DNA synthesis stimulated by bombesin and insulin in Swiss 3T3 cells. J Biol Chem 272:2509–14.

Wittenberg C, Sugimoto K, Reed SI (1990): G_1-specific cyclins of S. cerevisiae: cell cycle periodicity, regulation by mating pheromone, and association with the $p34^{CDC28}$ protein kinase. Cell 62:225–37.

Woods YL, Cohen P, Becker W, Jakes R, Goedert M, Wang X, Proud CG (2001): The kinase DYRK phosphorylates protein-synthesis initiation factor eIF2Bε at Ser^{539} and the microtubule-associated protein tau at Thr^{212}: Potential role for DYRK as a glycogen synthase kinase 3-priming kinase. Biochem J 355: 609–15.

Wormington M (1993): Poly(A) and translation: developmental control. Curr Opin Cell Biol 5:950–4.

Xu G, Kwon G, Marshall CA, Lin T-A, Lawrence JC, Jr., McDaniel ML (1998): Branched-chain amino acids are essential in the regulation of PHAS-I and p70 S6 kinase by pancreatic β-cells. J Biol Chem 273:28178–84.

Yan R, Rhoads RE (1995): Human protein synthesis initiation factor eIF-4γ is encoded by a single gene (EIF4G) that maps to chromosome 3q27-qter. Genomics 26:394–8.

Ye X, Fong P, Iizuka N, Choate D, Cavener DR (1997): Ultrabihorax and Antennapedia 5′-untranslated regions promote developmentally regulated internal translation initiation. Mol Cell Biol 17:1714–21.

Yenush L, Fernandez R, Myers MG, Jr., Grammer TC, Sun XJ, Blenis J, Pierce JH, Schlessinger J, White MF (1996): The *Drosophila* insulin receptor activates multiple signaling pathways but requires insulin receptor substrate proteins for DNA synthesis. Mol Cell Biol 16:2509–17.

Yu K, Toral-Barza L, Discafani C, Zhang WG, Skotnicki J, Frost P, Gibbons JJ (2001): mTOR, a novel target in breast cancer: the effect of CCI-779, an mTOR inhibitor, in preclinical models of breast cancer. Endocr Relat Cancer 8:249–258.

Zong Q, Schummer M, Hood L, Morris DR (1999): Messenger RNA translation state: The second dimension of high-throughput expression screening. Proc Natl Acad Sci USA 96:10632–36.

PART III

CHAPTER 13

TELOMERE STRUCTURE AND FUNCTION PROVIDES INSIGHTS INTO THE GENERATION OF GENOMIC INSTABILITY AND CARCINOGENESIS

COLLEEN FORDYCE and THEA D. TLSTY
Department of Pathology and UCSF Comprehensive Cancer Center, University of California at San Francisco, San Francisco, CA 94143-0511

INTRODUCTION

For several years we have suspected that telomeres and telomerase play an important role in the formation of malignancy. Recent studies have begun to elucidate this role and provide tools to use this information in the possible prognosis of disease. Exciting recent results suggest that individuals with short telomeres are susceptible to a host of disease states that go far beyond carcinogenesis. Improved methods for monitoring the status of telomeres in individual cells as well as in cell populations are changing our view of these important cellular structures.

TELOMERE COMPOSITION AND STRUCTURE

Telomeres are protein nucleic acid complexes that protect and stabilize the ends of linear chromosomes (Blackburn, 2000; Blackburn et al., 1995; de Lange et al., 1990; Greider and Blackburn, 1996). As such telomeres are an essential safeguard against genomic instability for any organism without a circular genome. In humans the nucleic acid component of the telomere is comprised of two domains including the telomere repeat, which is located at the extreme end of the chromosome, and the telomere-associated sequences (TAS), which are located proximally to the telomere repeat. The TAS represents a dynamic structure with many

Cell Cycle and Growth Control: Biomolecular Regulation and Cancer, Second Edition, Edited by Gary S. Stein and Arthur B. Pardee
ISBN 0-471-25071-6

internal repeats of differing length and function. In contrast to the TAS, the extreme ends of the telomeres consists of a hexanucleotide 5′TTAGGG3′ that is tandemly repeated 1000 to 2000 times (Moyzis et al., 1988). Unlike the rest of the chromosome, the 3′ end of the telomere has a single-stranded G-rich overhang. This overhang is believed to be produced primarily through the action of cell cycle specific exonucleases (Dionne and Wellinger, 1996; Shay and Bacchetti, 1997; Wright et al., 1997; Huffman et al., 2000).

In 1999 the cooperative efforts of the Griffith and de Lange laboratories revealed a new structure at the ends of human and mice chromosomes (Griffith et al., 1999). Using psoralen crosslinking, in combination with electron microscopy, it was possible to visualize a unique loop structure. This structure is created when the 3′ G-rich overhang folds back over the telomere repeat and intercalates into a portion of the double-stranded telomere, creating a D-loop. The invasion of the G overhang creates a larger telomere specific loop, or a T-loop. The T-loop is stabilized by a number of telomere binding proteins. Telomere repeat factor 1 (TRF1) has been shown to bend DNA (Bianchi et al., 1997) and is believed to help fold the telomere back. Acting synergistically with TRF1, TRF2, another telomere binding protein, promotes the intercalation of the 3′ G overhang into a double-stranded portion of the telomere (Griffith et al., 1999). Previous experiments demonstrate that loss of TRF1 or TRF2 alters telomere length and increases the frequency of fused chromosomes and alters cell cycle progression, respectively, behaviors which are consistent with their role in the T-loop model (Karlseder et al., 1999; van Steensel and de Lange, 1997; van Steensel et al., 1998; Broccoli et al., 1997). There are a number of other telomere binding proteins. These include, but are not limited to TIN2, tankyrase and telomerase that act either indirectly, through other telomere binding proteins, or directly to modify telomere length (Smith et al., 1998; Kim et al., 1999; Bailey et al., 1999).

In the 1960s a group of papers by Hayflick and colleagues demonstrated that cells in culture have a finite replicative capacity and therefore, postulated that cells have an intrinsic mechanism for recording cellular age (Hayflick, 1974, 2000). Olinkov and Watson independently proposed that the loss of telomeric DNA might be one mechanism by which cells record their age since some telomeric sequences are lost each time a cell divides (Olovnikov, 1973; Blackburn et al., 1995). DNA polymerase cannot replicate the ends of the linear DNA molecule, since lagging strand synthesis proceeds as a series of discrete steps, producing Okazaki fragments, each requiring an RNA primer. At the 3′ends of the chromosome there is no additional sequence to which the primer can anneal, thus DNA polymerase cannot fill in the gap between the final Okazaki fragment and the end of the chromosome. This difficulty is called the end replication problem. Without some mechanism to counteract the end replication problem, truncated telomeres are inherited by daughter cells, and the process repeats itself in the next generation, resulting in the gradual degradation of telomere repeats. The supposition that telomere loss is involved in cellular aging is supported by a number of investigations. These include the observations that reduced telomere length is associated with increasing age, predicts decreased replicative capacity of cells in culture, and is associated with decreased cell division in vivo and in vitro (Hayflick 1965, 1968, 1974, 1989, 2000; Daniel et al., 1968).

MALFUNCTIONING TELOMERES FACILITATE GENOMIC INSTABILITY

Seminal experiments by Barbara McClintock demonstrated that broken chromosomes are highly recombinogenic, fusing together to produce dicentric chromosomes. Such chromosomes are subject to additional recombination events once they have progressed through a round of mitosis, where they are pulled to opposite spindle poles and are literally torn at random sites. The "sticky" ends of broken chromosomes, which initiate the formation of dicentric chromosomes, also arise from loss of telomere function (de Lange, 1990; Saltman, 1993). These telomere-associated dicentric chromosomes are subject to the same breakage-bridge-fusion cycle described by McClintock (Lo et al., 2002).

Indeed, it has been shown that telomere repeats are absent or significantly reduced at the junctions of dicentric chromosomes (Wan et al., 1999), and that telomere dysfunction triggers extensive types of genomic instability including: DNA fragmentation, and regional amplification and deletion in vivo (Gisselsson, 2001; O'Hagan, 2002). In 1992 Counter and coworkers demonstrated that the genomic instability associated with telomere loss could be arrested by the introduction of telomerase, the enzyme that lengthens telomeres (Counter et al., 1992). Additional experiments utilizing a mouse model deficient in telomerase activity have further validated the relationship between telomere malfunction and genomic instability. As the telomeres in these knockout mice become progressively shorter, they acquire a striking amount of genomic instability and a predisposition to cancer (Blasco et al., 1997; Lee et al., 1998; Hande et al., 1999; Artandi, 2000). Once immortalized by exogenous telomerase, cells from these animals are tumorigenic (Blasco et al., 1997). Interestingly it has recently been shown that the shortened telomeres of the telomerase knockout mice can bind to DNA breaks induced by ionizing radiation thereby sterically inhibiting the repair of DNA damage (Latre et al., 2003). Collectively these data demonstrate that malfunctioning telomeres give rise to complex types of genomic instability both in vivo and in vitro.

In normal cells this telomere malfunction is recognized as a type of DNA damage and induces cell arrest or apoptosis. It is likely that p53 plays an essential role in sensing and responding to a malfunctioning telomere. For example, Saretzki and colleagues observed that treating cells with a telomere-like oligonucleotide induced a p53-dependent cell cycle arrest at Go_0. However, treatment with an oligonucleotide with identical composition, but scrambled sequence did not cause arrest (Saretzki et al., 1999). Similarly introduction of a dominant-negative TRF2 into a number of different cell lines was able to initiate p53 dependent apoptosis (Karlseder et al., 1999).

Investigations in the telomerase knockout mouse have further strengthened the relationship between p53, telomeres and cell cycle control. In male germ-line cells and lymphocytes from telomerase knockout mice, telomere loss elicits a strong apoptotic response. In mouse embryonic fibroblasts (MEFs) derived from the telomerase knockout mice, telomere loss causes a potent cell cycle block that occurs in concert with activation of p53 and p21. Loss of p53 rescues the cell cycle arrest in MEFs (Greenberg et al., 1999; Chin et al., 1999). Taken together, these experiments demonstrate that p53 forms part of a system that detects the loss of telomeres and the structural consequences of telomere erosion, subsequently signaling growth arrest or apoptosis.

It is important to note, however, that the exact component of the telomere required to initiate a DNA damage response is not clear. Titration of endogenous TRF2 by the dominant negative protein or telomere-like oligonucleotides may destabilize the T-loop, thereby exposing telomere DNA and initiating a DNA damage response. Alternatively, changes in the localization of key telomere-binding proteins may promote protein-protein interactions that ultimately result in the initiation of a DNA damage signal. Interestingly tumor-progenitor cells are able to overcome the DNA damage signals induced by telomere malfunction. In fact it seems likely that loss of p53 attenuates the severe effects of telomere malfunction in its earliest stages. However, as cells continue to divide past this growth barrier, sustained telomere loss and p53 malfunction coupled to the eventual up-regulation of telomerase can facilitate cellular transformation (Greenberg et al., 1999; Chin et al., 1999). Therefore telomere shortening may in some instances be a protective mechanism against cancer development and in other instances, when continued telomere erosion supplies the selective pressure needed to induce telomerase up-regulation, facilitate carcinogenesis.

TELOMERE MALFUNCTION: AN EARLY EVENT IN CARCINOGENESIS?

The relationship between extensive genomic instability and cancer is well documented. Given the observation that telomere malfunction results in the same complex types of genomic instability observed in cancer, it is logical to hypothesize that telomere malfunction may facilitate carcinogenesis. Since genomic instability is likely to be one of the early events in cancer development and progression, this hypothesis makes the prediction that telomere loss should likewise be an early event.

Recent studies have highlighted the role that telomeres may play in the earliest events in cancer formation. Normal human mammary epithelial cells in culture (nHMEC) grow for 15 to 20 population doublings before reaching a p16-dependent growth plateau. Several years ago, studies described a rare subpopulation of variant HMEC (vHMEC, also called postselection cells, M0 cells or postsenescent cells) that were able to escape from the proliferative barrier by silencing p16. These vHMEC grow for another 20 to 50 population doublings before reaching a second growth plateau termed agonescence (Romanov, 2001). As vHMEC proliferate, their telomeres become progressively shorter, at a rate that is comparable to that measured in isogenic fibroblasts. As the telomere sequences approach a critically short length, they acquire the types and amounts of genomic instability commonly seen in early events in cancer. Given these interesting properties of telomeric dysfunction and genomic instability, agonescent cells were studied for the identification of unique markers and the possibility that they may exist in vivo. Cells with the coincident expression of markers identified in vHMEC have been found in histologically normal areas of mammary tissue in reduction mammoplasties from women without signs of occult breast disease (Holst et al., 2003). These observations suggest that cells with the potential to acquire telomeric dysfunction exist in foci in normal breast tissue and identify these foci as candidates for precursor lesions.

These data are consistent with a number of recent investigations looking at changes in telomere length in cancer predisposing or precursor lesions. For example, it has been shown that chromosomal instability in ulcerative colitis is related to decreasing telomere length. Moreover patients whose telomeres were most reduced, were more likely to have dysplasia or cancer (O'Sullivan et al., 2002). This study by the Rabinovitch group is particularly provocative since it demonstrates a direct correlation with genomic instability, telomere loss and progression towards colon cancer. Additionally, it has been shown that telomere loss is observed in prostate tumorigenesis (Meker et al., 2002) and is a possible predictor for the development of hepatocellular carcinoma (Dsokawa et al., 1999). Similarly, a recent study by Wiemann an coworkers has suggested that telomere shortening is also a characteristic of liver cirrhosis, a disease which often predisposes patients to hepatocellular cancer (Wiemann et al., 2002).

TELOMERE LENGTH AND PROGNOSIS IN CANCER

It is known that the up-regulation of telomerase is a common and often essential step in cellular immortalization. However, the telomeres in most tumors appear to be shorter than those in surrounding tissue, implying that the up-regulation of telomerase is not sufficient to restore telomeres in tumor cells to their original length but rather to maintain them. Therefore the timing of telomerase up-regulation may play a critical role in tumorigenesis. If telomerase is up-regulated early, prior to extensive telomeric DNA loss, cells emerge with some genomic instability. If, however, telomerase is up-regulated later, after extensive telomere loss, the emerging population has a higher degree of genomic instability and presumably a greater probability of acquiring the necessary genetic programs to induce angiogenesis, invasion and metastasis.

There have been several investigations examining the relationship between telomere length, or its proxies, and outcome in cancer. Hiyama and colleagues conducted one of the earliest of these in 1992. They demonstrated that neuroblastomas with shorter telomeres were more likely to have high tumor stage, high fraction of cells in S phase and poor outcomes (Hiyama et al., 1992). Other studies in solid tumors, including breast and prostate cancer, have also shown associations between decreased telomeres and increased metastasis and decreased survival, respectively (Donaldson et al., 1999; Griffith et al., 1999). There have also been a number of investigations in hematological disorders. Where it has been shown that telomere loss is associated with an increased duration of aplastic anemia with persistent cytopenia further suggesting that telomere loss may be relevant to the pathology of myelodysplastic syndromes (Ball et al., 1998). Additional investigations have further elucidated the relationship between telomere malfunction and hematological disorders, in these studies, reduced telomeric DNA was associated with decreased survival in B-cell lymphocytic leukemia, chronic myeloid leukemia and chronic lymphocytic leukemia (Ohyashiki et al., 1994, 1999, 2000; Brummendorf et al., 2000; Hultdin et al., 2003; Wu et al., 2003).

In contrast to these results, there have been some investigations that suggest that telomere lengthening is associated with poor prognosis in non–small cell lung cancer and in head and neck cancer (Hirashima et al., 2000; Patel et al., 2002). In 2002 Ishibe and coworkers investigated the relationship between telomeres and prognosis in familial chronic lymphocytic leukemia. In opposition to the much larger study by Hultin and coworkers (2003), Ishibe and colleagues did not find a relationship between telomere length and survival (Ishibe et al., 2002). Many of these investigations have limitations including small sample size, highly selected patient populations and limited follow-up data. However, collectively they imply that telomere loss may have potential as a prognostic or predictive marker. Additional studies with larger patient populations and more stringent case selection has the potential of offering new and powerful insights into cancer development and progression as well as more accurate assessment of patient outcomes.

NEW METHODS FOR TELOMERE QUANTIFICATION

Classically telomere length is measured using a modification of southern blotting. In this approach genomic DNA is isolated from tissue and is treated with restriction endonucleases. These endonucleases cleave the genomic DNA internally of the telomere repeat to produce a telomere restriction fragment. The digested DNA is then separated by size using gel electophoresis and transferred to a membrane. Telomere restriction fragments are visualized following hybridization with a labeled telomere-specific probe. This methodology has a number of limitations. Since restriction sites for the endonucleases are located within the TAS, these measurements include more than just the telomere repeat itself. Additionally mean telomere restriction fragment length can vary depending on which restriction endonucleases are used. The necessity for large (μg) quantities of unfragmented DNA makes analysis of telomere restriction fragment length problematic on anything other than fresh, frozen or freshly fixed paraffin embedded tissues (de Lange et al., 1990; de Lange, 1994; Slagboom et al., 1994). Finally this technique measures telomere length in the population as a whole and is insensitive to heterogeneity within that population.

There are now a number of new methodologies that can be used to measure telomere content, a proxy for length. The first of these is based on slot blotting. In this application genomic DNA is purified from the tissue of interest, denatured, and fixed directly to a membrane using a vacuum. Much like telomere restriction fragment length analysis, the hybridization of labeled telomere-specific probe is utilized for quantification. Since the DNA is localized to a single dot or slot on the membrane, quantification is straightforward with data presented as a fraction or percentage of an internal control, typically DNA purified from human placenta. Unlike telomere restriction fragment length analysis, far less DNA is required (ηg quantities) and fragmentation does not appear to affect the assay. However, this method is highly dependent on the concentration of genomic DNA loaded into each well, since errors in DNA quantity would necessarily result in artificially high or low telomere content values. This limitation can be overcome by utilizing a second hybridization step where centromere DNA is also probed. In this approach telomere content is expressed as a ratio of telomere signal to centromere signal from the same sample. Typically tumors are heterogeneous, containing a mixture of tumor and normal epithelial cells as well as stromal and immune cells. Thus microdissection of tumor tissues is necessary if measurements of a single-cell type are desired (Bryant et al., 1997; Norwood and Dimitrov, 1998; Fordyce et al., 2002).

To gain further insight into the role of telomeres in disease, it is important to be able to analyze cells on an individual basis. Recent studies have provided for the analysis of single cells using two types of fluorescent in situ hybridization (FISH). In nonsolid tumors, telomere contents can be determined using flow cytometry. Termed flow-FISH, cells are heat denatured and then allowed to hybridize to a telomere-specific PNA probe labeled with a fluorchrome. Following hybridization, the excess probe is removed with several washes and telomere content is analyzed using a flow cytometer. In contrast to previous methods, this approach can be used to measure telomeres in specific cell subtypes and in small numbers of cells (Hultdin et al., 1998; Baerlocher et al., 2002). Recently this fluorescence-based approach has been adapted for use in solid tumors. Preparation of the samples is accomplished is a similar fashion as for flow-FISH. Freshly cut frozen or paraffin-embedded tissues are denatured and allowed to hybridize to a PNA telomere-specific probe. However, in contrast to flow-FISH, data are captured using a fluorescent microscope. Tissue sections are counterstained with DAPI to identify nuclei. In-house and commercial algorithms are then utilized for the quantification of these images. Typically such algorithms rely on the production of a nuclei mask and a series of threshold values to differentiate between background fluorescence and probe-specific signal (Meeker et al., 2002a, b; O'Sullivan et al., 2002). Since tissue architecture is preserved, it is possible to examine telomeres in specific cell types and in a small number of cells.

Telomeres can also be measured using quantitative PCR. In this approach telomere-specific primers were designed to specifically amplify telomeric repeats without generating primer dimer-derived products. Although this assay quantitates the telomere, the product of the assay is 76 base pairs, a length determined by the size of the telomere-specific primers. The number of 76 base pair products represents the number of possible sites available for primer binding and therefore is proportional to the total telomere length. These primers are used to compare the relative telomere length to a single copy gene from the same DNA sample. Quantification of the fluorescent signal is performed using the standard software supplied with quantitative PCR machines. Telomere content is expressed as a ratio of the number of PCR cycles required for the telomere-specific fluorescence to reach a threshold value to the number of PCR cycles required for the single-copy gene fluorescence to reach a threshold value (Cawthon, 2002; Cawthon et al., 2003). Perhaps more than any other, this method of telomere analysis has the greatest potential to be utilized in high-throughput investigations of large numbers of samples. However, unlike the other approaches, this method has not yet been externally validated and, similarly to southern blotting, the integrity of the DNA is essential. Hopefully the technical advances of these alternative methods of telomere measurement will continue to spur additional investigations into the relationship between telomere content and outcome in cancer. Studies that measure telomere length and structure in situ (as described below) will be particularly informative.

TELOMERE MALFUNCTION IN HISTOLOGICALLY NORMAL CELLS

Using the same quantitative PCR-based method for analysis of telomere content described previously, Cawthon and colleagues have recently reported that telomere loss in white blood cells from individuals 60 years and older was associated with increased mortality. In this study population atherosclerotic and infectious disease were the primary causes of mortality. Their observation was striking, since the white blood cells utilized for the investigation presumably were normal and variation in telomere content therefore reflects individual polymorphisms (Cawthon et al., 2003). This finding is consistent with the recent observation that a mutation in the RNA component of telomerase results in dyskeratosis congenita in which patients have characteristics of premature aging (Dokal, 2000; Vulliamy et al., 2001). It has also been shown that fibroblasts from progeria donors, who have accelerated rates of aging, had shorter telomeres than controls (Allsopp et al., 1992). Similar conclusions have been shown in cells from Downs Syndrome patients (Vaziri et al., 1993). Collectively these studies imply that shorter telomeres can predispose individuals to multiple age-related diseases, including cancer.

As previously stated, telomere malfunction can cause genomic instability, including loss of heterozygousity (LOH) (Chang et al., 2001; Gisselsson et al., 2001; Lo et al., 2002; O'Sullivan et al., 2002). Moreover telomere loss is likely an early event in carcinogenesis. It has been assumed that the field containing truncated telomeres was limited to tumor or tumor progenitor cells. However, the recent report by Cawthon and coworkers implies that reduced telomeres may extend beyond the tumor. A number of investigations have examined genomic instability in tumor-adjacent histologically normal cells. For instance, several investigations have shown that epithelial cells adjacent to tumors have multiple sites of LOH. In one instance, the sites of LOH between breast tumors and normal epithelial cells were similar, while in others (Deng et al., 1996), they were distinct (Lakhani et al., 1999; Forsti et al., 2001; Larson et al., 2002). Analogous studies have examined the relationship between genomic instability in ductal carcinoma in situ, breast tumors, and adjacent stromal cells. These studies have shown that stromal cells adjacent to epithelial lesions have LOH (Moinfar et al., 2000; Kurose et al., 2001). Euhus and coworkers investigated LOH in breast cells obtained with fine needle aspirate from 30 women. Approximately 50% of these women had some proliferative cytology, the degree of which was associated with the degree of LOH observed in the same tissue. When the patients' risk for breast cancer was accessed with the Gail model, it was shown that patients with more LOH were at higher risk (Euhus et al., 2002). Li and coworkers more fully examine the relationship between genomic instability and prognosis. In this study LOH at 13p11-26 was examined in normal epithelial cells adjacent to breast tumors. LOH at this locus was associated with a four- to fivefold increased risk of breast cancer recurrence when compared to patients without LOH at this site (Li et al., 2002).

Based on these data it is tempting to speculate that telomere malfunction occurs in histologically normal tissue where it is manifested as genomic instability. Thus individuals may harbor reservoirs of genetically unstable cells, which may have the ability to become tumors. The timing of such telomere shortening would influence the size of the affected region. An event that induces telomere loss early in the development would be propagated throughout a single organ or perhaps the body, and produce a wide field. In contrast, telomere loss mediated by some stochastic event in an adult would produce a much smaller field, and therefore is less likely to result in disease. In 1994 Odagiri and colleagues showed that reduced telomeres were associated with high-grade breast tumors. Strikingly they also report telomere loss in tumor-adjacent histologically normal tissue. Unfortunately, there are no data linking this reduction in telomeres in histologically normal tissue to either the tumor's characteristics or patient outcome (Odagiri et al., 1994). Recent advances in telomere content quantification make it possible to address these and other questions in a robust and critical fashion, and suggest that the relationship between telomere function and cancer has much to teach us.

REFERENCES

Allsopp RC, Vaziri H, Patterson C, Goldstein S, Younglai EV, Futcher AB, Greider CW, Harley CB (1992): Telomere length predicts replicative capacity of human fibroblasts. Proc Natl Acad Sci USA 89:10114–18.

Artandi SE, Chang S, Lee SL, Alson S, Gottlieb GJ, Chin L, DePinho RA (2000): Telomere dysfunction promotes non-reciprocal translocations and epithelial cancers in mice. Nature 406:641–5.

Baerlocher GM, Mak J, Tien T, Lansdorp PM (2002): Telomere length measurement by fluorescence in situ hybridization and flow cytometry: Tips and pitfalls. Cytometry 47:89–99.

Bailey SM, Meyne J, Chen DJ, Kurimasa A, Li GC, Lehnert BE, Goodwin EH (1999): DNA double-strand break repair proteins are required to cap the ends of mammalian chromosomes. Proc Natl Acad Sci USA 96:14899–904.

Ball SE, Gibson FM, Rizzo S, Tooze JA, Marsh JC, Gordon-Smith EC (1998): Progressive telomere shortening in aplastic anemia. Blood 91:3582–92.

Bianchi A, Smith S, Chong L, Elias P, de Lange T (1997): TRF1 is a dimer and bends telomeric DNA. EMBO J 16:1785–94.

Blackburn EH (2000): Telomere states and cell fates. Nature 408:53–6.

Blackburn EH, Greider CW (1995): Telomeres, Vol. 1, Plainveiw, NY: Cold Spring Harbor Laboratory Press, p 396.

Blasco MA, Lee HW, Hande MP, Samper E, Lansdorp PM, DePinho RA, Greider CW (1997): Telomere shortening and tumor formation by mouse cells lacking telomerase RNA. (See Comments.) Cell 91:25–34.

Broccoli D, Smogorzewska A, Chong L, de Lange T (1997): Human telomeres contain two distinct Myb-related proteins, TRF1 and TRF2. Nat Genet 17: 231–5.

Brummendorf TH, Holyoake TL, Rufer N, Barnett MJ, Schulzer M, Eaves CJ, Eaves AC, Lansdorp PM (2000): Prognostic implications of differences in telomere length between normal and malignant cells from patients with chronic myeloid leukemia measured by flow cytometry. Blood 95:1883–90.

Bryant JE, Hutchings KG, Moyzis RK, Griffith JK (1997): Measurement of telomeric DNA content in human tissues. Biotechniques 23:476–8.

Cawthon RM (2002): Telomere measurement by quantitative PCR. Nucl Acids Res 30:e47.

Cawthon RM, Smith KR, O'Brien E, Sivatchenko A, Kerber RA (2003): Association between telomere length in blood and mortality in people aged 60 years or older. Lancet 361:393–5.

Chang S, Khoo C, DePinho RA (2001): Modeling chromosomal instability and epithelial carcinogenesis in the telomerase-deficient mouse. Semin Cancer Biol 11:227–39.

Chin L, Artandi SE, Shen Q, Tam A, Lee SL, Gottlieb GJ, Greider CW, DePinho RA (1999): p53 deficiency rescues the adverse effects of telomere loss and cooperates with telomere dysfunction to accelerate carcinogenesis. Cell 97: 527–38.

Counter CM, Avilion AA, LeFeuvre CE, Stewart NG, Greider CW, Harley CB, Bacchetti S (1992): Telomere shortening associated with chromosome instability is arrested in immortal cells which express telomerase activity. EMBO J 11:1921–9.

Daniel CW, De Ome KB, Young JT, Blair PB, Faulkin LJ (1968): The in vivo life span of normal and preneoplastic mouse mammary glands: A serial transplantation study. Proc Natl Acad Sci USA 61:53–60.

de Lange T (1994): Activation of telomerase in a human tumor. Proc Natl Acad Sci USA 91:2882–5.

de Lange T, Shiue L, Myers RM, Cox DR, Naylor SL, Killery AM, Varmus HE (1990): Structure and variability of human chromosome ends. Mol Cell Biol 10:518–27.

Deng G, Lu Y, Zlotnikov G, Thor AD, Smith HS (1996): Loss of heterozygosity in normal tissue adjacent to breast carcinomas. Science 274:2057–9.

Dionne I, Wellinger RJ (1996): Cell cycle-regulated generation of single-stranded G-rich DNA in the absence of telomerase. Proc Natl Acad Sci USA 93: 13902–7.

Dokal I (2000): Dyskeratosis Congenit in all its forms. Br J Haematol 110:768–79.

Donaldson L, Fordyce C, Gilliland F, Smith A, Feddersen R, Joste N, Moyzis R, Griffith J (1999): Association between outcome and telomere DNA content in prostate cancer. J Urol 162:1788–92.

Euhus DM, Cler L, Shivapurkar N, Milchgrub S, Peters GN, Leitch AM, Heda S, Gazdar AF (2002): Loss of heterozygosity in benign breast epithelium in relation ot breast cancer risk. J Natl Cancer Inst 94:858–60.

Fordyce CA, Heaphy CM, Griffith JK (2002): Chemiluminescent Measurement of Telomere DNA Content in Biopsies. Biotechniques 33:144–8.

Forsti A, Louhelainen J, Soderberg M, Wijkstrom H, Hemminiki K (2001): Loss of heterozygosity in tumor-adjacent normal tissue of breast and bladder cancer. Eur J Cancer 37:1372–80.

Gisselsson D, Jonson T, Petersen A, Strombeck B, Dal Cin P, Hoglund M, Mitelman F, Martens F, Mandahl N (2001): Telomere dysfunction triggers extensive DNA fragmentation and evolution of complex chromsome abnormalities in human malignant tumors. Proc Natl Acad Sci USA 98:12683–8.

Greenberg RA, Chin L, Femino A, Lee KH, Gottlieb GJ, Singer RH, Greider CW, DePinho RA (1999): Short dysfunctional telomeres impair tumorigenesis in the INK4a(delta2/3) cancer-prone mouse. Cell 97:515–25.

Greider CW, Blackburn EH (1996): Telomeres, telomerase and cancer. Scientific American 274:92–7.

Griffith JD, Comeau L, Rosenfield S, Stansel RM, Bianchi A, Moss H, de Lange T (1999): Mammalian telomeres end in a large duplex loop. Cell 97:503–14.

Griffith JK, Bryant JE, Fordyce CA, Gilliland FD, Joste NE, Moyzis RK (1999): Reduced telomere DNA content is correlated with genomic instability and metastasis in invasive human breast carcinoma. Breast Cancer Res Treat 54: 59–64.

Hande MP, Samper E, Lansdorp P, Blasco MA (1999): Telomere length dynamics and chromosomal instability in cells derived from telomerase null mice. J Cell Biol 144:589–601.

Harley CB, Futcher AB, Greider CW (1990): Telomeres shorten during ageing of human fibroblasts. Nature 345:458–60.

Hayflick L (1965): Tissue cultures and mycoplasmas. Tex Rep Biol Med Suppl 23:285ff.

Hayflick L (1974): The longevity of cultured human cells. J Am Geriatr Soc 22: 1–12.

Hayflick L (1989): Antecedents of cell aging research. Expr Gerontol 24:355–65.

Hayflick L (2000): The illusion of cell immortality. Br J Cancer 83:841–6.

Hirashima T, Komiya T, Nitta T, Takada Y, Kobayashi M, Masuda N, Matui K, Takada M, Kikui M, Yasumitu T, Ohno A, Nakagawa K, Fukuoka M, Kawase I (2000): Prognostic significance of telomeric repeat length alterations in pathological stage I-IIIA non–small cell lung cancer. Anticancer Res 20: 2181–7.

Hiyama E, Hiyama K, Yokoyama T, Ichikawa T, Matsuura Y (1992): Length of telomeric repeats in neuroblastoma: Correlation with prognosis and other biological characteristics. Jap J Cancer Res 83:159–64.

Holst CR, Nuovo GJ, Esteller M, Chew K, Baylin SB, Herman JG, Tlsty TD (2003): Methylation of p16(INK4a) promoters occurs in vivo in histologically normal human mammary epithelia. Cancer Res 63:1596–601.

Huffman KE, Levene SD, Tesmer VM, Shay JW, Wright WE (2000): Telomere shortening is proportional to the size of the G-rich telomeric 3′-overhang. J Biol Chem 275:19719–22.

Hultdin M, Gronlund E, Norrback K, Eriksson-Lindstrom E, Just T, Roos G (1998): Telomere analysis by fluorescence in situ hybridization and flow cytometry. Nucl Acids Res 26:3651–6.

Hultdin M, Rosenquist R, Thunberg U, Tobin G, Norrback KF, Johnson A, Sundstrom C, Roos G (2003): Association between telomere length and V(H) gene mutation status in chronic lymphocytic leukaemia: Clinical and biological implications. Br J Cancer 88:593–8.

Ishibe N, Prieto D, Hosack DA, Lempicki RA, Goldin LR, Raffeld M, Marti GE, Caporaso NE (2002): Telomere length and heavy-chain mutation status in familial chronic lymphocytic leukemia. Leuk Res 26:791–4.

Isokawa O, Suda T, Aoyagi Y, Kawai H, Yokota T, Takahashi T, Tsukada K, Shimizu T, Mori S, Abe Y, Suzuki Y, Nomoto M, Mita Y, Yanagi M, Igarashi H, Asakura H (1999): Reduction of telomeric repeats as a possible predictor for development of hepatocellular carcinoma: Convenient evaluation by slot-blot analysis. Hepatology 30:408–12.

Karlseder J, Broccoli D, Dai Y, Hardy S, de Lange T (1999): p53- and ATM-dependent apoptosis induced by telomeres lacking TRF2. Science 283:1321–5.

Kim SH, Kaminker P, Campisi J (1999): TIN2, a new regulator of telomere length in human cells (See Comments.) Nat Genet 23:405–12.

Kurose K, Hoshaw-Woodard S, Adeyinka A, Lemeshow S, Watson PH, Eng C (2001): Genetic model of multi-step breast carcinogenesis involving the epithelium and stroma: Clues to tumour-microenvironment interactions. Hum Mol Genet 10:1907–13.

Lakhani SR, Chagger R, Davies S, Jones C, Collins N, Odel C, Stratton MR, O'Hare MJ (1999): Genetic alterations in "normal" luminal and myoepithelial cells of the breast. J Pathol 189:496–503.

Larson PS, de Las Morenas A, Bennett SR, Cupples A, Rosenberg CL (2002): Loss of heterozygosity or allele imbalance in histologically normal breast epithelium is distinct from loss of heterozygosity or allele imbalance in co-existing carcinomas. Am J Pathol 161:283–90.

Latre L, Tusell L, Martin M, Miro R, Egozcue J, Blasco MA, Genesca A (2003): Shortened telomeres join to DNA breaks interfering with their correct repair. Expr Cell Res 287:282–8.

Lee HW, Blasco MA, Gottlieb GJ, Horner JW, Greider CW, DePinho RA (1998): Essential role of mouse telomerase in highly proliferative organs. Nature 392: 569–74.

Li Z, Moore DH, Meng ZH, Ljung BM, Gray JW, Dairkee SH (2002): Increased risk of local recurrence is associated with allelic loss in normal lobules of breast cancer patients. Cancer Res 62:1000–3.

Lo AW, Sabatier L, Fouladi B, Pottier G, Ricoul M, Murnane JP (2002): DNA amplification by breakage/fusion/bridge cycles initiated by spontaeous telomere loss in a human cancer cell line. Neoplasia 4:531–8.

Meeker AK, Gage WR, Hicks JL, Simon I, Coffman JR, Platz EA, March GE, De Marzo AM (2002a): Telomere length assessment in human archival tissues: combined telomere fluorescence in situ hybridization and immunostaining. Am J Pathol 160:1259–68.

Meeker AK, Hicks JL, Platz EA, March GE, Bennett CJ, Delannoy MJ, De Marzo AM (2002b): Telomere shortening is an early somatic DNA alteration in human prostate tumorigenesis. Cancer Res 62:6405–9.

Moinfar F, Man YG, Arnould L, Bratthauer GL, Ratschek M, Tavassoli FA (2000): Concurrent and independent genetic alterations in the stromal and epithelial cells of mammary carcinoma: Implications for tumorigenesis. Cancer Res 60:2562–6.

Moyzis RK, Buckingham JM, Cram LS, Dani M, Deaven LL, Jones MD, Meyne J, Ratliff RL, Wu JR (1988): A highly conserved repetitive DNA sequence, (TTAGGG)n, present at the telomeres of human chromosomes. Proc Natl Acad Sci USA 85:6622–6.

Norwood D, Dimitrov DS (1998): Sensitive method for measuring telomere lengths by quantifying telomeric DNA content of whole cells. Biotechniques 25:1040–5.

O'Hagan R, Chang S, Maser R, Mohan R, Artandi S, Chin L, DePinho R (2002): Telomere dysfunction provokes regional amplification and deletion in cancer genomes. Cancer Cell 2:149.

O'Sullivan JN, Bronner MP, Brentnall TA, Finley JC, Shen WT, Emerson S, Emond MJ, Gollahon KA, Moskovitz AH, Crispin DA, Potter JD, Rabinovitch PS (2002): Chromosomal instability in ulcerative colitis is related to telomere shortening. Nat Genet 32:280–4.

Odagiri E, Kanada N, Jibiki K, Demura R, Aikawa E, Demura H (1994): Reduction of telomeric length and c-erbB-2 gene amplification in human breast cancer, fibroadenoma, and gynecomastia: Relationship to histologic grade and clinical parameters. Cancer 73:2978–84.

Ohyashiki JH, Iwama H, Yahata N, Ando K, Hayashi S, Shay JW, Ohyashiki K (1999): Telomere stability is frequently impaired in high-risk groups of patients with myelodysplastic syndromes. Clin Cancer Res 5:1155–60.

Ohyashiki JH, Ohyashiki K, Fujimura T, Kawakubo K, Shimamoto T, Iwabuchi A, Toyama K (1994): Telomere shortening associated with disease evolution patterns in myelodysplastic syndromes. Cancer Res 54:3557–60.

Ohyashiki K, Iwama H, Tauchi T, Shimamoto T, Hayashi S, Ando K, Kawakubo K, Ohyashiki JH (2000): Telomere dynamics and genetic instability in disease progression of chronic myeloid leukemia. Leuk Lymphoma 40:49–56.

Olovnikov AM (1973): A theory of marginotomy: The incomplete copying of template margin in enzymic synthesis of polynucleotides and biological significance of the phenomenon. J Theor Biol 41:181–90.

Patel MM, Parekh LJ, Jha FP, Sainger RN, Patel JB, Patel DD, Shah PM, Patel PS (2002): Clinical usefulness of telomerase activation and telomere length in head and neck cancer. Head Neck 24:1060–7.

Romanov SR, Kozakiewicz BK, Holst CR, Stampfer MR, Haupt LM, Tlsty TD (2001): Normal human mammary epithelial cells spontaneously escape senescence and acquire genomic changes. Nature 409:633–7.

Saltman D, Morgan R, Cleary ML, de Lange T (1993): Telomeric structure in cells with chromosome end associations. Chromosoma 102:121–8.

Saretzki G, Sitte N, Merkel U, Wurm RE, von Zglinicki T (1999): Telomere shortening triggers a p53-dependent cell cycle arrest via accumulation of G-rich single stranded DNA fragments. Oncogene 18:5148–58.

Shay JW, Bacchetti S (1997): A survey of telomerase activity in human cancer. Eur J Cancer 33:787–91.

Slagboom PE, Droog S, Boomsma DI (1994): Genetic determination of telomere size in humans: a twin study of three age groups. Am J Hum Genet 55:876–82.

Smith S, Giriat I, Schmitt A, de Lange T (1998): Tankyrase, a poly(ADP-ribose) polymerase at human telomeres (See Comments.) Science 282:1484–7.

van Steensel B, de Lange T (1997): Control of telomere length by the human telomeric protein TRF1. Nature 385:740–3.

van Steensel B, Smogorzewska A, de Lange T (1998): TRF2 protects human telomeres from end-to-end fusions. Cell 92:401–13.

Vaziri H, Schachter F, Uchida I, Wei L, Zhu X, Effros R, Cohen D, Harley CB (1993): Loss of telomeric DNA during aging of normal and trisomy 21 human lymphocytes. Am J Hum Genet 52:661–7.

Vulliamy T, Marrone A, Goldman F, Dearlove A, Bessler M, Mason P, Dokal I (2001): The RNA component of telomerase is mutated in autosomal dominant dyskeratosis congenita. Nature 413:432–4.

Wan TS, Martens UM, Poon SS, Tsao SW, Chan LC, Lansdorp PM (1999): Absence or low number of telomere repeats at junctions of dicentric chromosomes. Genes Chrom Cancer 24:83–6.

Wiemann SU, Satyanarayana A, Tsahuridu M, Tillmann HL, Zender L, Klempnauer J, Flemming P, Franco S, Blasco MA, Manns MP, Rudolph KL (2002): Hepatocyte telomere shortening and senescence are general markers of human liver cirrhosis. FASEB J 16:935–42.

Wright WE, Tesmer VM, Huffman KE, Levene SD, Shay JW (1997): Normal human chromosomes have long G-rich telomeric overhangs at one end. Genes Dev 11:2801–9.

Wu KD, Orme LM, Shaughnessy J, Jr, Jacobson J, Barlogie B, Moore MA (2003): Telomerase and telomere length in multiple myeloma: Correlations with disease heterogeneity, cytogenetic status, and overall survival. Blood 101: 4982–9.

CHAPTER 14

IMMORTALIZATION BY SV40 LARGE T ANTIGEN

ROWENA L. LOCK, SILVIA BENVENUTI, and PARMJIT S. JAT
Ludwig Institute for Cancer Research, Royal Free and University College School of Medicine, London, United Kingdom

INTRODUCTION

Simian virus 40 (SV40) is a member of the papovavirus family of DNA tumor viruses. It is a double-stranded DNA virus of rhesus monkey origin, with a circular genome of 5243 base pairs (Tooze, 1981). SV40 was first isolated in the late 1950s as a contaminating virus from rhesus macaque monkey renal cells that were being used to grow poliovirus for the early polio vaccine, and was later shown to be able to induce tumors in hamsters (reviewed in Hilleman, 1998).

SV40 carries out either of two types of infection. Infection of cells permissive for viral infection (rhesus monkey or African green monkey cells) results in a lytic infection that leads to full viral gene expression, synthesis of progeny particles, and eventually cell death. In contrast, nonpermissive cells (mouse) survive an infection and progeny particles are never released, but the early proteins are expressed albeit transiently. Semipermissive cells (hamster or human) lie between the two extremes, and a small fraction of infected cells permit replication of the virus. Infection of either hamster or rodent cells causes a proportion of them to become stably transformed. The resulting transformed cell lines are usually oncogenic when introduced into syngeneic animals. Lytic infection may be divided into two phases, early and late, defined by the onset of viral DNA replication. During the early phase of infection, viral genes from the early region are expressed; induction of a variety of cellular enzymes also occurs during this phase (Tooze, 1981). Replication of both cellular and viral DNA begins approximately 12 to 15 hours after infection and defines the end of the early phase. Viral DNA replication requires only one functional early gene product, the large T antigen

Cell Cycle and Growth Control: Biomolecular Regulation and Cancer, Second Edition, Edited by Gary S. Stein and Arthur B. Pardee
ISBN 0-471-25071-6

(Tegtmeyer, 1972; Tegtmeyer, 1975) and precedes extensive expression of the late region genes that encode the structural components of the virus particle. Although during the late phase expression of the late region predominates, the early region is still transcribed, albeit at a much lower level (Laub and Aloni, 1975).

The SV40 genome codes for seven gene products. The early genes are encoded by three alternatively spliced mRNAs that give rise to proteins that are identical for the amino-terminal 82 residues (referred to as "T/t common region"). These gene products are called large T antigen, small t antigen, and 17kT antigen (sometimes known as tiny t antigen). They are involved in many functions which include regulation of early and late gene transcription, up-regulation of transcription in the host cell, replication of viral DNA and virion assembly (reviewed in Conzen and Cole, 1995; Ali and DeCaprio, 2001; Grand, 2001; Sullivan and Pipas, 2002). These proteins were called T(umor) antigens because they were first identified when antisera from hamsters with SV40-induced tumors were used to search for antigens expressed in SV40 infected cells. SV40 also encodes the late gene products VP1, VP2, and VP3, which are generated from alternatively spliced mRNAs and alternative initiation codons and which are the structural proteins that form the viral capsid (see Fig. 14.1). SV40 also encodes the agno protein, a small highly basic polypeptide that is translated from the

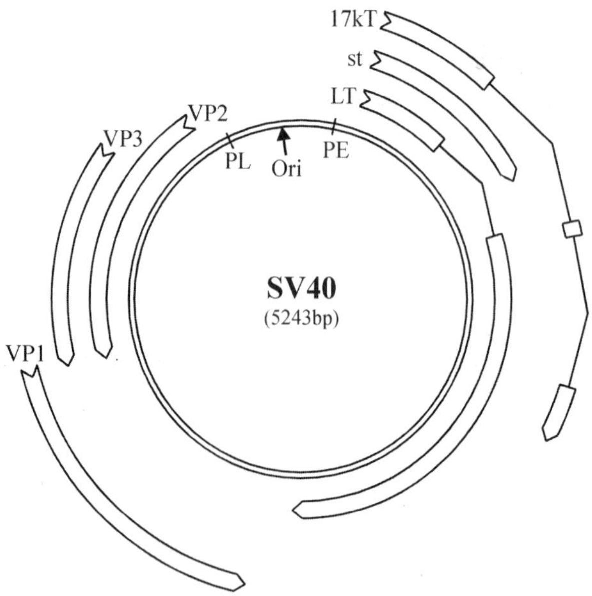

Figure 14.1. The genomic organization of the SV40 virus. SV40 virus has a circular double-stranded DNA genome, with promoters for early and late genes (PE and PL, respectively) located on either side of the origin of replication. The early genes encode the large T antigen (LT), small t antigen (st), and tiny t antigen (17kT) proteins, while the late genes encode the VP1, VP2, and VP3 proteins. The open reading frames that give rise to these proteins are indicated.

leader region of SV40 late mRNAs and may have a role in the assembly of the capsid and/or its release from the host cell (Jay, Nomura et al., 1981; Cole, 1996).

Large T antigen is largely responsible for the functions of the virus that induce the host cell to produce the enzymes that are necessary for replication of the viral genome. Introduction of large T antigen into primary rodent cells enables these cells to acquire an infinite proliferative potential (Jat and Sharp, 1989) but does not necessarily result in their transformation. Inactivation of large T in these immortal cells results in a rapid and irreversible loss of the proliferative potential in either G_1 or G_2 phases of the cell cycle, showing that the large T antigen is continuously required to maintain the proliferative state (Jat and Sharp, 1989; Gonos, Burns et al., 1996). However, introduction of large T antigen into immortalized cell lines can result in full transformation (reviewed in Ali and DeCaprio, 2001). The potency of large T antigen for inducing both immortalization and transformation of many cell types has led to its extensive use as a model system to study the critical steps for this process. These studies have provided a great insight into many cellular mechanisms, including the regulation of cell proliferation, senescence and apoptosis, in addition to the processes of immortalization and tumorigenesis. In this review we focus on the functions of large T antigen that specify its immortalization potential and the possible mechanism by which this may occur. However, before we describe the various functions, it is important to define immortalization and the various assays that are used to measure this activity.

IMMORTALIZATION

Immortalization has been defined as the ability to produce cell lines that can be serially cultivated indefinitely without the cells undergoing crisis. It can be tested in a variety of cell types but the critical property they must possess is that they must only be able to undergo a finite number of divisions as established cell lines have already undergone the changes necessary for immortalization (i.e., are already genetically abnormal). Immortalization is measured by the ability to obtain colonies upon transfection or infection of primary cultures and expansion of these colonies into cell lines that can be serially cultivated. The production of cell lines that can be serially cultivated is a critical component of this assay because it is possible to obtain an extension of life span without immortalization. The most common cell type used is embryonic fibroblasts, since they can be easily prepared and propagated in vitro. In contrast, transformation has been defined as the process that gives rise to oncogenically transformed cells. The assays that are commonly used to measure transformation potential are production of dense foci, foci that overgrow a monolayer, growth in semisolid medium, namely anchorage-independent growth, and growth in low serum concentrations or at high saturation densities. However, the ultimate test has to be whether tumors can be induced in either syngeneic or immunosuppressed animals. The closest correlate to tumors in experimental animals is anchorage-independent growth. Transformation ability does not have to be measured in primary cells. It is usually carried out in established cell lines. In fact, if primary cells are to be used, then there may be a requirement for the presence of an active immortalizing gene, because transforming genes such as activated *ras* induce premature senescence in the absence of an immortalizing gene.

Mouse embryo fibroblasts have the potential to undergo about 20 divisions before they cease dividing. However, they will readily undergo spontaneous immortalization, and thus great care needs to be taken when these cells are used for immortalization assays. Rat embryo fibroblasts undergo slightly less divisions but immortalize spontaneously at a much lower frequency. In marked contrast, human fibroblasts undergo many more divisions, approximately 50 to 70, defined as the Hayflick limit (Hayflick, 1961) prior to undergoing senescence, and have never been demonstrated to undergo spontaneous immortalization. The addition of an immortalizing gene, such as SV40 large T antigen, into human cells does not cause immortalization but can extend the in vitro life span of the cells, adding an extra 20 to 30 population doublings for fibroblasts (reviewed in Shay, Wright et al., 1991). At the end of the extended life span another proliferative decline known as "crisis" is observed. It has been proposed that senescence be defined as mortality stage I (M1), and that crisis be defined as mortality stage II (M2), to make a clear distinction between these two different states in the proliferative decline of human cells (Wright, Pereira-Smith et al., 1989). The inability of human cells to undergo spontaneous immortalization has led to the proposal that human cells have acquired extra control mechanisms to prevent tumor formation. This most likely corresponds to the shortening of telomeres in human somatic cells (reviewed in Sedivy 1998). In these cells the telomeres shorten each time the cells divide due to the end replication problem and the absence of functional telomerase activity in most human somatic cells. Telomerase activity can be reconstituted by ectopically expressing hTERT, the catalytic subunit of telomerase (Bodnar, Ouellette et al., 1998; Counter, Meyerson et al., 1998; Vaziri and Benchimol, 1998). In fact it has been proposed by some that reconstitution of telomerase activity is all that is required for immortalization of human cells (Bodnar, Ouellette et al., 1998; Vaziri and Benchimol, 1998; Yang, Chang et al., 1999; Ouellette, Liao et al., 2000). In contrast, rodent cells have much longer telomeres, and it is not possible to demonstrate telomere shortening upon serial passaging. Thus cellular senescence in rodent cells is not regulated by telomere-dependent mechanisms.

The loss of in vitro proliferative potential observed in rodent fibroblasts can be readily overcome by the action of any member of the family of immortalizing genes, including SV40 large T antigen (reviewed in Katakura, Alam et al., 1998). Furthermore it has been shown that these cells only become dependent on the immortalizing gene to continue dividing when their endogenous life span has elapsed; after this time the immortal cells are dependent on the immortalization gene to continue dividing (Ikram, Norton et al., 1994). The growth arrest is irreversible after about 72 hours and can occur in both G_1 and G_2 phases (Jat and Sharp, 1989; Gonos, Burns et al., 1996). The same result was obtained in other rodent cell types, namely myoblasts and neural cells. This property has led to large T antigen being one of the most frequently used reagents for generating immortal cell lines.

In contrast to researchers who have proposed that reconstitution of telomerase activity is all that is required for immortalization of human cells, others have found that additional activities are required. These can be provided by Human papilloma virus 16/18 E7 protein or the SV40 large T antigen (Kiyono, Foster et al., 1998; Hahn, Counter et al., 1999; O'Hare, Bond et al., 2001). Interestingly, as for rodent fibroblasts, human mammary fibroblasts immortalized with a combination of hTERT, and SV40 large T antigen only become dependent on large T antigen to maintain their growth when the endogenous life span has elapsed. Furthermore these cells depend on the T antigen to maintain their growth, even if telomerase activity has been constitutively reconstituted. Inactivation of T antigen resulted in a growth arrest in both G_1 and G_2 phases that was irreversible after about 7 days (O'Hare, Bond et al., 2001). As for rodent cells, similar results have been obtained with other cell types, namely microvascular endothelial cells, mammary epithelial cells, myoblasts, and prostate epithelial cells.

SV40 LARGE T ANTIGEN

Large T antigen is a multifunctional phosphoprotein of 708 amino acids that localizes mostly within the nucleus via a nuclear localization signal (NLS) in the N-terminal region of the protein (Soule and Butel, 1979; Kalderon, Richardson et al., 1984), although a very small fraction is bound to the plasma membrane (Santos and Butel, 1982). It has various biochemical activities (Fanning, 1992), including ATPase activity (Tjian and Robbins, 1979) and DNA and RNA helicase activity (Scheffner, Knippers et al., 1989). It can also bind to RNA covalently (Carroll, Samad et al., 1988), as well as to DNA both specifically and nonspecifically (Carroll, Hager et al., 1974). While these activities are necessary for replication of the viral DNA, none of them are required for the immortalization or transformation of rodent cells (Stringer, 1982; Manos and Gluzman, 1984; Peden, Spence et al., 1990). Large T antigen can also activate and repress transcription from various viral and cellular promoters (Alwine, Reed et al., 1977; Hansen, Tenen et al., 1981; Mitchell, Wang et al., 1987; Saffer, Jackson et al., 1990; Zhu, Rice et al., 1991; Gilinger and Alwine, 1993; Gruda, Zabolotny et al., 1993; Rice and Cole, 1993; Rushton, Jiang et al., 1997).

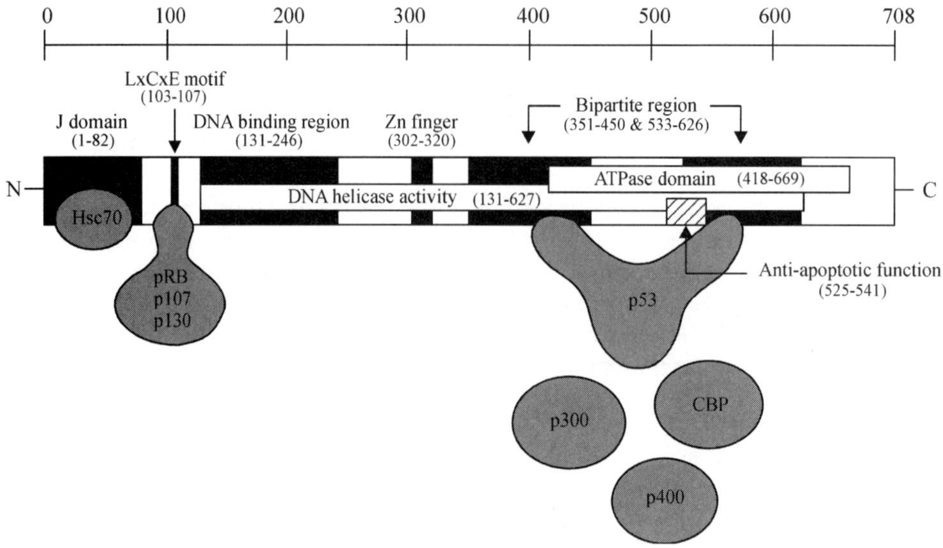

Figure 14.2. The functional domains of SV40 large T antigen. The locations of the domains of large T antigen that are responsible for interaction with host proteins are indicated, along with the regions responsible for other activities of the protein.

Large T antigen associates with a number of host cellular proteins, which include p53 (Lane and Crawford, 1979; Linzer and Levine, 1979), pRB (DeCaprio, Ludlow et al., 1988), p107 (Dyson, Buchkovich et al., 1989; Ewen, Ludlow et al., 1989), p130 (Hannon, Demetrick et al., 1993), DNA polymerase α (Smale and Tjian, 1986), heat shock protein Hsc70 (Sawai and Butel, 1989), p185 (Kohrman and Imperiale, 1992), CREB-binding protein (CBP), p300 (Avantaggiati, Carbone et al., 1996; Eckner, Ludlow et al., 1996), p400 (Lill, Tevethia et al., 1997) SUG1, a regulatory component of the proteasome (Grand, Turnell et al., 1999) and TBP, the TATA-binding protein (Martin, Subler et al., 1993) (see Fig. 14.2).

Mutational analyses of large T antigen have indicated that three regions of large T antigen are critical for its immortalizing activity (Conzen and Cole, 1995). These regions are the amino terminal 82 amino acids (the T/t common region), amino acids 102–114 and amino acids 350–560 (see Fig. 14.2). It should be noted, however, that while all three regions are required for immortalization, C3H10T1/2 cells (an established cell line) are still at least partially transformed by T antigen with mutations in the second or third regions, but mutations in the amino terminal 82 residues abrogate transformation of these cells. This illustrates the reduced requirement of large T antigen functions for the transformation of established lines compared with immortalization of primary cells. These regions of large T antigen are all involved in interactions with host proteins, and these interactions and their possible functions are discussed in detail below.

THE RETINOBLASTOMA FAMILY OF PROTEINS

The retinoblastoma protein (pRB) is the product of the retinoblastoma susceptibility gene, Rb1, which was the first tumor-suppressor gene to be identified in humans. Homozygous null mutations of the Rb1 gene cause tumors of the eye in children. pRB is essential for mouse development, as homozygous mutants of the gene are embryonic lethal in mice (Lee, Chang et al., 1992). It has also been shown that heterozygous mutations in the RB gene in mice result in almost 100% incidence of spontaneous pituitary tumors (Hu, Gutsmann et al., 1994).

Rb1 maps to human chromosome 13, band q14 (Dryja, Bruns et al., 1983) and pRB is localized in euchromatic areas with low DNA density in interphase nuclei. The RB protein is a ubiquitously expressed protein of 928 amino acids. During metaphase and anaphase pRB dissociates from condensing chromosomes and disperses throughout the cytoplasm but reassociates with decondensing chromatin during telophase (Szekely, Uzvolgyi et al., 1991). It contains at least three domains: an N-terminal domain, a central A/B (bipartite) pocket domain, and a C-terminal (C pocket) domain. The two other pRB-family members, p107 and p130, also possess the A/B pocket region, which is found to be the region with which most cellular and viral interacting proteins bind, and also the region in which most tumor-associated RB mutations occur (Hu, Dyson et al., 1990; Huang, Wang et al., 1990; Classon and Dyson, 2001). The RB gene, as well as genes involved in the RB regulatory pathway, are known to be mutated in a diverse range of human cancers (reviewed in Malumbres and Barbacid, 2001).

The pRB family of proteins function to control proliferation, differentiation, and apoptosis, by maintaining a tight control on cell cycle progression (see Fig. 14.3). Once a dividing cell has completed mitosis it enters G_1. At approximately two-thirds of the way through G_1, the cell reaches the "restriction (R) point" (Pardee, 1989), when it assesses the environmental conditions (e.g., the presence of mitogens) and decides whether to continue through G_1 and divide again, or to enter G_0 and become quiescent. If the decision is to continue cycling, the cell cycle proceeds unless an event such as DNA damage hinders its progress, and once R is passed, growth no longer depends on the presence of mitogens. pRB is phosphorylated and dephosphorylated during the cell cycle, and interestingly the site-specific phosphorylation of pRB appears to coincide with the transition of the cell through R (Weinberg, 1995; Bartek, Bartkova et al., 1996). It has been shown that multiple cyclin/cdk complexes target distinct functions of the pRB protein to progressively inactivate it. During G_0 and early G_1 phase of the cell cycle, pRB is largely in its active, underphosphorylated form. During mid G_1, pRB is phosphorylated by cyclin D/cdk(4/6), whereas at the restriction point R and in late G_1, it is further phosphorylated by the cyclin E/cdk2 complex (Adams, 2001). The inactive, hyperphosphorylated form predominates during G_2 and M phases of the cycle; pRB only becomes underphosphorylated again upon exit from M phase. pRB is also phosphorylated

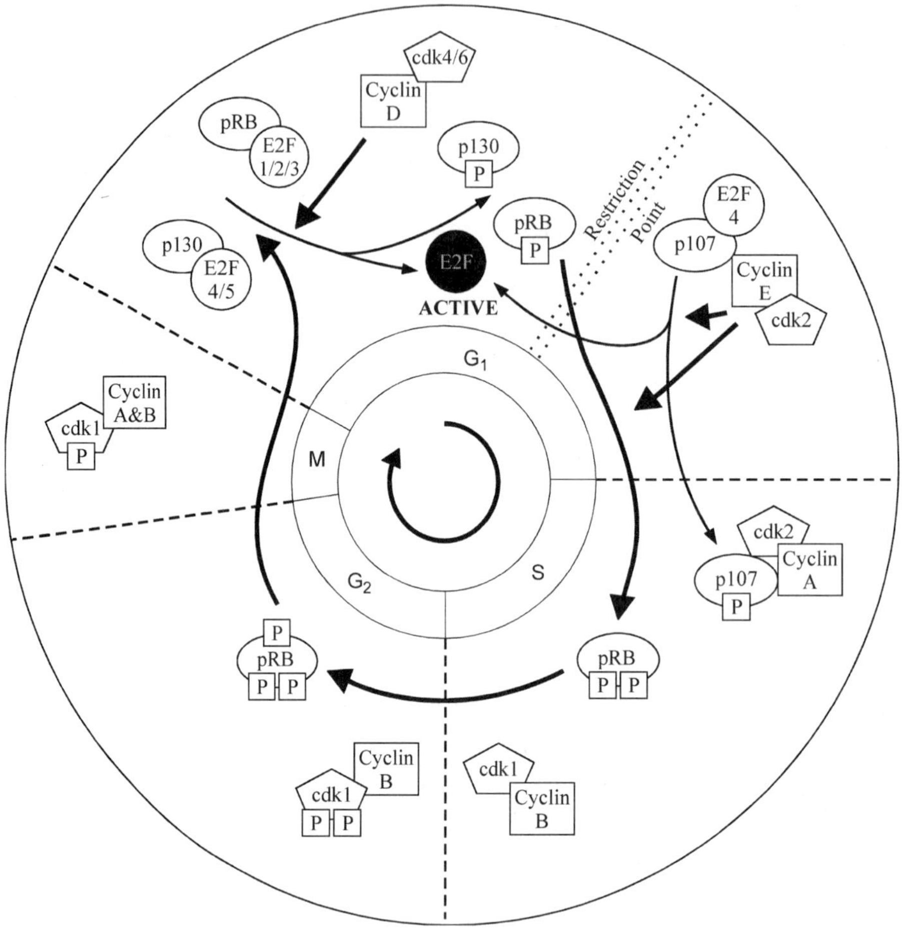

Figure 14.3. Role of the pRB family of proteins in cell cycle progression. The phosphorylation state of pRB and the related p107 and p130 proteins determines their ability to activate or repress E2F. Different cyclin/cdk complexes are responsible for phosphorylating these proteins at different stages of the cell cycle.

during S phase and at the G_2/M transition, perhaps by the cyclin B/cdk1 complex (previously known as cyclin B/p34^{cdc2}) (DeCaprio, Furukawa et al., 1992) prior to becoming dephosphorylated at M phase by PP1, a type 1 serine/threonine phosphatase (Nelson, Krucher et al., 1997), suggesting additional roles for pRB in the regulation of cell cycle progression besides its function at R.

pRB acts to prevent the cell cycle transition from G_1 to S phase partly via its interaction with E2F, a transcriptional activator for S phase-specific genes. It is only the active, underphosphorylated form of pRB that binds to E2F, thus inhibiting its transactivation activity. The interaction of pRB with E2F is regulated by phosphorylation of pRB by several cyclin-dependent kinases (CDKs), in a cell cycle dependent manner (Adams, 2001). Human cells have six E2F family members in total (E2F1–E2F6), with each (apart from E2F6) forming heterodimers with the two DP proteins (DP1 and DP2). DNA binding, transactivation and pRB-binding activities of E2F are enhanced by heterodimerization. The heterodimers bind to sequence specific DNA sites in the promoters of several genes required for entry into S phase, including cyclins A and E, dihydrofolate reductase, thymidylate synthetase, c-myc, and c-myb. When bound to the promoter, the E2F-DP heterodimers function to both activate transcription and to recruit members of the pRB family. Once bound to pRB family members, the role of E2F proteins changes from that of transcriptional activators, to transcriptional repressors. When complexed to E2F, pRB also recruits chromatin remodeling enzymes, such as histone deacetylase (HDAC), which further block the transactivating activity of E2Fs. The release of E2F from pRB upon hyperphosphorylation allows derepression of target genes and for E2F to induce expression of S phase genes required for DNA synthesis and transcriptional regulation. Many proteins that interact with pRB contain a LxCxE motif, and although the actual LxCxE binding site of pRB is entirely within the B region of the pocket, it is also dependent on other interactions with the A and B regions (Lee, Russo et al., 1998). E2F proteins do not have the LxCxE motif and are able to bind to the A/B domain of pRB concurrently with a LxCxE peptide, showing that the binding sites for these two types of protein are distinct and thus E2F can recruit complexes that contain both pRB and other proteins, such as those with the LxCxE motif, to a promoter. Even though the presence of the A/B domain is sufficient for pRB to bind E2F subunits such as E2F-1, additional interactions provided by the C-terminus of pRB are required for binding to functional E2F heterodimers (Qin, Chittenden et al., 1992; Hiebert, 1993). The C-terminal (C pocket) domain of pRB is very flexible and probably undergoes changes in conformation upon phosphorylation, allowing pRB to regulate a diverse range of cellular processes by interaction with many proteins including the c-Abl tyrosine kinase and MDM2 (Welch and Wang, 1993; Xiao, Chen et al., 1995; Whitaker, Su et al., 1998). The binding sites for c-Abl tyrosine kinase and MDM2 in the C-terminal domain of pRB are distinct from the E2F binding site. When c-Abl is complexed with pRB, its tyrosine kinase activity is blocked, but when pRB is hyperphosphorylated, active c-Abl is released. This interaction seems to be important for the growth suppression function of pRB (Whitaker, Su et al., 1998). pRB can also form a trimeric complex with p53 and MDM2, thereby blocking the anti-apoptotic activity of MDM2 by preventing the degradation of p53, and stabilizing the p53 protein (Hsieh, Chan et al., 1999). This indicates one of many levels of crosstalk between the pRB and p53 pathways.

The function of the N-terminal domain of pRB is still not known. This region contains several consensus cyclin-dependent kinase (CDK) phosphorylation sites, suggesting that activity of pRB may be regulated by phosphorylation of these sites during different stages of the cell cycle. The pRB family members p107 and p130 also contain multiple sites for phosphorylation by CDKs, including a large cluster of sites in the extreme C-terminus. In addition the spacer sequences which separate the A and B subdomains of the pocket in these proteins are longer than the analogous region in pRB and include a high-affinity binding site for cyclinA/cdk2 and cyclinE/cdk2, which may enable these proteins to act as CDK inhibitors (Classon and Dyson, 2001).

Large T antigen forms a stable complex with pRB, and this interaction is critical for immortalization because mutants that are defective for binding pRB have a reduced immortalization potential (DeCaprio, Ludlow et al., 1988; Powell, Darmon et al., 1999). Some transformation defective mutants of large T antigen are defective for binding to pRB, indicating the additional importance of this interaction for the transforming function of large T antigen (reviewed in Ali and DeCaprio 2001). All viral oncoproteins, including large T antigen, and many cellular proteins that bind to pRB, have in common the LxCxE motif, which is responsible for the interaction with the conserved pocket region of pRB. The LxCxE motif of large T antigen is located at residues 103–107, within the region of the protein that has homology to conserved region 2 (CR2) domain of adenovirus E1A and the human papilloma virus type 16 E7 protein. The CR2 domain of each of these proteins is responsible for their interactions with the pRB family of proteins (DeCaprio, Ludlow et al., 1988; Moran, 1988; Munger, Werness et al., 1989). Large T antigen only binds to the underphosphorylated growth suppressing form of pRB (Ludlow, DeCaprio et al., 1989; Ludlow, Shon et al., 1990), which normally binds E2F, resulting in its sequestration. This has a similar effect to the phosphorylation-mediated inactivation of pRB. In the presence of large T antigen, E2F is thus free to activate transcription, resulting in entry into S phase. Large T antigen also targets the other pRB family members, p107 and p130, similarly disrupting their interactions with E2F proteins, and additionally is able to perturb the phosphorylation status of p107 and p130 and stimulate p130 degradation (but not the degradation of p107) (Stubdal, Zalvide et al., 1996). It has been suggested that p107 and p130 are the critical targets of large T antigen for its immortalization function, and that pRB may not be so important for this activity (Srinivasan, McClellan et al., 1997). The N-terminal J domain of large T antigen (residues 1–82) is also involved in the interaction with pRB family members, and it has been suggested that a functional J domain as well as the LxCxE motif are required for a productive interaction with pRB (Zalvide, Stubdal et al., 1998). The phosphorylation state of the pRB family members p130 and p107 is affected by large T antigen, whereas there is no effect on the phosphorylation state of pRB. In cells expressing large T antigen, only the underphosphorylated form of p130 and p107 can be detected, and as for pRB, both the LxCxE motif, and the J domain are required for this effect. Dissociation of p130 from E2F-4 by large T antigen requires the J domain, Hsc70 protein, and ATP, suggesting that the J domain may play a function in coupling ATP hydrolysis by Hsc70 to the endothermic disruption of the p130-E2F-4 complex (Sullivan, Cantalupo et al., 2000).

It has been suggested that large T antigen disrupts the pRB-E2F complex via the recruitment of cellular proteins such as the Hsc70 chaperone in a manner similar to the p130-E2F-4 complex. This is proposed to occur by transfer of the pRB-E2F complex from large T antigen to the substrate binding site of Hsc70, and then to the ATPase domain of Hsc70 resulting in ATP hydrolysis and conformational changes in Hsc70. This leads to closure of the substrate binding cavity and subsequent dissociation of pRB and E2F separately from the complex. (Laufen, Mayer et al., 1999; Lee and Cho, 2002).

THE p53 PROTEIN

p53 was originally identified as a cellular protein that co-immunoprecipitated with large T antigen using antisera from tumor-bearing animals (Lane and Crawford, 1979; Linzer and Levine, 1979). Interaction of large T antigen with p53 leads to its stabilization. P53 mutants have been shown to compete with wild-type p53, and the resulting inhibition of the wild-type p53-dependent transcription is perceived as a dominant-negative effect (reviewed in Blagosklonny, 2000). Although it was originally believed that p53 was an oncogene, it was subsequently realized that wild-type p53 is a tumor suppressor, whereas mutant p53 is an immortalizing gene. p53 has proved to be one of the most commonly mutated genes in human cancer (Hollstein, Rice et al., 1994). It exhibits sequence-specific DNA binding and acts to increase the transcription of genes that cause growth arrest in G_1 and G_2, and induce apoptosis in response to genotoxic stress. Its function as a tumor suppressor and "guardian of the genome" (Lane, 1992) has been attributed to its negative regulation of cell proliferation in the presence of DNA damage. In senescent cells p53 is active (Atadja, Wong et al., 1995; Kulju and Lehman, 1995; Bond, Haughton et al., 1996), and introduction of wild-type p53 into p53-null cells can induce the cells to become senescent (Sugrue, Shin et al., 1997). p53 activity has also been shown to be critical for the maintenance of the senescent state, since microinjection of anti-p53 antibodies can induce DNA synthesis in senescent cells (Gire and Wynford-Thomas, 1998).

p53 is ubiquitously expressed in all cells but is not stable due to its interaction with MDM2, a ubiquitin E3 ligase shuttling protein that targets it for degradation by the proteasome (Momand, Wu et al., 2000). However, upon genotoxic stress such as irradiation, chemotoxin exposure, or unscheduled DNA synthesis induced by a virus, the normally short-lived p53 protein becomes stabilized, resulting in a corresponding increase in p53-mediated transcription (Maltzman and Czyzyk, 1984; Kastan, Onyekwere et al., 1991; Kuerbitz, Plunkett et al., 1992). The phosphorylation of residues within the N-terminus of p53 stabilize the protein by blocking the binding of MDM2; these modifications are also crucial for acetylation of C-terminal residues and these together result in the full p53-mediated response to genotoxic stress. Activation of p53 by ionizing radiation requires the ATM protein, which mediates the direct phosphorylation of serine 15 and indirectly mediates the phosphorylation of serine 20 by controlling activation of chk2 (Turenne, Paul et al., 2001). The ATM-Rad3-related protein ATR also regulates phosphorylation of serine 15 in response to UV DNA damage. Additional sites for DNA damage-induced phosphorylation at serines 33 and 37 may be regulated by ATR and other kinases (Tibbetts, Brumbaugh et al., 1999). When the C-terminus of p53 is modified, the ability of this region to negatively regulate sequence-specific DNA binding is inhibited, resulting in activation of transcription of downstream targets in response to DNA damage (Lill, Grossman et al., 1997; Appella and Anderson, 2001). This leads to a cascade of events that includes the transcriptional activation of the CDK inhibitor, $p21^{Cip1/Waf1/Sdi1}$, MDM2, as well as the apoptotic protein bax, cyclin G1, and several other factors. In addition p53 can also repress the transcription of other genes, including those encoding the anti-apoptotic protein bcl-2, pRB, the transcription factors c-*fos*, c-*jun*, and c-*myc* and the proliferating cell nuclear antigen (PCNA) (Ginsberg, Mechta et al., 1991; Mercer, Shields et al., 1991; Moberg, Tyndall et al., 1992; Shiio, Yamamoto et al., 1992; Miyashita, Krajewski et al., 1994). The basis for p53-mediated transcriptional repression is not well understood but has been proposed to be mediated via indirect interactions with other promoter-bound transcription factors (Ragimov, Krauskopf et al., 1993; Horikoshi, Usheva et al., 1995). It has more recently been demonstrated that p53 repression is dependent on interaction with the co-repressor Sin3a, which recruits HDACs (Murphy, Ahn et al., 1999). It may also occur via direct binding, as suggested by the finding that p53 can repress transcription by binding to a novel head-to-tail site within the MDR1 promoter (Johnson, Ince et al., 2001).

If a cell undergoing cell division gets damaged, p53 is activated resulting in induction of $p21^{Cip1/Waf1/Sdi1}$, which causes cell cycle arrest by the inhibition of the cell cycle-promoting CDKs. This allows repair of the damaged DNA, ensuring that mutations are not transmitted to the daughter cells. In G_1, $p21^{Cip1/Waf1/Sdi1}$ negatively regulates cyclinE/cdk2, and at high levels it also represses cyclinD/cdk4–6 kinase activities. These are the kinases that phosphorylate the pRB family of tumour suppressors during G_1, and so in the presence of high levels of $p21^{Cip1/Waf1/Sdi1}$, the pRB family members remain in their inhibitory, underphosphorylated forms, maintaining their repression of E2F, and thus cell cycle arrest. Elevated levels of $p21^{Cip1/Waf1/Sdi1}$ can also block DNA replication and promote DNA repair by competing for binding to PCNA (Cox, 1997; Warbrick, 1998), a component of the DNA replication machinery that is required for both replication and repair of DNA. Binding of $p21^{Cip1/Waf1/Sdi1}$ to PCNA impairs PCNA-dependent DNA replication, but PCNA-dependent nucleotide excision repair remains intact (Flores-Rozas, Kelman et al., 1994; Li, Waga et al., 1994; Waga, Hannon et al., 1994). Thus induction of $p21^{Cip1/Waf1/Sdi1}$ by p53 in response to DNA damage ensures that damage is repaired prior to DNA replication. Once DNA has been repaired, p53 levels are reduced thus lowering the levels of $p21^{Cip1/Waf1/Sdi1}$, and the CDKs become active again. It should be noted that the CDK inhibitor $p16^{INK4}$ also participates in G_1 arrest in response to DNA damage, and can carry out this function in a p53-null background (Robles and Adami, 1998; Shapiro, Edwards et al., 1998), indicating that other mechanisms besides p53 are also involved in the growth arrest. If the DNA damage is too extensive to be repaired, p53 induces exit from the cell cycle and cell death via apoptosis. p53 can also halt DNA replication during S phase of the cell cycle if damage has occurred, again either promoting arrest until the DNA is repaired or inducing apoptosis if the DNA is irreparable, indicating the pivotal role played by p53 as a signal integrator in the determination of cell fate. MDM2 acts to down-regulate p53, inducing its ubiquitin-mediated degradation, thus providing a negative feedback loop mechanism that restores the normal functions of the cell once the genotoxic stress has diminished (Momand, Wu et al., 2000). p300 may also be required for the degradation of p53 by MDM2 (Grossman, Perez et al., 1998). MDM2 not only regulates the stability of p53 but also inactivates its transcriptional activity by binding to and concealing its transactivation domain. Phosphorylation of p53 at theonine 18 or serine 20 in response to DNA damage causes p53 to dissociate from MDM-2, resulting in its stabilization (Kussie, Gorina et al., 1996). p53 can also induce a G_2 arrest, but the mechanism is less clear and may involve $p21^{Cip1/Waf1/Sdi1}$ and 14-3-3σ (North and Hainaut, 2000; Taylor and Stark, 2001). 14-3-3σ sequesters Cdc25C and cyclinB1/cdk1 in the cytoplasm, and helps maintain a G_2 arrest (Hermeking, Lengauer et al., 1997).

It is not yet fully understood how the interaction of large T antigen with p53 affects the functions of p53, but it is generally accepted that large T antigen inactivates p53 tumor-suppressor function by directly binding to the region of p53 involved in specific DNA binding, and mutations of p53 that disrupt its sequence-specific DNA binding also prevent binding to large T antigen. The region of large T antigen required for interaction with p53 is actually bipartite, consisting of amino acids 351–450 and 533–626; the C-terminal region of large T antigen beyond amino acid 627 can be removed without loss of immortalization, transformation, or tumorigenic activity (Tevethia, Pipas et al., 1988). Interestingly immortalization of mouse cells is very much dependent on the direct inactivation of p53 (Conzen and Cole, 1995), whereas rat cells can be immortalized by amino terminal fragments of large T antigen that cannot interact with p53 (Sompayrac and Danna, 1991; Powell, Darmon et al., 1999). However, it has been suggested that these mutants may be able to act downstream of p53 to inactivate the pathway, such as via pRB family binding (Michael-Michalovitz, Yehiely et al., 1991; Quartin, Cole et al., 1994; Rushton, Jiang et al., 1997).

The interaction of large T antigen with p53 prevents wild-type p53 from binding to DNA and acting as a transcriptional activator of genes that induce cell cycle arrest in response to genotoxic stress. The failure of p53 to monitor the genome in SV40-immortalized cells is thought to be the reason behind the increased occurrence of mutations in these cells (Ray, Peabody et al., 1990; Stewart and Bacchetti, 1991; Woods, LeFeuvre et al., 1994). However, interaction of large T antigen with p53 has also been shown to increase the half-life and steady state levels of p53 (Oren, Maltzman et al., 1981; Deppert, Haug et al., 1987). SV40 small t antigen plays a role in promoting the metabolic stabilization of p53 complexed to large T antigen, and it has been suggested that the stabilized p53 might have a positive function in SV40-mediated cellular transformation (Tiemann, Zerrahn et al., 1995; Zerrahn, Tiemann et al., 1996). The observation that MDM2 binds large T antigen directly in vitro and coprecipitates in a complex with p53 in vivo has led to the suggestion that large T antigen enhances the stability of p53 partly by entrapment of the p300/MDM2/p53 complex responsible for targeting its degradation (Brown, Deb et al., 1993; Henning, Rohaly et al., 1997; Grossman, Perez et al., 1998).

Several studies have shown that large T antigen can also prevent p53-dependent gene expression and growth suppression by a mechanism that does not require a direct interaction between the proteins, but does require large T/ pRB interaction. For example, an N-terminal fragment that consists of the J domain and pRB-binding motif, but no p53-binding domain was shown to block p53-dependent growth suppression (Michael-Michalovitz, Yehiely et al., 1991; Quartin, Cole et al., 1994). Also the growth suppression function of p53 cannot be blocked by large T antigen mutants affecting the pRB binding domain, even though these mutants are still capable of binding to and stabilizing p53. Moreover transient transfection assays have revealed that the J domain and pRB binding functions of large T antigen are essential for the prevention of p53-dependent gene expression (Rushton, Jiang et al., 1997). Small t antigen has been shown to play a role in the inactivation of p53. It is the C-terminal, cysteine-rich region of small t, rather than the J domain that is essential for this activity (Gjoerup, Chao et al., 2000). This region is also important for the inhibitory interaction of small t antigen with the cellular protein phosphatase complex, PP2A (Rundell and Parakati, 2001), perhaps implying a role for small t in the regulation of the phosphorylation state and inactivation of p53 via PP2A binding. It was also shown that either amino-terminal fragments of large T antigen encoding the J domain and pRB-binding domains, or small t antigen alone were capable of complementing large T antigen mutants defective for pRB-binding, allowing them to inhibit p53-dependent growth suppression (Gjoerup, Chao et al., 2000). The sequestration of p53 by large T antigen may also abrogate p53-mediated apoptosis (McCarthy, Symonds et al., 1994). In cell lines conditionally immortalized by a temperature-sensitive mutant of T antigen, its inactivation has in some studies resulted in growth arrest without significant cell death (Jat and Sharp, 1989), while in others its inactivation resulted in cell death by apoptosis, perhaps caused by a sudden release of a large amount of stable p53 (Yanai and Obinata, 1994).

THE J DOMAIN AND Hsc70

Chaperone proteins, such as the DnaK protein Hsc70, function to ensure the correct folding of proteins, and also to prevent the aggregation of proteins when the cell is under stress (Hartl, 1996). DnaJ and its J domain homologues are co-chaperone regulators, responsible for the stimulation of the ATPase activity of the DnaK proteins such as Hsc70, and for promoting their interactions with their substrates (Caplan, Cyr et al., 1993; Cheetham and Caplan, 1998). All J proteins contain a region of approximately 70 amino acids called the J domain that directly binds to Hsc70. The structure of the J domain consists of three or four alpha-helices in which helices II and III form an antiparallel finger-like projection held together by hydrophobic interactions (Jiang, Greener et al., 1997; Huang, Flanagan et al., 1999; Berjanskii, Riley et al., 2000; Kim, Ahn et al., 2001). These two helices are linked by a loop consisting of amino acids HPD, and this tripeptide motif directly contacts Hsc70 and has been found to be universally conserved in all J proteins. Some viruses, including bacteriophage λ, only make use of host-encoded chaperones, whereas others, including SV40, encode their own virus-specific chaperones. Many observations have led to the conclusion that large T antigen can function as a molecular chaperone (reviewed in Sullivan and Pipas, 2002).

First, the N-terminal region (residues 1–82) of large T antigen shows sequence similarity with the conserved residues of the type 3 DnaJ-like proteins including the absolutely conserved HPD tripeptide loop of the J domain (Kelley and Landry, 1994). The HPD motif present in the J domain of large T antigen is proximal to the pRB-binding region, supporting the model that large T antigen acts as a bridge to direct the action of Hsc70 on pRB-E2F complexes (Kim, Ahn et al., 2001). Second, it has been shown that the J domain of large T antigen is capable of functionally substituting for the J domains of *E. coli* DnaJ in a phage λ growth assay (Kelley and Georgopoulos, 1997). This complementation was not observed when the histidine of the conserved HPD residues was mutated, and mutations in the HPD loop were also found to render large T antigen defective for SV40 DNA replication, transformation and assembly (reviewed in Sullivan and Pipas 2002). Finally the J domain of large T antigen has also been shown to be capable of stimulating the ATPase activity of bovine Hsc70 and Ssa1p (cytosolic yeast Hsc70), binding to Hsc70, and promoting the release of a bound substrate from Ssa1p (Sawai and Butel, 1989; Srinivasan, McClellan et al., 1997). In addition the J domain of SV40 large T antigen has been shown to be required for efficient viral DNA replication in vivo (Campbell, Mullane et al., 1997).

Mutational analyses have shown that the J domain is essential for many different viral functions, including the replication of the viral DNA and for virion assembly in addition to its roles in transformation and immortalization. Transfection experiments in mouse embryo fibroblasts have shown that both the J domain and the pRB-binding region of large T antigen are required for p130 and p107 to stimulate growth, and immortalization (Christensen and Imperiale, 1995; Stubdal, Zalvide et al., 1997). Moreover the J domain was found to be required in *cis* with the pRB-binding region for immortalization and transformation (Srinivasan, McClellan et al., 1997), leading to the proposal of a model in which the pRB-transcription factor complexes are directly affected by the J domain chaperone activity. As described above, it is proposed that the J domain recruits Hsc70 to the pRB-complex (pRB-E2F, or perhaps a complex with other factors such as MyoD or c-Abl) (Brodsky and Pipas, 1998), induces the ATPase activity of Hsc70, and causes disruption of the complex, either directly by inducing a conformational change of pRB or E2F, or indirectly by recruiting other proteins to the complex. E2F is then free to activate gene transcription. This model is supported by the observation that the J domain is required in *cis* with the pRB-binding domain to up-regulate exogenous promoters that contain multiple E2F binding sites (Sheng, Denis et al., 1997; Zalvide, Stubdal et al., 1998). It is further supported by the detection of a p130-E2F-4 DNA binding complex in cellular lysates from cells not expressing large T antigen, or those expressing J domain mutants of large T antigen, whereas the complex was not observed in the presence of wild-type large T antigen (Harris, Christensen et al., 1998; Zalvide, Stubdal et al., 1998; Sullivan, Tremblay et al., 2000). Although it remains a possibility that the J domain disrupts the pRB-E2F complexes indirectly by driving the cells to cycle in a way that leads to the liberation of E2F, biochemical studies have clearly shown that T antigen can directly disrupt pRB-E2F complexes in vitro (Sullivan, Cantalupo et al., 2000). Large T antigen can also prevent apoptosis in a neural astrocyte precursor cell line when growth factors are removed. This function required both the pRB-binding and the J domains, providing further evidence that the J domain was necessary for abrogation of some of the many activities of pRB family members (Slinskey, Barnes et al., 1999).

In addition to these observations, it has also been demonstrated that large T antigen is capable of enhancing cell growth in the absence of its J domain, but requiring the pRB binding motif (Tevethia, Lacko et al., 1997). Also the pRB-binding domain, but not the J domain of large T antigen, is required to restore growth to a cell line which has been arrested by the conditional expression of p53 (Quartin, Cole et al., 1994; Gjoerup, Chao et al., 2000). We have also shown that a J domain mutant that was null for immortalization was active for maintenance of the immortalized state, indicating that J domain functions may be critical for initiating immortalization (Powell, Darmon et al., 1999). Therefore large T antigen can disrupt many of the activities of the pRB family of proteins involved in cell growth and death, and these functions are carried out by both J domain-dependent and independent mechanisms.

p300 AND THE CBP FAMILY OF PROTEINS

p300 was first identified as an adenovirus E1A interacting protein and has also been shown to bind to SV40 large T antigen (Avantaggiati,

Carbone et al., 1996; Eckner, Ludlow et al., 1996). p300, and the related proteins p400 and CBP (CREB-binding protein) are members of a family of transcriptional co-activators involved in coordinating the formation of specific transcription factor complexes (Struhl, 1998). They enhance the transcriptional activity of many proteins, including p53 and NFκB p65 subunit, either by directly interacting with transactivators or by modulating the chromatin structure via recruitment of histone acetyltransferases (Goodman and Smolik, 2000; Hottiger and Nabel, 2000). p300/CBP acetyltransferases promote the acetylation of p53, thus activating p53-mediated transcription; this activation of p53 by p300 is potentiated by ionizing radiation (Avantaggiati, Ogryzko et al., 1997).

In addition to their role in transcriptional regulation, these proteins may also be involved in cell cycle control and development. p300 and CBP may also have a tumor-suppressor function, as mice with a null mutation in one allele of CBP are known to develop haematological malignancies (reviewed in Janknecht, 2002), and various human carcinomas have been linked to p300 mutations (Muraoka, Konishi et al., 1996; Gayther, Batley et al., 2000). The binding site for p300 and CBP on large T antigen has been mapped to both the N-terminal and the C-terminal domains of the protein (Eckner, Ludlow et al., 1996; Lill, Tevethia et al., 1997). It is not clear if interaction of these proteins with large T antigen is direct or via p53, but it has been shown that large T antigen expression inactivates the transcriptional activation function of p300, and also changes the phosphorylation state of p300 (Avantaggiati, Carbone et al., 1996; Eckner, Ludlow et al., 1996). This inactivation of p300 leads to the inhibition of p53 transcriptional activation activity, further helping the virus to maintain an environment in which it can replicate in an uncontrolled manner.

p63 AND p73 FAMILY OF PROTEINS

p63 and p73 genes encode p53-related proteins that have been found to have both structural and functional similarities to p53 (reviewed in Levrero, De Laurenzi et al., 2000). They share a high amino acid sequence homology within the DNA binding region, suggesting a potential role in regulating transcription. Several findings have suggested that p63 and p73 do not play a similar role to p53 in the development of cancer. For example, while transgenic mice expressing mutant p53 or $p53^{-/-}$ mice are very prone to both spontaneous and induced tumors (Donehower, Harvey et al., 1992), $p63^{-/-}$ and $p73^{-/-}$ mice are not, instead displaying developmental defects (Mills, Zheng et al., 1999; Yang, Schweitzer et al., 1999; Yang, Walker et al., 2000). The p63 and p73 genes are expressed using alternative promoters and undergo differential splicing, to give rise to several different isoforms. The use of alternative promoters results in proteins with two different amino termini, TA forms that contain a transactivating domain, and ΔN forms that lack this domain. The differential splicing occurs mostly at the 3′ end (in the part of the sequence that is not present in p53), and results in isoforms with different carboxy termini, α (the longest form), β and γ (the shortest form) in the case of p63 (Yang, Kaghad et al., 1998), and α, β, γ, δ, ε, and ζ in the case of p73 (Kaghad, Bonnet et al., 1997; De Laurenzi, Costanzo et al., 1998; De Laurenzi, Catani et al., 1999; Ueda, Hijikata et al., 1999). Some of these splice forms (e.g., TAp63γ) can activate transcription from the $p21^{Cip1/Waf1/Sdi1}$ promoter and induce apoptosis (Ishida, Yamashita et al., 2000), while others (e.g., ΔNp63γ) function as endogenous inhibitors of p53 and p63 transcriptional activity (Yang, Kaghad et al., 1998). All splice forms of both p63 and p73 proteins have an extended carboxy terminal region that is absent in p53, and some of the isoforms contain a sterile alpha motif (SAM) domain previously identified in other proteins that regulate development (Thanos and Bowie, 1999). As mentioned above, in contrast to p53, both p63 and p73 deficient mice show developmental abnormalities, and it has been found that the human p63 gene is mutated in children who suffer from EEC (ectrodactly, ectodermal dysplasia and facial clefts), a condition similar to the phenotype of p63-deficient mice (Celli, Duijf et al., 1999).

We have found that p63α proteins are capable of interacting with large T antigen, albeit weakly, in a p53-independent manner, although this interaction was significantly enhanced in the presence of p53 (Djelloul, Tarunina et al., 2002). It remains to be determined if this interaction is direct or if it is mediated via another protein. This interaction is likely to be via the DNA binding domain that is highly conserved between p53 and all p63 family members, suggesting that they all have the potential to interact with large T antigen. It is suggested that the interaction between large T antigen and p63α proteins might modulate their activities in a similar way to the inhibition of p53 activity by large T antigen. We have proposed that differential expression of p63 splice forms in proliferating cells versus senescent cells could modulate the activity of p53, either by competition for p53 DNA binding sites or by directly interacting with DNA-bound p53 protein. Perhaps in young proliferating cells where ΔNp63 isoform levels are high, p53 activity is suppressed, whereas in senescent cells p53 activity is enhanced by the presence of the TAp63α and γ isoforms. The interaction of large T antigen with p63 proteins could perhaps affect their activities, and thus have an effect on expression of downstream targets.

CONCLUSIONS

We have described how the viral SV40 large T antigen immortalizes primary cells via its ability to perturb a variety of pathways in the cells (see Fig. 14.4.). These include the pRB and p53 pathways, as well as abrogation of the activities of the p300/CBP family of proteins, among others. The ability of large T antigen to overcome the finite proliferative potential of many cell types has led to its extensive use as a model system for immortalization studies, and its use in the generation of cell lines, in addi-

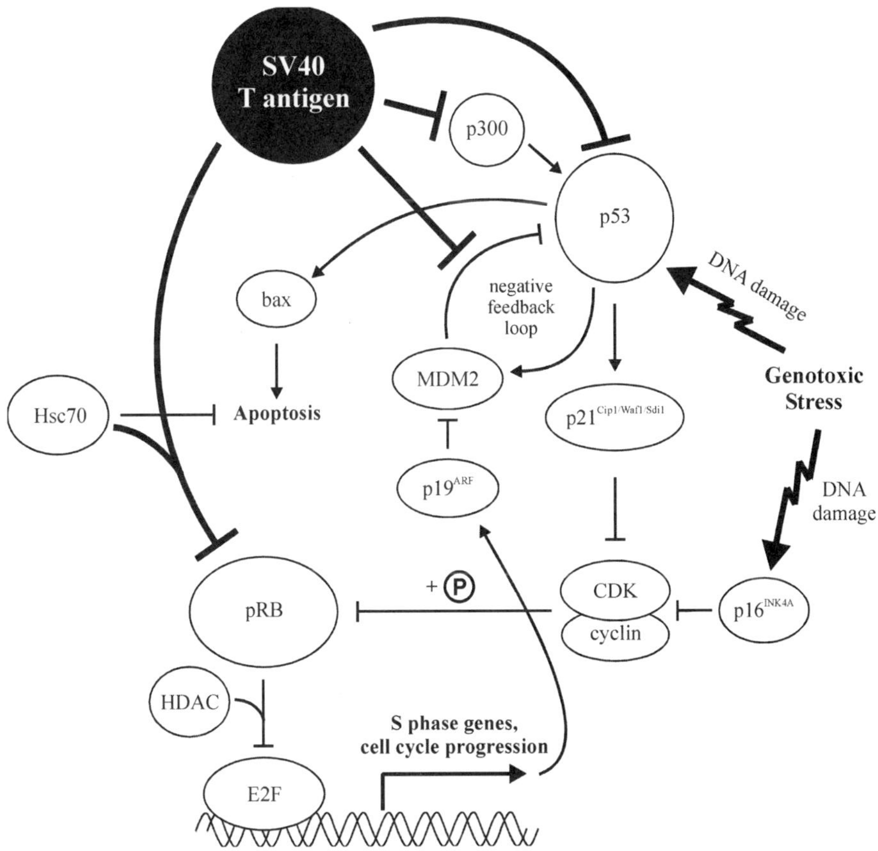

Figure 14.4. Cellular pathways affected by SV40 large T antigen. This figure summarizes some of the pathways that are involved in cell cycle progression, and shows the proteins that are affected by large T antigen.

tion to its use in allowing us to understand the process of transformation. The usefulness of large T antigen in these studies has been greatly enhanced by the availability of temperature sensitive mutants of the protein, enabling the senescence program of the cell to be induced at will via inactivation of large T antigen upon temperature shift to the non-permissive temperature. This has allowed insights to be gained into the wild-type function of these pathways. New interactions of large T antigen with host proteins continue to be identified, enabling a further understanding of how large T antigen causes deregulation of cell growth.

REFERENCES

Adams PD (2001): Regulation of the retinoblastoma tumor suppressor protein by cyclin/cdks. Biochim Biophys Acta 1471(3):M123–33.

Ali SH, DeCaprio JA (2001): Cellular transformation by SV40 large T antigen: Interaction with host proteins. Semin Cancer Biol 11(1):15–23.

Alwine JC, Reed SI, et al. (1977): Characterization of the autoregulation of simian virus 40 gene A. J Virol 24(1):22–7.

Appella E, Anderson CW (2001): Post-translational modifications and activation of p53 by genotoxic stresses. Eur J Biochem 268(10):2764–72.

Atadja P, Wong H, et al. (1995): Increased activity of p53 in senescing fibroblasts. Proc Natl Acad Sci USA 92(18):8348–52.

Avantaggiati ML, Carbone M, et al. (1996): The SV40 large T antigen and adenovirus E1a oncoproteins interact with distinct isoforms of the transcriptional co-activator, p300. EMBO J 15(9):2236–48.

Avantaggiati ML, Ogryzko V, et al. (1997): Recruitment of p300/CBP in p53-dependent signal pathways. Cell 89(7):1175–84.

Bartek J, Bartkova J, et al. (1996): The retinoblastoma protein pathway and the restriction point. Curr Opin Cell Biol 8(6):805–14.

Berjanskii MV, Riley MI, et al. (2000): NMR structure of the N-terminal J domain of murine polyomavirus T antigens: Implications for DnaJ-like domains and for mutations of T antigens. J Biol Chem 275(46):36094–103.

Blagosklonny MV (2000): p53 from complexity to simplicity: mutant p53 stabilization, gain-of-function, and dominant-negative effect. Faseb J 14(13): 1901–7.

Bodnar AG, Ouellette M, et al. (1998): Extension of life-span by introduction of telomerase into normal human cells. Science 279(5349):349–52.

Bond J, Haughton M, et al. (1996): Evidence that transcriptional activation by p53 plays a direct role in the induction of cellular senescence. Oncogene 13(10):2097–104.

Brodsky JL, Pipas JM (1998): Polyomavirus T antigens: Molecular chaperones for multiprotein complexes. J Virol 72(7):5329–34.

Brown DR, Deb S, et al. (1993): The tumor suppressor p53 and the oncoprotein simian virus 40 T antigen bind to overlapping domains on the MDM2 protein. Mol Cell Biol 13(11):6849–57.

Campbell KS, Mullane KP, et al. (1997): DnaJ/hsp40 chaperone domain of SV40 large T antigen promotes efficient viral DNA replication. Genes Dev 11(9):1098–110.

Caplan AJ, Cyr DM, et al. (1993): Eukaryotic homologues of *Escherichia coli* dnaJ: A diverse protein family that functions with hsp70 stress proteins. Mol Biol Cell 4(6):555–63.

Carroll RB, Hager L, et al. (1974): Simian virus 40 T antigen binds to DNA. Proc Natl Acad Sci USA 71(9):3754–7.

Carroll RB, Samad A, et al. (1988): RNA is covalently linked to SV40 large T antigen. Oncogene 2(5):437–44.

Celli J, Duijf P, et al. (1999): Heterozygous germline mutations in the p53 homolog p63 are the cause of EEC syndrome. Cell 99(2):143–53.

Cheetham ME, Caplan AJ (1998): Structure, function and evolution of DnaJ: Conservation and adaptation of chaperone function. Cell Stress Chaperones 3(1):28–36.

Christensen JB, Imperiale MJ (1995): Inactivation of the retinoblastoma susceptibility protein is not sufficient for the transforming function of the conserved region 2-like domain of simian virus 40 large T antigen. J Virol 69(6):3945–8.

Classon M, Dyson N (2001): p107 and p130: versatile proteins with interesting pockets. Expr Cell Res 264(1):135–47.

Cole CN (1996): In: BN Fields, DM Knipe, PM Howley (eds): Polyomaviridae: The Viruses and Their Replication. Philadelphia: Lippincott-Raven, pp 1997–2043.

Conzen SD, Cole CN (1995): The three transforming regions of SV40 T antigen are required for immortalization of primary mouse embryo fibroblasts. Oncogene 11(11):2295–302.

Counter CM, Meyerson M, et al. (1998): Telomerase activity is restored in human cells by ectopic expression of hTERT (hEST2), the catalytic subunit of telomerase. Oncogene 16(9):1217–22.

Cox LS (1997): Who binds wins: competition for PCNA rings out cell-cycle changes. Trends Cell Biol 7:493–8.

De Laurenzi V, Costanzo A, et al. (1998): Two new p73 splice variants, gamma and delta, with different transcriptional activity. J Expr Med 188(9):1763–8.

De Laurenzi VD, Catani MV, et al. (1999): Additional complexity in p73: Induction by mitogens in lymphoid cells and identification of two new splicing variants epsilon and zeta. Cell Death Differ 6(5):389–90.

DeCaprio JA, Furukawa Y, et al. (1992): The retinoblastoma-susceptibility gene product becomes phosphorylated in multiple stages during cell cycle entry and progression. Proc Natl Acad Sci USA 89(5):1795–8.

DeCaprio JA, Ludlow JW, et al. (1988): SV40 large tumor antigen forms a specific complex with the product of the retinoblastoma susceptibility gene. Cell 54(2):275–83.

Deppert W, Haug M, et al. (1987): Modulation of p53 protein expression during cellular transformation with simian virus 40. Mol Cell Biol 7(12):4453–63.

Djelloul S, Tarunina M, et al. (2002): Differential protein expression, DNA binding and interaction with SV40 large tumour antigen implicate the p63-family of proteins in replicative senescence. Oncogene 21(7):981–9.

Donehower LA, Harvey M, et al. (1992): Mice deficient for p53 are developmentally normal but susceptible to spontaneous tumours. Nature 356(6366): 215–21.

Dryja TP, Bruns GA, et al. (1983): Isolation of DNA fragments from chromosome 13. Retina 3(2):121–5.

Dyson N, Buchkovich K, et al. (1989): The cellular 107K protein that binds to adenovirus E1A also associates with the large T antigens of SV40 and JC virus. Cell 58(2):249–55.

Eckner R, Ludlow JW, et al. (1996): Association of p300 and CBP with simian virus 40 large T antigen. Mol Cell Biol 16(7):3454–64.

Ewen ME, Ludlow JW, et al. (1989): An N-terminal transformation-governing sequence of SV40 large T antigen contributes to the binding of both p110Rb and a second cellular protein, p120. Cell 58(2):257–67.

Fanning E (1992): Simian virus 40 large T antigen: The puzzle, the pieces, and the emerging picture. J Virol 66(3):1289–93.

Flores-Rozas H, Kelman Z, et al. (1994): Cdk-interacting protein 1 directly binds with proliferating cell nuclear antigen and inhibits DNA replication catalyzed by the DNA polymerase delta holoenzyme. Proc Natl Acad Sci USA 91(18): 8655–9.

Gayther SA, Batley SJ, et al. (2000): Mutations truncating the Ep300 acetylase in human cancers. Nat Genet 24(3):300–3.

Gilinger G, Alwine JC (1993): Transcriptional activation by simian virus 40 large T antigen: Requirements for simple promoter structures containing either TATA or initiator elements with variable upstream factor binding sites. J Virol 67(11):6682–8.

Ginsberg D, Mechta F, et al. (1991): Wild-type p53 can down-modulate the activity of various promoters. Proc Natl Acad Sci USA 88(22):9979–83.

Gire V, Wynford-Thomas D (1998): Reinitiation of DNA synthesis and cell division in senescent human fibroblasts by microinjection of anti-p53 antibodies. Mol Cell Biol 18(3):1611–21.

Gjoerup O, Chao H, et al. (2000): pRB-dependent, J domain-independent function of simian virus 40 large T antigen in override of p53 growth suppression. J Virol 74(2):864–74.

Gonos ES, Burns JS, et al. (1996): Rat embryo fibroblasts immortalized with simian virus 40 large T antigen undergo senescence upon its inactivation. Mol Cell Biol 16(9):5127–38.

Goodman RH, Smolik S (2000): CBP/p300 in cell growth, transformation, and development. Genes Dev 14(13):1553–77.

Grand RJ, Turnell AS, et al. (1999): Adenovirus early region 1A protein binds to mammalian SUG1-a regulatory component of the proteasome. Oncogene 18(2):449–58.

Grand RJA (2001): Viruses, Cell Transformation and Cancer. Amsterdam: Elsevier.

Grossman SR, Perez M, et al. (1998): p300/MDM2 complexes participate in MDM2-mediated p53 degradation. Mol Cell 2(4):405–15.

Gruda MC, Zabolotny JM, et al. (1993): Transcriptional activation by simian virus 40 large T antigen: Interactions with multiple components of the transcription complex. Mol Cell Biol 13(2):961–9.

Hahn WC, Counter CM, et al. (1999): Creation of human tumour cells with defined genetic elements. Nature 400(6743):464–8.

Hannon GJ, Demetrick D, et al. (1993): Isolation of the Rb-related p130 through its interaction with CDK2 and cyclins. Genes Dev 7(12A):2378–91.

Hansen U, Tenen DG, et al. (1981): T antigen repression of SV40 early transcription from two promoters. Cell 27(3, Pt 2):603–13.

Harris KF, Christensen JB, et al. (1998): Novel mechanisms of E2F induction by BK virus large-T antigen: Requirement of both the pRb-binding and the J domains. Mol Cell Biol 18(3):1746–56.

Hartl FU (1996): Molecular chaperones in cellular protein folding. Nature 381(6583):571–9.

Hayflick L, Moorhead PS (1961): The serial cultivation of human diploid cell strains. Expr Cell Res 25:585–621.

Henning W, Rohaly G, et al. (1997): MDM2 is a target of simian virus 40 in cellular transformation and during lytic infection. J Virol 71(10):7609–18.

Hermeking H, Lengauer C, et al. (1997): 14-3-3 sigma is a p53-regulated inhibitor of G2/M progression. Mol Cell 1(1):3–11.

Hiebert SW (1993): Regions of the retinoblastoma gene product required for its interaction with the E2F transcription factor are necessary for E2 promoter repression and pRb-mediated growth suppression. Mol Cell Biol 13(6): 3384–91.

Hilleman MR (1998): Discovery of simian virus 40 (SV40) and its relationship to poliomyelitis virus vaccines. Dev Biol Stand 94:183–90.

Hollstein M, Rice K, et al. (1994): Database of p53 gene somatic mutations in human tumors and cell lines. Nucl Acids Res 22(17):3551–5.

Horikoshi N, Usheva A, et al. (1995): Two domains of p53 interact with the TATA-binding protein, and the adenovirus 13S E1A protein disrupts the association, relieving p53-mediated transcriptional repression. Mol Cell Biol 15(1):227–34.

Hottiger MO, Nabel GJ (2000): Viral replication and the coactivators p300 and CBP. Trends Microbiol 8(12):560–5.

Hsieh JK, Chan FS, et al. (1999): RB regulates the stability and the apoptotic function of p53 via MDM2. Mol Cell 3(2):181–93.

Hu N, Gutsmann A, et al. (1994): Heterozygous Rb-1 delta 20/+ mice are predisposed to tumors of the pituitary gland with a nearly complete penetrance. Oncogene 9(4):1021–7.

Hu QJ, Dyson N, et al. (1990): The regions of the retinoblastoma protein needed for binding to adenovirus E1A or SV40 large T antigen are common sites for mutations. EMBO J 9(4):1147–55.

Huang K, Flanagan JM, et al. (1999): The influence of C-terminal extension on the structure of the "J-domain" in *E. coli* DnaJ. Protein Sci 8(1):203–14.

Huang S, Wang NP, et al. (1990): Two distinct and frequently mutated regions of retinoblastoma protein are required for binding to SV40 T antigen. EMBO J 9(6):1815–22.

Ikram Z, Norton T, et al. (1994): The biological clock that measures the mitotic life-span of mouse embryo fibroblasts continues to function in the presence of simian virus 40 large tumor antigen. Proc Natl Acad Sci USA 91(14): 6448–52.

Ishida S, Yamashita T, et al. (2000): Adenovirus-mediated transfer of p53-related genes induces apoptosis of human cancer cells. Jpn J Cancer Res 91(2): 174–80.

Janknecht R (2002): The versatile functions of the transcriptional coactivators p300 and CBP and their roles in disease. Histol Histopathol 17(2):657–68.

Jat PS, Sharp PA (1989): Cell lines established by a temperature-sensitive simian virus 40 large-T-antigen gene are growth restricted at the nonpermissive temperature. Mol Cell Biol 9(4):1672–81.

Jay G, Nomura S, et al. (1981): Identification of the SV40 agnogene product: a DNA binding protein. Nature 291(5813):346–9.

Jiang RF, Greener T, et al. (1997): Interaction of auxilin with the molecular chaperone, Hsc70. J Biol Chem 272(10):6141–5.

Johnson RA, Ince TA, et al. (2001): Transcriptional repression by p53 through direct binding to a novel DNA element. J Biol Chem 276(29):27716–20.

Kaghad M, Bonnet H, et al. (1997): Monoallelically expressed gene related to p53 at 1p36, a region frequently deleted in neuroblastoma and other human cancers. Cell 90(4):809–19.

Kalderon D, Richardson WD, et al. (1984): Sequence requirements for nuclear location of simian virus 40 large-T antigen. Nature 311(5981):33–8.

Kastan MB, Onyekwere O, et al. (1991): Participation of p53 protein in the cellular response to DNA damage. Cancer Res 51(23, Pt 1):6304–11.

Katakura Y, Alam S, et al. (1998): Immortalization by gene transfection. Methods Cell Biol 57:69–91.

Kelley WL, Georgopoulos C (1997): The T/t common exon of simian virus 40, JC, and BK polyomavirus T antigens can functionally replace the J-domain of the *Escherichia coli* DnaJ molecular chaperone. Proc Natl Acad Sci USA 94(8):3679–84.

Kelley WL, Landry SJ (1994): Chaperone power in a virus? Trends Biochem Sci 19(7):277–8.

Kim HY, Ahn BY, et al. (2001): Structural basis for the inactivation of retinoblastoma tumor suppressor by SV40 large T antigen. EMBO J 20(1–2):295–304.

Kiyono T, Foster SA, et al. (1998): Both Rb/p16INK4a inactivation and telomerase activity are required to immortalize human epithelial cells. Nature 396(6706):84–8.

Kohrman DC, Imperiale MJ (1992): Simian virus 40 large T antigen stably complexes with a 185-kilodalton host protein. J Virol 66(3):1752–60.

Kuerbitz SJ, Plunkett BS, et al. (1992): Wild-type p53 is a cell cycle checkpoint determinant following irradiation. Proc Natl Acad Sci USA 89(16):7491–5.

Kulju KS, Lehman JM (1995): Increased p53 protein associated with aging in human diploid fibroblasts. Expr Cell Res 217(2):336–45.

Kussie PH, Gorina S, et al. (1996): Structure of the MDM2 oncoprotein bound to the p53 tumor suppressor transactivation domain. Science 274(5289): 948–53.

Lane DP (1992): Cancer: p53, guardian of the genome. Nature 358(6381):15–6.

Lane DP, Crawford LV (1979): T antigen is bound to a host protein in SV40-transformed cells. Nature 278(5701):261–3.

Laub O, Aloni Y (1975): Transcription of simian virus 40. V. Regulation of simian virus 40 gene expression. J Virol 16(5):1171–83.

Laufen T, Mayer MP, et al. (1999): Mechanism of regulation of hsp70 chaperones by DnaJ cochaperones. Proc Natl Acad Sci USA 96(10):5452–7.

Lee C, Cho Y (2002): Interactions of SV40 large T antigen and other viral proteins with retinoblastoma tumour suppressor. Rev Med Virol 12(2):81–92.

Lee EY, Chang CY, et al. (1992): Mice deficient for Rb are nonviable and show defects in neurogenesis and haematopoiesis. Nature 359(6393):288–94.

Lee JO, Russo AA, et al. (1998): Structure of the retinoblastoma tumour-suppressor pocket domain bound to a peptide from HPV E7. Nature 391(6670):859–65.

Levrero M, De Laurenzi V, et al. (2000): The p53/p63/p73 family of transcription factors: Overlapping and distinct functions. J Cell Sci 113(Pt 10):1661–70.

Li R, Waga S, et al. (1994): Differential effects by the p21 CDK inhibitor on PCNA-dependent DNA replication and repair. Nature 371(6497):534–7.

Lill NL, Grossman SR, et al. (1997): Binding and modulation of p53 by p300/CBP coactivators. Nature 387(6635):823–7.

Lill NL, Tevethia MJ, et al. (1997): p300 family members associate with the carboxyl terminus of simian virus 40 large tumor antigen. J Virol 71(1):129–37.

Linzer DI, Levine AJ (1979): Characterization of a 54K dalton cellular SV40 tumor antigen present in SV40-transformed cells and uninfected embryonal carcinoma cells. Cell 17(1):43–52.

Ludlow JW, DeCaprio JA, et al. (1989): SV40 large T antigen binds preferentially to an underphosphorylated member of the retinoblastoma susceptibility gene product family. Cell 56(1):57–65.

Ludlow JW, Shon J, et al. (1990): The retinoblastoma susceptibility gene product undergoes cell cycle-dependent dephosphorylation and binding to and release from SV40 large T. Cell 60(3):387–96.

Maltzman W, Czyzyk L (1984): UV irradiation stimulates levels of p53 cellular tumor antigen in nontransformed mouse cells. Mol Cell Biol 4(9):1689–94.

Malumbres M, Barbacid M (2001): To cycle or not to cycle: A critical decision in cancer. Nat Rev Cancer 1(3):222–31.

Manos MM, Gluzman Y (1984): Simian virus 40 large T-antigen point mutants that are defective in viral DNA replication but competent in oncogenic transformation. Mol Cell Biol 4(6):1125–33.

Martin DW, Subler MA, et al. (1993): p53 and SV40 T antigen bind to the same region overlapping the conserved domain of the TATA-binding protein. Biochem Biophys Res Commun 195(1):428–34.

McCarthy SA, Symonds HS, et al. (1994): Regulation of apoptosis in transgenic mice by simian virus 40 T antigen-mediated inactivation of p53. Proc Natl Acad Sci USA 91(9):3979–83.

Mercer WE, Shields MT, et al. (1991): Growth suppression induced by wild-type p53 protein is accompanied by selective down-regulation of proliferating-cell nuclear antigen expression. Proc Natl Acad Sci USA 88(5):1958–62.

Michael-Michalovitz D, Yehiely F, et al. (1991): Simian virus 40 can overcome the antiproliferative effect of wild-type p53 in the absence of stable large T antigen-p53 binding. J Virol 65(8):4160–8.

Mills AA, Zheng B, et al. (1999): p63 is a p53 homologue required for limb and epidermal morphogenesis. Nature 398(6729):708–13.

Mitchell PJ, Wang C, et al. (1987): Positive and negative regulation of transcription in vitro: Enhancer-binding protein AP-2 is inhibited by SV40 T antigen. Cell 50(6):847–61.

Miyashita T, Krajewski S, et al. (1994): Tumor suppressor p53 is a regulator of bcl-2 and bax gene expression in vitro and in vivo. Oncogene 9(6):1799–805.

Moberg KH, Tyndall WA, et al. (1992): Wild-type murine p53 represses transcription from the murine c-myc promoter in a human glial cell line. J Cell Biochem 49(2):208–15.

Momand J, Wu HH, et al. (2000): MDM2—Master regulator of the p53 tumor suppressor protein. Gene 242(1–2):15–29.

Moran E (1988): A region of SV40 large T antigen can substitute for a transforming domain of the adenovirus E1A products. Nature 334(6178):168–70.

Munger K, Werness BA, et al. (1989): Complex formation of human papillomavirus E7 proteins with the retinoblastoma tumor suppressor gene product. EMBO J 8(13):4099–105.

Muraoka M, Konishi M, et al. (1996): p300 gene alterations in colorectal and gastric carcinomas. Oncogene 12(7):1565–9.

Murphy M, Ahn J, et al. (1999): Transcriptional repression by wild-type p53 utilizes histone deacetylases, mediated by interaction with mSin3a. Genes Dev 13(19):2490–501.

Nelson DA, Krucher NA, et al. (1997): High molecular weight protein phosphatase type 1 dephosphorylates the retinoblastoma protein. J Biol Chem 272(7):4528–35.

North S, Hainaut P (2000): p53 and cell-cycle control: A finger in every pie. Pathol Biol (Paris) 48(3):255–70.

O'Hare MJ, Bond J, et al. (2001): Conditional immortalization of freshly isolated human mammary fibroblasts and endothelial cells. Proc Natl Acad Sci USA 98(2):646–51.

Oren M, Maltzman W, et al. (1981): Post-translational regulation of the 54K cellular tumor antigen in normal and transformed cells. Mol Cell Biol 1(2):101–10.

Ouellette MM, Liao M, et al. (2000): Subsenescent telomere lengths in fibroblasts immortalized by limiting amounts of telomerase. J Biol Chem 275(14):10072–6.

Pardee AB (1989): G_1 events and regulation of cell proliferation. Science 246(4930):603–8.

Peden KW, Spence SL, et al. (1990): A DNA replication-positive mutant of simian virus 40 that is defective for transformation and the production of infectious virions. J Virol 64(6):2912–21.

Powell AJ, Darmon AJ, et al. (1999): Different functions are required for initiation and maintenance of immortalization of rat embryo fibroblasts by SV40 large T antigen. Oncogene 18(51):7343–50.

Qin XQ, Chittenden T, et al. (1992): Identification of a growth suppression domain within the retinoblastoma gene product. Genes Dev 6(6):953–64.

Quartin RS, Cole CN, et al. (1994): The amino-terminal functions of the simian virus 40 large T antigen are required to overcome wild-type p53-mediated growth arrest of cells. J Virol 68(3):1334–41.

Ragimov N, Krauskopf A, et al. (1993): Wild-type but not mutant p53 can repress transcription initiation in vitro by interfering with the binding of basal transcription factors to the TATA motif. Oncogene 8(5):1183–93.

Ray FA, Peabody DS, et al. (1990): SV40 T antigen alone drives karyotype instability that precedes neoplastic transformation of human diploid fibroblasts. J Cell Biochem 42(1):13–31.

Rice PW, Cole CN (1993): Efficient transcriptional activation of many simple modular promoters by simian virus 40 large T antigen. J Virol 67(11):6689–97.

Robles SJ, Adami GR (1998): Agents that cause DNA double strand breaks lead to p16INK4a enrichment and the premature senescence of normal fibroblasts. Oncogene 16(9):1113–23.

Rundell K, Parakati R (2001): The role of the SV40 ST antigen in cell growth promotion and transformation. Semin Cancer Biol 11(1):5–13.

Rushton JJ, Jiang D, et al. (1997): Simian virus 40 T antigen can regulate p53-mediated transcription independent of binding p53. J Virol 71(7):5620–3.

Saffer JD, Jackson SP, et al. (1990): SV40 stimulates expression of the transacting factor Sp1 at the mRNA level. Genes Dev 4(4):659–66.

Santos M, Butel JS (1982): Association of SV40 large tumor antigen and cellular proteins on the surface of SV40-transformed mouse cells. Virology 120(1):1–17.

Sawai ET, Butel JS (1989): Association of a cellular heat shock protein with simian virus 40 large T antigen in transformed cells. J Virol 63(9):3961–73.

Scheffner M, Knippers R, et al. (1989): RNA unwinding activity of SV40 large T antigen. Cell 57(6):955–63.

Sedivy JM (1998): Can ends justify the means? Telomeres and the mechanisms of replicative senescence and immortalization in mammalian cells. Proc Natl Acad Sci USA 95(16):9078–81.

Shapiro GI, Edwards CD, et al. (1998): p16INK4A participates in a G_1 arrest checkpoint in response to DNA damage. Mol Cell Biol 18(1):378–87.

Shay JW, Wright WE, et al. (1991): Defining the molecular mechanisms of human cell immortalization. Biochim Biophys Acta 1072(1):1–7.

Sheng Q, Denis D, et al. (1997): The DnaJ domain of polyomavirus large T antigen is required to regulate Rb family tumor suppressor function. J Virol 71(12):9410–6.

Shiio Y, Yamamoto T, et al. (1992): Negative regulation of Rb expression by the p53 gene product. Proc Natl Acad Sci USA 89(12):5206–10.

Slinskey A, Barnes D, et al. (1999): Simian virus 40 large T antigen J domain and Rb-binding motif are sufficient to block apoptosis induced by growth factor withdrawal in a neural stem cell line. J Virol 73(8):6791–9.

Smale ST, Tjian R (1986): T-antigen-DNA polymerase alpha complex implicated in simian virus 40 DNA replication. Mol Cell Biol 6(11):4077–87.

Sompayrac L, Danna KJ (1991): The amino-terminal 147 amino acids of SV40 large T antigen transform secondary rat embryo fibroblasts. Virology 181(1): 412–5.

Soule HR, Butel JS (1979): Subcellular Localization of simian virus 40 large tumor antigen. J Virol 30(2):523–32.

Srinivasan A, McClellan AJ, et al. (1997): The amino-terminal transforming region of simian virus 40 large T and small t antigens functions as a J domain. Mol Cell Biol 17(8):4761–73.

Stewart N, Bacchetti S (1991): Expression of SV40 large T antigen, but not small t antigen, is required for the induction of chromosomal aberrations in transformed human cells. Virology 180(1):49–57.

Stringer JR (1982): Mutant of simian virus 40 large T-antigen that is defective for viral DNA synthesis, but competent for transformation of cultured rat cells. J Virol 42(3):854–64.

Struhl K (1998): Histone acetylation and transcriptional regulatory mechanisms. Genes Dev 12(5):599–606.

Stubdal H, Zalvide J, et al. (1997): Inactivation of pRB-related proteins p130 and p107 mediated by the J domain of simian virus 40 large T antigen. Mol Cell Biol 17(9):4979–90.

Stubdal H, Zalvide J, et al. (1996): Simian virus 40 large T antigen alters the phosphorylation state of the RB-related proteins p130 and p107. J Virol 70(5): 2781–8.

Sugrue MM, Shin DY, et al. (1997): Wild-type p53 triggers a rapid senescence program in human tumor cells lacking functional p53. Proc Natl Acad Sci USA 94(18):9648–53.

Sullivan CS, Cantalupo P, et al. (2000): The molecular chaperone activity of simian virus 40 large T antigen is required to disrupt Rb-E2F family complexes by an ATP-dependent mechanism. Mol Cell Biol 20(17): 6233–43.

Sullivan CS, Pipas JM (2002): T antigens of simian virus 40: Molecular chaperones for viral replication and tumorigenesis. Microbiol Mol Biol Rev 66(2): 179–202.

Sullivan CS, Tremblay JD, et al. (2000): Species-specific elements in the large T-antigen J domain are required for cellular transformation and DNA replication by simian virus 40. Mol Cell Biol 20(15):5749–57.

Szekely L, Uzvolgyi E, et al. (1991): Subcellular localization of the retinoblastoma protein. Cell Growth Differ 2(6):287–95.

Taylor WR, Stark GR (2001): Regulation of the G2/M transition by p53. Oncogene 20(15):1803–15.

Tegtmeyer P (1972): Simian virus 40 deoxyribonucleic acid synthesis: The viral replicon. J Virol 10(4):591–8.

Tegtmeyer P (1975): Function of simian virus 40 gene A in transforming infection. J Virol 15(3):613–8.

Tevethia MJ, Lacko HA, et al. (1997): Adding an Rb-binding site to an N-terminally truncated simian virus 40 T antigen restores growth to high cell density, and the T common region in trans provides anchorage-independent growth and rapid growth in low serum concentrations. J Virol 71(3):1888–96.

Tevethia MJ, Pipas JM, et al. (1988): Requirements for immortalization of primary mouse embryo fibroblasts probed with mutants bearing deletions in the 3′ end of SV40 gene A. Virology 162(1):76–89.

Thanos CD, Bowie JU (1999): p53 Family members p63 and p73 are SAM domain-containing proteins. Protein Sci 8(8):1708–10.

Tibbetts RS, Brumbaugh KM, et al. (1999): A role for ATR in the DNA damage-induced phosphorylation of p53. Genes Dev 13(2):152–7.

Tiemann F, Zerrahn J, et al. (1995): Cooperation of simian virus 40 large and small T antigens in metabolic stabilization of tumor suppressor p53 during cellular transformation. J Virol 69(10):6115–21.

Tjian R, Robbins A (1979): Enzymatic activities associated with a purified simian virus 40 T antigen-related protein. Proc Natl Acad Sci USA 76(2):610–4.

Tooze J (1981): Molecular Biology of Tumor Viruses: DNA Tumor Viruses, 2nd ed. Cold Spring Harbor: Cold Spring Harbor Laboratory.

Turenne GA, Paul P, et al. (2001): Activation of p53 transcriptional activity requires ATM's kinase domain and multiple N-terminal serine residues of p53. Oncogene 20(37):5100–10.

Ueda Y, Hijikata M, et al. (1999): New p73 variants with altered C-terminal structures have varied transcriptional activities. Oncogene 18(35):4993–8.

Vaziri H, Benchimol S (1998): Reconstitution of telomerase activity in normal human cells leads to elongation of telomeres and extended replicative life span. Curr Biol 8(5):279–82.

Waga S, Hannon GJ, et al. (1994): The p21 inhibitor of cyclin-dependent kinases controls DNA replication by interaction with PCNA. Nature 369(6481):574–8.

Warbrick E (1998): PCNA binding through a conserved motif. Bioessays 20(3): 195–9.

Weinberg RA (1995): The retinoblastoma protein and cell cycle control. Cell 81(3):323–30.

Welch PJ, Wang JY (1993): A C-terminal protein-binding domain in the retinoblastoma protein regulates nuclear c-Abl tyrosine kinase in the cell cycle. Cell 75(4):779–90.

Whitaker LL, Su H, et al. (1998): Growth suppression by an E2F-binding-defective retinoblastoma protein (RB): Contribution from the RB C pocket. Mol Cell Biol 18(7):4032–42.

Woods C, LeFeuvre C, et al. (1994): Induction of genomic instability in SV40 transformed human cells: Sufficiency of the N-terminal 147 amino acids of large T antigen and role of pRB and p53. Oncogene 9(10):2943–50.

Wright WE, Pereira-Smith OM, et al. (1989): Reversible cellular senescence: implications for immortalization of normal human diploid fibroblasts. Mol Cell Biol 9(7):3088–92.

Xiao ZX, Chen J, et al. (1995): Interaction between the retinoblastoma protein and the oncoprotein MDM2. Nature 375(6533):694–8.

Yanai N, Obinata M (1994): Apoptosis is induced at nonpermissive temperature by a transient increase in p53 in cell lines immortalized with temperature-sensitive SV40 large T-antigen gene. Expr Cell Res 211(2):296–300.

Yang A, Kaghad M, et al. (1998): p63, a p53 homolog at 3q27–29, encodes multiple products with transactivating, death-inducing, and dominant-negative activities. Mol Cell 2(3):305–16.

Yang A, Schweitzer R, et al. (1999): p63 is essential for regenerative proliferation in limb, craniofacial and epithelial development. Nature 398(6729):714–8.

Yang A, Walker N, et al. (2000): p73-deficient mice have neurological, pheromonal and inflammatory defects but lack spontaneous tumours. Nature 404(6773):99–103.

Yang J, Chang E, et al. (1999): Human endothelial cell life extension by telomerase expression. J Biol Chem 274(37):26141–8.

Zalvide J, Stubdal H, et al. (1998): The J domain of simian virus 40 large T antigen is required to functionally inactivate RB family proteins. Mol Cell Biol 18(3): 1408–15.

Zerrahn J, Tiemann F, et al. (1996): Simian virus 40 small t antigen activates the carboxyl-terminal transforming p53-binding domain of large T antigen. J Virol 70(10):6781–9.

Zhu JY, Rice PW, et al. (1991): Mapping the transcriptional transactivation function of simian virus 40 large T antigen. J Virol 65(6):2778–90.

CHAPTER 15

APOPTOSIS SIGNALING IN NORMAL AND CANCER CELLS

SHULIN WANG and WAFIK S. EL-DEIRY
Howard Hughes Medical Institute, Departments of Medicine, Genetics, Pharmacology, and Cancer Center, University of Pennsylvania, Philadelphia, PA 19104

INTRODUCTION

The number of cells in an organ is determined by the rates of cell migration, cell division, and cell death (Raff, 1996). In the early 1970s the discovery of new patterns of cell death led to emergence of the concept of apoptosis (Kerr et al., 1972). Several terms have been used to describe the morphology and biochemistry of physiological cell death (Schweichel and Merker, 1973; Hengartner, 2000), all of which fit under the more global term "programmed cell death," which was originally defined as a series of events that culminate in the death of a cell (Lockshin and Williams, 1965). These genetically regulated programmed cell deaths are distinct from necrosis in response to an insult, which results in leaking of the cell contents and inflammation. During apoptosis there are remarkably arranged morphological and biochemical events, while necrosis is an unordered and accidental form of cell death (Wyllie et al., 1980). The concept claims that the cell death from pathophysiological stimuli is an evolutionarily conserved mechanism of cell elimination upon morphogenetic and homeostatic signals in animals and plants. Defects in apoptotic pathways are now thought to contribute to a number of human diseases, ranging from neurodegenerative disorders to malignancy (Barr and Tomei, 1994; Rathmell and Thompson, 2002) and drug resistance (Johnstone et al., 2002; Reed, 2003). In the last decade basic cancer research has produced remarkable advances in our understanding of cancer biology and cancer genetics. Among the most important advances is the realization that apoptosis and the genes that control it have a profound effect on the malignant phenotype. More recent highlights in apoptosis research include the discovery of death

Cell Cycle and Growth Control: Biomolecular Regulation and Cancer, Second Edition, Edited by Gary S. Stein and Arthur B. Pardee
ISBN 0-471-25071-6

receptors and their ligands (Nagata, 1996; Ashkenazi, 2002; Wu et al., 1997), definition of the separate apoptosis signaling pathways, and the identification of the components of the common cell death effector machinery and their biochemical functions (Cory and Adams, 2002; Thornberry and Lazebnik, 1998; Green and Reed, 1998; Shi, 2002). Understanding the molecular events that contribute to drug-induced apoptosis, and how tumors evade apoptotic death, provides a paradigm to explain the relationship between cancer genetics and treatment sensitivity and should enable a more rational approach to anticancer drug design and therapy.

MAJOR MEDIATORS OF APOPTOSIS

So far several functional groups of molecules have been identified and found to be involved in triggering and affecting the apoptotic process, including caspases, members of the Bcl-2 family of proteins, members of the tumor necrosis factor (TNF) receptor (TNF-R) superfamily, and the p53 family.

Caspases

A group of cysteine proteases, now called caspases, are central components of the machinery responsible for apoptosis. The critical involvement of a cysteine protease CED-3 in apoptosis was first discovered in the nematode *Caenorhabditis elegans* approximately 10 years ago (Yuan et al., 1993). Since then, compelling evidence has demonstrated that the mechanisms of apoptosis are evolutionarily conserved, and executed by a family of cysteine proteases that recognize tetrapeptide motifs and cleave their substrates on the carboxyl side of an aspartate residue (Alnemri et al., 1996). At least 14 caspases have been identified in mammals (Nicholson and Thornberry, 1997; Fig. 15.1), with their orthologues present in species ranging from the nematode to the dipteran *Drosophila melanogaster* and the lepidopteran *Spodoptera frugiperda*.

Caspases involved in apoptosis are generally divided into two categories, the initiator caspases, which include caspase-2, -8, -9, and -10, and the effector caspases, which include caspase-3, -6, and –7 (Fig. 15.1). All caspases are produced in cells as catalytically inactive zymogens and must undergo proteolytic activation during apoptosis. The activation of an effector caspase (e.g., caspase-3 or -7) is performed by an initiator caspase (e.g., caspase-9) through cleavage at a specific internal Asp residue that separates the large (p20) and small (p10) subunits (Fig. 15.1). The initiator caspases, however, are autoactivated. As the activation of an initiator caspase in cells inevitably triggers a cascade of downstream caspase activation, it is tightly regulated and often requires the assembly of a multicomponent complex under apoptotic conditions. Some caspases, particularly effector caspases, cleave and inactivate a broad spectrum of cellular targets such as DNA repair enzymes, lamin, gelsolin,

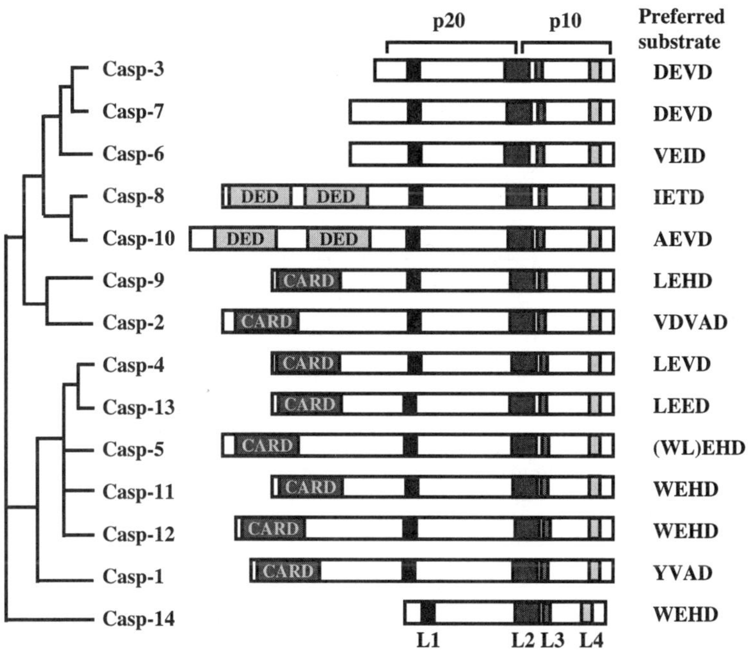

Figure 15.1. Mammalian caspase superfamily. The phylogenetic relationship of this family (*left*) appears to correct with their function in apoptosis and inflammation. Death effector domains (DED) and caspase recruitment domains (CARD) are shown, The four surface loops (L1–L4) that shape the catalytic groove are indicated. The large and small subunits, and their preferred substrate, are also indicated.

MDM2, and protein kinase C_{δ}, leading ultimately to cell death (Nicholson and Thornberry, 1997; Thornberry and Lazebnik, 1998).

Caspases are regulated by many cellular processes. For example, caspase are subject to transcriptional regulation and posttranslational modification (Earnshaw et al., 1999). In addition active caspases can be permanently eliminated by the ubiquitination-mediated proteasome degradation pathway (Huang et al., 2000, Suzuki et al., 2001). The enzymatic activity of caspases is subject to inhibition by the conserved IAP (inhibitor of apoptosis) family of proteins (Deveraux and Reed, 1999; Hay, 2000). Eight distinct mammalian IAPs, including XIAP, c-IAP1, c-IAP2, and ML-IAP/livin (Ashhab et al., 2001; Kasof and Gomes, 2001; Vucic et al., 2000) have been identified, and they target the initiator caspase, caspase-9, and the effector caspase-3 and -7 (Deveraux and Reed, 1999). These IAP proteins do not inhibit other caspases, such as caspase-6 or -8.

The Bcl-2 Family

The Bcl-2 (B-cell lymphoma 2) family of intracellular proteins is the central regulator of caspase activation, and its opposing factions of anti-

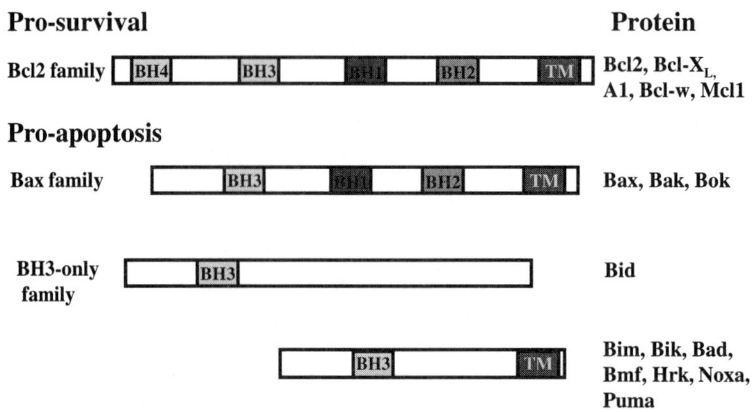

Figure 15.2. Structural comparision of the members of the Bcl-2 protein family. The Bcl-2 family of proteins can be divided into two subgroups, those that inhibit apoptosis and those that enhance it. Four highly conserved region (BH1–4) (Bcl-2 homology domains) are indicated. Most members have a carboxyl-terminal hydrobic domain that aid association with intracellular membranes, the exceptions being A1 and many of the BH3-only proteins (Bad, Bid, Noxa, Bmf, and Puma). TM is the transmembrane domain.

and pro-apoptotic members arbitrate the life-or-death switch (Cory and Adams, 2002). In mammals, Bcl-2 has at least 20 relatives, all of which share at least one conserved Bcl-2 homology (BH; Fig. 15.2). The Bcl-2 family includes four other anti-apoptotic proteins: Bcl-X_L, Bcl-w, A1 and Mcl1, and two groups of proteins that promote cell death: the Bax and the BH3-only domain families (Adams and Cory, 1998) (Fig. 15.2). Members of the Bax death family (Oltvai et al., 1993) have sequences that are similar to those in Bcl-2, especially the BH1, BH2, and BH3 regions, whereas other pro-apoptotic proteins have only the short BH3 motif, an interaction domain that is both necessary and sufficient for their killing action (Fig. 15.2). Both types of pro-apoptotic proteins are required to initiate apoptosis: the BH3-only proteins seem to act as damage sensors and direct antagonists of the pro-survival proteins, whereas the Bax-like proteins act further downstream, probably in mitochondrial disruption (Huang and Strasser, 2000).

Many members of the Bcl-2 family have a conserved C-terminal transmembrane region. This region localizes proteins to the outer leaflet of the nuclear envelop, the outer mitochondrial membrane, and the endoplasmic reticulum (Chen-Levy et al., 1989; Krajewski et al., 1993). Bcl-2 is an integral membrane protein, even in healthy cells, whereas Bcl-w and Bcl-X_L only become tightly associated with the membrane after a cytotoxic signal. This subcellular localization appears to be fixed for some family members, but others can change their subcellular localization. It has been reported that Bax is a cytosolic protein until the cell is triggered to undergo apoptosis, when it becomes redistributed to mitochondrial membranes (Nguyen et al., 1994; Gross et al., 1998). It is possible that such subcellular redistribution may be a mechanism of regulating the activity of Bcl-2 family members.

Remarkably little overall sequence similarity is needed for similar function among the pro-apoptotic Bcl-2 family members (Fig. 15.2). Because all of the pro-apoptotic proteins so far discovered have a BH3 region and not all of the anti-apoptotic proteins have this region, this may be a defining characteristic for the two subfamilies. For instance Bik/Nbk, Bid, Bim, and Hrk/DP5 have only the BH3 domain in common, yet all are pro-apoptotic when overexpressed, and all bind to anti-apoptotic Bcl-2 family members (Boyd et al., 1995; Wang et al., 1996). The limited common features of the pro-apoptotic members of the Bcl-2 family mean that each has some specific activity, such as interaction with specific upstream signaling molecules, and that their common function is to bind and antagonize the anti-apoptotic effects of their ligands. This implies that the biochemical effects of the Bcl-2 family on apoptosis are mediated through the anti-apoptotic members.

Bcl-2 might control the activation of several initiator caspases that act upstream or independently of any mitochondrial breach. For instance, Bcl-2 can control apoptosis from the ER, and caspase-12, which can process other caspases in the absence of Apaf-1 or cytochrome c, is activated by ER-regulated stress and by serum deprivation (Lee et al., Hacki et al., 2000; Rudner et al., 2001; Rao et al., 2002; Nakagawa et al., 2000; Kilic et al., 2002). The Bcl-2 like proteins presumably also function at the mitochondrion to prevent Bax/Bak oligomerization. In the absence of convincing evidence for physical interaction of these opposing factions under physiological conditions, indirect models must be considered. First, if Bcl-2 helps to gate a mitochondrial pore, as some findings indicate, engagement of Bcl-2 by a BH-3 only protein might allow the release of small molecules that provoke a conformation change in Bax/Bak (Tsujimoto and Shimizu, 2000). Second, if Bcl-2 and Bax/Bak compete for an unknown target on mitochondria, the ligation of Bcl-2 might free it, allowing Bax to bind and nucleate pore formation. Third, if Bcl-2 sequesters caspase activators, their release from Bcl-2 might allow an activated caspase to mediate Bax translocation, perhaps via cleavage of a Bid-like protein or an outer mitochondrial membrane protein, as suggested for caspase-2 (Guo et al., 2002). The Bax/Bak oligomers might not only produce pores in the mitochondrial outer membrane; they could also perturb the ER/nuclear membrane. Alternatively, Bax/Bak oligmers might serve, somehow, as a platform for activation of some upstream caspases.

The activity of different pro-apoptotic Bcl-2 family members is controlled by post-translational modifications. The activity of Bad can be regulated by Akt or protein kinase A-mediated phosphorylation (Zha et al., 1996; del Peso et al., 1997; Datta et al., 1997). Bid is activated through caspase-mediated proteolysis (Luo et al., 1998; Li et al., 1998), and Bim is regulated by binding to the Dynein motor complex (Puthalakath et al., 1999). It is therefore possible that these molecules act as sentinels of cellular damage at distinct sites and that different apoptotic stimuli induce cell death by activating distinct members of the Bcl-2 family. Post-translational modification is not limited to pro-apoptotic members of the Bcl-2 family. Bcl-2 and Bcl-X_L have been shown to be regulated by phosphorylation (Chang et al., 1997; Ito et al., 1997), and Bcl-X_L and Ced-9 can be caspase substrates (Clem et al., 1998; Xue and Horvitz, 1997).

The Tumor-Necrosis Factor Receptor Family

Tumor-necrosis factor (TNF) was discovered many years ago as a serum factor that was able to kill cancer cells in mice. The TNF receptor (TNFR) was shown to be expressed by mammalian cells years later, and led to the discovery of a superfamily of transmembrane proteins. The discoveries led to the identification of two gene families that include 18 ligands and 28 receptors, many of which are being targeted as anticancer therapies (Ashkenazi, 2002). Most TNFR-superfamily members function as transmembrane signal transducers that respond to ligand binding. Some of the receptors, however, do not signal death or other intracellular effects. Instead, they seem to act as decoys that compete for the interaction of cognate ligands with their signaling receptors (Ashkenazi and Dixit, 1998). The signaling members of the TNFR superfamily can be divided into two main subgroups on the basis of their cytoplasmic region (Fig. 15.3). One class of receptors, called the death receptor (DR), con-

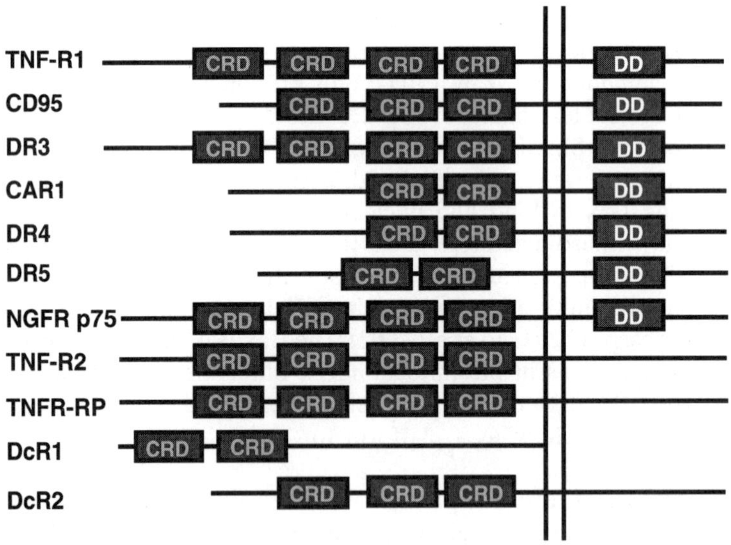

Figure 15.3. Structural comparision of the members of the TNF receptor family. The extracellular ligand-binding of the receptors are characterized by various numbers of cysteine-rich domain (CRD). Death receptors contain a death domain (DD) in their intracellular region, which is essential for apoptosis signaling. Decoy receptors DcR1 and DcR2 compete with DR4 and DR5 for binding to TRAIL.

tains a cytoplasmic death domain; the other class does not. Death domains mediate interaction of death receptors with death-domain-containing adaptor proteins. The adaptors contain additional sequence modules that mediate binding to intracellular effector enzymes. One of the adaptors, called FADD (Fas-associated death domain), activates specific caspases that initiate apoptosis (Kischkel et al., 2000; Sprick et al., 2000). Another adaptor, called TRADD (TNFR-associated death domain), stimulates protein kinases that control phosphorylation cascades, to induce transcription of immune-system modulation genes. Alternatively, TRADD can initiate apoptosis through FADD (Hsu et al., 1996a,b).

Much has also been learned about other death-inducing ligands and their receptors (Fig. 15.3). TRAIL/APO-2L (TNF-related apoptosis-inducing ligand; TRAIL) induces apoptosis preferentially in transformed and cancer cells, and in contrast to other death-inducing ligands, it is expressed in a wide range of tissues (Wiley et al., 1995). Several receptors have been identified for TRAIL: two death receptors, DR4 (TRAIL-R1) and DR5 (also called KILLER or TRAIL-R2), and two decoy receptors, DcR1 and DcR2 (also called TRAIL-R3 and TRAIL-R4, respectively) (Pan et al., 1997a, b; Sheridan et al., 1997; Wu et al., 1997). As a third decoy receptor, OPG, identified initially as a receptor for RANKL/OPGL, was shown later to bind APO2L/TRAIL (Emery et al., 1998). The physiological relevance of OPG as a receptor for APO2L/TRAIL remains unclear, however, because the affinity for this ligand at physiological temperatures is very low (Truneh et al., 2000). Although the main biological activity of APO2L/TRAIL seems to be the induction of apoptosis, the complete physiological role of this ligand is not fully understood. The relatively complex receptor interaction pattern of APO/TRAIL is unmatched not only in the TNF superfamily but also in other cytokine systems. Mouse gene knockout studies indicate that APO/TRAIL has an important role in anti-tumor surveillance by immune cells (Cretney et al., 2002).

After binding of FasL or APO2L/TRAIL, the death receptors Fas, DR4 or DR5 each assemble a death-inducing signaling complex (DISC): through adaptor protein FAS-associated death domain (FADD), they recruit and activate the apoptosis-initiating proteases caspase-8 and/or caspase-10. The physical proximity of the recruited enzymes in the ligand-induced signaling complex leads to their autoactivation by proteolysis, thereby triggering intracellular signaling cascades that induce specific cellular responses. After binding tumor-necrosis factor (TNF), TNF receptor1 (TNFR1) first recruits TNFR-associated death domain (TRADD) as a platform adaptor, and then assembles alternative signaling complexes through a second adaptor (Hsu et al., 1996a, b). One type of complex is the DISC that involves FADD and caspase-8 and triggers apoptosis in a manner similar to the other death receptors. Another complex involves the receptor-interacting protein (RIP), which links receptor stimulation to the inhibitor of κB-kinase (iKK) cascade, activating the nuclear factor of κB (NF-κB) transcription factor. A third complex involves TNFR-associated factor 2 (TRAF2), which couples receptor engagement to the JUN N-terminal kinase (JNK) cascade, stimulating the AP-1 transcription factor. Death receptor DR3, which was identified by virtue of its homology to TNFR1, assembles signaling complexes that are similar to those of TNFR1 (Chinnaiyan et al., 1996; Marsters et al., 1996; Kitson et al., 1996). DR6, a similarly discovered death receptor, seems to signal mostly through JNK/AP-1 pathway (Zhao et al., 2001).

p53 Family

The p53 tumor-suppressor protein, first identified in 1979, act as a major node in a complex network that evolved to sense diverse cellular stresses including DNA damage and hyperproliferative signals (Ko and Prives, 1996). Once stabilized and activated by genotoxic stress, p53 can either activate or repress a wide array of different gene targets, and thus can regulate cell cycle, cell death, DNA repair, angiogenesis, and other outcomes. p53 functions, including apoptosis, are thought to require its sequence-specific DNA binding and transcriptional activation activities, and a number of apoptosis-related genes are induced by p53 activation (Johnstone et al., 2002). The p53 tumor suppressor belongs to a small family of related proteins that includes two other members, p63 and p73. Although structurally and functionally related, p63 and p73 have distinct roles in normal development (Irwin and Kaelin, 2001), whereas p53 seems to have evolved in higher organisms to prevent tumor development.

p53 can induce expression of a wide array of death effectors (Fig. 15.4). p53 inducible genes that might contribute to the induction of both death-receptor and mitochondrial apoptotic pathways have been described, although studies so far indicate that the principal role of p53 is in the induction of the apoptotic cascade associated with mitochondrial release of factors such as cytochrome c and SMAC (Ryan et al., 2001). In addition to inducing genes that drive apoptosis, p53 can activate the expression of genes that inhibit survival signaling (Hoffman et al., 2001). The ability to engage various apoptotic pathways via several routes is likely to be particularly important for the tumor-suppressor activity of p53, as the selective pressure to loss pro-apoptotic gene function is extremely high during tumor development. In addition to the activation of apoptotic target genes, p53 can also repress gene expression and act independently of the regulation of transcription functions that have also been implicated in the induction of the full apoptotic response (Ryan et al., 2001). The transcriptional targets up-regulated by p53 that can induce apoptosis fall into at least three categories: death domain containing proteins including two proapoptotic death receptors (Herr and Debation, 2001), proteins that act at the levels of the mitochondria including three proapoptotic Bcl-2 family members, Bax, Bak, and Bid, and antiapoptotic Bcl-2, and a third group of proteins that lead to the generation of reactive oxygen species (Li et al., 1999) (Fig. 15.4).

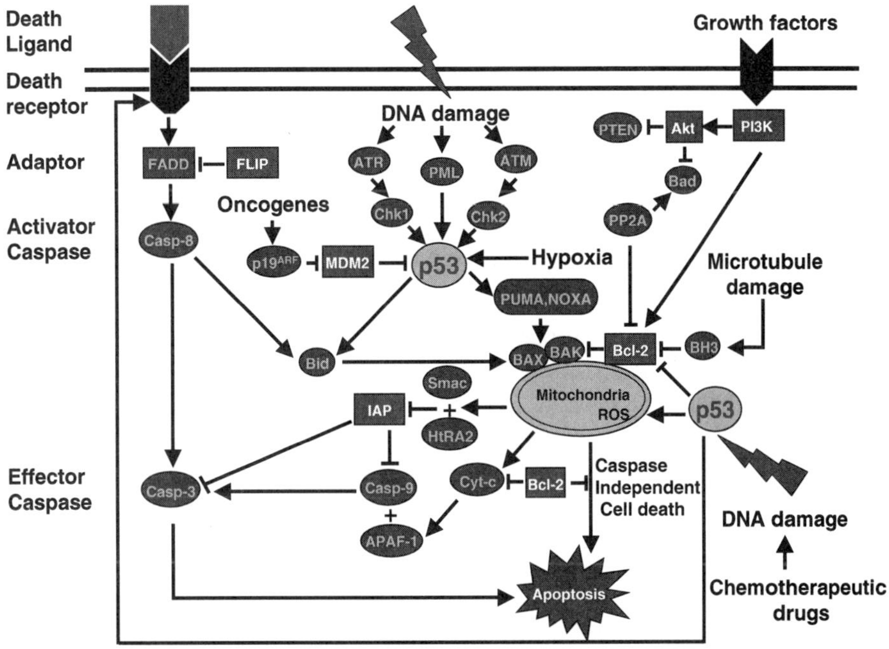

Figure 15.4. Crosstalk between apoptosis signaling pathways. A schematic diagram showing some of the known components of the intrinsic and death receptor apoptotic programs that may modulate tumor development and therapy. Components in red box inhibit apoptosis while those in blue circle promote apoptosis.

APOPTOSIS SIGNALING

Two main signaling pathways have been delineated to initiate the apoptotic suicide program in mammalian cells and both are utilized depending on the stimulus (Fig. 15.4). The first pathway is known as the mitochondrial pathway or the intrinsic pathway. The cell intrinsic pathway triggers apoptosis in response to DNA damage, defective cell cycle, detachment from the extracellular matrix, hypoxia, loss of survival factors, and other types of severe cell stresses. The pathway involves the activation of the pro-apoptotic arm of Bcl-2 gene superfamily, which in turn engages the mitochondria to cause the release of apoptogenic factors such as cytochrome c and SMAC/DIABLO into the cytosol (Adams and Cory, 1998; Green, 2000; Hunt and Evan, 2001). In the cytosol, cytochrome c binds the adaptor APAF-1, forming an "apoptosome" that activates the apoptosis-initiating protease caspase-9. In turn, caspase-9 activates "executioner" protease caspase-3, -6, and -7. SMAC/DIABLO promotes apoptosis by binding to inhibitor of apoptosis (IAP) proteins and preventing these factors from attenuating caspase activation (Du et al., 2000; Verhagen et al., 2000). Intrinsic stress, such as oncoproteins, direct DNA damage, hypoxia, and survival factor deprivation can activate the intrinsic apoptotic pathway. As a sensor of the cellular stress, p53 is a critical initiator of this pathway (Lowe and Lin, 2000; Fig. 15.4). For example, proteins that sense DNA damage, such as ATM and Chk2, phosphorylate and stabilize p53 directly, and inhibit MDM2-mediated ubiquitination of p53 (Khanna and Jackson, 2001). Mitogenic oncogenes can also activate p53 through a mechanism that is distinct from DNA damage, and can involve p19^{ARF}, the alternative reading frame product of the INK4α/ARF tumor suppressor locus. p19^{ARF}, in turn, binds and inactivates Mdm2, leading to p53 activation (Lowe and Lin, 2000; Sherr and Weber, 2000).

The cell extrinsic pathway triggers apoptosis in response to engagement of death receptors by their ligands (Fig. 15.4). This pathway stimulates the apoptotic caspase machinery independently of p53. The cell extrinsic pathway is initiated by members of the tumor-necrosis factor (TNF) superfamily (Fig. 15.3). Members of the TNF superfamily are type II membrane proteins with conserved C-terminal extracellular domains responsible for trimer formation. Several members of this family (TNFα, FasL, and TRAIL/APO2L) have been shown to induce apoptosis through binding their respective receptors, Fas/APO1 and DR4 (TRAIL-R1) and KILLER/DR5 (TRAIL-R2) respectively (Ashkenazi, 2002). Ligation of FasL or TRAIL to its receptor results in trimerization of the receptors and clustering of the receptor's intracellular death domain (DD), leading to the formation of the death inducing signaling complex (DISC). Trimerization of the death domains leads to the recruitment of an adaptor molecule, FADD and subsequent binding and activation of caspase-8 and -10. Activated caspase-8 and -10 then cleave caspase-3 which in turn leads to cleavage of the death substrates (Fig. 15.4). These two apoptosis signaling pathways communicate with each other. Caspase-8 has been shown to cleave the pro-apoptotic Bcl-2 family member Bid. The cleavage of Bid by caspase-8 and the translocation of truncated Bid to mitochondria to promote cytochrome c release through interaction with Bax and Bak provide a plausible mechanistic link between the extrinsic and intrinsic pathways (Green, 2000; Fig. 15.4). This apparently amplifies the apoptotic signal following death receptor activation, and different cell types may be more reliant on this amplification pathway than others (Fulda et al., 2001). Conversely, activators of the intrinsic pathway can sensitize cells to extrinsic death ligands.

The relative contribution of death receptor versus mitochondrial pathways in apoptosis may vary depending on the dose and kinetics but may also reflect the existence of two different cell types with respect to CD95 signaling. In type I cells, caspase-8 activation sufficient to kill cells occurs as a direct consequence of death receptor ligation, with activating downstream (effector) pro-caspases such as caspase-3. This step is independent of mitochondria and is not blocked by overexpression of Bcl-2. In contrast, type II (intrinsic) cell death is dependent on amplification of death receptor signals via the mitochondrial pathway controlled by Bcl-2. At the molecular level these two cell types differ principally in the amount of caspase 8 recruited to CD95 via FADD to form DISC. Whereas type I cells contain large amount of DISC in response to CD95 antibodies, type II cells do not and thus are dependent on stimulation of the intrinsic apoptotic pathway to undergo cell death. Mitochondria are activated in both type I and type II cells but are dispensable for the death of type I cells. Previous studies in cell lines and in vivo have demonstrated the existence of type I and type II cell lines in response to FasL (Scaffidi et al., 1998) or TRAIL (Burns and El-Deiry, 2001; Ozoren et al., 2000).

APOPTOSIS AND NORMAL PHYSIOLOGY

Organisms as different as worms and humans have conserved the genes that encode the core cell death machinery. Furthermore genetic studies using worm, fly, and mouse models indicate that physiological cell death is essential for normal development (Baehreck, 2002). Studies of developing animals illustrate many reasons for cells to die, for instance, during the formation and deletion of structures, to control cell numbers and to eliminate abnormal cells. Cell death is critical for animal development. In mammals some developmental cell deaths are autonomous, meaning that pro-apoptotic signals from neighboring cells are not needed for the demise of the doomed cell. In fact it is believed that in many of these deaths, neighboring cells provide survival signals and that cell death is initiated by withdrawal of growth factors and/or loss of cell attachment. These pathways to cell death are often referred to as death by neglect and can, in many instances, be blocked by anti-apoptotic members of the Bcl-2 protein family. Consistent with this, mice lacking Bcl-2 or Bcl-X_L exhibit specific defects in organogenesis. The activation of programmed cell death is tightly regulated and ecotopic activation can be catastrophic (Strasser et al., 2000). Several factors are involved in the activation process during development, including cell-lineage information, extracellular survival factors, steroid hormones, membrane-bound death receptors, and DNA-damaging agents such as radiation. Studies of cell lineage, survival factors and steroids have provided us with detailed mechanisms for how these signals activate cell demise during animal development (Baehrecke, 2002).

APOPTOSIS AND CANCER

Tumorigenesis is a multistep process in which mutations in key cellular genes produce a series of acquired capabilities that allow the developing cancer cells to grow unchecked in the absence of growth-stimulating signals, while overcoming growth-inhibitory signals and host immune responses. They also allow the tumor to replicate indefinitely, maintain an oxygen and nutrient supply, and invade adjacent and distant tissues. Importantly, the ability of cells to evade apoptosis is also an essential "hallmark of cancer."

Pro-survival Signaling and Cancer

Akt is emerging as one of the central mediators of tumorigenesis. Akt/PKB was first discovered as the cellular homologue of the viral oncogene, v-Akt (Bellacosa et al., 1991). Akt is now known as a family of closely related, highly conserved cellular homologues. In human, three closely related family members exist, Akt1, Akt2, and Akt3 (Testa and Bellacosa, 2001). There has been enormous interest in the mechanisms and cellular consequences of signal propagation from receptor tyrosine kinases to Akt. Akt activation involves both membrane translocation and phosphorylation. In response to a variety of signals including growth factors, insulin, IGF-1 or activated Ras, Akt is recruited to the membrane and activated. This occurs through the phosphatidylinositiol 3-kinase (PI3K) pathway. In response to growth factor signaling, PI3K is recruited to the plasma membranes where it phosphorylates membrane phosphoinositides generating 3′phosphorylated phosphoinositides primarily PI-3, 4, 5-P3 and PI-3, 4-P2. Akt then binds these phsopholipids and is phosphorylated and activated by PDK1 and a yet unidentified PDK2 (Balendran et al., 1999). Several pro-apoptotic substrates have been identified for Akt and each may have a role in mediating cell survival depending on the cellular context. The proapoptotic BH3-only Bcl-2 family member Bad is phosphorylated by Akt and then sequestered by 14-3-3 proteins (Datta et al., 1997; del Peso et al., 1997). Akt also appears to phosphorylate and inhibit caspase-9 (Cardone et al., 1998). However, the caspase-9 Akt phosphorylation site is not conserved in mice, and it remains unclear whether this is a critical substrate for promoting survival (Fujita et al., 1999). Akt can also disrupt the p53 pathway through phosphorylation of the negative p53 regulator, Mdm2. Upon phosphorylation, Mdm2 translocates into the nucleus where it can bind and degrade p53 (Mayo and Donner, 2001; Zhou et al., 2001). Finally Akt has been implicated in increasing the pro-survival NF-κB activity (Kane et al., 1999; Romashkova and Makarov, 1999). Akt has been shown to phosphorylate IκB kinase (IKK), which leads to degradation of IκB and translocation of NFκB to the nucleus (Ozes et al., 1999). Akt activation has also been implicated in TRAIL resistance (Asakuma et al., 2003). Amplification of Akt1 in gastric cancer and Akt2 in ovarian and pancreatic cancers has been observed at a significant rate (Testa and Bellacosa, 2001).

NFκB is a transcriptional factor and NFκB activity is required for the induction of more than 150 genes involved in cell growth, differentiation, development, apoptosis, and adaptive response to changes in cellular redox balance (Mercurio and Manning, 1999). A wide variety of external stimuli including cytokines, pathogens, stress, and chemotherapeutic agents can lead to the activation of NFκB (Baeuerle and Baltimore, 1996). These stimuli induce phosphorylation and subsequent degradation of IκB inhibitory proteins, thereby releasing NFκB proteins for translocation to the nucleus to function as transcription factors (Verma et al., 1995). NFκB transcription factors form heterodimer and homodimer complexes of related proteins that contain a Rel homology domain involved in specific DNA binding, protein dimerization, and nuclear import. The Rel proteins predominantly found in mammalian cells consist of two transcriptionally inactive forms, NFκB1 (p50) and NFκB2 (p52), and three transcriptionally active subunits known as RelA (p65), c-Rel, and RelB (Baeuerle and Baltimore, 1996). NFκB target genes that may provide anti-apoptotic function include the IAP family of caspase inhibitory proteins, TRAF1 and TRAF2, and c-FLIP, thought to suppress caspase 8 activation, the pro-survival Bcl-2 homologue proteins Bfl1/A1 and Bcl-X_L (Pahl, 1999; Wang et al., 1998). Whereas much evidence highlights an apparently paradoxical pro-apoptotic role for NFκB. The pro-apoptotic effect of NFκB may be due to the promoter activation of death receptors and death ligands such as CD95, CD95-L (Kasibhatla et al., 2000), and the TRAIL receptors DR4 and KILLER/DR5 (Ravi et al., 2001). Also crosstalk between NFκB and the caspase pathway has been found (Pahl, 1999; Wang et al., 1998). On the basis of these studies, it appears that opposite functions of NFκB may depend on the expression of its subunits where c-Rel and RelA functions as pro-apoptotic and anti-apoptotic proteins, respectively (Chen et al., 2003). There are many examples in which NFκB is activated in human neoplasia. Furthermore several cellular oncogenes such as Ras and Her2/Neu through activation of Akt and the BCR-ABL fusion protein can lead to NFκB activation. Elevated NFκB activity has been implicated in the development of many solid tumors, including breast and gastric carcinomas (Karin et al., 2002).

Another key survival factor family is the inhibitor of apoptosis family of IAP proteins that includes Survivin. The IAP family of endogenous inhibitors of caspase is conserved throughout animal evolution with homologies in viruses, yeast, flies, worms, mice, and human. All IAPs contain one to three baculovirus IAP repeat (BIR) domains that may be involved in the inhibition of caspase activity. The mammalian IAPs, X-IAP, c-IAP-1, c-IAP2, ML-IAP, and livin inhibit activated caspase-3 and -7, and the activation of caspase-9 mediated by Apaf-1-cytochrome c (Fig. 15.4), while survivin has been implicated in regulation of the cell cycle during mitosis (Deveraux and Reed, 1999). Because survivin, ML-IAP, and livin are overexpressed in many cancers but not in normal adult tissues, these molecules represent possible targets for development of drugs that selectively eliminate cancer cells (Altieri, 2003).

Dysregulation of the Intrinsic Pathway and Cancer

Disruption of the intrinsic apoptotic pathway is common in cancer. Indeed, the p53 tumor-suppressor gene is the most frequently mutated gene in human tumors, and loss of p53 function can both disable apoptosis and accelerate tumor development in transgenic mice (Ryan et al., 2001). Moreover functional mutations or altered expression of p53 downstream effectors (*PTEN, Bax, Bak, and Apaf-1*), or upstream regulators (ATM, Chk2, Mdm2, and $p19^{ARF}$) occurs in human tumors

(Schmitt et al., 1999). Given the importance of Bcl-2 family members in regulating the intrinsic apoptotic pathway, it is not surprising that these genes are altered in tumor samples. In fact Bcl-2 was first identified based on its translocation in follicular lymphoma, and is overexpressed in a variety of cancers (Reed, 1999). In addition mutations or altered expression of upstream regulators of Bcl-2 proteins are associated with cancer. For example, the Bad-kinase Akt is positively regulated by various oncoproteins and negatively regulated by PTEN tumor suppressor (Datta et al., 1999). Activation of Akt and loss of PTEN occur with high frequency in a variety of solid cancers, indicating the importance of this pathway in regulating tumorigenesis.

Inhibition of Postmitochondrial Death Processes and Cancer

While mutations in cancer cells often target regulators of the intrinsic apoptotic pathway such as p53 and the Bcl-2 family proteins, alterations that disrupt apoptosis downstream of the mitochondria have been reported. For example, siliencing of Apaf-1 occurs in metastatic melanoma, and overexpression of IAPs and heat shock protein (Hsp), which can inhibit caspase-9 activation, is commonly observed in human tumors (Deveraux and Reed, 1999). This indicates that downstream defects in apoptosis contribute to tumorigenesis.

Defects in Death Receptor-Induced Apoptosis and Cancer

Although tumorigenic disruptions in the death receptor pathway occur less frequently than the intrinsic pathway, changes accompanying tumorigenesis can alter the regulation of these pathways. Tumor cells are often resistant to death receptor-mediated apoptosis, and mutations or failure to up-regulate CD95 and CD95-L may result in apoptosis defects of tumor cells. In solid tumors and non-T-cell leukemia, tumor surveillance and immune escape may also be important. While cytotoxic lymphocytes predominantly kill tumor cells via the granule exocytosis pathway, CD95-L and TRAIL are also utilized. Thus inactivation of the death receptor pathway could allow escape from immune responses, and this provides a survival advantage to developing tumor cells. In fact loss of CD95 or TRAIL function can promote tumor growth and metastasis (Takeda et al., 2001).

APOPTOSIS AND CANCER THERAPY

Since the late 1950s the main strategy to treat cancer, besides surgery, has been radiotherapy or chemotherapy. These approaches work primarily by damaging proliferating cells at the level of DNA replication or cell division, and induce apoptotic cell suicide as a secondary response to the damage. Although this form of therapy has been successful for treatment of some tumors such as testicular cancer and certain leukemias, its success for the treatment of common epithelial tumors of the breast, colon, and lung has been less than spectacular. Over the last decade, our understanding of cellular damage responses and physiological cell death mechanisms has improved, leading in turn to new insights into drug-induced cell death. Drugs of differing structure and specificity induce the characteristic morphological changes associated with apoptosis, and it is now believed that the apoptotic pathways contribute to the cytotoxic action of most chemotherapeutic drugs.

Extrinsic Pathway

The extrinsic pathway is activated in vivo by TNF family ligands that engage DD-containing receptors, resulting in activation of the DED-containing caspases. Activation of the extrinsic pathway may be one of the potential mechanisms for activating tumor-specific apoptosis. Unfortunately, attempts to apply TNF as a biological agent for cancer treatment was stymied by the pro-inflammatory effects of this cytokine (Debs et al., 1990), while agonisitic antibodies that trigger the receptor Fas (CD95) are highly toxic to the liver (Ogasawara et al., 1993). The tumor-necrosis factor-related apoptosis-inducing ligand (TRAIL), is a member of the TNF superfamily, and recombinant soluble TRAIL induces apoptosis in cell lines from a broad spectrum of human cancers, including colon, lung, breast, prostate, pancreas, kidney, central nervous system, and thyroid cancer, as well as leukemia and multiple myeloma, indicating that this ligand might be useful for the treatment of many cancers (French and Tschopp, 1999; Walczak, et al., 1999). Most important, unlike FasL and TNFα whose severe systemic toxic side effects have precluded their clinical use, no systemic side effects in murine or nonhuman primates have been observed with TRAIL. Recent studies have raised questions of whether TRAIL exerts no systemic toxic effects on normal cells as human hepatocytes or kertinocytes were found to be sensitive to TRAIL. However, the cytotoxic effect may depend on the particular preparation of TRAIL used, and nontagged, zinc-bound recombinant TRAIL has been suggested to not be toxic toward normal human cell lines (Kelley et al., 2001; Lawrence et al., 2001). Initial studies in nonhuman primates, such as *cynomolgus* monkeys and chimpanzees, showed that short-term intravenous administration of nontagged, zinc-bound TRAIL was well tolerated even at high does (Kelley et al., 2001; Lawrence et al., 2001). Therefore TRAIL remains a promising therapeutic agent for a wide range of humor tumors and is currently being developed for clinical trials. Although TRAIL can kill tumor cells independently of mitochondria and the intrinsic pathway, in some types of cells it is clear that TRAIL-induced cell death can be inhibited by Bcl-2 family members such as Bax or Bcl-2 in some cell types. Therefore the expression level of Bcl-2 family member and the inhibitor of extrinsic pathway such as c-FLIP should be considered when using this as a potential therapeutic agent. Furthermore combinations of TRAIL and certain DNA-damaging drugs or radiotherapy have also been shown to have synergistic antitumor activity in several xenograft models, and some chemotherapeutic agents can re-sensitize TRAIL-resistant tumor cells to TRAIL (LeBlanc et al., 2002).

Intrinsic Pathway

Bcl-2 family proteins are central regulators of the intrinsic pathway. Overexpression of the antiapoptotic Bcl-2 or Bcl-X_L probably occurs in more than half of all cancers (Adams and Cory, 1998), rendering tumor cells resistant to myriad apoptotic stimuli, including most cytotoxic anticancer drugs. Several strategies have been designed to overcome the cytoprotective effects of Bcl-2 and Bcl-X_L in cancer cells by several groups. Nuclease-resistant antisense oligonucleotides targeting the Bcl-2 mRNA have advanced to phase III clinical trail for melanoma, myeloma, CLL, and AML, with phase II activity for a variety of solid tumors (Konopleva et al., 2000). Because Bcl-2 antisense enhances sensitivity to cytotoxic anticancer drugs in vitro and in xenograft models, most clinical trials combine the antisense agent with conventional chemotherapy (Chi et al., 2001). Strategies for attacking the Bcl-2 and Bcl-X_L proteins using small-molecule drugs have emerged from an understanding of how our own cells keep these antiapoptotic proteins under control. In this regard a large family of endogenous antagonists of bcl-2 and Bcl-X_L has been revealed (so-called BH3-only proteins) (Huang and Strasser, 2000). Several experiments using the synthetic BH3 peptides have confirmed the validity of strategies based on generating small-molecule drugs that occupy the BH3 binding site on Bcl-2 or Bcl-X_L, abrogating their cytoprotective functions (Letai et al., 2002). Other strategies for countering Bcl-2 and Bcl-X_L in cancer include inducing expression of opposing proapoptotic family members such as Bax with p53 adenovirus (in phase III trials) (Ferreira et al., 2002; Reed, 2003) or with Mda 7 (IL24) adenovirus gene therapy (complete phase I trials) (Cao et al., 2002) It might be also possible to activate the Bax and Bak proteins using drugs that mimic agonistic BH3 peptides (Letai et al., 2002).

Convergence Pathway

Ultimately the intrinsic and extrinsic pathways for caspase activation converge on downstream effector caspases. The capacity of caspase-9 to activate downstream caspases seems to be under the control of a regulator system based on endogenous inhibitors. In this regard the inhibitor of apoptosis proteins (IAPs) represent a family of evolutionarily conserved apoptosis suppressors. Pathological overexpression of IAPs occurs in many cancers. Recent attempts to use IAPs as targets for anticancer therapy have focused on Survivin and XIAP (preclinical). Antisense-mediated reductions in XIAP and Survivin can induce apoptosis in tumor cell lines in culture or sensitize cells to cytotoxic anticancer drugs, thus providing evidence that the pathological elevations of IAPs found in cancers are important for maintaining tumor cell survival and resistance to chemotherapy (Deveraux and Reed, 1999). Moreover the net effect of IAPs possibly depends on their interaction with regulatory molecules such as Smac/ DIABLO, HtrA2, and XIAP-associated factor-1, the antagonists of IAPs. Synthetic peptides that mimic SMAC and HtrA2 (preclinical) trigger apoptosis or sensitize tumor cell lines to apoptosis induced by cytotoxic anticancer drugs or TRAIL in vitro even in some tumor xenograft models (Fulda et al., 2002).

Other Regulators

Certain signaling proteins have emerged as potential drug targets for modulating apoptosis pathways, based on their ability to influence the expression or function of other proteins. The transcription factor p53 regulates the expression of multiple apoptosis-regulating genes that affect either the intrinsic and extrinsic pathways. Loss or mutation of p53 is a very common genetic abnormality in cancer. Initial phase I p53-based gene therapy trials suggested that p53 replacement could lead to an increase in apoptosis in tumor cells and surrounding cells through a bystander effect (Roth and Cristiano, 1997). As suggested by the preclinical data, it is likely that the combination of chemotherapy with the reintroduction of wild-type p53 may increase tumor cell killing. An alternative approach to target p53 is via compounds that stabilize its DNA-binding domain in the active conformation. Two compounds, CP-257042 and CP-31398, appear to stabilize p53 in its wild-type conformation (Foster et al., 1999). CP-31398 has been demonstrated to induce apoptosis or growth arrest and also exerts tumor suppressor effects in human xenograft models without systemic toxicity. Additional development is awaited to provide a better idea about the clinical potential of these compounds.

Similarly Akt directly phosphorylates multiple protein targets of relevance to apoptosis. Pathological elevations in Akt activity are a common occurrence in tumors, due to loss of tumor suppressor PTEN, hyperactivity of PI3K-activating growth factor receptors and oncoproteins, or other mechanisms (Testa and Bellacosa, 2001). Small-molecule drugs that occupy the ATP binding pocket of the catalytic domain of Akt thus represent a highly attractive approach to restoration of apoptosis sensitivity in cancers (Reed, 2003). NFκB transcription factor family can also induce expression of anti-apoptotic proteins that oppose the intrinsic (Bcl-X_L, Bfl-1), extrinsic (FLIP), and the convergence (cIAP2) pathways, as well as suppressing the expression of the death inducer Bax. Therefore agents that inhibit NFκB activation are desired, and are currently being pursued, based on strategies that seek to maintain levels of endogenous NFκB-inhibiting IκB family proteins (Orlowski and Baldwin, 2002). In addition to p53, Akt, and NFκB, multiple cancer-relevant proteins that operate upstream of these factors could to some extent be viewed as apoptosis modulators and can also be future targets for cancer therapy.

CONCLUSIONS AND PERSPECTIVE

It has been established that apoptotic cell death plays an essential role in normal development and functioning of multicellular organisms and that abnormalities in this process can cause diseases such as cancer. The molecular mechanisms that control and execute apoptotic cell death are coming into focus. Although there is much more to learn, our current understanding of apoptosis mechanisms and multiple novel strategies for restoring apoptosis in cancers provides new avenues for cancer diagnostics, prognosis and therapy. In the coming years, it seems likely that rational strategies to manipulate cell suicide programs will produce new therapies that are less toxic and mutagenic than current treatment regimens.

ACKNOWLEDGMENTS

W. S. El-Deiry is an Assistant Investigator of the Howard Hughes Medical Institute.

REFERENCES

Adams JM, Cory S (1998): The Bcl-2 protein family: Arbiters of cell survival. Science 281(5381):1322–6.

Alnemri ES, Livingston DJ, Nicholson DW, Salvesen G, Thornberry NA, Wong WW, Yuan J (1996): Human ICE/CED-3 protease nomenclature. Cell 87(2):171.

Altieri DC (2003): Validating survivin as a cancer therapeutic target. Nat Rev Cancer 3(1):46–54.

Asakuma J, Sumitomo M, Asano T, Asano T, Hayakawa M (2003): Selective Akt inactivation and tumor necrosis actor-related apoptosis-inducing ligand sensitization of renal cancer cells by low concentrations of paclitaxel. Cancer Res 63(6):1365–70.

Ashhab Y, Alian A, Polliack A, Panet A, Ben Yehuda D (2000): Two splicing variants of a new inhibitor of apoptosis gene with different biological properties and tissue distribution pattern. FEBS Lett 495(1–2):56–60.

Ashkenazi A (2002): Targeting death and decoy receptors of the tumour-necrosis factor superfamily. Nat Rev Cancer 2(6):420–30.

Ashkenazi A, Dixit VM (1998): Death receptors: signaling and modulation. Science 281(5381):1305–8.

Baehrecke EH (2002): How death shapes life during development. Nat Rev Mol Cell Biol 3(10):779–87.

Baeuerle PA, Baltimore D (1996): NF-kappa B: ten years after. Cell 87(1):13–20.

Barr PJ, Tomei LD (1994): Apoptosis and its role in human disease. Biotechnology 12(5):487–93.

Balendran A, Casamayor A, Deak M, Paterson A, Gaffney P, Currie R, Downes CP, Alessi DR (1999): PDK1 acquires PDK2 activity in the presence of a synthetic peptide derived from the carboxyl terminus of PRK2. Curr Biol 9(8):393–404.

Bellacosa A, Testa JR, Staal SP, Tsichlis PN (1991): A retroviral oncogene, akt, encoding a serine-threonine kinase containing an SH2-like region. Science 254(5029):274–7.

Boyd JM, Gallo GJ, Elangovan B, Houghton AB, Malstrom S, Avery BJ, Ebb RG, Subramanian T, Chittenden T, Lutz RJ, et al. (1995): Bik, a novel death-inducing protein shares a distinct sequence motif with Bcl-2 family proteins and interacts with viral and cellular survival-promoting proteins. Oncogene 11(9):1921–8.

Burns TF, El-Deiry WS (2001): Identification of inhibitors of TRAIL-induced death (ITIDs) in the TRAIL-sensitive colon carcinoma cell line SW480 using a genetic approach. J Biol Chem 276(41):37879–86.

Cao XX, Mohuiddin I, Chada S, Mhashilkar AM, Ozvaran MK, McConkey DJ, Miller SD, Daniel JC, Smythe WR (2002): Adenoviral transfer of mda-7 leads to BAX up-regulation and apoptosis in mesothelioma cells, and is abrogated by over-expression of BCL-XL. Mol Med 8(12):869–76.

Cardone MH, Roy N, Stennicke HR, Salvesen GS, Franke TF, Stanbridge E, Frisch S, Reed JC (1998): Regulation of cell death protease caspase-9 by phosphorylation. Science 282(5392):1318–21.

Chang BS, Minn AJ, Muchmore SW, Fesik SW, Thompson CB (1997): Identification of a novel regulatory domain in Bcl-$X_{(L)}$ and Bcl-2. EMBO J 16(5):968–77.

Chen-Levy Z, Nourse J, Cleary ML (1989): The bcl-2 candidate proto-oncogene product is a 24-kilodalton integral-membrane protein highly expressed in lymphoid cell lines and lymphomas carrying the t (14; 18) translocation. Mol Cell Biol 9(2):701–10.

Chen X, Kandasamy K, Srivastava RK (2003): Differential roles of RelA (p65) and c-Rel subunits of nuclear factor kappa B in tumor necrosis factor-related apoptosis-inducing ligand signaling. Cancer Res 63(5):1059–66.

Chi KN, Gleave ME, Klasa R, Murray N, Bryce C, Lopes de Menezes DE, D'Aloisio S, Tolcher AW (2001): A phase I dose-finding study of combined treatment with an antisense Bcl-2 oligonucleotide (Genasense) and mitoxantrone in patients with metastatic hormone-refractory prostate cancer. Clin Cancer Res 7(12):3920–7.

Chinnaiyan AM, O'Rourke K, Yu GL, Lyons RH, Garg M, Duan DR, Xing L, Gentz R, Ni J, Dixit VM (1996): Signal transduction by DR3, a death domain-containing receptor related to TNFR-1 and CD95. Science 274(5289):990–2.

Clem RJ, Cheng EH, Karp CL, Kirsch DG, Ueno K, Takahashi A, Kastan MB, Griffin DE, Earnshaw WC, Veliuona MA, Hardwick JM (1998): Modulation of cell death by Bcl-x_L through caspase interaction. Proc Natl Acad Sci USA 95(2):554–9.

Cory S, Adams JM (2002): The Bcl2 family: Regulators of the cellular life-or-death switch. Nat Rev Cancer 2(9):647–56.

Cretney E, Takeda K, Yagita H, Glaccum M, Peschon JJ, Smyth MJ (2002): Increased susceptibility to tumor initiation and metastasis in TNF-related apoptosis-inducing ligand-deficient mice. J Immunol 168(3):1356–61.

Datta SR, Brunet A, Greenberg ME (1999): Cellular survival: A play in three Akts. Genes Dev 13(22):2905–27.

Datta SR, Dudek H, Tao X, Masters S, Fu H, Gotoh Y, Greenberg ME (1997): Akt phosphorylation of BAD couples survival signals to the cell-intrinsic death machinery. Cell 91(2):231–41.

Debs RJ, Fuchs HJ, Philip R, Brunette EN, Duzgunes N, Shellito JE, Liggitt D, Patton JR (1990): Immunomodulatory and toxic effects of free and liposome-encapsulated tumor necrosis factor alpha in rats. Cancer Res 50(2): 375–80.

del Peso L, Gonzalez-Garcia M, Page C, Herrera R, Nunez G (1997): Interleukin-3-induced phosphorylation of BAD through the protein kinase Akt. Science 278(5338):687–9.

Deveraux QL, Reed JC (1999): IAP family proteins—Suppressors of apoptosis. Genes Dev 13(3):239–52.

Du C, Fang M, Li Y, Li L, Wang X (2000): Smac, a mitochondrial protein that promotes cytochrome c-dependent caspase activation by eliminating IAP inhibition. Cell 102(1):33–42.

Earnshaw WC, Martins LM, Kaufmann SH (1999): Mammalian caspases: Structure, activation, substrates, and functions during apoptosis. An Rev Biochem 68:383–424.

Emery JG, McDonnell P, Burke MB, Deen KC, Lyn S, Silverman C, Dul E, Appelbaum ER, Eichman C, DiPrinzio R, Dodds RA, James IE, Rosenberg M, Lee JC, Young PR (1998): Osteoprotegerin is a receptor for the cytotoxic ligand TRAIL. J Biol Chem 273(23):14363–7.

Ferreira CG, Epping M, Kruyt FA, Giaccone G (2002): Apoptosis: Target of cancer therapy. Clin Cancer Res 8(7):2024–34.

Foster BA, Coffey HA, Morin MJ, Rastinejad F (1999): Pharmacological rescue of mutant p53 conformation and function. Science 286(5449):2507–10.

French LE, Tschopp J (1999): The TRAIL to selective tumor death. Nat Med 5(2):146–7.

Fujita E, Jinbo A, Matuzaki H, Konishi H, Kikkawa U, Momoi T (1999): Akt phosphorylation site found in human caspase-9 is absent in mouse caspase-9. Biochem Biophys Res Commun 264(2):550–5.

Fulda S, Meyer E, Friesen C, Susin SA, Kroemer G, Debatin KM (2001): Cell type specific involvement of death receptor and mitochondrial pathways in drug-induced apoptosis. Oncogene 20(9):1063–75.

Fulda S, Wick W, Weller M, Debatin KM (2002): Smac agonists sensitize for Apo2L/TRAIL- or anticancer drug-induced apoptosis and induce regression of malignant glioma in vivo. Nat Med 8(8):808–15.

Green DR, Reed JC (1998): Mitochondria and apoptosis. Science 281(5381): 1309–12.

Hay BA (2000): Understanding IAP function and regulation: A view from *Drosophila*. Cell Death Differ 7(11):1045–56.

Green DR (2000): Apoptotic pathways: paper wraps stone blunts scissors. Cell 102(1):1–4.

Gross A, Jockel J, Wei MC, Korsmeyer SJ (1998): Enforced dimerization of BAX results in its translocation, mitochondrial dysfunction and apoptosis. EMBO J 17(14):3878–85.

Guo Y, Srinivasula SM, Druilhe A, Fernandes-Alnemri T, Alnemri ES (2002): Caspase-2 induces apoptosis by releasing proapoptotic proteins from mitochondria. J Biol Chem 277(16):13430–7.

Hacki J, Egger L, Monney L, Conus S, Rosse T, Fellay I, Borner C (2000): Apoptotic crosstalk between the endoplasmic reticulum and mitochondria controlled by Bcl-2. Oncogene 19(19):2286–95.

Hengartner MO (2000): The biochemistry of apoptosis. Nature 407:770–6.

Herr I, Debatin KM (2001): Cellular stress response and apoptosis in cancer therapy. Blood 98(9):2603–14.

Hoffman WH, Biade S, Zilfou JT, Chen J, Murphy M (2001): Transcriptional repression of the anti-apoptotic survivin gene by wild type p53. J Biol Chem 277(5):3247–57.

Hsu H, Huang J, Shu HB, Baichwal V, Goeddel DV (1996a): TNF-dependent recruitment of the protein kinase RIP to the TNF receptor-1 signaling complex. Immunity 4(4):387–96.

Hsu H, Shu HB, Pan MG, Goeddel DV (1996b): TRADD-TRAF2 and TRADD-FADD interactions define two distinct TNF receptor 1 signal transduction pathways. Cell 84(2):299–308.

Huang H, Joazeiro CA, Bonfoco E, Kamada S, Leverson JD, Hunter T (2000): The inhibitor of apoptosis, cIAP2, functions as a ubiquitin-protein ligase and promotes in vitro monoubiquitination of caspases 3 and 7. J Biol Chem 275(35):26661–4.

Huang DC, Strasser A (2000): BH3-only proteins-essential initiators of apoptotic cell death. Cell 103(6):839–42.

Hunt A, Evan G (2001): Apoptosis: Till death us do part. Science 293(5536):1784–5.

Ito T, Deng X, Carr B, May WS (1997): Bcl-2 phosphorylation required for anti-apoptosis function. J Biol Chem 272(18):11671–3.

Irwin MS, Kaelin WG (2001): p53 family update: p73 and p63 develop their own identities. Cell Growth Differ 12(7):337–49.

Johnstone RW, Ruefli AA, Lowe SW (2002): Apoptosis: A link between cancer genetics and chemotherapy. Cell 108(2):153–64.

Kane LP, Shapiro VS, Stokoe D, Weiss A (1999): Induction of NF-kappaB by the Akt/PKB kinase. Curr Biol 9(11):601–4.

Karin M, Cao Y, Greten FR, Li ZW (2002): NF-kappaB in cancer: from innocent bystander to major culprit. Nat Rev Cancer 2(4):301–10.

Kasibhatla S, Brunner T, Genestier L, Echeverri F, Mahboubi A, Green DR (1998): DNA damaging agents induce expression of Fas ligand and subsequent apoptosis in T lymphocytes via the activation of NF-kappa B and AP-1. Mol Cell 1(4):543–51.

Kasof GM, Gomes BC (2001): Livin, a novel inhibitor of apoptosis protein family member. J Biol Chem 276(5):3238–46.

Kelley SK, Harris LA, Xie D, Deforge L, Totpal K, Bussiere J, Fox JA (2001): Preclinical studies to predict the disposition of Apo2L/tumor necrosis factor-related apoptosis-inducing ligand in humans: characterization of in vivo efficacy, pharmacokinetics, and safety. J Pharmacol Expr Ther 299(1):31–8.

Kerr JF, Wyllie AH, Currie AR (1972): Apoptosis: a basic biological phenomenon with wide-ranging implications in tissue kinetics. Br J Cancer 26:239–57.

Khanna KK, Jackson SP (2001): DNA double-strand breaks: signaling, repair and the cancer connection. Nat Genet 27(3):247–54.

Kilic M, Schafer R, Hoppe J, Kagerhuber U (2002): Formation of noncanonical high molecular weight caspase-3 and -6 complexes and activation of caspase-

12 during serum starvation induced apoptosis in AKR-2B mouse fibroblasts. Cell Death Differ 9(2):125–37.

Kischkel FC, Lawrence DA, Chuntharapai A, Schow P, Kim KJ, Ashkenazi A (2000): Apo2L/TRAIL-dependent recruitment of endogenous FADD and caspase-8 to death receptors 4 and 5. Immunity 12(6):611–20.

Kitson J, Raven T, Jiang YP, Goeddel DV, Giles KM, Pun KT, Grinham CJ, Brown R, Farrow SN (1996): A death-domain-containing receptor that mediates apoptosis. Nature 384(6607):372–5.

Ko LJ, Prives C (1996): p53: puzzle and paradigm. Genes Dev 10(9):1054–72.

Konopleva M, Tari AM, Estrov Z, Harris D, Xie Z, Zhao S, Lopez-Berestein G, Andreeff M (2000): Liposomal Bcl-2 antisense oligonucleotides enhance proliferation, sensitize acute myeloid leukemia to cytosine-arabinoside, and induce apoptosis independent of other antiapoptotic proteins. Blood 95(12):3929–38.

Krajewski S, Tanaka S, Takayama S, Schibler MJ, Fenton W, Reed JC (1993): Investigation of the subcellular distribution of the bcl-2 oncoprotein: residence in the nuclear envelope, endoplasmic reticulum, and outer mitochondrial membranes. Cancer Res 53(19):4701–14.

Lawrence D, Shahrokh Z, Marsters S, Achilles K, Shih D, Mounho B, Hillan K, Totpal K, DeForge L, Schow P, Hooley J, Sherwood S, Pai R, Leung S, Khan L, Gliniak B, Bussiere J, Smith CA, Strom SS, Kelley S, Fox JA, Thomas D, Ashkenazi A (2001): Differential hepatocyte toxicity of recombinant Apo2L/TRAIL versions. Nat Med 7(4):383–5.

LeBlanc H, Lawrence D, Varfolomeev E, Totpal K, Morlan J, Schow P, Fong S, Schwall R, Sinicropi D, Ashkenazi A (2002): Tumor-cell resistance to death receptor-induced apoptosis through mutational inactivation of the proapoptotic Bcl-2 homolog Bax. Nat Med 8(3):274–81.

Lee ST, Hoeflich KP, Wasfy GW, Woodgett JR, Leber B, Andrews DW, Hedley DW, Penn LZ (1999): Bcl-2 targeted to the endoplasmic reticulum can inhibit apoptosis induced by Myc but not etoposide in Rat-1 fibroblasts. Oncogene 18(23):3520–8.

Letai A, Bassik MC, Walensky LD, Sorcinelli MD, Weiler S, Korsmeyer SJ (2002): Distinct BH3 domains either sensitize or activate mitochondrial apoptosis, serving as prototype cancer therapeutics. Cancer Cell 2(3):183–92.

Li H, Zhu H, Xu CJ, Yuan J (1998): Cleavage of BID by caspase 8 mediates the mitochondrial damage in the Fas pathway of apoptosis. Cell 94(4):491–501.

Li PF, Dietz R, von Harsdorf R (1999): p53 regulates mitochondrial membrane potential through reactive oxygen species and induces cytochrome c-independent apoptosis blocked by Bcl-2. EMBO J 18(21):6027–36.

Lockshin RA, Williams CM (1965): Program cell death. I. Cytology of degeneration in the intersegmental muscles of the pernyi silkmoth. J Insect Physiol 11:123–33.

Lowe SW, Lin AW (2000): Apoptosis in cancer. Carcinogenesis 21(3):485–95.

Luo X, Budihardjo I, Zou H, Slaughter C, Wang X (1998): Bid, a Bcl2 interacting protein, mediates cytochrome c release from mitochondria in response to activation of cell surface death receptors. Cell 94(4):481–90.

Marsters SA, Sheridan JP, Donahue CJ, Pitti RM, Gray CL, Goddard AD, Bauer KD, Ashkenazi A (1996): Apo-3, a new member of the tumor necrosis factor receptor family, contains a death domain and activates apoptosis and NF-kappaB. Curr Biol 6(12):1669–76.

Mayo LD, Donner DB (2001): A phosphatidylinositol 3-kinase/Akt pathway promotes translocation of Mdm2 from the cytoplasm to the nucleus. Proc Natl Acad Sci USA 98(20):11598–603.

Mercurio F, Manning AM (1999): Multiple signals converging on NF-kappaB. Curr Opin Cell Biol 11(2):226–32.

Nakagawa T, Zhu H, Morishima N, Li E, Xu J, Yankner BA, Yuan J (2000): Caspase-12 mediates endoplasmic-reticulum-specific apoptosis and cytotoxicity by amyloid-beta. Nature 403(6765):98–103.

Nagata S (1996): Fas ligand and immune evasion. Nat Med 2(12):1306–7.

Nicholson DW, Thornberry NA (1997): Caspases: killer proteases. Trends Biochem Sci 22(8):299–306.

Nguyen M, Branton PE, Walton PA, Oltvai ZN, Korsmeyer SJ, Shore GC (1994): Role of membrane anchor domain of Bcl-2 in suppression of apoptosis caused by E1B-defective adenovirus. J Biol Chem 269(24):16521–4.

Ogasawara J, Watanabe-Fukunaga R, Adachi M, Matsuzawa A, Kasugai T, Kitamura Y, Itoh N, Suda T, Nagata S (1993): Nature 364(6440):806–9.

Oltvai ZN, Milliman CL, Korsmeyer SJ (1993): Bcl-2 heterodimerizes in vivo with a conserved homolog, Bax, that accelerates programmed cell death. Cell 74(4):609–19.

Orlowski RZ, Baldwin AS, Jr. (2002): NF-kappaB as a therapeutic target in cancer. Trends Mol Med 8(8):385–9.

Ozes ON, Mayo LD, Gustin JA, Pfeffer SR, Pfeffer LM, Donner DB (1999): NF-kappaB activation by tumour necrosis factor requires the Akt serine-threonine kinase. Nature 401(6748):82–5.

Ozoren N, Kim K, Burns TF, Dicker DT, Moscioni AD, El-Deiry WS (2000): The caspase 9 inhibitor Z-LEHD-FMK protects human liver cells while permitting death of cancer cells exposed to tumor necrosis factor-related apoptosis-inducing ligand. Cancer Res 60(22):6259–65.

Pahl HL (1999): Activators and target genes of Rel/NF-kappaB transcription factors. Oncogene 18(49):6853–66.

Pan G, Ni J, Wei YF, Yu G, Gentz R, Dixit VM (1997a): An antagonist decoy receptor and a death domain-containing receptor for TRAIL. Science 277(5327):815–18.

Pan G, O'Rourke K, Chinnaiyan AM, Gentz R, Ebner R, Ni J, Dixit VM (1997b): The receptor for the cytotoxic ligand TRAIL. Science 276(5309):111–13.

Puthalakath H, Huang DC, O'Reilly LA, King SM, Strasser A (1999): The proapoptotic activity of the Bcl-2 family member Bim is regulated by interaction with the dynein motor complex. Mol Cell 3(3):287–96.

Raff MC (1996): Size control: The regulation of cell numbers in animal development. Cell 86(2):173–5.

Rao RV, Castro-Obregon S, Frankowski H, Schuler M, Stoka V, del Rio G, Bredesen DE, Ellerby HM (2002): Coupling endoplasmic reticulum stress to the cell death program: An Apaf-1-independent intrinsic pathway. J Biol Chem 277(24):21836–42.

Rathmell JC, Thompson CB (2002): Pathways of apoptosis in lymphocyte development, homeostasis, and disease. Cell Suppl 109:S97–107.

Ravi R, Bedi GC, Engstrom LW, Zeng Q, Mookerjee B, Gelinas C, Fuchs EJ, Bedi A (2001): Regulation of death receptor expression and TRAIL/Apo2L-induced apoptosis by NF-kappaB. Nat Cell Biol 3(4):409–16.

Reed JC (2003): Apoptosis-targeted therapies for cancer. Cancer Cell 3(1):17–22.

Romashkova JA, Makarov SS (1999): NF-kappaB is a target of AKT in anti-apoptotic PDGF signalling. Nature 401(6748):86–90.

Roth JA, Cristiano RJ (1997): Gene therapy for cancer: what have we done and where are we going? J Natl Cancer Inst 89(1):21–39.

Rudner J, Lepple-Wienhues A, Budach W, Berschauer J, Friedrich B, Wesselborg S, Schulze-Osthoff K, Belka C (2001): Wild-type, mitochondrial and ER-restricted Bcl-2 inhibit DNA damage-induced apoptosis but do not affect death receptor-induced apoptosis. J Cell Sci 114:4161–72.

Ryan KM, Phillips AC, Vousden KH (2001): Regulation and function of the p53 tumor suppressor protein. Curr Opin Cell Biol 13(3):332–7.

Scaffidi C, Fulda S, Srinivasan A, Friesen C, Li F, Tomaselli KJ, Debatin KM, Krammer PH, Peter ME (1998): Two CD95 (APO-1/Fas) signaling pathways. EMBO J 17(6):1675–87.

Schmitt CA, McCurrach ME, de Stanchina E, Wallace-Brodeur RR, Lowe SW (1999): INK4a/ARF mutations accelerate lymphomagenesis and promote chemoresistance by disabling p53. Genes Dev 13(20):2670–7.

Schweichel JU, Merker HJ (1973): The morphology of various types of cell death in prenatal tissues. Tetratology 7:253–66.

Sheridan JP, Marsters SA, Pitti RM, Gurney A, Skubatch M, Baldwin D, Ramakrishnan L, Gray CL, Baker K, Wood WI, Goddard AD, Godowski P, Ashkenazi A (1997): Control of TRAIL-induced apoptosis by a family of signaling and decoy receptors. Science 277(5327):818–21.

Sherr CJ, Weber JD (2000): The ARF/p53 pathway. Curr Opin Genet Dev 10(1): 94–9.

Shi Y (2002): Mechanisms of caspase activation and inhibition during apoptosis. Mol Cell 9(3):459–70.

Sprick MR, Weigand MA, Rieser E, Rauch CT, Juo P, Blenis J, Krammer PH, Walczak H (2000): FADD/MORT1 and caspase-8 are recruited to TRAIL receptors 1 and 2 and are essential for apoptosis mediated by TRAIL receptor 2. Immunity 12(6):599–609.

Strasser A, O'Connor L, Dixit VM (2000): Apoptosis signaling. An Rev Biochem 69:217–45.

Suzuki Y, Nakabayashi Y, Takahashi R (2001): Ubiquitin-protein ligase activity of X-linked inhibitor of apoptosis protein promotes proteasomal degradation of caspase-3 and enhances its anti-apoptotic effect in Fas-induced cell death. Proc Natl Acad Sci USA 98(15):8662–7.

Takeda K, Hayakawa Y, Smyth MJ, Kayagaki N, Yamaguchi N, Kakuta S, Iwakura Y, Yagita H, Okumura K (2001): Involvement of tumor necrosis factor-related apoptosis-inducing ligand in surveillance of tumor metastasis by liver natural killer cells. Nat Med 7(1):94–100.

Testa JR, Bellacosa A (2001): AKT plays a central role in tumorigenesis. Proc Natl Acad Sci USA 98(20):10983–5.

Thornberry NA, Lazebnik Y (1998): Caspases: Enemies within. Science 281(5381): 1312–16.

Truneh A, Sharma S, Silverman C, Khandekar S, Reddy MP, Deen KC, McLaughlin MM, Srinivasula SM, Livi GP, Marshall LA, Alnemri ES, Williams WV, Doyle ML (2000): Temperature-sensitive differential affinity of TRAIL for its receptors. DR5 is the highest affinity receptor. J Biol Chem 275(30):23319–25.

Tsujimoto Y, Shimizu S (2000): VDAC regulation by the Bcl-2 family of proteins. Cell Death Differ 7(12):1174–81.

Verhagen AM, Ekert PG, Pakusch M, Silke J, Connolly LM, Reid GE, Moritz RL, Simpson RJ, Vaux DL (2000): Identification of DIABLO, a mammalian protein that promotes apoptosis by binding to and antagonizing IAP proteins. Cell 102(1):43–53.

Verma IM, Stevenson JK, Schwarz EM, Van Antwerp D, Miyamoto S (1995): Rel/NF-kappa B/I kappa B family: Intimate tales of association and dissociation. Genes Dev 9(22):2723–35.

Vucic D, Stennicke HR, Pisabarro MT, Salvesen GS, Dixit VM (2000): ML-IAP, a novel inhibitor of apoptosis that is preferentially expressed in human melanomas. Curr Biol 10(21):1359–66.

Walczak H, Miller RE, Ariail K, Gliniak B, Griffith TS, Kubin M, Chin W, Jones J, Woodward A, Le T, Smith C, Smolak P, Goodwin RG, Rauch CT, Schuh JC, Lynch DH (1999): Tumoricidal activity of tumor necrosis factor-related apoptosis-inducing ligand in vivo. Nat Med 5(2):157–63.

Wang CY, Mayo MW, Korneluk RG, Goeddel DV, Baldwin AS Jr (1998): NF-kappaB antiapoptosis: Induction of TRAF1 and TRAF2 and c-IAP1 and c-IAP2 to suppress caspase-8 activation. Science 281(5383):1680–3.

Wang K, Yin XM, Chao DT, Milliman CL, Korsmeyer SJ (1996): BID: A novel BH3 domain-only death agonist. Genes Dev 10(22):2859–69.

Wiley SR, Schooley K, Smolak PJ, Din WS, Huang CP, Nicholl JK, Sutherland GR, Smith TD, Rauch C, Smith CA, et al. (1995): Identification and characterization of a new member of the TNF family that induces apoptosis. Immunity 3(6):673–82.

Wu GS, et al. (1997): KILLER/DR5 is a DNA damage-inducible p53-regulated death receptor gene. Nature Genet 17(2):141–3.

Wyllie AH, Kerr JF, Currie AR (1980): Cell death: The significance of apoptosis. Int Rev Cytol 68:251–306.

Xue D, Horvitz HR (1997): Caenorhabditis elegans CED-9 protein is a bifunctional cell-death inhibitor. Nature 390(6657):305–8.

Yuan J, Shaham S, Ledoux S, Ellis HM, Horvitz HR (1993): The *C. elegans* cell death gene ced-3 encodes a protein similar to mammalian interleukin-1 beta-converting enzyme. Cell 75(4):641–52.

Zha J, Harada H, Yang E, Jockel J, Korsmeyer SJ (1996): Serine phosphorylation of death agonist BAD in response to survival factor results in binding to 14-3-3 not BCL-$X_{(L)}$. Cell 87(4):619–28.

Zhao H, Yan M, Wang H, Erickson S, Grewal IS, Dixit VM (2001): Impaired c-Jun amino terminal kinase activity and T cell differentiation in death receptor 6-deficient mice. J Expr Med 194(10):1441–8.

Zhou BP, Liao Y, Xia W, Zou Y, Spohn B, Hung MC (2001): HER-2/neu induces p53 ubiquitination via Akt-mediated MDM2 phosphorylation. Nat Cell Biol 3(11):973–82.

PART IV

CHAPTER 16

MUTAGENESIS, MUTATIONS, AND DNA REPAIR

ROGER D. JOHNSON
Departments of Cancer Biology and Cell Biology, University of Massachusetts Medical School, Worcester, MA 01605

INTRODUCTION

Evolution is defined as the change in the genetic composition of a population during successive generations. The forces that drive this process are genetic variation among individuals, natural selection, fitness, and probability. Hence the survival of an individual is related to its environment, the phenotypes it exhibits, and chance. Individuals in a population and their progeny that exhibit traits that enable them to better compete for resources, avoid predators, and reproduce are likely to dominate the population in coming generations. On the macro level these processes have been verified by such classical observations as the adaptation of bird beaks to exploit various food sources or the alteration in the color of gypsy moths in postindustrial England from gray to black to camouflage them from predators.

These same principles of evolution, when adapted to a cellular milieu, provide important insights into tumor formation and cancer. Under normal circumstances the cellular composition of tissues and organs is defined and maintained by elaborate intra- and extracellular regulatory mechanisms that control cell differentiation, proliferation, and death (see Fig. 16.1 and Chapters 1 and 15). Despite these elaborate, and sometimes redundant mechanisms, tissue composition and function can be drastically compromised when a single cell becomes tumorigenic and acquires traits that enable it to evade normal growth control constraints. In this situation the tumorigenic cell will have acquired one or more phenotypes that allow it to proliferate when it should be quiescent (analogous to increasing its reproductive fitness) and/or allow it to ignore signals that would result in its death (analogous to acquiring camouflage to avoid predation). Mutations that activate proto-oncogenes (Chapters

Cell Cycle and Growth Control: Biomolecular Regulation and Cancer, Second Edition, Edited by Gary S. Stein and Arthur B. Pardee
ISBN 0-471-25071-6

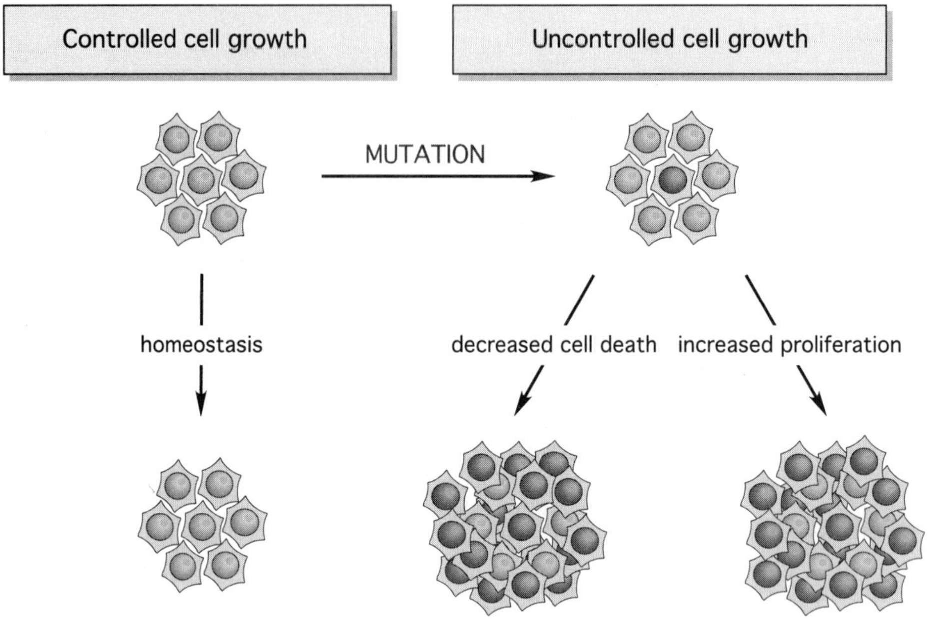

Figure 16.1. Uncontrolled cell growth leads to malignancy. Under normal circumstances the composition of tissues is tightly controlled by intra- and extracellular mechanisms leading to homeostasis. Mutation can allow a cell to evade normal growth control constraints, usually by increasing its ability to proliferate or preventing its death. These types of mutations can lead to tumor formation.

4 and 5), inactivate cell cycle inhibitory genes (Chapter 7), or alter genes that control apoptosis (chapter 14) can directly contribute to tumorigenic transformation. However, recent experiments demonstrate that some tumorigenic mutations occur in genes that do not directly control proliferation or apoptosis. Many of these genes function in DNA repair pathways that reduce mutation formation. Hence mutations in these genes indirectly lead to tumor formation by increasing genetic and/or chromosomal instability. This chapter will focus on the major DNA repair and maintenance pathways and their relevance to cancer.

DNA DAMAGE AND REPAIR

The diverse biological processes that DNA participates in and its unique chemical structure invariably leads to DNA damage. Some DNA lesions are endogenous, caused by a unique confluence of chemistry and biology (Fig. 16.2). Under physiological conditions some chemical bonds within the DNA structure are prone to degrade spontaneously (Lindahl, 1993). Two chemical alterations, base hydrolysis and deamination, occur at a high enough frequency in DNA to pose a serious challenge to genetic integrity. In addition to these spontaneous reactions, cellular metabolism

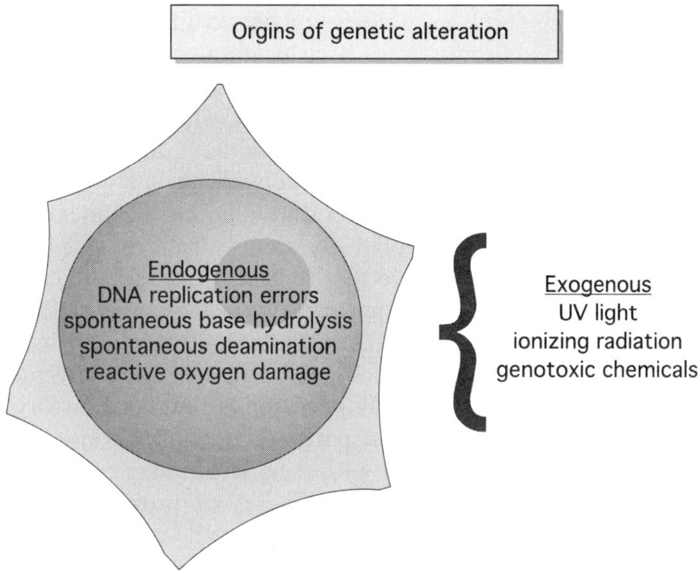

Figure 16.2. Endogenous and exogenous sources of genetic alteration. Numerous events can lead to DNA damage. Some sources of DNA damage, such as DNA replication errors, spontaneous chemical alteration of the DNA structure, and chemicals that can cause DNA alterations, are endogenous, emanating from within the cell. Other sources of DNA damage are endogenous, emanating from outside of the cell. Unless DNA damage is accurately repaired it can lead to genetic alterations.

generates reactive oxygen species (ROS), mainly hydroxyl radicals, and singlet oxygen, which can damage and distort DNA structure (Breen and Murphy, 1995). In contrast to these mechanisms that chemically alter DNA structure, a more subtle source of genetic alterations is DNA metabolism, specifically DNA replication. Insertions, deletions, and misincorporated bases can result from errors in DNA replication and lead to altered genetic content. While the sources of some DNA lesions emanate from within the cell, the sources of other lesions emanate from outside of the cell. The exogenous sources of DNA damage are frequently radiation or UV light, but they can also be chemical compounds that physically alter DNA structure. Whatever the cause, unless DNA damage is repaired, it leads to mutation.

Numerous genetic alterations have been identified that are associated with tumorigenesis. These alterations fall into four basic categories: nucleotide sequence changes, alterations in chromosome number, chromosome rearrangements, and gene amplifications. Nucleotide sequence changes alter the sequence of a gene or a regulatory element that controls transcription of a gene. These changes usually alter one to several nucleotides, with the modifications leading to altered gene function or expression. Alterations in chromosome number involve gain or loss of whole chromosomes and can be found in most tumor types. Chromosome translocations result from the fusion of two different chromosomes to form a chimeric chromosome. Translocation can be categorized as reciprocal, where a part of one chromosome is exchanged for a part of a different chromosome, or nonreciprocal, where one chromosome receives a region of a second chromosome without donating a region to that same chromosome. Gene amplifications result in a given region of a chromosome being duplicated several to hundreds of times.

The fact that the genome is relatively stable reflects the efficiency with which DNA repair pathways recognize and repair DNA damage. In human cells five major DNA repair pathways have been defined: mismatch repair, nucleotide-excision repair, base-excision repair, nonhomologous end-joining, and homologous recombination. Because the need to repair DNA damage has persisted throughout history, these repair pathways are, for the most part, evolutionarily conserved from prokaryotes to eukaryotes. Much of our understanding of the basics of these repairs pathways comes from studies of prokaryotic systems; however, evolution has led to some significant alterations in these repair pathways to accommodate for the different genomic composition and challenges of mammalian cells and multicellular organisms. In general, defects in any of these DNA repair pathways lead to cancer susceptibility.

MISMATCH REPAIR

The mismatch repair system (MMR) was originally identified in bacteria and was shown to play an important role in preventing genetic instability by repairing DNA replication errors (Radman and Wagner, 1986). Many features of this system are evolutionarily conserved from bacteria to humans, facilitating our understanding of the importance of MMR in cancer avoidance. The primary function of the mismatch repair system is to eliminate base-base mismatches and insertion/deletion loops (Fig. 16.3B) that result from DNA polymerase misincorporation or slippage during DNA replication. Inactivating the MMR pathway results in

---

Figure 16.3. Mismatch repair substrates and repair mechanism. Nucleotide that are not paired correctly and nucleotides that are not paired at all are substrates for cellular mismatch repair mechanisms. **(A)** Schematic representation of DNA substrates that would be corrected by the mismatch repair mechanism including base-pair mismatches and insertion and deletion loops. **(B)** Steps involved in mismatch repair. **(I)** Mismatch repair proteins recognize different types DNA lesions. **(II)** Base-base mismatches and one base-pair insertion/deletion loops (IDL) are recognized by the MSH2/MSH6 heterodimer. IDL containing one or more base pairs are recognized by the MSH2/MSH3 heterodimer. **(III)** The MSH heterodimer recruits a MLH heterodimer, usually MLH1/PMS2, which helps propagate mismatch repair. **(IV)** Protein-protein interactions can then recruit an exonuclease, such as EXO1, which can remove the strand containing the mismatch. **(V)** Replication factors resynthesize the DNA strand that has been removed.

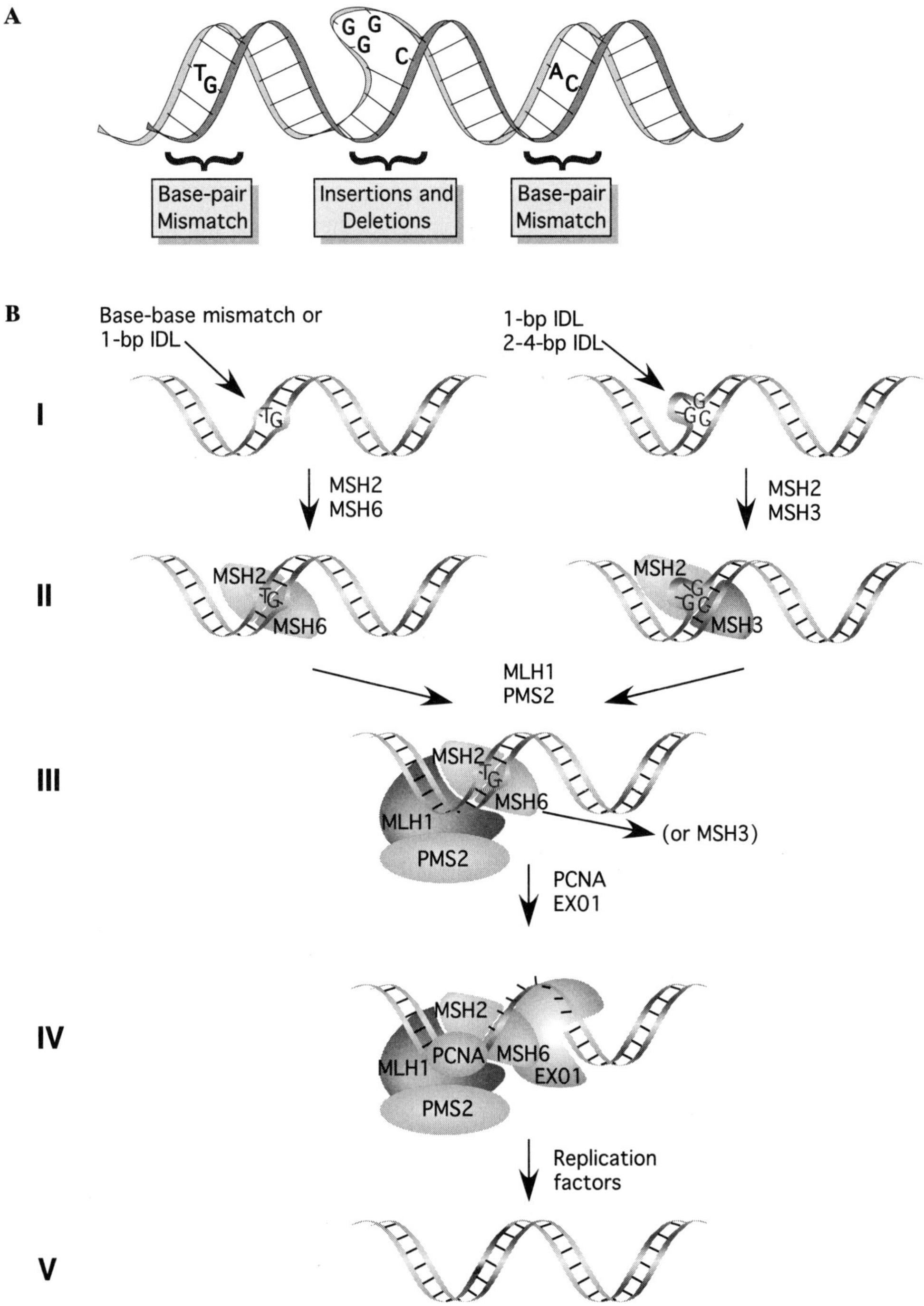
A
TG
GGG
C
AC
Base-pair Mismatch
Insertions and Deletions
Base-pair Mismatch
B
Base-base mismatch or 1-bp IDL
1-bp IDL 2-4-bp IDL
I
TG
GGG
MSH2 MSH6
MSH2 MSH3
II
MSH2
MSH6
MSH3
MLH1 PMS2
III
MLH1
PMS2
(or MSH3)
PCNA EX01
IV
PCNA
EX01
Replication factors
V

a profound DNA repair defect and a coincident increase in the mutation frequency. Microsatellites, regions of mono-, di-, tri-, and tetranucleotide repeats, are particularly susceptible to mutation by defects in the MMR pathway. Microsatellite instability (MSI) was often observed in colorectal cancers, leading to the hypothesis that defects in the MMR pathway in humans could promote cancers. The ensuing genetic analysis of MSI^+ colorectal tumors confirmed that defects in the MMR pathway could lead to a cancer-prone syndrome called hereditary nonpolyposis colorectal cancer (HNPCC; see Fishel et al., 1993; Leach et al., 1993). Subsequently analysis of other tumor types identified MMR gene defects in sporadic tumors, suggesting an even broader role for MMR in tumor suppression (Claij and te Riele, 1999).

Repair Mechanism

In *E. coli* the DNA mismatch repair system was originally identified by the isolation of strains that had genetic defects that elevated spontaneous mutation frequencies. Subsequent analysis verified that three genes, *mutS, mutL*, and *mutH*, are dedicated to mismatch repair (Lahue et al., 1989). In addition in vivo reconstitution of mismatch repair has demonstrated that a number of nonspecialized proteins, including MutU/UvrD, several exonucleases, DNA polymerase III, and DNA ligase, are required for efficient mismatch repair (Modrich and Lahue, 1996). MutS carries out the initial step in mismatch repair, recognition of mispaired or unpaired nucleotides within a DNA duplex. To catalyze this process MutS forms an asymmetric homodimer that must bind both ATP and a DNA mismatch simultaneously in order to stimulate other repair proteins and propagate mismatch repair (Galio et al., 1999; Lamers et al., 2000; Obmolova et al., 2000). Strand discrimination between the template and the newly synthesized DNA strand in accomplished through the function of MutH. MutH binds hemimethylated GATC sequences and cleaves the unmethylated strand, thereby distinguishing the strand that will be removed during mismatch repair (Welsh et al., 1987). MutL acts as an intermediate to both couple and stimulate the activities of MutS and MutH. First, a homodimer of MutL interacts with MutS bound at a mismatch and stimulates the ATP hydrolysis-dependent translocation of MutS. This activity enables MutS to search for the nearest hemimethylated GATC site occupied by MutH (Ban et al., 1999; Ban and Yang, 1998). Second, MutL, complexed with MutS and mismatched DNA, can stimulate the endonuclease activity of MutH. Finally MutL is required to load the MutU/UvrD helicase at the site of the DNA nick, which promotes unwinding of the DNA duplex and subsequent exonucleolytic removal of the fragment containing the mismatch (Dao and Modrich, 1998; Yamaguchi et al., 1998). Because the nick catalyzed by MutH can lay on either side of the mismatch, either a 5′–3′ or a 3′–5′ exonucleases can effect removal of the mismatch. At least two 5′–3′ single-strand specific exonucleases (RecJ and ExoVII) and two 3′–5′ single-strand specific exonucleases (ExoI and ExoX) are required for excision, as mismatch repair, in vivo and in vitro, is impeded only when all four exonucleases are absent (Burdett et al., 2001). After removal of the large fragment of DNA containing the mismatch, single-strand DNA binding protein, DNA polymerase III holoenzyme and DNA ligase, resynthesizes and seals the gap generated during mismatch repair.

In eukaryotes, multiple MutS and MutL homologues exist, but no MutH or MutU/UvrD homologues have been identified, reflecting both conservation and divergence from the *E. coli* mismatch repair process (Fig. 16.3B; Harfe and Jinks-Robertson, 2000). Base-base mismatches and insertion/deletion loops are identified by several of the eukaryotic MutS homologues (MSH2, 3, and 6). Unlike *E. coli*, different MSH complexes identify distinct MMR substrates. Heterodimers of MSH2/MSH6, also called MutSα, specialize in identifying mismatches and single-base loops, while heterodimers of MSH2/MSH3, also called MutSβ, specialize in identifying larger insertion/deletion loops (Acharya et al., 1996; Genschel et al., 1998). Other MutS homologues (MSH1, 4, and 5) have also been identified, but their involvement in nuclear MMR is unlikely. MSH1 appears to be involved in mitochondrial genome maintenance (Chi and Kolodner, 1994). Interestingly MSH4 and MSH5 have completely lost the ability to participate in the normal mismatch repair process but appear to function solely in meiosis, promoting crossover and chromosome synapsis (de Vries et al., 1999; Edelmann et al., 1999).

Four MutL homologues have been identified in humans, MLH1, MLH3, PMS1, and PMS2 (Harfe and Jinks-Robertson, 2000). The names for some of these genes differ between humans and yeast and have been the source of considerable confusion in the field. This chapter will only use the names for the human genes. As with the MutS homologues, heterodimers of the MutL homologues function in MMR in humans. Three different heterodimeric MutL complexes have been identified, all of which contain MLH1. Based on genetic and biochemical studies in yeast and humans cells, the heterodimer of MLH1 and PMS2, also called MutLα, is the most important form of MutL homologues in MMR (Kolodner, 1996).

In humans the later steps of MMR, strand discrimination, excision, and resynthesis are poorly understood; however, MutS, in cooperation with MutL, can recruit other factors to the site of mismatch repair, consistent with the notion that a higher order complex catalyzes these steps. Several lines of evidence suggest that MMR is intimately coupled with DNA replication, providing possible explanations for how the later steps of MMR are catalyzed. Proliferating cell nuclear antigen (PCNA) is able to bind both MutSα and MutSβ and has been proposed to guide the mismatch repair proteins to free DNA termini, potentially playing a role in strand discrimination (Clark et al., 2000; Flores-Rozas et al., 2000). In addition to PCNA, genetic and biochemical experiments suggest a role for the replication factors DNA polymerases δ and ε in MMR. The 3′ to 5′ proofreading functions of δ and ε have been implicated in the excision step, while the DNA polymerase activity of δ and possibly ε have been implicated in the resynthesis step (Longley et al., 1997). Although the coupling of MMR to DNA replication functions is clearly an appealing model for how strand discrimination, excision, and resynthesis is accomplished, more proof is need to validate this hypothesis.

Cancer Susceptibility

Hereditary nonpolyposis colon cancer (HNPCC) is an autosomal dominant cancer predisposition syndrome that is defined by three or more family members being diagnosed with colorectal cancer in two or more generations, with at least one of the affected members being diagnosed before the age of 50 (Lynch et al., 1991). HNPCC accounts for approximately 5% of all cases of colon cancer, with predisposed individuals from HNPCC families having an extremely high lifetime risk of developing colorectal cancer (80%), endometrial cancer (50%), as well as various other cancers (15%). Collectively this tumor predisposition is known as the HNPCC tumor spectrum. The genetic nature of HNPCC led scientists to search for tumor susceptibility loci in these patients. Linkage analysis of large kindreds identified two chromosomal regions, on chromosome 2 and chromosome 3, that were frequently lost in tumors derived from HNPCC patients (Lindblom et al., 1993; Peltomaki et al., 1993). Moreover many of these tumors displayed genetic instability of microsatellites and nucleotide repeat regions, reminiscent of the mutator phenotype observed in bacteria or yeast harboring a defect in a mismatch repair gene (Peinado et al., 1992). The confluence of these observations led to the hypothesis that germ-line mutations in human mismatch repair genes could lead to tumor susceptibility. In 1993 and 1994 scientists confirmed this hypothesis by identifying mutations in *hMSH2*, the human homologue of the *E. coli mutS* gene, and *hMLH1*, the human homologue of *E. coli mutL*, as the genetic defects on chromosome 2 and chromosome 3 respectively, that predisposed HNPCC patients to developing colorectal cancer (Bronner et al., 1994; Fishel et al., 1993; Leach et al., 1993; Papadopoulos et al., 1994).

The majority of HNPCC patients are heterozygous for recessive, germ-line mutation in one of the human mismatch repair genes. To date, more that 400 different mutations in mismatch repair genes have been shown to be associated with HNPCC. Defects in four of the mismatch repair genes account for approximately 80% of all HNPCC cases. Of these cases nearly 50% affect hMLH1, about 40% affect hMSH2, and about 10% affect hMSH6, with a small percentage affecting hPMS2 (see Web resources section). Mutations in other mismatch repair genes may predispose individuals to HNPCC but any such link has not been verified. Because HNPCC patients are heterozygous for a mutant and normal allele of a mismatch repair gene, normal tissue from these patients displays no overt defects in mismatch repair. Tumor development is nearly always associated with inactivation of the normal allele, which can occur by loss of heterozygosity, somatic mutation or transcription down regulation by promoter hypermethylation.

hMLH1 and Cancer

The *hMLH1* gene is located on chromosome 3p21.3 and consists of 19 exons. Defects in *hMLH1* account for approximately 50% of mutations found in HNPCC patients. Most of these alterations create frameshift or splice site mutations that result in truncated gene products. Tumor formation in individuals containing an *hMLH1* mutation is nearly always associated with inactivation of the remaining normal allele. Only 30% to 35% of tumors from *hMLH1* patients show loss or somatic mutation of the wild-type allele (Hemminki et al., 1994). The remaining tumors reduce or abolish *hMLH1* gene expression by promoter hypermethylation (Kuismanen et al., 2000). Promoter hypermethylation of *hMLH1* also appears to be an important event in sporadic colorectal cancers. Many sporadic colorectal cancers display enhanced microsatellite instability. Biallelic hypermethylation of the *hMLH1* promoter accounts for approximately 90% of these cases (Cunningham et al., 1998; Thibodeau et al., 1998). Thus defects in *hMLH1* contribute not only to hereditary colorectal cancer but also to sporadic colorectal cancer.

hMSH2 and Cancer

The *hMSH2* gene contains 15 exons and is located on chromosome 2p22–p21. Approximately 40% of HNPCC kindreds harbor mutations in *hMSH2*. Most of the mutations are unique and lead to premature termination of the hMSH2 gene product. A smaller number of alterations lead to point mutations that lead to total or partial loss of hMSH2 function. Unlike *hMLH1*, inactivation of the normal *hMSH2* allele almost always occurs by loss of heterozygosity or somatic mutation (Leach et al., 1993).

Genetic Alterations

Numerous lines of evidence indicate repeated sequences are highly susceptible to misalignment during replication. If unrepaired, these replication errors lead to nucleotide insertion and deletion mutations. The mismatch repair system is responsible for identifying and correcting any such errors. Hence cells that are defective for the mismatch repair system show an enormous increase in the mutation rate. Many of these repeated sequences in the mammalian genome are located in noncoding or nonregulatory regions, however; a number of human genes exist that contain mononucleotide tracts, making them potential targets for alteration in mismatch repair deficient cells. Many of these genes are altered in MSI^+ tumors. These genes include important growth regulatory and signal transduction genes (TGF-β-RII, IGFIIR, and PTEN see; Markowitz et al., 1995; Shin et al., 2001; Souza et al., 1996), proapoptotic genes (BAX and caspase-5; see Rampino et al., 1997; Schwartz et al., 1999), transcription factors (E2F4 and TDF-4; see Duval et al., 2000; Ikeda et al., 1998), and DNA repair genes (MSH6, MSH3, MLH3,

MED-1, RAD50, DNA-PK$_{cs}$, and BLM; see Bader et al., 1999; Duval et al., 2001; Loukola et al., 2000; Malkhosyan et al., 1996; Ohmiya et al., 2001; Riccio et al., 1999; Yin et al., 1997). Whether these genes are true targets for inactivating frameshift mutations that contribute to tumor initiation or progression or only reflect the high mutation rate associated with mismatch repair defects is an important issue in understanding MSI-associated carcinogenesis. One example, TGFβ-RII, effectively illustrates the connection between mismatch repair deficiency, mutation, and carcinogenesis.

Several aspects of TGFβ-RII biology and structure make it a prime candidate for involvement in colorectal cancers with microsatellite instability. TGFβ-RII encodes a receptor for TGFβ and has been shown to negatively regulate the growth of colonic epithelial cells. Moreover the TGFβ-RII open reading frame contains a tract of 10 adenines within its coding region, making it a prime candidate for frameshift mutations as a result of defective mismatch repair function. Studies of the TGFβ-RII have confirmed that it is mutated at an extremely high frequency (~85%) in colorectal cancer with microsatellite instability, confirming the hypothesis that TGFβ-RII inactivation is an important step in human colorectal tumorigenesis (Takenoshita et al., 1997). The importance of the tract of adenines in TGFβ-RII in mismatch repair deficient tumors is illustrated by comparing tumors in MSH2-deficient humans and mice. Unexpectedly the majority of MSH2-knockout mice succumb to lymphoid tumors at an early age, rather than to colorectal cancer (de Wind et al., 1995; Reitmair et al., 1995). In order to determine whether the difference in tumor spectrum was due to differences in the TGFβ-RII, the coding region of the mouse TGFβ-RII gene was determined. Unlike the human gene, which contained a tract of 10 adenines, the corresponding mouse gene sequence was AAAAGAAAAG (Jacob and Praz, 2002). The first A to G transition is silent, while the second A to G transition is conservative, changing a lysine residue to arginine. The poly-A tract in the mouse coding region, since two guanine residues interrupt it, is likely to be more resistant to polymerase slippage and defective mismatch repair-induced mutagenesis. Thus the difference in the tumor spectrum between mismatch repair deficient mice and humans may be due to a lower mutation frequency of the mouse TGFβ-RII gene.

NUCLEOTIDE EXCISION REPAIR

Many exogenous DNA damaging agents, such as ultraviolet light, are capable of chemically modifying DNA bases. Frequently these alterations distort the normal shape of the DNA helix by creating inappropriate chemical bonds between bases or covalently adding bulky chemical groups. These types of lesions can lead to mutations when a DNA polymerase replicates across the damaged base and inserts the wrong nucleotide. Nucleotide excision repair provides an important defense against a large variety of structurally unrelated DNA alterations that distort the DNA helix (Fig. 16.4A). The importance of this repair pathway in preventing cancer is exemplified by the human syndrome xeroderma pigmentosum (XP). XP patients are extremely sensitive to sunlight and have a 1000-fold higher chance of developing melanoma than the population in general (Kraemer et al., 1994). The cellular basis for this is a defect in nucleotide excision repair. Cells from these patients fail to remove modified bases, including pyrimidine dimers, that result from exposure to sunlight. As a consequence these patients frequently develop *skin* cancers in areas that are exposed to light.

Repair Mechanism

Nucleotide excision repair in mammalian cells is a complex biochemical process that requires approximately 30 proteins, is divided into several distinct steps, and can be divided into two subpathways. One pathway, called transcription coupled repair (TCR), deals specifically with lesions that arrest or impede RNA polymerase II that is transcribing an expressed gene. The other pathway, called global genomic repair (GGR), functions to remove DNA damage in nontranscribed regions of the genome. While the first step, identification of a DNA lesion, is the same for both of the subpathways, the proteins that carry out this step are different (Fig. 16.4B). In GGR a complex of XPC-HHRAD23B and the UV DNA damage binding protein UV-DDB are responsible for identifying lesions that need to be repaired (Hwang et al., 1999; Sugasawa et al., 1998; Tang and Chu, 2002). For many lesions recognition by XPC-HHRAD23B is sufficient to initiate repair, while for other lesions, such as cyclobutane pyrimidine dimers, UV-DDB is also required. UV-DDB is thought to enhance the DNA helix distortion caused by the cyclobutane pyrimidine dimer to more efficiently recruit the XPC-HHRAD23B complex. For TCR the ability of the lesion to block RNA polymerase II seems to be critical. The RNA polymerase II complex is thought to include proteins, which, although not necessarily components of the normal transcription machinery, are recruited to sites of arrested transcription. Among these are two proteins, CSA and CSB, that are thought to help displace the stalled RNA polymerase and then help recruit other nucleotide excision repair factors (Le Page et al., 2000).

The subsequent stages of GGR and TCR are thought to be similar if not identical. After identification of the damaged DNA, several other factors are recruited to this site. These include TFIIH, a subcomplex of RNA polymerase II that contains two helicases, XPB and XPD. These helicase unwind a small, localized region of approximately 30 base pairs that contains the DNA damage. XPA probably confirms the presence of DNA damage by probing for abnormal backbone structure (Buschta-Hedayat et al., 1999). If XPA is absent or cannot find any DNA damage nucleotide excision repair is aborted. RPA, the single-stranded DNA binding protein that is essential for DNA replication, also acts at this stage, likely stabilizing the open DNA helix and facilitating the action of XPA.

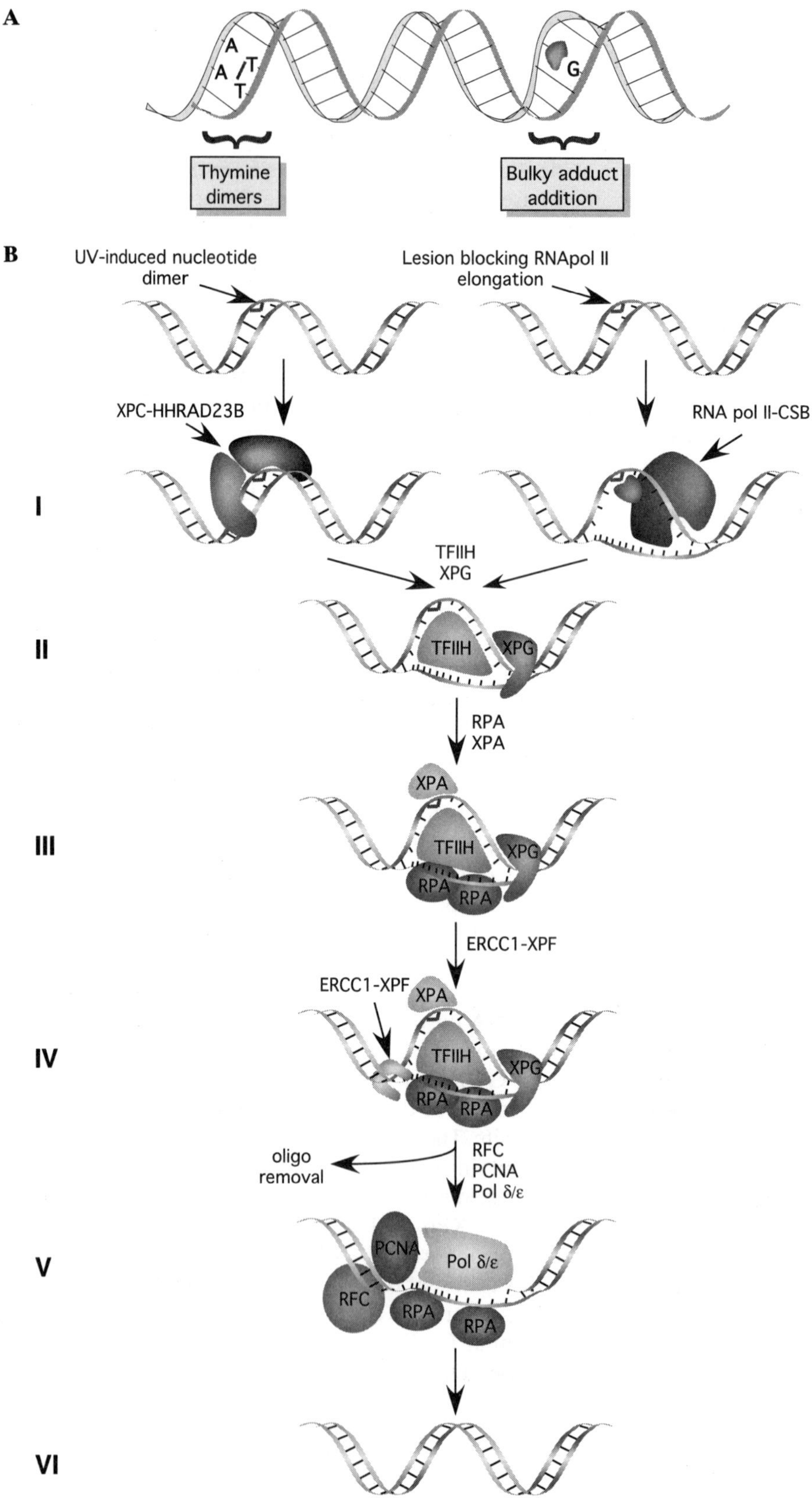
A
A
A
T
T
G
Thymine dimers
Bulky adduct addition
B
UV-induced nucleotide dimer
Lesion blocking RNApol II elongation
XPC-HHRAD23B
RNA pol II-CSB
I
TFIIH
XPG
II
TFIIH
XPG
RPA
XPA
XPA
III
TFIIH
XPG
RPA
RPA
ERCC1-XPF
ERCC1-XPF
XPA
IV
TFIIH
XPG
RPA
RPA
oligo removal
RFC
PCNA
Pol δ/ε
PCNA
Pol δ/ε
V
RFC
RPA
RPA
VI

After identification of the DNA damage and opening of the DNA helix containing the damage, endonucleases affect removal of a single-stranded oligonucleotide containing the DNA damage (Sugasawa et al., 2001). One of the factors, XPG, is recruited early during the repair process but functions at the cleavage step to make a single-stranded DNA break 3′ of the DNA damage at the junction of the single- and double-stranded DNA. The other endonuclease is a complex of XRCC1 and XPF, which cleaves the DNA 5′ to the DNA damage. This generates a single-stranded oligonucleotide of approximately 30 base pairs containing the DNA damage. Currently it is unknown if the fragment is actively or passively removed from the DNA repair site.

The final step in nucleotide excision repair is filling in the gap created by removal of the oligonucleotide containing the DNA damage. Repair DNA synthesis requires either DNA polymerase δ or ε, and several accessory factors including RFC, PCNA, and RPA. After repair DNA synthesis, the DNA break is then sealed by DNA ligase.

Cancer Susceptibility

Xeroderma pigmentosum is an autosomal recessive cancer predisposition syndrome that was originally described in 1874 by Moritz Karposi and Ferdinand Hebra. XP is characterized by an acute predisposition to skin cancers, photosensitivity, a moderately elevated frequency of internal tumors, and a predisposition to neurodegeneration. The molecular basis of XP was not known until the 1960s, when two independent laboratories demonstrated that cells from XP patients were defective for nucleotide excision repair (Cleaver, 1968; Epstein et al., 1970). Complementation analysis by cell fusions has led to the identification of seven different complementation groups, designated XP-A to XP-G. The genes

---

Figure 16.4. Nucleotide excision repair substrates and mechanism. Nucleotide excision repair recognized numerous types of DNA lesions, with the most prevalent being thymine dimers and the addition of bulky chemical adducts. **(A)** Schematic representation of DNA lesions recognized by the nucleotide excision repair mechanism. **(B)** Two subpathways exist in nucleotide excision repair, global genome repair, and transcription coupled repair. The difference between these two subpathways lies in the molecules that initially recognize the DNA lesion and initiate repair. **(I)** In global genome repair, the XPC-HHRAD23B complex recognizes the lesion. In transcription coupled repair, the CSA and CSB proteins are involved in recognizing lesions that stall RNA polymerase II transcription. **(II)** The following steps in nucleotide excision repair are likely the same for global genome repair, and transcription coupled repair, starting with recruitment of TFIIH and XPG. **(III)** Next XPA is recruited to ensure the presence of a DNA lesion and RPA helps to stabilize the opened DNA helix. **(IV)** Recruitment of the ERCC1-XPF complex allows cleavage on both sides of the region containing the DNA damage. ERCC1-XPF cleaves 5′ of the DNA damage while XPG cleaves 3′ of the DNA damage. **(V, VI)** DNA replication factors resynthesize the DNA that has been removed and ligate the broken strand. (See color insert.)

responsible for the genetic defects in each of the complementation groups have now been cloned and analyzed in patients, confirming the link between a genetic defect in one of these genes and XP. Moreover, because nucleotide excision repair has been reconstituted in vitro, the physical involvement of the protein products from these genes has been biochemically verified. Hence tumor susceptibility in these patients results from an accumulation of mutations due to inappropriately repaired DNA damage. Support for this comes from the analysis of p53 in XP patients. In these experiments skin taken from areas that had not been exposed to sunlight contained no mutations in p53. In skin exposed to sunlight numerous mutations were found, with nearly all of them being UV-specific C to T or CC to TT transitions at dipyrimidine sites (Williams et al., 1998).

One interesting observation concerning nucleotide excision repair genes is that mutations in some of the genes do not lead to tumor susceptibility. Cockayne syndrome is an autosomal recessive disorder that is caused by defects in the *CSA* or *CSB* genes (Henning et al., 1995; Lehmann, 1982; Tanaka et al., 1981; Troelstra et al., 1992). Symptoms in CS patients vary greatly, with most patients exhibiting neurological abnormalities and sun sensitivity; however, CS patients are not predisposed to cancer. Skin fibroblasts from these patients are, as expected, more sensitive to UV-induced killing than normal fibroblasts, however; a detailed analysis of these cells demonstrated that they were defective for TCR but were normal for GGR. Currently it is not known how defects in CSA or CSB lead to CS.

Tricothiodystrophy (TTD) shares many of the same symptoms as CS, but has the additional symptom of brittle hair and nails (Price et al., 1980). Remarkably, mutations in XPB and XPD can give rise to all three diseases, XP, CS, and TTD. The discovery that TFIIH is required for both transcription and nucleotide excision repair and that XPB and XPD are essential components of TFIIH has provided some answers to these observations. The simplest explanation for how this might occur is that some mutations in XPB or XPD affect different cellular functions. For example, XP would result from mutations in XPB, or XPD led to defective nucleotide excision repair but permitted normal TFIIH function. In contrast, mutations in XPB or XPD that affected transcription by TFIIH but did not impair nucleotide excision repair would lead to CS or TTD.

BASE EXCISION REPAIR

The most common types of DNA damage are nucleotide base lesions caused by spontaneous deamination or methylating and oxidizing agents. Reactive oxygen species, such as hydroxyl radicals, can react with DNA and alter base and sugar structures or lead to abasic sites. Spontaneous deamination has been estimated to affect approximately 10,000 bases per cell per day (Evans et al., 2000). Base excision repair (BER) is a major

DNA repair pathway protecting mammalian cells against single-base DNA damage (Fig. 16.5A). Base excision repair is mediated through at least two subpathways, one involving single nucleotide removal and the other involving removal of longer DNA patches of 2 to 15 nucleotides. Although BER is responsible for much of the cellular DNA repair, defects in BER have not been associated with cancer susceptibility in humans. This may stem from the fact that inactivation of some BER proteins leads to embryonic lethality, while inactivation of other proteins leads to a minimal phenotype (Tebbs et al., 1999; Xanthoudakis et al., 1996). The most compelling evidence for the involvement of BER in cancer prevention comes from epidemiological studies of polymorphisms of *OGG1* and *XRCC1* and their association with cancer (Goode et al., 2002). Mouse models for one BER gene, poly (ADP-ribose) polymerase also called PARP, have directly demonstrated that loss of PARP, coupled with p53 inactivation, decreases viability and widens the spectrum of tumors observed in $p53^{-/-}$ mice (Tong et al., 2001b). These studies have emphasized how genome instability caused by defects in BER can lead to cancer susceptibility.

Repair Mechanism

The BER pathway can repair numerous types of single nucleotide alterations, ranging from altered nucleotide bases and abasic sites to single-stranded DNA breaks. These types of lesions, and how the different repair subpathways intersect, are shown in Figure 16.5. The identification of altered nucleotide bases within the DNA helix activates one of the core BER subpathways. Numerous DNA glycoslyases, such as *ANPG* and *OGG1*, recognize a relatively narrow spectrum of lesions and initiate the core BER subpathway by flipping the altered nucleotide out of the DNA helix and then catalyzing cleave of the base-sugar bond (Boiteux and Radicella, 2000). This generates an apurinic/apyrimidinic site (AP site). DNA glycolsylases, such as ANPG, are monofunctional and remove only the damaged base, while DNA glycoslyases, such as OGG1, are bifunctional, removing not only the damaged base but also generating an incision in the DNA phosphate backbone 3′ of the damaged base.

The next step in the BER pathway is catalyzed by the apurinic/apyrimidinic endonuclease APE1. Biochemically, APE1 cleaves the phosphodiester backbone 5′ of the AP site. Following cleavage by a monofunctional glycosylase APE1 creates a single-stranded DNA break with a 5′-terminal baseless sugar, while APE1 activity coupled with a bifunctional glycosylase leads to complete removal of the AP site (Mol et al., 1995). APE1 may also be an important regulator of BER. Several protein-protein interactions have been identified between APE1 and other proteins involved in BER, suggesting that APE1 may help regulate later steps in the BER pathway. APE1 protein-protein interactions include interactions with XRCC1, a protein that stimulates APE1 activity and interacts with other BER proteins; PCNA and DNA polymerase

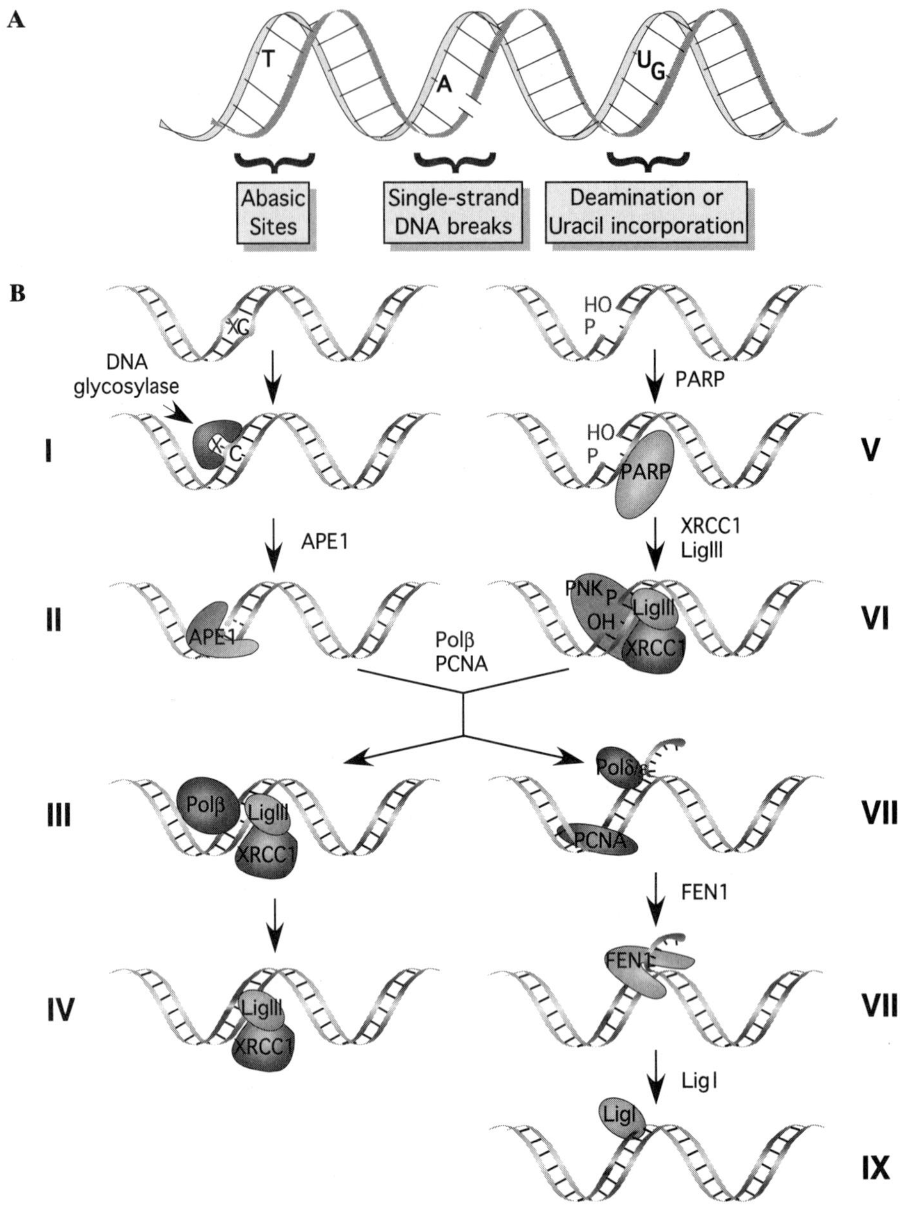

Figure 16.5. Base excision repair substrates and mechanism. Base excision repair recognizes numerous single base alterations. Several of the alterations are depicted in **(A)**. **(B)** Two mechanisms of base excision repair exist. One mechanism removes a single nucleotide while the other mechanism removes 2 to 15 nucleotides. **(I)** Base excision repair is usually initiated by a DNA glycosylase that removes the damaged base. **(II)** APE1 cleaves the DNA backbone, which allows DNA polymerase β **(III)** to resynthesize the base that has been removed. **(IV)** The XRCC1-ligase III complex then seals the remaining nick. **(V)** Alternatively, the single-stranded nick can be recognized by PARP, which can then recruit **(VI)** polynucleotide kinase (PNK) and XRCC1-ligase III. **(VII)** Recruitmentof PCNA and polymerase δ (or ε) can lead to a short patch of DNA being synthesized, displacing a single-stranded DNA flap. **(VIII)** FEN1 can then cleave the flap allowing ligase I **(IX)** to seal the remaining nick.

β, involved in repair DNA synthesis of the DNA break; and FEN1, an endonuclease that can remove DNA flaps (Caldecott, 2001; Dianova et al., 2001; Kubota et al., 1996; Ranalli et al., 2002). These interactions suggest that APE1 may help to coordinate the cellular response to DNA breaks. Recent studies suggest that after APE1 activity the other proteins involved in BER may recognize the DNA product-protein complex, rather than binding to an intermediate that is free in solution, ensuring the correct processing of the single-stranded DNA break, and protecting it from other processing reactions.

After a single-stranded DNA break has been generated, either short-patch or long-patch repair can heal the DNA break. In mammalian cells short-patch repair is the dominant repair mechanism. In this mechanism DNA polymerase β performs a one-nucleotide gap-filling reaction, and if necessary, DNA polymerase β can also remove the 5-terminal baseless sugar by its DNA lyase activity. This results in a ligatable nick that is sealed by the ligase activity of the XRCC1-DNA ligase IIIα complex (Whitehouse et al., 2001). XRCC1, by virtue of its ability to interact with DNA polymerase β and DNA ligase IIIα, may function as a scaffold protein, facilitating the final steps of the short-patch repair pathway. In the long-patch repair pathway three DNA polymerases (β, δ, or ε) can perform the synthesis step. When DNA polymerases δ or ε is used, PCNA is also required. DNA synthesis results in a patch of 2 to 15 nucleotides being added to the 3′ end of the single-stranded DNA break, leading to disassociation of the 5′ end of the single-stranded DNA break and the formation of a DNA flap. FEN1 then removes the displaced DNA flap and DNA ligase I seals the break.

The BER pathway is also important for repairing single-stranded DNA breaks that may arise randomly throughout the genome. In contrast to the BER subpathway that uses DNA glycoslyases to initiate repair the single-stranded break, the repair subpathway uses a different surveillance mechanism to initiate repair. The initial protein to interact with unscheduled single-stranded DNA breaks is PARP. PARP is capable of binding unusual DNA structures, including DNA single-stranded breaks, which stimulates its enzymatic activity. Once activated, PARP catalyzes the attachment of ADP ribose units to target proteins (Tong et al., 2001a). The exact role that this activity plays in DNA repair is unknown but may function in recruiting and stimulating other repair enzymes or may modulate local chromatin structure in order to facilitate DNA repair. In addition to its enzymatic activity PARP interacts with XRCC1, a scaffold protein for BER (Caldecott et al., 1996; Masson et al., 1998).

The next step in single-stranded break repair probably depends on the type of damage located at the break. XRCC1 is capable of interacting with APE1, which may be needed to process unligatable DNA ends. XRCC1 also interacts with polynucleotide kinase, an enzyme that possesses both a 3′-DNA phosphatase activity and a 5′-DNA kinase activity (Whitehouse et al., 2001). Once the break is processed, it can be funneled into either the short-patch or long-patch BER subpathways.

Cancer Susceptibility

Defective BER in *E. coli*, yeast, and mammalian cells leads to an elevated frequency of spontaneous mutations. One of the major lesions found in BER-defective cells are point mutations, such as have been found to inactivate some tumor-suppressor genes. Although this circumstantial evidence points to a link between defective BER and cancer, definitive proof remains elusive. Part of this may be due to the fact that inactivation of BER genes, such as APE1 and XRCC1, leads to embryonic lethality (Tebbs et al., 1999; Xanthoudakis et al., 1996). No appropriate mouse model systems have been designed to effectively test the hypothesis that defects in APE1 or XRCC1 leads to tumor susceptibility. Another possibility is that inactivation of BER genes that do not lead to embryonic lethality, such as the DNA glycosylases, do not lead to cancer susceptibility because of partial redundancy and an overlap with transcription coupled repair.

The strongest evidence that links defective BER with cancer susceptibility comes from studies of polymorphisms of *OGG1* and *XRCC1* with cancer susceptibility (Goode et al., 2002). The human *OGG1* gene maps to chromosome 3, a chromosome that is frequently deleted in various types of human cancers. Recent studies have shown a high frequency of loss of heterozygosity of the *OGG1* gene in several types of human cancers, which is associated with retention of a mutated *OGG1* gene. All of the mutations so far detected in cancers cells are point mutations resulting in OGG1 proteins that have amino acid substitutions. Functional assays have confirmed that some of these mutant proteins have reduced BER activities. One of the mutations has an altered damaged base recognition activity, which leads to a hypermutation phenotype. Some of these polymorphisms were consistently associated with a small, but significant, increase in cancer risk, suggesting that defects in *OGG1* may contribute to cancer susceptibility.

Epidemiological studies of *XRCC1* polymorphisms and cancer are more difficult to interpret than the studies of *OGG1*. One *XRCC1* polymorphism (R914W) is consistently linked to a reduced risk of cancer, while another allele (R399Q) showed associations with increased and decreased cancer susceptibility depending on cancer type. Such conflicting data make it impossible to determine the overall role of *XRCC1* in tumor suppression.

While studies linking *OGG1* and *XRCC1* to cancer susceptibility are not definitive, studies demonstrating that *PARP1* suppresses tumor formation are. Several observations have led to the proposal that PARP1 is important for BER. As mentioned above, PARP1 is the first protein recruited to the site of single-stranded DNA breaks and interacts with XRCC1. Cells defective for PARP1 are hypersensitive to ionizing radiation, all of which suggests a role for PARP1 in DNA repair. Several other observations have challenged the validity of this conclusion. While inactivation of PARP1 leads to radiation sensitivity, biochemical assays of these cells indicates that DNA repair proceeds normally, suggesting that PARP1 is dispensable for BER.

Despite conflicting observations concerning the role of PARP1 in BER, accumulating evidence shows that it is important for genome stability. PARP1 knockout cells exhibit an increased frequency of spontaneous and DNA damage-induced sister-chromatid recombination, aneuploidy, chromosomal fragmentation, and chromosomal fusions; however, this genome instability is not sufficient to lead to tumor susceptibility (de Murcia et al., 1997; Simbulan-Rosenthal et al., 1999; Wang et al., 1997). Tumor susceptibility in animals defective for PARP1 is only revealed upon inactivation of p53 or Ku80 (Tong et al., 2001a; Tong et al., 2001b). A deficiency in PARP1 widens the tumor spectrum of mice deficient for p53, resulting in an increased frequency of tumor formation in mammary gland, lung, prostate, skin, and brain. Inactivation of PARP1 and Ku80 leads to embryonic lethality, while inactivation of either gene by itself does not, suggesting an important redundancy in these activities for embryo development. Inactivation of either of these genes does not lead to tumor susceptibility; however, haploinsufficieny of Ku80 in a $PARP1^{-/-}$ background promotes the development of hepatocellular carcinoma. Taken together, these data suggest that defects in PARP1 can be combined with defects in other DNA repair or signaling genes to promote tumor formation.

NONHOMOLOGOUS END-JOINING

DNA double-strand breaks are considered to be one of the most lethal forms of DNA damage. In mammalian cells there are two major DNA repair pathways responsible for repairing DNA double-strand breaks, nonhomologous end-joining, and homologous recombination (Fig. 16.6A) (Liang et al., 1998). Nonhomologous end-joining, as the name implies, repairs DNA double-strand breaks irrespective of sequence. Initial experiments found that nonhomologous end-joining could repair DNA double-strand breaks quite efficiently in somatic mammalian cells. Subsequent analyses revealed that nonhomologous end-joining was important not only as a general DNA repair pathway but also for V(D)J recombination in developing B and T lymphocytes (Taccioli et al., 1993). A hallmark of defective nonhomologous end-joining is genome instability. Cell lines derived from somatic cells or mice defective for factors that promote nonhomologous end-joining manifest an increased frequency of chromosomal aberrations such as translocations and amplifications (Ferguson et al., 2000; Gu et al., 1997). How defects in nonhomologous end-joining can promote tumor development is best understood in mouse models that harbor deletions of genes that promote nonhomologous end-joining. In these animals induction of V(D)J recombination promotes oncogenic translocations and amplifications that lead primarily to pro-B cell lymphomas (Zhu et al., 2002). These studies have been

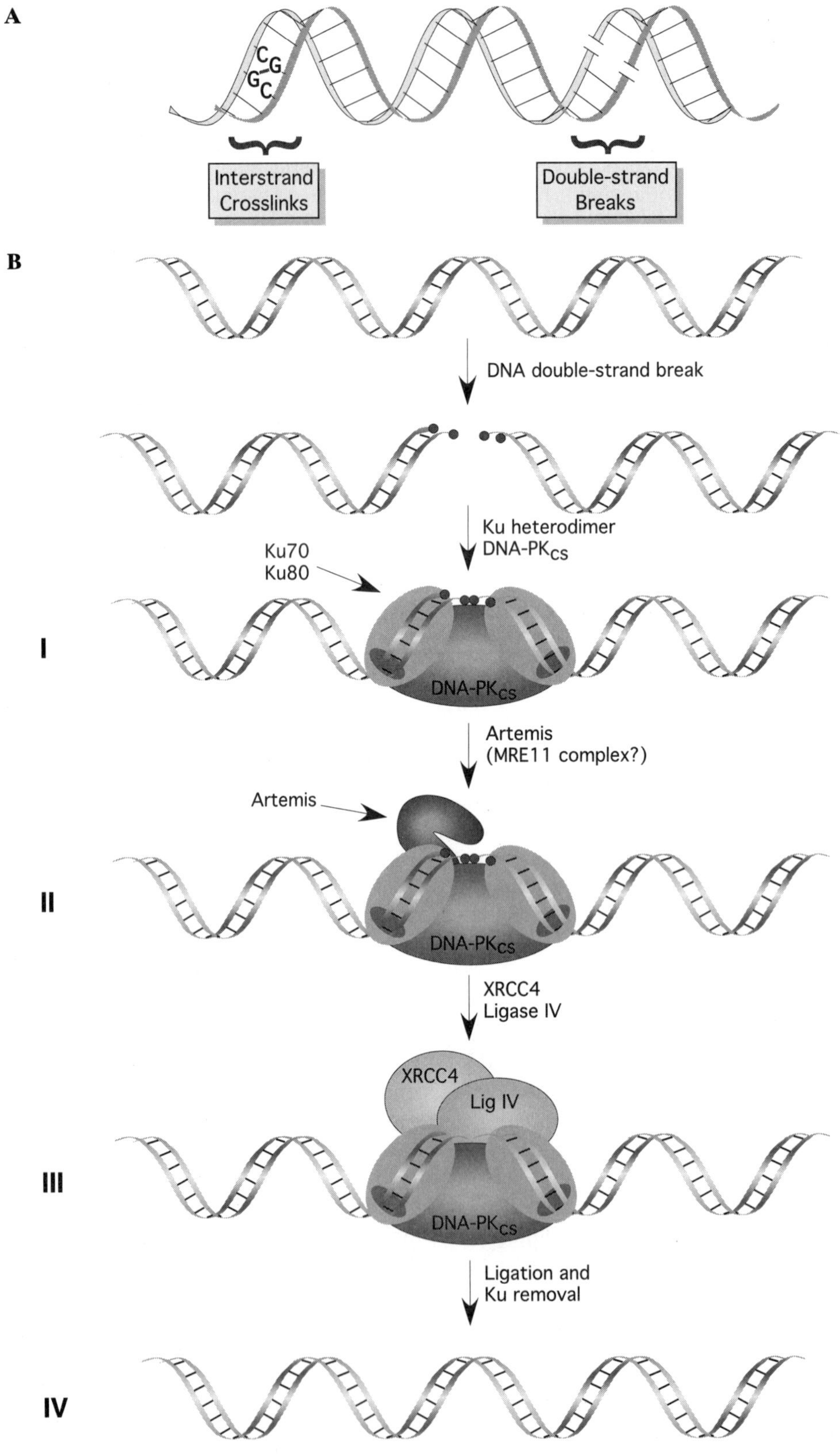
A
C
G
G
C
Interstrand Crosslinks
Double-strand Breaks
B
DNA double-strand break
Ku heterodimer
DNA-PK$_{CS}$
Ku70
Ku80
I
DNA-PK$_{CS}$
Artemis
(MRE11 complex?)
Artemis
II
DNA-PK$_{CS}$
XRCC4
Ligase IV
XRCC4
Lig IV
III
DNA-PK$_{CS}$
Ligation and
Ku removal
IV

critical in understanding how defects in nonhomologous end-joining may contribute to human malignancies.

Repair Mechanism

Nonhomologous end-joining is believed to be the predominant form of DNA double-strand break repair in mammalian cells, particularly during the G_0 and G_1 phases of the cell cycle when sister chromatids do not exist. Genetic and biochemical evidence indicate that nonhomologous end-joining is catalyzed by at least five factors, the Ku heterodimer, the catalytic subunit of the DNA-dependent protein kinase (DNA-PK_{cs}), and the XRCC4/DNA ligase IV protein complex (Lieber, 1999). Because DNA double-strand breaks frequently produce overhanging ends that cannot be ligated, processing of the ends prior to ligation is required. The factor(s) that carry out this step are most likely Artemis or the RAD50/MRE11/NBS1 complex (the MRE11 complex; Ma et al., 2002; Paull, 2001). How these factors function in repairing DNA double-strand breaks is depicted in Figure 16.6B.

The first step in nonhomologous end-joining is carries out by Ku, a heterodimer of 70 and 86 kDa subunits. Ku binds specifically to the ends of DNA double-strand breaks and is unable to bind closed circular DNA (Dynan and Yoo, 1998; Ramsden and Gellert, 1998). While a broken DNA end is essential for Ku binding, neither the sequence nor the structure (e.g., blunt, recessed, or hairpinned) influences binding. The mode of Ku binding to DNA ends was elegantly determined by J. Goldberg and his colleagues when they solved the crystal structure of the Ku heterodimer bound to DNA (Walker et al., 2001). The structure of the Ku heterodimer resembles a ring, with positively charged residues on the inside of the ring that promotes nonspecific DNA binding. The shape of the ring is asymmetric, with one side of the ring forming a large platform and groove that can accommodate a stretch of approximately 14 DNA base pairs. The other side of the ring forms a very thin "bridge" which covers 1–2 DNA base pairs and completely encircles the DNA double helix. The structure of the Ku heterodimer and its interaction with DNA make it obvious why Ku requires a DNA end for binding and suggests

---

Figure 16.6. Recombinational repair substrates and nonhomologous end-joining. Certain DNA lesions require recombinational repair. **(A)** DNA double-strand breaks can be repaired by either nonhomologous end-joining or homologous recombination. Interstrand crosslinks are primarily repaired by homologous recombination. **(B)** Schematic representation of the nonhomologous end-joining mechanism. **(I)** DNA-PK consisting of the Ku heterodimer and DNA-PK_{cs} recognizes DNA double-strand breaks. **(II)** The ends of the break are then processed by Artemis, making them compatible for DNA ligation. **(III)** The complex of DNA ligase IV and XRCC4 is recruited to the DNA break through interactions with other nonhomologous end-joining proteins and ligates the DNA break. **(IV)** The Ku heterodimer is then removed from the DNA through an unknown mechanism.

how DNA is held in place by Ku, yet still remains accessible to factors that process and ligate the DNA ends.

The Ku heterodimer is an essential part of a DNA-dependent protein kinase (DNA-PK) that promotes nonhomologous end-joining. In addition to Ku, DNA-PK also contains a catalytic subunit, DNA-PK_{cs}. While DNA-PK is essential for nonhomologous end-joining, its specific function(s) remains unclear. Structurally DNA-PK_{cs} is most related to phosphatidyl-inositol 3-kinase proteins such as ATM and ATR, suggesting a role for DNA-PK in damage signaling (Hartley et al., 1995). In support of this idea mutations within the kinase domain of DNA-PK_{cs} fail to complement DNA repair or V(D)J recombination defects in DNA-PK_{cs}-defective cell lines, indicating that the kinase activity is essential for DNA-PK function (Kurimasa et al., 1999). There is also evidence that suggests that the kinase activity of DNA-PK is required to alter or enhance the activity of other proteins that are directly involved in repairing the break. Some evidence suggests that DNA-PK may perform a structural role by functioning as a scaffold for assembly and coordination of the repair enzymes at DNA breaks. Whether it is one or all of these functions, the activity of DNA-PK_{cs} is important for nonhomologous end-joining.

DNA double-strand breaks induced by damaging agents are frequently complex lesions that cannot simply be ligated. Because of this nonhomologous end-joining frequently requires nuclelolytic processing of the DNA ends so that they are compatible for ligation. The factor(s) that carries out this step in nonhomologous end-joining is controversial, but accumulating evidence suggests that the nuclease Artemis catalyzes this process. Artemis was originally identified as a factor involved in V(D)J recombination of B and T cells. Mutations in Artemis lead to radiosensitivity and inherited severe combined immunodeficiency (SCID), similar to mutations in Ku, DNA-PK_{cs}, XRCC4, or DNA ligase IV (Moshous et al., 2001). Biochemical analysis revealed that Artemis possesses a single-strand specific 5′ to 3′ exonuclease activity (Ma et al., 2002). Artemis can form a complex with and be phosphorylated by DNA-PK_{cs}. When phosphorylated Artemis acquires an endonucleolytic activity on 5′ and 3′ overhangs as well as hairpins. It appears that phosphorylation and complex formation with DNA-PK_{cs} alters the enzymatic activity of Artemis, enabling it to open hairpins in V(D)J recombination or process incompatible DNA ends in nonhomologous end-joining.

Artemis may not be the only factor that processes the ends of double-strand breaks in nonhomologous end-joining. In yeast, which do not contain an Artemis homologue, processing of DNA breaks in nonhomologous end-joining is likely carried out by a multiprotein complex consisting of Rad50, Mre11, and Xrs2 called the Mre11 complex. Disruption of yeast *RAD50*, *MRE11*, or *XRS2* in either wild-type, *yku70*, or *lig4* mutant backgrounds revealed that these genes function epistatically in the yeast nonhomologous end-joining pathway (Boulton and Jackson, 1998; Milne et al., 1996). Although the function of the Mre11 complex in nonhomologous end-joining is unclear, structural and biochemical data suggest that the Mre11 complex may process DNA ends prior to ligation and/or bring together DNA ends enhancing DNA ligation. Rad50 and Mre11 share homology with the *E. coli* ATP-dependent exonuclease proteins SbcC and SbcD respectively. Biochemical characterization of Mre11 confirmed that it is a nuclease, capable of processing numerous types of DNA ends (Paull and Gellert, 1999). In general, the DNA processing activities of Mre11 are stimulated by Rad50 binding and ATP, suggesting possible mechanisms for regulating Mre11 activity.

Sequence analysis and crystallographic studies indicate that Rad50 is a member of a class of proteins known as structural maintenance of chromosome (SMC) proteins, which are involved in chromosome condensation and sister chromatid cohesion in eukaryotes (Hirano, 2002). As with other SMC proteins, yeast Rad50 forms an antiparallel homodimer with a flexible hinge region at one end and a head region at the other end (Hopfner et al., 2000). Several lines of evidence indicate that Mre11 and Xrs2 bind to the head region. In order to facilitate end-joining, it has been proposed that two Rad50 molecules could interact through their hinge regions, mimicking the configuration of other SMC proteins. Subsequent interactions with Mre11 and Xrs2, bound to DNA ends could then facilitate the juxtaposition of the ends and stimulate end-joining.

The Mre11 complex is conserved from yeast to human. While human homologues of Rad50 and Mre11 have been identified, no sequence homologue of Xrs2 has been found. In human cells MRE11 and RAD50 form a complex with NBS1, the protein whose deficiency leads to Nijmegen breakage syndrome (Carney et al., 1998). Recent studies indicate that NBS1, while lacking sequence similarity, is a functional homologue of Xrs2. Like the yeast Mre11 complex, the human MRE11 complex also exhibits nuclease activity, consistent with a putative role in processing of DNA breaks prior to ligation. While the human MRE11 complex may participate in nonhomologous end-joining that is probably not its only role. The MRE11 complex has also been implicated in homologous recombination, replication, meiosis, DNA damage signaling, and telomere maintenance.

The final step in nonhomologous end-joining is ligation. In nonhomologous end-joining this is carried out by the DNA ligase IV/XRCC4 complex. DNA ligase IV is one of three genetically distinct ligases in mammalian cells. XRCC4 interacts with ligase IV to form heteromultimers. The interaction between XRCC4 and ligase IV stabilizes the complex in vivo, and stimulates its ligase activity in vitro (Grawunder et al., 1997). Moreover interactions between DNA-PK and ligase IV/XRCC4 seem to stimulate nonhomologous end-joining. Ku has been shown to recruit ligase IV/XRCC4 to DNA ends via protein-protein interactions, while XRCC4 has been shown to stimulate the assembly of the DNA-PK complex on DNA ends (Nick McElhinny et al., 2000). These interactions indicate a coordinated response in repairing DNA breaks by nonhomologous end-joining.

Defects in Nonhomologous End-Joining and Genome Instability

One of the phenotypes that define cells defective for nonhomologous end-joining is an extreme sensitivity to ionizing radiation and compounds that induce DNA double-strand breaks. Other experiments indicated that these same cells are defective for V(D)J recombination, suggesting that the core defect in these cells was an inability to repair DNA double-strand breaks. Cytological analysis of cells defective for nonhomologous end-joining revealed that these cells had a significant increase in chromosome and chromatid breaks. Subsequent analyses by spectral karyotyping determined that cells defective for nonhomologous end-joining showed an extreme predisposition to spontaneously rearrange chromosomes, frequently resulting in nonreciprocal and complex translocations (Ferguson and Alt, 2001). Thus the inability to repair DNA double-strand breaks by nonhomologous end-joining leads to genome instability, a phenotype frequently found in human tumors.

Cancer Susceptibility

Studies of patients or animal models harboring mutations in genes involved in nonhomologous end-joining show that these defects lead to SCID syndromes. In humans, mutations in DNA ligase IV have been identified in patients exhibiting developmental delay and immunodeficiency (O'Driscoll et al., 2001). In addition a different study identified a group of SCID patients that displayed radiosensitivity. Defects in these patients were localized to the Artemis gene (Moshous et al., 2001). While defects in nonhomologous end-joining genes are most frequently associated with SCID, one patient, containing a mutation in DNA ligase IV, was found to have leukemia, suggesting a link between nonhomologous end-joining and cancer. If the MRE11 complex is indeed involved in nonhomologous end-joining the link between defects in nonhomologous end-joining and cancer is strengthened. One component of the human MRE11 complex is NBS1 or nibrin. Defects in NBS1 lead to Nijmegen breakage syndrome, an AT-like disorder that is characterized by developmental defects, radiosensitivity and predisposition to cancer (Varon et al., 1998). Mutations in the MRE11 gene were recently found in patients with a different AT-like disorder, AT-LD, which is virtually indistinguishable from ataxia-telangiectasia and leads to cancer susceptibility (Stewart et al., 1999).

Lessons from Animal Models with Mutations in Nonhomologous End-Joining Genes

While the link between nonhomologous end-joining defects and cancer in humans is conjectural, the link between nonhomologous end-joining defects and cancer in mice is certain. Mice that are deficient for Ku, DNA-PK_{cs}, XRCC4, or Lig4 in combination with p53 deficiency succumb to pro-B-cell lymphoma at an early age (Difilippantonio et al., 2000;

Frank et al., 2000; Gao et al., 2000; Lim et al., 2000). In each case lymphoma development is dependent on the presence of the nonhomologous end-joining defect and the presence of functional RAG1 and RAG2, suggesting that the inability to repair the RAG1/RAG2-induced DNA double-strand break in V(D)J recombination is the initiating event in these cancers. In nearly all instances the presence of a nonreciprocal translocation containing the centromere region of chromosome 12 and the telomeric region of chromosome 15 was observed in these lymphomas. In addition to the nonreciprocal translocation these lymphomas harbored a complex chromosome rearrangement containing the centromere region of chromosome 15, a portion of chromosome 12, and a portion of another chromosome that varied from tumor to tumor.

Cytological and molecular analysis of these translocations revealed several important similarities between these tumors. First, southern blot analysis indicated that these tumors harbored rearrangements and amplifications of IgH and the protooncogene *c-myc* sequences, consistent with the idea that the initiating lesion in these translocations is an inappropriately repaired break during V(D)J recombination. Second, the amplified *c-myc* and IgH region was always located on the complex chromosomal rearrangement containing the centromere of chromosome 15. Third, the junction point for these translocations nearly always contains a region of microhomology that has been proposed to help mediate the DNA repair process in the absence of functional nonhomologous end-joining factors. These observations have led to a model proposed by Fred Alt and colleagues for how defective nonhomologous end-joining can lead to gene amplifications and complex chromosome translocations (Fig. 16.7) (Zhu et al., 2002).

The first step in these complex translocations and amplifications is the DNA double-strand break initiated by the RAG1/RAG2 endonuclease. Because these cells are defective for nonhomologous end-joining and V(D)J recombination the break cannot be repaired normally and must be repaired by an alternate mechanism. In these tumors repair involves an illegitimate recombination event between chromosome 12 and chromosome 15. Sequence data from the junction sites suggests that this illegitimate repair mechanism involves regions of microhomology. The resulting rearrangement is a triradial containing three chromosomal arms. Progression through S phase and replication of the triradial results in three products, an unaltered chromosome 15, a nonreciprocal translocation between chromosome 12 and chromosome 15, and a dicentric chromosome containing the IgH region of chromosome 12 and the *c-myc* gene from chromosome 15. Two of these products, the unaltered chromosome 15 and the nonreciprocal translocation between chromosome 12 and chromosome 15, are stable but the dicentric chromosome is unstable.

During cell division the centromeres on the dicentric chromosome can be pulled toward opposite poles, resulting in breakage of the chromosome. Because the broken chromosome it not capped by a telomere, it remains unstable and is able to recombine with other chromosomes. The

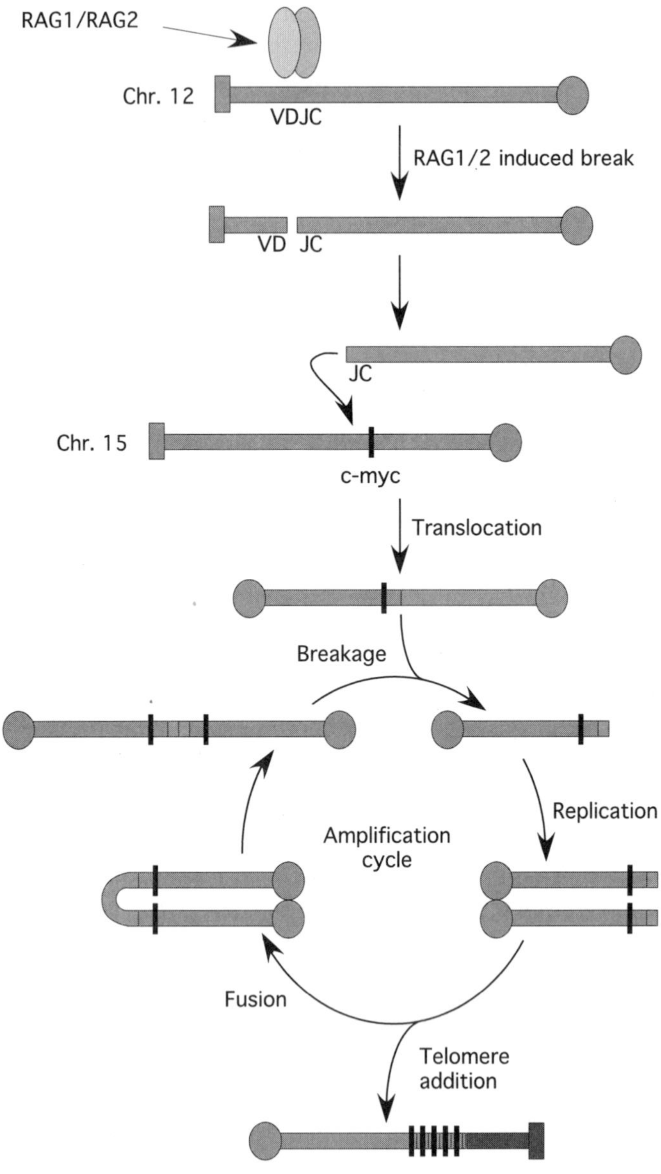

Figure 16.7. Oncogenic translocation and amplification in nonhomologous end-joining mice. Chromosomal instability is initiated when the RAG1/RAG2 endonuclease cleaves the V(D)J locus to initiate V(D)J recombination. Because these cells are defective for nonhomologous end-joining, V(D)J recombination cannot be completed and the DNA double-strand break is repaired by an alternate repair mechanism that uses microhomologies. Frequently this inappropriate repair mechanism leads to translocations between chromosome 12 and chromosome 15. One of the products from this translocation is a dicentric chromosome containing the *c-myc* oncogene. The dicentric chromosome is broken during cell division, resulting in a chromosome that does not contain a telomere. The broken chromosome is duplicated during DNA replication but the sister chromatids frequently fuse, resulting in a dicentric chromosome containing two copies of the *c-myc* oncogene. This cycle continues until the broken chromosome is lost or is healed by the fortuitous addition of a telomere. The telomere sequence usually comes from a different chromosome.

broken chromosome can then enter into a cycle called the breakage-bridge-fusion cycle first described by Barbara McClintock in 1941. In this cycle the chromosome without a telomere is replicated during S phase and, because the broken end is recombinogenic, frequently fuses with its sister chromatid. This results in a dicentric chromosome that is then broken during subsequent cell divisions to reinitiate the cycle. The cycle does not end until the broken chromosome is either lost from the cell or obtains a telomere to cap the end of the chromosome. The end product is a complex translocation containing an amplification of the IgH and *c-myc* genes.

HOMOLOGOUS RECOMBINATION

For years the importance of homologous recombination for repairing DNA double-strand breaks in somatic mammalian cells was discounted because of very active and efficient nonhomologous end-joining mechanisms. Recent experiments have shown that homologous recombination is essential for cellular proliferation, providing a repair mechanism for DNA double-strand breaks arising during DNA replication. One feature of homologous recombination that distinguishes it from nonhomologous end-joining is that it is nonmutagenic, using either a homologous chromosome or, when available, the sister chromatid to direct repair DNA synthesis. As with nonhomologous end-joining, defects in homologous recombination lead to genome instability including chromosomal translocations and aneuploidy, events associated with carcinogenesis. The significance of homologous recombination in the prevention of cancer became evident with the finding that the *BRCA2*, a tumor suppressor gene that, when mutated, is responsible for approximately 20% of all hereditary breast cancer cases, promotes the homologous recombination process (Wooster and Weber, 2003). Since then defects in other homologous recombination proteins have also been implicated in tumor suppression (Thompson and Schild, 2002).

Repair Mechanism

A working model for DNA double-strand break repair via homologous recombination is presented in Figure 16.8. The appearance of DNA double-strand breaks, such as following exposure to DNA damaging agents or during DNA replication, initiates a complex cascade of events that culminates in the formation of DNA crossovers called Holliday junctions. Because the pathway uses an undamaged homologous duplex to direct repair DNA synthesis and restore lost information at the site of the DNA break, this process is, for the most part, error free. Initially the ends of the break are resected by an endonuclease, leaving 3′ single-stranded DNA overhangs. Because of its nuclease activity, the MRE11 complex has been suggested to provide this activity. Support for a role of the MRE11 complex in homologous recombination comes from using

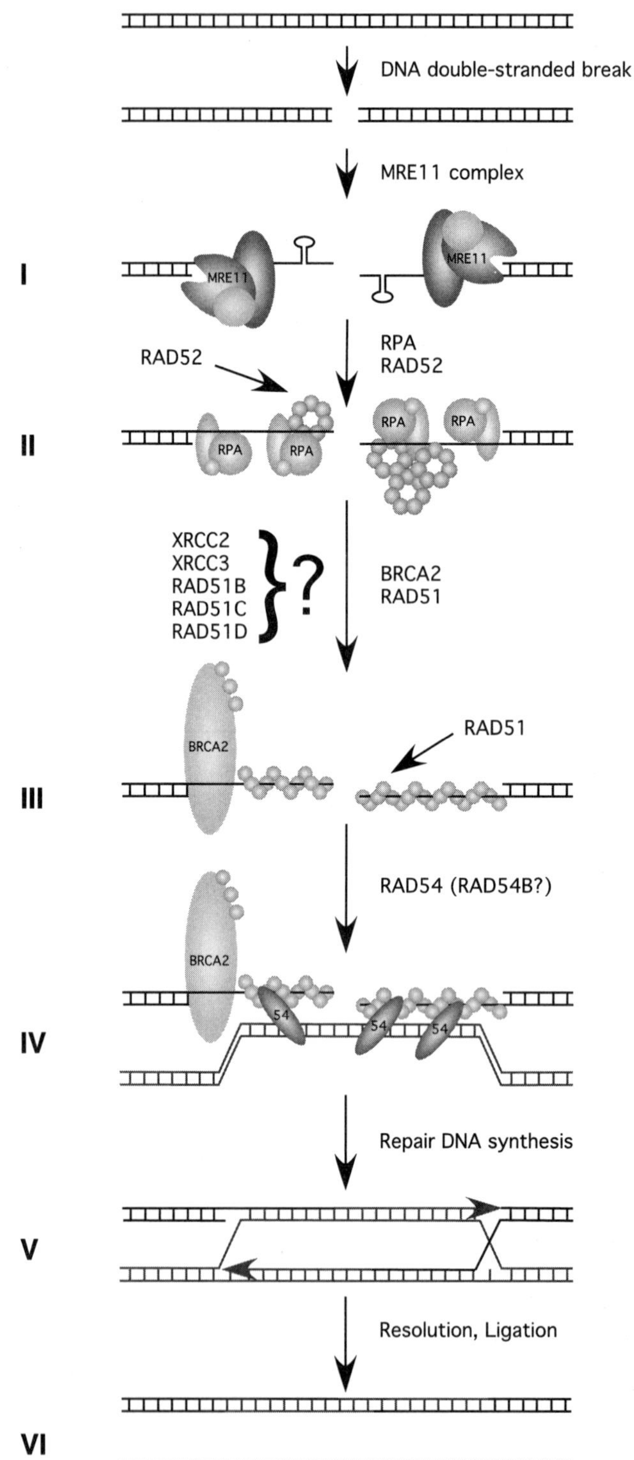
DNA double-stranded break
MRE11 complex
I
MRE11
MRE11
RPA
RAD52
RAD52
II
RPA
RPA
RPA
RPA
XRCC2
XRCC3
RAD51B
RAD51C
RAD51D
?
BRCA2
RAD51
BRCA2
RAD51
III
RAD54 (RAD54B?)
BRCA2
54
54
54
IV
Repair DNA synthesis
V
Resolution, Ligation
VI

vertebrate cells that have been deleted for NBS1, one of the members of the MRE11 complex. Inactivation of NBS1 leads to ionizing radiation sensitivity and defects in sister chromatid recombination and double-strand break-induced homologous recombination (Tauchi et al., 2002). While nonhomologous end-joining and homologous recombination are essentially independent, data seem to suggest that the MRE11 complex can participate in both pathways.

Once formed, the single-stranded DNA overhangs need to be protected from nucleases and prepared for subsequent step in the pathway. Depending on the specific DNA sequence, single-stranded DNA can form secondary structures that are inhibitory for subsequent homologous recombination steps. RPA, a trimeric protein complex that binds single-stranded DNA and can remove secondary structure performs this function (Baumann and West, 1997; Sung, 1994). In addition to RPA, RAD52 may also act at this step. RAD52 forms clearly defined heptameric ring-shaped oligomers that in turn form higher order complexes. RAD52 can interact with RPA, single-stranded DNA, and RAD51, suggesting that RAD52 may facilitate the transition from RPA binding the single-stranded overhang to RAD51 forming a nucleoprotein filament on the single-stranded overhang (Benson et al., 1998; Lloyd et al., 2002; Stasiak et al., 2000).

Although RAD52 likely assists RAD51 filament formation, it is unlikely to be the only protein involved in this process in vivo. Following DNA damage and during S phase RAD51 forms discrete subnuclear foci, presumably forming nucleoprotein filaments at the sites of DNA damage. Genetically the presence of BRCA1, BRCA2, and the RAD51 paralogues (XRCC2, XRCC3, RAD51B, RAD51C, and RAD51D) is essential for RAD51 focus formation (Takata et al., 2001; Tarsounas et al., 2003). Moreover recent evidence suggests that the BRCA2 protein plays an important regulatory role in this pathway through its interactions with RAD51 and single-stranded DNA. Biochemical and crystallographic studies suggest that BRCA2 participates directly in localizing RAD51 to the sites of DNA damage and may assist in loading RAD51 onto single-stranded DNA (Davies et al., 2001; Pellegrini et al., 2002). Genetic studies also suggest that the RAD51 paralogues participate at

---

Figure 16.8. Schematic representation of the homologous recombination mechanism. Homologous recombination is initiated by a DNA double-strand break. The break is processed by the MRE11 complex **(I)**, leading to the production of a 3′ single-stranded DNA end. **(II)** RPA, a single-stranded DNA-binding protein and RAD52 help protect the single-stranded DNA end and help recruit other DNA repair factors. **(III)** RAD51 is recruited to the DNA double-strand break through an unknown mechanism that likely involves its interaction with RAD52, BRCA2, and possibly other homologous recombination proteins. **(IV)** RAD54, a chromatin remodeling factor, assists RAD51 in strand invasion of the repair template. **(V)** The undamaged template is used to direct DNA repair synthesis of the damaged strand. **(VI)** Holliday junction intermediates are resolved to separate the DNA strands, and ligase then seals any remaining nicks. (See color insert.)

this step; however, the specific mechanism by which they participate is unknown.

After RAD51 nucleoprotein filaments are formed the next step in the repair mechanism is strand exchange. This process requires search for a homologous sequence, such as a sister chromatid, and then strand invasion. This reaction is influenced by other homologous recombination factors such as RAD54. RAD54 is a member of the *SWI2/SNF2* family of ATP-dependent chromatin remodeling enzymes. The biochemical activities of RAD54 suggest that it removes or alters the chromatin structure of the repair template, facilitating strand exchange (Dronkert et al., 2000; Sigurdsson et al., 2002). In addition to RAD54 there is a second RAD54 homologue in human cells, called RAD54B, which may participate (Miyagawa et al., 2002; Tanaka et al., 2000).

The final step in homologous recombination is resolution of recombination intermediates. In *E. coli* resolution of recombination intermediates, four-stranded DNA structures called Holliday junctions, is carried out by the RuvA, RuvB, and RuvC complex (Eggleston et al., 1997). The RuvA and RuvB proteins catalyze branch migration of the Holliday junction until it reaches the sequence $5'\text{-}{}^{A}/_{T}TT^{G}/_{C}\text{-}3'$, where the RuvC endonuclease symmetrically nicks the Holliday junction to promote resolution of the recombining DNA. A biochemical activity that promotes a similar activity has been identified in mammalian cells, but the proteins responsible for this activity have yet to be identified.

Cancer Susceptibility

Defects in homologous recombination undermine DNA double-strand break repair and lead to chromosomal instability. Karyotypic studies have shown that most cancers harbor chromosomal alterations such as chromosomal rearrangements and aneuploidy, events frequently found in homologous recombination defective cells. Although this strongly suggests that defects in homologous recombination may promote tumor formation, this association was only proved with the discovery that BRCA1 and BRCA2 promote homologous recombination.

BRCA1 Function and Cancer

Currently more than 190,000 women in the United States and almost 1.2 million women worldwide are diagnosed with breast cancer every year. For years scientists have known that individuals with a family history of breast cancer are at much higher risk of developing breast cancer, indicating that genetics plays an important role in breast cancer. In 1990 chromosome 17q21 was identified as the location of a breast cancer susceptibility gene, and in 1994 several groups used positional cloning to isolate the gene, now termed *BRCA1* (Hall et al., 1990; Miki et al., 1994). *BRCA1* encoded an extremely large protein of 1863 amino acids, but because it contained little homology to any known genes, the protein sequence provided little information concerning its biological function.

Since then BRCA1 has been under intense scientific scrutiny in an attempt to determine its cellular functions. BRCA1 physically interacts with numerous other proteins. These interactions have implicated BRCA1 in transcriptional regulation, transcription and transcription-coupled repair, ubiquitin ligation, X chromosome inactivation, DNA damage sensing and signaling, and resection of DNA double-strand breaks (Anderson et al., 1998; Ganesan et al., 2002; Hashizume et al., 2001; Li et al., 2000; Morris et al., 2002; Moynahan et al., 1999; Scully et al., 1997; Xu et al., 1999; Zhong et al., 1999).

While it is possible that loss of any of these functions could promote cancer, this section will focus on the role of BRCA1 in DNA repair. Genetically defects in *BRCA1* impair double-strand break-induced homologous recombination and lead to genome instability (Moynahan et al., 1999). Results from biochemical and cellular experiments suggest two possible functions for BRCA1 in DNA repair, regulation of DNA double-strand break processing, and DNA damage signaling. During S phase and following exposure to DNA damaging agents, BRCA1 is one of the earliest proteins to migrate to the sites of DNA breaks, and may help localize several different proteins to these site. BRCA1 physically interacts with SWI/SNF, a complex of proteins that remodels chromatin, and BACH1, a novel DNA helicase (Bochar et al., 2000; Cantor et al., 2001). These interactions have been suggested to be important in modifying the local DNA topology surrounding the DNA break, making it more accessible to other repair proteins. BRCA1 can also interact with the MRE11 complex. As indicated above the MRE11 complex is likely involved in processing of DNA double-strand breaks to form 3′ single-stranded DNA ends. Under certain conditions BRCA1 can inhibit this activity, suggesting that BRCA1 may also help regulate the length of single-stranded DNA ends produced at the break site. However, BRCA1 may also play a role in DNA damage signaling. After DNA damage BRCA1 can be phosphorylated by three different kinases, ATM, ATR, and CHK2. ATM and CHK2 phosphorylate BRCA1 following exposure to ionizing radiation, and ATR phosphorylates, BRCA1 following UV light exposure (Cortez et al., 1999; Gatei et al., 2001; Lee et al., 2000). BRCA1 has been implicated in several checkpoint events. Following exposure to ionizing radiation, BRCA1 defective cells fail to arrest DNA synthesis and fail to arrest in G_2. These observations suggest an important role for BRCA1 during early events in DNA double-strand break repair.

Mutations in BRCA1 lead to an autosomal dominant inheritance pattern of breast cancer as well as an increased frequency of ovarian cancer. After identification of BRCA1, numerous disease-associated mutations were identified, with most resulting in loss or truncation of the protein product. Most disease-associated mutations of BRCA1 can be accessed at the Breast Cancer Information Core (see Web resources section). Mutations of BRCA1 are highly penetrant, with female carriers exhibiting up to a 90% lifetime risk of developing breast cancer. While inheritance of one mutant BRCA1 allele is sufficient to cause cancer predisposition, inactivation of the second allele is essential for tumor formation, demonstrating that BRCA1 functions as a typical tumor-suppressor gene.

BRCA2 Function and Cancer

Like *BRCA1*, *BRCA2* is a highly penetrant breast cancer susceptibility gene thought to be involved in homologous recombination. BRCA2 is located on chromosome 13q12–13 and leads to an autosomal dominant predisposition to breast cancer (Tavtigian et al., 1996; Wooster et al., 1995). Mutations in BRCA2 confer a lifetime risk of breast cancer of between 30% to 85%, depending on the specific mutation. Over the last several years genetic, cellular, biochemical, and crystallographic evidence has shown that BRCA2 likely plays a direct role in DNA double-strand break repair by homologous recombination. Genetically, cell lines defective for BRCA2 function are impaired for DNA double-strand break-induced homologous recombination and have significantly increased frequencies of genome instability (Moynahan et al., 2001; Patel et al., 1998). During S phase and following exposure to DNA-damaging agents RAD51, the strand exchange protein in mammalian cells, and BRCA2 colocalize in subnuclear foci, suggesting that BRCA2 and RAD51 physically interact to promote DNA repair. This has been confirmed by biochemical and yeast two-hybrid analyses that demonstrate a direct interaction between BRCA2 and RAD51. BRCA2 contains at least six domains, called BRC repeats, that can interact with RAD51. The specific molecular interactions between BRCA2 and RAD51 have been identified by crystallization of a chimeric protein containing BRCA2 and RAD51 (Pellegrini et al., 2002). In addition to its interaction with RAD51, BRCA2 can bind to single-stranded DNA, which has led to the hypothesis that BRCA2 stimulates homologous recombination by helping to localize RAD51 to single-stranded DNA generated at the sites of DNA damage (Yang et al., 2002). Without BRCA2 function, it is likely that DNA damage is directed to other error-prone repair pathways that leads to genome instability and promotes tumor formation.

OTHER HOMOLOGOUS RECOMBINATION GENES AND CANCER

Although the evidence that BRCA1 and BRCA2 suppress tumor formation is definitive, evidence implicating other homologous recombination genes is ambiguous. To date no cancer susceptibility syndromes have been linked to defects in RAD51, RAD52, RAD54, XRCC2, XRCC3, RAD51B, RAD51C, or RAD51D. The data implicating RAD51 in tumor suppression are particularly confusing. Two examinations of breast and metastatic brain tumors did not find that mutations in RAD51 occurred at a higher frequency than in control samples (Kuschel et al., 2002; Schmutte et al., 1999), while one report found an association between a RAD51 mutation and bilateral breast cancer (Kato et al., 2000). Some reports have suggested that overexpression of RAD51 is frequently associated with tumorigenesis, with overexpression of RAD51 being correlated with histological grading of sporadic invasive ductal breast cancer (Maacke et al., 2000; Raderschall et al., 2002). These results suggest that overexpression of RAD51 may be an important factor in tumor formation and progression. However, the validity of this hypothesis was brought into question by a different report that found that a significant number of breast carcinomas had a reduced level of RAD51 (Yoshikawa et al., 2000). Although RAD51 has been the most heavily scrutinized for involvement in tumorigenesis, epidemiological studies of polymorphisms in other homologous recombination genes have been analyzed to determine if they are associated with cancer. For the most part these studies have shown no correlation or a weak correlation between the polymorphism and cancer. One exception involves a translocation within the RAD51B gene and uterine leiomyomas (Schoenmakers et al., 1999; Takahashi et al., 2001). Mouse models for these homologous recombination genes have done little to help elucidate their importance in tumor suppression. These genes fall into two classes. One class, consisting of RAD51, XRCC2, RAD51B, and RAD51D, leads to embryonic lethality (Deans et al., 2000; Lim and Hasty, 1996; Pittman and Schimenti, 2000; Shu et al., 1999; Tsuzuki et al., 1996); therefore their involvement in tumor suppression cannot be determined. Genes in the other class, consisting of RAD52 and RAD54, are viable but do not lead to an obvious susceptibility to cancer (Essers et al., 1997; Rijkers et al., 1998).

Fanconi Anemia

Fanconi anemia is an extremely rare autosomal recessive disorder, affecting 1 in 100,000 live births, that is characterized by progressive bone marrow failure and susceptibility to cancers such as acute myeloblastic leukemia (AML). Cells from fanconi anemia patients are hypersensitive to DNA-damaging agents, particularly crosslinking agents and chemicals that generate DNA double-stranded breaks, and have elevated levels of chromosomal instability. Complementation studies of cells from fanconi anemia patients revealed that there are at least eight complementation groups. So far, seven fanconi anemia genes have been cloned with six of these genes, FANC-A, C, D2, E, F, and G, being unique (Meyn, 1997). The other gene that can lead to fanconi anemia is BRCA2 (Howlett et al., 2002). Moreover BRCA1 has been shown to colocalize with FANC-D2 and directly interact with FANC-A. These finding strongly suggest that the fanconi anemia genes are involved in homologous recombination; however, their specific functions in this repair pathway are unknown.

RecQ Helicases

To date five human helicase have been identified that contain significant sequence homology to the *E. coli* RecQ gene. Three of these helicases,

BLM, WRN, and RECQ4, are responsible for rare cancer susceptibility disorders termed Bloom's syndrome, Werner's syndrome, and Rothmund-Thompson syndrome, respectively (Hickson, 2003). Cells from Bloom's syndrome patients exhibit a high frequency of sister-chromatid recombination and chromosomal breaks, and the BLM helicase has been shown to physically interact with the RAD51 recombinase, suggesting that BLM may be involved in homologous recombination (Bischof et al., 2001; Wu et al., 2001). However, BLM also interacts with TOPOIIIα and MLH1, raising the possibility that BLM is involved in several DNA repair pathways (Ababou et al., 2000; Langland et al., 2001). Werner's syndrome patients usually develop normally until the onset of puberty but fail to thrive thereafter, with death occurring due to cancer or vascular disease. Typically Werner's syndrome patients are of short stature and look as if they have aged prematurely. Cells from Werner's syndrome patients exhibit elevated mutation frequencies, particularly deletions. The WRN protein physically interacts with numerous proteins, including proteins involved in replication, nonhomologous end-joining, and base excision repair, suggesting that WRN may act in more than one DNA repair pathway (Brosh et al., 2001; Cooper et al., 2000). Rothmund-Thompson patients are cancer prone, with cancer usually occurring before the age of 25. The role of RECQ4 in DNA repair remains poorly understood.

CONCLUDING REMARKS

It is now widely accepted that cancer results from an accumulation of mutations in genes that control cell proliferation and cell death. While mutations are sometimes inherited, frequently they result from unrepaired or inappropriately repaired DNA damage. Given that cells are continuously exposed to endogenous and exogenous DNA-damaging agents, it is somewhat surprising that cancer is so infrequent. This undoubtedly represents the efficiency and coordination of DNA damage recognition, signaling, and repair pathways. Over the last few decades the importance of these DNA repair pathways in preventing cancer has become evident. Numerous DNA repair defects, which increase the frequency of genetic and chromosomal instability, have now been documented to predispose individuals to cancer susceptibility. Hope for the future lies in a better understanding of these repair pathways, leading to rational and effective therapies that can combat cancers due to DNA repair defects.

WEB RESOURCES

http://www.nfdht.nl/ International collaborative group on hereditary nonpolyposis colorectal cancer *http://research.nhri.nih.gov/bic/* National human genome research institute—breast cancer information

http://xps.org Xeroderma pigmentosum society

http://www.ncbi.nlm.nih.gov/disease/Cockayne.html National center for biomedical information—cockayne syndrome

http://www.cancerindex.org/geneweb/NBS1.htm Nijmegen breakage syndrome

REFERENCES

Ababou M, Dutertre S, Lecluse Y, Onclercq R, Chatton B, Amor-Gueret M (2000): ATM-dependent phosphorylation and accumulation of endogenous BLM protein in response to ionizing radiation. Oncogene 19:5955–63.

Acharya S, Wilson T, Gradia S, Kane MF, Guerrette S, Marsischky GT, Kolodner R, Fishel R (1996): hMSH2 forms specific mispair-binding complexes with hMSH3 and hMSH6. Proc Natl Acad Sci USA 93:13629–34.

Anderson SF, Schlegel BP, Nakajima T, Wolpin ES, Parvin JD (1998): BRCA1 protein is linked to the RNA polymerase II holoenzyme complex via RNA helicase A. Nat Genet 19:254–6.

Bader S, Walker M, Hendrich B, Bird A, Bird C, Hooper M, Wyllie A (1999): Somatic frameshift mutations in the MBD4 gene of sporadic colon cancers with mismatch repair deficiency. Oncogene 18:8044–7.

Ban C, Junop M, Yang W (1999): Transformation of MutL by ATP binding and hydrolysis: A switch in DNA mismatch repair. Cell 97:85–97.

Ban C, Yang W (1998): Crystal structure and ATPase activity of MutL: Implications for DNA repair and mutagenesis. Cell 95:541–52.

Baumann P, West SC (1997): The human Rad51 protein: polarity of strand transfer and stimulation by hRP-A. EMBO J 16:5198–206.

Benson FE, Baumann P, West SC (1998): Synergistic actions of Rad51 and Rad52 in recombination and DNA repair. Nature 391:401–4.

Bischof O, Kim SH, Irving J, Beresten S, Ellis NA, Campisi J (2001): Regulation and localization of the Bloom syndrome protein in response to DNA damage. J Cell Biol 153:367–80.

Bochar DA, Wang L, Beniya H, Kinev A, Xue Y, Lane WS, Wang W, Kashanchi F, Shiekhattar R (2000): BRCA1 is associated with a human SWI/SNF-related complex: Linking chromatin remodeling to breast cancer. Cell 102:257–65.

Boiteux S, Radicella JP (2000): The human OGG1 gene: Structure, functions, and its implication in the process of carcinogenesis. Arch Biochem Biophys 377:1–8.

Boulton SJ, Jackson SP (1998): Components of the Ku-dependent nonhomologous end-joining pathway are involved in telomeric length maintenance and telomeric silencing. EMBO J 17:1819–28.

Breen AP, Murphy JA (1995): Reactions of oxyl radicals with DNA. Free Radic Biol Med 18:1033–77.

Bronner CE, Baker SM, Morrison PT, Warren G, Smith LG, Lescoe MK, Kane M, Earabino C, Lipford J, Lindblom A, et al. (1994): Mutation in the DNA mismatch repair gene homologue hMLH1 is associated with hereditary nonpolyposis colon cancer. Nature 368:258–61.

Brosh RM, Jr, von Kobbe C, Sommers JA, Karmakar P, Opresko PL, Piotrowski J, Dianova I, Dianov GL, Bohr VA (2001): Werner syndrome protein interacts with human flap endonuclease 1 and stimulates its cleavage activity. EMBO J 20:5791–801.

Burdett V, Baitinger C, Viswanathan M, Lovett ST, Modrich P (2001): In vivo requirement for RecJ, ExoVII, ExoI, and ExoX in methyl-directed mismatch repair. Proc Natl Acad Sci USA 98:6765–70.

Buschta-Hedayat N, Buterin T, Hess MT, Missura M, Naegeli H (1999): Recognition of nonhybridizing base pairs during nucleotide excision repair of DNA. Proc Natl Acad Sci USA 96:6090–5.

Caldecott KW (2001): Mammalian DNA single-strand break repair: an X-ra(y)ted affair. Bioessays 23:447–55.

Caldecott KW, Aoufouchi S, Johnson P, Shall S (1996): XRCC1 polypeptide interacts with DNA polymerase beta and possibly poly (ADP-ribose) polymerase, and DNA ligase III is a novel molecular "nick-sensor" in vitro. Nucl Acids Res 24:4387–94.

Cantor SB, Bell DW, Ganesan S, Kass EM, Drapkin R, Grossman S, Wahrer DC, Sgroi DC, Lane WS, Haber DA, Livingston DM (2001): BACH1, a novel helicase-like protein, interacts directly with BRCA1 and contributes to its DNA repair function. Cell 105:149–60.

Carney JP, Maser RS, Olivares H, Davis EM, Le Beau M, Yates JR, 3rd, Hays L, Morgan WF, Petrini JH (1998): The hMre11/hRad50 protein complex and Nijmegen breakage syndrome: Linkage of double-strand break repair to the cellular DNA damage response. Cell 93:477–86.

Chi NW, Kolodner RD (1994): The effect of DNA mismatches on the ATPase activity of MSH1, a protein in yeast mitochondria that recognizes DNA mismatches. J Biol Chem 269:29993–7.

Claij N, te Riele H (1999): Microsatellite instability in human cancer: A prognostic marker for chemotherapy? Expr Cell Res 246:1–10.

Clark AB, Valle F, Drotschmann K, Gary RK, Kunkel TA (2000): Functional interaction of proliferating cell nuclear antigen with MSH2-MSH6 and MSH2-MSH3 complexes. J Biol Chem 275:36498–501.

Cleaver JE (1968): Defective repair replication of DNA in xeroderma pigmentosum. Nature 218:652–6.

Cooper MP, Machwe A, Orren DK, Brosh RM, Ramsden D, Bohr VA (2000): Ku complex interacts with and stimulates the Werner protein. Genes Dev 14: 907–12.

Cortez D, Wang Y, Qin J, Elledge SJ (1999): Requirement of ATM-dependent phosphorylation of brca1 in the DNA damage response to double-strand breaks. Science 286:1162–6.

Cunningham JM, Christensen ER, Tester DJ, Kim CY, Roche PC, Burgart LJ, Thibodeau SN (1998): Hypermethylation of the hMLH1 promoter in colon cancer with microsatellite instability. Cancer Res 58:3455–60.

Dao V, Modrich P (1998): Mismatch-, MutS-, MutL-, and helicase II-dependent unwinding from the single-strand break of an incised heteroduplex. J Biol Chem 273:9202–7.

Davies AA, Masson JY, McIlwraith MJ, Stasiak AZ, Stasiak A, Venkitaraman AR, West SC (2001): Role of BRCA2 in control of the RAD51 recombination and DNA repair protein. Mol Cell:7:273–82.

de Murcia JM, Niedergang C, Trucco C, Ricoul M, Dutrillaux B, Mark M, Oliver FJ, Masson M, Dierich A, LeMeur M, Walztinger C, Chambon P, de Murcia G (1997): Requirement of poly(ADP-ribose) polymerase in recovery from DNA damage in mice and in cells. Proc Natl Acad Sci USA 94:7303–7.

de Vries SS, Baart EB, Dekker M, Siezen A, de Rooij DG, de Boer P, te Riele H (1999): Mouse MutS-like protein Msh5 is required for proper chromosome synapsis in male and female meiosis. Genes Dev 13:523–31.

de Wind N, Dekker M, Berns A, Radman M, te Riele H (1995): Inactivation of the mouse Msh2 gene results in mismatch repair deficiency, methylation tolerance, hyperrecombination, and predisposition to cancer. Cell 82:321–30.

Deans B, Griffin CS, Maconochie M, Thacker J (2000): *Xrcc2* is required for genetic stability, embryonic neurogenesis and viability in mice. EMBO J 19: 6675–85.

Dianova II, Bohr VA, Dianov GL (2001): Interaction of human AP endonuclease 1 with flap endonuclease 1 and proliferating cell nuclear antigen involved in long-patch base excision repair. Biochemistry 40:12639–44.

Difilippantonio MJ, Zhu J, Chen HT, Meffre E, Nussenzweig MC, Max EE, Ried T, Nussenzweig A (2000): DNA repair protein Ku80 suppresses chromosomal aberrations and malignant transformation. Nature 404:510–14.

Dronkert ML, Beverloo HB, Johnson RD, Hoeijmakers JH, Jasin M, Kanaar R (2000): Mouse RAD54 affects DNA double-strand break repair and sister chromatid exchange. Mol Cell Biol 20:3147–56.

Duval A, Rolland S, Compoint A, Tubacher E, Iacopetta B, Thomas G, Hamelin R (2001): Evolution of instability at coding and non-coding repeat sequences in human MSI-H colorectal cancers. Hum Mol Genet 10:513–18.

Duval A, Rolland S, Tubacher E, Bui H, Thomas G, Hamelin R (2000): The human T-cell transcription factor-4 gene: Structure, extensive characterization of alternative splicings, and mutational analysis in colorectal cancer cell lines. Cancer Res 60:3872–9.

Dynan WS, Yoo S (1998): Interaction of Ku protein and DNA-dependent protein kinase catalytic subunit with nucleic acids. Nucl Acids Res 26:1551–9.

Edelmann W, Cohen PE, Kneitz B, Winand N, Lia M, Heyer J, Kolodner R, Pollard JW, Kucherlapati R (1999): Mammalian MutS homologue 5 is required for chromosome pairing in meiosis. Nat Genet 21:123–7.

Eggleston AK, Mitchell AH, West SC (1997): In vitro reconstitution of the late steps of genetic recombination in *E. coli*. Cell 89:607–17.

Epstein JH, Fukuyama K, Reed WB, Epstein WL (1970): Defect in DNA synthesis in skin of patients with xeroderma pigmentosum demonstrated in vivo. Science 168:1477–8.

Essers J, Hendriks RW, Swagemakers SM, Troelstra C, de Wit J, Bootsma D, Hoeijmakers JH, Kanaar R (1997): Disruption of mouse RAD54 reduces ionizing radiation resistance and homologous recombination. Cell 89:195–204.

Evans AR, Limp-Foster M, Kelley MR (2000): Going APE over ref-1. Mutat Res 461:83–108.

Ferguson DO, Alt FW (2001): DNA double strand break repair and chromosomal translocation: Lessons from animal models. Oncogene 20:5572–9.

Ferguson DO, Sekiguchi JM, Chang S, Frank KM, Gao Y, DePinho RA, Alt FW (2000): The nonhomologous end-joining pathway of DNA repair is required for genomic stability and the suppression of translocations. Proc Natl Acad Sci USA 97:6630–3.

Fishel R, Lescoe MK, Rao MR, Copeland NG, Jenkins NA, Garber J, Kane M, Kolodner R (1993): The human mutator gene homolog MSH2 and its association with hereditary nonpolyposis colon cancer. Cell 75:1027–38.

Flores-Rozas H, Clark D, Kolodner RD (2000): Proliferating cell nuclear antigen and Msh2p-Msh6p interact to form an active mispair recognition complex. Nat Genet 26:375–8.

Frank KM, Sharpless NE, Gao Y, Sekiguchi JM, Ferguson DO, Zhu C, Manis JP, Horner J, DePinho RA, Alt FW (2000): DNA ligase IV deficiency in mice leads to defective neurogenesis and embryonic lethality via the p53 pathway. Mol Cell:5:993–1002.

Galio L, Bouquet C, Brooks P (1999): ATP hydrolysis-dependent formation of a dynamic ternary nucleoprotein complex with MutS and MutL. Nucl Acids Res 27:2325–31.

Ganesan S, Silver DP, Greenberg RA, Avni D, Drapkin R, Miron A, Mok SC, Randrianarison V, Brodie S, Salstrom J, Rasmussen TP, Klimke A, Marrese C, Marahrens Y, Deng CX, Feunteun J, Livingston DM (2002): BRCA1 supports XIST RNA concentration on the inactive X chromosome. Cell 111:393–405.

Gao Y, Ferguson DO, Xie W, Manis JP, Sekiguchi J, Frank KM, Chaudhuri J, Horner J, DePinho RA, Alt FW (2000): Interplay of p53 and DNA-repair protein XRCC4 in tumorigenesis, genomic stability and development. Nature 404:897–900.

Gatei M, Zhou BB, Hobson K, Scott S, Young D, Khanna KK (2001): Ataxia telangiectasia mutated (ATM) kinase and ATM and Rad3 related kinase mediate phosphorylation of Brca1 at distinct and overlapping sites: In vivo assessment using phospho-specific antibodies. J Biol Chem 276:17276–80.

Genschel J, Littman SJ, Drummond JT, Modrich P (1998): Isolation of MutSbeta from human cells and comparison of the mismatch repair specificities of MutSbeta and MutSalpha. J Biol Chem 273:19895–901.

Goode EL, Ulrich CM, Potter JD (2002): Polymorphisms in DNA repair genes and associations with cancer risk. Cancer Epidemiol Biomarkers Prev 11: 1513–30.

Grawunder U, Wilm M, Wu X, Kulesza P, Wilson TE, Mann M, Lieber MR (1997): Activity of DNA ligase IV stimulated by complex formation with XRCC4 protein in mammalian cells. Nature 388:492–5.

Gu Y, Jin S, Gao Y, Weaver DT, Alt FW (1997): Ku70-deficient embryonic stem cells have increased ionizing radiosensitivity, defective DNA end-binding activity, and inability to support V(D)J recombination. Proc Natl Acad Sci USA 94:8076–81.

Hall JM, Lee MK, Newman B, Morrow JE, Anderson LA, Huey B, King MC (1990): Linkage of early-onset familial breast cancer to chromosome 17q21. Science 250:1684–9.

Harfe BD, Jinks-Robertson S (2000): DNA mismatch repair and genetic instability. An Rev Genet 34:359–99.

Hartley KO, Gell D, Smith GC, Zhang H, Divecha N, Connelly MA, Admon A, Lees-Miller SP, Anderson CW, Jackson SP (1995): DNA-dependent protein kinase catalytic subunit: a relative of phosphatidylinositol 3-kinase and the ataxia telangiectasia gene product. Cell 82:849–56.

Hashizume R, Fukuda M, Maeda I, Nishikawa H, Oyake D, Yabuki Y, Ogata H, Ohta T (2001): The RING heterodimer BRCA1-BARD1 is a ubiquitin ligase inactivated by a breast cancer-derived mutation. J Biol Chem 276:14537–40.

Hemminki A, Peltomaki P, Mecklin JP, Jarvinen H, Salovaara R, Nystrom-Lahti M, de la Chapelle A, Aaltonen LA (1994): Loss of the wild type MLH1 gene is a feature of hereditary nonpolyposis colorectal cancer. Nat Genet 8:405–10.

Henning KA, Li L, Iyer N, McDaniel LD, Reagan MS, Legerski R, Schultz RA, Stefanini M, Lehmann AR, Mayne LV, et al. (1995): The Cockayne syndrome group A gene encodes a WD repeat protein that interacts with CSB protein and a subunit of RNA polymerase II TFIIH. Cell 82:555–64.

Hickson ID (2003): RecQ helicases: caretakers of the genome. Nat Rev Cancer 3:169–78.

Hirano T (2002): The ABCs of SMC proteins: Two-armed ATPases for chromosome condensation, cohesion, and repair. Genes Dev 16:399–14.

Hopfner KP, Karcher A, Shin DS, Craig L, Arthur LM, Carney JP, Tainer JA (2000): Structural biology of Rad50 ATPase: ATP-driven conformational control in DNA double-strand break repair and the ABC-ATPase superfamily. Cell 101:789–800.

Howlett NG, Taniguchi T, Olson S, Cox B, Waisfisz Q, De Die-Smulders C, Persky N, Grompe M, Joenje H, Pals G, Ikeda H, Fox EA, D'Andrea AD (2002): Biallelic inactivation of BRCA2 in Fanconi anemia. Science 297:606–9.

Hwang BJ, Ford JM, Hanawalt PC, Chu G (1999): Expression of the p48 xeroderma pigmentosum gene is p53-dependent and is involved in global genomic repair. Proc Natl Acad Sci USA 96:424–8.

Ikeda M, Orimo H, Moriyama H, Nakajima E, Matsubara N, Mibu R, Tanaka N, Shimada T, Kimura A, Shimizu K (1998): Close correlation between mutations of E2F4 and hMSH3 genes in colorectal cancers with microsatellite instability. Cancer Res 58:594–8.

Jacob S, Praz F (2002): DNA mismatch repair defects: Role in colorectal carcinogenesis. Biochimie 84:27–47.

Kato M, Yano K, Matsuo F, Saito H, Katagiri T, Kurumizaka H, Yoshimoto M, Kasumi F, Akiyama F, Sakamoto G, Nagawa H, Nakamura Y, Miki Y (2000): Identification of Rad51 alteration in patients with bilateral breast cancer. J Hum Genet 45:133–7.

Kolodner R (1996): Biochemistry and genetics of eukaryotic mismatch repair. Genes Dev 10:1433–42.

Kraemer KH, Lee MM, Andrews AD, Lambert WC (1994): The role of sunlight and DNA repair in melanoma and nonmelanoma skin cancer: The xeroderma pigmentosum paradigm. Arch Dermatol 130:1018–21.

Kubota Y, Nash RA, Klungland A, Schar P, Barnes DE, Lindahl T (1996): Reconstitution of DNA base excision-repair with purified human proteins: Interaction between DNA polymerase beta and the XRCC1 protein. Embo J 15: 6662–70.

Kuismanen SA, Holmberg MT, Salovaara R, de la Chapelle A, Peltomaki P (2000): Genetic and epigenetic modification of MLH1 accounts for a major share of microsatellite-unstable colorectal cancers. Am J Pathol 156:1773–9.

Kurimasa A, Kumano S, Boubnov NV, Story MD, Tung CS, Peterson SR, Chen DJ (1999): Requirement for the kinase activity of human DNA-dependent protein kinase catalytic subunit in DNA strand break rejoining. Mol Cell Biol 19:3877–84.

Kuschel B, Auranen A, McBride S, Novik KL, Antoniou A, Lipscombe JM, Day NE, Easton DF, Ponder BA, Pharoah PD, Dunning A (2002): Variants in DNA double-strand break repair genes and breast cancer susceptibility. Hum Mol Genet 11:1399–407.

Lahue RS, Au KG, Modrich P (1989): DNA mismatch correction in a defined system. Science 245:160–4.

Lamers MH, Perrakis A, Enzlin JH, Winterwerp HH, de Wind N, Sixma TK (2000): The crystal structure of DNA mismatch repair protein MutS binding to a G × T mismatch. Nature 407:711–17.

Langland G, Kordich J, Creaney J, Goss KH, Lillard-Wetherell K, Bebenek K, Kunkel TA, Groden J (2001): The Bloom's syndrome protein (BLM) interacts with MLH1 but is not required for DNA mismatch repair. J Biol Chem 276:30031–5.

Le Page F, Kwoh EE, Avrutskaya A, Gentil A, Leadon SA, Sarasin A, Cooper PK (2000): Transcription-coupled repair of 8-oxoguanine: Requirement for XPG, TFIIH, and CSB and implications for Cockayne syndrome. Cell 101: 159–71.

Leach FS, Nicolaides NC, Papadopoulos N, Liu B, Jen J, Parsons R, Peltomaki P, Sistonen P, Aaltonen LA, Nystrom-Lahti M, et al. (1993): Mutations of a mutS homolog in hereditary nonpolyposis colorectal cancer. Cell 75:1215–25.

Lee JS, Collins KM, Brown AL, Lee CH, Chung JH (2000): hCds1-mediated phosphorylation of BRCA1 regulates the DNA damage response. Nature 404: 201–4.

Lehmann AR (1982): Three complementation groups in Cockayne syndrome. Mutat Res 106:347–56.

Li S, Ting NS, Zheng L, Chen PL, Ziv Y, Shiloh Y, Lee EY, Lee WH (2000): Functional link of BRCA1 and ataxia telangiectasia gene product in DNA damage response. Nature 406:210–15.

Liang F, Han M, Romanienko PJ, Jasin M (1998): Homology-directed repair is a major double-strand break repair pathway in mammalian cells. Proc Natl Acad Sci USA 95:5172–7.

Lieber MR (1999): The biochemistry and biological significance of nonhomologous DNA end joining: An essential repair process in multicellular eukaryotes. Genes Cells 4:77–85.

Lim D-S, Hasty P (1996): A mutation in mouse rad51 results in an early embryonic lethal that is suppressed by a mutation in p53. Mol Cell Biol 16: 7133–43.

Lim DS, Vogel H, Willerford DM, Sands AT, Platt KA, Hasty P (2000): Analysis of ku80-mutant mice and cells with deficient levels of p53. Mol Cell Biol 20: 3772–80.

Lindahl T (1993): Instability and decay of the primary structure of DNA. Nature 362:709–15.

Lindblom A, Tannergard P, Werelius B, Nordenskjold M (1993): Genetic mapping of a second locus predisposing to hereditary non-polyposis colon cancer. Nat Genet 5:279–82.

Lloyd JA, Forget AL, Knight KL (2002): Correlation of biochemical properties with the oligomeric state of human rad52 protein. J Biol Chem 277:46172–8.

Longley MJ, Pierce AJ, Modrich P (1997): DNA polymerase delta is required for human mismatch repair in vitro. J Biol Chem 272:10917–21.

Loukola A, Vilkki S, Singh J, Launonen V, Aaltonen LA (2000): Germline and somatic mutation analysis of MLH3 in MSI-positive colorectal cancer. Am J Pathol 157:347–52.

Lynch HT, Lanspa S, Smyrk T, Boman B, Watson P, Lynch J (1991): Hereditary nonpolyposis colorectal cancer (Lynch syndromes I & II). Genetics, pathology, natural history, and cancer control, Part I. Cancer Genet Cytogenet 53: 143–60.

Ma Y, Pannicke U, Schwarz K, Lieber MR (2002): Hairpin opening and overhang processing by an Artemis/DNA-dependent protein kinase complex in nonhomologous end joining and V(D)J recombination. Cell 108:781–94.

Maacke H, Opitz S, Jost K, Hamdorf W, Henning W, Kruger S, Feller AC, Lopens A, Diedrich K, Schwinger E, Sturzbecher HW (2000): Overexpression of wild-type Rad51 correlates with histological grading of invasive ductal breast cancer. Int J Cancer 88:907–13.

Malkhosyan S, Rampino N, Yamamoto H, Perucho M (1996): Frameshift mutator mutations. Nature 382:499–500.

Markowitz S, Wang J, Myeroff L, Parsons R, Sun L, Lutterbaugh J, Fan RS, Zborowska E, Kinzler KW, Vogelstein B, et al. (1995): Inactivation of the type II TGF-beta receptor in colon cancer cells with microsatellite instability. Science 268:1336–8.

Masson M, Niedergang C, Schreiber V, Muller S, Menissier-de Murcia J, de Murcia G (1998): XRCC1 is specifically associated with poly(ADP-ribose) polymerase and negatively regulates its activity following DNA damage. Mol Cell Biol 18:3563–71.

Meyn MS (1997): Chromosome instability syndromes: lessons for carcinogenesis. Curr Top Microbiol Immunol 221:71–148.

Miki Y, Swensen J, Shattuck-Eidens D, Futreal PA, Harshman K, Tavtigian S, Liu Q, Cochran C, Bennett LM, Ding W, et al. (1994): A strong candidate for the breast and ovarian cancer susceptibility gene BRCA1. Science 266:66–71.

Milne GT, Jin S, Shannon KB, Weaver DT (1996): Mutations in two Ku homologs define a DNA end-joining repair pathway in *Saccharomyces cerevisiae*. Mol Cell Biol 16:4189–98.

Miyagawa K, Tsuruga T, Kinomura A, Usui K, Katsura M, Tashiro S, Mishima H, Tanaka K (2002): A role for RAD54B in homologous recombination in human cells. EMBO J 21:175–80.

Modrich P, Lahue R (1996): Mismatch repair in replication fidelity, genetic recombination, and cancer biology. An Rev Biochem 65:101–33.

Mol CD, Kuo CF, Thayer MM, Cunningham RP, Tainer JA (1995): Structure and function of the multifunctional DNA-repair enzyme exonuclease III. Nature 374:381–6.

Morris JR, Keep NH, Solomon E (2002): Identification of residues required for the interaction of BARD1 with BRCA1. J Biol Chem 277:9382–6.

Moshous D, Callebaut I, de Chasseval R, Corneo B, Cavazzana-Calvo M, Le Deist F, Tezcan I, Sanal O, Bertrand Y, Philippe N, Fischer A, de Villartay JP (2001): Artemis, a novel DNA double-strand break repair/V(D)J recombination protein, is mutated in human severe combined immune deficiency. Cell 105:177–86.

Moynahan ME, Chiu JW, Koller BH, Jasin M (1999): Brca1 controls homology-directed DNA repair. Mol Cell 4:511–18.

Moynahan ME, Pierce AJ, Jasin M (2001): BRCA2 is required for homology-directed repair of chromosomal breaks. Mol Cell 7, in press.

Nick McElhinny SA, Snowden CM, McCarville J, Ramsden DA (2000): Ku recruits the XRCC4-ligase IV complex to DNA ends. Mol Cell Biol 20: 2996–3003.

O'Driscoll M, Cerosaletti KM, Girard PM, Dai Y, Stumm M, Kysela B, Hirsch B, Gennery A, Palmer SE, Seidel J, Gatti RA, Varon R, Oettinger MA, Neitzel H, Jeggo PA, Concannon P (2001): DNA ligase IV mutations identified in patients exhibiting developmental delay and immunodeficiency. Mol Cell 8:1175–85.

Obmolova G, Ban C, Hsieh P, Yang W (2000): Crystal structures of mismatch repair protein MutS and its complex with a substrate DNA. Nature 407: 703–10.

Ohmiya N, Matsumoto S, Yamamoto H, Baranovskaya S, Malkhosyan SR, Perucho M (2001): Germline and somatic mutations in hMSH6 and hMSH3 in gastrointestinal cancers of the microsatellite mutator phenotype. Gene 272: 301–13.

Papadopoulos N, Nicolaides NC, Wei YF, Ruben SM, Carter KC, Rosen CA, Haseltine WA, Fleischmann RD, Fraser CM, Adams MD, et al. (1994): Mutation of a mutL homolog in hereditary colon cancer. Science 263:1625–9.

Patel KJ, Yu VP, Lee H, Corcoran A, Thistlethwaite FC, Evans MJ, Colledge WH, Friedman LS, Ponder BA, Venkitaraman AR (1998): Involvement of Brca2 in DNA repair. Mol Cell 1:347–57.

Paull TT (2001): New glimpses of an old machine. Cell 107:563–5.

Paull TT, Gellert M (1999): Nbs1 potentiates ATP-driven DNA unwinding and endonuclease cleavage by the Mre11/Rad50 complex. Genes Dev 13:1276–88.

Peinado MA, Malkhosyan S, Velazquez A, Perucho M (1992): Isolation and characterization of allelic losses and gains in colorectal tumors by arbitrarily primed polymerase chain reaction. Proc Natl Acad Sci USA 89:10065–9.

Pellegrini L, Yu DS, Lo T, Anand S, Lee M, Blundell TL, Venkitaraman AR (2002): Insights into DNA recombination from the structure of a RAD51-BRCA2 complex. Nature 420:287–93.

Peltomaki P, Aaltonen LA, Sistonen P, Pylkkanen L, Mecklin JP, Jarvinen H, Green JS, Jass JR, Weber JL, Leach FS, et al. (1993): Genetic mapping of a locus predisposing to human colorectal cancer. Science 260:810–12.

Pittman DL, Schimenti JC (2000): Midgestation lethality in mice deficient for the RecA-related gene, Rad51d/Rad51l3. Genesis J Genet Dev 26:167–73.

Price VH, Odom RB, Ward WH, Jones FT (1980): Trichothiodystrophy: Sulfur-deficient brittle hair as a marker for a neuroectodermal symptom complex. Arch Dermatol 116:1375–84.

Raderschall E, Stout K, Freier S, Suckow V, Schweiger S, Haaf T (2002): Elevated levels of Rad51 recombination protein in tumor cells. Cancer Res 62:219–25.

Radman M, Wagner R (1986): Mismatch repair in *Escherichia coli*. An Rev Genet 20:523–38.

Rampino N, Yamamoto H, Ionov Y, Li Y, Sawai H, Reed JC, Perucho M (1997): Somatic frameshift mutations in the BAX gene in colon cancers of the microsatellite mutator phenotype. Science 275:967–9.

Ramsden DA, Gellert M (1998): Ku protein stimulates DNA end joining by mammalian DNA ligases: A direct role for Ku in repair of DNA double-strand breaks. EMBO J 17:609–14.

Ranalli TA, Tom S, Bambara RA (2002): AP endonuclease 1 coordinates flap endonuclease 1 and DNA ligase I activity in long patch base excision repair. J Biol Chem 277:41715–24.

Reitmair AH, Schmits R, Ewel A, Bapat B, Redston M, Mitri A, Waterhouse P, Mittrucker HW, Wakeham A, Liu B, et al. (1995): MSH2 deficient mice are viable and susceptible to lymphoid tumours. Nat Genet 11:64–70.

Riccio A, Aaltonen LA, Godwin AK, Loukola A, Percesepe A, Salovaara R, Masciullo V, Genuardi M, Paravatou-Petsotas M, Bassi DE, Ruggeri BA, Klein-Szanto AJ, Testa JR, Neri G, Bellacosa A (1999): The DNA repair gene MBD4 (MED1) is mutated in human carcinomas with microsatellite instability. Nat Genet 23:266–8.

Rijkers T, Van Den Ouweland J, Morolli B, Rolink AG, Baarends WM, Van Sloun PP, Lohman PH, Pastink A (1998): Targeted inactivation of mouse RAD52 reduces homologous recombination but not resistance to ionizing radiation. Mol Cell Biol 18:6423–9.

Schmutte C, Tombline G, Rhiem K, Sadoff MM, Schmutzler R, von Deimling A, Fishel R (1999): Characterization of the human Rad51 genomic locus and examination of tumors with 15q14–15 loss of heterozygosity (LOH). Cancer Res 59:4564–9.

Schoenmakers EF, Huysmans C, Van de Ven WJ (1999): Allelic knockout of novel splice variants of human recombination repair gene RAD51B in t(12;14) uterine leiomyomas. Cancer Res 59:19–23.

Schwartz S, Jr, Yamamoto H, Navarro M, Maestro M, Reventos J, Perucho M (1999): Frameshift mutations at mononucleotide repeats in caspase-5 and other target genes in endometrial and gastrointestinal cancer of the microsatellite mutator phenotype. Cancer Res 59:2995–3002.

Scully R, Anderson SF, Chao DM, Wei W, Ye L, Young RA, Livingston DM, Parvin JD (1997): BRCA1 is a component of the RNA polymerase II holoenzyme. Proc Natl Acad Sci USA 94:5605–10.

Shin KH, Park YJ, Park JG (2001): PTEN gene mutations in colorectal cancers displaying microsatellite instability. Cancer Lett 174:189–94.

Shu Z, Smith S, Wang L, Rice MC, Kmiec EB (1999): Disruption of muREC2/RAD51L1 in mice results in early embryonic lethality which can Be partially rescued in a p53(-/-) background. Mol Cell Biol 19:8686–93.

Sigurdsson S, Van Komen S, Petukhova G, Sung P (2002): Homologous DNA pairing by human recombination factors Rad51 and Rad54. J Biol Chem 277:42790–4.

Simbulan-Rosenthal CM, Haddad BR, Rosenthal DS, Weaver Z, Coleman A, Luo R, Young HM, Wang ZQ, Ried T, Smulson ME (1999): Chromosomal aberrations in PARP(-/-) mice: Genome stabilization in immortalized cells by reintroduction of poly(ADP-ribose) polymerase cDNA. Proc Natl Acad Sci USA 96:13191–6.

Souza RF, Appel R, Yin J, Wang S, Smolinski KN, Abraham JM, Zou TT, Shi YQ, Lei J, Cottrell J, Cymes K, Biden K, Simms L, Leggett B, Lynch PM, Frazier M, Powell SM, Harpaz N, Sugimura H, Young J, Meltzer SJ (1996): Microsatellite instability in the insulin-like growth factor II receptor gene in gastrointestinal tumours. Nat Genet 14:255–7.

Stasiak AZ, Larquet E, Stasiak A, Muller S, Engel A, Van Dyck E, West SC, Egelman EH (2000): The human Rad52 protein exists as a heptameric ring. Curr Biol 10:337–40.

Stewart GS, Maser RS, Stankovic T, Bressan DA, Kaplan MI, Jaspers NG, Raams A, Byrd PJ, Petrini JH, Taylor AM (1999): The DNA double-strand break repair gene hMRE11 is mutated in individuals with an ataxia-telangiectasia-like disorder. Cell 99:577–87.

Sugasawa K, Ng JM, Masutani C, Iwai S, van der Spek PJ, Eker AP, Hanaoka F, Bootsma D, Hoeijmakers JH (1998): Xeroderma pigmentosum group C protein complex is the initiator of global genome nucleotide excision repair. Mol Cell 2:223–32.

Sugasawa K, Okamoto T, Shimizu Y, Masutani C, Iwai S, Hanaoka F (2001): A multistep damage recognition mechanism for global genomic nucleotide excision repair. Genes Dev 15:507–21.

Sung P (1994): Catalysis of ATP-dependent homologous DNA pairing and strand exchange by yeast RAD51 protein. Science 265:1241–3.

Taccioli GE, Rathbun G, Oltz E, Stamato T, Jeggo PA, Alt FW (1993): Impairment of V(D)J recombination in double-strand break repair mutants. Science 260:207–10.

Takahashi T, Nagai N, Oda H, Ohama K, Kamada N, Miyagawa K (2001): Evidence for RAD51L1/HMGIC fusion in the pathogenesis of uterine leiomyoma. Genes Chromosomes Cancer 30:196–201.

Takata M, Sasaki MS, Tachiiri S, Fukushima T, Sonoda E, Schild D, Thompson LH, Takeda S (2001): Chromosome instability and defective recombinational repair in knockout mutants of the five Rad51 paralogs. Mol Cell Biol 21: 2858–66.

Takenoshita S, Tani M, Nagashima M, Hagiwara K, Bennett WP, Yokota J, Harris CC (1997): Mutation analysis of coding sequences of the entire transforming growth factor beta type II receptor gene in sporadic human colon cancer using genomic DNA and intron primers. Oncogene 14:1255–8.

Tanaka K, Hiramoto T, Fukuda T, Miyagawa K (2000): A novel human rad54 homologue, Rad54B, associates with Rad51. J Biol Chem 275:26316–21.

Tanaka K, Kawai K, Kumahara Y, Ikenaga M, Okada Y (1981): Genetic complementation groups in cockayne syndrome. Somatic Cell Genet 7:445–55.

Tang J, Chu G (2002): Xeroderma pigmentosum complementation group E and UV-damaged DNA-binding protein. DNA Repair (Amst) 1:601–16.

Tarsounas M, Davies D, West SC (2003): BRCA2-dependent and independent formation of RAD51 nuclear foci. Oncogene 22:1115–23.

Tauchi H, Kobayashi J, Morishima K, van Gent DC, Shiraishi T, Verkaik NS, van Heems D, Ito E, Nakamura A, Sonoda E, Takata M, Takeda S, Matsuura S, Komatsu K (2002): Nbs1 is essential for DNA repair by homologous recombination in higher vertebrate cells. Nature 420:93–8.

Tavtigian SV, Simard J, Rommens J, Couch F, Shattuck-Eidens D, Neuhausen S, Merajver S, Thorlacius S, Offit K, Stoppa-Lyonnet D, Belanger C, Bell R, Berry S, Bogden R, Chen Q, Davis T, Dumont M, Frye C, Hattier T, Jammulapati S, Janecki T, Jiang P, Kehrer R, Leblanc JF, Goldgar DE, et al. (1996): The complete BRCA2 gene and mutations in chromosome 13q-linked kindreds. Nat Genet 12:333–7.

Tebbs RS, Flannery ML, Meneses JJ, Hartmann A, Tucker JD, Thompson LH, Cleaver JE, Pedersen RA (1999): Requirement for the Xrcc1 DNA base excision repair gene during early mouse development. Dev Biol 208:513–29.

Thibodeau SN, French AJ, Cunningham JM, Tester D, Burgart LJ, Roche PC, McDonnell SK, Schaid DJ, Vockley CW, Michels VV, Farr GH, Jr, O'Connell MJ (1998): Microsatellite instability in colorectal cancer: Different mutator phenotypes and the principal involvement of hMLH1. Cancer Res 58: 1713–18.

Thompson LH, Schild D (2002): Recombinational DNA repair and human disease. Mutat Res 509:49–78.

Tong WM, Cortes U, Wang ZQ (2001a): Poly(ADP-ribose) polymerase: A guardian angel protecting the genome and suppressing tumorigenesis. Biochim Biophys Acta 1552:27–37.

Tong WM, Hande MP, Lansdorp PM, Wang ZQ (2001b): DNA strand break-sensing molecule poly(ADP-Ribose) polymerase cooperates with p53 in telomere function, chromosome stability, and tumor suppression. Mol Cell Biol 21:4046–54.

Troelstra C, van Gool A, de Wit J, Vermeulen W, Bootsma D, Hoeijmakers JH (1992): ERCC6, a member of a subfamily of putative helicases, is involved in Cockayne's syndrome and preferential repair of active genes. Cell 71:939–53.

Tsuzuki T, Fujii Y, Sakumi K, Tominaga Y, Nakao K, Sekiguchi M, Matsushiro A, Yoshimura Y, Morita T (1996): Targeted disruption of the Rad51 gene leads to lethality in embryonic mice. Proc Natl Acad Sci USA 93:6236–40.

Varon R, Vissinga C, Platzer M, Cerosaletti KM, Chrzanowska KH, Saar K, Beckmann G, Seemanova E, Cooper PR, Nowak NJ, Stumm M, Weemaes CM, Gatti RA, Wilson RK, Digweed M, Rosenthal A, Sperling K, Concannon P, Reis A (1998): Nibrin, a novel DNA double-strand break repair protein, is mutated in Nijmegen breakage syndrome. Cell 93:467–76.

Walker JR, Corpina RA, Goldberg J (2001): Structure of the Ku heterodimer bound to DNA and its implications for double-strand break repair. Nature 412:607–14.

Wang ZQ, Stingl L, Morrison C, Jantsch M, Los M, Schulze-Osthoff K, Wagner EF (1997): PARP is important for genomic stability but dispensable in apoptosis. Genes Dev 11:2347–58.

Welsh KM, Lu AL, Clark S, Modrich P (1987): Isolation and characterization of the Escherichia coli mutH gene product. J Biol Chem 262:15624–9.

Whitehouse CJ, Taylor RM, Thistlethwaite A, Zhang H, Karimi-Busheri F, Lasko DD, Weinfeld M, Caldecott KW (2001): XRCC1 stimulates human polynucleotide kinase activity at damaged DNA termini and accelerates DNA single-strand break repair. Cell 104:107–17.

Williams C, Ponten F, Ahmadian A, Ren ZP, Ling G, Rollman O, Ljung A, Jaspers NG, Uhlen M, Lundeberg J, Ponten J (1998): Clones of normal keratinocytes and a variety of simultaneously present epidermal neoplastic lesions contain a multitude of p53 gene mutations in a xeroderma pigmentosum patient. Cancer Res 58:2449–55.

Wooster R, Bignell G, Lancaster J, Swift S, Seal S, Mangion J, Collins N, Gregory S, Gumbs C, Micklem G (1995): Identification of the breast cancer susceptibility gene BRCA2. Nature 378:789–92.

Wooster R, Weber BL (2003): Breast and ovarian cancer. N Engl J Med 348: 2339–47.

Wu L, Davies SL, Levitt NC, Hickson ID (2001): Potential role for the BLM helicase in recombinational repair via a conserved interaction with RAD51. J Biol Chem 276:19375–81.

Xanthoudakis S, Smeyne RJ, Wallace JD, Curran T (1996): The redox/DNA repair protein, Ref-1, is essential for early embryonic development in mice. Proc Natl Acad Sci USA 93:8919–23.

Xu X, Weaver Z, Linke SP, Li C, Gotay J, Wang XW, Harris CC, Ried T, Deng CX (1999): Centrosome amplification and a defective G_2-M cell cycle checkpoint induce genetic instability in BRCA1 exon 11 isoform-deficient cells. Mol Cell 3:389–95.

Yamaguchi M, Dao V, Modrich P (1998): MutS and MutL activate DNA helicase II in a mismatch-dependent manner. J Biol Chem 273:9197–201.

Yang H, Jeffrey PD, Miller J, Kinnucan E, Sun Y, Thoma NH, Zheng N, Chen PL, Lee WH, Pavletich NP (2002): BRCA2 function in DNA binding and recombination from a BRCA2-DSS1-ssDNA structure. Science 297:1837–48.

Yin J, Kong D, Wang S, Zou TT, Souza RF, Smolinski KN, Lynch PM, Hamilton SR, Sugimura H, Powell SM, Young J, Abraham JM, Meltzer SJ (1997): Mutation of hMSH3 and hMSH6 mismatch repair genes in genetically unstable human colorectal and gastric carcinomas. Hum Mutat 10:474–8.

Yoshikawa K, Ogawa T, Baer R, Hemmi H, Honda K, Yamauchi A, Inamoto T, Ko K, Yazumi S, Motoda H, Kodama H, Noguchi S, Gazdar AF, Yamaoka Y, Takahashi R (2000): Abnormal expression of BRCA1 and BRCA1-interactive DNA-repair proteins in breast carcinomas. Int J Cancer 88:28–36.

Zhong Q, Chen CF, Li S, Chen Y, Wang CC, Xiao J, Chen PL, Sharp ZD, Lee WH (1999): Association of BRCA1 with the hRad50-hMre11-p95 complex and the DNA damage response. Science 285:747–50.

Zhu C, Mills KD, Ferguson DO, Lee C, Manis J, Fleming J, Gao Y, Morton CC, Alt FW (2002): Unrepaired DNA breaks in p53-deficient cells lead to oncogenic gene amplification subsequent to translocations. Cell 109:811–21.

CHAPTER 17

ONCOGENES

STACEY J. BAKER and E. PREMKUMAR REDDY
Fels Institute for Cancer Research and Molecular Biology, Temple University School of Medicine, Philadelphia, PA 19140

DISCOVERY OF VIRAL AND CELLULAR ONCOGENES

Oncogenes were discovered through the study of transforming retroviruses that produced tumors in animals. Several decades ago it was observed that chickens predominantly die of cancer, suggesting that they are genetically predisposed to this disease. Examination of the tumors derived from these animals led to the identification of transforming retroviruses. In 1911 Peyton Rous at the Rockefeller Institute in New York isolated the first retrovirus from a spontaneously forming chicken sarcoma (Rous, 1911). He established the viral etiology of the tumor by demonstrating that tumor-derived extracts could be filtered through bacteria-retaining membranes and still have the ability to induce tumors in animals injected with the filtered extracts. The virus isolated by Rous has been named the Rous sarcoma virus (RSV), in honor of its discoverer.

Interestingly RSV and other transforming viruses were found to contain RNA as their genetic material, and hence they were termed RNA tumor viruses, or retroviruses. Later studies revealed that extracts from animal tumors often contain two types of RNA tumor viruses, termed acute transforming viruses and leukemia viruses. These viruses differed form each other in a number of properties, as listed in Table 17.1. The most important biological difference between these two classes of viruses was manifested in their ability to induce tumors in vivo as a function of time. While acute transforming viruses could induce tumors in susceptible hosts within a very short period of time (1–2 weeks), the leukemia viruses required several months to years to induce tumors. A second important difference between the two classes of viruses seemed to be manifested in their ability to transform cells in vitro. The acute transforming viruses could readily transform cells in culture while the leukemia viruses failed to do so. Finally, several of the acute

Cell Cycle and Growth Control: Biomolecular Regulation and Cancer, Second Edition, Edited by Gary S. Stein and Arthur B. Pardee
ISBN 0-471-25071-6

TABLE 17.1. Retroviruses or Type C RNA Tumor Viruses

Properties	Chronic Leukemia Viruses	Acute Transforming Viruses
Induce	Lymphomas	Sarcomas, carcinomas and hematopoietic tumors
Tumor latency period	Months to years	Weeks
Transformation of cells in culture	No	Fibroblasts and/or hematopoietic cells
Replication competent	Yes	Generally not
Mechanism of transformation	Integration next to oncogene	Discrete cell-derived gene within viral genome (oncogene)

transforming viruses (with the exception of Rous sarcoma virus) were replication incompetent, while leukemia viruses appeared to replicate readily in vitro and in vivo.

Comparison of the structures of the acute transforming and leukemia viral genomes revealed that the acute transforming viruses contained an additional segment of genomic RNA that was not present in the leukemia viruses. More important, deletion of this sequence abolished the transforming ability of these viruses, suggesting that a protein encoded by this unique piece of genetic material was responsible for the induction of tumors. The first of these genes was originally found in the Rous sarcoma virus, and was therefore designated as the *src* oncogene. Structural analyses of several of the acute transforming viruses revealed that very often acute transforming viruses suffered deletions in their env, pol, and/or gag sequences, which explained their inability to replicate in vitro or in vivo (Fig. 17.1). The only exception to this rule is the Rous sarcoma virus, which contains all of the three structural genes of the avian leukosis virus in addition to the *src* gene, and thus constitutes the only replication competent acute transforming virus.

The discovery that oncogenes of acute transforming viruses were in fact derived form normal cellular DNA, and that this genetic information is transduced by acute transforming viruses via genetic recombination, was a landmark in the field of cancer research. An analysis of normal cellular DNA using a probe derived from the *src* oncogene (derived from RSV) revealed the presence of endogenous oncogene-related sequences in normal chicken DNA (Stehelin et al., 1976). Following this lead, retrovirologists have isolated more than 200 different tumor-producing acute transforming viruses from animal tumors of different species. Table 17.2 provides a list of some of the acute transforming viruses and their transduced oncogenes. An examination of this table shows that several of the acute transforming viruses isolated from different animal species contained the same oncogene. It also suggests that there are only a finite number of transforming genes and that these genes often undergo recombination with replicating retroviruses leading to the formation of acute transforming viruses. Since the first identification of the v-*src* oncogene, a number of approaches have been used to identify genes that are

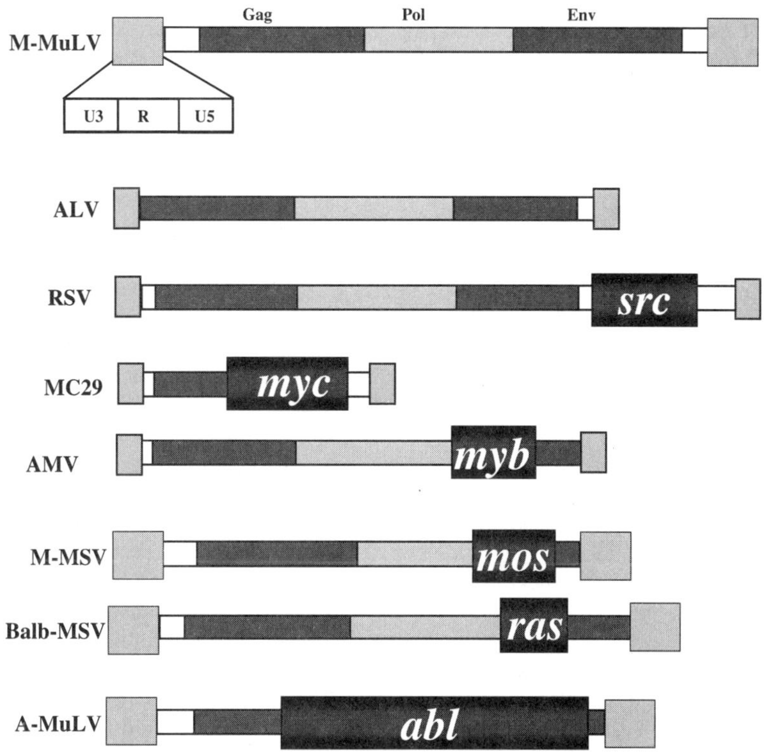

Figure 17.1. Comparison of the genomic structures M-MuLV, ALV, and selected acute trnsforming retroviruses.

responsible for the altered growth properties of tumor cells. Therefore the term oncogene is now broadly used to include any gene whose expression is associated with enhanced growth of tumor cells.

MECHANISMS OF RETROVIRAL-MEDIATED ACTIVATION OF ONCOGENES

Since this discovery of the *src* oncogene, the number of cellular homologues to retrovirus-associated oncogenes numbers approximately 50. These studies clearly demonstrated that all retroviral oncogenes are derived form normal cellular DNA. A detailed comparison of the structure of the viral oncogenes with their cellular counterparts revealed that several of the viral genes often represent mutant versions of their cellular homologues. Many protooncogenes are highly conserved through evolution, as some of them can be readily detected even in yeast, suggesting an important role for these genes in cell proliferation and survival.

The observation that retroviral oncogenes are derived from normal cell DNA but readily transform cells in culture posed an important

TABLE 17.2. Selected Retroviral Transforming Genes

Gene	Virus	Animal Origin	Protein	Function
abl	A-MuLV	Mouse	p120	Protein kinase
crk	ASV CT10	Chicken	p47	Adaptor protein
erb	AEV	Chicken	p45 + p75	Receptor
fes	ST-FeSV	Cat	p85	Protein kinase
	GA-FeSV	Cat	p85	Protein kinase
fgr	GR-FeSV	Cat	p70	Protein kinase
fms	MS-FeSV	Cat	p170	Protein kinase
fos	FBJ MSV	Mouse	p55	Transcription factor
fps	FSV	Chicken	p140	Protein kinase
	PRCII	Chicken	p105	Protein kinase
	UR1	Chicken	p150	Protein kinase
	16L	Chicken	p142	Protein kinase
jun	ASV17	Chicken	p55	Transcription factor
mos	Mo-MSV	Mouse	p37	Protein kinase
myc	MC29	Chicken	p110	Transcription factor
	CMII	Chicken	p90	Transcription factor
	MH2	Chicken	p100	Transcription factor
	OK10	Chicken	p200	Transcription factor
raf	3611MSV	Mouse	p75	Protein kinase
	MH2	Chicken	p100	Protein kinase
ras	Ki-MSV	Rat	p21	G-protein
	Ha-MSV	Rat	p21	G-protein
	BALB-MSV	Mouse	p21	G-protein
rel	AEV	Turkey	p56	Transcription factor
ros	UR2	Chicken	p68	Protein kinase
sis	SSV	Monkey	p28	Growth factor
ski	SK	Chicken	p110	Transcription factor
src	RSV	Chicken	p60	Protein kinase
	B77	Chicken	p60	Protein kinase
	rASV	Chicken, Quail	p60	Protein kinase
yes	Y73	Chicken	p90	Protein kinase
	ESC	Chicken	p80	Protein kinase

question: How can the same gene, when expressed in the context of normal cell growth, have no deleterious effect but, when expressed as an integral component of retrovirus, readily induce transformation? A detailed structural analysis of the v- (virus) and c- (cell) oncogenes revealed that retroviral oncogenes represent gain-of-function mutants of c-oncogenes. This is exemplified by the analysis of five retroviral oncogenes, v-*src*, v-*abl*, v-*ras*, v-*myb*, and v-*myc* and their normal cellular homologues, c-*src*, c-*abl*, c-*ras*, c-*myb*, and c-*myc*.

c-*src*

An analysis of the c-*src* gene shows that it encodes a protein that contains four well-defined structural domains, termed the unique, SH3, SH2, and tyrosine kinase (SH1) domains (reviewed in Brown and Cooper,

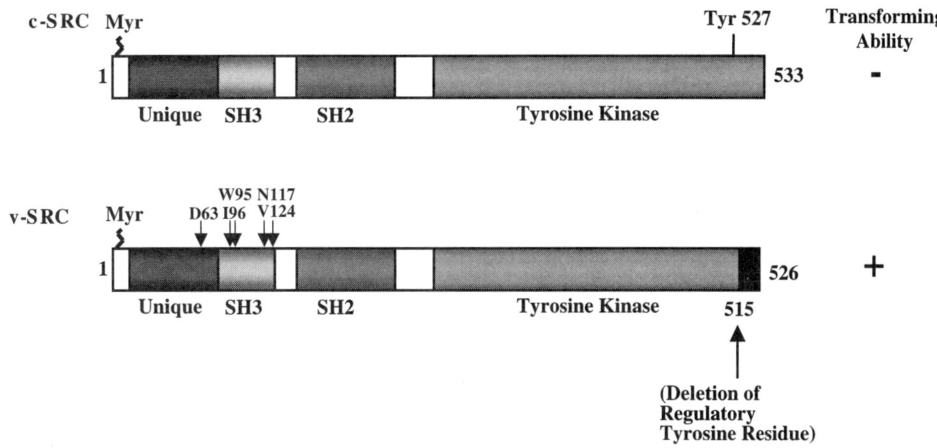

Figure 17.2. Activation of the Src oncoprotein. c-*src* encodes a tyrosine kinase whose activity is regulated by tyrosine phosphorylation. In a normal, resting cell, Src is phosphorylated on a C-terminal negative-regulatory tyrosine residue (by Csk) and has a low or undetectable level of catalytic activity. The protein is dephosphorylated and activated following stimulation with growth factors and cytokines. As a consequence of transduction by RSV, the v-*src* oncogene has a deletion in its 3' end and does not encode a negative-regulatory tyrosine residue. As a result, its kinase activity cannot be regulated and remains constitutively active, thereby causing transformation. v-*src* also contains a number of additional point mutations that are thought to enhance its oncogenic activity; however, none of these mutations by themselves have been shown to activate c-Src. (See color insert.)

1996). The kinase domain, which is located at the C-terminus of the protein, contains a negative regulatory tyrosine residue (Y527 in human Src) that is phosphorylated by a related kinase, Csk, in an unstimulated resting cell (Cooper et al., 1986; Okada and Nakagawa, 1989; reviewed in Frame, 2002) (Fig. 17.2). This phsophorylated state results in a folded, secondary structure that renders the kinase inactive, whereby the phosphorylated residue is bound to the SH2 domain in an inactive, or "closed," conformation. Other interactions between the SH3 domain and the linker sequences that are located between the SH2 and kinase domains also aid in maintaining the inactive state of the Src protein (Superti-Furga and Gonfloni, 1997; Gonfloni et al., 2000). Mitogenic stimulation of cells results in the dephosphorylation of this tyrosine, which subsequently results in enhanced kinase activity of the protein. In the v-Src protein, this negative regulatory region at the C-terminus is deleted, resulting in constitutive activation of its kinase activity due to the inability of this protein to achieve a closed conformation (reviewed in Brown and Cooper, 1996).

c-abl

Abelson murine leukemia virus (A-MuLV) is a replication defective retrovirus that was isolated after inoculation of Moloney murine leukemia virus (M-MuLV) into prednisolone-treated BALB/c mice (Abelson and Rabson, 1970). A-MuLV induces B-cell lymphomas in vivo and transforms both lymphoid and fibroblastic cells in vitro. Sequence analysis of the proviral genome of A-MuLV revealed that this virus encodes a single protein that is a fusion product of the viral-derived gag and cellular-derived *abl* sequences (Reddy et al., 1983). Analysis of c-*abl* sequences revealed that like the *src* gene, the *abl* gene codes for a tyrosine kinase that also contains the unique, SH3, SH2, and tyrosine kinase domains (Oppi et al., 1987). However, unlike the *src* gene, the *abl* gene product contains an additional C-terminal proline-rich domain whose function is thought to be that of an adaptor protein. Immediately following this domain is a putative DNA binding domain (Kipreos and Wang, 1992). In addition, the ABL protein does not contain a negative regulatory tyrosine at its C-terminus. Instead, the kinase activity of the c-Abl protein appears to be negatively regulated by its SH3 domain, which is deleted from the v-*abl* gene product. Interestingly the v-*abl* gene product was also found to contain a point mutation in its C-terminal sequence, which appears to enhance its tyrosine kinase and transforming activities (Shore et al., 1990). Thus both the v-*src* and v-*abl* genes have alterations in their regulatory sequences that result in the constitutive activation of their tyrosine kinase activities, which correlates with their transforming abilities (Fig 17.3).

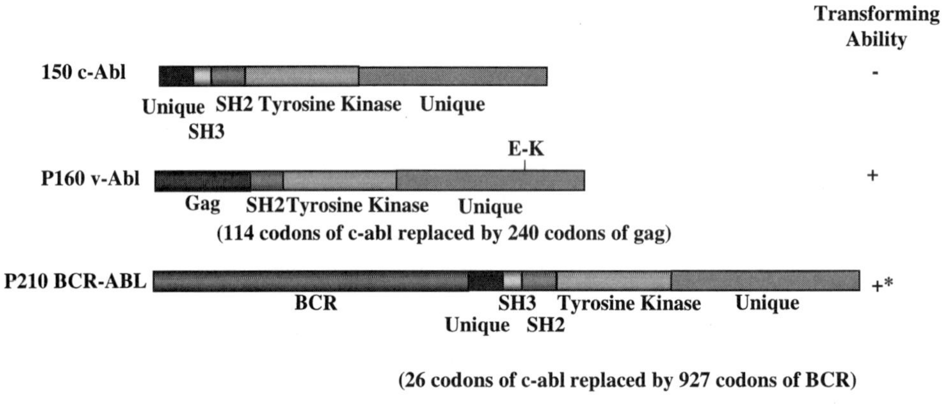

Figure 17.3. Activation of the Abl oncoprotein. c-*abl* encodes a tyrosine kinase, whose activity is regulated by phosphorylation and conformation via association with its own SH3 domain. This negative regulation is lost in v-Abl due to the substitution of the c-Abl SH3 domain with viral-derived gag sequences. v-Abl also has an amino acid substitution within its C-terminus that has been shown to enhance its tyrosine kinase activity. The *BCR-ABL* gene, which is formed as a consequence of a reciprocal chromosomal translocation (9; 22) encodes a protein, which has a similar structure where the N-terminal region of c-Abl protein is replaced by the N-terminal sequences of the BCR protein. (See color insert.)

c-ras Genes

The *ras* gene has been transduced by three rodent retroviruses (Harvey sarcoma virus, Balb MSV, and Kirsten sarcoma virus) and has been extensively studied because of its role in growth factor-associated signal transduction pathways (Barbacid, 1987). The viral *ras* genes encode 21 Kd proteins that belong to the G-protein superfamily. The proportion of GTP-bound p21 Ras is tightly regulated in normal cells by feedback mechanisms and represents less than 5% of the total Ras protein. A comparison of v- and c-*ras* revealed that the viral oncogenes contain two point mutations, one in codon 12 and a second in codon 59, both of which seem to impair the intrinsic GTPase activity of the mutant proteins and render them resistant to negative regulation (Barbacid, 1987; Bollag and McCormick, 1991). As a result the v-Ras proteins remain in a constitutively activated (or GTP-bound) state, which contributes to their oncogenic activity.

c-myb

The avian myeloblastosis virus, isolated in 1941 from a chicken tumor, is an acute transforming virus that causes myeloblastic leukemia in chickens and transforms myelomonocytic cells in vitro (Hall, 1941). Like most acute transforming viruses, AMV is replication defective, having arisen by recombination between a nondefective leukemia virus and chicken *myb* sequences. The *c-myb* gene encodes a nuclear transcription factor that appears to be essential for the proliferation of lymphoid, myeloid and erythoblastoid cells. Sequence analysis of c-*myb* revealed that the encoded protein contains an amino terminal DNA-binding domain, a central transactivation domain and a C-terminal negative regulatory domain (reviewed in Baluda and Reddy, 1994; Weston and Bishop, 1989; Oh and Reddy, 1999). A comparison of v-*myb* and c-*myb* sequences showed that the viral oncogene arose as a result of deletions in the 5′ and 3′ portions of the coding sequences, which results in the deletion of a portion of the DNA-binding domain and the entire negative regulatory domain. While the deletion within the DNA-binding domain does not seem to affect the DNA-binding ability of the viral protein, the deletion of the C-terminal negative regulatory domain appears to result in enhanced transcriptional transactivation activity by the v-Myb protein, which in turn appears to enhance the proliferative activity of virus-infected myeloid cells (Oh and Reddy, 1999) (Fig 17.4).

c-myc

The *myc* oncogene was originally identified as the transforming gene of the MC29 acute transforming virus that was isolated from a chicken with spontaneous myelocytomatosis (Vanov et al., 1964). The *myc*-related sequences were subsequently identified in three other independently derived chicken retroviral isolates, called CMII, OK10, and MH2. A

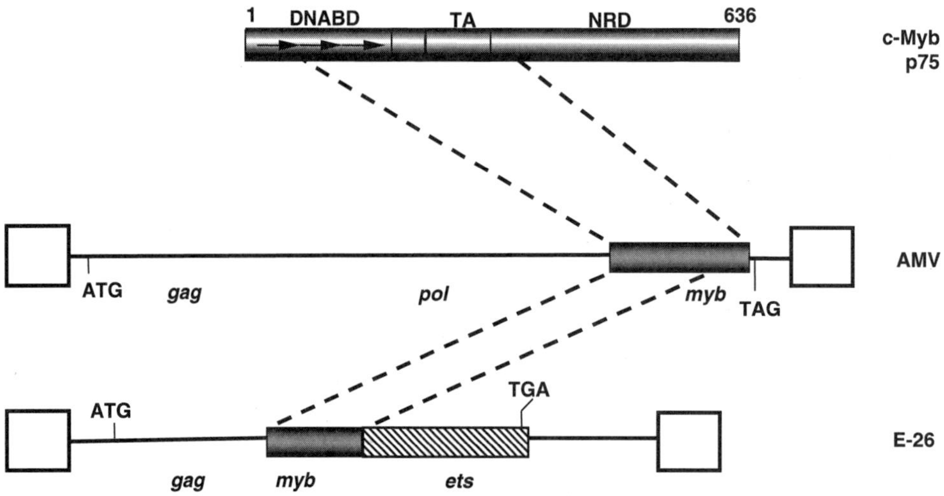

Figure 17.4. Structural comparison of the coding regions of normal c-*myb* with oncogenic forms encoded by the AMV and E26 viruses. Long terminal repeats are indicted by rectangles. The cellular-derived portion of the *myb* sequences is indicated by black boxes.

comparison of the MC29-derived v-*myc* and chicken c-*myc* sequences showed that the viral oncogene contains the entire coding sequence of the chicken c-*myc* gene fused to the viral gag sequences, thus producing a gag-*myc* fusion protein (Reddy et al., 1983; Watson et al., 1983). While the fusion of the *myc* sequences to the gag gene appears to result in elevated levels of Myc protein expression, deletion analysis of the proviral genome suggests that elevated expression of *myc* sequences alone is adequate for the transforming activity of this oncogenic virus. Thus, unlike with other oncogenic viruses, structural alterations seem to play a less important role in the oncogenic activity of this gene, and overexpression of the encoded protein product alone appears to be adequate for inducing transformation of appropriate target cells.

LEUKEMIA VIRUS ACTIVATION OF ONCOGENES

The study of acute transforming viruses has generated a wealth of knowledge regarding the mechanisms associated with oncogene activation, but the mechanisms associated with the transformation by leukemia viruses such as ALV or Mo-MuLV were not addressed by these studies. As noted in Table 17.1, leukemia viruses lack a cellular-derived oncogene but nevertheless induce leukemias, albeit over prolonged periods of time. The long latency period associated with the development of leukemias suggest that in addition to virus infection, other genetic events are required for the induction of transformation by this group of viruses. The first clues into the mechanisms associated with leukemia virus induced transformation came from studies with ALV-induced bursal lymphomas in chickens (Hayward et al., 1981) and Mo-MuLV-induced leukemias in mice (Cuypers et al., 1984).

An analysis of the chicken lymphomas induced by ALV demonstrated that these tumors contained integrated copies of the proviral genome. Most interestingly, these integration sites were often located within the c-*myc* proto-oncogene (Hayward et al., 1981). A detailed analysis of the c-*myc* locus from these tumors indicated that in normal cells, the c-*myc* gene contained three exons, of which exon 1 contained the 5′ non-coding sequences and exons 2 and 3 contained the entire coding region of the gene (Fig. 17.5). In ALV-induced bursal lymphomas this locus was disrupted by the insertion of the proviral genome within intron 1 of the c-*myc* gene (Fig. 17.5). Several tumors analyzed showed that following the integration of the provirus, the viral genome itself often underwent rearrangements, resulting in the deletion of most of the structural genes as well as one of the LTR sequences. This caused the positioning of a viral LTR immediately upstream of exons 2 and 3 of the c-*myc* gene, placing c-*myc* under the transcriptional control of the viral LTR. Since the viral LTR is designed to express high levels of RNA, the lymphoid cells containing the viral LTR integration expressed abnormally high levels of *myc* RNA and protein, which in combination with other genetic alterations led to the formation of the tumor. *c-myc* expression can also be deregulated by a similar mechanism due to chromosomal translocation (see section on Chromosomal Abnormalities and Oncogenes).

Mo-MuLV has been shown to induce thymic lymphomas when injected into newborn mice (Cuypers et al., 1984) and myeloid leukemias when injected into pristane-primed BALB/C mice (Shen-Ong et al., 1984). Analysis of the thymic lymphomas induced by this virus revealed that the integration of the provirus often occurs into two common loci. As with ALV-induced lymphomas, many of the Mo-MuLV-induced lymphomas exhibited the integration of the proviral genome into the c-*myc* locus and therefore induced deregulated expression of this gene (Cuypers et al., 1984). Other tumors exhibited integration of the provirus in a new locus, which was termed as *pim-1* (proviral integration MuLV) (Cuypers et al., 1984). *pim-1* encodes a kinase that acts as a transforming gene when overexpressed in lymphoid cells. Interestingly a number of tumors showed integration of the provirus in both the c-*myc* and *pim-1* loci, and therefore have elevated transcriptional rates of both genes.

Analysis of myeloid tumors induced by Mo-MuLV revealed that the viral integrations in these tumors often occur in the c-*myb* locus (Shen-Ong et al., 1984), the oncogene originally identified in the avian myeloblastosis virus (AMV). The mouse c-*myb* gene contains 15 exons, and very often, viral integrations were found to occur either in intron 3 or intron 9 resulting in either amino-terminal truncations or a C-terminal truncation of the protein after splicing. Interestingly the aberrant mRNAs produced by these tumors exhibit deletions similar to those observed in the v-*myb* oncogene (Tantravahi et al., 1996). Thus it appears that the activation of the *myb* gene is invariably accompanied by

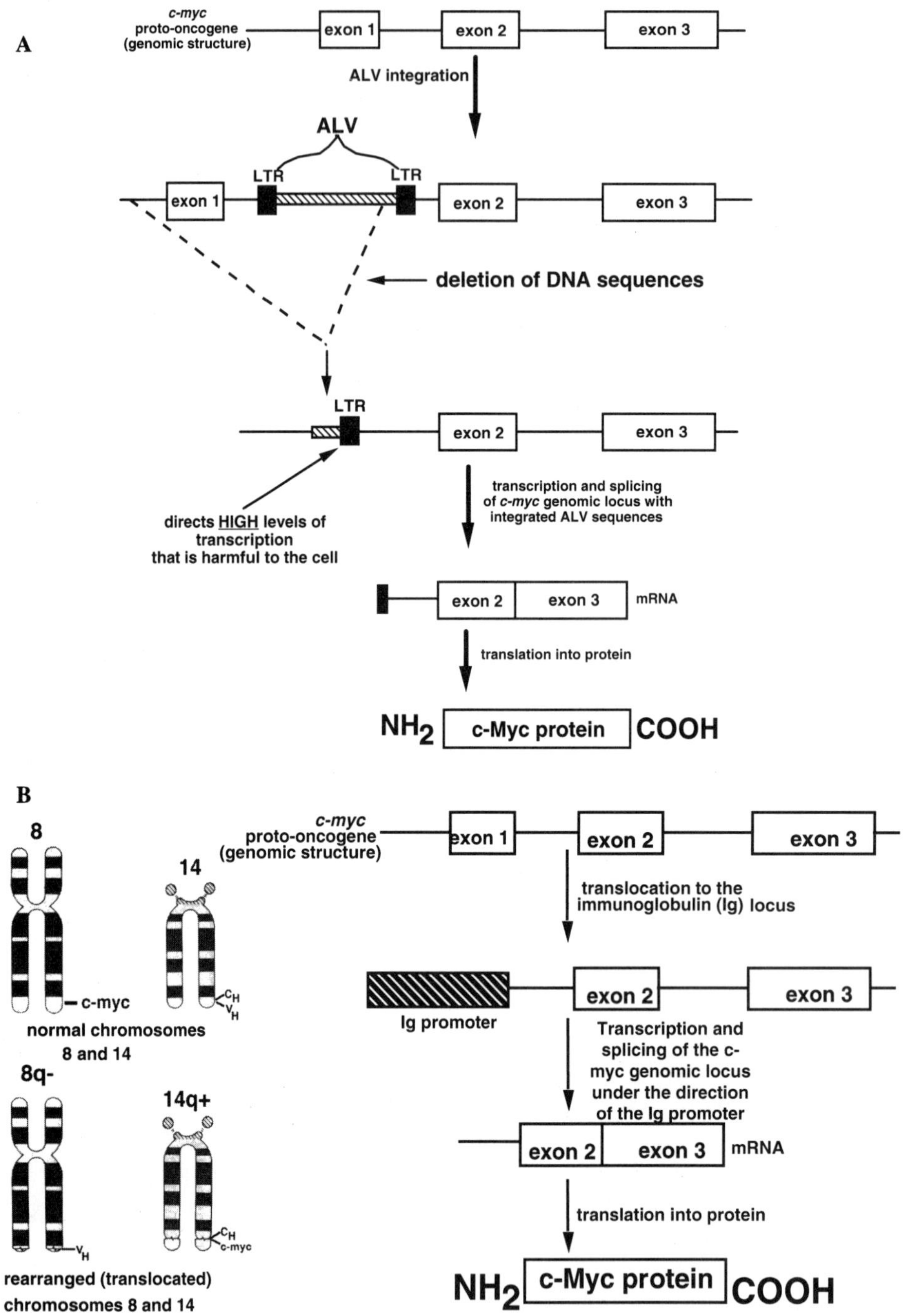
A
c-myc proto-oncogene (genomic structure)
exon 1
exon 2
exon 3
ALV integration
ALV
LTR
LTR
exon 1
exon 2
exon 3
deletion of DNA sequences
LTR
exon 2
exon 3
directs HIGH levels of transcription that is harmful to the cell
transcription and splicing of c-myc genomic locus with integrated ALV sequences
exon 2
exon 3
mRNA
translation into protein
NH2
c-Myc protein
COOH
B
8
14
c-myc
CH
VH
normal chromosomes 8 and 14
8q-
14q+
VH
CH
c-myc
rearranged (translocated) chromosomes 8 and 14
c-myc proto-oncogene (genomic structure)
exon 1
exon 2
exon 3
translocation to the immunoglobulin (Ig) locus
Ig promoter
exon 2
exon 3
Transcription and splicing of the c-myc genomic locus under the direction of the Ig promoter
exon 2
exon 3
mRNA
translation into protein
NH2
c-Myc protein
COOH

deletions in the coding regions. This finding lends support to the hypothesis that these structural alterations result in a gain of function for this oncoprotein, which in turn results in its oncogenicity.

c-akt

The AKT8 retrovirus was isolated from a spontaneous T-cell lymphoma obtained from an AKR mouse. Sequencing of the proviral genome revealed the presence of cellular derived sequences, termed *v-akt*, which arose due to recombination between viral gag sequences and the 5′ untranslated region of *c-akt* (Staal, 1987). The normal Akt (PKB) protein encodes a 55 kD protein with an N-terminal pleckstrin homology (PH) domain, a central serine-threonine kinase domain that shares similarities with protein kinase C (PKC) and PKA, and a C-terminal regulatory domain (reviewed in Nicholson and Anderson, 2002). In a normal cell, Akt is activated by phosphatidyl-inositol-3 kinase (PI3-K)-dependent kinases that phosphorylate the protein's PH domain and direct it to the plasma membrane in response to extracellular stimuli (Burgering and Coffer, 1995; Franke et al., 1005). Because v-Akt is a GAG-ONC fusion protein, the myristylation signal present in the GAG sequence permanently anchors the viral protein to the plasma membrane, where it remains constitutively active. v-Akt and constitutively active mutants of c-Akt have been shown to suppress apoptosis, suggesting that this is a mechanism by which these proteins induce transformation (Bellacosa et al., 1991, 1993; Ahmed et al., 1993).

MOUSE MAMMARY TUMOR VIRUS (MMTV) INDUCED ACTIVATION OF ONCOGENES

Similar to tumors induced by the leukemia viruses, MMTV-induced tumors arise due to insertional mutagenesis and subsequent viral-mediated transcriptional control of genes located near the integration

Figure 17.5. Similarities in the activation of the c-*myc* oncogene in ALV-induced lymphomas and Burkitt's lymphomas. (**A**) In a normal cell the three exons of the c-*myc* gene are spliced into an mRNA that encodes the c-Myc protein (only exons 2 and 3 are coding exons). When ALV integrates into the c-*myc* locus (usually between exons 1 and 2), the LTR sequences drive high levels of c-Myc expression, thereby promoting trasformation. ALV integration is often accompanied by deletions within the viral sequences, leaving one of the LTRs intact. (**B**) In certain cases of Burkitt's lymphoma, a portion of chromosome 8 that contains the c-*myc* locus translocates to chromosome 14 at a site immediately following the immunoglobulin (Ig) heavy chain locus (C_H). This translocation results in the placement of the c-*myc* locus downstream of the Ig regulatory sequences. Because the Ig promoter positively regulates high levels of transcription in B lumphocytes, the c-*myc* gene (instead of Ig) is overexpressed in the cell, resulting in oncogenesis.

site(s). MMTV was initially isolated from the milk of certain inbred strains of mice with unusually high incidences of mammary tumors (reviewed in Peters, 1988). These mice initially develop hormone-dependent hyperplastic alveolar nodules, which, along with the primary tumor itself, are induced by pregnancy. Although these tumors regress after parturition, newly formed hormone-dependent tumors arise after repeated pregnancies (reviewed in Peters, 1988), supporting the notion that MMTV is a latent oncogenic retrovirus. Detailed molecular analysis of MMTV proviral integration sites in mammary tumors has lead to the discovery of several protooncogenes. These genes, termed *int* genes, are not expressed in normal mammary tissue and have therefore been isolated using MMTV-derived sequences as probes to clone genes present at the sites of viral integration. The *int* name is practically all that they share in common, since most *int* genes are not structurally related; however, all *int* genes can be categorized into two main groups: the *wnt-1/int-4* and the *int-2/hst2* families (reviewed in Tekmal et al., 1997). Wnt proteins belong to a family of secreted glycoproteins that regulate many cellular processes, including differentiation, migration and proliferation. *Int-2*, on the other hand, is a member of the fibroblast growth factor (FGF) family (reviewed in Tekmal et al., 1997). In some tumors, MMTV proviruses have been observed at each end of both loci, and the position of the proviruses have indicated that MMTV induces increased levels of transcription of both genes (Dickson et al., 1990; reviewed in Tekmal et al., 1997). In other tumors, viral integration actually disrupts the structure of the RNA transcript. *Int-3*, the homologue of the *Drosophila melanogaster Notch* gene, is deregulated in certain mouse strains infected with MMTV. Viral integration occurs at the middle of the locus, whereby one LTR terminates transcription while the other LTR acts as a promoter and activates transcription of the 3′ end of the gene (reviewed in Tekmal et al., 1997). The end result is a truncated transcript whose expression is driven by MMTV regulatory sequences. Because the defective Int-3 protein contains a deletion in the extracellular domain, and overexpression of the cytoplasmic domain results in enhanced signaling, the end result is the formation of tumors in the mammary gland (Jhappan et al., 1992).

ISOLATION OF DOMINANT TRANSFORMING GENES FROM HUMAN TUMORS

While the study of retrovirus-associated oncogenes led to the identification of a number of oncogenes, the link between these genes and human cancer was made through the identification of dominant transforming genes found in human tumor DNAs. The development of the NIH/3T3 cell transformation assay system has contributed immensely to the identification of the transforming genes present in human tumors. When the DNA from different human tumors was transfected and assayed for the induction of foci, about 20% of human tumors contained sequences that could transform NIH/3T3 cells (Cooper, 1982). Molecular cloning and characterization of these sequences resulted in the identification of several new oncogenes. One of the first human oncogenes identified from the T24 human bladder carcinoma cell line was found to be related to the viral *ras* gene, previously found to be present in the Harvey sarcoma virus that was isolated from a rat tumor (Santos et al., 1982; Der et al., 1982). A detailed characterization of the genetic lesion associated with the transforming gene of T24 bladder carcinoma cell line showed that a point mutation in codon 12 (similar to that seen in the v-*ras* oncogene) results in its oncogenic activation (Tabin et al., 1982; Reddy et al., 1982).

The human genome contains three closely related *ras* genes, all of which code for proteins of 21 kD and exhibit a high degree of homology (Fig. 17.6) (Barbacid, 1987). These three genes have been termed as H-*ras*, K-*ras* and N-*ras*. Of these, the H-*ras* gene is homologous to the v-*ras* genes derived from the Harvey sarcoma virus and BALB-MSV, while the K-*ras* gene is homologous to the oncogene encoded by the Kirsten sarcoma virus. The N-*ras* gene was originally isolated from a neuroblastoma following transfection of the tumor DNA into NIH/3T3 cells and analyzing the nature of the sequences that were responsible for the induction of transformed foci. A detailed analysis of the tumor-derived oncogenic *ras* genes showed that almost all of these genes contained mutations either in the 12th or the 61st codon, resulting in the constitutive activation of the protein products, a mechanism similar to that seen in their viral counterparts.

ret

The ret receptor tyrosine kinase was originally identified as an oncogene that was cloned following the transfection of NIH/3T3 cells with DNA isolated from a T-cell lymphoma (Takajashi et al., 1985). The low transformation efficiency (i.e., the formation of a single focus during the first round of transfection) and the absence of rearrangement in the primary tumor suggested that the *ret* oncogene arose due to recombination of normal DNA during transfection (reviewed in Eva, 1988). This recombination resulted in the formation of a fusion protein, whereby the Ret cytoplasmic domain was fused with a portion of the Rfp zinc finger protein, thereby causing Ret to be constitutively active (reviewed in Hunter, 1997). Since its initial discovery, mutations in the *ret* gene have been identified in several human cancers. In most, if not all cases, the protein's cytoplasmic domain is fused to an N-terminus, which causes constitutive dimerization and activation. Ret is also mutated in hereditary multiple endocrine neoplasia (MEN) 2A and 2B. Although the genetic defects that are associated with this disease are point mutations, they have been shown to facilitate receptor dimerization (causing constitutive activation) and even alter substrate specificity (reviewed in Hunter, 1997).

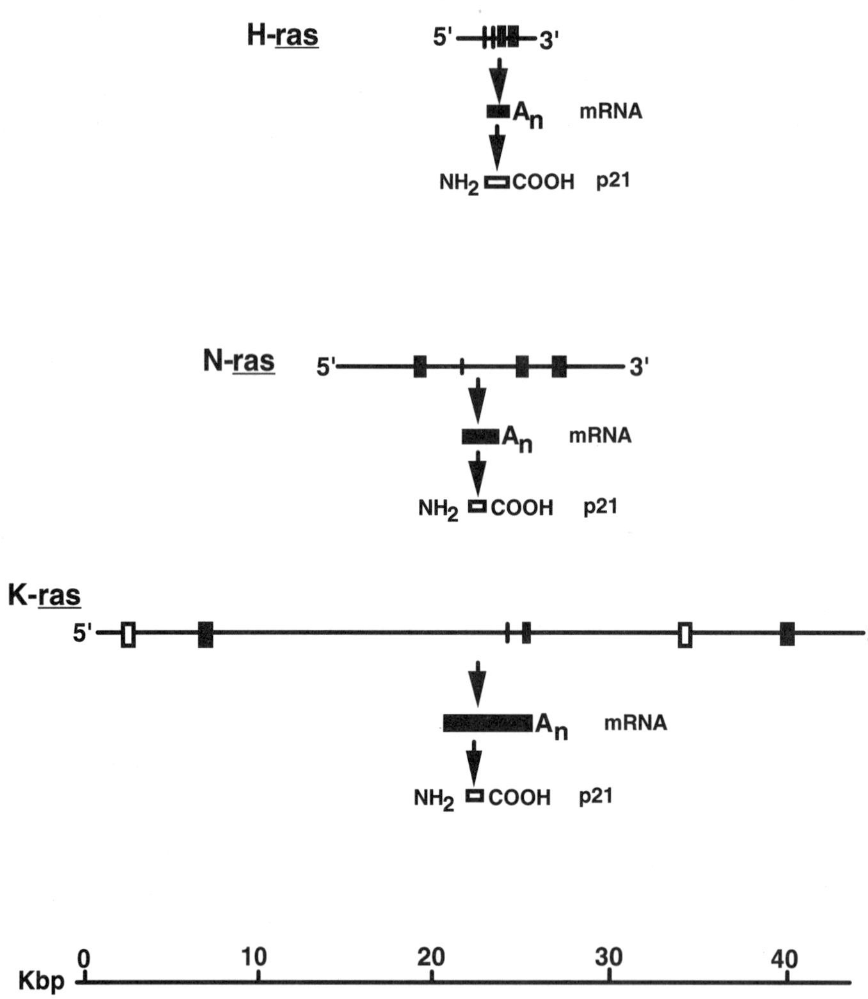

Figure 17.6. Comparison of the genomic structures of the *ras* protooncogenes. Although all *ras* genes encode four exons, K-*ras* encodes a fifth exon that is alternatively spliced during transcription. Although the length of each *ras* locus and transcript varies significantly, the size of each protein is identical.

CHROMOSOMAL ABNORMALITIES AND ONCOGENES

For many years it was known that tumor cells contain nonrandom chromosomal abnormalities. Although the hypothesis that such abnormalities might play a role in cancer was put forth in 1890 by Von Hansenmann (Von Hansemann, 1890), the first supporting experimental evidence was provided by Nowell and Hungerford in 1960, when they discovered the chronic myelogenous leukemia (CML)-associated Philadelphia chromosome (Nowell and Hungerford, 1960). This disease provides a paradigm of the genetic lesions associated with cancer because more than 90% of the cases involve this translocation. Over the past decade a number of chromosomal abnormalities in various tumor types were studied at the molecular level. All the evidence accumulated so far suggests that these chromosomal breakpoints occur within or in the vicinity of an oncogene, resulting in its activation. Gene-mapping studies have established that the c-*abl* oncogene is located on chromosome 9q34, the location where the breakpoint has been shown to occur in the Philadelphia chromosome. During the formation of the Philadelphia chromosome, a portion of the c-*abl* gene is translocated to chromosome 22 and is fused to a portion of the *bcr* gene, which itself is disrupted during the translocation process (Figs. 17.3 and 17.7). This process results in the generation of a new gene, termed *bcr-abl*, with enhanced oncogenic activity whose expression leads the development of leukemia (Groffen et al., 1984; Shitvelman et al., 1985). It is interesting to note that the BCR-ABL protein is structurally very similar to the GAG-ABL protein encoded by the Abelson murine leukemia virus; both GAG-ABL and BCR-ABL exhibit deletions in the N-terminal domain of c-Abl that result in high levels of tyrosine kinase activity, which is essential for their transforming activities.

A second example of a tumor-associated chromosomal abnormality is observed in Burkitt's lymphoma and involves a translocation between chromosomes 8 and 14 (Taub et al., 1982; Dalla-Favera et al., 1983). This translocation results in the juxtaposition of the *c-myc* oncogene with the regulatory sequences of the immunoglobulin (Ig) genes (Fig. 17.5). Because the Ig loci normally direct high levels of transcription in B lymphocytes, the *c-myc* oncogene is transcriptionally regulated by the Ig promoter elements. The end result is a B cell that abnormally produces extremely high levels of *c-myc* mRNA and protein (instead of antibody), which induces a malignant phenotype.

A third type of chromosomal abnormality is associated with the abnormal amplification of certain regions of chromosomes that contain oncogenic sequences. The amplified DNA in these tumors can exist as a tandem copy of a set of genes on a given chromosome, or it could exist as double-minute chromosomes. Double-minute chromosomes present themselves as homogeneously staining regions (HSR) that are distributed over the entire field in a chromosome preparation. A variety of human neuroblastomas have been found to contain amplification of the N-*myc* gene (Alitalo et al., 1983), while amplification of the *erb*B-2 gene is associated with several solid tumors (Dougall et al., 1994). Interestingly the prognosis of these tumors appears to correlate with their level of oncogene amplification, where poor prognosis is associated with greater amplification of a given oncogenic sequence.

ONCOGENES ARE OFTEN TARGETED BY CHEMICAL CARCINOGENS

Chemical carcinogens have a very broad range of structures, and the diversity in their structures clouded early studies of possible mechanisms

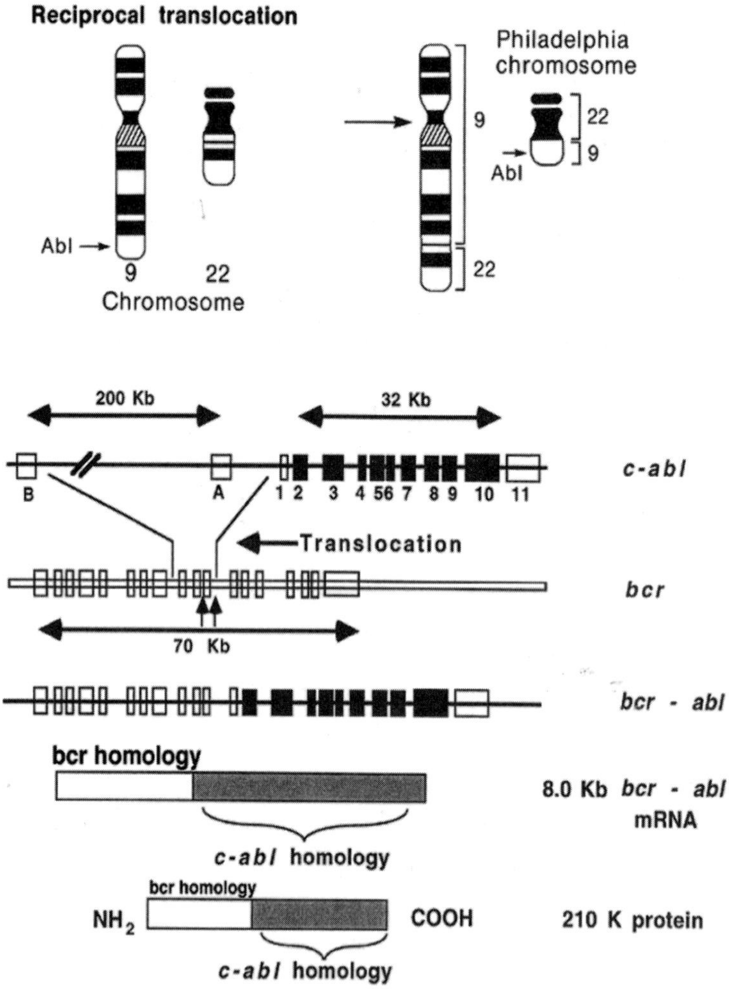

Figure 17.7. Chromosomal transclocation generates the BCR-ABL oncoprotein. The c-*abl* gene is located in chromosome 9 and encodes a tyrosine kinase. In certain cases of CML a portion of chromosome 9, which contains the c-*abl* locus, translocates to chromosome 22 at the breakpoint cluster region (*bcr*) locus, thereby generating the BCR-ABL oncoprotein. Because the translocation deletes the negative regulatory regions of c-Abl, the fusion protein has constitutive tyrosine kinase activity.

associated with their mechanism of action. Rapid progress in this field was made when it was discovered that most chemical carcinogens do not act directly but undergo modifications by cytochrome p450 proteins within the cell, thereby generating a new group of chemicals that act as carcinogens (Conney, 1982). There are two types of chemical carcinogens, one called direct-acting carcinogens, which, as the name implies, act directly without any metabolic activation, and the other, called indirect-acting carcinogens, which require metabolic activation for them to succeed as carcinogens. The terms pro-carcinogen and ultimate carcinogen have been used to describe the pre- and postprocessing states of indirect-acting carcinogens. Once inside the cell, both types of carcinogens seem to interact with a wide variety of cellular components, including DNA, RNA, and proteins. It is now believed that the most important of these interactions is with DNA, a process that results in the introduction of mutations into its sequence. In fact the finding that most carcinogens are mutagens allowed Bruce Ames to develop a number of assays using bacterial systems, which allowed a rapid analysis of various environmental agents for their possible oncogenic activity (Ames, 1979).

While it was becoming clear that chemical carcinogens might exert their influence by acting as mutagens, the identity of their targets remained unknown for quite some time. The first clues regarding their targets came from DNA transfection experiments using NIH/3T3 cells, where it could be shown that DNA from chemically transformed cell lines contained activated *ras* genes. These studies suggested that protooncogenes might be the crucial targets of the chemical carcinogens (Sukumar et al., 1983). The concept that protooncogenes are indeed the targets of chemical carcinogens received the much needed experimental support by the demonstration that in a majority of rat mammary tumors induced with a single dose of *N*-nitroso-*N*-methylurea MNU during sexual development, the *ras* gene undergoes a point mutation that converts this protooncogene into a potent transforming gene (Sukumar et al., 1983). Following this discovery, a number of chemical carcinogens were shown to act on protooncogenes, by converting them into dominant transforming genes. For instance, other models in which neuroblastomas or gliomas were induced by either ethyl-nitrourea (ENU) or MNU were found to reproducibly contain an activated the *neu* oncogene (Barbacid, 1986).

FUNCTION OF ONCOPROTEINS

The first clues regarding the normal function of oncogenes came from the observation that the *v-sis* oncogene product of the simian sarcoma virus is highly related to the β-chain of the platelet-derived growth factor receptor (PDGFR; reviewed by Aaronson, 1991). A second discovery that linked oncogenes with growth factor signaling was the finding that the Erb-B2 oncogene, encoded by the avian erythroblastosis virus (Table 17.2), is an activated form of EGF receptor (reviewed by Aaronson, 1991). These two observations suggested for the first time that oncogenes might encode growth factors, their receptors, or signal transducing proteins. In normal cells the primary event leading to a cellular mitotic response is the binding and activation of cellular receptors by growth factors such as PDGF or epidermal growth factor (EGF; Egan et al., 1993; McCorkick, 1984). Structural analysis of these receptors shows that they contain a tripartite structure consisting of an external ligand-binding domain, a transmembrane domain, and an intracellular tyrosine kinase domain, whereby ligand binding induces receptor dimerization and activation of kinase activity. Activation of kinase activity leads to the intermolecular phosphorylation of multiple tyrosine residues located in the intracellular portion of the receptor, which act as docking sites for a number of signal-transducing proteins, some of which are devoid of catalytic activity. One such protein, growth factor receptor bound protein 2 (Grb2) interacts with RTKs via its SH2 domain (Lowenstein et al., 1992). Grb-2 also contains a SH3 domain and associates with proline-rich regions of Ras guanine exchange factors (GEFs), such as son of sevenless (SOS), via this SH3 domain (Rozakis-Adcock et al., 1993; Egan et al., 1993; Olivier et al., 1993). Ras GEFs catalyze the exchange of Ras-bound GDP for GTP, thereby promoting activation of Ras proteins. Although the duration of Ras activation can vary, depending on the nature of the stimulus, activated Ras is rapidly deactivated by GAPs that stimulate GTP hydrolysis (reviewed in Macaluso et al., 2002). Point mutations acquired by the Ras protein (in tumor cells) result in the down-regulation of its GTPase activity leading to its constitutive activation. Activated Ras proteins have many downstream targets, including PI-3 kinase, GEFs for Ral proteins, MEKK, and the Raf serine/threonine kinase, which is probably the most widely studied Ras effector (reviewed in Macaluso et al., 2002) (Fig. 17.8).

The *v-raf* oncogene was originally identified as the transforming gene of the murine sarcoma virus 3611 (Rapp et al., 1983). In a normal cell, Ras associates with cellular Raf when it is bound to another protein, 14-3-3. This association prompts a conformational change in Raf that facilitates its localization at the plasma membrane and the subsequent activation of phosphorylation cascades (Tzivion et al., 1998; McPherson et al., 1999). Recent studies have shown that members of the *raf* gene family are frequently mutated in certain tumors, again leading to constitutive activation of this pathway (Davies et al., 2002; reviewed in Mercer and Pritchard, 2003). Activation of *ras* and *raf* constitute short-lived signals, which lead to the activation of longer-lasting kinase cascades that eventually relay the signals to the nucleus. This relay system consists of MAPK cascades that are evolutionary conserved and are required for a wide variety of biological responses. In vertebrates, multiple isoforms of MAPK have been identified and categorized into three subfamilies: the extracellular signal-regulated kinases (ERKs), p38mapk, and the Jun N-terminal kinases (JNKs) or stress-activated protein kinases (reviewed in Johnson and Lapadt, 2002). All of these kinases are regulated by phosphorylation. Activation of MAP-kinases requires phosphorylation of both a threonine and a tyrosine residue, which are separated by a single amino acid. The kinase that phosphorylates both amino acids, termed MAP-kinase-kinase (MAPKK), is itself activated by serine/threonine phosphorylation. This phosphorylation event is catalyzed by MAP-kinase-kinase-kinase (MAPKKK), which is activated by Raf kinases (reviewed in Johnson and Lapadt, 2002). The "classical" ERKs, ERK1/$p44^{mapk}$, and ERK2/$p42^{mapk}$ signal downstream of Raf-1 and MEK1, and together signal in response to a variety of extracellular stimuli (reviewed

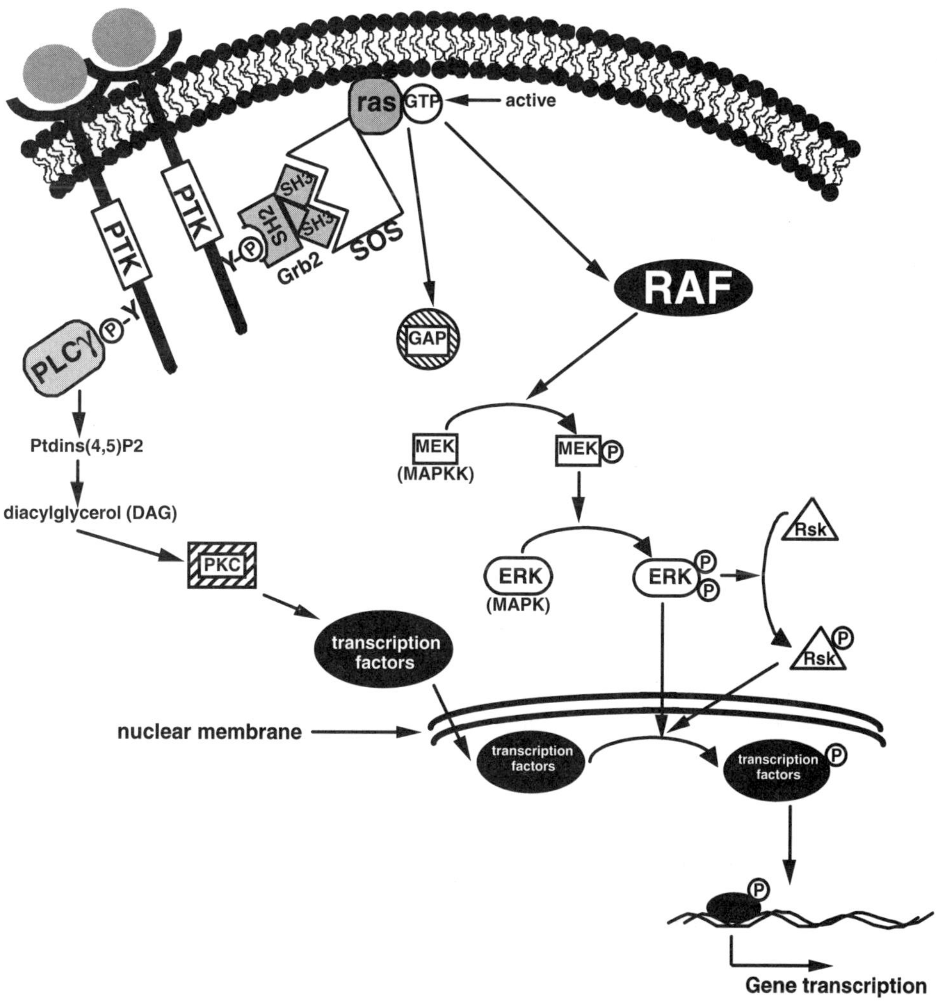

Figure 17.8. Schematic representation of receptor tyrosine kinase (RTK) signaling via the MAP kinase pathway. Upon growth factor stimulation, RTKS dimerize and undergo autophosphorylation on multiple tyrosine residues. These phosphorylated residues now serve as docking sites for adaptor molecules, such as the Grb2-Sos complex. Because adaptor proteins tether multiple proteins to a single signaling pathway, the RTK:Grb2:Sos complex triggers the activation of Ras via the exchange of GDP for GTP. Ras subsequently initiates an orderly phosphorylation cascade of Raf, Mek, Erk, and Rsk, which is essential for proliferation. The phosphorylation of multiple transcription factors enables them to bind to DNA and induce transcription.

in Johnson and Lapadt, 2002; Davis, 1993) (Fig. 17.8). MAPK signaling pathways have been implicated in control of numerous cellular responses including proliferation, differentiation, and apoptosis. These response pathways are partly triggered via the translocation of activated MAPK into the nucleus (Chen et al., 1992), where it phosphorylates and activates transcription factors such as Sap1 (Dalton and Treisman, 1992), c-Jun (Rozek and Pfizer, 1993), and ATF2 (Gille et al., 1992; Rozek and Pfizer, 1993). These transcription factors stimulate the expression of the immediate–early response oncogenes, including *c-fos* (Gille et al., 1992) and *c-jun* (Rozek and Pfizer, 1993), which comprise activator protein-1 (AP-1) (Shaulian and Karin, 2002).

The Ribosomal S6 kinases (RSK) family of serine/threonine kinases are activated at the end of the ERK-MAPK pathway (reviewed in Ballif and Blenis, 2001), whereby they are phosphorylated by ERK (Fig. 17.8) and subsequently translocate to the nucleus and phosphorylate nuclear substrates, including, c-Fos (Chen et al., 1993), CREB (Xing et al., 1996), CAAT enhancer-binding protein (C/EBP; Buck et al, 1999), and the estrogen receptor (ER; Joel et al., 1998). RSK may also have a role in chromatin remodeling via phosphorylation of histone H3 (Sassone-Corsi et al., 1999). The diversity of these substrates suggests that RSK regulates a variety of cellular functions.

Growth Factors and Activation of c-Src

As stated earlier, activation of growth factor receptors leads to the intermolecular phosphorylation of multiple tyrosine residues located in the intracellular portion of the receptor, which act as docking sites for a number of signal-transducing proteins. One such protein, c-Src, associates with the PDGFR via its SH2 domain, whereby it binds to specific phosphotyrosine residues located in the receptor's juxtamembrane region. Src's association with PDGFR activates its catalytic activity, and several studies have demonstrated that Src is required for PDGF-induced mitogenesis and that the PDGFR itself is a substrate for Src (Mori et al., 1993). Src has been shown to directly phosphorylate downstream targets, such as signal transducers and activators of transcription (STAT)-3 (reviewed in Biscardi et al., 2000). Furthermore dominant-negative forms of STAT proteins can inhibit v-Src-mediated transformation, c-Src-dependent proliferation in response to PDGF (Bowman et al., 2001) as well as other growth factors, including EGF. Stimulation of cells with EGF has also been shown to activate Src activity by a mechanism that is similar to that of PDGF. Src-overexpressing fibroblasts proliferate more rapidly in response to EGF (Luttrell et al., 1988) and as with PDGF-induced signaling, dominant-negative forms of Src can inhibit EGF-induced mitogenesis (Roche et al., 1995). Overexpression of both EGFR and Src in certain cell lines can also synergistically induce DNA synthesis and promote growth in soft agar, both of which are dependent on the physical association of Src with the EGFR (Maa et al., 1995). Src can also directly phosphorylate the EGFR on specific tyrosine residues, thereby creating docking sites for itself (via its SH2 domain) and other signaling molecules such as PI-3 kinase (Strover et al., 1995), as well as triggering the phosphorylation of STATs 1, 3, and 5 and other downstream proteins (reviewed in Biscardi et al., 2000). Other transcription factors, including c-Myc, have been proposed to function downstream of Src and can even reverse the phenotype of cells expressing dominant-negative Src (reviewed in Courtneidge, 2002). This effect, however, may be indirect as c-Myc has been identified as a target of STAT3 (a substrate of Src) in v-src transformed fibroblasts (Bowman et al., 2001).

Growth Factors and the Phosphatidylinolitol-3 Kinase/Akt Pathway

One of the proteins that binds to the phosphotyrosine moiety of activated growth factor receptors is PI-3 kinase. The first clue into the importance of PI-3 kinases in oncogenesis, RTK signaling and signal transduction in general came from avian sarcoma virus 16 (ASV16) (Chang et al., 1997; reviewed in Vogt et al., 1999). PI-3 kinase is comprised of two subunits, the p85 regulatory subunit and the p110 catalytic subunit, and catalyzes the formation of 3′ phosphoinositides, phosphatidylinositol 3-phosphate, phosphatidylinositol 3,4-biphosphate, and phosphatidylinositol 3,4,5-triphosphate, which act as second messengers (reviewed in Roymans and Siegers, 2001). p110 activity is tightly regulated by interaction with p85. However, p3k, the viral homologue of the p110 subunit, encodes a GAG-ONC fusion protein that does not need to interact with p85 for regulation of enzymatic activity and is therefore unresponsive to upstream signals from growth factors such as PDGF and EGF (reviewed in Vogt et al., 1999). PI-3 kinase also regulates the activity of Akt in a variety of cell systems. Akt is recruited to the plasma membrane where it is phosphorylated on a key threonine residue by phosphoinositide-dependent kinase-1 (PDK1), resulting in partial activation of the protein (reviewed in Nicholson and Anderson, 2001). The list of Akt substrates is growing and includes prosurvival proteins such as Bad (del Paso et al., 1997), Ask1 (Kim et al., 2001), and procaspase-9 (Cardone et al., 1998), supporting many reports that demonstrated a role for Akt during apoptosis. Other Akt substrates, such as Raf (Zimmermann and Moelling, 1999; Guan et al., 2000), also act downstream of Ras proteins in RTK-mediated mitogenesis.

Initial clues into the importance of Akt in cell proliferation and survival came from studies with v-Akt. Ectopic expression of v-Akt can prolong survival (Songyang et al., 1997; Eves et al., 1998). It was later determined that Akt regulates and prevents apoptosis via phosphorylation of Bad, a Bcl-2 family member. In response to growth factor or cytokine stimulation, such as that of IL-3, Bad is phosphorylated by Akt. This phosphorylation results in its cytosolic sequestration by the 14-3-3 tau isoform and its inactivation, as the phosphorylated form is unable to associate with antiapoptotic Bcl-2 proteins (Zha et al., 1996; del Paso et al., 1997) (see section on Oncogenes and Apoptosis). A similar phosphorylation is also catalyzed by Raf-1 (Wang et al., 1996), reinforcing the role of the Ras/Raf/Akt pathway in the regulation of cell proliferation and survival.

Oncogenes and Apoptosis

Approximately 85% of follicular lymphomas carry the t(14;18) translocation in which the *bcl-2* (B-cell leukemia/lymphoma-2) gene, which is located on chromosome 18, is translocated to chromosome 14 (Tsujimoto et al., 1985). Because the breakpoint on chromosome 14 occurs at the Ig heavy chain locus, high levels of *bcl-2* gene expression are observed in cells containing this translocation. The fact that the *bcl-2* gene was disrupted during the translocation suggested that its protein product played a role in cell survival, and this prompted the hypothesis that it functioned as an oncogene. However, forced expression of Bcl-2 did not induce transformation nor did it cause cells to proliferate indefinitely. It was not until the genetics of *Caenorhabditis elegans* revealed that CED-9 was a homologue of Bcl-2 and that it played a role in cell survival that Bcl-2 was identified as a regulator of apoptosis (Hengartner and Horvitz, 1994).

Today bcl-2 is also known as the original member of a large and currently expanding family of proteins which, to date, consists of numerous members (reviewed in Reed, 1998; Strasser et al., 2000). Although these proteins are grouped together and are classified as such due to the presence of at least one of four conserved BH (bcl-2 homology) domains, BH1, BH2, BH3, and BH4, the most potent inducers of apoptosis are those proteins that contain only the BH3 domain (reviewed in Reed, 1998; Bouillet and Strasser, 2002; Borner, 2003). Accordingly proteins such as Bad and Bid convey the strongest death signals while other proteins such as Bax, Bak, and Bcl-x_s, which possess one or more additional BH domains, are somewhat weaker among the pro-apoptotic family members (reviewed in Reed, 1998; Bouillet and Strasser, 2002; Borner, 2003). The ability of these proteins to signal an apoptotic response has been shown to be dependent on the ratio of these proteins to those that are anti-apoptotic (including Bcl-2, Bcl-x_L) as well as their ability to homo- and heterodimerize with each other. (Dimerization among the family members is actually a characteristic of this protein family; Oltvai et al., 1993; Oltvai and Korsmeyer, 1994.) The membrane localization of Bcl-2 family proteins, combined with the determination of the three-dimensional structure of certain family members has lead to the notion that these proteins may function as pores or channels within the membrane. This has prompted the current hypothesis that they regulate mitochondrial membrane integrity and cytochrome c release (Munchmore et al., 1996). Accordingly recent studies have demonstrated that Bcl-2, Bcl-x_L, and bax possess pore-forming capacities, in vitro and that ectopic expression of Bcl-2 or Bcl-X_L can inhibit the release of cytochrome c from mitochondria. Overexpression of the pro-apoptotic Bax protein, on the other hand, resulted in the release of cytochrome c (Manon et al., 1997). Although the mechanism by which these proteins regulate this phenomenon is not well understood, it is thought that Bcl-2 proteins are regulators of the mitochondrial permeability transition pore and/or that they regulate other apoptotic molecules.

Subversion of Apoptotic Pathways by Oncogenes in Hematopoietic Cells

Cells of hematopoietic lineages predominantly transduce their growth signals via cytokine receptors. Molecular cloning of these receptors and subsequent structure-function studies has revealed that unlike growth-factor receptors, several cytokine receptors lack a cytoplasmic kinase domain. Nevertheless, interaction of a cytokine with its receptor rapidly induces tyrosine phosphorylation of the receptors and cellular proteins, suggesting that these receptors transmit their signals through tyrosine kinases. During the past decade, new evidence has emerged to indicate that most cytokines transmit their signals via a new family of tyrosine kinases termed Janus Kinases (JAKs) as well as Src kinases (Ihle, 1995; Chaturvedi et al., 1998).

The JAK family differs markedly from other classes of tyrosine kinases by the presence of an additional pseudokinase domain. To date, this family consists of four members: JAK1, JAK2, JAK3, and TYK2. Although JAKs do not contain the SH2 and SH3 domains that characterize Src family tyrosine kinases, they do possess seven highly conserved JAK homology (JH) domains upstream of the kinase domains. These kinases, either alone or in conjunction with each other, appear to be responsible for effects mediated by several cytokines and neurokines including interleukins, interferons, erythropoietin, prolactin, growth hormone, oncostatin M (OSM), and ciliary neurotrophic factor (CNTF; reviewed in Rane and Reddy, 2002).

Current models suggest that in the case of single chain cytokine receptors, such as EPO, growth hormone, prolactin and GCSF, the interaction of cytokines with their receptors induces receptor dimerization. Dimerization increases the affinity of the cytoplasmic domain of the receptor for JAK kinases, resulting in a ligand-dependent increase of a complex containing the receptors and JAKs (Fig. 17.9) (Ihle, 1995). This association seems to require the membrane-proximal cytoplasmic domain of the receptor and results in the activation of the JAKs through an event associated with tyrosine phosphorylation. The activated kinases appear to then phosphorylate the receptors as well as cellular substrates, which regulate a wide variety of responses in cells. A surge of recent studies has also demonstrated that cytokine signaling is associated with the phosphorylation of STAT proteins (Darnell, 1997). In a resting cell these proteins remain unphosphorylated and are localized in the cytoplasm. Stimulation of cells with cytokines results in the phosphorylation of STATs, which seems to be mediated by both Src kinases and JAKs. The phosphorylated STAT proteins then migrate to the nucleus, where they bind to promoters that contain STAT-responsive elements (Fig. 17.9).

It is now well established that several oncogenic tyrosine kinases have profound effects on the cytokine dependence of hematopoietic cells and that the underlying mechanism is associated with the constitutive activation of STAT proteins. In a normal 32Dcl3 myeloblastic cell, stimulation with IL-3 results in the phosphorylation of JAK1 and 2, activation

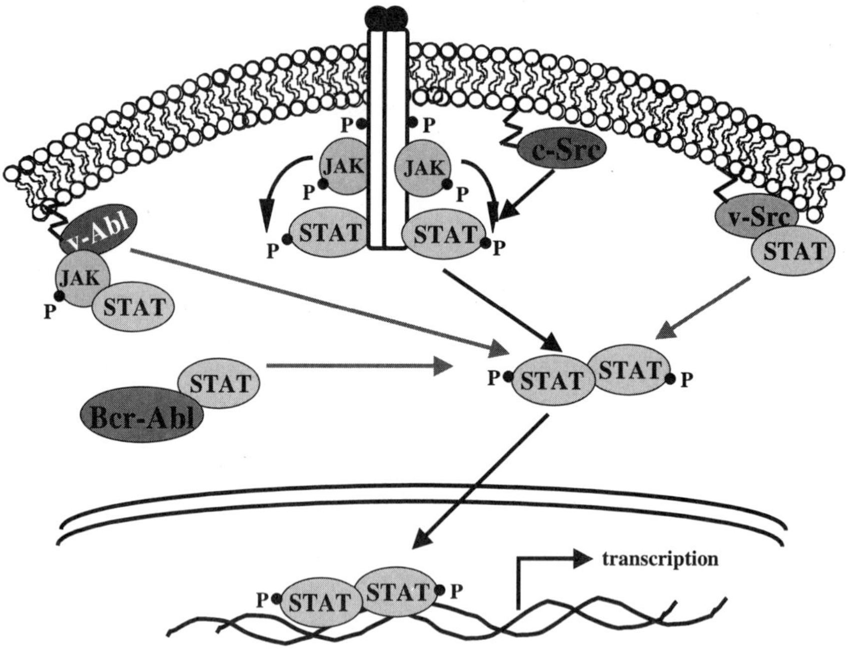

Figure 17.9. Activation of STAT proteins in response to cytokine stimulation. Cytokine binding induces receptor dimerization and the activation and phosphorylation of two families of tyrosine kinases, termed JAKs and Src kinases. These kinases subsequently phosphorylate a family of transcription factors, termed STATs. Once phosphorylated, these proteins dimerize and translocate to the nucleus where they bind to DNA and transactivate transcription of genes whose promoters contain STAT-responsive elements. Oncoproteins such as v-Src, v-Abl, and Bcr-Abl have the ability to activate STAT proteins directly, effectively bypassing the need for cytokines and causing constitutive activation of this pathway. The gray arrows indicate the oncogenic pathways that result in STAT phosphorylation.

of Src kinase activity, and nuclear translocation of STATs 1, 3, and 5. In v-Src transformed 32Dcl3 meyloblastic cells, the v-Src protein appears to mediate constitutive activation of STATs 1, 3, and 5, leading to IL-3 independent proliferation of these cells, suggesting that v-Src mimics the signaling pathways which are normally triggered by cytokine stimulation (Chaturvedi et al., 1997).

Similarly A-MuLV primarily induces cytokine-independent tumors of pre-B lymphocytic origin that are dependent on the kinase activity of v-Abl (Rosenberg, 1994). The mechanism that underlies the cytokine independence of these cells is also the constitutive activation of STAT proteins, namely STATs 5 and 6 (reviewed in Danial and Rothman, 2000). In these cells constitutive STAT activation is further correlated with the kinase activities of JAKs 1 and 3 (Danial et al., 1995), which are also dependent on the activity of v-Abl. Because JAK proteins can physically associate with v-Abl, (Danial et al., 1998) suggesting that they may be v-Abl substrates, it is possible that JAK kinases may relay the signal between v-Abl and STAT proteins (Fig. 17.9). It is also possible that v-Abl may directly phosphorylate STAT proteins, as is seen in the case of v-Src and BCR-ABL (Chaturvedi et al., 1997; reviewed in Danial and Rothman, 2000) (Fig. 17.9).

STAT5 is constitutively activated in BCR-ABL expressing cell lines and in some cells isolated from acute lymphocytic leukemia (ALL) patients (Chai et al., 1997). However, in cells isolated from some CML patients, STAT1 is the predominantly activated STAT. BCR-ABL, despite its homology to v-ABL, has not been shown to physically associate with JAKs, although JAK1 is constitutively activated in some BCR-ABL$^+$ cells (reviewed in Danial and Rothman, 2000). BCR-ABL, in an analogous situation to v-Src and STAT3 (Chaturvedi et al., 1997), has been shown to physically associate with STAT5 via its SH3 and SH2 domains, suggesting that it can directly activate STAT proteins (Nieborowska-Skorska et al., 1999). This observation also explains the inability of v-Abl to associate with STATs, since this oncoprotein lacks an SH3 domain.

Subversion of Cell Cycle Checkpoints by Oncogenes

Most cells, unless they have received a stimulus to proliferate or differentiate, remain in a resting state, termed G_0. However, when additional cells are required, as is frequently the case in the hematopoietic system, extracellular stimuli trigger cells to enter the G_1 phase of the cell cycle and become committed to cell division. It is at a late point in the G_1 phase that a potentially dividing cell reaches the "restriction point" (Pardee, 1974). Provided that conditions are conducive to proliferation, the cell proceeds past this checkpoint and enters S phase. If these conditions have not been met, then the cell will initiate apoptotic pathways. An absolute prerequisite for cell growth is the duplication of its genetic material, which occurs during the S phase. Once the DNA has been replicated, the cellenters a second resting phase called G_2, where it "ascertains" whether this process has been correctly executed during the second checkpoint, and provided that it has, the cell enters the mitotic phase where it completes final cell division. Oncoproteins as well as malignant cells can effectively override one or both intrinsic checkpoints.

The phosphorylation/inactivation of the retinoblastoma (Rb) tumor-suppressor protein is perhaps the most widely studied event in restriction point regulation. Cyclin-dependent kinases (CDK), CDK4, and CDK6, in conjunction with D-type cyclins, and CDK2, in conjunction with cyclins E or A, phosphorylate Rb proteins. Rb proteins suppress proliferation by associating with E2F transcription factors and preventing the transactivation E2F-responsive genes. Because these targets include proteins that control entry into S phase and DNA replication, active, unphosphorylated Rb effectively prevents cell cycle progression.

Phosphorylation/inactivation of Rb by CDKs prevents the association with E2F family proteins, thereby allowing the cell to proceed through S phase. CDKs are regulated by several mechanisms, including association with cyclin kinase inhibitors (CKIs). One class of CKIs includes $p21^{waf1/cip1}$, $p27^{kip1}$, and p57, all of which associate with G_1 cyclin/CDK complexes and act as CDK2 inhibitors. A second class of CKIs includes $p16^{INK4A}$, $p15^{INK4B}$, $p18^{INK4C}$, and $p19^{INK4D}$ and inhibits CDK4 and CDK6. Forced expression of CKIs induces G_1 arrest, and these proteins have therefore been shown to play key roles in growth suppression (reviewed in Donjerkovic and Scott, 2000).

Cell cycle restriction points, which function to prevent oncogenesis, are effectively overridden by transforming oncoproteins. For example, *v-src* has been shown to suppress the expression of the $p27^{Kip1}$ cyclin-dependent kinase inhibitor, resulting in a shorter G_1 phase, and to render cells unable to enter G_0 when they are deprived of growth factors (Johnson et al., 1998; Riley et al., 2001). The ability of v-*src* to cause cells to loose the ability to enter and remain in G_0 is further demonstrated by the use of a temperature-sensitive mutant in quiescent cells that, upon shift to the nonpermissive temperature, activates cyclin D/CDK4, 6, cyclin E/CDK2, and cyclin A/CDK2 (Riley et al., 2001) (Fig. 17.10). The net result is a cell that is forced to proceed through G_1 and enter S phase. These effects on the cell cycle require proteins that are downstream of Src. For example, induction of PI-3 kinase also triggers entrance into S phase and the activation of CDKs 4 and 2, and the use of PI-3 kinase inhibitors has been shown to result in decreased expression of p27 (Kippel et al., 1998). Other studies have also shown that Akt can modulate cell cycle progression by a similar mechanism. p27 expression is regulated in part by forkhead transcription factors that are substrates for Akt. Phosphorylated AFX is cytosolic, and in certain cell types it cannot upregulate p27 expression under conditions of sustained Akt activity. One study has also provided evidence that p27 is a substrate for Akt and that phospho-p27 is retained in the cytoplasm where it is unable to inhibit transition through the cell cycle (Collado et al., 2000; Graff et al., 2000).

p27 function is also deregulated as a result of constitutive Myc expression, whereby p27 is sequestered by cyclin D2-CDK4 complexes. Although the exact mechanisms by which Myc overrides G_1 arrest are not well defined, both cyclin D2 and CDK4 have been identified as Myc target genes (Bouchard et al., 1999; Hermeking et al., 2000). Myc is also capable of repressing the $p21^{waf1/cip1}$ promoter (Gartel et al., 2001), thereby forcing cell cycle progression through G_1. The ability of Myc to force cells through G_1 however, is insufficient to induce transformation of primary fibroblasts, nor is it sufficient to promote factor-independent proliferation of hematopoietic cells (Hume et al., 1998). However, co-transfection of *myc* and *ras* does induce transformation (Land et al., 1986), and this has been attributed to the ability of both oncogenes to regulate rate-limiting steps of the cell cycle via their synergistic actions on CDK function.

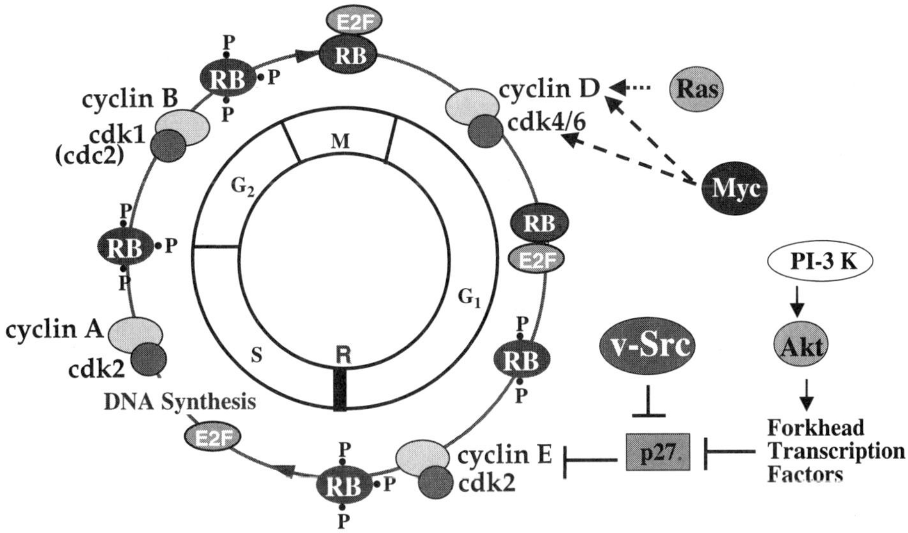

Figure 17.10. Modulation of the cell cycle by oncogenes. The cell cycle is divided into four phases, termed G_1, S, G_2, and M. Stimulation with growth factors induces proliferation and triggers the cell to enter the G_1 phase. This event requires the expression and assembly of different cyclin/CDK complexes that are formed at specific stages of the cell cycle and mediate cell cycle progression through S phase and mitosis. Of these, CDKs 4 and 6 mediate the progression through G_1, while CDK2 mediates the transition through S phase. Mitosis is then initiated by the CDK1-cyclin B complex. An important target of cyclin/CDK holoenzymes is pRb. pRb is phosphorylated in a cell cycle-dependent manner and hypophosphorylated forms of Rb constitute the active forms of this protein. pRb is hypophosphorylated in quiescent cells, and this form of pRb binds several cellular proteins and its phosphorylation results in the release of these associated proteins. One of these proteins is E2F-1, which appears to positively activate the transcription of genes whose products are required for S phase progression. Extracellular stimuli induce the cells to enter the G_1 phase of the cell cycle and become committed to cell division. It is at a point in late G_1 that a potentially dividing cell reaches the "restriction point," a time when the cell must determine whether the conditions are suitable for continued proliferation. Provided that conditions are conducive to proliferation, the cell proceeds past this checkpoint. Research in the past two decades has shown that malignant cells override this checkpoint through activation of oncogenes and disruption of growth suppressor gene pathways. Thus activation of the *ras* and *myc* genes has been found to activate the cyclin D/CDK4/6 complex, while the activation of v-Src and Akt pathways appear to subvert the negative regulation of cyclin E/CDK2 complexes.

Oncogenic Ras expression, like v-Src, can result in factor-independent cell cycle progression and proliferation (Mavilio et al., 1989). Stimulation of cells in G_0 with growth factors causes Ras activity to peak in two phases: the first is observed during early G_1 and is associated with activation of Raf, while the second is associated with the activation of PI-3 kinase and Akt and occurs in the middle of G_1 (Gille and Downward,

1999). Ras, like PI-3 kinase and Akt, forces progression through G_1 and downregulation of p27. Ras activity can also cause upregulation of cyclin D1 expression, although PI-3 kinase has also been shown to play a role in this process. Several more recent studies have also provided a link between Ras and Rb, whereby inhibition of Ras activity causes the accumulation of hypophosphorylated Rb and the induction of G_1 arrest (reviewed in Macaluso et al., 2002) (Fig. 17.10).

It is of interest to note that all of the viral homologues of protooncogenes whose protein products play a role in regulating the cell cycle in response to growth factor stimulation appear to do so by a similar mechanism, namely rapidly driving the cell through G_1 (Fig. 17.10). These findings indicate that overriding intrinsic controls of the cell cycle was key to the evolutionary success of retroviruses as transforming agents.

CONCLUSION

In just two decades we have come a long way in our understanding of the molecular mechanisms associated with cell growth, differentiation, and oncogenesis. It is becoming increasingly clear that aberrations in oncogenes, which act as positive regulators of growth, and tumor-suppressor genes, which act as negative regulators of growth, contribute to the development of the neoplastic state.

Two phenomena in the field of oncogene research are especially interesting. The first is that multiple transforming retroviruses have transduced identical oncoproteins. The second is that many oncoproteins, despite their structural unrelatedness, inappropriately activate identical signaling pathways in order to induce transformation. From these similarities and overlaps we have yet to define the precise pathways that lead to controlled proliferation versus uncontrolled growth and tumorigenesis. It is already apparent that cell proliferation and differentiation are interrelated and that mechanisms that lead to a block in cell differentiation result in increased proliferation of cells and ultimately into neoplasia. It is not unreasonable to hope that a more detailed understanding of the signal transduction pathways that are associated with cell growth and differentiation will provide us with clues that allow us to override the deregulated proliferation and block in differentiation that are seen in tumor cells. This is hinted by recent studies that show that Gleevec®/STI571/imatinib mesylate, a tyrosine kinase inhibitor, selectively inhibits the catalytic activity of BCR-ABL, thereby allowing the selective killing of CML tumor cells (reviewed in Druker, 2002). The prospect of development of additional therapeutic approaches that target signaling pathways that are unique to tumor cells for the treatment of cancer appears to be thus promising in the immediate future. The identification of oncogenes and delineation of their function clearly has had a major impact on the development of such approaches.

REFERENCES

Aaronson SA (1991): Growth factors and cancer. Science 254:1146–53.

Abelson HT, Rabstein LS (1970): Liposarcoma: Virus-induced thymic-independent disease in mice. Cancer Res 30:2213–22.

Ahmed NN, Franke TF, Bellacosa A, Datta K, Gonzalez-Portal ME, Taguchi T, Testa JR, Tsichlis PN (1993): The proteins encoded by c-*akt* and v-*akt* differ in post-translational modification, subcellular localization and oncogenic potential. Oncogene 8:1957–63.

Alitalo K, Schwab M, Lin CC, Varmus HE, Bishop JM (1983): Homogeneously staining chromosomal regions contain amplified copies of an abundantly expressed cellular oncogene (c-*myc*) in malignant neuroendocrine cells from a human colon carcinoma. Proc Nat Acad Sci USA 80:1707–11.

Ames BN (1979): Identifying environmental chemicals causing mutations and cancer. Science 204:587–93.

Baluda MA, Reddy EP (1994): Anatomy of an integrated avian myeloblastosis provirus: structure and function. Oncogene 9:2761–74.

Ballif BA, Blenis J (2001): molecular mechanisms mediating mammalian mitogen-activated protein kinase (MAPK) kinase (MEK)-MAPK cell survival signals. Cell Growth Diff 12:397–408.

Barbacid M (1987): Ras genes. Ann Rev Biochem 56:779–827.

Bellacosa A, Franke TF, Gonzalez-Portal ME, Datta K, Taguchi T, Gardner J, Cheng JQ, Testa JR, Tsichlis PN (1993): Structure, expression and chromosomal mapping of c-*akt*: Relationship to v-*akt* and its implications. Oncogene 8:745–54.

Bellacosa A, Testa JR, Staal SP, Tsichlis PN (1991): A retroviral oncogene, *akt*, encoding a serine-threonine kinase containing an SH2-like region. Science 254:274–7.

Biscardi JS, Ishizawar RC, Silva CM and Parsons Sj (2000): Tyrosine kinase signaling in breast cancer: Epidermal growth factor receptor and c-Src interactions in breast cancer. Breast Cancer Res 2:203–10.

Bollag G, McCormick F (1991): Regulators and effectors of ras proteins. An Rev Cell Biol 7:601–32.

Borner, C (2003): The Bcl-2 protein family: Sensor and checkpoints for life-or-death decisions. Mol Immunol 39:615–47.

Bouchard C, Thieke K, Maier A, Saffrich R, Hanley-Hyde J, Ansorge W, Reed S, Sicinski P, Bartek J, Eilers M (1999): Direct induction of cyclin D2 by Myc contributes to cell cycle progression and sequestration of p27. EMBO J 18:5321–33.

Bouillet P, Strasser A (2002): BH3-only proteins—Evolutionarily conserved pro-apoptotic Bcl-2 family members essential for initiating programmed cell death. J Cell Sci 115:1567–74.

Bowman T, Broome MA, Sinibali D, Wharton W, Pledger WJ, Sedivy JM, Irby R, Yeatman T, Courtneidge SA, Jove R (2001): Stat3-mediated Myc expression is required for Src transformation and PDGF-induced mitogenesis. Proc Nat Acad Sci USA 98:7319–24.

Brown MT, Cooper JA (1996): Regulation, substrates and functions of src. Biochem Biophys Acta 1287:121–49.

Buck M, Poli V, van der Geer P, Chojkier M, Hunter T (1999): Phosphorylation of rat serine 105 or mouse threonine 217 in C/EBP beta is required for hepatocyte proliferation induced by TGF alpha. Mol Cell 4:1087–92.

Burgering BM, Coffer PJ (1995): Protein kinase B (c-Akt) in phosphatidylinositol-3-OH kinase signal transduction. Nature 376:599–602.

Cardone MH, Roy N, Stennicke HR, Salvesen GS, Franke TF, Stanbridge E, Frisch S, Reed JC (1998): Regulation of cell death protease caspase-9 by phosphorylation. Science 282:1318–21.

Chai SK, Nichols GL, Rothman P (1997): Constitutive activation of JAKs and STATs in BCR-Abl-expressing cell lines and peripheral blood cells derived from leukemic patients. J Immunol 159:4720–8.

Chang HW, Aoki M, Fruman D, Auger KR, Bellacosa A, Tsichlis PN, Cantley LC, Roberts TM, Vogt PK (1997): Transformation of chicken cells by the gene encoding the catalytic subunit of PI 3-kinase. Science 276:1848–50.

Chaturvedi P, Reddy MV, Reddy EP (1998): Src kinases and not JAKs activate STATs during IL-3 induced myeloid cell proliferation. Oncogene 16:1749–58.

Chaturvedi P, Sharma S, Reddy EP (1987): Abrogation of interleukin-3 dependence of myeloid cells by the v-src oncogene requires SH2 and SH3 domains which specify activation of STATs. Mol Cell Biol 17:3295–304.

Chen RH, Sarnecki C, Blenis J (1992): Nuclear localization and regulation of erk and rsk-encoded protein kinases. Mol Cell Biol 12:915–27.

Chen R, Abate C, Blenis J (1993): Phosphorylation of the c-Fos transrepression domain by mitogen-activated protein kinase and 90-kDa ribosomal S6 kinase. Proc Natl Acad Sci USA 90:10952–6.

Collado M, Medema RH, Garcia-Cao I, Dubuisson ML, Barradas M, Glassford J, Rivas C, Burgering BM, Serrano M, Lam EW (2000): Inhibition of the phosphoinositide 3-kinase pathway induces a senescence-like arrest mediated by p27Kip1. J Biol Chem 275:21960–8.

Conney AH (1982): Induction of microsomal enzymes by foreign chemicals and carcinogenesis by polycyclic aromatic hydrocarbons: G. H. A. Clowes Memorial Lecture. Cancer Res 42:4875–82.

Cooper GM (1982): Cellular transforming genes. Science 218:801–6.

Cooper JA, Gould KL, Cartwright CA, Hunter T (1986): Tyr527 is phosphorylated in pp60c-src: Implications for regulation. Science 231:1431–4.

Courtneidge SA (2002): Role of Src in signal trnasduction pathways. Biochem Soc Trans 30:11–17.

Cuypers HT, Selton G, Quint W, Zijlstra M, Robanus-Maandag E, Boelens W, Van Wezenbeek P, Meleif C, Berns A (1984): Murine leukemia virus-induced T-cell lymphomagenesis: Integration of proviruses in a distinct chromosomal region. Cell 37:141–50.

Dalla-Favera R, Martinitti S, Gallo RC, Erikson J, Croce C (1983): Translocation and rearrangements of the c-myc oncogene locus in human undifferentiated B-cell lymphomas. Science 219:963–7.

Dalton S, Treisman R (1992): Characterization of Sap-1, a protei recruited by serum response factor to c-*fos* serum response element. Cell 68:607–12.

Danial NN, Losman JA, Lu T, Yip N, Krishnan K, Krolewski J, Goff SP, Wang JY, Rothman PB (1998): Direct interaction of Jak1 and v-Abl is required for v-Abl-induced activation of STATs and proliferation. Mol Cell Biol 18: 6795–804.

Danial NN, Pernis A, Rothman PB (1995): Jak-STAT signaling induced by the v-abl oncogene. Science 269:1875–7.

Danial NN, Rothman P (2000): JAK-STAT signaling activated by Abl oncogenes. Oncogene 10:2523–31.

Darnell JE, Jr. (1997): STATs and gene regulation. Science 277:1630–5.

Davies H, Bignell GR, Cox C, Stephens P, Edkins S, Clegg S, Teague J, Woffendin H, Garnett MJ, Bottomley W, Davis N, Dicks E, Ewing R, Floyd Y, Gray K, Hall S, Hawes R, Hughes J, Kosmidou V, Menzies A, Mould C, Parker A, Stevens C, Watt S, Hooper S, Wilson R, Jayatilake H, Gusterson BA, Cooper C, Shipley J, Hargrave D, Pritchard-Jones K, Maitland N, Chenevix-Trench G, Riggins GJ, Bigner DD, Palmieri G, Cossu A, Flanagan A, Nicholson A, Ho JW, Leung SY, Yuen ST, Weber BL, Seigler HF, Darrow TL, Paterson H, Marais R, Marshall CJ, Wooster R, Stratton MR, Futreal PA (2002): Mutations of the BRAF gene in human cancer. Nature 417:949–54.

Davis RJ (1993): The mitogen activated protein kinase signal transduction pathway. J Biol Chem 268:14553–6.

del Paso L, Gonzalez G, Page C, Herrera R, Nunez G (1997): Interleukin-3 induced phosphorylation of BAD through the protein kinase AKT. Science 278:687–9.

Der C, Krontiris TG, Cooper GM (1982): Transforming genes of human bladder and lung carcinoma cell lines are homologous to the ras genes of Harvey and Kirsten sarcoma viruses. Proc Nat Acad Sci USA 79:3637–40.

Dickson C, Smith R, Brookes G, Peters G (1990): Proviral insertions in the int-2 gene can generate multiple anomalous transcripts but leave the protein-coding domain intact. J Virol 64:784–93.

Donjerkovic D, Scott DW (2000): Regulation of the G_1 phase of the mammalian cell cycle. Cell Res 10:1–16.

Dougall WC, Qian X, Peterson NC, Miller MJ, Samanta A, Green MI (1994): The neu-oncogene: Signal transduction pathways, transformation mechanisms and evolving therapies. Oncogene 9:2109–23.

Druker BJ (2002): Inhibition of the Bcr-Abl tyrosine kinase as a therapeutic strategy for CML. Oncogene 21:8541–6.

Egan SE, Giddings BW, Brooks MW, Buday L, Sizeland AM, Weinberg RA (1993): Association of Sos Ras exchange protein with Grb2 is implicated in tyrosine kinase signal transduction and transformation. Nature 363:45–51.

Eva A (1988): New human oncogenes. In: EP Reddy, AM Skalka, T Curran (eds): The Oncogene Handbook. Amsterdam: Elsevier, pp 515–26.

Eves EM, Xiong W, Bellacosa A, Kennedy SG, Tsichlis PN, Rosner MR, Hay N (1998): Akt, a target of phosphatidylinositol 3-kinase, inhibits apoptosis in a differentiating neuronal cell line. Mol Cell Biol 18:2143–52.

Frame MC (2002): Src in Cancer: Deregulation and consequences for cell behaviour. Biochem Biophys Acta 1602:114–30.

Franke TF, Yang SI, Chan TO, Datta K, Kazlauskas A, Morrison DK, Kaplan DR, Tsichlis PN (1995): The protein kinase encoded by the Akt proto-oncogene is a target of the PDGF-activated phosphatidylinositol 3-kinase. Cell 5: 727–36.

Gartel AL, Ye X, Goufman E, Shianov P, Hay N, Najmabadi F, Tyner AL (2001): Myc represses the p21(WAF1/CIP1) promoter and interacts with Sp1/Sp3. Proc Natl Acad Sci USA 98:4510–15.

Gille H, Downward J (1999): Multiple ras effector pathways contribute to G(1) cell cycle progression. J Biol Chem 274:22033–40.

Gille H, Sharrocks AD, Shaw PE (1992): Phosphorylation of transcription factor p62TCF by MAP kinase stimulates ternary complex formation at the c-*fos* promoter. Nature 358:414–17.

Gonfloni S, Weijland A, Kretzschmar J, Superti-Furga G. (2000): Crosstalk between the catalytic and regulatory domains allows bidirectional regulation of Src. Nat Struct Biol 4:281–6.

Graff JR, Konicek BW, McNulty AM, Wang Z, Houck K, Allen S, Paul JD, Hbaiu A, Goode RG, Sandusky GE, Vessella RL, Neubauer BL (2000): Increased AKT activity contributes to prostate cancer progression by dramatically accelerating prostate tumor growth and diminishing p27Kip1 expression. J Biol Chem 275:24500–5.

Groffen J, Stephenson JR, Heisterkamp N, De Klein A, Bartam CB, Grosveld G (1984): Philadelphia chromosomal breakpoints are clustered within a limited region, bcr, on chromosome 22. Cell 36:93–9.

Guan KL, Figueroa C, Brtva TR, Zhu T, Taylor J, Barber TD, Vojtek AB (2000): Negative regulation of the serine/threonine kinase B-Raf by Akt. J Biol Chem 275:27354–9.

Hall WJ, Bean CW, Pollard M (1941): Transmission of fowl lekosis through chick embryos and adult chicks. Am J Vet Res 2:272–9.

Hayward WS, Neel BG, Astrin SM (1981): Activation of a cellular onc gene by promoter insertion in ALV-induced lymphoid leukosis. Nature 290:475–80.

Hengartner MO, Horvitz HR (1994): *C. elegans* cell survival gene ced-9 encodes a functional homolog of the mammalian proto-oncogene bcl-2. Cell 76:665–76.

Hermeking H, Rago C, Schuhmacher M, Li Q, Barrett JF, Obaya AJ, O'Connell BC, Mateyak MK, Tam W, Kohlhuber F, Dang CV, Sedivy JM, Eick D, Vogelstein B, Kinzler KW (2000): Identification of CDK4 as a target of c-MYC. Proc Natl Acad Sci USA 97:2229–34.

Hume CR, Nocka KH, Sorrentino V, Lee JS, Fleissner E (1998): Constitutive c-*myc* expression enhances the response of murine mast cells to IL-3, but does not eliminate their requirement for growth factors. Oncogene 2:223–6.

Hunter T (1997): Oncoprotein Networks. Cell 88:333–46.

Ihle JN (1995): The Janus protein tyrosine kinase family and its role in cytokine signaling. Adv Immunol 60:1–35.

Jhappan C, Gallahan D, Stahle C, Chu E, Smith GH (1992): Expression of an activated notch-related int-3 transgene interferes with cell differentiation and induces neoplastic transformation in mammary and salivary glands. Genes Dev 6:345–55.

Joel PB, Smith J, Sturgill TW, Fisher TL, Blenis J, Lannigan DA (1998): pp90rsk1 regulates estrogen receptor-mediated transcription through phosphorylation of Ser-167. Mol Cell Biol 18:1978–84.

Johnson D, Frame MC, Wyke JA (1998): Expression of the v-Src oncoprotein in fibroblasts disrupts normal regulation of the CDK inhibitor p27 and inhibits quiescence. Oncogene 16:2017–28.

Johnson GL, Lapadat R (2002): Mitogen-activated protein kinase pathways mediated by ERK, JNK, and p38 protein kinases. Science 298:1911–12.

Kim AH, Khursigara G, Sun X, Franke TF, Chao MV (2001): Akt phosphorylates and negatively regulates apoptosis signal-regulating kinase 1. Mol Cell Biol 21:892–901.

Klippel A, Escobedo MA, Wachowicz MS, Apell G, Brown TW, Giedlin MA, Kavanaugh WM, Williams LT (1998): Activation of phosphatidylinositol 3-kinase is sufficient for cell cycle entry and promotes cellular changes characteristic of oncogenic transformation. Mol Cell Biol 18:5699–711.

Kipreos ET, Wang JY (1992): Cell cycle-regulated binding of c-Abl tyrosine kinase to DNA. Science 256:382–5.

Land H, Chen AC, Morgenstern JP, Parada LF, Weinberg RA (1986): Behavior of *myc* and *ras* oncogenes in transformation of rat embryo fibroblasts. Mol Cell Biol 6:1917–25.

Lowenstein EJ, Daly RJ, Batzer AG, Li W, Margolis B, Lammers R, Ullrich A, Skolnik EY, Bar-Sagi D, Schlessinger J (1992): The SH2 and SH3 domain-containing protein GRB2 links receptor tyrosine kinases to ras signaling. Cell 70:431–42.

Luttrell DK, Luttrell LM, Parsons SJ (1988): Augmented mitogenic responsiveness to epidermal growth factor in murine fibroblasts that overexpress pp60c-src. Mol Cell Biol 8:497–501.

Maa MC, Leu Th, McCarley DJ, Schatzman RC, Parsons SJ (1995): Potentiation of EGF-mediated oncogenesis by c-Src: Implications for the etiology of multiple human cancers. Proc Nat Acad Sci USA 92:6981–5.

Macaluso M, Russo G, Cinti C, Bazan V, Gebbia N, Russo A (2002): Ras family genes: An interesting link between cell cycle and cancer. J Cell Physiol 192: 125–30.

Manon S, Chaudhuri B, Guerin M (1997): Release of cytochrome c and decrease of cytochrome c oxidase in Bax-expressing yeast cells, and prevention of these effects by coexpression of Bcl-xL. FEBS Lett 415:29–32.

Mavilio F, Kreider BL, Valtieri M, Naso G, Shirsat N, Venturelli D, Reddy EP, Rovera G (1989): Alteration of growth and differentiation factors response by Kirsten and Harvey sarcoma viruses in the IL-3-dependent murine hematopoietic cell line 32D C13(G). Oncogene 4:301–8.

McCormick F (1994): Activators and effectors of ras p21 proteins. Curr Opin Genet Dev 4:71–6.

McPherson RA, Harding A, Roy S, Lane A, Hancock JF (1999): Interactions of c-Taf-1 with phosphatidylserine and 14-3-3. Oncogene 18:3862–9.

Mercer KE, Pritchard CA (2003): Raf proteins and cancer: B-Raf is identified as a mutational target. Biochim Biophys Acta 1653:25–40.

Mori S, Ronnstrand L, Yokote K, Engstron A, Courtneidge SA, Claesson-Welsh L, Heldin CH (1993): Identification of two juxtamembrane autophosphorylation sites in the PDGF beta-receptor; involvement in the interaction with Src family tyrosine kinases. EMBO J 12:2257–64.

Munchmore SW, Sattler M, Liang H, Meadows RP, Harlan JE, Yoon HS, Nettesheim D, Chang BS, Thompson CB, Wong SL, Ng SL, Fesik SW (1996): X-ray and NMR structure f human Bcl-xL, an inhibitor of programmed cell death. Nature 381:335–41.

Nicholson KM, Anderson NG (2002): The protein kinase B/Akt signaling pathway in human malignancy. Cell Signal 14:381–95.

Nieborowska-Skorska M, Wasik MA, Slupianek A, Salomoni P, Kitamura T, Calabretta B, Skorski T (1999): Signal transducer and activator of transcription (STAT)5 activation by BCR/ABL is dependent on intact Src homology (SH)3 and SH2 domains of BCR/ABL and is required for leukemogenesis. J Expr Med (189):122912–42.

Nowell PC, Hungerford DA (1960): A minute chromosome in human granulocytic leukemia. Science 132:1497–9.

Oh I, Reddy EP (1999): The *myb* gene family in growth, differentiation and apoptosis. Oncogene 18:3017–33.

Okada M, Nakagawa H (1989): A protein tyrosine kinase involved in regulation of pp60c-src function. J Biol Chem 264:20886–93.

Olivier JP, Raabe T, Henkemeyer M, Dickson B, Mbamalu G, Margolis B, Schlessinger J, Hafen E, Pawson T (1993): A *Drosophila* SH2-SH3 adaptor protein implicated in coupling the sevenless tyrosine kinas e to an activator of Ras guanine nucleotide exchange, Sos. Cell 73:179–91.

Oltvai ZN, Korsmeyer SJ (1994): Checkpoints of dueling dimers foil death wishes. Cell 79:189–92.

Oltvai ZN, Millman CL, Korsmeyer SJ (1993): Bcl-2 heterodimerizes in vivo with a conserved homolog, Bax, that accelerates programmed cell death. Cell 74: 609–19.

Oppi C, Shore SK, Reddy EP (1987): Nucleotide sequence of testis-derived c-*abl* cDNAs: Implications for testis-specific transcription and *abl* oncogene activation. Proc Nat Acad Sci USA 84:8200–4.

Pardee AB (1974): A restriction point for control of normal animal cell proliferation. Proc Natl Acad Sci USA 71:1286–90.

Patel G, Kreider B, Rovera G, Reddy EP (1993): v-*myb* blocks granulocyte colony-stimulating factor-induced myeloid cell differentiation but not proliferation. Mol Cell Biol 13:2269–76.

Peters G (1988): The int oncogenes. In: EP Reddy, AM Skalka, T Curran (eds): The Oncogene Handbook. Amsterdam: Elsevier, pp 487–94.

Rane SG, Reddy EP (2002): JAKs, STATs and Src kinases in hematopoiesis. Oncogene 21:3334–58.

Rapp UR, Goldsborough MD, Mark GE, Bonner TI, Groffen J, Reynolds FH, Jr, Stephenson J (1983): Structural and biological activity of v-*raf*, a unique oncogene transduced by a retrovitus. Proc Nat Acad Sci USA 80:4218–22.

Reddy EP, Reynolds RK, Santos E, Barbacid M (1982): A point mutation is responsible for the acquisition of transforming properties by the T24 human bladder carcinoma oncogene. Nature 300:149–52.

Reddy EP, Reynolds RK, Watson DK, Schultz RA, Lautenberger J, Papas TS (1983): Nucleotide sequence analysis of the proviral genome of avian myelocytomatosis virus (MC29). Proc Nat Acad Sci USA 80:2500–4.

Reddy EP, Smith MJ, Srinivasan A (1983): Nucleotide sequence of Abelson murine leukemia virus genome: Structural similarity of its transforming gene product to other onc gene products with tyrosine-specific kinase activity. Proc Nat Acad Sci USA 80:3623–7.

Reed JC (1998): Bcl-2 family proteins. Oncogene 17:3225–36.

Riley D, Carragher NO, Frame MC, Wyke JA (2001): The mechanism of cell cycle regulation by v-Src. Oncogene 20:5941–50.

Rous P (1911): A sarcoma of the fowl transmissible by an agent separable from tumor cells. J Expr Med 13:397–411.

Roche S, Koegl M, Barone MV, Roussel MF, Courtneidge SA (1995): DNA synthesis induced by some but not all growth factors requires Src family tyrosine kinases. Mol Cell Biol 15:1102–9.

Rosenberg N (1994): Abl-mediated transformation, immunoglobulin gene rearrangements and arrest of B lymphocyte differentiation. Semin Cancer Biol 5:95–102.

Rozakis-Adcock M, Fernley R, Wade J, Pawson T, Bowtell D (1993): The SH2 and SH3 domains of mammalian Grb2 couple the EGF receptor to the Ras activator mSos1. Nature 363:83–5.

Rozek D, Pfizer GP (1993): In vitro protein-DNA interaction at the c-jun promoter: Preformed complexes mediate the UV response. Mol Cell Biol 13: 5490–9.

Santos E, Tronick SR, Aaronson SA, Pulciani S, Barbacid M (1982): T24 human bladder carcinoma oncogene is an activated form of the normal human homologue of BALB- and Harvey-MSV transforming genes. Nature 298: 343–7.

Sassone-Corsi P, Mizzen CA, Cheung P, Crosio C, Monaco L, Jacquot S, Hanauer A, Allis CD (1999): Requirement of Rsk-2 for epidermal growth factor-activated phosphorylation of histone H3. Science 285:886–91.

Shaulian E, Karin M (2002): AP-1 as a regulator of cell life and death. Nat Cell Biol 4:E131–6.

Shen-Ong GLC, Potter M, Mushinski JF, Reddy EP (1984): Activation of the c-*myb* locus by viral insertional mutagenesis in plasmacytoid lymphosarcomas. Science 226:1077–80.

Shore SK, Bogart SL, Reddy EP (1990): Activation of murine c-abl protooncogene: Effect of a point mutation on oncogenic activation. Proc Nat Acad Sci USA 87:6502–6.

Shtivelman E, Lifshitz RP, Gale RP, Canaani E (1985): Fused transcript of abl and bcr genes in chronic myelogenous leukaemia. Nature 315:550–3.

Songyang Z, Baltimore D, Cantley LC, Kaplan DR, Franke TF (1997): Interleukin 3-dependent survival by the Akt protein kinase. Proc Natl Acad Sci USA 94:11345–50.

Staal SP (1987): Molecular cloning of the akt oncogene and its human homologues AKT1 and AKT2: Amplification of AKT1 in a primary human gastric adenocarcinoma. Proc Natl Acad Sci USA 14:5034–7.

Stehelin D, Varmus HE, Bishop JM, Vogt PK (1976): DNA related to the transforming gene(s) of avian sarcoma viruses is present in normal avian DNA. Nature 260:170–3.

Strasser A, O'Connor LM, Dixit V (2000): Apoptosis signaling. An Rev Biochem 69:217–45.

Strover DR, Becker M, Liebetanz J, Lydon SB (1995): Src phosphorylation of the epidermal growth factor receptor at novel sites mediates interaction with Src and p85. J Biol Chem 270:15591–7.

Sukumar S, Notario V, Martin-Zanca D, Barbacid M (1983): Induction of mammary carcinomas in rats by nitroso-methylurea involves malignant activation of H-ras-1 locus by single point mutations. Nature 306:658–61.

Superti-Furga G, Gonfloni S (1997): A crystal milestone: The structure of regulated Src. Bioessays 6:447–50.

Tabin CJ, Bradley SM, Bargmann CI, Weinberg RA, Papageorge AG, Scolnick EM, Dhar R, Lowy DR, Chang EH (1982): Mechanism of activation of a human oncogene. Nature 300:143–9.

Takahashi M, Ritz J, Cooper MG (1985): Activation of a novel human transforming gene, ret, by DNA rearrangement. Cell 42:581–8.

Tantravahi R, Dudek H, Patel G, Reddy EP (1996): Murine myeloid leukemic cells with disrupted myb loci show splicing anomalies that account for heterogeneous sizes in myb proteins. Oncogene 13:1187–96.

Taub R, Kirsh I, Morton C, Lenoir G, Swan D, Tronick S, Aaronson S, Leder P (1982): Translocation of the c-*myc* gene into the immunoglobulin heavy chain locus in human Burkitt lymphoma and murine plasmacytoma cells. Proc Nat Acad Sci USA 79:7837–41.

Tekmal RR, Keshava N (1997): Role of MMTV integration locus cellular genes in breast cancer. Frontiers Biosci 2:519–26.

Tsujimoto Y, Gorham J, Cossman J, Jaffe E, Croce CM (1985): The t(14; 18) chromosome translocations involved in B cell neoplasms result form mistakes in VDJ joining. Science 229:1390–3.

Tzivion G, Luo Z, Avruch J (1998): A dimeric 14-3-3 protein is an essential cofactor for Raf kinase activity. Nature 394:88–92.

Vogt PK, Aoki M, Bottoli I, Chang H, Fu S, Hecht A, Iacovoni JS, Jiang B, Kruse U (1999): A random walk in oncogene space: The quest for targets. Cell Growth Diff 10:777–84.

Vanov IX, Mladenov Z, Nedyalkov S, Todorov TG, Yakimov M (1964): Bull Inst Pathol Comp Anim 10:5–38.

Von Hansemann D (1890): Uber asymmetrische Zellteilung in epithel Krebsen und der en biologische Bedeutung. Virchows Arch Pathol Anatom Physiol 119:299–326.

Wang HG, Rapp UR, Reed JC (1996): Bcl-2 targets the protein kinase Raf-1 to mitochondria. Cell 87:629–38.

Watson DK, Reddy EP, Duesberg PH, Papas TS (1983): Nucleotide sequence analysis of the chicken c-myc gene reveals homologous and unique coding regions by comparison with the transforming gene of avian myelocytomatosis virus MC29, delta gag-myc. Proc Nat Acad Sci USA 8:2164–50.

Weston K, Bishop JM (1989): Transcriptional activation by the v-*myb* oncogene and its cellular progenitor, c-myb. Cell 58:85–93.

Xing JD, Ginty D, Greenberg ME (1996): Coupling of the RAS-MAPK pathway to gene activation by RSK2, a growth factor-regulated CREB kinase. Science 273:959–63.

Zha J, Harada H, Yang E, Jockel J, Korsmeyer SJ (1996): Serine phosphorylation of death agonist BAD in response to survival factor results in binding to 14-3-3-not Bcl-X_L. Cell 87:619–28.

Zimmerman S, Moelling K (1999): Phosphorylation and regulation of Raf by Akt (protein kinase B). Science 286:1741–4.

CHAPTER 18

ROLE OF THE RETINOBLASTOMA FAMILY IN CELL CYCLE PROGRESSION AND GROWTH CONTROL

VALERIA MASCIULLO[1] and ANTONIO GIORDANO[2]
[1]Departments of Pathology, Anatomy, and Cell Biology, Thomas Jefferson University, Philadelphia, PA 19107
[2]Sbarro Institute for Cancer Research and Molecular Medicine, Center for Biotechnology, Temple University, Philadelphia, PA 19122

INTRODUCTION

The retinoblastoma gene family has three members, namely the product of the retinoblastoma gene (pRb), which is one of the most studied tumor-suppressor genes, and two related proteins, pRb2/p130 and p107, which often are collectively called the pocket proteins. pRb was originally identified as the gene that, when mutated in the germ line, results in the development of retinoblastoma (Friend et al., 1986). Inactivation of both copies of the Rb gene is associated with development of retinoblastoma in humans. Rb became of even greater interest to researchers when it was discovered that it was a specific target of small DNA tumor viruses, such as adenovirus E1A, SV40 large T antigen, and human papillomavirus E7 proteins. Because these oncoproteins are found to bind directly to pRb, the critical growth suppressive properties of the retinoblastoma gene, it suggested that these viruses force infected cells into DNA synthesis (Whyte et al., 1988). Further studies on these viral oncoproteins show that they are able to bind in the same region as two other cellular proteins, pRb2/p130 and p107 (Dyson et al., 1989; Li et al., 1993). Subsequent isolation of cDNAs encoding p107 (Ewen et al., 1991) and pRb2/p130 (Mayol et al., 1993) show that these proteins are not only structurally very similar to pRb but share many biochemical and functional properties.

Cell Cycle and Growth Control: Biomolecular Regulation and Cancer, Second Edition, Edited by Gary S. Stein and Arthur B. Pardee
ISBN 0-471-25071-6

Structural Characteristics of the Retinoblastoma Family

pRb/p105, pRb2/p130, and p107 are structurally and biochemically very similar proteins that are composed of an N-terminus region, a C-terminus region, and a pocket region. Rb family members share large sequence homology, expecially in the pocket region which is composed of two conserved functional domains (A and B) separated by a spacer region (J), which differs substantially among the three Rb proteins, pRb/p105, pRb2/p130, and p107. The N- and C-terminus regions share less sequence homology, suggesting that these fragments may endow specific functions for each family member.

The Pocket Region. The pocket domain, which confers a peculiar steric conformation to Rb family members, is mainly responsible for the specific protein interactions and their functional activity. In this region, pRb2/p130 and p107 are much more closely related to each other (50% of identity) than they are to pRb/p105 (30–35% identity). p107 and pRb2/p130 also share a motif in the spacer region that is not present in the pRb sequence, enabling them to form a strong binding site to cyclin A/CDK2 and to cyclin E/CDK2 complexes. A distinct kinase inhibitor domain specifically inhibiting cdk2 kinase activity in vitro also has been identified in the spacer region of pRb2/p130, suggesting that this member of the family can modulate kinase activity by itself.

The A and B pocket domains are targeted by proteins that carry a LxCxE peptide sequence such as oncoproteins from several DNA tumor viruses, the D-type cyclins, and histone deacetylase (HDACs). Simian virus 40 (SV40) large tumor antigen (T-Ag), JC virus T-Ag, adenovirus E1A, and human papillomavirus E7 oncoproteins all associate with Rb family members, and this binding is essential for transformation. It was shown that T-Ag carrying mutations in the LxCxE domain is unable to transform Rb-deficient mouse embryo fibroblasts, suggesting that the disrupted binding to Rb proteins renders the oncoprotein transformation defective. Conversely, it was shown that mutations within the A subdomain of pRb/p105 affect its binding to LxCxE viral oncoproteins at the B subdomain, suggesting that the A subdomain is able to affect the conformation and/or the accessibility of the B subdomain.

HDACs comprise a family of seven proteins, of which HDAC1 is the functional binding partner of pRb/p105 (Brehm et al., 1998; Magnaghi-Jaulin et al., 1998) and pRb2/p130 (Stiegler et al., 1998). Both Rb family members need the LxCxE-binding motif present in the pocket region (residues 379–792 for pRb) as well as a functional C-terminus region in order to form a stable complex with HDAC1 in vitro as well in vivo. E2Fs, pRb/p105, pRb2/p130, and HDAC1 were found as a complex in mammalian cells because the binding sites for E2F and HDAC1 are distinct from each other. Viral proteins are able to disrupt the interaction between pRb/p105 and HDAC1 in vitro by binding to Rb through a LxCxE motif (Lee et al., 1998), suggesting that the interaction between Rb family members and HDAC1 may be fundamental for linking chromatin remodeling to cell cycle control, and thus a main point of attack for oncogenic processes (Brehm and Kouzarides, 1999). The importance of the pocket domain for tumor suppression by pRb is highlighted by knowing that all mutations or deletions identified in cancer patients map in this region.

The pocket region itself also is a target of LxCxE-lacking proteins such as E2F transcriptional family of proteins, which are physiologically relevant Rb-binding proteins. The pocket region is required for Rb family members to repress polymerase III (Pol III) transcription both in vivo and in vitro. The binding of E7 oncoprotein in this region disrupts the interaction between the Rb proteins and a general Pol III factor called TFIIIB, inducing Pol III transcription in vivo. More recently (Ciarmatori et al., 2001) it has been shown that the pocket region of pRb2/p130 and pRb shares the ability to repress Pol-I transcription by binding and inhibiting the Pol-I specific factor UBF (upstream binding factor), whereas p107 seems ineffective in this system (see below).

The C-Terminus. The carboxy-terminus region differs in length from pRb/p105 and the two other family members p107 and pRb2/p130, the latter two being quite similar in amino acid sequence. This region contains the nuclear localization signal (NLS) and the domains involved in the binding to HDAC1 and cyclin/cdks complexes (see below). Mutations of the COOH terminus (exon 21 and/or 22) of pRb2/p130 have been found in several malignancies, including lung (Claudio et al., 2000b), Burkitt's lymphoma (Cinti et al., 2000b), and nasopharyngeal cancer (Claudio et al., 2000a).

The nuclear shuttling of pRb/p105 and pRb2/p130 has been reported to be under the control of a bipartite NLS consisting of basic residues in two clusters separated by a stretch of amino acids that frequently includes proline residues (Zacksenhaus et al., 1993; Cinti et al., 2000a). The NLS has been well-characterized in pRb/p105 and pRb2/p130 and provides an additional function for Rb family members as carriers for the E2Fs transcription factors. Two amino acid substitutions into these sites, or deletion of the NLS, may confine proteins to the cytoplasm, strongly impairing Rb protein function in cell growth (Cinti et al., 2000a). Recently (Chestukhin et al., 2002) two independent NLSs that could target reporter proteins to the nucleus were identified in the C-terminus of p130. Mutation of the C-terminus pRb2/p130 NLSs separately or together did not disrupt the nuclear localization of pRb2/p130 nor its ability to bind to and repress E2F activity. In contrast, deletion of the C-terminus of pRb2/p130 resulted in loss of binding to E2F-4 and the ability to induce growth arrest. This finding suggests that the functional importance of the p130 C-terminus is in its binding to E2F4 rather than in translocating pRb2/p130 in the nucleus as previously reported. In addition to the C-terminus NLSs, the intact pocket domain of pRb2/p130 itself was sufficient for nuclear translocation.

The N-Terminus. Little is known about the function of the NH2 terminus of Rb family members. B-myb recognizes an N-terminus p107 region, which overlaps the larger cyclin-binding domain. The binding of cyclin E/A to the major cyclin binding domain occurs at the N-terminus of pRb2/p130 (Hansen et al., 2001). Recently an N-terminus domain of pRb2/p130 that serves as a cytoplasmic retention signal also has been identified (Chestukhin et al., 2002).

Phosphorylation of Rb Family Members. The phosphorylation status of each Rb family member is essential for its functional regulation (see below) and is regulated by binding with specific cyclin/cdk complexes at the N-terminus, at the pocket, and at the C-terminus domain of the proteins (Hansen et al., 2001). An accurate analysis of Rb structure reveals the existence of several distinct phosphorylation sites; however, the exact number of serine and threonine residues of pRb that can be phosphorylated during the G_1 phase is not yet clear. Research from several groups over the last decade has identified between 10 to 12 of the 16 candidate CDK consensus sites of pRb as targets of cyclin D/cdk4 and /or cyclin E/cdk2 (Lees et al., 1991; Knudsen and Wang, 1996; Mittnacht, 1998). Recently 22 serine/threonine residues of pRb2/p130 phosphorylated in vivo have been mapped, and a few more remain to be identified, suggesting that the regulation of pRb2/p130 by phosphorylation is significantly more complex than that of pRb. Of the 22 phosphorylation sites, three localize close to the border between the N-terminal part and the A-pocket, six localize in the spacer region, another six localize in the C-terminal region, and seven sites localize in the B-pocket, within a sequence of pRb2/p130 not conserved in pRb and p107. The latter region, also known as the loop region, is probably linked to some unique function of pRb2/p130 (Canhoto et al., 2000), and it contains the only sites whose mutation affects phosphorylation of other residues. Only 3 of the 22 mapped residues are conserved in pRb, while 10 are conserved in p107, and 12 sites appear unique to pRb2/p130 (Hansen et al., 2001). Multiple kinase activities converge to phosphorylate pRb2/p130, including cdk4(6), cdk2, and unidentified non-cdk kinases. pRb2/p130 phosphorylation events are initiated by non-cdk kinases in early G_1 followed by cyclin D-cdk4(6), cyclin E-cdk2, and finally by non-cdk kinases, which require pre-phosphorylation of pRb2/p130 by cdks.

Functional Characteristics of the Retinoblastoma Family

Regulation of Cell Cycle Transition from G_1 through S Phase. The control of the G_1 through S phase transition in the cell cycle is an important checkpoint in regulating cell proliferation. Cell cycle progression is controlled by a precise sequence of events that determines the cell's decision to either continue proliferating or withdraw from the cycle and re-enter a quiescent state. This is controlled by the temporally regulated expression of cyclins, a large number of cell cycle phase-specific proteins.

pRb/p105, pRb2/p130, and p107 are critical targets for cyclins and cdks, and their phosphorylation is required for progression of the cell through G_1 and S phase. Underphosphorylated Rb family members are negative regulators of the cell cycle because they arrest cells during the

G_1 phase by repressing transcription of genes required for G_1 to S phase transition. Progression of cells through G_1 and S phase requires inactivation of Rb proteins by phosphorylation. Phosphorylation status of these proteins is regulated by CKIs binding to and inhibiting the cyclin/cdk complexes (see below).

The three Rb family members differ in expression patterns detected at various stages of the cell cycle. pRb/p105 is abundant at all times with only slight variations in expression levels but significant differences in its phosphorylation status. p107 expression is lost in cells withdrawn from the cell cycle but is high throughout the proliferative cell cycle (Stiegler et al., 1998). p107 is predominant during late G_1 phase through G_2/M, and its expression is strictly regulated by its E2F-dependent promoter. However, accumulation of p107 in late G_1 and S phase, and its inability to bind to the transcriptional repressor HBP1, indicate a contradictory role as a negative cell cycle regulator A (Nevins, 1998).

pRb2/p130 is highly expressed in nonproliferating cells. Although pRb is expressed in quiescent cells, the dominant presence of the pRb2/p130/E2F-4 complex is a good indicator of cells in a G_0 state. The promoter activity of pRb2/p130 does not fully explain the dramatic increase of this protein, which could be due to enhanced protein stability in arrested cells. Protein stability of pRb2/p130 was found to be regulated by 26S proteasome degradation after hyperphosphorylation in G_1 phase (Nevins, 1998). The hyperphosphorylated form of pRb2 is believed to be degraded by the ubiquitin-proteasome pathway because its level drops dramatically as cells exit G_0 and progress through G_1, and because it can be stabilized by proteasome inhibitors (Grana X. et al., 1998). pRb2/p130 was shown to be ubiquinated in vitro when phosphorylated on serine 672, and mutation of this residue to alanine increased the protein stability in vivo (Tedesco et al., 2002), providing additional evidence for its degradation in vivo by the ubiquitin-proteasome pathway.

The three pocket proteins are regulated by phosphorylation events. As a consequence of its phosphorylation and dephosphorylation events, pRb is considered the guardian of the G_1-S phase transition at the restriction point gate. Before G_1 transition starts, pRb is underphosphorylated, thus becoming active in repressing cell cycle progression. As the cells progress into G_1/S phase, pRb becomes hyperphosphorylated, causing the inactivation of its growth inhibitory function. pRb then maintains this hyperphosphorylated state throughout the cell cycle, becoming underphosphorylated once again at the end of the M phase (Weinberg, 1995).

The best-characterized targets of Rb family proteins are the E2F family of transcription factors, which have binding sites in the promoters of cell cycle genes important for progression of cells from G_1 to S phase. The Rb proteins repress gene transcription by directly binding to the transactivation domain of E2F and by binding to the promoter of these genes as a complex with E2F. The activity of E2F transcription factors also is regulated through phosphorylation by cyclin E/cdk2, which increases the transcriptional activity of E2F in G_1 phase, and by cyclin A/cdk2, which inhibits E2F binding to DNA in S phase. The Rb family members have different affinities for different E2F members. pRb complexes preferentially bind to E2F1-3 but also E2F-4 (Lees et al., 1993), while only p107/E2F-4 complexes were described in vivo (Beijersbergen et al., 1994). E2F-4 associates with pRb2/p130 during G0 and switches to p107 and pRb as the cells re-enter the G_1 phase (Moberg et al., 1996).

The recent generation (Sage et al., 2000) of triple knock out mouse embryonic fibroblasts (TKO MEFs) shows that these cells have a shorter cell cycle than wild type, single or double knockout control cells. TKO cells are resistant to G_1 arrest following DNA damage, contact inhibition or serum starvation. TKO cells do not undergo senescence in culture and show some features of transformation.

Interaction with Cyclin/CDK Complexes. Phosphorylation of retinoblastoma family members occurs as a consequence of the binding of cyclin/CDKs complexes with the C-terminus domain of the pocket proteins (Knudsen and Wang, 1996). During the G_1 phase of the cell cycle, CDKs phosphorylate all Rb-related members, which in turn release E2Fs from the complexes, thus inducing DNA synthesis (MacLachlan et al., 1995).

Several cyclin-dependent kinases are implicated in phosphorylation of Rb family members. Cyclin D1/CDK4-6 complexes perform the main phosphorylation of pRb and p107 and are negatively regulated by the cyclin-dependent kinase inhibitors (CKIs) (Bejiersbergen et al., 1995; Mittnacht, 1998; Paggi and Giordano, 2001). In vitro studies indicate that cyclin D3/CDK4 complexes are able to use pRb2/p130 as a substrate. pRb2/p130 also associates with cyclin D3 both in vitro and in vivo, suggesting that this complex is responsible for phosphorylation of pRb2/p130 (Dong et al., 1998).

pRb2/p130 and p107 family members, by contrast to pRb, are able to stably bind cyclin A/CDK2 and cyclin E/CDK2 complexes (Faha et al., 1992; Hannon et al., 1993; Li et al., 1993; Claudio et al., 1996). The functional consequences of this binding remain unclear. This could represent a simple mechanism by which pRb2/p130 is efficiently phosphorylated by the CDKs, or, this interaction could serve to target the kinase for another function (Hauser et al., 1997). It remains possible that p107 and pRb2/p130 function as cdk inhibitors either by sequestering cdk's away from other substrates or by using the N-terminal domain to reduce the activity of the associated kinases (Zhu et al., 1995). Two studies have shown that pRb2/p130 and p107 contain a p21-like kinase inhibitory domain that has been shown to inhibit CDK2 kinase activity both in vitro and in vivo for p107 (Zhu et al., 1995; Woo et al., 1997) and in vitro for pRb2/p130 (De Luca et al., 1997). We also identified a pRb2/p130 spacer region as inhibiting CDK2-dependent kinase activity, which correlates with the diminished CDK2 kinase activity observed in quiescent cells (De Luca et al., 1997a).

To explore the effects of pRb2/p130 induction on cyclin, CDKs and CKIs, we set up a tetracycline regulated system to control the expression of pRb2/p130 in JCV-induced hamster brain tumor cells (Howard et al.,

2000). In vivo induction of pRb2/p130 specifically inhibits cyclin A and cyclin E associated kinase activity and, in so doing, increases p27 levels. The last effect is probably because of inhibition of targeted proteolysis of p27 by cyclin E-CDK2 phosphorylation of p27. pRb2/p130 also decreases cyclin A while increasing cyclin E at both the transcriptional and protein level of expression. We hypothesize that cyclin E protein levels remain high in the presence of pRb2/p130 expression because cyclin E-associated kinase activity is known to be the rate-limiting factor involved in G_1-S phase transition (Koff et al., 1992). Both pRb2/p130 and cyclin E may exhibit negative feedback regulation of each other. pRb2/p130 is capable of maintaining cyclin E expression both at transcriptional level and by stabilizing the short half-life of cyclin E (Lew et al., 1991; Koff et al., 1992). Cyclin E is maintained in an inactive form by pRb2/p130 inhibition of cyclin E-associated kinase activity, both directly and by induction of the CKI p27. When the cell finally prepares for DNA replication, pRb2/p130 is inactivated, most likely by phosphorylation. This event diminishes inhibition of cyclin E/CDK2 kinase activity, which enhances pRb2/p130 phosphorylation and inactivation, as well as the phosphorylation and degradation of p27, allowing the cells to progress through the G_1/S phase transition. At this point cyclin E is degraded without the protection of pRb2/p130 which prevents endoreduplication, maintains DNA fidelity, and allows the cells to renew (Spruck et al., 1999).

Interaction with p53. Loss of function of both the p53 pathway and the retinoblastoma protein (pRb) pathway plays a significant role in the development of most human cancers. Loss of pRb results in lack of differentiation, hyperproliferation, and apoptosis, while loss of p53 desensitizes cells to checkpoint signals, including apoptosis (Yamasaki et al., 1998). Over the past 10 years, mouse genetics and gene expression profiling have led to major advances in understanding how the p53 and pRb pathways cooperate in regulating apoptosis, and thus developing tumors. The major amount of evidence clarifying the cooperation between p53 and pRb comes from several studies in gene knockout mice. Germ-line mutations in Rb or p53 predispose mice to tumor formation (Lee et al., 1992; Jacks et al., 1992; Clarke et al., 1992; Donehower et al., 1992). $Rb^{-/-}$ mice are not viable because of defects in hematopoiesis and neuronal cell death. A large amount of apoptotic cells are present in the lens fiber in the early embryos. There are no apoptotic cells in the abnormally developed lenses in the early embryos of the $Rb^{-/-}$, $p53^{-/-}$ mice (Morgenbesser et al., 1994; Williams et al., 1994), suggesting a cooperation between pRb and p53. Lack of functional pRb results in a failure of terminal differentiation possibly leading to hyperproliferation, while p53 induces apoptosis in the abnormally differentiated cells. Thus loss of both p53 and pRb/p105 may lead to tumorigenesis. However, compared with the viable $Rb^{+/-}$ mice and the $p53^{-/-}$ mice, different tumors were found in the $Rb^{+/-}$, $p53^{-/-}$ mice, suggesting that tumor development in some tissues requires loss of both Rb and p53 genes.

pRb/p105 acts as a transcriptional repressor by targeting the E2F transcription factors whose functions are required for entry into S phase. Recently (Lomazzi et al., 2002) showed that although E2F1 alone is not sufficient to induce S phase in diploid mice and human fibroblasts, increased E2F1 activity can result in S phase entry in diploid fibroblasts, in which the p53-mediated G_1 checkpoint is suppressed, or in primary mouse fibroblasts lacking pRb. These results indicate an overlapping role for p53 or pRb at the G_1 checkpoint in repressing E2F-induced S phase entry.

Interaction with Chromatin. The retinoblastoma family of proteins suppress cell growth by regulating E2F-dependent mRNA transcription, rRNA and tRNA transcription, and HDAC1 recruitment and chromatin packaging. Rb can actively repress transcription through several mechanisms: (1) by interfering with the assembly of the general transcription apparatus near the transcription-initiation site (i.e., binding to TAFII250, a component of the TFIID basal-transcription complex; Shao et al., 1995: (2) by interfering with the function of other DNA-bound transcriptional activators such as PU.1, MYC, and ELF1, by disrupting their interaction with the transcription machinery (Weintraub et al., 1995); and (3) by inducing the formation of a repressive chromatin structure that surrounds the promoter. The chromatin structure can be modified through the acetylation and deacetylation of lysine residues in the N-terminal tails of histones. Histones are basic proteins that form complexes with DNA to keep it well organized in loose or condensed nucleosomes. Whereas histone acetylation is believed to weaken the interaction between histones and DNA, histone deacetylation leads to formation of a repressive chromatin conformation (Grunstein, 1997), thus limiting the accessibility of transcription factor-binding sites.

Several groups have reported that pRb/p105 can bind to at least two members of the HDAC family, HDAC1 and HDAC2 (Brehm et al., 1998; Luo et al., 1998; Magnaghi-Jaulin et al., 1998). HDAC1 preferentially binds to the active, hypophosphorylated form of Rb (Luo et al., 1998). E2F1, Rb, and HDAC1 can form a trimeric complex, and Rb can recruit histone deacetylase activity in vitro (Brehm et al., 1998; Magnaghi-Jaulin et al., 1998). We demonstrated that pRb2/p130 also binds to HDAC1 (Stiegler et al., 1998), thus increasing the ability of pRb2/p130 to inhibit transcription of the E2F-dependent cyclin A promoter in vivo. p107 is also able to interact physically with HDAC1 in vivo through an LXCEX-like motif, similar to that used by viral transforming proteins to bind and inactivate pocket proteins. Indeed, p107 is able to interact simultaneously with HDAC1 and E2F4 (Ferreira et al., 1998).

These data suggest a novel mechanism of Rb family mediated repression (Fig. 18.1). A complex containing an hypophosphorylated Rb family member, HDAC, and E2F forms early in G_1 phase. Then HDAC, by deacetylating the nucleosomes surrounding the promoter, blocks access of transcription factors to their binding sites. During the S phase, Rb phosphorylation leads to dissociation of the E2F-RB-HDAC repressor

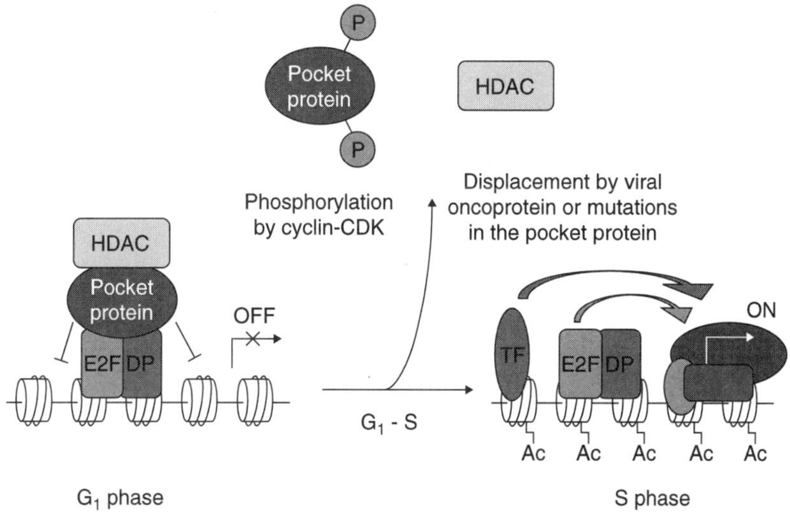

Figure 18.1. Model of pocket protein transcriptional repression during the cell cycle. To block transcription the pocket protein recruits, the histone deacetylase (HDAC) to the promoter of S phase-specific genes through a heterodimer is comprised of one E2F and one differentiation-regulated transcription factor polypeptide (DP). HDAC induce the formation of a repressive chromatin structure. During G_1, phosphorylation of the pocket protein by the cyclin-dependent kinases cause the release of HDAC. The hypoacetylated chromatin state is no longer maintained, and transcription factors (TFs) gain access to their binding sites and activate transcription. Mutations or viral inactivation of one of the pocket proteins can alterate this process, leading to cellular transformation.

complex, and E2F becomes available for transcription by interacting with basal elements of the transcription machinery or with histone acetyltransferases, such as CBP (Trouche and Kouzarides, 1996). Transcription of the genes encoding p107 and cyclin E is up-regulated in Rb-expressing osteosarcoma cells and in mouse embryo fibroblasts, respectively, when HDACs are inhibited by treatment of cells with TSA (Luo et al., 1998). However, Rb-induced repression of several synthetic promoters showed that not all of them are sensitive to treatment with the potent histone-deacetylase inhibitor trichostatin A (TSA). These data raise questions about the possibility that other mechanisms are involved in Rb-repressed transcription, such as the repression of transcriptional activators bound close to the E2F-Rb repressor complex.

Chromatin structure appears to be different in $Rb^{+/+}$ and $Rb^{-/-}$ MEFs, while chromatin from $Rb^{-/-}$ cells shows an open conformation (Herrera et al., 1996a). This effect could be attributable to the Rb ability to repress cyclin E transcription and consequently histone-H1 phosphorylation (Herrera et al., 1996a) or, alternatively, to Rb interaction with chromatin-modifying complexes other than HDAC complexes. It was shown recently that pRb has the potential to interact with different classes of chromatin-modifying complexes, such as SWI/SNF complex (Cairns,

1998), that are involved in transcriptional regulation. We reported in a recent study that pRb2/p130, p107-E2F4, and pRb2/p130-HDAC1 complexes all are inner nuclear-matrix-associated proteins, and that they localize to sites different from those of the pRb/p105 (Zini et al., 2001). A significantly larger amount of pRb2/p130, p107, E2F4, and their complexes were found in the interchromatin regions as compared with the heterochromatin regions. Because active transcriptional sites are in the less condensed interchromatin regions, it is not surprising that both Rb proteins and E2Fs, possibly associated with HDAC1, are more numerous at this location. We also labeled pRb2/p130, p107, E2F4, and their complexes in the nucleoli, where the rRNA genes and the DNA polymerase I (Pol I) are located (Zini et al., 2001). Both pRb/p105 and pRb2/p130 act as regulators of Pol I activity by inhibiting the UBF-mediated transcription. It has been shown that pRb/p105 competes with the hystone acetyltransferase CREB-binding protein (CBP) in recruiting UBF. The presence of both E2F4 and pRb2/p130-E2F4 complexes in the nucleoli could be due to a competition between E2F4 and UBF for binding with pRB2/p130. This hypothesis is supported further by the fact that both transcription factors share a binding site in the pRbs pocket region (Hannan et al., 2000). It is therefore conceivable that pRb/p105 and pRb2/p130 family members through their binding to HDACs and E2F transcription factors can control both processes of chromatin acetylation/deacetylation and rDNA transcription/repression. In agreement with this hypothesis, we recently demonstrated the ability of pRb2/p130 to repress the transcription of estrogen receptor α (ER- α) in a complex with chromatin-modifying enzymes (Macaluso et al., 2003). The role of p107 in the transcription remains controversial (Voit et al., 1997; Hannan et al., 2000; Pelletier et al., 2000). It is possible, however, that its role in the nucleolus is more to maintain chromatin condensation than to suppress rRNA transcription.

GROWTH SUPPRESSIVE PROPERTIES

The retinoblastoma family of proteins shows cell growth suppressive properties that are cell-type specific. The human osteosarcoma cell line SaOs-2 is growth arrested in the G_0/G_1 phase of the cell cycle by each of the retinoblastoma proteins (Claudio et al., 1994). In addition overexpression of pRb2/p130 in the glioblastoma cell line T98G can confer growth arrest, while pRb and p107 have no such effect (Claudio et al., 1994). On the other hand, C33A cells, which are derived from human cervical carcinoma, are insensitive to pRb/p105-mediated growth suppression but are sensitive to either p107 or pRb2/p130 overexpression (Zhu et al., 1993).

Hematopoietic cells are able to progress through the cell cycle in response to cytokines despite the presence of pRb/p105, whereas their growth is strongly blocked by ectopic expression of pRb2/p130 (Hoshikawa et al., 1998). Enforced expression of pRb2/p130, but not pRb, inhibits the myeloid cell proliferation that is concomitantly associated with granulocytic differentiation (Mori et al., 1999). In primary human T cells and in serum-starved mouse fibroblasts, the quiescent status is maintained by pRb2/p130 rather than by pRb/p105. On the other hand, T lymphocytes proliferate normally in culture and possess a normal cell-mediated immune function in vivo in pRb2/p130-deficient mice (Mulligan et al., 1998). This suggests that, besides having overlapping roles, some fundamental differences exist in the molecular pathways activated by each of the retinoblastoma family member in growth control.

Rb FAMILY AND DEVELOPMENT

The most convincing evidence about the role of Rb family members in development comes from an analysis of knockout mice. In contrast to mice homozygously deleted for Rb (Lee et al., 1992), which die between days 13 and 15 of gestation with defects in erythroid, neuronal, and lens development, mice deleted for p107 or pRb2/p130 develop normally. It is likely that either pRb2/p130 or p107 can compensate for each other in maintaining a normal phenotype because of overlapping functions. This relationship is supported by data showing that mice lacking both p107 and pRb2/p130 die shortly after birth and acquire developmental defects different from the pRb-deleted mice (Cobrinik et al., 1996). It has been observed recently that loss of Rb leads to excessive proliferation of trophoblast cells with severe disruption of the normal placental architecture and function, suggesting a key function for Rb in extra-embryonic cell lineages. The rescue of Rb-null pups by conditional knockout strategies abolished most of the neurological and erythroid abnormalities responsible for the embryonic lethality of $Rb^{-/-}$ mice, which were carried to term but died soon after birth (Wu et al., 2003).

Embryos deficient in pRb/p107 or pRb/pRb2 exhibit a phenotype that is similar to that of their $Rb^{-/-}$ littermates. However, they die two days earlier and show accelerated apoptosis in the erythroid compartment or in the nervous system (Lee et al., 1996; Lipinsky et al., 1999). This acceleration of the mutant phenotype suggests that p107 and pRb2/p130 can partially compensate for the Rb role during development. On a mixed 129/SvxC57BL/6 genetic background, $p130^{-/-}$, $p107^{-/-}$ mice die just after birth with defects in bone formation and anomalies in chondrocyte proliferation (Cobrinik et al., 1996). In addition the generation of chimeric animals by embryionic stem (ES) cells lacking of both pRb and p107 (Robanus-Maandag et al., 1998) reveals that $pRb^{-/-}/p107^{-/-}$ cells are not capable of forming most adult tissues, while $Rb^{-/-}$ cells of $Rb^{-/-}$:$Rb^{+/+}$ chimeras can be shown to contribute to most adult organs. These data suggest a functional overlap between family members but do not clarify the full role of each pocket protein in development.

The genetic background also seems important in defining the role of each pocket protein in development. Two different studies (Le Couter,

1998a, b) performed in Balb/c strain showed embryonic lethality at days 11 to 13 in pRb2/p130$^{-/-}$ embryos, while p107$^{-/-}$ mice were viable but had a severe postnatal growth deficiency and overproliferation of some compartments. These defects were rescued by a single cross with C57BL/6 mice, suggesting that different pathways may modify the penetrance of null mutation phenotypes of p107 and pRb2/p130.

However, interpreting the Rb knockout phenotypes remains complicated because of functional compensation and genetic redundancy among different Rb family members. Moreover every phenotype results from a coordinated and intricate network of biological processes, such as cell cycle regulation, specific gene expression, and apoptosis—all events regulated by pocket proteins but by largely unknown definition.

Rb FAMILY AND APOPTOSIS

pRb not only is a growth suppressor but also is an anti-apoptotic factor. pRb knockout mice have defects in the haematopoietic system and have impaired development of the central and the peripheral nervous systems because of massive cell death. Inappropriate cell death is present also in the liver, lens, and skeletal muscle precursors (Jacks et al., 1992).

In vitro experiments confirm results obtained with transgenic mice. TGF-β1 causes cell death by suppressing pRb expression and phosphorylation (Fan et al., 1996). IFNγ-induced apoptosis is blocked in pRb-positive cells (Berry et al., 1996). Restoring the function of pRb in a pRb-null osteosarcoma cell line inhibits irradiation-induced apoptosis (Haas-Kogan et al., 1995). The presence of HPV in HeLa cells abolishes functional p53 and pRb. In this cell line, ectopic expression of p53 leads to cell death, but the lethal effect of p53 can be reversed by cotransfecting pRb (Haupt et al., 1995). Additional support for a protective role of pRb against apoptosis comes from the observation that pRb is the target of caspases. pRb is cleaved in different apoptotic systems such as tumor necrosis factor-, staurosporine-, Fas- and cytosine arabinoside-induced apoptosis (An et al., 1996; Dou et al., 1997; Janicke et al., 1996).

The role of pRb in apoptosis has yet to be clarified. Studies conducted on E2F-1 may help to clarify the anti-apoptotic function of pRb. Transgenic mice null for E2F-1 show inappropriate cell proliferation in the thymus and lymph nodes. Considering E2F-1's role in stimulating cell cycle progression, the expected effect of inactivating E2F-1 should be hypoproliferation. The explanation for this surprising result may be that E2F-1, in addition to promoting cell proliferation, controls apoptotic pathways. E2F-1 mutants unable to interact with pRb show an even greater ability to induce apoptosis. This suggests that pRb has the ability to regulate E2F-1 function during the cell cycle but also can inhibit the apoptotic pathway that E2F-1 stimulates. In fact E2F-1 cannot induce apoptosis when pRb is co-expressed (Fan et al., 1996). The current data indicate that pRb inhibits apoptosis by avoiding improper activation of E2F-1.

Few data are currently available about the role of the other two Rb family members, pRb2/p130 and p107, in apoptosis. It is probable that these two proteins have roles in apoptosis because they share other characteristics with the better known pRb. Data supporting a p107 anti-apoptotic role come from a study conducted in mice. The liver and the central nervous system of $pRb^{-/-}$; $p107^{-/-}$ embryos show more extensive apoptosis than do $pRb^{-/-}$ embryos. In addition the double-knockout animals die two days prior to the single mutant (Lee et al., 1996). Conversely, we recently showed (Pucci et al., 2002) that pRb2/p130 is able to promote γ-irradiation-induced apoptosis by down-regulating bcl-2 and upregulating p73 in the SaOs2 osteosarcoma cell line. The pro-apoptotic activity of pRb2 was not associated with its ability to arrest cells in G0/G1 phase, thus suggesting that pRb2 inhibits tumor progression not only by arresting cell cycle but also by inducing cell death.

These observations are given credence by the fact that in 42 human retinoblastomas expression of pRb2/p130 is inversely correlated with the apoptotic index (Bellan et al., 2002) and that in the osteosarcoma cell line Hos p73 transcription is activated by E2F-1-pRb2/p130–p300 complexes (La Sala et al., 2003).

Rb FAMILY AND DIFFERENTIATION

A role for Rb family members in differentiation is supported by several experiments conducted in cell culture systems (Lipinsky and Jacks, 1999). Rb family members interact with several differentiation-specific transcription factors, such as Myo D, NF-IL6, and HBP1 (Tevosian et al., 1997; Wang et al., 1997; Yee et al., 1998; Carnac et al., 2000) (Table 18.1). Rb proteins are highly expressed in some terminal differentiated cells, such as neurons and skeletal muscles where pRb2/p130 also is highly expressed, while p107 shows higher levels in breast and prostate epithelial cells (Baldi et al., 1997).

pRb is implicated in adipogenesis, myogenesis (Zacksenhaus et al., 1996; Novitch et al., 1996 and 1999), and hematopoiesis (Condorelli et al., 1995; Condorelli and Giordano, 1997). Conversely, p107 or pRb2/p130 can have either redundant or even antagonistic functions in some differentiation programs. pRb associates with members of the MyoD bHLH family of transcription factors, and cells lacking pRb are unable to undergo myogenic activation (Gu et al., 1993). However, the myogenic process can be partially rescued by overexpressing p107 in the same system (Schneider et al., 1994). In contrast, pRb2/p130 specifically accumulates in a subset of myoblasts named "reserve cells," that are quiescent and undifferentiated but do have the capacity to self-renew and to give rise to differentiated myoblasts (Carnac et al., 2000). In addition muscle cells overexpressing pRb2/p130 have reduced levels of the muscle-promoting factor MyoD. It is therefore conceivable that pRb2/p130 is essential for maintaining a pool of reserve cells during terminal differentiation.

TABLE 18.1. Transcription Factors That Associate with Rb Family Members

Transcription Factor	Rb Protein	Biological Function	References
hBRM/ hBRG1	pRb, pRb2, p107	Chromatin remodeling	Strober et al. (1996)
CtIp	pRb, pRb2	Chromatin remodeling	Meloni et al. (1999)
HDAC	pRb, pRb2, p107	Chromatin remodeling	Luo et al. (1998), Stiegler et al. (1998), Ferreira et al. (1998)
Id2	pRb, pRb2, p107	Cell cycle regulation	Lasorella et al. (2000)
Myo D	pRb, p107	Muscle differentiation	Gu et al. (1993), Schneider et al. (1994), Zacksenhaus et al. (1996)
HBP1	pRb, pRb2	Muscle differentiation	Tevosian et al. (1997), Shih et al. (1998)
C/EBPS	pRb	Adipogenesis	Chen et al. (1996a)
NF-IL6 (C/EBPβ)	pRb	Monocyte/macrophage differentiation	Chen et al. (1996b)
c-*myc*, N-*myc*	pRb, p107	Cell proliferation and growth	Rustgi et al. (1991), Beijersbergen et al. (1994)
Sp1	pRb, p107		Kim et al. (1992), Datta et al. (1995)
TFIIIB	pRb, pRb2, p107	RNA Pol III transcription	Larminie et al. (1997), Sutcliffe et al. (1999)
E2F1–E2F3	pRb	Cell cycle regulation	Qin et al. (1995)
E2F4	pRb, pRb2, p107	Cell cycle regulation	Vairo et al. (1995), Ginsberg et al. (1994) Moberg et al. (1996)
E2F5	pRb2	Cell cycle regulation	Hijmans et al. (1995)

In murine fibroblasts, pRb activates C/EBPS, a family of transcription factors involved in adipocyte differentiation. Its presence is required for fibroblast conversion in adipocyte differentiation following C-EBP transfection (Chen et al., 1996). However, roles of pRb2/p130 and p107 in adipogenesis remain controversial. In fact, although p107 and pRb2/p130 are up-regulated during the different phases of adipocyte differentiation, cells lacking p107 or both p107 and pRb2/p130 differentiate better than wild-type cells (Classon et al., 2000). These data suggest that Rb family members can have opposite functions when analyzed singularly, and therefore their role in different systems is difficult to establish.

pRb2/p130 also accumulates selectively during the in vitro differentiation of myeloid progenitor cells into granulocytes in response to granulocyte-colony stimulating factor (Mori et al., 1999). In vitro studies on DMSO-induced differentiation of N1E-115 murine neuroblastoma cells show that pRb2/p130 is strongly upregulated while pRb/p105 and p107 are markedly decreased (Raschella et al., 1998). pRb2/p130 also accumulates selectively during in vitro differentiation of the myeloid progenitor cell, 32Dcl3, into a granulocyte in response to granulocyte-colony stimulating factor (G-CSF) (Mori et al., 1999). We demonstrated recently that overexpression of pRb2/p130 by adenoviral-mediated gene transfer is able to induce astrocyte differentiation (Galderisi et al., 2001).

The effect of Rb family mutations has been examined in mouse embryonic fibroblasts (MEFs) in culture. $Rb^{-/-}$ and $p107^{-/-}$; $Rb2^{-/-}$ fibroblasts (Herrera et al., 1996b) each have mild defects in cell cycle regulation. However, pRb but not p107 or pRb2/p130 is required for terminal differentiation and for maintenance of postmitotic state of MEFs induced to differentiate in the myogenic lineage (Novitch et al., 1996).

It has been recently shown that ES cells that lack all three Rb family members have an impaired differentiation capacity (Dannenberg et al., 2000). However, their ability to form tissue in chimeric animals remains to be tested.

In conclusion, the cellular processes that regulate differentiation seem to depend on the ability of Rb family members to balance proliferation and apoptosis. While much progresses has been made in understanding how pocket proteins regulate these two processes, the mechanism that determines their role in differentiation remains unknown and requires further investigations.

Rb FAMILY AND ANGIOGENESIS

Tumor progression and metastasis require persistent new blood vessel growth. Manipulating the "angiogenic switch" also is an important feature that a tumor must acquire for successful growth. This switch is regulated by the highly balanced activities of angiogenetic and antiangiogenetic molecules (Folkman, 1990). A large body of direct evidence incriminates vascular endothelial growth factor (VEGF), a potent tumor-secreted angiogenic factor, and its opposed antiangiogenic molecule Thrombospondin-1 (TSP-1) as central mediators of tumor-induced angiogenesis in vivo. The down-regulation of VEGF inhibits tumor formation in vivo. On the other hand, decreased expression of TSP-1, a p53 and pRb/p105 regulated angiogenesis inhibitor, has been observed in several human tumors, while its enhanced expression has been reported to suppress tumor growth and metastasis in some cell types. The angiogenic process is associated with inactivation of either the p53 or pRb/p105 tumor-suppressor genes, which are unable to up-regulate the expression of TSP-1, or with inactivation of p16 cyclin-dependent kinase inhibitor, which is able to down-regulate VEGF expression. A recent study showed that p53 and pRb/p105 are activated when endothelial cells are treated with the antiangiogenic agent TNP-470. We recently showed that pRb2/p130 also is involved in angiogenesis. Its enhanced expression by adenoviral mediated gene transfer in nude mice downregulates

VEGF expression, leading to the inhibition of tumor formation (Claudio et al., 2001). No study results currently are available on the role of p107 in angiogenesis.

DEREGULATION OF Rb FAMILY MEMBERS IN CANCER AND THERAPEUTIC PERSPECTIVES

pRb/p105

The retinoblastoma protein pRb encodes a 105 kDa nuclear phosphoprotein and was the first tumor-suppressor gene to be cloned (Lee et al., 1987). The Rb gene is located at chromosome 13q14. Homozygous genetic inactivation of pRb is found in several tumors and occurs by many different types of mutation including deletion in its promoter (Friend et al., 1986; Paggi et al., 1996). Studies on pRb/p105 show that its function is impaired in a high percentage of human tumors (Knudson et al., 1976 and Knudson, 1978). pRb/p105 is also essential for determining cell cycle arrest, differentiation, and apoptosis.

Over the past 10 years a large number of studies have examined the diagnostic and prognostic significance of pRb in human cancer, and almost all of these studies report decreased pRb expression in aggressive tumors. pRb expression also is reported to be an independent prognostic factor in several tumors including acute myelogenous leukemia (Kornblau et al., 1998), non-small cell lung carcinoma (Xu et al., 1994), papillary thyroid carcinoma (Omura et al., 1997), bladder (Cordon Cardo et al., 1992), prostate (Theodorescu et al., 1997), and ovarian cancer (Dong et al., 1997).

Several studies recently suggested that pRb sits in a key regulatory pathway, the so-called pRb/cyclin D1/cdk4/p16 pathway, characterized by the same end results. For example, irregular activation of E2F transcription factor is obtained by substituting events, such as overexpression of cyclin D1 or cdk4, and by functional inactivation of pRb or p16. Each of these alterations can be because of different mechanisms: by affecting the gene directly, such as deletions and mutations in p16 and Rb, or by chromosomal rearrangements of cyclin D1 and cdk4, or by affecting expression of the gene in the case of cyclin D1 overexpression because of constitutive activation by signaling pathways from activated growth factor receptors. Expression of the cyclin-dependent kinase inhibitor p16 is also under the control of E2F. This feedback mechanism ensures the turning off of cyclin D1/cdk4 at the end of G_1 phase. p16 inhibits cyclin D1/cdk4 activity by competitive binding at cdk4 at the same site where cdk4 binds to cyclin D1. As a result overstimulation of E2F, for instance, by mutation of Rb, leads to an enhanced p16, which replaces cyclin D1 in the complex with cdk4. Cyclin D1 then becomes available for proteosomal degradation, which explains why tumor cells harboring Rb mutations have low protein levels of cyclin D1.

Recent studies of primary MEFs (Bruce et al., 2000) show that p107 or pRb2/p130 are required, in addition to pRb, for p16 to arrest these cells in the G_1 phase of the cell cycle, thus supporting the idea that deregulation of cyclinD1/cdk4 does not only affect the activity of pRb but also affects all three family members.

pRb2/p130

Supporting involvement of the pRb2/p130 gene in human cancer as a tumor suppressor is the fact that it maps to human chromosome 16q12.2, an area in which deletions have been found in several tumors including breast, ovarian, hepatic, and prostate cancer (Yeung et al., 1993). Several studies of lung carcinomas (Baldi et al., 1996 and 1997; Caputi et al., 2002), endometrial cancer (Susini et al., 1998), choroidal melanomas (Massaro-Giordano et al., 1999), and prostate cancer (Claudio et al., 2002) show the loss of pRb2/p130, is correlated with unfavorable clinical outcomes. Genomic mutations of Rb2 gene also have been found in several cell lines, such as lymphoid (Cinti et al., 2000a) and nasopharyngeal, in samples of primary nasopharyngeal carcinomas (Claudio et al., 2000a), in lung tumors (Claudio et al., 2000b), and in Burkitt's lymphoma (Cinti et al., 2000b). These mutations include either splice acceptor site changes or mutations that prevent nuclear localization of pRb2, both instances resulting in a nonfunctional protein in tumor cells. A recent study, however, contradicts these findings in a similar set of tumor samples (Modi et al., 2000).

To investigate the putative tumor suppressor activity of pRb2/p130, we set up a tetracycline regulated gene expression system to control the expression of the encoded protein Rb2/p130 in JC-virus induced hamster brain tumor cells, and to study the effects of pRb2/p130 on the growth of such tumor cells in nude mice. pRb2/p130 induction resulted in a 69% reduction in the final tumor mass in nude mice and was able to overcome cellular transformation mediated by T antigen (Howard et al., 1998). We also observed that ectopic expression of pRb2/p130 in nude mice suppresses the tumorigenicity of the c-erbB-2 overexpressing SKOV-3 ovarian tumor cell line (Pupa et al., 1999).

Retrovirus mediated delivery of wild type pRb2/p130 to the lung tumor cell line H23 potently inhibited tumorigenesis in vivo and in vitro, as shown by the dramatic growth arrest observed in a colony assay and by the suppression of anchorage-independent growth potential and tumor formation in nude mice. The tumors transduced with the pRb2/p130 retrovirus diminished in size, and the reduction in tumor growth after pRb2/p130 transduction was statistically significant compared with controls. Microarray analysis of *RB2/p130* adenoviral transduction on the H23 lung adenocarcinoma cell line identified several down-regulated genes that were classified into 24 categories (Table 18.2) on the basis of a well-documented and established biological or pathological function of the encoded protein (Russo et al., 2003).

p107

No evidence exists at the moment that p107 is a tumor-suppressor gene. p107 maps in a chromosomal region that rarely shows cytogenetical

TABLE 18.2. Classification of RB2/p130 Down-regulated Genes by Category

Category	Number of Genes
ATPase/GTPase/ATP binding/GTP binding	2
Calcium/potassium/sodium/iron binding protein	1
Cell cycle/cyclins	11
Cell surface/antigen	8
Chromosome/chromatin/histone	3
Cytokines and growth factors	4
Cytoskeleton/microtubules/microfilaments/motility	3
Differentiation/development	9
Diseases	10
DNA binding/damage/recombination	3
G protein/regulators of G protein signaling	3
Hydrolase/hydrolysis/hydrolyzes	3
Kinases	14
Lipoproteins/lipids	1
Membrane trafficking	2
Mitochondrial proteins	2
Nuclear receptors/receptors	12
Oncogenes	4
Phosphatase/proteases/peptidase	3
Signal transduction	7
Synthetase/synthase	2
Transcription/transcription factor	6
Transporters	1
Transferases	2

alterations in tumor cells. However, it was shown in SaOs-2 cells, which lack a functional pRb, that p107 has growth suppressive activities (Zhu et al., 1993). In contrast to pRb2/p130, levels of p107 are often elevated in actively cycling cells as well as in tumor samples, and to date there is only one characterized rearrangements in the p107 locus, an intragenic deletion of the p107 gene in a B cell lymphoma line (Ichimura et al., 2000).

To date no tumor phenotype has been observed either in p107 or pRb2/p130 knockout mice or in $p107^{+/+}/pRb2^{-/-}$ or $p107^{-/-}/pRb2^{-/-}$ mice (doubly deficient mice are not viable). However, in vivo experiments do not exclude the possibility that p107 and pRb2 might cooperate in tumor suppression under certain circumstances, such as the example of p107 being able to act as a tumor suppressor in the context of pRb deficiency (Robanus-Maandag et al., 1998). The biological importance of p107 and pRb2/p130 appears also to be dependent on the genetic background (see above). Thus studies in chimeric animals harboring mutations of different pathways commonly involved in cancer might define the role of these two pocket proteins in tumorigenesis.

Cyclin-Rbs-E2F pathways are potential targets for pharmacological manipulation. The development of small molecules that specifically inhibit certain cyclin/cdk complexes or that can affect E2F-dependent transcription, as well as peptides that can mimic the functional regions of Rb proteins inactivated in specific tumors, are currently under investigation.

CONCLUSIONS

Considerable progress has been made in defining the biochemical and molecular mechanisms underlying growth control and tumor suppression by Rb family members. It is evident that all three pocket proteins have important and interdependent roles in the fundamental pathways governing normal cell growth, differentiation, embryogenesis, and apoptosis. Despite similarities among these three proteins, it is clear that individual family members have unique properties. However, the functional overlap between pRb, pRb2/p130 and p107 makes it extremely difficult to characterize the precise biological role of each pocket protein at present. Additional studies in multiple genetic systems will be essential to determine the roles of individual retinoblastoma family members in different cellular processes and to clarify the significance of their functional loss in cancer.

ACKNOWLEDGMENTS

We apologize to collegues whose work could not be cited due to space limitations. We thank Dr. J. Gartland for editing the manuscript. V.M. is supported by a fellowship from a training grant from the National Cancer Institute (PHS 5 T32 CA09137). This work was supported by Sbarro Health Research Organization and NIH grants.

REFERENCES

An B, Dou QP (1996): Cleavage of retinoblastoma protein during apoptosis: An interleukin 1 beta-converting enzyme-like protease as candidate. Cancer Res 56(3):438–42.

Baldi A, Esposito V, De Luca A, Howard CM, Mazzarella G, Baldi F, Caputi M, Giordano A (1996): Differential expression of the retinoblastoma gene family members pRb/p105, p107, and pRb2/p130 in lung cancer. Clin Cancer Res 2(7):1239–45.

Baldi A, Esposito V, De Luca A, Fu Y, Meoli I, Giordano GG, Caputi M, Baldi F, Giordano A (1997): Differential expression of Rb2/p130 and p107 in normal human tissues and in primary lung cancer. Clin Cancer Res 3(10):1691–7.

Beijersbergen RL, Kerkhoven RM, Zhu L, Carlee L, Voorhoeve PM, Bernards R (1994): E2F-4, a new member of the E2F gene family, has oncogenic activity and associates with p107 in vivo. Genes Dev 8(22):2680–90.

Beijersbergen RL, Hijmans EM, Zhu L, Bernards R (1994): Interaction of c-Myc with the pRb-related protein p107 results in inhibition of c-Myc-mediated transactivation. EMBO J 13(17):4080–6.

Beijersbergen RL, Carlee L, Kerkhoven RM, Bernards R (1995): Regulation of the retinoblastoma protein-related p107 by G_1 cyclin complexes. Genes Dev 9(11):1340–53.

Bellan C, De Falco G, Tosi GM, Lazzi S, Ferrari F, Morbini G, Bartolomei S, Toti P, Mangiavacchi P, Cevenini G, Trimarchi C, Cinti C, Giordano A, Leoncini L, Tosi P, Cottier H (2002): Missing expression of pRb2/p130 in human retinoblastomas is associated with reduced apoptosis and lesser differentiation. Invest Ophthalmol Vis Sci 43(12):3602–8.

Berry DE, Lu Y, Schmidt B, Fallon PG, O'Connell C, Hu SX, Xu HJ, Blanck G (1996): Retinoblastoma protein inhibits IFN-gamma induced apoptosis. Oncogene 12(8):1809–19.

Brehm A, Miska EA, McCance DJ, Reid JL, Bannister AJ, Kouzarides T (1998): Retinoblastoma protein recruits histone deacetylase to repress transcription. Nature 391(6667):597–601.

Brehm A, Kouzarides T (1999): Retinoblastoma protein meets chromatin. Trends Biochem Sci 24(4):142–5.

Bruce JL, Hurford RK Jr, Classon M, Koh J, Dyson N (2000): Requirements for cell cycle arrest by p16INK4a. Mol Cell 6(3):737–42.

Cairns BR (1998): Chromatin remodeling machines: similar motors, ulterior motives. Trends Biochem Sci 23(1):20–5.

Canhoto AJ, Chestukhin A, Litovchick L, DeCaprio JA (2000): Phosphorylation of the retinoblastoma-related protein p130 in growth-arrested cells. Oncogene 19(44):5116–22.

Caputi M, Groeger AM, Esposito V, De Luca A, Masciullo V, Mancini A, Baldi F, Wolner E, Giordano A (2002): Loss of pRb2/p130 expression is associated with unfavorable clinical outcome in lung cancer. Clin Cancer Res 8(12): 3850–6.

Carnac G, Fajas L, L'honore A, Sardet C, Lamb NJ, Fernandez A (2000): The retinoblastoma-like protein p130 is involved in the determination of reserve cells in differentiating myoblasts. Curr Biol 10(9):543–6.

Chen PL, Riley DJ, Chen Y, Lee WH (1996a): Retinoblastoma protein positively regulates terminal adipocyte differentiation through direct interaction with C/EBPs. Genes Dev 10(21):2794–804.

Chen PL, Riley DJ, Chen-Kiang S, Lee WH (1996b): Retinoblastoma protein directly interacts with and activates the transcription factor NF-IL6. Proc Natl Acad Sci USA 93(1):465–9.

Chestukhin A, Litovchick L, Rudich K, DeCaprio JA (2002): Nucleocytoplasmic shuttling of p130/RBL2: Novel regulatory mechanism. Mol Cell Biol 22(2): 453–68.

Ciarmatori S, Scott PH, Sutcliffe JE, McLees A, Alzuherri HM, Dannenberg JH, te Riele H, Grummt I, Voit R, White RJ (2001): Overlapping functions of the pRb family in the regulation of rRNA synthesis. Mol Cell Biol 21(17):5806–14.

Cinti C, Claudio PP, Howard CM, Neri LM, Fu Y, Leoncini L, Tosi GM, Maraldi NM, Giordano A (2000a): Genetic alterations disrupting the nuclear localization of the retinoblastoma-related gene RB2/p130 in human tumor cell lines and primary tumors. Cancer Res 60(2):383–9.

Cinti C, Leoncini L, Nyongo A, Ferrari F, Lazzi S, Bellan C, Vatti R, Zamparelli A, Cevenini G, Tosi GM, Claudio PP, Maraldi NM, Tosi P, Giordano A (2000b): Genetic alterations of the retinoblastoma-related gene RB2/p130 identify different pathogenetic mechanisms in and among Burkitt's lymphoma subtypes. Am J Pathol 156(3):751–60.

Clarke AR, Maandag ER, van Roon M, van der Lugt NM, van der Valk M, Hooper ML, Berns A, te Riele H (1992): Requirement for a functional Rb-1 gene in murine development. Nature 359(6393):328–30.

Classon M, Kennedy BK, Mulloy R, Harlow E (2000): Opposing roles of pRB and p107 in adipocyte differentiation. Proc Natl Acad Sci USA 97(20): 10826–31.

Claudio PP, Howard CM, Baldi A, De Luca A, Fu Y, Condorelli G, Sun Y, Colburn N, Calabretta B, Giordano A (1994): p130/pRb2 has growth suppressive properties similar to yet distinctive from those of retinoblastoma family members pRb and p107. Cancer Res 54(21):5556–60.

Claudio PP, De Luca A, Howard CM, Baldi A, Firpo EJ, Koff A, Paggi MG, Giordano A (1996): Functional analysis of pRb2/p130 interaction with cyclins. Cancer Res 56(9):2003–8.

Claudio PP, Howard CM, Fu Y, Cinti C, Califano L, Micheli P, Mercer EW, Caputi M, Giordano A (2000a): Mutations in the retinoblastoma-related gene RB2/p130 in primary nasopharyngeal carcinoma. Cancer Res 60(1):8–12.

Claudio PP, Howard CM, Pacilio C, Cinti C, Romano G, Minimo C, Maraldi NM, Minna JD, Gelbert L, Leoncini L, Tosi GM, Micheli P, Caputi M, Giordano GG, Giordano A (2000b): Mutations in the retinoblastoma-related gene RB2/p130 in lung tumors and suppression of tumor growth in vivo by retrovirus-mediated gene transfer. Cancer Res 60(2):372–82.

Claudio PP, Stiegler P, Howard CM, Bellan C, Minimo C, Tosi GM, Rak J, Kovatich A, De Fazio P, Micheli P, Caputi M, Leoncini L, Kerbel R, Giordano GG, Giordano A (2001): RB2/p130 gene-enhanced expression down-regulates vascular endothelial growth factor expression and inhibits angiogenesis in vivo. Cancer Res 61(2):462–8.

Claudio PP, Zamparelli A, Garcia FU, Claudio L, Ammirati G, Farina A, Bovicelli A, Russo G, Giordano GG, McGinnis DE, Giordano A, Cardi G (2002): Expression of cell-cycle-regulated proteins pRb2/p130, p107, p27(kip1), p53, mdm-2, and Ki-67 (MIB-1) in prostatic gland adenocarcinoma. Clin Cancer Res 8(6):1808–15.

Cobrinik D, Lee MH, Hannon G, Mulligan G, Bronson RT, Dyson N, Harlow E, Beach D, Weinberg RA, Jacks T (1996): Shared role of the pRB-related p130 and p107 proteins in limb development. Genes Dev 10(13):1633–44.

Condorelli GL, Testa U, Valtieri M, Vitelli L, De Luca A, Barberi T, Montesoro E, Campisi S, Giordano A, Peschle C (1995): Modulation of retinoblastoma gene in normal adult hematopoiesis: Peak expression and functional role in advanced erythroid differentiation. Proc Natl Acad Sci USA 92(11):4808–12.

Condorelli G, Giordano A (1997): Synergistic role of E1A-binding proteins and tissue-specific transcription factors in differentiation. J Cell Biochem 67(4):423–31.

Cordon-Cardo C, Wartinger D, Petrylak D, Dalbagni G, Fair WR, Fuks Z, Reuter VE (1992): Altered expression of the retinoblastoma gene product: Prognostic indicator in bladder cancer. J Natl Cancer Inst 84(16):1251–6.

Dannenberg JH, van Rossum A, Schuijff L, te Riele H (2000): Ablation of the retinoblastoma gene family deregulates G(1) control causing immortalization and increased cell turnover under growth-restricting conditions. Genes Dev 14(23):3051–64.

Datta PK, Raychaudhuri P, Bagchi S (1995): Association of p107 with Sp1: Genetically separable regions of p107 are involved in regulation of E2F- and Sp1-dependent transcription. Mol Cell Biol 15(10):5444–52.

De Luca A, MacLachlan TK, Bagella L, Dean C, Howard CM, Claudio PP, Baldi A, Khalili K, Giordano A (1997a): A unique domain of pRb2/p130 acts as an inhibitor of Cdk2 kinase activity. J Biol Chem 272(34):20971–4.

Donehower LA, Harvey M, Slagle BL, McArthur MJ, Montgomery CA Jr, Butel JS, Bradley A (1992): Mice deficient for p53 are developmentally normal but susceptible to spontaneous tumours. Nature 356(6366):215–21.

Dong F, Cress WD Jr, Agrawal D, Pledger WJ (1998): The role of cyclin D3-dependent kinase in the phosphorylation of p130 in mouse BALB/c 3T3 fibroblasts. J Biol Chem 273(11):6190–5.

Dong Y, Walsh MD, McGuckin MA, Cummings MC, Gabrielli BG, Wright GR, Hurst T, Khoo SK, Parsons PG (1997): Reduced expression of retinoblastoma gene product (pRB) and high expression of p53 are associated with poor prognosis in ovarian cancer. Int J Cancer 74(4):407–15.

Dou QP, An B, Antoku K, Johnson DE (1997): Fas stimulation induces RB dephosphorylation and proteolysis that is blocked by inhibitors of the ICE protease family. J Cell Biochem 64(4):586–94.

Dyson N, Buchkovich K, Whyte P, Harlow E (1989): The cellular 107K protein that binds to adenovirus E1A also associates with the large T antigens of SV40 and JC virus. Cell 58(2):249–55.

Ewen ME, Xing YG, Lawrence JB, Livingston DM (1991): Molecular cloning, chromosomal mapping, and expression of the cDNA for p107, a retinoblastoma gene product-related protein. Cell 66(6):1155–64.

Faha B, Ewen ME, Tsai LH, Livingston DM, Harlow E (1992): Interaction between human cyclin A and adenovirus E1A-associated p107 protein. Science 255(5040):87–90.

Fan G, Ma X, Kren BT, Steer CJ (1996): The retinoblastoma gene product inhibits TGF-beta1 induced apoptosis in primary rat hepatocytes and human HuH-7 hepatoma cells. Oncogene 12(9):1909–19.

Ferreira R, Magnaghi-Jaulin L, Robin P, Harel-Bellan A, Trouche D (1998): The three members of the pocket proteins family share the ability to repress E2F activity through recruitment of a histone deacetylase. Proc Natl Acad Sci USA 95(18):10493–8.

Folkman J (1990): What is the evidence that tumors are angiogenesis dependent? J Natl Cancer Inst 82(1):4–6.

Friend SH, Bernards R, Rogelj S, Weinberg RA, Rapaport JM, Albert DM, Dryja TP (1986): A human DNA segment with properties of the gene that predisposes to retinoblastoma and osteosarcoma. Nature 323(6089):643–6.

Galderisi U, Melone MA, Jori FP, Piegari E, Di Bernardo G, Cipollaro M, Cascino A, Peluso G, Claudio PP, Giordano A (2001): pRb2/p130 gene overexpression induces astrocyte differentiation. Mol Cell Neurosci 17(3):415–25.

Ginsberg D, Vairo G, Chittenden T, Xiao ZX, Xu G, Wydner KL, DeCaprio JA, Lawrence JB, Livingston DM (1994): E2F-4, a new member of the E2F transcription factor family, interacts with p107. Genes Dev 8(22):2665–79.

Grana X, Garriga J, Mayol X (1998): Role of the retinoblastoma protein family, pRB, p107 and p130 in the negative control of cell growth. Oncogene 17(25): 3365–83.

Grunstein M (1997): Histone acetylation in chromatin structure and transcription. Nature 389(6649):349–52.

Gu W, Schneider JW, Condorelli G, Kaushal S, Mahdavi V, Nadal-Ginard B (1993): Interaction of myogenic factors and the retinoblastoma protein mediates muscle cell commitment and differentiation. Cell 72(3):309–24.

Haas-Kogan DA, Kogan SC, Levi D, Dazin P, T'Ang A, Fung YK, Israel MA (1995): Inhibition of apoptosis by the retinoblastoma gene product. EMBO J 14(3):461–72.

Hannan KM, Hannan RD, Smith SD, Jefferson LS, Lun M, Rothblum LI (2000): Rb and p130 regulate RNA polymerase I transcription: Rb disrupts the interaction between UBF and SL-1. Oncogene 19(43):4988–99.

Hannon GJ, Demetrick D, Beach D (1993): Isolation of the Rb-related p130 through its interaction with CDK2 and cyclins. Genes Dev 7(12A):2378–91.

Hansen K, Farkas T, Lukas J, Holm K, Ronnstrand L, Bartek J (2001): Phosphorylation-dependent and -independent functions of p130 cooperate to evoke a sustained G_1 block. EMBO J 20(3):422–32.

Haupt Y, Rowan S, Oren M (1995): p53-mediated apoptosis in HeLa cells can be overcome by excess pRB. Oncogene 10(8):1563–71.

Hauser PJ, Agrawal D, Chu B, Pledger WJ (1997): p107 and p130 associated cyclin A has altered substrate specificity. J Biol Chem 272(36):22954–9.

Herrera RE, Chen F, Weinberg RA (1996a): Increased histone H1 phosphorylation and relaxed chromatin structure in Rb-deficient fibroblasts. Proc Natl Acad Sci USA 93(21):11510–5.

Herrera RE, Sah VP, Williams BO, Makela TP, Weinberg RA, Jacks T (1996b): Altered cell cycle kinetics, gene expression, and G_1 restriction point regulation in Rb-deficient fibroblasts. Mol Cell Biol 16(5):2402–7.

Hijmans EM, Voorhoeve PM, Beijersbergen RL, van't Veer LJ, Bernards R (1995): E2F-5, a new E2F family member that interacts with p130 in vivo. Mol Cell Biol 15(6):3082–9.

Hoshikawa Y, Mori A, Amimoto K, Iwabe K, Hatakeyama M (1998): Control of retinoblastoma protein-independent hematopoietic cell cycle by the pRB-related p130. Proc Natl Acad Sci USA 95(15):8574–9.

Howard CM, Claudio PP, Gallia GL, Gordon J, Giordano GG, Hauck WW, Khalili K, Giordano A (1998): Retinoblastoma-related protein pRb2/p130 and suppression of tumor growth in vivo. J Natl Cancer Inst 90(19):1451–60.

Howard CM, Claudio PP, De Luca A, Stiegler P, Jori FP, Safdar NM, Caputi M, Khalili K, Giordano A (2000): Inducible pRb2/p130 expression and growth-suppressive mechanisms: Evidence of a pRb2/p130, p27Kip1, and cyclin E negative feedback regulatory loop. Cancer Res 60(10):2737–44.

Ichimura K, Hanafusa H, Takimoto H, Ohgama Y, Akagi T, Shimizu K (2000): Structure of the human retinoblastoma-related p107 gene and its intragenic deletion in a B-cell lymphoma cell line. Gene 251(1):37–43.

Jacks T, Fazeli A, Schmitt EM, Bronson RT, Goodell MA, Weinberg RA (1992): Effects of an Rb mutation in the mouse. Nature 359(6393):295–300.

Janicke RU, Walker PA, Lin XY, Porter AG (1996): Specific cleavage of the retinoblastoma protein by an ICE-like protease in apoptosis. EMBO J 15(24): 6969–78.

Kim SJ, Onwuta US, Lee YI, Li R, Botchan MR, Robbins PD.(1992): The retinoblastoma gene product regulates Sp1-mediated transcription. Mol Cell Biol 12(6):2455–63.

Knudsen ES, Wang JY (1996): Differential regulation of retinoblastoma protein function by specific Cdk phosphorylation sites. J Biol Chem 271(14):8313–20.

Knudson AG Jr, Meadows AT, Nichols WW, Hill R (1976): Chromosomal deletion and retinoblastoma. N Engl J Med 295(20):1120–3.

Knudson AG Jr (1978): Retinoblastoma: a prototypic hereditary neoplasm. Semin Oncol 5(1):57–60.

Koff A, Giordano A, Desai D, Yamashita K, Harper JW, Elledge S, Nishimoto T, Morgan DO, Franza BR, Roberts JM (1992): Formation and activation of a cyclin E-cdk2 complex during the G_1 phase of the human cell cycle. Science 257(5077):1689–94.

Kornblau SM, Andreeff M, Hu SX, Xu HJ, Patel S, Theriault A, Koller C, Kantarjian H, Estey E, Deisseroth AB, Benedict WF (1998): Low and maximally phosphorylated levels of the retinoblastoma protein confer poor prognosis in newly diagnosed acute myelogenous leukemia: A prospective study. Clin Cancer Res 4(8):1955–63.

Larminie CG, Cairns CA, Mital R, Martin K, Kouzarides T, Jackson SP, White RJ (1997): Mechanistic analysis of RNA polymerase III regulation by the retinoblastoma protein. EMBO J 16(8):2061–71.

La Sala D, Macaluso M, Trimarchi C, Giordano A, Cinti C (2003): Triggering of p73-dependent apoptosis in osteosarcoma is under the control of E2Fs-pRb2/p130 complexes. Oncogene (in press).

Lasorella A, Noseda M, Beyna M, Yokota Y, Iavarone A (2000): Id2 is a retinoblastoma protein target and mediates signalling by Myc oncoproteins. Nature 407(6804):592–8.

LeCouter JE, Kablar B, Hardy WR, Ying C, Megeney LA, May LL, Rudnicki MA (1998a): Strain-dependent myeloid hyperplasia, growth deficiency, and accelerated cell cycle in mice lacking the Rb-related p107 gene. Mol Cell Biol 18(12):7455–65.

LeCouter JE, Kablar B, Whyte PF, Ying C, Rudnicki MA (1998b): Strain-dependent embryonic lethality in mice lacking the retinoblastoma-related p130 gene. Development 125(23):4669–79.

Lee EY, Chang CY, Hu N, Wang YC, Lai CC, Herrup K, Lee WH, Bradley A (1992): Mice deficient for Rb are nonviable and show defects in neurogenesis and haematopoiesis. Nature 359(6393):288–94.

Lee JO, Russo AA, Pavletich NP (1998): Structure of the retinoblastoma tumor-suppressor pocket domain bound to a peptide from HPV E7. Nature 391(6670):859–65.

Lee MH, Williams BO, Mulligan G, Mukai S, Bronson RT, Dyson N, Harlow E, Jacks T (1996): Targeted disruption of p107: functional overlap between p107 and Rb. Genes Dev 10(13):1621–32.

Lee WH, Bookstein R, Hong F, Young LJ, Shew JY, Lee EY (1987): Human retinoblastoma susceptibility gene: cloning, identification, and sequence. Science 235(4794):1394–9.

Lees JA, Buchkovich KJ, Marshak DR, Anderson CW, Harlow E (1991): The retinoblastoma protein is phosphorylated on multiple sites by human cdc2. EMBO J 10(13):4279–90.

Lees JA, Saito M, Vidal M, Valentine M, Look T, Harlow E, Dyson N, Helin K (1993): The retinoblastoma protein binds to a family of E2F transcription factors. Mol Cell Biol 13(12):7813–25.

Lew DJ, Dulic V, Reed SI (1991): Isolation of three novel human cyclins by rescue of G_1 cyclin (Cln) function in yeast. Cell 66(6):1197–206.

Li Y, Graham C, Lacy S, Duncan AM, Whyte P (1993): The adenovirus E1A-associated 130-kD protein is encoded by a member of the retinoblastoma gene family and physically interacts with cyclins A and E. Genes Dev 7(12A): 2366–77.

Lipinski MM, Jacks T (1999): The retinoblastoma gene family in differentiation and development. Oncogene 18(55):7873–82.

Lomazzi M, Moroni MC, Jensen MR, Frittoli E, Helin K (2002): Suppression of the p53- or pRB-mediated G_1 checkpoint is required for E2F-induced S-phase entry. Nat Genet 31(2):190–4.

Luo RX, Postigo AA, Dean DC (1998): Rb interacts with histone deacetylase to repress transcription. Cell 92(4):463–73.

Macaluso M, Cinti C, Russo G, Russo A, Giordano A (2003): pRb2/p130-E2F4/5-HDAC1-SUV39H1-p300 and pRb2/p130-E2F4/5-HDAC1-SUV39H1-DNMT1 multimolecular complexes mediate the transcription of estrogen receptor-α in breast cancer. Oncogene 22(23):3511–7.

MacLachlan TK, Sang N, Giordano A (1995): Cyclins, cyclin-dependent kinases and cdk inhibitors: Implications in cell cycle control and cancer. Crit Rev Eukaryot Gene Expr 5(2):127–56.

Magnaghi-Jaulin L, Groisman R, Naguibneva I, Robin P, Lorain S, Le Villain JP, Troalen F, Trouche D, Harel-Bellan A (1998): Retinoblastoma protein represses transcription by recruiting a histone deacetylase. Nature 391(6667):601–5.

Massaro-Giordano M, Baldi G, De Luca A, Baldi A, Giordano A (1999): Differential expression of the retinoblastoma gene family members in choroidal melanoma: Prognostic significance. Clin Cancer Res 5(6):1455–8.

Mayol X, Grana X, Baldi A, Sang N, Hu Q, Giordano A (1993): Cloning of a new member of the retinoblastoma gene family (pRb2) which binds to the E1A transforming domain. Oncogene 8(9):2561–6.

Meloni AR, Smith EJ, Nevins JR (1999): A mechanism for Rb/p130-mediated transcription repression involving recruitment of the CtBP corepressor. Proc Natl Acad Sci USA 96(17):9574–9.

Mittnacht S (1998): Control of pRB phosphorylation. Curr Opin Genet Dev 8(1):21–7.

Moberg K, Starz MA, Lees JA (1996): E2F-4 switches from p130 to p107 and pRB in response to cell cycle reentry. Mol Cell Biol 16(4):1436–49.

Modi S, Kubo A, Oie H, Coxon AB, Rehmatulla A, Kaye FJ (2000): Protein expression of the RB-related gene family and SV40 large T antigen in mesothelioma and lung cancer. Oncogene 19(40):4632–9.

Morgenbesser SD, Williams BO, Jacks T, DePinho RA (1994): p53-dependent apoptosis produced by Rb-deficiency in the developing mouse lens. Nature 371(6492):72–4.

Mori A, Higashi H, Hoshikawa Y, Imamura M, Asaka M, Hatakeyama M (1999): Granulocytic differentiation of myeloid progenitor cells by p130, the retinoblastoma tumor suppressor homologue. Oncogene 18(46):6209–21.

Mulligan GJ, Wong J, Jacks T (1998): p130 is dispensable in peripheral T lymphocytes: Evidence for functional compensation by p107 and pRB. Mol Cell Biol 18(1):206–20.

Nevins JR (1998): Toward an understanding of the functional complexity of the E2F and retinoblastoma families. Cell Growth Differ 9(8):585–93.

Novitch BG, Mulligan GJ, Jacks T, Lassar AB (1996): Skeletal muscle cells lacking the retinoblastoma protein display defects in muscle gene expression and accumulate in S and G_2 phases of the cell cycle. J Cell Biol 135(2):441–56.

Novitch BG, Spicer DB, Kim PS, Cheung WL, Lassar AB (1999): pRb is required for MEF2-dependent gene expression as well as cell-cycle arrest during skeletal muscle differentiation. Curr Biol 9(9):449–59.

Omura K, Nagasato A, Kanehira E, Kinsen H, Amaya S, Kimura K, Kajita T, Nozaki Y, Mizukami Y, Nonomura A, Watanabe (1997): Retinoblastoma protein and proliferating-cell nuclear antigen expression as predictors of recurrence in well-differentiated papillary thyroid carcinoma. J Clin Oncol 15(12): 3458–63.

Paggi MG, Baldi A, Bonetto F, Giordano A (1996): Retinoblastoma protein family in cell cycle and cancer: A review. J Cell Biochem 62(3):418–30.

Paggi MG, Giordano A (2001): Who is the boss in the retinoblastoma family? The point of view of Rb2/p130, the little brother. Cancer Res 61(12):4651–4.

Pelletier G, Stefanovsky VY, Faubladier M, Hirschler-Laszkiewicz I, Savard J, Rothblum LI, Cote J, Moss T (2000): Competitive recruitment of CBP and Rb-HDAC regulates UBF acetylation and ribosomal transcription. Mol Cell 6(5):1059–66.

Pucci B, Claudio PP, Masciullo V, Bellincampi L, Terrinoni A, Khalili K, Melino G, Giordano A (2002): pRb2/p130 promotes radiation-induced cell death in the glioblastoma cell line HJC12 by p73 upregulation and Bcl-2 downregulation. Oncogene 21(38):5897–905.

Pupa SM, Howard CM, Invernizzi AM, De Vecchi R, Giani C, Claudio PP, Colnaghi MI, Giordano A, Menard S (1999): Ectopic expression of pRb2/p130 suppresses the tumorigenicity of the c-erbB-2-overexpressing SKOV3 tumor cell line. Oncogene 18(3):651–6.

Qin XQ, Livingston DM, Ewen M, Sellers WR, Arany Z, Kaelin WG Jr (1995): The transcription factor E2F-1 is a downstream target of RB action. Mol Cell Biol 15(2):742–55.

Raschella G, Tanno B, Bonetto F, Negroni A, Claudio PP, Baldi A, Amendola R, Calabretta B, Giordano A, Paggi MG (1998): The RB-related gene Rb2/p130 in neuroblastoma differentiation and in B-myb promoter down-regulation. Cell Death Differ 5(5):401–7.

Robanus-Maandag E, Dekker M, van der Valk M, Carrozza ML, Jeanny JC, Dannenberg JH, Berns A, te Riele H (1998): p107 is a suppressor of retinoblastoma development in pRb-deficient mice. Genes Dev 12(11):1599–1609.

Russo G, Claudio PP, Fu Y, Stiegler P, Yu ZL, Macaluso M, Giordano A (2003): RB2/p130 target genes in non small lung cancer cells identified by microarray analysis. Oncogene 22(44):6959–69.

Rustgi AK, Dyson N, Bernards R (1991): Amino-terminal domains of c-myc and N-myc proteins mediate binding to the retinoblastoma gene product. Nature 352(6335):541–4.

Sage J, Mulligan GJ, Attardi LD, Miller A, Chen S, Williams B, Theodorou E, Jacks T (2000): Targeted disruption of the three Rb-related genes leads to loss of G(1) control and immortalization. Genes Dev 14(23):3037–50.

Schneider JW, Gu W, Zhu L, Mahdavi V, Nadal-Ginard B (1994): Reversal of terminal differentiation mediated by p107 in Rb-/- muscle cells. Science 264(5164):1467–71.

Shao Z, Ruppert S, Robbins PD (1995): The retinoblastoma-susceptibility gene product binds directly to the human TATA-binding protein-associated factor TAFII250. Proc Natl Acad Sci USA 92(8):3115–9.

Shih HH, Tevosian SG, Yee AS (1998): Regulation of differentiation by HBP1, a target of the retinoblastoma protein. Mol Cell Biol 18(8):4732–43.

Spruck CH, Won KA, Reed SI (1999): Deregulated cyclin E induces chromosome instability. Nature 401(6750):297–300.

Stiegler P, De Luca A, Bagella L, Giordano A (1998): The COOH-terminal region of pRb2/p130 binds to histone deacetylase 1 (HDAC1), enhancing transcriptional repression of the E2F-dependent cyclin A promoter. Cancer Res 58(22):5049–52.

Stiegler P, Kasten M, Giordano A (1998): The RB family of cell cycle regulatory factors. J Cell Biochem Suppl 30–31:30–6.

Strober BE, Dunaief JL, Guha, Goff SP (1996): Functional interactions between the hBRM/hBRG1 transcriptional activators and the pRB family of proteins. Mol Cell Biol 16(4):1576–83.

Susini T, Baldi F, Howard CM, Baldi A, Taddei G, Massi D, Rapi S, Savino L, Massi G, Giordano A (1998): Expression of the retinoblastoma-related gene Rb2/p130 correlates with clinical outcome in endometrial cancer. J Clin Oncol 16(3):1085–93.

Sutcliffe JE, Cairns CA, McLees A, Allison SJ, Tosh K, White RJ (1999): RNA polymerase III transcription factor IIIB is a target for repression by pocket proteins p107 and p130. Mol Cell Biol 19(6):4255–61.

Tedesco D, Lukas J, Reed SI (2002): The pRb-related protein p130 is regulated by phosphorylation-dependent proteolysis via the protein-ubiquitin ligase SCF (Skp2). Genes Dev 16(22):2946–57.

Tevosian SG, Shih HH, Mendelson KG, Sheppard KA, Paulson KE, Yee AS (1997): HBP1: a HMG box transcriptional repressor that is targeted by the retinoblastoma family. Genes Dev 11(3):383–96.

Theodorescu D, Broder SR, Boyd JC, Mills SE, Frierson HF Jr (1997): P53, bcl-2 and retinoblastoma proteins as long-term prognostic markers in localized carcinoma of the prostate. J Urol 158(1):131–7.

Trouche D, Kouzarides T (1996): E2F1 and E1A(12S) have a homologous activation domain regulated by RB and CBP. Proc Natl Acad Sci USA 93(4): 1439–42.

Vairo G, Livingston DM, Ginsberg D (1995): Functional interaction between E2F-4 and p130: Evidence for distinct mechanisms underlying growth suppression by different retinoblastoma protein family members. Genes Dev 9(7):869–81.

Voit R, Schafer K, Grummt I (1997): Mechanism of repression of RNA polymerase I transcription by the retinoblastoma protein. Mol Cell Biol 17(8): 4230–7.

Wang J, Guo K, Wills KN, Walsh K (1997): Rb functions to inhibit apoptosis during myocyte differentiation. Cancer Res 57(3):351–4.

Weinberg RA (1995): The retinoblastoma protein and cell cycle control. Cell 81(3):323–30.

Weintraub SJ, Chow KN, Luo RX, Zhang SH, He S, Dean DC (1995): Mechanism of active transcriptional repression by the retinoblastoma protein. Nature 375(6534):812–5.

Whyte P, Buchkovich KJ, Horowitz JM, Friend SH, Raybuck M, Weinberg RA, Harlow E (1988): Association between an oncogene and an anti-oncogene: The adenovirus E1A proteins bind to the retinoblastoma gene product. Nature 334(6178):124–9.

Williams BO, Remington L, Albert DM, Mukai S, Bronson RT, Jacks T (1994): Cooperative tumorigenic effects of germline mutations in Rb and p53. Nat Genet 7(4):480–4.

Woo MS, Sanchez I, Dynlacht BD (1997): p130 and p107 use a conserved domain to inhibit cellular cyclin-dependent kinase activity. Mol Cell Biol 17(7): 3566–79.

Wu L, De Bruin A, Saavedra HI, Starovic M, Trimboli A, Yang Y, Opavska J, Wilson P, Thompson JC, Ostrowski MC, Rosol TJ, Woollett LA, Weinstein M, Cross JC, Robinson ML, Leone G (2003): Extra-embryonic function of Rb is essential for embryonic development and viability. Nature 421(6926):942–7.

Xu HJ, Quinlan DC, Davidson AG, Hu SX, Summers CL, Li J, Benedict WF (1994): Altered retinoblastoma protein expression and prognosis in early-stage non–small-cell lung carcinoma. J Natl Cancer Inst 86(9):695–9.

Yamasaki L, Bronson R, Williams BO, Dyson NJ, Harlow E, Jacks T (1998): Loss of E2F-1 reduces tumorigenesis and extends the lifespan of Rb1(+/–)mice. Nat Genet 18(4):360–4.

Yee AS, Shih HH, Tevosian SG (1998): New perspectives on retinoblastoma family functions in differentiation. Front Biosci 3:532–47.

Yeung RS, Bell DW, Testa JR, Mayol X, Baldi A, Grana X, Klinga-Levan K, Knudson AG, Giordano A (1993): The retinoblastoma-related gene, RB2, maps to human chromosome 16q12 and rat chromosome 19. Oncogene 8(12): 3465–8.

Zacksenhaus E, Bremner R, Phillips RA, Gallie BL (1993): A bipartite nuclear localization signal in the retinoblastoma gene product and its importance for biological activity. Mol Cell Biol 13(8):4588–99.

Zacksenhaus E, Jiang Z, Chung D, Marth JD, Phillips RA, Gallie BL (1996): pRb controls proliferation, differentiation, and death of skeletal muscle cells and other lineages during embryogenesis. Genes Dev 10(23):3051–64.

Zini N, Trimarchi C, Claudio PP, Stiegler P, Marinelli F, Maltarello MC, La Sala D, De Falco G, Russo G, Ammirati G, Maraldi NM, Giordano A, Cinti C (2001): pRb2/p130 and p107 control cell growth by multiple strategies and in association with different compartments within the nucleus. J Cell Physiol 189(1):34–44.

Zhu L, van den Heuvel S, Helin K, Fattaey A, Ewen M, Livingston D, Dyson N, Harlow E (1993): Inhibition of cell proliferation by p107, a relative of the retinoblastoma protein. Genes Dev 7(7A):1111–25.

Zhu L, Harlow E, Dynlacht BD (1995): p107 uses a p21CIP1-related domain to bind cyclin/cdk2 and regulate interactions with E2F. Genes Dev 9(14): 1740–52.

CHAPTER 19

p53 TUMOR-SUPPRESSOR GENES

FAITH A. ZAMAMIRI-DAVIS and GERARD P. ZAMBETTI
Department of Biochemistry, St. Jude Children's Research Hospital, Memphis, TN 38105

INTRODUCTION

The p53 tumor-suppressor protein stepped onto the cell cycle scene 25 years ago and has taken center stage in both basic and clinical research fields ever since. A large portion of p53's mass appeal may be attributed to its additional starring role in the regulation of apoptosis. Playing such a prominent part in controlling cell growth and survival has earned this tumor suppressor several titles, ranging from "Guardian of the Genome" to "Master Executioner." The main appeal of p53 lies in its range—the amazing ability to have a hand in every stage of a cell's life from development to death. An early Saturday Night Live skit says it best, "It's a floor wax *and* a dessert topping!" The p53 tumor suppressor seems to do it all. While we may not enjoy p53 with our Ben & Jerry's per se, it is indispensable in detecting and removing irregular cells that lead to cancer. All allegory aside, the role of p53 in cell cycle regulation and cancer prevention is truly a matter of life or death. Defining the factors that cause p53 to choose between arresting a damaged cell in G_1 to allow repair of the defects or to initiate the cellular death program is fundamental to understanding cancer biology.

Ironically, a large amount of experimental evidence generated in the years immediately following the identification of p53 in 1979 (DeLeo, Shiku et al., 1977; DeLeo, Jay et al., 1979; Lane and Crawford, 1979; Linzer, Maltzman et al., 1979) prompted its initial classification as an oncogene. A full decade passed before it was realized that the exact opposite was true, and p53 was shown to be a bona-fide tumor suppressor (Baker, Fearon et al., 1989; Finlay, Hinds et al., 1989; Baker, 2003). Once p53 was classified as a clinically relevant tumor suppressor, the field

Cell Cycle and Growth Control: Biomolecular Regulation and Cancer, Second Edition, Edited by Gary S. Stein and Arthur B. Pardee
ISBN 0-471-25071-6

exploded and an onslaught of studies poured in, resulting in more than 25,000 published manuscripts to date. With this intense focus some not so worthy studies have found their way into the literature. However, many landmark discoveries have been made, and these findings serve to drive and direct the field. The study of p53 has become increasingly complex and has branched out into subspecialties. A recent issue of the journal *Cell Death and Differentiation* was dedicated to summarizing the past 25 years of p53 research and serves as an excellent resource for catching up with the wealth of information concerning p53 (Bourdon, Laurenzi et al., 2003). Throughout this chapter we will highlight developments in the arrest/apoptosis venue, along with additional "classics" of p53 research that have withstood the test of time. This is not intended to be an all-inclusive review; rather, we will focus on what we believe to be the current status, controversies and "coming events" in the field.

CLASSIFICATION OF A TUMOR SUPRESSOR: ALL IN THE FAMILY?

The p53 tumor suppressor gained additional family members in 1997, with the identification of the homologue p73 (Kaghad, Bonnet et al., 1997), and one year later, p63 (Yang, Kaghad et al., 1998). Similar to p53, p63 and p73 contain the three functional domains typical of a transcription factor: an acidic amino-terminal transactivation (TA) domain; a central core DNA-binding domain (DBD); and a carboxy-terminal oligomerization domain (OD) (Fig. 19.1) (Levrero, De Laurenzi et al., 2000; Cadwell and Zambetti, 2001; Benard, Douc-Rasy et al., 2003). All three family members bind p53 DNA consensus sites, transactivate p53-responsive genes, and induce cell cycle arrest or apoptosis (Benard, Douc-Rasy et al., 2003).

Under normal conditions p53 expression is quite low, and upon cellular stress, such as DNA damage, p53 protein is stabilized allowing for the protein to accumulate. The mechanism for p53 induction is almost exclusively regulated at the post-translational level, although some degree of translational control may also contribute to its regulation. There is little if any regulation of p53 expression at the transcriptional level. By contrast, the consistent expression of multiple protein isoforms of p63 and p73, due to differential mRNA splicing and alternative promoter usage, are within themselves a means by which these family members set themselves apart from p53 (Pozniak, Radinovic et al., 2000; Yang and McKeon, 2000) (Fig. 19.2) (Levrero, De Laurenzi et al., 2000; Benard, Douc-Rasy et al., 2003). The P1 promoters of p63 and p73 yield isoforms that contain a transactivating (TA) domain with p53 homology; while the downstream P2 promoters give rise to N-terminal-truncated (ΔN) forms lacking a TA domain and possessing biological properties opposite to those of the P1-controlled p63/p73 TA isoforms (Table 19.1) (Levrero, De Laurenzi et al., 2000). This unusual gene regulation suggests a "two-genes in one" concept for the p53 family members, since TA

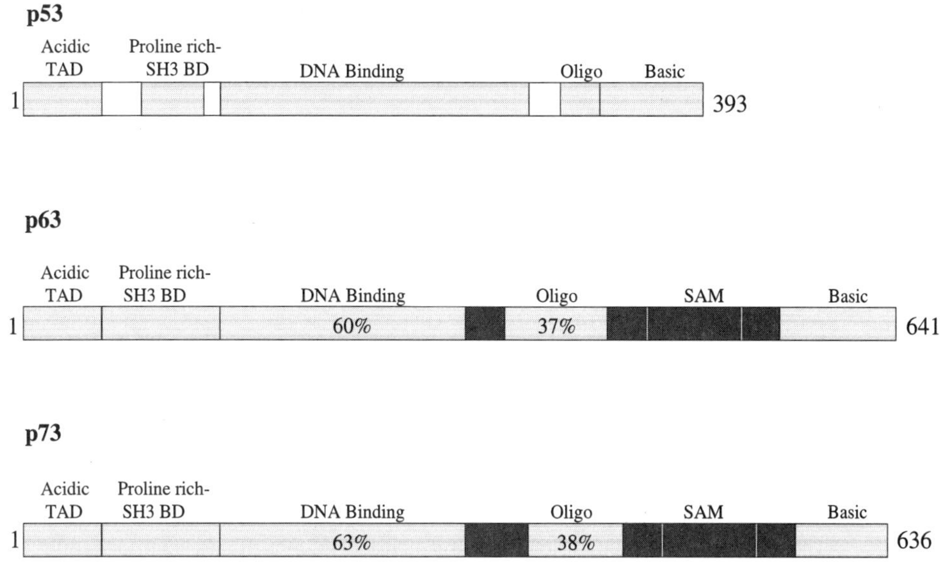

Figure 19.1. Gene organization of Human p53 family members. TAD = transactivation domain; SH3 BD= PXXP SH3 binding domain; Oligo = oligomerization domain; SAM = sterile A motif. Sequence homology for p63 and p73 DNA binding and oligomerization domains is shown by percent similarity to p53. Dark shade identifies regions unique to p63 and p73.

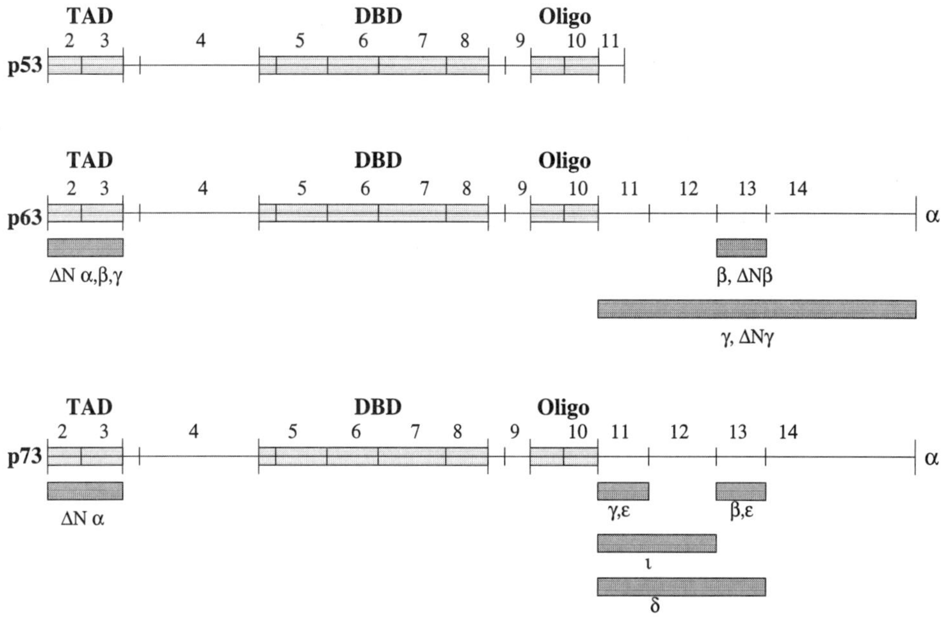

Figure 19.2. Alternative splicing of Human p53 family members. Dark shaded areas designate the following functional domains: TAD, Transactivation Domain; DBD = DNA Binding Domain; Oligo = Oligomerization Domain. Exons 2–11 (p53) and 2–14 (p63, p73) are numbered (exon 1 is non-coding and not represented). Light shaded boxes represent exons removed by alternative splicing.

TABLE 19.1. Summary of p63 and p73 Splice Variants

	Exons	Exons Spliced Out	Length
p53	11	—	393
p63α	14	—	641
β		13	516
γ		11,12,13,14	448
ΔNα		2,3	586
ΔNβ		2,3,13	461
ΔNγ		2,3,11,12,13,14	393
p73α	14	—	636
β		13	499
γ		11	475
δ		11,12,13	404
ε		11,13	555
ι		11,12	540
ΔNα		2,3	—

Note: Isoforms resulting from alternative splicing and their resulting characteristics are summarized. Note that the ΔN isoforms lack the transactivation domain (TAD).

and ΔN isoforms can act in vitro as tumor suppressors and oncogenes, respectively (Yang and McKeon, 2000).

While p53, p63, and p73 share remarkable sequence homology, as well as significant structural and functional similarity, there is increasing evidence from gene-targeting studies that each member functions in very different processes. p53-knockout mice invariably develop spontaneous tumors and succumb to their disease usually within the first 6 months of life (Donehower, Harvey et al., 1992; Jacks, Remington et al., 1994). The fact that deletion of only the p53 tumor suppressor, out of approximately 30–40,000 mouse genes, causes tumors in 100% of the animals is truly remarkable and underscores the importance of p53 in protecting against cancer. It should also be emphasized that although p53 does not appear to be involved in apoptotic processes that normally occur during development, its function is crucial for DNA damage and stress-induced cell death (Clarke, Purdie et al., 1993; Lowe, Ruley et al., 1993).

On the other hand, both p63 and p73 null mice are not prone to spontaneous tumor formation but are associated with multiple developmental defects (Yang, Schweitzer et al., 1999; Yang and McKeon, 2000; Yang, Walker et al., 2000). p63 knockout mice are born alive with severe malformation of limbs and epithelia (Mills, Zheng et al., 1999; Yang, Schweitzer et al., 1999). These phenotypic disorders have also been confirmed by the presence of human p63 mutations associated with syndromes of limb and ectodermal development, such as cleft lip/palate, split hand/foot malformation (SHFM), and limb-mammary syndrome (LMS) (Celli, Duijf et al., 1999; Yang, Schweitzer et al., 1999; van Bokhoven, Hamel et al., 2001).

The first phenotypic traits observed in p73-deficient mice have been neuronal defects, due to the predominant expression of p73 in normal developing murine brain and sympathetic neurons (Yang, Walker et al., 2000). The implicated isoform expressed in these cells, ΔNp73, was also shown to inhibit neuronal apoptosis by blocking the pro-apoptotic function of p53 (Pozniak, Radinovic et al., 2000). Interestingly p73-knockout males are infertile due to possible perturbations in olfactory and hormonal pathways and not to overt reproductive tract defects. A more recent study takes it a step further by demonstrating that additional ΔN isoforms of p73 also inactivate another tumor suppressor, Rb, by increased phosphorylation. This effect of p73 on Rb is independent of its established ability to inhibit p53, and suggests a unique contribution of p73 to a tumorigenic phenotype (Stiewe, Stanelle et al., 2003). Indeed, initially p73, which is localized on human chromosome 1p32, was thought to be the tumor suppressor that is frequently targeted in neuroblastoma. However, it is important to keep in mind that $p73^{-/-}$ animals are not tumor prone, and no evidence exists that p73 is altered either genetically or epigenetically in human cancers.

To date it appears that p53 stands alone among its family as the critical suppressor of tumorigenesis. The reviews by Lohrum (Lohrum and Vousden, 2000), Moll (Moll, Erster et al., 2001), and Benard (Benard, Douc-Rasy et al., 2003) are excellent summaries of the intricacies, interactions, and implications associated with the diversity among this tumor suppressor family of proteins. For the purpose of this text, we will focus the remainder of our attention on the founder of this tumor suppressor family, p53.

p53 FUNCTION: THE TERMINATOR OR DIE ANOTHER DAY?

In its most basic sense, p53 functions as a sequence-specific transcription factor that coordinates an elaborate cellular stress response by binding to consensus sites of target genes. The products of p53 target genes, whose numbers increase almost daily in the literature, make a significant contribution to the biological functions of p53. The culmination of p53 activity in a given cell context may be apoptotic death or any combination of the following: inhibition of cell cycle progression, DNA repair, senescence or differentiation (Oren, 2003). The factors that initiate the p53 response and influence the final outcome are often specific to the exact nature, duration and context of the cellular stress. This aspect of p53 activation and function will be the focus of the following section.

Cell Cycle Control

Activation of p53 often results in cell cycle arrest, presumably to allow for DNA repair before replication or mitosis. A primary mechanism by which p53 negatively controls the cell cycle is through the transcriptional activation of $p21^{WAF1/CIP1}$ (el-Deiry, Tokino et al., 1993; Prives and Hall, 1999) In normal cells, p21 inhibits cyclin-dependent kinases, resulting in the prevention of phosphorylation and subsequent inactivation of Rb. The final result is the arrest of cell cycle progression in G_1/S phase. In support of this role, mouse embryonic fibroblasts (MEFs) derived from *p21*$^{-/-}$ mice exhibit a markedly impaired ability to arrest in G_1 following DNA damage (Deng, Zhang et al., 1995). Additionally, cells isolated from *p53*-knockout mice fail to induce p21 following UV or γ-IR treatment and are defective in cell cycle arrest after DNA damage.

p53 induces additional targets effecting cell cycle arrest, such as *Gadd45*α (Kastan, Zhan et al., 1992; Wang, Zhan et al., 1999). The *Gadd45* gene product interacts with other cell cycle-related proteins including Cdc2 and proliferating cell nuclear antigen (PCNA), leading to G_2/M arrest and DNA repair (Zhan, Antinore et al., 1999). Consistent with these in vitro findings, lymphocytes from *Gadd45*α-knockout mice fail to arrest in G_2/M after ultraviolet (UV) irradiation and display genomic instability. Similarly another p53 target, *14-3-3*σ, negatively regulates G_2/M progression following IR and somatic human *14-3-3*σ-knockout cells are consequently defective in IR-induced cell cycle arrest and undergo mitotic cell death (Hermeking, Lengauer et al., 1997). Whether p53 suppresses tumorigenesis through the regulation of these cell cycle regulators remains to be determined.

Transrepression of target genes by p53 also contributes to the regulation of cell proliferation and survival. The c-Myc protooncogene product is a potent promoter of cell cycle progression and is necessary and sufficient to drive quiescent cells into S phase (Baudino and Cleveland, 2001). Wild-type p53 represses c-*Myc* expression most likely through a transcriptional mechanism (Moberg, Tyndall et al., 1992; Levy, Yonish-Rouach et al., 1993). Therefore loss of p53 either through mutation or deletion could be considered a double-edged sword that results in a faulty break system with the accelerator locked in the on position. Consequently c-Myc, when overexpressed due to p53 inactivation, promotes aggressive cell growth, genomic instability, and transformation (Yin, Grove et al., 1999). It is therefore not surprising that p53 mutations and deregulated c-Myc expression are common events in human cancer.

Two endogenous genes involved in microtubule polymerization, *MAP4* and *stathmin*, are also repressed by wild-type p53 (Murphy, Ahn et al., 1999). Here the mechanism is somewhat better defined and the transrepression of *MAP4* by p53 involves the mammalian transcriptional co-repressor, mSin3a, which mediates the interaction of p53 with histone deacetylases (HDACs). Although p53 and Sin3a can be isolated in complex with the *MAP4* promoter by chromatin immunoprecipitation assays (ChIPs), there is no known DNA sequence element that confers p53-transrepression activity. Transrepression of p53 targets is an intense field of study with important recent developments that impact not only cell cycle control by p53 but also the initiation of apoptosis. The role of transrepression in modulating cell survival will be addressed further in the apoptosis section below.

Cellular Senescence

Cellular senescence is the ability of cultured cells to undergo a limited number of passages before ceasing to divide. Hence it is also referred to as "replicative senescence." Hayflick and colleagues made the initial observation that normal human embryonic or newborn fibroblast populations will divide approximately 60 to 80 times in culture before ceasing to reproduce (Hayflick, 1961). This number of doublings is appropriately termed, the Hayflick limit, and it differs significantly with that of murine cells. Fibroblast cultures that do not adhere to the Hayflick principle and continue to divide infinitely are said to be immortal, in comparison with their mortal senescent counterparts. Mouse embryonic fibroblasts (MEFs) are more likely to become immortalized than human fibroblasts; however, human fibroblasts in culture demonstrate longer proliferative potential. What accounts for the differences in these two systems? A review by Sherr and DePinho (2000) suggest that both extrinsic and intrinsic signals contribute to the expression of a common set of pro-senescent cell cycle regulators, including the CDK inhibitors $p16^{Ink4a}$ and $p21^{Cip1}$ as well as p53 and its inducer, $p19^{Arf}$ (Arf). It is the differing responses to such factors as cell culture shock (extrinsic) and telomere integrity (intrinsic) that result in the variation of senescent or immortal behavior observed in cultures of human and mouse origin.

The role of p53 in the proliferative capacity of fibroblast cultures is unquestionable. Nearly 80% of immortal MEFs contain mutant p53 alleles. Furthermore, MEFS derived from p53 or ARF null mice do not senesce and will continue to replicate indefinitely (Sherr and DePinho, 2000). Deletion of the p53 target, $p21^{Cip1}$ or the Rb tumor suppressor is not sufficient in preventing growth arrest (Sherr and DePinho, 2000), implicating a role for additional p53 targets. In addition normal primary MEFs will undergo premature senescence following exposure to oncogenic stimuli, such as the *Ras* oncogenes (Serrano, Lin et al., 1997); while the same stimuli will result in immortal transformation of cultures lacking p53 or Arf (Kamijo, Zindy et al., 1997). Taken together, loss of p53 function either directly through mutation or via alterations of regulating genes, such as Arf (loss of function) and/or Mdm2 (gain of function), in murine cell cultures is considered essential for overcoming normal replicative limits.

The case for p53 in human fibroblast proliferation is more complicated. Loss of p53 (or Rb) lengthens the replicative capacity of human cultures, but unlike murine counterparts, loss of p53 alone is not sufficient to immortalize these cells (Wright and Shay, 1992). Studies have shown that this may be due to shorter telomere length in human cells. Over time and many replications, telomeres become shortened and may induce DNA damage responses like p53 activation, which would contribute to the onset of senescence. This may be bypassed if the p53 checkpoint is compromised by oncoproteins or mutations. In this case cells may continue to proliferate, but only to the extent that their chromosomes will allow, since telomerase is not normally expressed in human cells. When telomeres become completely eroded, human cells exhibit severe genetic instability, which leads to a state of "crisis" and eventually cell death. A series of experiments using enforced expression of TERT (telomerase reverse transcriptase), the catalytic subunit of telomerase, demonstrated that the activation of this enzyme, and thus maintenance of telomere length and chromosomal stability can overcome senescence and allow the fibroblasts to replicate infinitely (Bodnar, Ouellette et al., 1998). It should also be noted that the majority of immortalized human cultures are derived from established, telomerase-expressing neoplasms. Most of these tumors will also have lost p53 or Rb function, allowing the life span of the cells to be extended to the point of telomerase activation.

In summary, it appears as though human cells in culture are more effective than murine cells in their ability to detect and repair DNA damage before reaching the point of no return. Their extended proliferative capacity and resistance to transformation in comparison with mouse cultures seem to reflect the differences in life span of these two species.

Initiation of Apoptotic Pathways

Recent studies have significantly advanced our understanding of how p53 functions in tumor suppression. However, the mechanism by which p53 controls cell death is not as clearly understood (or agreed upon) as its role in cell cycle regulation. Molecules associated with both the death receptor- and mitochondria-mediated apoptotic pathways have been implicated in the p53 pathway during cellular stress (Vousden, 2000; Sax and El-Deiry, 2003). Indeed, the list of p53-regulated survival and death genes is already extensive and still growing. Many of these candidates are identified by differential gene expression approaches in cells undergoing p53-mediated apoptosis. These targets are then "validated" by overexpressing the candidate genes at nonphysiological levels, which consequently induces cell death. In a few circumstances the candidates are tested in more rigorous detail by interfering with their expression through antisense or RNAi strategies. Although these approaches are considered acceptable practice, the biological contribution of p53 targets to apoptosis and tumor suppression awaits more physiologically relevant analyses, such as knockout mouse models. Several of the more interesting targets that may be involved in p53-induced cell death are highlighted in this section (Table 19.2) (Cadwell and Zambetti, 2001; Oren, 2003).

Two proapoptotic members of the tumor necrosis factor receptor (TNFR) superfamily, Fas/Apo1 and Killer/DR5, are regulated in a p53-dependent manner in response to chemothereapeutic drugs (Wu, Burns et al., 1997; Muller, Wilder et al., 1998; Wu, Burns et al., 1999; Sax and El-Deiry, 2003). The activation of these signaling pathways through the up-regulation of these receptors by p53 is thought to play a role in sensitizing cells to DNA damage. However, thymocytes from Fas-deficient mice, while resistant to death induced by ligation of Fas or CD3, are susceptible to DNA damage-induced apoptosis, indicating that other targets are required for p53-mediated cell death, at least under these defined conditions. One such candidate is the membrane-bound protein PERP (p53 apoptosis effector related to PMP-22). A subtractive hybridization

TABLE 19.2. Key p53 Target Genes in Apoptosis and Cell Cycle Arrest

Cell cycle/DNA repair		
↑ p21	↑ DcR2	
↑ Gadd45α	↑ HB-EGF	
↑ 14-3-3σ	↓ c-Myc	
Apoptosis		
↑ Bax	↑ PIDD	↑ Apaf-1
↑ Killer/DR5	↑ PERP	↑ Zac-1
↑ Fas/ApoI	↑ Puma	↓ Bcl-2
↑ Noxa	↑ PIGs[a]	↓ MAP4

[a] For example, Ei24.

strategy, in which G_1-arrested MEF cDNA populations were subtracted from apoptotic MEF cDNA populations, identified PERP as a p53 target gene that induces apoptosis. Overexpression of PERP is able to induce cell death, which can in turn be blocked by the overexpression of anti-apoptotic protein Bcl-2. Although the complete PERP signaling pathway that leads to cell death is not yet defined, it is clear that the transcriptional induction of this protein by p53 is able to initiate an apoptotic pathway (Attardi, Reczek et al., 2000).

Recently a p53 cytoplasmic protein containing a death domain was identified and appropriately named p53-induced protein with a death domain (PIDD; Lin, Ma et al., 2000). When compared to wild-type MEFs, p53-deficient MEFs have much lower levels of PIDD, and the interference of PIDD expression by antisense strategies attenuates p53-mediated cell death (Lin, Ma et al., 2000; Sax and El-Deiry, 2003). The precise mechanism by which PIDD induces apoptosis is still not known.

The regulation of the extrinsic (receptor-mediated) pathway at the most upstream level, the plasma membrane, provides a link between DNA damage and the initiation of a cascade that may work to induce apoptosis directly, or alternatively, through the disruption of mitochondria (Sax and El-Deiry, 2003). Mitochondrial dysfunction, usually typified by the release of cytochrome c and activation of Apaf-1 and Caspase 9, initiates the intrinsic apoptotic pathway. This pathway is closely regulated by the balance between prosurvival and proapoptotic Bcl-2 family members (Adams and Cory, 1998). One of the earliest p53 target genes associated with mitochondrial-mediated apoptosis is *Bax*, a proapoptotic Bcl-2 family member (Miyashita and Reed, 1995). *Bax*, through its binding to the survival factors Bcl-2 and Bcl-X_L shifts the balance of cell viability in favor of apoptosis (Miyashita and Reed, 1995; Adams and Cory, 1998). Some questions arise to the authenticity of *Bax* as a bona fide target: (1) the p53-consensus sites within the *Bax* promoter are not conserved between mouse and human (Schmidt, Korner et al., 1999), (2) Bax protein levels are not consistently induced during p53 activation (McCurrach, Connor et al., 1997), and (3) thymocytes isolated from wild-type and *Bax*$^{-/-}$ mice, but not p53-knockout animals, undergo massive cell death following ionizing irradiation (Knudson, Tung et al., 1995). The latter observation clearly demonstrates a strict requirement for p53 in DNA damage-induced death of primary thymocytes and that Bax is dispensable. It may be that Bax plays a significant role in stress-induced death but is redundant with other pro-apoptotic factors such as Bak. Evidence supporting this possibility is provided by studies using cells derived from *Bax/Bak* double-knockout mice, which are remarkably resistant to many apoptotic signals (Lindsten, Ross et al., 2000; Wei, Zong et al., 2001).

If p53 is required for DNA damage-induced death of primary cells independently of Bax, what other downstream targets may mediate this apoptotic response? Several interesting p53-regulated genes have been recently identified that are strongly pro-apoptotic, including two Bcl-2 family members, PUMA (p53 upregulated mediator of apoptosis) and Noxa (Nakano and Vousden, 2001; Yu, Zhang et al., 2001). Both of these proteins belong to the subset of Bcl-2 related pro-apoptotic factors that consists of only a BH3 domain. Noxa expression is induced by p53 in mouse embryo fibroblasts and thymocytes following DNA damage. Through its BH3 domain, Noxa interacts with pro-survival Bcl-2 proteins, thereby altering mitochondrial membrane permeability, which leads to cell death (Oda, Ohki et al., 2000). Its physiological contribution to apoptosis, however, awaits the development of a knockout mouse model.

Puma was originally identified as a p53 pro-apoptotic target by differential gene expression approaches and in a yeast two-hybrid screen for Bcl-2 binding proteins. Interestingly Puma expression is regulated by diverse death signals, including those that are p53 dependent (DNA damage) and independent (e.g., glucocorticoids or serum deprivation). Inhibition of PUMA expression by antisense attenuates p53 apoptotic responses, while its overexpression potently suppresses colony formation in human tumor cell lines, presumably due to PUMA-mediated apoptosis (Nakano and Vousden, 2001; Yu, Zhang et al., 2001). Recently we generated the first Puma knockout mouse, whose phenotype confirms a critical role for the protein in both p53-dependent and independent apoptotic pathways, the implications of which are discussed below (unpublished data).

Another intrinsic target induced by p53 is p53-regulated apoptosis-inducing factor (p53AIP1). This protein is not a member of the Bcl-2 family, but it is localized to the mitochondria following DNA damage, subsequent to phosphorylation of p53 on serine-46 (Oda, Arakawa et al., 2000). These findings present yet another paradox to the field; specifically, that phosphorylation of serine-46 is required for p53-mediated cell death. Post-translational modifications intuitively play a critical role in the induction and activation of p53 in response to DNA damage (see below). Many of these modification sites are conserved throughout evolution, presumably to maintain important functional activities. However, serine-46 of human p53 is one such site that is not conserved in mice or other rodents, raising the question of whether this modification is really required for p53's apoptotic activity.

The p53 tumor suppressor can also transcriptionally activate a subset of genes referred to as *PIGs* (p53 inducible genes), which are involved in the production of reactive oxygen species (ROS). In turn, ROS damages the mitochondria and induces apoptosis (Polyak, Xia et al., 1997). In particular, expression of *PIG3*, which is highly related to a plant NADPH oxidoreductase, is strongly induced by p53 but does not appear to be sufficient for generating ROS or inducing cell death. It has been proposed that the induction of multiple *PIGs* may be required in concert to initiate an apoptotic response.

The absolute requirement for p53 to function as a DNA binding protein that activates transcription and expression of downstream target genes to induce apoptosis has been controversial. This issue stems in part from studies demonstrating that p53 fragments (amino acids 1–214 or 319–393) containing only the amino or carboxy terminus can elicit some degree of cell death, despite their inability to bind DNA (Haupt, Rowan et al., 1995; Wang, Vermeulen et al., 1996). In addition, recent studies support a direct role for p53 at the mitochondrial surface in the induction of the apoptotic response (Dumont, Leu et al., 2003; Mihara, Erster et al., 2003). These studies should be considered with some reservation as they relied either on the overexpression of p53 peptides (Haupt, Rowan et al., 1995; Wang, Vermeulen et al., 1996) or forced mitochondrial targeting strategies (Dumont, Leu et al., 2003; Mihara, Erster et al., 2003). Furthermore the region that is reportedly involved in mitochondrial complex formation is localized to the same core region of p53 involved in sequence-specific DNA binding (Manfredi, 2003; Mihara, Erster et al., 2003). Hence any disruption of the mitochondrial targeting region will simultaneously impair the p53 transactivation response, making it impossible to rule out a transcription-dependent role for p53 in cell death. This mindset is further supported by the finding that while the arginine-72 polymorphic variants (amino acid 72 is either Arg or Pro) demonstrate increased localization to the mitochondria, they also show a striking transcriptional upregulation of the human version of PERP, a previously identified proapoptotic p53 inducible gene (see above) (Dumont, Leu et al., 2003; Manfredi, 2003). These concerns raise the important point that it is imperative to address p53 functions in a physiological manner. Indeed, cells that are genetically engineered to express a transactivation-defective mutant p53 protein from the endogenous locus as a knock-in allele are defective in both growth arrest and apoptosis in response to cell stress (Weber and Zambetti, 2003).

An excellent case in point is the recent development of *Puma*$^{-/-}$ mice by our lab (G.Z., data submitted). Puma is required for hematopoietic cell death triggered by diverse death signals, including ionizing radiation (IR), deregulated c-Myc expression and cytokine withdrawal. Puma is also essential in vivo for IR-induced death in the thymus and throughout the developing nervous system. Most important, the lack of neuronal cell death in the *Puma*-knockouts following DNA damage rivals that observed in p53-null mice, indicating that Puma accounts for nearly all of the apoptotic activity attributed to p53. These findings establish Puma as the principal mediator of p53-dependent cell death and highlight the importance of p53 as a transcription factor in this process. Moreover these findings demonstrate that endogenous wild-type p53 is not sufficient to induce efficient cell death when Puma is absent, thus arguing against mitochondrial localization and similar models that evoke transcriptionally independent mechanisms for p53-mediated cell death.

p53 REGULATION: STRESS MANAGEMENT 101

p53 responds to numerous forms of stress including DNA damage, hypoxia, oncogene activation, nucleotide depletion, and hyperproliferation. Depending on the nature of the stress and cell context, p53 can efficiently inhibit cell growth by inducing cell cycle arrest or apoptosis (Vousden, 2000). The factors that influence this cell fate decision are not yet defined, although several intriguing models have been proposed (Weber and Zambetti, 2003). p53 function must be maintained at sufficiently low levels to permit normal growth and development and yet retain the capacity for rapid induction during stress (e.g., inappropriate cell growth due to oncogenic events) to prevent tumorigenesis. A recent review by Vousden summarizes the major mechanisms controlling p53 function: the regulation of p53 protein levels, control of the localization of the p53 protein, and modulation of the activity of p53, particularly its ability to function as a transcription factor (Vousden, 2002). The p53 protein is subject to several post-transcriptional modifications and numerous interactions with other cell proteins. The contributions of these modifications and interactions to the regulation of p53 stability, location, and activity are just beginning to be understood. An excellent recent review by Appella and Anderson summarizes how these modifications regulate p53 function (Appella and Anderson, 2001).

In the absence of stress, p53 is maintained at low steady state levels. The limited p53 molecules that do exist in this state are not very effective as transcriptional activators but nevertheless contribute to the maintenance of basal levels of target genes. It is well established that Mdm2 (murine double minute-2; HDM2 in humans) is responsible for the majority of negative regulation of p53, even in the absence of stress (see reviews in Momand and Zambetti, 1997; Daujat, Neel et al., 2001; Michael and Oren, 2002). In basic terms, Mdm2 binds to p53 and renders it inactive (Momand, Zambetti et al., 1992). There are at least three molecular mechanisms by which this occurs. First, Mdm2 physically interferes with the transcriptional activity of p53 by binding its N-terminal transactivation domain, which blocks critical interactions with transcriptional components, such as transcription activating factors (TAFs), that are necessary for p53-dependent regulation of gene expression (Oliner, Pietenpol et al., 1993). Second, Mdm2 shuttles p53 out of the nucleus, thus ensuring that p53 cannot transcriptionally activate its target genes

(Roth, Dobbelstein et al., 1998). Third, Mdm2 "terminates" p53 through its E3 ubiquitin ligase activity, which targets p53 for degradation by the 26S proteosome (Grossman, Deato et al., 2003). The functional interaction between p53 and Mdm2 is absolutely essential and represents an intricate balance that is designed to regulate itself (Barak, Juven et al., 1993; Wu, Bayle et al., 1993). The Mdm2 gene contains two adjacent p53 binding sites, and is a bona-fide target of p53, hence establishing a negative regulatory feedback loop. This loop serves as the receiving end for incoming stress signals, which collectively determine the state of p53 activity under a given set of conditions and cell types. Thus elevated levels of Mdm2 will interfere with the activity of p53 and conversely, signals that render p53 immune to Mdm2 binding (i.e., phosphorylation of N-terminal residues; see below) will stabilize and activate p53 (Oren, 2003).

The genetic proof that Mdm2 regulates p53 comes from elegant knockout mouse studies (Jones, Roe et al., 1995; Montes de Oca Luna, Wagner et al., 1995). Deletion of Mdm2 causes early embryonic lethality around days E4.5–E6.5, and this is due to p53 blocking cell proliferation and/or inducing rampant cell death. Remarkably the embryonic lethality associated with deletion of Mdm2 is fully eliminated by crossing the Mdm2 knockout allele onto a p53-deficient background, and Mdm2/p53 double-knockout mice are born at near Mendelian frequency. Moreover the phenotype of the double-knockout mice is essentially the same as p53-null animals, indicating that Mdm2 plays a critical role in negatively regulating p53. These findings are clinically relevant, as Mdm2 is amplified and overexpressed in nearly one-third of human soft tissue sarcomas and osteosarcomas (Oliner, Pietenpol et al., 1993). These tumors express wild-type p53, which is rendered nonfunctional due to the elevated levels of Mdm2. Overexpression of Mdm2 also occurs in other tumor types to varying extents (for review, see Momand and Zambetti, 1997).

In addition to ubiquitination by Mdm2, p53 undergoes multiple post-translational modifications that positively influence its activity (Fig. 19.3A) (Buschmann, Adler et al., 2000; Appella and Anderson, 2001; Cadwell and Zambetti, 2001). A growing number of protein kinases have been discovered that phosphorylate p53, including ATM, ATR, Chk-1, and Chk-2 (Appella and Anderson, 2001). Phosphorylation of p53 by these kinases following DNA damage is required for the efficient release of p53 from Mdm2, thus stabilizing p53 and activating its tumor-suppressor functions. Perhaps the best understood modification of p53 is phosphorylation of Serine-15 (S15) by ATM. Here S15 is conserved throughout evolution, and this site becomes modified only after DNA damage or certain other forms of cell stress, such as hypoxia. Patients with germ-line mutations in ATM are highly sensitive to DNA damage and are tumor prone due in part to a profound inability to activate p53. If p53 is not efficiently phosphorylated at S15 during DNA damage, such as in ATM patients, its tumor-suppressor function is compromised. It has been proposed that phosphorylation of S15 sets off a cascade of modifi-

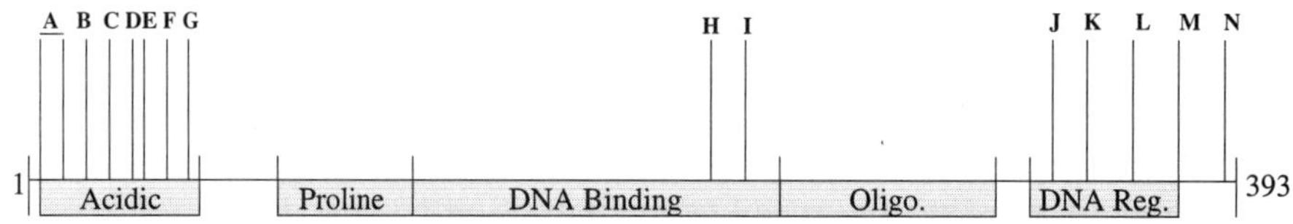

A. Phosphorylation

A. CKI- S6, 9
B. ATM, ATR, DNA-PK- S15
C. CHK1,2- S20
D. CAK- S33
E. ATR- S37
F. HIPK2- S46
G. TAF_{II} 50- T55
H. CDK2, CDC2- S315
J. PKC- S371
L. PKC- S378
N. CKII- S386

B. Acetylation

I. pCAF- K320
K. p300- K373
M. p300- K382

Figure 19.3. Post-translational modification of p53. Phosphorylation sites and their respective kinases are shown in (**A**). Acetylation targets are summarized in (**B**).

cations that prevent Mdm2 from binding as well as activating its DNA binding activity (Sakaguchi, Herrera et al., 1998).

A recently identified homeodomain-interacting protein kinase-2 (HIPK2) was shown to regulate the phosphorylation of p53 at serine 46 (S46), in conjunction with wild-type p53-inducible phosphatase (Wip1/PPM1D) (Fiscella, Zhang et al., 1997; D'Orazi, Cecchinelli et al., 2002). Mutation of this serine selectively impairs the induction of a proapoptotic target gene, p53AIP1, resulting in compromised p53-mediated apoptosis, while the expression of $p21^{Cip1}$ remains unaffected (Oda, Arakawa et al., 2000). This suggests that post-translational modifications of p53 may play a role in directing p53 to specific target genes that will ultimately determine cell fate (Weber and Zambetti, 2003). However, as discussed above, modification of S46 has not been critically challenged in a physiological manner. Since S46 is not conserved in rodents, it may be possible to create a somatic knockin mutation in human cell lines, similar to the strategy used for generating the p53-deficient HCT116 colon cancer cells (Bunz, Dutriaux et al., 1998). Alternatively, Hollstein and coworkers have developed a "humanized p53" knockin mouse model, referred to as Hupki mice, in which the endogenous mouse p53 gene has been swapped with the corresponding human fragment (Luo, Yang et al., 2001). The resultant Hupki mice faithfully express human p53 and are phenotypically normal. They have adopted the human p53 as their own and are not tumor prone. Moreover the signaling pathway that leads to S46 phosphorylation is intact (Clarke and Hollstein, 2003). These animals should provide an ideal opportunity for mutating S46 and testing its consequence on p53-dependent cell death and tumor suppression.

Acetylation is also thought to regulate p53 function by activating its DNA binding activity (Fig. 19.3B) (Buschmann, Adler et al., 2000; Appella and Anderson, 2001; Cadwell and Zambetti, 2001). Factors such as CBP and PCAF selectively acetylate several key lysines in the C-terminus of p53 (e.g., K320, K373, K381, and K382), which enhances p53 DNA binding in vitro. Consistent with a functional role for these lysines in p53 activity, mutation of these sites markedly impairs the ability of p53 to function as a transactivator or negative growth regulator in transient transfection-based assays. Somewhat surprisingly, ChIP analysis demonstrated that acetylation of p53 did not increase its binding to the $p21^{CIP1}$ promoter but instead facilitated co-activator recruitment and histone acetylation (Barlev, Liu et al., 2001). These are clearly convincing and intriguing results that warrant further study. It will be interesting to challenge the role of these modification sites in a knockin mutant mouse model. The reviews by Prives (Prives and Manley, 2001), Xu (Xu, 2003) and Melchior (Melchior and Hengst, 2002) are excellent resources for studying the contribution of acetylation as well as other modifications in the regulation of p53 tumor-suppressor activity.

In addition to post-translational modifications, p53 can be activated by upstream regulators, such as $p19^{Arf}$ ($p14^{ARF}$ in humans). The p19Arf (alternative reading frame) tumor suppressor is encoded by the *Ink4a* locus, which also encodes $p16^{Ink4a}$. This is truly fascinating, since the existence of a single gene encoding two completely unrelated proteins is unprecedented in higher eukaryotes. What is perhaps the most important and interesting aspect of *Ink4a* is that p16 and $p19^{Arf}$ play pivotal roles in activating the two predominant tumor-suppressor pathways, Rb and p53, respectively (p16 is addressed in Chapter 18 by Masciullo and Giordano). In light of these critical interactions, tumors frequently target *Ink4a* either by deletion or by silencing, and a general inverse correlation seems to exist between Arf and p53 expression: Arf levels are high in tumors with mutant p53, and conversely, tumors with wild-type p53 do not express Arf. Loss of Arf expression cripples the p53 pathway, and Arf null tumors can therefore tolerate wild-type p53. Normally Arf functions as a sensor of inappropriate cell proliferation initiated by deregulated oncogenes. Enforced expression of c-Myc, E2F1, or E1A induces Arf mRNA and protein through unknown mechanisms. The Arf protein binds and sequesters Mdm2 to the nucleolus, thus physically segregating Mdm2 from p53 (Weber, Taylor et al., 1999; Lohrum, Ashcroft et al., 2000). Moreover Arf antagonizes the E3 ubiquitin ligase activity of Mdm2, allowing for p53 to become stabilized (Honda, Tanaka et al., 1997; Honda and Yasuda, 1999). Therefore Arf induces p53 expression and function through the inactivation of Mdm2 at multiple levels. In turn, when p53 levels reach a certain threshold, it represses Arf expression, which then allows for Mdm2 to down-regulate the response (Robertson and Jones, 1998). How Arf responds to oncogenic signals and what role it plays in other biological processes (e.g., RNA processing and p53-

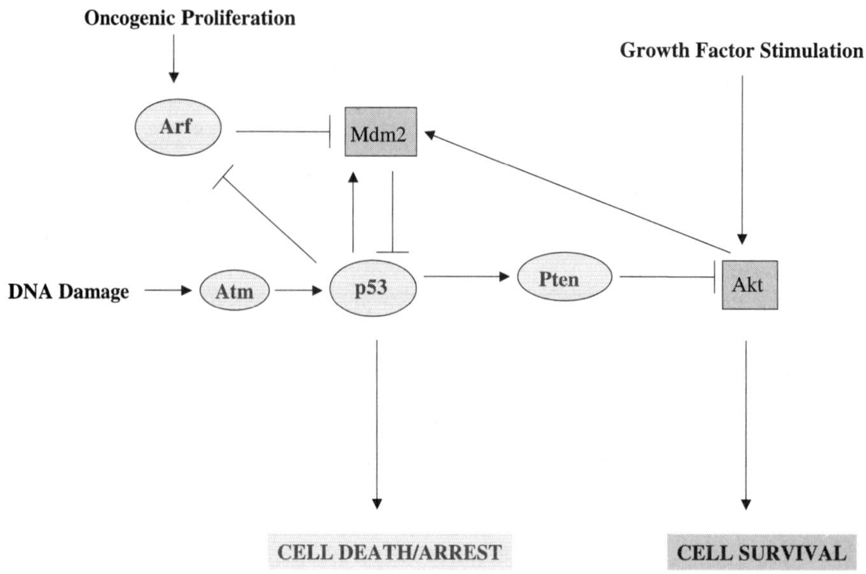

Figure 19.4. Key elements in the p53 signaling pathway. Hyperproliferation induces Arf, which antagonizes Mdm2 and leads to the activation of p53 and cell cycle arrest and/or death. DNA damage activates PI3-like kinases (e.g., Atm), which phosphorylate p53 such that Mdm2 is unable to bind. Alternatively, activation of Akt leads to the phosphorylation of Mdm2 and consequent repression of p53, which favors cell growth and survival. By contrast, the expression of Pten (a downstream target of p53) blocks Akt signaling and promotes apoptosis.

independent tumor-suppressor pathways) are important questions that remain to be answered (Lowe and Sherr, 2003).

In a parallel signal transduction pathway, mitogens activate the Akt/PKB serine-threonine kinase, resulting in the phosphorylation of Mdm2 at serines 166 and 186 (Mayo and Donner, 2001). Phosphorylation of Mdm2 at these sites promotes nuclear localization and consequently the repression of p53, and the induction of cell proliferation. The PTEN tumor suppressor antagonizes AKT and consequently blocks phosphorylation of Mdm2. This allows for Mdm2 to accumulate within the cytoplasm during p53 activation. Since human PTEN is a downstream target of wild-type p53, PTEN expression will also be induced, resulting in a further accumulation of Mdm2. The balance between PTEN, p53, and MDM2 is considered important to allow damaged cells the opportunity to repair the defects or to delete irrevocably damaged cells by apoptosis (Mayo and Donner, 2002) (Fig. 19.4) (Oren, 2003).

Arf and PTEN autogulatory loops are just two hallmark examples that only scratch the surface of existing multi-tiered interactions regulating p53 activity. It is important to keep in mind that each of the tumor suppressors described above, as well as the Mdm2 protein, possess the ability to interact with targets other than p53. Thus these loops will impact many additional processes, which will contribute to the biological complexity associated with p53 (Weber, Jeffers et al., 2000; Daujat, Neel et al., 2001; Eymin, Leduc et al., 2003). In summary of this section, p53-mediated tumor suppression via cell cycle control and apoptosis is undoubtedly linked to its transcriptional regulation, both positively and negatively, of downstream target genes.

p53 AND TUMORIGENESIS: BORN TO BE WILD-TYPE

Multiple lines of evidence highlight the importance of wild-type p53 in tumor suppression: (1) p53 is the most frequently mutated gene identified in human cancer (>50% in all cancers); (2) Li-Fraumeni syndrome (LFS) is a genetic disease attributed to germ-line p53 mutations (Malkin, Li et al., 1990), and people affected by this syndrome typically develop malignant tumors of widespread tissue types by early adulthood, upon mutation or deletion of the normal p53 allele (reviewed in Varley, 2003); and (3) p53 knockout mice develop normally but are remarkably predisposed to developing lymphoma and a broad spectrum of other cancers, whose malignancies result in death by approximately 4 to 6 months of age (Donehower, Harvey et al., 1992; Jacks, Remington et al., 1994). Driving home the point that p53 is clinically relevant to human cancer, more than 18,000 mutations have been detected in 150 different cancer types. These findings are organized in a comprehensive database (*www.iarc.fr/p53/*) that also serves as an excellent resource for learning the latest findings on p53 (Hollstein, Rice et al., 1994; Beroud and Soussi, 2003).

Mutation of p53 occurs by deletion, insertion, truncation, or point mutation, and tumors frequently undergo loss of heterozygosity (LOH) in which the wild-type allele is deleted. Most of the p53 mutations (85%) result in a single amino acid substitution and the production of a missense protein (Hollstein, Rice et al., 1994). The vast majority (>90%) of these mutations occur within the sequence-specific DNA core-binding domain, and about 50% alter codons 175, 248, 249, 273, or 282 (Levine, 1997), frequently referred to as "hot spot" mutations (Fig. 19.5) (Hainaut, Hernandez et al., 1998; Hernandez-Boussard, Rodriguez-Tome et al., 1999). If p53 fails to properly bind to DNA, it cannot regulate target gene expression and therefore is unable to function in tumor suppression. The crystal structure of p53 (Fig. 19.6) (Cho, Gorina et al., 1994; Berman, Westbrook et al., 2000) elegantly illustrates how wild-type p53 binds to DNA and provides insight into how the "hot spot" mutants either corrupt the conformation of the DNA binding domain (e.g., R175H) or directly disrupt the contact points (e.g., R248W and R273H) (Cho, Gorina et al., 1994).

Mutation-independent mechanisms also exist to functionally inactivate p53 during tumorigenesis. As mentioned in previous sections, p53 activity is tightly regulated via specific interactions with other proteins. It is this interdependence of tumor suppression pathways that leads to tumor formation in which wild-type *p53* alleles are tolerated, yet unable

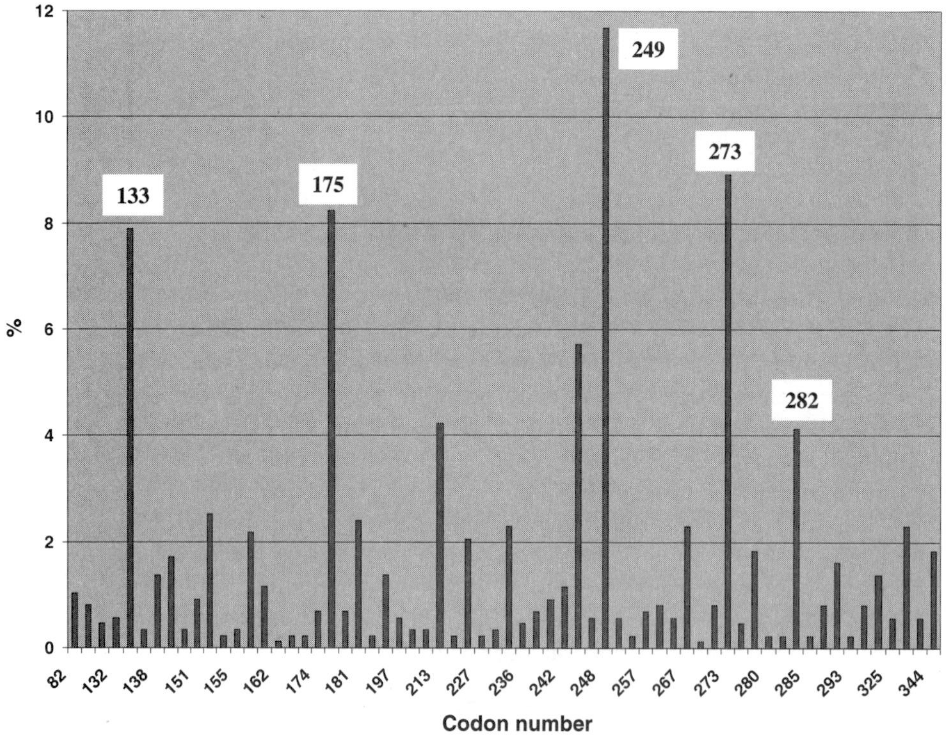

133
175
249
273
282
%
Codon number

A

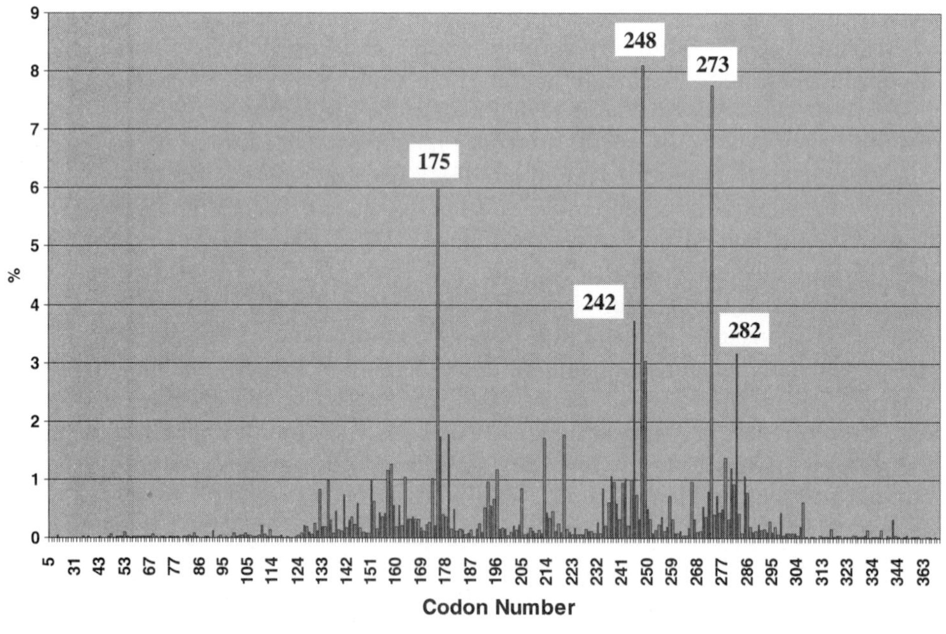

175
248
273
242
282
%
Codon Number

B

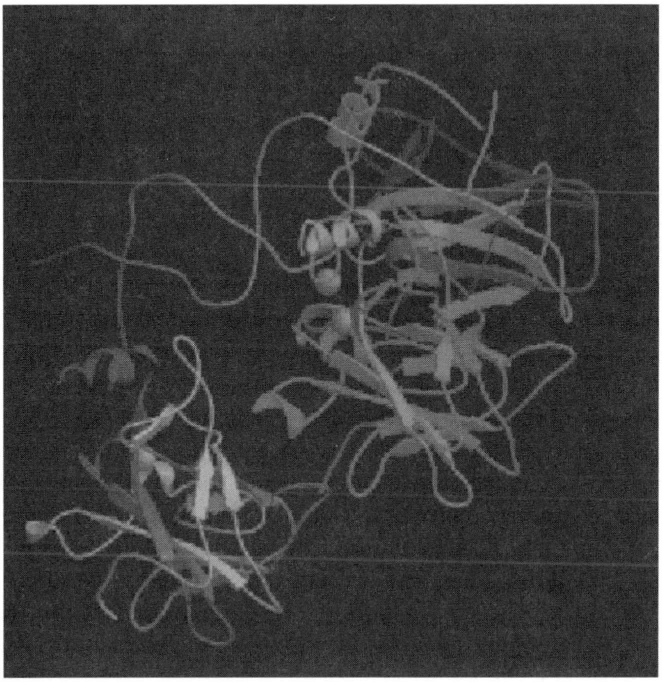

Figure 19.6. Structure of wild-type p53 bound to DNA. Protein Data Bank ID: 1TUP (see Web Resources). (See color insert.)

to function effectively. Key examples are tumors in which *ARF/INK4A* is deleted or silenced (e.g., colorectal, gastric, and uterine tumors) or *MDM2* is amplified and overexpressed (~30% soft tissue and osteosarcomas) (Esteller, Cordon-Cardo et al., 2001). Furthermore the causative agents of cervical carcinoma, human papillomavirus (HPV) serotypes 16 and 18, express the viral E6 oncoprotein that targets p53 for ubiquitin-dependent protein turnover (Scheffner, Werness et al., 1990; Werness, Levine et al., 1990). Although these tumors often retain wild-type p53 alleles, there is little functional p53 protein expressed in these cells due to inactivation by E6. Interruption of p53 downstream signaling through silencing of proapoptotic factors, such as Apaf-1, has also been

Figure 19.5. p53 "Hot spot" Mutations. (**A**) Codon distribution of missense and nonsense germ-line mutations (n = 874) IARC TP53 database, version R8. Codons 133, 175, 248, 273, and 282 are located within the DNA-binding domain and are most frequently the targets of inherited point mutations. (**B**) Codon distribution of missense and nonsense somatic mutations (n = 15,082) identified from about 150 distinct tumor types. Similar to the germ-line mutations, somatic hot spots occur at codons 175, 245, 248, 273, and 282. Figures are based on the most current data available from the IARC p53 mutation database (see Web Resources), version 8 (June 2003).

observed in malignant melanoma with wild-type p53 (Soengas, Capodieci et al., 2001).

Last, p53 can be functionally inactivated due to alterations in its subcellular localization. Approximately 95% of neuroblastomas express wild-type p53 protein, which often accumulates to high levels within the cytoplasm. A newly recognized factor, Parc (p53-associated parkin-like cytoplasmic protein), has been implicated as the anchor that sequesters p53 within the cytoplasm (Kastan and Zambetti, 2003; Nikolaev, Li et al., 2003). Interference with nuclear localization of p53 as seen in neuroblastomas, as well as inflammatory breast carcinomas (Moll, Riou et al., 1992), intuitively impairs its ability to regulate downstream targets and to suppress tumorigenesis. The identification of such mutation-independent mechanisms for p53 inactivation provides the framework for developing strategies that could one day be used in treating patients with these tumors (see below). Taken together, it appears that most, if not all tumors, are defective in p53 tumor-suppressor function, either by mutation of the p53 gene or by corruption of the p53 pathway.

Specifically, mutation of p53 contributes to tumorigenesis in at least three possible ways: (1) Mutation of p53 results in a loss of wild-type p53 tumor-suppressor function (e.g., inability to activate downstream targets); (2) mutant p53 can bind to and inhibit wild-type p53, and possibly p63 and p73 family members, in a dominant-negative manner; and (3) mutant p53 exhibits gain of function activities, which confer a selective growth and survival advantage to tumors. Each of these mechanisms is well documented, and we will highlight several models of particular interest in assessing the consequences of p53 mutations on tumorigenesis.

Mouse models utilizing targeted p53 alleles are being applied to emulate the genetic defects that give rise to human cancers, and the generation of these mice, and their respective phenotypes, were recently reviewed by Parant and Lozano (2003). It is important to consider that although the p53 knockout mice were especially revealing in demonstrating the importance of p53 as a tumor suppressor, the most common p53 alteration in both somatic and inherited cancers is a missense mutation. Therefore mice expressing endogenous mutant p53 alleles would be considered more relevant to modeling human cancers. The first mouse model to achieve this goal contains a single nucleotide mutation at codon 172 (equivalent to the amino acid 175 hot spot mutation in humans), resulting in an Arg to His substitution (Liu, McDonnell et al., 2000). Mice heterozygous for the mutant 172H allele differed from $p53^{+/-}$ mice in tumor spectrum and, more important, the frequency of metastasis. These data indicate clear differences between p53 missense mutations and a null allele in vivo during tumorigenesis, and suggest that the 172 mutant expresses a gain of function that promotes metastasis (Parant and Lozano, 2003). Additional models are currently being developed to test other p53 mutations, including those that are commonly found in human cancers, as well as those that will challenge suspected functional residues, such as those sites that are modified during cell stress (see above).

Of special relevance to human mutant p53 phenotypes is our discovery of a novel p53 germ-line mutation that is specifically associated with adrenal cortical tumors (ACT) (Ribeiro, Sandrini et al., 2001). Pediatric ACT is very rare, yet in southern Brazil the incidence is significantly elevated, 10 to 15 times higher than worldwide estimates. Pediatric ACT is almost always diagnostic of a germ-line p53 mutation, which is usually associated with Li-Fraumeni syndrome (LFS). LFS is characterized by a highly penetrant tumor phenotype with carriers developing cancer, sometimes multiple forms, as children or young adults. Although the children from southern Brazil were not from tumor prone families and only susceptible to ACT, we examined the p53 status of 45 ACT patients and their families. Remarkably all but one of these children had an identical germ-line point mutation in p53, resulting in an Arg to His substitution at amino acid 337, which lies outside the DNA binding domain. Interestingly the mutation occurs in the oligomerization domain, corresponding exactly to the residue that participates in a stabilizing salt bridge. This inherited mutant allele exists in unrelated families and polymorphic marker analyses demonstrated that at least some mutant alleles arose independently, thus eliminating a founder effect. Applying multiple approaches (e.g., DNA binding, promoter-reporters, colony reduction, apoptosis, and gain-of-function assays) to studying the biological consequence of the R337H to p53 function failed to reveal any defects, at least in tissue culture cell-based models. By contrast, the analysis of primary ACT samples demonstrated that the wild-type allele was deleted (typical of tumor-suppressor genes) and that the missense p53 protein was highly expressed in the nucleus of these tumor cells. If p53 was functional in these tumors, it should either block cell growth and/or induce apoptosis, which was clearly not the case as these tumors can reach a size of 1 kg or more. The most convincing piece of this puzzle comes from the finding that the inheritance of the R337H mutation increases the risk of developing ACT, and no other tumor types, by 300,000-fold. Therefore this inherited R337H p53 mutation represents a low-penetrance p53 allele that contributes in a tissue-specific manner to the development of pediatric ACT (Ribeiro, Sandrini et al., 2001).

Additional clues to understanding how the R337H mutation selectively promotes ACT are provided by structural analyses of the mutant protein. In wild-type p53, Arg-337 forms a stabilizing salt bridge, and this interaction relies on the biochemical properties of arginine, which will be positively charged within the physiological pH range. We predicted that histidine would be less efficiently protonated and therefore less likely to be competent for participating in the salt bridge. Indeed, the mutant tetramerization domain (p53tet-R337H) is considerably less stable than the wild-type domain (p53tet-wt) and completely unfolds at slightly basic pH values (between 7 and 8), which is well within physiological limits. The sensitivity of the R337H mutant strictly correlates with the protonation state of the His 337 residue and its ability to form the stabilizing salt bridge. These results identify the ACT-associated R337H missense protein as the first mutant of p53 that displays a pH-sensitive molecular defect. We speculate that subtle differences in the environment of the adrenal gland compromises R337H function, which leads to the development of ACT (DiGiammarino, Lee et al., 2002). Finally we propose that the significantly higher propensity of the R337H mutant to form amyloid-like fibers, compared to wild-type p53, provides a possible mechanism for the observed nuclear accumulation of p53 in ACT cells (Lee, Galea et al., 2003).

These findings also lead to the notion that other familial tumor syndromes could be related to novel germ-line p53 mutations. Whereas mutations that disrupt DNA contact points or alter the overall structure of p53 so that it no longer binds to DNA will predispose the carrier to LFS, subtle types of mutations, such as the R337H missense protein, could favor the development of specific tumor types. It is therefore imperative that future tumor studies considering a role for p53 in tumorigenesis examine the entire coding region of p53, not just the DNA binding domain. Additional considerations include splicing mutations, which have been documented in LFS patients, as well as other epigenetic mechanisms (e.g., any factor that could regulate p53 activity, such as Mdm2, ATM, and Chk2).

THERAPEUTIC TARGETS IN THE p53 PATHWAY: SHOW ME THE MONEY!

It seems appropriate to conclude this chapter by highlighting the somewhat obvious therapeutic potential that lies within the p53 pathway. The main therapeutic strategies may be divided into two main categories: those that target tumors with wild-type p53 and those expressing mutant p53, which is the majority of the cases (Lane and Lain, 2002).

As expected, treating tumors that possess wild-type p53 will need to focus on activating the endogenous p53 gene within the tumor. This approach will depend on how the signaling pathway is disrupted and will need to be tailored accordingly. Such strategies will need to target factors such as Mdm2, Arf, HPV-E6, and the newly discovered cytoplasmic retention protein Parc. Intuitively, tumors with wild-type p53 could have the advantage of bypassing the reintroduction of genes into each tumor cell; rather, a small molecule that could disrupt the binding of Mdm2, E6, or Parc to p53 protein could release p53 to block tumor growth through cell cycle arrest or apoptosis. This is not a trivial goal and regrettably no specific therapies involving these types of approaches have yet been uncovered.

Initial attempts in treating tumors that express mutant p53 utilized gene therapy strategies in which a wild-type p53 gene is introduced back into tumor cells. Indeed, some encouraging results have been generated from phase I and II clinical trials using adenovirus vectors to express wild-type p53 in treating some tumor types (e.g., non–small cell lung cancer, and head and neck tumors), especially in combination with radiation or chemotherapy (Swisher, Roth et al., 2003) (Swisher and Roth,

2002). An interesting twist on this theme is the generation of a "smart virus" that kills only tumor cells by virtue of their inherent defect in p53 function. Adenovirus must inactivate p53, which is carried out by the viral E1B-55K protein, to successfully replicate and lyse its host cell. By engineering the virus so that it lacks E1B, referred to as the Onyx-015 adenovirus, it can infect but not replicate in normal cells, since it cannot inactivate wild-type p53. In tumor cells where p53 is mutated or inactivated by other means, the virus is competent for replication and will kill the tumor cell, hence the notion of it being "smart" (Bischoff, Kirn et al., 1996). Here, too, clinical trials show promising results (Khuri, Nemunaitis et al., 2000; Nemunaitis, Khuri et al., 2001). However, not all clinical studies have been so encouraging, and multiple complications exist that can impede the efficiency of this method, such as gene silencing, incomplete spreading of virus throughout the tumor, and inactivation of the vaccine by the immune system (Zeimet and Marth, 2003). Nevertheless, this approach remains an encouraging approach for treating tumors that have otherwise failed conventional treatment.

An alternative to the strategies above is to rescue the function of mutant p53. Since approximately half of all cancers express mutant p53 protein, the development of small molecules to reactivate p53 would be a logical approach that has great potential. However, there are many different point mutations of p53 found in human cancers, and these mutations can have very different effects on its structure and activity. The task of developing a drug to combat any one of them is daunting; nonetheless, progress is being made. A screen of reported p53-binding peptides has identified and characterized a panel of small peptides that can stabilize the core domain of mutant p53 proteins (Friedler, Hansson et al., 2002). Moreover second site mutations were identified that restore wild-type p53 transactivation and apoptotic functions to tumor-derived p53 mutants (Venkatachalam, Shi et al., 1998). Restoration of normal tumor-suppressor functions to mutant p53 establishes a proof-of-principle that at least certain mutants could be rescued using small molecules. Bullock and Fersht provide a detailed examination of this complicated, yet exciting strategy (Bullock and Fersht, 2001).

Recently this goal has been taken to the next level and a small compound, called Prima1, was identified by Wiman and colleagues in a drug screen looking for targets that can restore p53 transcriptional activity to a comprehensive panel of different missense p53 proteins (Bykov, Issaeva et al., 2002). Prima-1 appears to restore a more normal protein conformation to mutant p53s, thereby activating their DNA binding, transactivation, and growth suppression functions. The mechanism by which the drug reactivates p53 is not yet known, and in actuality is quite puzzling, since it "repairs" both DNA contact mutants (wild-type p53 structure) and conformation mutants (denatured structure). Nevertheless, as a colleague from a preeminent drug company once said, "How it works is not so important, all that matters is the drug works." In a follow-up confirmatory study conducted in our own lab, Prima-1 restored tumor-suppressor function to human mutant p53 proteins (Fig. 19.7)

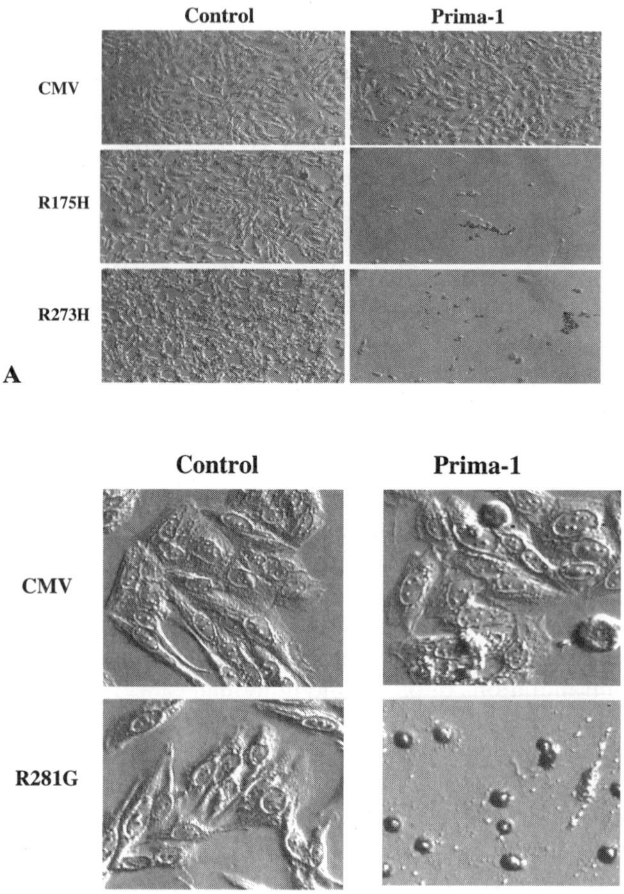

Figure 19.7. Prima-1 reactivates mutant p53. (**A**) Restoration of wild-type p53 activity to mutant p53 by Prima-1 in mouse 10(3) cells. Murine (10)3 fibroblasts lacking endogenous p53 were engineered to express only the selectable marker (CMV) or either the human mutant p53-R175H or R273H. Cells were grown under normal culture conditions (control) or treated with Prima-1 (10 μM) for 48 hours and stained for morphological analysis. Note that cells lacking p53 maintained viability after Prima-1 treatment (upper right panel), whereas cells expressing mutant p53 underwent apoptosis (middle and lower right panels) (unpublished data). (**B**) Restoration of wild-type p53 activity to mutant p53 by Prima-1 in Saos-2 cells. Human osteosarcoma Saos-2 cells lacking endogenous p53 were engineered to express only the selectable marker (CMV) or human mutant p53-R281G. Cells were grown under normal culture conditions (Control) or treated with Prima-1 (75 μM) for 48 hours and stained for morphological analysis. Note that cells lacking p53 maintained viability after Prima-1 treatment (upper right panel) whereas cells expressing mutant p53 underwent apoptosis (lower right panel) (unpublished data). (See color insert.)

which is quite exciting, especially in light of the prevalence of p53 mutations in human cancer. This is not to say that there are no inherent problems associated with the drug. Prima-1 at slightly higher doses appears to be toxic to cells that have no p53, indicating that the window of efficacy may be a limitation. However, preliminary data by the Wiman group argues that Prima-1 works well in vivo, at least in a xenograft human cancer model (Bykov, Issaeva et al., 2002). Perhaps second-generation compounds derived from the parental structure of Prima-1 will be less toxic and even more efficacious in treating tumor cells with mutant p53.

These are truly exciting times. Encouraging results from clinical trials continue to provide the impetus for designing drugs to correct other genetic abnormalities that occur in human cancers, and no other alteration is as common as mutation of the p53 tumor-suppressor gene. The wealth of knowledge generated by 25 years of research fuels the exploration for drugs that will rescue mutant p53 or that can block Mdm2 and other rational targets from inhibiting wild-type p53. Although these efforts are in an early stage, the rapid pace at which the field is still evolving provides hope that the best in p53 discoveries may be yet to come.

WEB RESOURCES

IARC p53 Mutation Database: *http://www.iarc.fr/p53*
The Protein Data Bank (PDB): *http://www.pdb.org/*

ACKNOWLEDGMENTS

We especially thank Wayne and Daveen Speer for their heartfelt dedication to St. Jude Children's Research Hospital (SJCRH). Their generosity and support is truly appreciated by the children and their families, physicians, scientists and staff of SJCRH. Through the Speer's commitment and that of all supporters of the American and Lebanese Syrian Associated Charities (ALSAC), we hope to fulfill Danny Thomas's dream that "No child should die in the dawn of life."

REFERENCES

Adams JM, Cory S (1998): The Bcl-2 protein family: Arbiters of cell survival. Science 281(5381):1322–6.

Appella E, Anderson CW (2001): Post-translational modifications and activation of p53 by genotoxic stresses. Eur J Biochem 268(10):2764–72.

Attardi LD, Reczek EE, et al. (2000): PERP, an apoptosis-associated target of p53, is a novel member of the PMP-22/gas3 family. Genes Dev 14(6):704–18.

Baker SJ (2003): Redefining p53 entering the tumor suppressor era. Cell Cycle 2(1):7–8.

Baker SJ, Fearon ER, et al. (1989): Chromosome 17 deletions and p53 gene mutations in colorectal carcinomas. Science 244(4901):217–21.

Barak Y, Juven T, et al. (1993): mdm2 expression is induced by wild type p53 activity. EMBO J 12(2):461–8.

Barlev NA, Liu L, et al. (2001): Acetylation of p53 activates transcription through recruitment of coactivators/histone acetyltransferases. Mol Cell 8(6):1243–54.

Baudino TA, Cleveland JL (2001): The Max network gone mad. Mol Cell Biol 21(3):691–702.

Benard J, Douc-Rasy S, et al. (2003): TP53 family members and human cancers. Hum Mutat 21(3):182–91.

Berman HM, Westbrook J, et al. (2000): The protein data bank. Nucl Acids Res 28(1):235–42.

Beroud C, Soussi T (2003): The UMD-p53 database: New mutations and analysis tools. Hum Mutat 21(3):176–81.

Bischoff JR, Kirn DH, et al. (1996): An adenovirus mutant that replicates selectively in p53-deficient human tumor cells. Science 274(5286):373–6.

Bodnar AG, Ouellette M, et al. (1998): Extension of life-span by introduction of telomerase into normal human cells. Science 279(5349):349–52.

Bourdon JC, Laurenzi VD, et al. (2003): p53: 25 years of research and more questions to answer. Cell Death Differ 10(4):397–9.

Bullock AN, Fersht AR (2001): Rescuing the function of mutant p53. Nat Rev Cancer 1(1):68–76.

Bunz F, Dutriaux A, et al. (1998): Requirement for p53 and p21 to sustain G_2 arrest after DNA damage. Science 282(5393):1497–501.

Buschmann T, Adler V, et al. (2000): p53 phosphorylation and association with murine double minute 2, c-Jun NH2-terminal kinase, p14ARF, and p300/CBP during the cell cycle and after exposure to ultraviolet irradiation. Cancer Res 60(4):896–900.

Bykov VJ, Issaeva N, et al. (2002): Restoration of the tumor suppressor function to mutant p53 by a low-molecular-weight compound. Nat Med 8(3):282–8.

Cadwell C, Zambetti GP (2001): The effects of wild-type p53 tumor suppressor activity and mutant p53 gain-of-function on cell growth. Gene 277(1–2):15–30.

Celli J, Duijf P, et al. (1999): Heterozygous germline mutations in the p53 homolog p63 are the cause of EEC syndrome. Cell 99(2):143–53.

Cho Y, Gorina S, et al. (1994): Crystal structure of a p53 tumor suppressor-DNA complex: Understanding tumorigenic mutations. Science 265(5170):346–55.

Clarke AR, Hollstein M (2003): Mouse models with modified p53 sequences to study cancer and ageing. Cell Death Differ 10(4):443–50.

Clarke AR, Purdie CA, et al. (1993): Thymocyte apoptosis induced by p53-dependent and independent pathways. Nature 362(6423):849–52.

Daujat S, Neel H, et al. (2001): MDM2: Life without p53. Trends Genet 17(8): 459–64.

DeLeo AB, Jay G, et al. (1979): Detection of a transformation-related antigen in chemically induced sarcomas and other transformed cells of the mouse. Proc Natl Acad Sci USA 76(5):2420–4.

DeLeo AB, Shiku H, et al. (1977): Cell surface antigens of chemically induced sarcomas of the mouse. I. Murine leukemia virus-related antigens and alloantigens on cultured fibroblasts and sarcoma cells: description of a unique antigen on BALB/c Meth A sarcoma. J Expr Med 146(3):720–34.

Deng C, Zhang P, et al. (1995): Mice lacking p21CIP1/WAF1 undergo normal development, but are defective in G_1 checkpoint control. Cell 82(4):675–84.

DiGiammarino EL, Lee AS, et al. (2002): A novel mechanism of tumorigenesis involving pH-dependent destabilization of a mutant p53 tetramer. Nat Struct Biol 9(1):12–6.

Donehower LA, Harvey M, et al. (1992): Mice deficient for p53 are developmentally normal but susceptible to spontaneous tumours. Nature 356(6366):215–21.

D'Orazi G, Cecchinelli B, et al. (2002): Homeodomain-interacting protein kinase-2 phosphorylates p53 at Ser 46 and mediates apoptosis. Nat Cell Biol 4(1): 11–19.

Dumont P, Leu JI, et al. (2003): The codon 72 polymorphic variants of p53 have markedly different apoptotic potential. Nat Genet 33(3):357–65.

el-Deiry WS, Tokino T, et al. (1993): WAF1, a potential mediator of p53 tumor suppression. Cell 75(4):817–25.

Esteller M, Cordon-Cardo C, et al. (2001): p14ARF silencing by promoter hypermethylation mediates abnormal intracellular localization of MDM2. Cancer Res 61(7):2816–21.

Eymin B, Leduc C, et al. (2003): p14ARF induces G_2 arrest and apoptosis independently of p53 leading to regression of tumours established in nude mice. Oncogene 22(12):1822–35.

Finlay CA, Hinds PW, et al. (1989): The p53 proto-oncogene can act as a suppressor of transformation. Cell 57(7):1083–93.

Fiscella M, Zhang H, et al. (1997): Wip1, a novel human protein phosphatase that is induced in response to ionizing radiation in a p53-dependent manner. Proc Natl Acad Sci USA 94(12):6048–53.

Friedler A, Hansson LO, et al. (2002): A peptide that binds and stabilizes p53 core domain: Chaperone strategy for rescue of oncogenic mutants. Proc Natl Acad Sci USA 99(2):937–42.

Grossman SR, Deato ME, et al. (2003): Polyubiquitination of p53 by a ubiquitin ligase activity of p300. Science 300(5617):342–4.

Hainaut P, Hernandez T, et al. (1998): IARC Database of p53 gene mutations in human tumors and cell lines: Updated compilation, revised formats and new visualisation tools. Nucl Acids Res 26(1):205–13.

Haupt Y, Rowan S, et al. (1995): Induction of apoptosis in HeLa cells by trans-activation-deficient p53. Genes Dev 9(17):2170–83.

Hayflick L (1961): The serial cultivation of human diploid cell strains. Exp Cell Res 25:585–621.

Hermeking H, Lengauer C, et al. (1997): 14-3-3 sigma is a p53-regulated inhibitor of G2/M progression. Mol Cell 1(1):3–11.

Hernandez-Boussard T, Rodriguez-Tome P, et al. (1999): IARC p53 mutation database: A relational database to compile and analyze p53 mutations in human tumors and cell lines. International Agency for Research on Cancer. Hum Mutat 14(1):1–8.

Hollstein M, Rice K, et al. (1994): Database of p53 gene somatic mutations in human tumors and cell lines. Nucl Acids Res 22(17):3551–5.

Honda R, Tanaka H, et al. (1997): Oncoprotein MDM2 is a ubiquitin ligase E3 for tumor suppressor p53. FEBS Lett 420(1):25–7.

Honda R, Yasuda H (1999): Association of p19(ARF) with Mdm2 inhibits ubiquitin ligase activity of Mdm2 for tumor suppressor p53. EMBO J 18(1):22–7.

Jacks T, Remington L, et al. (1994): Tumor spectrum analysis in p53-mutant mice. Curr Biol 4(1):1–7.

Jones SN, Roe AE, et al. (1995): Rescue of embryonic lethality in Mdm2-deficient mice by absence of p53. Nature 378(6553):206–8.

Kaghad M, Bonnet H, et al. (1997): Monoallelically expressed gene related to p53 at 1p36, a region frequently deleted in neuroblastoma and other human cancers. Cell 90(4):809–19.

Kamijo T, Zindy F, et al. (1997): Tumor suppression at the mouse INK4a locus mediated by the alternative reading frame product p19ARF. Cell 91(5): 649–59.

Kastan MB, Zambetti GP (2003): Parc-ing p53 in the cytoplasm. Cell 112(1):1–2.

Kastan MB, Zhan Q, et al. (1992): A mammalian cell cycle checkpoint pathway utilizing p53 and GADD45 is defective in ataxia-telangiectasia. Cell 71(4): 587–97.

Khuri FR, Nemunaitis J, et al. (2000): A controlled trial of intratumoral ONYX-015, a selectively-replicating adenovirus, in combination with cisplatin and 5-fluorouracil in patients with recurrent head and neck cancer. Nat Med 6(8):879–85.

Knudson CM, Tung KS, et al. (1995): Bax-deficient mice with lymphoid hyperplasia and male germ cell death. Science 270(5233):96–9.

Lane DP, Crawford LV (1979): T antigen is bound to a host protein in SV40-transformed cells. Nature 278(5701):261–3.

Lane DP, Lain S (2002): Therapeutic exploitation of the p53 pathway. Trends Mol Med 8(4 Suppl): S38–42.

Lee AS, Galea C, et al. (2003): Reversible amyloid formation by the p53 tetramerization domain and a cancer-associated mutant. J Mol Biol 327(3):699–709.

Levine AJ (1997): p53, the cellular gatekeeper for growth and division. Cell 88(3):323–31.

Levrero M, De Laurenzi V, et al. (2000): The p53/p63/p73 family of transcription factors: overlapping and distinct functions. J Cell Sci 113 (Pt 10):1661–70.

Levy N, Yonish-Rouach E, et al. (1993): Complementation by wild-type p53 of interleukin-6 effects on M1 cells: Induction of cell cycle exit and cooperativity with c-myc suppression. Mol Cell Biol 13(12):7942–52.

Lin Y, Ma W, et al. (2000): Pidd, a new death-domain-containing protein, is induced by p53 and promotes apoptosis. Nat Genet 26(1):122–7.

Lindsten T, Ross AJ, et al. (2000): The combined functions of proapoptotic Bcl-2 family members bak and bax are essential for normal development of multiple tissues. Mol Cell 6(6):1389–99.

Linzer DI, Maltzman W, et al. (1979): The SV40 A gene product is required for the production of a 54,000 MW cellular tumor antigen. Virology 98(2):308–18.

Liu G, McDonnell TJ, et al. (2000): High metastatic potential in mice inheriting a targeted p53 missense mutation. Proc Natl Acad Sci USA 97(8):4174–9.

Lohrum MA, Ashcroft M, et al. (2000): Identification of a cryptic nucleolar-localization signal in MDM2. Nat Cell Biol 2(3):179–81.

Lohrum MA, Vousden KH (2000): Regulation and function of the p53-related proteins: same family, different rules. Trends Cell Biol 10(5):197–202.

Lowe SW, Ruley HE, et al. (1993): p53-dependent apoptosis modulates the cytotoxicity of anticancer agents. Cell 74(6):957–67.

Lowe SW, Sherr CJ (2003): Tumor suppression by Ink4a-Arf: Progress and puzzles. Curr Opin Genet Dev 13(1):77–83.

Luo JL, Yang Q, et al. (2001): Knock-in mice with a chimeric human/murine p53 gene develop normally and show wild-type p53 responses to DNA damaging agents: a new biomedical research tool. Oncogene 20(3):320–8.

Malkin D, Li FP, et al. (1990): Germ line p53 mutations in a familial syndrome of breast cancer, sarcomas, and other neoplasms. Science 250(4985):1233–8.

Manfredi JJ (2003): p53 and apoptosis: it's not just in the nucleus anymore. Mol Cell 11(3):552–4.

Mayo LD, Donner DB (2001): A phosphatidylinositol 3-kinase/Akt pathway promotes translocation of Mdm2 from the cytoplasm to the nucleus. Proc Natl Acad Sci USA 98(20):11598–603.

Mayo LD, Donner DB (2002): The PTEN, Mdm2, p53 tumor suppressor-oncoprotein network. Trends Biochem Sci 27(9):462–7.

McCurrach ME, Connor TM, et al. (1997): bax-deficiency promotes drug resistance and oncogenic transformation by attenuating p53-dependent apoptosis. Proc Natl Acad Sci USA 94(6):2345–9.

Melchior F, Hengst L (2002): SUMO-1 and p53. Cell Cycle 1(4):245–9.

Michael D, Oren M (2002): The p53 and Mdm2 families in cancer. Curr Opin Genet Dev 12(1):53–9.

Mihara M, Erster S, et al. (2003): p53 has a direct apoptogenic role at the mitochondria. Mol Cell 11(3):577–90.

Mills AA, Zheng B, et al. (1999): p63 is a p53 homologue required for limb and epidermal morphogenesis. Nature 398(6729):708–13.

Miyashita T, Reed JC (1995): Tumor suppressor p53 is a direct transcriptional activator of the human bax gene. Cell 80(2):293–9.

Moberg KH, Tyndall WA, et al. (1992): Wild-type murine p53 represses transcription from the murine c-myc promoter in a human glial cell line. J Cell Biochem 49(2):208–15.

Moll UM, Erster S, et al. (2001): p53, p63 and p73—Solos, alliances and feuds among family members. Biochim Biophys Acta 1552(2):47–59.

Moll UM, Riou G, et al. (1992): Two distinct mechanisms alter p53 in breast cancer: Mutation and nuclear exclusion. Proc Natl Acad Sci USA 89(15): 7262–6.

Momand J, Zambetti GP (1997): Mdm-2: big brother of p53. J Cell Biochem 64(3):343–52.

Momand J, Zambetti GP, et al. (1992): The mdm-2 oncogene product forms a complex with the p53 protein and inhibits p53-mediated transactivation. Cell 69(7):1237–45.

Montes de Oca Luna R, Wagner DS, et al. (1995): Rescue of early embryonic lethality in mdm2-deficient mice by deletion of p53. Nature 378(6553):203–6.

Muller M, Wilder S, et al. (1998): p53 activates the CD95 (APO-1/Fas) gene in response to DNA damage by anticancer drugs. J Expr Med 188(11):2033–45.

Murphy M, Ahn J, et al. (1999): Transcriptional repression by wild-type p53 utilizes histone deacetylases, mediated by interaction with mSin3a. Genes Dev 13(19):2490–501.

Nakano K, Vousden KH (2001): PUMA, a novel proapoptotic gene, is induced by p53. Mol Cell 7(3):683–94.

Nemunaitis J, Khuri F, et al. (2001): Phase II trial of intratumoral administration of ONYX-015, a replication-selective adenovirus, in patients with refractory head and neck cancer. J Clin Oncol 19(2):289–98.

Nikolaev AY, Li M, et al. (2003): Parc: A cytoplasmic anchor for p53. Cell 112(1):29–40.

Oda E, Ohki R, et al. (2000): Noxa, a BH3-only member of the Bcl-2 family and candidate mediator of p53-induced apoptosis. Science 288(5468):1053–8.

Oda K, Arakawa H, et al. (2000): p53AIP1, a potential mediator of p53-dependent apoptosis, and its regulation by Ser-46-phosphorylated p53. Cell 102(6):849–62.

Oliner JD, Pietenpol JA, et al. (1993): Oncoprotein MDM2 conceals the activation domain of tumour suppressor p53. Nature 362(6423):857–60.

Oren M (2003): Decision making by p53: Life, death and cancer. Cell Death Differ 10(4):431–42.

Parant JM, Lozano G (2003): Disrupting TP53 in mouse models of human cancers. Hum Mutat 21(3):321–6.

Polyak K, Xia Y, et al. (1997): A model for p53-induced apoptosis. Nature 389(6648):300–5.

Pozniak CD, Radinovic S, et al. (2000): An anti-apoptotic role for the p53 family member, p73, during developmental neuron death. Science 289(5477):304–6.

Prives C, Hall PA (1999): The p53 pathway. J Pathol 187(1):112–26.

Prives C, Manley JL (2001): Why is p53 acetylated? Cell 107(7):815–8.

Ribeiro RC, Sandrini F, et al. (2001): An inherited p53 mutation that contributes in a tissue-specific manner to pediatric adrenal cortical carcinoma. Proc Natl Acad Sci USA 98(16):9330–5.

Robertson KD, Jones PA (1998): The human ARF cell cycle regulatory gene promoter is a CpG island which can be silenced by DNA methylation and down-regulated by wild-type p53. Mol Cell Biol 18(11):6457–73.

Roth J, Dobbelstein M, et al. (1998): Nucleo-cytoplasmic shuttling of the hdm2 oncoprotein regulates the levels of the p53 protein via a pathway used by the human immunodeficiency virus rev protein. EMBO J 17(2):554–64.

Sakaguchi K, Herrera JE, et al. (1998): DNA damage activates p53 through a phosphorylation-acetylation cascade. Genes Dev 12(18):2831–41.

Sax JK, El-Deiry WS (2003): p53 downstream targets and chemosensitivity. Cell Death Differ 10(4):413–7.

Scheffner M, Werness BA, et al. (1990): The E6 oncoprotein encoded by human papillomavirus types 16 and 18 promotes the degradation of p53. Cell 63(6): 1129–36.

Schmidt T, Korner K, et al. (1999): The activity of the murine Bax promoter is regulated by Sp1/3 and E-box binding proteins but not by p53. Cell Death Differ 6(9):873–82.

Serrano M, Lin AW, et al. (1997): Oncogenic ras provokes premature cell senescence associated with accumulation of p53 and p16INK4a. Cell 88(5):593–602.

Sherr CJ, DePinho RA (2000): Cellular senescence: mitotic clock or culture shock? Cell 102(4):407–10.

Soengas MS, Capodieci P, et al. (2001): Inactivation of the apoptosis effector Apaf-1 in malignant melanoma. Nature 409(6817):207–11.

Stiewe T, Stanelle J, et al. (2003): Inactivation of retinoblastoma (RB) tumor suppressor by oncogenic isoforms of the p53 family member p73. J Biol Chem 278(16):14230–6.

Swisher SG, Roth JA (2002): p53 Gene therapy for lung cancer. Curr Oncol Rep 4(4):334–40.

Swisher SG, Roth JA, et al. (2003): Induction of p53-regulated genes and tumor regression in lung cancer patients after intratumoral delivery of adenoviral p53 (INGN 201) and radiation therapy. Clin Cancer Res 9(1):93–101.

van Bokhoven H, Hamel BC, et al. (2001): p63 Gene mutations in eec syndrome, limb-mammary syndrome, and isolated split hand-split foot malformation suggest a genotype-phenotype correlation. Am J Hum Genet 69(3):481–92.

Varley JM (2003): Germline TP53 mutations and Li-Fraumeni syndrome. Hum Mutat 21(3):313–20.

Venkatachalam S, Shi YP, et al. (1998): Retention of wild-type p53 in tumors from p53 heterozygous mice: reduction of p53 dosage can promote cancer formation. EMBO J 17(16):4657–67.

Vousden KH (2000): p53: Death star. Cell 103(5):691–4.

Vousden KH (2002): Activation of the p53 tumor suppressor protein. Biochim Biophys Acta 1602(1):47–59.

Wang XW, Vermeulen W, et al. (1996): The XPB and XPD DNA helicases are components of the p53-mediated apoptosis pathway. Genes Dev 10(10): 1219–32.

Wang XW, Zhan Q, et al. (1999): GADD45 induction of a G_2/M cell cycle checkpoint. Proc Natl Acad Sci USA 96(7):3706–11.

Weber JD, Jeffers JR, et al. (2000): p53-independent functions of the p19(ARF) tumor suppressor. Genes Dev 14(18):2358–65.

Weber JD, Taylor LJ, et al. (1999): Nucleolar Arf sequesters Mdm2 and activates p53. Nat Cell Biol 1(1):20–6.

Weber JD, Zambetti GP (2003): Renewing the debate over the p53 apoptotic response. Cell Death Differ 10(4):409–12.

Wei MC, Zong WX, et al. (2001): Proapoptotic BAX and BAK: A requisite gateway to mitochondrial dysfunction and death. Science 292(5517):727–30.

Werness BA, Levine AJ, et al. (1990): Association of human papillomavirus types 16 and 18 E6 proteins with p53. Science 248(4951):76–9.

Wright WE, Shay JW (1992): The two-stage mechanism controlling cellular senescence and immortalization. Expr Gerontol 27(4):383–9.

Wu GS, Burns TF, et al. (1997): KILLER/DR5 is a DNA damage-inducible p53-regulated death receptor gene. Nat Genet 17(2):141–3.

Wu GS, Burns TF, et al. (1999): Molecular cloning and functional analysis of the mouse homologue of the KILLER/DR5 tumor necrosis factor-related apoptosis-inducing ligand (TRAIL) death receptor. Cancer Res 59(12): 2770–5.

Wu X, Bayle JH, et al. (1993): The p53-mdm-2 autoregulatory feedback loop. Genes Dev 7(7A):1126–32.

Xu Y (2003): Regulation of p53 responses by post-translational modifications. Cell Death Differ 10(4):400–3.

Yang A, Kaghad M, et al. (1998): p63, a p53 homolog at 3q27–29, encodes multiple products with transactivating, death-inducing, and dominant-negative activities. Mol Cell 2(3):305–16.

Yang A, McKeon F (2000): P63 and P73: P53 mimics, menaces and more. Nat Rev Mol Cell Biol 1(3):199–207.

Yang A, Schweitzer R, et al. (1999): p63 is essential for regenerative proliferation in limb, craniofacial and epithelial development. Nature 398(6729):714–8.

Yang A, Walker N, et al. (2000): p73-deficient mice have neurological, pheromonal and inflammatory defects but lack spontaneous tumours. Nature 404(6773):99–103.

Yin XY, Grove L, et al. (1999): C-myc overexpression and p53 loss cooperate to promote genomic instability. Oncogene 18(5):1177–84.

Yu J, Zhang L, et al. (2001): PUMA induces the rapid apoptosis of colorectal cancer cells. Mol Cell 7(3):673–82.

Zeimet AG, Marth C (2003): Why did p53 gene therapy fail in ovarian cancer? Lancet Oncol 4(7):415–22.

Zhan Q, Antinore MJ, et al. (1999): Association with Cdc2 and inhibition of Cdc2/Cyclin B1 kinase activity by the p53-regulated protein Gadd45. Oncogene 18(18):2892–900.

PART V

CHAPTER 20

CELL CYCLE AND GROWTH CONTROL: CURRENT CLINICAL APPLICATIONS

MICHAEL DEININGER
Oregon Health and Science University Center for Hematologic Malignancies Portland, OR 97201

INTRODUCTION

Impairment or loss of growth control is characteristic of many disease states. Conceptually, a distinction must be made between situations, where loss of growth control is the central problem, as in cancer, and situations, where increased or decreased proliferation is the consequence of another initiating event. Examples for the latter include proliferative retinopathy in patients with diabetes or proliferation of mesangial cells in certain types of glomerulonephritis. Less frequently, impaired proliferation may also be relevant. Neurodegenerative diseases like Parkinson's or disorders with tissue hypoplasia such as Sudeck's atrophy of the radius are examples where reduced proliferation is the pivotal problem.

This chapter focuses on the clinical aspects and applications that have arisen from the study of the cell cycle and its regulation. Much room is given to drug development, particularly in relation to agents that directly interfere with the cell cycle machinery. At present, the treatment of malignant diseases is by far the most important application of cell cycle-directed therapy, but other areas are also developing. Another topic of interest emerges from the observation that certain cell cycle-related proteins may have additional functions in physiological or pathological circumstances. Examples include the phosphorylation by cyclin-dependent kinases (Cdks) of certain proteins involved in Alzheimers's disease. Apart from direct therapeutic applications, cell cycle-related parameters, such as the expression of genes involved in cell cycle regulation or the cell cycle status of malignant cells prior to therapy, have been used to aid prognostication.

Cell Cycle and Growth Control: Biomolecular Regulation and Cancer, Second Edition, Edited by Gary S. Stein and Arthur B. Pardee
ISBN 0-471-25071-6

CELL CYCLE AND CANCER THERAPY

General Considerations

Cancer has been defined as a cell cycle disease. The G_1 phase of the cell cycle before the restriction point is particularly vulnerable to oncogenic events (Sherr, 1996). Disruption of the function of the retinoblastoma susceptibility gene product (pRb) by a variety of mechanisms is demonstrable in most malignancies (Wall, 1996). Basic science has led to a greatly improved understanding of the biochemical pathways that regulate cell cycle progression under physiological conditions. Frequently human malignancies have served as experiments of nature that draw attention to the particular importance of individual proteins involved in cell cycle regulation. Examples include cyclin D1, which was originally identified in parathyroid tumors with rearrangements of the 11q13 chromosomal band (Motokura et al., 1991), and the retinoblastoma susceptibility gene, whose deletion is causal to familial retinoblastoma (Friend et al., 1986).

Drug therapy of malignant diseases has classically relied on cytotoxic agents that are toxic to normal as well as neoplastic tissues. Although their mechanisms of action are diverse, most of these agents act in a cell cycle specific fashion or influence the cell cycle of the malignant cells, usually in an indirect manner. Attempts have been made to exploit the cell cycle specificity of some cytotoxic agents in order to improve therapeutic efficacy. These conventional chemotherapeutic agents will be considered first.

More recently specific agents have been introduced into the clinic. Some of these agents, such as all-*trans*-retinoic acid for the treatment of acute promyelocytic leukemia have been found empirically, before their specific targets were known (Flynn et al., 1983). Others, such as the anti-CD19 antibody rituximab for the treatment of B-cell lymphoma (Maloney et al., 1997) or the Abl-specific tyrosine kinase inhibitor imatinib (Gleevec™) (Druker et al., 1996) were rationally designed to target specific features of the malignant cells. This field of molecularly targeted therapy is rapidly evolving at present. Although many of these compounds have profound effects on cell cycle regulation, their effects are by definition secondary and not related to direct interference with the cell cycle machinery. However, from the perspective of successful drug development, important lessons can be learned that might have implications for cell cycle-targeted compounds. Thus the development of imatinib will be discussed in some detail in the second part.

Last, there is an increasing number of compounds that directly target cell cycle regulators, particularly cyclin-dependent kinases. With few exceptions these agents have not yet been tested clinically. Some are not suitable for clinical use, because of toxicity or unfavorable pharmacokinetic properties, while others are just about to move into clinical trials. From the point of exploiting abnormalities of cell cycle regulation for cancer therapy, these compounds are the most interesting group, and will thus be reviewed in detail.

Conventional Cytotoxic Agents

Three major groups of conventional cytotoxic agents can be distinguished according to their mechanism of action. Some of these agents exhibit various degrees of cell cycle phase specificity (Table 20.1).

1. Drugs that target DNA directly, mainly the alkylating agents, have little cell cycle specificity. They generate free radicals that cause damage to DNA, regardless of the cell cycle position of the cell. Examples include busulfan, cyclophosphamide, and *cis*-platinum.
2. Drugs that interfere with DNA synthesis. This group comprises antimetabolites such as methotrexate (an inhibitor of dihydrofolate reductase), which depletes the cells of intermediates required for nucleotide synthesis, and nucleoside analoga such as 1-beta-D-arabinofuranosylcytosine (cytarabine), which inhibits DNA polymerase. Other agents interfere with different steps of the replication process. Examples include topotecan and etoposide, which inhibit topoisomerases I and II, respectively.
3. Drugs that interfere with the mitotic spindle by binding or modifying tubulin. Vincristin and paclitaxel belong to this group.

There are agents with yet other mechanisms of action such as asparaginase which depletes the body from the amino acid L-asparagine and 5-fluorouracil, an inhibitor of RNA synthesis.

Most malignancies are treated with a combination of two or more cytotoxic agents rather than monotherapy. This concept is based on three main considerations:

1. Use of noncrossresistant drugs should increase tumor cell kill and/or avoid the emergence of resistance.
2. Use of drugs with different toxicity profiles should allow for higher dose intensity.
3. The individual drugs may act synergistically.

Many combinations of conventional cytotoxic agents have been tested in vitro and in clinical trials. In several instances this has led to major advances. Polychemotherapy is capable of curing most patients with germ cell tumors (Einhorn, 2002) and Hodgkin's disease (Josting et al., 2000), and most children with acute lymphoblastic leukemia (Rubnitz and Pui, 2003). Results are slightly less favorable in the case of Non–Hodgkin's lymphoma and acute myelogenous leukemia (AML). In sharp contrast, most solid tumors in adults are incurable with current chemotherapy.

With growing understanding of cell cycle regulation and apoptosis, the sequence and timing of drug administration has received more attention. In the most straightforward scenario, it would appear unwise to use a drug that induces a G_1 or G_2 arrest before an S phase specific agent is

TABLE 20.1. Cell Cycle Specific Activity of Conventional Cytotoxic Agents

		Activity			
	Mechanism	S Phase	G_2/M Phase	G_1	Clinical Use (examples)
Asparaginase	L-asparagine depletion	+	+	+	Acute lymphoblastic leukemia
Bleomycine	Free radical damage to DNA	+	+	+	Hodgkin's disease, testicular cancer
Busulfan	DNA alkylation	+	+	+	Chronic myelogenous leukemia
Chlorambucil	DNA alkylation	+	+	+	Chronic lymphocytic leukemia
Cisplatin	DNA crosslinks	+	++	+++	Germ cell tumors
Cyclophosphamide	DNA alkylation	+	+	+	Non–Hodgkin's lymphoma
Cytarabine	DNA synthesis inhibitor	+	–	–	Acute myelogenous leukemia
Daunorubicin	DNA intercalation; topoisomerase II inhibition	+	–	–	Acute myelogenous leukemia
Etoposide	Topoisomerase II inhibition	+	–	–	Germ cell tumors
Fludarabin	DNA synthesis inhibitor	+	–	–	Non–Hodgkin's lymphoma
Gemcitabine	DNA synthesis inhibitor	+	–	–	Gastric cancer
Hydoxyurea	DNA synthesis inhibitor	+	–	–	Chronic myelogenous leukemia
Melphalan	DNA alkylation	+	+	+	Multiple myeloma
Mercaptopurine	DNA synthesis inhibitor	+	–	–	Acute lymphoblastic leukemia
Methotrexate	DNA synthesis inhibitor	+	–	–	Osteosarcoma
Mitomycin	DNA crosslinking	+	+	+	Breast cancer
Paclitaxel	Inhibition of microtubuli depolymerisation	–	+	–	Ovarian cancer
Topotecan	Topoisomerase I inhibitor	+	–	–	Ovarian cancer
Vincristin	Tubulin binding	–	+	–	Lymphoma

TABLE 20.2. ATP-Competitive Inhibitors of Cyclin-Dependent Kinases

Compound	Class	Cdk1/ CyclinB	Cdk2/ Cyclins A, E	Cdk4/ Cyclin D	Cdk5/p25	In vitro Anti-tumor Trials	Clinical Activity	Reference
Flavopiridol	Flavone	0.4	0.1 (A)	0.4	Not known	Yes	Yes	De Azevedo et al. (1996)
Staurosporine	Indolocarbazole	0.006	0.007	<10	Not known	Yes	No	Lawrie et al. (1997)
Olomoucine	Purine	7	7	>1000	3	Yes	No	Schulze-Gahmen et al. (1995)
Roscovitine	Purine	0.45	0.7	>100	0.16	Yes	Yes	Schulze-Gahmen et al. (1995)
Indirubin	Indolinone	10	2.2 (A) 7.5 (E)	12	5.5	Yes	Yes	Hardcastle et al. (2002)
Purvalanol B	Purine	0.006	0.006 (A) 0.009 (E)	>10	0.006	Yes	No	Gray et al. (1998)
CINK4	Triaminopyrimidine	>100	>50 (A) >50 (E)	1.5	25	Yes	No	Soni et al. (2001)
Hymenialdisine	Pyrroloazepine	0.022	0.07 (A) 0.04 (E)	0.6	0.028		No	Meijer et al. (2000)
SU9516	Indolinone	0.04	0.022 (A)	0.2	Not known	Yes	No	Lane et al. (2001)
Compound 26a	Diarylurea	0.12	0.078	0.042	Not known		No	Honma et al. (2001a,b)
Compound 15b	Diarylurea	1.8	0.44 (A)	0.0023	Not known	Yes	No	Honma et al. (2001a,b)
Alsterpaullone	Benzazepinone	0.035	0.015 (A) 0.2 (E)	>10	0.04	Yes	No	Schultz et al. (1999)

administered. This issue has been studied extensively for the combination of cytarabine, an S phase specific agent, with other drugs, such as daunorubicin or methotrexate. Daunorubicin is a DNA intercalating agent and topoisomerase II inhibitor; methotrexate inhibits dihydrofolate reductase, depriving the cells of C_1 units needed for purine biosynthesis. Daunorubicin and cytarabine form the backbone of therapy for AML. As early as 1974 it was noted that cytarabine followed by daunorubicin, but not the reverse sequence, produced synergistic cytotoxicity toward cancer cell lines in vitro (Rao et al., 1975; Edelstein et al., 1974). It was initially thought that the sequence dependence of synergism was a direct consequence of cell cycle effects of cytarabine, which synchronized the cells in S phase, rendering them susceptible to the action of daunorubicin. However, subsequent investigations found that, for maximal synergism to occur, synchronization of cells in S phase was not required. Rather than on the S phase fraction, synergy was more dependent on the interval between exposure to cytarabine and exposure to daunorubicin (Ritch et al., 1981). This indicates that synergism of chemotherapeutic agents depends on complex interaction; the cell cycle position of a cell may not be the only and most important variable that determines sensitivity. However, regardless of the precise mechanism, in vitro data may nonetheless have clinical importance. For example, cytarabine followed by mitoxantrone (a drug very similar to daunorubicin) 6 hours later was used in one study of relapsed or refractory patients with AML; there was a 35% response rate in patients who had been refractory to simultaneous administration of cytarabine and daunorubicin before (Paciucci et al., 1997). Thus drug resistance may be overcome by optimizing drug scheduling, although improved efficacy may not necessarily be related to the cell cycle position of the leukemic cells.

Another example for the importance of sequence of drug administration is paclitaxel in combination with various other cytotoxic agents. Paclitaxel inhibits the depolymerization of microtubuli and is thus an M phase specific agent. Induction of apoptosis by paclitaxel depends on the activity of cyclin B/Cdk1 (Yu et al., 1998). In agreement with this observation, treatment of cancer cell lines with paclitaxel followed by the Cdk inhibitor flavopiridol (see below) was synergistic, while the reverse sequence of exposure was antagonistic (Motwani et al., 1999). Similarly, if exposure to paclitaxel precedes treatment with methotrexate, the combined effects are antagonistic. Presumably this is due to the G_2/M arrest that is induced by paclitaxel (Kano et al., 1998); by contrast, if methotrexate is followed by placlitaxel, synergism results (Yeh et al., 1994).

The applicability of in vitro studies to clinical trial design is limited by the fact that many other factors complicate the issue. Important variables are pharmacokinetics and the development of drug resistance by somatic mutation of tumor cells. A number of mathematical models have been developed that range from simple models that focus on one particular aspect (e.g., genetic drug resistance vs. cell kinetic resistance) to complex models that integrate multiple parameters (e.g., multiple drugs, evolution of resistance, and cell cycle kinetics, recently reviewed in Gardner, 2002). The hope is that in the future such models will help to improve prediction of therapeutic response in individual patients.

Conventional cytotoxic agents do not specifically target cell cycle regulatory proteins. However, the cell cycle and apoptosis machinery are crucial for their cytotoxic effects to occur. Current thinking holds that malignant cells exhibit depressed cell cycle checkpoints, resulting in the accumulation of genetic damage (Zhou and Elledge, 2000). Nonetheless, pathways that link checkpoints with apoptotic responses are maintained in cancer cells. It is thought that checkpoint-mediated apoptosis is activated as a result of DNA damage, accounting for the therapeutic effects as well as for the side effects of conventional chemotherapy (Rich et al., 2000). The intimate relation between the integrity of cell cycle checkpoints and DNA repair versus apoptosis implies that it may even be possible to exploit the malignant cell's depressed checkpoint regulation more directly, without first inflicting DNA damage. This approach has been termed activated checkpoint therapy (Li et al., 2003).

Cell Cycle Target-Specific Therapies

General Considerations. Targeted therapy implies that an agent affects a pathway that is specific to the malignant cell (Fig. 20.1). The best examples are Bcr-Abl, the oncogenic tyrosine kinase responsible for chronic myelogenous leukemia (CML, Daley et al., 1990) and the PML-retinoic acid receptor alpha (PML-RARA) fusion protein that causes acute

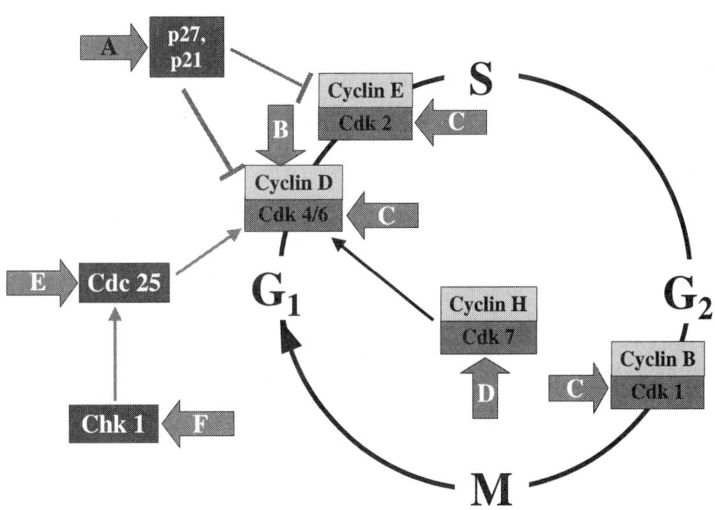

Figure 20.1. Mechanisms of pharmacological interference with cell cycle traverse. (**A**) up-regulation of physiological Cdk inhibitors; (**B**) down-regulation of cyclins required for Cdk activation; (**C**) direct (small molecule) inhibition of Cdks; (**D**) inhibition of Cdk-activating kinases; (**E**) direct inhibition of Cdk-activating phosphatases; (**F**) inhibition of cdc25 phosphorylation.

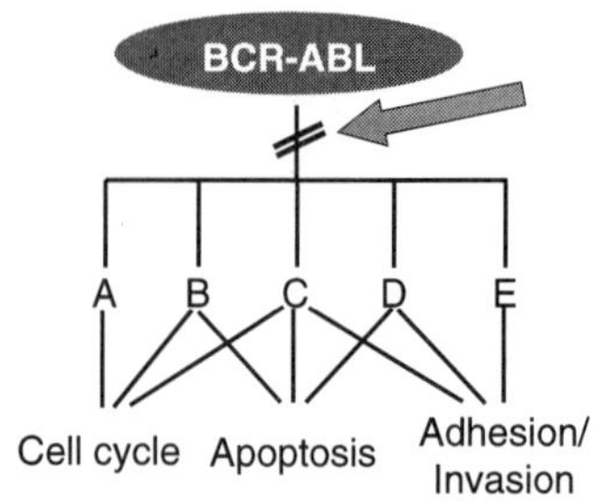

Figure 20.2A. Targeting the oncogeneic process at its origin. Bcr-Abl activates multiple signaling pathways, most of which are dependent on its kinase activity. Malignant transformation is critically dependent on kinase activity, which represents the ideal therapeutic target. Thus, repression of kinase activity (thick arrow) interrupts oncogenic signal transduction at its origin.

promyelocytic leukemia (de The et al., 1990). All-*trans* retinoic acid (ATRA) as a treatment for acute promyelocytic leukemia was found empirically. By contrast, imatinib for CML is the result of a systematic approach.

Imatinib and ATRA target the oncogenic process at its origin (Fig. 20.2A). For obvious reasons, this approach will be more specific than interfering with processes further downstream in the signaling cascades. In the latter case, physiological signals using the same effector molecules will also be inhibited, which will potentially lead to more side effects. In terms of therapeutic efficacy, targeting downstream effectors after the initial oncogenic signal has branched out is less promising. In this regard the example of Bcr-Abl signaling is instructive. As a potent tyrosine kinase, Bcr-Abl phosphorylates multiple substrates and activates numerous signaling pathways (Deininger et al., 2000). Using a variety of strategies, efforts were made to identify downstream effectors of Bcr-Abl that are essential for malignant transformation. With the availability of knockout mice, it became possible to test the importance of individual effector molecules more directly. It was found that the Bcr-Abl-driven leukemogenesis is not impaired in mice with disruption of the IL-3, GM-CSF (Li et al., 2001), STAT5 (Sexl et al., 2000), CRKL (Hemmeryckx et al., 2002), or SHIP genes (Jiang et al., 2003). For all these downstream mediators, there was in vitro evidence for an important or essential role in *BCR-ABL*-positive leukemia. The studies in the knockout animals clearly show, however, that pharmacological targeting of these proteins would not necessarily be effective.

Compared to CML, at least in the chronic phase of the disease, the genetic makeup of most other malignancies is far more complicated, and the cooperation of multiple genetic lesions may be responsible for their malignant phenotype. Nonetheless, these independent events may feed into common pathways downstream. Thus, in case of malignancies with a multifactorial pathogenesis, attacking downstream effectors may be more feasible than interfering with multiple pathways

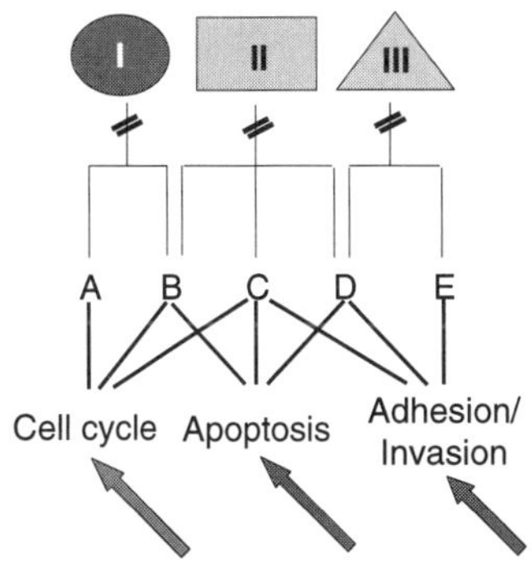

Figure 20.2B. In malignancies with complex pathogenetic make-up, multiple independent genetic lesions (I, II, III) may cooperate, and there may be no single upstream pathway that is absolutely required for malignant transformation. Thus, targeting common pathways further downstream (arrows), such as the cell cycle, may be more easily achievable than interfering with multiple activating lesions upstream.

upstream. An obvious and attractive target is the cell cycle machinery (Fig. 20.1B).

Direct Inhibitors of Cell Cycle Traverse. Cyclin-dependent kinase inhibitors in a narrower sense directly block the enzymatic activity of Cdks, in most cases by competing with ATP. It is thought that the human genome encodes for approximately 1000 protein kinases that can be classified into four major groups (Hanks, 1995). Cdks belong to the CMGC group, which also contains the MAP (mitogen-activated protein kinase), glycogen synthase kinase 3 (GSK3), casein kinase, and Cdk-like families. Given the similarity of these kinases, designing specific inhibitors represents a major challenge. Not surprisingly, many compounds that were originally thought to be specific Cdk inhibitors turned out to be also potent inhibitors of GSK3 (Leclerc et al., 2001). Most of the early compounds have been found empirically or during screening of natural sources for agents with Cdk inhibitory activity. Many of these agents have a broad spectrum of activity toward various Cdks. More recently efforts were made to design of inhibitors that preferentially target specific Cdks over others. One obstacle is that until now Cdk2 is the only member, whose three-dimensional structure has been resolved in isolation, while the structures of Cdk4 and Cdk6 have been determined only in complex with Cdk inhibitors (e.g., $p16^{INK4}$), which may affect structure and ATP binding (Brotherton et al., 1998). Thus homology models based on Cdk2 are being exploited for the structure-based development of inhibitors of other Cdks. Almost all Cdk inhibitors developed thus far are ATP-competitive; however, substrate-competitive compounds or mimetics of physiological Cdk inhibitors, such as p27, can also be envisaged and may offer greater specificity.

The biological effects of Cdk inhibitors are much more complex than the simple inhibition of cell cycle traverse. One reason is that many of the compounds inhibit more than one Cdk as well as other targets. Furthermore cell cycle regulation and apoptosis are intimately linked. Cell cycle dysregulation is a very potent stimulus for induction of apoptosis (King and Cidlowski, 1995). Key components of the cell cycle machinery such as the retinoblastoma gene product (Kaelin, 1999), the E2F transcription factor (Qin et al., 1994), p27 (St. Croix et al., 1996), and cyclin D1 (Niu et al., 2001) have been shown to influence the apoptotic response. In addition cell cycle checkpoint function is impaired in many cancer cells (Li et al., 2003). It is thought that this contributes to the genetic instability of malignant cells and favors the accumulation of mutations. On the other hand, in contrast to normal cells, cancer cells may be more likely to activate an apoptotic program rather than efficient DNA repair in response to genotoxic stress, a property that is key to the (limited) selectivity of unspecific cytotoxic agents. Specific inhibition of checkpoint function may be exploited therapeutically: rather than to arrest and allow for repair after exposure to DNA-damaging agents, malignant cells would progress into mitosis, resulting in mitotic catastrophe and cell death.

Flavopiridol. The anti-proliferative potential of quercetin, a carcinogenic flavonoid, had been known since the mid-1970s (Suolinna et al., 1975). It was originally thought that this effect was related to inhibition of glycolysis. Later it was discovered that genistein, another flavonoid, was a potent and relatively selective tyrosine kinase inhibitor that competitively blocked ATP binding (Akiyama et al., 1987). Genistein inhibits the proliferation of leukemia cell lines (Tohda et al., 1991) and selectively spares normal hematopoietic progenitor cells versus progenitors derived from chronic myeloid leukemia (CML, Carlo Stella et al., 1996). However, for clinical applications, the specificity of these compounds is not sufficient. Another semisynthetic flavonoid, L86-8275, later renamed flavopiridol, had originally been characterized as an inhibitor of various protein kinases, including the epidermal growth factor receptor (EGFR) and protein kinase A. When tested against the NCI screening panel of cancer cell lines, it effectively inhibited the growth of a wide variety of tumor cell lines, although with various lag times (Kaur et al., 1992). Further studies into its mechanism of action revealed that it blocked cells both in G_1/S and G_2/M cell cycle transitions. This pattern suggested that flavopiridol may act as a Cdk inhibitor; in fact subsequent studies revealed potent inhibition of Cdk1 (Losiewicz et al., 1994), 2 and 4 (Carlson et al., 1996). Of particular importance with regard to clinical applications is that fact that flavopiridol is a potent inducer of apoptosis in certain cell lines, particularly at higher concentrations, and in cells of hematopoietic origin (Konig et al., 1997; Parker et al., 1998; Byrd et al., 1998; Patel et al., 1998). Both extrinsic (Achenbach et al., 2000) and intrinsic (Decker et al., 2001) pathways of apoptosis may be involved. Since disruption of the cell cycle is a potent pro-apoptotic stimulus, these findings could be explained on these grounds only. However, there is evidence that flavopiridol has additional biochemical effects that may contribute to its cytotoxic potential. For example, it was recently reported that flavopiridol potently inhibits glycogen phosphorylase, which may partially explain its cytotoxicity (Kaiser et al., 2001). Moreover the cell cycle effects are not exclusively mediated by Cdk inhibition but also by modulation of cell cycle related proteins such as cyclin D1 (Carlson et al., 1999). Yet other effects include the inhibition of angiogenesis (Melillo et al., 1999).

In 1998 the first phase I clinical trial was reported; it included 76 patients with refractory neoplasms, mainly solid tumors (Senderowicz et al., 1998). The drug was administered as a 72-hour continuous infusion every two weeks. Dose-limiting toxicity was secretory diarrhea that defined a maximum tolerated dose of 50 mg/m^2; with antidiarrheal medication, dose escalation to 78 mg/m^2 was possible. Other remarkable side effects were hypotension as well as a complex of symptoms that was termed "proinflammatory syndrome." Mean plasma concentrations were 271 and 344 nM, respectively, which are both within the range of 200 to 400 nM that induced Cdk and cell cycle inhibition in vitro. Clinical responses were observed in 4/71 evaluable patients, the best response being a partial remission in a patient with renal cancer. Stable disease was seen in an additional 10 patients. A second phase I trial with an identical dosing schedule determined the maximun tolerated dose as 40 mg/m^2, with mean steady state plasma concentrations of 417 nM (Thomas et al., 2002). In this study the cell cycle profile of peripheral blood lymphocytes drawn at the time of flavopiridol administration was determined in an attempt to monitor a biological end point. However, no changes were observed. Notably, one patient with metastatic gastric cancer achieved a complete remission that has been sustained for four years.

The clinical activity seen in the phase I studies sparked a series of phase II trials that used the 72-hour administration regime. Although some antitumor activity was observed, the overall results of these studies were rather disappointing. For example, in one study in minimally pretreated patients with renal carcinoma, only 6% partial responses and no complete responses were seen (Stadler et al., 2000). Similarly there were no major objective responses in a trial in patients with advanced gastric cancer (Schwartz et al., 2001), non–small cell lung cancer (Shapiro et al., 2001), and refractory mantle cell lymphoma (Lin et al., 2002). From these and similar studies it was concluded that the 72-hour dosing schedule was not effective. In order to achieve higher peak plasma concentrations, 1-hour infusions for 1 to 5 days every three weeks were tested in another phase I trial of patients with advanced tumors (Tan et al., 2002). Median peak plasma concentrations of up to 1700 nM were reached. Stable disease was observed in 12/50 patients, but no objective responses were reported. Subsequently several phase II studies that used the 1-hour dosing regimen were started. In patients with mantle cell lymphoma, 41% of whom had no prior chemotherapy, partial responses were seen in only 12% (Connors et al., 2001). In another study in metastatic melanoma, no objective responses were seen (Burdette-Radoux et al., 2002). The conclusion from all these trials is that single agent flavopiridol has limited clinical activity. Although induction of apoptosis was demonstrable in vitro, disease stabilization but rarely remissions were seen in patients. Current efforts aim at combining flavopiridol with conventional chemotherapeutic agents. Objective responses were seen in patients treated with sequential paclitaxel-flavopiridol (Schwartz et al., 2002), sequential paclitaxel-cisplatin-flavopiridol (Shah et al., 2003), or sequential paclitaxel-carboplatin-flavopiridol (Gries et al., 2003). In the latter study, 6/18 patients with previously untreated non–small cell lung cancer (NSCLC) achieved partial or complete remissions.

Another clinical trial that combines flavopiridol with imatinib in imatinib-resistant CML will start recruiting patients soon. In vitro studies showed powerful synergism of the two agents (Yu et al., 2002).

It is not clear why the clinical efficacy of flavopiridol is inferior to its potent in vitro effects on tumor cell lines and in animal models. Two aspects appear to be important. First, there is no evidence that a pharmacodynamic end point such as the inhibition of Cdk2 activity or, simpler, cell cycle arrest was actually achieved in the tumor tissues during exposure to the drug. Since, in the case of solid tumors, such tissue is not readily available, two studies monitored peripheral blood lymphocytes as a surrogate (Stadler et al., 2000; Schwartz et al., 2001) but failed to detect any effects on cell cycle profiles or apoptosis. It is somewhat questionable if lymphocytes, which are predominantly noncycling, are an adequate readout. Future strategies to improve the results will much depend on knowing if the pharmacodynamic end point was actually achieved in the target tissue. Second, the fact the one patient with metastatic gastric carcinoma achieved and maintained CR is an indication that there are tumors with a molecular design that renders them extremely sensitive to flavopiridol. Identifying this molecular design and individualizing therapy will be critical for the success of any molecularly targeted therapy in heterogeneous malignancies.

Staurosporine and 7-Hydroxystaurosporine (UCN-01). Staurosporine was discovered when extracts of streptomyces sp. Were screened for compounds with protein kinase C (PKC) inhibitory activity (Omura et al., 1977). The discovery that PKC is activated in some human malignancies (O'Brian et al., 1989; Benzil et al., 1992) stimulated the development of staurosporine analogues, the most prominent examples of which are 7-hydroxystaurosporine (UCN-01) and CGP41251. However, abnormally low rather than high PKC expression has also been described in certain malignancies (Kopp et al., 1991), indicating a cell-type specific role for PKC. Thus PKC is of questionable value as a drug target (reviewed in Gescher, 2000), and growth inhibition by staurosporine and analogues is likely mediated by mechanisms other than inhibition of PKC.

Extensive studies into the mechanism of action of UCN-01 revealed a very complex picture. This compound inhibits Cdks 2, 4, and 6 at concentrations of approximately 50 nM, and cell cycle progression is blocked in G_1 (Kawakami et al., 1996). However, there is also accumulation of Cdk inhibitors like p27 that is independent of the G_1 arrest, suggesting the additional mechanisms are involved in mediating the biological effects (Nishi et al., 1998). In B-cell chronic lymphocytic leukemia cell lines, UCN-01 leads to down-regulation of various anti-apoptotic proteins. The precise mechanism underlying these effects is not known (Kitada et al., 2000).

The most intriguing activity of UCN-01 is its ability to abrogate the accumulation of cells in G_2 after DNA damage. As a result of DNA damage, normal cells activate a cell cycle checkpoint and arrest in G_2, a response that involves p53 dependent and independent mechanisms. It was found that in cells treated with UCN-01, this G_2 arrest was abrogated, an effect that was more prominent in cells with defective p53 (Wang et al., 1996). Further studies showed that this effect, although eventually mediated by Cdk1, was in fact the result of an event further upstream (Yu et al., 1998) that was subsequently identified as ATP-competitive inhibition of checkpoint kinase 1 (chk1) (Graves et al., 2000; Zhao et al., 2002). Under physiological conditions, Chk1 phosphorylates cdc25, excluding it from the nucleus and preventing it from dephosphorylating Cdk1. As a result Cdk1 remains inactive, and the cells are arrested in G_2. Conversely, inhibition of chk1 releases this cell cycle block. In cancer cell lines this is associated with apoptosis. Thus UCN-01 may potentiate the cytotoxicity of drugs that cause DNA damage, such as *cis*-platinum (Wang et al., 1996) or ionizing radiation (Xiao et al., 2002). Blocking Chk1 activity may deprive p53-deficient cancer cells of an essential mechanism to repair DNA damage. Instead, the cells proceed into M and undergo apoptosis as a result of mitotic catastrophe (Dixon and Norbury, 2002). Of note, methylxanthines like caffein, and pentoxyfyllin have similar effects on the G_2 arrest after DNA damage. This effect may be the consequence of inhibition of Ataxia teleangiectasia mutated (ATM), a protein kinase upstream of Chk1 (Zhou et al., 2000).

Most studies indicate that the effects of UCN-01 are primarily mediated by abrogation of the G_2 checkpoint. Leukemia cells treated with cytostatic concentrations of gemcitabine, a purine analogue, accumulate in the S phase. After withdrawal of gemcitabine, in the presence of non-toxic concentrations of UCN-01, the cells undergo apoptosis, without resuming DNA synthesis and without arresting in G_2 (Shi et al., 2001).

UCN-01 is currently being tested in clinical trials. Results of the first phase I trial in 47 patients with refractory neoplasms have recently been reported (Sausville et al., 2001). Dose-limiting toxicities were hyperglycemia with resultant metabolic acidosis, pulmonary dysfunction, nausea, vomiting, and hypotension. Unexpectedly, the plasma half-life of turned out to be approximately 30 days (Senderowicz et al., 2000), necessitating an extension of the interval between cycles. The recommended phase 1 dose is 45 mg/m^2, 3 days continuous infusion, every two weeks. The long plasma half-life had not been anticipated from preclinical studies and was shown to be the result of specific binding to human alpha 1 acidic glycoprotein (Fuse et al., 1999). The phase I trial also monitored phosphorylation of adducin, a PKC substrate and abrogation of G_2 checkpoint function. It was demonstrated that UCN-01 influenced these biochemical and biological end points. With one partial remission in a patient with melanoma, the clinical efficacy was very modest. As in the case of flavopiridol, UCN-01 may be more effective in combination with conventional cytotoxic agents or with other signal transduction inhibitors such as inhibitors of the MAP kinase pathway (Decker et al., 2001). Combination treatment of U937 leukemia cells with phorbol esthers (PMA) and UCN-01 led to increased apoptosis, probably as a result of PMA-induced inhibition of growth arrest (Rahmani and Grant, 2002).

Olomoucine and Roscovitine. Olomoucine, a compound isolated from marine invertebrates, has been found to inhibit Cdkl/cyclin B, Cdk2/cyclin A and E kinases, Cdk5, and the ERK1/MAP-kinase in the low micromolar dose range (Vesely et al., 1994; Havlicek et al., 1997). The Cdk4/cyclin D1 and Cdk6/cyclin D3 kinases are not significantly sensitive to olomoucine (IC50 values greater than 1mM and 150 microM, respectively), nor are many other kinases tested. Modifications of olomoucine led to the synthesis of rocovitine, which displays enhanced activity against Cdk1. Both agents are ATP-competitive inhibitors. Consistent with the inhibition of Cdk1 and Cdk2, they block both G_1/S and G_2/M transitions. Induction of apoptosis was observed in tumor biopsies of a dog with a spontaneous melanoma that had received a course in intravenous olomoucine (Hajduch et al., 1997). R-Roscovitine (CYC202) has good oral bioavailability and is currently being tested in phase I trials. Until now, no responses have been observed, but stable disease was seen in several patients with refractory solid tumors (Pierga et al., 2003).

Indirubin. Indirubin, the red-colored isomer of indigo, was identified as the active component of a herbal mixture that had been used to treat CML in traditional Chinese medicine. Indirubin acts as an ATP-competitive inhibitor of Cdk1-cyclin B, Cdk2-cyclin A, Cdk2-cyclin E, and Cdk5/p35 at micromolar concentrations, while most other kinases are unaffected. Various derivatives have been synthesized that exhibit increased activity towards Cdks (Hoessel et al., 1999). More recently activity against GSK-3 was also demonstrated (Leclerc et al., 2001). Tumor cell lines treated with indirubin preferentially arrest in G_2/M. Since indirubin also inhibits DNA synthesis, it is not clear if all its biological activities are attributable to inhibition of Cdks. Unlike many other Cdk inhibitors, the compound has actually been tested in clinical trials in China, and activity was demonstrated in patients with CML (reviewed in Han, 1994).

CINK4. CINK4, a triaminopyrimidine derivative, was identified during a high-throughput screen to identify small molecules that inhibit pRb phosphorylation by Cdk4 (Soni et al., 2001). Unlike other compounds, it has considerable specificity towards Cdk4/cyclin D1 (IC50 = 1.5 μM) and Cdk6/cyclin D1 (IC50 = 5.6 μM) compared to Cdk1, 2 and 5, and also to Cdk4/cyclin D2 and Cdk6/cyclin D2, which all show IC50 values of ≥25 μM. Tumor cell lines treated with CINK4 were arrested in G_1, regardless of their p16 status, and this was accompanied by decreased phosphorylation of serines 780 and 795 of pRb, two residues that are specific targets of Cdk4 (Connell-Crowley et al., 1997). As predicted, CINK4 was inactive in cells that lacked functional pRb. Modest activity against a human colon carcinoma line was shown in a murine xenograft model. Thus far no clinical trials have been reported.

Hymenialdisine. Hymenialdisine was discovered when marine invertebrates were screened for Cdk inhibitors (Meijer et al., 2000). This compound inhibits Cdk1, 2, and 5 at nanomolar concentrations. Casein kinase 1 and GSK3 are inhibited at similar doses, and several other kinases are affected at approximately 5 to 10 times higher concentrations. Thus the biological effects of this agent may not be related only to inhibition of Cdks. One intriguing feature of hymenialdisine is the fact that it inhibits Cdk5 in addition to GSK3; both kinases are known to phosphorylate tau proteins, the main constituents of the neurofibrillary tangles that are among the key neuroanatomical features of Alzheimer's disease (Grundke-Iqbal et al., 1986). In addition it has recently been demonstrated that GSK3 is responsible for the formation of amyloid-beta peptides, another hallmark of Alzheimer's disease (Phiel et al., 2003). Thus inhibition of Cdk5 and GSK3 with hymenialdisine may simultaneously block two important pathways in Alzheimer's disease.

CGP74514A. CGP74514A, a trisubstituted purine derivative that is structurally related to olomoucine, preferentially inhibits Cdk1. At low concentrations, leukemia cells arrest at G_2/M. Higher concentrations induce apoptosis, an effect that can be counteracted by overexpression of anti-apoptotic proteins like Bcl-2. The effects of CGP74514A are diverse; apart from inhibition of Cdk1 activity there is activation of various signaling cascades such as the MAP kinase and JNK. Thus it is impossible to attribute the biological effects solely to Cdk inhibition (Dai et al., 2002).

Paullones. The COMPARE algorithm was developed to detect similarities in the patterns of activity of compounds tested against the NCI panel of tumor cell lines (Zaharevitz et al., 1999). Using the pattern of flavopiridol as a template, a new group of potent Cdk inhibitors was discovered. The lead compound, kenpaullone, inhibits Cdk1/cyclin B, Cdk2/cyclin A, and Cdk5/p25 at concentrations of less than 1 μM and Cdk2/cyclin E at 7.5 μM. Inhibition of Erk1, Erk2 casein kinase 1, Raf, and Src occurs at cocentrations between 9 and 38 μM, while the IC_{50} values of a series of other kinases were >100 μM. Kenpaullone has only modest antitumor activity, as assessed by the NCI cell line screen. A series of derivatives was subsequently synthesized, and a compound termed Alsterpaullone was discovered with much more potent antitumor activity than the parental compound (Schultz et al., 1999). Alsterpaullone retained potent Cdk1 inhibitory activity. However, several other derivatives that are weak inhibitors of Cdk1 have nonetheless potent antitumor activity, while one compound with potent Cdk1 inhibition was virtually ineffective. These observations indicate that the antiproliferative activity of paullones is not mediated by inhibition of Cdk1. Alsterpaullone was chosen for further development. As yet clinical trials have not been initiated.

Indolinone Derivatives. SU9516, a 3-substituted indolinone derivative, has relative selectivity for Cdk2 (IC50 = 22 nM), while the values for Cdk1 and Cdk4 are 1.8- and 9-fold higher, respectively (Lane et al., 2001). The cell cycle effects in tumor cell lines were partially dependent on the cell type, with one line arresting in G_1 and another predominantly in G_2/M. In both instances, caspase activation and apoptosis were demonstrable. Recent data indicate that SU9516 has additional affects such as downregulation of cyclin D1 and Cdk2 (Yu et al., 2002). To the best of our knowledge, clinical trials have not yet been initiated.

Other Cdk Inhibitors. The list of Cdk inhibitors is growing rapidly. Active compounds not yet mentioned include butyrolactone, a Cdk1/2 inhibitor isolated from Aspergillus species (Kitagawa et al., 1993) and Purvanalol A and B. These compounds preferentially inhibit Cdk1/2 at nanomolar concentrations (Gray et al., 1998; Chang et al., 1999). PD 0183812, another recently described compound, has selectivity toward Cdk4 and 6 (Fry et al., 2001).

Many of the compounds discovered thus far, including those that are being tested in clinical trials, are relatively unselective. Thus current efforts are directed at developing more selective inhibitors that target one specific Cdk. Cdk4 has received particular attention, as a result of the high frequency of p16 inactivation in human tumors (Hall and Peters, 1996). Restoration of p16 function would seem a logical approach to the treatment of such malignancies. Using the structure of Cdk2 bound to various inhibitors as the starting point, Honma et al. (2001a) were able to identify four novel lead compounds, one of them diarylurea. Further development led to a compound with selectivity over Cdk1/2 (780-fold/190-fold) but also many other kinases (>430-fold) (Honma et al., 2001b).

Further Perspectives for Clinical Development of Cdk Inhibitors

Can Cdk Inhibitors Be as Successful as Imatinib? Imatinib, a specific ATP-competitive inhibitor of the Abl, Kit and platelet-derived growth factor receptor tyrosine kinases, has set the gold standard for successful targeted therapy. Imatinib is extremely effective for two malignant disorders: CML, where the causal lesion is the Bcr-Abl tyrosine kinase (Druker et al., 2001; O'Brien et al., 2003), and gastrointestinal stromal tumors (GISTs), which carry activating mutations of Kit (van Oosterom et al., 2001). Several conclusions can be drawn from the clinical development of imatinib that may explain why Cdk inhibitors were less successful.

Defining the Right Target. Murine leukemia models indicate that Bcr-Abl is capable of inducing leukemia, without the need for cooperating lesions (Daley et al., 1990; Huettner et al., 2000). Thus the choice of the right target in this setting is straightforward. Similarly mutated Kit is key to the pathogenesis of GISTs (Blanke et al., 2001). By contrast, if a molecular target is expressed by the tumor cells but not involved in the pathogenesis of the disease, specific inhibition is not efficacious. This was strikingly exemplified when the use of imatinib was extended to acute myeloid leukemia, where the leukemic cells frequently express Kit. Only isolated responses were seen (Kindler et al., 2003), while the results as a whole were disappointing.

Establishing the right target in most other tumors will be much more difficult, since multiple lesions are likely to cooperate. It may be more promising to target lesions that occur early in disease development. The fact that CML in advanced phase still responds to imatinib shows that the initial oncogenic event remains crucial to the pathogenesis (Druker et al., 2001). There are relatively few examples where mutations to cell cycle related genes are the initiating event. One exception is familial cutaneous melanoma where germ-line mutations of p16 or Cdk4 are regularly found, suggesting that disruption of the p16/Cdk4 interaction is a crucial event (Platz et al., 2000). In further support of this concept knockin mice with a Cdk4 mutation found in melanoma families that interferes with p16 binding are prone to develop melanoma (Sotillo et al., 2001). This suggests that restoration of Cdk inhibition with a specific chemical inhibitor may reverse the malignant phenotype in these tumors. Defining the patients in whom a given target such as Cdk4 is indeed the bottleneck will be of critical importance for the success of specific therapy.

The consequence is that enrollment into trials with such agents must be based on biochemical characteristics; otherwise, active compounds might be lost for further development, because they are active only in a small fraction of patients. As long as such biochemical characterization is not possible, Cdk inhibitors with a broad spectrum of activity may hold more promise.

The example of the paullones illustrates another important aspect. It was originally assumed that the antitumor effects of these compounds was mediated by inhibition of Cdk1. However, synthesis of a series of compounds led to the discovery that the most active Cdk1 inhibitor has lost most antiproliferative activity, while another compound with minimal anti-Cdk1 activity was a potent antiproliferative agent. Thus use of Cdk1 activity as a pharmacodynamic end point in a clinical trials would have been misleading.

Defining the Right Patient Population for Study. In CML there is a more than 90% correlation between the morphological diagnosis and the presence of the BCR-ABL translocation (Deininget et al., 2000). Thus molecular stratification was easy to achieve when patients were entered into the imatinib studies. It is conceivable, for example, that the morphological diagnosis of gastric carcinoma covers a range of conditions with diverse molecular mechanisms underlying tumor growth. If a specific molecular therapy was tested in such a cohort, individual patients may respond well, while there is no effect in the majority. The overall low response rate may then prevent further clinical development of such agents. In a recent study the "tyrosine kinome" was sequenced in a panel of colon cancer cell lines. Mutations that might lead to activation were found in several different tyrosine kinases (Bardelli et al., 2003). If the functional relevance of these mutations is confirmed, this would imply that targeted therapy should be tailored according to the responsible kinase in each individual patient. The same may hold true for inhibitors of specific Cdks.

Targeting Gain of Function Rather Than Loss of Function. The transforming capacity of Bcr-Abl is correlated with its tyrosine kinase activity (Lugo et al., 1990). No cooperating loss of function lesions have been demonstrated in the early (chronic) phase of the disease. This is naturally a more favorable starting point than, for example, the inactivation of the p53 tumor-suppressor gene. Given the much more complex pathogenesis of solid tumors, where loss of function lesions cooperate with gain of function lesions, it is evident that no single compound can be expected to be equally effective as imatinib.

Physiological Function of the Target Is Not Essential. There was considerable concern that inhibition of Abl along with Bcr-Abl may lead to side effects. Although mice with homozygous disruption of the ABL gene locus are viable, there is a very high neonatal mortality, and the animals have a number of other defects (Tybulewicz et al., 1991). Of even more concern, disruption of both ABL and the ABL-related gene, ARG, a tyrosine kinase that is also inhibited by imatinib, is embryonically lethal (Koleske et al., 1998). Disruption of platelet-derived growth factor A or B is also lethal (Soriano, 1994, 1997), and KIT mutant mice have severe hematological defects (Geissler et al., 1988). Surprisingly the side effects of imatinib are mostly minor, although the drug is given over prolonged periods of time. This suggests that the activity of the target kinases may be crucial only during embryogenesis or that enough residual activity persists to maintain the physiological function. Compared to imatinib, side effects of flavopiridol have been more severe, and the maximum tolerated dose could easily be established. The relative weaker specificity of this compound as well as the crucial physiological function of its targets may be responsible.

Defining the Right Dose. In the trials of imatinib in CML, no dose-limiting toxicity was observed, and thus the maximum tolerated dose (MTD) could not be established. This is in contrast to conventional agents, where side effects are usually dose limiting before the maximal therapeutic effect has been achieved (Deininger et al., 2003). This indicates that definition of a biochemical endpoint to guide dosing is crucial. In the case of CML, inhibition of phosphorylation of Crkl, a major Bcr-Abl substrate, was used (Druker et al., 2001). Similarly, in the phase I trial of UCN-01, PKC activity was measured as a phamacodynamic end point (Sausville et al., 2001). However, unlike Crkl phosphorylation, this is only a surrogate marker, since PKC is most likely not the biologically relevant target. Moreover imatinib activity could be measured in the target tissue, while this not usually possible in the case of solid tumors.

Overall, although there are plenty of ways to explain why Cdk inhibitors were much less successful than imatinib, it still remains somewhat mysterious why many of the compounds are so dysproportionately more active in vitro than in vivo.

Indirect Inhibitors of Cell Cycle Traverse. Many compounds affect components of the cell cycle machinery indirectly. Differentiation-inducing agents such as Phorbol esters and histone deacetylase inhibitors induce p21, various compounds down-regulate cyclin D1, and yet others, such as inhibitors of the 26s proteasome, up-regulate p27 (Grant and Roberts, 2003). As a result cell cycle progression is halted. Virtually any compound with relevant antitumor activity will affect the cell cycle machinery in some way. Compounds with indirect cell cycle effects will not be considered in detail, with the exception of inhibitors of the 26s proteasome that have recently received much attention due to their therapeutic efficacy.

It is estimated that approximately 80% of cellular proteins are degraded via the 26s proteasome, implying a fundamental role for this pathway in cellular metabolism (Bochtler et al., 1999). Degradation by the proteasome contributes to regulation of abundance of numerous proteins involved in cell cycle regulation such as cyclins (A, B, D, and E) and Cdk inhibitors. Another important group are transcription factors such as p53, Myc, Fos, and inhibitory protein such as IkB (Schenkein, 2002). Consistent with this central role, continuous inhibition of the proteasome is not compatible with life.

The variety of proteins that are degraded by the proteasome might suggest that the effects of inhibitors are rather nonspecific, affecting normal no less than malignant cells. However, it turned out that transformed cells are much more vulnerable to proteasome block than normal cells. It is conceivable that normal cells may be capable of activating checkpoint mechanisms that hold proliferation in the face of the profound disruption in the turnover of cell cycle regulatory proteins that is induced by proteasome inhibition. Thus proliferation is only resumed when proteasome function has been restored, while malignant cells continue to proliferate and are thus driven into apoptosis. While the precise mechanisms underlying this differential response are not well understood (Goldberg and Rock, 2002), there are some defined changes in transformed cells that are reversed by proteasome inhibition. One example is p27, a Cdk inhibitor that is down-regulated in many tumors.

Conversely, proteasome inhibition restores p27 expression and induces a cell cycle block in leukemia cells expressing the Bcr-Abl oncoprotein (Jonuleit et al., 2000). Another important proteasome target is IκB. This protein binds the transcription factor NFκB, preventing it from localizing to the nucleus. NFκB regulates the expression of a number of genes, including cytokines, adhesion proteins and anti-apoptotic proteins. Dysregulation of NFκB activity is a frequent finding in cancer cells (Richmond, 2002).

The first proteasome inhibitor to be tested in clinical trials was PS-341, a dipedityl boronic acid now named Bortezomib. This compound has activity against a broad range of tumor cells in vitro as well as in xenograft tumor models (Adams et al., 1999). By contrast, normal bone marrow mononuclear cells are unaffected at concentrations of PS-341 that induce growth arrest and apoptosis in malignant cells (Hideshima et al., 2001). A number of phase I studies with PS-341 have been initiated and partly completed. Toxicity was manageable and included low grade fever, weakness, diarrhea and peripheral neuropathy. Objective responses were seen in a variety of tumor types (Orlowski et al., 2001; Aghajanian et al., 2002). Since PS-341 is cleared from the plasma very rapidly, a bioassay was developed that allowed for determination of residual proteasome activity in peripheral blood mononuclear cells and helped guide dose-escalation schemes. Stimulated by the promising results from the phase I studies, several phase II trials were initiated. In these studies activity was demonstrated in some patients with solid tumors; however, the most striking results were seen in refractory multiple myeloma. In 202 myeloma patients enrolled in phase I and II protocols, the overall response rate was 35%, with 4% achieving a complete response (reviewed in Adams, 2003). These results were encouraging enough to trigger an international phase III trial. In addition in vitro data suggest synergism between PS-341 and conventional cytotoxic drugs (Mitsiades et al., 2003). PS-341 has recently been approved by the FDA for the treatment of refractory multiple myeloma.

Given the complex cellular consequences of proteasome inhibition, it is impossible to ascribe their effects on malignant cells solely to cell cycle effects.

Cell Cycle and Drug Resistance

Sensitizing Leukemia Cells to S-Phase Specific Drugs. For several decades 1-beta-D-arabinofuranosylcytosine (cytarabine) has formed the backbone of chemotherapy for acute myeloid leukemia. Early on it was recognized that cytarabine specifically acts on cells in S phase (Skipper et al., 1967). Thus, when myeloid growth factors became available, several studies noticed that the sensitivity of leukemic blasts to cytarabine could be increased by prior stimulation with various growth factors, mainly G-SCF, GM-CSF, and IL-3 (Miyauchi et al., 1989; Lista et al., 1990; Tafuri and Andreeff, 1990; te Boekhorst et al., 1993). Studies in AML patients prior to and after short-term administration of G-CSF showed significant increases in the blast count and proportion of cells in S phase, suggesting that the desired effects were indeed induced in vivo (Baer et al., 1996; Bai et al., 1999). Since the initial reports, a large number of clinical trials have been published. The design of these studies varied considerably in terms of timing, overall length of exposure to the various cytokines, and the patient populations studied. The consensus interpretation of all these studies is that patients may benefit with respect to the duration of neutropenia and incidence of infectious complications. However, there was no consistent benefit with respect to complete remission rates and progression-free survival. In fact a decreased rate of complete remission was noted in one relatively small study (Zittoun et al., 1996). Thus a recent review concluded that clinical studies that have attempted to exploit possible potentiation of chemotherapeutic activity by recruitment of leukemic cells into the cell cycle have generally been disappointing (Richardson et al., 2000). It is not clear why the well-founded in vitro concept did not work in vivo, although pronounced effects were demonstrable in patient samples analyzed ex vivo (Baer et al., 1996; Bai et al., 1999). Possible explanations are that relapse originates from a population of cells that is either not recruitable by growth factors or that has additional features conferring resistance.

Sensitizing Leukemia Cells to Immunotherapy. Recruitment of resting leukemic cells into cycle may also be important for an effective immune response. It was recently shown that treatment of AML-193 leukemia cells with interferon-alpha or GM-CSF recruited quiescent G_0 cells into cycle. Treatment of these cells with an activating anti-CD95 (Fas) antibody led to a complete depletion of cells in G_1, while cells in G_0 were absolutely, and cells in G_2/M were relatively, protected. This indicates that the cell cycle status of the target cells may also influence their sensitivity to immune responses (Jedema et al., 2003). Similar observations were made for TRAIL-induced apoptosis in SW480 colon cancer and H460 lung cancer cell lines (Jin et al., 2002). It is not known if the cell cycle effects of interferon-alpha occur in vivo, and if they contribute to the enhanced antileukemic effect in patients who receive alpha-interferon in addition to donor lymphocytes for relapse of leukemia after allogeneic stem cell transplantation.

Persistence of Leukemic Cells in CML. In CML the issue of minimal residual disease was until recently confined to patients after allogeneic transplants. There is compelling evidence that in these cases an allogeneic or autologous T cell response is essential for disease eradication (Apperley et al., 1986; Kolb et al., 1990; Mollderm et al., 2000). Unlike previously available drug treatment, imatinib induces complete cytogenetic remissions in 75% of newly diagnosed patients in first chronic phase (O'Brien et al., 2003). However, residual disease remains detectable in most cases if sensitive techniques such as RT-PCR are employed (Hughes et al., 2002). Two groups reported the existence of a quiescent population of CML cells that are resistant to imatinib

(Graham et al., 2002; Holtz et al., 2002). These cells are in $G_{1/0}$, express BCR-ABL mRNA but are not sensitive to imatinib in vitro, which appears to affect predominantly cycling cells. It is conceivable that these quiescent cells may be responsible for the persistence of minimal residual disease in vivo. If their drug resistance is a direct consequence of their cell cycle status, or if they have additional features (e.g., increased expression of the drug efflux protein MDR1), is not known. Whatever the precise mechanism, this finding is surprising, since most BCR-ABL-positive cell lines undergo apoptosis upon treatment with imatinib (Druker et al., 1996; Deininger et al., 1997; Gambacorti-Passerini et al., 1997). Since relapses have occurred in patients with complete cytogenetic responses, targeting these residual cells may be of great importance for the long-term prognosis of patients on imatinib.

DIAGNOSTIC AND PROGNOSTIC USE OF CELL CYCLE PARAMETERS

Functional Parameters

Attempts were made to utilize growth characteristics of malignant cells for prediction of response to treatment and outcome. An early study in AML found that the response to high-dose cytarabine was dependent on the percentage of cells in S phase: no patient with less than 6% phase in S phase (as assessed by 3H thymidine incorporation) entered remission (Preisler et al., 1984). Another study showed that a longer total cell cycle time was associated with longer remission duration in AML patients treated with cytarabine/daunorubicin (Raza et al., 1990), suggesting that the leukemic cells in these cases may not be able to re-grow between cycles of chemotherapy. Over the years a number of additional reports appeared that correlated response to treatment with cell cycle parameters (Kaaijk et al., 2003). However, although easy to perform, cell cycle analysis as a prognostic toll has not been introduced into clinical routine. This is also a reflection of the fact that cytogenetic and/or molecular markers have been discovered that are extremely powerful predictors of prognosis (Giles et al., 2002).

Expression of Cell Cycle-Related Genes

Several cell cycle regulatory genes were found to have prognostic significance. Low expression of p27 is adversely associated with prognosis in many solid tumors (Huang et al., 2002; Khoo et al., 2002; Nitti et al., 2002; Kirla et al., 2003). Since the E3 ubiquitin ligase Skp2 regulates p27 abundancy (Mamillapalli et al., 2001), it is not surprising that high levels of Skp2 have also been linked with adverse outcome (Kudo et al., 2001). High levels of cyclin E are also an adverse prognostic factor in diverse malignancies (Dong et al., 2000; Fukuse et al., 2000; Dosaka et al., 2001; Sui et al., 2001), most importantly, in breast cancer (Keyomarsi et al.,

1994, 2002). High expression of cyclin D2 predicts unfavorable outcome in gastric cancer (Takano et al., 1999). It is likely that parameters that are easy to determine such as the expression of cyclins in histological sections by immunohistochemistry will aid treatment stratification in the future. It is obvious that prognostic scores that use a limited number of cell cycle-related genes will be inferior to systems that include more parameters. In studies that employed genome-wide microarray technology to identify prognostic signatures, cell cycle-related genes were only a very small part of the signature genes, for example in breast cancer (van de Vijver et al., 2002). The hope would be that expression profiling will allow not only for more accurate prognostication but also for a more rational approach to therapy. On the other hand, given the complexity of cell cycle regulation, one may doubt, for example, that overexpression of a specific Cdk would predict response to a specific inhibitor of this kinase.

TARGETING CELL CYCLE COMPONENTS IN NONMALIGNANT DISORDERS

Drugs that interfere with cell cycle proteins have been developed mainly for the treatment of malignant diseases, while other applications have evolved more slowly. Still related to cancer treatment is the aspect that Cdk inhibitors may protect normal cells from the effects of cytotoxic chemotherapy. In one study normal mammary epithelial cells and breast cancer cells were treated with staurosporine followed by cytotoxic drugs. Normal epithelial cells arrested in $G_{0/1}$ upon exposure to staurosporine, while breast cancer cells continued to proliferate. Subsequent treatment with doxorubicin or camptothecin selectively killed breast cancer cells, while the normal cells resumed proliferation after the drugs had been washed out. In this setting, normal cells tolerated doses of camptothecin that were more than 100-fold higher than those needed for in vitro tumor eradication (Chen et al., 2000). Another example is the prevention of chemotherapy-induced alopecia by a topically applied Cdk inhibitor of the indolinone class that prevents progression of cells into S phase.

The use of Cdk inhibitors may however stretch much wider than oncology. Of particular interest are neurodegenerative diseases like Alzheimer's and ischemic brain injury. In Alzheimer's disease there is aberrant phosphorylation of tau protein, the major constituent of neurofibrils, by Cdk5 and GSK3. Thus inhibition of these kinases may have therapeutic benefit (recently reviewed in Lau et al., 2002). Cdk4/cyclin D1 activity increases after ischemic stroke. In a rat model of focal cerebral ischemia, administration of flavopiridol prior to ischemia and throughout the reperfusion period reduced neuronal apoptosis by 80% (Osuga et al., 2000). Other benign conditions where Cdk inhibitors may have a role in reducing pathologically increased proliferation include atherosclerosis and proliferative nephropathy (Clough, 2002). Until now, no clinical trials have been carried out.

CONCLUSION

Understanding the principles of cell cycle regulation is key to understanding the pathogenesis of cancer but also of other disease states with imbalanced proliferation. Basic scientific knowledge has increased at a very rapid pace, and the enormous complexity of the cell cycle and its interactions with the apoptotic machinery and other vital cell functions is being realized. The challenge is now to move this knowledge from the bench to the bedside. Naturally cancer as a "cell cycle disease" is in the center of the interest. Early on attempts concentrated on exploiting the relative cell cycle phase specificity of conventional cytotoxic agents to increase their therapeutic efficacy by rational choice of sequence of administration. It turned out that compared to the cell line models used in vitro, drug interactions in vivo were much more complex, cell cycle status being only one of a number of parameters that determine response. Nonetheless, the impression remains that drug scheduling may not yet be fully exploited as a means of optimizing response. However, compounds that directly interfere with cell cycle regulators, most important, Cdks, have recently attracted more attention. Many of these agents are extremely powerful inducers of cell cycle arrest and apoptosis in vitro. Several compounds have been developed into drugs and tested in clinical trials, but activity was generally modest. The reasons for these rather sobering results are probably manifold. The hope is that with adequate changes to trial design, results will improve. Perhaps the biggest challenge is to shift from a pathological to a molecular stratification of malignancies that would eventually allow matching the right drug with the right patient. Combination with conventional agents or other signal transduction inhibitors also holds promise. Cell cycle specific therapy for nonmalignant disorders has received less attention, but prevention of ischemic cell damage or Alzheimer's disease may be interesting areas in the future. Last, cell cycle regulatory genes are valuable as prognostic parameters in certain types of cancer. Giving the rapid accumulation of knowledge, many more clinical applications are likely to emerge in the future.

REFRENCES

Achenbach TV, Muller R, Slater EP (2000): Bcl-2 independence of flavopiridol-induced apoptosis: Mitochondrial depolarization in the absence of cytochrome c release. J Biol Chem 275:32089–97.

Adams J, Palombella VJ, Sausville EA, et al. (1999): Proteasome inhibitors: A novel class of potent and effective antitumor agents. Cancer Res 59:2615–22.

Adams J (2003): Potential for proteasome inhibition in the treatment of cancer. Drug Discov Today 8:307–15.

Aghajanian C, Soignet S, Dizon DS, et al. (2002): A phase I trial of the novel proteasome inhibitor PS341 in advanced solid tumor malignancies. Clin Cancer Res 8:2505–11.

Akiyama T, Ishida J, Nakagawa S, et al. (1987): Genistein, a specific inhibitor of tyrosine-specific protein kinases. J Biol Chem 262:5592–5.

Apperley JF, Jones L, Hale G, et al. (1986): Bone marrow transplantation for patients with chronic myeloid leukaemia: T-cell depletion with Campath-1 reduces the incidence of graft-versus-host disease but may increase the risk of leukaemic relapse. Bone Marrow Transplant 1:53–66.

Baer MR, Bernstein SH, Brunetto VL, et al. (1996): Biological effects of recombinant human granulocyte colony-stimulating factor in patients with untreated acute myeloid leukemia. Blood 87:1484–94.

Bai A, Kojima H, Hori M, et al. (1999): Priming with G-CSF effectively enhances low-dose Ara-C-induced in vivo apoptosis in myeloid leukemia cells. Expr Hematol 27:259–65.

Bardelli A, Parsons DW, Silliman N, et al. (2003): Mutational analysis of the tyrosine kinome in colorectal cancers. Science 300:949.

Benzil DL, Finkelstein SD, Epstein MH, Finch PW (1992): Expression pattern of alpha-protein kinase C in human astrocytomas indicates a role in malignant progression. Cancer Res 52:2951–6.

Blanke CD, Eisenberg BL, Heinrich MC (2001): Gastrointestinal stromal tumors. Curr Treat Options Oncol 2:485–91.

Bochtler M, Ditzel L, Groll M, Hartmann C, Huber R (1999): The proteasome. An Rev Biophys Biomol Struct 28:295–317.

Brotherton DH, Dhanaraj V, Wick S, et al. (1998): Crystal structure of the complex of the cyclin D-dependent kinase Cdk6 bound to the cell-cycle inhibitor p19INK4d. Nature 395:244–50.

Burdette-Radoux S, Tozer RG, Lohmann R, et al. (2002): NCIC CTG phase II study of flavopiridol in patients with previously untreated metastatic malignant melanoma (Ind.137). Proc Am Soc Clin Oncol 21:346a.

Byrd JC, Shinn C, Waselenko JK, et al. (1998): Flavopiridol induces apoptosis in chronic lymphocytic leukemia cells via activation of caspase-3 without evidence of bcl-2 modulation or dependence on functional p53. Blood 92: 3804–16.

Carlo Stella C, Dotti G, Mangoni L, et al. (1996): Selection of myeloid progenitors lacking BCR/ABL mRNA in chronic myelogenous leukemia patients after in vitro treatment with the tyrosine kinase inhibitor genistein. Blood 88:3091–100.

Carlson BA, Dubay MM, Sausville EA, Brizuela L, Worland PJ (1996): Flavopiridol induces G_1 arrest with inhibition of cyclin-dependent kinase (CDK) 2 and CDK4 in human breast carcinoma cells. Cancer Res 56:2973–8.

Carlson B, Lahusen T, Singh S, et al. (1999): Down-regulation of cyclin D1 by transcriptional repression in MCF-7 human breast carcinoma cells induced by flavopiridol. Cancer Res 59:4634–41.

Chang YT, Gray NS, Rosania GR, et al. (1999): Synthesis and application of functionally diverse 2,6,9-trisubstituted purine libraries as CDK inhibitors. Chem Biol 6:361–75.

Chen X, Lowe M, Herliczek T, et al. (2000): Protection of normal proliferating cells against chemotherapy by staurosporine-mediated, selective, and reversible G(1) arrest. J Natl Cancer Inst 92:1999–2008.

Clough J (2002): CDK inhibitor shows promise for inflammatory kidney disease. Drug Discov Today 7:789–90.

Connell-Crowley L, Harper JW, Goodrich DW (1997): Cyclin D1/Cdk4 regulates retinoblastoma protein-mediated cell cycle arrest by site-specific phosphorylation. Mol Biol Cell 8:287–301.

Connors JM, Kouroukis C, Belch A, et al. (2001): Flavopiridol for mantle cell lymphoma: Moderate activity and frequent disease stabilization. Blood 98: 807a.

Dai Y, Dent P, Grant S (2002): Induction of apoptosis in human leukemia cells by the CDK1 inhibitor CGP74514A. Cell Cycle 1:143–52.

Daley GQ, Van Etten RA, Baltimore D (1990): Induction of chronic myelogenous leukemia in mice by the P210bcr/abl gene of the Philadelphia chromosome. Science 247:824–30.

Davis ST, Benson BG, Bramson HN, et al. (2001): Prevention of chemotherapy-induced alopecia in rats by CDK inhibitors. Science 291:134–7.

de The H, Chomienne C, Lanotte M, Degos L, Dejean A (1990): The t(15;17) translocation of acute promyelocytic leukaemia fuses the retinoic acid receptor alpha gene to a novel transcribed locus. Nature 347:558–61.

Decker RH, Dai Y, Grant S (2001): The cyclin-dependent kinase inhibitor flavopiridol induces apoptosis in human leukemia cells (U937) through the mitochondrial rather than the receptor-mediated pathway. Cell Death Differ 8:715–24.

Deininger M, Goldman JM, Lydon NB, Melo JV (1997): The tyrosine kinase inhibitor CGP57148B selectively inhibits the growth of BCR-ABL positive cells. Blood 90:3691–8.

Deininger MW, Goldman JM, Melo JV (2000): The molecular biology of chronic myeloid leukemia. Blood 96:3343–56.

Deininger MW, O'Brien SG, Ford JM, Druker BJ (2003): Practical management of patients with chronic myeloid leukemia receiving imatinib. J Clin Oncol 21:1637–47.

Dixon H, Norbury CJ (2002): Therapeutic exploitation of checkpoint defects in cancer cells lacking p53 function. Cell Cycle 1:362–8.

Dong Y, Sui L, Tai Y, Sugimoto K, Hirao T, Tokuda M (2000): Prognostic significance of cyclin E overexpression in laryngeal squamous cell carcinomas. Clin Cancer Res 6:4253–8.

Dosaka-Akita H, Hommura F, Mishina T, et al. (2001): A risk-stratification model of non–small cell lung cancers using cyclin E, Ki-67, and ras p21: Different roles of G_1 cyclins in cell proliferation and prognosis. Cancer Res 61:2500–4.

Druker BJ, Sawyers CL, Kantarjian H, et al. (2001): Activity of a specific inhibitor of the BCR-ABL tyrosine kinase in the blast crisis of chronic myeloid leukemia and acute lymphoblastic leukemia with the Philadelphia chromosome. N Engl J Med 344:1038–42.

Druker BJ, Talpaz M, Resta DJ, et al. (2001): Efficacy and safety of a specific inhibitor of the BCR-ABL tyrosine kinase in chronic myeloid leukemia. N Engl J Med 344:1031–7.

Druker BJ, Tamura S, Buchdunger E, et al. (1996): Effects of a selective inhibitor of the Abl tyrosine kinase on the growth of Bcr-Abl positive cells. Nat Med 2:561–6.

Edelstein M, Vietti T, Valeriote F (1974): Schedule-dependent synergism for the combination of 1-beta-D-arabinofuranosylcytosine and daunorubicin. Cancer Res 34:293–7.

Einhorn LH (2002): Chemotherapeutic and surgical strategies for germ cell tumors. Chest Surg Clin N Am 12:695–706.

Flynn PJ, Miller WJ, Weisdorf DJ, Arthur DC, Brunning R, Branda RF (1983): Retinoic acid treatment of acute promyelocytic leukemia: in vitro and in vivo observations. Blood 62:1211–7.

Friend SH, Bernards R, Rogelj S, et al. (1986): A human DNA segment with properties of the gene that predisposes to retinoblastoma and osteosarcoma. Nature 323:643–6.

Fry DW, Bedford DC, Harvey PH, et al. (2001): Cell cycle and biochemical effects of PD 0183812. A potent inhibitor of the cyclin D-dependent kinases CDK4 and CDK6. J Biol Chem 276:16617–23.

Fukuse T, Hirata T, Naiki H, Hitomi S, Wada H (2000): Prognostic significance of cyclin E overexpression in resected non–small cell lung cancer. Cancer Res 60:242–4.

Fuse E, Tanii H, Takai K, et al. (1999): Altered pharmacokinetics of a novel anticancer drug, UCN-01, caused by specific high affinity binding to alpha1-acid glycoprotein in humans. Cancer Res 59:1054–60.

Gambacorti-Passerini C, le Coutre P, Mologni L, et al. (1997): Inhibition of the ABL kinase activity blocks the proliferation of BCR/ABL+ leukemic cells and induces apoptosis. Blood Cells Mol Dis 23:380–94.

Gardner SN (2002): Cell cycle phase-specific chemotherapy: Computational methods for guiding treatment. Cell Cycle 1:369–74.

Geissler EN, Ryan MA, Housman DE (1988): The dominant-white spotting (W) locus of the mouse encodes the c-kit proto-oncogene. Cell 55:185–92.

Gescher A (2000): Staurosporine analog. Crit Rev Oncol Hematol 34:127–35.

Giles FJ, Keating A, Goldstone AH, Avivi I, Willman CL, Kantarjian HM (2002): Acute myeloid leukemia. Hematology (Am Soc Hematol Educ Program) 73–110.

Goldberg AL, Rock K (2002): Not just research tools—Proteasome inhibitors offer therapeutic promise. Nat Med 8:338–40.

Graham SM, Jorgensen HG, Allan E, et al. (2002): Primitive, quiescent, Philadelphia-positive stem cells from patients with chronic myeloid leukemia are insensitive to STI571 in vitro. Blood 99:319–25.

Grant S, Roberts JD (2003): The use of cyclin-dependent kinase inhibitors alone or in combination with established cytotoxic drugs in cancer chemotherapy. Drug Resist Updat 6:15–26.

Graves PR, Yu L, Schwarz JK, et al. (2000): The Chk1 protein kinase and the Cdc25C regulatory pathways are targets of the anticancer agent UCN-01. J Biol Chem 275:5600–5.

Gray NS, Wodicka L, Thunnissen AM, et al. (1998): Exploiting chemical libraries, structure, and genomics in the search for kinase inhibitors. Science 281:533–8.

Gries JM, Kasimis B, Schwarzenberger P, et al. (2003): Phase I study of HMR1275 (flavopiridol) in non–small cell lung cancer (NSCLC) patients after 24 hr IV administration in combination with paclitaxel and carboplatin. Proc Am Soc Clin Oncol (in press).

Grundke-Iqbal I, Iqbal K, Quinlan M, Tung YC, Zaidi MS, Wisniewski HM (1986): Microtubule-associated protein tau: A component of Alzheimer paired helical filaments. J Biol Chem 261:6084–9.

Hajduch M, Kolar Z, Novotny R, et al. (1997): Induction of apoptosis and regression of spontaneous dog melanoma following in vivo application of synthetic cyclin-dependent kinase inhibitor olomoucine. Anticancer Drugs 8:1007–13.

Hall M, Peters G (1996): Genetic alterations of cyclins, cyclin-dependent kinases, and Cdk inhibitors in human cancer. Adv Cancer Res 68:67–108.

Han R (1994): Highlight on the studies of anticancer drugs derived from plants in China. Stem Cells 12:53–63.

Hanks SK, Hunter T (1995): Protein kinases 6. The eukaryotic protein kinase superfamily: Kinase (catalytic) domain structure and classification. FASEB J 9:576–96.

Havlicek L, Hanus J, Vesely J, et al. (1997): Cytokinin-derived cyclin-dependent kinase inhibitors: synthesis and cdc2 inhibitory activity of olomoucine and related compounds. J Med Chem 40:408–12.

Hemmeryckx B, Reichert A, Watanabe M, et al. (2002): BCR/ABL P190 transgenic mice develop leukemia in the absence of Crkl. Oncogene 21:3225–31.

Hideshima T, Richardson P, Chauhan D, et al. (2001): The proteasome inhibitor PS-341 inhibits growth, induces apoptosis, and overcomes drug resistance in human multiple myeloma cells. Cancer Res 61:3071–6.

Hoessel R, Leclerc S, Endicott JA, et al. (1999): Indirubin, the active constituent of a Chinese antileukaemia medicine, inhibits cyclin-dependent kinases. Nat Cell Biol 1:60–7.

Holtz MS, Slovak ML, Zhang F, Sawyers CL, Forman SJ, Bhatia R (2002): Imatinib mesylate (STI571) inhibits growth of primitive malignant progenitors in chronic myelogenous leukemia through reversal of abnormally increased proliferation. Blood 99:3792–800.

Honma T, Hayashi K, Aoyama T, et al. (2001a): Structure-based generation of a new class of potent Cdk4 inhibitors: new de novo design strategy and library design. J Med Chem 44:4615–27.

Honma T, Yoshizumi T, Hashimoto N, et al. (2001b): A novel approach for the development of selective Cdk4 inhibitors: library design based on locations of Cdk4 specific amino acid residues. J Med Chem 44:4628–40.

Huang LW, Chao SL, Hwang JL, Chou YY (2002): Down-regulation of p27 is associated with malignant transformation and aggressive phenotype of cervical neoplasms. Gynecol Oncol 85:524–8.

Huettner CS, Zhang P, Van Etten RA, Tenen DG (2000): Reversibility of acute B-cell leukaemia induced by BCR-ABL1. Nat Genet 24:57–60.

Hughes T, Kaeda J, Branford S, et al. (2002): Molecular responses to imatinib (STI571) or interferon+Ara-C as initial therapy for CML; results in the IRIS study. Blood 100:93a.

Jedema I, Barge RM, Willemze R, Falkenburg JH (2003): High susceptibility of human leukemic cells to Fas-induced apoptosis is restricted to G_1 phase of the cell cycle and can be increased by interferon treatment. Leukemia 17: 576–84.

Jiang X, Stuible M, Li A, et al. (2003): Leukemogenesis by BCR-ABL transduction of primitive SHIP null cells is unaltered and SHIP is not downregulated in CD34+ CML cells. Blood 100:743a.

Jin Z, Dicker DT, El Deiry WS (2002): Enhanced sensitivity of G_1 arrested human cancer cells suggests a novel therapeutic strategy using a combination of simvastatin and TRAIL. Cell Cycle 1:82–9.

Jonuleit T, van der KH, Miething C, et al. (2000): Bcr-Abl kinase down-regulates cyclin-dependent kinase inhibitor p27 in human and murine cell lines. Blood 96:1933–9.

Josting A, Wolf J, Diehl V (2000): Hodgkin disease: prognostic factors and treatment strategies. Curr Opin Oncol. 12:403–11.

Kaaijk P, Kaspers GJ, Van Wering ER, et al. (2003): Cell proliferation is related to in vitro drug resistance in childhood acute leukaemia. Br J Cancer 88: 775–81.

Kaelin WG Jr (1999): Functions of the retinoblastoma protein. Bioessays 21: 950–8.

Kaiser A, Nishi K, Gorin FA, Walsh DA, Bradbury EM, Schnier JB (2001): The cyclin-dependent kinase (CDK) inhibitor flavopiridol inhibits glycogen phosphorylase. Arch Biochem Biophys 386:179–87.

Kano Y, Akutsu M, Tsunoda S, Furuta M, Yazawa Y, Ando J (1998): Schedule-dependent synergism and antagonism between paclitaxel and methotrexate in human carcinoma cell lines. Oncol Res 10:347–54.

Kaur G, Stetler-Stevenson M, Sebers S, et al. (1992): Growth inhibition with reversible cell cycle arrest of carcinoma cells by flavone L86–8275. J Natl Cancer Inst 84:1736–40.

Kawakami K, Futami H, Takahara J, Yamaguchi K (1996): UCN-01, 7-hydroxyl-staurosporine, inhibits kinase activity of cyclin-dependent kinases and reduces the phosphorylation of the retinoblastoma susceptibility gene product in A549 human lung cancer cell line. Biochem Biophys Res Commun 219:778–3.

Keyomarsi K, O'Leary N, Molnar G, Lees E, Fingert HJ, Pardee AB(1994): Cyclin E, a potential prognostic marker for breast cancer. Cancer Res 54:380–5.

Keyomarsi K, Tucker SL, Buchholz TA, et al. (2002): Cyclin E and survival in patients with breast cancer. N Engl J Med 347:1566–75.

Khoo ML, Beasley NJ, Ezzat S, Freeman JL, Asa SL (2002): Overexpression of cyclin D1 and underexpression of p27 predict lymph node metastases in papillary thyroid carcinoma. J Clin Endocrinol Metab 87:1814–8.

Kindler T, Breitenbuecher F, Marx A, et al. (2003): Sustained complete hematologic remission after administration of the tyrosine kinase inhibitor imatinib mesylate in a patient with refractory, secondary AML. Blood 101:2960–2.

King KL, Cidlowski JA (1995): Cell cycle and apoptosis: Common pathways to life and death. J Cell Biochem 58:175–80.

Kirla RM, Haapasalo HK, Kalimo H, Salminen EK (2003): Low expression of p27 indicates a poor prognosis in patients with high-grade astrocytomas. Cancer 97:644–8.

Kitada S, Zapata JM, Andreeff M, Reed JC (2000): Protein kinase inhibitors flavopiridol and 7-hydroxy-staurosporine down-regulate antiapoptosis proteins in B-cell chronic lymphocytic leukemia. Blood 96:393–7.

Kitagawa M, Okabe T, Ogino H, et al. (1993): Butyrolactone I, a selective inhibitor of Cdk2 and cdc2 kinase. Oncogene 8:2425–32.

Kolb HJ, Mittermuller J, Clemm C, et al. (1990): Donor leukocyte transfusions for treatment of recurrent chronic myelogenous leukemia in marrow transplant patients. Blood 76:2462–5.

Koleske AJ, Gifford AM, Scott ML, et al. (1998): Essential roles for the Abl and Arg tyrosine kinases in neurulation. Neuron 21:1259–72.

Konig A, Schwartz GK, Mohammad RM, Al Katib A, Gabrilove JL (1997): The novel cyclin-dependent kinase inhibitor flavopiridol downregulates Bcl-2 and induces growth arrest and apoptosis in chronic B-cell leukemia lines. Blood 90:4307–12.

Kopp R, Noelke B, Sauter G, Schildberg FW, Paumgartner G, Pfeiffer A (1991): Altered protein kinase C activity in biopsies of human colonic adenomas and carcinomas. Cancer Res 51:205–10.

Kudo Y, Kitajima S, Sato S, Miyauchi M, Ogawa I, Takata T (2001): High expression of S-phase kinase-interacting protein 2, human F-box protein, correlates with poor prognosis in oral squamous cell carcinomas. Cancer Res 61:7044–7.

Lane ME, Yu B, Rice A, et al. (2001): A novel Cdk2-selective inhibitor, SU9516, induces apoptosis in colon carcinoma cells. Cancer Res 61:6170–7.

Lau LF, Seymour PA, Sanner MA, Schachter JB (2002): Cdk5 as a drug target for the treatment of Alzheimer's disease. J Mol Neurosci 19:267–73.

Leclerc S, Garnier M, Hoessel R, et al. (2001): Indirubins inhibit glycogen synthase kinase-3 beta and CDK5/p25, two protein kinases involved in abnormal tau phosphorylation in Alzheimer's disease: A property common to most cyclin-dependent kinase inhibitors? J Biol Chem 276:251–60.

Li S, Gillessen S, Tomasson MH, Dranoff G, Gilliland DG, Van Etten RA (2001): Interleukin 3 and granulocyte-macrophage colony-stimulating factor are not required for induction of chronic myeloid leukemia-like myeloproliferative disease in mice by BCR/ABL. Blood 97:1442–50.

Li Y, Sun X, LaMont JT, Pardee AB, Li CJ (2003): Selective killing of cancer cells by beta -lapachone: direct checkpoint activation as a strategy against cancer. Proc Natl Acad Sci USA 100:2674–8.

Lin TS, Howard OM, Neuberg DS, Kim HH, Shipp MA (2002): Seventy-two hour continuous infusion flavopiridol in relapsed and refractory mantle cell lymphoma. Leuk Lymphoma 43:793–7.

Lista P, Brizzi MF, Rossi M, Resegotti L, Clark SC, Pegoraro L (1990): Different sensitivity of normal and leukaemic progenitor cells to Ara-C and IL-3 combined treatment. Br J Haematol 76:21–6.

Losiewicz MD, Carlson BA, Kaur G, Sausville EA, Worland PJ (1994): Potent inhibition of CDC2 kinase activity by the flavonoid L86–8275. Biochem Biophys Res Commun 201:589–95.

Lugo TG, Pendergast AM, Muller AJ, Witte ON (1990): Tyrosine kinase activity and transformation potency of bcr-abl oncogene products. Science 247: 1079–82.

Maloney DG, Grillo-Lopez AJ, White CA, et al. (1997): IDEC-C2B8 (Rituximab) anti-CD20 monoclonal antibody therapy in patients with relapsed low-grade non-Hodgkin's lymphoma. Blood 90:2188–95.

Mamillapalli R, Gavrilova N, Mihaylova VT, et al. (2001): PTEN regulates the ubiquitin-dependent degradation of the CDK inhibitor p27(KIP1) through the ubiquitin E3 ligase SCF(SKP2). Curr Biol 11:263–7.

Meijer L, Thunnissen AM, White AW, et al. (2000): Inhibition of cyclin-dependent kinases, GSK-3beta and CK1 by hymenialdisine, a marine sponge constituent. Chem Biol 7:51–63.

Melillo G, Sausville EA, Cloud K, Lahusen T, Varesio L, Senderowicz AM (1999): Flavopiridol, a protein kinase inhibitor, down-regulates hypoxic induction of vascular endothelial growth factor expression in human monocytes. Cancer Res 59:5433–7.

Mitsiades N, Mitsiades CS, Richardson PG, et al. (2003): The proteasome inhibitor PS-341 potentiates sensitivity of multiple myeloma cells to conventional chemotherapeutic agents: Therapeutic applications. Blood 101:2377–80.

Miyauchi J, Kelleher CA, Wang C, Minkin S, McCulloch EA (1989): Growth factors influence the sensitivity of leukemic stem cells to cytosine arabinoside in culture. Blood 73:1272–8.

Molldrem JJ, Lee PP, Wang C, et al. (2000): Evidence that specific T lymphocytes may participate in the elimination of chronic myelogenous leukemia. Nat Med 6:1018–23.

Motokura T, Bloom T, Kim HG, et al. (1991): A novel cyclin encoded by a bcl1-linked candidate oncogene. Nature 350:512–15.

Motwani M, Delohery TM, Schwartz GK (1999): Sequential dependent enhancement of caspase activation and apoptosis by flavopiridol on paclitaxel-treated human gastric and breast cancer cells. Clin Cancer Res 5:1876–83.

Nishi K, Schnier JB, Bradbury EM (1998): The accumulation of cyclin-dependent kinase inhibitor p27kip1 is a primary response to staurosporine and independent of G_1 cell cycle arrest. Expr Cell Res 243:222–31.

Nitti D, Belluco C, Mammano E, et al. (2002): Low level of p27(Kip1) protein expression in gastric adenocarcinoma is associated with disease progression and poor outcome. J Surg Oncol 81:167–75.

Niu MY, Menard M, Reed JC, Krajewski S, Pratt MA (2001): Ectopic expression of cyclin D1 amplifies a retinoic acid-induced mitochondrial death pathway in breast cancer cells. Oncogene 20:3506–18.

O'Brian C, Vogel VG, Singletary SE, Ward NE (1989): Elevated protein kinase C expression in human breast tumor biopsies relative to normal breast tissue. Cancer Res 49:3215–7.

O'Brien SG, Guilhot F, Larson RA, et al. (2003): Imatinib compared with interferon and low-dose cytarabine for newly diagnosed chronic-phase chronic myeloid leukemia. N Engl J Med 348:994–1004.

Omura S, Iwai Y, Hirano A, et al. (1977): A new alkaloid AM-2282 OF Streptomyces origin. Taxonomy, fermentation, isolation and preliminary characterization. J Antibiot (Tokyo) 30:275–82.

Orlowski RZ, Stinchcombe TE, Mitchell BS, et al. (2002): Phase I trial of the proteasome inhibitor PS-341 in patients with refractory hematologic malignancies. J Clin Oncol 20:4420–7.

Osuga H, Osuga S, Wang F, et al. (2000): Cyclin-dependent kinases as a therapeutic target for stroke. Proc Natl Acad Sci USA 97:10254–9.

Paciucci PA, Cuttner J, Gottlieb A, Davis RB, Martelo O, Holland JF (1997): Sequential mitoxantrone, daunorubicin, and cytosine arabinoside for patients with newly diagnosed acute myelocytic leukemia. Am J Hematol 56:214–18.

Parker BW, Kaur G, Nieves-Neira W, et al. (1998): Early induction of apoptosis in hematopoietic cell lines after exposure to flavopiridol. Blood 91:458–65.

Patel V, Senderowicz AM, Pinto D Jr, et al. (1998): Flavopiridol, a novel cyclin-dependent kinase inhibitor, suppresses the growth of head and neck squamous cell carcinomas by inducing apoptosis. J Clin Invest 102:1674–81.

Phiel CJ, Wilson CA, Lee VM, Klein PS (2003): GSK-3alpha regulates production of Alzheimer's disease amyloid-beta peptides. Nature 423:435–9.

Pierga JY, Faivre S, Vera K, et al. (2003): A phase I and pharmacokinetic (PK) trial of CYC202, a novel oral cyclin-dependent kinase (CDK) inhibitor, in patients (pts) with advanced solid tumors. Proc Am Soc Clin Oncol 22:210.

Platz A, Ringborg U, Hansson J (2000): Hereditary cutaneous melanoma. Semin Cancer Biol 10:319–26.

Preisler HD, Azarnia N, Raza A, et al. (1984): Relationship between the per cent of marrow cells in S phase and the outcome of remission-induction therapy for acute nonlymphocytic leukaemia. Br J Haematol 56:399–407.

Qin XQ, Livingston DM, Kaelin WG Jr, Adams PD (1994): Deregulated transcription factor E2F-1 expression leads to S-phase entry and p53-mediated apoptosis. Proc Natl Acad Sci USA 91:10918–22.

Rahmani M, Grant S (2002): UCN-01 (7-hydroxystauorsporine) blocks PMA-induced maturation and reciprocally promotes apoptosis in human myelomonocytic leukemia cells (U937). Cell Cycle 1:273–81.

Rao PN, Mahagaokar S, Freireich EJ, Loo TL, Gottlieb JA (1975): Schedule-dependent synergistic cytotoxicity of arabinofuranosylcytosine with adriamycin or 3,6-bis(5-chloro-2-piperidinyl)-2,5-piperazinedione in cultured cells. Cancer Res 35:2996–3000.

Raza A, Preisler HD, Day R, et al. (1990): Direct relationship between remission duration in acute myeloid leukemia and cell cycle kinetics: A leukemia intergroup study. Blood 76:2191–7.

Rich T, Allen RL (2000): Wyllie AH. Defying death after DNA damage. Nature 407:777–83.

Richardson DS, Newland AC (2000): Current perspectives on the use of growth factors in the therapy of acute myeloid leukaemia; malignancy. Hematol 5:189–203.

Richmond A (2002): Nf-kappa B, chemokine gene transcription and tumour growth. Nat Rev Immunol 2:664–74.

Ritch PS, Occhipinti SJ, Cunningham RE, Shackney SE (1981): Schedule-dependent synergism of combinations of hydroxyurea with adriamycin and 1-beta-D-arabinofuranosylcytosine with adriamycin. Cancer Res 41:3881–4.

Rubnitz JE, Pui CH (2003): Recent advances in the treatment and understanding of childhood acute lymphoblastic leukaemia. Cancer Treat Rev 29:31–44.

Sausville EA, Arbuck SG, Messmann R, et al. (2001): Phase I trial of 72-hour continuous infusion UCN-01 in patients with refractory neoplasms. J Clin Oncol 19:2319–33.

Schenkein D (2002): Proteasome inhibitors in the treatment of B-cell malignancies. Clin Lymphoma 3:49–55.

Schultz C, Link A, Leost M, et al. (1999): Paullones, a series of cyclin-dependent kinase inhibitors: synthesis, evaluation of CDK1/cyclin B inhibition, and in vitro antitumor activity. J Med Chem 42:2909–19.

Schwartz GK, Ilson D, Saltz L, et al. (2001): Phase II study of the cyclin-dependent kinase inhibitor flavopiridol administered to patients with advanced gastric carcinoma. J Clin Oncol 19:1985–92.

Schwartz GK, O'Reilly E, Ilson D, et al. (2002): Phase I study of the cyclin-dependent kinase inhibitor flavopiridol in combination with paclitaxel in patients with advanced solid tumors. J Clin Oncol 20:2157–70.

Senderowicz AM, Headlee D, Stinson SF, et al. (1998): Phase I trial of continuous infusion flavopiridol, a novel cyclin-dependent kinase inhibitor, in patients with refractory neoplasms. J Clin Oncol 16:2986–99.

Senderowicz AM, Sausville EA (2000): Preclinical and clinical development of cyclin-dependent kinase modulators. J Natl Cancer Inst 92:376–87.

Sexl V, Piekorz R, Moriggl R, et al. (2000): Stat5a/b contribute to interleukin 7-induced B-cell precursor expansion, but abl- and bcr/abl-induced transformation are independent of stat5. Blood 96:2277–83.

Shah M, Kaubisch A, O'Reilly E (2003): A phase IB clinical trial of the sequence dependent combination of paclitaxel and cisplatin with the cyclin dependent kinase inhibitor flavopiridol in patients with advanced solid tumors. Proc Am Soc Clin Oncol 42:542.

Shapiro GI, Supko JG, Patterson A, et al. (2001): A phase II trial of the cyclin-dependent kinase inhibitor flavopiridol in patients with previ-ously untreated stage IV non–small cell lung cancer. Clin Cancer Res 7:1590–9.

Sherr CJ (1996): Cancer cell cycles. Science 274:1672–7.

Shi Z, Azuma A, Sampath D, Li YX, Huang P, Plunkett W (2001): S-Phase arrest by nucleoside analogues and abrogation of survival without cell cycle progression by 7-hydroxystaurosporine. Cancer Res 61:1065–72.

Skipper HE, Schabel FM Jr, Wilcox WS (1967): Experimental evaluation of potential anticancer agents. XXI. Scheduling of arabinosylcytosine to take advantage of its S-phase specificity against leukemia cells. Cancer Chemother Rep 51:125–65.

Soni R, O'Reilly T, Furet P, et al. (2001): Selective in vivo and in vitro effects of a small molecule inhibitor of cyclin-dependent kinase 4. J Natl Cancer Inst 93:436–46.

Soriano P (1994): Abnormal kidney development and hematological disorders in PDGF beta-receptor mutant mice. Genes Dev 8:1888–96.

Soriano P (1997): The PDGF alpha receptor is required for neural crest cell development and for normal patterning of the somites. Development 124: 2691–700.

Sotillo R, Garcia JF, Ortega S, et al. (2001): Invasive melanoma in Cdk4-targeted mice. Proc Natl Acad Sci USA 98:13312–17.

St Croix B, Florenes VA, Rak JW, et al. (1996): Impact of the cyclin-dependent kinase inhibitor p27Kip1 on resistance of tumor cells to anticancer agents. Nat Med 2:1204–10.

Stadler WM, Vogelzang NJ, Amato R, et al. (2000): Flavopiridol, a novel cyclin-dependent kinase inhibitor, in metastatic renal cancer: A University of Chicago Phase II Consortium study. J Clin Oncol 18:371–5.

Sui L, Dong Y, Ohno M, et al. (2001): Implication of malignancy and prognosis of p27(kip1), Cyclin E, and Cdk2 expression in epithelial ovarian tumors. Gynecol Oncol 83:56–63.

Suolinna EM, Buchsbaum RN, Racker E (1975): The effect of flavonoids on aerobic glycolysis and growth of tumor cells. Cancer Res 35:1865–72.

Tafuri A, Andreeff M (1990): Kinetic rationale for cytokine-induced recruitment of myeloblastic leukemia followed by cycle-specific chemotherapy in vitro. Leukemia 4:826–34.

Takano Y, Kato Y, Masuda M, Ohshima Y, Okayasu I (1999): Cyclin D2, but not cyclin D1, overexpression closely correlates with gastric cancer progression and prognosis. J Pathol 189:194–200.

Tan AR, Headlee D, Messmann R, et al. (2002): Phase I clinical and pharmacokinetic study of flavopiridol administered as a daily 1-hour infusion in patients with advanced neoplasms. J Clin Oncol 20:4074–82.

te Boekhorst PA, Lowenberg B, Vlastuin M, Sonneveld P (1993): Enhanced chemosensitivity of clonogenic blasts from patients with acute myeloid leukemia by G-CSF, IL-3 or GM-CSF stimulation. Leukemia 7:1191–8.

Thomas JP, Tutsch KD, Cleary JF, et al. (2002): Phase I clinical and pharmacokinetic trial of the cyclin-dependent kinase inhibitor flavopiridol. Cancer Chemother Pharmacol 50:465–72.

Tohda S, Nara N, Imai Y, Aoki N (1991): Effect of protein kinase inhibitors on the proliferation of leukemic cells stimulated by granulocyte colony-stimulating factor, granulocyte-macrophage colony-stimulating factor or interleukin-3. Leukemia 5:813–14.

Tybulewicz VL, Crawford CE, Jackson PK, Bronson RT, Mulligan RC (1991): Neonatal lethality and lymphopenia in mice with a homozygous disruption of the c-abl proto-oncogene. Cell 65:1153–63.

van de Vijver MJ, He YD, van't Veer LJ, et al. (2002): A gene-expression signature as a predictor of survival in breast cancer. N Engl J Med 347:1999–2009.

van Oosterom AT, Judson I, Verweij J, et al. (2001): Safety and efficacy of imatinib (STI571) in metastatic gastrointestinal stromal tumours: A phase I study. Lancet 358:1421–3.

Vesely J, Havlicek L, Strnad M, et al. (1994): Inhibition of cyclin-dependent kinases by purine analogues. Eur J Biochem 224:771–86.

Wang Q, Fan S, Eastman A, Worland PJ, Sausville EA, O'Connor PM (1996): UCN-01: A potent abrogator of G_2 checkpoint function in cancer cells with disrupted p53. J Natl Cancer Inst 88:956–65.

Xiao HH, Makeyev Y, Butler J, Vikram B, Franklin WA (2002): 7-Hydroxystaurosporine (UCN-01) preferentially sensitizes cells with a disrupted TP53 to gamma radiation in lung cancer cell lines. Radiat Res 158:84–93.

Yeh YA, Olah E, Wendel JJ, Sledge GW Jr, Weber G (1994): Synergistic action of taxol with tiazofurin and methotrexate in human breast cancer cells: schedule-dependence. Life Sci 54:L431–5.

Yu B, Lane ME, Wadler S (2002): SU9516, a cyclin-dependent kinase 2 inhibitor, promotes accumulation of high molecular weight E2F complexes in human colon carcinoma cells. Biochem Pharmacol 64:1091–100.

Yu C, Krystal G, Dent P, Grant S (2002): Flavopiridol potentiates STI571-induced mitochondrial damage and apoptosis in BCR-ABL-positive human leukemia cells. Clin Cancer Res 8:2976–84.

Yu D, Jing T, Liu B, et al (1998): Overexpression of ErbB2 blocks Taxol-induced apoptosis by upregulation of p21Cip1, which inhibits p34Cdc2 kinase. Mol Cell 2:581–91.

Yu L, Orlandi L, Wang P, et al. (1998): UCN-01 abrogates G_2 arrest through a Cdc2-dependent pathway that is associated with inactivation of the Wee1Hu kinase and activation of the Cdc25C phosphatase. J Biol Chem 273:33455–64.

Zaharevitz DW, Gussio R, Leost M, et al. (1999): Discovery and initial characterization of the paullones, a novel class of small-molecule inhibitors of cyclin-dependent kinases. Cancer Res 59:2566–9.

Zhao B, Bower MJ, McDevitt PJ, et al. (2002): Structural basis for Chk1 inhibition by UCN-01. J Biol Chem 277:46609–15.

Zhou BB, Chaturvedi P, Spring K, et al. (2000): Caffeine abolishes the mammalian G(2)/M DNA damage checkpoint by inhibiting ataxia-telangiectasia-mutated kinase activity. J Biol Chem 275:10342–8.

Zhou BB, Elledge SJ (2000): The DNA damage response: Putting checkpoints in perspective. Nature 408:433–9.

Zittoun R, Suciu S, Mandelli F, et al. (1996): Granulocyte-macrophage colony-stimulating factor associated with induction treatment of acute myelogenous leukemia: A randomized trial by the European Organization for Research and Treatment of Cancer Leukemia Cooperative Group. J Clin Oncol 14: 2150–9.

PART VI

CHAPTER 21

MISREGULATED FATE—CANCER

ARTHUR B. PARDEE
Dana-Farber Cancer Institute, Boston, MA 02115

In the Middle of difficulty lies opportunity.
Albert Einstein

CANCER VERSUS NORMAL CELLS

The aim of this chapter is to survey the basis of the aberrant properties of cancer cells as changed from the normal ones described in this book. Molecular mechanisms are generally qualitatively the same in normal and cancer cells; their changes are often quantitative. The basis of cancer is chaotic control, which modifies all cell fates—proliferation, differentiation, senescence, migration, programmed death—and their underlying mechanisms in cancers.

All the possible fates of a cell and their underlying mechanisms are modified in cancers. On the one hand, these differences provide growth and survival benefits to cancer cells, and on the other hand, they suggest targets for therapy. The cancer cell is an outlaw whose properties are determined by numerous perturbations of the elegant and complex metabolic and regulatory mechanisms that maintain homeostasis of a normal cell with the whole organism. Molecular mechanisms are generally qualitatively the same in normal and cancer cells; their changes are often quantitative. Very numerous critical processes can be defectively regulated, almost all of which may be altered in some cancers. These include, among others, controls of (1) genetic stability, (2) environmental interactions, (3) proliferation, (4) differentiation, (5) senescence, (6) location, (7) apoptosis, and (8) damage repair.

What genetic and biochemical differences have been found between normal and cancer cells, and how are these related to discordant cell behavior? Cancer provides "experiments of nature" that give us insights about genes and their functions. "The pathological illuminates the normal." Comparisons of the information in this chapter with that in pre-

Cell Cycle and Growth Control: Biomolecular Regulation and Cancer, Second Edition, Edited by Gary S. Stein and Arthur B. Pardee
ISBN 0-471-25071-6

vious chapters about normal cells are valuable. One also needs to ask questions in the context of actions and effects of external agents, including mutagens and anticancer drugs. Early steps of carcinogenesis create cells with imbalanced regulation. A new balance is produced in later stages through mutations of additional genes. Cancer therapy fundamentally requires a differential activity in tumor versus normal cells as a target. This difference can be provided by the altered homeostasis of cancer cells relative to normal cells.

Cancer research is making great progress, but it is ongoing and much remains to be discovered and integrated. The primary objective here is to provide a snapshot in time, to help the reader discover further information. Motifs are discussed that I find promising for understanding cancer, and that hopefully will be useful to develop therapies. Points of view are suggested rather than attempting to provide the book-length project of providing complete coverage.

Justice cannot be done here to historical landmarks of basic cancer research, the concepts and experiments on which the subject is founded. Excellent overviews of molecular and cellular biology of cancer are found in the principal comprehensive texts on cancer. A few arbitrarily chosen highlights of concepts include (1) the early hypothesis of a deranged energy metabolism that uncouples glycolysis and oxidative phosphorylation, (2) mutagenesis by chemicals and radiation, (3) the two-hit mutational basis of hereditary cancer, (4) discovery of oncogenes, (5) tumor suppressors, and (6) dominance of normal versus tumor phenotypes.

Effects of chemotherapeutic agents are noted because some property differentially expressed by cancer versus normal cells must underlie an even partly effective therapy. Such differences can provide potential therapeutic windows. A therapeutic result whose basis is known supports the proposed underlying difference. Otherwise, a clue is provided that deserves investigation. However, cancer therapy with drugs and drug resistance, in themselves very important, are not subjects of this chapter (for clinical applications, see Chapter 20 by Deininger).

This chapter cannot be exhaustive. Much is omitted because of the sheer mass of information, limitations of space, and of gaps in the writer's knowledge. For example, about 200 reviews on tumor suppressors appeared in PubMed in the past year, there were 6600 abstracts for the 2003 Annual Meeting of AACR, and relevant publications on a single molecule can run into the hundreds, as in a recent critical review on NF-κB activation in cancer. Therefore only selected topics can be be summarized and some central differences emphasized. Information about proceses in normal cells, with references, figures, and tables, can be found in the other chapters.

No references are provided here. A few germinal and overview articles are included in Chapter 1. This is in part because of the mass of data alluded to in this chapter. It is principally because at this time each reader by searching on the Internet (PubMed) can easily find numerous reviews that provide the most recent overviews and primary information on any subject of individual interest. The database can furthermore be mined to discover connections by combining definitive sets of key terms, or "terms + review" for more breadth, an approach we named "conceptual biology." And searches often uncover unexpected treasures. They provide interactive learning experiences with little effort, in contrast to the passive reading of text. Finally I apologize for not directly listing the very many individuals to whom credit for these researches is due.

MUTATIONS IN CANCER

Cancer arises from genetic changes that accumulate as a tumor grows Regulatory genes of every sort are mutated, creating positively functioning oncogenes and repressing tumor-suppressor genes, and eventually causing neoplastic transformation of rare cells. Spontaneously arising genetic changes in somatic cells are responsible for most cancers, though heritable mutations have roles in a minority, and a few cancers develop from uptake of viral genes. Several novel techniques have been developed for discovering genes expressed differently in cancer versus normal cells. Roles of the implicated genes can be investigated by blocking their expression with use of techniques such as dominant negative gene action, antisense RNA, or the new RNAi technique and observing resulting changes of phenotype.

Mutants

Controlled cell replication depends on an intricate balance between multiple regulators including oncogenes, tumor-suppressor genes, and other cell cycle-associated proteins. Deregulation of this machinery switches a normal cell to a cancerous cell. A most striking difference of cancer cells as compared to normal cells is their numerous mutations, as seen from increasing alterations of chromosomal structure and number as tumors progress. Cancer arises from accumulated genetic changes as a tumor mass grows from a mutated single cell. The accumulation of mutations in genes that maintain this homeostasis eventually causes neoplastic transformation in rare cells. Human breast tumors were shown to be comprised of diverse populations of differently mutated cells. A minority of these could form tumors in immunosuppressed mice, and these could be distinguished by their expressed surface markers.

In humans, at least four to six mutations are required to reach this state. These stepwise genetic and epigenetic alterations in cancer development include localized mutations, chromosome rearrangements, and viral integration-mediated genetic alterations. Identified mutations are changes in nucleotide sequence that include point mutations, amplifications, deletions that confer loss of heterozygosity (LOH), microsatellite instability, and gross chromosomal rearrangments. Some changes regulate transcription, while others operate in signal transduction pathways that are involved in processes of cell division, differentiation, DNA repair, apoptosis, drug resistance, and the like. Newer techniques for observing gross chromosomal changes include spectral karyotyping (SKY) and comparative genomic hybridization (CGH).

Mutated genes involved in cancer are broadly classified as oncogenes and tumor suppressors (see Chapter 19 by Zamamiri-Dans and Zambetti). Positively functioning oncogenes are activated and the repressing tumor-suppressor genes are inactivated in cancers. These respectively permit ever more rapid growth and avoid growth-arrest mechanisms. Tumor suppressors and oncogenes are regulated and misregulated by the same strategies as are other proteins. A landmark is the discovery that mutated *ras* is an oncogene. Transfection of mutant *ras* into mouse fibroblast line 3T3 made these cells tumorigenic, and therefore mutation of this gene was proposed to be the basis of cancer. This one-step proposal was challenged on the basis of many data demonstrating that cancer must arise from several mutations, and led to the demonstration that these 3T3 cells already contain other mutated oncogenic and tumor-suppressor genes such as *myc*, and that *ras* provides only the final step. Ras-Myc collaboration is essential for maintaining the balance between excessive proliferation, on the one hand, and apoptosis, on the other. This research led to the concepts of both oncogenes and tumor-suppressor genes.

Normal cells have been converted to tumor forming cells by insertion of only four genes: *ras*, *SV40 small T* (which affects protein phosphatase 2A), *large T* (which inactivates p53 and pRb), and *telomerase* (which stabilizes telomeres). As models, RIP-TAG transgenic mice transfected with SV40 form tumors, as also do K14 mice carrying the human papilloma virus 16 gene. But numerous mutations are required for development of cancers in vivo. As is often noted here, regulatory genes of every sort are mutated; for example, p53, p16, DCC, and DPC4/SMAD4 are frequently altered in prostate cancers. Some genes that are overexpressed in cancers include VEGF, COX-2, Her-2/neu, c-Myc, and Rad51. Crucial molecular events include derangement of the Wnt- and the transforming growth factor beta (TGF-β) signaling pathways, which exert a synergistic effect on the cell cycle. With loss of p53 function as well, several checks and balances are disrupted that pave the way to gross chromosomal aberrations and aneuploidy.

Defective Control of Mutation

Mutations accumulate as a tumor progresses. Whether the mutations appear in a definite sequential order and whether there is usually an initiating event are not established. One model proposes that a definite sequence of mutations underlies colorectal cancer, but others propose that the order varies. The numerous genetic changes in cancer raise the question of the mutational mechanism that creates them. The spontaneous rate of mutation acting over a long time has been calculated to be sufficient. To the contrary, the normal mutation rate has been proposed to be insufficient to be responsible for the thousands of random mutations appearing in one cancer cell. A major role is suggested for an initial mutation that causes a high rate of further mutations. Such an accelerated rate could result from mutation of genes that preserve genetic stability in normal cells, for example, spontaneous mutations arise owing to errors in DNA replication and their imperfect repair. Such a mutator phenotyope would appear early in tumor progression, and even in benign tumors. In turn, the mutator phenotype would initiate a cascade of further mutations, some of which would confer a selective advantage that expands the cell into a population. For example, in follicular lymphoma the cells survive that do not undergo programmed cell death because they overexpress anti-apoptotic Bcl-2 (see below). Certainly mutations are increased by extracellular mutagens. Carcinogens damage DNA, and thereby create mutations. Important practical examples of carcinogenesis are chemicals in tobacco smoke that cause lung cancer, and sunlight for skin cancer. There are several stages of carcinogenesis. Carcinogens modify DNA, and tumor promoters then cause hyperplasia and clonal expansion. TNF-α is a major promoter on mouse skin, acting upon inflammation through NF-κB and TGF-β.

A current question is the relative importance of localized mutations versus chromosomal rearrangements, both of which are frequent in cancers. DNA base changes, gene amplifications, and deletions have direct effects, resulting in modifications, increases, and eliminations of gene products, respectively. Point mutations are common, for example a single base is changed at a definite position in the *ras* oncogene in many cancer cells. A point mutation in a coding region can change functional properties of the product, and in a noncoding sequence it can quantitatively modify the level of a gene's expression.

Chromosomal changes are also frequent in cancer. A very early observation is that karyotypes are rarely normal in tumor cells, which exhibit multiple abnormalities of both number and structure and cells become increasingly aneuploid as the tumor progresses. Direct evidence for a genetic basis of cancer came from identification of tumor-specific translocations in leukemias and lymphomas, for example, the Philadelphia chromosome in chronic myelogenous leukemia (CML) is a 9/22 translocation. And in acute promyelocytic leukemia the retinoic acid receptor RARα fuses with the PML gene. Chromosomal rearrangements are also proposed to be responsible for mutations that change drug sensitivity. Mitosis is the most dramatic and potentially dangerous event in the cell cycle, when one copy of every chromosome is irreversibly segregated to each daughter cell. Defects in the checkpoints that normally maintain the fidelity of mitosis can lead to chromosomal instability and cancer, through defective segregation that produces aneuploid cells with consequent misexpression of genes. Thus both local DNA modifications and chromosomal rearrangements are fundamental in cancer progression.

Hereditary Cancer

A relatively small proportion of human cancer is hereditary. But genetic predisposition imposes a high risk for development of one or several kinds of cancer. More than 30 mutant genes for hereditary cancers have been cloned. One inherited allele is usually mutated, and some random later somatic mutation of the other allele causes onset of clonal expansion to form a tumor. A hereditary mutation in the familial adenomatous polyposis coli (FAP) gene is responsible for the disease, and it is rate limiting in most sporadic colorectal cancers. Mutation in the second allele gradually activates a series of molecular and histological changes that alter oncogenes or tumor-suppressor genes; the Wnt transduction pathway is constitutively activated. BRCA-1 and -2 are high penetrance herditary genes whose mutations are found in patients with familial breast and ovarian cancers, in addition to other mutated genes. BRCA1 is germ-line mutate in 50% of inherited breast cancer patients. BRCA1 and 2 form multiprotein complexes that have functions in repair, and are is localized in nuclear DNA damage foci. The mutations found in familial BRCA1/2 and other ovarian cancers are specific to tumors of a particular type, and are associated with differences in differentiation and stage. They are thought to be involved in defective DNA repair. An association with breast cancer was found for 13 polymorphisms in 10 genes described in one breast cancer study. A number of chromosomal regions have been identified that have frequent deletions. The haploinsufficiency hypothesis proposes that one such one allele may be useful for identifying gene targets of subsequent chromosomal deletions. In addition there is functional and/or genetic evidence supporting roles in cancer of genes in these regions. *PTEN* is currently the most frequently mutated gene in prostate cancer, and *KLF5* most frequently has hemizygous deletion and loss of expression. Such genes may aid development of biomarkers and therapeutic regimens.

Cancer Viruses

Several viruses have been associated with human cancers. Their inserted genes modify specific cellular processes. They are expressed as oncogenes involved in transformation and immortalization, and are required for the progression toward malignant cancer. As suspected 30 years ago, human papilloma viruses (HPV) are agents of cervical cancer. HPV-type specific DNAs have been found in almost all cervical cancers. HPV-type E16, E6, and E7 oncoproteins independently induce chromosome instability by inactivating p53 and pRb, respectively. As another example, infection with human polyoma virus does not kill the cell under some circumstances, but its T-antigens result in the cell becoming transformed and tumorigenic, mainly by action of middle T, which assembles a large multi-protein complex at the cell membrane whose tyrosine kinase stimulates p21(ras), P13K and PLC gamma-1 activity, and the MAP kinase cascade. HIV Tat antagonizes the expression and apoptotic function of p53. Preneoplastic changes in gene expression in hepatocellular carcinomas (HCC) result from the actions of hepatitis B and C viruses. Subsequent tumor progression follows by a multistep process involving DNA methylation, point mutations, or loss of heterozygosity (LOH) in selected cellular genes. These changes are often distinct in different HCC nodules, suggesting molecular heterogeneity. Multiple and perhaps redundant negative growth regulatory pathways appear to protect cells against transformation by these viruses.

Gene Expression

Mutations both change the expression levels of genes and produce modified proteins, which results in alterations of the downstream biochemistry, physiology, and cell structure (see Chapter 2). The various differentiated normal cells and individual cancer cells express different genes as mRNAs and then as proteins. Thousands of these changes develop randomly in every cancer cell, though many may not have any role in the disease. Some mRNAs are misexpressed because the gene that codes for them is mutated (class I), but the majority (class II) do not arise from their gene being altered at the DNA level, but indirectly due to defective transciption controls produced by a primary mutation of another gene. That is, mutation of one gene modifies expression of other target genes. For example, a mutated kinase can alter the activity of a transcription factor or of another enzyme. As a class II example, expressions of many genes in colon cancers are increased by overexpression of Pinl. Pinl binds to β-catenin, thereby decreasing its nuclear exclusion and degradation through an interaction with FAP protein. Mutations of different genes can change activity of the same target, for example, of *ras* expression. These are related by differently modifying the same upstream regulatory or biochemical pathway. Furthermore expression levels are changed by environmental agents. One must consider the cancer problem to involve interacting networks of reactions, some of which are redundently controlled and are therefore not readily evident.

Gene misexpressions in cancer versus normal cells identify mRNAs and proteins whose functions can provide information about critical events in tumors. Several novel approaches have been developed for discovering genes expressed differently as mRNAs in cancer versus normal cells. A few such differences as of tumor-suppressor maspin expression, were found by applying the laborious technique of subtractive hybridization. Newer methods, especially differential display, have rapidly led to discoveries of numerous new as well as known mRNAs whose expressions are modified by effects of oncogenes, growth factors, ligands, drugs, dominant negatives, and so on. Array methods now make possible rapid searches for differential expressions of genes. They permit rapid screening for finding pattern changes of tens of thousands of known mRNAs. Genes for metastasis were discovered with microarrays, and they can be present in cancer cells prior to clinical evidence. A combined display-array technique has been applied to identify mRNA differences in small blood samples from normal individuals and breast cancer patients. Methods for identifying protein changes (proteomics) are available using two-dimensional gel patterns.

These results suggest that the diverse pathways exert broadly overlapping effects on tumor production. Global expression monitoring was used to explore the relationships between receptor tyrosine kinase-activated signaling pathways and transcriptional induction of known immediate early genes (IEGs). Activation of the PDGF-β receptor induced 66 genes in fibroblasts. Mutant receptors lacking binding sites for activation of the PLCgamma, PI3K, SHP2, and Ras-GAP pathways retained partial ability to induce 64 of these IEGs. Removal of the Grb2-hinding site further broadly decreased induction. PDGF-β receptor and fibroblast growth factor receptor 1 each induced essentially identical IEGs. Roles of the implicated genes can then be investigated by blocking their expression with techniques such as dominant negative gene action, antisense RNA, and the new RNAi technique in order to observe the resulting changes of phenotype.

GENERAL THEMES OF DYSREGULATION

Summary

Proliferation of each individual cell in a multicellular eukaryote must be closely coordinated with the whole if the organism is to survive. The cell cycle is the sequence of events that produces two cells from one. Progression through the cycle is arrested at specific loci (checkpoints) by conditions, such as inadequate extracellular conditions for growth stimulation, or by stress, such as is caused by DNA damage. Defective checkpoint control is found in cancers. The organism's defense to remove potentially dangerous cells is programmed cell death (apoptosis), a normal physiological process that functions thoroughout life. Since growth of a tumor depends on an imbalance of proliferation versus cell death, this mechanism is comparable in importance to proliferation. The processes are intertwined; excessive proliferation appears to signal apoptosis, which is usually preceded by transient arrest of proliferation. Other oncogenic changes eliminate apoptosis, a process benefical to tumor growth, and those mutant cells that preferentially survive can proliferate rapidly.

General Concepts

We summarize some regulatiory principles here, before discussing defective regulations in cancer. Proliferation of each individual cell in a multicellular eukaryote must be closely coordinated with the whole if the organism is to survive. This is in contrast to rapid growth which is the survival strategy of single-cell organisms such as yeasts and bacteria. As a result orthologues of their regulators have to be investigated in mammalian systems to discover their roles in cancer. Signals from the whole organism, delivered via blood or through contacts with adjacent cells, determine the behavior of normal cells. Defects of these controls are major factors in cancer.

Proliferation of cells is a major process that is increased in cancers. The cell cycle is the sequence of events that produce two cells from one. Fail-safe mechanisms protect cells in extreme situations. For example, yeast forms spores under adverse growth conditions. Progress through the cycle is arrested at specific loci in the cycle (checkpoints) by conditions such as conditions inadequate for growth stimulation. Cells damaged by stress such as that caused by DNA damage arrest at checkpoints, which provides them with time for adjustment and repair of their lesions. Failing this, they undergo irreversible growth arrest (senescence) or programmed cell death (apoptosis). Checkpoint controls, senescence, and apoptosis are major factors in cancer because they are an organism's defense to remove potentially dangerous cells (see Chapter 3). They are frequently defective in cancer cells.

The net growth of a tumer depends on an imbalance of cell proliferation over death. A tumor could exceed body mass in less than two months if its cells simply doubled daily and did not die. But cancer cells do die in large numbers, thereby slowing tumor growth. Elimination of apoptosis is benefical to the tumor, and those mutant cells that preferentially survive may proliferate rapidly. Because cancer cells are intrinsically fragile, many die during the growth of a tumor, and so additional mutations that decrease apoptosis are essential for their survival. Having counteracted their intrinsic instability, cells are selected during tumor progression. Apoptosis, as for most central processes, therefore is perturbed as cancers develop. Some oncogenic changes promote apoptosis, which are thus comparable in importance to those controlling proliferation. Proliferation and programmed cell death are intertwined. Excessive proliferation appears to signal apoptosis, which is usually preceded by transient arrest of proliferation.

Cellular events are the consequences of not one but several interacting extra- or intracellular molecular signals. Many anticancer agents decrease proliferation as well as increase apoptosis. Such balances between opposing natural forces are fundamental; for every action there is a reaction, a phenomenon long recognized in China as the yin-yang principle. Elaborate regulatory systems are required to maintain homeostasis and viability of normal cells, and they cause apoptosis of cancer and of other cells whose balance is irreversibly perturbed. Conflicting molecular signals that cause defective control of growth could trigger apoptosis. The highly innovative Clash hypothesis proposes that unbalanced competitive simultanous molecular signalings for proliferation and for growth arrest create a discord that activates apoptosis mechanisms. Mutations that block apoptosis are important for cancer cells in other ways. As a consequence of defects in apoptotic mechanisms, DNA lesions persist in cancer cells and create mutations. Cells that have been mutated to resist apoptosis are harder to kill by cytotoxic chemotherapy, since most of these agents would otherwise induce apoptosis. The failure to die contributes to persistence of DNA damage, which is mutagenic when incompletely repaired. The resulting deregulation leads to unlimited, self-sufficient cell growth and ultimately generates invasive and lethal metastatic tumors: "cancer goes from bad to worse." Therapeutic techniques are being developed to overcome this barrier raised by anti-apoptotic mechanisms, based on the increasing knowledge regarding underlying mechanisms.

LEVELS OF REGULATION

Summary

Proliferation control is modified at many levels in cancer cells. Biochemical regulatory pathways involve initiation by growth factors. Cascades of phosphorylations catalyzed by kinases then transmit these extracellular signals through the cell membrane and on to the nucleus, where they induce transcriptions of specific genes that produce sets of messenger RNAs that code for the protein machinery for proliferation. Transcription provides one major mode of regulating cell functions. There is, however, more to life than transcription. Control is exerted upon every process of molecular biology and biochemistry, and mutations in cancers modify the controls normally exerted upon all of these many steps.

Proliferation

Controls in cancer cells are modified at many levels, transcriptional and post-transcriptional. Cascades of phosphorylations catalyzed by kinases transmit extracellular signals through the plasma membrane and on to the nucleus. These signal transduction pathways induce transcriptions of specific genes to produce the biochemical machinery for proliferation. In the classical mechanism derived from bacterial β-galactosidase regulation, the repressor protein, a transcription factor, when bound to a specific DNA sequence in the promoter, represses expression of the target gene. Transcription is activated by noncovalent binding of a β-galactoside to another (allosteric) site of the repressor. In eukaryotes the process is much more complex. Major steps can include kinase-catalyzed entry into the nucleus of a transcription factor protein, its binding to a specific DNA promoter sequence, to which other proteins are added. Covalent modification of these proteins, such as by their phosphorylation, can further modify transcription. Binding of a transcription factor complex to the RNA polymerase complex activates synthesis of a heterogeneous nuclear RNA corresponding in sequence to a strand of the downstream DNA. This RNA is processed by splicing that removes the introns, usually in several ways to produce sets of messenger RNAs whose 3′-ends are then poly-adenylated and 5′-ends are capped. These mRNAs are exported to the cytoplasm where they provide the sequence information for translation, the linking of amino acids into proteins catalyzed by ribosomes. Mutations in cancers modify all of these steps. Protein phosphorylations resulting from altered kinase activities are remarkably frequently altered in cancers. These activate some enzymes and inactivate others, by changing affinities to cooperating molecules, their locations and proteolytic degradations. Why are there so many levels of regulation? Time frames of the mechanisms differ greatly. Non-covalent regulations act within minutes, whereas effects of DNA methylations can persist for a lifetime.

Most of the subsequent molecular biological and biochemical procesess are catalyzed by multiprotein complexes rather than by free enzymes. These complexes include the transcription complexes for synthesis of RNA, translation complexes (ribosomes) for protein synthesis, replitase complexes for DNA synthesis, DNA repair complexes, apoptosomes for apoptosis, proteasomes for protein degradation, and scaffolds that line up kinase cascades. These complexes can be very large. They contain not only enzymes but proteins that bind small regulatory molecules, such as the G proteins than bind activating GTP or inactivating GDP, and also scaffold proteins.

DNA Methylation and Histone Acetylation

In addition to transcription an important mechanism for regulating gene expression in eukaryotes is by changes caused by covalent modifications. The important mechanism is methylation of DNA. This methylation modifies nearby histones, the main building blocks of nucleosomes around which DNA is wrapped, and thereby modify chromosome structure (see Chapters 2 and 13 on cell architecture and histone modifications.) The mechanism is that protein MeCP2 specifically binds to methylated DNA sites, where it binds deacetylases (HDACs) that remove acetyls from amino termini on nearby histones. The major consequence is to increase the fraction of deacetylated histones that are locally bound to DNA. Epigenetics is a term for such altered gene expression without changes in DNA nucleotide sequence. Abnormally imprinted cells are susceptible to epigenetic modification.

Cytosines in "CpG islands" are the prime targets for hypermethylation of DNA. Up to 30% of cytosines are methylated in DNA of higher eukaryotes. Their essential function is to silence genes. These base sequences are associated with at least half of all cellular genes that are usually methylation-free in normal cells; this methylation is almost solely on genes subject to genomic imprinting, as on the inactivated female X-chromosome; only the non-imprinted allele is transcribed. The pattern varies in different tissues. Dense methylation of cytosine residues within these islands causes strong and heritable transcriptional silencing. As a simplified picture, DNA methyltransferases (DnMT1) catalyze transfer of methyl from the donor methylene-tetrahydrofolate to the 5′position of cytosine in CpGs of promoter DNA. DnMT1 is overexpressed in many tumors. Ras up-regulates methylation by inducing DnMT1, most likely via EGF1 activated transcription. There is also a global demethylation in cancers.

DNA methylation is frequently changed in cancers. Aberrant hypermethylation of the CpG islands associated with tumor-suppressor genes has been proposed to contribute to carcinogenesis and tumor progression. Differential methylation hybridization assays showed that poorly differentiated tumors become increasingly hypermethylated. In cancer cells numerous methylation-dependent gene expression differences were found by expression genetics. Tumor-related changes of gene expression are caused not only by global CpG hypermethylation, especially of CpG islands, but also by more general demethylation; there was a 10× increased mutation rate in Dnmt1-minus cancer cells. Furthermore S-adenosylmethonine is a methyl donor, and metabolism of methionine is altered in cancers; tumor cells requires methionine for growth, whereas normal cells can methylate homocysteine to form methionine. Methylation-induced gene silencing in cancer cells correlates with methylation, altered chromatin structure, and transcriptional accessibility. Questions arise about underlying mechanisms of these changes, and the precise consequences of DNA methylation in developmental biology and cancer. These need to be clarified before the importance of genomic methylation patterns can be understood.

Key genes are inactivated by highly methylated promoters in cancer cells. Repression of pRb is created by methylation at a single site. The Rb family of proteins function is linked to chromatin remodeling enzymes, and not only regulates exit of cells from both G_1 and S phase of the cell cycle. A suppression of p53 in cancers is caused by inactivation of ARF through methylation of its promoter. As another example, BRCA1 transcription is repressed in sporadic breast cancers. The repair gene hMLH1 is frequently hypermethylated in colon cancers, an early event that causes mutations. And many tumor-suppressor genes, such as the cdk inhibitor p16, and protooncogenes, for example, *raf*, have hypermethylated promoters in cancers. The *ras* oncogene acting via Rb/E2F and Dnmt-1 causes epigenetic changes by methylation, such as inactivation of p16, which activates tumor cell growth.

Epigenetic changes are more readily reversed by chemical agents than are mutations. Many reviews have appeared on reactivating genes silenced by methylation. 5-Aza-2′-deoxycytidine is a specific inhibitor of cytosine DNA methyltransferase that inhibits methylation. Such agents potentially have numerous effects against cancer. Zebularine, a DNA methylation inhibitor that covalently binds to DNA methyltransferases, inhibits CpG island methylation in the p16 gene promoter. It has antitumor activity in a mouse system. Numerous results link retinoids and retinoic acid receptors to methylation and cancer. The retinoic acid receptor RARβ-2 is involved in cancer therapy as a proapoptotic tumor suppressor. It is lost in breast cancers by methylation of its promoter, and 5-aza-2′-deoxycytidine reactivates it. Tumor-suppressor gene-1 (DRG-1) expression is controled by several differentiating agents, including retinoids. Its transfection up-regulates several differentiation markers, and its overexpression decreases invasion in culture and metastasis in mice. Its expression is decreased in metastatic colon cancers, in which it is lost early due to promoter hypermethylation. Retinoids are, however, oxidizable-reducible compounds that can also have effects that are independent of their receptors. Glutathione (GSH) is oxidizable-reducible, and its conjugation with electrophilic compounds catalyzed by glutathione–S-transferase detoxifies them, thereby protecting DNA. This transferase is a marker for detection of prostate cancer.

Acetylation of histones is catalyzed by histone acetyltransferases (HATs), and net acetylation is a balance between acetylase and deacetylase (HDAC), which activate rather than silence transcription of nearby genes. These enzymes are important in DNA functions, including transcription, replication, and repair. They are being increasingly associated with normal and tumor cell processes such as proliferation and differentiation. p300 and CBP are transcriptional coactivators that bind to transcription factors and transcription machinery, providing a scaffold for integration, and their acetyltransferase activity increases acetylation and thereby modifies chromation structure. Their genes are mutated in tumors, and their functions are modifed by small molecules. When retinoic acid binds to nuclear receptor dimers RAR-RXR, they interact with histone acetylase and SRC/p300 coactivators. But in the absence of the ligand they form a repressing complex with histone deacetylase (HDAC) Sin3 and corepressors SMRT or Nco-R. These opposite activities provide another example of biological regulation by the yin-yang principle. Histone deacetylase inhibitors such as SAHA are being applied for cancer therapy. In addition to the better known acetylation, covalent histone modifications including phosphorylation, methylation, and ubiquitination change transcriptional regulation. Histones are modified on their N-terminal tails by methylation, causing permanent negative inhibition of transcription.

Post-transcriptional Regulations

Much current emphasis is on transcriptional regulations that alter gene expression to produce hnRNAs. This interest arose from exciting developments with oncogenes and tumor-suppressor genes. But signaling pathways are regulated at both molecular genetic and biochemical levels, and controls are exerted upon each of the many steps from a gene to a functional protein. These include RNA processing with alternative splicing, degradation, translation, covalent modifications, and degradation of proteins that include transcription factors, histones, and enzymes. As examples, TPP promotes degradation of transcripts of TNF-α. Translation is inhibited by TIA-1 binding to mRNA prior to attachment of the large ribosomal subunit. The functional consequence of binding of thymidylate synthase protein to its own mRNA, and also to p53 mRNA, provides an example among many of post-transcriptional regulation. mTOR is a mammalian protein kinase with several roles in translation, whose activity is needed for passage from G_1 to S. It is the target of rapamycin, a macrolide natural product with activity against tumor xenografts.

After transcription, properties of proteins are changed by a variety of covalent additions, for example, the acetylation of histones described above. As another example, ubiquitination of Smad7 is prevented by actylation of its lysines, thereby blocking subsequent proteosomal degradation. Changes of phosphorylation are remarkably frequent in cancers, resulting from altered kinase activities. Phosphates add negative charges that alter the protein's structure and thereby modify its properties, activating some enzymes and inactivating others, altering regulations by changing affinity of cooperating ligands and of other proteins, locations, or proteolytic degradation by subsequent ubiquitination and protosome activity. Proteins are also altered by other substitutions such as by methylations.

Proteasomes

Degradation of key proteins is increasingly found to be a major mechanism for regulation of cell growth and apoptosis. This proteolysis is initiated by an ATP-dependent phosphorylation of the target protein, which then covalently binds one or several molecules of the small protein ubiquitin by a sequence of three enzymatic steps that activate and bind ubiquitin to its target. A large family of E3 ubiquitin ligases confers specificity for ubiquitin binding. For example, skp2 ligates ubiquitins to the p27 protein. The proteosome, a multi-protein proteolysis machine, then recognizes and hydrolyzes this product. A dynamic balance is created by the converse activity of a family of more than 60 deubiqitinating (DUB) enzymes. Among the key proteins degraded by proteosomes are p53, cyclins, NF-κB, and E2F-1.

Proteosomes have roles in many diseases, and therapeutic applications are ongoing. Several proteasome inhibitors have activity against tumors, including some cancers that are resistant to conventional chemotherapy. Proteasome inhibitors can stabilize p53, activate stress kinases, and induce heat shock proteins (Hsp). High expression of oncogenic c-Myc makes cancer cells more susceptible to the apoptosis induced by proteasome inhibitors. This varies between cell type, and some drugs exhibit selectivity by preferentially inducing proteolysis in transformed or proliferating cells. A mechanism by which proteasome inhibitors increase the sensitivity of cancer cells to apoptosis is by inhibiting anti-apoptotic NF-κB activation, which is elevated in many tumor cells. The proteosome inhibitor PS-341 blocks degradation of the inhibitor protein IkB, and thereby activation of NF-κB and modulation of apoptosis-related genes. It can inhibit growth of carcinomas and melanomas in mice. These drugs also can prevent angiogenesis and metastasis in vivo. They are being tested clinically against multiple myeloma.

Compartmentalization

Cell architecture has a very major role in regulation and and cancer. (See Chapter 2 on cell archicture.) Many regulations function by relocalization of proteins to or from the intracell compartment. They may function, for example, in the nucleus where they control gene expression, or be removed to the nucleolus. As an important example, enzymes related to DNA synthesis are transported into the nucleus at the G_1/S transition. Another example is the action of kinases in the cytoplasm, as they release IκB from NF-κB, which permits the latter's transport into the nucleus where it interacts with DNA and activates transcriptions. Stimulation of PKCδ causes its translocation to mitochondria and nucleus.

The multiple mechanisms that import and export proteins and RNAs between cellular compartments are a subject of intensive investigation. Transport across the nuclear membrane is through pores. Carrier proteins that specifically bind to target proteins have signaling sequences that bind to organelle-specific receptors. In response to specific signals they interact with several very large translocon complexes such as the nuclear pore complex. These carry the proteins through the proper membrane. There are three major classes of nuclear transport receptors; the largest class is named importins/exportins. Another class transports the small GTPase Ran, an evolutionarily conserved member of the Ras superfamily, and its binding partners including the guanine nucleotide exchange factor and several related factors. This highly active system provides energy for and regulates the first class. The third class mediates export of mRNAs from the nucleus to cytoplasm. Several drugs that inhibit nuclear transport are under investigation. Furthermore, inside the nucleus there is dynamic compartmentation (see Chapter 2 by Braastad et al.).

QUIESCENCE VERSUS PROLIFERATION

Summary

Normal cells are usually quiescent in adults and are only occasionally stimulated to proliferate. When transferred to suboptimal external conditions in culture, they complete their cycle and then enter an arrested state (G_0). If adequate conditions are restored, G_0 cells reenter the cell cycle and transit through the cycle's G_1 phase and resume growth. The frequency of initiation of cycling from quiescence is the major factor that determines growth. Cancer cells are not as readily arrested by inadequate growth conditions. Normal cells, but not tumor cells, also cease growth when their culture becomes confluent; proliferation is then inhibited by surface interactions between cells in which adhesion proteins function as primary sensors. Also adequate supplies of extracellular molecules including oxygen, ions, glucose, essential amino acids, and vitamins are essential for proliferation of both normal and cancer cells, In vivo these nutrients must be supplied through the blood to permit growth. Solid tumors grow within an environment that is characterized by an abnormal vasculature, which is needed to provide sufficient oxygen and nutrients. Hydrophilic biomolecules, including amino acids, nuclosides, and vitamins, must be transported through the cell membrane, and their transport systems are modified in cancer cells. Several growth factor proteins are communicators between an organism's cells. Their interactions with specific cell surface receptors activate major intracell signaling paths. Some cancers aberrantly overexpress growth factors, thereby stimulating them and nearby cells, and the overexpression of receptors accelerates cancer progression. Estrogen, androgen, and retinoids also carry signals for growth and differentiation through the organism. In contrast to the pathways of growth factor proteins, these lipophilic molecules diffuse into the nucleus where they bind specifically to receptor proteins and activate transcriptions. Cancer is characterized by inappropriate growth in space as well as in time. Whereas normal solid tissues cells are localized, tumors eventually become metastatic, spreading and growing in secondary locations. Cancer cells adhere weakly to adjacent cells because their surfaces are modified by mutated extracellular proteases, thereby permitting their detachment as a prelude to metastasis.

Quiescence

Normal cells are usually quiescent in adults and are only occasionally stimulated to proliferate, for example, when contact is lost after adjacent cells die or after wounding. Nerve cells do not proliferate. Some cells including those in skin, bone marrow, and colon proliferate at a rate sufficient to replace aged dying cells. Only a fraction of cancer cells are proliferating in vivo at any time. Proliferation is counterbalanced by cell death; many rapidly growing cancer cells die. Some clinically applied drugs such as 5F-uracil and methotrexate are active against DNA synthesis and Taxol is active mainly in M phase, and therefore these compounds are selective against proliferating cells (see Chapter 20 by Deininger). The frequency of initiation of cycling from quiescence is the major factor that determines growth of a mass of cells, rather than duration of the cycle which is fairly constant. But some kinds of normal cells initiate growth in vivo more frequently than do cancer cells, which creates a major problem with those chemotherapies that are based on cell cycle events, because both growing normal as well as tumor cells are killed.

Normal cells in culture complete their cycle and then enter an arrested state (G_0) when transferred to suboptimal conditions, such as upon reaching confluence or at low concentration of a growth factor or essential nutrient. Cancer cells are not as readily arrested by growth conditions that are inadequate for normal cells. Their proliferative advantage arises from their ability to bypass quiescence.

Proliferation

G_0 cells reenter the cycle and resume growth in G_1 if adequate conditions are restored without too long a delay. These G_1 phase cells have the same amount of DNA as do G_0 cells. But they are not the same biochemically, and many protein and RNA differences are found, for example, by array technologies. As specific examples, the Rb-related p130 protein is present in G_0 cells, and it disappears in G_1 phase. Conversely, Ki-67 is expressed in G_1 but not in quiescent G_0 cells. It often serves as a marker to distinguish between these states and is a prognostic and diagnostic tool; 3500 published articles featuring it demonstrate major interest.

Extracellular Regulation

Whether or not a cell initiates proliferation is determined by external conditions that affect the cell machinery that functions in the G_0 to G_1 transition (named competence) and in late G_1. Adequate supplies of extracelluiar molecules, including oxygen, ions, glucose, amino acids, and vitamins are essential for proliferation of both normal and cancer cells. In vivo, nutrients must be supplied through the blood to permit growth and to prevent necrosis. Asparagine is essential for survival of some cancers; its depletion by asparaginase is applied clinically. Furthermore adequate, but low, caloric intake decreases cancer growth and increases survival of animals. Tumors are often hypoxic, and their oxygenation status is a parameter of prognosis; hypoxia independently indicates poor outcome of prostate, head and neck, and cervical cancers. It affects genomic stability, apoptosis, angiogenesis, and metastasis. And hypoxia limits effectiveness of radiotherapy and most chemotherapy. However, this property within the unique tumor microenvironment can provide the basis for tumor therapies by bioreductive drugs such as tirapamazine. These are specifically toxic to hypoxic cells, and also depend on hypoxia-specific gene delivery systems.

Solid human tumors grow within an environment that is characterized by an abnormal vasculature that is needed to provide sufficient oxygen and nutrients (see Chapter 10 on angiogenesis). They generate new blood vessels to provide adequate amounts of these substances. This angiogenesis requires stimulating the proliferation of cells that form the blood vessels. A question is whether these are normal endothelial or cancer cells. Furthermore angiogenesis facilitates escape of potentially metastatic cells from tumors. Vascular endothelial growth factor (VEGF) is a major activator of angiogenesis; it stimulates *ras* plus *myc* signaling. Insulin-like binding protein-3 (IGFBP-3) was shown by array technology to be overexpressed in endothelial cells of mouse breast tumor vessels, and may be a marker for angiogenesls. Thrombospondin 1, 2 inhibits angiogenesis. The importance of supplying oxygen and nutrients is illustrated by current interest in inhibitors of angiogenesis, compounds that inhibit creation of new vasculature and thereby prevent adequate nutrition for tumor growth. They are being tested alone and in combinations as chemotherapeutic and chemopreventive agents. As an example, Avasatin (anti-VEGF) in combination with chemotherapy showed positive survival benefits for colon cancer patients. Thalidomide is a tumor-specific anti-angiogenic agent, especially active against myeloma.

Nutrients are supplied in the medium to cells in culture. An inadequate supply stops proliferation, and the cells return to G_0, for example, after deprivation of an essential amino acid such as isoleucine. Transport across the cell membrane and into the cell is the first step of small molecule utilization. The proteins that are involved in transport of numerous low molecular weight compounds are changed in cancers. The SLC26 family of anion exchangers that transport sulfate, bicarbonate, and chloride may be modified, as can transporters of cations including H^+, Na^+, and K^+. Mg^{++} is reported to have a major general role in metabolism, perhaps owing to its complexing ATP and other negatively charged phosphorylated compounds. A variety of active transport systems carry biologically important ions into and out of cells. A major example is Ca^{++}, which is involved in a multitude of intracellular signals that control numerous diverse processes including cell proliferation and differentiation. Calcium binds to the ubiquitous proteins calmodulin (CaM) and calcineurin, which, in response to intracell Ca^{++} concentration, regulate transcription through changing phosphorylation of transcription factors. Thapsigargin, produced by the "death carrot," blocks Ca^{++} transport and kills cells. Neoplastic cells have a high requirement for iron because enzymes such as ribonuclotide reductase for DNA synthesis and the cytochromes for energy production contain iron. They overexpress transferrin receptor 1 and very rapidly internalize Fe^{+++} carried by transferrin protein. Other molecules that also have roles in iron metabolism and cellular proliferation include transferrin receptor 2 (TfR2), a transferrin receptor-like protein that is inducible by estrogen, melanotransferrin, ceruloplasmin, and ferritin.

Mammalian cells utilize sugar as a major substrate for energy production. Glucose is transported into the cell by facilitative glucose transporters (GLUT), whose isoforms are cell specific and controlled by hormones and environment. The majority of cancers overexpress the GLUTs present in their tissue of origin and also others that are not normally present in these tissues. Various growth factors and their signaling pathway kinases such as Akt modulate GLUT through their effects on insulin action,

Hydrophilic biomolecules including amino acids, nuclosides, and vitamins are actively transported through the cell membrane, and their transport systems are modified in cancer cells. For example, uridine concentration in tissues is tightly controlled by its transport mechanism. This uptake is balanced by uridine phosphorylase, whose activity is higher in human tumor specimens than in paired normal tissue. Its expression is directly regulated by the tumor suppressor gene p53. Nucleoside transport inhibitors can exert differential effects on tumor vs. normal cells. Dipyridamole and *p*-nitrobenzylthioinosine have indicated anticancer efficacy in combination with NB 1011, a novel anticancer agent that targets tumor cells expressing high levels of thymidylate synthase, but they do not synergistically kill normal cells. The antibiotic WS-5995 blocks nucleoside transport and decreases viability of L1210 leukemia cells.

Extracellular Structures

Proliferation is regulated by both a cell surface's interactions with soluble growth factors that ligate to specific membrane receptors and with other cells and molecules in the extracellular matrix (ECM). In vivo, cells are bound to ECM, and in culture they deposit matrix molecules onto their substratum (see Chapter 9 by Rizki and Bissell on extracellular matrix). The ECM is comprised of four major classes of proteins: collagens, proteoglycans, glycoproteins, and elastin. It regulates growth in G_1, although somewhat differently than the stimulation by growth factors, activating integrins and signaling through the Ras, Raf, MEK, and ERK kinase cascade. Also growth ceases when normal cells come into contact as their culture becomes confluent. This density-dependent inhibition on contact can be via Ras signaling. To grow in culture, normal cells must attach to a suitable surface such as specially chemically prepared or protein-coated plastic. If they are detached, they stop growing and undergo a programmed cell death called anoikis.

In contrast, cancer cells continue to proliferate at confluence and unattached in suspension. They are generally less adhesive than normal cells, they deposit less ECM, and their growth becomes independent of ECM. Their growth into colonies, when suspended in soft agar, is a classic test for tumorigenic transformation. Malignant cells circumvent anchorage dependence by actions of oncoproteins that modify the signaling pathways. Loosened matrix adhesion contributes to the ability of tumor cells to leave their original position, enter the circulation, and then adhere to remote endothelia and there establish metastatic colonies.

Nontumor cells surround the cancer cells in a tumor; malignant cells comprise only about 1% of the tumor mass in Hodgkins disease, the bulk being stromal cells. Four other cells in stroma interact with tumor cells. Endothelial cells are stimulated by growth factors VEGF and FGF produced by the tumor and are angiogenic but are not major direct contributors to tumor growth. Several kinds of immuno-inflammatory cells supply MMP-9 and gelatinase B, but not urokinase, and activate VEGF, which with its receptor VEGFR2 enhances angiogenesis and early tumor growth. These proteins are not active in invasion. Pericytes contribute somewhat by stimulating MAP kinase. And a few percent of fibroblasts stimulated growth of PC3 prostate cancer cells, through contact, cytokine release, and activation of anti-apoptotic NF-κB. Drugs that affect normal fibroblasts can modulate tumor growth, and are under investigation. Tumor-stromal cell interactions are reciprocal; both undergo permanent changes after interaction. These findings support utilization of mixed cell and three-dimensional culture methods for investigations of cancer. Drugs that inhibit VEGFR are being developed for potential therapy.

Adhesion molecules on the cell surface (CAMs) include integrins and cadherins. Integrins are transmembrane proteins that function as primary sensors of the extracellular environment. They interact with ECM proteins, such as fibronectin, to initiate signaling pathways that regulate cell migration, growth, and survival. Integrins, focal adhesion kinase (FAK), and the adapter protein Shc activate the anchorage-dependent process. ECM and integrins are also important for angiogenesis because the endothelial cells depend on alpha v integrins for survival. Integrins are altered in cancers, and compounds that block their activity suggest anticancer strategies. For example, integrin $\alpha 6\beta 4$ is constitutively active in many tumor cells. The $\alpha 6\beta 4$ and $\alpha 3\ \beta 1$ integrins ligate laminin-5 (Ln-5), a component of ECM that regulates cell adhesion, migration, and morphogenesis. Ln-5 chains have tissue-specific patterns in tumors, and are often up-regulated in gliomas, gastric carcinomas, and squamous carcinomas but are down-regulated in prostate and basal cell carcinomas and markedly so in breast tumor cell lines when compared with various normal breast epithelial cells. Ln-5 is often lower in cell lines derived from early-stage breast tumors, and it is absent in lines derived from later-stage tumors. These expression changes could not be attributed to large-scale mutations or gene rearrangements.

E-cadherins are transmembrane proteins that regulate adhesion of adjacent cell surfaces. Their function is highly dependent on interactions of their cytoplasmic domain with intracellular catenin proteins. Alterations in cadherin-catenin complexes have a major role in adhesion defects in carcinomas. β-Catenin is upregulated in cancer as a result of inactivating mutations in the APC and AXTN tumor-suppressor proteins and by gain-of-function mutations in β-catenin itself. However, this β-catenin deregulation appears to have consequences from inactivation of E-cadherin or α-catenin. Germline mutation of the E-cadherin gene is the basis of hereditary diffuse gastric cancer.

Thrombospondin-1 (TSP-1) is a glycoprotein that influences cellular phenotype and the structure of the ECM that is important in tissue remodeling associated with neoplasia and angiogenesis. Mutations in oncogenes and tumor-suppressor genes are frequently associated with its decreased expression. TSP-1 produced by stromal fibroblasts, endothelial cells and immune cells suppresses tumor progression, and inhibits angiogenesis through direct effects on endothelial cell migration and survival, as well as indirectly affecting growth factor activity. TSP-1 in the microenvironment can suppress tumor cell growth through activation of transforming growth factor beta (TGF-β).

Carbohydrates on the cell surface have for many years been implicated in tumor development. Highly metastatic cells frequently have altered binding of lectin proteins to their surface carbohydrates. The most consistent change in cell surface saccharide expression during tumor progression in vivo was an increase in specific binding to *N*-acetyl-D-galactosamine of soybean agglutinin (SBA), one of five lectins tested. Experimental liver metastasis, SBA binding, and hepatocyte clumping (rosetting) were significantly directly correlated. Preincubation of the tumor cells but not hepatocytes with D-galactosamine inhibited rosetting of hepatocytes and homotypic tumor cell binding, suggesting that saccharide-specific, lectin-like receptors on tumor cells have a role in liver metastasis binding. This change may provide reduction in susceptibility of tumor cells to natural antibodies, natural killer cells, and activated macrophages, and increased tumorigenicity and metastasis.

Growth Factors

These proteins function as major regulatory communicators between an organism's cells. Interactions of the several growth factors with their specific cell surface receptors activate major intracell signaling paths. The numerous growth factors are reviewed in other chapters. Altered functions in cancers of some of them are summarized here. Some cancers aberrantly overexpress growth factors, thereby causing autocrine stimulation and also paracrine stimulation to nearby cells. Epidermal growth factor (EGF) is the paradigmatic growth factor. It binds to a specific receptor (EGFR/HER1/ERB1) on the cell surface and dimerizes, which activates its intracellular tyrosine kinase. This enzyme initiates a kinase cascade that includes protein kinase C and activates transcriptions of growth stimulatory proteins including cyclin D. EGFR is amplified or overexpressed in some cancers and is structurally altered in others. Some cancers are independent of this growth factor because the kinase is constitutively activated; their truncation eliminates EGF binding. Another member of this receptor family is HER2/NEU/ERB2, which is activated by heregulin/neuregulin and whose kinase cascade includes MAPK and Akt. Its gene is amplified and overexpressed in 25% to 30% of breast cancers, making the cells independent of stimulation by estrogen (estrogen receptor negative, ER^-). Overexpression of the HER2 gene is associated with aggressive clinical behavior.

The paths of cell growth stimulated by EGFRs provide targets for differential therapy. Clinically effective therapeutics have been demonstrated with monoclonal antibodies such as Herceptin against HER2 overexpressing breast cancers, and by inhibitors targeted to the EGF receptor family tyrosine kinases. Iressa is such a selective tyrosine kinase inhibitor, to which about 10% of patients respond. It acts synergistically with taxol derivatives. Conversely, a potential therapy designed to protect normal cells during therapy is suggested by first arresting nontumorigenic EGF-dependent (MCF-10A) cells in G_0/G_1 by withdrawing EGF or by providing the EGF receptor blocking drug AG1478. These pretreatments prevented the lethality of subsequently provided paclitaxel which functions at G2/M.

Hepatocyte growth factor/scatter factor (Hgf/Sf) activates both mobility and cell proliferation. It is a protease that is potent in invasion and angiogenesis by activating the urokinase plasminogen activator network, and this increases cell motility by proteolytic dissolution of the ECM. Hgf/Sf is synthesized by mesenchymal cells, and its receptor on epithelial cells is Met, a tyrosine kinase. This provides an example of cell-cell interactions. Met is activated by mutations in most solid tumors, both hereditary and spontaneous, although this is rare in breast cancers.

Insulin-like growth factors IGF-I and IGF-II, their receptors IGFR-I and IGFR-II, and six IGFR binding proteins have major roles in proliferation, apoptosis, and differentiation. IGF-I transcription is controlled by tumor suppressors p53, WT, and BRCA1. The binding of IGF-1 to its receptor IGFR-I is principally antiapoptotic but also is mitogenic, more dramatically in vivo than in culture. It signals via PI-3K /Akt, although other pathways including MAPK are involved. IGF-I is inhibited by TNFα, and inhibitors that block digestion of IGF-I by cysteine proteases prevent IGF-I cycling into and out of cells. The prostate specific antigen (PSA), a protease, also cleaves it. IGFs and their receptors are frequently mutated or deleted in cancers. IGF-IR loss in breast cancers contributes to higher growth rate and decreases apoptosis.

IGF-II is also frequently lost or mutated in cancers, consistent with its tumor-suppressor role; 90% of gastrointestinal tumors were mutated in IGF-II or TGF-β. Its activity is complex, binding to both IGFR-I and IGFR-II. IGF-II degradation is activated by IGFR-II. IGFR-II is paternally controlled and imprinted in mice. The tumor-suppressor activity is perhaps activated by retinoic acid, which binds at a second site on IGFR-II and up-regulates it. Retinoic acid also strongly modifies intracell distribution of lysosomal mannose 6 phosphate (M6P) enzymes, such as cathepsins B and D, involved in cancer growth. These lysosomal proteins bind at a third site on IGFR-II. Secretion of procathepsin L increases metastatic potential due to defective function of IGFR-II (M6P) in the tumor cells. The proteins proteolytically cleave and inactivate TGF-β, thereby increasing proliferation and decreasing apoptosis. These effects suggest a basis for cancer therapy with retinoic acid.

Six IGF binding proteins limit availability of IGFs for binding to their receptors by attaching to and stabilizing them. IGFBP inhibit proliferation in cooperation with TGF-β. IGFBP-3 is a major binding protein, which, by binding to IGF-1, is growth inhibitory and proapoptotic. This interaction is frequently disregulated in cancers. IGFBP-3 is cleaved after it binds to the immortalizing and antiapoptotic E7 protein of human papilloma virus 16. IGFBP activity is controlled by several molecules, including RXRα plus retinoic acid which stimulates apoptosis, IGF-I transcription, and TNF-α action. These proteins move to the nucleus when retinoic acid or its analogues are provided. There they bind to response elements (RxRE) and regulate dependent transcriptions that induce apoptosis. IGFBP-5 activates survival against the apoptosis caused by ceramides.

Another proposed function of IGF-I and IGF-II in mediating cancer cell growth is regulation of transport of key amino acids across the cell. After glutamate and leucine transport were significantly increased by glutamine deprivation, they were blocked by a monoclonal antibody that recognizes IGF-IR and that also increases IGFR-I protein on the cell surface. This antibody significantly decreased proliferation and DNA and protein biosynthesis of neuroblastoma cells, in both control and glutamine-deprived media.

The Wnt pathway is involved in cell-cell and cell-matrix signaling. Wnt signaling illustrates two major regulatory mechanisms, namely translocation between intracell compartments and control of specific protein degradation. Wnt binds to its receptor Frizzled-2, which releases the transcriptional coactivator β-catenin from binding with adenomatous polyposis coli (APC) protein on an Axin scaffold. β-catenin then translocates to the nucleus, where it generates an active complex with T cell Tcf/lymphoid enhancer factor that regulates transcriptions associated with cancer that include cyclin D1 and c-*myc*. Regulation is also influenced by modifying activity of the Tcf proteins through complexing with transcriptional repressors. This pathway activates release of stored Ca^{++} and it signals by decreasing cGMP, which is blocked by inhibiting cGMP specific phosphodiesterases. Normal cells closely regulate β-catenin degradation by ubiquitination and proteasome activity, which is inhibited via a GSK3 phosphorylation mechanism. Agents that inhibit β-catenin signaling include TGF-β, retinoic acid and vitamin D. The β-catenin-related gene E-cadherin may be repressed in vivo by the SLUG zinc-finger protein in breast cancer.

Cadherin as well as catenin levels are changed in human cancers. Mutations that constitutively activate the transcriptional response of the Wnt and Hedgehog pathways are frequently associated with specific human cancers such as colon carcinoma and melanoma. Genetic aberrations of several components of the β-catenin pathways potentially contribute to colorectal carcinogenesis. This role was first suggested by β-catenin protein's association with APC, and by dysregulation of its expression at all stages of the adenoma-carcinoma sequence. This deregulation is achieved via mutation of APC, β-catenin frizzled, TCF-4 or Axin. Other components release β-catenin from the adherens complex and/or encourage translocation to the nucleus; phosphoryation leads to its degradation. The transcription promotes the expression of numerous genes important in colorectal carcinoma.

The transforming growth factor beta (TGF-β) family has many functions including both inhibition and stimulation of cell proliferation, differentiation, migration, and modulation of immune functions. It is primarily a potent growth inhibitor with tumor-suppressing activity, and it can protect various normal cells from antineoplastic drugs including 5-F-uracil, both in culture and in vivo. Its effects are in part through control of ECM synthesis and degradation. TGF-β binds to its receptor whose serine/threonine kinases phosphorylate intracellular membrane-bound R-Smad regulatory proteins. These then form complexes with Co-Smads, which translocate into the nucleus and regulate the transcription of target genes. Smads normally block growth stimulation by inducing the cyclin/Cdk inhibitor p15. A third class (I-Smads) that are also induced by TGF-β inhibit these signals. Thus Smads comprise an autoinhibitory pathway. They function as components of a network regulated by other signaling pathways, and Smads modulate transcriptions that are targets of other signaling pathways. Protein kinase C (PKC) phosphorylates Smads, which changes their binding affinity to DNA and blocks TGF-β activity. Smads are mutated and inactivated, for example, in pancreatic and colorectal cancers.

TGF-β is antitumorigenic in early cancers, but its inhibitory action is lost through several mechanisms including inactivation of its receptor or downstream partners. Cancers are often resistant to the inhibitory effects of TGF-β because of perturbation of its signaling pathway, such as by Ras activation or by mutational loss of signaling components. But TGF-β can support growth of later cancers. Carcinomas often secrete excess TGF-β1, and they respond by enhanced invasion and metastasis. Blocking TFG-β stimulates apoptosis. Hyaluronidase that degrades extracell matrix carbohydrates counteracts suppressing effects of the TGF-β signaling complex on TNF and up-regulates apoptosis through p53.

The two platelet-derived growth factors (PDGF α and β) are important stimulators of fibroblast proliferation. The *sis* oncogene of simian sarcoma virus is related to them. They dimerize, and their binding to two tyrosine kinase receptors initiates proliferation and migration during wound healing. This system is frequently mutated in sarcomas, and PDGF is overproduced.

Chemokines comprise four chemically different families of small peptides that bind to several receptors on the cell membrane. They activate G-protein downstream signal pathways, such as PI3 kinase –AKT-NEMO kinase, that strongly activate dependent transcriptions by phosphorylating the p65 subunit of transcription factor NF-κB. They are activated by cytokines, including tumor necrosis factor (TNF) and IL-1, and are inhibited by TGF-β. Chemokine genes are induced constitutively in melanomas due to altered kinase signals that modulate NF-κB activation, thereby stimulating tumorigenesis and metastasis.

Lipophilic molecules such as estrogen, androgen, and retinoids also carry signals for growth and differentiation through the organism. In contrast to the pathways stimulated by growth factor proteins that bind to receptors on the cell membrane and activate proliferation through kinase cascades, they bind to carrier proteins in blood, diffuse through the membranes of target cells and into the nucleus. There they bind specifically to receptor proteins that interact with promoter sites and activate transcriptions. Estradiol is a very important molecule for signaling growth of many breast and ovarian cancers. It activates transcriptions in hormone-dependent target cells by binding to their nuclear receptors. These are low in normal breast cells, but mutations cause overexpression. For comparison, about two-thirds of human breast tumors is estrogen responsive ER^+, a third is ER^-, and a tenth is Her^{2+}. Androgen is similarly critical in prostate cancer cell proliferation, apoptosis, and differentiation.

INTRACELL SIGNALING

Summary

Kinase cascades are central to the transmission of proliferative signals from surface growth factor receptors to nucleus. The tyrosine kinases of these receptors are overexpressed in many tumors. Ras protein provides a major link beetween activated receptors in the cell membrane and downstream effector kinases. It is also involved in inducing senescence and cell death, suggesting that alternatively it can activate anti-oncogenic pathways. Point mutated *Ras* genes at one of three locations are frequently found, in one-third of tumors. Mitogen-activated tyrosine kinases turn on downstream cascades of serine-threonine protein kinases. These modulate targets including transcription factors, regulators, enzymes, and structural proteins that are involved in proliferation, gene expression, metabolism, mitosis, cell movement, and apoptotic cell death. Mitogen-activated protein kinase (MAPK) is a major cascade (Ras/Raf/MEK/ERK) that is stimulated by growth factors and also by steroids. The signaling pathways are complex in both their branching and interacting structures and in their downstream consequences. Kinases are perhaps the most investigated specific molecular targets for current anticancer strategies.

Ras

Extracellular factors initiate intracellular kinase cascades that transmit signals from the cell membrane to the nucleus, where they stimulate transcriptions. These primarily initiate entry into and transit through the cell cycle's G_1 phase. The major signaling networks that control and link cell proliferation and death have often been summarized (see Chapter 3 by Ford et al. on cell cycle cascades). As discussed here, their major molecules, including Ras, cyclin-dependent kinases, Myc, and pRb, are all mutated in cancers.

Ras provides a principal example of signaling from the intracell membrane, following activation by tyrosine phosphorylation of receptors. The Ras superfamily of about 28 members are small monomeric proteins whose association with the cytoplasmic membrane is through their post-translational modification by covalent prenyl binding, catalyzed by three main classes of prenyltransferases. Statins that inhibit the farnesyl/prenyl biosynthetic pathway at mevalonate synthesis block cell cycling in early G_1 phase. These inhibitors indicate important roles of Ras and other prenylated proteins in regulation of cell cycle progression through both G_0/G_1 and G_2/M, depending on the cell line. They also inhibit proteasomes and are being investigated as anticancer agents.

Diverse cell-signaling pathways and mechanisms activate Ras, including tyrosine kinases, G-protein-coupled receptors, adhesion molecules, second messengers, and various protein-interacting molecules that relocate and elevate intracellular Ras-GTP, and also through interactions with Grb, Sos, and scaffolding proteins such as Gab/Dos that hold them together. Ras-dependent kinase pathways are activated when Ras binds GTP, and are inactive when it hydrolyzes GTP, and GDP is then bound. Guanine nucleotide exchange factors (GEFs) catalyze the dissociation of GDP from inactive Ras proteins, to which GTP then binds and induces a conformational change that permits interaction with downstream effector kinases.

The Ras pathway also depends on regulation of transport mechanisms that determine intracell concentrations of ions. It is sensitive to changes in intracellular Ca^{++}. G protein-coupled receptor functioning is related to the G protein-coupled receptor kinases, composed of six members whose activity is selectively regulated by calcium binding to different calcium sensor proteins. Ni^{++} is a carcinogen, perhaps because it competes with Ca^{++}. An inhibitory effect of Zn^{++} is suggested because Ras-mediated signaling in *C. elegans* requires the protein CDF-1, which decreases the intracellular concentration of this ion.

Ras genes that are point mutated at one of three locations are frequently found in one-third of various tumors; H-RasV^{12} is the most frequent. This mutation produces a constitutively active protein that locks in GTP. Oncogenic *ras* are implicated through multiple functions in several human malignancies. They cause cell cycle deregulation, moderate responses to several mitogens and differentiation factors, and alter enzymatic activities that enhance downstream signals for cell proliferation and transformation. They transform immortal rodent cells to a tumorigenic state, in part by constitutively transmitting mitogenic signals through the MAPK cascade. Ras-activated MEK permanently arrests primary murine and human fibroblasts, but it forces uncontrolled mitogenesis and transformation in cells lacking either p53 or INK4a. Oncogenic *ras* activates the Arf-p53 program to suppress wild-type primary murine keratinocyte transformation from becoming tumorigenic. These cells arrest proliferation with features of senescence and terminal differentiation. Ras did not promote cell cycle arrest of Arf-null keratinocytes, induce differentiation markers, nor activate p53, and these cells rapidly formed carcinomas in vivo. Ras mutations precede p53 and INK4a/ARF mutations during the progression toward malignancy of chemically induced skin carcinomas in mice. BRAF is a Ras-regulated kinase that was activated by mutation in the majority of melamomas tested.

Transfection of wild-type *ras* reverses the oncogenic phenotype of transformed cells, indicating that it is a tumor suppressor, and expression of wild-type *ras* genes in several human malignancies is associated with good prognosis. *Ras* is also involved in inducing of senescence and apoptosis, suggesting that it alternatively activates anti-oncogenc pathways. The relationships of *ras* activation, apoptosis, cellular proliferation, and cancer are complex, activating at least three overlapping and cooperative kinase cascades Raf/MEK/ERK, Ral GTPase, and PI3K/Akt, which inhibits cytochrome c release. Ras also decreases the inhibitory association of p27 with cyclin E/Cdk2. Ras further activates oncogenic sphingosine kinase, and sphingolipids are mediators of cell death induced through cytochrome c release by agents that include Fas and ionizing radiation. Ceramide is 50% lower in colon cancer than in normal colon, and ceramidase inhibitors (B13) that increase ceramide were apoptotic and prevented tumor growth.

Among numerous *ras*-related proteins, Rho GTPase, a branch of the ras family, promotes malignant transformation and metastasis, and increasing evidence implicates its mutation in cancer. The guanosine phosphate binding (G) protein coupled receptor (GPCR) family of extracellular signal regulators have a role in prostate cancer. Expression of some GPCRs and GPCR ligands is elevated in prostate cancer cells and also in adjacent stromal tissue; this enhances proliferation and decreases apoptosis. Stimulation of GPCRs by lysophosphatidic acid or bradykinin induces proliferation of prostate cells, and also increases survival via activation of antiapoptotic NF-κB. RERG is a *ras*-related growth inhibitory gene that is regulated by estrogen and is often lost in breast cancers.

Understanding effects of *ras* requires elucidation of the downstream signal transduction pathway that it controls. Reverse genetic approaches can trace these pathways, and they also provide an example of gene discovery. A novel oncogenic Ras target, *mob-5*, was identified by differential display. *Mob-5* expression could be induced by oncogenic Ha-Ras and Ki-Ras, but not by activation of normal *ras*. It is a 23-kDa cytokine-like secreted protein. Inhibitors of both Ha-Ras and mitogen-activated protein kinase kinase completely abolished *mob-5* expression with concomitant loss of the transformation phenotype. Its structure, chromosomal localization, and cytokine-like properties identified it as Interleukin 24 (IL-24), a secreted member of the IL-10 family of cytokines involved in the immune system. A putative Mob-5 receptor was identified on the surface of oncogenic *ras* transformed cells. Mob-5 ligands two heterodimeric receptors that bind it with similar kinetics; either one activates signal transducers and transcription. Mob-5 with its receptor may be an autocrine loop coordinately activated by oncogenic Ras.

Mda-7, now identified as Mob-5, was independently discovered by subtraction hybridization, when it increased during terminal differentiation of human melanoma cells. It appears to mediate induction of the DNA damage inducible (GADD) family of genes via the p38 MAPK pathway. Infection of melanoma cells and several human cancer cells but not normal cells with a replication-incompetent adenovirus carrying *mda-7* caused growth arrest and which correlated with induction of apoptosis. This Mda-7 infection specifically increased the phosphorylation of p38 MAPK and heat shock protein 27. A selective inhibitor of the p38 mitogen-activated protein kinase (MAPK) pathway inhibited this apoptosis and induction of GADD genes, and effectively blocked down-regulation of the antiapoptotic protein Bcl-2. Antisense, and also an adenovirus that expresses a dominant negative mutant of p38 MAPK, had similar effects.

The hyperactivation of *ras* provides opportunites for therapy. Farnesyl transferase inhibitors are being applied. Caco-2 cells transfected with an activated *K-ras* expressed higher levels of cyclooxygenase (COX-2) mRNA and protein than parental cells, and more quickly formed tumors in mice. Sulindac is a nonsteroidal anti-inflammatory drug that inhibits COX enzyme activities. *Ras*-dependent signaling is suppressed by sulindac analogues (Exisulind and CP461) developed as inhibitors of cyclic GMP phosphodiesterases, thereby increasing cGMP which activates and induces cGMP-dependent protein kinase. This results in increased phosphorylation of β-catenin and enhanced apoptosis of colon tumors. Sulindac, its sulfide, and sulfone (Exisulind) metabolites are apoptotic and anticarcinogenic in experimental models, and show promise against precancers and cancers without affecting normal cells. Exisulind also inhibits EGF-induced phosphorylations of ERK-1/2 and proapoptotic Bad in colon cancer. Exisulind prevented colorectal polyp formation for 24 months in patients with familial adenomatous polyposis (FAP). It also inhibited the rise of prostate-specific antigen (PSA) after radical prostatectomy, and was well tolerated by most patients. Preclinical data indicate its additive or synergistic antineoplastic effects with other drugs. Also by screening 30,000 compounds for their toxicity to mutant *ras*, a new cytidine analogue that inhibits cancer was found.

Kinase Cascades

There are more than 100 known protein tyrosine kinases. They are modified in many ways in cancers; only a few are mentioned here. As an example, Src kinase is a critical signal transducer that modulates a wide variety of cellular functions by phosphorylating protein tyrosines. Elevated expression and/or activity of Src is implicated in cancer development, enhancing tumor growth and stimulating migratory and invasive activities of normally relatively nonmotile cells. Src is activated by a variety of mechanisms, including heterotrimeric guanine-nucleotide-binding proteins. These translocate Src to the inner surface of the plasma membrane, where covalent binding of myristate mediates its attachment. There its kinase activity initiates signal transduction pathways that increase cell adhesions.

Point mutations and rearrangements of the RET protein tyrosine kinase convert it to a dominant transforming oncogene. Gain of function of this kinase is associated with human cancer. Mutations and rearrangements both increase tyrosine kinase activity of RET and downstream signaling. Its germ-line point mutations are responsible for multiple endocrine neoplasias. Somatic gene rearrangements connect the tyrosine kinase domain of RET to heterologous partners in papillary carcinomas of the thyroid.

Cascades of serine-threonine protein kinases are activated by downstream signaling of mitogen-tyrosine kinases. They modulate targets including transcription factors, regulators, enzymes, and structural proteins that are involved in gene expression. The MAPK signal transduction cascade follows stimulation by both growth factors and steroid hormones. Three subfamilies of MAPKs are active as three-step phospho-relays, for example, growth factor/receptor, RAS GTP activator, c-Raf1, MKK1, ERK1, and p90RSK. ERKs are involved in cell division, JNKs in transcription and in apoptosis, and p38s in environmental stress, among their other functions. A scaffold of MAP kinases is localized to microtubules. It binds a dozen other kinases at its different sites via protein-protein interactions, and mediates cell fates including growth, proliferation, and survival through modulating apoptosis-related proteins Bad and Bcl-2. Its mechanisms ultimately alter gene expression, and its negative regulators include phosphatases.

MAPK signaling pathways are complex in both their branching and interacting structures and in their downstream consequences. An example involves the transcription factor encoded by the immediate early growth response gene *Egr1*, which is rapidly induced by growth factors to create a proliferative signal. It has a role in progression of growing tumors through generating a hypoxic signal; angiogenesis is stimulated and survival is improved. Induction of *Egr1* is generally transient, though it may be sustained in some prostate tumor cell lines and tumors, and it is often absent or decreased in breast, lung, and brain tumors. Its re-expression aids tumor cell survival by producing anti-apoptotic activity. A contradiction is that *Egr1* is required for apoptosis after stress of both normal and tumor cells. How these diverse effects can be achieved is not clear. Many of the kinases become oncogenic through mutations.

Approximately half of breast tumors express MAP kinase more activated than in the surrounding benign tissues, and this activity is higher in primary tumors of node positive than in node negative patients. This kinase up-regulation does not appear to arise from mutations of *ras* but results from enhanced growth factor activity. The MAP kinase pathway that is dependent on a particular *Raf* to regulate proliferation, arrest, and apoptosis through Bad and Bcl-2 is frequently mutated in cancers. The pathway that involves ERK-1 and -2 is highly relevant for human breast cancers. Its major regulators are growth factors acting through tyrosine kinase receptors. Estradiol, progesterone, and testosterone also activate MAP kinase, as do ligands acting through G protein receptors. Cell proliferation and anchorage-independent growth of a squamous carcinoma cell line transfected with activated Ha-Ras were inhibited when ERK pathways were blocked by a MEK inhibitor.

Phosphoinositol 3-kinase (PI3K) activation is catalyzed by EGF family receptors. This enzyme forms inositol 3,4,5-triphosphate from inositol 4,5-diphosphate, which is produced by phospholipase C's hydrolysis of phosphatidyl inositol 4,5-diphosphate. It activates protein kinase B (Akt) by releasing stored Ca^{++} acting in combination with diacylglycerol the other product of phospholipase C action. The tumor promoter phorbol myristate acetate acts similarly. PI3K is proposed to function early in MAPK activation. It interacts with the Ras/mitogen-activated protein kinase pathway. Akt, which is also activated by the her2/neu receptor, is the focal point of many signal transduction pathways that control multiple major processes. It binds several regulators including heat shock protein 90 and Cdc37. There is a massive literature on the role of the PI3K pathway in cancer.

PTEN has an action opposite of PI3K. It is a 3-phosphoinositide phosphatase that removes the phosphate added by PI3K. The balance of these reactions may coordinate G protein-coupled signaling pathways during eukaryotic proliferation and chemotaxis. These competing activities provide an excellent example of the yin-yang effect, so frequently found in biological regulations. PTEN is a tumor-suppressor gene whose chromosome region 10q23.3 is frequently deleted or mutated in many cancers. Epigenetics and perhaps inappropriate subcellular compartmentalization are mechanisms for PTEN silencing. Several syndromes have germ-line loss-of-function PTEN mutations. Somatic intragenic PTEN mutations are rare in primary epithelial thyroid tumors, although hemizygous deletion occurr in 10% to 20% of thyroid adenomas and carcinomas.

Therapies Directed Against Kinases

Modern anticancer strategies are being designed against specific molecular targets, with the goal of sparing normal non-neoplastic tissues. Choosing a tumor-specific molecular target for therapy is, however, difficult. And for proof of an inhibitor's function, it is necessary to determine a role for a target enzyme by its specific modulation with RNAi, dominant negative, knockout, and so on. Kinases are perhaps the most investigated targets. An important candidate for therapeutic intervention is cyclin-dependent kinase 2 (cdk2). It phosphorylates retinoblastoma protein (pRb), which is an inhibitor of proliferation in the transition from G_1 to S phase. Cdk2 also plays a critical role in the transition through S phase and at the S to G_2 transition. As examples, UCN-01 and flavopyridol are kinase inhibitors with a variety of effects; they are in clinical trials, and derivatives are being developed. But these compounds and most inhibitors of cyclin-dependent kinases do not act specifically. Because they are chemically modified ATPs, they may inhibit multiple processes. More selective Cdk2 inhibitors are in development. Inhibitory oligonucleotides that block gene expressions rather than enzymes are also being investigated; GEM-231 inhibits overexpression of a subunit of protein kinase A that is associated with many cancers.

Gleevec (STI-571) is the first selective tyrosine-kinase inhibitor to be approved for the treatment of a cancer. Its specificity is based on overexpression of Abl tyrosine kinase in the blast crisis of chronic myeloid leukemia and acute lymphoblastic leukemia with the Philadelphia chromosome, where a translocation puts Abl kinase under control of the Bcr promoter. But remission in the clinic was observed in after only 2 to 6 months, due to a threonine to isoleucine mutation at the ATP binding site of Abl kinase. This line of investigation is continuing with identification of the directly acting small molecule tyrosine kinase inhibitor PKC412, a candidate agent for treating acute myeloblastic leukemia (AML), in which constitutively activating mutated FLT3 receptors are found in 35% of patients. PKC412 is a potent inhibitor that selectively induces G_1 arrest and apoptosis of Ba/F3 leukemia cell lines expressing mutant FLT3. Cell lines made resistant to PKC412 overexpressed mutant FLT3, confirming that FLT3 is the target of this drug. Leukemia was prevented by PKC412 in Balb/c mice transplanted with marrow cells transduced with a FLT3-ITD expressing retrovirus. Inhibitors of are being developed for treatment of diseases.

Overexpressed EGFR, hyperactive *ras* and other overproduced targets act though PKC to stimulate proliferation and invasion and inhibit apoptosis. Inhibition of PKC is thus the basis of therapies. PKC-inhibiting agents given to patients with refractory high-grade malignant gliomas have led to some clinical responses. A new selective therapy is based on the elevated NF-κB found in some ER-cancers. Inhibiting PKC with the drug Go6796 decreases the activation of downstream NF-κB and reverses its anti-apoptotic activity. Thus interference with PKC activity is a novel form of experimental cancer treatment that may both restrain hyperproliferation and invasion and create apoptosis, without the toxicity of classical agents.

Myc

Myc is a multifunctional protein that acts by multiple mechanisms. This transcription factor is the cellular homologue of the avian myelocytic leukemia virus gene. c-Myc is tightly regulated in normal cells. A transient block of Myc causes permanent differentiation. It is central to progression through G_1 by activating E2F1 near the G_1/S transition, and it also increases cyclin D/cdk4. But it appears not to be required for proliferation because its complete removal only slows cell growth. Myc perturbs the balance of cell growth by affecting both proliferation and apoptosis, at least in part owing to the proteins that interact with it. Excess Myc allows fibroblasts to proliferate in 0.1% serum, but it prevents a net cell increase via the balance of proliferation vs. apoptosis. Its activation in adult mature beta cells induces uniform proliferation, but this oncogenic potential also is masked by apoptosis. Upon suppression of apoptosis by coexpression of Bcl-xL, c-Myc rapidly triggers progression into angiogenic invasive tumors. Subsequent c-Myc deactivation causes rapid regression associated with vascular degeneration and beta cell apoptosis. It also induces apoptosis in response to various negative conditions including limited growth factors. This pathway is through p19ARF/mdm2/p53 and cytochrome c release.

Numerous binding proteins with potential impacts on Myc function have been found. Interactions of these with Myc could determine the cell's response. Serum greatly increases its brief half-life, due to blocked proteosomal degradation. Myc forms a heterodimer with Max that binds specifically to E-box DNA sequences, where it forms a complex with several other proteins. Mad-Max heterodimers compete with Myc-Max to inhibit binding at these sites. Transcriptional repression by c-Myc may involve the zinc-finger factor mMiz-1, and may expain how Myc interferes with cell cycle arrest after DNA danage and other conditions, such as when the APC gene is mutated. A growth factor that represses c-Myc is TGF-β, whose rapid signaling is mediated by nuclear translocation of a complex composed of Smad3 plus E2F-1 or E2F-4. The Smad complex has been suggested as a chemotherapeutic target, because a mechanism of TGF-β inactivation is via the inhibitory interaction of c-Myc with

Smads. Functional interactions of Myc with Ras are involved at several levels.

The activated *Myc* oncogene deregulates both cell growth and death checkpoints, and thereby can rapidly accelerate carcinogenesis. *Myc* is over expressed in nearly all cancers; it is implicated in a large number of human and animal solid tumors and also leukemias. c-*Myc* amplification is frequently found in invasive breast cancer, in which it appears early and is involved in a more aggressive phenotype that has been associated with poor disease outcome. *Myc* is the prototype for oncogene activation by chromosomal translocation, and integration of oncogenic viruses frequently target the *Myc* locus, causing its structural or functional alterations.

Activation of Transcription by Small Molecules

Certain lipophilic compounds pass directly into the nucleus and there activate transcriptions by binding to their receptors. Ligation of a small molecule to its nuclear receptor protein is a process of transcription activation simpler than and different from activation by growth factors and kinase cascades. A single factor can activate transcription of many downstream genes; retinoic acid is reported to directly or indirectly regulate expressions of 532 genes. Mutations in cancers alter retinoic acid functioning. A well-known example is acute promyelocytic leukemia in which translocation replaces a part of the gene for retinoic acid receptor RARα with a PML sequence; this creates a dominant negative mutation against differentiation. And RARβ is down-regulated in mammary carcinoma cell lines. Effects like these of small molecules provide connections between transcription and phenotype, and also suggest potential targets for therapy. For example, retinoic analogues are being developed for chemoprevention and against tumor growth.

Estradiol (E2) is very important for growth of many breast and ovarian cancers, as mentioned above. MAP kinase pathways can crosstalk with ER-induced transcription as well as by directly modifying the cell cycle. Estrogen analogues such as tamoxifen are applied clinically against tumors of this class, and they are effective against some. Many breast cancers that are not responsive to antiestrogens have undergone other mutations, either intially or subsequently by ER up-regulating mutations. Some of these tumors are mutated to increase their EGF or her2/neu receptors, and are treated with antibodies or drugs that block EGF-receptor-dependent reactions (see above).

Androgen is critical in prostate cell proliferation, apoptosis, and differentiation. It acts during passage through G_1 phase, by mechanisms under investigation that appear to be different from those of estrogen. A modest hereditary predisposition to prostate cancer is connected to the length of a CAG repeat in the coding sequence of the androgen receptor (AR) gene. Many prostate cancers initially require androgen-dependent transcriptions for growth, and they undergo apoptosis after androgen is decreased. Therefore patients are often treated with drugs such as flutamide or by surgery to decrease their androgen. But these tumors soon become androgen independent because of AR mutations. Mutations were found in ARs of patients who received androgen blockade, and the mutant cells were strongly stimulated by flutamide. These cells soon produce bone marrow metastases.

Metastasis

Cancer is characterized by inappropriate growth in space, as well as in time. Whereas normal solid tissues cells are localized, tumors eventually become metastatic, spreading and growing in secondary locations. They are then usually incurable with current treatments (see Chapter 21 on metastasis). Interference with tumor cell attachment by integrin-binding peptides has been shown to be an effective antimetastatic strategy in animal experiments.

Almost all metastases appear in only four tissues—bone, liver, brain, and lung. Specific tissue affinities may underlie the tendency of some tumors to metastasize preferentially to certain tissues. Metastasis to bone is blocked by bisphosphonate, which inhibits release of osteoclast produced growth factor.

Numerous genes that regulate breast cancer metastasis are being discovered, and clues are emerging concerning their mechanisms of action. Some are metastasis activator genes (*ras, MEK1, mta1*, proteinases, adhesion molecules, chemoattractants, receptors, *autotaxin*, *PKC*, *S100A4*, *RhoC*, *osteopontin*) and others (*Nm23, E-cadherin, TIMPs, KiSS1, Kai1, Maspin, MKK4, BRMS1*) are metastasis suppressors. Gene *nm23* may suppress early steps of carcinogenesis in human bladder cancer. All-*trans* retinoic acid could suppress metastasis by up-regulating *nm23-H1*, a process that is opposed by EGF; down-regulation may increase c-*erbB/neu*. Among the involved genes, activation of *Smad2* expression alone induces migration. Elevated *H-ras* is required for nuclear accumulation of *Smad2*, and sequential elevation of both is essential for the epithelial-mesenchymal transition. P53 also has a role in metastasis-related gene expression. Metastasis-associated protein (MTA1) that is involved in chromatin remodeling by histone deacylation is frequently higher in prostate cancer.

Metastasis is a multistep process that involves cell surfaces and is misregulated by defective expression of genes. Primary tumor cells must escape into the blood, their viability must be maintained during dissemination to distant sites, and then invasion and re-establishment at these sites is necessary. Multiple genetic changes are therefore required to establish metastasis. These mutations are generally believed to arise at the end of progression of normal human cells as differentiated squamous carcinoma, move to a motile invasive stage, make an overt change from epithelial to mesenchymal cell type, and finally produce rare metastatic cells. But this sequence is now debated as mutations seem to appear early. A biopsy subset from primary human tumors has recently, by microarray analysis, identified metastatic signatures of 17 mis-expressed genes.

Cancer cells adhere weakly to adjacent cells because their surfaces are altered (see above), thereby permitting their detachment as a prelude to metastasis. Extracellular proteases modify ECM during several tumor-related processes, including metastasis and angiogenesis. Matrix metalloproteases (MMP) are essential for normal ECM remodeling. Their production by both cancer and normal stromal fibroblast cells is critical for metastatic spread of tumors. Breast cancer cells in culture release EMMPRIN, which in turn promotes the release from normal fibroblasts of pro-MMP2, providing an example of interaction between cancer and stromal cells. The generation of MMP2 is mediated by activation of phospholipase A2 and 5-lipoxygenase. Increased mRNA activity and secretion of MMP-12 by statins is reported. These compounds are inhibitors of the cholesterol biosynthetic pathway; they block cells in G_1 phase and have anticancer activity. One target of MMP is RECK, the membrane-anchored product of a metastasis/angiogenesis tumor-suppressor gene that is downregulated in tumor cells and tumors. RECK transcription is down-regulated by Ras via SP1 transcription factor, and DNA methylation may be involved. Transfection of it produces flat revertants of *K-ras* transformed fibroblasts. RECK suppresses invasion and angiogenesis by regulating MMP-2, 9, and MT1-MMP, both by supresssing pro-MMP-9 secretion and directly inhibiting the enzyme's activation. Matrix metalloprotease inhibitors are being developed as potential anticancer agents.

Serine proteases also are involved in motility and invasion of breast tumors. Maspin is a mammary serine protease inhibitor protein (serpin) that is a tumor-suppressor gene expressed in normal human mammary epithelial cells and involved in normal breast development. Maspin inhibits plasminogen activators and modulates cell surface integrins. It is active on the cell surface. Maspin inhibits motility and invasion of cells in culture, osteolysis, angiogenesis, and tumor growth and metastasis in nude mice. p53 may induce maspin expression by transcriptional activation, a control that is at least in part by promoter hypermethylation. And tyrosines of recombinant maspin protein are phosphorylated in vitro by the kinase domain of the epidermal growth factor, which could have a functional role. Maspin is down-regulated during progression of breast cancer. The clinical significance of maspin expression, and its correlation with p53 protein expression in human breast cancer patients has not been elucidated. Paraffin-embedded carcinoma tissues from 168 female patients diagnosed with invasive ductal carcinoma were investigated by immunoreactivity with antibodies to maspin and p53. Tumors that were scored positive significantly correlated with more advanced tumors and shorter relapse-free and overall survival. These results seem to be contrary to previous reports defining maspin as a tumor-suppressor gene. However, an inverse relationship was observed between maspin and estrogen receptor or progesterone receptor status. Also results with 58 cases of ductal carcinoma in situ suggest that maspin expression may initially be down-regulated and then up-regulated during malignant progression. Maspin is potentially a breast cancer marker, and its immunohistochemical detection in carcinoma cells may select for breast cancer patients with an aggressive phenotype. These proteases could be targets for therapy.

THE CELL CYCLE

Summary

Restoration of optimal conditions to G_0 cells signals transcriptions that initiate entry into the G_1 phase and semisynchronous cycling. Cyclins D, E, A, and B are a set of sequentially produced key regulatory proteins that activate cyclin dependent kinases (cdks). Inhibitory proteins counteract the stimulation of these kinases by cyclins. The interactions of cyclins and inhibitors provides another excellent example of the yin-yang principle in biology. The cyclin/cdk activities culminate at the restriction point (R) in late G_1 phase, a key event that controls cell proliferation shortly before DNA synthesis is initiated. In normal cells it is activated only if extracellular conditions are optimal for cyclin production. A major concept is that proliferation control of cancer cells is lost by defective regulation of their R point. At R the cdks regulated by cyclin D and then E phosphorylate the retinoblastoma (pRb) protein. The *Rb* gene is mutated in one-third of human cancers, and is frequently functionally inactive in a variety of tumors. Phosphorylation of pRb releases bound factor E2F-1, which transcribes numerous genes whose products are essential for DNA synthesis. E2F activity is thus necessary for entry into S phase. Many of the enzymes involved in both DNA and deoxynucleotide synthesis assemble on DNA into a Replitase replication complex present in S but not in G_1 cells. As well as DNA synthesis, histone syntheses and chromosome assembly are activated in S phase, and mechanisms for coordination of these processes require activation of many genes.

Proliferation

Restoration of optimal conditions to G_0 cells signal transcriptions that initiate entry into the G_1 phase and semi-synchronous cycling. The cell cycle is a sequence of molecular biological, biochemical, and morphological events that produces two cells from one. Its successive steps, G_1, S, G_2, and M, have been worked out in detail (see Chapter 2 by Braastad et al. on the cycle). In contrast to durations of G_1 which vary by over threefold in individual cells in a culture, lengths of S, G_2, and M phases are relatively invarient and are similar in normal and cancer cells. Both randomly cycling normal and cancer cells become arrested at specific locations in the cycle after they are treated with certain drugs. Lovastatin blocks them in early G_1; other compounds including high thymidine, mimosine, and DNA synthesis inhibitors arrest cells when they reach the G_1/S interface; and colchicine and taxol, which affect microtubules, block cell entry into mitosis. The addition of one of these agents, followed by its removal after cycling has finished, is the basis for forced synchonization of a culture that starts growth of all the cells from the arrested cycle position.

Cyclins D, E, A, and B are a set of sequentially produced allosteric regulatory proteins that activate cyclin-dependent kinases (cdks). For example, cyclin D positively controls cdks 4 and 6, and cyclin E activates cdk2; these are essential for passage through G_1 phase. The cyclins are key regulatory molecules. Because they are unstable, rapid general protein synthesis is necessary for them to accumulate to their critical functional levels. Dynamic steady states such as this, created by synthesis versus degradation, provide a mechanism for highly responsive regulation. Proliferation control is relaxed in tumor cells by mutations that modify the balance of such cycle regulatory genes. For example, cyclin D is overexpressed by gene amplification in many tumors. Overexpression and truncation of cyclin E increases with cancer stage, and it dramatically predicts inability to cure breast cancers. Suppression of mammary tumor growth in cyclin D1 deficient mice can be compensated by expression of subsequently acting cyclin E.

Cdk inhibitory proteins (CKIs) counteract stimulation of the kinases by cyclins. CKI families are the KIPs p21 (responding to DNA damage), p27 (to G_1 arrest), and p57. The INKS are p15, p16, p18, and p19 (see Chapter 7 by Carneiro and Koff on cycle inhibitory proteins). This interaction of activating cyclins and their inhibitors provides another excellent example of the yin-yang principle, where positive and negative activities are balanced. In contrast to cyclins, mutations modulating CKIs are rare, although *INK4*, which encodes p16INK4, p15INK4b, and p19ARF, is very frequently down-regulated in human cancers, often by DNA methylation. Two other major inhibitors of the cdk/cyclins are p21 and p27. They are activated in response to antimitogenic signals or DNA damage, and are proposed to have additional roles that depend on cellular localizations of their targets. p27 is inactivated by phosphorylation catalyzed by cyclin E/cdk2, and is removed by proteosomes. They can be misregulated in cancers by mutation of only one gene or by cytoplasmic relocalization (of p27). Cyclin-Cdks are further modified by both phosphorylations and dephosphorylations. Cdc25 is a dual-specificity phosphatase that catalyzes activation of the Cdks, thereby causing initiation and progression of successive phases of the cell cycle. Kinase cascades that activate this phosphatase are also central in initiating the DNA damage created checkpoints (see below). Multiple links are emerging between defects in these checkpoints, genomic instability, and oncogenesis.

Restriction Point

These processes culminate in the restriction point (R), a key event that controls cell proliferation. This process is located in late G_1 phase, shortly before DNA synthesis is initiated. If R is not passed, when extracellular conditions are suboptimal for proliferation, the cells reversibly revert to

G_0. After cells pass beyond R, their passage through the rest of the cycle is irreversible, and they proceed through S, G_2, and M phases and cell division independent of external stimuli. When cells then enter a new G_1, they must again pass the R point. The useful name "checkpoint" has been given to several similar processes that arrest cell cycling under adverse conditions, principally after DNA is damaged (see below). A main concept is that lost proliferation control of cancer cells is based on defective regulation of their R point. Their mutations avoid the nutritional, growth factor, and other requirements for rapid synthesis of cyclins D and E that must accumulate to permit G_1 phase transit. These rapidly turning over essential cyclins require increased ribosomes in cancers to keep their synthesis ahead of degradation, as is reflected by changed nucleolar organizer regions (NOR). Drugs are being developed that modify the restriction point mechanism (see kinase inhibitors). Cdk inhibitors in clinical trial include flavopiridol, UCN-01 (7-OH-staurosporine), and paullones. Another example is roscovitine (CYC202) an inhibitor whose action is strongest against cyclin E/Cdk2.

At R, cdks phosphorylate the retinoblastoma (pRb) family of proteins, pRb, p107, and p130. These are central regulators that when hypophosphorylated restrict cell proliferation, inhibit apoptosis, and promote differentiation (see Chapter 17 by Baker and Premkumar Reddy on pRb protein). At least 110 proteins have been reported to associate with pRb. This raises questions such as how many functions pRb possesses, and which of these contribute to tumor suppression or development. Principal attention is on negative regulation of transcription factor E2F-1 with which hypophosphorylated pRb combines and inhibits. The phosphorylation of pRb by cdks, regulated by first cyclin D and then E, activates E2F-1 and permits its transcription of numerous genes essential for DNA synthesis. pRb also can inhibit progression through S phase by targeting cyclin A/cdk2. The *Rb* gene is mutated in one-third of human cancers, and it is frequently functionally inactive in a variety of tumors. A heritable loss of one *Rb* allele followed by somatic mutation of the second is the basis of herditary retinoblastoma. SV40 viruses and human papilloma viruses are oncogenic in part because they inactivate pRb. Conversely, constitutively active pRb inhibits E2F activity and the stimulation of cyclin E transcription. But existing cyclin E and its dependent functions including centrosome duplication are not affected.

Completion of the Cycle

Cells do not require growth factors after they have passed beyond the R point. Internal controls, about which less is known, determine timing of the subsequent successive cell cycle events. Completion of one molecular process has been proposed to initiate the next. E2F-1 is necessary for entry into S phase. After release from pRb, the E2F protein family combines with DP proteins and activates transcriptions that produce enzymes involved in synthesis of deoxynucleotides and DNA. E2F activated transcription, with ORC1 and CDC6, initiate DNA synthesis. Novel target genes involved in E2F-1 regulated cellular functions such as cell cycle control, DNA replication, angiogenesis, invasion, metastasis, and apoptosis have been found by biochemistry and by expression analysis with cDNA microarrays and RT-PCR. As examples, MdmX protein binds to and decreases ability of E2F-1 to bind specifically to DNA, thereby decreasing transcription. E2F-1 also binds p53 and thereby affects apoptosis (see below). E2F protein and its coactivator CBP are elevated in non–small cell lung cancers, as is E2F mRNA. This correlated with higher proliferation rather than decreased apoptosis, and with deregulated pRB/p53/mdm2.

Histone syntheses as well as DNA synthesis and chromosome assembly are activated in S phase, and mechanisms for coordination of these processes are necessary (see the Chapter 2 by Braastad et al. on cell architecture). Close coupling between initiation of DNA and histone syntheses requires activation of many genes including 14 that encode histone H4. The initiation of histone synthesis is controlled by a complex signaling system that is independent of E2F. It is activated by cyclin E/cdk2, via NPAT protein, whose gene maps at the ataxia telangiectasia (ATM) locus, which activates HiNF-P/MIZF that binds to a conserved II locus in the promoter DNA of H4 and activates transcription. PCNA and chromatin assembly factor-1 (Caf-1) are involved in the subsequent assembly on DNA of histones into nucleosomes, to form chromatin. The anaphase-promoting complex is an E3 ligase that targets mitosis related proteins cdc2 and securin (Pds1), which permits activity of the protease separase.

Many of the enzymes involved in both DNA and deoxynucleotide synthesis assemble on DNA into a Replitase replication complex, which is present in S but not G_1 cells (see Chapter 5 by Prem-Veer Reddy et al. on S phase). These enzymes move from cytoplasm to nucleus at G_1/S, providing an example of regulation by intracell localization. How this relocalization relates to intercompartment transport mechanisms is unknown. Is it triggered by E2F, cyclin E, or cyclin A, all of which are active in S as well as in G_1 phase? They are directly involved in assembly and stability of origin of replication complexes (ORC), which are associated with chromatin throughout the cycle. Cdc6p, replication licensing factor Cdt1, MCM, and other proteins assemble with ORC in G_1 to turn on DNA synthesis at specific replication origins. Cdc6 is degraded in early S, which prevents reinitiation of DNA synthesis during one cycle. It is down-regulated in prostate cancer. The transcription repressor Mad3 that antagonizes Myc activity is induced by E2F specifically in S phase.

G_2 and M phases do not appear to he drastically modified in cancer versus normal cells. Their responses to DNA damage and other stresses, however, differ (see below). Aurora kinase A is transcribed late in the cell cycle and is involved in both meiosis and mitosis. It is overexpressed in cancers, and it is transforming and causes aneuploidy. The anaphase-promoting complex is an E3 ligase that targets mitosis-related proteins cdc2 and securin (Pds1), which permits activity of the protease separase.

As examples of the many therapies based on the cell cycle, K vitamins with short thio-ethanol side chains at the 2 or 3 position of the core naphthoquinone are growth inhibitors of various tumor cell lines in vitro. The analogue Cpd 5 represents a novel class of drugs with selective antagonism to phosphotyrosine phosphatase activity; several Cdc25 substrates remain tyrosine phosphorylated and thereby are inhibitory. The growth inhibitory effects are correlated with ERK1/2 phosphorylation. Cpd 5 inhibited both normal liver regeneration and hepatoma growth in vivo, and DNA synthesis also was inhibited in several hepatocyte culture systems.

GROWTH TERMINATION

Summary

A hallmark of cancer is defective differentiation (anaplasia), a part of which is underlying regulatory changes that allow escape from the terminal growth arrest normally associated with differentiation. The growth-terminating senescence response of normal cells and their limited proliferative potential may have evolved to suppress tumorigenicity. Intimately implicated in senescence are telomeres, long repetitive DNA sequences that form structures at the ends of chromosomes. They become shorter at each progressive cell division, and eventually when they become critically short, they trigger either replicative senescence or apoptosis. A major mechanism for overcoming senescence in cancer cells is activation of telomerase, an enzyme that increases telomere length. Its activity is high in essentially all major types of cancer; 90% of human tumors express telomerase. It is not detectable in most normal human cells. Senescence of mammalian cells also is modified by factors other than telomere shortening, such as damage or stress.

Differentiation and Arrest of Proliferation

Differentiation and proliferation mechanisms are interconnected (see also Chapter 11 on metamorphosis). Differentiation is highly specifically programmed in different cells. Proliferation arrest is followed by appearance of products of differentiation-related gene. Defective differentiation (anaplasia) is a hallmark of cancer, in which underlying regulatory changes both alter diffentiation and cause escape from terminal growth arrest. The cells can be blocked at intermediate stages of differentiation that still permit proliferation, as in leukemia. Tumor cells have many characteristics similar to normal embryonic cells, including differentiation changes, rapid proliferation, angiogenesis, and patterns of gene expression such as for *myo D*, which regulates both processes. Mutations in the homeotic differentiation master genes modulate adult cell differentiation programs. The homeobox gene *HSIX* that is involved in differentiation is overexpressed in breast cancers, where it overcomes G_2/M checkpoint arrest after DNA damage. Retinoic acids have a major role in differentiation, activating or inactivating various HOX genes. Incorrect activation of the sonic hedgehog-Gli transcription signaling pathway is found in some tumors, including those of brain and skin. Another example of altered gene expression in differentiation is A33, an antigen expressed in most colon cancers. Four genes of the inhibitor of differentiation (Id) family are involved in differentiation, and they are overexpressed in aggressive breast tumors, which correlates with stimulated proliferation, invasion, and metastasis as well as blocked differentiation. The Id proteins might negatively regulate function of basic helix-loop-helix transcription factors. They increase expression of matrix metalloproteinase. Also Id binds to important cell cycle regulatory proteins, the pRb and Ets-family transcription factors. Mutations initiating cancers in adults have been proposed to arise upon differentiation of stem cells. If so, their asymmetric division to produce a stem cell and a diffentiating cell cells raises the question of which progeny cell might be cancer-related. Genes involved in stem cell differentiation include p53 and PTEN and p21. Differentiation under the influences of proteinaceous factors of multipotent stem cells to hematopoietic cells which differ dramatically in properties and structures provides a prime model.

Several drugs that reactivate differentiation are proposed to have therapeutic utility, selectively arresting or killing cancer cells. Histone deacetylation modifies chromatin structure (see above), and it decreases expression of many genes involved in differentiation. By inhibiting histone deacetylase, the suberoyl hydroxamic acid SAHA induces differentiation of breast cancer cells and stops their growth. It has entered clinical trials. Differentiation has also been proposed to put cancer cells into a pro-apoptotic state. Apoptosis of HL-60 tumor cells was increased when an antineoplastic drug was followed with a differentiating agent, retinoic acid or *n*-butyrate. Butyrate caused arrest in G_1 and apoptosis of breast cancer cell lines MCF-7, MCF-7ras, T47-D and BT-20, and in G_2/M of MDA-MB-231 cells. Whereas p53 was not involved, butyrate increased Fas and Fas ligand levels of MCF-7 cells, and this was decreased by anti-Fas antibody, which indicates apoptosis mediated by Fas signaling. These agents can increase apoptotic activities caused by other treatments.

Retinoic acids modulate growth and are important mediators of differentiation of both normal and malignant cells. They affect transcriptions through ligation to six nuclear retinoic acid receptors (RAR and RXR) that function as dimers, and they interact with a large number of coactivators and repressors. RXR with its peroxisome proliferator receptor partners (PPAP) has similar functions. AP2 alpha is a retinoic acid inducible tumor suppressor whose activity is mediated through a direct interaction with p53, which potentiates transcription of the proliferation inhibitor p21. An example of silencing is by hypermethylation of the cellular retinol binding protein-1 gene. Effects on cellular proliferation and migration have made retinoic acid (RA) useful in the treatment of several types of cancer. Acute promyelocytic leukemia (PML) is a special case of effective clinical treatment with all-trans RA. Treatment of prostate neoplasia is in progress. But RA is toxic at high concentrations, and it is a strong teratogen. Numerous RA analogues therefore have been synthesized and tested in searches for greater specificity. Some of these induce apoptosis in a wide variety of malignant cells and show potential for treatment. Although potential targets have been identified, their mechanism(s) of action and receptors are unclear. CD437/AHPN is a RARγ agonist that forms nonspecific adducts with DNA and specifically affect differentiation and causes apoptosis. When AHPN binds to it, the orphan receptor nur77 translocates to mitochondria and causes apoptosis of prostate cancers. Combination with histone deacetylase inhibitors or other agents may increase efficacy of retinoids against other malignancies.

Aging, Senescence, and Immortalization

Chronological age and cancer are interrelated. The incidence of cancer rises exponentially with age in humans and other mammalian species. For example, the frequency of cancer of the large intestine is 1/million at age 30 and increases 400-fold at 80. This increase could result from accumulation of oncogenic mutations in epithelial cells of renewable tissues, which produce most age-related cancers, in synergy with age-related pro-oncogenic changes in the tissue milieu (ECM). Senescent cells secrete factors that can disrupt tissue architecture and/or stimulate nearby cells to proliferate. In particular, senescent stromal fibroblasts create a pro-oncogenic tissue environment that promotes the development of epithelial cancers.

Defects in DNA repair are involved in this age-related increase of cancer incidence. Individuals with the inherited human premature aging disorders Werner's, Rothmund-Thomson, and especially Bloom's syndrome develop many types of cancer at an early age. The basis of these diseases is hereditary mutations in the RecQ family of ATP-dependent helicases that catalyze DNA unwinding. These defects diminish precision of repair and elevate the level of recombination. A recQ-like protein is involved in repair of double-strand breaks in *Drosophila*, and Werner protein (WRN) and Bloom syndrome protein (BLM) may similarly be involved in mammalian cells. These helicases are activated via the WNT path, and replin and pontin are co-proteins. Werner's syndrome involves genomic instability and premature senescence. WRN co-localizes and interacts through its RecQ-conserved C-terminal region with the critical telomere-binding protein TRF2, which stimulates helicase activities and causes defects associated with telomere maintenance including accelerated erosion. TRF2 also has high affinity in vitro for BLM. These helicases in combination with replication Protein A unwind long telomeric duplex regions that are pre-bound by TRF2. Organismal aging also may be influenced by p53. Several mice with aging-related phenotypes have mutations that increase p53 signaling. A mutant mouse line that appears to have an enhanced p53 response has a shorter life span and has phenotypes associated with early aging, and it also is more cancer resistant. This aging phenotype is consistent with a model in which aging is driven in part by a gradual depletion of stem cell functional capacity. Several kinds of evidence suggest that the senescence response and limited proliferative potential of normal human cells may have evolved to suppress tumorigenicity. Senescence irreversibly arrests proliferation in response to stimuli that could otherwise initiate neoplastic transformation, in which overcoming senescence is a key early event. Bypass pathways arise in cancers through mutations.

Senescence of normal cells is observed in tissue culture; after numerous cycles cell proliferation becomes slower and eventually it ceases, by 50 doublings (see Chapter 13 and 14 on senescence). Most senescent mammary epithelial cells die as they approach the growth plateau, but rare cells undergo a crisis, termed agonescence. Even agnosecent human mammary epithelial cells do not immortalize spontaneously. They can, however, be induced to immortalize by expressing telomerase. These cells can be maintained as viable cultures, termed cell lines.

This cellular senescence is thought to contribute to organismal aging, but the connection is not clearly established. Problems include the differences between conditions in vivo versus in culture, including absence of stromal cells and tumor-stromal interactions. Mixed cell and three-dimensional culture techniques are being utilized. Expression of some genes and also extrinsic factors involved in growth control modulate senescence. Epithelial cells and fibroblasts that have undergone cellular senescence accumulate in tissues during aging. The subepithelial layer (stroma) composed of extracellular matrix and several cell types essential for epithelial function is maintained, remodeled, and repaired by resident fibroblasts. Senescent human fibroblasts stimulated proliferation in culture of premalignant and malignant but not normal epithelial cells, even when only 10% of the fibroblasts were senescent. This effect was due, at least in part, to soluble and insoluble factors secreted by the senescent cells. It was equally strong when senescence was produced by multiple replications, oncogenic *ras*, p14(ARF), or hydrogen peroxide. Senescent, much more than pre-senescent, fibroblasts in mice caused premalignant and malignant epithelial cells to form tumors.

Gene products that stop proliferation by inhibiting cdk activities have roles in senescence. Primary fibroblasts derived from a melanoma-prone family had a finite life span but were not arrested by oncogenic *ras*. These cells were homozygous for an intragenic deletion in the CDKN2A tumor-suppressor locus, which encodes p16(INK4a) and p14(ARF), two proteins that have been implicated in senescence. They were p16(INK4a) deficient but expressed a frameshift protein with functions of p14(ARF). In normal human fibroblasts, ARF was not demonstrably induced by *ras*, indicating differences in CDKN2A-dependent senescence control in various cell types. These findings suggest interaction of intracellular mechanisms for cell proliferation and DNA repair.

Ras is initially mitogenic in primary cells but it eventually induces premature senescence. An irreversible senescence-like cell cycle arrest is produced in murine and human fibroblasts by coexpression of oncogenic *ras* and enhanced p53 levels. But extremely high p53 levels without oncogenic *ras* did not induce senescence. Inactivation by RNAi of either p53 or p16 prevented Ras-induced arrest in rodent cells, and E1A caused a similar effect in human cells. This operates via the MEK/MAP kinase cascade, and p19(ARF) is required. Furthermore activated MEK forced uncontrolled mitogenesis and transformation in cells lacking either p53 or INK4a. This opposite response of normal and immortalized cells to activation of the MAPK cascade implies that premature senescence is a mechanism that limits transformation caused by excessive *rag* signaling.

Telomeres have been intimately implicated in senescence (see Chapter 12 by Rhoads). They cap the ends of chromosomes; each is composed of hundreds of repeats of a sequence of 6 bases, 5′-TTAGGG-3′, with associated proteins. The telomeric protein TRF2 primarily protects human chromosome ends by capping them, which involves formation of the telomeric loop, a higher order structure at the end. This closed loop binds several proteins, including tankyrases, DNA-PK, and H-TERT, which is the reverse transcriptase part of telomerase. TRF-1 is a small protein that with other proteins controls telomere length.

Telomeres become shorter at each round of DNA synthesis at progressive cell divisions because the DNA replicating enzymes cannot duplicate them completely. They eventually trigger either replicative senescence or apoptosis when telomere length becomes critically short. This process has been suggested to be a mitotic clock that counts the number of duplications in a cell's history. Telomeres distinguish natural chromosome ends from damaged DNA breaks, and maintain the stability of eukaryotic genomes by allowing cellular repair mechanisms to act specifically on the latter. Aged cells accumulate chromosome abnormalities, probably at least in part because their chromosome ends become unprotected by telomere attrition. This loss permits end-to-end ligations that cause rearrangements during mitosis, and thereby massive cell death, and also genome instability, which greatly increases the probability that additional mutations will be created. Thus, although cellular senescence suppresses tumorigenesis early in life, it may promote cancer in aged organisms by destabilizing chromosomes. Senescence of mammalian cells also is modified by factors other than telomere shortening, such as damage or stress. Low oxygen permits many more doublings of mouse fibroblasts, but not of human cells. In cultured cells, senescence in response to a variety of cellular stresses is associated with elevated p53 activity, and lowered p53 may have the opposite effect.

A major mechanism for overcoming senescence is activation of telomerase, a ribonucleoprotein enzyme that increases telomerase length by adding many TTAGGGs to chromosome ends. Telomerase is not detectable in most normal human cells, except for those that go through many cycles such as activated T cells and stem keratinocytes. The block to its activation is extremely stringent in normal cells. But transduction of its catalytic subunit, h-TERT, avoids agonescence and creates immortality. Conversion is the gradual process that leads to telomerase activation, telomere length stabilization, decreased p57(KIP2) expression, and increased ability to grow uniformly well in the presence or absence of TGF-β. Conversion may represent a rate-limiting step in immortal transformation of cells with active p53. But the events that reactivate telomerase during carcinogenesis are not known. Overcoming agonescence can be separated from activation of telomerase in cells immortalized by a chemical carcinogen.

Telomerase activity is high in essentially all major types of cancer. Its promoter is not changed. But it is sequestered in nucleoli of cancer cells, and is released into nucleoplasm at G_2/M after DNA damage and after SV40 virus infection. The large T antigen gene of SV40 virus that immortalizes rodent cells and bypasses proliferation arrest does not immortalize human cells. It increases the life span of human primary hone cells for only a few passages before cells' net growth ceases and they enter crisis. When human osteoblast precursors and marrow stromal cells transformed with the SV40 T antigen were reconstituted with telomerase, the cells became immortal. Like pre-crisis cells, they were able to differentiate despite chromosomal abnormalities. A temperature-sensitive SV40 T antigen gene transfected into mice permits isolation from any tissue of cells that can be cultured indefinitely at the permissive temperature. These cells function nearly normally under nonpermissive conditions.

Telomerase activity has a significant role in the development and potentially in treatment of human cancer. Understanding telomere biology of normal versus cancer cells may lead to clinically effective telomerase-based therapies. A mouse lymphoma model points to an unsuspected role of drug-induced senescence. This approach is predicted to be different in important ways from traditional cytotoxic drug therapies.

Bmi-1, an oncogene that cooperates with Myc, can overcome senescence, extend replicative life span, and immortalize mammary epithelial cells. Bmi-1 was overexpressed in immortal cell lines, several breast cancer cell lines, and leukemia and in some non–small cell lung cancers. It activates hTERT transcription and telomerase activity, induces immortalization, and represses p16/p19ARF. Bmi-1 was not overexpressed in hTERT-immortalized cells, suggesting that it functions upstream of hTERT. Its activity appeared to require the RING finger and a conserved helix-turn-helix-turn domain in the hTERT promoter, independently of c-Myc binding sequences. Other involved genes are Mad, which blocks Myc/Max functions, TGF-β which represses h-TERT through SIP1, and menin another activator of hTERT.

A NAD^+ dependent deactylase in yeast named Sir2 is involved in responses to DNA damage and chromosome stability, gene silencing, and decreased cell aging. Sir2 produces nicotinamide from NAD^+, and nicotinamide feedback inhibits the enzyme. Environmental stresses transcribe

Pnc1, which hydrolyzes nicotinamide and thereby releases Sir2 activity. A similar system probably functions in humans.

CELL DEATH

Summary

Programmed cell death (apoptosis) is as important as is proliferation for cancer. Apoptosis functions thoroughout life, eliminating cells after they are no longer needed. Commitment to apoptosis and the ability to prevent apoptosis involves closely interconnected interactions of complex survival and death pathways, in which many genes and molecules of cell proliferation, such as the mitogen-activated protein kinase (MAPK) family signaling pathways, also function. The transforming growth factor TGF-β induces apoptosis in both G_1 and S phases, in addition to inhibiting entry into S phase. Tumor-suppressor gene p53 has a major role in preventing cancer development by first arresting growth of DNA-damaged potential tumor cells and then killing them by apoptosis. Cancer cells must develop a variety of defenses against apoptosis, such as loss of p53 activity which is found in half of all human cancers. Furthermore many tumors carry mutations that prevent full activation of p53. Protein Mdm2 suppresses cell death by activating degradation of p53. Increased Mdm2 thus has oncogenic consequences similar to the mutations that inactivate p53. However, p53-negative cells can undergo apoptosis when activated by stress or drugs, which suggests that other molecules can to some degree substitute for p53. An antitumorigenic safeguard mechanism independent of p53 might depend on p73, a protein related but differing structurally from p53.

Apoptosis

Apoptosis is a normal physiological process that functions thoroughout life, eliminating cells after they are no longer needed, such as during differentiation, or when they become aged, such as for white blood cells following a few months of functioning. For example, normal prostate cells rapidly undergo apoptosis if deprived of androgen (see Chapter 15 by Wang and El-Deiry on apoptosis and necrosis). A definite intracell signaling program is activated when apoptosis is stimulated. It cleanly eliminates specific cells with minimal effects on the microenvironment and nearby cells. In contrast, necrosis is a mechanism of cell death that causes inflammation of surrounding tissues due to released components of dead cells. The conditions and maintenance of cellular energy pools can determine which mode of cell death ensues. High concentrations of benzamide riboside predominantly induce necrosis, which correlates with depletion of ATP and dATP and DNA strand breaks. Replenishment of the ATP pool by addition of adenosine prevents necrosis and favors apoptosis.

A variety of signals intiate the apoptotic cascade. Some main events are aberrant growth stimulation, as initiated by Myc or E2F-1 via p19ARF/Mdm2/p53, and DNA damage signaled via ATM/ATR and chk2. In addition a "murder," in contrast to "suicide" pathway is initiated, for example, by T cells through the Fas-Fas ligand mechanism that activates the programmed cell death mechanism in target cells. An intracell localization change permits interaction of Fas with its ligand Fas-L. Apoptosis in the absence of protein synthesis involves increased extracellular Fas/FasL interaction as well as activation of FLICE a Fas/APO-1-specific protease. This Fas extracellularly activated pathway and the intracell pathways converge to activate a series of proteases named caspases (see below).

Commitment to apoptosis and the ability to prevent apoptosis involves closely interconnected interactions of complex survival and death pathways in which many genes and molecules of cell proliferation also function. As an example of upstream-signaling molecules, *ras* is involved in apoptosis as well as in proliferation. It activates Akt, which has several anti-apoptotic effects including translocation of mdm2 to the nucleus where it targets p53 for destruction. Continued production of mutant murine *K-ras4b(G12D)* was necessary to maintain the viability of lung adenocarcinomas. Lesions appeared by seven days after *K-ras* was conditionally expressed in transgenic mice, and after two months their lungs contained adenomas and adenocarcinomas. Removal *of K-ras* caused apoptosis and regression by three days. Similarly tumors that appeared histologically more malignant arose within one month in animals deficient in either the Arf or p53 gene, and they also regressed rapidly when Arf was removed. The influence of Ras on cell survival or death is proposed to depend on relative activations of its target proteins. The Nore1-Mst1 protein complex is a new downstream target for the Ras GTPases, which reveals a mechanism by which *ras* influences survival versus apoptosis.

Mitogen-activated protein kinase (MAPK) family signaling pathways and apoptotic regulatory machinery are connected. MAPK/ERK kinase (MEK) inhibitors can block survival and increase cytotoxicity after some drug treatments. Activities of the major MAPK subgroups differ, depending on conditions that include the cell line and antitumor agent. JNK kinase and p38 have pro-apoptotic functions. *myc* activates proliferation and also in excess activates apoptosis both by increasing p19ARF, which blocks Mdm2 and thereby stabilizes apoptotic p53, and by mitochondrial destabilization and release of cytochrome c, which cooperates with proapoptotic members of the Bcl-2 family. Fibroblasts lacking *bak* are susceptible to c-Myc-induced apoptosis whereas Bax-deficient fihroblasts are resistant, and apoptosis can be restored by ectopic expression of Bax. However, c-Myc activation had no detectable effects on Bax expression, localization, or conformation. This role of *myc* is at least partly prevented by *ras*, which regulates Myc stability. Thus neoplastic lesions can be both induced and maintained in vtvo by interlocking molecular processes.

TGF-β1 induces apoptosis in both G_1 and S phases, as well as inhibiting entry into S phase. The preapoptotic changes are in part reversible upon its removal, and the majority of cells then rapidly enter S phase. This apoptosis is associated with a marked increase in activity of E2F (see below). Both binding of E2F to DNA and formation of E2F-responsive reporter constructs were increased, and the formation of an E2F-DP-1 complex was altered, but no significant changes were observed in E2F mRNA and protein levels. This apoptosis was inhibited by overexpression of pRb, an effect that might be removed when Rb is replaced by the E2F-1 partner DP-1. Phosphorylation of DP-1 by cyclin A/cdk2 releases E2F-1, and failure of this reaction causes S phase arrest and apoptosis. E2F-DP-2 exhibited little change in the preapoptotic cells.

p53 and Apoptosis

Tumor-suppressor gene p53 is often referred to as the guardian of the genome (see Chapter 18 by Masciullo and Gordano on p53). It has a major role in preventing cancer development by arresting growth of DNA-damaged potential tumor cells and then killing them. It also functions in regulating lethality of many antineoplastic drugs. p53 is the most often mutated gene in human cancers, because eliminating programmed cell death is important for cancer cell survival. Loss of its activity is found in about half of all human cancers. Furthermore many tumors carry mutations that prevent full activation of wild-type p53. The result is in either case a defect in the ability to induce an apoptotic p53 response. Furthermore those tumor cells that lack p53 are generally not arrested in G_1 but continue to cycle until they reach the G_2/M checkpoint, in contrast to p53 positive cells.

Growth arrest and apoptosis are both stimulated by p53, which alone is not sufficient to specify between these two fates. Rb is involved in this choice, as a necessary effector in p53-mediated growth arrest that inhibits E2F and nuclear c-Abl tyrosine kinase. Rb also binds mdm2 and thus regulates p53 activity. p53 is phosphorylated and is hyperacetylated via p300 by stresses including DNA damage, which increases its DNA-binding activity and is necessary for responses to it. Sir2 deacetylates p53, which inhibits p21 expression. 14-3-3 protein has a role in activation of p53.

Genomic integrity is compromised by impaired p53 function, and this increases mutation rates of other genes and contributes to tumor progression. Mutated p53 can modulate the mechanism of mutation. Loss of p53 function increases mutations resulting from nonhomologous recombination. Human lymphoblastoid cells with impaired p53 function increased both spontaneous and induced mutant frequencies at the autosomal thymidine kinase locus.

Activated p53 in turn transcriptionally activates some pro-apoptotic family members, including GADD45 and NF-κB-related BH3-containing NOXA, and thereby apoptosis. Transcription-independent, pro-apoptotic activities of p53 have also been described. The *PUMA* (p53 upregulated modulator of apoptosis) gene is induced in cells following p53 activation. It encodes two BH3 domain-containing proteins, PUMA-α and PUMA-β that have similar activities. Exogenous expression of PUMA caused much faster apoptosis of colorectal cancer cells than resulted from exogenous expression of p53. Thus PUMA may be a direct mediator of p53-associated apoptosis. Antisense inhibition of PUMA expression reduces the apoptotic response to p53. PUMA is likely to play a role in mediating p53-induced cell death through the cytochrome c/Apaf-1-dependent pathway. It binds to Bcl-2 and Bcl-X(L), and it localizes to mitochondria to induce cytochrome c release. Proapoptotic members of the Bcl-2 family such as Bid, Bax, and Bad proteins bind to and block antiapoptotic Bcl-2 and Bcl-xL that stabilize the mitochondrial membrane. The Bcl-2 family includes more than 10 proapoptotic BH-3 domain only proteins such as NOXA and PUMA. Proteolytie modification of these key regulatory molecules involved in apoptotic and survival pathways is a feature of the control of programmed cell death. Four molecules of the family (BID, Bcl-2, Bax, Bcl-xL) are cleaved during apoptosis, as are XIAP and RIP and TRAF1, two proteins involved in NF-κB activation, and MEKK1, a molecule involved in a protein kinase stress signaling cascade that contributes to apoptosis and NF-κB activation. These cleavage products can inactivate an existing function or produce a new function. Many slow growing tumors overexpress Bcl-2, and tumor cells often have elevated antiapoptic (Bcl-2, Bcl-xL) versus proapoptotic (Bax, Bak) proteins. Conformational changes and mitochondrial targeting of proapoptotic Bax may be the downstream consequence of the apoptosis cascade that was caused by the kinase inhibitor flavopiridol.

Cancer cells develop a variety of defenses against apoptosis. These modifications can provide targets for chemotherapies designed to destroy an anti-apoptotic mechanism unique to cancer cells. This therapeutic approach differs from the classical ones directed against cell proliferation. The best known example of inactivation of apoptosis is by mutation of p53. Thus a selective strategy is to raise or restore p53 of tumor cells. Geldanamycin modifies p53 structure and selectively destabilizes and conformationally alters mutated p53, suggesting its application to proapoptic therapy. A novel therapeutic modality is proposed with peptides, derived from the C-terminus of p53, that reactivate mutant p53 proteins. They induced rapid apoptosis in breast cancer cell lines defective in p53 but were not toxic to nonmalignant human cell lines containing wild-type p53. Binding of the peptide to a N-terminal regulatory region of p53 was suggested. Another approach is with the E1B gene-attenuated adenovirus ONYX-015, which selectively causes apoptosis of those tumor cells in which absence of p53 in permits virus survival and lethalty. This antitumor effect can be augmented by standard chemotherapeutic agents. The mechanism and clinical potential are under study. Another adeno-associated virus kills p53-defective cancer cells, but it causes only growth arrest in G_2 of normal cells.

Several therapeutic strategies have been proposed to decrease the host toxicity of therapy-related apoptotic treatment, by previously chemically lowering normal cells' p53 and thereby protecting them. Pfithrin-alpha is a small molecule that reversibly blocks p53-dependent transcription and thereby apoptosis; it protected mice against lethal effects of an anticancer treatment. This is similar in intent to selectively blocking normal cells but not tumor cells at the restriction point prior to application of a cycle-specific agent (see above).

Mdm2 suppresses the ability of p53 to inhibit cellular proliferation and to cause cell death. This protein functions primarily as an E3 ubiquitin-protein ligase that induces degradation of p53 by the 26S proteasome, thereby blocking apoptosis. Increasing Mdm2 thus has consequences similar to mutations that inactivate p53, a property that underlies Mdm2's oncogenic potential. It is amplified in human sarcomas, and overexpressed in 30% to 70% of breast tumors. More than 40 different splice variants of Mdm transcripts have been identified in both tumors and normal tissues, and the majority of these variants do not contain the sequence encoding the p53-binding site. The potential functions of these variants are associated with tumor progression and prognosis. In addition they interact with full-length Mdm2 protein. Mdm2 is sequestered in the nucleolus by p19ARF binding, which thereby increases apoptosis due to p53. Some tumors have lost both ARF and p53. Mdm2 also has p53-independent functions that target other proteins, and that might be counteracted by surveillance mechanisms that need to be inactivated for Mdm2 oncogenic activity.

Functional diversity and selectivity of a protein depends, in general, on its associations with numerous partner proteins, and is closely controlled by complex post-translational modifications. The p53-Mdm2 paradigm represents one of the best-studied relationships between a tumor-suppressor gene and an oncogene. This kind of interaction is involved in numerous cellular biological systems. Not only does Mdm2 decrease p53 stability, but also p53 down-regulates Mdm2 expression. These determine the stability and activity of both p53 and Mdm2. Regulatory metabolic loops delicately balance p53. As an example, after p53 is activated, it rapidly shuts itself off by inducing Mdm2. But it later induces PTEN, which, by inhibiting activation of Akt, retains mdm2 in the cytoplasm and promotes p53 function. The consequence is that there is p53-dependent apoptosis of badly damaged cells after a time lag. In one, Akt is antiapoptotic, by phosphorylating and inactivating propapoptotic Bad. Increasing Akt activity also moves Mdm2 into the nucleus where it causes loss of p53, and this in turn increases Akt, which further decreases p53. As another example, elevated β-catenin inactivates Mdm2, which increases p53, and this in turn degrades β-catenin. These balances are distorted in cancers.

p73

Some p53-negative cells and tumors undergo apoptosis when they are activated by stress or drugs. This suggests that other molecules can to some degree substitute for p53. p73 and p63 are two proteins that are related to but differ structurally from p53. Their alternatively spliced forms create a complex network of proteins, and the gradient of regulatory functions among them ranges from proliferation and apoptosis to development. Knockouts of p63 and p73, hut not of p53, cause abnormalities in mice that strongly suggest the involvement of p63 and p73 in development. Inactivation in human cancers of p73 by either mutation or DNA methylation was not observed, unlike p53. Deregulation of E2F-1 activates p73 in p53-negative cells, leading to transcription of genes that are normally p53 responsive and to apoptosis. Intratumoral injection of an adenovirus vector expressing E2F-1 in combination with gemcitabine strongly induced apoptosis and decreased the size of p53-independent pancreatic cancers. These effects were directly correlated with induction of p73, suggesting that the E2F-1/p73 pathway plays a critical role. Disruption of p73 function in a tumor-derived p53 mutant reduced E2F-1-mediated apoptosis. Activation of p73 might thus provide an antitumorigenic safeguard mechanism that is independent of p53.

The stimuli and molecular mechanisms regulating pro-apoptotic activity of p73 are not yet well understood, and could be related to either p14ARF or p53 mutation or both. c-Abl stabilizes the p73 protein and activates its pro-apoptotic function in S/G_2 cells, but not in G_1 cells because of their Rb action. P73 can bind mdmX, mdm2, p300/CAF, and adenovirus E4-orf6 proteins, and activates promoters including p21, bax, mdm2, gadd45, cyclin G, IGFBP-3, and 14-3-3 sigma, in which Cdc25 may be involved.

E2F-1 and Apoptosis

Many genes induced by this protein initiate and enhance progression through S phase. But it is very frequently overexpressed in tumor cells, which is an indicator of aberrant proliferation. Its overexpression provides control by apoptosis, thus activating pathways that protect the organism from potential cancers. This pair of conflicting processes provides an example supporting the hypopthesis that excess of a molecule needed for proliferation signals senescence or apoptosis. The quantity of E2F-1 is normally closely regulated, balanced by mechanisms of activation, inactivation, and proteosomal degradation. Underphosphophorylated pRb inhibits activation of E2F-1 by directly binding to it and recruiting HDACs. E2F, pRb, p107, p130, and p107-cyclin A-cdk2 are pRb-associated proteins that have been identified in more than four E2F complexes in preapoptotic cells. Activation of E2F-1 is primarily by pRb phosphorylation, which controls transit through the G_1 restriction point, and so E2F-1 activation can be deregulated through modifications of cyclin/cdk activities in cancers. Apoptosis follows pRb's degradation by caspases; a caspase-resistant Rb protein desensitizes cells to apoptosis. E2F1-dependent transcriptions are repressed when pRb is bound by prohibitin, a potential tumor suppressor recruiting n-CoR and HDAC1. This mechanism differs from the repression by underphosphorylated Rb alone.

In S phase, binding to DNA and degradation of E2F is normally regulated by its phosphorylation on specific amino acids, which is catalyzed by cyclin A/cdk2. Then ubiquitination by a specific E3 ligase p45 (skp2), in which Mdm2 has a role, is followed by proteolysis. Mutations that affect these reactions enhance the ability of E2F to induce apoptosis. E2F-1 thus has major roles in both S phase proliferation and apoptosis, which is consistent with its S phase specificity and its necessary binding to DNA.

E2F-1 activates p53 phosphorylation in p53 positive cells and thereby stabilizes it, which activates apoptosis. Phosphatidylinositol 3-kinase is proposed to be involved, since caffeine inhibits this activity, overrides the S phase checkpoint, and abolishes E2F-dependent apoptosis. E2F-1 also specifically activates the ARF promoter and the increased P19ARF blocks Mdm2, thereby stabilizing p53 and as a consequence increases apoptosis. Furthermore P19ARF binds to and activates proteosomal degradation of E2F-1. And E2F-1-2-3 are sequestered in the nucleolus by p19ARF. E2F/ARF thus provides a possible negative feedback loop, like the one for the p53 and Mdm2 interaction. P14ARF (mouse p14ARF = human p19ARF) is the alternate product of the INK4A/ARF locus, and as a tumor suppressor it is frequently inactivated, as in lung cancers. Homozygous deletions of p14ARF were detected in 12 of 53 human cell lines and 16 of 8 primary lung cell carcinomas. A loss of p14ARF could be functionally equivalent to a p53 mutation in being anti-apoptotic and oncogenic. Since E2F-1 can signal p53 phosphorylation in the absence of p14ARF, which is lost in some tumors, these processes may interact as a network rather than as a simple linear pathway. The downstream mechanism of induction of apoptosis initiated by E2F also involves modulation of activity of the cytochrome c promoter. Another target is Apaf-1; elevated E2F-1 upregulated transcription of Apaf-1 and also activated caspase 9, but it did not release cytochrome c into the cytoplasm. E2F-1 also inhibits survival signals, whose loss by cancer cells would be oncogenic.

Therapies are being developed based on agents that modify the E2F-1 apoptosis pathway. Mutations of E2F-1 that block binding of cyclin A/cdk inhibit protosomal degradation and thereby increase the amount of E2F-1. This result led to designing short peptides that compete with E2F-1 for interaction with cyclin A/cdk complexes, and thereby selectively and specifically kill tumor cells. The novel retinoid CD437/AHPN causes dissociation of E3 ligase from E2F-1 and decreases E2F-1 ubiquitination. It increased E2F-1 of prostate carcinoma cells and caused their S phase arrest subsequent apoptosis. Naphthoquinone NA activated release of E2F-1 from Rb, induced p73 mRNA and protein, and stimulated expression of the p73-induced downstream genes *p21* and *Bax* in p53-independent HeLa cells, causing their apoptosis; overexpression of pRb was inhibitory.

Differences in cell cycle checkpoint controls between cancer and normal cells offer additional potentials to selectively target cancer cells for apoptosis, especially since oncogenic pathways converge on checkpoints despite molecular complexity and heterogeneity of individual cancers. Direct checkpoint activators are thus proposed as selective therapeutic agents. The drug β-lapachone activates the S phase checkpoint in p53-independent cells. It produces apoptosis after sustained elevations of the level and activity of E2F-1, proliferation block in S, production of pro-apoptotic products, release of mitochondrial cytochrome c, and cleavage of PARP. This drug depletes NAD and ATP, through reactions that are inhibited by dicoumarol. β-Lapachone is proposed to activate redox cycling by involving diaphorase-sensitive NAD(P)H-quinone oxidoereductase, which depletes ATP and NADH, modifes Ca^{++} pools and causes apoptosis. Cytosolic Ca^{++} is increased and activates the Ca^{++}-dependent protease u-calpain, which translocates to the nucleus and stimulates apoptosis; caspases are not involved in this novel mechanism. These events are inhibited by the intracell chelator BAPTA-AM. This lethality also involves increased E2F-1. β-lapachone has a strong potential for therapy, based on its potent tumor-specific activity in xenografled human tumor cell models, and shows selective apoptosis of cancer cells but not normal cells. These distinctive properties are not shared by most anticancer agents.

Restoring Apoptosis to Tumor Cells

In addition to the inactivation of p53 discussed above, various molecular alterations diminish the apoptotic tendency of tumor cells. The differential created by development in tumor cells of an anti-apoptotic response can be the basis for therapies that provide tumor specificity of drugs that restore the masked apoptotic activity. Also such a pro-apoptotic drug can be applied in combination with conventional antitumor agents that mediate their effects by stimulating apoptosis, despite having diverse primary mechanisms of action. Apoptosis is restored in tumor cells by post-translational deamidation of Bcl-xL, which prevents its binding to proapoptotic BH3-containing proteins. This deamidation is specific to tumor cells because in them underphosphorylated pRb, which blocks the reaction, is absent and so Bcl-xL remains inhibitory to apoptosis. Also chaperone heat shock proteins (HSP) affect cell death and might provide targets of this kind. Geldanamycin, herbimycin A, and radicicol that bind Hsp90 activate ubiquitination and proteasomal degradation, and thereby destabilize associated proteins including v-Src, Bcr-Abl, Raf-1, ErbB2, and some growth factor and steroid receptors. Depending on the cellular context, this process can cause growth arrest, differentiation, and apoptosis, or it can prevent apoptosis. HSP70 is anti-apoptotic and is overexpressed in many tumor cells. It interacts directly with key apoptotic molecules, including cytochrome c and Apaf-1. Increase of HSP70 during tumor progression may suppress a transformation-associated death program that is unusual in being independent of caspases, Bcl-2, and so on. Preventing HSP70 synthesis with antisense caused massive apoptosis of breast cancer cells but not of normal cells. Although not cancer specific, as for most chemotherapy, compounds that interact with them can preferentially kill certain tumor cells. HSP-active agents are undergoing clinical trials.

MECHANISMS OF APOPTOSIS

Summary

Their involvement in apoptosis provides a new role for mitochondria. The principal function of these intracell organelles is to produce energy by oxidizing nutrients. Reactive oxygen species (ROS) are rare by products of this oxidation, cause DNA damage, and are mutagenic. Mitochondrial mutations may be involved in normal aging and degenerative diseases, and they accumulate in cancers. In apoptosis the outer mitochondrial membrane undergoes a permeability transition. Cytochrome c is released into the cytoplasm through the open permeability pores. It recruits a protein complex that binds and activates the caspase proteinases involved in apoptosis. Pro- and anti-apoptotic Bcl-2 family proteins activate or prevent this permeabilization, respectively, by interacting with the permeahilty transition pore complex. Bcl-2 is elevated in some tumors. Apoptosis is delicately balanced by activities of pro- and anticaspase molecules such as the inhibitors of apoptosis family that include survivin and X-linked IAP, and are frequently upregulated in cancer cells. The NF-κB family of transcription factors has a major role through its anti-apoptotic activity. It is present at high levels in a large fraction of human and rodent mammary tumors, promoting both cell survival and proliferation. DNA-damaging agents and chemotherapy activate NF-κB in some tumors, and thereby decrease efficacy of their treatments. Blocking the activation of anti-apoptotic NF-κB provides a basis for therapy.

Mitochondria and Apoptosis

Involvement in apoptosis provides a new role for mitochondria. Signals for apoptosis, including those initiated by DNA damage, lead to mitochondria. To summarize, under apoptotic conditions the outer mitochondrial membrane undergoes a permeability transition that releases apoptotic proteins cytochrome c and Smac/DIABLO through channels and into the cytoplasm. Note the remarkable dual functions of one protein, cytochrome c, in both oxidative phosphorylation and apoptosis. Pro- and anti-apoptotic Bci-2 family proteins activate or prevent this permeabilization, respectively, by interacting with the permeabilty transition pore complex of mitochondria, and they are the basis for several potential therapeutic drug activities.

The principal function of these intracell organelles is to produce ATP through oxidative phosphorylation, energized by H^+ transport across the potential of the double membrane that separates their interior from the cell cytoplasm. This membrane potential concentrates positively charged molecules, for example, the positively charged fluorescent dye rhodamine 123. Alterations of mitochondrial location, shape, and organization can be observed with this specific probe, such as those induced by colchicine treatment. AMP-activated protein kinase is a key regulator of this energy metabolism.

Reactive oxygen species (ROS) are by-products of oxidative phosphorylation in mitochondria. They include superoxide, which is converted to hydrogen peroxide and a hydroxyl radical. Superoxide also reacts with nitric oxide to form peroxynitrite, another strong oxidant. These are formed under conditions of oxidative stress, such as in ischemia-reoxygenation, and they cause cell injury. ROS damage DNA and are mutagenic, so mitochondrial function is altered as a consequence.

Another major property of mitochondria is their storage of Ca^{++}. This ion leaves mitochondria by one or a combination of pathways upon stimulation, by oxidants, for example. Proteases, nucleases, and phospholipases are activated by Ca^{++} efflux, and then pathways of apoptosis are initiated. Several genes have been suggested to control Ca^{++} efflux, and apoptosis by mechanisms, such as by overexpression of Bcl-2 at the membrane.

Mitochondria are organelles that contain their own DNA (mtDNA) of small size (15,569 bp in humans), coding for 13 enzymes that are involved in mitochondrial respiration and 4 RNAs. Most mitochondrial proteins are, however, coded by nuclear genes. mtDNA is dependent on the nuclear genome for transcription, translation, replication, and repair, but precise mechanisms of how the two genomes interact and integrate with each other are poorly understood. mtDNA is maternally inherited. Individual normal cells contain abundant mitochondrial point mutations that had been clonally expanded. Each cell contains hundreds of morphologically and functionally heterogeneous mitochondria. Mechanisms for efficient homogenization of mitochondrial genomes within individual cells are proposed, but they are likely to be different between tissues.

mtDNA is proposed to be involved in carcinogenesis because of its 10× higher susceptibility to mutation and its limited repair mechanisms. It lacks introns, and so most of these mutations are in coding sequences. To become relevant in terms of pathology, a mutation must generally affect at least between 50% and 70% of mtDNA molecules in a tissue. To reach this level, mutated mitochondria that are deficient in oxidative phosphorylation or apoptosis may be preferentially replicated and so produce a clone of cells with this characteristic. This type of amplification of mutant mtDNA has recently been shown in colon cancer cell lines. Mitochondrial mutations may be involved in normal aging and degenerative diseases, and accumulate in a cancer. Mutated mtDNA is found in cells from a variety of cancers. These mutations appear early; they are found in primary prostate cancers and their paired PIN lesions, and can be detected in body fluids of some early stage patients. Twenty mtDNA mutations were detected in the tumors of three prostate cancer patients and in the PIN lesion from one patient. Identical mutations were also detected in matched urine and plasma samples obtained from all three patients. They are less common in prostate adenocarcinoma.

Mitochondrial proteins are involved in several cancer functions. Alterations in expression of mtDNA-encoded polypeptides required for oxidative phosphorylation and cellular ATP generation may be a general characteristic of cancer cells. These mutations could modify apoptosis or increase the production of ROS, and thereby accelerate tumor progression. Methods have been devised to detect mitochondrial changes associated with apoptosis. The PRDX3 gene is required to maintain normal mitochondrial function, encoding a mitochondrial protein of the peroxiredoxin gene family. It is a target of c-*Myc* that is decreased in c-$Myc^{-/-}$ cells, and is important for *Myc*-mediated proliferation, transformation, and apoptosis after glucose withdrawal. Specific fluorescent probes demonstrate that PRDX3 is essential for maintaining mitochondrial mass and membrane potential in transformed rat and human cells. Another mitochondrial protein involved in apoptosis is dynamin-related Protein 1. Furthermore mitochondria, plasma membrane microdomains and lysosomes compartment ceramide whose deregulated functions are involved in apoptosis and cancer.

Anticancer drugs targeted against mitochondria are being developed. A clinically applied example is Tirapazamine, a bioreductive drug that is selectively toxic toward hypoxic cells. At high doses mitochondria metabolize it to a DNA-damaging compound. Several compounds, including retinoids, act directly upon mitochondria to cause apoptosis. PolyADP-ribose polymerase-1 is produced following DNA damage (see below), and it activates apoptosis possibly by depleting NAD and thereby limiting oxidative phosphorylation and ATP production. 3-Br-pyruvate which blocks ATP production by both glycolysis and oxidative phosphorylation when given by arterial delivery is specifically effective against liver tumors in rabbits.

Downstream Events—Caspases

The cytochrome c released from mitochondria into the cytoplasm recruits the scaffolding adapter protein Apaf-1 and procaspase-9 in the presence of dATP to form 1000 kD apoptosomes. This complex binds and activates cascades of caspases, proteases that form signaling pathways for apoptosis and that are active in metastatic cancers. Caspases are normally inactive and require proteolytic cleavage for their activation, starting with caspase-9, which in turn cleaves and activates effector caspases-3 and -7 that execute the death program by cleaving their substrates. Additionally released proteins include Smac/DIABLO, which cooperatively represses proteins that inhibit caspase activation (IAP). But Smac/DIABLO was reported not to affect apoptosis in vivo. Downstream effects include cleavage of numerous proteins, changed localization of membrane phospholipids, then fragmentation of DNA, and eventually loss of cell viability. A novel alternative mechanism to caspases is proteolysis by activated Ca^{++}-dependent protease u-calpain.

Apoptosis is delicately balanced by activities of pro- and anti-apoptotic molecules, and the latter are frequently up-regulated in cancer cells. The inhibitors of apoptosis (IAP) family are E3 ubiquitin ligases that include survivin and X-linked IAP (XIAP). They inhibit apoptosis by binding to and interfering with activation of caspases-3 and -7. Survivin is a recently described protein that has also been implicated in both the control of cell proliferation and the regulation of cell life span. It is overexpressed in most human cancers. The IAP proteins are regulated by the MAPK, PI3K pathways and are destroyed by proteosome activity. Another inhibitor is AKT, which can block apoptosis by inhibiting caspases-9 and -3, even in the presence of released cytochrome c.

NE-κB and Anti-apoptosis

The NE-κB family of dimeric transcription factors is involved in numerous major cell processes. It has a major role in cancer through its anti-apoptotic activity, and in addition it activates cyclin D and thereby proliferation. NF-κB is expressed at a low level in all normal cells, with the exception of B cells, but is inactive and sequestered in the cytoplasm by the specific inhibitory IκB protein. It is present at high levels in a large fraction of human and rodent mammary tumors, promoting both cell survival and proliferation. NF-κB's constitutive activation is an early event in rats treated with the carcinogen DMBA; it increased by 3 weeks and prior to the detection of tumors after 7 to 9 weeks. Exposure of human mammary epithelial cells in culture to a carcinogen also up-regulated NE-κB prior to malignant transformation. Constitutively active NE-κB was found in all tested multiple myelomas and their cell lines. The enzyme IκB kinase (IKK) phosphorylates IκB, causing its dissociation from NF-κB, ubiquitination, and degradation by proteasomes. This removal of IκB exposes nuclear localization signals on NF-κB, which then translocates to the nucleus where it activates transcriptions of tumorigenic, anti-apoptotic, and angiogenic genes. IKK is a 900,000 kD multiprotein complex containing two kinases, α and β, and an essential modulator scaffold protein IKKγ (NEMO). Its activation and deactivation are precisely regulated. An IκBα deubiqitinating (DUB) enzyme CYLD associates with IKKγ, blocks proteolysis, and thereby NF-κB activation. An IκBα super-repressor increased chemosenstivity of pancreatic carcinoma cells. Inhibition of IκK by an IκKb dominant negative protein demonstrated that activated Akt requires IKK to efficiently stimulate the transactivation domain of the p65 subunit of NF-κB. Inhibition of endogenous Akt activity was associated with a loss of NE-κB transcriptional activity and sensitized cells to H-Ras(V12)-induced apoptosis. And Akt-transformed cells required NF-κB to suppress apoptosis induced by etoposide. In contrast, Akt stimulates NF-κB predominantly by upregulating the transactivation potential of Rel A/p65, unlike activated *ras* which can activate both DNA binding and the transcriptional activity of NF-κB by parallel pathways. A different mechanism of IκBa degradation is by calpain (a protease activated by kinase CK2) that is active in many primary breast cancers.

Activation of NF-κB can be stimulated in cancers by overexpressed growth factors such as EGF and IGF-1, their receptors, cytokines such as TNF, and oncoproteins such as *H-ras*. As an example of stimulation by growth factors, heregulin is an autocrine activator of NF-κB, and it has a major role in breast cancer cell lines derived from transgenic mice that overexpress specific oncogenic growth factors. Constitutive NF-κB activation by heregulin correlated with phosphorylation of the Erb3 receptor and was inhibited by Tryphostin AG1517. In contrast, Emodin blocked phosphorylation of her2/neu by heregulin but failed to decrease NF-κB activation. Extracellular matrix also is important for NF-κB activity. Activators initiate kinase cascades that phosphorylate and activate the IKK kinase complex. Constitutive activation of NF-κB in cancer cells is reported to be by Ras/MEKK activity. But several different kinases have been reported to be involved, perhaps because different cells and media were investigated. Seven pancreatic cancer cell lines with *K-ras* mutated at codon 12 had different pathways of Akt phosphorylation, as shown by analysis of mutations and sensitivities to serum and inhibitors. But they expressed similar Akt-dependent downstream mRNA patterns. Akt and NIK activate IKK-α, whereas MEKK1 and PKC isoforms activate IKK-β. PI3K-Akt or PKC provides a survival signal by increasing NF-κB. Protein kinase A and IκB kinase are reported to be involved as downstream signaling molecules. Specific inhibitors of these kinases but not of ERK/MAP kinase or protein kinase C decreased heregulin-mediated NE-κB activation.

Pro-apoptotic Therapy

Cells that overexpress NF-κB are frequently resistant to conventional chemotherapy, presumably because apoptosis is blocked. DNA-damaging agents and chemotherapeutic agents activate NF-κB in some tumors, and thus can decrease the efficacy of treatment. The anti-apoptotic activity is through induction by NF-κB of TRAF1, TRAF2, CIP1, and CIAP2 and suppressing caspase-8 activation. Diverse agents act at multiple levels to inhibit the NF-κB signal transduction pathway in tumor cells and create apoptosis; stimulators of apoptosis are therefore being developed, some of which should soon reach the clinic. Many are related to endogenous inhibitors, such as Bcl-2, and IAP proteins, such as survivin and XIAP. Ubiquitination and proteosomal degradation of IκB is one attractive approach. Several IKK inhibitors are being tested, including aspirin and prostaglandin A1. The nontoxic natural product curcumin blocks NF-κB production by inhibiting phosphorylation of IκB by IKKs, and thereby it decreases COX2, which is overexpressed in colon cancers; it prevented colon cancer in tumor models. A synthetic retinoid 4-HPR suppresses NF-κB and induces apoptosis in some tumors, inhibition of ERK kinase increases this apoptosis by c-jun terminal kinase and caspase-3 activation. The IκB-α phosphorylation inhibitory compound Bay 117082 increases apoptosis of resistant cell lines. Calagualine, a methanolic extract of a fern, blocks activation of NF-κB by TNF-α or phorbol ester, by inhibiting the phosphorylation of IKKβ by NIK. A new class of retinoid antagonist anticancer agents (MX781) inhibits NF-κB activation by binding to IκB, and induces apoptosis independently of retinoid receptors, among the many effects. Inhibitors of proteosome activity such as MG132 decrease NF-κB and induced apoptosis in colon cancers. The inhibitor PS-341 (Velcade) blocks NF-κB activation, and has been approved for the treatment of multiple myeloma. Readhering colon cancer cells were extremely sensitive to NF-κB inhibitors.

TNF-α regulates both pro- and anti-apoptotic responses, for which NF-κB appears to be the critical determinant. TNF-α alone was not cytotoxic to A549 human lung carcinoma cells, but expression of a dominant negative IκKb resulted in apoptosis after TNF-α exposure. An inhibition of anti-apoptotic TNF-α receptor signaling was linked to E2F-1 by an increase of mRNA of NF-κB inducing kinase (NIK, MAP3K14), which interacts with TRAE-2 a direct target of E2F-l, and through expression of Bcl-xL. NF-κB blocks apoptosis by TNF-α, but it is also activated by TNE-α. Concurrent inhibition of upstream effectors in TNF-mediated NF-κB activation, the Racl and IκK pathways, sensitizes lung cancer cells to apoptosis induced by TNF-α. The critical determinant of the anti-apoptotic response in epithelial cells infected with adenoviral constructs carrying dominant negative mutants of Racl and IκK or constitutively active mutant of Racl appears to be activation of NF-κB. Dominant negative Racl further sensitized these cells. A menadione analogue naphthoquinone (NA) causes apoptosis of HeLa cells treated with TNF-α, by decreasing the anti-apoptotic effect of NF-κB. The mechanism is through cleavage of the p65 subunit with caspase-3, activated by the cytochrome c/caspase pathway.

These studies suggest practicality of a target-directed chemotherapy against EGF-responsive breast cancers, based on blocking NF-κB and thereby reactivating apoptosis. As proof of this principle, Go6976 is potentially a clinically applicable drug shown to block ER-breast tumor growth in mice and even cause disappearance of the established tumors without harm to the animals. It is a staurosporine derivative that inhibits protein kinase C, inhibits EGF-induced activation, nuclear translocation, and DNA-binding activity of NF-κB in ER-breast cancer cells, and causes apoptosis. Apoptotic stimuli translocate protein kinase C to nucleus, mitochondria, and Golgi where it is phosphorylated and interacts with several important apoptotic proteins. cDNA microarray analysis revealed that expression changes of a subset of apoptosis-related genes that are altered by EGF treatment were reversed by Go6976. The related multi-kinase inhibitor flavopiridol, given with or followed by TNF or with the ligand TRAIL, inhibited NF-κB-dependent transcription and rapidly caused apoptosis of several carcinoma lines. These studies suggest practicality of a target-directed chemotherapy for EGF-responsive breast cancers, based on blocking NF-κB activation and thereby reactivating apoptosis. Also Go6976, like caffeine, overcomes the G_2/M checkpoint. And tumors with high NE-κB are generally resistant to chemotherapy.

STRESSED CELLS

Summary

Damage to DNA is mutagenic, and so is a most important stress for producing cancer. Furthermore currently applied anticancer treatments with X rays and many drugs function by damaging DNA and therefore are carcinogenic. Four very different outcomes of DNA damage are recovery, mutation, senescence, and death, with the frequencies depending on severity of the insult, the time available for its repair, and the target cell (normal or cancer). DNA damage activates genes and mechanisms involved in repair of the lesions. Their intial step is recognition of a lesion by interaction of receptor molecules with the damaged DNA. This initiates kinase cascades similar to proliferation signaling. If repair is not soon completed, cells undergo apoptosis, and thereby their possible carcinogenic effect is removed.

DNA Damage

The stress most important for cancer is damage to DNA because it is mutagenic. As discussed initially, DNA alterations can arise intracellularly from errors of replication or mitosis. Furthermore, as mentioned above, currently applied anticancer treatments with X rays and many drugs damage DNA. Exposure to extracellular DNA-damaging agents creates a variety of lesions that include thymine dimers from ultraviolet irradiation, single and double-strand breaks from ionizing radiation, chemical adducts as with benzopyrene, and methylation by alkylating agents. These are repaired by different enzymes. Five major DNA repair pathways are involved: homologous recombinational repair (HRR), non-homologous end joining (NHEJ), nucleotide excision repair (NER), base excision repair (BER), and mismatch repair (MMR). Two repair pathways act upon progression of replication forks. Recovery, mutation, senescence, or death are four very different outcomes of DNA damage; their frequencies depend on severity of the insult, the time available for repair, and the target cell (normal or cancer). Cells undergo apoptosis if repair is not soon completed, and the possible carcinogenic effect is thereby removed.

Repair of DNA damage involves four steps: (1) damage recognition, (2) transduction of signaling, (3) cell cycle inhibition, and (4) DNA repair. The initial step is recognition of a lesion by interaction of the damaged DNA with receptor molecules. Sensors of damage that bind to ionizing irradiation-damaged DNA are ATM and ATR, which are PI3K-related Ataxia telangiectasia kinases. ATM recognizes double-strand breaks, and ATR other kinds of DNA damage. They phosphorylate various damage-related proteins and activate cell cycle arrest by blocking cdc2 with p21 or cdc25A. DNA damage activates repair genes involved in repair of the lesions. Kinase cascades similar to proliferation signaling, and culminating with JNK and p38 kinases, are initiated. ATM phosphorylates checkpoint kinase Chk2 (also named Cds1). Downstream targets are the cdc25 family of phosphatases whose action blocks Cdk1/cyclin B and entry into mitosis. Similarly ATR acting via Chkl destabilizes Cdc25A and blocks progression through S phase by inhibition of Cdk2-cyclin E. Nuclear clustrin is induced by DNA damage and translocates to the nucleus. Along with DNA damage sensor Ku70, it forms a dimeric multifunctional protein with Ku80 that binds to double-strand DNA ends and to the ends of telomeres; in excess, it triggers apoptosis. It has DNA-dependent ATPase and helicase activities, and it is the regulatory subunit of DNA-dependent protein kinase (DNA-PK), a nuclear serine/threonine kinase that is activated by irradiation-generated DNA damage, is involved in repair and proliferation arrest, and phosphorylates many major enzymes and transcription factors. The DNA-PKA C-terminal peptide of Ku80 disrupts the complex and blocks its activity. Ku and NF-κB interact, suggesting a link between damage and apoptosis. Therefore DNA damage responses in cancers could be changed because their NF-κB is frequently elevated.

Another nick sensor is Poly (ADP-ribose) polymerase (PARP1), an abundant nuclear enzyme that is activated as a very early event following damage. When it binds to DNA strand breaks, it catalyzes synthesis from NAD of long (100+) ADP-rihose polymers on nearby proteins including PARP itself, and especially on core histones. This molecule opens chromatin, functioning as a switch to activate DNA repair, increase rates of RNA elongation, and block binding of transcription factors involved in proliferation. High PARP is correlated with genetic stability in cancers, and drugs that modulate PARP's function are being investigated. PARP-1 interacts with other repair-related protein systems including Ku and NF-κB involved in maintaining genomic stability. The binding of NF-κB to DNA was inversely correlated to PARP activity in mutant L1210 cells, and elevated PARP decreased high levels of NF-κB complexes to nearly normal levels. Conversely, binding of PARP1 to damaged DNA was inhibited by NF-κB, with subunits of which it forms a NAD-dependent complex that stimulates transcriptions involved in inflammatory processes. This interaction could be one basis for the increased apoptosis following DNA damage. The Sir complex may have similar properties in yeast. Also PARP binding mediates repression of retinoic acid-thyroid hormone receptor activities.

Repair Checkpoints

A cell has a limited time to repair its damaged DNA because permanent changes may result if repair is not completed before the damage becomes irreversible owing to its replication in S phase or its defective distribution in M (see Mutation above). Checkpoint responses to DNA damage temporarily arrest the cell and provide more time for repair. Cycling proceeds if and when repair is completed, and the cell resumes replication. Checkpoints and repair thus cooperate to help prevent genomic instability and preserve integrity of the genome. Their coordination is important following chemo or radiation therapy, which create extensive DNA damage. Quiescent cells have more time for repair prior to cycling than do cycling cells, so they are more resistant to DNA damage, such as from chemotherapy. Effects of repair infidelity can be diminished by blocking repair with a drug that increases death resulting from cell damage. Application of β-lapachone, a topoisomerase I inhibitor, caused far greater lethality of damaged DNA, and this was specific to cancer cells.

Damagc chcckpoints act in all phases of the cell cycle, in addition to the G_1 restriction point created by inadequate mitogens or by defectively controlled proliferation (see above). The G_1 damage checkpoint initiated by DNA damage activates distinct pathways that converge on the same components that regulate G_1 phase transit. A kinase cascade triggered by stress and heat shock stimulates MAPKAP kinase-2 and phosphorylation of small heat shock proteins. Stresses stabilize and activate p53 protein, which initially results in p21 and arrest in G_1 phase, which is followed by apoptosis or senescence if repair is not soon accomplished. p53-independent G_1 arrest requires p16(INK4A), which prevents Rb phosphorylation and activation of E2F1. Rb-deficient cells are hypersensitive to apoptosis induced by DNA damage. They cannot undergo G_1, mid-S, or G_2 arrest following DNA damage, although they can activate a reversible G_2 checkpoint. Rb thus has a critical role in detemining cell fate following DNA damage.

Damage in other phases of the cell cycle activates mechanisms that differ from the G_1 phase checkpoint. A checkpoint arises after DNA is damaged in S phase. The cells react by arresting proliferation, activating DNA repair, and eventually apoptosis which involves over expressed E2F-1. The Mrel 1 complex associates with the E2F family specifically in S phase and is involved in the S phase checkpoint. p53 is phosphorylated and is hyperacetylated via p300 by stresses, including DNA damage, and thus increases its DNA binding activity and is necessary for responses to it. p73 is involved in this DNA damage response, and it triggers cell death through a different pathway than p53, as discussed above. In response to ionizing radiation, its increased phosphorylation by cAbl kinase activates a proapoptotic pathway that is the same as is initiated by Myc. Proteins that cause cycle arrest are phosphorylated, including p53, p21, Chk1, and BRCA1. A role in repair of Chk2 is suggested by the finding that it interacts with Mus81, the Holliday junction-resolving activity molecule. A p53 binding protein (53BP1) is a mediator of the S and G_2 DNA damage checkpoints, as shown by its inactivation with siRNA.

Histone synthesis as well as DNA synthesis must be coordinated; otherwise, cells stop at the S phase checkpoint, their DNA is damaged, and they eventually undergo apoptosis. Hydroxyurea, which inhibits ribonucleotide reductase and thereby blocks DNA synthesis, destabilizes Cdc25A and stops cycling, acting similarly to DNA damage and through

ATR and Chk1. Cytosine arabinoside also trigger a concerted repression of histone synthesis, results indicating that histone synthesis depends on continued DNA synthesis. Conversely, ectopic expression of HIRA, a repressor of histone gene transcription, also blocks DNA synthesis and produces nucleosome-deficient DNA that is sensitive to micrococcal nuclease. These effects of uncoupling of DNA and histone synthesis during S phase are independent of cyclin/cdk2 activity, because the response to inhibition is not accompanied by prolonged arrest of cyclin A/cdk2 or E/cdk2 activity.

The G_2/M damage checkpoint arrests entry into and passage through M phase (see Chapter 6 by Sluder et al. on mitosis and cytokinesis). It depends on both phosphorylations and dephosphorylations on different amino acids of cdc2 and its subsequent proteolysis, which can be blocked by an inhibitory peptide. Exposure of cells to the retinoid analogue CD437 results in enhanced association of the cyclin B E3 ligase APC with and rapid proteolysis of cyclin B. CD437 also modulates the expression of cyclin B through simultaneous inhibition of the E2F-1 E3 ligase.

Because their G_1 checkpoint is not effective, DNA-damaged p53-negative tumor cells are arrested prior to entry into mitosis, rather than in G_1, like normal cells. The mechanism depends on multi-phosphorylation of Rad 9 by ATR, which activates several transducers including BRCA1, chk1, chk2, and Rad 53 and blocks cdc25C, thereby preventing cdc2/cyclin B activation and causing G_2 arrest. Effects of bypassing this checkpoint are illustrated by application of caffeine or its analog pentoxifylline. A DNA-damaging drug applied at a moderate dose causes cultured tumor cells to arrest at the G_2 checkpoint. After a few hours these cells recover and resume proliferation. But, when also given pentoxifylline, the damaged cells rapidly pass through mitosis and enter G_1 phase, their chromosomes are severely disrupted, and they die. Furthermore treatment of mice with an alkylating agent followed by pentoxifylline kills implanted cancer cells without adverse effects on the host. The kinase C inhibitor Go6976 also has this G_2 bypass effect on cells in culture, and at lower concentrations.

If damage is not repaired in M phase, cells with excessive lesions are arrested at the spindle checkpoint in metaphase and die. The presence of improperly attached kinetochores is detected through attachment-sensitive phosphorylations, to which Mad2 protein binds and prevents activation by cdc20 protein of the anaphase-promoting complex (APC). Still-defective cells that have passed this checkpoint are tetraploid, and they are arrested in G_1 by a mechanism that involves both p53 and pRb. DNA damage can also start endocycles of DNA replication from G_2 arrest, and the resulting cells in some cases repair the DNA and produce mitotic progeny. In contrast, undamaged M phase cells do not undergo cytokinesis when treated with cytochalasin B, and they become tetraploid but pass normally through their next cycle.

Chromosomal instability has for decades been a recognized feature of human tumors. Defective checkpoints in cancer cells cause chromosome aberrations, a driving force of tumorigenesis. An insight into potential mechanisms arises from increased centrosome numbers in cells that have become genetically unstable. Efficient coordination of centrosome duplication and DNA synthesis requires proteins that associate with and regulate Gadd45a transcription. Deletion of Gadd45a produces centrosome amplification, abnormal mitosis, and aneuploidy. Chromosome instability could provide a basis for specific vulnerability of cancer cells to treatments. The goal is to identify compounds that directly target the mitotic error mechanisms at the heart of this process.

Damage and Death

DNA repair and senescence or apoptosis are sequentially related processes essential for genome integrity. A single DNA-double-strand break has been estimated to be sufficient to cause a normal human cell to senescence. Moderate damage can be repaired during G_2/M arrest, and the cells then proceed through mitosis. The cells undergo apoptosis if repair is not soon completed, and thereby the possible carcinogenic effect is removed. Upon failure of repair an apoptotic cascade activates p53 in normal cells. But for p53-negative tumor cells the damage frequently does not lead to death, which results in mutations that cause tumor progression. If DNA damage is excessive, proteins that contribute to cellular survival by functioning in DNA repair become executioners. The molecular basis underlying the decision between these time-dependent processes is not understood. It is presumably achieved by the time-dependent coordination of activities of proliferative cyclin-dependent kinases, checkpoints, and repair biochemistry. IGF-I mediated Akt activation in primary human fibroblasts postpones the onset of ultraviolet light-induced apoptosis by providing more time for removal of cyclobutane dimers. One such mechanism could involve effects of PARP, whose repair activity is limited to a few minutes by its inactivation, release from DNA by self-modification, and cleavage by apoptotic caspases. And the poly(ADP-ribose) bound to PARP itself and other target proteins is soon removed by a specific glycohydrolase, thereby reversing the initial protective effects. Apoptosis subsequently might be activated through interactions with NF-κB and also by depletion of NAD by PARP activity.

DEVELOPMENTS IN THE NEAR FUTURE

Summary

Great progress has been made in the past few decades in unraveling molecular mechanisms underlying cancer. Based on these discoveries, future understanding and practical applications should be rapidly achieved. The directions of progress are, however, impossible to predict or manage; a truly novel insight or technique can suddenly open new vistas.

Basic Science

The biological and molecular differences between normal cells and the variety of cancer cells are evidently remarkably numerous. Some are beginning to be investigated, but much remains to be learned about these deranged pathways and the fundamental cellular processes they control. The improved techniques being developed for discovering novel differences of gene and protein expressions and their functions between cell types or under altered conditions will accelerate current progress. Validation of proposed connections between causes and effects, such as linking genes or drugs with proliferation and apoptosis requires novel tools. Degradation of a specific mRNA by action of the RNAi system has great promise in this regard.

For basic-clinical interface studies, heterogeneity of tumor biopsy samples that contain both normal and diverse tumor cells creates problems of analysis. Other problems are the differences between conditions in vivo versus in culture. These include absence of stromal cells and tumor-stromal interactions. Some differential effects found in cultures might be related to growth of cancer cells versus quiescence of most normal cells in vivo. Mixed cell and three-dimensional techniques are being utilized.

The complexity of the multiple interacting networks of pathways that are modifed in cancers requires an integrative bioinformatic approach, especially because of feedbacks and yin-yang steady state equilibria, of which examples are cyclins/inhibitors and +/− apoptotic Bcl-2 family members. Positive and negative autoregulatory and feedback loop processes include (1) PARP automodification and tuning with time, (2) p14ARF-Mdm2-p53 interactions, and (3) elevated β-catenin inactivation of Mdm2 that accumulates p53, which in turn feedback down-regulates β-catenin, and elevated Akt potentiation of Mdm2, which blocks p53 and turns down Akt in several ways. These delicate balances are modified in cancers. And the molecular basis of successive timing and of altenative outcomes of biological processes such as growth cessation versus apoptosis need explanations.

Clinincal Applications

Improved methods for treating cancers are sorely needed and are being created as based on new knowledge. Classical chemotherapy depends on the application of poisons that interrupt some vital process, for example, by damaging DNA. The theoretical basis of these treatments is minimal. Major problems that limits efficacy of chemotherapy include toxicity of drugs to normal cells versus tumor cells (a low therapeutic index). Progress in finding novel targets for agents directed against the disease will hopefully develop from differences between normal and tumor cells, such as those outlined here. In current therapy, drugs are applied in combinations because single-drug “magic bullet” therapies do not cure most cancers. Knowledge is essential on which interacting drugs with different modes of action including pro-apoptosis to apply, and this is being discovered. A promising example is the greatly enhanced tumor-specific lethality of β-lapachone with Taxol. New methods for drug synthesis and discovery are emerging. Development of appropriate and rapid assays will accelerate them.

The heterogeneous malignant cells in a tumor can respond differently to a therapy. Their genetic and functional changes continue to increase as a cancer progresses; earlier treatments should be superior for this reason as well as for others such as development of drug resistance. Methods for earlier cancer detection are being developed, based on molecular biology. These will complement mammography, Pap smears, and prostate-specific antigen to improve early finding of warning signs, and will result in better diagnostics and sooner treatment. Finally, testing a new treatment is a very difficult project both clinically and financially, and its administrative, and legal problems, if greatly simplified and improved, would be for the benefit of patients.

INDEX

Cell Cycle and Growth Control: Biomolecular Regulation and Cancer, Second Edition, Edited by Gary S. Stein and Arthur B. Pardee
ISBN 0-471-25071-6